Stem Cell Technologies
Basics and Applications

Stem Cell Technologies
Basics and Applications

Chief Editor
Kaushik D. Deb
Advanced NeuroScience Allies (ANSA) Pvt. Ltd.
Bangalore

Associate Editor
Satish M. Totey
Advanced NeuroScience Allies (ANSA) Pvt. Ltd.
Bangalore

McGraw-Hill

New York Chicago San Francisco Lisbon London
Madrid Mexico City Milan New Delhi San Juan
Seoul Singapore Sydney Toronto

The **McGraw·Hill** Companies

Copyright © 2010 by Tata McGraw-Hill Education Private Limited. All rights reserved. Printed in the United States of America. Except as permitted under the United States Copyright Act of 1976, no part of this publication may be reproduced or distributed in any form or by any means, electronic, mechanical, photocopying, recording, or otherwise, or stored in a data base or retrieval system, without the prior written permission of the publisher. Program listings (if any) may be entered, stored, and executed in a computer system, but they may not be reproduced for publication.

1 2 3 4 5 6 7 8 9 0   DOC/DOC   1 6 5 4 3 2 1 0

ISBN 978-0-07-163572-1
MHID 0-07-163572-6

The sponsoring editor for this book was Taisuke Soda and the production supervisor was Pamela A. Pelton. The art director for the cover was Jeff Weeks.

Printed and bound by RR Donnelley.

This book was previously published as *Stem Cells: Basics and Applications* by Tata McGraw-Hill Education Private Limited, New Delhi, India, copyright © 2009.

McGraw-Hill books are available at special quantity discounts to use as premiums and sales promotions, or for use in corporate training programs. To contact a representative, please e-mail us at bulksales@mcgraw-hill.com.

This book is printed on acid-free paper.

Information contained in this work has been obtained by The McGraw-Hill Companies, Inc. ("McGraw-Hill") from sources believed to be reliable. However, neither McGraw-Hill nor its authors guarantee the accuracy or completeness of any information published herein, and neither McGraw-Hill nor its authors shall be responsible for any errors, omissions, or damages arising out of use of this information. This work is published with the understanding that McGraw-Hill and its authors are supplying information but are not attempting to render engineering or other professional services. If such services are required, the assistance of an appropriate professional should be sought.

About the Editors

Dr. Kaushik D. Deb is General Manager, Operations and Research, at Advanced NeuroScience Allies, Pvt. Ltd., a research and development company focusing on different areas of neuroscience and allied research. His current research interests are in the fields of directed cell trafficking technologies and applications of nanotechnology and tissue engineering in regenerative medicine.

Dr. Satish M. Totey is Co-Founder, President, and Chief Executive Officer of Advanced NeuroScience Allies, Pvt. Ltd. He is among a small number of researchers who pioneered the isolation and characterization of pluripotent stem cells from human embryos, which provided an important framework for the development of human embryonic stem cells. Dr. Totey has derived several human embryonic stem cell lines, some of which have been listed at the NIH stem cell registry. He has also developed a method for large-scale clinical scale-up of mesenchymal stem cells and successfully done several pilot studies and clinical trials in various degenerative diseases. Dr. Totey has more than 100 peer-reviewed research papers in international journals and 19 national and international patents to his credit.

PRAISE FOR *STEM CELL TECHNOLOGIES*

Stem Cell Technologies: Basics and Applications will be of significant interest to researchers, faculty members teaching biology, medicine or embryology as well as graduate and post-graduate students. The practical aspects of stem cell research will also be of interest to those involved in stem cell banks and therapies. The area of my particular interest covered in this book is isolation and characterisation of germ and spermatogonial stem cells. I belive that experts in the area of human and animal reproduction will find discussions of derivation of gametes from embryonic stem cells highly relevant and revealing. I hope that the book finds a wide audience and fosters interest in stem cell biology and applications.

Dmitri Dozortsev, MD, PhD
Center for Women's Medicine
Reproductive Laboratories Director,
Houston, Texas, USA

Stem Cell Technologies: Basics and Applications will be of great interest to undergraduate and post-graduate students as well as scientists who use stem cell technologies for different purposes.

The book covers the basic concepts on isolation, characterisation, culture and differentiation of stem cells, as well as their possible future application in improving human health. The book has brought together scientists from different countries and with different backgrounds, allowing different opinions and perspectives on a relatively new area of biology. Interestingly, the book also has a historical perspective on medicine, covering major milestones from the discovery of stem cells to the evolution of regenerative medicine. It will increase our current knowledge about stem cells derived from different organisms, such as mammal, zebra fish and avian.

Alysson R. Muotri, PhD
Assistant Professor,
Department of Pediatrics/Cellular & Molecular Medicine
University of California
San Diego, USA

Stem Cell Technologies: Basics and Applications presents a comprehensive overview of the 2009-2010 state-of-the-art in the use of stem cells in regenerative medicine, drug discovery, genetics of human development, and the development of cancers. With a virtual explosion in the number and range of papers published on this subject, this book gives its readership an in-depth view of the nature of stem cells from embryonic to umbilical cord to amniotic to adult. My own research with adult-derived stem cells counters the view that adult stem cells have very little plasticity in their ability to differentiate among all tissues of the body. However, I do agree with Prof. David Williams that all avenues of stem cell research, be it embryonic, umbilical cord, amniotic, hematopoietic, spermatogonial, dental pulp, and adult-derived stem cells, should be investigated to their fullest extent to gain insights into all possibilities of harnessing regenerative medicine. We are currently engaged in research on repairing damaged tissues, curing both somatic and genetic diseases, understanding the role of stem cells in the development of cancers, and developing are ethnic-based pharmaceuticals, utilising autologous, allogeneic, and xenogeneic stem cells.

Henry E. Young, PhD
Professor of Anatomy, Division of Basic Medical Sciences,
Professor, Department of Anesthesiology,
Professor, Department of Pediatrics,
Professor, Department of Obstetrics & Gynecology,
Director of Embryology, Gross Anatomy, and Gross Anatomical Dissection,
Director of Embryology, Histology, and
Gross Anatomical Dissection for Certified Registered Nurse, Anesthetists,
Director of Surgical Anatomy,
Director, Adult Stem Cell Research Laboratory,
Mercer University School of Medicine,
Macon, Georgia, USA

Contents

List of Contributors — xix
Foreword — xxv
Preface — xxvii

1 Stem Cells and Regenerative Medicine—The Evolving Story — 1
   Introduction *1*
   The Evolution of Regenerative Medicine *2*
   Tissue Engineering *3*
   Nanotechnology *3*
   Stem Cells in Regenerative Medicine *4*
   Adult Stem Cells *7*
   Concluding Remarks *10*
   Acknowledgements *11*
   References *11*

PART 1 — Embryonic Stem Cells

2 Zebrafish Stem Cells — 21
   Introduction *21*
   Protocol for the Initiation and Maintenance of Zebrafish Primary Embryo Cell Cultures *23*
   Derivation of Zebrafish ESC Cultures *25*
   Germ-Line Chimera Production from ES Cell Cultures *26*
   Homologous Recombination in the Zebrafish ES Cell Cultures *27*
   Derivation of Zebrafish PGC Culture *31*
   PGC Isolation *32*

Frequency of Zebrafish Germ-Line Chimera Production from PGC Cultures *34*
Conclusion 35
Acknowledgements 35
References 36

3 Medaka Stem Cells   40

Introduction *40*
Medaka ES Cells *41*
Derivation of Medaka ES Cell Cultures *41*
Characteristics of Medaka ES Cells *43*
Chimera Formation from Medaka ES Cells *44*
Medaka Spermatogonial Stem Cells *46*
Application and Future Directions of Medaka Stem Cells *49*
Acknowledgement 51
References 51

4 Chicken Embryonic Stem Cells as Another Non-Mammalian Embryonic Stem Cell Type   56

Introduction: Totipotency and Pluripotency in Animal Cells *56*
Mouse Embryonic Stem Cells as ESC Archetype *57*
Pluripotency and Isolation of Embryonic Stem Cells in Birds *58*
Biochemical Characterisation *60*
In Vitro Differentiation Potentialities *60*
Genetic Modification *61*
In Vivo Chimerism *62*
Germ-line Competency and In Vivo Contribution of cESC *62*
Molecular Characterisation of Chicken Pluripotency: OCT4 Is also Present in Chicken *65*
Presence of Oct4 in Non-Mammalian Species *66*
The Nanog Homeodomain Gene Is also Present in Chicken *67*
Industrial Interest of Avian Stem Cells *69*
Conclusion *70*
References *70*

5 Derivation, Culture and Genetic Modification of Mouse Embryonic Stem Cells   76

Introduction *76*
Derivation of Mouse Embryonic Stem Cells *77*

Culture of Mouse Embryonic Stem Cells 79
Properties of Mouse Embryonic Stem Cells 81
Genetic Modification of Mouse Embryonic Stem Cells 88
Factors Affecting Homologous Recombination 97
Future Applications of Mouse Embryonic Stem Cell Technologies 103
References 106

6 Derivation of Human Embryonic Stem Cells from Blastocysts — 120
Introduction 120
Cell Differentiation in Embryos 131
References 135

7 Developing a Serum-free and a Novel Reporter System for hESCs — 138
Introduction 138
Current hESC Culture Systems 139
Feeder-free and Serum-free Defined Culture Systems for hESCs 140
Conclusion 149
References 149

8 Epigenetics and Embryonic Stem Cells — 152
Introduction 152
Molecular Mechanisms in Epigenetic: Tagging of Histones 155
PcG and Stem Cell Maintenance 159
Molecular Targets of PcG Protein 160
Conclusion 160
Acknowledgements 161
References 162

9 Embryonic Stem Cells as a Therapy for Diabetes — 169
Introduction 169
Alternative β-Cell Replacement Therapies 172
Non-Ontogeny-Based Differentiation of ESCs into Insulin-Producing β-Cells 173
Limitations in Non-Ontogeny Based Differentiation Strategies 175
The Generation of Definitive Endoderm from Pluripotent Cells 179
Development of the Primitive Gut Tube from Definitive Endoderm 182
Production of PDX-1-Expressing Pancreatic Progenitor Cells 183
Differentiation of Pancreatic Progenitor Cells to Endocrine Cells 186
Endocrine Cell Specification to Mature Insulin-Producing β-Cells 188

Other Approaches to Embryonic Stem Cell Differentiation
into Insulin-Producing β-Cells *189*
The Future *190*
Conclusion 192
Acknowledgements 193
References 193

10 Manipulation of Biomaterial Interface by Biomimetic Strategies: Possibilities of Exploring Stem Cells — 200

Introduction *200*
Stem Cells: Origin and Exploration for Tissue Engineering
and Regenerative Medicine *203*
Inflammation at an Implant Site *202*
Stem Cell Behavior in Relation to Inflammatory Response *203*
Stem Cells and Chronic Inflammation *205*
Stem Cells and Controlled Differentiation *205*
Stem Cells and Engraftment *205*
Stem Cells and Angiogenesis *206*
Stem Cell Behavior in Relation to Immune Response *206*
Modulating Immune Response a Lesson from Fetal Engraftment *207*
Surface the Culprit *208*
Stem Cell Interaction with Biomaterials *208*
Biomimetic Approaches in Surface Modification using Dense Thin Solid Films *209*
Optimization of Heterogeneous Biomimetic Lipid Composition *210*
Stem Cell Adhesion and Proliferation Studies *210*
Conclusion 211
Acknowledgements 211
References 211

11 Immunomodulatory Aspects of Human Embryonic and Adult Stem Cells — 215

Introduction *215*
Transplantation Immunology: Rejection Issues *216*
Transplanting Adult Stem Cells *221*
Mesenchymal Stem Cells *222*
Transplanting Allogeneic Embryonic Stem Cells *224*
Strategies for Generating Autologous ES Cells *226*
Conclusion 237
References 238

12 Human Embryonic Stem Cells as a Model System to Study Human Genetics   244

Introduction *244*
Human Embryonic Stem Cells *245*
Differentiating hESCs as a Model System to Study Embryogenesis *245*
Derivation of Mutant hESC Lines *247*
Nuclear-Transfer-Derived Embryos *251*
Transcription-Factor-Induced Pluripotent Stem Cells *251*
Potential Applications and Future Goals *252*
Summary 257
References 257

13 Germ Cells and Artificial Gametes Production from Embryonic Stem Cells   262

Introduction *262*
Gametes Derived from Embryonic Stem Cells *270*
Are Artificial Gametes Functional? *275*
Benefits of ESC-Derived Gametes *276*
Conclusion 276
Acknowledgements 277
References 277

14 Using Embryonic Stem Cells to Introduce Mutations into the Mouse Germ Line: Animal Models and Future Applications   281

Introduction *281*
Escs Can be Genetically Manipulated *282*
Using Escs to Study Embryonic Development via Random Transgenesis *283*
Escs as Cellular Models in Developmental Biology and Pathology *290*
Conclusion 291
References 292

15 Application of Human Embryonic Stem Cells in Drug Discovery   293

Introduction *293*
Isolation, Propagation and Characterization of hESCs *295*
Spontaneous Differentiation of hESCs *296*
Drawbacks with hESCs in Regenerative Medicine *297*
hESCs in Drug Discovery and Development *297*
Developmental Toxicity *299*

Measurement Criteria *300*
Differential Gene Expression during EB Development *300*
Cytotoxicity *300*
Role of Genetically Modified hESCs in toxicity screening *300*
Application for Cardiomyocytes Derived from hESCs *301*
Cardiomyocytes: Derivation and Characteristics *302*
Cardiomyocytes: Safety Pharmacology and Discovery of Novel Drug Targets *302*
Application for Hepatocytes Derived from hESCs *304*
Hepatocytes: Derivation and Characteristics *304*
Hepatocytes: Studies of Drug Metabolism and Drug-Induced Hepatotoxicity *306*
Conclusion 307
References 308

PART 2 Adult Stem Cells

16 Recent Advancements on Tissue-Resident Adult Stem/Progenitor Cell Biology and Their Therapeutic Applications in Regenerative Medicine and Cancer Therapies — 319

Introduction *319*
Phenotypic and Functional Properties of Adult Stem/Progenitor Cells *322*
Implication of Adult StemProgenitor Cell Dysfunctions in Disease Development *326*
Tissue-Resident Adult Stem/Progenitor Cell Types and their Therapeutic Applications *327*
Conclusion 339
Abbreviations 340
Acknowledgements 340
References 340

17 Mesenchymal Stem Cells and their Therapeutic Potential in the Nervous System — 358

Introduction *358*
Potential Therapeutic Applications of MSCs *365*
MSC-based Therapy in the Treatment of Osseous and Adipose Tissue Defects *366*
MSC-based Therapy in the Treatment of Neurological Deficits *367*

Mechanisms by which Transplanted MSCS Elicit Recovery
in the Damaged Nervous System *370*
In Vitro 'Trans'-Differentiation of MSCs into Neural Cell Types *370*
In Vitro 'Trans'-Differentiation of into Neural Cell Types—An Artifact of Cell Culture *372*
In Vivo 'Trans'-Differentiation of MSCs into Neural Cell Types *375*
Spontaneous Cell Fusion of MSCs with Host Neural Cells *377*
MSCs as Trophic Mediators of Neural Differentiation *377*
Secretion of Neurotrophic Factors by MSCs *378*
Secretion of Non-Neurotrophic Factors by MSCs *380*
Analysis of MSC-derived Soluble Factors *In Vitro:*
A Conditioned Media Approach *381*
References *383*

18 Mesenchymal Stem Cells: Biology, Culture Optimization and Potential Clinical Use — 391

Introduction *391*
Historical Background *392*
MSC are Present in all the Tissues *392*
MSC Niche *393*
Optimal Culture Conditions and Up-Scaling *395*
Phenotypic Properties of MSC *398*
Immunogenicity of MSC *400*
Immunomodulatory Properties of MSC *401*
MSC Facilitate Tissue Repair *401*
MSCs for Clinical Application *403*
Conclusion *404*
References *405*

19 Mesenchymal Stem Cells: Culture, Characterization and Therapeutic Applications in Neurodegenerative Diseases — 411

Introduction *411*
Isolation and Culture of Rat MSCs *412*
Characterization of MSCs *413*
Transplantation of MSCs for CNS Repairs *413*
Effects of MSCs Transplantation in the Central Nervous System *418*
Discussion *421*
Conclusion *421*

20 Dental Pulp Stem Cells and Perspectives of Future Application in Cell Therapy — 420

Introduction *426*
Isolation of Stem Cells from Dental Pulp *427*
Tissue Processing, Expansion in Culture and Dental Pulp Cell Banking *432*
In Vitro Differentiation Assays *434*
In Vivo Differentiation Assays *434*
Conclusion *437*
Glossary *438*
Acknowledgements 438
References 438

21 Hematopoietic Stem Cells and Their Dynamics — 442

Introduction *442*
Evidence for Hematopoietic Stem Cells *442*
Properties of Hematopoietic Stem Cells *443*
Sources of Hematopoietic Stem Cells *445*
The Stem Cell Niche *446*
How Many Stem Cells *447*
Expansion of the HSC During Human Ontogeny *448*
Stem Cell Aging/Clonal Succession *449*
Stochastic Dynamics within the HSC Pool *450*
Symmetry of Stem Cell Replication and Fitness *454*
Stem Cell Pool, Longevity and Cancer *456*
Inherited and Acquired Hematopoietic Stem Cell Disorders *458*
Acquired Hematopoietic Stem Cells Disorders with CML and PNH as Prototypes *459*
Future Developments *462*
Acknowledgements 464
References 464

22 Spermatogonial Stem Cells — 473

Introduction *473*
Basic Principles and Jargons Related to Spermatogonial Stem Cells *474*
Testicular Compartment and Cellular Interaction *475*
Proliferation and Differentiation of SSCs *476*

Acknowledgements 422
References 422

Regulation of SSCs Development *479*
Current Status of the Research *483*
Markers Express on SSCs during Their Development *484*
Future Applications and Caution *489*
Conclusions *490*
References *490*

23 Epigenetics as a Cause for the Eloquent Silence of Stem Cells — 505

Introduction *505*
What Is Epigenetics? *505*
Stark Examples of Epigenetic Phenomena in Biology *507*
Markers of Epigenetic Imprint *511*
Cellular Memory Modules *516*
Epigenetic Regulators of Stem Cell Pluripotency *529*
Epigenetics: A Trail from Stem Cells to Cancer *534*
Acknowledgements *537*
References *537*

24 Cancer Stem Cells: The Enemy within Cancers — 551

Introduction to Cancer *551*
Clonality in Cancer *554*
Cancer Heterogeneity *555*
Stochastic and Hierarchy Models *559*
Stem Cells and Hierarchy *559*
Cancer Stem Cells *561*
Techniques used to Isolate CSCs *562*
Techniques used to Assess CSCs *574*
Identification and Isolation of CSCs from Solid Tumours *580*
CSCs and Metastasis *583*
Cell of Origin of CSC *585*
Therapeutic Implications of CSCs *586*
Future Directions *592*
Definitions of Important Terms *593*
References *595*

25 Exploring the Potential of Adult Stem Cells: Why Umbilical Cord Blood? — 603

Introduction *603*
Umbilical Cord Blood: An Important Source of Stem Cells *603*

Umbilical Cord Blood Stem Cells and Transplants *604*
General and Ethical Topics about UCB Banking *605*
Umbilical Cord Blood as a Source of Mesenchymal Stem Cells *606*
Umbilical Cord Derived Stem Cells for Tissue Therapy: Current and Perspectives *608*
References *613*

26 Cell Therapy and Tissue Engineering Strategies in Regenerative Medicine — 617

Introduction *617*
The Basic Components of Tissue Engineering *618*
Tissue Engineering of Specific Tissues and Organs *625*
Other Emerging Technologies *633*
Conclusion *636*
Acknowledgement *637*
References *637*

27 Biomaterials for Enhancing Corneas and Spinal Cord Regeneration — 645

Introduction *645*
Models of the Nervous System *648*
Corneal Substitutes *652*
Spinal Cord Repair and Regeneration *658*
Conclusions *661*
References *662*

Index *669*

List of Contributors

Alexandre Kerkis
Applied Genetics, Veterinary Activities Ltd.
(Genética Aplicada),
São Paulo,
BRAZIL.

Amitha Jaykumar
Gunasheela Institute for Research in
Reproduction, Bangalore,
INDIA

Ann P. Chidgey
Monash Immunology and Stem Cell Laboratories
(MISCL),
Level 3, Building-75, Monash University,
Wellington Road, Clayton
Melbourne, Victoria 3800
AUSTRALIA
ann.chidgey@med.monash.edu.au

Anna Michalska
Monash Immunology and Stem Cells
Laboratories,
Monash University,
Wellington Road, Clayton,
Victoria 3800,
AUSTRALIA

Annapoorni Rangarajan
Indian Institute of Science,
Bangalore 560 012,
INDIA
anu@mrdg.iisc.ernet.in

Anthony Atala
The W. Boyce Professor and Chair,
Department of Urology
Director, Wake Forest Institute for
Regenerative Medicine
Wake Forest University School of Medicine
USA
aatala@wfubmc.edu

Antonietta Giudice
Monash Immunology and Stem Cells
Laboratories,
Monash University,
Wellington Road, Clayton,
Victoria 3800,
AUSTRALIA

Anurag N. Paranjape
Indian Institute of Science,
Bangalore 560 012,
INDIA

Arne Traulsen
Max Planck Institute for Evolutionary Biology,
24306 Ploen,
GERMANY

B. Pain
CNRS, UMR 6247, Université de Clermont, Inserm,
U931, Facult'e de Medecine, 28,
place Henri Dunant,
Clermont-Ferrand, F-63001,
FRANCE
bepain@u-clermont1.fr

Benu Sethi
Department of Chemical and Biological
Engineering,
Univ. of Ottawa, 161 Louis-Pasteur,
Ottawa, ON K1N 6N5,
CANADA

Bernard E. Tuch.
Diabetes Transplant Unit, Prince of Wales
Hospital, High Street, Randwick, NSW, 2031,
AUSTRALIA
b.tuch@unsw.edu.au

Bhaskar Thyagarajan,
Invitrogen Corporation
5781 Van Allen Way, Carlsbad, CA 92008
3175 Staley Road, Grand Island, NY 14072,
USA

Chandra P Sharma
Biosurface Technology division, BMT Wing,
Sree Chitira Tirunal Institute for Medical Science
and Technology, Thiruvananthapuram,
Kerala,
INDIA
sharmacp@sctimst.ac.in

Chuah Chong Boon,
Stempeutics Research Malaysia Sdn Bhd
LG-E-2A, Technology Park Malaysia, Enterprise-4
Bukit Jalil 57000,
Kuala Lumpur,
MALAYSIA

Daniel J. Maltman
School of Biological and Biomedical Sciences,
Durham University,
South Road,
Durham DH1 3LE,
UK

Devaveena Dey
Indian Institute of Science,
Bangalore 560 012,
INDIA

David Dingli
Division of Hematology,
Mayo Clinic College of Medicine,
200, First Street SW, Rochester, MN 55905,
USA
dingli.david@mayo.edu

Devika Gunasheela
Gunasheela Institute for Research in
Reproduction, Bangalore,
INDIA

Eiges Rachel
Stem Cell Research Laboratory,
Medical Genetics Unit,
Shaare Zedek Medical Center, Jerusalem,
ISRAEL,
rachela@szmc.org.il

Eitan Lunenfeld
Department of Obstetrics and Gynaecology
Soroka University Medical Center
Ben-Gurion University of the Negev
Beer-Sheva,
ISRAEL

F Lavial
Stem Cell Biology,
Institute of Reproductive and Developmental
Biology, Faculty of Medicine,
Imperial College London, Hammersmith Hospital,
Du Cane Road,
London,
UK

Henry Sathananthan,
Monash Immunology and Stem Cell Laboratory,
Monash University,
Melbourne,
AUSTRALIA

Huseyin Sumer
NH&MRC Peter Doherty Biomedical Research
Fellow Centre for Reproduction & Development
Monash Institute of Medical Research
27-31 Wright Street
Clayton, VIC 3168,
AUSTRALIA
Huseyin.Sumer@med.monash.edu.au

Irina Karkis
Laborarory of Genetics, Butantan Institute,
São Paulo, Brazil
Av. Vital Brazil 1500, 05503-900, São Paulo, SP,
Brazil
E-mail: ikerkis@butantan.gov.br

Jayant G Mehta
Sub-Fertility Unit, Queens Hospital
Rom Valley way, Romford
Essex RM7 OAG,
UK
jayantgmehta@gmail.com

Jennifer C.Y. Wong,
Diabetes Transplant Unit,
Prince of Wales Hospital,
High Street, Randwick, NSW, 2031,
AUSTRALIA

Joanna Lim
Monash Immunology and Stem Cells Laboratories,
Monash University,
Wellington Road,
Clayton, Victoria 3800,
AUSTRALIA

Joanne M. Hackett
University of Ottawa Eye Inst, 501
Smyth Road,
Ottawa, ON K1H 8L6,
CANADA

Jonathan D. Chesnut,
Invitrogen Corporation
5781 Van Allen Way, Carlsbad, CA 92008
3175 Staley Road, Grand Island, NY 14072,
USA

Jorge M. Pacheco
ATP-Group, CFTC & Departamento de Física da
Faculdade de Ciências,
P-1649-003 Lisboa Codex,
PORTUGAL

K.G. Vijayendran,
Stempeutics Research Malaysia Sdn Bhd
LG-E-2A, Technology Park Malaysia, Enterprise-4
Bukit Jalil 57000, Kuala Lumpur,
MALAYSIA

K. Kaladhar
Biosurface Technology division, BMT Wing,
Sree Chitira Tirunal Institute for Medical Science
and Technology, Thiruvananthapuram,
Kerala,
INDIA

Karen L. Jones,
Centre for Reproduction & Development
Monash Institute of Medical Research
27-31 Wright Street
Clayton, VIC 3168
AUSTRALIA
Karen.Jones@med.monash.edu.au

Katherine Wagner
Invitrogen Corporation
5781 Van Allen Way, Carlsbad, CA 92008
3175 Staley Road, Grand Island, NY 14072,
USA

Kaushik D. Deb
General Manager (Operations and Research)
Advanced NeuroScience Allies (ANSA) Pvt. Ltd.
#560, 2nd Floor, 9th A Main, Indiranagar,
Bangalore 560038
Karnataka, INDIA

Lianchun Fan
Department of Animal Sciences,
Purdue University,
West Lafayette, IN 47907,
USA

M. Naseer
Indian Institute of Science,
Bangalore 560 012,
INDIA

Mahendra S. Rao
Invitrogen Corporation
5781 Van Allen Way, Carlsbad, CA 92008
3175 Staley Road, Grand Island, NY 14072,
USA

Mahmoud AbuElhija
The Shraga Segal Department of
Microbiology and Immunology
Ben-Gurion University of the Negev
Beer-Sheva,
ISRAEL

Mahmoud Huleihel
The Shraga Segal Department of
Microbiology and Immunology
Ben-Gurion University of the Negev
Beer-Sheva, ISRAEL
huleihel@bgu.ac.il

Mariane Secco
Human Genome Research Center,
Department of Genetic and Evolutive Biology,
University of São Paulo,
Rua do Matao, 106 – Cidade Universitária –
São Paulo – SP, 05508-090,
BRAZIL
marianesecco@usp.br

MaryLynn Tilkins
Invitrogen Corporation
5781 Van Allen Way, Carlsbad, CA 92008
3175 Staley Road, Grand Island, NY 14072,
USA

May Griffith
University of Ottawa Eye Inst,
501 Smyth Road, Ottawa, ON K1H 8L6, Canada,
Dept. of Cellular and Molecular Medicine,
University of Ottawa,
451 Smyth Road,
Ottawa, ON K1H 8M5,
CANADA

Mayana Zatz
Human Genome Research Center,
Department of Genetic and Evolutive Biology,
University of São Paulo,
Rua do Matao, 106 – Cidade Universitária –
São Paulo – SP,
BRAZIL, 05508-090
mayazatz@usp.br

Meera Saxena
Indian Institute of Science,
Bangalore 560 012,
INDIA

Mehrdad A. Rafat
University of Ottawa Eye Inst,
501 Smyth Road,
Ottawa, ON K1H 8L6,
CANADA

Meisheng Yi
Department of Biological Sciences,
National University of Singapore,
10 Kent Ridge Crescent,
SINGAPORE 119260
dbshyh@nus.edu.sg);

Mohan C. Vemuri
Invitrogen Corporation
5781 Van Allen Way, Carlsbad, CA 92008
3175 Staley Road, Grand Island, NY 14072,
USA

Murielle Mimeault
Department of Biochemistry and Molecular Biology,
Eppley Institute for Research in Cancer and
Allied Diseases,
University of Nebraska Medical Center,
Omaha, Nebraska 68198-5870,
USA,
mmimeault@unmc.edu

Natalie Seach
Monash Immunology and Stem Cell
Laboratories (MISCL),
Level 3, Building-75, Monash University,
Wellington Road, Clayton 3800,
Melbourne, Victoria,
AUSTRALIA

Paul Collodi
Department of Animal Sciences,
Purdue University,
West Lafayette,
IN 47907,
USA

Paul J Verma
Stem Cell Biology
Monash Institute of Medical Research
27-31 Wright Street
Clayton, VIC 3168,
AUSTRALIA
Paul.Verma@med.monash.edu.au

Pollyanna A. Tat,
Centre for Reproduction & Development
Monash Institute of Medical Research
27-31 Wright Street
Clayton, VIC 3168
AUSTRALIA
Pollyanna.Tat@med.monash.edu.au

Rebecca Lim
Monash Immunology and Stem Cell
Laboratories (MISCL),
Level 3, Building-75,
Monash University,
Wellington Road, Clayton 3800,
Melbourne, Victoria,
AUSTRALIA

Rajarshi Pal
Manipal Institute of Regenerative Medicine
Manipal University Branch Campus
10 Service Road, Domlur Layout
Bangalore- 560 071,
INDIA

Rakhi Pal,
Stempeutics Research Private Limited
9th Floor, Manipal Hospital, Bangalore-570017,
INDIA

Richard L. Boyd
Monash Immunology and Stem Cell
Laboratories (MISCL),
Level 3, Building-75,
Monash University, Wellington Road, Clayton
3800, Melbourne, Victoria,
AUSTRALIA

Samuel SW Tay
Department of Anatomy
Yong Loo Lin School of Medicine
National University of Singapore
#4 Medical Drive, Blk MD10
SINGAPORE 117597

Satish M. Totey
Advanced NeuroScience Allies (ANSA) Pvt. Ltd.
#560, 2nd Floor, 9th A Main, Indiranagar,
Bangalore 560038
Karnataka,
INDIA

Shipra Bhatia
Dr. B.R. Ambedkar Centre for Biomedical
Research, University of Delhi
Delhi-110007,
INDIA

Sonja E. Lobo
Laboratory of Genetics, Butantan Institute &
Department of Genetics and Morphology,
Federal University of São Paulo São Paulo,
BRAZIL.
E-mail: sonjael@terra.com.br

Soojung Shin,
Invitrogen Corporation
5781 Van Allen Way, Carlsbad, CA 92008
3175 Staley Road, Grand Island, NY 14072,
USA

Stefan A. Pryzborski
School of Biological and Biomedical Sciences,
Durham University,
South Road,
Durham DH1 3LE,
UK,
stefan.przyborski@durham.ac.uk

Steven A. Hardy,
School of Biological and Biomedical Sciences,
Durham University,
South Road, Durham DH1 3LE,
UK

Steven Y. Gao
Diabetes Transplant Unit,
Prince of Wales Hospital
High Street, Randwick,
NSW, 2031,
AUSTRALIA

Sulochana Gunasheela,
Gunasheela Institute for Research in
Reproduction,
Bangalore,
INDIA

Surinder K. Batra
Department of Biochemistry and
Molecular Biology,
Eppley Institute for Research in Cancer and
Allied Diseases,
University of Nebraska Medical Center,
Omaha, Nebraska 68198-5870,
USA,
sbatra@unmc.edu

Swapnil Totey
Stempeutics Research Malaysia Sdn Bhd
LG-E-2A, Technology Park Malaysia, Enterprise-4
Bukit Jalil 57000,
Kuala Lumpur,
MALAYSIA

Ten-Tsao Wong
Department of Animal Sciences,
Purdue University,
West Lafayette, IN 47907,
USA
wong20@purdue.edu

Thais M.C. Lavagnolli
Department of Genetics and Morphology,
Federal University of São Paulo & Laboratory of
Genetics, Butantan Institute, São Paulo,
BRAZIL.
tlavagnolli@gmail.com

Vani Brahmachari
Dr. B.R. Ambedkar Centre for Biomedical
Research, University of Delhi
Delhi-110007,
INDIA

Viviane Abreu Nunes
Stem Cells Lab
Human Genome Research Center - University of
São Paulo
BRAZIL
Email: vanunes@ib.usp.br

Xudong Cao
Department of Chemical and Biological
Engineering,
University of Ottawa,
161 Louis-Pasteur,
Ottawa, ON K1N 6N5,
CANADA

Yunhan Hong
State Key Laboratory of Freshwater Ecology and
Biotechnology,
Institute of Hydrobiology,
Wuhan 430072,
CHINA

Foreword

Interestingly the history of medicine combines a story of the evolution of knowledge about the human body, the diseases which affect it and the therapies intended to deal with these diseases, with a number of remarkable events that have themselves moved the subject forward with quantum leap. The discovery of antibiotics superimposed on the evolution of knowledge of infectious diseases, the development of open heart surgical techniques that revolutionised the treatment of heart disease and the introduction of functional imaging devices to markedly improve diagnostic techniques—all come to mind here. Perhaps, after a few decades medical historians, will look back to the turn of this century, and pronounce that the ability to harvest and manipulate stem cells for the purposes of regeneration of tissue in adult humans was one of those major, remarkable events.

We can already see the potential of regenerative medicine. From my own perspective, as that of a biomedical engineer attempting to treat diseased and traumatised tissue through replacement by synthetic structures, the great strides that we have made over half a century are likely to be overshadowed by the gains that will be made by various technologies of regenerative medicine. There will always be significant limitations to what we can expect to achieve through the replacement of tissues by non-living synthetic materials. Although we have had success with joint replacements, prosthetic heart valves and arteries, implantable electronic devices such as pacemakers and defibrillators, and intraocular lenses, such success will be limited to mechanical and physical functions. How much better would it be to improve biological and physiological functions, through the regeneration of the affected tissues? This is what regenerative medicine is about—a combination of cell therapy, gene therapy and tissue engineering. Bearing in mind that as adults we have very limited ability to regenerate our tissues, this is by no means, a trivial task, which essentially has to involve the sourcing of appropriate cells and persuading them, through various manipulations, to using new functional tissues.

The nature of the cells for these purposes has been, and will continue to be, a controversial issue from a variety of scientific, clinical and ethical positions, with debates about the respective qualities of autologous, allogeneic and xenogeneic sources; fully differentiated cells, progenitor and stem cells; and adult or embryonic sources of cells. Without doubt, however, a major factor in this whole debate is the suitability of stem cells for these functions. Although stem cells have been known for some time, the biology of stem cells and their manipulation for therapeutic purposes have become the subject of intense research only in the last decade. It is highly appropriate at this time, therefore, that an analysis of the current state of knowledge about these cells and the technologies associated with these applications is carried out, and set forth in a book dedicated to this purpose. I congratulate the editors of this volume for their contributions and am sure that it will serve a valuable purpose in establishing stem cell science and technology at the forefront of twenty-first century medicine.

PROFESSOR DAVID WILLIAMS, FREng.,
Emeritus Professor, University of Liverpool, UK
Professor and Director of International Affairs,
Wake Forest Institute of Regenerative Medicine, North Carolina, USA
Visiting Professor, Christian Barnard Department of Cardiothoracic Surgery,
Cape Town, South Africa,
Visiting Professor, Graduate School of Biomedical Engineering,
University of New South Wales, Australia
Guest Professor, Tsinghua University, Beijing and Visiting Professor,
Shanghai Jiao Tong Medical University, China
Editor-in-Chief, *Biomaterials*

PREFACE

Currently, stem cells research is generating considerable excitement and debate, which is clearly reflected in the ever-increasing number of scientific publications in peer-reviewed journals and clinical trials using stem cells. There is a growing consensus amongst the policy analysts and scientists alike that India is likely to play a key role in the scientific, clinical and commercial development of stem cell research. Although at a nascent stage, stem cell research and therapies is a rapidly growing field in India and the market is estimated to be of around US $ 600 million. The global market for stem cell therapy is expected to be worth around US $ 20 billion by 2010.

Although significant progress has been made in understanding their mechanisms and unraveling their potential for therapeutic application, widespread use of stem cell-based therapies is still in a nascent stage but can one day benefit a vast number of patients. At present, stem cells from multiple sources have been identified and characterised and this list is increasing. Embryonic stem cells, stem cells from bone marrow, adipose tissues, umbilical cord, dental pulp and several other locations can offer a potential source of allograft or auto graft in future. Recent establishment of induced pluripotent stem cells may revolutionise drug screening and toxicity.

Translational research in the field of stem cells is advancing at a rapid pace, mostly in the area of adult stem cells. However, these successes to some extent are limited by the problems of reproducibility, poor therapeutic outcome and lack of understanding of the mechanisms needed for clinical advances. It is, therefore, pertinent to research into all possible sources of stem cells at this point—embryonic as well as adult—as it is not clear which approach would lead to successful therapies for different diseases.

While putting together this book, we were attentive to its role in education and clinical research around the world, including India. Our primary goal was to maintain the unique components that make this book a first line resource—comprehensive yet succinct and clearly written chapters addressing all major areas of stem cell biology and their potential application in therapies. Many renowned experts have contributed to this book and we are greatly indebted to our contributing authors for giving us their valuable time while preparing these chapters. As the authors have described some of their own findings, the book also serves as a first-hand reference for researchers. The book covers the entire cross-section of stem cell biology and will be valuable for researchers, students and beginners.

Each chapter covers the basic as well as advanced material on each topic. It is assumed that students or readers using this book would have previously acquired a basic biological vocabulary. The book has been divided in two parts. *Part One* covers embryonic stem cells, the most potent stem cells known so far. This section also contains a chapter on the emerging field of embryonic stem cell-based drug screening platforms. *Part Two* deals with the multipotent adult stem cells from different tissue types. This section also covers some unique concepts such as cancer stem cells and tissue engineering-based approaches for designing the right microenvironments for tissue regeneration. An attempt has also been made to present a very balanced outlook on the therapeutic and technological advances in each area. We have made all efforts to ensure that this book provides an up-to-date, accurate and a complete resource for the topics covered. Nonetheless, the field of stem cell is multifaceted and rapidly evolving, and there is a remarkable scope for improvement.

KAUSHIK D. DEB
SATISH M. TOTEY

Stem Cell Technologies
Basics and Applications

1

STEM CELLS AND REGENERATIVE MEDICINE—THE EVOLVING STORY

Ann P. Chidgey*, Natalie Seach, Rebecca Lim, Richard L. Boyd

Monash Immunology and Stem Cell Laboratories (MISCL), Monash University, Melbourne, Victoria, Australia

INTRODUCTION

The continual quest of man to improve health care is not new—it is just the technologies that change. Over 2500 years ago, The Chinese developed substances to fight "infections" reaching through to quinine for malaria in the 17th century to 1877 when Pasteur and Koch found that an airborne bacillus had "antibiosis" actions, to Ehrlich's "antibiotic" for syphilis in 1909. But it was not until 1928 with Fleming's discovery of penicillin that the concept of antibiotics was popularised, ultimately culminating in a Nobel prize shared with Chain and Florey in 1939. The hallmark of defined antibiotics, coupled with major improvements in general sanitation, hygiene and nutrition, was the trigger to major improvements in human health and well-being. One of the ironies accompanying the increased lifespan of the population from approximately 40 years at the turn of 1900 to almost 80 years now is an unprecedented aging population and the onset of many types of degenerative diseases.

Globally, the aging society is not content with suffering, their impatience driving further expectations of better treatments and even cures. But this is counter balanced by the financially conservative nature of governments and their available input into health improvement. How do they triage the clinical needs before them? One of the issues is the time (10–20 years) and cost (hundreds of millions of dollars) involved to make the transition from "bench to bedside". Current estimates are that there are over 70 forms of intractable, degenerative conditions affecting hundreds of millions

---

* *ann.chidgey@med.monash.edu.au*

of people globally. It is this social and clinical burden that has been the catalyst to a new breakthrough technology, stem cell based therapies—a new revolution in medicine. Indeed this is best exemplified by the formation of the Californian Institute for Regenerative Medicine (http://www.cirm.ca.gov)—a $3 billion enterprise that is leading the world with the funding of stem cell research and its clinical applications.

The chapters in this book highlight the explosion in this area of research. For the first time, so many diseases have been covered under the same therapeutic umbrella—that provided by stem cells as the new foundation to regenerative medicine.

THE EVOLUTION OF REGENERATIVE MEDICINE

William Heseltine, Chairman and CEO of Human Genome Sciences (Rockville, MD), about 10 years ago described "regenerative medicine" as *"the broad range of disciplines . . . working towards a common goal of replacing or repairing damaged or diseased tissue."* This section discusses the history and evolution of regenerative medicine and its associated paradigm shifts, as well as the interplay of factors that shaped the pace and direction of its development.

The earliest concept of tissue regeneration dates as far back as the ancient Greek mythologies. Hercules cut off the head of the monster Hydra and was shocked to observe two heads growing back in its place; Prometheus was chained to a rock where daily his liver was eaten by an eagle, but was renewed every night. In nature, newts are able to regenerate severed limbs and a number of lower vertebrates can regenerate central nervous system neurons following axotomy. Although humans are not capable of such dramatic regeneration, however on a daily basis the integrity of all cells, tissues and organs is homeostatically maintained by continual replacement from residual pockets of specific stem and progenitor cells. In many forms of tissue injury, adult stem cells both native to the organ as well as those distal to the site play a role in the repair and regeneration of the tissue. In the case of the skin, wound healing involves local cutaneous cells to reconstitute the epidermis as well as distant bone marrow derived cells, and the adjacent uninjured dermal mesenchyme to reconstitute the dermal fibroblast population [1]. In liver resection, up to two-thirds of the liver can be removed and remaining hepatocytes undergo replication to restore original organ size and function; however, under severe and sustained injury, intra-hepatic stem cells from the Canals of Hering become activated and take on the task of regeneration [2, 3].

While there have been suggestions of bone autografts for centuries and many sporadic attempts at transplantation surgery occurred up until the 1900s, the modern-day concept of transplantation as a paradigm for regenerative medicine did not become popularised until December 1967 with the first successful heart transplant performed by Christiaan Barnard. Interestingly, there were two main barriers to success—appropriate surgical procedures and immune rejection. The former is now stunning in its precision, while the latter still poses a major challenge.

Stem cell based therapies and their clinical applications to regenerative medicine have as their origins the world's first successful bone marrow transplant (BMT) in 1968, wherein whole bone marrow was transplanted into a patient and the stem cells contained within the graft successfully replaced all haematopoietic cells [4]. While today's BMT benefits from extensive human leukocyte antigen (HLA)-typing, earlier unsuccessful attempts at BMT for the treatment of aplastic anaemia by Osgood in 1939 [5] preceded our understanding of the immunological factors involved in functional engraftment and graft-versus-host disease (GVHD). Early theories on the mechanisms of rejection were proposed by Ehrlich in 1906 and Carrel in 1910 [6, 7], who suggested that this rejection was due to malnutrition and physiological disturbances in donor tissues. As the ABO blood group system and immunological nature of allograft rejection in humans were discovered, more light was shed on the restrictions of transplantation, and hence the strategies required improving successful engraftment and long-term acceptance. However, despite the fact that it is now over 80 years since the first definition of major histocompatibility complex (MHC) antigens [8], the immunological barriers to successful, non-self transplants, including those of stem cells, remain a major challenge.

TISSUE ENGINEERING

Bioengineering is increasingly being recognised as an important component in the field of regenerative medicine. In the later part of the century, scientists faced yet another paradigm shift as tissue engineering was born from collaborative efforts by four independent laboratories at the Massachusetts Institute of Technology (MIT), which resulted in a number of products that are still available today [9]. Concurrent research at the Harvard Medical School to develop a bio-artificial pancreas [10, 11] led to a variety of biohybrid organ projects including Amcyte's and Novocell's Phase I/II trials of microencapsulated allogeneic islets cells. In 1993, *Science* published the first major review on tissue engineering [12] and by 1998, the US FDA approved the world's first allogeneic bioengineered product, Apligraf®, which was marketed as a living skin equivalent.

As was the case for monoclonal antibodies and genetic engineering, initial claims for bioengineering were over-simplified and over-played. Suggestions that human organs could be grown in a petri dish resulted in unrealistic public expectations and media hype. The peak of this was perhaps in 2000 when *Time* magazine named tissue engineers as "The Hottest Job" for the future. There is, however, a re-incarnation of tissue engineering and its integration into creating artificial niches for stem cell growth.

NANOTECHNOLOGY

The concepts of nanotechnology were first introduced by Richard Feynman in 1959 [13] and further defined by Taniguchi in 1974 [14]. Today's interest in nanomedicine is based on the application of nanotechnology tools to the development of structures at the molecular level, which then allows for the improvement of interactions between synthetic materials and biological entities. Within the genre of regenerative medicine, nanoparticle research has mainly addressed the development of entrapment and delivery systems for genetic material, therapeutic agents, biomolecules and as a reinforcing- or bioactivity-enhancement phase for polymeric matrices in 3D scaffolds. The advances in nanotechnology facilitate the synthesis of an extracellular matrix (ECM) that can promote and influence cell growth, cell mobility and adhesion—much more than mimicking a cell's natural environment. Micro- and nano-topography have been shown to be crucial cues for cell behaviour including differentiation [15, 16]. In the case of tissues such as tendons, nerves, corneal stroma and intervertebral disc regeneration, cell orientation using contact guidance afforded by nanoscaffolds is essential to achieve functional tissue [17, 18]. The capability of nanomedicine to produce nanostructures, which can mimic natural tissues, as well as nanoparticles in delivery systems has elicited research interest in this field. One of the challenges scientists are facing today is developing artificial nanocarriers that can pass tight junctions, blood–brain barriers and capillaries but with the efficiency and specificity similar to that of viruses [19].

STEM CELLS IN REGENERATIVE MEDICINE

Embryonic Stem Cell Therapy

Recently, human embryonic stem cells (hESCs) have been touted as heralding the dawn of a revolutionary age of regenerative medicine [20, 21]. Following the "discovery" of hESCs in 1998 [22], and their capacity for diverse differentiation [23], our understanding of the biology of these cells, their physiological requirements and the parameters of their potential has increased enormously. Protocols have been developed to maintain hESCs in their undifferentiated state, and panels of primitive markers related to *"stemness"* have been described to identify, sort and isolate the cells. The main characteristic that separates hESCs from all other types of cells is that they are truly pluripotent, that is, they can potentially give rise to virtually all cell types in the human body and, therefore, can provide exciting new therapies for tissue regeneration. Unlike haematopoietic and other adult stem cells that cannot undergo long-term self-renewal *in vitro* [24], hESCs appear to be capable of infinite self-renewal and can, therefore, expand to very large numbers [25].

Some of society's most devastating diseases, such as diabetes, kidney failure, lung disorders, spinal injury, stroke, neurodegenerative disorders, haematological disorders and heart failure, for example, may potentially be treated with defined cell populations able to repair or replace the damaged tissue. Whilst generating specific cell types from ESCs *in vitro* poses many challenges, the activation or inhibition of specific factors required to induce the differentiation of ESCs through the many developmental stages towards lineage-specific progenitors and mature cells, are slowly being unravelled [26]. Despite the enormous potential of ESCs, and the fact that many ethical issues have been overcome, their clinical utility has been tempered by safety concerns. These have arisen directly from their characteristic propensity for teratoma formation (tumours derived from all three germ layers) and whether therapeutic ES products do indeed faithfully recapitulate the normal cells and tissues they are designed to repair or replace. This does not preclude the value of ESCs for generating invaluable cell lines as diagnostics for drug testing. The issue of immune rejection will be a challenge with traditional ESCs.

Tolerance to Stem Cell Therapeutics

The immunogenicity of ESC transplants has been the topic of much debate. The discovery that ESCs expressed only low levels of MHC proteins—few MHC Class I and virtually no MHC Class II [27]—led to the possibility that ESCs might be "immune privileged", that is, they can elude the host immune system. However, compelling evidence of immune rejection has conclusively shown recently that this is not the case. Using a sophisticated non-invasive molecular imaging technique, Swijnenburg and colleagues [28] unequivocally demonstrated immune rejection of hESC within 10 days of transplantation into immune competent mice. Furthermore, upon re-injection of hESC, the already primed immune system rejected them within 2–4 days. This study suggests that patients will reject allogeneic hESC transplants, as they would for any other allogeneic solid organ transplant.

It was initially perceived that the environment in which ESCs would be transplanted into may induce the differentiation of these cells into the surrounding tissue type, with the caveat that aged niches may have reduced capacity to stimulate stem cell regeneration [29]. In this respect, Nussbaum and colleagues [30] confirmed that adult tissues may lack the cues to induce ESCs to form mature cell types. In carefully controlled experiments, syngeneic undifferentiated mouse ESCs were transplanted into normal and infarcted adult hearts. The damaged cardiac tissue lacked the required inductive cues for ESC differentiation into cardiomyocytes and teratomas formed. Furthermore, while the grafts were accepted in immunocompromised and syngeneic hosts, they were eventually rejected in allogeneic hosts [30].

Inducing ESCs to differentiate *in vitro* into the appropriate cell types prior to transplantation would avoid the issue of teratoma formation, provided there were no lingering undifferentiated cells. However, it is highly probable that the differentiated progeny of ESCs will have increased expression of MHC proteins (referred to as HLA), even more so in inflammatory environments, which stimulate the production of cytokines such as IFNg. This would increase their likelihood of immune rejection in an allogeneic setting. Clearly efficient tolerance strategies will need to be addressed to avoid life-long administration of immunosuppressive treatments, which are associated with high morbidity and loss of quality of life, with rejection ultimately inevitable.

Several alternatives to generalised immune suppression have been proposed in order to overcome immune rejection [26, 31, 32]. These include the accessibility of enough ESC lines to provide histocompatibility matching to the genetically diverse population, to engineered ESCs that suppress HLA expression or secrete immunosuppressive molecules. Each has its associated difficulties and risks. It was estimated that 150 different cell lines will be required to provide a full match at HLA-A, -B and -DR types for 20% of recipients, a beneficial match for 38% with one HLA-A or one HLA-B mismatch only, and one HLA-DR match for 84% [33]; immunosuppressive treatment will, therefore, be required even after that.

Another approach—using the same rules that establish and maintain self-tolerance, and has many examples of proof of concept—is to re-educate the immune system to accept donor tissue by creating haematopoietic chimeras. Patients receiving a bone marrow, or haematopoietic stem cell (HSC), transplant prior to solid organ transplants from the same donor were taken off immunosuppressive treatments once donor engraftment was achieved (for example, [34–39]). Whilst successes in this approach have been described for over a decade, it is surprising that it has not been incorporated more often in clinical transplantation therapy, at least in young recipients who have a more robust immune regenerative potential. Thus, by deriving HSCs from the same hESC line that was used to develop the cell therapeutic, central tolerance could be established, provided the hESC-derived HSCs readily engraft into the bone marrow and then migrate as appropriate progenitors to the thymus. Here they convert to not only T-cells but also dendritic cells, which can deliver tolerogenic signals to the developing thymocytes, eliminating or silencing any that may be donor (or self-) reactive.

One limiting factor to such tolerance-inducing approach is the age of the recipient. This is of paramount importance as the vast majority of people requiring regenerative medicine-based therapies will be adults through to the elderly. However, the thymus—the function of which is critical for central tolerance—atrophies with age such that by mid-life it has less than 10% of its maximal functional tissue [40]. The bone marrow niche may also be altered with age, although this phenomenon is not as clear-cut as that evident in the thymus, with the possibility of both intrinsic and extrinsic effects altering HSC and B-cell developmental potential with age [41–47]. The realisation that damage from pre-conditioning regimes required for HSC transplants may further compromise the function of the thymus and bone marrow in the elderly has led to increased attention towards developing strategies to enhance T-cell reconstitution. Such strategies include regenerating the bone marrow [48] and thymic niches [49, 50], adoptive transfer of T-cell precursors [51] and induction of regulatory T-cells (Treg) [52–54]. Changes in the immune-endocrine axis with age have provided some insight into the mechanisms of thymic involution and these are currently being investigated as potential therapeutics for thymic regeneration. Blocking the suppressive effects of sex steroids by administering a luteinising hormone releasing hormone (LHRH) agonist has had dramatic results in regenerating lymphopoiesis and immune reconstitution in aged animal models and following chemotherapy-induced damage [55–60] and is currently showing positive results in clinical trials [61]. Growth factors, such as growth hormone, keratinocyte growth factor, interleukin-7 and Fms-like tyrosine kinase 3, have also shown potential as regenerative agents [62].

Strategies to produce patient-specific "embryonic" stem cells are being developed to avoid rejection, such as induced pluripotential stem (iPS) cell lines, generated by the dedifferentiation or reprogramming of patient somatic cells [63–66]. The reprogramming requires integration of retroviruses encoding specific transcription factors. The risks in using viral vectors are presently too great for clinical application; however, developing technologies for transient gene transfer or the transfection of proteins rather than genes may overcome this and the possibility of immune rejection, if viral antigens are present.

ADULT STEM CELLS

Although BMT has now been a very successful therapy for many decades, recently, in part instigated by the ethical and safety issues with ESCs, adult stem cells have become a major focus of interest. It is now clear that every tissue and organ in this body has its own reservoir of stem cells that provide the "engine room" for homeostatic maintenance of the body. The aging process no doubt reflects a numerical or functional degradation in these stem cells—hence major efforts being made to identify and target adult stem cells for regenerative medicine.

Indeed pre-clinical studies are already being translated into the clinic. In particular, bone marrow derived stem/progenitor cells may also migrate to distant extramedullary peripheral sites after severe tissue damage and participate in repair, remodelling and the regeneration process [67–70]. This unique feature makes them a relevant source of immature cells for tissue repair based on the body's own regenerative capabilities [71]. The tissue regeneration mediated via adult stem/progenitor cells is usually accompanied by changes in the local environment orchestrated by growth factor and cytokine-initiated cascades including EGF-EGFR, Wnt/β-catenin, Notch, BMP, SDF-1-CXCR4 signalling pathways [72–74]. Additionally, adult stem cells are being studied as a novel means to deliver gene therapy, for example anti-angiogenic or cytotoxic agents can be directed to specific tumoural sites as a treatment for aggressive and metastatic tumours, which are unamenable to traditional therapy [75–78].

Two adult stem cells, namely the mesenchymal stem cell (MSC) and the amnion epithelial stem cell, have emerged recently in the literature as cells with marked potential in the realm of immunomodulatory and regenerative medicine. MSCs (also referred to as mesenchymal stromal cells) were originally isolated from the bone marrow but are now found in numerous tissues, including adipose, skin, umbilical cord and placental membranes. Isolated from their milieu on the basis of adherence to plastic and negative expression of haematopoietic markers, MSCs demonstrate clear multipotency for differentiation into cells of mesenchymal lineage such as adipocytes, chondrocytes and osteoclasts and are currently being used clinically for the treatment of genetic bone disorders such as *Osteogenesis imperfecta*, as well as mechanical bone and cartilage injuries (refer to Table 1.1). More controversially, MSCs have also reported trans-differentiation into cells of both endodermal [79, 80] and ectodermal germ layers [79, 81], indicating potential for the treatment of neurological conditions such as stroke as well as therapeutic applications in both pancreatic, renal and liver function [82]. An *in vivo* study has recently reported improved neurological function in a cohort of patients injected with autologous MSC for the treatment of multiple system atrophy [83]. In addition, multiple clinical trials are underway to assess the effect of site-specific injection of MSC in patients with end-stage liver disease as well as a broad range of cardiac myopathies (refer Table 1.1).

MSCs also function as potent regulators of the immune system, suppressing immune responses both *in vitro* and *in vivo* [84]. This remarkable property has, most notably, led to its clinical use in patients undergoing myeloablation for HSC transplant. Numerous studies have now demonstrated the powerful effect of MSC infusion in reducing the incidence and/or severity of GVHD in patients receiving allogeneic HSC transplantation, as well as improvement in donor stem cell engraftment and function [85–92]. The immunosuppressive properties of MSC are currently being evaluated as potential therapy for other immune disorders including Crohn's disease, systemic lupus erythematosus and Type 1 diabetes (refer to Table 1.1).

Amnion epithelial cells (AECs) are isolated from the amnionic membrane of discarded placental tissues. Placental amnion has been widely used in medical history for its pro-epithelial and anti-inflammatory properties, in particular for the treatment of both chemical and thermal burns and corneal defects [93]. In more recent times, AECs have been isolated as single cells from the epithelial layer of the amniotic membrane and have been shown to retain pluripotent properties similar to ESCs. Unlike ESCs, however, human AECs do not form teratomas when injected into SCID mice and thus represent a safer therapeutic option for the treatment of a wide range of tissue-related disorders [94, 95]. Ongoing research indicates that AECs retain the ability to differentiate into tissue cell types from all three germ layers. Currently, most of the work pertaining to AEC has occurred in pre-clinical animal models and demonstrates promising outcomes in a wide range of disorders including liver disease [95–97] and neural disorders [98–101]. A clinical trial is now underway to assess the effect of transplanted, culture-derived AEC in repairing damaged ocular surfaces (Clinicaltrials.gov identifier NCT00344708). Recent studies have also indicated immunosuppressive properties of AEC, similar to that observed for MSC [102]. Importantly, both MSC and AEC appear to be immune privileged and thus are not encumbered by the problems of immune recognition, allowing clinical MSC transplants of both autologous and "off-the-shelf" allogeneic therapeutic products. The ease of isolation and harvest of these cells types, such as bone marrow or adipose biopsies for MSC isolation and discarded, full-term placental tissue for AEC extraction, are both relatively non-invasive and herald minor ethical considerations. Both AEC and MSC represent a viable and promising source of cell therapy for regenerative medicine.

Table 1.1 *Clinical Use of Mesenchymal Stem Cells—Published and Ongoing Clinical Trials*

Disease/condition	Intervention	Patient number/ clinical phase	Reference/clinical trial identifier (Clinicaltrials.gov)
Graft versus host disease	MSC infusion (I.V.), HLA-identical	1 patient	Le Blanc et al., 2004 [89]
	MSC infusion (I.V.), HLA-identical, haploidentical, HLA-mismatched	8 patients	Ringden et al., 2006 [92]
	MSC infusion, HLA-mismatched	2 children	Fang et al., 2007 [103]
	MSC infusion (I.V.), HLA-identical, haploidentical, HLA-mismatched	Phase II	Le Blanc et al., 2008 [88]
	MSC infusion, HLA-mismatched	1 child	Ball et al., 2008 [104]
	MSC infusion (I.V.), allogeneic	Phase I	NCT00361049
	MSC infusion (I.V.), allogeneic	Phase I/II	NCT00314483
	MSC infusion (I.V.), allogeneic	Phase I/II	NCT00447460
	MSC infusion (I.V.), Umbilical cord-derived	Phase I/II	NCT00749164
	MSC infusion (I.V.), OTI-010	Phase II	NCT00081055

(Contd.)

Table 1.1 (Contd.)

	MSC infusion (I.V.), Prochymal™	Phase II	NCT00504803
	MSC infusion (I.V.), Prochymal™	Phase II	NCT00603330
	MSC infusion (I.V.), Prochymal™	Phase II	NCT00476762
	MSC infusion (I.V.), Prochymal™	Phase II	NCT00136903
	MSC infusion (I.V.), Prochymal™	Phase II	NCT00284986
	MSC infusion (I.V.), Prochymal™	Phase III	NCT00366145
Haemopoietic engraftment			
Breast cancer	MSC infusion (I.V.), autologous	28 patients	Koc et al., 2000 [86]
Acute myelogenous leukaemia	MSC infusion (I.V.), HLA-mismatched	1 patient	Lee et al., 2002 [91]
Hematologic malignancy	MSC infusion (I.V.), HLA-identical	phase I	Lazarus et al., 2005 [87]
Leukaemia, aplastic anaemia, severe combined immunodeficiency (SCID)	HLA identical, haploidentical	7 patients	Le Blanc, 2007 [90]
Cardiac myopathy (total 7)			
Myocardial infarction	MSC infusion, Provacel™	Phase I	NCT00114452
Congestive heart failure	Intramyocardial MSC injection, autologous	Phase I/II	NCT00644410
Myocardial infarction	Intramyocardial MSC injection, autologous	Phase I/II	NCT00587990
Chronic myocardial ischemia	Intramyocardial MSC injection, autologous	Phase I/II	NCT00260338
Acute myocardial infarction	Transendocardial injection, allogeneic	Phase I/II	NCT00555828
Heart failure	Transendocardial injection, allogeneic	Phase II	NCT00721045
Heart failure	Intramyocardial MSC injection, autologous	Phase II	NCT00418418
Osteogenesis Imperfecta	MSC infusion (I.V.), allogeneic	6 children	Horwitz et al., 2002 [105]
	In-uterine transplantation of allogeneic MSC	1 patient	Le Blanc et al., 2005 [106]
	Bone marrow cell infusion (CD3 depleted), allogeneic	Pilot	NCT00187018
Chrohn's disease	MSC infusion (I.V.), Prochymal™	Phase II	NCT00294112
	MSC infusion (I.V.), Prochymal™	Phase III	NCT00543374
	MSC infusion (I.V.), Prochymal™	Phase III	NCT00482092
	MSC infusion (I.V.), Prochymal™	Phase III	NCT00609232

(Contd.)

Table 1.1 (Contd.)

Diabetes (type1)	-Cotransplantation of islet (allograft) and MSC infusion-autologous	Phase I/II	NCT00646724
	-MSC infusion (I.V.) in recently diagnosed type I diabetics, Prochymal™	Phase II	NCT00690066
Chronic obstructive pulmonary disease	MSC infusion (I.V.), Prochymal™	Phase II	NCT00683722
Multiple sclerosis	MSC infusion, autologous	Phase I/II	NCT00395200
End stage liver disease	MSC (differentiated into progenitor hepatocytes) infusion (portal vein), autologous	Phase I/II	NCT00420134
Systemic Lupis Erythematosus	MSC infusion (I.V.), autologous	Phase I/II	NCT00659217
	MSC infusion (I.V.), allogeneic	Phase I/II	NCT00698191
Decompensated cirrhosis	MSC infusion (I.V.), autologous	Phase II	NCT00476060
Tibial fracture	Local MSC implantation, autologous	Phase I/II	NCT00250302
Partial medial meniscectomy	Intra-articular MSC injection, autologous	Phase I/II	NCT00702741
Adult periodontitis	Implant of scaffold including MSC and ostoclasts	Phase I/II	NCT00221130
Multiple system atrophy	MSC infusion (intra-arterial and I.V.), autologous	18 patients	Lee et al., 2007 [83]

Note: I.V., intravenous.

CONCLUDING

From the earliest applications of regenerative medicine in BMT nearly five decades ago, this dynamic area of research has grown to include both adult and embryonic stem cells, fusing the continually evolving molecular biology with cutting edge sciences such as nanotechnology and tissue engineering. Thus regenerative medicine in this day and age incorporates both the transplantation of cells or synthesised material into the body, as well as aiding the body's natural regenerative capacity. It is not surprising that this field of science has expanded exponentially with our evolving understanding of signalling pathways vital to the differentiation of stem cells and better grasp of nanotechnology in order to build more complex scaffolds to include growth factors, ECM and other proteins. The prospect of *ex vivo* cell, tissue and organ genesis is conceivable.

Our ability to create better animal models for studying regeneration and graft acceptance has also played a vital role in the evolution of regenerative medicine. Tail amputations performed on the zebrafish model shed light on the role of Wnt/β-catenin in limb regeneration, and the production of "humanised" mice allows the testing of potential therapeutics in a system that mimics the human haematopoietic-lymphoid system, and although this does not exclude the need for large animal studies, they are vital in the speeding up of preclinical evaluation of novel agents.

While the potential for hESCs and adult stem cells in regenerative medicine is obvious, it is important to keep in mind the challenges that need to be overcome:

(i) allogenic transplantation of adult stem cells, hESCs and hESC-derived progenitors will still require donor tolerance induction by its recipient;

(ii) hESCs can form teratomas, for which grafts must be free of all undifferentiated cells prior to transplantation; and

(iii) the microenvironment of the grafted tissue must be conducive to the survival of the graft because the graft will be exposed to all the host factors that cause tissue damage in the first place.

Indeed the new revolution in regenerative medicine that the age of stem cells has instigated is tantalisingly close. It will be fascinating to revisit the contents of this excellent book in 5 and 10 years time—how accurate it was and what amazing discoveries are still in store!

ACKNOWLEDGEMENTS

We acknowledge generous research support from the Australian Stem Cell Centre, the National Health & Medical Research Council of Australia and Norwood Immunology.

REFERENCES

1. Fathke C., L. Wilson, J. Hutter, V. Kapoor, A. Smith, A. Hocking, F. Isik (2004). Contribution of bone marrow-derived cells to skin: collagen deposition and wound repair. *Stem Cells.* **22**: 812–822.
2. Chen Y.K., X.X. Zhao, J.G. Li, S. Lang, Y.M. Wang (2006). Ductular proliferation in liver tissues with severe chronic hepatitis B: an immunohistochemical study. *World J. Gastroenterol.* **12**: 1443–1446.
3. Koenig S., I. Probst, H. Becker, P. Krause (2006). Zonal hierarchy of differentiation markers and nestin expression during oval cell mediated rat liver regeneration. *Histochem Cell Biol.* **126**: 723–734.
4. McCann S.R. (2003). In: *The History of Bone Marrow Transplantation.* N.S. Hakim, V.E. Papalois (Eds.) Organ & Cell Transplantation. Imperial College Press, London, UK.
5. Osgood E.E., M.C. Riddle, T.J. Matthew (1939). Aplastic anemia treated with daily transfusions and intravenous marrow; case report. *Ann. Intern. Med.* **13**: 357–367.
6. Baar H.S. (1963). From Ehrlich-Pirquet to Medawar and Burnet; A revolution in immunology. *J. Main. Med. Assoc.* **54**: 209–214.
7. Carrel A. (1910). The replantation of the kidney and the spleen. *J Exp Med* **12**(2): 146–150.
8. Gorer P.A. (1937). The genetic and antigenic basis of tumour transplantation. *J. Pathol. Bacteriol.* **44**: 691–697.
9. Bell E., B. Ivarsson, C. Merrill (1979). Production of a tissue-like structure by contraction of collagen lattices by human fibroblasts of different proliferative potential *in vitro. Proc. Natl. Acad. Sci. USA.* **76**: 1274–1278.

10. Chick W.L., A.A. Like, V. Lauris, P.M. Galletti, P.D. Richardson, G. Panol, T.W. Mix, C.K. Colton (1975). A hybrid artificial pancreas. *Trans. Am. Soc. Artif. Intern. Organs.* **21**: 8–15.
11. Chick W.L., J.J. Perna, V. Lauris, D. Low, P.M. Galletti, G. Panol, A.D. Whittemore, A.A. Like, C.K. Colton, M.J. Lysaght (1977). Artificial pancreas using living beta cells: effects on glucose homeostasis in diabetic rats. *Science.* **197**: 780–782.
12. Langer R., J.P. Vacanti (1993). Tissue engineering. *Science.* **260**: 920–926.
13. Feynman R. (1959). There's Plenty of Room at the Bottom. *American Physical Society Caltech*, CA.
14. Taniguchi N., "On the Basic Concept of 'Nano-Technology'," *Proc. Intl. Conf. Prod. Eng. Tokyo, Part II, Japan Society of Precision Engineering*, 1974.
15. Boyan B.D., L.F. Bonewald, E.P. Paschalis, C.H. Lohmann, J. Rosser, D.L. Cochran, D.D. Dean, Z. Schwartz, A.L. Boskey (2002). Osteoblast-mediated mineral deposition in culture is dependent on surface microtopography. *Calcif Tissue Int.* **71**: 519–529.
16. Zinger O., G. Zhao, Z. Schwartz, J. Simpson, M. Wieland, D. Landolt, B. Boyan (2005). Differential regulation of osteoblasts by substrate microstructural features. *Biomaterials.* **26**: 1837–1847.
17. Gomez N., Y. Lu, S. Chen, C.E. Schmidt (2007). Immobilized nerve growth factor and microtopography have distinct effects on polarisation versus axon elongation in hippocampal cells in culture. *Biomaterials.* **28**: 271–284.
18. Teixeira A.I., G.A. McKie, J.D. Foley, P.J. Bertics, P.F. Nealey, C.J. Murphy (2006). The effect of environmental factors on the response of human corneal epithelial cells to nanoscale substrate topography. *Biomaterials.* **27**: 3945–3954.
19. Mastrobattista E., M.A. van der Aa, W.E. Hennink, D.J. Crommelin (2006). Artificial viruses: A nanotechnological approach to gene delivery. *Nat Rev Drug Discov.* **5**: 115–121.
20. Braude P., S.L. Minger, R.M. Warwick (2005). *Stem cell therapy: Hope or hype? BMJ* **330**: 1159–1160.
21. Kitzinger J., C. Williams (2005). Forecasting science futures: Legitimising hope and calming fears in the embryo stem cell debate. *Soc Sci Med.* **61**: 731–740.
22. Thomson J.A., J. Itskovitz-Eldor, S.S. Shapiro, M.A. Waknitz, J.J. Swiergiel, V.S. Marshall, J.M. Jones (1998). Embryonic stem cell lines derived from human blastocysts. *Science.* **282**: 1145–1147.
23. Reubinoff B.E., M.F. Pera, C.Y. Fong, A. Trounson, A. Bongso (2000). Embryonic stem cell lines from human blastocysts: Somatic differentiation *in vitro. Nat. Biotechnol.* **18**: 399–404.
24. Wang L., P. Menendez, C. Cerdan, M. Bhatia (2005). Hematopoietic development from human embryonic stem cell lines. *Exp Hematol.* **33**: 987–996.
25. Avery S., K. Inniss, H. Moore (2006). The regulation of self-renewal in human embryonic stem cells. *Stem Cells Dev.* **15**: 729–740.
26. Murry C.E., G. Keller (2008). Differentiation of embryonic stem cells to clinically relevant populations: lessons from embryonic development. *Cell.* **132**: 661–680.
27. Drukker M., G. Katz, A. Urbach, M. Schuldiner, G. Markel, J. Itskovitz-Eldor, B. Reubinoff, O. Mandelboim, N. Benvenisty (2002). Characterisation of the expression of MHC proteins in human embryonic stem cells. *Proc. Natl. Acad. Sci. USA.* **99**: 9864–9869.
28. Swijnenburg R.J., S. Schrepfer, J.A. Govaert, F. Cao, K. Ransohoff, A.Y. Sheikh, M. Haddad, A.J. Connolly, M.M. Davis, R.C. Robbins, J.C. Wu (2008). Immunosuppressive therapy mitigates immunological rejection of human embryonic stem cell xenografts. *Proc. Natl. Acad. Sci. USA.* **105**: 12991–12996.

29. Carlson M.E., I.M. Conboy (2007). Loss of stem cell regenerative capacity within aged niches. *Aging. Cell.* **6**: 371–382.
30. Nussbaum J., E. Minami, M.A. Laflamme, J.A. Virag, C.B. Ware, A. Masino, V. Muskheli, L. Pabon, H. Reinecke, C.E. Murry (2007). Transplantation of undifferentiated murine embryonic stem cells in the heart: Teratoma formation and immune response. *Faseb. J* **21**: 1345–1357.
31. Cabrera C.M., F. Cobo, A. Nieto, A. Concha (2006). Strategies for preventing immunologic rejection of transplanted human embryonic stem cells. *Cytotherapy.* **8**: 517–518.
32. Drukker M., N. Benvenisty (2004). The immunogenicity of human embryonic stem-derived cells. *Trends Biotechnol.* **22**: 136–141.
33. Taylor C.J., E.M. Bolton, S. Pocock, L.D. Sharples, R.A. Pedersen, J.A. Bradley (2005). Banking on human embryonic stem cells: estimating the number of donor cell lines needed for HLA matching. *Lancet.* **366**: 2019–2025.
34. Alexander S.I., N. Smith, M. Hu, D. Verran, A. Shun, S. Dornery, A. Smith, B. Webster, P.J. Shaw, A. Lammi, M.O. Stormon (2008). Chimerism and tolerance in a recipient of a deceased-donor liver transplant. *N. Engl. J. Med.* **358**: 369–374.
35. Helg C., B. Chapuis, J.F. Bolle, P. Morel, D. Salomon, E. Roux, V. Antonioli, M. Jeannet, M. Leski (1994). Renal transplantation without immunosuppression in a host with tolerance induced by allogeneic bone marrow transplantation. *Transplantation.* **58**: 1420–1422.
36. Kadry Z., B. Mullhaupt, E.L. Renner, P. Bauerfeind, U. Schanz, B.C. Pestalozzi, G. Studer, R. Zinkernagel, P.A. Clavien (2003). Living donor liver transplantation and tolerance: A potential strategy in cholangiocarcinoma. *Transplantation.* **76**: 1003–1006.
37. Kawai T., A.B. Cosimi, T.R. Spitzer, N. Tolkoff-Rubin, M. Suthanthiran, S.L. Saidman, J. Shaffer, F.I. Preffer, R. Ding, V. Sharma, J.A. Fishman, B. Dey, D.S. Ko, M. Hertl, N.B. Goes, W. Wong, W.W. Williams, Jr., R.B. Colvin, M. Sykes, D.H. Sachs (2008). HLA-mismatched renal transplantation without maintenance immunosuppression. *N. Engl. J. Med.* **358**: 353–361.
38. Sayegh M.H., N.A. Fine, J.L. Smith, H.G. Rennke, E.L. Milford, N.L. Tilney (1991). Immunologic tolerance to renal allografts after bone marrow transplants from the same donors. *Ann. Intern. Med.* **114**: 954–955.
39. Svendsen U.G., S. Aggestrup, C. Heilmann, N. Jacobsen, C. Koch, B. Larsen, A. Svejgaard, B. Thisted, G. Petterson (1995). Transplantation of a lobe of lung from mother to child following previous transplantation with maternal bone marrow. *Eur. Respir. J.* **8**: 334–337.
40. Flores K.G., J. Li, G.D. Sempowski, B.F. Haynes, L.P. Hale (1999). Analysis of the human thymic perivascular space during aging. *J. Clin. Invest.* **104**: 1031–1039.
41. Liang Y., G. Van Zant, S.J. Szilvassy (2005). Effects of aging on the homing and engraftment of murine hematopoietic stem and progenitor cells. *Blood.* **106**: 1479–1487.
42. Rossi D.J., D. Bryder, J.M. Zahn, H. Ahlenius, R. Sonu, A.J. Wagers, I.L. Weissman (2005). Cell intrinsic alterations underlie hematopoietic stem cell aging. *Proc. Natl. Acad. Sci. USA.* **102**: 9194–9199.
43. Scadden D.T. (2006). The stem-cell niche as an entity of action. *Nature.* **441**: 1075–1079.
44. Stephan R.P., C.R. Reilly, P.L. Witte (1998). Impaired ability of bone marrow stromal cells to support B-lymphopoiesis with age. *Blood.* **91**: 75–88.
45. Tsuboi I., K. Morimoto, Y. Hirabayashi, G.X. Li, S. Aizawa, K.J. Mori, J. Kanno, T. Inoue (2004). Senescent B lymphopoiesis is balanced in suppressive homeostasis: decrease in interleukin-7 and

transforming growth factor-beta levels in stromal cells of senescence-accelerated mice. *Exp. Biol. Med. (Maywood)* **229**: 494–502.

46. Xing Z., M.A. Ryan, D. Daria, K.J. Nattamai, G. Van Zant, L. Wang, Y. Zheng, H. Geiger (2006). Increased hematopoietic stem cell mobilisation in aged mice. *Blood*. **108**: 2190–2197.
47. Yilmaz O.H., M.J. Kiel, S.J. Morrison (2006). SLAM family markers are conserved among hematopoietic stem cells from old and reconstituted mice and markedly increase their purity. *Blood*. **107**: 924–930.
48. Adams G.B., R.P. Martin, I.R. Alley, K.T. Chabner, K.S. Cohen, L.M. Calvi, H.M. Kronenberg, D.T. Scadden (2007). Therapeutic targeting of a stem cell niche. *Nat. Biotechnol*. **25**: 238–243.
49. Chidgey A.P., D. Layton, A. Trounson, R.L. Boyd (2008). Tolerance strategies for stem-cell-based therapies. *Nature*. **453**: 330–337.
50. van den Brink M.R., O. Alpdogan, R.L. Boyd (2004). Strategies to enhance T-cell reconstitution in immunocompromised patients. *Nat. Rev. Immunol*. **4**: 856–867.
51. Zakrzewski J.L., A.A. Kochman, S.X. Lu, T.H. Terwey, T.D. Kim, V.M. Hubbard, S.J. Muriglan, D. Suh, O.M. Smith, J. Grubin, N. Patel, A. Chow, J. Cabrera-Perez, R. Radhakrishnan, A. Diab, M.A. Perales, G. Rizzuto, E. Menet, E.G. Pamer, G. Heller, J.C. Zuniga-Pflucker, O. Alpdogan, M.R. van den Brink (2006). Adoptive transfer of T-cell precursors enhances T-cell reconstitution after allogeneic hematopoietic stem cell transplantation. *Nat. Med*. **12**: 1039–1047.
52. Steiner D., N. Brunicki, B.R. Blazar, E. Bachar-Lustig, Y. Reisner (2006). Tolerance induction by third-party "off-the-shelf" CD4+CD25+ Treg cells. *Exp. Hematol*. **34**: 66–71.
53. Tu W., Y.L. Lau, J. Zheng, Y. Liu, P.L. Chan, H. Mao, K. Dionis, P. Schneider, D.B. Lewis (2008). Efficient generation of human alloantigen-specific CD4+ regulatory T cells from naive precursors by CD40-activated B cells. *Blood*. **112**: 2554–2562.
54. Zhang X., M. Li, D. Lian, X. Zheng, Z.X. Zhang, T.E. Ichim, X. Xia, X. Huang, C. Vladau, M. Suzuki, B. Garcia, A.M. Jevnikar, W.P. Min (2008). Generation of therapeutic dendritic cells and regulatory T cells for preventing allogeneic cardiac graft rejection. *Clin. Immunol*. **127**: 313–321.
55. Goldberg G.L., O. Alpdogan, S.J. Muriglan, M.V. Hammett, M.K. Milton, J.M. Eng, V.M. Hubbard, A. Kochman, L.M. Willis, A.S. Greenberg, K.H. Tjoe, J.S. Sutherland, A. Chidgey, M.R. van den Brink, R.L. Boyd (2007). Enhanced immune reconstitution by sex steroid ablation following allogeneic hemopoietic stem cell transplantation. *J. Immunol*. **178**: 7473–7484.
56. Goldberg G.L., J. Dudakov, J.J. Reiseger, N. Seach, T. Ueno, K. Vlahos, M.V. Hammett, L. Young, T.S.P. Heng, R.L. Boyd, A.P. Chidgey (2009). Sex steroid ablation enhances immune reconstitution following cytotoxic antineoplastic therapy in young mice. *J. Pathol*.
57. Goldberg G.L., J.S. Sutherland, M.V. Hammet, M.K. Milton, T.S. Heng, A.P. Chidgey, R.L. Boyd (2005). Sex steroid ablation enhances lymphoid recovery following autologous hematopoietic stem cell transplantation. *Transplantation*. **80**: 1604–1613.
58. Greenstein B.D., F.T. Fitzpatrick, M.D. Kendall, M.J. Wheeler (1987). Regeneration of the thymus in old male rats treated with a stable analogue of LHRH. *J Endocrinol*. **112**: 345–350.
59. Marchetti B., V. Guarcello, M.C. Morale, G. Bartoloni, Z. Farinella, S. Cordaro, U. Scapagnini (1989). Luteinizing hormone-releasing hormone-binding sites in the rat thymus: characteristics and biological function. *Endocrinology*. **125**: 1025–1036.
60. Sutherland J.S., G.L. Goldberg, M.V. Hammett, A.P. Uldrich, S.P. Berzins, T.S. Heng, B.R. Blazar, J.L. Millar, M.A. Malin, A.P. Chidgey, R.L. Boyd (2005). Activation of thymic regeneration in mice and humans following androgen blockade. *J Immunol*. **175**: 2741–2753.

61. Sutherland J.S., L. Spyroglou, J.L. Muirhead, T.S. Heng, A. Prieto-Hinojosa, H.M. Prince, A.P. Chidgey, A.P. Schwarer, R.L. Boyd (2008). Enhanced Immune System Regeneration in Humans Following Allogeneic or Autologous Hemopoietic Stem Cell Transplantation by Temporary Sex Steroid Blockade. *Clin Cancer Res.* **14**: 1138–1149.

62. Chidgey A., J. Dudakov, N. Seach, R. Boyd (2007). Impact of niche aging on thymic regeneration and immune reconstitution. *Semin Immunol.* **19**: 331–340.

63. Nakagawa M., M. Koyanagi, K. Tanabe, K. Takahashi, T. Ichisaka, T. Aoi, K. Okita, Y. Mochiduki, N. Takizawa, S. Yamanaka (2008). Generation of induced pluripotent stem cells without Myc from mouse and human fibroblasts. *Nat. Biotechnol.* **26**: 101–106.

64. Park I.H., R. Zhao, J.A. West, A. Yabuuchi, H. Huo, T.A. Ince, P.H. Lerou, M.W. Lensch, G.Q. Daley (2008). Reprogramming of human somatic cells to pluripotency with defined factors. *Nature.* **451**: 141–146.

65. Takahashi K., K. Tanabe, M. Ohnuki, M. Narita, T. Ichisaka, K. Tomoda, S. Yamanaka (2007). Induction of pluripotent stem cells from adult human fibroblasts by defined factors. *Cell.* **131**: 861–872.

66. Yu J., M.A. Vodyanik, K. Smuga-Otto, J. Antosiewicz-Bourget, J.L. Frane, S. Tian, J. Nie, G.A. Jonsdottir, V. Ruotti, R. Stewart, Slukvin, II, J.A. Thomson (2007). Induced pluripotent stem cell lines derived from human somatic cells. *Science.* **318**: 1917–1920.

67. Braga L.M., K. Rosa, B. Rodrigues, C. Malfitano, M. Camassola, P. Chagastelles, S. Lacchini, P. Fiorino, K. De Angelis, B.D. Schaan, M.C. Irigoyen, N.B. Nardi (2008). Systemic delivery of adult stem cells improves cardiac function in spontaneously hypertensive rats. *Clin. Exp. Pharmacol. Physiol.* **35**: 113–119.

68. Endres M., K. Neumann, T. Haupl, C. Erggelet, J. Ringe, M. Sittinger, C. Kaps (2007). Synovial fluid recruits human mesenchymal progenitors from subchondral spongious bone marrow. *J. Orthop. Res.* **25**: 1299–1307.

69. Hatch H.M., D. Zheng, M.L. Jorgensen, B.E. Petersen (2002). SDF-1alpha/CXCR4: a mechanism for hepatic oval cell activation and bone marrow stem cell recruitment to the injured liver of rats. *Cloning Stem Cells.* **4**: 339–351.

70. Kopp H.G., C.A. Ramos, S. Rafii (2006). Contribution of endothelial progenitors and proangiogenic hematopoietic cells to vascularisation of tumour and ischemic tissue. Curr Opin Hematol **13**: 175–181.

71. Mimeault M., S.K. Batra (2008). Recent progress on tissue-resident adult stem cell biology and their therapeutic implications. *Stem Cell Rev.* **4**: 27–49.

72. Brack A.S., I.M. Conboy, M.J. Conboy, J. Shen, T.A. Rando (2008). A temporal switch from notch to Wnt signaling in muscle stem cells is necessary for normal adult myogenesis. *Cell Stem Cell.* **2**: 50–59.

73. Gurtner G.C., S. Werner, Y. Barrandon, M.T. Longaker (2008). Wound repair and regeneration. *Nature.* **453**: 314–321.

74. Sun X., X. Fu, Z. Sheng (2007). Cutaneous stem cells: something new and something borrowed. *Wound Repair Regen.* **15**: 775–785.

75. Hall B., J. Dembinski, A.K. Sasser, M. Studeny, M. Andreeff, F. Marini (2007). Mesenchymal stem cells in cancer: tumour-associated fibroblasts and cell-based delivery vehicles. *Int. J. Hematol.* **86**: 8–16.

76. Kucerova L., V. Altanerova, M. Matuskova, S. Tyciakova, C. Altaner (2007). Adipose tissue-derived human mesenchymal stem cells mediated prodrug cancer gene therapy. *Cancer Res.* **67**: 6304–6313.

77. Mapara K.Y., C.B. Stevenson, R.C. Thompson, M. Ehtesham (2007). Stem cells as vehicles for the treatment of brain cancer. *Neurosurg Clin. N. Am.* **18**: 71–80, ix.
78. Yu J.J., X. Sun, X. Yuan, J.W. Lee, E.Y. Snyder, J.S. Yu (2006). Immunomodulatory neural stem cells for brain tumour therapy. *Expert. Opin. Biol. Ther.* **6**: 1255–1262.
79. Jiang Y., B.N. Jahagirdar, R.L. Reinhardt, R.E. Schwartz, C.D. Keene, X.R. Ortiz-Gonzalez, M. Reyes, T. Lenvik, T. Lund, M. Blackstad, J. Du, S. Aldrich, A. Lisberg, W.C. Low, D.A. Largaespada, C.M. Verfaillie (2002). Pluripotency of mesenchymal stem cells derived from adult marrow. *Nature.* **418**: 41–49.
80. Wang G., B.A. Bunnell, R.G. Painter, B.C. Quiniones, S. Tom, N.A. Lanson, Jr., J.L. Spees, D. Bertucci, A. Peister, D.J. Weiss, V.G. Valentine, D.J. Prockop, J.K. Kolls (2005). Adult stem cells from bone marrow stroma differentiate into airway epithelial cells: potential therapy for cystic fibrosis. *Proc. Natl. Acad. Sci. USA.* **102**: 186–191.
81. Kopen G.C., D.J. Prockop, D.G. Phinney (1999). Marrow stromal cells migrate throughout forebrain and cerebellum, and they differentiate into astrocytes after injection into neonatal mouse brains. *Proc. Natl. Acad. Sci. USA.* **96**: 10711–10716.
82. Brooke G., M. Cook, C. Blair, R. Han, C. Heazlewood, B. Jones, M. Kambouris, K. Kollar, S. McTaggart, R. Pelekanos, A. Rice, T. Rossetti, K. Atkinson (2007). Therapeutic applications of mesenchymal stromal cells. *Sem. Cell. Dev. Biol.* **18**: 846–858.
83. Lee P.H., J.W. Kim, O.Y. Bang, Y.H. Ahn, I.S. Joo, K. Huh (2007). Autologous Mesenchymal Stem Cell Therapy Delays the Progression of Neurological Deficits in Patients With Multiple System Atrophy. *Clin. Pharmacol. Ther.* **83**: 723–730.
84. Blanc K. Le, O. Ringden (2007). Immunomodulation by mesenchymal stem cells and clinical experience. *J. Intern Med.* **262**: 509–525.
85. Fouillard L., A. Chapel, D. Bories, S. Bouchet, J.M. Costa, H. Rouard, P. Herve, P. Gourmelon, D. Thierry, M. Lopez, N.C. Gorin (2007). Infusion of allogeneic-related HLA mismatched mesenchymal stem cells for the treatment of incomplete engraftment following autologous haematopoietic stem cell transplantation. *Leukemia.* **21**: 568–570.
86. Koc O.N., S.L. Gerson, B.W. Cooper, S.M. Dyhouse, S.E. Haynesworth, A.I. Caplan, H.M. Lazarus (2000). Rapid hematopoietic recovery after coinfusion of autologous-blood stem cells and culture-expanded marrow mesenchymal stem cells in advanced breast cancer patients receiving high-dose chamotherapy. *J. Clin. Oncol.* **18**: 307–316.
87. Lazarus H.M., O.N. Koc, S.M. Devine, P. Curtin, R.T. Maziarz, H.K. Holland, E.J. Shpall, P. McCarthy, K. Atkinson, B.W. Cooper, S.L. Gerson, M.J. Laughlin, F.R. Loberiza, Jr., A.B. Moseley, A. Bacigalupo (2005). Cotransplantation of HLA-identical sibling culture-expanded mesenchymal stem cells and hematopoietic stem cells in hematologic malignancy patients. *Biol. Blood. Marrow. Transplant.* **11**: 389–398.
88. Blanc K. Le, F. Frassoni, L. Ball, F. Locatelli, H. Roelofs, I. Lewis, E. Lanino, B. Sundberg, M.E. Bernardo, M. Remberger, G. Dini, R.M. Egeler, A. Bacigalupo, W. Fibbe, O. Ringden (2008). Mesenchymal stem cells for treatment of steroid-resistant, severe, acute graft-versus-host disease: A phase II study. *Lancet.* **371**: 1579–1586.
89. Blanc K. Le, I. Rasmusson, B. Sundberg, C. Gotherstrom, M. Hassan, M. Uzunel, O. Ringden (2004). Treatment of severe acute graft-versus-host disease with third party haploidentical mesenchymal stem cells. *Lancet.* **363**: 1439–1441.

90. Blanc K. Le, H. Samuelsson, B. Gustafsson, M. Remberger, B. Sundberg, J. Arvidson, P. Ljungman, H. Lonnies, S. Nava, O. Ringden (2007). Transplantation of mesenchymal stem cells to enhance engraftment of hematopoietic stem cells. *Leukemia.* **21**: 1733–1738.

91. Lee S.T., J.H. Jang, J.W. Cheong, J.S. Kim, H.Y. Maemg, J.S. Hahn, Y.W. Ko, Y.H. Min (2002). Treatment of high-risk acute myelogenous leukaemia by myeloablative chemoradiotherapy followed by co-infusion of T cell-depleted haematopoietic stem cells and culture-expanded marrow mesenchymal stem cells from a related donor with one fully mismatched human leucocyte antigen haplotype. *Br. J. Haematol.* **118**: 1128–1131.

92. Ringden O., M. Uzunel, I. Rasmusson, M. Remberger, B. Sundberg, H. Lonnies, H.U. Marschall, A. Dlugosz, A. Szakos, Z. Hassan, B. Omazic, J. Aschan, L. Barkholt, K. Le Blanc (2006). Mesenchymal stem cells for treatment of therapy-resistant graft-versus-host disease. *Transplantation.* 81: 1390–1397.

93. Sippel K.C., M.J. J., F.C. S. (2001). Amniotic membrane surgery. *Curr Opin Ophthalmol.* **12**: 269–281.

94. Ilancheran S., A. Michalska, G. Peh, E.M. Wallace, M. Pera, U. Manuelpillai (2007). Stem cells derived from human Foetal membranes display multilineage differentiation potential. *Biol. Reprod.* **77**: 577–588.

95. Miki T., T. Lehmann, H. Cai, D.B. Stolz, S.C. Strom (2005). Stem cell characteristics of amniotic epithelial cells. *Stem Cells.* **23**: 1549–1559.

96. Sakuragawa N., S. Enosawa, T. Ishii, R. Thangavel, T. Tashiro, T. Okuyama, S. Suzuki (2000). Human amniotic epithelial cells are promising transgene carriers for allogeneic cell transplantation into liver. *J. Hum. Genet.* **45**: 171–176.

97. Takashima S., H. Ise, P. Zhao, T. Akaike, T. Nikaido (2004). Human amniotic epithelial cells possess hepatocyte-like characteristics and functions. *Cell Struct. Funct.* **29**: 73–84.

98. Elwan M.A., N. Sakuragawa (1997). Evidence for synthesis and release of catecholamines by human amniotic epithelial cells. *Neuroreport.* **8**: 3435–3438.

99. Kakishita K., M.A. Elwan, N. Nakao, T. Itakura, N. Sakuragawa (2000). Human amniotic epithelial cells produce dopamine and survive after implantation into the striatum of a rat model of Parkinson's disease: A potential source of donor for transplantation therapy. *Exp. Neurol.* **165**: 27–34.

100. Kakishita K., N. Nakao, N. Sakuragawa, T. Itakura (2003). Implantation of human amniotic epithelial cells prevents the degeneration of nigral dopamine neurons in rats with 6-hydroxydopamine lesions. *Brain. Res.* **980**: 48–56.

101. Sakuragawa N., H. Misawa, K. Ohsugi, K. Kakishita, T. Ishii, R. Thangavel, J. Tohyama, M. Elwan, Y. Yokoyama, O. Okuda, H. Arai, I. Ogino, K. Sato (1997). Evidence for active acetylcholine metabolism in human amniotic epithelial cells: Applicable to intracerebral allografting for neurologic disease. *Neurosci. Lett.* **232**: 53–56.

102. Banas R.A., C. Trumpower, C. Bentlejewski, V. Marshall, G. Sing, A. Zeevi (2008). Immunogenicity and immunomodulatory effects of amnion-derived multipotent progenitor cells. *Hum. Immunol.* **69**: 321–328.

103. Fang B., Y. Song, Q. Lin, Y. Zhang, Y. Cao, R.C. Zhao, Y. Ma (2007). Human adipose tissue-derived mesenchymal stromal cells as salvage therapy for treatment of severe refractory acute graft-vs.-host disease in two children. *Pediatr. Transplant.* **11**: 814–817.

104. Ball L., R. Bredius, A. Lankester, J. Schweizer, M. van den Heuvel-Eibrink, H. Escher, W. Fibbe, M. Egeler (2008). Third party mesenchymal stromal cell infusions fail to induce tissue repair despite

successful control of severe grade IV acute graft-versus-host disease in a child with juvenile myelo-monocytic leukemia. *Leukemia.* **22**: 1256–1257.

105. Horwitz E.M., P.L. Gordon, W.K. Koo, J.C. Marx, M.D. Neel, R.Y. McNall, L. Muul, T. Hofmann (2002). Isolated allogeneic bone marrow-derived mesenchymal cells engraft and stimulate growth in children with osteogenesis imperfecta: Implications for cell therapy of bone. *Proc. Natl. Acad. Sci. USA.* **99**: 8932–8937.

106. Blanc K. Le, C. Gotherstrom, O. Ringden, M. Hassan, R. McMahon, E. Horwitz, G. Anneren, O. Axelsson, J. Nunn, U. Ewald, S. Norden-Lindeberg, M. Jansson, A. Dalton, E. Astrom, M. Westgren (2005). Foetal mesenchymal stem-cell engraftment in bone after in utero transplantation in a patient with severe osteogenesis imperfecta. *Transplantation.* **79**: 1607–1614.

PART 1

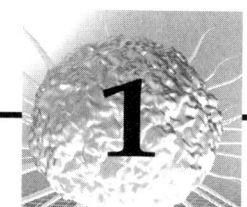

Embryonic Stem Cells

2. Zebrafish Stem Cell
3. Medaka Stem Cells
4. Chicken Embryonic Stem Cells as Another Non-mammalian Embryonic Stem Cell Type
5. Derivation, Culture and Genetic Modification of Mouse Embryonic Stem Cells
6. Derivation of Human Embryonic Stem Cells from Blastocysts
7. Developing a Serum-Free Platform and a Novel Reporter System for hESCs
8. Epigenetics and Embryonic Stem Cells
9. Embryonic Stem Cells as a Therapy for Diabetes
10. Manipulation of Biomaterial Interface by Biomimetic Strategies: Possibilities of Exploring Stem Cells
11. Immunomodulatory Aspects of Human Embryonic and Adult Stem Cells
12. Human Embryonic Stem Cells as a Model System to Study Human Genetics
13. Germ Cells and Artificial Gametes Production from Embryonic Stem Cells
14. Using Embryonic Stem Cells to Introduce Mutations into the Mouse Germ Line: Animal Models and Future Applications
15. Application of Human Embryonic Stem Cells in Drug Discovery

2
ZEBRAFISH STEM CELLS

*Ten-Tsao Wong, Lianchun Fan, Paul Collodi**
Department of Animal Sciences, Purdue University, West Lafayette, IN 47907

INTRODUCTION

This chapter discusses about the derivation, selection, maintenance and the application of zebrafish embryonic stem (ES) cell and primordial germ cell (PGC) cultures. A key component of the cell culture system is the feeder cells derived from the rainbow trout spleen cell line (RTS34st), which is required to maintain pluripotency and germ-line competency of the ES cells and PGCs. Pluripotency of the stem cells is further enhanced by using RTS34st feeders that were engineered to express recombinant zebrafish kit ligand a (Kitlga). Addition of Kitlga to the cultures increases the number of passages during which the stem cells maintain characteristics of pluripotency such as a high level of alkaline phosphatase activity and differentiated embryoid body formation. For gene-targeting experiments, both the ES cell and PGC cultures can incorporate plasmid DNA in a targeted fashion by homologous recombination. The zebrafish PGC cultures continue to express germ cell markers after multiple passages and they possess a greater capacity to contribute to the germ line than ES cells following transplantation into host embryos. These latter characteristics should make the PGC cultures a valuable tool for developing gene-targeting approach with zebrafish.

Due to its many favourable characteristics, the zebrafish has become an established model for *in vivo* studies of vertebrate development. The zebrafish is easy to maintain in the laboratory, has a short generation time and produces a large number of embryos throughout the year. The transparent embryos develop *ex vivo* within a relatively short period of time, making it convenient to use them for a wide range of experimental manipulations [1]. In addition to studies of embryogenesis, neurobiology [2–5], and tissue regeneration [6–8], more recently the zebrafish has

---

* pcollodi@purdue.edu

also served as a model for investigating genetic diseases [9, 10] including cancer [11, 12] and congenital heart defects [13–15], as well as a tool for drug discovery [16–18]. For genetic studies, more than two thousand mutations that perturb normal zebrafish development have been generated through large-scale mutagenesis screens [19–21]. Together with chemical- or radiation-induced mutagenesis, positional cloning strategies have been applied to recover some of the disrupted genes [22, 23]. Also, retrovirus- or transposon-mediated insertional mutagenesis methods have been established to facilitate the isolation of the mutated genes [24–27]. In addition to methods for random mutagenesis, other molecular tools and methods such as high-density genetic maps, expressed sequence tags (ESTs) [28, 29], and commercially available large insert yeast, bacterial, and P1 artificial chromosome libraries [30] that are important for genomics research are available with the zebrafish.

Targeted inhibition of gene expression is commonly accomplished in zebrafish through the antisense phosphoramidite morpholino oligomers (MOs) to block translation or processing of the target mRNA [31]. This approach is applied to efficiently knock-down gene expression during embryogenesis; however, the transient nature of the inhibition makes the approach less suitable for targeting genes expressed late in development or in the adult [32]. Another genetic approach that has been applied to zebrafish is referred to as target-induced local lesions in genomes (TILLING) that involves the identification of mutations at a specific locus by high-throughput screening of DNA from a large number of randomly generated mutant fish lines [33]. Very recently targeted mutations have been introduced in zebrafish using zinc finger nucleases (ZFNs). ZFNs introduce gene disruptions by non-homologous end-joining re-ligation that results in a gene lesion by frameshift [34, 35]. As this approach becomes widely accessible, it should provide a powerful method for generating knockout lines of zebrafish by direct embryo manipulation. Since ZFNs have also been used to target genes by homologous recombination in cell culture [36, 37], it is feasible to combine this approach with stem cell mediated germ-line chimera production to generate both knockout and knockin lines of fish.

In mice, the use of pluripotent ES cell cultures to produce knockout mutants has provided a valuable approach to the study of gene function during embryogenesis and growth [32, 38]. ES cells that have undergone the targeting event are selected and grown in culture, and the genetic alteration is transferred to the germ line of a host embryo when the cultured cells are transplanted into the embryo [39]. As the sequence of the zebrafish genome is nearly complete (*Danio rerio* sequence project: www.sanger.ac.uk/Projects/ D_rerio/), the ability to employ this sequence information for introducing targeted mutations would greatly facilitate studies of gene function in this model system. A gene-targeting strategy involving homologous recombination in pluripotent cells combined with the zinc finger approach would provide *in vitro* models of zebrafish gene expression and regulation, and a method for the production of knockout/knockin lines of fish.

To establish a zebrafish gene-targeting approach by homologous recombination, methods were developed to derive zebrafish ES cell lines that remain germ-line competent for multiple passages in culture [40–42]. The extremely low efficiency of germ-line transmission and germ-line chimera formation using the ES cell cultures, however, led us to develop methods for the derivation of PGC cultures that are initiated from late-stage embryos. Since the PGCs are committed to the germ-cell lineage early during embryogenesis, it was postulated that cell cultures initiated from a homogeneous population of PGCs isolated from later-stage embryos would have the capacity to generate germ-line chimeras at a higher frequency than ES cells following transplantation into a host embryo. Transplantation studies conducted so far indicate that the PGCs do contribute to the germ-line following their introduction into a host embryo (as discussed further). In order to perform the gene-targeting approach, inactivation of the targeted gene is accomplished in the stem cell culture by the targeted insertion of foreign DNA into the coding region of the gene by homologous recombination [43]. Cells that have undergone the targeting event are selected and grown in culture, and the genetic alteration is transferred to the germ line of a host embryo when the cultured cells are transplanted into the embryo, where they contribute to the germ-cell lineage (Fig. 2.1). We have demonstrated that the zebrafish ES cell and PGC cultures can undergo homologous recombination at a frequency that is sufficient to allow the selection of targeted colonies for expansion. The major obstacle with implementing this approach to date, however, has been to isolate colonies of homologous recombinants that retain the capacity to generate germ-line chimeras when transplanted into a host embryo. In the following sections we describe the procedures for zebrafish ES cell and PGC isolation, culture and genetic manipulation.

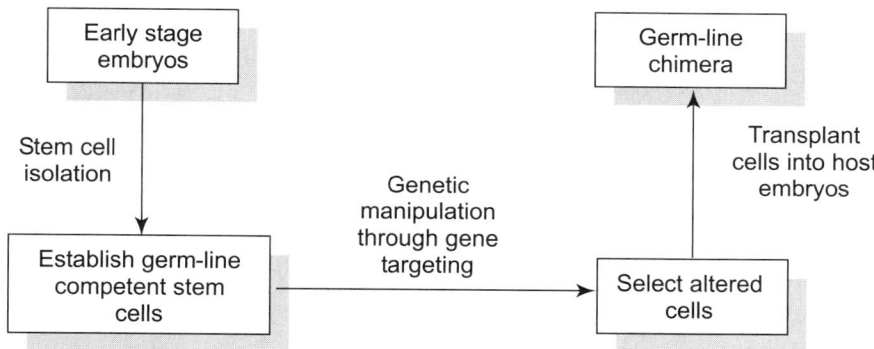

▶ Fig. 2.1 *Cell-mediated gene targeting using germ-line competent stem cell cultures. Gene inactivation is accomplished in the stem cell cultures by the targeted insertion of foreign DNA into the coding region of the gene by homologous recombination. The altered cells are selected in culture and reintroduced into a host embryo where they colonize the germ line. The resulting chimera is used to generate a knockout or knockin line.*

PROTOCOL FOR THE INITIATION AND MAINTENANCE OF ZEBRAFISH PRIMARY EMBRYO CELL CULTURES

This section describes the general protocol for initiating primary cultures from zebrafish embryos for either ES cell or PGC cultures. The following sections will discuss specific methods for each cell type. Zebrafish embryos at the desired stage of development are collected, rinsed several times with water to remove debris, and transferred to a Petri dish (Fig. 2.2). The embryos are further divided into groups of approximately 50 individuals, and transferred into a 1.5 mL microcentrifuge tube that has the bottom removed and replaced with fine mesh. Each tube containing embryos sitting on the net is submerged into 70% ethanol for 10 s and then rinsed immediately by submerging the embryos into a beaker of sterile egg water. After removing the dead individuals, the embryos are rinsed three times with LDF medium (50% Leibowitz's L-15, 35% Dulbecco's Modified Eagle's, and 15% Ham's F12 media, GIBCO-BRL) and then treated with bleach solution (0.5% Chlorox bleach in water) for 2 min, followed immediately by a rinse in LDF. The bleach treatment and rinse steps are repeated two additional times.

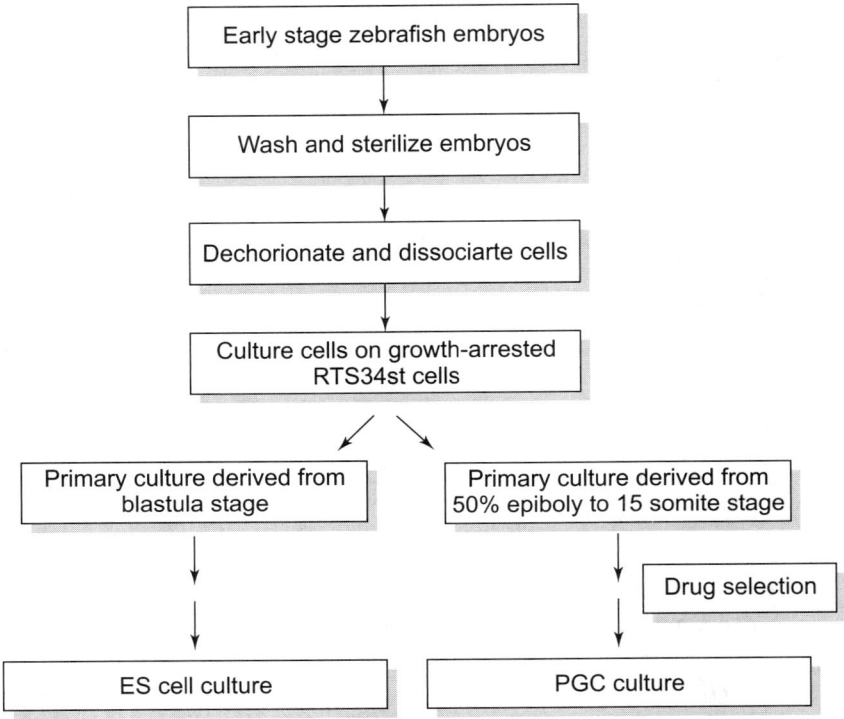

▶ **Fig. 2.2** *General protocol for the derivation of zebrafish ES cell and PGC culture initiated from zebrafish embryos.*

Following the final bleach treatment, the embryos are rinsed three times more with LDF medium. The chorions are removed by incubating each group of embryos in pronase (0.5 mg/mL) dissolved in Hank's solution for approximately 15 min or until the chorions begin to break apart. The suspended embryos are gently swirled in the dish to release them from the digested chorion. Then the dechorionated embryos are rinsed with LDF medium and treated with trypsin/ethylenediaminetetraacetic acid (EDTA) solution (0.2% trypsin and 1 µM EDTA in phosphate-buffered saline (PBS)) for about 2 min to dissociate the cells. The trypsin is stopped by adding approximately 300 µL of fetal bovine serum (FBS) and the cells are collected by centrifugation (800 g, 5 min). The cell pellet obtained from each group of approximately 50 embryos is re-suspended in 1.8 mL LDF medium and transferred to a single well of a six-well tissue culture plate containing a confluent monolayer of growth-arrested RTS34st cells [44]. The plate is then left undisturbed for 30 min to enable the embryo cells to attach to the RTS34st monolayer. After the cells have attached, the following supplements are added at the final concentration indicated—FBS (5%), zebrafish embryo extract (40 µg protein/mL), trout plasma (1%, East Coast Biologics, North Berwick, ME), bovine insulin (10 µg/mL, Sigma, St Louise, MO), human epidermal growth factor (EGF, 50 ng/mL, Invitrogen, CA), human basic fibroblast growth factor (bFGF, 30 ng/mL, Invitrogen, CA), and RTS34st-conditioned medium (30%). The culture is incubated at 22°C and the cell aggregates increase in size as the cells proliferate. As the cell aggregates grow larger, they should continue to possess a homogeneous appearance and be composed of tightly adherent cells. Although zebrafish cell cultures are normally propagated at 26–28°C, the embryo cell cultures are maintained at 22°C to accommodate the feeder layer of trout spleen cells.

To passage the primary culture, the cells from each well are harvested by adding 2 mL of trypsin/EDTA solution and incubating for approximately 30 s. The cells are dislodged from the culture surface by gentle pipetting and the resulting suspension of cell aggregates is partially dissociated by pipetting up and down several times before transferring the suspension to a 15 mL centrifuge tube. The action of the trypsin is stopped by adding 0.2 mL of FBS to the cell suspension. The cells are collected by centrifugation (800 g, 5 min) and the cell pellet is re-suspended into 3.6 mL of LDF medium. A portion of the cell suspension (1.8 mL) is added to each of two wells of a six-well plate containing a confluent monolayer of growth-arrested RTS34st cells, and the other supplements (listed earlier) are added to each well after the cells have attached (approximately 30 min). The plate is incubated at 22°C and the medium is changed every 5 days, and fresh growth-arrested feeder cells are added every 2 weeks. The cultures can be cryopreserved when they begin to grow as a monolayer (about passage 4) and a portion of the culture can be frozen at each passage.

DERIVATION OF ZEBRAFISH ESC CULTURES

Primary ES cell cultures can be initiated from zebrafish embryos at either the blastula- or germ-ring stage of development and maintained on a feeder layer of growth-arrested RTS34st (Fig. 2.3A). Since the blastula consists entirely of non-differentiated cells, it is the optimal stage to initiate the ES cell cultures. The primary cultures initiated from blastulas consist of multiple dense cell aggregates that possess a homogeneous appearance and lack any morphological indication of differentiation. As the cells proliferate, the aggregates continue to increase in size without losing their homogeneous, dense appearance (Fig. 2.3B). In contrast, because cell differentiation has begun to occur during gastrulation, the majority of cell aggregates in cultures initiated from the germ-ring embryos will not appear homogeneous and will contain recognizable differentiated cell types, including pigmented melanocytes, neural cells, and fibroblasts. As such, it is necessary to manually select the individual cell aggregates that possess an ES-like morphology for further passage from the primary germ-ring cultures to establish pluripotent ES cell cultures. Using these methods, continuously growing ES cell lines have been maintained for more than 200 population doublings in conditions that were previously shown to promote cell growth and preserve pluripotency of the cells [42]. The cell culture conditions consist of LDF basal nutrient medium supplemented with human EGF, human bFGF, bovine insulin, FBS, trout serum, and a feeder layer of growth-arrested RTS34st. The growth-arrested RTS34st feeder cells secrete factor(s) that maintain the ES cells in a germ-line competent condition. ES cells co-cultured in the presence of the RTS34st feeder layer exhibit several *in vitro* characteristics associated with pluripotency, including alkaline phosphatase activity, recognition by the SSEA-1 antibody, and the capacity to form differentiated embryoid bodies in suspension culture. A small percentage of the zebrafish ES cells also maintain the capacity to contribute to the germ-cell lineage of a recipient embryo when they are grown on a feeder layer for at least six passages (6 weeks) in culture [43, 45].

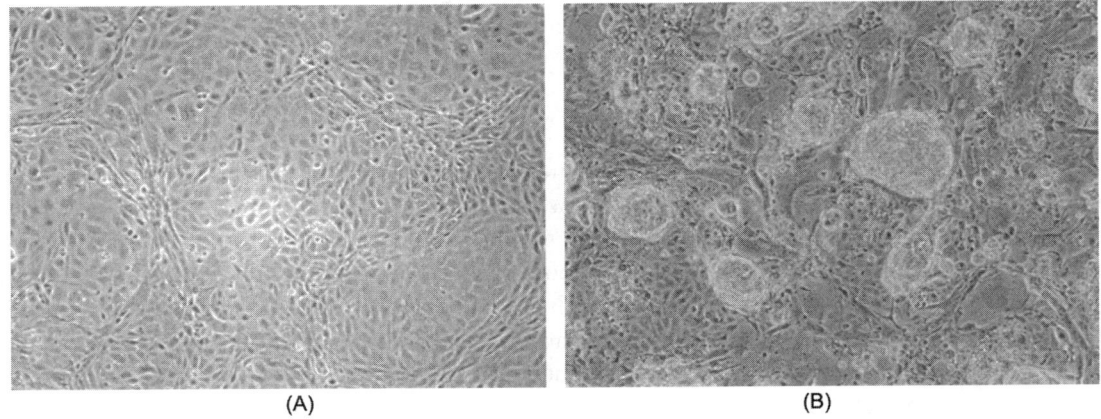

▶ Fig. 2.3 *(A) Feeder cells derived from the rainbow trout spleen cell line RTS34st. (B) Zebrafish blastula derived culture showing ES-like morphology with aggregates of embryo cells possessing a homogeneous appearance.*

GERM-LINE CHIMERA PRODUCTION FROM ES CELL CULTURES

To evaluate the ES cells' ability to contribute to the germ line of a host embryo, cultures were initiated from a transgenic line of fish that expresses green fluorescence protein (GFP) throughout the body and possesses wild-type pigmentation [46]. At each passage, the cultured ES cells were injected into host embryos of the GASSI strain that lacks melanocyte pigmentation. Cultured ES cells used for microinjection are harvested by trypsinisation, collected by centrifugation, washed, and suspended in serum-free culture medium. The cells are loaded into a needle formed from a drawn-out Pasteur pipette and are immediately injected into the host embryos. The recipient embryos for the cell transplants are prepared as follows. Synchronously developing blastula-stage (1000 cells) GASSI embryos are dechorionated in pronase, rinsed in water, and placed in a depression made in agarose in a 60-mm Petri dish. Approximately 50–100 cultured cells are injected into the cell mass of each recipient blastula by using a dissecting microscope and a hand-held pipette aid containing a drawn-out Pasteur pipette. After injection, the embryos are transferred into a Petri dish containing water (26°C) and 3 days later the potential germ-line chimeras are identified by the presence of GFP+ cells in the region of the gonad. To confirm germ-line chimerism, the potential germ-line chimeras are raised to sexual maturity and bred with non-injected GASSI mates and the resulting F1 generation is examined for the expression of GFP throughout the body that is derived from the injected cells. The frequency of germ-line transmission is low when using the ES cells, and it continues to decrease as the cells are maintained for multiple passages in culture. Comparison of cultures at the first passage with those at passage 5 revealed that the frequency of germ-line chimera production decreased from approximately 5% to less than 1% of the total number of injected embryos that survived till sexual maturity. At passage 5, two germ-line chimeras were identified from approximately 250 fish screened [45].

HOMOLOGOUS RECOMBINATION IN THE ZEBRAFISH ES CELL CULTURES

In order to be useful for gene-targeting experiments, in addition to contributing to the germ cell lineage of a host embryo, the ES cells must also incorporate plasmid DNA in a targeted fashion by homologous recombination. We have examined the frequency of homologous recombination in the zebrafish ES cells by targeting the *no tail* gene (*ntl*) since a mutation at this locus produces an obvious and well-characterized phenotype [47, 48]. To target the inactivation of *ntl* in the ES cell cultures, targeting plasmids were designed to utilize two different selection strategies for isolating cell colonies that incorporated the targeting plasmid by homologous recombination [43]. A neo expression cassette was used as a positive selection marker in both strategies. The neo cassette was flanked by a 2.45-kb *ntl* 5′ homologous region, which included the *ntl* promoter region and extended into the first exon. Neo was also flanked by a 3.8-kb *ntl* 3′ homologous region that spanned exons 1 through 9. In the first selection strategy, the gene encoding the red fluorescent protein (*DsRed*) was added outside of the homologous region which served as a visual selection maker (pTG-DsRed-Neo-ntl, Fig. 2.4A). Using this construct, the cells that undergo random insertion are *DsRed* positive, while cells that incorporate the targeting plasmid into their genome by homologous recombination lose the *DsRed*. In the second selection strategy, the diphtheria toxin A-chain gene (*dt*) is inserted into the targeting plasmid outside of the homologous region and serves as the negative selection marker instead of DsRed (pTG-DT-Neo-ntl, Fig. 2.4B). Using this selection method, the cells that undergo random insertion are eliminated by the expression of DT. Together with G418 selection, the candidates (Neo+/DsRed- or DT-) can be selected and identified for further confirmation.

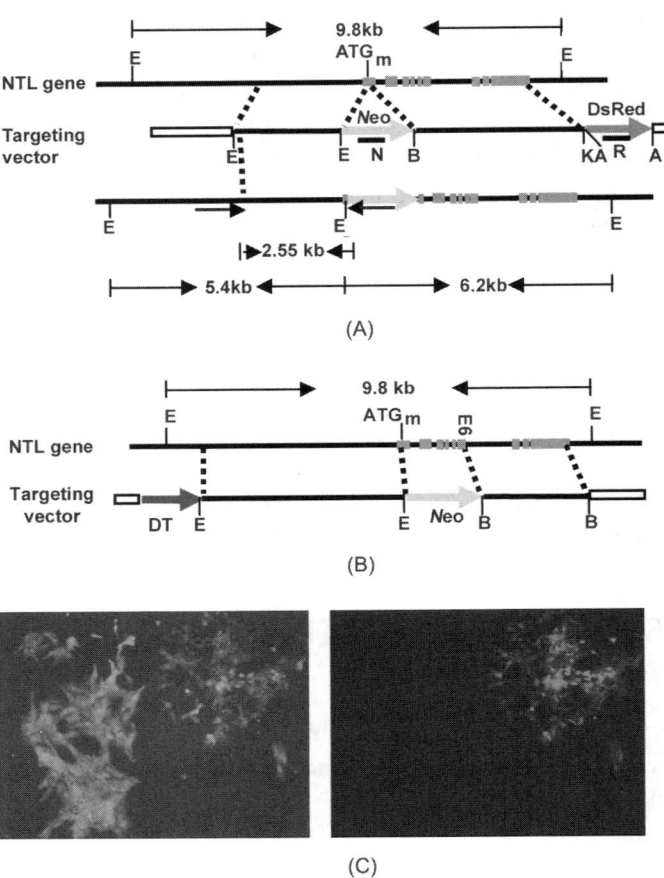

▶ **Fig. 2.4** *(A) pTG-Neo-DsRed-ntl targeting plasmid. Diagram shows the linearised vector (middle) along with the endogenous gene before (top) and after (bottom) homologous recombination with the targeting plasmid. (B) pTG-Neo-DT-ntl targeting plasmid construction. Diagram shows the endogeneous gene (top) before homologous recombination with the targeting vector (bottom) The solid arrows represent the DNA fragments shown on southern blots after EcoRI (marked as E) digestion in Fig. 2.5 (C) Photomicrograph showing the same two G418-resistant colonies following pTG-Neo-DsRed-ntl electroporation. The colonies are shown through red and green filters. The DsRed⁻ colony (arrow) is a homologous recombinant. Figure 2.4 (C) is taken from Fig. 2 of Ref. [43] and used with kind permission of Springer Science and Business Media. (For colour figure see Plate 1).*

To introduce the targeting plasmid by electroporation, early passage zebrafish ES cells are harvested with trypsin/EDTA, washed twice with PBS, and 1–2 million cells are suspended in 0.8 mL PBS containing 50 µg of linearized and sterile targeting plasmid. The electroporation is performed in a 0.4 cm electroporation cuvette using the following conditions: 950 μF, 300 V with a time constant of 11.6 ms. Immediately after electroporation, the cells are seeded into dishes that contain a monolayer of growth-arrested RTS34st feeder cells; G418 (500 μg/mL) is added 24 h after electroporation. Colony visualisation can be aided by using zebrafish ES cell cultures derived from transgenic *actin-GFP* fish that stably express the fluorescent protein (Fig. 2.4C). Following G418 selection, all of the colonies are visible by fluorescence microscopy for GFP expression, while the colonies containing random plasmid insertions can be distinguished from the potential homologous recombinants by the presence or absence of DsRed expression, respectively (when pTG-DsRed-Neo-ntl targeting plasmid is used) (Fig. 2.4C). About 5–6 weeks after electroporation, the G418–resistant DsRed⁻ colonies can be manually selected from the dishes, expanded into flasks, and examined by polymerase chain reaction (PCR) for the presence of a junction sequence created by the targeted insertion of the plasmid (Fig. 2.5A). The PCR analysis is conducted by using primers recognizing endogenous targeted gene sequence that is located outside of the 5′ or 3′ homologous region with primers that recognize *neo* sequence introduced by the vector. Using this approach, the amplified PCR product is only obtained if the targeting plasmid is incorporated by homologous recombination into the target gene. Confirmation of the targeting event is obtained by Southern blot analysis of the PCR positive colonies. When the blot is hybridized with a probe that recognizes a region of the targeted gene that lies outside of the 3′ homologous arms, two fragments corresponding to the wild-type and targeted alleles are identified (Fig. 2.5B). When the blot is hybridized with a probe that recognizes a region of the targeted gene located within the 5′ homologous arms included on the plasmid, three DNA fragments corresponding to the targeted allele, the wild-type allele, and the plasmid that inserted randomly were visualized in those cells that

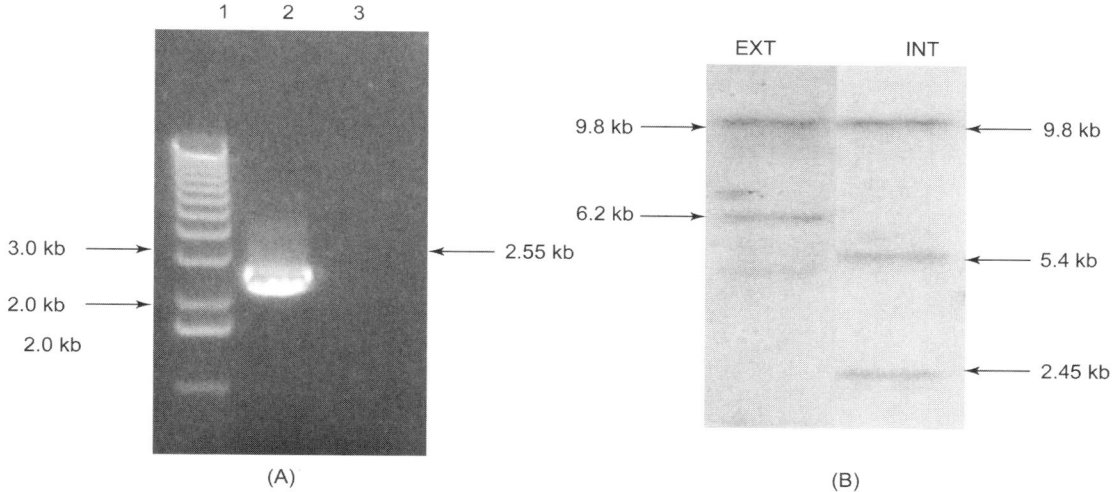

▶ **Fig. 2.5** *Targeted incorporation of pTG-Neo-DsRed-ntl into zebrafish ES cells. (A) PCR analysis of DNA isolated from targeted cells showing 2.55 kb junction fragment created by insertion of pTG-Neo-DsRed-ntl. (B) Southern blot of DNA isolated from a DsRed⁺ clone containing both targeted and random insertions.*

possessed both homologous recombination and random insertion of targeting plasmid (Fig. 2.5B).

The blot was hybridized with a probe recognizing *ntl* sequences located external (EXT) of the 3' homologous region. Two fragments corresponding to the wild-type and targeted alleles (9.8 kb and 6.2 kb, respectively) were detected in colonies that incorporated the targeting plasmid by homologous recombination. In those colonies that had undergone both homologous recombination and random insertion of the targeting plasmid, three DNA fragments were visualized (5.4 kb fragment corresponding to the targeted allele, 9.8-kb fragment corresponding to the wild-type allele, and 2.5 kb fragment corresponding to the randomly inserted plasmid) when hybridisation is conducted with the probe recognizing *ntl* sequences located internal (INT) to the 5' homologous region. Taken from Fig. 2.3 of Ref. [43] and used with kind permission of Springer Science and Business Media.

In addition to *ntl*, we are also working to inactivate *hag* by homologous recombination in the stem cell cultures. Hag is an ortholog of mouse Dactylin (*Dac*) and encodes an F-box/WD40-repeat protein, which is involved in stripe pattern formation. Hag mutations generated by proviral insertion were dominant and resulted in obvious pigmentation pattern anomalies [49]. We are comparing the use of a direct PCR screen of neo-resistant colonies with visual marker expression and positive/negative selection to isolate individual colonies of cells that possess an inactivated *hag* gene. The *hag*-targeting plasmid has been designed to contain puromycin along with the DsRed fusion

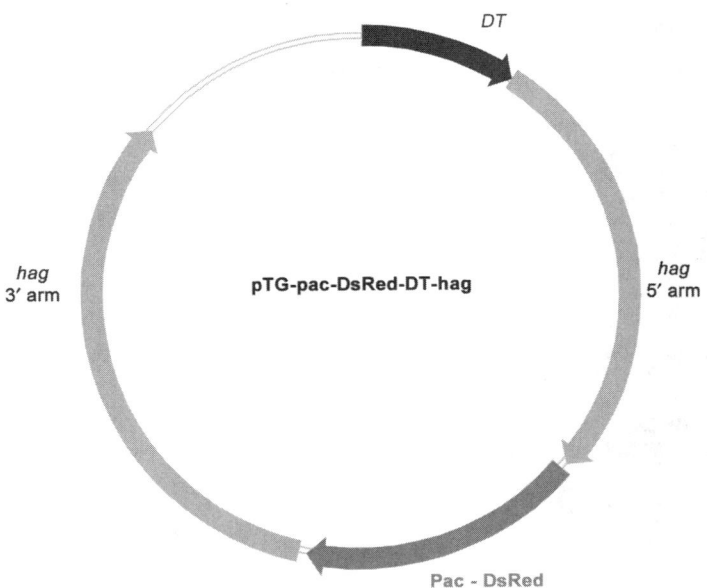

▶ **Fig. 2.6** *pTG-pac-DsRed-DT-hag. In addition to puromycin (pac) selection, DsRed is used to visualise the colonies that had undergone targeted plasmid insertion by homologous recombination. (For colour figure see Plate 2)*

expression cassette (pTG-pac-DsRed-DT-hag) so that it can be used for both visual and positivenegative selection approaches (Fig. 2.6). The plasmid contains a Pac-DsRed expression cassette flanked by 4.3 kb of 5′ *hag* homologous sequence and a 4.1 kb 3′ *hag* homologous arm along with *dt* driven by a constitutively active promoter located outside of the homologous region. This modification will also enable us to visualize cells (DsRed$^+$) that incorporate plasmid by homologous recombination.

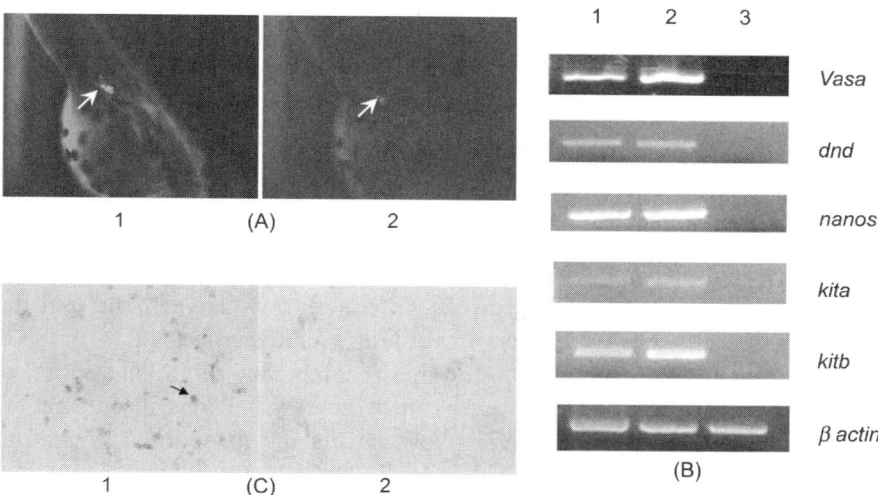

▶ **Fig. 2.7** *Derivation of homogeneous multi-passage PGC cultures. (A) a vasa-GFP embryo injected with pac-DsRed-nanos1-3′UTR mRNA, showing co-expression of green (1) and red (2) fluorescence specifically in PGCs. (B) Lane 1, RT-PCR detection of PGC-specific gene expression in passage 8 cells; lane 2, RNA isolated from 1-day-old embryo (positive control); lane 3, muscle RNA (negative control). (C) In situ hybridization analysis of vasa expression in 3 month PGC cultures showing that a large percentage of the cells continue to express the gene—(1) antisense and (2) sense probes. Taken from fig. 7 of Ref. [59] and used with kind permission of Mary Ann Liebert, Inc. (For colour figure see Plate 3)*

DERIVATION OF ZEBRAFISH PGC CULTURE

In addition to the use of pluripotent ES cell cultures, germ-line contribution can be accomplished by microinjecting host embryos with cultured PGCs—the embryonic germ (EG) cell precursor [50]. Zebrafish PGCs are specified early in development by the incorporation of maternally derived germ plasm [51, 52]. After specification, the PGCs proliferate and migrate to the developing gonad and eventually give rise to spermatogonia or oogonia following sexual differentiation [53]. Mammalian and avian PGCs that are isolated from the developing gonad and maintained in culture behave like pluripotent ES cells that possess an ES-like morphology, express pluripotency markers, form embryoid bodies in suspension culture, and contribute to multiple tissues following transplantation into a host embryo [50, 54, 55]. The PGC-derived pluripotent cells, referred to as EG cells, have been used to generate transgenic mice [56] and chickens [57] by germ-line chimera production similar to ES cells.

In order to develop a PGC-mediated transgenic approach in zebrafish, we have established a culture condition for zebrafish PGCs. Successful culture of EG cells from mouse, chick, and pig requires feeder layers and the addition of Kitlga, FGF, and leukemia inhibitory factor (LIF) [50, 55, 56, 58]. Consistent with these studies, our results have shown that growth of zebrafish PGCs in culture also requires the presence of a feeder layer, FGF, and Kitlga. Our data showed that the addition of zebrafish Kitlga increased the number of PGC colonies by 107.5% and the total number of PGCs by 87.4% after 16 days in culture. In the presence of zebrafish Kitlga, the average size of the PGC colonies is also larger than the culture without Kitlga. The addition of zebrafish Kitlgb, however, has no significant effect on the growth of PGCs *in vitro* [59]. The growth-enhancing effect of Kitlga on zebrafish PGCs in culture is surprising since studies have shown that the PGCs do not express the kit receptor in the embryo [60]. In contrast, we have found that the zebrafish PGCs begin to express kit receptor by 15 days in culture, which is consistent with the observed *in vitro* response to Kitlga [59]. We have also found that recombinant zebrafish stromal cell derived factor 1 (Sdf1) enhances the survival or the growth of zebrafish PGCs in culture. In mice and zebrafish, SDF1/Sdf1 serves as a chemokine that controls the directional migration of PGCs Zebrafish have two *sdf1* genes that share approximately 70% identity (cDNA accession #s: *sdf1a*, AY147915; *sdf1b*, AY347314). Consistent with our results showing an effect of Sdf1 on zebrafish PGCs in culture, mouse SDF1 has been shown to promote PGC survival *in vivo* [62]. We have produced feeder cell lines that express recombinant Sdf1a and Sdf1b that we are using for PGC culture. Our results revealed that in the presence of Sdf1a or Sdf1b, the total number of PGCs present in 16-day cultures increased by 34.4% and 66.8% and number of PGC colonies also increased by 17.8% and 48.8%, respectively. Recombinant Sdf1b is more effective than Sdf1a in promoting zebrafish PGC growth. Furthermore, the use of the combined feeder cells of zebrafish Kitlga and Sdf1b plus their mixed conditioned medium increased the total number of PGCs by 16.5% and number of PGC colonies by 14.1% compared to cultures grown in the presence of zebrafish Kitlga alone [59].

PGC ISOLATION

Although the results demonstrate that recombinant zebrafish Kitlga can stimulate PGC colony formation in primary cultures, the number of PGCs present in these short-term cultures represented a small proportion of the total number of cells present in the culture. A long-term culture consisting of a homogeneous population of proliferating PGCs was needed to characterize the effects of Kitlga and other growth factors on *in vitro* growth and behaviour of PGCs and to develop a PGC-mediated transgenic approach. Hence, we applied a drug selection strategy to establish multi-passage PGC/EG cell cultures. To accomplish this, cDNA encoding puromycin–DsRed fusion protein with the nanos-3' UTR to direct expression in the PGC lineage (*Pac-DsRed-nanos*-3' UTR) was constructed under the control of T7 promoter. Using this cDNA as template, mRNA was synthesized (T7 mMESSAGE kit, Ambion, Austin, TX) and approximately 20–40 pg of the mRNA was injected into one to four cell-stage embryos from *vasa-GFP* fish. Two days after injection, the expression of both red and green fluorescence was examined by fluorescence microscopy. Co-expression of both fluorescent proteins in the same cells indicated that the mRNA was specifically incorporated and expressed in the PGCs (Fig. 2.7A). These embryos were collected, dissociated in trypsin, and cell cultures were initiated on a feeder layer of growth-arrested RTS34st–Kitlga, as described above. Puromycin (5 µg/ml final) was added to the medium after 24 hours. Puromycin selection was maintained on the cultures as long as the cells continued to express DsRed fluorescence indicating that the injected mRNA was still present. PGC cultures derived from the injected embryos exhibited puromycin-resistance and stable DsRed expression up to 12 to 14 days in culture. After drug selection, most of the somatic cells were eliminated from the cultures and a relatively pure population of PGCs remained and continued to proliferate for multiple passages. At passage 8 (3 month old), RT-PCR analysis using RNA extracted from drug-selected PGC cultures revealed that the cells continued to exhibit strong expression of PGC-specific markers including *vasa*, *dnd* and *nanos* (Fig. 2.7B). *In situ* hybridisation with anti *vasa* probe indicated that a large proportion of the late passage cells still expressed *vasa* (Fig. 2.7C).

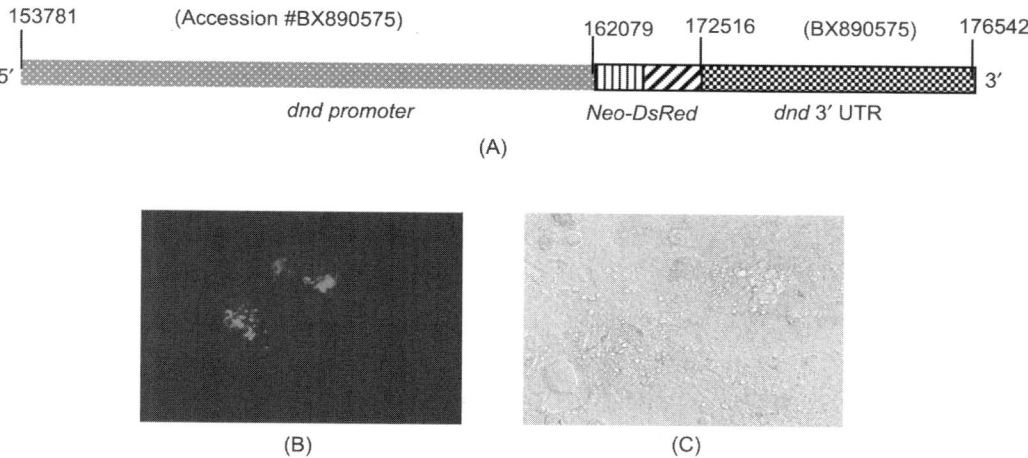

▶ **Fig. 2.8** *Diagram of pdnd-Neo-DsRed plasmid constructs. (A) The expression of Neo–DsRed fusion protein is controlled by the PGC-specific promoter dnd and its 3'UTR. (B) The plasmid was injected into one-cell embryos, and later, the embryos were used to initiate cultures. Following G418 selection, the DsRed-expressing colonies were identified. (C) The same view as in (B), under bright field. (For colour figure see Plate 3).*

The second strategy that we are pursuing to isolate homogeneous cultures of PGCs is by a long-term drug selection. We designed a plasmid (p*dnd-Neo-DsRed*) that encodes a Neo–DsRed fusion protein under the control of the *deadend* promoter and its 3'-UTR (Fig. 2.8). Zebrafish *deadend* has been shown to be specifically expressed in the PGCs through at least day-5 post-fertilisation and is required for germ cell formation in the embryo [65]. To test the construct, embryos at the one-cell stage were injected with p*dnd-Neo-DsRed* and allowed to develop to the 12–15-somite stage before they were dissociated and the cells used to initiate cultures using LDF medium in the presence of growth-arrested RTS34 feeder cells that express zebrafish Kitlga. Two days later, G418 (500 μg/mL) was added to the cultures and fresh growth-arrested feeder cells were added every 2 weeks. After approximately 6 weeks, drug-resistant and DsRed$^+$ colonies were identified (Fig. 2.8). Presently, the DsRed$^+$ and G418-resistant colonies have been expanded into six-well plates and have been growing for more than 4 months. The results demonstrate that the cloned *dnd* promoter is able to drive stable expression of the fusion proteins since the cells continue to proliferate under G418 selection and express DsRed. After 4 months under G418 selection, the cultures continue to express PGC markers including *nanos*, *dnd*, and *vasa* (data not shown). These results are similar to those obtained with EG cultures derived from *Pac-DsRed-nanos*-3'UTR mRNA injection (Fig. 2.7B).

FREQUENCY OF ZEBRAFISH GERM-LINE CHIMERA PRODUCTION FROM PGC CULTURES

To evaluate the PGCs' ability to contribute to the germ line *in vivo*, the cultured cells were injected into the wild-type or GASSI recipient embryos. Three days after injection, potential germ-line chimeras were identified by the presence of DsRed$^+$ cells in the region of the gonad. Recipient embryos were injected with one of the PGC cultures that had been growing *in vitro* for 4 months. The PGC culture was derived by introducing the pdnd-Neo-DsRed construct into one cell-stage embryos from the *actin-GFP* transgenic line of fish and initiating cultures from the founder transgenic embryos in the presence of G418. The resulting PGC culture expressed GFP constitutively along with the Neo–DsRed fusion protein under the control of the *dnd* promoter. From three separate experiments involving the injection of the cultured PGCs into recipient embryos, a total of 322 chimeric larvae were obtained and 30 were found to have DsRed-positive cells in the gonad. If the DsRed-positive larvae are considered to be potential germ-line chimeras, this frequency (9%) is much higher than that obtained with cultured zebrafish ES cells (<1%). The region of the gonad comprised of DsRed-positive cells continued to increase in size when the chimeric larvae were observed over a period of 2 weeks indicating that the transplanted PGC were proliferating in the host. The potential germ-line chimeras were raised to sexual maturity and each was bred with a non-chimeric mate. Twenty-four fish out of 30 potential founders survived to sexual maturity and 9 were identified as germ-line chimeric fish by the transmission of *Neo–DsRed* sequence to the F1 embryos. PCR analysis was conducted on DNA isolated from groups of approximately 50 F1 embryos using primers designed to amplify a *Neo-DsRed* junction fragment donated by the transplanted PGCs. Thirteen out of a total of 57 batches of embryos (23%) screened contained *Neo-DsRed* sequence. The results revealed that germ-line chimeras were produced from cultured PGCs maintained up to 4 months. Although the inheritance of *dnd-Neo-DsRed* was detected in F1 embryos, we did not obtain any larvae that carried the plasmid sequence and continued to survive to sexual maturity. These results indicate that the embryos generated from transplanted PGCs may not be healthy enough to survive and reach sexual maturity. Our attempts to circumvent this problem are discussed in the Summary.

CONCLUSION

The zebrafish has tremendous potential as a model for the elucidation and characterisation of gene function and the dissection of genetic pathways. To fully realize this potential, it will be important to develop gene-targeted strategies that will complement the other genetic tools currently available for use with the zebrafish. A stem cell based approach for introducing targeted mutations will greatly facilitate the ability to assign function to specific genes and elucidate their role in development or a disease process. Successful employment of this approach will require the derivation of embryo cell lines that are able to generate viable germ cells *in vivo*. Toward this end, our laboratory and others have been working to derive fish embryo cell cultures that are suitable for use in the production of transgenic and knockout lines of fish. We have established cultures of PGCs initiated from zebrafish embryos, and optimized conditions that promote the *in vitro* growth and survival of zebrafish PGCs for at least 4 months in culture. The PGCs continue to express late markers of differentiated germ cells in culture and possess a greater capacity to contribute to the germ line following transplantation into a host embryo compared to ES cells. To date, none of the F1 embryos that are derived from the transplanted PGCs have survived to sexual maturity indicating that the cultured PGCs may accumulate genetic aberrations during the 4-month period in culture that prevents them from producing normal embryos. To resolve this problem, we are working to identify genetically normal PGC colonies by karyotype analysis before the cells are used for transplantation to host embryos. Also, we are working to optimize the culture conditions to include other growth factors that have been shown to promote the survival and growth of genetically normal mammalian PGCs. Some of the zebrafish factors that we have cloned and are investigating include recombinant LIF [50], gonadal soma derived growth factor [66], and growth arrest specific 6 [67]. Additionally, other germ cell specific promoters such as *nanos, vasa, kop,* and *ziwi* are being used to drive expression of the drug-resistant and fluorescence protein genes in the transgenic embryos. PGC cultures will be isolated from each of these transgenic lines and evaluated for germ-line chimera production.

ACKNOWLEDGEMENTS

This work was supported by grants from NIH (R01GM069384) and USDA (2005-35206-15261).

REFERENCES

1. Fishman M.C. (2001). Genomics. Zebrafish—the canonical vertebrate. *Science.* **294**(5545): 1290–1291.
2. Fetcho J.R. and K.S. Liu (1998). Zebrafish as a model system for studying neuronal circuits and behaviour. *Ann. N Y Acad. Sci.* **860**: 333–345.
3. Key B. and C.A. Devine (2003). Zebrafish as an experimental model: Strategies for developmental and molecular neurobiology studies. *Methods Cell Sci.* **25**(1–2): 1–6.
4. Mueller T., P. Vernier and M.F. Wullimann (2004). The adult central nervous cholinergic system of a neurogenetic model animal, the zebrafish Danio rerio. *Brain Res.* **1011**(2): 156–169.
5. Lohr J.L. and H.J. Yost (2000). Vertebrate model systems in the study of early heart development: Xenopus and zebrafish. *Am. J. Med. Genet.* **97**(4): 248–257.
6. Poss K.D., L.G. Wilson and M.T. Keating (2002). Heart regeneration in zebrafish. *Science.* **298**(5601): 2188–2190.
7. Keating M.T. (2004). Genetic approaches to disease and regeneration. *Philos. Trans. R. Soc. Lond. B. Biol. Sci.* **359**(1445): 795–798.
8. Lepilina A., A.N. Coon, K. Kikuchi, J.E. Holdway, R.W. Roberts, C.G. Burns and K.D. Poss (2006). A dynamic epicardial injury response supports progenitor cell activity during zebrafish heart regeneration. *Cell.* **127**(3): 607–619.
9. Ackermann G.E. and B.H. Paw (2003). Zebrafish: A genetic model for vertebrate organogenesis and human disorders. *Front. Biosci.* **8**: d1227–1253.
10. Bassett D.I. and P.D. Currie (2003). The zebrafish as a model for muscular dystrophy and congenital myopathy. *Hum. Mol. Genet. 12 Spec.* **2**: R265–270.
11. Amatruda J.F., J.L. Shepard, H.M. Stern and L.I. Zon (2002). Zebrafish as a cancer model system. *Cancer Cell.* **1**(3): 229–231.
12. Stern H.M. and L.I. Zon (2003). Cancer genetics and drug discovery in the zebrafish. *Nat Rev Cancer.* **3**(7): 533–539.
13. Khuchua Z., Z. Yue, L. Batts and A.W. Strauss (2006). A zebrafish model of human Barth syndrome reveals the essential role of tafazzin in cardiac development and function. *Circ. Res.* **99**(2): 201–208.
14. Raya A., A. Consiglio, Y. Kawakami, C. Rodriguez-Esteban and J.C. Izpisua-Belmonte (2004). The zebrafish as a model of heart regeneration. *Cloning Stem Cells.* **6**(4): 345–351.
15. Ticho B.S., D.Y. Stainier, M.C. Fishman and R.E. Breitbart (1996). Three zebrafish MEF2 genes delineate somitic and cardiac muscle development in wild-type and mutant embryos. *Mech. Dev.* **59**(2): 205–218.
16. Langheinrich U. (2003). Zebrafish: A new model on the pharmaceutical catwalk. *Bioessays.* **25**(9): 904–912.
17. Langheinrich U., E. Hennen, G. Stott and G. Vacun (2002). Zebrafish as a model organism for the identification and characterisation of drugs and genes affecting p53 signaling. *Curr. Biol.* **12**(23): 2023–2028.
18. Zon L.I. and R.T. Peterson (2005). In vivo drug discovery in the zebrafish. *Nat. Rev. Drug Discov.* **4**(1): 35–44.
19. Currie P.D. (1996). Zebrafish genetics: Mutant cornucopia. *Curr. Biol.* **6**(12): 1548–1552.
20. F.J. van Eeden, M. Granato, J. Odenthal and P. Haffter (1999). Developmental mutant screens in the zebrafish. *Methods Cell Biol.* **60**: 21–41.

21. Holder N. and A. McMahon (1996). Genes from zebrafish screens. *Nature.* **384**(6609): 515–516.
22. Zhang J., W.S. Talbot and A.F. Schier (1998). Positional cloning identifies zebrafish one-eyed pinhead as a permissive EGF-related ligand required during gastrulation. *Cell.* **92**(2): 241–251.
23. Brownlie A., A. Donovan, S.J. Pratt, B.H. Paw, A.C. Oates, C. Brugnara, H.E. Witkowska, S. Sassa and L.I. Zon (1998). Positional cloning of the zebrafish sauternes gene: A model for congenital sideroblastic anaemia. *Nat. Genet.* **20**(3): 244–250.
24. Gaiano N., A. Amsterdam, K. Kawakami, M. Allende, T. Becker and N. Hopkins (1996). Insertional mutagenesis and rapid cloning of essential genes in zebrafish. *Nature.* **383**(6603): 829–832.
25. Amsterdam A., S. Burgess, G. Golling, W. Chen, Z. Sun, K. Townsend, S. Farrington, M. Haldi and N. Hopkins (1999). A large-scale insertional mutagenesis screen in zebrafish. *Genes Dev.* **13**(20): 2713–2724.
26. Golling G., A. Amsterdam, Z. Sun, M. Antonelli, E. Maldonado, W. Chen, S. Burgess, M. Haldi, K. Artzt, S. Farrington, S.Y. Lin, R.M. Nissen and N. Hopkins (2002). Insertional mutagenesis in zebrafish rapidly identifies genes essential for early vertebrate development. *Nat. Genet.* **31**(2): 135–140.
27. Ivics Z., Z. Izsvak and P.B. Hackett (1999). Genetic applications of transposons and other repetitive elements in zebrafish. *Methods Cell Biol.* **60**: 99–131.
28. Woods I.G., P.D. Kelly, F. Chu, P. Ngo-Hazelett, Y.L. Yan, H. Huang, J.H. Postlethwait and W.S. Talbot. (2000). A comparative map of the zebrafish genome. *Genome Res.* **10**(12): 1903–1914.
29. Kelly P.D., F. Chu, I.G. Woods, P. Ngo-Hazelett, T. Cardozo, H. Huang, F. Kimm, L. Liao, Y.L. Yan, Y. Zhou, S.L. Johnson, R. Abagyan, A.F. Schier, J.H. Postlethwait and W.S. Talbot (2000). Genetic linkage mapping of zebrafish genes and ESTs. *Genome Res.* **10**(4): 558–567.
30. Amemiya C.T., T.P. Zhong, G.A. Silverman, M.C. Fishman and L.I. Zon (1999). Zebrafish YAC, BAC, and PAC genomic libraries. Methods *Cell Biol.* **60**: 235–258.
31. Nasevicius A. and S.C. Ekker (2000). Effective targeted gene 'knockdown' in zebrafish. *Nat. Genet.* **26**(2): 216–220.
32. Ekker S.C. and J.D. Larson (2001). Morphant technology in model developmental systems. *Genesis.* **30**(3): 89–93.
33. Zhang W., R.R. Behringer and E.N. Olson (1995). Inactivation of the myogenic bHLH gene MRF4 results in up-regulation of myogenin and rib anomalies. *Genes Dev.* **9**(11): 1388–1399.
34. Meng X., M.B. Noyes, L.J. Zhu, N.D. Lawson and S.A. Wolfe (2008). Targeted gene inactivation in zebrafish using engineered zinc-finger nucleases. *Nat. Biotechnol.* **26**(6): 695–701.
35. Doyon Y., J.M. McCammon, J.C. Miller, F. Faraji, C. Ngo, G.E. Katibah, R. Amora, T.D. Hocking, L. Zhang, E.J. Rebar, P.D. Gregory, F.D. Urnov and S.L. Amacher (2008). Heritable targeted gene disruption in zebrafish using designed zinc-finger nucleases. *Nat. Biotechnol.* **26**(6): 702–708.
36. Urnov F.D., J.C. Miller, Y.L. Lee, C.M. Beausejour, J.M. Rock, S. Augustus, A.C. Jamieson, M.H. Porteus, P.D. Gregory and M.C. Holmes (2005). Highly efficient endogenous human gene correction using designed zinc-finger nucleases. *Nature.* **435**(7042): 646–651.
37. Moehle E.A., J.M. Rock, Y.L. Lee, Y. Jouvenot, R.C. DeKelver, P.D. Gregory, F.D. Urnov and M.C. Holmes (2007). Targeted gene addition into a specified location in the human genome using designed zinc finger nucleases. *Proc. Natl. Acad. Sci. USA.* **104**(9): 3055–3060.
38. Wienholds E., S. Schulte-Merker, B. Walderich and R.H. Plasterk (2002). Target-selected inactivation of the zebrafish rag1 gene. *Science.* **297**(5578): 99–102.

39. Minami T., J.A. Kuivenhoven, V. Evans, T. Kodama, R.D. Rosenberg and W.C. Aird (2003). Ets motifs are necessary for endothelial cell-specific expression of a 723-bp Tie-2 promoter/enhancer in Hprt targeted transgenic mice. Arterioscler Thromb. *Vasc. Biol.* **23**(11): 2041–2047.
40. Fan L., J. Crodian and P. Collodi (2004). Culture of embryonic stem cell lines from zebrafish. *Methods Cell Biol.* **76**: 151–160.
41. Fan L., X. Liu, A. Alestrom, P. Alestrom, J. Crodian and P. Collodi (2004). Zebrafish embryo cells remain pluripotent and germ-line competent for multiple passages in culture. *Zebrafish.* **1**: 22–26.
42. Ma C., L. Fan, R. Ganassin, N. Bols and P. Collodi (2001). Production of zebrafish germ-line chimeras from embryo cell cultures. *Proc. Natl. Acad. Sci. USA.* **98**(5): 2461–2466.
43. Fan L., J. Moon, J. Crodian and P. Collodi (2006). Homologous recombination in zebrafish ES cells. *Transgenic Res.* **15**(1): 21–30.
44. Ganassin R.C. and N.C. Bols (1999). A stromal cell line from rainbow trout spleen, RTS34ST, that supports the growth of rainbow trout macrophages and produces conditioned medium with mitogenic effects on leukocytes. *in vitro Cell Dev. Biol. Anim.* **35**(2): 80–86.
45. Fan L., J. Crodian and P. Collodi (2004). Production of germ line chimeras for targeted gene disruption. *Methods Cell Biol.* **77**: 113–119.
46. Higashijima S., H. Okamoto, N. Ueno, Y. Hotta and G. Eguchi (1997). High-frequency generation of transgenic zebrafish which reliably express GFP in whole muscles or the whole body by using promoters of zebrafish origin. *Dev. Biol.* **192**(2): 289–299.
47. Schulte-Merker S., R.K. Ho, B.G. Herrmann and C. Nusslein-Volhard. (1992). The protein product of the zebrafish homologue of the mouse T gene is expressed in nuclei of the germ ring and the notochord of the early embryo. *Development.* **116**(4): 1021–1032.
48. Schulte-Merker S., F.J. van Eeden, M.E. Halpern, C.B. Kimmel and C. Nusslein-Volhard (1994). no tail (ntl) is the zebrafish homologue of the mouse T (Brachyury) gene. *Development.* **120**(4): 1009–1015.
49. Kawakami K., A. Amsterdam, N. Shimoda, T. Becker, J. Mugg, A. Shima and N. Hopkins (2000). Proviral insertions in the zebrafish hagoromo gene, encoding an F-box/WD40-repeat protein, cause stripe pattern anomalies. *Curr. Biol.* **10**(8): 463–466.
50. Matsui Y., K. Zsebo and B.L. Hogan (1992). Derivation of pluripotential embryonic stem cells from murine primordial germ cells in culture. *Cell.* **70**(5): 841–847.
51. Braat A.K., T. Zandbergen, S. van de Water, H.J. Goos and D. Zivkovic (1999). Characterisation of zebrafish primordial germ cells: Morphology and early distribution of vasa RNA. *Dev. Dyn.* **216**(2): 153–167.
52. Raz E. (2003). Primordial germ-cell development: The zebrafish perspective. *Nat. Rev. Genet.* **4**(9): 690–700.
53. Starz-Gaiano M. and R. Lehmann (2001). Moving towards the next generation. *Mech. Dev.* **105**(1–2): 5–18.
54. Park T.S. and J.Y. Han (2000). Derivation and characterisation of pluripotent embryonic germ cells in chicken. *Mol. Reprod. Dev.* **56**(4): 475–482.
55. Park T.S., Y.H. Hong, S.C. Kwon, J.M. Lim and J.Y. Han (2003). Birth of germline chimeras by transfer of chicken embryonic germ (EG) cells into recipient embryos. *Mol. Reprod. Dev.* **65**(4): 389–395.
56. Piedrahita J.A., K. Moore, B. Oetama, C.K. Lee, N. Scales, J. Ramsoondar, F.W. Bazer and T. Ott (1998). Generation of transgenic porcine chimeras using primordial germ cell-derived colonies. *Biol. Reprod.* **58**(5): 1321–1329.

57. van de Lavoir M.C., J.H. Diamond, P.A. Leighton, C. Mather-Love, B.S. Heyer, R. Bradshaw, A. Kerchner, L.T. Hooi, T.M. Gessaro, S.E. Swanberg, M.E. Delany and R.J. Etches (2006). Germline transmission of genetically modified primordial germ cells. *Nature*. **441**(7094): 766–769.
58. Durcova-Hills G., I.R. Adams, S.C. Barton, M.A. Surani and A. McLaren (2006). The role of exogenous fibroblast growth factor-2 on the reprogramming of primordial germ cells into pluripotent stem cells. *Stem Cells*. **24**(6): 1441–1449.
59. Fan L., J. Moon, T.T. Wong, J. Crodian and P. Collodi (2008). Zebrafish primordial germ cell cultures derived from vasa: RFP transgenic embryos. *Stem Cells Dev.* **17**(3): 585–97.
60. Parichy D.M., J.F. Rawls, S.J. Pratt, T.T. Whitfield and S.L. Johnson (1999). Zebrafish sparse corresponds to an orthologue of c-kit and is required for the morphogenesis of a subpopulation of melanocytes, but is not essential for haematopoiesis or primordial germ cell development. *Development*. **126**(15): 3425–3436.
61. Doitsidou M., M. Reichman-Fried, J. Stebler, M. Koprunner, J. Dorries, D. Meyer, C.V. Esguerra, T. Leung and E. Raz (2002). Guidance of primordial germ cell migration by the chemokine SDF-1. *Cell*. **111**(5): 647–659.
62. Molyneaux K.A., H. Zinszner, P.S. Kunwar, K. Schaible, J. Stebler, M.J. Sunshine, W. O'Brien, E. Raz, D. Littman, C. Wylie and R. Lehmann (2003). The chemokine SDF1/CXCL12 and its receptor CXCR4 regulate mouse germ cell migration and survival. *Development*. **130**(18): 4279–4286.
63. Raz E. (2004). Guidance of primordial germ cell migration. *Curr. Opin. Cell Biol.* **16**(2): 169–173.
64. Schier A.F. (2003). Chemokine signaling: rules of attraction. *Curr. Biol.* **13**(5): R192–194.
65. Weidinger G., J. Stebler, K. Slanchev, K. Dumstrei, C. Wise, R. Lovell-Badge, C. Thisse, B. Thisse and E. Raz (2003). Dead end, a novel vertebrate germ plasm component, is required for zebrafish primordial germ cell migration and survival. *Curr. Biol.* **13**(16): 1429–1434.
66. Sawatari E., S. Shikina, T. Takeuchi and G. Yoshizaki (2007). A novel transforming growth factor-beta superfamily member expressed in gonadal somatic cells enhances primordial germ cell and spermatogonial proliferation in rainbow trout (Oncorhynchus mykiss). *Dev. Biol.* **301**(1): 266–275.
67. Matsubara N., Y. Takahashi, Y. Nishina, Y. Mukouyama, M. Yanagisawa, T. Watanabe, T. Nakano, K. Nomura, H. Arita, Y. Nishimune, M. Obinata and Y. Matsui (1996). A receptor tyrosine kinase, Sky, and its ligand Gas 6 are expressed in gonads and support primordial germ cell growth or survival in culture. *Dev. Biol.* **180**(2): 499–510.

3

MEDAKA STEM CELLS

Meisheng Yi[1,*], *Yunhan Hong*[1,2]

[1] *Department of Biological Sciences, National University of Singapore, Singapore 119260*
[2] *State Key Laboratory of Freshwater Ecology and Biotechnology, Institute of Hydrobiology, Wuhan 430072, China*

INTRODUCTION

Medaka, like zebrafish, is a small fish, which has high laboratory use as it daily produces eggs that are easily controllable by light cycles. This fish represents a unique lower vertebrate. Medaka has been the first vertebrate besides mouse that has given rise to pluripotent embryonic stem (ES) cell lines. Medaka ES cells retain a diploid karyotype and developmental pluripotency. They can be directed to differentiate into particular cell types. Medaka has also been the first animal that has generated SG3—a normal adult spermatogonial (SG) stem cell line capable of test-tube sperm production. Medaka is the most distantly related vertebrate to mammals and its stem cell lines provide an ideal reference to mammalian ES cells for stemness. Medaka stem cell lines offer an excellent system for experimental analyses *in vitro* and *in vivo*, thanks to its easy embryology and transparency.

In animals, a new life starts with a zygote—the product of fertilization between an egg and a sperm. This zygote divides into a large number of cells, with diverse cell types through differentiation—this process consequently gives rise to over 200 different types of cells making up a vertebrate body. In this process of embryonic development, stem cells play an essential role in keeping a balance between the stem cell pool and terminally functional cells. A stem cell is characterised by two fundamental properties—self-renewal and pluripotency. Self-renewing divisions produce progeny stem cells identical to parental cells, and pluripotency is the developmental potential to produce many—if not all—types of specialized cells by differentiation.

---

* *dbsym@nus.edu.sg*

In this regard, a zygote represents a totipotent stem cell. Stem cells are also present in a wide variety of adult tissues, such as germ stem cells in the gonad and somatic stem cells including hematopoietic stem cells in the bone marrow, neural stem cells in the nervous system, and so on.

Stem cells can be maintained under defined conditions to develop into stable cell lines. The first embryonic stem (ES) cell lines were established from the inner cell mass of mouse blastocyst embryos in 1981 [1, 2]. Attempts toward ES cell cultures have been made in other mammalian species including the hamster [3], sheep [4], mink [5], rat [6], pig [7], bovine [8], and monkey [9], and non-mammalian species including chicken [10] and fish (as discussed further). Human ES cell lines have been derived successfully 10 years ago [11, 12]. Today, ES cells provide an excellent model system for analyzing early mammalian development *in vitro*, and human ES cells hold great promise for cell therapy.

Fish are among the first vertebrate in which ES cell cultures were attempted. Pioneer research was done in two small laboratory fish species—medaka (*Oryzias latipes*) and zebrafish (*Danio rerio*). In addition, ES-like cell cultures have also been reported in marine fish species including the gilthead seabream [13], red seabream [14], sea perch [15], and Asian sea bass [16]. In Chapter 2, Collodi has discussed the research on zebrafish ES cells, while in this chapter we will focus on medaka ES cells and germ stem cells from the adult testis.

MEDAKA ES CELLS

Work towards establishing embryo-derived cell lines, and ultimately ES cell lines, in fish started nearly 20 years ago. Various methods used to develop mouse ES cell cultures have been tested on fish. However, as in mammalian species other than the mouse, the results obtained with fish have generally been unsatisfactory. However, an exception is medaka, which we have used as a model for establishing ES cell technology in fish.

DERIVATION OF MEDAKA ES CELL CULTURES

A major challenge in ES cell cultures is to inhibit spontaneous differentiation. In mouse, this was achieved by cultivating the inner-cell mass cells of blastocysts on a layer of feeder cells [1] or in a conditioned medium [2]. In medaka, Wakamatsu et al. [17] applied the feeder layer technique. They used primary cultures from blastula- and gastrula-stage embryos of the medaka strain HNI as feeder cells and developed a medium formulation, which included basic fibroblast growth factor (bFGF) and fish serum from the common carp (*Cyprinus carpio*), besides other major supplements common to media for mouse ES cell cultures. These culture conditions enabled the establishment of a pluripotent cell line, OLES1, from a blastula embryo of strain HNI. OLES1 exhibited stable growth, ES-like morphology, high alkaline phosphatase (AP) activity (a marker for undifferentiated ES cells) and the potential to be induced by retinoic acid to differentiate into several cell types. This work is quite valuable with respect to ES cell derivation in medaka.

Nichols et al. [18] successfully derived mouse ES cell lines in the presence of leukemia-inhibiting factor (LIF) without using feeder cells. Initially, our labs have attempted both feeder and feeder-free culture conditions. Feeder cells that are routinely used to maintain mouse ES cells include mouse embryonic fibroblasts and the STO line. However, none was found to be able to prevent medaka mid-blastula-derived cells from differentiation (Hong, unpublished). Therefore, we focused our efforts on, and succeeded in, the development of the feeder cell free culture conditions for medaka ES cell cultures [19]. Under these conditions, medaka mid-blastula embryo (MBE) cells are dissociated and grown on a gelatin-coated surface. We have obtained three medaka ES cell lines—MES1, MES2, and MES3 [20].

The first ES medium (ESM1) is formulated on the basis of Dulbecco's modified Eagle medium (DMEM) supplemented with growth factors, medaka embryo extract, and fish serum. Analyses of growth responses of MBE cell cultures and MES1 line reveal that LIF has no effect, whereas bFGF, fish serum, and medaka embryo extract are essential for MBE initiation and maintenance [19]. Decrease in concentration of growth factors and removal of LIF result in the formulation of media ESM2–ESM4 (Table 3.1). We usually use ESM3 for primary cultures and ESM4 for cell-line maintenance. LIF is essential for mouse ES cells but not for human ES cells [21]. The LIF we tested was the recombinant human LIF. It is still under debate whether the human LIF operates in fish or not. Recent reports on putative fish LIF [22] and its receptor [23] have revealed a considerable divergence in LIF protein sequence.

Table 3.1 *Compositions of Medaka ES Cell Culture Media*

Medium components	ESM1	ESM2	ESM3	ESM4
Glutamine (mM)	2	2	2	2
Penecillin-Streptomycin (μg-unit/mL)	100	100	100	100
Non-protein supplements				
Non-essential amino acid (mM)	1	1	1	1
Na Pyruvate (mM)	1	1	1	1
Na-Selenite (nM)	2	2	2	2
2-mercaptoethanol (μM)	50	100	100	100
Phenylthiourea (mM)	0	1	1	1
Protein supplements				
Foetal bovine serum (%)	15	15	15	15
LIF, human recombinant (ng/mL)	10	2	0	0
bFGF human recombinant (ng/mL)	10	2	10	2
Fish serum (%)	1	1	1	1
Fish embryo extract (FEE) (embryo/mL)	1	0.4	1	0.4

Note: Basal medium: DMEM (4.5 g/L glucose), Hepes (20 mM), adjust to pH 7.7 and filtrate. Components are added sequentially in the order in the table.

Phenylthiourea is added to prevent the appearance of pigmented cells in primary cultures.

The successful derivation of ES cell lines is strongly dependent not only on proper culture media and conditions such as cell density [20] but also on the genetic background that has a profound effect on the derivation of medaka MBE cell cultures. For example, out of 12 medaka strains tested, only certain strains such as HNI, HB32C, and HB12A reliably give rise to chimera-competent ES cell cultures, whereas other strains, such as Sakura, Kaga and Yokote, do not [24a]. A refractory genetic background usually has two problems against ES cell culture: the first is poor adhesion, or no adhesion at all, of MBE cells to substrata; and the second is spontaneous differentiation during early days of culture.

Is the feeder-free condition we developed for the medaka widely applicable for derivation of pluripotent cells from fish blastula embryos? As early as in 1980, the first fish embryonic cell line was derived from blastulae of the goldfish and their nuclei were later used for the successful production of an adult nuclear transplant ([24b] and reference therein). Interestingly, this cell line was established using standard conditions for adult tissue cultures, indicating that derivation of cell lines from early embryos is easy in this fish and a feeder layer is not necessary here. In our laboratories, four other species with different relatedness to the medaka were tested—*O. minutillus, O. curvinotus*, and *O. mekongenesis* allow for easy initiation and maintenance of MBE cell cultures, whereas *O. celebensis* does not. By using similar culture conditions, we also obtained one MBE cell line from a particular source of zebrafish. This feeder-free system has been successfully used for the derivation of ES-like cells even from several marine fish species including the gilthead seabream [13], red seabream [14], sea perch [15], and Asian sea bass (sea perch) [14]. Because the medaka, zebrafish, and the marine fish are distantly related and live in either freshwater or seawater habitats, it appears that the ease with which MBE cell cultures can be obtained in fish has little to do with the phylogenetic relatedness. This is analogous to the situation in mammals. For instance, the rat, rabbit, and hamster are very closely related to the mouse. They are, however, no better than other mammalian species for ES cell derivation by using the techniques established in the mouse. It is noteworthy that this feeder-free system is also suitable for germ cell cultures in medaka (discussed further).

CHARACTERISTICS OF MEDAKA ES CELLS

Among the three medaka ES cell lines, MES1 has been extensively characterized *in vitro* and *in vivo*. *In vitro* this cell line shows all the known features of mouse ES cells. These include stable growth, ES cell morphology (small size with relatively large nuclei and prominent nucleoli, and round or polygonal shape (Fig. 3.1E) and high AP activity (Fig. 3.1F)), a normal diploid karyotype, and the ability to form embryoid body-like structures in suspension culture. MES1 has the potential to be induced to differentiate under defined conditions into various cell types including melanin-synthesizing pigment cells, contracting muscle cells, nerve cells, and fibroblasts. MES1 also has the ability for clonal growth forming compacted cell colonies. All descendants from a single colony of cells uniformly displaying the ES cell morphology were able to give rise not only to ES cells but also to the same spectrum of differentiated derivatives as the parental line. This property is retained after long-term cultivation (>400 passages during more than 10 years of culture) and is not abolished by cryostorage.

CHIMERA FORMATION FROM MEDAKA ES CELLS

Chimera formation is a stringent criterion for testing the pluripotency of putative ES cells. To establish conditions for chimera production, the procedure for transplanting non-cultivated medaka blastomeres obtained from the deep layer [25, 26] was modified for blastomeres previously dispersed from whole blastulae. This led to the efficient production of chimeras (>90%) displaying donor-derived wild-type melanocytes in as early as 2 days of embryonic development [24]. In this case of uncultivated donor blastomeres, transplantation of as many cells as possible had little effect on the survival and development of host embryos, which should be due to full physiological compatibility between the donor and recipient blastomeres. Interestingly, although different donor strains gave rise to a similar frequency of pigmented chimeras, they showed various patterns of chimeric pigmentation in terms of compartmental distribution of donor-derived melanocytes.

A procedure to transplant medaka MBE cell cultures was established. Introduction of too many cultured MBE cells (>100) severely affects the viability and chimera frequency. This phenomenon is common to all cultures that have been maintained over 3 days *in vitro*, regardless of the donor strain. When up to 100 cultured MBE cells were transferred into each recipient, pigmented chimeras were obtained. The frequency of chimeras and the degree of chimerism were high when early MBE cultures were used as donors. Although there was a stepwise decline in the efficiency of chimera formation after prolonged cultivation of donor MBE cells, in general the chimera-competence was retained in MBE cells following cultivation up to 70 days [24]. In zebrafish, MBE cells produced a frequency of 37% for pigmented chimeras following 2 days of culture [27] and 15% for PCR-detectable chimeras following 14 days of culture [28].

When MES1 cells were transplanted into albino blastulae, 90% of the recipients developed into chimeric fry as revealed by genotype-specific PCR-assays [29]. Moreover, pigmented chimeras were obtained at approximately 5% in numerous independent experiments. To investigate whether MES1 cells were able to contribute not only to the pigment cell lineage but also to other cell lineages, they were labelled by transfection with a construct expressing the green fluorescent protein (GFP) from a strong constitutive, ubiquitously active promoter. Two days after transfection, 7% of the cells were GFP-positive. These transfected cultures were used for transplantation to host blastulae and the resulting embryos were examined up to the hatching stage. More than 90% of these blastulae developed into GFP-positive fry. These chimeras contained from one to more than 50 GFP-expressing MES1 cells that were found in one to several different areas (Fig. 3.1G). The GFP-positive cells contributed to all major organ systems of all three germ layers, for example, epithelial cells in undulating fins, contracting muscles in hearts, and extraembryonic cells in the yolk sac [29]. Furthermore, when uniform populations of cells stably expressing GFP were obtained by long-term drug selection and used for transplantation, we obtain 100% chimera formation [30]. Stable GFP-expressing transfectants also allowed following the behaviour of MES1 cells in more detail. It turned out that the fate of MES1 cells depends largely on their distribution.

When in regions where the future heart, blood, or fins will form, MES1 cells differentiated into corresponding cell types and participated in these organs. This may reflect that MES1 cells are totipotent *in vivo* and they express this totipotency by responding to various local signals of the developing embryo.

The degree of chimerism—the donor contribution in the MES1-derived chimeras—was generally low (2–10%). This phenomenon was most obvious from the small number (6%) of pigmented chimeras and the degree of chimerism in pigmented organs, which is significantly lower compared to mouse ES cells (see Ref. [31] for example) or to chimeras formed from non-cultivated medaka blastomeres [24, 25]. A series of transplantation experiments using a large number of independent MBE cell cultures has proven that this is not specific to the MES1 line but common to cultured fish MBE cells [24]. The reduced degree of chimerism obtained with MBE cell cultures compared to blastomeres may be ascribed to possible barriers to the physiological and genetic compatibility between the donor and host.

On average, a total of 100 MES1 donor cells were introduced into each host blastula that consists of 1000 cells. Thus, providing the transferred donors could behave like normal blastomeres, a degree of chimerism of approx 10% should be observed. Based on pigmentation that detects the genuine contribution of donors in a physiologically functional cell lineage, we generally obtained a degree of 1–2%: 1–2 melanocytes in chimeras versus 84 melanocytes in an embryo of the wild-type donor strain. Considering a frequency of 5–6% for pigmented chimeras, the degree of chimerism for all MES1-transplanted embryos will be around 0.1%. Thus, there is a 100-fold difference between expected and observed degrees of chimerism. This indicates that cultivated donors are weaker than endogenous blastomeres in terms of propagation. In parallel, cultured cells differ remarkably from blastomeres of host embryos in cell cycle length: 33–48 h for MES1 under different culture conditions versus 30 min for cleavage until the mid-blastula stage [32]. The difference is approximately 70–100-fold, which is comparable to the 100-fold difference in the degree of chimerism. The lengthy cell cycling time may be a major reason for the low degree of chimerism. We concluded here that using weaker host blastulae could improve chimera production. Joly et al. [33] have enhanced germ-line transmission of non-cultivated medaka blastomeres by using gamma-irradiated host embryos. Similar conditions were adopted for chimera formation from MES1. Extensive transplantation experiments showed a dramatic enhancement in both chimera frequency and degree of chimerism.

In mice, the combination of donor and host strains strongly affects chimera frequency and degree of chimerism [31, 34, 35]. A similar phenomenon has also been observed in fish, as medaka ES-like cells from different donor strains showed highly variable efficiencies in chimera formation [24]. Of the two albino strains we have used as hosts, i^1 is superior to i^3 in the formation of pigmented chimeras from MES1 and other MBE cell cultures: >5% in i^1 but <1% in i^3 and a wide distribution of donor-derived melanocytes in i^1 but restricted distribution to the yolk sac (extraembryonic structure) in i^3. Interestingly, when gamma-irradiated blastulae were used as the hosts, i^1 were unable to produce pigmented chimeras, while i^3 produced a considerably enhanced formation of pigmented chimeras in which the donor-derived melanocytes were found predominantly in the embryonic body. This suggests that the donor-host genetic compatibility can be experimentally modulated. Although we do not know how the donor–host compatibility operates and is modulated by irradiation, this observation provides a clue to adjust this compatibility in the future. Of particular importance, germ-line transmission depends heavily not only on the totipotency of ES cells but also on the host genetic background. For example, some C57BL/6 ES lines produce germ-line chimeras in embryos of inbred BalbC but not C3H [31, 34]. Availability of many different medaka strains and populations allows extensive examination of various combinations. There is a high probability that intensive investigation will provide an optimal donor–host combination.

One of the major applications of fish ES cells is to achieve cell culture-mediated germ-line transmission. Production of germ-line chimeras from long-term cultured ES cells has not yet been achieved in fish. Besides a proper combination of donor and host strains as mentioned earlier, other factors have to be considered. A critical question rises concerning the possibility for ES cells to enter the developing germ-line. Zebrafish has been shown to have preformed germ-line, where primordial germ cell (PGC) specification is determined during cleavage stages (before ES cells can be isolated and transplanted) by maternally supplied germ plasm [36, 37]. Shinomiya et al. [38] reported that the medaka vasa—a germ cell marker—exhibited a different expression during early embryogenesis from zebrafish. Herpin et al. [39] have reported that medaka PGCs are formed autonomously. A detailed analysis of the timing and mode of PGC formation in medaka is required to see if the germ-line in this organism is accessible to ES cell contribution.

MEDAKA SPERMATOGONIAL STEM CELLS

Overview

Male germ stem cells spermatogonia (SG) are the foundation of spermatogenesis in male animals. Throughout adult life of most animals, male germ cells in the testis produce sperm that transmit genetic information between generations. These cells originate from PGCs, which are segregated from somatic cells early in development and migrate to the embryonic gonad, where they become gonocytes. At the onset of spermatogenesis, testicular gonocytes or prospermatogonia resume proliferation and become undifferentiated type-A SG—the male germ stem cells that self-renew themselves to maintain the SG stem cell pool or differentiate through meiosis into fertile sperm. Because SG stem cells are the only stem cells in the body that can transmit genetic information to the next generation, they have become a prime option for gene modification. Nevertheless, because of their small population and the lack of identification methods, SG stem cells have been extremely difficult to study. Brinster and Zimmermann [40] developed a germ cell transplantation technique that provided the first functional assay for SG stem cell. The development of the germ cell transplantation technique gave rise to the possibility of controlling animal reproduction by genetic manipulation of the germ line in vertebrates. However, the ability to manipulate SG cells was still limited because of the lack of long-term cultivation of normal SG cells into a stable cell line. In mice, the number of spermatogenic cells decreases by 50% and 90% after 2 and 7 days of culture, respectively [41, 42]. Only by immortalization male germ cell lines have been established. One cell line was derived from 6-week-old mice and was able to generate meiotic cells at 6 months of culture [43], but lost this ability subsequently [44]. Similarly, an SG cell line has been obtained by telomerase immortalization from 6-day-old mice [45]. In the cattle, SGs from 5-month-old testes have been cocultured with Sertoli-like cells for 100 days [46]. Kanatsu-Shinohara et al [47] cultured the mouse testicular cells for up to 5 months in the presence of growth factors and the cells have been shown to maintain their SG potential. They also developed a long-term culture of mouse male germ-line stem cells under serum- or feeder-free conditions [48], and recently obtained a long-term culture of hamster SG cells which had the ability to enter meiosis in the recipient testis, although the haploid germ cells derived from the donor cells were undetectable [49].

In lower vertebrates such as fish, primary culture system of spermatogenic cells has been obtained in eel [50], medaka [51], and zebrafish [52]. In medaka, we have obtained the first normal SG cell line SG3 capable of producing sperm *in vitro* [53], demonstrating the ability of adult SG to produce stable stem cell lines without immortalization or transformation. Indeed, Guan et al. [54] succeeded in SG stem cells from adult mouse testis.

Derivation of Medaka Spermatogonial Stem Cells

SG3 originated from an adult testis. It expresses several germ cell markers, can undergo meiosis, and can eventually produce mobile sperm in culture. In this culture system, sperm are continuously generated, and activated in culture medium immediately after maturation, excluding the possibility to test their fertility. Therefore, SG3 will provide a model system to determine whether adult SG cells can produce alternate for germ-line transmission by producing fertile sperm in culture toward artificial insemination. Note that the conditions for deriving SG3 line is similar to that used for medaka ES cell cultures. Furthermore, SG3 and MES1 are similar in growth and cellular phenotype (Fig. 3.1H).

The procedure for developing SG3 is similar to that for medaka ES cells [19]. The testis from a mature medaka of albino strain i^3 was dissected and minced with a fine scissors. Testicular fragments were trypsinized, and dissociated cells were seeded into a gelatin-coated plate. Sperm and spermatids were removed by medium change. A subset of attached cells underwent stable growth and eventually developed into SG3 line consisting of pure SG cells. SG3 cells are polygonal at low densities, and become round or oval at high densities. They display the typical features of SG cells—a diploid karyotype, a smooth contour, sparse cytoplasm, and relatively large nuclei with prominent nucleoli and positive AP-staining. SG3 has the ability of clonal growth and undergo self-renewing divisions as MES1.

Characteristics of Medaka SG Stem Cells

SG3 resembles medaka ES cells like MES1 in growth, morphology, and colony formation. We used several germ cell marker molecules to characterize the germ cell identity of SG3 [53]. These include *dazl*, *piwi*, *vasa* and *c-kit*. *Dazl* encodes an RNA-binding protein of the deleted in-azoospermia (*DAZ*) gene family and is expressed in SG and spermatocytes of medaka [55]. *Piwi*

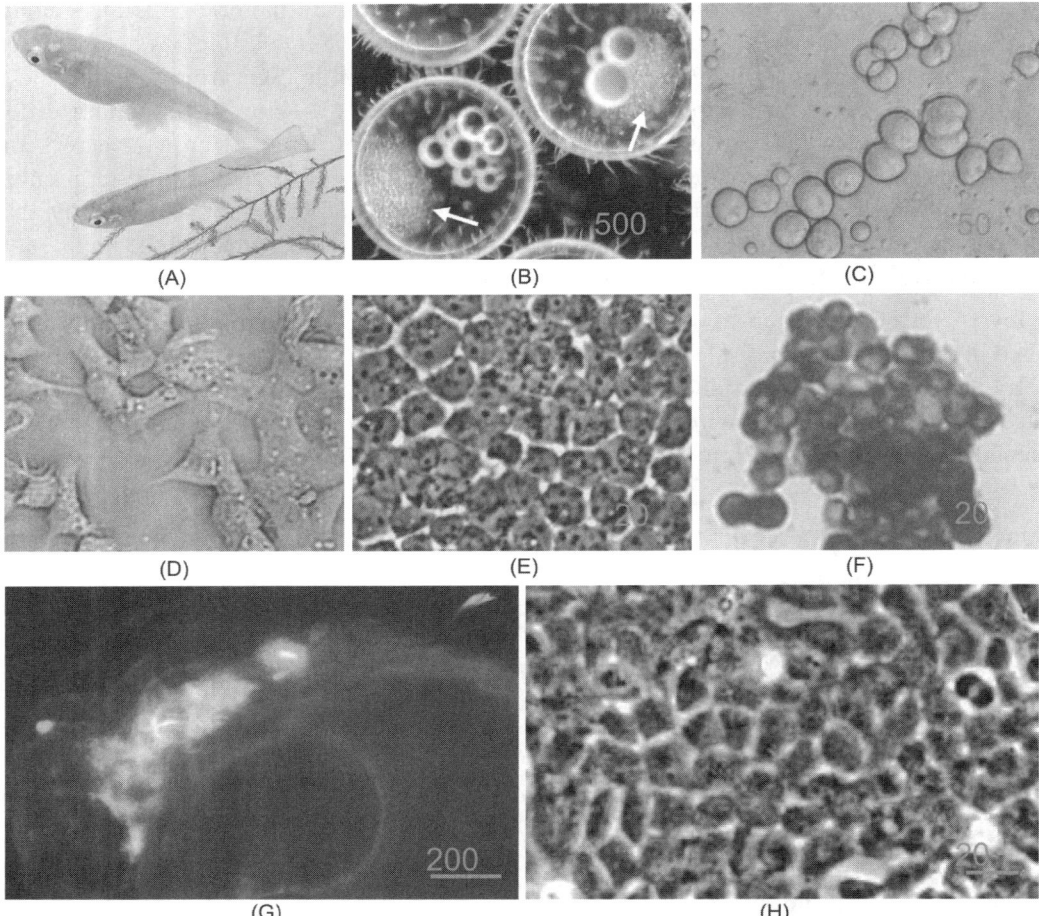

► **Fig. 3.1** *Medaka and its stem cells. (A) Adult fish. Female fish (top), with eggs attached to the belly, and male (bottom). (B) Medaka embryos at mid-blastula stage showing the blastoderms (arrows). (C) Freshly isolated cells from blastoderms. (D) Primary culture of MBE cells. (E and F) MES1 and its AP-staining. (G) Chimera from MES1 cells expressing GFP from a transgene. (H) SG3 cells derived from adult testis. (For colour figure see Plate 4).*

expression in the adult medaka testis colocalizes with *vasa* (Hong, unpublished data). *c-kit* is a tyrosine kinase receptor that, in the adult mouse testis, is highly expressed by SG and spermatocytes [56]. SG3 transcribes all four of these genes, and thus appears to comprise male germ cells including SG.

In the adult testis, type-A SG undergo a series of differentiating divisions to produce type-B SG, which become primary spermatocytes with a tetraploid DNA content and go through meiosis I to generate secondary spermatocytes with a haploid metaphase but a diploid DNA content, while the second spermatocytes in turn go through meiosis II to produce spermatids with a haploid DNA content. Usually, the four round spermatids from a primary spermatocyte are connected together by cytoplasmic bridges to form a tetrad. We applied meiosis and its progression to further define SG3. When SG3 cells were maintained at full confluence without subculture for more than 2 weeks, cell detachment occurred under this condition. Newly detached cells usually formed clusters of four round cells like tetrads. In the medaka, single tetrads are present in freshly isolated cells from mature testes and also in primary cultures of spermatocytes [51]. Detached single cells took a round shape and often displayed a cytoplasmic extension, resembling elongating spermatids. The detached cells have the ability to form sperm in vitro. After 1 day of suspension culture in bacteriological dishes, some of the detached cells generated sperm with a forming residual body. When red fluorescent protein RFP-labelled SG3 cells were mixed in suspension with stromal cells of the cell-line RTG (a cell line from embryonic rainbow trout gonad), chimeric aggregates formed between SG3 and RTG cells within 2 h after cell seeding. At day 2–3 of coculture, the round spermatids, elongating spermatids, and RFP-positive motile sperm were sequentially observed [53]. Thus, SG3 cells appear to be capable of meiosis and spermatogenesis. Furthermore, SG3 can produce a haploid karyotype at a low frequency of 1% and express the protamine mRNA, a marker for the meiotic products spermatids, indicating that SG3 can complete meiosis *in vitro*.

Most importantly, SG3 in the above coculture system with RTG can produce a substantial proportion of haploid cells as analysed by flow cytometry. In accordance with this, we obtained meiotic products at various stages of spermiogenesis, including round spermatids in tetrad, elongating spermatids and ultimately motile sperm. Although it remains to be determined whether the SG3-derived test-tube sperm have fertilizing ability, our results demonstrate that adult medaka SG can give rise to a stable cell line that retains all the properties of a male germ stem cell for self-renewal and differentiation even under culture conditions.

APPLICATION AND FUTURE DIRECTIONS OF MEDAKA STEM CELLS

Human stem cells hold great promise in cell-based therapy of human diseases. And medaka stem cells accelerate the development of genetic engineering by gene targeting and an *in vitro* model system for experimentally analyzing the processes and mechanisms of stem cell renewal and differentiation.

Gene Targeting

In mouse, gene targeting is currently the most powerful tool for introducing defined genetic modifications into specific sites of the genome with ES cells [57, 58] or even SG stem cells [59, 60], thus allowing the elucidation of the physiological functions of the genes under study. However, this approach has so far been limited to mouse.

Transfection and drug selection in the carp cell line *endothelial progenitor cells* (EPC) demonstrated that these selectable marker genes work as well [61], indicating the possibility for use in fish. We have optimized conditions for gene transfer and drug selection in MES1 as a prelude for gene targeting in fish [62]. MES1 gave rise to a moderate to high transfection efficiency by the calcium phosphate co-precipitation (5%), commercial reagents Fugene (11%), GeneJuice (21%), and electroporation (>30%). Transient gene transfer and chloramphenicol acetyltransferase (CAT) reporter assay revealed that several enhancers/promoters and their combinations including CMV, RSV, and ST (the SV40 virus early gene enhancer linked to the thymidine kinase promoter) were suitable regulatory sequences to drive transgene expression in MES1. Expression of genes for neo, hyg, or pac can confer resistance to G418, hygromycin, or puromycin for positive selection, while the HSV-tk generates sensitivity to gancyclovir (Gc) for negative selection. The positive–negative selection (PNS) procedure that is widely used for gene targeting in mouse ES cells works well in MES1. Importantly, MES1 cells after gene transfer and long-term drug selection, retain the developmental pluripotency, as it can undergo induced differentiation *in vitro* and contribute to various tissues and organs during chimeric embryogenesis.

The key to the gene replacement approach in fish is the presence of homologous recombination (HR) activity in fish ES cells. MES1 appears to possess reasonable HR activity as evidenced by the appearance of sister-chromatid exchanges following sister-chromatid differential staining. The medaka *p53* gene was isolated and found to be a single copy gene in the medaka genome [63]. It was used to prepare HR vectors [64]. MBE cell cultures, including the MES1 line, were transfected by electroporation with the vectors. Five independent electroporated cell pools, before and after dual drug selections, were subjected to screening by PCR. All the five pools yielded a PCR product with a size expected for the correct HR event. Furthermore, drug-selected pools produced a significantly stronger signal than non-selected ones, indicating the effectiveness of PNS in medaka ES cells. Most importantly, southern blot analysis revealed that four of the five drug-selected pools produced a detectable band of the expected size. Clonal expansion of these pools under PNS conditions led to the formation of single cell colonies, which were isolated and expanded to cell clones. All this is indicating the probability of gene targeting in medaka. Extensive research is required in this area to obtain targeted ES clones that remain the pluripotency and, more importantly, to achieve germ-line transmission.

Medaka Stem Cells for Directed Differentiation and Transplantation Biology

Two major tasks for stem-cell-based therapy are the production of homogenous populations of specialized cells and the functional integration of these stem cell derivatives under physiological condition upon transplantation into patients. In these contexts, medaka and its stem cells provide a unique model system *in vitro* and *in vivo*. *In vitro*, MES1 can be directed into particular cell lineages. For example, when transfected with the melanocyte-specific isoform of the microphtalmia transcription factor (MITF), MES1 specifically develops into melanocytes by morphology, melanosome motility, and gene expression patterns of completely differentiated pigment cells [65]. This work demonstrated that Mitf expression is sufficient to direct the proper differentiation of medaka ES cells into melanocytes, and that MES1 cells represent a suitable differentiation model. *In vivo*, medaka offers external embryology and a unique see-through vertebrate where internal organs can easily be visualized in living animals [66]. This transparency throughout life will allows for tracing the fate and behaviour of transplanted stem cells and their differentiation derivative. Future work is needed to establish reliable protocols for cell transplantation into adult fish for transplantation biology.

ACKNOWLEDGEMENTS

This work was supported by grants from the Chinese Ministry of Sciences and Technology (High Tech-973 Program 2004CB117406), National University of Singapore (R-154-000-153-720), Biomedical Research Council of Singapore (R-154-000-204-305), and Ministry of Education of Singapore (R154-000-285-112).

REFERENCES

1. Evans M.J. and M.H. Kaufman (1981). Establishment in culture of pluripotential cells from mouse embryos. *Nature*. **292**: 154–156.
2. Martin G.R. (1981). Isolation of a pluripotent cell line from mouse embryo cultures in medium conditioned by teratocarcinoma stem cells. *Proc Natl Acad Sci USA*. **78**: 7634–7638.
3. Doetschman T., P. Williams and N. Maeda (1988). Establishment of hamster blastocyst-derived embryonic stem (ES) cells. *Dev. Biol.* **127**: 224–227.
4. Piedrahita J.A., G.B. Anderson and R.H. BondDurant (1990). On the isolation of embryonic stem cells: Comparative behaviour of murine, porcine and ovine embryos. *Theriogenology*. **34**: 879–901.
5. Sukoyan M.A., S.Y. Vatolin, A.N. Golubitsa, A.I. Zhelezova, L.A. Semenova and O.L. Serov (1993). Embryonic stem cells derived from morulae, inner cell mass and blastocysts of mink: Comparisons of their pluripotencies. *Mol Reprod Dev*. **36**: 148–158.
6. Iannaccone P.M., G.U. Taborn, R.L. Garton, M.D. Caplice and D.R. Brenin (1994). Pluripotent embryonic stems cells from the rat are capable of producing chimera. *Dev. Biol.* **163**: 288–292.
7. Wheeler M.B. (1994). Development and validation of swuine embryonic stem cells: *A review*. *Reprod Fertil Dev*. **6**: 563–568.
8. First N.L., M.M. Sims, S.P. Park and M.J. Kent-Fisrt (1994). Systems for production of calves from cultured bovine embryonic cells. *Reprod. Fertil. Dev.* **6**: 553–562.
9. Thomson J.A., J.K. Kalishman, T. Golos, M. Urning, C.P. Harris, R.A. Becker and J.P. Hearn (1995). Isolation of a primate embryonic stem cell line. *Proc. Natl. Acad. Sci. USA*. **92**: 7844–7848.
10. Pain B., M.E. Clark, M. Shen, H. Nakazama, M. Sakurai, J. Samarut and R.J. Etches (1996). Long-term in vitro culture and characterisation of avian embryonic stem cells with multiple morphogenetic potentialities. *Development*. **122**: 2339–2348.
11. Thomson J.A., J. Itskovitz-Eldor, S.S. Shapiro, M.A. Waknitz, J.J. Swiergiel, V.S. Marshall and J.M. Jones (1998). Embryonic stem cell lines derived from human blastocysts. *Science*. **282**: 1145–1147.
12. Reubinoff B.E., M.F. Pera, C.Y. Fong, A. Trounson and A. Bongso (2000). Embryonic stem cell lines from human blastocysts: Somatic differentiation in vitro. *Nat Biotechnol*. **18**: 399–404.

13. Béjar J., Y. Hong and M.C. Alvarez (2002). An ES-like cell line from the marine fish Sparus aurata: Characterization and chimera production. *Transgenic Res.* **11**: 279–289.
14. Chen S.L., H. Ye, Q. Sha and C.Y. Xhi (2003). Derivation of a pluripotent embryonic cell line from red sea bream blastulas. *J Fish Biol.* **63**: 795–805.
15. Chen S.L., Z.X. Sha, H.Q. Ye, Y. Liu, Y.S. Tian, Y. Hong and Q.S. Tang (2007). Pluripotency and chimera competence of an embryonic stem cell line from the sea perch (*Lateolabrax japonicus*). *Mar Biotechnol.* (NY) **9**(1): 82–91.
16. Parameswaran V., R. Sukla, R. Bhonde and A.S. Hameed (2007). Development of a pluripotent ES-like cell line from Asian sea bass (*Lates calcarifer*)—an oviparous stem cell line mimicking viviparous ES cells. *Mar Biotechnol.* (NY) **9**(6): 766–775.
17. Wakamatsu Y., K. Ozato and T. Sasado (1994). Establishment of a pluripotent cell line derived from a medaka (*Oryzias latipes*) blastula embryo. *Mol Mar Biol Biotechnol.* **3**: 185–191.
18. Nichols J., E.P. Evans and A.G. Smith (1990). Establishment of germ-line-competent embryonic stem (ES) cells using differentiation inhibiting activity. *Development.* **110**(4): 1341–13418.
19. Hong Y. and M. Schartl (1996). Establishment and growth responses of early medakafish (*Oryzias latipes*) embryonic cells in feeder layer-free cultures. *Mol. Mar. Biol. Biotechnol.* **5**: 93–104.
20. Hong Y., C. Winkler and M. Schartl (1996). Pluripotency and differentiation of embryonic stem cell lines from the medakafish (*Oryzias latipes*). *Mech. Dev.* **60**: 33–44.
21. Bojani M. and H.R. Scholer (2005). Regulatory networks in embryo-derived pluripotent stem cells. *Nat. Rev. Mol. Cell Bio.* **6**(11): 872–884.
22. Abe T., T. Mikekado, S. Haga, Y. Kisara, K. Watanabe, T. Kurokawa and T. Suzuki (2007). cDNA cloning, and mRNA localization of a zebrafish ortholog of leukemia inhibitory factor. *Comp Biochem Physiol B. Biochem Mol. Biol.* **147**(1): 38–44.
23. Hanington P.C., S.A. Patten, L.M. Reaume, A.J. Waskiewicz, M. Belosevic and D.W. Ali (2008). Analysis of leukemia inhibitory factor and leukemia inhibitory factor receptor in embryonic and adult zebrafish (*Danio rerio*). *Dev. Biol.* **314**(2): 250–260.
24a. Hong Y., C. Winkler and M. Schartl (1998b). Efficiency of cell culture derivation from blastula embryos and of chimera formation in the medaka (*Oryzias latipes*) depends on donor genotype and passage number. *Dev. Genes. Evol.* **208**: 595–602.
24b. Chen H., Y. Yi, M. Chen and X. Yang (1986). Studies on the developmental potentiality of cultured cell nuclei of fish. Acta. Hydrobiologica. *Sinica.* **10**: 1–7.
25. Wakamatsu Y., K. Ozato, H. Hashimoto, M. Kinoshita, M. Sakaguchi, T. Iwamatsu, Y. Hyodo-Taguchi and H. Tomita (1993). Generation of germ-line chimeras in medaka (*Oryzias latipes*). *Mol Mar. Biol. Biotechnol.* **2**: 325–332.
26. Ando S. and Y Wakamatsu (1995). Production of chimeric medaka (*Oryzias latipes*). *Fish Biol J Medak.* **7**: 65–68.
27. Bradford C.S., L. Sun and D.W. Barnes (1994). Basic fibroblast growth factor stimulates proliferation and suppresses melanogenesis in cell cultures derived from early zebrafish embryos. *Mol. Mar. Biol. Biotechnol.* **3**(2): 78–86.
28. Sun L., C.S. Bradford, C. Ghosh, P. Collodi and D.W. Barnes (1995b). ES-like cell cultures derived from early zebrafish embryos. *Mol. Mar. Biol. Biotechnol.* **4**: 193–199.

29. Hong Y., C. Winkler and M. Schartl (1998a). Production of medakafish chimeras from a stable embryonic stem cell line. *Proc. Natl. Acad. Sci. USA*. **95**: 3679–3684.
30. Hong Y., C. Winkler, T. Liu, G. Chai and M. Schartl (2004a). Activation of the mouse Oct4 promoter in medaka embryonic stem cells and its use for ablation of spontaneous differentiation. *Mech. Dev.* **121**: 933–943.
31. Kawase E., H. Suemori, H. Takahashi, K. Okazaki and N. Nakatsuji (1994). Strain difference in establishment of mouse embryonic stem (ES) cell lines. *Int. J. Dev. Biol.* **38**: 385–390.
32. Iwamatsu T. (2004). Stages of normal development in the medaka *Oryzias latipes*. *Mech. Dev.* **121**: 605–619.
33. Joly J.S., C. Kress, M. Vandeputte, F. Bourrat and D. Chourrout (1999). Irradiation of fish embryos prior to blastomere transfer boosts the colonisation of their gonads by donor-derived gametes. *Mol. Reprod. Dev.* **53**: 394–397.
34. Schwartzberg P.L., S.P. Goff and E.J. Robertson (1989). Germ-line transmission of a *c-abl* mutation produced by targeted gene disruption in ES cells. *Science*. **246**(4931): 799–803.
35. Lerdermann B. and K. Burki (1991). Establishment of a germ-line competent C57BL/6 embryonic stem cell line. *Exp. Cell. Res.* **197**: 254–258.
36. Yoon C., K. Kawakami and N. Hopkins (1997). Zebrafish vasa homologue RNA is located to the cleavage planes of 2- and 4-cell-stage embryos and its expression in the primordial germ cells. *Development*. **124**: 3157–3166.
37. Raz E. (2003). Primordial germ-cell development: The zebrafish perspective. *Nat. Rev. Genet.* **4**(9): 690–700.
38. Shinomiya A., M. Tanaka, T. Kobayashi, Y. Nagahama and S. Hamaguchi (2000). The vasa-like gene, *olvas*, identifies the migration path of primordial germ cells during embryonic body formation stage in the medaka, *Oryzias latipes*. *Dev. Growth. Differ.* **42**: 317–326.
39. Herpin A., S. Rohr, D. Riedel, N. Kluever, E. Raz and M. Schartl (2007). Specification of primordial germ cells in medaka (*Oryzias latipes*). *BMC. Dev. Biol.* **7**: 3.
40. Brinster R.L. and J.W. Zmmermann (1994). Spermatogonesis following male germ-cell transplantation. *Proc. Natl. Acad. Sci. USA*. **91**: 11298–11302.
41. Nagano M., C.J. Brinster, K.E. Orwig, B.Y. Ryu, M.R. Avarbock and R.L. Brinster (2001). Transgenic mice produced by retroviral transduction of male germ-line stem cells. *Proc. Natl. Acad Sci. USA* **98**: 13090–13095.
42. Brinster R.L. (2002). Germline Stem Cell Transplantation and Transgenesis. *Science*. **296**: 2174–2176.
43. Hofmann M.C., R.A. Hess, E. Goldberg and J.L. Millan (1994) Immortalized germ cells undergo meiosis in vitro. *Proc. Natl. Acad. Sci. USA*. **91**: 5533–5537.
44. Wolkowicz M.J., S.A. Coonrod, P.P. Reddi, J.L. Millan, M.C. Hofmann and J.C. Herr (1996). Refinement of the differentiated phenotype of the spermatogenic cell line GC-2spd(ts). *Biol. Reprod.* **55**: 923–932.
45. Feng K.X., Y. Chen, L. Dettin, R.A. Pera, J.C. Herr, E. Goldberg and M. Dym (2002). Generation and in Vitro Differentiation of a Spermatogonial Cell Line. *Science*. **297**: 392–395.
46. Izadyar F., K. Den Ouden, L.B. Creemers, G. Posthuma, M. Parvinen and D.G. De Rooij (2003). Proliferation and differentiation of bovine type A spermatogonia during long-term culture. *Biol. Reprod.* **68**: 272–281.

47. Kanatsu-Shinohara M., N. Ogonuki, K. Inoue, H. Miki, A. Ogura, S. Toyokuni and T. Shinohara (2003). Long-term proliferation in culture and germline transmission of mouse male germline stem cells. *Biol. Reprod.* **69**: 612–616.

48. Kanatsu-Shinohara M., H. Miki, K. Inoue, N. Ogonuki, S. Toyokuni, A. Ogura and T. Shinohara (2005). Long-term culture of mouse male germline stem cells under serum-or feeder-free conditions. *Biol. Reprod.* **72**(4): 985–891.

49. Kanatsu-Shinohara M. , T. Muneto, J. Lee, M. Takenaka, S. Chuma, N. Nakatsuji, T. Horiuchi and T. Shinohara (2008). Long-term culture of male germline stem cells from hamster testes. *Biol. Reprod.* **78**: 611–617.

50. Miura T., K. Yamauchi, H. Takahashi and Y. Nagahama (1991). Hormonal induction of all stages of spermatogenesis in vitro in the male Japanese eel (*Anguilla japonica*). *Proc. Natl. Acad. Sci USA.* **88**: 5774–5778.

51. Sakai A., M. Tamura, M. Matsumoto, J. Katowgi, A. Watanabe and K. Onitake (1997). Establishment of in vitro spermatogenesis from spermatocytes in the medaka, *Oryzias latipes. Dev. Growth. Differ.* **39**: 337–344.

52. Sakai N. (2002). Transmeiotic differentiation of zebrafish germ cells into functional sperm in culture. *Development.* **129**: 3359–3365.

53. Hong Y., T. Liu, H. Zhao, H. Xu, W. Wang, R. Liu, T. Chen, J. Deng and J. Gui (2004b). Establishment of a normal medakafish spermatogonial cell line capable of sperm production in vitro. *Proc Natl Acad. Sci. USA.* **101**: 8011–8016.

54. Guan K., K. Nayernia, L.S. Maier, S. Wagner, R. Dressel, J.H. Lee, J. Nolte, F. Wolf, M. Li, W. Engel and G. Hasenfuss (2006). Pluripotency of spermatogonial stem cells from adult mouse testis. *Nature.* **440**(7088): 1199–1203.

55. Xu H.Y., M.Y. Li, J.F. Gui and Y.H. Hong (2007). Coning and expression of medaka dazl during embryogenesis and gematogenesis. *Gene. Expr. Patterns.* **7**: 332–338.

56. Sette C., S. Dolci, R. Geremia and P. Rossi (2000). The role of stem cell factor and of alternative *c-kit* gene products in the establishment, maintenance and function of germ cells. *Int. J. Dev. Biol.* **44**: 599–608.

57. Thomas K.R. and M.R. Capecchi (1987). Site-directed mutagenesis by gene targeting in mouse embryo-derived stem cells. *Cell.* **51**: 501–512.

58. Müller U. (1999). Ten years of gene targeting: Targeted mouse mutants from vector design to phenotype analysis. *Mech. Dev.* **82**: 3–21.

59. Kanatsu-Shinohara M., M. Ikawa, M. Takehashi, N. Ogonuki, H. Miki, K. Inoue, Y. Kazuki, J. Lee, S. Toyokuni, M. Oshimura, A. Ogura and T. Shinohara (2006). Production of knockout mice by random or targeted mutagenesis in spermatogonial stem cells. *Proc. Natl. Acad. Sci. USA.* **103**(21): 8018–8023.

60. Takehashi M., M. Kanatsu-Shinohara, H. Miki, J. Lee, Y. Kazuki, K. Inoue, N. Ogonuki, S. Toyokuni, M. Oshimura, A. Ogura and T. Shinohara (2007). Production of knockout mice by gene targeting in multipotent germline stem cells. *Dev. Biol.* **312**(1): 344–352.

61. Chen S., Y. Hong and M. Schartl (2002a). Development of a positive-negative selection procedure for gene targeting in fish cells. *Aquaculture.* **214**: 67–79.

62. Hong Y., S. Chen, J. Gui and M. Schartl (2004c). Retention of the developmental pluripotency in medaka embryonic stem cells after gene transfer and long-term drug selection for gene targeting in fish. *Transgenic. Res.* **13**(1): 41–50.
63. Chen S., Y. Hong, S. Scherer and M. Schartl (2001). Lack of UV light inducibility of the medakafish (*Oryzias latipes*) tumor suppressor gene *p53*. *Gene.* **264**: 197–203.
64. Chen S., Y. Hong and M. Schartl (2002b). Cloning, structural analysis and construction of homologous recombination vector of *p53* gene in medaka fish (*Oryzias latipes*). *Acta. Zool. Sin.* **48**: 519–526.
65. Béjar J., T. Hong and M. Schart (2003). Mitf expression is sufficient to direct differentiation of medaka blastula derived stem cells to melanocytes. *Development.* **130**: 6545–6553.
66. Wakamatsu Y., S. Pristyazhnyuk, M. Kinoshita, M. Tanaka and K. Ozato (2001). The see-through medaka: A fish model that is transparent throughout life. *Proc. Natl. Acad. Sci. USA.* **98**(18): 10046–10050.

4

CHICKEN EMBRYONIC STEM CELLS AS ANOTHER NON-MAMMALIAN EMBRYONIC STEM CELL TYPE

Lavial F[1], B. Pain[2,3],*

[1]*Institut de Génomique Fonctionnelle de Lyon, Université de Lyon, CNRS, INRA Ecole Normale Supérieure de Lyon, France*

[2]*CNRS, UMR 6247, Université de Clermont, Inserm, U931 Clermont-Ferrand, France*

[3]*Département de Physiologie Animale, INRA*

INTRODUCTION: TOTIPOTENCY AND PLURIPOTENCY IN ANIMAL CELLS

Pioneering works performed both in invertebrates (Sea urchin) and vertebrates (Newt) demonstrated as early as the last decade of the 19th Century about the existence of totipotent and pluripotent cells in early embryos. Indeed, H. Driesch (1891, Die Stockbildung bei den Hydroidpolypen und ihre theoretische Bedeutung. Biol Zentralbl 11:14–21) demonstrated that a small sea urchin blastoderm dice, isolated at 2 or 4 cells, will develop and form independently two animals, as for the zygote, except that they will be smaller, comparatively, to the parental zygote. This phenomenon is known as the blastomere totipotency.

Speeman replicated these results in 1902 with another species, the newt, with blastomeres taken at a 2-cell stage (Biography of Hans Spemann, The Nobel Foundation. 1 November 1999, http://www.nobel.se/laureates/medicine-1935-1-bio.html). He also demonstrated that a newt larva could be split into two equal parts at 4, 8 or 16 blastomeres, and it could regenerate into two complete embryos. At this stage, the blastomeres are considered as pluripotent, i.e. they are able to

---

* *bepain@u-clermont1.fr*

participate in the development of different embryonic tissues but not autonomous in reconstituting a whole organism.

The development of mammalian embryos takes place exclusively in the mother's body. It is divided into three phases—the pre-implantation period during which the embryo goes through the oviduct to arrive in the uterus; the uterine implantation phase and the post-implantation during which the embryo adheres to the mothers uterine tissue and develops through the placenta formation, allowing exchanges with the mother's body to continue its development.

After fertilisation, the egg, which becomes a diploid cell, enters into division and forms a small cluster of cells called the morula. Eventually, it compacts and the most outer cells form the trophectoderm annexes that give rise to extra-embryonic tissues, while the inner cell mass (ICM), derived from the inner part of the morula, will form the future embryo. During this expansion process, a cavity will be formed, known as the blastocoel cavity. The ICM then forms two derivatives—the epiblast (Epi) that will form the embryo and the primitive endoderm (PrE) that will produce to extra-embryonic tissues. At this stage, the embryo is called a blastocyst. The implantation then takes place and the embryo continues its development. Depending on the species, the kinetics of the differentiation of these early tissues is not the same.

MOUSE EMBRYONIC STEM CELLS AS ESC ARCHETYPE

It was in 1981, more than 25 years ago now, that the first embryonic stem cells of mice (mESC) were isolated from cells of the ICM) of a E3.5–E4.5 dpc (embryonic days postcoitum) blastocyst of the line 129 [1, 2].This discovery was recently honoured by the Nobel prize to Sir M. Evans in 2006 with O. Smithies and M. Capecchi for their demonstration of ESC genetic modification.

These embryonic stem cells (mESC) are maintained undifferentiated under specific cultured condition on a feeder of irradiated mouse embryonic fibroblasts (MEFs). mESC are small cells (8–12 μM) with a high nuclear cytoplasmic ratio that grow in aggregate. These cells were progressively and extensively characterised—they exhibit the same surface markers that the mouse embryonic carcinoma (mEC) cells show (SSEA1, ECMA7, EMA1, etc....), an alkaline phosphatase, activity, an endogenous telomerase activity and large properties for *in vitro* differentiation. Progressively, the culture conditions have been improved including the use of feeder cells supplemented by MEF-conditioned medium [3], before the identification of leukemia inhibitory factor (LIF) cytokine as the cytokine of choice to grow mESC. Different cytokines from the gp130 family were then demonstrated to support self-renewal of the mESC, including CNTF, IL-6, etc. [4]. However, in mouse, the inactivation of the LIF cytokine is not lethal for early and late embryonic development suggesting that the gp130 is not the only signalling pathway and that other molecules can maintain ICM cells [5–7]. At the molecular level, these mESCs express a number of important transcription factors, such as Oct4, Nanog and Sox2, whose respective functions start to be identified in details as discussed in several reviews.

Using the mouse model, a general definition of ESC may be formulated: these are cells capable, even after a long period of culture:

- to self-renew indefinitely *in vitro*, without major alteration of their phenotype or their karyotype,
- to differentiate *in vitro* into many cell types representative of the three embryonic layers—ectoderm, mesoderm and endoderm,
- to participate *in vivo* in the formation of every cell in the embryo, including germ-line when they are re-injected into an early embryo, and
- to form tumours (teratocarcinomas) when injected into an adult organism.

Note that it is possible to establish populations of cells from pluripotent cells in the early embryo (ES cells), but also, under certain conditions, from the cells of the germ-line (EG cells), which retain similarities with embryonic cells [8].

In 1981, Evans and Kaufman have used markers produced from mESCs to define the embryonic stage from which ESCs could be potentially derived [1]. Recently, in a similar approach, authors have succeeded in isolating pluripotent stem cells from mouse epiblast tissue taken at E6.5 dpc. Maintained in the presence of feeder and of ActivinA and FGF2 but in the absence of LIF, these cells were found to proliferate actively [9, 10]. At the morphological level, these so-called "EpiSC" are more similar to the hESC than to the mESC. Despite a great potential of differentiation both *in vitro* with EB and *in vivo* with teratoma formation, these EpiSC are unable to form chimeras either through direct injection into blastocysts or using the morula aggregation protocol. Moreover, these cells are expressing genes such as *Dax1*, *Stella*, *PiwiL2*, *Stra8* and *Dazl*, which are also expressed in hESC and mESC [10].

Beside these different ESCs, stem cells were also derived from the trophectoderm and primitive endoderm, as example of extra-embryonic lineages. Trophectodermal stem (TS) cells can be derived from trophectoderm of E3.5-dpc-old embryos in the presence of feeder, heparin and FGF4 [11, 12]. The TS cells express specific transcription factors different from those of mESC such as Cdx2, Eomes and Hand1, but these cells do not express Oct4. Cells with close characteristics of the extra-embryonic endoderm (ExE) can be isolated from blastocysts at E3.5 in the growing conditions of the TS cells, although fibroblast growth factor (FGF) signalling is not required to maintain them in culture [13]. These cells, known as XEN cells, present a peculiar morphology and express the transcription factors also found in the ExE such as Gata4, Gata6, Sox7, Hnf4 and FoxA2, but again, they do not express Oct4.

With regard to this large range of stem cells obtained in mouse model, an important step for both fundamental and applied research is to identify what would be the equivalent cell type in chicken, if they exist and if culture condition would allow us to derive and amplify them for various purposes.

PLURIPOTENCY AND ISOLATION OF EMBRYONIC STEM CELLS IN BIRDS

In chicken, the egg is laid 20–23 h after fertilisation. The early embryonic development that occurs in the oviduct is arbitrarily divided into 14 stages, numbered in Roman numerals (stage I–XIV), according to Eyal-Giladi and Kochav [14]. The fertilized cell undergoes rapid division to be laid at the so-called Stage X embryo, which is already composed of 20,000–50,000 cells, called blastodermal cells (BCs). Morphologically, this Stage X embryo can be divided into area opaca at the peripheral part and the area pellucida representing the central part of the embryo. The entire embryo and some of the extra-embryonic tissues are derived from the epiblast, which is separated from the yolk by the subgerminal cavity. Hypoblast is then set up by delamination from the epiblast. Hypoblast cells are in direct contact with the yolk and divisions are not completed.

Once incubated, the epiblast cells undergo a number of morphogenetic movements enabling the establishment of the primitive streak, the future axis of the embryo. The embryonic development after laying is arbitrarily divided into 46 stages mentioned by Arabic numerals (stages 1–46) according to Hamburger and Hamilton, established in 1951 [15a].

An important step was achieved by Spratt and Haas [15b]. They demonstrated the pluripotent nature of the BCs by dividing a blastoderm into four equal parts, each of which was therefore able to produce a normal embryo. Furthermore, it was demonstrated that BCs could be injected and colonise a recipient embryo [16a]. From cell clusters derived from freshly laid non-incubated embryo, chimeras (3 on 239) were obtained, based on coloured feathers. However, no alive chicks were hatched. No new results were recorded until 1990 when Pettite and colleagues [16b] slightly modified the process of obtaining chimeras, mainly by dissociating the BCs using trypsin as dissociating enzyme. They were able to obtain alive chimeras from stage X BCs, in particular a germ-line chimera, i.e. the chick born would be able to transmit to his progeny, the genotype of the injected cells, demonstrating for the first time the germ-line competency of the injected cells and, therefore, the germ-line competency of the stage X BCs. Slight irradiation of the recipient embryos increased both the yield of somatic chimerism to 15–20% as well as the germ-line chimerism around 3% [17]. This somatic and germ-line colonisations demonstrated that Stage X embryonic cells were pluripotent.

Based on the above findings, we decided to isolate and grow *in vitro* the cells that harbour ESC features. We were the first to establish an *in vitro* culture of chicken ESCs (cESC) [18]. The chicken BCs (cBC) were derived from a fertile non-incubated Stage X embryo. Once taken, the cBCs are maintained as their murine counterparts on a feeder of irradiated embryonic fibroblasts. It has to be noticed that different fibroblasts were tested such as primary MEF, established STO line as well as avian cells including primary chicken embryonic fibroblasts (CEF), quail established QT6 line, etc. Among the different tested feeders, the STO provided the most steady and reproducible results once cBCs were platted on it. A cocktail of cytokines initially composed of recombinant LIF, interleukin 11 (IL11), stem cell factor (SCF), insulin growth factor 1 (IGF1) and basic FGF (bFGF) ensure their initial proliferation in a non-differentiated status. Since, more simple combinations have been established especially by the use of the IL6 and IL6Rs as a key component of the gp130 signalling pathway. Moreover, the identification of the avian LIF allows the maintenance of cESC in the sole presence of this cytokine with serum [19]. However, it was not established whether this cytokine is able to provide cESC culture competent for either somatic and/or germ-line contribution.

It has to be mentioned that the genetic background of cESC isolation is not so well characterised. Indeed, different strains were tested for their ability to provide cESC in culture. We observed differences in the establishment efficiency including variations in egg- and meat-producing strains. However, as relationships between the different strains of chicken—from a genetic point of view—are not so well established, it was rather difficult to assess the ability to obtain cESC from one or more characters.

BIOCHEMICAL CHARACTERISATION

Different markers are routinely used to characterise and control the culture conditions. In particular classical antibodies including SSEA-1, SSEA3 and SSEA4 are used. These antibodies raised against mESCs were demonstrated to recognised different ESCs from various species. We were the first to describe a cross-reactivity of these antibodies with chicken cells derived from cultured BCs. In particular, reactivity towards both SSEA1 and SSEA3 is a chicken-specificity as mESCs are positive for SSEA1 labelling but not for SSEA4, and hESC are in contrast negative for SSEA1 and positive for SSEA4. The presence of intense alkaline phosphatase activity is also a common and rapid way to look for the good culture conditions.

We also demonstrated that highly proliferative cESCs exhibit a high endogenous telomerase activity [18]. After this first demonstration of a telomerase activity in chicken cells, the chicken telomerase gene was cloned [20, 21]. In contrast, as soon as the cESCs are induced to differentiate, this endogenous expression is rapidly down-regulated [22].

IN VITRO DIFFERENTIATION POTENTIALITIES

As for their murine counterparts, different methods were developed to demonstrate and control the *in vitro* differentiation potentialities of the cESCs. One approach is to plate cESCs in a culture medium containing no cytokines and growth factors needed to sustain cell proliferation. In particular, the absence of one of the gp130 family cytokines (LIF, IL-6, CNTF, GPA, IL-11) leads to a slowdown in the proliferation and a progressive loss of the pluripotency markers. However, the differentiation of cells submitted to this process is not uniform and different cell types will be obtained depending on the initial platting density, the expected autocrine and paracrine cell production and the influence of the cell on each other. The heterogeneity of the cells obtained through this non-controlled process can be estimated by detecting the presence of early lineages markers, such as genes the *Brachyury* and *Goosecoid* genes, specific to mesendoderm lineage, the *Sox1*, *Sox2*, and *Pax6* genes, markers of neurectoderm lineage and the *Hnf3* marker specific to the endoderm lineage. List of primers for QRT-PCR detection of markers specific for each embryonic lineage is available [22].

A second approach involves the production of embryoid bodies (EBs) by seeding the cells in a dish that is not culture-treated. The cells are trypsinised and platted under non-adherent suspension either in a large volume under slow agitation or in a small volume in hanging drops. All these differentiations are performed in a serum-depleted medium (from 0.5% to 5% serum, for example) and in the absence of growth factors and cytokines required for normal un-differentiated proliferation. By preventing the cell adhesion and basolateral polarisation, ESCs differentiates by adopting a three-dimensional structure that mimics the embryonic lineages. Once formed after 2–5 days, EBs can be platted again to obtain more fully differentiated cells in the presence of specific growth factors or used directly for further characterisation.

A third approach is to use chemical inducers to reinforce the cESC differentiation. By chemical induction, it implies using any non-peptide chemical molecule, whether natural or obtained through the chemical synthesis. For example, DMSO (diméthylsulfoxide) is often used as a general inducer generating a mixed differentiated cell population with several cell types. In contrast, retinoic acid usually allows a more rapid differentiation. These different inductions were used as an illustration to demonstrate that the chicken Oct4 and Nanog gene expressions were down-regulated following induction of cESC differentiation [22].

A forth approach is to silence a specific gene by using conditional shRNA. By applying the system used in mouse [23], we developed this innovative approach in chicken cells by using a conditional floxed U6 promoter to direct the expression of shRNA against different genes including pluripotent-associated genes. By inducing their expression, we obtained cESC induced to differentiate into a cell type that is still under investigation but that did not proliferate as their parental cells. Moreover, gene expression profile clearly demonstrated that such inactivation leads to the loss of pluripotent markers. Briefly, an expression vector comprising the shRNA and carrying the Neomycine resistance cassette is introduced into cESC as described. Once selection is applied, resistant clones emerged, expanded and submitted to a second round of transfection with an expression vector allowing the expression of the CRE-ERT2 recombinase [24]. Following a second selection (e.g. hygromycine selection), amplified clones were submitted to the action of tamoxifen to induce the recombinase activity. Adding tamoxifen induces the excision of the floxed resistance cassette and allows the shRNA expression leading to a strong phenotypic change observed as soon as 48 h and confirmed 96 h after drug addition [22].

Another approach is to use enrichment protocols that use of genetic selection process positive. For this, expression of a resistance gene to a drug (neomycin, hygromycine, zéomycine, blasticidine, etc.) or of a phenotypic marker for the cell sorting including fluorescent proteins such as green fluorescent protein (GFP) is placed under the control of a tissue-specific promoter. Once induced to differentiate, the cells expressing this resistance and/or this marker can be enriched by this particular way. Another way is to over-express the cDNA of interest that can induce the differentiation in a given lineage. This over-expression can be performed using vectors allowing random insertion, but also with a knock in strategy in a defined locus of a specific lineage. Retroviruses are also useful to infect a large number of cells. The last three strategies are presently under development in our group.

GENETIC MODIFICATION

We have previously shown that cESCs could be genetically modified by various vectors including simple expression vectors [25], gene trap vectors [26] and homologous recombination vectors (unpublished data, patent Application FRN°. 01/15111). For simple expression vector, different promoters were active in the non-differentiated as well as differentiated cESCs. In particular, strong promoters such as viral cytomegalovirus (CMV) promoter or chicken β-actin-derived promoter such as the CAG (a combination of chicken beta-actin promoter and cytomegalovirus immediate-early enhancer) promoter are often used to over-express transgene in the proliferating cESCs. Beside, tests of retroviruses have been demonstrated to be efficient directly on these cells (unpublished data) as well as directly on BCs present in fertile embryo [27a].

IN VIVO CHIMERISM

To test the potential influence to the development of mESCs, microinjection of these cells into a blastocyst host has been made for many years [27b]. Cells are dissociated using proteolytic enzymes and directly injected into the blastocoelic cavity of E3.5 dpc embryo. In another classical approach, several embryos are merged to get chimeras. For example, the aggregation of two morulas—each of which is composed of eight cells—can produce viable embryo composed of cells from the two genetic pools [27c]. These experiments showed that cells at the morula stage can integrate a number of information to ensure proper development of all tissues of the embryo demonstrating their pluripotency. However, this technique by aggregation cannot be extended to blastocyst-derived stage.

In chicken, the structure of the recipient blastoderm is slightly different. One of the advantages is that cells can be directly injected into blastoderm embryo already in place and that no foster mother is required! Different studies have demonstrated the ability of the cBCs to participate in the morphogenesis and the germ-line of the embryos. We have demonstrated that cultured cESC can produce high-level somatic chimeras and sometimes germ-line chimeras. We and other have also found that this germ-line competency is rapidly lost when cells are cultured for a long period. This results suggest that the germinal competent cells are either lost during the culture process or the culture conditions are unable to maintain them leading in both cases to a loss of germ-line competency. Molecular mechanisms governing this germ-line competency and experimental ways of maintaining it is currently under investigation. In particular, it would have been interesting to study the ability of germ-line competent cells to proliferate without loosing their features [18, 28, 29].

In all cases, injection into recipient embryo is usually performed at Stage X embryo, i.e. non-incubated embryo when cells are not completely induced. Another approach is to inject directly into the blood stream of an incubated embryo. This way could provide a better homing for germ cells as these last cells present the property of migrating through blood stream to reach the gonads.

GERM-LINE COMPETENCY AND *IN VIVO* CONTRIBUTION OF CESC

Relationship between ESCs and germ-line competency is still an open field of investigation. In particular, it is hypothesised that the germ-line determination is based on the pre-formation model in chicken. Then, it is still unknown whether the cESCs can keep long-term germinal competency once maintained for long-term *in vitro* culture. Indeed, germ cells retain the ability to form a new individual and must therefore not respond to differentiation signals from surrounding cells during development. Schematically, establishment of the germ-line is associated with the suppression of somatic differentiation in many organisations. Two main ways of germ-line formation are identified: the pre-formation and the induction models [30].

In *Caenorhabditis elegans* and *Drosophila melanogaster* at the pluripotent zygote stage, precursors of the germ-line called primordial germ cells (PGCs) are specified by maternal components present in the cytoplasm. Asymetrical division induces this cell determination, and different genes including *Pie1* for *C. elegans* and *Gcl* for *D. melanogaster* are among the factors controlling the transcriptional repression leading to this complex germ cell [31].

In mice, the PGCs appear under inductive model. Indeed, some cells in the proximal epiblast will receive signals, mainly BMP4, BMP8b from the extra-embryonic ectoderm and the visceral endoderm visceral. These primed cells will become competent to form germ cells, highlighted by the expression of the fragilis early marker. A key event of the determinism of PGC is the expression of transcriptional repressor Blimp1, which presents a histone methyltransferase activity. Indeed, some of the cells expressing fragilis become Blimp1 positive and are rapidly engaged into germinal differentiation while adjacent cells are engaged into somatic differentiation [32, 33]. It seems that Blimp1 represses the expression of *Hox* genes and therefore a program of somatic differentiation. Moreover, after their specification, the PGCs undergo a strong epigenetic reprogramming associated with the loss of dimethylation H3-K9 dimethylation, an increase of H3-K27 trimethylation as also observed in *C. elegans* and a loss of DNA methylation.

Even if there are differences between models, a number of mechanisms are kept at the establishment of the germ-line, in particular the transcription repression of transcription and a strong chromatin remodelling [34].

In chicken, the debate remains open to know whether the germ cells occur following a pre-determined process. Indeed, from the Stage X embryo (EG&K), clusters of cells expressing surface antigens such as SSEA-1 and EMA-1 and also some specific germ cell markers such as *Cvh* (chicken vasa homolog) and *Dazl* are detected as early as this non-incubated stage. These markers, as well as the presence of strongly positive cells for alkaline phosphatase, are also detected in the germinal crescent after 18 h of incubation [35–37].

Experimental ablations of the germinal crescent at 7–10 (H&H) show that this embryonic structure is the sole source of PGC of the chicken embryo. One of the most distinctive feature of chicken germ cells is their transient presence in embryonic blood as circulating PGCs during few hours (from 48 h to 55 h of development) of Stage 14 to Stage 17 (H&H) embryos just before their arrival in the undifferentiated gonads. Only the cells that come to these gonadal structures will differentiate into germ cells. The other cells enter into apoptosis as they did not find favourable proliferation conditions.

In mouse, different cells with pluripotent-associated properties were isolated from germ cells. Initially identified by their positivity towards phosphatase alkaline, the cells are detected as early as E7 dpc at the junction between epiblast and extra-embryonic [38]. Until E12.5, they migrate and proliferate to form a population of 25,000 cells. At that time, the PGC can be isolated from the genital ridge of E8.5–E12.5 and a combination of SCF, LIF and bFGF allowed the growth of PGCs that become EG cells (embryonic germinal cells) [39, 40]. The EG cells present very similar characteristics as the mESCs as they form teratomas and the potential to form chimears have been demonstrated [41–43]. However they present a difference in term of methylation of some loci such as *Igf2R*.

Using another combination of growth factor and cytokine, mainly GDNF, EGF, LIF and bFGF, germ stem cells (GSCs) were established from newborn mouse testes [44, 45]. These cells were stem cells but different from mESC and mEG cells in term of differentiation potentialities. More recently, pluripotent GSCs were derived from adult mouse testes (maGSC) [46] using a genetic enrichment procedure. Spermatogonial stem cells (SSCs) were isolated by the presence of GFP under the control of the Stra8 promoter, known to be expressed in spermatogonia [47].

In chicken, different attempts have been made to isolate and grow germ-derived cells. Isolated during this developmental window, PGCs can be maintained under non-differentiated conditions in the presence of cytokines such as LIF, SCF and bFGF [48]. According to the authors, these cells are round and do not attach to the substrate. They form germ chimeras at a high frequency when injected into Stage 13–15 (H&H) embryos, but do not participate in the somatic, even if they are injected into early Stage X (EG & K) embryos. Moreover, when transferred into a medium with a decreased level of serum and the absence of FGF, these cells turned into adherent cells presenting EG characters. However, surprisingly, these cells do not contribute anymore to germ-line even if they are able to contribute to somatic tissues [48].

When taken directly in the developing gonads from a Stage 28 (H&H) embryo, the germ cells can be established *in vitro* under culture conditions including the use of LIF, SCF, FGF2, IGF-1 and IL-11 highly similar to those conditions we developed for cESC [18]. This protocol led to the establishment of EG-like cells from scattered embryonic gonadal cells from a 5.5-day incubated embryo [49, 50]. The establishment is a two-step process including a first phase of colony formation of gonadal PGC (gPGC) directly on stroma cells derived from the platted gonads that progressively turn into more homogenous culture when these proliferating PGCs are passed to a second phase on irradiated feeder cells. These cells have the ability to differentiate both *in vitro* by forming EBs and *in vivo* by contributing to somatic chimeras when injected into Stage X (EG & K) embryos. Injection experiments in Stage 17 (H&H) embryos demonstrated that these cells can form germ-line chimeras. However, as observed with the cESC, this ability is also rapidly lost when continuing culture, greatly limiting their use for biotechnological approaches [49].

In contrast to the PGC, the germ cells taken from gonads and also to the initial cBC, we have indeed found that chicken cESC are expressing reduced level of the *Cvh* marker and no expression at all of *Dazl* gene [22]. In contrast, they over-express other pluripotency-associated genes such as *Oct4* and *Nanog* (cf. *infra*).

In chicken a key question remains opened regarding the presence of gene controlling the germinal competency of these different stem cells and the relationship between PGC and cESC. We have started investigating in detail this molecular relationship.

MOLECULAR CHARACTERISATION OF CHICKEN PLURIPOTENCY: *OCT4* IS ALSO PRESENT IN CHICKEN

In mESC, a large number of genes are now described as regulators of the pluripotent state. These genes are usually expressed both in the early embryo and in the pluripotent cells, but only a few exceptions exist such as *Sox2* [50] or *Zfx* [51]. *Oct4*, *Nanog* and *Sox2* are among the key factors for pluripotency maintenance. Their function is now well described and expression levels are used as genuine sensors to assess the pluripotency status of a cell.

Oct4, also called *Pou5f1* or *Oct3/4*, belongs to the family of transcription factors POU domain to Class V, which includes genes *Pit* (P), *Oct* (O) and *Unc* (U). In mouse, Oct4 is a protein of 352 amino acids (AA), which contains a domain POU DNA-binding (150AA) led by two transactivation domains in N- and C-terminal, with virtually redundant functions and required to maintain undifferentiated ES cells [52]. The POU domain is composed of two distinct parts connected by a hinge region, one of 75 specific regions AA POU (POUs) and the other 60 AA consisting of a homeodomain (POUh). This area can recognise the consensus sequence ATGCAAAT DNA sequence, which is sometimes found degenerated and involved in modulating expression of target genes.

The transcript *Oct4* was cloned in 1990 from a cDNA library of cells mEC [53]. In mice, the expression profile is restricted to totipotent and pluripotent cells. The level of expression of *Oct4* is actually stronger in the oocyte, the cells of the morula and the ICM of the blastocyst. The expression is then rapidly restricted to the epiblast at E4.5–E5, although a sharp but transient expression is detected in the primitive endoderm. It is then maintained in the PGCs at E7.5 and in the germ-line later [53, 54]. *In vitro*, mESC as well as mEC and newly isolated EpiSC express *Oct4* [9]. *Oct4* is expressed from the zygotic stage to all cells in the morula. Then, its expression starts to be restricted to cells from the ICM of the blastocyst, while trophectoderm cells express other transcription factors such as Cdx2 and Eomes [55]. A physical interaction between Oct4 and Cdx2 was described [56] leading to a determination of each cell type by a mechanism still to be discovered. Cdx2 is known to form a complex with Oct4 protein that recruits repressors and leads to the inhibition of *Oct4* expression and its targets. Moreover, genes positively regulated usually determine their own expression [57].

Mutant embryos in which the *Oct4* gene is inhibited stop developing at the blastocyst stage [58]. Their analysis revealed that cells in the ICM do not maintain their identity, differentiate into trophectoderm cells and stop proliferating. One of the primary targets is probably the *GFG4* gene whose expression is completely abolished in these mutants. *In vitro*, reduced level of Oct4 induces the mESC to differentiate into trophectoderm cells expressing *Hand1* or *Cdx2* genes as elegantly demonstrated by a conditional *Oct4* inhibition [56, 59, 60]. Over-expression of *Oct4* leads to a differentiation into a primitive endoderm cell type with the induction of gene expression such as *Gata4* and *Gata6* [59].

Regarding the germ cells, *Oct4* is also expressed by cells of the germ-line and their derivatives *in vitro* as EG cells or mGSC [45, 46]. The function of *Oct4* in these cells could not be studied because the inactivation of the gene leads to a lethal phenotype to the blastocyst stage. An inactivation specific to the germ cells was performed using Lox sequences placed in the locus of *Oct4* gene and the Cre recombinase under the control of the tissue non-specific alkaline phosphatase (TNAP), which is expressed only in germ cells at the early development [61]. A severe phenotype of infertility is observed in the male and female mutants. During the development, the mutant PGCs for *Oct4* enter into apoptosis at E10.5 dpc prior to their entry into the gonad. *Oct4* is therefore necessary for the survival of post-migratory PGCs. The initial E7.5 dpc pool of PGC is not affected by the inactivation of Oct4, especially when intense proliferative phase occurs because deletion has not yet taken place via the Cre recombinase or *Oct4* is necessary when the PGC have migrated into proliferation intense phase [62].

The *Oct4* gene has been identified in a large number of mammalian species including mice, humans, pigs, cattle, rabbits, etc. [63]. The analysis of promoters shows the existence of highly conserved regions [64] and inactivation of OCT4 siRNA in human ESCs also induces the differentiation into trophectoderm cells, suggesting a conserved function between mouse and human at least.

PRESENCE OF *OCT4* IN NON-MAMMALIAN SPECIES

It has been shown that the mouse Oct4 promoter could be activated in zebrafish embryonic cells and in early embryos suggesting a common activation mechanism [65]. The *Spg/Pou2* gene was therefore also claimed to be an orthologue of *Oct4* [66, 67]. The homology is based on the protein alignments and the expression in the early embryo. However, this gene expression is also detected in neurectodermic derivatives [67]. The zygotic mutant fails to form neurectoderm tissues, and the use of morpholinos shows that *Spg/Pou2* is needed to maintain the proliferative embryonic character of the cells in early embryo and is required for endoderm formation [68]. This gene also has a role in the early establishment of the dorsoventral axis [69a] in non-mammalian species. Even if *Spg/Pou2* expression is detected in the ovaries by RT-PCR [66], the identity of expressing cells in the gonad is still unknown. Note that the zygotic mutants have no germ-line defects as the PGCs are present and they express the Vasa marker as expected at the end of the gastrulation. In addition, experiments in which mutant embryos were rescued by injecting Spg/Pou2 mRNA at one-cell stage results in fertile embryos, indicating that Spg/Pou2 is not required for subsequent differentiation of germ cells, in contrast to the mouse model [68]. It is, therefore, possible that other POU factors are engaged in this germ-line specific function in zebrafish.

It is also interesting to analyse the situation in *Xenopus*. Three genes encoding transcription factors with POU domain of Class V have been identified in the *Xenopus—Xoct25, Xoct60* and *Xoct91*. These genes are expressed in a sequential manner in the early embryo until neurulation [69b]. *Xoct91* maintains the embryonic cells in an undifferentiated status and makes them competent for neurectodermal differentiation in response to a FGF signal [70]. The use of morpholinos against the three transcripts induced defects in the neural system [71]. Note that the depletion of these POU Class V proteins (Xoct25, Xoct60 and Xoct91) led to an expansion of the expression of *Xcad3*, the orthologue of the *Cdx2* gene in mice. This situation reminds the balance between *Oct4* and *Cdx2* observed in the mouse ES cells [56]. In addition, this depletion leads to an earlier expression of endodermal differentiation genes such as *Sox17* during gastrulation suggesting that the expression of these PouV proteins restricts and can control early entry into differentiation [71].

In chicken, there was no data regarding the presence of genes involved in the maintenance of pluripotency during early embryonic development. A report [72] hypothesised the absence of the *Oct4* orthologue in the chicken genome.

In order to identify some of the major players involved in the maintenance of cESCs and to understand their function, a differential hybridisation strategy was launched between proliferative cESC maintained in undifferentiated status; these same cells induce to differentiate by forming chicken EBs (cEB) for 48 h. Among a large number of sequences, some still under analysis, a particular one presented a strong homology with a POU domain. Following extensive characterisation, we were able to establish that this new gene was the avian homologue of the mammalian *Oct4*. Indeed, by cloning and sequencing the complete associated cDNA, we identified a new gene called *cPouV* coding for a 295 AA protein. The phylogenetic analyses have confirmed that the protein sequence identified was closer to that of mammalian *Oct4* orthologues and avian paralogues (*Oct6, Oct2* chicken), suggesting that this protein is the avian counterpart of *Oct4*. Located on chicken chromosome 17, the syntenic relationship is only partially conserved between mammals and the other species in the highly particular Oct4–PouV related locus.

Expression level was particularly high in proliferative cESC. As soon as we were inducing these cells in differentiation, whatever be the kind, we observed a strong and rapid decrease of its expression level as described in detail in Ref. [22].

THE *NANOG* HOMEODOMAIN GENE IS ALSO PRESENT IN CHICKEN

Beside the *cPouV* gene, we also identified another key factor of the pluripotency in chicken—the *Nanog* gene. The Nanog transcription factor was isolated more recently by two different approaches. The first was to develop a functional screening by stable transfection of mESCs to identify genes that impact their pluripotency. In this screening, expression of *Nanog* is thus able to maintain the mESC cells in an undifferentiated status without the presence of cytokines [73, 74]. The second method was based on *in silico* approach using the digital differential display (http://www.ncbi.nlm.nih.gov/UniGene/info_ddd.shtml), which allows them to identify specific transcripts from EST databases derived from different tissues. By subtracting a number of EST libraries from those derived from mESC, they identified 20 genes known as *Ecat* (ESC-associated transcripts) and confirmed that 9 of them were expressed specifically in mESCs. *Nanog* was one among these genes [75].

The murine protein Nanog is composed of 305 amino acids and belongs to the NKX homeodomain protein family. The protein can bind to the DNA sequence consensus (C/G) (G/A) (C/G) C (G/C) ATTAN (G/C). The rest of the protein had no significant homology with other factors, but transactivation tests in yeast demonstrated that the N- and C-terminal parts of the murine Nanog protein present a transactivation activity [76].

Similarly to *Oct4*, *Nanog* is expressed *in vitro* by mESC, mEC and EpiSC cells [9, 73]. *In vivo* expression is detectable in the compact morula, then restricted to certain cells of the ICM cells and then to the epiblast [77]. No signal was detected in the primitive endoderm. The *Nanog* expression is also detected in germ cells at E7.75, throughout their migration to the gonads [73]. After reaching the gonad, the expression of *Nanog* extinguished in male and female germ cells at the time of their entry into meiosis [78]. In mouse, inactivation of Nanog is lethal embryonic. At E5.5 dpc, embryos present disrupted extra-embryonic structures; the epiblast and the extra-embryonic ectoderm are not present. At E3.5 dpc, the ICM of the mutant embryos is indistinguishable from that of control embryos. If the ICM is maintained *in vitro*, cells do not proliferate like a control embryo but present a parietal endoderm-like morphology [74, 75]. Nanog is therefore important for maintaining the epiblast identity. *In vitro*, mESC, in which the two alleles of Nanog are invalidated, present a different morphology and express lower levels of pluripotency genes as *Oct4* and *Rex1* but an induction of primitive endoderm markers (Gata4, Gata6), of parietal endoderm markers (LamininB1, Dab2) and of visceral endoderm marker (Bmp2). The exogenous expression of *Nanog* allows maintaining mESC in an undifferentiated status, even in the absence of cytokines [73–75]. The recent data on reprogramming approaches also suggest a key role for establishment of pluripotency-associated status [79].

In chicken, we cloned the *Nanog* gene composed of 310 AA from a cDNA of 930 bp. The homology is around 40% with the other mammalian Nanog cDNAs and on less than 30% for the other Nkx members. *In vivo* expression profile analysis indicated a complex and dynamic profile especially during gastrulation period. Later on, *Nanog* gene is mainly restricted to neural tube and germ cells. This expression is also decreasing once the germinal cells reach the gonads.

The third factor of pluripotency, *Sox2*, is also expressed in avian cells. It has been also demonstrated that the transcription factor Sox2 is involved in the maintenance of pluripotency as well. The *Sox2* gene belongs to the very large Sox family of transcription factors (SRY-related HMG box), which includes more than 20 members in mice [80]—it belongs to the B1 subgroup with *Sox1* and *Sox3*. In contrast to the *Oct4* gene, expression profile is less specific to the pluripotent ESC. Indeed, the Sox2 transcript is detected in some morula cells at E2.5, and then becomes restricted to the ICM cells at the blastocyst stage. The expression persists in the epiblast, and then narrows again at the front of neurectoderm presumptive to E7–7.5 [50]. However, another site of expression of *Sox2* is the extra-embryonic ectoderm at E6.5. At E9.5, Sox2 transcript is detected in the brain, neural tube and the endoderm of the primitive intestine [50]. Interestingly, it is also expressed in both male and female germ cells. *In vitro*, *Sox2* is expressed not only in mESC, mEC and EpiSC [9, 81a], but also in TS cells [50].

Inactivation of the gene *Sox2* phenotype results in embryonic lethality. The ICM is formed, in contrast to the *Oct4* mutant embryo [58], but the epiblasts cells do not maintain their identity as observed for the Nanog mutant. *In vitro* culture of these mutant blastocysts confirms that the trophectoderm is present at the blastocyst stage, but that the expression of *Oct4* and *FGF4* epiblastic markers is gradually lost leading to a trophectoderm differentiation. This late phenotype could be due to the presence of the maternal protein, detected until the blastocyst stage and of other *Sox* factors such as *Sox3* that could play a redundant role [50].

As Sox2 forms a complex with Oct4 and this complex can bind to DNA, many genes are co-regulated by this gene and *Nanog* is among the direct targets of such complex [81b, 82, 83]. The study of the function of *Sox2* in mESC was particularly complex because of redundancy with other factors Sox also expressed in these cells. Following conditional inactivation can be considered as a powerful factor stabilising the level of expression of *Oct4* [52, 84].

In chicken, we indeed identified the *Sox2* but also the *Sox3* gene as expressed in non-differentiated cESCs. By using QRT-PCR analysis, we were able to quantify the expression level of these two genes in control conditions and during induction of differentiation.

Regarding the respective data presented in mouse and in chicken, it appears immediately that few data are proportionally obtained in chicken, in particular regarding the gene network present or absent in these cESC.

INDUSTRIAL INTEREST OF AVIAN STEM CELLS

cESC have been successively identified, characterised and established in long-term *in vitro* culture. They have been successfully genetically modified by various kinds of vectors. Different patents have been filled on these different steps by our laboratory and others leading to a potential use of these cells for various biotechnological and industrial applications.

Different aspects could be mentioned—first, the production of molecules of therapeutic interest in eggs is one of the most promising potential applications of the genetically modified cESCs. There is a great expectation from such production as cost of these products could be highly competitive and they can produce large quantities of complex molecules. Interestingly, the glycosylations of protein in birds and in chicken in particular are very similar to those observed in human. This important point can be considered as an additional competitive advantage for this kind of protein production. Besides, sanitary conditions to maintain high-quality animals are already well-mastered worldwide as eggs are already produced in these standards for vaccines production.

Second, these cells could be used to produce chicken lines resistant to diseases. The aim is to provide tools for basic researches on avian resistance/susceptibility to infectious disease and mainly to supply the chicken market with a line resistant to disease. The focus is on the agricultural market in countries where breeding is in development stage, including new EU members and emerging countries. In this case, regulations on the use of genetic modifications are not defined in a similar way between different countries. Production of animals resistant to diseases responds to a need and a demand for obtaining animals and products for human consumption whose health status must constantly move forward. Food safety is still a general public health problem in many places. In addition, using chemicals and veterinary treatment is becoming increasingly questionable on the part of consumers and producers for the reasons of cost and environmental impact.

Third, the *in vitro* use of these cESCs is also an important axis of industrial challenge for vaccine production. Indeed, a large number of vaccines are presently produced on eggs after direct infection of embryos by viruses of interest such as the flu virus and other highly pathogenic viruses. Therefore, these cells, either under proliferative non-differentiated conditions or following induction towards a specific phenotype, could be seen as a good substrate for virus production to be transformed in vaccines. As an example, these cESCs start to be tested at the industrial level and some developments are presently taking place in international vaccine companies as well as in biotechnologies societies.

In the coming years, using these remarkable cells, a good control of genetic modification and a well-mastered production of efficient germ-line chimeras can develop new modes for producing proteins and vaccines for both human and animals.

CONCLUSION

ESCs are mainly studied in mammals and, particularly, in mouse. However, we demonstrated that cESCs can be identified and amplified *in vitro* with specific ESC features. Despite the lack of complete molecular characterisation, we indeed already identified two of the key factors controlling chicken pluripotency. These first steps are presently actively pursued and developing their industrial application is the present scientific and industrial challenge.

To conclude, we are indeed convinced that the cESCs represent a unique model to study and bring a better understanding in the stem cell physiology in a non-mammalian species. All these reasons as discussed in this chapter have inspired our sustained research on these remarkable cells.

REFERENCES

1. Evans M.J., M.H. Kaufman (1981). Establishment in culture of pluripotential cells from mouse embryos. *Nature.* **292**: 154–156.
2. Martin G.R. (1981). Isolation of a pluripotent cell line from early mouse embryos cultured in medium conditioned by teratocarcinoma stem cells. *Proc. Natl. Acad. Sci. USA* **78**: 7634–7638.
3. Smith A.G., M.L. Hooper (1987). Buffalo rat liver cells produce a diffusible activity which inhibits the differentiation of murine embryonal carcinoma and embryonic stem cells. *Dev. Biol.* **121**: 1–9.
4. Yoshida K., I. Chambers, J. Nichols, A. Smith, M. Saito, K. Yasukawa, M. Shoyab, T. Taga, T. Kishimoto (1994). Maintenance of the pluripotential phenotype of embryonic stem cells through direct activation of gp130 signalling pathways. *Mech. Dev.* **45**: 63–71.
5. Ware C.B., M.C. Horowitz, B.R. Renshaw, J.S. Hunt, D. Liggitt, S.A. Koblar, B.C. Gliniak, H.J. McKenna, T. Papayannopoulou, B. Thoma, et al. (1995). Targeted disruption of the low-affinity leukemia inhibitory factor receptor gene causes placental, skeletal, neural and metabolic defects and results in perinatal death. *Development.* **121**: 1283–1299.
6. Yoshida K., T. Taga, M. Saito, S. Suematsu, A. Kumanogoh, T. Tanaka, H. Fujiwara, M. Hirata, T. Yamagami, T. Nakahata, T. Hirabayashi, Y. Yoneda, K. Tanaka, W.Z. Wang, C. Mori, K. Shiota, N. Yoshida, T. Kishimoto (1996). Targeted disruption of gp130, a common signal transducer for the interleukin 6 family of cytokines, leads to myocardial and haematological disorders. *Proc. Natl. Acad. Sci. USA.* **93**: 407–411.
7. Kunath T., M.K. Saba-El-Leil, M. Almousailleakh, J. Wray, S. Meloche, A. Smith (2007). FGF stimulation of the Erk1/2 signalling cascade triggers transition of pluripotent embryonic stem cells from self-renewal to lineage commitment. *Development.* **134**: 2895–2902.

8. Chambers I., A. Smith (2004). Self-renewal of teratocarcinoma and embryonic stem cells. *Oncogene.* **23**: 7150–7160.

9. Brons, I.G., L.E. Smithers, M.W. Trotter, P. Rugg-Gunn, B. Sun, Chuva de Sousa S.M. Lopes, S.K. Howlett, A. Clarkson, L. Ahrlund-Richter, R.A. Pedersen, et al. (2007). Derivation of pluripotent epiblast stem cells from mammalian embryos. *Nature.* **448**: 191–195.

10. Tesar, P.J., J.G. Chenoweth, F.A. Brook, T.J. Davies, E.P. Evans, D.L. Mack, R.L. Gardner, and R.D McKay (2007). New cell lines from mouse epiblast share defining features with human embryonic stem cells. *Nature.* **448**: 196–199.

11. Tanaka S., T. Kunath, A.K. Hadjantonakis, A. Nagy, J. Rossant (1998). Promotion of trophoblast stem cell proliferation by FGF4. *Science.* **282**: 2072–2075.

12. Uy G.D., K.M. Downs, R.L. Gardner (2002). Inhibition of trophoblast stem cell potential in chorionic ectoderm coincides with occlusion of the ectoplacental cavity in the mouse. *Development.* **29**: 3913–3924.

13. Kunath T., D. Arnaud, G.D. Uy, I. Okamoto, C. Chureau, Y. Yamanaka, E. Heard, R.L. Gardner, P. Avner, J. Rossant (2005). Imprinted X-inactivation in extra-embryonic endoderm cell lines from mouse blastocysts. *Development.* **132**: 1649–1661.

14. Eyal-Giladi H., S. Kochav (1976). From cleavage to primitive streak formation: a complementary normal table and a new look at the first stages of the development of the chick. I. General morphology. *Dev. Biol.* **49**: 321–337.

15a. Hamburger V., H.L. Hamilton (1951). A series of normal stages in the development of the chick embryo. *J. Embryol. Exp. Morphol.* **88**: 49–92

15b. Spratt, N.T. and H. Haas (1960). Integrative mechanisms in development of the chick blastoderm. I. Regulative potentiality of separated parts./.exp. *Zool.* **145**: 97–138.

16a. Marzullo G. (1970). Production of chick chimaeras. *Nature.* **225**: 72–73.

16b. Petitte J.N., M.E. Clark, G. Liu, Verrinder A.M. Gibbins, R.J. Etches (1990). Production of somatic and germline chimeras in the chicken by transfer of early blastodermal cells. *Development.* **108**(1): 185–189.

17. Carsience, R.S., M.E. Clark, A.M. Verrinder Gibbins, and R.J. Etches (1993). Germline chimeric chickens from dispersed donor blastodermal cells and compromised recipient embryos. *Development.* **117**: 669–675.

18. Pain B., M.E. Clark, M. Shen, H. Nakazawa, M. Sakurai, J. Samarut, R.J. Etches (1996). Long-term *in vitro* culture and characterisation of avian embryonic stem cells with multiple morphogenetic potentialities. *Development.* **122**: 2339–2348.

19. Horiuchi H., A. Tategaki, Y. Yamashita, H. Hisamatsu, M. Ogawa, T. Noguchi, M. Aosasa, T. Kawashima, S. Akita, N. Nishimichi, N. Mitsui, S. Furusawa, H. Matsuda (2004). Chicken leukemia inhibitory factor maintains chicken embryonic stem cells in the undifferentiated state. *J. Biol. Chem.* **279**: 24514–24520.

20. Delany M.E., L.M. Daniels (2004). The chicken telomerase reverse transcriptase (chTERT): Molecular and cytogenetic characterisation with a comparative analysis. *Gene.* **15**: 339: 61–69.

21. Swanberg S.E., W.S. Payne, H.D. Hunt, J.B. Dodgson, M.E. Delany (2004). Telomerase activity and differential expression of telomerase genes and c-myc in chicken cells *in vitro*. *Dev. Dyn.* **231**: 14–21.

22. Lavial, F., H. Acloque, F. Bertocchini, D.J. Macleod, S. Boast, E. Bachelard, G. Montillet, S. Thenot, H.M. Sang, C.D. Stern, et al. (2007). The Oct4 homologue PouV and Nanog regulate pluripotency in chicken embryonic stem cells. *Development.* **134**: 3549–3563.

23. Coumoul X., W. Li, R.H. Wang, C. Deng (2004). Inducible suppression of Fgfr2 and Survivin in ES cells using a combination of the RNA interference (RNAi) and the Cre-LoxP system. *Nucleic. Acids. Res.* **32**: e85.
24. Feil R., J. Wagner, D. Metzger, P. Chambon (1997). Regulation of Cre recombinase activity by mutated estrogen receptor ligand-binding domains. *Biochem Biophys Res Commun* **237**: 752–757.
25. Pain B., P. Chenevier, J. Samarut (1999). Chicken embryonic stem cells and transgenic strategies. *Cells Tissues. Organs.* **165**: 212–219.
26. H. Acloque, V. Risson, A.M. Birot, R. Kunita, B. Pain, J. Samarut (2001). Identification of a new gene family specifically expressed in chicken embryonic stem cells and early embryo. *Mech. Dev.* **103**: 79–91.
27a. McGrew M.J., A. Sherman, F.M. Ellard, S.G. Lillico, H.J. Gilhooley, A.J. Kingsman, K.A. Mitrophanous, H. Sang (2004). Efficient production of germline transgenic chickens using lentiviral vectors. *EMBO. Rep.* **5**: 728–733.
27b. Gardner, R.L. (1968). Mouse chimeras obtained by the injection of cells into the blastocyst. *Nature.* **220**(5167): 596–597.
27c. Tarkowski A.K. (1961). True hermaphroditism in chimaeric mice. *J. Embryol. Exp. Morphol.* **12**: 735–757.
28. Petitte J.N., G. Liu, Z. Yang (2004). Avian pluripotent stem cells. *Mech. Dev.* **121**: 1159–1168.
29. Van de Lavoir, M.C. C. Mather-Love, P. Leighton, J.H. Diamond, B.S. Heyer, R. Roberts, L. Zhu, P. Winters-Digiacinto, A. Kerchner, T. Gessaro, S. Swanberg, M.E. Delany, R.J. Etches (2006b). High-grade transgenic somatic chimeras from chicken embryonic stem cells. *Mech. Dev.* **123**: 31–41.
30. Extavour, C.G., and M. Akam (2003). Mechanisms of germ cell specification across the metazoans: Epigenesis and preformation. *Development.* **130**: 5869–5884.
31. Strome S., R. Lehmann (2007). Germ versus soma decisions: lessons from flies and worms. *Science.* **316**: 392–393.
32. Saitou M., S.C. Barton, M.A. Surani (2002). A molecular programme for the specification of germ cell fate in mice. *Nature.* **418**: 293–300.
33. Hayashi K., S.M. de Sousa Lopes, M.A. Surani (2007). Germ cell specification in mice. *Science.* **316**: 394–396.
34. Surani M.A., K. Hayashi, P. Hajkova (2007). Genetic and epigenetic regulators of pluripotency. *Cell.* **128**: 747–762.
35. Urven L.E., C.A. Erickson, U.K. Abbott, J.R. McCarrey (1988). Analysis of germ line development in the chick embryo using an anti-mouse EC cell antibody. *Development.* **103**: 299–304.
36. Karagenc L., Y. Cinnamon, M. Ginsburg, and J.N. Petitte (1996). Origin of primordial germ cells in the prestreak chick embryo. *Developmental genetics.* **19**: 290–301.
37. Tsunekawa N., M. Naito, Y. Sakai, T. Nishida, and T. Noce (2000). Isolation of chicken vasa homolog gene and tracing the origin of primordial germ cells. *Development.* **127**: 2741–2750.
38. Lawson K.A., W.J. Hage (1994). Clonal analysis of the origin of primordial germ cells in the mouse. *Ciba Found Symp* **182**: 68–84; *discussion.* 84–91.
39. Godin I., R. Deed, J. Cooke, K. Zsebo, M. Dexter, C.C. Wylie (1991). Effects of the steel gene product on mouse primordial germ cells in culture. *Nature.* **352**: 807–809.
40. Matsui Y., D. Toksoz, S. Nishikawa, S. Nishikawa, D. Williams, K. Zsebo, B.L. Hogan (1991). Effect of Steel factor and leukaemia inhibitory factor on murine primordial germ cells in culture. *Nature.* **353**: 750–752.

41. Resnick J.L., L.S. Bixler, L. Cheng, P.J Donovan (1992). Long-term proliferation of mouse primordial germ cells in culture. *Nature*. **359**: 550–551.

42. Matsui Y., Zsebo K., B.L. Hogan (1992). Derivation of pluripotential embryonic stem cells from murine primordial germ cells in culture. *Cell*. **70**: 841–847.

43. Labosky P.A., D.P. Barlow, B.L. Hogan (1994). Mouse embryonic germ (EG) cell lines: Transmission through the germline and differences in the methylation imprint of insulin-like growth factor 2 receptor (*Igf2r*) gene compared with embryonic stem (ES) cell lines. *Development*. **120**: 3197–3204.

44. Kanatsu-Shinohara M., N. Ogonuki, T. Iwano, J. Lee, Y. Kazuki, K. Inoue, H. Miki, M. Takehashi, S. Toyokuni, Y. Shinkai, M. Oshimura, F. Ishino, A. Ogura, T. Shinohara (2005). Genetic and epigenetic properties of mouse male germline stem cells during long-term culture. *Development*. **132**: 4155–4163.

45. Kanatsu-Shinohara M., T. Shinohara (2007). Culture and genetic modification of mouse germline stem cells. *Ann. NYAcad. Sci.* **1120**: 59–71.

46. Guan K., K. Nayernia, L.S. Maier, S. Wagner, R. Dressel, J.H. Lee, J. Nolte, F. Wolf, M. Li, W. Engel, G. Hasenfuss (2006). Pluripotency of spermatogonial stem cells from adult mouse testis. *Nature*. **440**: 1199–1203.

47. Oulad-Abdelghani, M., P. Bouillet, D. Decimo, A. Gansmuller, S. Heyberger, P. Dolle, S. Bronner, Y. Lutz, and P. Chambon (1996). Characterisation of a premeiotic germ cell-specific cytoplasmic protein encoded by Stra8, a novel retinoic acid-responsive gene. *J. Cell. Biol.* **135**: 469–477.

48. van de Lavoir, M.C., J.H. Diamond, P.A. Leighton, C. Mather-Love, B.S. Heyer, R. Bradshaw, A. Kerchner, L.T. Hooi, L.T. Gessaro, S.E. Swanberg, et al. (2006a). Germline transmission of genetically modified primordial germ cells. *Nature*. **441**: 766–769.

49. Park T.S., Hong Y.H., Kwon S.C., Lim J.M., J.Y. Han (2003). Birth of germline chimeras by transfer of chicken embryonic germ (EG) cells into recipient embryos. *Mol. Reprod. Dev.* **65**: 389–395.

50. Avilion A.A., S.K. Nicolis, L.H. Pevny, L. Perez, N. Vivian, R. Lovell-Badge (2003). Multipotent cell lineages in early mouse development depend on SOX2 function. *Genes. Dev.* **17**: 126–140.

    Park T.S., J.Y. Han (2000). Derivation and characterisation of pluripotent embryonic germ cells in chicken. *Mol. Reprod. Dev.* **56**: 475–482.

51. Galan-Caridad J.M., S. Harel, T.L. Arenzana, Z.E. Hou, F.K. Doetsch, L.A. Mirny, B. Reizis (2007). Zfx controls the self-renewal of embryonic and haematopoietic stem cells. *Cell*. **129**: 345–357.

52. Tomioka M., M. Nishimoto, S. Miyagi, T. Katayanagi, N. Fukui, H. Niwa, M. Muramatsu, A. Okuda (2002). Identification of Sox-2 regulatory region which is under the control of Oct-3/4-Sox-2 complex. *Nucleic. Acids. Res.* **30**: 3202–3013.

53. Schöler H.R., S. Ruppert, N. Suzuki, K. Chowdhury, P. Gruss (1990). New type of POU domain in germ line-specific protein Oct-4. *Nature*. **344**: 435–439.

54. Yeom Y.I., G. Fuhrmann, C.E. Ovitt, A. Brehm, K. Ohbo, M. Gross, K. Hubner, H.R. Scholer (1996). Germline regulatory element of Oct4 specific for the totipotent cycle of embryonal cells. *Development*. **122**: 881–894.

55. Russ A.P., S. Wattler, W.H. Colledge, S.A. Aparicio, M.B. Carlton, J.J. Pearce, S.C. Barton, M.A. Surani, K. Ryan, M.C. Nehls, V. Wilson, M.J. Evans (2000). Eomesodermin is required for mouse trophoblast development and mesoderm formation. *Nature*. **404**: 95–99.

56. Niwa H., Y. Toyooka, D. Shimosato, D. Strumpf, K. Takahashi, R. Yagi, J. Rossant (2005). Interaction between Oct3/4 and Cdx2 determines trophectoderm differentiation. *Cell*. **123**: 917–929.

57. Okumura-Nakanishi S., M. Saito, H. Niwa, F. Ishikawa (2005). Oct-3/4 and Sox2 regulate Oct-3/4 gene in embryonic stem cells. *J. Biol. Chem.* **280**: 5307–5317.
58. Nichols J., B. Zevnik, K. Anastassiadis, H. Niwa, D. Klewe-Nebenius, I. Chambers, H. Scholer, A. Smith (1998). Formation of pluripotent stem cells in the mammalian embryo depends on the POU transcription factor Oct4. *Cell.* **95**: 379–391.
59. Niwa H., J. Miyazaki, A.G. Smith (2000). Quantitative expression of Oct-3/4 defines differentiation, dedifferentiation or self-renewal of ES cells. *Nat. Genet.* **24**: 372–376.
60. Tolkunova E., F. Cavaleri, R. Reinbold, L.K. Christenson, H. Schöler, A. Tomolin (2006). The Caudal-related protein Cdx2 promotes trophoblast differentiation of mouse embryonic stem cells. *Stem Cells.* **24**: 139–144.
61. Ramos-Mejia V., D. Escalante-Alcalde, T. Kunath, L. Ramirez, M. Gertsenstein, A. Nagy, H. Lomeli (2005). Phenotypic analyses of mouse embryos with ubiquitous expression of Oct4: Effects on mid-hindbrain patterning and gene expression. *Dev. Dyn.* **232**: 180–190.
62. Kehler J., E. Tolkunova, B. Koschorz, M. Pesce, L. Gentile, M. Boiani, H. Lomeli, A. Nagy, K.J. McLaughlin, H.R. Scholer, A. Tomilin (2004). Oct4 is required for primordial germ cell survival. *EMBO. Rep.* **5**: 1078–1083.
63. Shi J.J., D.H. Cai, H.Z. Sheng (2007). Cloning and characterisation of the rabbit POU5F1 gene. *DNA Seq.* **19**(1): 56–61.
64. Nordhoff. V., K. Hubner, A. Bauer, I. Orlova, A. Malapetsa, H.R. Scholer (2001). Comparative analysis of human, bovine, and murine Oct4 upstream promoter sequences. *Mamm. Genome.* **12**: 309–317.
65. Hong Y., C. Winkler, T. Liu, G. Chai, M. Schartl (2004). Activation of the mouse Oct4 promoter in medaka embryonic stem cells and its use for ablation of spontaneous differentiation. *Mech. Dev.* **121**: 933–943.
66. Takeda H., T. Matsuzaki, T. Oki, T. Miyagawa, H. Amanuma (1994). A novel POU domain gene, zebrafish pou2: Expression and roles of two alternatively spliced twin products in early development. *Genes. Dev.* **8**: 45–59.
67. Burgess S., G. Reim, W. Chen, N. Hopkins, M. Brand (2002). The zebrafish spiel-ohne-grenzen (*spg*) gene encodes the POU domain protein Pou2 related to mammalian Oct4 and is essential for formation of the midbrain and hindbrain, and for pre-gastrula morphogenesis. *Development.* **129**: 905–916.
68. Lunde K., H.G. Belting, W. Driever (2004). Zebrafish PouVf1/pou2, homolog of mammalian Oct4, functions in the endoderm specification cascade. *Curr. Biol.* **14**: 48–55.
69a. Reim G., M. Brand (2002). Spiel-ohne-grenzen/pou2 mediates regional competence to respond to Fgf8 during zebrafish early neural development. *Development.* **129**: 917–933.
69b. Hinkley C.S., J.F. Martin, D. Leibham, M. Perry (1992). Sequential expression of multiple POU proteins during amphibian early development. *Mol. Cell. Biol.* **12**(2): 638–649.
70. Snir M., R. Ofir, S. Elias, D. Frank (2006). Xenopus laevis POU91 protein, an Oct3/4 homologue, regulates competence transitions from mesoderm to neural cell fates. *EMBO. J.* **25**: 3664–3674.
71. Morrison G.M., J.M. Brickman (2006). Conserved roles for Oct4 homologues in maintaining multipotency during early vertebrate development. *Development.* **133**: 2011–2022.
72. Soodeen-Karamath S., A.M. Gibbins (2001). Apparent absence of Oct 3/4 from the chicken genome. *Mol. Reprod. Dev.* **58**: 137–148.

73. Chambers I., D. Colby, M. Robertson, J. Nichols, S. Lee, S. Tweedie, A. Smith (2003). Functional expression cloning of Nanog, a pluripotency sustaining factor in embryonic stem cells. *Cell.* **113**: 643–655.
74. Chambers I., J. Silva, D. Colby, J. Nichols, B. Nijmeijer, M. Robertson, J. Vrana, K. Jones, L. Grotewold, A. Smith (2007). Nanog safeguards pluripotency and mediates germline development. *Nature.* **450**: 1230–1234.
75. Mitsui K., Y. Tokuzawa, H. Itoh, K. Segawa, M. Murakami, K. Takahashi, M. Maruyama, M. Maeda, S. Yamanaka (2003). The homeoprotein Nanog is required for maintenance of pluripotency in mouse epiblast and ES cells. *Cell.* **113**: 631–642.
76. Pan G., D. Pei (2005). The stem cell pluripotency factor NANOG activates transcription with two unusually potent subdomains at its C terminus. *J. Biol. Chem.* **280**: 1401–1407.
77. Chazaud C., Y. Yamanaka, T. Pawson, J. Rossant (2006). Early lineage segregation between epiblast and primitive endoderm in mouse blastocysts through the Grb2-MAPK pathway. *Dev. Cell.* **10**: 615–624.
78. Yamaguchi S., H. Kimura, M. Tada, N. Nakatsuji, T. Tada (2005). Nanog expression in mouse germ cell development. *Gene. Expr. Patterns.* **5**: 639–646.
79. Wernig, M., A. Meissner, R. Foreman, T. Brambrink, M. Ku, K. Hochedlinger, B.E. Bernstein, and R. Jaenisch (2007). In vitro reprogramming of fibroblasts into a pluripotent ES-cell-like state. *Nature.* **448**: 318–324.
80. Lefebvre V., B. Dumitriu, A. Penzo-Méndez, Y. Han, B. Pallavi (2007). Control of cell fate and differentiation by Sry-related high-mobility-group box (Sox) transcription factors. *Int J Biochem Cell. Biol.* **39**: 2195–2214.
81a. Zhao S., J. Nichols, A.G. Smith, M. Li (2004). SoxB transcription factors specify neuroectodermal lineage choice in ES cells. *Mol. Cell. Neurosci.* **27**: 332–342.
81b. Reményi A., K. Lins, L.J. Nissen, R. Reinbold, H.R. Schöler, M. Wilmanns (2003). Crystal structure of a POU/HMG/DNA ternary complex suggests differential assembly of Oct4 and Sox2 on two enhancers. *Genes. Dev.* **17**(16): 2048–2059.
82. Chew J.L., Y.H. Loh, W. Zhang, X. Chen, W.L. Tam, L.S. Yeap, P. Li, Y.S. Ang, B. Lim, P. Robson, H.H. Ng (2005). Reciprocal transcriptional regulation of PouVf1 and Sox2 via the Oct4/Sox2 complex in embryonic stem cells. *Mol. Cell. Biol.* **25**: 6031–6046.
83. Rodda D.J., J.L. Chew, L.H. Lim, Y.H. Loh, B. Wang, H.H. Ng, P. Robson (2005). Transcriptional regulation of Nanog by OCT4 and SOX2. *J. Biol. Chem.* **280**: 24731–24737.
84. Masui S., Y. Nakatake, Y. Toyooka, D. Shimosato, R. Yagi, K. Takahashi, H. Okochi, A. Okuda, R. Matoba, A.A. Sharov, M.S. Ko, H. Niwa (2007). Pluripotency governed by Sox2 via regulation of Oct3/4 expression in mouse embryonic stem cells. *Nat. Cell. Biol.* **9**: 625–635.

5

DERIVATION, CULTURE AND GENETIC MODIFICATION OF MOUSE EMBRYONIC STEM CELLS

Antonietta Giudice*, Joanna Lim, Anna Michalska

Monash Immunology and Stem Cells Laboratories, Monash University, Wellington Road, Clayton, Victoria 3800, Australia

INTRODUCTION

Mouse embryonic stem cells (mESCs) are permanent cell lines isolated from the inner cell mass (ICM) of blastocyst-stage embryos [1,2]. The essential features of mESC are their capacity to self-renew and pluripotency. *In vitro*, mESC can be grown indefinitely in the undifferentiated state and, under appropriate conditions, they can differentiate into many cell types representative of the three primary germ layers: the ectoderm, the endoderm, and the mesoderm [3]. This unique characteristic of mESC and their close similarity to pluripotent cells of the embryo, from which they are derived, makes them invaluable to study cell differentiation and development, provides a model of embryogenesis, as well as generates cells suitable for drug development and toxicology testing. The pluripotent nature of mESC is demonstrated by their ability to contribute to development *in vivo* [4]. When introduced into a host blastocyst, mESC recolonize the developing embryo and contribute to all tissues of chimeric animal, including functional gametes. This in turn permits the germ-line transmission of the mESC genotype to future offspring.

The ability to grow mESC *in vitro* makes them accessible to genetic manipulation and permits subsequent characterization and selection before such cells are used for the generation of chimeric mice [5–7]. In combination with the ongoing and rapid developments in gene discoveries,

---

*Anna.Michalska@med.monash.edu.au

sequencing, and manipulation, the value of mESC has been established as a powerful tool enabling targeted (site-directed) mutagenesis in mESC.

The importance of this field of research has been recognized by awarding the 2007 Nobel Prize in Physiology or Medicine to Professors Mario Capecchi, Martin Evans, and Oliver Smithies for their seminal discoveries, made in the 1980s. Martin Evans discovered pluripotent mESC and demonstrated that following genetic modification in culture they can generate genetically modified offspring. Mario Capecchi and Oliver Smithies, independently of each other, discovered the principle of gene targeting by homologous recombination in cells and developed methods to generate genetically modified 'knockout' mice.

Accordingly, genetically modified mESC and resulting mutant mice have been hallmarks of functional genomics, enabling the generation of a wealth of information on gene expression, regulation, function and behaviour *in vivo*, spatially, at the cell or organ level and temporally, during specific stages of development. In addition, the generation of modified mESC and mutant mouse models has improved our understanding of mechanisms of human diseases and served as a basis for therapeutic developments. It is not surprising that the list of 'customized' mutant mice is growing rapidly for a range of applications as these mice remain to be one of the most valuable tools in understanding mammalian biology in a complex and relevant environment.

In this chapter, we attempt to provide an overview of the current status of knowledge in derivation, culture, and genetic modification of mESC, and their application to generate mESC-derived mutant mice.

DERIVATION OF MOUSE EMBRYONIC STEM CELLS

Derivation, maintenance, and characterization of mESC evolved from earlier study on embryonal carcinoma cells (ECC)—the malignant stem cells of germ cell tumors called teratocarcinomas that occur spontaneously in testes [8] or ovaries [9] of certain strains of mice (reviewed in Ref. [10]). A number of ECC lines have been established over time (e.g. [11–14] and the full list is available in [15]). Several of these lines remain pluripotent following extended culture. After injection into extra-uterine sites of host mice, these ECC can form benign teratomas composed of various somatic tissues, and malignant teratocarcinomas that, in addition to mature tissues, contain a population of undifferentiated ECC. More importantly, clonal cell lines derived from a single ECC can also give derivatives of three germ layers following injection into a suitable host. Note that ECC are prone to developing karyotypic abnormalities before establishment as cultured cell lines [16].

Although ECC can differentiate *in vitro* into derivatives of all three embryonic germ layers (ectoderm, endoderm, and mesoderm) either spontaneously or following drug inducement, they vary in the degree of differentiation, with some lines showing a limited potential for differentiation [17, 18]. Once injected into a mouse blastocyst, ECC can participate in embryonic development and contribute to various tissues in chimeric mice [19, 20]. However, this often results in a restricted pattern of differentiation, in that rarely germ-line transmission occurs and frequent development of ECC-derived tumors is observed [16, 21–23]. Mouse ECC can also be induced experimentally by transplanting embryos into extra-uterine sites [24, 25]. A number of studies have demonstrated that ECC share many morphological, biochemical, and immunological characteristics with normal cells of the early embryo (reviewed in Ref. [26]). These characteristics and behaviour of the ECC served as a model to isolate comparable cells from mouse embryos.

The isolation of pluripotent mESC lines was first reported by two independent groups almost 27 years ago, in 1981 [1, 2]. Two different approaches were used in these original experiments. Evans and Kaufman [1] used blastocysts in which normal implantation was delayed by ovariectomy [27]. Following attachment of such 'delayed' blastocysts to feeder cells, the ICM outgrowths were picked, dissociated, and passaged. In contrast, Martin [2] used ICM isolated from blastocysts by immunosurgery–a procedure where trophoblast cells were selectively killed by exposure to mouse antiserum followed by treatment with complement [28].

There are a number of reports on establishing pluripotent mESC from single blastomeres obtained from early cleavage-stage mouse embryos [29–32]. The most efficient and the most commonly used method of mESC derivation is from the preimplantation embryo at the blastocyst stage (Fig. 5.1) [33]. At this stage, the blastocyst is composed of two morphologically distinct cell types: trophectoderm and pluripotent ICM, which following implantation will give rise to placenta and all foetal and some extraembryonic tissues, respectively. Although in the initial experiments Evans and Kaufman, and Martin [1, 2] used 'delayed' blastocysts or isolated ICM, it was subsequently shown that the mESC isolation from the intact, expanded blastocyst is as efficient. In this approach, the whole blastocyst (3.5 days post-fertilization) is allowed to attach to either fibroblast feeders or gelatin-coated plastic. After 4–6 days of culture, when the blastocyst reaches a stage equivalent to the early post-implantation embryo, the ICM outgrowths are disaggregated and replated onto fresh feeder cells. Following 2–5 days of culture, several types of primary colonies normally appear. These can be classified on the basis of morphology, and colonies with the characteristics of undifferentiated cells are individually picked and expanded further through culture [33].

More recently, stem cells were derived from the epiblast of the early post-implantation embryo [34]. These cells, termed EpiSC (post-implantation epiblast-derived stem cells), have a wide developmental potency as determined by *in vitro* and *in vivo* differentiation. However, following blastocyst injection or morula aggregation, these cells do not contribute to chimeric mice possibly due to lack of compatibility with the pre-implantation embryo environment.

Another source of embryos from which mESC can be derived are parthenogenetically activated oocytes. Parthenogenesis is the process by which embryonic development is initiated without male contribution. In mouse, a number of experimental methods were developed to initiate parthenogenesis [35]. Such embryos are capable of normal development during pre-implantation stages but only limited development occurs following transfer to surrogate mothers and implantation [36, 37]. Although parthenogenetic ESC (pESC) can be derived with a relatively high efficiency, these cells have a limited potential to differentiate as compared to fertilized embryo-derived mESC, due to the epigenetic disorders arising from the lack of male genome [38, 39].

Pluripotent mESC have also been isolated directly from primordial germ cells, a small population of epiblast-derived cells that migrate to genital ridges during development and eventually give rise to sperm or oocytes [40-42]. The established cells, termed embryonic germ cells (EGC), are morphologically indistinguishable from blastocyst-derived ESC. They express markers typical of pluripotent mESC, differentiate efficiently *in vitro* and *in vivo*, and, following injection into a host blastocyst, give rise to germ-line chimeric mice [40, 43].

Somatic cell nuclear transfer (SCNT) offers yet another method for producing embryos from which mESC can be derived (Fig. 5.1) [44, 45]. In this approach, the nucleus from an adult somatic cell is transferred to an unfertilized oocyte devoid of maternal genetic material and subsequent blastocysts, which develop under the control of the transferred nucleus and are used for derivation of ESC [46]. These cells are referred to as NT-ESC to distinguish them from cells derived from fertilized blastocysts. The NT-ESC display the characteristic morphology and expression of ESC marker genes, are capable of extensive differentiation into all three embryonic germ layers *in vitro* and *in vivo*, and also contribute to germ-line chimeric mice [45, 47–49].

Recently, ESC-like cells (induced pluripotent stem (iPS) cells) have been derived from terminally differentiated somatic cells without the use of SCNT-derived embryos (Fig. 5.1). Refer to *Future Applications of Mouse Embryonic Stem Cell Technologies* for a more detailed discussion.

The efficiency of the isolation of mESC mainly depends on the strain of mice. The majority of currently available mESC lines has been derived from the inbred strain 129—the strain characterized by a high incidence of spontaneous testicular teratomas or teratocarcinomas, and from which a large number of ECC lines were derived [25]. The reported success rate of establishing mESC lines from the permissive 129 strain line is between 10% and 30% [33]. Mouse ESC have also been derived from the C57BL/6 strain but with a lower efficiency [50–54]. Isolation of germ-line transmitting mESC lines from only a few other inbred strains of mice have also been reported ([55] and references within). These lines, in addition to low efficiency of isolation, can be difficult to culture *in vitro* and generate chimeras less efficiently than the permissive 129 strain.

CULTURE OF MOUSE EMBRYONIC STEM CELLS

Mouse ESC must be cultured in optimal culture conditions that maintain their normal karyotype, and preserve their pluripotency and self-renewal capacity. Detailed protocols have been published describing the routine propagation of mESC [33,56,57]. Mouse ESC can be cultured in a number of conventional tissue culture media, such as Dulbecco's modified minimal essential medium (DMEM) or a mix of DMEM and Ham's F12 (DMEM/F12), supplemented with 10–20% of either foetal calf serum (FCS) or serum replacement (SR), non-essential amino acids (NEAA), and 2-mercaptoethanol. It has to be pointed out that the quality of FCS is critical and a number of batches must be tested to select a suitable product. To prevent differentiation, the mESC are co-cultured on fibroblast feeder cells in the presence of the cytokine, leukemia inhibitory factor (LIF). Mouse ESC are usually propagated every 2–3 days. Colonies are enzymatically dissociated with trypsin to obtain a single-cell suspension that is plated onto fresh feeder cells.

The first mESC were isolated on inactivated STO fibroblasts—a permanent cell line derived from SIM strain of mice that are Thioguanine- and Ouabain-resistant [58]. Other established embryonic fibroblast cell lines, such as C3H 10T1/2 and BALB-3T3/A31, were also shown to have the capacity to maintain undifferentiated mESC in culture [59, 60]. However, the most commonly used feeder cells are mouse embryonic fibroblasts (MEF) isolated from mid-gestation mouse foetuses. These cells are easy to isolate and propagate in culture and can be expanded to large numbers according to standard protocols [61]. Feeder cells are produced by inhibition of cell division with mitomycin C- or Y-irradiation. Following treatment, the feeder cells remain metabolically active, and cells produce extracellular matrices and a number of soluble factors that are important for proliferation and maintenance of undifferentiated ESC.

The main factor secreted by mouse feeder cells that is responsible for supporting mESC is the cytokine LIF, which not only allows mESC to grow but also to be derived *de novo* in the absence of feeders in a medium containing LIF [62–64]. LIF acts by activating the JAK/Stat3 signaling pathway, and activated Stat3 seems to be sufficient for the maintenance of undifferentiated mESC [65, 66]. However, LIF alone does not support clonal growth of mESC in feeder- and serum-free conditions. It has been postulated that other factors, provided by feeder cells or serum, such as bone morphogenic protein (Bmp) and Wnt act synergistically with LIF to maintain mESC self-renewal [67, 68].

In an effort to develop chemically defined culture conditions for mESC a number of feeder- and serum-free protocols have been proposed [68–71]. One such medium formulation consists of DMEM/F12 supplemented with N2 and B27 (commercially available growth supplements used to enrich serum-free media), Bmp4 and LIF ([68]; commercially available ESGRO complete medium). The N2 and B27 components improve viability of mESC, while Bmp, in combination with LIF, promotes mESC self-renewal. This action of Bmp is mediated through the Smad transcription pathway that in turn activates the inhibitor of DNA-binding (Id) proteins resulting in the suppression of neural differentiation. Mouse ESC grown under these conditions can be expanded from single cells, display typical characteristics of pluripotent stem cells (including multilineage differentiation), and contribute to generation of germ-line chimeras. New mESC lines have been isolated from 129 and C57BL/6 strains of mice using this medium [72].

Recently, a modified formulation of the above medium and novel culture conditions that allow scale-up mESC expansion have been published [69]. The medium, termed ESN2, consists of DMEM/F12, N2, LIF, and basic fibroblast growth factor (bFGF). The omission of exogenous Bmp is probably compensated by endogenous Bmp since mESC are cultured in suspension in nonlimiting density conditions. The cells are grown as small spheres and harvested by filtration, thus removing any differentiated adherent cells. The extended growth in these conditions does not seem to affect the pluripotency of mESC. The cells express a number of stem cell markers, differentiate *in vitro* to neurons and cardiomyocytes, and contribute to chimeric mice; however, their germ-line transmission has not been tested [69].

Current experimental approaches and techniques aimed at drug discovery through high-throughput molecular screening require a large number of ESC. This demand has been addressed in the past few years through the development of bioreactor technology where large-scale expansion of mESC is carried out in suspension culture [73, 74]. This has been possible because mESC can be grown in the presence of LIF without MEF. However, mESC are anchorage-dependent and do not grow as a single-cell suspension; it is necessary to grow them as aggregates [75–77] or on microcarriers [76, 78, 79]. These suspension bioreactor techniques allow the expansion of some mESC lines up to 100-fold, with the resultant cells expressing markers associated with pluripotency, forming embryoid bodies (EB), and differentiating into three germ-line lineages *in vitro* and in teratomas *in vivo*. However, their contribution to chimeric mice remains to be demonstrated.

Although mESC can be grown conveniently in a medium supplemented with LIF or in more defined media formulations, it is still recommended to generate stock mESC on feeder layers to prevent culture deterioration and to maintain pluripotency [80].

PROPERTIES OF MOUSE EMBRYONIC STEM CELLS

Undifferentiated mESC have a very characteristic morphology—the cells are small (approximately 12–15 μm in diameter), with a large, clear nucleus containing prominent nucleoli, and very little cytoplasm (a high nuclear-to-cytoplasm ratio). They form tight, three-dimensional, dome-shaped colonies with sharp edges (Fig. 5.2). Any flattening of a colony, development of a rough appearance, or formation of fibroblast- or endoderm-like outgrowths are indicative of differentiation that will lead to the loss of pluripotency. When maintained under the proper conditions, mESC have unlimited self-renewal capacity and an mESC line can be cloned from a single cell. This is a very important feature of mESC that allows selection and expansion of genetically manipulated clones.

Mouse ESC are characterized by rapid cell division with a proliferation rate of approximately 10–11 hours and with a short G1 cell cycle phase [81, 82]. Mouse ESC also display a high level of mTert gene expression and corresponding telomerase activity, both of which are rapidly downregulated during differentiation [83]. The elevated activity of telomerase in mESC correlates with the ability of stem cells to divide indefinitely in culture [84].

The widely accepted standard markers for undifferentiated mESC are the surface marker stage-specific embryonic antigen-1 antigen (SSEA-1; [85]), the intracellular enzyme alkaline phosphatase (AP; [86]), and the nuclear transcription factor octamer 4 (Oct4; [87]; refer to section *Molecular basis of pluripotency*) (Fig. 5.2). E-cadherin (E-cad), a cell adhesion molecule, is also associated with (but not restricted to) the undifferentiated state of mESC [88]. Loss of pluripotency, following spontaneous or induced differentiation, is associated with the progressive loss of expression of these markers [89].

Mouse ESC cultured under optimal conditions maintain a normal (euploid) karyotype and have a modal chromosomal number of 40 (19 pairs of autosomes and one pair of sex chromosomes). The majority of existing mESC lines are derived from male embryos and are 40XY. The XY chromosome component appears to be stable in culture compared to XX lines that have a high incidence of spontaneous deletion of one of the active X chromosomes [90,91]. As with any cell

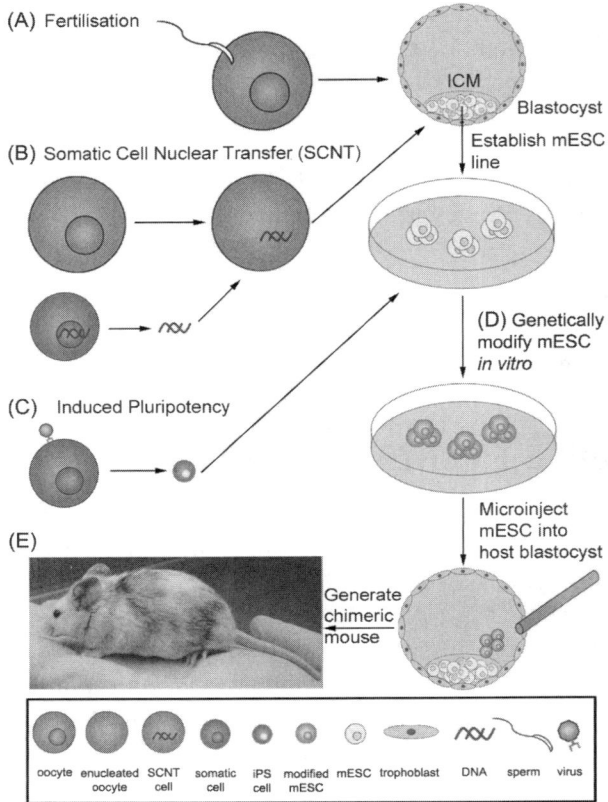

▶ **Fig. 5.1** *Isolation of mESC lines and generation of chimeric mice.*
Mouse ESC can be derived from a number of sources. (A) The most common approach is to isolate the mESC from a blastocyst that results from fertilization of an oocyte by sperm. At this stage (3.5 days after fertilization), the blastocyst consists of an ICM and trophectoderm. Mouse ESC lines are established from the ICM. (B) mES-like cells can also be derived following SCNT, the process when the genome from a somatic cell is transferred into an enucleated oocyte. (C) Induced pluripotent stem (iPS) cell lines that display ESC-like characteristics can be generated by the dedifferentiation of a somatic cell, by introducing a set of defined factors by retroviral transduction. (D) The genome of mESC can be manipulated in vitro, and upon introduction into a host blastocyst, the modified mESC are able to contribute to all cells of a chimeric mouse. (E) A photograph of a chimera generated from the injection of mESC derived from a pigmented strain of mouse into an albino host blastocyst. Chimeras are bred with wild type mice to produce a transgenic mouse line carrying the genetic modification initially introduced to mESC. (For colour figure see Plate 5).

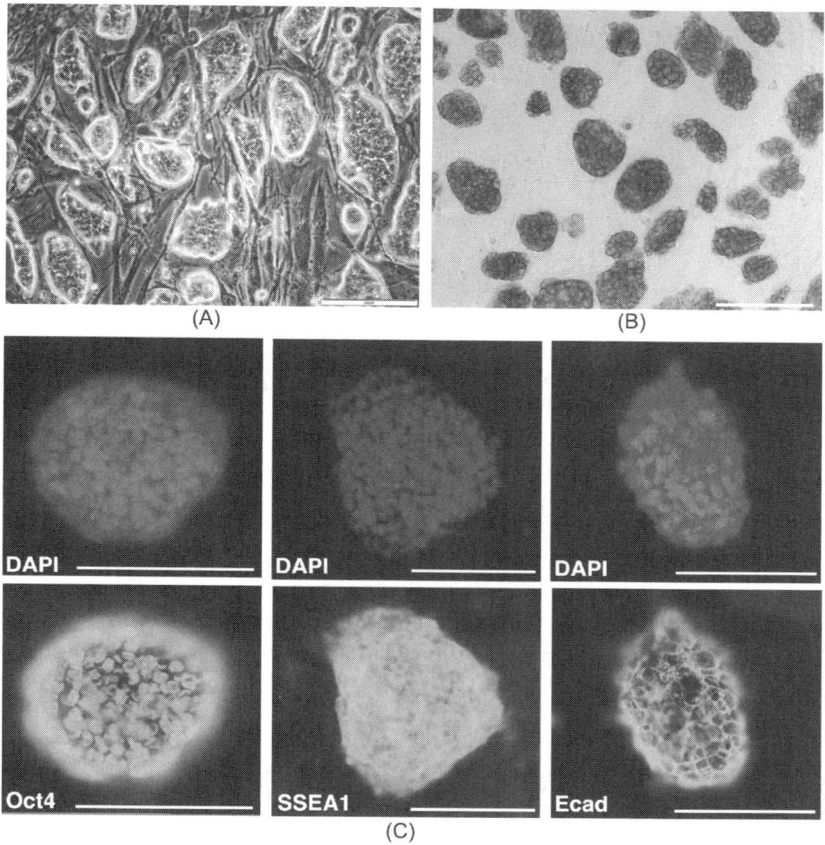

▶ Fig. 5.2 *Characteristics of mESC:*
Morphology and the expression of pluripotent markers of mESC cultured on a fibroblast feeder cells and in the presence of LIF. (A) Morphology of mESC colonies. (B) Expression of alkaline phosphatase. (C) Expression of nuclear (Oct4) and cell surface (SSEA-1, Ecad) markers. Images in (C): immunofluorescence microscopy; blue, DAPI staining of nuclei; green, respective markers. Scale bar: (a) and (b) 200 μm; (c) 100 μm. (For colour figure see Plate 6).

types grown continuously *in vitro*, the long-term culture of mESC has been associated with karyotypic changes and after 20 passages they can rapidly become aneuploid, most typically with the loss or gain of an entire chromosome [92]. There is a linear correlation between the proportion of euploid cells and their ability to form functional germ cells in chimeras. Clones with less than 50% of euploid metaphases fail to contribute to all tissues of chimeras, including functional germ cells. One of the most common chromosomal abnormalities in mESC is trisomy 8, detected in over

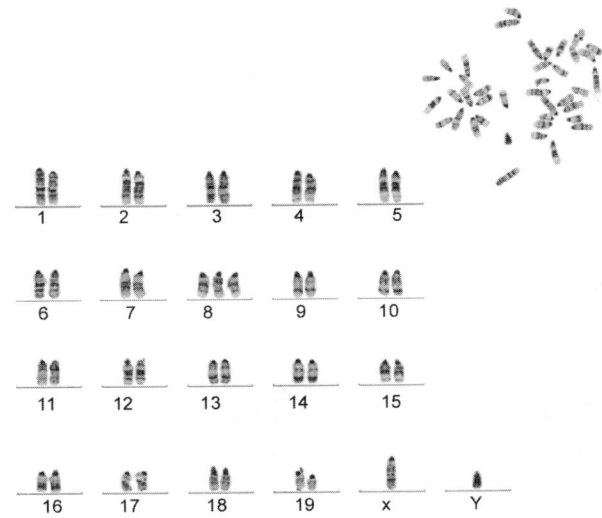

▶ **Fig. 5.3** *Karyotype of abnormal mESC.*
Giemsa-banded metaphase spread and karyotype analysis of a male mESC showing XY sex chromosome complement and an additional chromosome 8. Trisomy 8 is a common abnormality that appears following long-term culture. Image provided by Lucy Gugasyan, Cytogenetics Unit, Monash Medical Centre, Australia.

80% of all analyzed aneuploid cell lines (Fig. 5.3) [93, 94]. These cells have a higher growth rate compared to cells with the normal karyotype and can overgrow the normal cells in five passages. Since the ability to contribute to the germ-line is the most important feature of mESC, and since genetic manipulation procedures sometimes require an extensive culture period, karyotype analysis is a prerequisite for ensuring the normalcy of ESC. This could consist of a simple chromosome count to determine modal distribution number and ideally should be followed by Giemsa (G)-banding that allows examination of chromosomal integrity.

Molecular Basis of Pluripotency

The molecular mechanisms regulating the self-renewal of pluripotent mESC are poorly understood. Currently available evidence suggests that the pluripotency of mESC is regulated by a network of key transcription factors [95, 96]. Of these, the best characterized is Oct4 (also known as Oct3/4 or POU5F1), a member of the POU family of homeodomain transcription factors, that is specifically expressed in preimplantation embryos, epiblast and germ cells, and also in undifferentiated ESC [66, 87, 97]. The expression of Oct4 diminishes upon differentiation and loss of pluripotency. Oct4 is required not only for self-renewal of mESC but it also appears to inhibit differentiation. However, precisely defined amounts of Oct4 are required to maintain mESC pluripotency. Repression of Oct4 below 50% of normal expression causes mESC differentiation into trophectoderm, while less that twofold increase in expression leads to differentiation toward the extraembryonic endoderm and mesoderm [98]. Interestingly, in the presence of forced constitutive expression of Oct4, LIF is still required to maintain the self-renewal of mESC, indicating that Oct4 is not a 'master' gene for pluripotency [98].

Nanog is another divergent homeoprotein, which is expressed in the ICM of a blastocyst and mESC, and appears to play an essential role in the maintenance of pluripotency [99, 100] [101]. The expression of Nanog is restricted to undifferentiated cells and is downregulated during differentiation. Like Oct4, Nanog appears to be involved in both self-renewal and inhibition of differentiation of mESC. Disruption of Nanog expression in mESC results in differentiation into the extraembryonic endoderm lineages, while its over-expression can maintain self-renewal of mESC independent of LIF/Stat3.

Sox2 (sex determining region Y-box 2), a member of the high-mobility group (HMG) family proteins, is yet another transcription factor required for efficient self-renewal of mESC. Sox2 is co-expressed with Oct4 in the ICM of the preimplantation embryo, mESC and germ cells, and occupies many gene targets with Oct4 [102].

Another transcription factor implicated in the maintenance of pluripotency of the ICM and mESC that interacts with Oct4 is the fork-head family transcription factor FoxD3 (forkhead-box D3; [103]). However, similar to Sox2, the expression of FoxD3 is not restricted to pluripotent mESC and can be detected in somatic stem cells during later development.

Novel screening strategies, such as microarray analysis and genome-wide chromatin-immunoprecipitation (ChIP) array analysis, identified additional candidate genes and their products involved in the maintenance of pluripotency of mESC and provided evidence of a complex interconnected transcriptional network regulating the self-renewal of mESC [104–106].

Differentiation Potential of Mouse Embryonic Stem Cells

One of the most striking features of pluripotent mESC is their ability to differentiate into multiple cell and tissue types *in vitro* and *in vivo*. Because of this characteristic, mESC can be used for the analysis of gene function, pharmacological screening, and potential development of cell therapy, as well as for designing suitable models of differentiation and gene expression in early development.

Differentiation *in vitro*

Many protocols have been developed to differentiate mESC *in vitro*. One of the simplest is to allow cells to differentiate spontaneously following prolonged, monolayer adherent cultures in the absence of LIF and feeder layers. However, the routine method used is to grow the single mESC under nonadherent conditions without LIF or feeders. Under these conditions, the cells form three-dimensional spherical aggregates called EB because they resemble tissues of an early embryo [107]. Formation of EB can be achieved by three basic methods: suspension culture in liquid or semisolid, methylcellulose-containing media, and culture in hanging drops [108]. The pattern of differentiation of mESC as EB parallels that of the early embryo, and the structure of EB differs depending on the length of the culture [107]. After 2–3 days of culture, simple EB are formed that resemble a morula-stage embryo. From 3 to 8 days of culture, cystic EB develop that resemble blastocyst- and egg-cylinder-stage embryo. Cystic EB consist originally of two layers: inner ectoderm (representing pluripotent tissue) and outer endoderm, followed by the appearance of mesoderm as the culture progresses. After 8–10 days, EB expand to large cystic structures that are analogous to the visceral yolk sac of the postimplantation embryo. When the EB are replated in monolayer cultures, a range of specialized cell types grows out of them including contracting cardiomyocytes, neuroectodermal, and hematopoietic cells ([109] and references within). By exposing differentiating mESC and EB to specific extracellular factors known to influence lineage choices in developing embryos, differentiation can be directed into specific tissue types [110–114].

There are two other methods used to induce mESC differentiation *in vitro* that do not rely on the formation of EB. One is the co-culture with feeder cell lines where direct interaction with specific stromal cells promotes mESC differentiation. Examples of the most commonly used stromal feeders include murine stromal cell lines PA6 and MS5, both effective inducers of neural differentiation [115–117], and OP9 cell line that directs mESC differentiation toward hematopoietic lineages [118]. Another technique is to differentiate mESC in a monolayer on specific extracellular matrix proteins type IV collagen or simply on gelatin [119, 120].

Differentiation of mESC to a broad spectrum of functional cell and tissue types using the above techniques has been achieved, and the reader is directed to a number of recent reviews summarizing this subject [3, 109, 121–123].

Differentiation *in vivo*

One of the standard methods to assess full differentiation capacity of mESC is to evaluate their potential to form teratomas [1, 2]. This is done by transplanting the mESC to extra-uterine site (such as under the kidney or testicular capsule) of severe combined immunodeficient (SCID) mice. After 7 to 10 weeks, complex teratomas develop that consist of a variety of differentiated cell types, such as epithelia, neural, and connective tissues, as well as undifferentiated stem cells, which are derived from the ectoderm, the endoderm, and the mesoderm, thus demonstrating the pluripotency of the mESC (Fig. 5.4).

The most important feature of pluripotent mESC is their ability to colonize all foetal tissues, including the germ-line when the cells are introduced into a host embryo. This is considered the 'gold standard' of mESC pluripotency and has been instrumental in the production of mutant chimeric mice from genetically modified mESC. Chimeric mice have been produced by either co-culture of eight-cell embryos on a lawn of ESC [124], aggregation between two eight-cell embryos [125, 126], or injection into morula-stage embryos [127]; however, the most commonly used method is the injection of ESC into a blastocyst-stage embryo [4, 125]. In this approach, between 10 and 15 mESC are injected into the blastocoel cavity using a micropipet and micromanipulators, and such blastocysts are then transferred into pseudo-pregnant foster female mice. Mouse ESC participate fully in normal development giving rise to chimeric mice.

By selecting distinguishable coat colour alleles of the donor mESC and the host blastocyst, chimeras can be easily identified by differential hair pigmentation (Fig. 5.1). A good indication of a

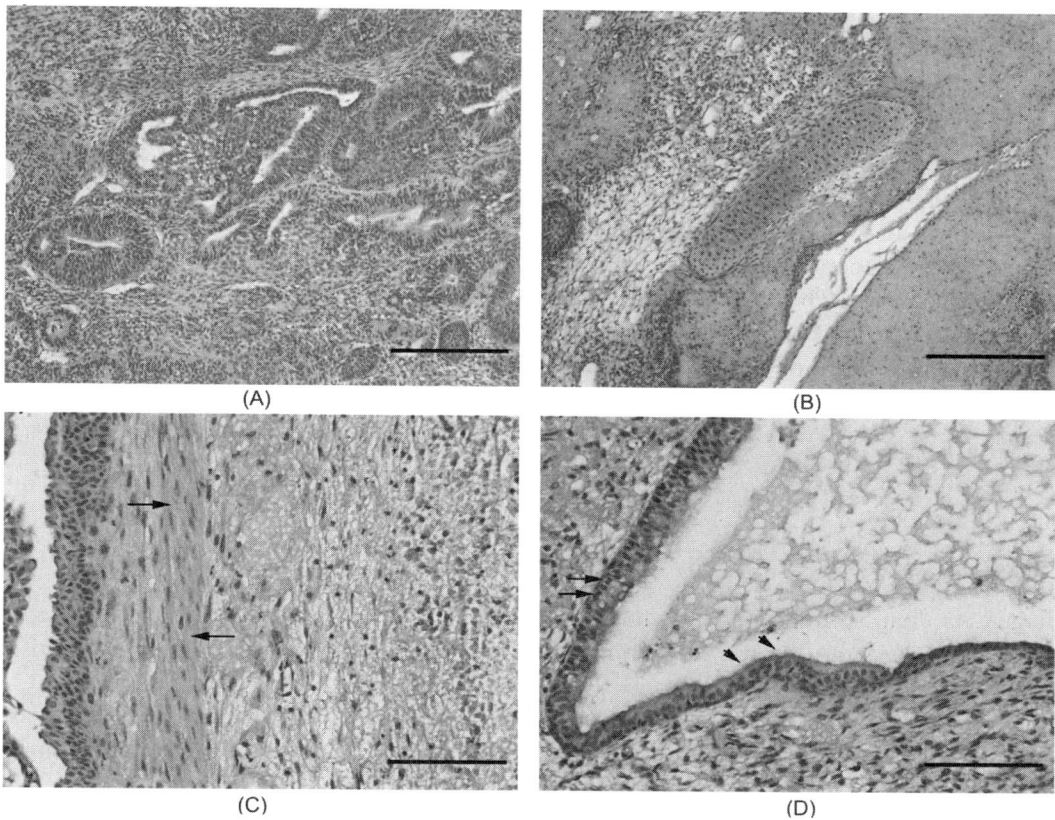

▶ **Fig. 5.4** *In vivo differentiation of mESC in teratomas.*
Haematoxylin and eosin staining of paraffin sections of a 7-week teratoma, showing cells of all three germ-line lineages. (A) Neural epithelium (ectoderm: (B) Cartilage, (C) Mesoderm and/or ectoderm: Smooth muscle cells; (arrows, mesoderm). (D) Columnar epithelium with goblet cells (arrows) and cilia (arrow heads; endoderm). Scale bar: (a) and (b) 100 μm; (c) and (d) 50 μm. (For colour figure see Plate 7).

germ-line capability of mESC is the high rate of male chimeras. This results from the conversion of female embryos to male embryos by the XY genotype of the mESC. As discussed earlier, most lines are 40XY and when randomly injected into a male or a female blastocyst, they can convert indifferent gonad of an XX recipient embryo into testicular development thus giving rise to a phenotypic male chimera [4, 5].

Chimeric mice are bred with wild-type mice. In the case of genetically modified mice, half of the ESC-derived first-generation offspring will carry the introduced gene as heterozygotes. The interbreeding of heterozygous siblings yields homozygous animals, which then can be tested for the effects of the gene alteration. In this conventional approach to generating a gene-targeted mouse using genetically modified ESC, transmission through the germ-line is slow and laborious. It can take up to 12 months from the identification of the targeted heterozygous mESC clone to obtain a homozygous gene-targeted mouse. Not only does breeding take a long time, but also all produced mice have to be genotyped at each stage.

The production of gene-targeted mice can be dramatically shortened by application of a technology where the mutated mESC clones are aggregated with, or injected into, tetraploid embryos [126, 128, 129]. In this system, tetraploid embryos with a double number of chromosomes are generated experimentally by, for example, electrofusion of the blastomeres of two-cell stage embryo. They are restricted in their developmental potential and give rise mostly to primitive endoderm and trophectoderm, which will form extra-embryonic tissues, but not the embryo proper. The combination of tetraploid embryos with diploid mESC results in the development of entirely ESC-derived foetuses. Thus, the generation of a homozygous mutant mouse can be achieved within 8 weeks. However, the efficiency of ESC-tetraploid aggregation technology is still limited due to the low frequency of recovering live ESC mice.

GENETIC MODIFICATION OF MOUSE EMBRYONIC STEM CELLS

Stable genetic modification of mESC can be achieved by chemical mutagenesis and by integration of exogenous DNA into the mESC genome. Chemical mutagens such as N-ethyl-N-nitrosourea (ENU) or methanesulfonic acid ethyl ester (EMS) cause point mutations, while ICR19—an intercalating agent that commonly targets stretches of guanine—results in frameshift mutations. These mutations can be difficult to detect and require large-scale screening to identify; however, they do permit the generation of subtle mutations, which might be difficult to accomplish through the alternative approach. The chemical mutagenesis approach will not be discussed further in this chapter as it has been recently reviewed elsewhere [130]. Random and targeted genomic alterations can also be introduced following the delivery of exogenous DNA. The various delivery methods include electroporation, nucleofection, chemical-mediated approaches (lipofection and cationic-polymer based), and viral transduction. Different methods may be best suited for different applications [131]. The method of choice will also depend on cost, time, level of effort required, and the availability of equipment, reagents, and expertise [132]. The following section will provide an overview of the current status of genome engineering of mESC via the DNA delivery approach, including examples of some of its applications.

Random Integration

Random integration involves insertion of exogenous DNA at any location in the genome. It represents a less time-consuming and cumbersome approach compared with the alternative gene-targeting approach, especially when viral vectors (typically retroviral or lentiviral) are employed. A transgenic construct contains a promoter fragment, with or without an enhancer(s), driving expression of a cDNA encoding a gene of interest, followed by polyadenylation (poly A) signals. Depending on the promoter used, it is possible to induce gene expression constitutively, or in a spatial (tissue-dependent) and temporal (developmental stage-dependent) manner. Often a positive selection marker is also included in the vector to facilitate elimination of clones not harboring the construct. Transgenes have been used for a variety of applications in mESC. They can be used to mediate specific gene knockdown by RNA interference (RNAi; discussed further) and for ectopic gene expression to investigate perturbations in gene expression [133, 134]. Transgenes can also be used to investigate the intracellular localization of gene products [135] and protein-protein interactions [136, 137], provided that there are mechanisms to detect the recombinant protein. Indeed, tags such as green fluorescent protein (GFP), hemagglutin (HA), and glutathione-S-transferase (GST) are often incorporated in the protein for this purpose. When mESC are subjected to the appropriate culture conditions, transgenes can be used to force them to adopt specific characteristics [135, 138–140]. Importantly, transgene expression can be regulated in an inducible fashion to enable controlled differentiation (see *Inducible Systems for Regulating Expression of Ectopic Sequences*). The transgenic approach can also be used to introduce reporter genes or selectable markers into mESC for labeling and enrichment of specific derivatives [140–142], and for gene identification (see *Gene Trapping*).

Bacterial artificial chromosomes (BAC)—capable of bearing up to ~350 kilobases (kb) of genomic sequences—and BAC-related vectors can also be introduced into mESC to achieve ectopic expression of a gene of interest. This includes a BAC in which a gene of interest has been modified by introducing a reporter gene. In this system the expression of the reporter gene is regulated by the control elements within BAC [106,143,144]. Given the presence of a large amount of homology to a region in the genome, BAC can undergo homologous recombination with the corresponding endogenous sequence, resulting in the generation of gene-targeted clones, as well as random integrants [145, 146].

The limitations of this random integration approach and methods to overcome them have been described [131]. Briefly, this approach has the potential to cause functional disruption or oncogenic activation of endogenous genes and can result in differential expression of the transgene depending on the integration site (referred to as position effects) and copy number. Furthermore, for the promoter-driven approach, promoter fragments do not always harbor sufficient regulatory sequences for directing accurate expression. In the case of randomly integrated BAC transgenes, these position effects are less likely. The large size of the BAC genomic sequences should contain all relevant gene regulatory elements that will allow for a more precise gene-specific expression. BAC sequences may, however, influence expression of adjacent endogenous genes, and can result in ectopic expression of other genes residing within the BAC. All these drawbacks require analysis of multiple cell lines to determine transgene-specific effects.

Other approaches can be employed to overcome or diminish some or all of the problems associated with the transgenic approach. These include the use of the phiC31 integrase technology, which appears to facilitate the integration at specific genomic 'hot spots' [147], thus serving to minimize insertional mutagenesis. Recombinase-mediated cassette exchange (RMCE) can also be used with Cre-, Flp- or phiC31-based systems to insert sequences into already targeted sites that are transcriptionally active and not functionally affected by alteration, such as the Rosa26 locus [148, 149]. In both instances, the expression of an additional transgene—the recombinase—is required. Even though the recombinase only needs to be transiently expressed, the potential for genomic uptake of the expression vector still exists. Alternatively, gene targeting can be used for site-specific integration, making it possible to resolve all complications of the transgenic approach.

Gene Trapping

Gene trapping is a term used to describe a random insertion approach applied to mESC for identification of genes. This method uses a selection marker (antibiotic resistance or reporter gene) that is often either promoterless or possesses a promoter but lacks poly A signals. The selection marker expression becomes activated upon integration within a gene, provided that, in the first instance, it is expressed in mESC. Other modifications, such as internal ribosome entry sites (IRES) or splice acceptor sites, might also be included to improve the efficiency of transgene expression when not inserted within or in frame with exonic sequences. Often, gene-trapped mESC lines result in gene knockouts; however, since the insertion site is random, the resulting mutants are not always completely null. As such, gene trapping can result in the generation of truncated native proteins that might be fused to the selection marker, and normal (if the gene is subject to alternative splicing and the exon containing the genetic modification is normally spliced out of a message in an isoform) and aberrantly spliced versions. Recently, all major groups in the gene-trap community have formed a collaboration to centralize their resources [150]. They have created repositories for their >135,000 well-characterized gene-trap mESC lines. These lines are available to the scientific community and are accessible via the International Gene Trap Consortium (IGTC; http://www.genetrap.org).

Viral Delivery Methods

Stable viral delivery into mESC is commonly achieved through the use of lentiviral vectors derived from HIV-1. These vectors represent efficient carriers to deliver DNA into cells and do so randomly. They are also very effective due to their resistance to silencing, unlike other retroviruses. Lentiviral vectors have been used to stably express transgenes in mESC after long-term culture, after *in vitro* (in EBs) and *in vivo* (in teratomas) differentiation, in chimeric mice generated from them and in the germ-line [151, 152]. These vectors are also effective because multiple copies tend to be inserted into the genome of a single cell. The copy number can be easily modified by varying the viral dose to reduce the undesirable consequences of random integration. Lentiviral transduction of gene-specific short hairpin (sh) RNA in mESC is the current preferred method of achieving gene knockdown by RNAi in these cells and mice [104, 153–155] (reviewed in [152]; see *RNA Interference*).

Adenoviral (Ad) vectors, which exist as self-replicating or non-replicating entities (episomes) separate from the host genome, can be employed to achieve transient gene expression. This is because these vectors become diluted and eventually lost as the cell undergoes cell division. Adenoviral vectors have been applied to achieve transgene expression in mESC [156]. Mouse ESC transduced with Ad vectors carrying the LacZ gene driven by one of four different promoters, identified the EF-1α promoter to be the most superior [157]. In the same study, the Ad vector system demonstrated utility in directing differentiation of mESC. It allowed the evaluation of ectopic expression of a number of transgenes, including the Oct4 transgene, which resulted in enhanced differentiation toward the three germ layers.

Gene Targeting (Homologous Recombination)

Gene targeting simply means the introduction of planned, site-specific genomic alterations through homologous recombination. Provided that there is sufficient sequence information about a particular gene of interest, virtually any modification is possible in the attempt to address a specific biological question. Indeed, extensive gene-sequencing efforts and the availability of freely accessible sequence databases for a number of inbred mouse strains have made this feasible. Homologous recombination between injected foreign DNA was first documented in mammalian cells where it was shown to be responsible for the insertion of multiple copies of plasmid DNA at a single genomic location (concatemerization) [158]. Subsequently, it was demonstrated in cells transfected with either of two non-overlapping mutants of a *neomycin phosphotransferase* (*neo*) resistance marker that could not yield stable transfectants in the presence of an antibiotic G418 unless cotransfected [159]. In 1985, homologous recombination between sequences contained within a plasmid and that of an endogenous gene was shown for the first time in human cells [160]. This led to the earliest modifications of mESC at the *hypoxanthine phosphoribosyltransferase* (*Hprt*) locus [6,7,161] and derivation of gene-targeted mice [162]. These same principles and techniques have been used for the generation of a variety of genetically modified mESC and mice for almost 20 years [163,164]. Indeed, vast improvements in the gene-targeting technologies have been made since [165–168].

In order to introduce a specific mutation into the mouse genome, it is essential to generate a suitable targeting vector that contains a sufficient amount of homologous sequences to enable it to localize to the desired region of the genome and undergo homologous recombination. A targeting vector is comprised of two homology arms that flank a positive selection marker, typically neo thus conferring resistance to the antibiotic G418 (and foreign sequences in the case of knockins), and a bacterial vector backbone. Recently, fluorescent markers have also been used for selection [169]. The basic design of a gene targeting vector is illustrated in Fig. 5.5.

Two types of targeting vectors have been used, namely replacement and insertion vectors (Fig. 5.6). Replacement vectors result in the exchange of homologous sequence, including the intervening foreign sequence, in the vector with homologous sequence in the genome. Insertion vectors have been used less frequently and result in the insertion of the entire vector into the target genomic site. The vectors are structurally identical except for the position of the linearization site. Replacement vectors are linearized within the vector backbone, whereas insertion vectors are linearized within one of the homology arms.

The positive-negative selection strategy was developed to reduce the number of random integrants and enrich for gene-targeted events [161]. This approach dramatically improves the efficiency of obtaining correctly targeted clones as it allows elimination of numerous random integrants (which are much more common) through negative selection. The negative selection

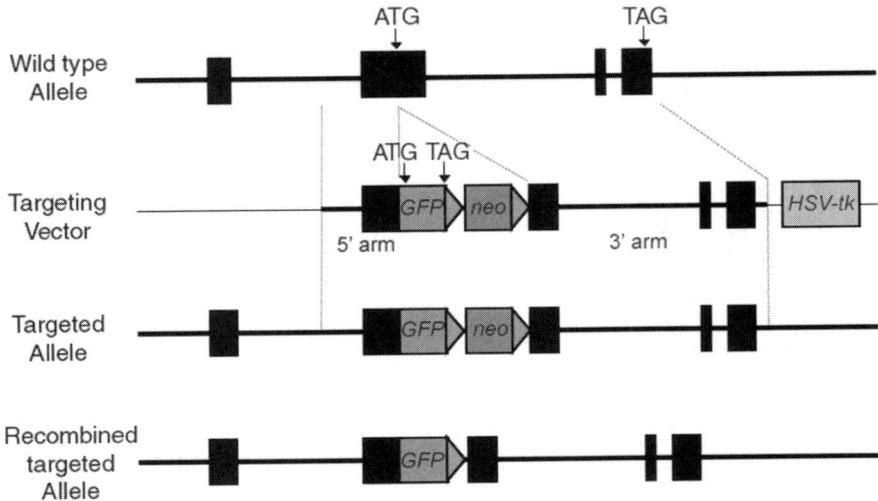

▶ **Fig. 5.5** *Basic knockin gene-targeting strategy.*
The approach shown targets the 5' end of a gene. Reporter gene expression is controlled by the regulatory elements of the endogenous gene. Hence, sequence deletions should be avoided. Exons are represented as black boxes. The 5' and 3' homology arms flank the reporter gene (GFP) and the selectable marker (neo). The neo shown is floxed, i.e. flanked by loxP or Frt sites (triangles) so that it can be removed after gene targeting. Correctly targeted mESC are identified by positive–negative selection, followed by molecular analysis. ESC that have integrated the neo are identified by positive selection with G418. Negative selection with ganciclovir is used to eliminate cells that have incorporated the HSV-tk. Cells harboring the HSV-tk represent random integrants, as the HSV-tk lies in the vector backbone (thin line) and cannot be involved in homologous recombination. The selection marker is removed following exposure to Cre or Flp recombinase, usually by transient transfection with an expression vector. A single recombination site remains in the modified gene. (For colour figure see Plate 8).

cassette—usually the *Herpes Simplex Virus thymidine kinase* (*HSV-tk*) gene-is included in the targeting vector but is not flanked by homologous sequences, so it cannot participate in homologous recombination. Random integration usually involves insertion of the entire linearized vector; therefore the *HSV-tk* becomes integrated in the genome. An antiviral drug ganciclovir is toxic to these cells, hence serves as a means to eradicate them.

Conventional Knockouts

The most common application of gene targeting has been in loss-of-function studies (generation of knockouts or nulls) in mice. In this approach, mice are generated from targeted mESC with the functionally mutated gene present in every cell of the body, thus enabling the evaluation of the overall biological significance of the gene. The positive selection marker is not only used to assist in

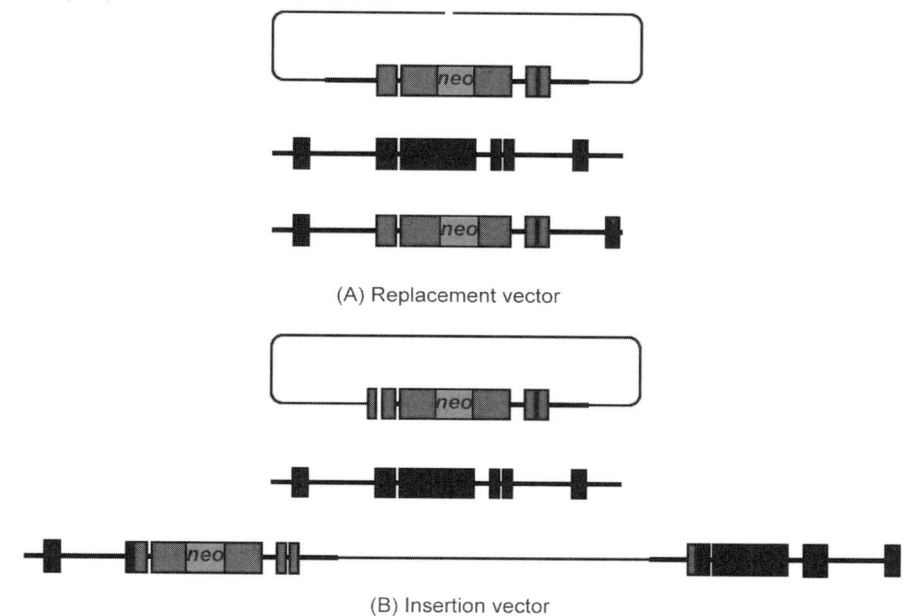

➤ **Fig. 5.6** *Schematic representation of the replacement and insertion targeting vectors and their resulting integration at a specific gene locus.*
Exons are shown as filled boxes—those in blue have been derived from the vector whereas those in black are pre-existing in the genome. Thick lines represent intronic sequences, and thin lines represent the vector backbone. The linearization sites have been shown as discontinuities in the vectors. (A) Targeted integration with the replacement vector results in the exchange of the homologous sequence present in the genome with homologous sequences in the vector plus neo, since it is continuous with the homology region. The vector can be linearized at any unique site in the vector backbone. (B) Targeted integration with insertion vectors results in the integration of the entire vector sequence at the specific gene locus. The vector is linearized at an appropriate site in the homology region. (For colour figure see Plate 9).

the identification of ESC with integrated extrachromosomal DNA by allowing them to survive when exposed to antibiotic selection, but also to disrupt the target gene either by replacement of or by insertion into genomic sequences. The marker is often incorporated into the 5′ end of the gene, usually at, or immediately downstream of the start codon, to circumvent the production of a stable truncated version of the native protein that can still have functionality—although different to the unmodified protein. A targeting strategy is often aimed at one of the loci of a gene of interest. The other allele can be targeted with a separate targeting construct that contains a different positive selection marker [170]. Knockout mESC lines (targeted at both loci) can also be generated with a single construct, by increasing the concentration of the antibiotic used for selection on heterozygote targeted clones [170, 171]. Knockout mice are obtained by cross-breeding heterozygous animals carrying one modified allele. Alternatively, they can be generated by aggregation of knockout ESC lines with tetraploid embryos [128] (see *Derivation of Mouse Embryonic Stem Cells*).

The phenotypes of different knockout mice are highly variable, ranging from failure to develop beyond foetal stage to no obvious effects. Information on mutant mouse phenotypes and associated genes can be accessed at the Mouse Genome Informatics: Phenotypes, Alleles & Disease Models (http://www.informatics.jax.org/phenotypes.html). Occasionally a phenotype becomes apparent only after mice have been challenged by stressful conditions or when they are exposed to different surroundings or stimuli. Sometimes it is possible to miss the detection of a phenotype as the knockout might only have a very subtle effect that requires specialized methods and expertise to identify, or current phenotyping protocols used are not suitable for its detection. It has been suggested that ablating the function of some genes might not result in a readily detectable phenotype due to genetic robustness or existence of alternative 'back-up' mechanisms [172]. The genetic background of the targeted mutation can have a major influence on the penetrance of the phenotype, reflecting the effects of 'modifier genes' [173, 174]. For example, a behavioural difference was detected between two different strains of the *Fmr1−/−* mouse—a model of the human Fragile X mental retardation syndrome [175]. Knockouts in mice of a mixed genetic background (C57BL/6 and 129/OlaHsd) demonstrated a learning difficulty in a specific visuospatial task; however, no abnormality was detected on the pure C57BL/6 strain. In another example, *Nogo*-A−/− mice generated in both the 129X1/SvJ and C57BL/6 strains demonstrated enhanced regeneration in a model of neural injury; however, the extent of regeneration was greater for the former [176]. This increased regenerative capacity was confirmed in an *in vitro* assay, showing increased neurite formation from dorsal root ganglia of the 129X1/SvJ strain. In the case of mixed genetic backgrounds, one must be extremely cautious in interpreting the phenotype, ensuring that the effects are due to the target gene modification rather than genes that are closely linked to the targeted locus.

The International Mouse Knockout Consortium was recently established with the aim to generate mESC knockouts for every gene (including conditional versions and knockins of reporter genes) through targeted and random gene-trapping approaches, and store them at various repositories [177, 178]. These mutated mESC lines are available to the scientific community and the corresponding mice or cryopreserved embryos may also be obtained. Another goal of the consortium is to perform standard phenotyping on the mice and compile information that is readily accessible. However, since it is a costly and onerous task, more specialized phenotyping of lower priority genes is expected to be undertaken by researchers.

Conditional Knockouts

Knocking out a gene often causes extensive abnormalities that disrupt development and the resulting embryo or foetus dies, thus prohibiting studies of the gene function in adult mice. Additionally, when genes are expressed in multiple tissues, a total knockout does not allow for the analysis of tissue-specific gene function. A way to circumvent this problem is the generation of conditional knockout mice where deletion of the gene can occur in specific cells or at a particular age of the mouse. This approach utilizes genetic constructs that control the deletion of the gene of interest [179], such as the bacteriophage P1-derived *loxP-Cre or the Saccharomyces cerevisiae derived Frt-Flp* systems [180]. Specifically, these systems involve site-specific recombinases Cre or Flp that recognize short (34 base pair), specific DNA sequences loxP and Frt, respectively, and mediate recombination between them. Various modifications are possible, including deletions, insertions, inversions, or translocations depending on how the sites are arranged. For the conditional knockout approach, the sites are situated in the gene flanking an important exon, typically the exon that contains the start codon (and sometimes the positive selection marker) or all or most of the coding exons provided that they are not too far apart, in the targeting vector. The sites are orientated in the same direction and within intronic sequences away from sequences that might be involved in splicing or gene regulation so that their presence does not affect the normal functioning of the gene. How these systems are used for excision of genomic sequences is illustrated in Fig. 5.5. Mice harboring the *loxP* or *Frt* flanked exon are phenotypically normal; however, a knockout can be achieved in the progeny when crossed with Cre or Flp transgenic mice. Knockouts specific to cell and tissue type, or developmental stage, can be generated depending on the promoter driving expression of the recombinase. For example, a conditional knockout of the oxytocin receptor (*Oxtr*)—a gene that is normally expressed in multiple tissues—was generated in the forebrain [181]. Mice harboring loxP sites flanking both copies of the coding sequence of the gene were bred with transgenic mice expressing Cre recombinase driven by the Ca^{2+}/calmodulin-dependent protein kinase IIa promoter. These mice differed from the more routine total knockout mice as they were able to lactate. Both examples demonstrated occurrence of differences in social recognition behaviour. Viral vectors can also be used to deliver the recombinase transgene *in vivo*. For example, a conditional knockout of hepatocyte growth/scatter factor (HGF/SF) was generated, whereby the gene was deleted postnatally only in the liver [182]. This was achieved by intravenous administration of recombinant adenoviral Cre recombinase expression vector in mice carrying *loxP* sites flanking exon 5 at both *HGF/SF* loci. Unlike total knockouts, which are not viable, mice lacking HGF/SF in the liver exhibited no phenotypic differences compared to control mice treated with an equivalent vector containing *lacZ* instead of the recombinase and, therefore, expressing normal levels of HGF/SF. However, a reduced capacity to regenerate hepatocytes was observed in these mice after carbon tetrachloride induced liver injury. Timed excision of DNA sequences is also possible through exogenous administration of specific effector molecules such as RU486, and tamoxifen [183,184]. These molecules can either induce expression of the Cre by activating its promoter through direct effects on a chimeric regulator (i.e. a synthetic transcription factor that is fused to the effector-binding domain enabling responsiveness to the effector), or render a modified version of the Cre protein functional by enabling it to translocate to the nucleus where its action is required.

Knockins

Knockins represent targeted events that have resulted in the introduction of a cDNA coding sequence into a gene of interest so that its expression is controlled by the promoter and regulatory elements of the target gene. Such cDNA sequences often encode reporter genes such as fluorescent proteins (e.g. GFP, DsRed) or proteins that can enzymatically produce chemiluminescence (e.g. β-galactosidase, AP). These are particularly useful for studying the expression profiles of specific genes during various stages of development and in the adult in an *in vivo* context. Recently, our laboratory generated a GFP reporter knockin of the mouse *Pdx1* gene—a marker of early pancreatic organogenesis [185]. This study enabled a very detailed spatial characterization of Pdx1 expression in chimeric and heterozyogote embryos based on fluorescence as GFP and Pdx1 expression were co-expressed [185, 186] (Fig. 5.7). In addition, knockins can be used to determine lineage-specific molecular markers that are associated with GFP expression in a subset of cells in a developing embryo. Reporter knockins can also serve as very useful tools to optimize the *in vitro* directed differentiation of ESC toward specific cell fates and enable their purification [131, 140]. Markers conferring antibiotic resistance have also been used for the purpose of cell enrichment [140, 187]. Integration of the GFP reporter gene into the mouse *Olig2* gene enabled the isolation of a subset of cells following directed differentiation to ESC-derived neural lineage cells that comprised neurons, oligodendrocytes and astrocytes [188]. Most of the GFP-expressing cells had a morphological resemblance to oligodendrocytes and expressed specific markers (O4, O1, and Olig2) of this lineage. GFP expression was used to mark putative cells of the pancreatic lineage during *in vitro* directed differentiation of the Pdx1 knockin ESC line [185]. It facilitated the optimization of conditions to increase yields of the desired cell types and enabled their purification by fluorescence-activated cell sorting (FACS). Characterization of the GFP-expressing cells by reverse transcription-polymerase chain reaction (RT-PCR) revealed expression of *Pdx1*

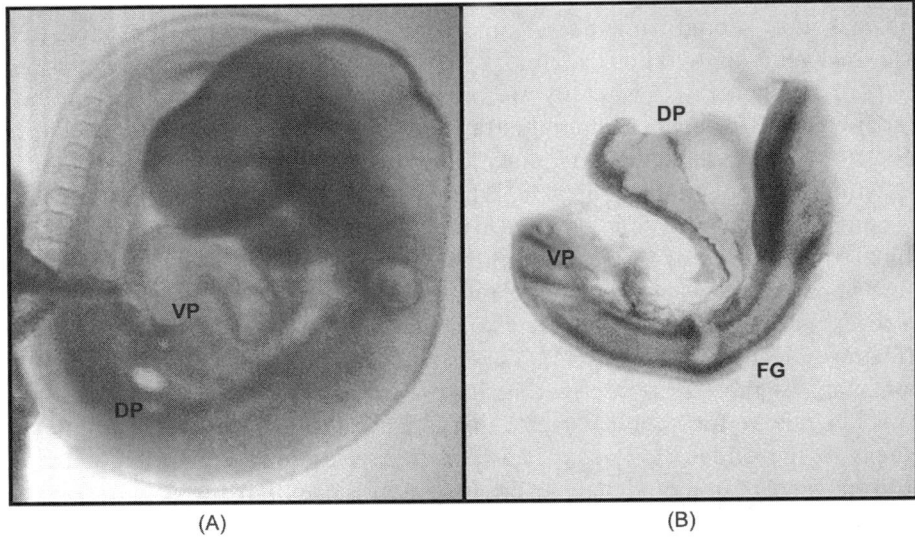

(A)   (B)

▶ **Fig. 5.7** *GFP expression in Pdx1-GFP reporter knockin mouse embryo.*
GFP expression marks the expression of Pdx in (A) pancreas of whole embryo at day E9.5 and in (B) the dissected foregut region from day E13.5 embryo. VP, ventral pancreas; DP, dorsal pancreas; FG, foregut. Image provided by Suzanne Micallef, Monash Immunology & Stem Cell Laboratories, Australia. (For colour figure see Plate 10).

and several endoderm markers, indicating that the cells were indeed committed to the pancreatic lineage.

Knockins have also been used to produce humanized mouse models by replacing the mouse version of a particular gene with the human orthologue, sometimes a disease-causing version. These models appeal greatly to the pharmaceutical industry as they can facilitate the development of potential therapeutic agents, and allow a more accurate assessment of the actions of these drugs in an *in vivo* system in the presence of relevant human proteins. For example, the generation of humanized complement 5a receptor (C5aR) mice enabled the generation of highly specific monoclonal antibodies to the receptor, following immunization of wild-type mice with knockin-derived neutrophils expressing high levels of the transgenic protein [189]. The antibodies were shown to have therapeutic effects by preventing and reversing the effects of inflammatory arthritis when administered to knockin mice, by impairing or preventing C5a-mediated activation of the receptor. Such molecules are expected to be effective in treating equivalent and other human diseases where C5aR is implicated.

Knockins often result in the disruption of one copy of an endogenous gene, i.e. in a knockout. This is problematic for genes that are expressed by only one set of chromosomes or where the gene dosage effects are important (so-called mono-allelic expression or haploinsufficiency). A method to circumvent this is to target the 3' untranslated region (UTR) of the gene with the inclusion of an IRES upstream of the cDNA. This enables the production of a bicistronic mRNA from which two protein products can be derived—the endogenous gene product and the foreign gene product. However, the efficiency of translation of the second coding sequence is variable and lower than that of the first [190]. Hence, expression of the reporter gene might not adequately reflect expression of the native gene. Furthermore, IRES can have unexpected phenotypic effects due to influences on endogenous gene expression [191].

FACTORS AFFECTING HOMOLOGOUS RECOMBINATION

Variations that affect the efficiency of gene targeting include the combined length of the homology arms, sequence composition, and chromatin status of the target genomic locus, method of delivery of the targeting vector, and the stage of the cell cycle. Generally, the longer the homology arms, the higher the efficiency of homologous recombination [192, 193]. This is highly dependant on the specificity of the homologous sequences to the target locus, that is, whether they are comprised of repetitive elements. The RepeatMasker program (http://www.repeatmasker.org/) allows putative homology arms to be scanned for repetitive sequences, providing 'a readout' of the types of elements present and helping one determine the suitability of fragments as components of targeting vectors. The use of isogenic DNA (i.e. derived from the same mouse strain) has been shown to significantly increase targeting efficiencies [194–196]. Current methods of obtaining homologous sequences include long-range PCR from genomic DNA or BAC clones carrying the gene of interest [197]. BAC libraries for a variety of mouse strains, including 129Sv and C57BL/6, exist and are available (http://bacpac.chori.org/libraries.php-disp=t) [198]. Such clones can also be used to isolate arms by sub-cloning using restriction endonuclease digestion or BAC recombineering such as Red/ET (http://www.genebridges.com/) [197]. Compared with conventional cloning involving ligations of PCR products and fragments generated by restriction digestion, recombineering technology allows more rapid generation of targeting constructs and enables cloning of larger fragments that are error free.

Electroporation is the routinely used method to introduce the targeting vector into mESC. It is anticipated that the new electroporation-based technology, termed nucleofection, which results in direct delivery of the DNA into the nucleus, could enable greater targeting efficiency. Studies performed in synchronized rat fibroblasts indicate that homologous recombination is cell-cycle dependent, which occurs mainly in early to mid S-phase [199].

Zinc finger nucleases (ZFN) and triple-helix forming oligonucleotide (TFO)-linked nucleases are synthetic site-specific enzymes that have been used to promote gene targeting. This is based on the observation that the rate of homologous recombination is increased when a double-strand break (DSB) is introduced within or in close proximity to a target site [166, 167, 200, 201]. Homing endonucleases, naturally occurring nucleases derived from microorganisms, can also be applied for this purpose; however, it is required that their recognition sequence is initially introduced into the foreign genome [201]. Artificial versions of these endonucleases are being created through mutagenesis with the aim of producing types that can recognize other sequences [166]. It is unlikely, however, that they will enable the generation of sufficient variants for widespread application. TFO-linked nucleases are less versatile than ZFN since they recognize polypurine/polypyrimidine stretches in the major groove of the DNA double helix [201]. ZFN are chimeric proteins comprised of the non-specific endonuclease *FokI* cleavage domain fused to a series of usually three to four zinc finger motifs [166, 167, 200, 201]. Each motif is capable of recognizing and binding to a nucleotide triplet; however, motifs for all possible triplet combinations are yet to be identified. Hence, the sequential arrangement of multiple interconnected motifs should allow DSB to be introduced at almost any specific location in the genome. Since ZFN function as heterodimers, with each component recognizing one of the two DNA strands, expression vectors for both proteins must be introduced into cells. Current efforts are focusing on reducing toxicity of ZFN by improving their specificity, as they can produce off-target cleavage through the action of homodimers or even monomers, resulting in mutations. This includes identifying ways to make their effects transient [202]. Surprisingly, there have been no reports on the use of ZFN to create genetically modified mESC; however, they have been successfully applied to cell types that have proven difficult to target through conventional approaches, including human ESC and somatic cells, and the genomes of other species [166, 200].

RNA Interference

RNAi is a conserved gene-silencing mechanism first discovered in the nematode worm *Caenorhabditis elegans* [203]. It is a cellular process that inhibits gene expression at the post-transcriptional and transcriptional level using exogenous double-stranded RNA (dsRNA) and the cell's own machinery. The dsRNA is cleaved into smaller 21–23 nucleotide RNA duplexes, known as small interfering RNA (siRNA), by the RNase III enzyme Dicer [204, 205]. One strand

Plate 1

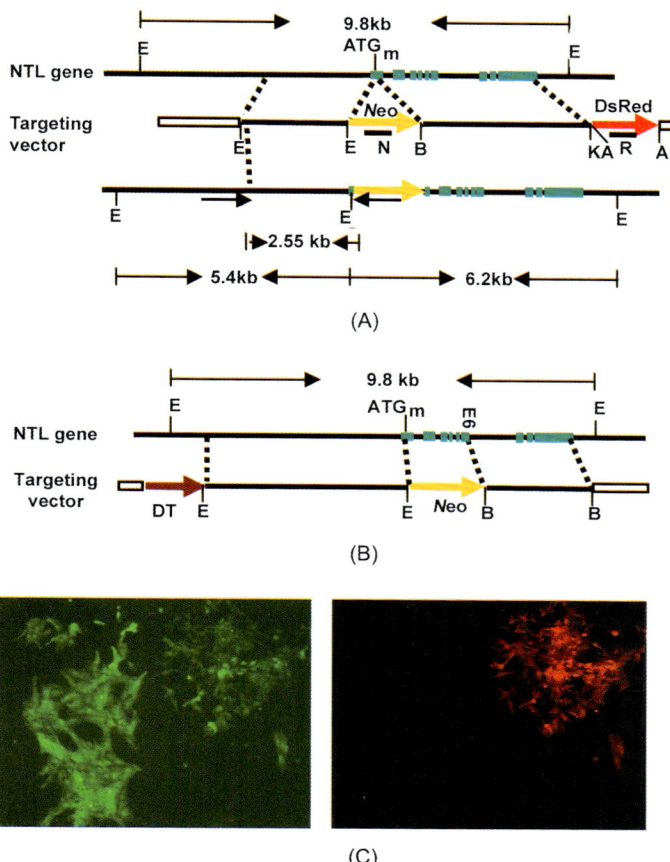

➤ Fig. 2.4(A) pTG-Neo-DsRed-ntl targeting plasmid. Diagram shows the linearised vector (middle) along with the endogenous gene before (top) and after (bottom) homologous recombination with the targeting plasmid. (B) pTG-Neo=DT-ntl forgeting plasmid construction. Diagram shows the endogeneous gene (top) before homologous recombination with the targeting ector (Bottom). The solid arrows represent the DNA fragments shown on southern blots after EcoRI (marked as E) digestion in Fig. 2.5 (C). Photomicrograph showing the same two G418-resistant colonies following pTG-Neo-DsRed-ntl electroporation. The colonies are shown through red and green filters. The DsRed⁻ colony (arrow) is a homologous recombinant. Figure 2.4 (C) is taken from fig. 2 of Ref. [43] and used with kind permission of Springer Science and Business Media.

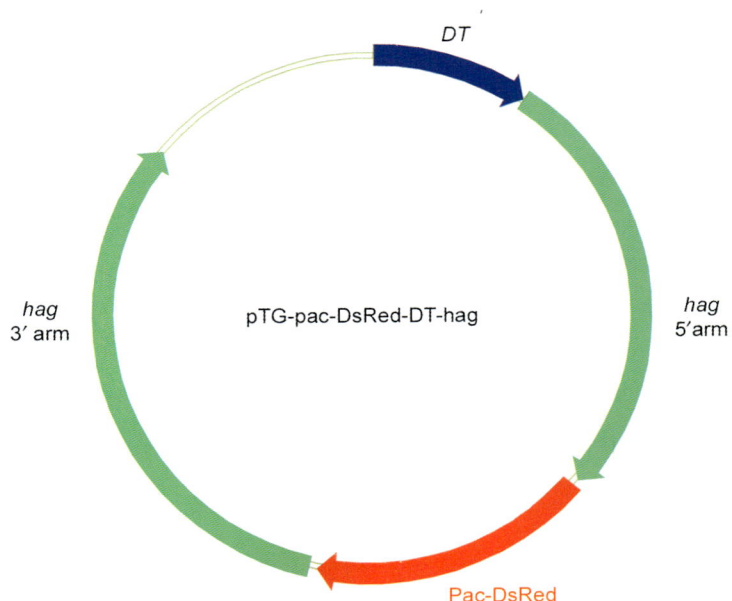

▶ *Fig. 2.6 pTG-pac-DsRed-DT-hag. In addition to puromycin (pac) selection, DsRed is used to visualise the colonies that had undergone targeted plasmid insertion by homologous recombination.*

Plate 3

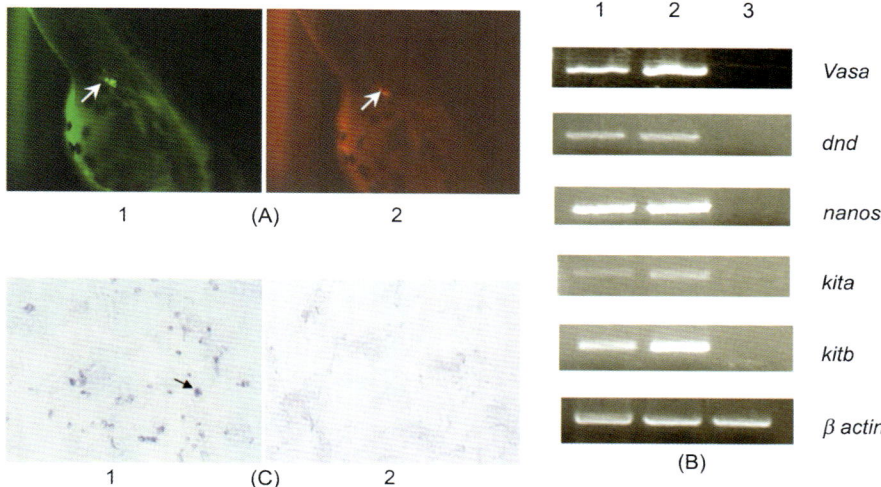

► *Fig. 2.7 Derivation of homogeneous multi-passage PGC cultures. (A) a vasa-GFP embryo injected with pac-DsRed-nanos1-3'UTR mRNA, showing co-expression of green (1) and red (2) fluorescence specifically in PGCs. (B) Lane 1, RT-PCR detection of PGC-specific gene expression in passage 8 cells; lane 2, RNA isolated from 1-day-old embryo (positive control); lane 3, muscle RNA (negative control). (C) In situ hybridization analysis of vasa expression in 3 month PGC cultures showing that a large percentage of the cells continue to express the gene—(1) antisense and (2) sense probes. Taken from fig. 7 of Ref. [59] and used with kind permission of Mary Ann Liebert, Inc.*

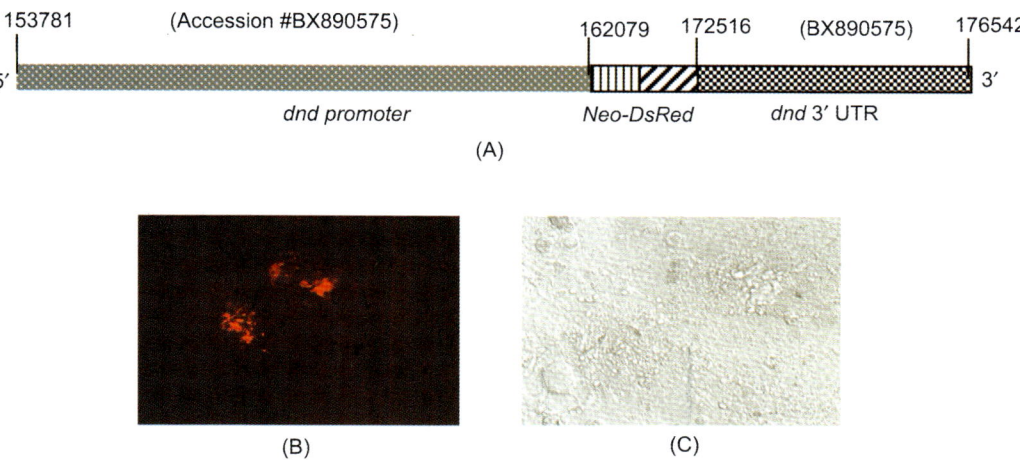

► *Fig. 2.8 Diagram of pdnd-Neo-DsRed plasmid constructs. (A) The expression of Neo–DsRed fusion protein is controlled by the PGC-specific promoter dnd and its 3'UTR. (B) The plasmid was injected into one-cell embryos, and later, the embryos were used to initiate cultures. Following G418 selection, the DsRed-expressing colonies were identified. (C) The same view as in (B), under bright field.*

Plate 4

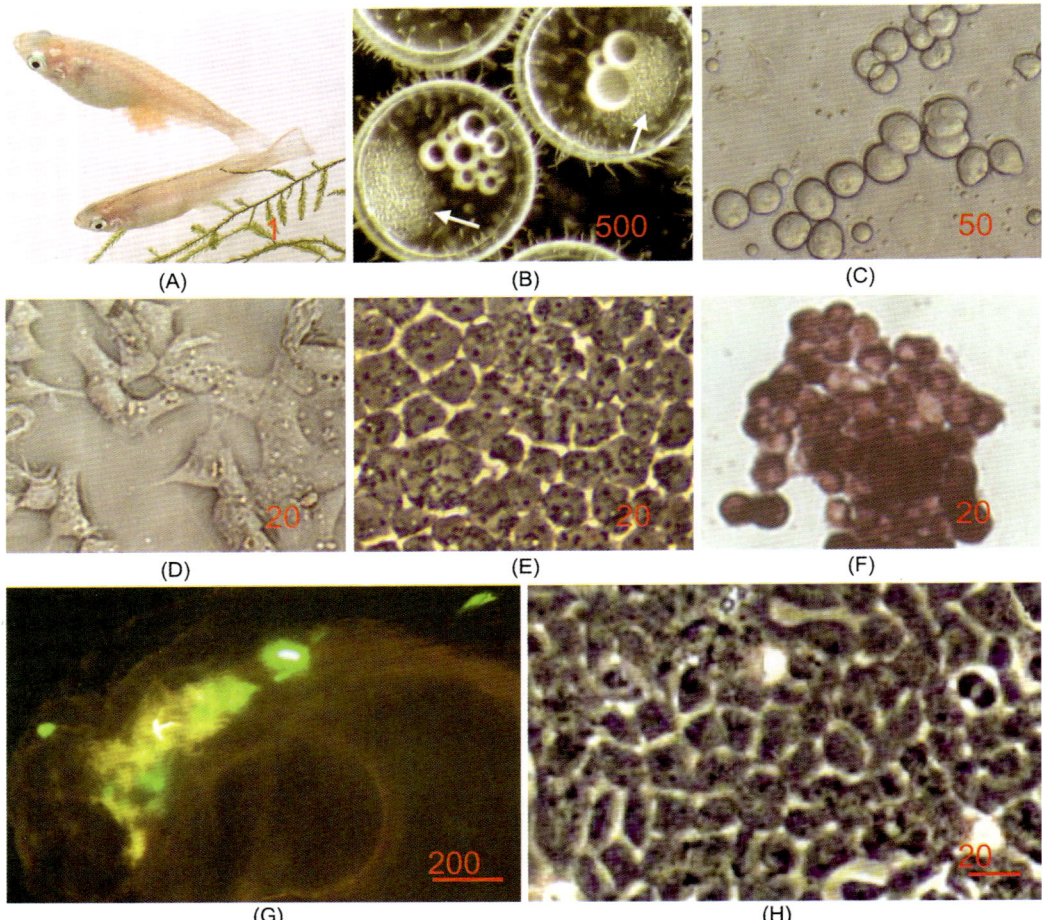

▶ *Fig. 3.1 Medaka and its stem cells. (A) Adult fish. Female fish (top), with eggs attached to the belly, and male (bottom). (B) Medaka embryos at mid-blastula stage showing the blastoderms (arrows). (C) Freshly isolated cells from blastoderms. (D) Primary culture of MBE cells. (E) and (F) MES1 and its AP-staining. (G) Chimera from MES1 cells expressing GFP from a transgene. (H) SG3 cells derived from adult testis.*

Plate 5

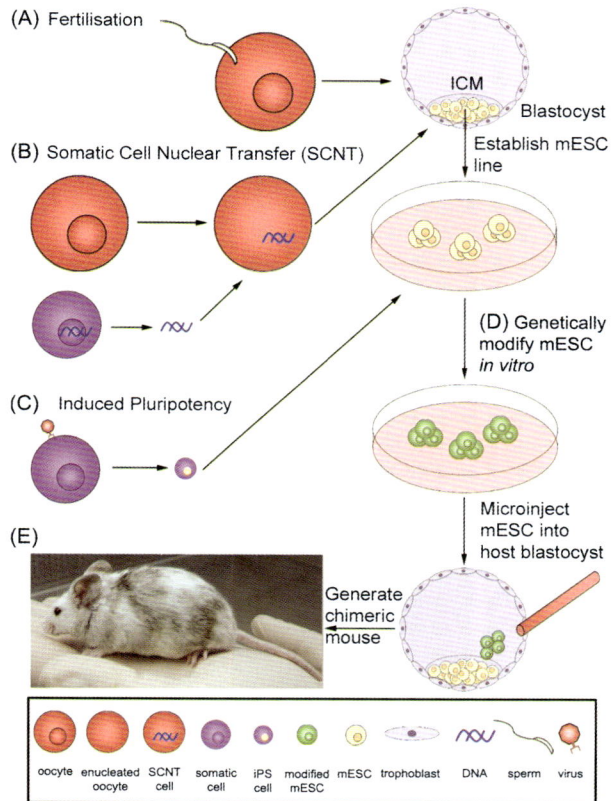

▶ *Fig. 5.1 Isolation of mESC lines and generation of chimeric mice.*

Mouse ESC can be derived from a number of sources. (A) The most common approach is to isolate the mESC from a blastocyst that results from fertilization of an oocyte by sperm. At this stage (3.5 days after fertilization), the blastocyst consists of an ICM and trophectoderm. Mouse ESC lines are established from the ICM. (B) mES-like cells can also be derived following SCNT, the process when the genome from a somatic cell is transferred into an enucleated oocyte. (C) Induced pluripotent stem (iPS) cell lines that display ESC-like characteristics can be generated by the dedifferentiation of a somatic cell, by introducing a set of defined factors by retroviral transduction. (D) The genome of mESC can be manipulated in vitro, and upon introduction into a host blastocyst, the modified mESC are able to contribute to all cells of a chimeric mouse. (E) A photograph of a chimera generated from the injection of mESC derived from a pigmented strain of mouse into an albino host blastocyst. Chimeras are bred with wild type mice to produce a transgenic mouse line carrying the genetic modification initially introduced to mESC.

Plate 6

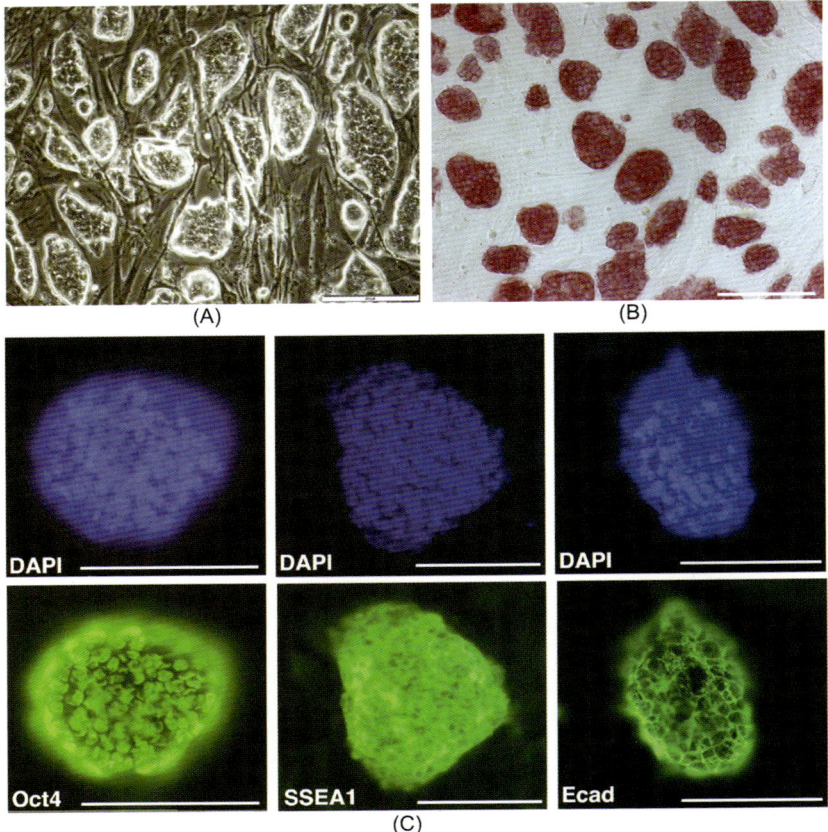

▶ *Fig. 5.2 Characteristics of mESC:*

Morphology and the expression of pluripotent markers of mESC cultured on a fibroblast feeder cells and in the presence of LIF. (A) Morphology of mESC colonies. (B) Expression of alkaline phosphatase. (C) Expression of nuclear (Oct4) and cell surface (SSEA-1, Ecad) markers. Images in (C): immunofluorescence microscopy; blue, DAPI staining of nuclei; green, respective markers. Scale bar: (a) and (b) 200 μm; (c) 100 μm.

Plate 7

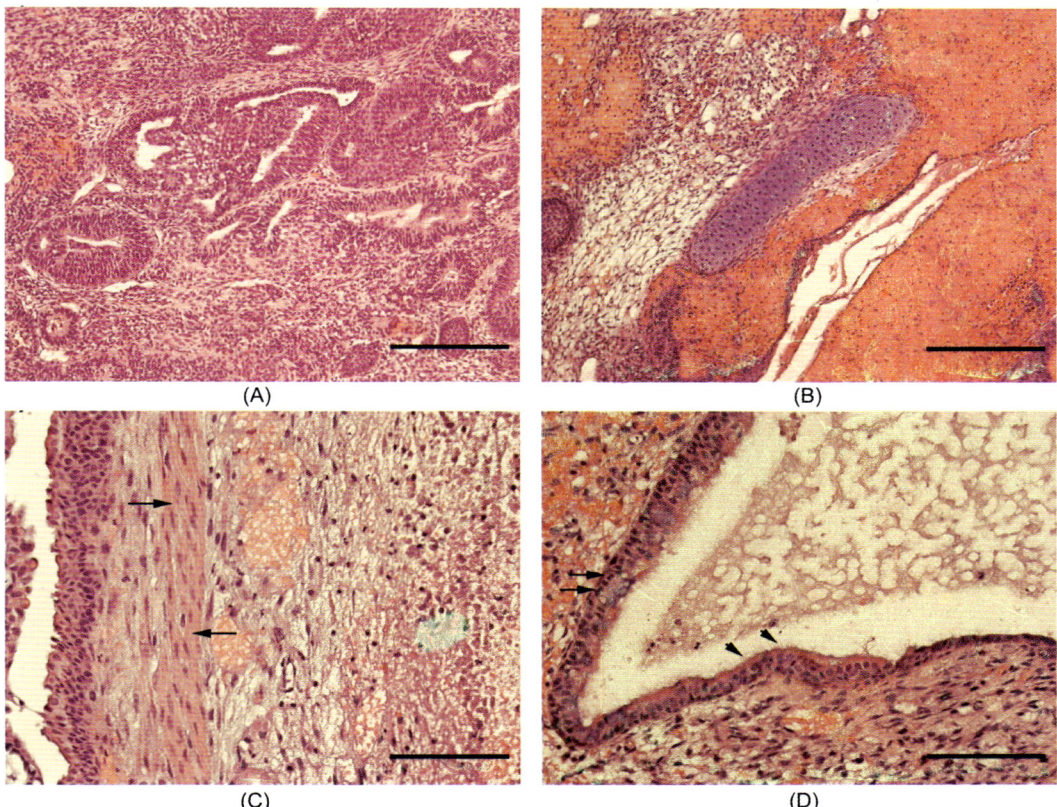

➤ Fig. 5.4 *In vivo differentiation of mESC in teratomas.*

Haematoxylin and eosin staining of paraffin sections of a 7-week teratoma, showing cells of all three germ-line lineages. (A) Neural epithelium (ectoderm), (B) Cartilage, (C) Mesoderm and/or ectoderm: Smooth muscle cells; (arrows, mesoderm). (D) Columnar epithelium with goblet cells; (arrows) and cilia (arrow heads; endoderm). Scale bar: (a) and (b) 100 μm; (c) and (d) 50 μm.

Plate 8

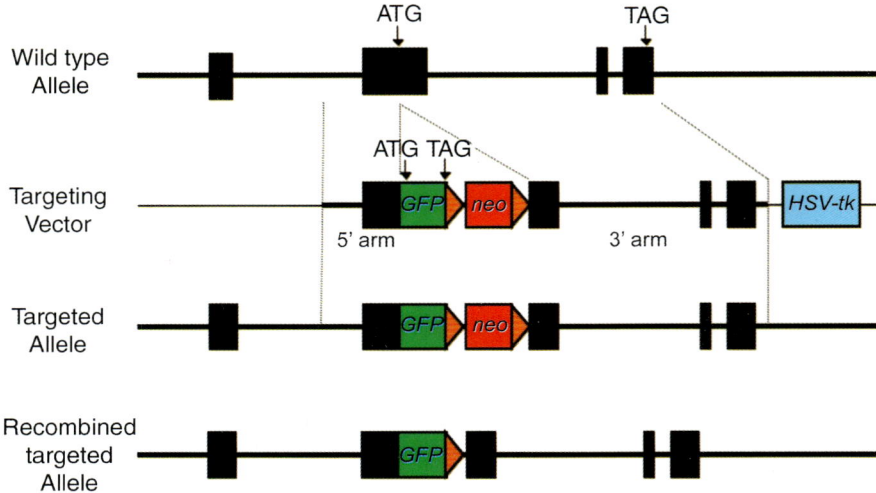

➤ *Fig. 5.5 Basic knockin gene-targeting strategy.*

The approach shown targets the 5' end of a gene. Reporter gene expression is controlled by the regulatory elements of the endogenous gene. Hence, sequence deletions should be avoided. Exons are represented as black boxes. The 5' and 3' homology arms flank the reporter gene (GFP) and the selectable marker (neo). The neo shown is floxed, i.e. flanked by loxP or Frt sites (triangles) so that it can be removed after gene targeting. Correctly targeted mESC are identified by positive–negative selection, followed by molecular analysis. ESC that have integrated the neo are identified by positive selection with G418. Negative selection with ganciclovir is used to eliminate cells that have incorporated the HSV-tk. Cells harboring the HSV-tk represent random integrants, as the HSV-tk lies in the vector backbone (thin line) and cannot be involved in homologous recombination. The selection marker is removed following exposure to Cre or Flp recombinase, usually by transient transfection with an expression vector. A single recombination site remains in the modified gene.

Plate 9

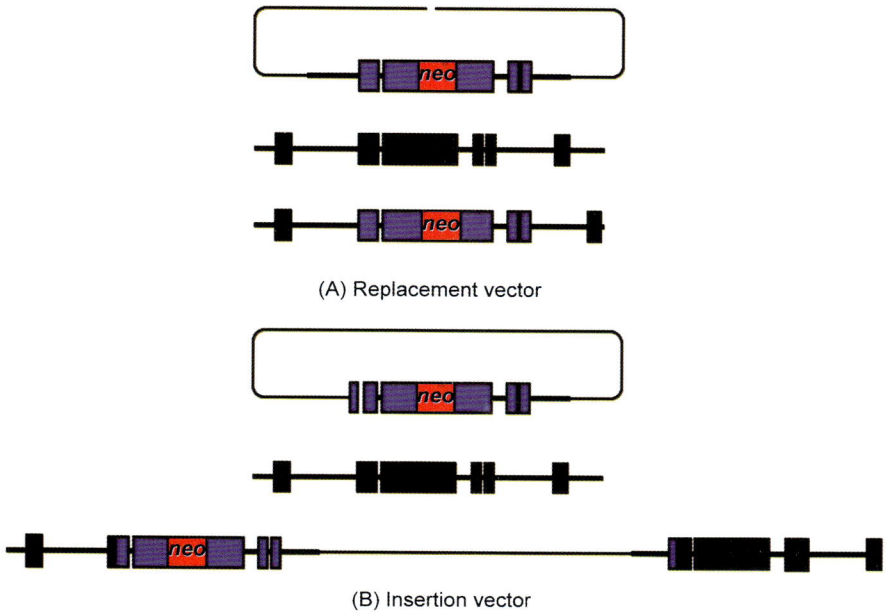

➤ *Fig. 5.6 Schematic representation of the replacement and insertion targeting vectors and their resulting integration at a specific gene locus.*

Exons are shown as filled boxes—those in blue have been derived from the vector whereas those in black are pre-existing in the genome. Thick lines represent intronic sequences, and thin lines represent the vector backbone. The linearization sites have been shown as discontinuities in the vectors. (A) Targeted integration with the replacement vector results in the exchange of the homologous sequence present in the genome with homologous sequences in the vector plus neo, since it is continuous with the homology region. The vector can be linearized at any unique site in the vector backbone. (B) Targeted integration with insertion vectors results in the integration of the entire vector sequence at the specific gene locus. The vector is linearized at an appropriate site in the homology region.

Plate 10

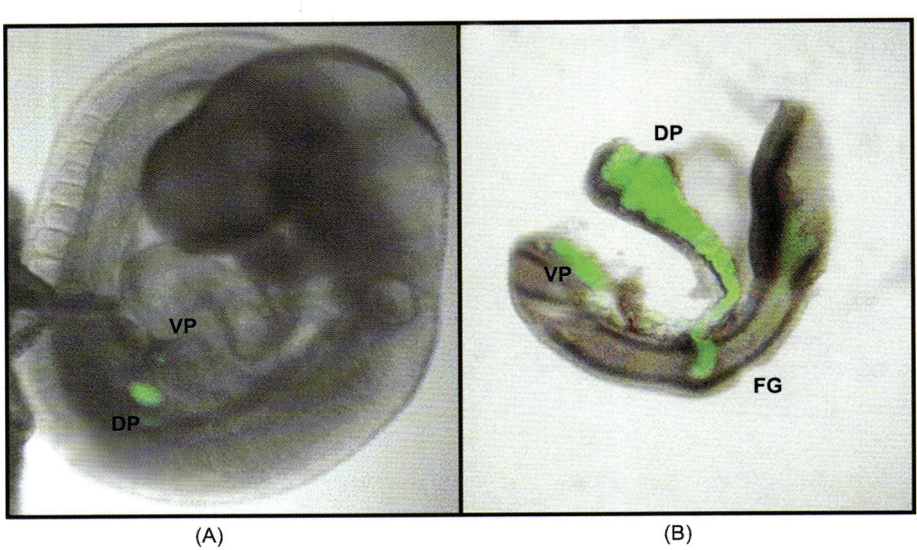

➤ *Fig. 5.7 GFP expression in Pdx1-GFP reporter knockin mouse embryo.*

GFP expression marks the expression of Pdx in (A) pancreas of whole embryo at day E9.5 and in (B) the dissected foregut region from day E13.5 embryo. VP, ventral pancreas; DP, dorsal pancreas; FG, foregut. Image provided by Suzanne Micallef, Monash Immunology & Stem Cell Laboratories, Australia.

Plate 11

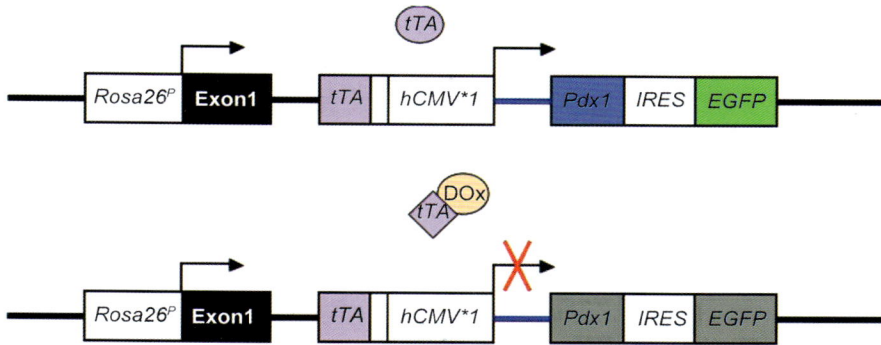

▶ *Fig. 5.8 Integration of the tet-off system into the Rosa26 locus to enable inducible regulation of ectopic Pdx1 expression.*

A schematic representation of the single modified Rosa26 allele in mESC. Note: Not all modifications are included. The Rosa26 promoter is designated as Rosa26^P. Intronic sequences are shown as a solid line. The blue line represents the rabbit β-globin second intron. A splice acceptor site (not shown) was included immediately upstream from the tTA coding sequence to prevent it from being lost through RNA processing of transcripts arising from the Rosa26 promoter. An insulator sequence (empty box) was included to suppress potential interactions between the adjacent expression units. The arrows represent transcription. Actively expressed genes are indicated in colour. A poly A signal (not shown) was included downstream of the EGFP coding sequence. tTA is constitutively expressed from the Rosa26 promoter. In the absence of Dox, the tTA protein activates the hCMV*1 promoter, enabling co-expression of the Pdx1 and EGFP genes. In the presence of Dox, the hCMV*1 promoter is inactivated as the Dox binds to tTA, preventing it from binding to the promoter. Figure adapted from Ref. [220].

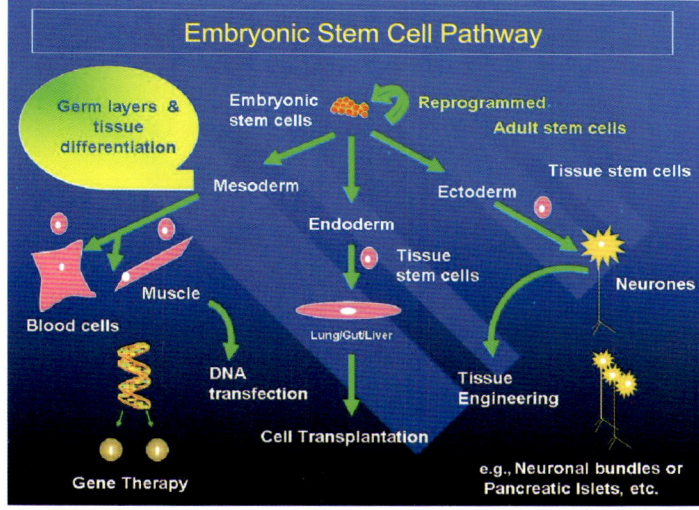

Courtesy: A. Trounson

▶ *Fig. 6.1   Embryonic stem cell pathway: Cells to therapy*

Plate 12

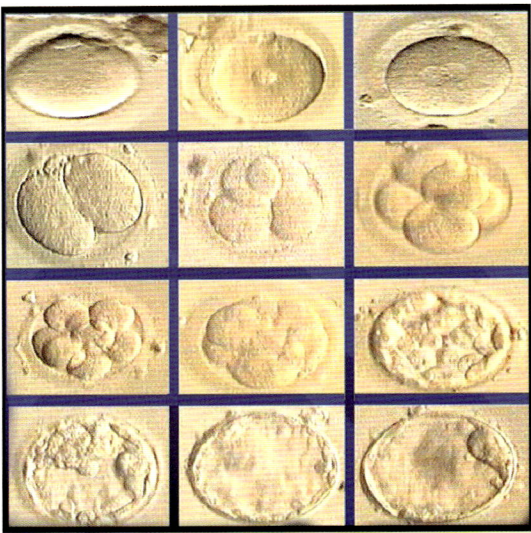

► *Fig. 6.2 Embryo development—egg to blastocyst. Mature egg, unfertilised egg, fertilised ovum, two-cell, four-cell, five-cell, eight-cell, compaction, morula, early blastocyst, hatching blastocyst (early), expanded blastocyst (day 6); ✗400 (from Ref. [29]).*

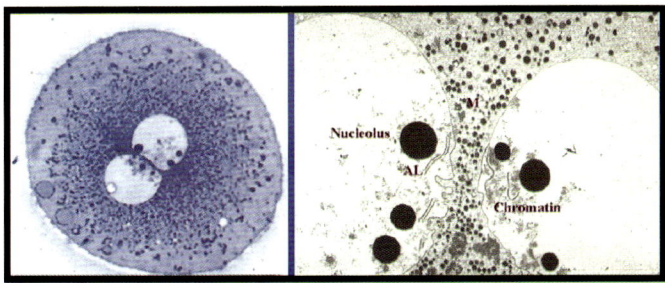

► *Fig. 6.3 Normal fertilized ovum (LM & TEM). These bipronuclears, after monospermic fertilization, seem normal. Note the alignment of nucleoli adjacent to apposing pronuclear membranes. What is more significant is the alignment of chromatin, associated with nucleoli, which would condense to form the male and female chromosomes at syngamy; ✗35,700 (from Ref. [30]).*

Plate 13

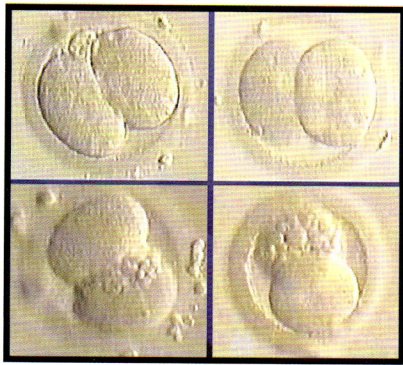

➤ *Fig. 6.4 Two-cell embryos—normal and fragmented (video). Both normal and abnormal embryos are evident. Fragments appear over the cleavage furrow or a whole cell can fragment totally (probably apoptosis); ✗400 (from Ref. [29]).*

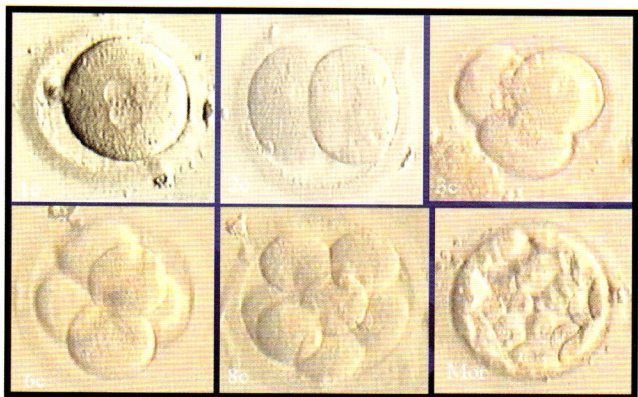

➤ *Fig. 6.5 Normal embryos—fertilised ovum to morula (video). The cleaving embryos have equal blastomeres and minimal cytoplasmic fragmentation, except the three-cell embryo; ✗400 (from Ref. [29]).*

Plate 14

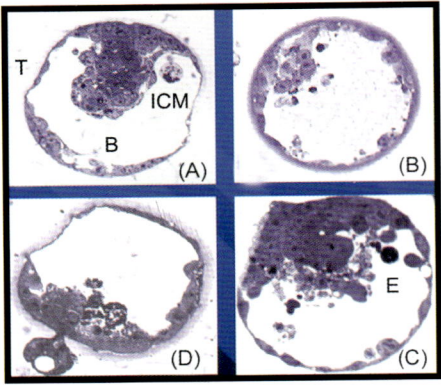

➤ *Fig. 6.6 Normal and abnormal blastocysts* **(LM).** *(A) Normal blastocyst with trophoblast* **(T)**, *inner-cell mass* **(ICM)** *and blastocoele* **[B]**. *(B) Disorganised ICM. (C) Disorganised endoderm* **[E]**. *(D) Failed hatching—degenerating.* ✕*400 (from Ref. [11]).*

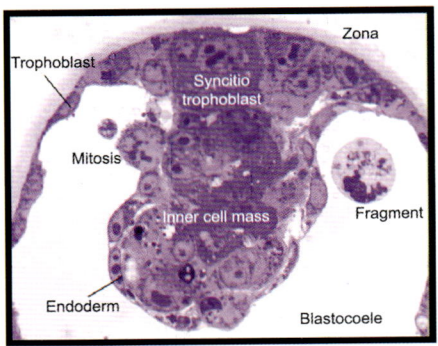

➤ *Fig. 6.7 Human expanded blastocyst—day 6. Embryonic stem cells are derived from the inner-cell mass of blastocysts. Most cell types of the blastocyst, including endoderm, are shown in this section;* ✕*1000 (From Ref. [25]).*

Plate 15

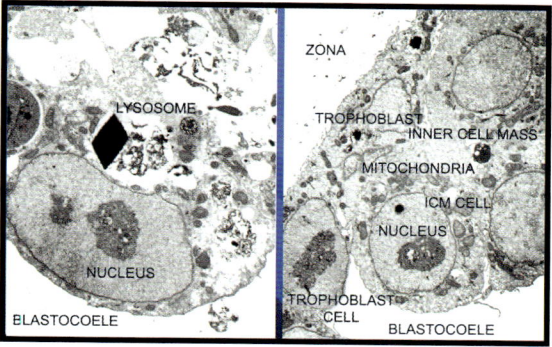

➤ *Fig. 6.8 Human blastocyst—endoderm cell and ICM/trophoblast junction (TEM). The endoderm cell is typically phagocytic. The ICM cell has a reticulated nucleus and mitochondria like all the other cells. ✗6800 and ✗850, respectively (from Ref. [11]).*

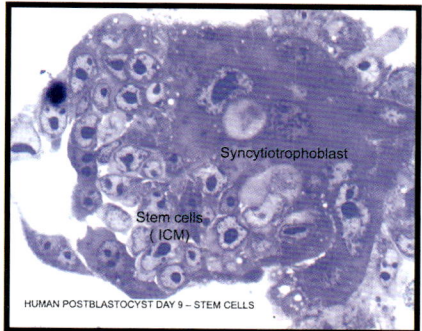

➤ *Fig. 6.9 Human day-9 blastocyst in culture-stem cells (LM). The ESCs have proliferated within the ICM and show a more simplified structure than the ICM. Note the amniotic-like cavity (left), ✗1000 (from Ref. [25]).*

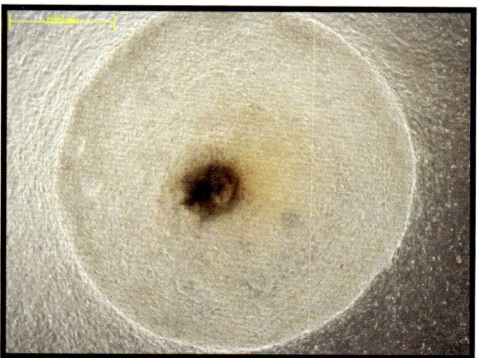

➤ *Fig. 6.10 Saucer-shaped ESC colony (P142). The colony is flat and has grown from the centre toward the periphery. Cultured on mouse embryonic fibroblasts. ✗1000 (from Ref. [25]).*

Plate 16

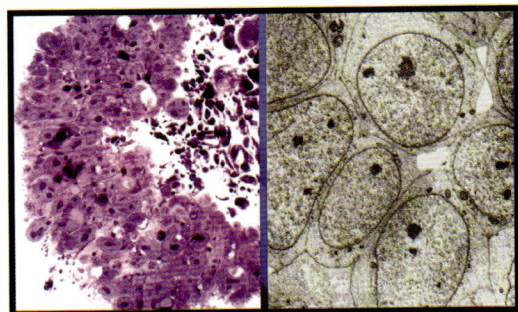

➤ *Fig. 6.11 ESCs in a colony—passage 35 (LM & TEM). The ESC have large nuclei with reticulated nucleoli and scanty cytoplasm with few mitochondria. Differentiating cells are seen within the colony with mouse fibroblasts. ✗1000 and ✗5250 for the left and right panel, respectively (from Ref. [25]).*

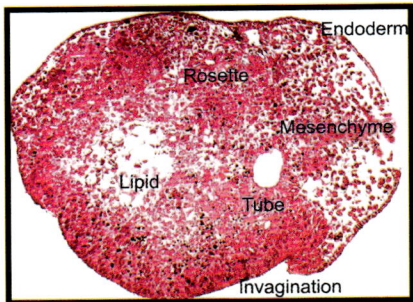

➤ *Fig. 6.12 Embryoid body (section) showing spontaneous differentiation (P130). Cells representing all three germ layers, including rosettes, tubular and capillary-like formations were seen in EB. Epithelium was cuboidal, columnar or stratified. Note the cells at surface invagination. ✗100 (from Ref. [25]).*

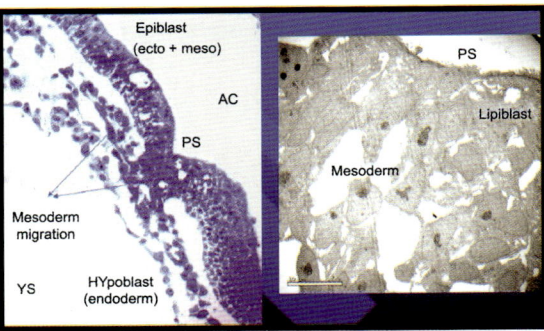

➤ *Fig. 6.13 Primitive streak, ectopic embryo showing three germ layers (LM & TEM). Sections of embryonic disc in the third week showing migration of mesodermal cells through the primitive streak establishing the three germ layers—ectoderm, mesoderm and endoderm. AC, amniotic cavity; PS, primitive streak; YS, yolk sac. ✗1000 and ✗4000 for the left and right panel, respectively, (from Ref. [26]).*

(passenger strand) of the siRNA duplex is degraded, while the other stand (guide strand) binds to multiple proteins to form the RNA-induced silencing complex (RISC). Most somatic mammalian cells are unable to tolerate long dsRNA as it triggers the interferon (IFN) response, causing non-specific destruction of RNA through activation of RNase L and inhibition of post-translational mechanisms, ultimately resulting in cell death [206, 207]. For this reason, short hairpin RNA (shRNA), which are single stranded but contain a dsRNA domain, or siRNA, are more commonly used. Stable RNAi is achieved by delivering shRNA usually driven by an RNA polymerse-III dependent promoters, such as U6 or H1, in a vector (plasmid or more frequently lentiviral), whereas siRNA is used to achieve transient effects. Within the cell, shRNA is also processed into siRNA and incorporated into the RISC. RISC mediates enzymatic cleavage of specific RNA transcripts (including message, non-coding, and viral), which are homologous to the guide strand, resulting in destruction of the transcript, and hence decreased protein production in the case of mRNA [206, 208]. RNAi-based gene silencing approaches have been widely used in loss-of-function genetic studies in mESC and mice derived from them [153, 155, 207, 209, 210]. Recently, it was used to demonstrate functional redundancy between *Klf2*, *Klf4* and *Klf5* in mESC [153]. RNAi-mediated silencing of all three *Klfs* promoted differentiation of mESC, whereas knockdown of each individual member, or two of the members, had no observable consequences. This method offers a quicker, less tedious, and cost-effective approach compared to gene knockouts. In addition, gene silencing can be achieved to variable extend and in an inducible fashion, depending on the effector and promoter used. However, a complete loss of gene expression is not achieved via this approach.

RNAi can be used to influence cell fate. Recently, RNAi mediated by lentiviral transduction of specific shRNAs was used to demonstrate the importance of seven genes for the self-renewal of mESC *in vitro* [104]. The downregulation of each gene resulted in differentiation of mESC toward different lineages, reflecting their different roles in the regulation of cell identity or specification. The transient downregulation of *Shp2* and *PPARgamma* by siRNA was able to impair the *in vitro* differentiation of mESC toward the hematopoietic lineage and promote their differentiation toward the osteoblastic lineage, respectively [211, 212]. Endogenous RNAi triggers in the form of microRNA (miRNA) are known to exist. These are short, approximately 22 nucleotide, RNA derived from the processing of larger primary RNA (pri-miRNA). A single miRNA type is thought to have a wide range of targets, hence enabling it to simultaneously affect the expression of numerous genes. Identification of the miRNA profiles of various cell types could, therefore, lead to the identification of key molecules, which when overexpressed as shRNA or introduced as synthetic siRNA in mESC or their derivatives, force differentiation toward specific phenotypes. Indeed this approach has been used to enhance the recovery of mesoderm and cardiomyocytes from mESC [154].

Several parameters are known to influence the efficiency of RNAi in mESC and other mammalian cells. The most important being the design of the RNAi effector molecules and their delivery. Indeed it has been demonstrated that variable levels of downregulation can be achieved for a particular transcript, depending on the sequence of the siRNA [213]. Empirical guidelines and algorithms, including web-based programs (http://biotools.swmed.edu/siRNA; http://

www.ncbi.nlm.nih.gov/genome/RNAi/) have been proposed to aid in designing potent RNAi effectors [213, 214, 215]. At present, however, effector success can only be realized through experimentation with the appropriate controls. In this context, an important control is to repeat the experiment with another siRNA that targets the same transcript but is homologous to a different region. The same or similar results should be obtained if the effects are transcript-specific. Scientific databases containing validated sequences are also available (http://www.ncbi.nlm.nih.gov/genome/RNAi/), making the task of obtaining specific and effective siRNAs or shRNAs less onerous. To remove the need to identify a potent effector, which can be a time-consuming process, it has been suggested that multiple siRNAs be utilized simultaneously for each target gene [215]. New developments in this field that enable rapid large-scale identification of effective inhibitors, phenotypic effects of RNAi or miRNA expression profiles to identify candidate effectors for lineage specification include the emergence of RNAi libraries [216, 217] and miRNA arrays [218].

Inducible Systems for Regulating Expression of Ectopic Sequences

A number of inducible systems have been developed to control expression of transgenes and shRNA. The *loxP-Cre-/FRT/*-Flp-based and tetracycline (tet) response (tet-on and tet-off) systems are the most commonly used. The site-specific recombinase approach results in permanent modifications. Once the recombinase is expressed, a genetic deletion occurs resulting in loss or activation of gene expression. As described earlier, excision can be controlled by regulating Cre transgene expression according to the promoter used or by activating modified Cre protein by exposure to the appropriate exogenous substances. The alternative inducible approach produces reversible changes. It relies on the addition of tet, doxycycline (Dox), or related compounds (membrane permeable and capable of entering the nucleus) to the culture medium of cells, or the drinking water of mice. Within the nucleus, these exogenous molecules regulate transgene expression by binding to a synthetic transactivator either tTA (for the tet-off system) or rtTA (for the tet-on system), causing conformational changes that prevent or permit, respectively, binding of the transcription factor to the transgenic promoter. Withdrawal of the exogenous molecule enables transgene expression to return to its previous state, though this is expected to vary depending on whether the cell's phenotype has been altered. This is because different promoters (even constitutive ones) display differential expression in different cell types. Improvements have been made to these tet systems to reduce leakiness of the promoter controlling the inducible gene, which involve additional components. For detailed reviews on these and some of the other approaches, we direct the reader to Refs. [165, 168, 184]. Some instances where these inducible systems are employed include:

1. In dissecting the spatial and temporal role of endogenous genes *in vivo*. This is described earlier, in the section Conditional Knockouts.
2. To observe the effects of genetic modifications (transgenes, knockins, shRNA) in the absence of selection markers, which could interfere with expression.
3. In lineage tracing studies where activation of a promoter driving expression of a recombinase is used to achieve constitutive expression of a reporter gene.

All points listed here involve permanent removal of sequences, although Point 1 can also be investigated using a reversible approach, such as in the context of gene knockdown via the RNAi approach.

4. To enable greater control during *in vitro* directed differentiation. Constitutive expression of transgenes or shRNA could affect the cell's ability to become mature, or in the case of a progenitor, limit its differentiation potential by preventing it from generating its usual range of specialized progeny.

5. In controlling transgene and shRNA expression that can be toxic.

Points 3 and 4 could involve permanent removal of the genetic modification or reversible gene regulation strategies.

It was proposed that the ideal inducible system for analyzing gene function and controlling cell fate and behaviour should be reversible to enable experimental mESC and animals to serve as their own controls [184]. This removes the problems of gene linkage effects due to a mixed genetic background and position effects for randomly integrated transgenes.

A number of limitations have been reported with these inducible systems, and reviewed recently [165,184,219]. Some of these limitations are as follows:

- Occasional incomplete excision of sequences using the recombinase approach, resulting in mice that are mosaic in the target tissues composed of cells with and without the deletion.
- In the case of Cre, tTA, and rtTA, inappropriate expression can occur due to position effects and the use of a poorly defined promoter, resulting in deletions at undesired sites or less stringent control of inducible gene expression. This problem can be overcome by using a suitable site-specific integration approach.
- General toxicity and adverse health issues have been observed for Cre recombinase and the tet systems. For Cre, this is possibly due to the existence of 'pseudo' *loxP* sites in the mammalian genome.

The LacO/LacI^R system derived from the bacteria *Escherichia coli* has been nominated as potentially being the best method for achieving reversible gene regulation without toxic effects [184]. In this system, transgene expression can be controlled in mammalian cells by including *LacO* binding sites in the promoter. In the presence of the transcriptional repressor LacI^R, transgene expression is significantly reduced; however, exposure to the compound isopropyl β-D-1-thiogalactopyranoside (IPTG) results in its activation. IPTG interacts with LacI^R causing it to uncouple from the promoter. The issue of integration as described earlier also applies for the LacI^R transgene.

The example shown in Fig. 5.8 describes the use of a tetracycline response system for inducible gene regulation in driving specific differentiation of mESC. Homologous recombination was used to insert the tet-off system into the *Rosa26* locus to enable inducible expression of ectopic Pdx1 *in vitro* (Fig. 5.8) [220]. The *Rosa 26* locus is routinely used for the insertion of transgenes and various reporter constructs into mouse genome as it can be targeted with high efficiency in mESC and is expressed in most adult tissues [221,222]. The Rosa26 promoter was used to drive

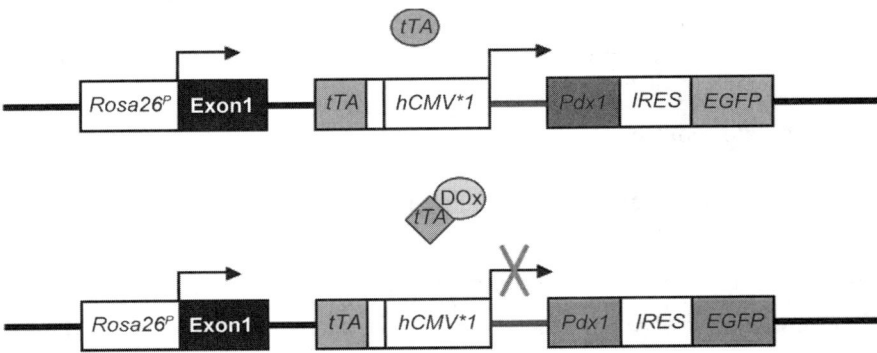

▶ **Fig. 5.8** *Integration of the tet-off system into the Rosa26 locus to enable inducible regulation of ectopic Pdx1 expression.*
*A schematic representation of the single modified Rosa26 allele in mESC. Note: Not all modifications are included. The Rosa26 promoter is designated as $Rosa26^P$. Intronic sequences are shown as a solid line. The blue line represents the rabbit β-globin second intron. A splice acceptor site (not shown) was included immediately upstream from the tTA coding sequence to prevent it from being lost through RNA processing of transcripts arising from the Rosa26 promoter. An insulator sequence (empty box) was included to suppress potential interactions between the adjacent expression units. The arrows represent transcription. Actively expressed genes are indicated in colour. A poly A signal (not shown) was included downstream of the EGFP coding sequence. tTA is constitutively expressed from the Rosa26 promoter. In the absence of Dox, the tTA protein activates the hCMV*1 promoter, enabling co-expression of the Pdx1 and EGFP genes. In the presence of Dox, the hCMV*1 promoter is inactivated as the Dox binds to tTA, preventing it from binding to the promoter. Figure adapted from Ref. [220]. (For colour figure see Plate 11).*

expression of tTA. The *CMV*1* promoter, which is responsive to tTA and used to drive expression of the Pdx1 transgene, was also incorporated into the Rosa26 locus, but downstream of the tTA coding sequence. Hence, the Pdx1 transgene expression status depends on the presence or absence of Dox. Dox 'switches-off' gene expression by binding to tTA, thus preventing it from binding DNA and activating the *CMV*1* promoter. Mouse ESC cultured in the absence of Dox expressed the Pdx1 transgene, and following *in vitro* differentiation the expression of numerous pancreas-specific genes was detected. However, when Dox was present in the culture medium during the differentiation process, most of these pancreas markers were not detected. It was speculated that the ectopic expression of Pdx1 might be responsible for producing pancreatic progenitors but prevented maturation of these cells as not all markers of the endocrine pancreas were detected. It remains to be tested whether addition of Dox to the pancreatic-like derivatives permits more terminal differentiation.

FUTURE APPLICATIONS OF MOUSE EMBRYONIC STEM CELL TECHNOLOGIES

New concepts and methodologies are constantly and rapidly evolving in the ESC field. Such methodologies enable the analysis of global gene expression profiles using microarray [216] and miRNA array approaches [218], and protein-DNA interactions through ChIP microarrays [223]. RNAi libraries have also been generated as a high-throughput screening approach [216, 217] that may be applied to ESC to identify strategies to direct specific differentiation. In this section, we focus on some of the other technologies that have recently received significant attention and have the potential to result in major scientific, medical, pharmaceutical, and industrial advancements.

Large-scale chemical library screening is being used to identify small molecules that influence cell fate decisions of ESC and their derivatives [224–226]. It is anticipated that such libraries can lead to the development of chemically defined culture conditions that enable the derivation of homogenous cell populations of interest—an important requirement, especially in regard to medical applications. These chemical libraries consist of over 100,000 compounds and have been generated using combinatorial chemistry and naturally occurring scaffolds. Examples of some successful applications are available on-line (http://www.scripps.edu/chem/ding/). One novel molecule, pluripotin, was shown to maintain pluripotency of mESC when they were cultured in a chemically defined medium without LIF, serum, and feeders [227]. These libraries will be useful for rapid identification of molecules that promote mESC differentiation to specific lineages when used in conjunction with lineage- or cell-specific ESC reporter lines. The generation of fluorescent reporter mESC lines is valuable for deriving pure cell types of interest. As the reporter gene expression can be easily monitored while the cells are alive and identifies a specific cell population of interest, these reporter lines are tools to develop optimal conditions for mESC differentiation. Indeed, this approach has already enabled the discovery of compounds such as TWS119 and Cardiogenol C, which promote differentiation of mESC to neurons and cardiomyocytes, respectively [228, 229].

Using small molecules libraries can also facilitate identification of potential therapeutic factors. An automated screening approach using motor neurons derived from mESC that have a spinal muscular atrophy (SMA) mutation is being used to identify compounds that rescue the diseased motor neurons from premature death [225]. Such factors could serve as candidate drugs for treatment of human diseases. To date, little effort has been made in this approach of using ESC carrying disease mutations to create the relevant cell types affected by disease by *in vitro* differentiation, and then using them to identify potential therapeutics (http://network.nature.com/boston/news/articles/2007/11/19/in-pursuit-of-a-dream-drug-screen). However, with the generation of iPS cells (as discussed further), this is expected to become a very promising approach in future, especially since customized human 'ESC-like' cells can be generated for each patient.

Numerous studies have demonstrated functionality of ESC derivatives in mouse models of human disease [140, 230]. Although there is still conjecture as to the degree of immunogenicity of human ESC, the *in vivo* analysis of these cells and their immediate therapeutic derivatives has been best studied in mice with a compromised immune system. The most common models used to prevent rejection of the graft, when the strain of the donor was not identical (allogeneic) to the recipient, are the nonobese diabetic (NOD), SCID, and nude mice. Given that ESC express some histocompatibility antigens, a major impediment to the application of allogeneic ESC-derived cells for human therapy is the immune barrier provided by patients. While common immunosuppressive regimes may overcome this rejection to some degree, there is a high degree of morbidity caused by generalized immunosuppression. To reduce this problem, human leukocyte antigen (HLA)-typed ESC banking has been proposed to provide an adequate match for most of the population in a particular geographic area [231, 232]. Each region could establish banks to cater for their own genetic diversity. So far, estimates of the numbers of cell lines required have been determined for two countries, United Kingdom and Japan. While these have been relatively small in each case (150 for the UK and 170 for Japan), some evidence suggests that hESC differ in their differentiation potential, probably reflecting genetic variability of the human population from whom hESC are derived [233].Therefore, it might be necessary to increase a number of HLA-matched ESC lines to facilitate the derivation of a wide range of cells for disease treatment.

An alternative approach is to induce immune tolerance in the recipient, specific to the donor transplant. This is currently the subject of investigation in our laboratory, where a novel strategy involving mESC is being tested in a mouse model of thymic regeneration (J. Lim, A. Giudice, R. Boyd; personal communication). The thymus is a crucial organ for generating all the T-cells required for immune defense against infections, but it is also responsible for establishing self-tolerance through deletion of potentially auto-reactive T-cells prior to their exit from the thymus. It also produces a class of T-cells called T-regulatory cells that further 'police' against self-reactivity. Resident thymic epithelial cells (TEC) play a major role in these processes. Hence, introduction of donor cells (e.g. as haemopoietic stem cells—HSC) into the host thymus would be expected to generate a new immune system tolerant to both the donor and host. While this has been demonstrated in young animals, the process is severely compromised in adults because the thymus undergoes profound atrophy with age post-puberty [234]. This results in a reduced output of T-cells into the periphery and consequently reduced immune competence in the elderly. Therefore, in older patients, strategies to regenerate the thymus should be incorporated into donor immune tolerance induction approaches. One avenue of regenerating the thymus and restoring immune function could be through the direct intra-thymic delivery of ESC-derived thymic epithelial progenitor cells (TEPC), resulting in a chimeric thymus [234]. Together with HSC derived from the same human ESC donor as the TEPC, which would provide a continual supply of immature T-cells and other progenitor cells required for the maturation process, tolerance to other donor-derived ESC derivatives could be induced. This could enable the application of human ESC for various cell-based therapies, provided that the HSC and TEPC transplants can functionally integrate into the bone marrow and thymus, respectively. We are currently in the process of generating mESC reporter lines that will aid in the development of efficient protocols to derive cells of the thymic lineage (J.Lim, A.Giudice, R.Boyd; personal communication).

Recently, a ground-breaking development by Yamanaka et al., resulted in derivation of iPS cells [235]. In this study, a combination of four transcription factors—Oct4, Sox2, c-Myc and Klf4, normally expressed in ESCs—was introduced into mouse fibroblasts by retroviral transduction. Ectopic expression of these factors resulted in reprogramming of somatic cells to ES-like iPS cells (Fig. 5.1). Following the success in mouse, reprogramming of various human somatic cells into iPS cells—using the same retroviral or a lentiviral transgene delivery approach—was recently reported [236–242]. The iPS cells resemble ESC in many regards, including morphology, growth, global gene expression profiles and methylation status of some promoters, and differentiation potential *in vitro*, and in teratoma formation *in vivo*. In addition, mouse iPS cells have also contributed to all tissues, including the germ-line, of chimeric mice. The iPS cells have received much attention since they apply stem cells in a therapeutic setting more accessible by showing that it is possible to produce a source of autologous cells for transplantation using a more direct method than that compared to generation of NT-ESC lines. Of course, alternative methods for induction of pluripotency must be identified as viral delivery is fraught with problems, including those associated with random integration and immunogenicity of viral elements. In the meantime, mESC can be used to define optimal conditions for maintenance as well as the efficient derivation of pure specific cell types. Embryonic stem cells will be also required as a reference to validate pluripotency of therapeutically safe (free of genetic modifications) iPS cells.

Gene modification technology developed in mESC has proven effective in iPS cells. Gene targeting has been used to correct the sickle cell mutation in iPS cells generated from mice with the disease [243]. Therapeutic application of hematopoietic progenitors derived from these gene-corrected cells resulted in curing the disease in mice. This demonstrates how lessons learnt from mESC can be extended to iPS cells using combined gene- and cell-based approaches to treat disease. This is not the first time that reprogrammed somatic cells resembling mESC have been used in this way to treat a genetic disease. NT-ESC generated from somatic cells of the SCID Rag2 knockout mice were targeted to correct one of the alleles [244]. It was possible to generate mice with corrected immune competence from the genetically corrected NT-ESC. Transplantation of bone marrow cells derived from these animals or hematopoietic precursors derived by *in vitro* differentiation of the corrected NT-ESC was shown to be capable of restoring almost complete immune function to the Rag2 knockout mouse.

The ability to genetically modify mESC and generate live mice from them has impacted significantly on our understanding of gene function *in vivo* in normal and diseased states. These studies have provided significant insights into the roles of human genes. In addition, mESC research has facilitated the development of technologies that enable a wide range of genetic manipulations and phenotypes to be obtained. This work has 'paved the way' for genetic modification of hESC and other stem cells, including 'manufactured' cell types such as iPS and NT-ESC. Despite rapid advances in hESC technology, mESC models are still beneficial and required for further exploration and discovery of new technologies. This is mainly due to the relative ease in handling, propagation, and genetic manipulation of these cells, and the ability to derive information that is relevant to the human system. Most importantly, they can continue to be used in an *in vivo* context to generate live animals for gene analysis, to test cell and molecular-based therapies and to demonstrate proof-of-concept for human applications.

REFERENCES

1. Evans M.J., M.H. Kaufman (1981). Establishment in culture of pluripotential cells from mouse embryos. *Nature*. **292**: 154–156.
2. Martin G.R. (1981). Isolation of a pluripotent cell line from early mouse embryos cultured in medium conditioned by teratocarcinoma stem cells. *Proc. Natl. Acad. Sci. USA*. **78**: 7634–7638.
3. Keller G. (2005). Embryonic stem cell differentiation: Emergence of a new era in biology and medicine. *Genes. Dev.* **19**: 1129–1155.
4. Bradley A., M. Evans, M.H. Kaufman, E. Robertson (1984). Formation of germ-line chimaeras from embryo-derived teratocarcinoma cell lines. *Nature*. **309**: 255–256.
5. Robertson E., A. Bradley, M. Kuehn, M. Evans (1986). Germ-line transmission of genes introduced into cultured pluripotential cells by retroviral vector. *Nature*. **323**: 445–448.
6. Thomas K.R., M.R. Capecchi (1987). Site-directed mutagenesis by gene targeting in mouse embryo-derived stem cells. *Cell*. **51**: 503–5012.
7. Doetschman, T., R.G. Gregg, N. Maeda, M.L. Hooper, D.W. Melton, S. Thompson, O. Smithies (1987). Targeted correction of a mutant HPRT gene in mouse embryonic stem cells. *Nature*. **330**: 576–578.
8. Stevens, L.C. (1973). A new inbred subline of mice (129-terSv) with a high incidence of spontaneous congenital testicular teratomas. *J. Natl. Cancer. Inst.* **50**: 235–242.
9. Stevens, L.C. and D.S. Varnum (1974). The development of teratomas from parthenogenetically activated ovarian mouse eggs. *Dev. Biol.* **37**: 369–380.
10. Solter, D. (2006). From teratocarcinomas to embryonic stem cells and beyond: A history of embryonic stem cell research. *Nat. Rev. Genet.* **7**: 319–327.
11. Kahan, B.W, B. Ephrussi (1970). Developmental potentialities of clonal *in vitro* cultures of mouse testicular teratoma. *J. Natl. Cancer. Inst.* **44**: 1015–1036.
12. Rosenthal, M.D., R.M. Wishnow, G.H. Sato (1970). *In vitro* growth and differetiation of clonal populations of multipotential mouse clls derived from a transplantable testicular teratocarcinoma. *J. Natl. Cancer. Inst.* **44**: 1001–1014.
13. Evans, M.J. (1972). The isolation and properties of a clonal tissue culture strain of pluripotent mouse teratoma cells. *J. Embryol. Exp. Morphol.* **28**: 163–176.
14. Jami, J., E. Ritz (1974). Multipotentiality of single cells of transplantable teratocarcinomas derived from mouse embryo grafts. *J. Natl. Cancer. Inst.* **52**: 1547–1552.
15. Silver, L.M., G.R. Martin, S. Strickland, (eds.) (1983). Teratocarcinoma Stem Cells. Cold Spring Harbor Laboratory Press, Cold Spring Harbor.
16. Papaioannou, V.E., M.W. McBurney, R.L. Gardner, M.J. Evans (1975). Fate of teratocarcinoma cells injected into early mouse embryos. *Nature*. **258**: 70–73.
17. Kleinsmith, L.J., G.B. Pierce, Jr. (1964). Multipotentiality of Single Embryonal Carcinoma Cells. *Cancer. Res.* **24**: 1544–1551.
18. Strickland S., V. Mahdavi (1978). The induction of differentiation in teratocarcinoma stem cells by retinoic acid. *Cell*. **15**: 393–403.
19. Brinster, R.L. (1974). The effect of cells transferred into the mouse blastocyst on subsequent development. *J. Exp. Med.* **140**: 1049–1056.

20. Illmensee, K., B. Mintz (1976). Totipotency and normal differentiation of single teratocarcinoma cells cloned by injection into blastocysts. *Proc. Natl. Acad. Sci. USA.* **73**: 549–553.
21. Mintz, B. and K. Illmensee (1975). Normal genetically mosaic mice produced from malignant teratocarcinoma cells. *Proc. Natl. Acad. Sci. USA.* **72**: 3585–3589.
22. Papaioannou, V.E., R.L. Gardner, M.W. McBurney, C. Babinet, M.J. Evans (1978). Participation of cultured teratocarcinoma cells in mouse embryogenesis. *J. Embryol. Exp. Morphol.* **44**: 93–104.
23. Rossant, J., M.W. McBurney (1982). The developmental potential of a euploid male teratocarcinoma cell line after blastocyst injection. *J. Embryol. Exp. Morphol.* **70**: 99–112.
24. Solter, D., N. Skreb, I. Damjanov (1970). Extrauterine growth of mouse egg-cylinders results in malignant teratoma. *Nature.* **227**: 503–504.
25. Stevens, L.C. (1970). The development of transplantable teratocarcinomas from intratesticular grafts of pre- and post-implantation mouse embryos. *Dev. Biol.* **21**: 364–382.
26. Damjanov, I. (2005). The road from teratocarcinoma to human embryonic stem cells. *Stem Cell. Rev.* **1**: 273–276.
27. Bergstrom, S. (1978). Experimentally delayed implantation. In: Methods in Mammalian Reproduction (J.C. Daniel, ed.), 419–435. *Academic Press Inc.*, London and New York.
28. Solter, D., B.B. Knowles (1975). Immunosurgery of mouse blastocyst. *Proc Natl Acad Sci USA.* **72**: 5099–5102.
29. Chung, Y., I. Klimanskaya, S. Becker, J. Marh, S.J. Lu, J. Johnson, L. Meisner, R. Lanza (2006). Embryonic and extraembryonic stem cell lines derived from single mouse blastomeres. *Nature.* **439**: 216–219.
30. Delhaise, F., V. Bralion, N. Schuurbiers, F. Dessy (1996). Establishment of an embryonic stem cell line from 8-cell stage mouse embryos. *Eur J Morphol.* **34**: 237–243.
31. Tesar, P.J. (2005). Derivation of germ-line-competent embryonic stem cell lines from preblastocyst mouse embryos. *Proc. Natl. Acad. Sci. USA.* **102**: 8239–8244.
32. Wakayama, S., T. Hikichi, R. Suetsugu, Y. Sakaide, H.T. Bui, E. Mizutani, T. Wakayama (2007). Efficient establishment of mouse embryonic stem cell lines from single blastomeres and polar bodies. *Stem Cells.* **25**: 986–993.
33. Robertson, E.J. (1987). Embryo-derived stem cells. In: Teratocarcinomas and Embryonic Stem Cells, A Practical Approach (E.J. Robertson, ed.), 71–112. *IRL Press*, Oxford, Washington DC.
34. Tesar, P.J., J.G. Chenoweth, F.A. Brook, T.J. Davies, E.P. Evans, D.L. Mack, R.L. Gardner, R.D. McKay (2007). New cell lines from mouse epiblast share defining features with human embryonic stem cells. *Nature.* **448**: 196–199.
35. Kaufman, M.H (1983). Early Mammalian Development: Parthenogenetic Studies. Cambridge University Press, Cambridge.
36. Surani, M.A., S.C. Barton (1983). Development of gynogenetic eggs in the mouse: Implications for parthenogenetic embryos. *Science.* **222**: 1034–1036.
37. Surani, M.A., S.C. Barton, M.L. Norris (1984). Development of reconstituted mouse eggs suggests imprinting of the genome during gametogenesis. *Nature.* **308**: 548–550.
38. Hikichi, T. et al. (2007). Differentiation potential of parthenogenetic embryonic stem cells is improved by nuclear transfer. *Stem Cells.* **25**: 46–53.

39. Shao, H., Z. Wei, L. Wang, L. Wen, B. Duan, L. Mang, S. Bou (2007). Generation and characterisation of mouse parthenogenetic embryonic stem cells containing genomes from non-growing and fully grown oocytes. *Cell Biol. Int.* **31**: 1336–1344.
40. Labosky, P.A., D.P. Barlow, B.L. Hogan (1994). Mouse embryonic germ (EG) cell lines: Transmission through the germline and differences in the methylation imprint of insulin-like growth factor 2 receptor (Igf2r) gene compared with embryonic stem (ES) cell lines. *Development.* **120**: 3197–3204.
41. Matsui, Y., K. Zsebo, B.L. Hogan (1992). Derivation of pluripotential embryonic stem cells from murine primordial germ cells in culture. *Cell.* **70**: 841–847.
42. Resnick, J.L., L.S. Bixler, L. Cheng, P.J. Donovan (1992). Long-term proliferation of mouse primordial germ cells in culture. *Nature.* **359**: 550–551.
43. Stewart, C.L., I. Gadi, H. Bhatt (1994). Stem cells from primordial germ cells can reenter the germ line. *Dev. Biol.* **161**: 626–628.
44. Kawase, E., Y. Yamazaki, T. Yagi, R. Yanagimachi, R.A. Pedersen (2000). Mouse embryonic stem (ES) cell lines established from neuronal cell-derived cloned blastocysts. *Genesis.* **28**: 156–163.
45. Munsie, M.J., A.E. Michalska, C.M. O' Brien, A.O. Trounson, M.F. Pera, P.S. Mountford (2000). Isolation of pluripotent embryonic stem cells from reprogrammed adult mouse somatic cell nuclei. *Curr. Biol.* **10**: 989–992.
46. Markoulaki, S., A. Meissner, R. Jaenisch (2008). Somatic cell nuclear transfer and derivation of embryonic stem cells in the mouse. *Methods.* **45**: 101–114.
47. Brambrink, T., K. Hochedlinger, G. Bell, R. Jaenisch (2006). ES cells derived from cloned and fertilized blastocysts are transcriptionally and functionally indistinguishable. *Proc. Natl. Acad. Sci. USA.* **103**: 933–938.
48. Wakayama, S. et al. (2006). Equivalency of nuclear transfer-derived embryonic stem cells to those derived from fertilized mouse blastocysts. *Stem. Cells.* **24**: 2023–2033.
49. Wakayama, T., V. Tabar, I. Rodriguez, A.C. Perry, L. Studer, P. Mombaerts (2001). Differentiation of embryonic stem cell lines generated from adult somatic cells by nuclear transfer. *Science.* **292**: 740–743.
50. Auerbach, W., J.H. Dunmore, V. Fairchild-Huntress, Q. Fang, A.B. Auerbach, D. Huszar, A.L. Joyner (2000). Establishment and chimera analysis of 129/SvEv- and C57BL/6-derived mouse embryonic stem cell lines. *Biotechniques.* **29**: 1024–1028: 1030: 1032.
51. Keskintepe, L., K. Norris, G. Pacholczyk, S.M. Dederscheck, A. Eroglu (2007). Derivation and comparison of C57BL/6 embryonic stem cells to a widely used 129 embryonic stem cell line. *Transgenic. Res.* **16**: 751–758.
52. Kontgen, F., G. Suss, C. Stewart, M. Steinmetz, H. Bluethmann (1993). Targeted disruption of the MHC class II Aa gene in C57BL/6 mice. *Int. Immunol.* **5**: 957–964.
53. Ledermann, B., K. Burki (1991). Establishment of a germ-line competent C57BL/6 embryonic stem cell line. *Exp. Cell Res.* **197**: 254–258.
54. Ware, C.B., L.A. Siverts, A.M. Nelson, J.F. Morton, W.C. Ladiges (2003). Utility of a C57BL/6 ES line versus 129 ES lines for targeted mutations in mice. *Transgenic. Res.* **12**: 743–746.
55. Gardner, R.L. (2004). Pluripotential stem cells from vertebrate embryos: Present perspective and future challenges. In: Handbook of Stem Cells (R. Lanza, J. Gearhart, B. Hogan, D. Melton, R. Pedersen, J. Thomson, M. West,eds.), pp. 15–26. Elsevier Academic Press, Oxford.

56. Nagy, A., M. Gertsenstein, K. Vintersten, R. Behringer (2003). Manipulating the Mouse Embryo: A Laboratory manual (eds.). Cold Spring Harbor Laboratory Press, Cold Spring Harbor.
57. Tremml, G., M. Singer, and R. Malavarca (2008). Culture of mouse embryonic stem cells. In: Current Protocols in Stem Cell Biology (M. Bhatia, A. Elefanty, S.J. Fisher, R. Patient, T. Schlaeger, and E. Snyder, eds.), IC.4.1–IC.4.19. John Wiley & Sons Inc., New York.
58. Ware, L.M. and A.A. Axelrad (1972). Inherited resistance to N- and B-tropic murine leukemia viruses *in vitro*: Evidence that congenic mouse strains SIM and SIM.R differ at the Fv-1 locus. *Virology*. **50**: 339–348.
59. Ogiso, Y., A. Kume, Y. Nishimune, and A. Matsushiro (1982). Reversible and irreversible stages in the transition of cell surface marker during the differentiation of pluripotent teratocarcinoma cell induced with retinoic acid. *Exp. Cell Res*. **137**: 365–372.
60. Rathjen, P.D., S. Toth, A. Willis, J.K. Heath, and A.G. Smith (1990). Differentiation inhibiting activity is produced in matrix-associated and diffusible forms that are generated by alternate promoter usage. *Cell*. **62**: 1105–1114.
61. Michalska, A.E (2008). Isolation and propagation of mouse embryonic fibroblasts and preparation of mouse embryonic feeder layer cells. In: Current Protocols in Stem Cell Biology (M. Bhatia, A. Elefanty, S.J. Fisher, R. Patient, T. Schlaeger, and E. Snyder, eds.), IC.3.1–IC.3.16. John Wiley & Sons, Inc., New York.
62. Nichols, J., E.P. Evans, and A.G. Smith (1990). Establishment of germ-line-competent embryonic stem (ES) cells using differentiation inhibiting activity. *Development*. **110**: 1341–1348.
63. Smith, A.G., J.K. Heath, D.D. Donaldson, G.G. Wong, J. Moreau, M. Stahl, and D. Rogers (1988). Inhibition of pluripotential embryonic stem cell differentiation by purified polypeptides. *Nature*. **336**: 688–690.
64. Williams, R.L. et al (1988). Myeloid leukaemia inhibitory factor maintains the developmental potential of embryonic stem cells. *Nature*. **336**: 684–687.
65. Matsuda, T., T. Nakamura, K. Nakao, T. Arai, M. Katsuki, T. Heike, and T. Yokota (1999). STAT3 activation is sufficient to maintain an undifferentiated state of mouse embryonic stem cells. *Embo. J*. **18**: 4261–4269.
66. Niwa, H., T. Burdon, I. Chambers, and A. Smith (1998). Self-renewal of pluripotent embryonic stem cells is mediated via activation of STAT3. *Genes. Dev*. **12**: 2048–2060.
67. Ogawa, K., R. Nishinakamura, Y. Iwamatsu, D. Shimosato, and H. Niwa (2006). Synergistic action of Wnt and LIF in maintaining pluripotency of mouse ES cells. *Biochem. Biophys. Res. Commun*. **343**: 159–166.
68. Ying, Q.L., J. Nichols, I. Chambers, and A. Smith (2003). BMP induction of Id proteins suppresses differentiation and sustains embryonic stem cell self-renewal in collaboration with STAT3. *Cell*. **115**: 281–292.
69. Andang, M., A. Moliner, C.A. Doege, C.F. Ibanez, and P. Ernfors (2008). Optimized mouse ES cell culture system by suspension growth in a fully defined medium. *Nat. Protoc*. **3**: 1013–1017.
70. Furue, M. et al (2005). Leukemia inhibitory factor as an anti-apoptotic mitogen for pluripotent mouse embryonic stem cells in a serum-free medium without feeder cells. *In vitro Cell Dev. Biol. Anim*. **41**: 19–28.
71. Hayashi, Y. et al (2007). Integrins regulate mouse embryonic stem cell self-renewal. *Stem Cells*. **25**: 3005–3015.

72. Nichols, J. and Q.L. Ying (2006). Derivation and propagation of embryonic stem cells in serum- and feeder-free culture. *Methods Mol. Biol.* **329**: 91–98.
73. King, J.A. and W.M. Miller (2007). Bioreactor development for stem cell expansion and controlled differentiation. *Curr. Opin. Chem. Biol.* **11**: 394–398.
74. Thomson, H. (2007). Bioprocessing of embryonic stem cells for drug discovery. *Trends Biotechnol.* **25**: 224–230.
75. Cormier, J.T., zur N.I. Nieden, D.E. Rancourt, and M.S. Kallos (2006). Expansion of undifferentiated murine embryonic stem cells as aggregates in suspension culture bioreactors. *Tissue Eng.* **12**: 3233–3245.
76. Fok, E.Y. and P.W. Zandstra (2005). Shear-controlled single-step mouse embryonic stem cell expansion and embryoid body-based differentiation. *Stem Cells.* **23**: 1333–1342.
77. zur Nieden, N.I., J.T. Cormier, D.E. Rancourt and M.S. Kallos (2007). Embryonic stem cells remain highly pluripotent following long term expansion as aggregates in suspension bioreactors. *J. Biotechnol.* **129**: 421–432.
78. Abranches, E., E. Bekman, D. Henrique, and J.M. Cabral (2007). Expansion of mouse embryonic stem cells on microcarriers. *Biotechnol. Bioeng.* **96**: 1211–1221.
79. Fernandes, A.M., T.G. Fernandes, M.M. Diogo, C.L. da Silva, D. Henrique and J.M. Cabral (2007). Mouse embryonic stem cell expansion in a microcarrier-based stirred culture system. *J. Biotechnol.* **132**: 227–236.
80. Evans, M. (2004). Isolation and maintenance of murine embryonic stem cells. In: Handbook of Stem Cells (R. Lanza, J. Gearhart, H. Hogan, D. Melton, R. Pedersen, J. Thomson, M. West, eds.). 413–417. Elsevier academic Press, Amsterdam.
81. Savatier, P., S. Huang, L. Szekely, K.G. Wiman, and J. Samarut (1994). Contrasting patterns of retinoblastoma protein expression in mouse embryonic stem cells and embryonic fibroblasts. *Oncogene.* **9**: 809–818.
82. Stead, E. et al. (2002). Pluripotent cell division cycles are driven by ectopic Cdk2: cyclin A/E and E2F activities. *Oncogene.* **21**: 8320–8333.
83. Armstrong, L., M. Lako, J. Lincoln, P.M. Cairns, and N. Hole (2000). mTert expression correlates with telomerase activity during the differentiation of murine embryonic stem cells. *Mech. Dev.* **97**: 109–116.
84. Tam, W.L., Y.S. Ang, and B. Lim (2007). The molecular basis of ageing in stem cells. *Mech Ageing Dev.* **128**: 137–148.
85. Solter, D. and B.B. Knowles (1978). Monoclonal antibody defining a stage-specific mouse embryonic antigen (SSEA-1). *Proc. Natl. Acad. Sci. USA.* **75**: 5565–5569.
86. Berstine, E.G., M.L. Hooper, S. Grandchamp and B. Ephrussi (1973). Alkaline phosphatase activity in mouse teratoma. *Proc. Natl. Acad. Sci. USA.* **70**: 3899–3903.
87. Scholer, H.R., S. Ruppert, N. Suzuki, K. Chowdhury, and P. Gruss (1990). New type of POU domain in germ line-specific protein Oct-4. *Nature.* **344**: 435–439.
88. Viswanathan, S., T. Benatar, S. Rose-John, D.A. Lauffenburger and P.W. Zandstra (2002). Ligand/receptor signaling threshold (LIST) model accounts for gp130-mediated embryonic stem cell self-renewal responses to LIF and HIL-6. *Stem Cells.* **20**: 119–138.
89. Palmqvist, L., C.H. Glover, L. Hsu, M. Lu, B. Bossen, J.M. Piret, R.K. Humphries and C.D. Helgason (2005). Correlation of murine embryonic stem cell gene expression profiles with functional measures of pluripotency. *Stem Cells.* **23**: 663–680.

90. Rastan, S. and E.J. Robertson (1985). X-chromosome deletions in embryo-derived (EK) cell lines associated with lack of X-chromosome inactivation. *J. Embryol. Exp. Morphol.* **90**: 379–388.
91. Robertson, E.J., M.J. Evans and M.H. Kaufman (1983). X-chromosome instability in pluripotential stem cell lines derived from parthenogenetic embryos. *J. Embryol. Exp. Morphol.* **74**: 297–309.
92. Longo, L., A. Bygrave, F.G. Grosveld and P.P. Pandolfi (1997). The chromosome make-up of mouse embryonic stem cells is predictive of somatic and germ cell chimaerism. *Transgenic Res.* **6**: 321–328.
93. Liu, X., H. Wu, J. Loring, S. Hormuzdi, C.M. Disteche, P. Bornstein, and R. Jaenisch (1997). Trisomy eight in ES cells is a common potential problem in gene targeting and interferes with germ line transmission. *Dev. Dyn.* **209**: 85–91.
94. Sugawara, A., K. Goto, Y. Sotomaru, T. Sofuni and T. Ito (2006). Current status of chromosomal abnormalities in mouse embryonic stem cell lines used in Japan. *Comp. Med.* **56**: 31–34.
95. Liu, N., M. Lu, X. Tian, and Z. Han (2007). Molecular mechanisms involved in self-renewal and pluripotency of embryonic stem cells. *J. Cell Physiol.* **211**: 279–286.
96. Niwa, H (2007). How is pluripotency determined and maintained? *Development.* **134**: 635–646.
97. Nichols, J., B. Zevnik, K. Anastassiadis, H. Niwa, D. Klewe-Nebenius, I. Chambers, H. Scholer and A. Smith. (1998). Formation of pluripotent stem cells in the mammalian embryo depends on the POU transcription factor Oct4. *Cell.* **95**: 379–391.
98. Niwa, H. (2000). (Molecular mechanism for cell-fate determination in ES cells). *Tanpakushitsu Kakusan Koso.* **45**: 2047–4055.
99. Chambers, I., D. Colby, M. Robertson, J. Nichols, S. Lee, S. Tweedie and A. Smith (2003). Functional expression cloning of Nanog, a pluripotency sustaining factor in embryonic stem cells. *Cell.* **113**: 643–655.
100. Mitsui, K. et al. (2003). The homeoprotein Nanog is required for maintenance of pluripotency in mouse epiblast and ES cells. *Cell.* **113**: 631–642.
101. Pan, G. and J.A. Thomson (2007). Nanog and transcriptional networks in embryonic stem cell pluripotency. *Cell Res.* **17**: 42–49.
102. Avilion, A.A., S.K. Nicolis, L.H. Pevny, L. Perez, N. Vivian and R. Lovell-Badge (2003). Multipotent cell lineages in early mouse development depend on SOX2 function. *Genes Dev.* **17**: 126–140.
103. Sutton, J. et al. (1996). Genesis, a winged helix transcriptional repressor with expression restricted to embryonic stem cells. *J Biol Chem.* **271**: 23126–23133.
104. Ivanova, N. et al. (2006). Dissecting self-renewal in stem cells with RNA interference. *Nature.* **442**: 533–538.
105. Kim, J., J. Chu, X. Shen, J. Wang, and S.H. Orkin. (2008). An extended transcriptional network for pluripotency of embryonic stem cells. *Cell.* **132**: 1049–1061.
106. Wang, S., C. Hu, and J. Zhu. (2007). Transcriptional silencing of a novel hTERT reporter locus during *in vitro* differentiation of mouse embryonic stem cells. *Mol Biol Cell.* **18**: 669–677.
107. Doetschman, T.C., H. Eistetter, M. Katz, W. Schmidt, and R. Kemler (1985). The *in vitro* development of blastocyst-derived embryonic stem cell lines: Formation of visceral yolk sac, blood islands and myocardium. *J Embryol Exp Morphol.* **87**: 27–45.
108. Kurosawa, H. (2007). Methods for inducing embryoid body formation: *In vitro* differentiation system of embryonic stem cells. *J Biosci Bioeng.* **103**: 389–398.
109. Wobus, A.M. and Boheler, K.R. (2005). Embryonic stem cells: Prospects for developmental biology and cell therapy. *Physiol Rev.* **85**: 635–678.

110. Gottlieb, D.I. (2002). Large-scale sources of neural stem cells. *Annu Rev Neurosci.* **25**: 381–407.
111. Guan, K., Chang, H., Rolletschek, A. and Wobus, A.M. (2001). Embryonic stem cell-derived neurogenesis. Retinoic acid induction and lineage selection of neuronal cells. *Cell Tissue Res.* **305**: 171–176.
112. Kubo, A., Shinozaki, K., Shannon, J.M., Kouskoff, V., Kennedy, M., Woo, S., Fehling, H.J. and Keller, G. (2004). Development of definitive endoderm from embryonic stem cells in culture. *Development.* **131**: 1651–1662.
113. Tada, S. et al. (2005). Characterisation of mesendoderm: A diverging point of the definitive endoderm and mesoderm in embryonic stem cell differentiation culture. *Development.* **132**: 4363–4374.
114. Yasunaga, M. et al. (2005). Induction and monitoring of definitive and visceral endoderm differentiation of mouse ES cells. *Nat Biotechnol.* **23**: 1542–1550.
115. Barberi, T. et al. (2003). Neural subtype specification of fertilization and nuclear transfer embryonic stem cells and application in parkinsonian mice. *Nat Biotechnol.* **21**: 1200–1207.
116. Kawasaki, H., K. Mizuseki, S. Nishikawa, S. Kaneko, Y. Kuwana, S. Nakanishi, S.I. Nishikawa and Y. Sasai (2000). Induction of midbrain dopaminergic neurons from ES cells by stromal cell-derived inducing activity. *Neuron.* **28**: 31–40.
117. Perrier, A.L., V. Tabar, T. Barberi, M.E. Rubio, J. Bruses, N. Topf, N.L. Harrison and L. Studer (2004). Derivation of midbrain dopamine neurons from human embryonic stem cells. *Proc. Natl. Acad. Sci. USA.* **101**: 12543–12548.
118. Nakano, T., H. Kodama, and T. Honjo (1994). Generation of lymphohematopoietic cells from embryonic stem cells in culture. *Science.* **265**: 1098–1101.
119. Nishikawa, S.I., S. Nishikawa, M. Hirashima, N. Matsuyoshi and H. Kodama, (1998). Progressive lineage analysis by cell sorting and culture identifies FLK1+VE-cadherin+ cells at a diverging point of endothelial and haemopoietic lineages. *Development.* **125**: 1747–1757.
120. Ying, Q.L., Stavridis, M., Griffiths, D., Li, M. and Smith, A. (2003). Conversion of embryonic stem cells into neuroectodermal precursors in adherent monoculture. *Nat. Biotechnol.* **21**: 183–186.
121. Keller, G.M. (1995). *In vitro* differentiation of embryonic stem cells. *Curr. Opin. Cell. Biol.* **7**: 862–869.
122. Murry, C.E. and G. Keller (2008). Differentiation of embryonic stem cells to clinically relevant populations: lessons from embryonic development. *Cell.* **132**: 661–680.
123. Spence, J.R. and J.M. Wells (2007). Translational embryology: Using embryonic principles to generate pancreatic endocrine cells from embryonic stem cells. *Dev. Dyn.* **236**: 3218–3227.
124. Wood, S.A., N.D. Allen, J. Rossant, A. Auerbach and A. Nagy (1993). Non-injection methods for the production of embryonic stem cell-embryo chimaeras. *Nature.* **365**: 87–89.
125. Bradley, A. (1987). Production and analysis of chimaeric mice. In: Teratocarcinomas and Embryonic Stem Cells, A Practical Approach (Robertson, E., ed.), pp. 113–151. IRL Press, Oxford, Washington DC.
126. Nagy, A., E. Gocza, E.M. Diaz, V.R. Prideaux, E. Ivanyi, M. Markkula and J. Rossant (1990). Embryonic stem cells alone are able to support foetal development in the mouse. *Development.* **110**: 815–821.
127. Lallemand, Y. and P. Brulet (1990). An in situ assessment of the routes and extents of colonisation of the mouse embryo by embryonic stem cells and their descendants. *Development.* **110**: 1241–1248.
128. Nagy, A., J. Rossant, R. Nagy, W. J.C. Abramow-Newerly and Roder (1993). Derivation of completely cell culture-derived mice from early-passage embryonic stem cells. *Proc. Natl. Acad. Sci. USA.* **90**: 8424–8428.

129. Wang, Z.Q., F. Kiefer, P. Urbanek, and E.F. Wagner (1997). Generation of completely embryonic stem cell-derived mutant mice using tetraploid blastocyst injection. *Mech Dev.* **62**: 137–145.
130. Becker, S., M. Hrabé de Angelis, and J. Beckers (2006). Use of chemical mutagenesis in mouse embryonic stem cells. In: Embryonic Stem Cell Protocols: Volume 1: Isolation and Characterization (Turksen, K., ed.), pp. 397–407. Humana Press Inc, Totowa, New Jersey.
131. Giudice, A. and A. Trounson (2008). Genetic modification of human embryonic stem cells for derivation of target cells. *Cell Stem Cell.* **2**: 422–433.
132. Kobayashi, N., J.D. Rivas-Carrillo, A. Soto-Gutierrez, T. Fukazawa, Y. Chen, N. Navarro-Alvarez, and N. Tanaka. 2005). Gene delivery to embryonic stem cells. *Birth Defects Res. C. Embryo. Today.* **75**: 10–18.
133. Hanahan, D., E.F. Wagner, and R.D. Palmiter (2007). The origins of oncomice: A history of the first transgenic mice genetically engineered to develop cancer. *Genes Dev.* **21**: 2258–2270.
134. Gotz, J. and L.M. Ittner (2008). Animal models of Alzheimer's disease and frontotemporal dementia. *Nat Rev Neurosci.* **9**: 532–544.
135. Li, X., J.W. Xiong, C.S. Shelley, H. Park. and M.A. Arnaout (2006). The transcription factor ZBP-89 controls generation of the hematopoietic lineage in zebrafish and mouse embryonic stem cells. *Development.* **133**: 3641–3650.
136. Torres, J. and F.M. Watt (2008). Nanog maintains pluripotency of mouse embryonic stem cells by inhibiting NFkappaB and cooperating with Stat3. *Nat Cell Biol.* **10**: 194–201.
137. Mizushima, N. et al. (2003). Mouse Apg16L, a novel WD-repeat protein, targets to the autophagic isolation membrane with the Apg12-Apg5 conjugate. *J. Cell Sci.* **116**: 1679–1688.
138. Blyszczuk, P., J. Czyz, G. Kania, M. Wagner, U. Roll, L. St-Onge, and A.M. Wobus (2003). Expression of Pax4 in embryonic stem cells promotes differentiation of nestin-positive progenitor and insulin-producing cells. *Proc. Natl. Acad. Sci. USA.* **100**: 998–1003.
139. Kim, J.H. et al. (2002). Dopamine neurons derived from embryonic stem cells function in an animal model of Parkinson's disease. *Nature.* **418**: 50–56.
140. Lai, Y., I. Drobinskaya, E. Kolossov, C. Chen and T. Linn (2008). Genetic modification of cells for transplantation. *Adv. Drug. Deliv. Rev.* **60**: 146–159.
141. Ishii, T. et al. (2005). *In vitro* differentiation and maturation of mouse embryonic stem cells into hepatocytes. *Exp. Cell. Res.* **309**: 68–77.
142. Drobinskaya, I., T. Linn, T. Saric, R.G. Bretzel, H. Bohlen, J. Hescheler and E. Kolossov (2008). Scalable Selection of Hepatocyte- and Hepatocyte Precursor-like Cells from Culture of Differentiating Transgenically Modified Murine ES Cells. *Stem. Cells.* **26**(9): 2245–2256.
143. Yang, X.W., P. Model and N. Heintz (1997). Homologous recombination based modification in Escherichia coli and germline transmission in transgenic mice of a bacterial artificial chromosome. *Nat. Biotechnol.* **15**: 859–865.
144. Tomishima, M.J., A.K. Hadjantonakis, S. Gong, and L. Studer (2007). Production of green fluorescent protein transgenic embryonic stem cells using the GENSAT bacterial artificial chromosome library. *Stem Cells.* **25**: 39–45.
145. Valenzuela, D.M. et al. (2003). High-throughput engineering of the mouse genome coupled with high-resolution expression analysis. *Nat. Biotechnol.* **21**: 652–659.
146. Yang, Y. and B. Seed (2003). Site-specific gene targeting in mouse embryonic stem cells with intact bacterial artificial chromosomes. *Nat. Biotechnol.* **21**: 447–451.

147. Chalberg, T.W., J.L. Portlock, E.C. Olivares, B. Thyagarajan, P.J. Kirby, R.T. Hillman, J. Hoelters, and M.P. Calos (2006). Integration specificity of phage phiC31 integrase in the human genome. *J. Mol Biol.* **357**: 28–48.

148. Hitz, C., W. Wurst, and R. Kuhn (2007). Conditional brain-specific knockdown of MAPK using Cre/loxP regulated RNA interference. *Nucleic Acids Res.* **35**: e90.

149. Wirth, D., L. Gama-Norton, P. Riemer, U. Sandhu, R. Schucht. and H. Hauser (2007). Road to precision: Recombinase-based targeting technologies for genome engineering. *Curr. Opin. Biotechnol.* **18**: 411–419.

150. Nord, A.S. et al. (2006). The International Gene Trap Consortium Website: A portal to all publicly available gene trap cell lines in mouse. *Nucleic. Acids. Res.* **34**: D642–648.

151. Pfeifer, A., M. Ikawa, Y. Dayn, and I.M. Verma (2002). Transgenesis by lentiviral vectors: Lack of gene silencing in mammalian embryonic stem cells and preimplantation embryos. *Proc. Natl. Acad. Sci. USA.* **99**: 2140–2145.

152. Pfeifer, A. (2006). Lentiviral transgenesis-A versatile tool for basic research and gene therapy. *Curr. Gene. Ther.* **6**: 535–542.

153. Jiang, J. et al. (2008). A core Klf circuitry regulates self-renewal of embryonic stem cells. *Nat. Cell. Biol.* **10**: 353–360.

154. Ivey, K.N. et al. (2008). MicroRNA regulation of cell lineages in mouse and human embryonic stem cells. *Cell. Stem Cell* **2**: 219–229.

155. Pfeifer, A., S. Eigenbrod, S. Al-Khadra, A. Hofmann, G. Mitteregger, M. Moser, U. Bertsch and H. Kretzschmar (2006). Lentivector-mediated RNAi efficiently suppresses prion protein and prolongs survival of scrapie-infected mice. *J. Clin. Invest.* **116**: 3204–3210.

156. Kawabata, K., F. Sakurai, N. Koizumi, T. Hayakawa and H. Mizuguchi (2006). Adenovirus vector-mediated gene transfer into stem cells. *Mol. Pharm.* **3**: 95–103.

157. Kawabata, K., F. Sakurai, T. Yamaguchi, T. Hayakawa, and H. Mizuguchi (2005). Efficient gene transfer into mouse embryonic stem cells with adenovirus vectors. *Mol. Ther.* **12**: 547–554.

158. Folger, K.R., E.A. Wong, G. Wahl, and M.R. Capecchi (1982). Patterns of integration of DNA microinjected into cultured mammalian cells: Evidence for homologous recombination between injected plasmid DNA molecules. *Mol Cell Biol.* **2**: 1372–1387.

159. Kucherlapati, R.S., E.M. Eves, K.Y. Song, B.S. Morse and O. Smithies (1984). Homologous recombination between plasmids in mammalian cells can be enhanced by treatment of input DNA. *Proc Natl Acad Sci. USA.* **81**: 3153–3157.

160. Smithies, O., R.G. Gregg, S.S. Boggs, M.A. Koralewski and R.S. Kucherlapati (1985). Insertion of DNA sequences into the human chromosomal beta-globin locus by homologous recombination. *Nature.* **317**: 230–234.

161. Mansour, S.L., K.R. Thomas, and M.R. Capecchi (1988). Disruption of the proto-oncogene int-2 in mouse embryo-derived stem cells: A general strategy for targeting mutations to non-selectable genes. *Nature.* **336**: 348–352.

162. Koller, B.H. et al. (1989). Germ-line transmission of a planned alteration made in a hypoxanthine phosphoribosyltransferase gene by homologous recombination in embryonic stem cells. *Proc. Natl. Acad. Sci. USA.* **86**: 8927–8931.

163. Capecchi, M.R. (2005). Gene targeting in mice: Functional analysis of the mammalian genome for the twenty-first century. *Nat. Rev. Genet.* **6**: 507–512.

164. Manis, J.P. (2007). Knock out, knock in, knock down-Genetically manipulated mice and the Nobel Prize. *N. Engl. J. Med.* **357**: 2426–2429.
165. Bockamp, E., R. Sprengel, L. Eshkind, T. Lehmann, J.M. Braun, F. Emmrich and J.G. Hengstler (2008). Conditional transgenic mouse models: From the basics to genome-wide sets of knockouts and current studies of tissue regeneration. *Regen. Med.* **3**: 217–235.
166. Porteus, M.H. and D. Carroll (2005). Gene targeting using zinc finger nucleases. *Nat. Biotechnol.* **23**: 967–973.
167. Pingoud, A. and G.H. Silva (2007). Precision genome surgery. *Nat. Biotechnol.* **25**: 743–744.
168. Bockamp, E., M. Maringer, C. Spangenberg, S. Fees, S. Fraser, L. Eshkind, F. Oesch and B. Zabel (2002). Of mice and models: Improved animal models for biomedical research. *Physiol. Genomics.* **11**: 115–132.
169. Hatada, S., L.W. Arnold, T. Hatada, J.E. Cowhig, Jr., D. Ciavatta and O. Smithies (2005). Isolating gene-corrected stem cells without drug selection. *Proc. Natl. Acad. Sci. USA.* **102**: 16357–16361.
170. Mortensen, R.M. (1993). Double knockouts. Production of mutant cell lines in cardiovascular research. *Hypertension.* **22**: 646–651.
171. Mortensen, R.M., D.A. Conner, S. Chao, A.A. Geisterfer-Lowrance and J.G. Seidman (1992). Production of homozygous mutant ES cells with a single targeting construct. *Mol. Cell. Biol.* **12**: 2391–2395.
172. Barbaric, I., G. Miller and T.N. Dear (2007). Appearances can be deceiving: Phenotypes of knockout mice. *Brief Funct Genomic Proteomic.* **6**: 91–103.
173. Wolfer, D.P., W.E. Crusio and H.P. Lipp (2002). Knockout mice: Simple solutions to the problems of genetic background and flanking genes. *Trends Neurosci.* **25**: 336–340.
174. Montagutelli, X. (2000). Effect of the genetic background on the phenotype of mouse mutations. *J. Am. Soc. Nephrol 11. Suppl.* **16**: S101–105.
175. Dobkin, C., A. Rabe, R. Dumas, El Idrissi, A., H. Haubenstock and W.T. Brown (2000). Fmr1 knockout mouse has a distinctive strain-specific learning impairment. *Neuroscience.* **100**: 423–429.
176. Dimou, L. et al. (2006). Nogo-A-deficient mice reveal strain-dependent differences in axonal regeneration. *J. Neurosci.* **26**: 5591–5603.
177. Collins, F.S., J. Rossant and W. Wurst (2007). A mouse for all reasons. *Cell* **128**: 9–13.
178. Collins, F.S., R.H. Finnell, J. Rossant, and W. Wurst (2007). A new partner for the international knockout mouse consortium. *Cell.* **129**: 235.
179. Gu, H., J.D. Marth, P.C. Orban, H. Mossmann, and K. Rajewsky (1994). Deletion of a DNA polymerase beta gene segment in T cells using cell type-specific gene targeting. *Science.* **265**: 103–106.
180. Branda, C.S. and S.M. Dymecki (2004). Talking about a revolution: The impact of site-specific recombinases on genetic analyses in mice. *Dev. Cell.* **6**: 7–28.
181. Lee, H.J., H.K. Caldwell, A.H. Macbeth, S.G. Tolu and W.S. Young, III (2008). A conditional knockout mouse line of the oxytocin receptor. *Endocrinology.* **149**: 3256–3263.
182. Phaneuf, D., A.D. Moscioni, C. LeClair, S.E. Raper, and J.M. Wilson (2004). Generation of a mouse expressing a conditional knockout of the hepatocyte growth factor gene: Demonstration of impaired liver regeneration. *DNA Cell Biol.* **23**: 592–603.
183. Tsai, S.Y., B.W. O'Malley, F.J. DeMayo, Y. Wang, and S.S. Chua (1998). A novel RU486 inducible system for the activation and repression of genes. *Adv. Drug. Deliv. Rev.* **30**: 23–31.

184. Matthaei, K.I. (2007). Genetically manipulated mice: A powerful tool with unsuspected caveats. *J. Physiol.* **582**: 481–488.
185. Micallef, S.J., M.E. Janes, K. Knezevic, R.P. Davis, A.G. Elefanty, and E.G. Stanley (2005). Retinoic acid induces Pdx1-positive endoderm in differentiating mouse embryonic stem cells. *Diabetes.* **54**: 301–305.
186. Holland, A.M., S.J. Micallef, X. Li, A.G. Elefanty, and E.G. Stanley (2006). A mouse carrying the green fluorescent protein gene targeted to the Pdx1 locus facilitates the study of pancreas development and function. *Genesis.* **44**: 304–307.
187. Li, M., L. Pevny, R. Lovell-Badge and A. Smith (1998). Generation of purified neural precursors from embryonic stem cells by lineage selection. *Curr. Biol.* **8**: 971–974.
188. Xian, H.Q., E. McNichols, A. St Clair and D.I. Gottlieb (2003). A subset of ES-cell-derived neural cells marked by gene targeting. *Stem. Cells.* **21**: 41–49.
189. Lee, H. et al. (2006). Human C5aR knock-in mice facilitate the production and assessment of anti-inflammatory monoclonal antibodies. *Nat. Biotechnol.* **24**: 1279–1284.
190. Mizuguchi, H., Z. Xu, A. Ishii-Watabe, E. Uchida and T. Hayakawa (2000). IRES-dependent second gene expression is significantly lower than cap-dependent first gene expression in a bicistronic vector. *Mol. Ther.* **1**: 376–382.
191. Miquerol, L., B.L. Langille and A. Nagy (2000). Embryonic development is disrupted by modest increases in vascular endothelial growth factor gene expression. *Development.* **127**: 3941–3946.
192. Hasty, P., J. Rivera-Perez and A. Bradley (1991). The length of homology required for gene targeting in embryonic stem cells. *Mol. Cell. Biol.* **11**: 5586–5591.
193. Thomas, K.R., C. Deng and M.R. Capecchi (1992). High-fidelity gene targeting in embryonic stem cells by using sequence replacement vectors. *Mol. Cell. Biol.* **12**: 2919–2923.
194. van Deursen, J. and B. Wieringa (1992). Targeting of the creatine kinase M gene in embryonic stem cells using isogenic and nonisogenic vectors. *Nucleic. Acids. Res.* **20**: 3815–3820.
195. te Riele, H., E.R. Maandag and A. Berns (1992). Highly efficient gene targeting in embryonic stem cells through homologous recombination with isogenic DNA constructs. *Proc. Natl. Acad. Sci. USA.* **89**: 5128–5132.
196. Deng, C. and M.R. Capecchi (1992). Reexamination of gene targeting frequency as a function of the extent of homology between the targeting vector and the target locus. *Mol. Cell. Biol.* **12**: 3365–3371.
197. Wilson, T.J., D. Truman, A. Giudice and P. Hertzog (2007). Rapid generation of gene-targeting constructs. In: PCR: Methods Express (S. Hughes, A. Moody, eds.), 173–196. Scion Publishing Ltd, Oxfordshire.
198. Adams, D.J. et al. (2005). A genome-wide, end-sequenced 129Sv BAC library resource for targeting vector construction. *Genomics.* **86**: 753–758.
199. Wong, E.A. and M.R. Capecchi (1987). Homologous recombination between coinjected DNA sequences peaks in early to mid-S phase. *Mol. Cell Biol.* **7**: 2294–2295.
200. Wu, J., K. Kandavelou and S. Chandrasegaran (2007). Custom-designed zinc finger nucleases: what is next- *Cell Mol. Life Sci.* **64**: 2933–2944.
201. Carroll, D. (2004). Using nucleases to stimulate homologous recombination. In: Genetic Recombination: Reviews and Protocols (A.S. Waldman, ed.), pp. 195–207. Humana Press Inc, Totowa, New Jersey.

202. Lombardo, A. et al. (2007). Gene editing in human stem cells using zinc finger nucleases and integrase-defective lentiviral vector delivery. *Nat Biotechnol.* **25**: 1298–1306.
203. Fire, A., S. Xu, M.K. Montgomery, S.A. Kostas, S.E. Driver and C.C. Mello (1998). Potent and specific genetic interference by double-stranded RNA in Caenorhabditis elegans. *Nature.* **391**: 806–811.
204. Bernstein, E., A.A. Caudy, S.M. Hammond and G.J. Hannon (2001). Role for a bidentate ribonuclease in the initiation step of RNA interference. *Nature.* **409**: 363–366.
205. Elbashir, S.M., W. Lendeckel and T. Tuschl (2001). RNA interference is mediated by 21- and 22-nucleotide RNAs. *Genes. Dev.* **15**: 188–200.
206. Martin, S.E. and N.J. Caplen (2007). Applications of RNA interference in mammalian systems. *Annu Rev Genomics Hum. Genet.* **8**: 81–108.
207. Ding, L. and F. Buchholz (2006). RNAi in embryonic stem cells. *Stem Cell Rev.* **2**: 11–18.
208. Hammond, S.M., E. Bernstein, D. Beach and G.J. Hannon (2000). An RNA-directed nuclease mediates post-transcriptional gene silencing in Drosophila cells. *Nature.* **404**: 293–296.
209. Kunath, T., G. Gish, H. Lickert, N. Jones, T. Pawson and J. Rossant (2003). Transgenic RNA interference in ES cell-derived embryos recapitulates a genetic null phenotype. *Nat. Biotechnol.* **21**: 559–561.
210. Xia, X.G., H. Zhou and Z. Xu (2006). Transgenic RNAi: Accelerating and expanding reverse genetics in mammals. *Transgenic Res.* **15**: 271–275.
211. Zou, G.M., R.J. Chan, W.C. Shelley and M.C. Yoder (2006). Reduction of Shp-2 expression by small interfering RNA reduces murine embryonic stem cell-derived *in vitro* hematopoietic differentiation. *Stem Cells.* **24**: 587–594.
212. Yamashita, A., T. Takada, K. Nemoto, G. Yamamoto and R. Torii (2006). Transient suppression of PPARgamma directed ES cells into an osteoblastic lineage. *FEBS Lett.* **580**: 4121–4125.
213. Reynolds, A., D. Leake, Q. Boese, S. Scaringe, W.S. Marshall and A. Khvorova (2004). Rational siRNA design for RNA interference. *Nat. Biotechnol.* **22**: 326–330.
214. Shah, J.K., H.R. Garner, M.A. White, D.S. Shames and J.D. Minna (2007). sIR: siRNA Information Resource, a web-based tool for siRNA sequence design and analysis and an open access siRNA database. *BMC. Bioinformatics.* **8**: 178.
215. Mittal, V. (2004). Improving the efficiency of RNA interference in mammals. *Nat. Rev. Genet.* **5**: 355–365.
216. Cheng, J.C., E.M. Horwitz, S.L. Karsten, L. Shoemaker, H.I. Kornblum, P. Malik and K.M. Sakamoto (2007). Report on the workshop "New Technologies in Stem Cell Research," Society for Pediatric Research, San Francisco, California, April 29: 2006. *Stem Cells.* **25**: 1070–1088.
217. Clark, J. and S. Ding (2006). Generation of RNAi Libraries for High-Throughput Screens. *J Biomed Biotechnol.* **2006**: 45716.
218. Lakshmipathy, U., B. Love, C. Adams, B. Thyagarajan and J.D. Chesnut (2007). Micro RNA profiling: an easy and rapid method to screen and characterize stem cell populations. *Methods Mol. Biol.* **407**: 97–114.
219. Schmidt-Supprian, M. and K. Rajewsky (2007). Vagaries of conditional gene targeting. *Nat. Immunol.* **8**: 665–668.
220. Miyazaki, S., E. Yamato and J. Miyazaki (2004). Regulated expression of pdx-1 promotes *in vitro* differentiation of insulin-producing cells from embryonic stem cells. *Diabetes.* **53**: 1030–1037.

221. Friedrich, G. and P. Soriano (1991). Promoter traps in embryonic stem cells: a genetic screen to identify and mutate developmental genes in mice. *Genes. Dev.* **5**: 1513–1523.
222. Soriano, P. (1999). Generalized lacZ expression with the ROSA26 Cre reporter strain. *Nat. Genet.* **21**: 70–71.
223. Aiba, K., M.G. Carter, R. Matoba and M.S. Ko (2006). Genomic approaches to early embryogenesis and stem cell biology. *Semin. Reprod. Med.* **24**: 330–339.
224. Chen, S., S. Hilcove and S. Ding (2006). Exploring stem cell biology with small molecules. *Mol. Biosyst.* **2**: 18–24.
225. Rubin, L.L. (2008). Stem cells and drug discovery: the beginning of a new era. *Cell.* **132**: 549–552.
226. Xu, Y., Y. Shi and S. Ding (2008). A chemical approach to stem-cell biology and regenerative medicine. *Nature.* **453**: 338–344.
227. Chen, S. et al. (2006). Self-renewal of embryonic stem cells by a small molecule. *Proc. Natl. Acad. Sci. USA.* **103**: 17266–17271.
228. Ding, S., T.Y. Wu, A. Brinker, E.C. Peters, W. Hur, N.S. Gray and P.G. Schultz (2003). Synthetic small molecules that control stem cell fate. *Proc. Natl. Acad. Sci. USA.* **100**: 7632–7637.
229. Wu, X., S. Ding, Q. Ding, N.S. Gray and P.G. Schultz (2004). Small molecules that induce cardiomyogenesis in embryonic stem cells. *J. Am. Chem. Soc.* **126**: 1590–1591.
230. Doss, M.X., C.I. Koehler, C. Gissel, J. Hescheler and A. Sachinidis (2004). Embryonic stem cells: A promising tool for cell replacement therapy. *J. Cell Mol. Med.* **8**: 465–473.
231. Taylor, C.J., E.M. Bolton, S. Pocock, L.D. Sharples, R.A. Pedersen and J.A. Bradley (2005). Banking on human embryonic stem cells: estimating the number of donor cell lines needed for HLA matching. *Lancet.* **366**: 2019–2025.
232. Nakajima, F., K. Tokunaga, and N. Nakatsuji (2007). Human leukocyte antigen matching estimations in a hypothetical bank of human embryonic stem cell lines in the Japanese population for use in cell transplantation therapy. *Stem Cells.* **25**: 983–985.
233. Osafune, K. et al. (2008). Marked differences in differentiation propensity among human embryonic stem cell lines. *Nat. Biotechnol.* **26**: 313–315.
234. Seach, N., D. Layton, J. Lim, A. Chidgey and R. Boyd (2007). Thymic generation and regeneration: a new paradigm for establishing clinical tolerance of stem cell-based therapies. *Curr. Opin. Biotechnol.* **18**: 441–447.
235. Takahashi, K. and S. Yamanaka (2006). Induction of pluripotent stem cells from mouse embryonic and adult fibroblast cultures by defined factors. *Cell.* **126**: 663–676.
236. Okita, K., T. Ichisaka and S. Yamanaka (2007). Generation of germline-competent induced pluripotent stem cells. *Nature.* **448**: 313–317.
237. Wernig, M., A. Meissner, R. Foreman, T. Brambrink, M. Ku, K. Hochedlinger, B.E. Bernstein, and R. Jaenisch (2007). *In vitro* reprogramming of fibroblasts into a pluripotent ES-cell-like state. *Nature.* **448**: 318–324.
238. Maherali, N. et al. (2007). Directly reprogrammed fibroblasts show global epigenetic remodeling and widespread tissue contribution. *Cell Stem Cell.* **1**: 55–70.
239. Takahashi, K., K. Tanabe, M. Ohnuki, M. Narita, T. Ichisaka, K. Tomoda, and S. Yamanaka (2007). Induction of pluripotent stem cells from adult human fibroblasts by defined factors. *Cell.* **131**: 861–872.

240. Yu, J. et al. (2007). Induced pluripotent stem cell lines derived from human somatic cells. *Science.* **318**: 1917–1920.
241. Wernig, M., A. Meissner, J.P. Cassady, and R. Jaenisch (2008). c-Myc is dispensable for direct reprogramming of mouse fibroblasts. *Cell Stem Cell.* **2**: 10–12.
242. Nakagawa, M. et al. (2008). Generation of induced pluripotent stem cells without Myc from mouse and human fibroblasts. *Nat. Biotechnol.* **26**: 101–106.
243. Hanna, J. et al. (2007). Treatment of sickle cell anemia mouse model with iPS cells generated from autologous skin. *Science.* **318**: 1920–1923.
244. Rideout, W.M., III, K. Hochedlinger, M. Kyba, G.Q. Daley and R. Jaenisch (2002). Correction of a genetic defect by nuclear transplantation and combined cell and gene therapy. *Cell.* **109**: 17–27.

6

DERIVATION OF HUMAN EMBRYONIC STEM CELLS FROM BLASTOCYSTS

Henry Sathananthan[1]*, Sulochana Gunasheela[2], Devika Gunasheela[2], Amitha Jaykumar[2]

[1]*Monash Immunology and Stem Cell Laboratory, Melbourne, Australia*
[2]*Gunasheela Institute for Research in Reproduction, Bangalore, India*

INTRODUCTION

Human embryonic stem cells (hESC), derived from the inner-cell mass (ICM) of blastocysts and grown up to 5–6 days in culture, have been recognized to show the capacity to generate cells of all the three primary embryonic germ layers, namely the ectoderm, endoderm, and mesoderm (Fig. 6.1). These could form more than 210 cell types in the adult body. Being pluripotent, they are superior to adult progenitor stem cells which form only a limited number of cell types. Because of their unlimited capacity of self-renewal, the ESCs have been regarded as a source of tissue for replacement of cells degenerated by injury or disease. On the other hand, adult stem cells and cord blood stem cells (non--ESC) have been used for treating blood and immune system related diseases, cancers, juvenile diabetes, Parkinson's disease, blindness, and spinal cord injuries. In view of this, there is a great demand for super-numerary human blastocysts available in IVF laboratories. Some of these are donated by parents out of altruistic sentiments or to promote academic research for cell therapy.

The ESCs are usually derived from the ICM of human blastocysts [1–5]. A blastocyst is a day 5/6 preimplantation embryo with an ICM, which becomes the embryo proper and an outer trophoblast, which becomes the placenta. The ICM is separated from the trophoblast by immunosurgery [6], or can be effectively isolated by mechanical means [7]. The ICM is grown on human or animal feeder cells and the ensuing ESCs are continuously propagated further on fresh

**Henry Sathananthan@med.monash.edu.au*

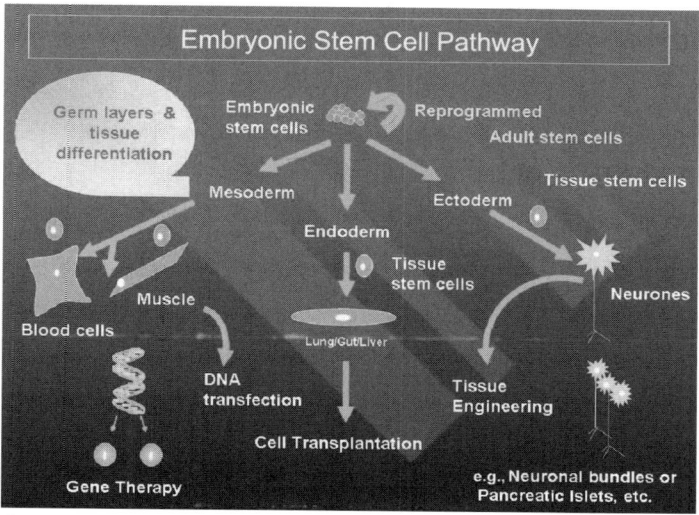

Courtesy: A. Trounson

▶ **Fig. 6.1** *Embryonic stem cell pathway: cells to therapy. (For colour figure see Plate 11)*

feeder cells in the presence of special nourishing media, renewed every 7-9 days until a cell line is established [4,5]. The ESCs derived are diploid, pluripotent, self-renewing cells that propagate indefinitely and have the potential to give rise to any tissue of the human body. Recently, attempts were made to derive ESC from blastomeres, biopsied from eight-cell embryos to overcome ethical concerns, which was not so successful [8,9].

Embryo and Blastocyst Culture

The most important requirement in generating viable ESC colonies or embryoid bodies (EBs) *in vitro* is the production of morphologically normal embryos in the IVF laboratory [10,11]. The success rate of generating viable ESC lines from blastocysts is variable [12], which must be improved. Both high-quality oocytes and normal sperm are required for the production of good embryos in IVF. See reviews for assessment of gametes and embryos [13,14].

The embryos need to develop normally to blastocysts, in space and time, without arresting or slowing down at any cleavage stage (Fig. 6.2 and Table 6.2). This is now possible in most established ART centers reporting pregnancy rates of 20-30% of embryos implanted. Both IVF and intracytoplasmic sperm injection (ICSI) embryos could be used to produce blastocysts, monitored on digital video using an inverted microscope with superior optics. Excess siblings of embryos that result in a pregnancy on embryo transfer on day 2 or 3 are optimal for ESC culture after blastocyst culture and freezing. There is no sense in using abnormal embryos for ESC biotechnology and consequent cell therapy.

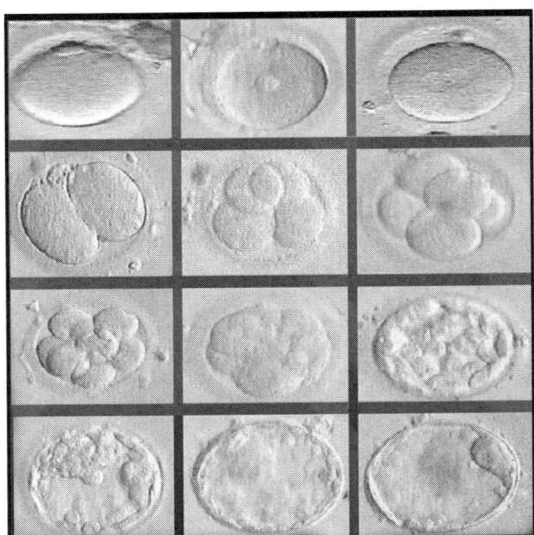

▶ **Fig. 6.2** *Embryo development—egg to blastocyst. Mature egg, unfertilised egg, fertilised ovum, two-cell, four-cell, five-cell, eight-cell, compaction, morula, early blastocyst, hatching blastocyst (early), expanded blastocyst (day 6) ↔400 (from Ref. [29]). (For colour figure see Plate 12)*

IVF and Embryo Culture Protocols

Basically blastocyst produced in the IVF laboratories are those derived from preimplantation embryos. Human embryos are produced by IVF after harvesting mature oocytes aspirated from ovarian follicles. The ovaries are stimulated by various stimulation protocols depending upon the age group of the patients, the body mass index (BMI), and other conditions relevant to safety in induction of ovulation.

In the majority of patients, downregulation of the pituitary gland was achieved with Gonadotropin-releasing hormone agonist (GnRH agonist) (Luprolide acetate, Sun Pharmaceuticals Ind. Ltd., India), injected in a dose of 0.5mg/day, for the first 12–13 days starting from Day 3 of the cycle until the E2 level was less than 50 pg/mL. Stimulation was then started with recombinant Follicle-stimulating hormone (FSH) (Gonal F, Serono, Switzerland) at a dose of 150 or 225 IU depending on the patient's BMI and Human Menopausal Gonadotrophin (HMG) (IVFM, L.G. Life Sciences, South Korea) was added as and when required. The dose of FSH and HMG were continuously adapted according to the serum concentrations of E2 and LH and to the dynamics of ovarian follicular growth.

Some of the patients were also stimulated directly with recombinant FSH (Gonal F, Serono, Switzerland) for 5 days from Day 3 to Day 7; GnRH antagonist Orgalutron 0.25 mg/day (Organon Ltd., Dublin, Ireland) was interposed from 6th day of stimulation along with continued administration of recombinant FSH. Throughout the course of induction of ovulation, serial follicular scans were performed using ultrasound machine (Aloka, Trivitron Medical Systems, India) with vaginal probe (UST 981 TB probe—90° with 5 MHz). Triggering of final maturation of oocytes was done by administering 10,000 IU of HCG IM (Ovutrig HP, VHB Life Sciences, Germany), when at least two follicles of 18 mm were seen on ultrasound along with four to five supporting follicles of approximately 15 mm.

In the long agonist protocol, the last injection of agonist was given one day before the administration of HCG, whereas in the antagonist protocol the last injection of antagonist was administered on the morning of HCG.

Oocyte Retrieval

Oocyte retrieval was performed per vagina 34 h after the dose of HCG using a single lumen Ova-Stiff ovum aspiration needle (Cook, Australia) with the help of a vaginal probe and suction machine (Labotech, Germany), with a negative pressure maintained at 100 mmHg.

Mature (metaphase II) oocytes were graded as 1, 2, and 3 depending on the cumulus quantity and cumulus corona expansions. MII oocytes were incubated at 37°C in fertilization medium (Quinns, Sage Bio Pharma, Pasadena, USA) in CO_2 incubator (Heracell, Germany, equilibrated with 5% CO_2 in air) for 3-6 h. Then they were freed off the cumulus oophorus and cell of the corona radiata, and were then transferred to a dish containing droplets of fertilization media covered with oil kept in CO_2 (5% CO_2 in air) incubator with temperature maintained at 37°C.

Intracytoplasmic Sperm Injection

Injection dishes were prepared using Falcon 3001 plastic culture dishes (Becton Dickinson, Plymouth, UK). The oocytes were placed in the micro drops of flushing medium (Quinns, Sage Bio Pharma, Pasadena, USA), previously equilibrated with 5% CO_2 in air and covered with pre-equilibrated tissue culture oil (Quinns, Sage Bio Pharma, Pasadena, USA). The dishes were maintained on a heated stage (37°C) of a Nikon (Tokyo, Japan) Diaphot inverted research microscope during all subsequent manipulations. The microscope was equipped with two Narashige micro manipulators (Tokyo, Japan). During the Intra Cytoplesmic Sperum Injection (ICSI) procedure, spermatozoa were placed and identified in a viscous poly vinyl pyriludin (PVP) solution (Quinns, Sage Bio Pharma, Pasadena, USA) and immobilized thereafter. Sperm immobilization was done by slashing the tail with injection needle. Oocytes were held by negative suction through a holding pipette (TPC, Thebarton, South Australia), and an immobilized spermatozoan was injected into each oocyte by injection pipette (TPC, Thebarton, South Australia). After the ICSI procedure, the injected oocytes were transferred to the Day 0 transfer dish and incubated at 37° C at 5% CO_2 for 16–18 h.

Fertilisation and Embryo Culture

Fertilization was assessed 16–18 h after the ICSI process. Zygotes were scored as having good morphology or poor morphology based on the assessment of the number and alignment of nucleoli in the pronuclei. They were then transferred to a dish containing droplets of cleavage medium (Quinns medium) 20 h later. The nature and rate of cleavage were assessed on the second day of oocyte insemination. Embryos were graded from grade IV to grade I based on the regularity of the blastomere and presence or absence of fragmentation at 48 h after ICSI. On Day 3, embryos were assessed for further division and graded as eight-cell, early compaction, compacting and compacted.

They were then transferred into blastocyst medium (Quinns, Sage Bio Pharma, Pasadena, USA), which is a modification of human tubal fluid (HTF). Modifications to the original formulation include the addition of an elevated concentration of glucose, selected non-essential and essential amino acids, alanyl-glutamine as a stable source of glutamine, MEM vitamin mix, increased levels of magnesium, and calcium lactate. On Day 5, the blastocysts were examined and graded as early cavitating blastocysts, expanding blastocysts, expanded blastocysts and hatching blastocysts. On Day 6, the blastocysts were loaded into eppendorf tubes containing 1 mL of equilibrated blastocyst medium sealed with parafilm paper and sent to laboratory for further growth and ESC culture. The blastocysts were fresh and not frozen (see Table 6.1).

Table 6.1 *Normal Embryonic Growth from Day 2–6*

Day	Embryo	Appearance/hours
D1	Fertilised ovum	2 PN (12 h) and syngamy (18–24 h)
D2	Cleaving embryo	2–6 cells: rounded blastomeres
D3	Cleaving embryo	8–10 cells or rounded blastomeres
	Compacting embryo	Blastomeres show evidence of adhesion
D4	Compacted morula	Blastomeres show increased adhesion
	Early cavitating	Beginning of blastocoele formation
D5	Early blastocyst	Blastocoele formed
	Mid-blastocyst	ICM, trophoblast and blastocyst clearly seen. Trophoblast expanding, zona thinning out.
	Expanding blastocyst	Embryo growing, blastocoele much increased
D6/7	Late blastocyst	Expanded~150–200 cells; diameter~215 μm
	Hatching blastocyst	Trophoblast hatching out of zona
	Hatched blastocyst	Trophoblast and ICM hatched out of empty zona

Modified from Ref. [18].

Table 6.2 *Embryo Grading for Embryo Transfer in the Laboratory*

Grades	Description
Grade-I	Blastomeres of equal size and no cytoplasmic fragmentation
Grade-II	Blastomeres of equal size and minor cytoplasmic fragmentation (<10%)
Grade-III	Blastomeres of unequal size and variable fragmentation
Grade-IV	Blastomeres of equal or unequal size and significant fragmentation (>10%)
Grade-V	Few blastomeres of any size and severe fragmentation (50%)
Grades I and II have a greater potential of establishing a clinical pregnancy	

Modified from Ref. [17].

Derivation of hESC from Blastocysts

The blastocyst ICM has about 20–40 viable cells for stem cell culture, depending on the size of the ICM. These cells evidently have fewer chromosomal and nuclear aberrations [10,11] compared to early cleavage embryos. Further, transfer of blastocysts in assisted reproductive technology (ART) has yielded higher live birth rates approaching 50% of embryos transferred. Fragmentation is common in early embryos, and these enucleated fragments are largely located in the perivitelline space (PVS) outside the trophoblast in blastocysts, to be discarded at hatching. Fragments are also found within the blastocoele, which later becomes the primary yolk sac cavity. Apoptosis, or programmed cell death, is also evident within the ICM, where abnormal cells seem to be eliminated by phagocytosis by primitive endodermal (hypoblast) cells that later proliferate and line the extra-embryonic yolk sac in late blastocysts [11]. Hence there are intrinsic mechanisms of eliminating fragments and abnormal cells when embryos grow to blastocysts. The hardiest of embryos will survive to blastocysts [10], while abnormal embryos may cease to develop or fragment at early stages, ensuring the selection of a more viable population of blastocysts for ART or ESC culture. Thus derivation of ESC from ICMs has a clear advantage over blastomeres of early embryos. Further, ICM cells are already committed to form the embryo proper (embryoblast), while the trophoblast will form the extra-embryonic membranes and placenta, the initial step in cell differentiation in early embryogenesis [15]. Endoderm has already split from the ICM in blastocysts on Day 6, will form the primitive gut, and line the primitive yolk sac. Isolation of ICM will include some of these endoderm cells. The pros and cons of blastocyst culture for ART have been discussed earlier [16].

Embryo Assessment

Live embryos are assessed in the laboratory by morphologically using an inverted microscope, preferably with Hoffman optics. These images of embryos could be enhanced and stored on video using a digital camera for future reference and assessment. An embryo that develops to the timetable shown in Table 6.1 is likely to be more viable than the one which shows slow growth or arrested development [14,17].

Assessment of Fertilisation

First, it is important to verify normal, monospermic fertilisation at the pronuclear stage when two pronuclei—male and female—are demonstrable in the central cytoplasm. The position and alignment of nucleoli within pronuclei are now assessed to predict normal development (Fig. 6.3). More importantly, these nucleoli are associated with chromatin that will condense to form the maternal and paternal chromosomes at syngamy. This chromatin cannot be seen with the laboratory microscope. Dispermy or digyny will result in tripronuclear, triploid embryos which should be discarded [14].

A normal bipronuclear (2PN) ovum shows the following:
- Two pronuclei—male and female—associated within the central ooplasm

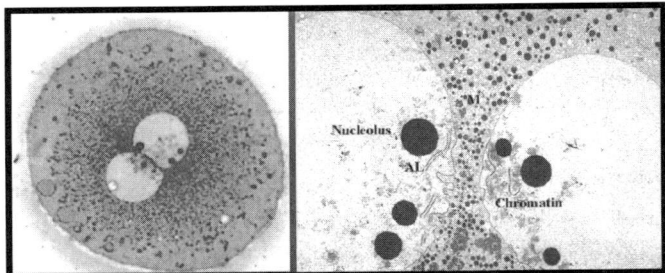

▶ **Fig. 6.3** *Normal fertilised ovum (LM & TEM). These bipronuclears, after monospermic fertilisation, seem normal. Note the alignment of nucleoli adjacent to apposing pronuclear membranes. What is more significant is the alignment of chromatin, associated with nucleoli, which would condense to form the male and female chromosomes at syngamy; ↔35,700 (from Ref. [30]). (For colour figure see Plate 12).*

- Two polar bodies—PB1 with chromosomes and PB2 with a nucleus
- Nucleoli aligned close to apposing PN nuclear envelopes
- No CG or few CG beneath oolemma after IVF
- Crowding of organelles, mostly mitochondria, around PN

(delayed CG exocytosis has been observed after ICSI at the two-cell stage)

An abnormal dispermic pronuclear ovum will show the following:

- Three pronuclei (two male and one female)
- Two polar bodies (PB1 and PB2) in PVS
- A triploid chromosome complement 69XXY or 69XXX
- Two male centrosomes and two sperm asters
- A bipolar or tripolar spindle at syngamy

(may cleave into two or three cells, show normal cleavage and develop to term.)

Cleavage-stage Embryo Assessment

The most important morphological parameters to assess cleavage embryo normality in the laboratory are blastomere appearance, fragmentation, and multinucleation. Those with equal blastomeres, minimal cytoplasmic fragmentation, and few multinucleated cells have a better prospect of implantation. The fate of each embryo could be monitored right up to blastocyst hatching (Figs. 6.4 and 6.5). Extensively fragmented embryos, common in arrested or slow-developing embryos, should be discarded. Fragmentation, regarded as an apoptotic phenomenon, goes hand in hand with multinucleation and micronucleation expressed by chromosomal or genetic abnormalities [14,19]. Aneuploidy, haploidy, polyploidy, and mosaicism are the chief causes of

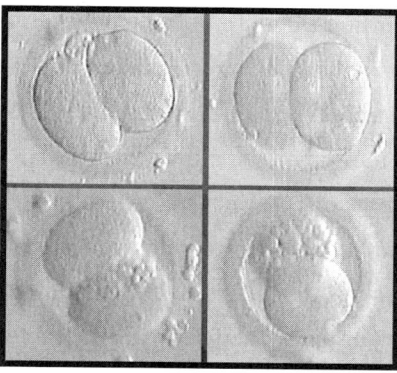

► Fig. 6.4 *Two-cell embryos—normal and fragmented (video). Both normal and abnormal embryos are evident. Fragments appear over the cleavage furrow or a whole cell can fragment totally (probably apoptosis); ↔400 (from Ref. [29]). (For colour figure see Plate 13).*

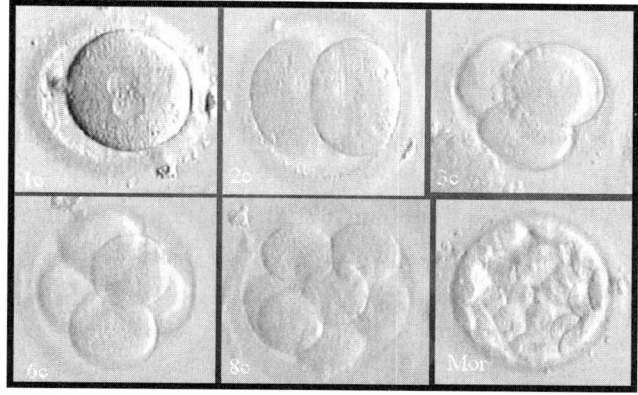

► Fig. 6.5 *Normal embryos—fertilised ovum to morula (video). The cleaving embryos have equal blastomeres and minimal cytoplasmic fragmentation, except the three-cell embryo; ↔400 (from Ref. [29]). (For colour figure see Plate 13).*

early embryonic loss approaching 60% in ART. Of course, these abnormalities could be detected in ESC cultures by karyotyping, as applied here.

Normal embryos have the following features:

- Rounded equal blastomeres, except when cells are dividing
- Blastomeres with well-defined outlines-cell membranes
- Cells with centralized, single nuclei
- No fragments or minimal fragmentation (>10%)

(embryos should develop according to the time frame without delay—Table 6.1)

Abnormal embryos show the following characteristics:

- Extensive cytoplasmic fragmentation of blastomeres (30–50%)
- Spontaneous fragmentation of whole blastomeres (probably apoptosis)
- Unequal or fused blastomeres of varying sizes
- Multinucleation of blastomeres—polyploidy, mosaicism
- Micronuclei in blastomeres beside normal nuclei-aneuploidy

Arrested or degenerating embryos show the following characteristics:

- Dark granular blastomeres with aggregation of organelles
- Extensive vacuolation of blastomeres—increases density
- Cells with eccentrically located nuclei
- Multinucleated cells, many fragments and unequal cells
- Lack of compaction of cells in later embryos and morulae

Blastocyst Assessment

Blastocysts cultured *in vitro* also need to be assessed for their viability and suitability for generating ESC. Blastocysts are classified according to their age, growth, activity and morphology [11]:

1. Early blastocysts (Day 5) show cavitation and formation of the blastocyst
2. Mid-blastocysts (Day 5/6) show growth with increasing cell number (thick zona)
3. Expanding blastocysts (Day 6) have a large blastocoele, distinct ICM, stretched trophoblast, and a thin zona
4. Hatching blastocysts (Day 6/7) show emergence of the embryo and a breached zona
5. Hatched blastocysts (Day 7) are expanded or contracted with no zona.

As blastocysts grow, their total cell numbers increase to about 150–250, determined by DAPI staining [10]. Optimally, an expanding, normal blastocyst should be used for isolation of ICM cells (Figs. 6.6–6.8). Hatched blastocysts have a clump of multinucleated syncitiotrophoblast at the embryo pole, which invade the ICM and may interfere with its isolation. Some characteristics of blastocysts are discussed further.

Normal blastocysts have a

- distinct trophoblast, ICM, and blastocoele,
- well-defined, compacted ICM with many cells,
- trophoblast forming a continuous epithelium,
- large fluid-filled blastocoele and

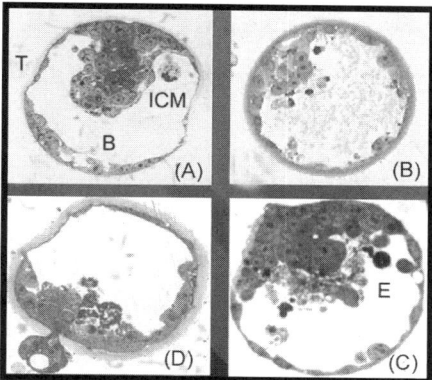

➤ Fig. 6.6 *Normal and abnormal blastocysts (LM). (A) Normal blastocyst with trophoblast (T), inner-cell mass (ICM) and blastocoele [B]. (B) Disorganised ICM. (C) Disorganised endoderm [E]. (D) Failed hatching—degenerating. ↔400 (from Ref. [11]). (For colour figure see Plate 14).*

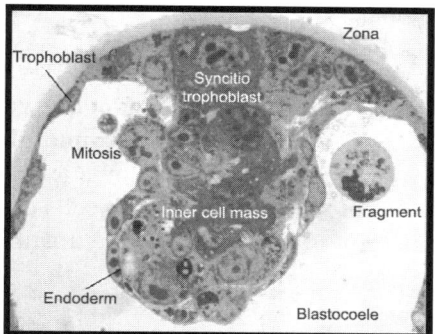

➤ Fig. 6.7 *Human expanded blastocyst—day 6. Embryonic stem cells are derived from the inner-cell mass of blastocysts. Most cell types of the blastocyst, including endoderm, are shown in this section; ↔1000 (From Ref. [25]). (For colour figure see Plate 14).*

- few early cleavage stage fragments in the blastocoele or PVS.

Abnormal blastocysts
- have no ICM or have a small or dispersed ICM,
- fail to expand and hatch on Day 6/7—are moribund or dead,
- arrest in development and eventually degenerate,
- have many early stage fragments in PVS—interferes with hatching and
- show many multinucleated cells in ICM, trophoblast and endoderm.

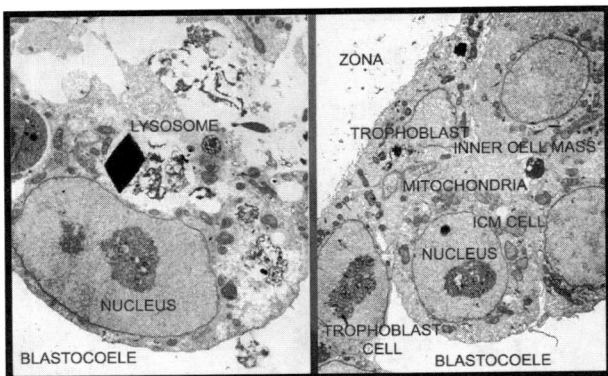

▶ **Fig. 6.8** *Human blastocyst—endoderm cell and ICM/trophoblast junction (TEM). The endoderm cell is typically phagocytic. The ICM cell has a reticulated nucleus and mitochondria like all the other cells. ↔6800 and ↔850, respectively (from Ref. [11]). (For colour figure see Plate 15).*

ICM Cell Structure

The ICM cell is a far more differentiated cell than an early cleavage blastomere [11], displaying reticulated nucleoli within its nucleus, tubular mitochondria with well-defined cristae, rough endoplasmic reticulum, Golgi complexes, lysosomes, and lipid globules (Fig. 6.8). Mitotic cells show typical centrosomes with two centrioles, like all somatic cells. The cells also show typical desmosomes and gap junctions. Apoptotic and phagocytic cells are also evident and occasionally clear cells resembling primordial germ cells. Overall their structure conforms to that of a protein synthesizing cell.

Post-blastocysts: Clues to the Origins of ESC

We extended our study to hatched, post-blastocysts by prolonged culture to Day 9. This is believed to be the stage when ICM cells are transformed into ESC. Some embryos retain their original organization (Fig. 6.9), while others become cystic [20] somewhat resembling EBs. The ICM cells proliferate and acquire a more simplified structure—a vital clue to the origins of ESC in human embryos.

ESC Structure and De-differentiation

Compared to ICM cells from which they are derived, ESC have a simplified structure, a phenomenon we attributed to de-differentiation [21]. The ESCs have large nuclei and scanty cytoplasm with few organelles, like mitochondria and centrosomes, both in colonies and EBs (Figs. 6.10–6.12). Cells may show primitive cell junctions, like desmosomes. It is remarkable that embryonic cells differentiate back and forth during their genesis. Now we know that even adult

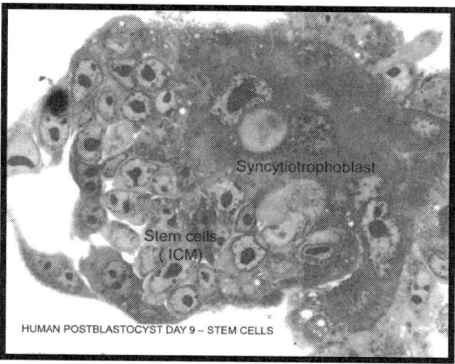

▶ **Fig. 6.9** *Human day-9 blastocyst in culture-stem cells (LM). The ESCs have proliferated within the ICM and show a more simplified structure than the ICM. Note the amniotic-like cavity (left); ↔1000 (from Ref. [25]). (For colour figure see Plate 15).*

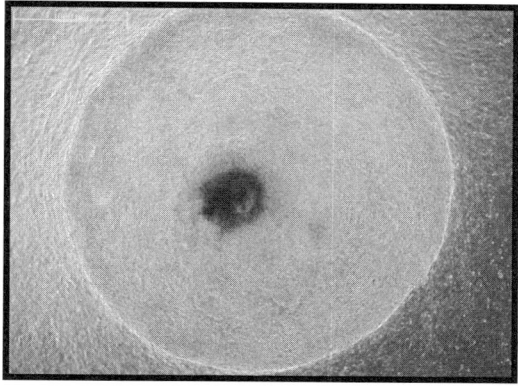

▶ **Fig. 6.10** *Saucer-shaped ESC colony (P142). The colony is flat and has grown from the centre toward the periphery. Cultured on mouse embryonic fibroblasts. ↔1000 (from Ref. [25]). (For colour figure see Plate 15).*

somatic cells can revert back to stemness by genetic manipulation—induced pluripotency [22,23]. However, it is also known that various human ESC lines have differential propensity to differentiate into one of the three germ lineages [24].

CELL DIFFERENTIATION IN EMBRYOS

In blastocysts, the first cell differentiation has occurred delineating an outer trophoblast layer and an ICM. A layer of endoderm cells (hypoblast) has also delaminated from the ICM on Day 6/7—the first event in gastrulation or germ layer formation. Thus in week 2, the embryo is a bilaminar disc

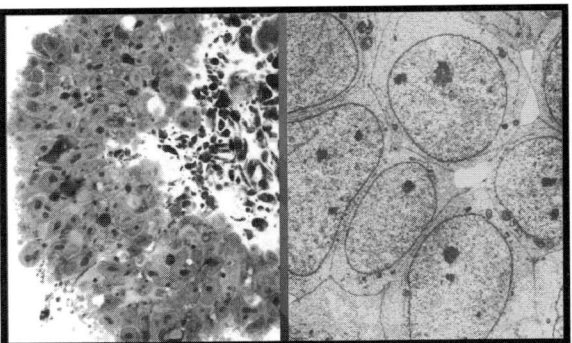

▶ **Fig. 6.11** *ESCs in a colony—passage 35 (LM & TEM). The ESC have large nuclei with reticulated nucleoli and scanty cytoplasm with few mitochondria. Differentiating cells are seen within the colony with mouse fibroblasts. ↔1000 and ↔5250 for left and right panel, respectively (from Ref. [25]). (For colour figure see Plate 16).*

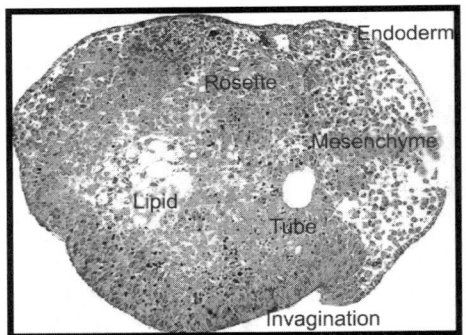

▶ **Fig. 6.12** *Embryoid body (section) showing spontaneous differentiation (P130). Cells representing all three germ layers, including rosettes, tubular and capillary-like formations were seen in EB. Epithelium was cuboidal, columnar or stratified. Note the cells at surface invagination. ↔100 (from Ref. [25]). (For colour figure see Plate 16).*

composed of epiblast (future ectoderm and mesoderm) and hypoblast. The events of cell differentiation are more dramatic in week 3, when the embryo becomes trilaminar, since mesoderm (and endoderm) cells migrate beneath the epiblast through a primitive streak, establishing all three germ layers—ectoderm, mesoderm and endoderm (Fig.6.13). This is the completion of gastrulation—the most important morphogenetic event in embryogenesis, when the right cell is put into its right place at the right time. If something goes wrong at this stage, there will be serious repercussions in tissue and organ formation later on. We have recently studied the mechanics of this process in ectopic human embryos using TEM [25,26]. The cells of germ layers represent the blue prints of all tissues and organs of the human body (Fig.6.1). They include epidermal, neural,

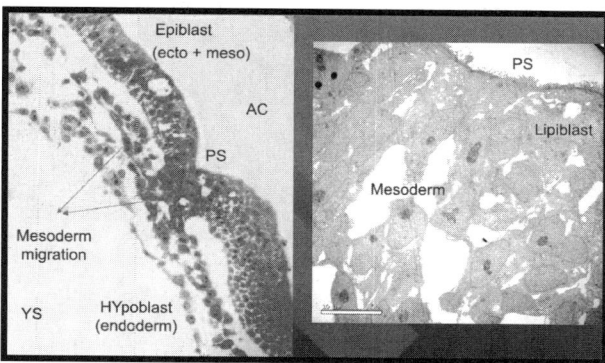

▶ Fig. 6.13: *Primitive streak, ectopic embryo showing three germ layers (LM & TEM). Sections of embryonic disc in the third week showing migration of mesodermal cells through the primitive streak establishing the three germ layers—ectoderm, mesoderm and endoderm. AC, amniotic cavity; PS, primitive streak; YS, yolk sac. ↔1000 and ↔4000 for left and right panel, respectively (from Ref. [26]). (For colour figure see Plate 16).*

mesenchymal, and endodermal stem cells at the next level of cell differentiation. These stem cells, as well as tissue cells such as fibroblasts, muscle, neural rosettes, epithelia, goblet-like cells, have been identified in ESC colonies and EBs (as discussed further), paralleling the events of differentiation that occur in weeks 3 and 4 in the embryo, which is not surprising [21, 27]. Further, in embryos there are also the rudiments of the yolk sac, amnion, chorion and allantois representing the extraembryonic membranes established in weeks 3/4. We have gone a little further in demonstrating the events of tissue differentiation in bovine uterine embryos ranging from blastocysts to Day 21 embryos, which follow a similar pattern of early cleavage observed in humans [28]. There is little doubt that the embryonic genes are activated after the eight-cell stage in humans and these influence preimplantation development and differentiation to blastocysts and beyond, reflected by increased complexity in cell structure. Recently, the morula (approximately 32 cells) has been used to generate ESCs [30]. The reader is advised to refer any embryology text book to review the cells differentiation process *in vivo* [15]. There is also a useful website of the famous Carnegie collection of human embryos at all stages with serial sections www.virtualhumanembryo.isuhsc.edu and in the website of early embryogenesis www.sathembryoart.com.

Spontaneous Differentiation of Cells in ESC Colonies and EBs

Spontaneous differentiation is an intrinsic feature of ESC and varies in different colonies and EBs depending on cell lines and culture conditions (Figs. 6.12, 6.14–6.16). All ESCs do not retain their stemness or pluripotency and start differentiating into various cell types [21, 25, 27]. Cells at the onset of differentiation are more like ICM cells synthesizing proteins. Differentiated cells resemble goblet cells with secretions, probably endodermal, while others had phagocytic vacuoles, like primitive endoderm of blastocysts. Cells at the surface were epithelioid with terminal cell junctions

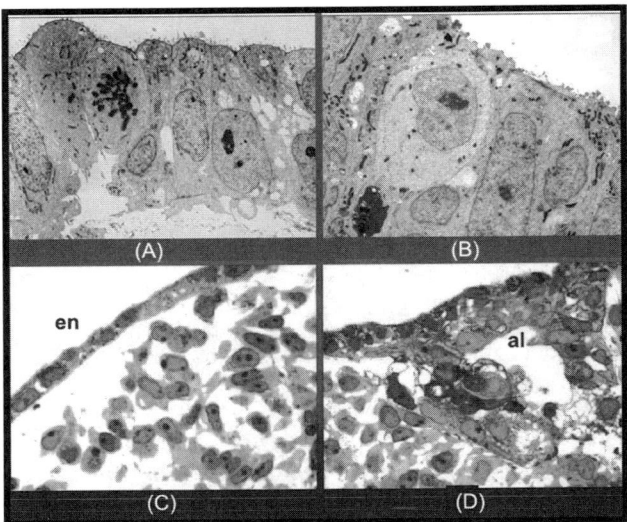

▶ **Fig. 6.14** *Spontaneous differentiation of ESC (LM and TEM). (A) Epithelioid layer with dividing cell. (B) Epthelioid layer with clear germ cell; ↔3400. (C) Mesenchyme stem cells and endoderm (en). (D) Lung cells in alveolus (al); ↔1000 (from Ref. [25]). (For colour figure see Plate 17).*

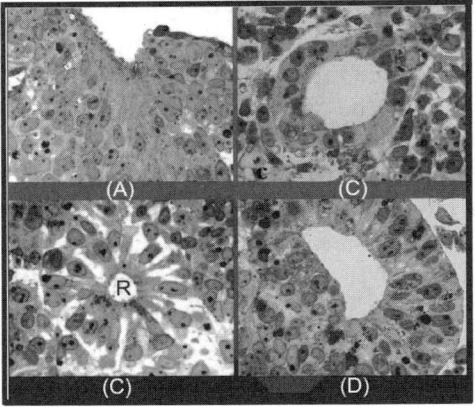

▶ **Fig. 6.15** *Spontaneous cell differentiation in EBs (LM). (A) Surface invagination in outer ectoderm. (B) Neural rosette [R] with neural stem cells. (C) Tubular formation with cuboidal cells. (D) Tubular formation like neural tube. ↔1000 (from Ref. [25]). (For colour figure see Plate 17).*

and polarized centrosomes, but had no basal lamina. Those beneath the surface were largely mesenchyme stem cells, differentiated fibroblasts with matrix and fibrils (connective tissue), and spindle-shaped cells (muscle) and branched cells (cardiac muscle) with myofilaments. There were

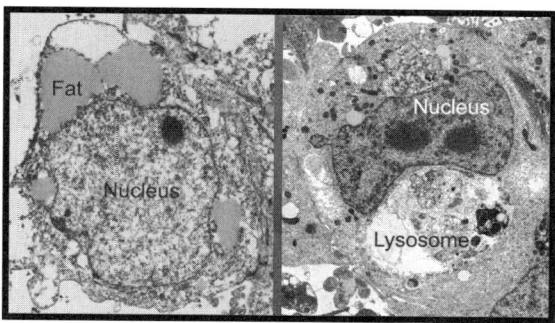

▶ **Fig. 6.16** *ESC colony—lipid and phagocytic cell (P35). TEM: The lipid cell has globules of fat and the phagocyte resembles an endoderm cell of blastocysts. Note the lysosome with cellular debris. ↔10,200 (from Ref. [25]). (For colour figure see Plate 18)*

also lipid cells (adipocytes) and clear cells resembling primordial germ cells. Others formed rosettes possibly neurogenic (neural stem cells) or tubular profiles like neural tubes, probably ectodermal. There were also endothelial formations like capillaries and alveoli with possible pneumocytes. Most cell types have been documented by TEM [21,25,27] and by fluorescent markers in colonies and EBs [6,12]. Optimally, fluorescent microscopy should be combined with TEM-00 the golden standard for the characterization of stem cells–during differentiation. Fluorescent microscopy alone does not reveal cell structure, unless paraffin or epoxy resin sections of sibling colonies or EBs are used in parallel. This needs to be followed by chromosomal and genetic studies. Thus the differentiated cells represent ectoderm, mesoderm, and endoderm lineages recapitulating the events that occur in weeks 3 and 4 of development. It appears that cells work in concert and interact with one other as in embryos, but are not organized as such in space and time. There is no pattern or axis formation characteristics of normal embryos.

REFERENCES

1. Bongso A., C. Y. Fong, S.C. Ng et al. (1994). Isolation and culture of inner cell mass cells from human blastocysts. *Hum. Reprod.* **9**: 2110–2117.
2. Hoffman L.M., M.K. Carpenter (2005). Characterization and culture of human embryonic stem cells. *Nat. Biotechnol.* **23**: 699–708.
3. Trounson A. (2006). The production and directed differentiation of human embryonic stem cells. *Endocr. Rev.* **27**: 208–219.
4. Richards M., C.Y. Fong, W.K. Chan, et al. (2002). Human feeders support prolonged undifferentiated growth of human inner cell masses and embryonic stem cells. *Nat. Biotechnol.* **20**: 933–936.
5. Ludwig T.E., V. Bergendahl, M.E. Levenstein, et al. (2006). Feeder-independent culture of human embryonic stem cells. *Nat. Methods.* **3**: 637–644.
6. Thomson J.A., J. Itskovitz-Eldor, S. Sander, et al. (1998). Embryonic stem cell lines derived from human blastocysts. *Science.* **282**: 1145–1147.

7. Ström S., J. Inzunza, K.H. Grinnemo, et al. (2007). Mechanical isolation of the inner cell mass is effective in derivation of new human embryonic stem cell lines. *Hum. Reprod.* **22**: 3051–3058.
8. Fong C-Y., M. Richards, A. Bongso (2006). Unsuccessful derivation of human embryonic stem cell lines from pairs of human blastomeres. Reprod. *Biomed. Online.* **13**: 295–300.
9. Klimanskaya I., Y. Chung, S. Becker, et al. (2006). Human embryonic stem cell lines derived from single blastomeres. *Nature doi, 1038.nature 05142.* 1–4. (Letter).
10. Bongso A. (1999). *Handbook on Blastocyst Culture*, pp. 93. Sydney Press Indusprint, Singapore.
11. Sathananthan A.H., S. Gunasheela, J. Menezes (2003). Critical evaluation of human blastocysts for assisted reproduction techniques and embryonic stem cell biotechnology. *Reprod. Biomed. Online.* **7**: 219–227.
12. Bongso A., E.H. Lee (eds.) (2005). Stem Cells From Bench to Bedside, p 565. *World Scientific, Singapore.*
13. Sathananthan A.H., S.C. Ng, A. Bongso, et al. (1993). Visual Atlas of Early Human Development for Assisted Reproductive Technology, 209. National University Hospital and Serono, Singapore.
14. Sathananthan A.H., S. Gunasheela (2007). Human oocyte and embryo assessment for ART. In: Human Preimplantation Embryo Selection, first ed. (Elder K., Cohen J., eds.), pp. 1–14. *Informa Healthcare*, UK.
15. Larsen W.J. (1998). *Essentials of Human Embryology*, 394. Churchill Livingstone, New York.
16. Sathananthan A.H. (2006). Blastocyst culture: Fact or fiction. In: Contemporary Perspectives on Assisted Reproduction (Allahbadia G.N., Merchant R., eds.), pp. 135–143. *Elsevier*, India.
17. Veeck L. (1999). An Atlas of Human Gametes and Conceptuses, pp. 215. Parthenon, London.
18. Gunasheela S. (2005). The A-Z Encyclopedia on Male and Female Infertility, pp. 146. *Jaypee*, New Delhi.
19. Munne S. (2006). Chromosome abnormalities and their relationship to morphology and development of human embryos. *Reprod. Biomed. Online.* **12**: 234–253.
20. Fong C.Y., A.H. Sathananthan, A. Bongso, et al. (2004). Nine-day old human embryo cultured *in vitro*: A clue to the origins of embryonic stem cells. *Reprod. BioMed. Online.* **9**: 321–325.
21. Sathananthan H., M. Pera, Trounson A. (2002). The fine structure of human embryonic stem cells. *Reprod. Biomed. Online.* **4**: 56–61.
22. Takahashi K., S. Yamanaka (2006). Induction of pluripotent stem cells from mouse embryonic and adult fibroblast cultures by defined factors. **Cell 126**: 663–676.
23. Yu J., M.A. Vodyanik, K. Smuga-Otto, et al. (2007). Induced pluripotent stem cell lines derived from human somatic cells. *Science.* **318**: 1917–1920.
24. Osafune K., L. Caron, M. Borowiak, et al. (2008). Marked differences in differentiation propensity among human embryonic stem cell lines. *Nat. Biotechnol.* **26**: 313–315.
25. Sathananthan A.H. (ed.) (2007). Embryonic Stem Cells (DVD): A multi-author production of digital images from Monash Immunology & Stem Cell Laboratories. Presented at the Monash booth at the 5th ISSCR Annual Meeting, *Cairns*, Australia.
26. Sathananthan A.H., K. Selvaraj (2007). Human germ layers: Clues to the origins of embryonic stem cells. *World Tubes Conference*, Kolkatta, India.
27. Sathananthan A.H., A. Trounson (2007). *In vitro* differentiation of human embryonic stem cells. In: *Human Embryonic Stem Cells: A Practical Handbook* (Sullivan S., Cowan C., Eggan K., eds.), pp. 149–168. Wiley Press, NJ.

28. Sathananthan A.H. (ed.) (2007). Bovine embryogenesis (DVD): A multiauthor, international, visual presentation of gametes, embryos, *in vivo* and *in vitro*, for ART, ESC and therapeutic cloning. Monash Institute of Medical Research, Melbourne, Australia.
29. Sathananthan A.H., J. Menezes, S. Gunasheela (2003). Video of human embryos (DVD).
30. Strelchenko N., Y. Verlinsky (2006). *Embryonic stem cells from Morula Methods in Enzymology.* **418**: 93–108.

DEVELOPING A SERUM-FREE PLATFORM AND A NOVEL REPORTER SYSTEM FOR hESCs

Jonathan D. Chesnut[1], Bhaskar Thyagarajan[1], Soojung Shin[1], Katherine Wagner[2], MaryLynn Tilkins[2], Mahendra S. Rao[1], Mohan C. Vemuri[2, *]

[1]*Invitrogen Corporation, Van Allen Way, Carlsbad, CA 92008*
[2]*Staley Road, Grand Island, NY 14072*

INTRODUCTION

Human embryonic stem cells are cultured on mouse embryonic fibroblasts or co-cultured on human foreskin fibroblast feeder layers, and are often grown in xenogenic conditions in culture media that contains bovine serum, or serum replacement, and on extracellular matrix, such as Matrigel™. Cell-based therapies require totally defined and controlled differentiation conditions. Considerable progress has been made over the past 2–3 years to modify culture conditions to develop xeno-free, clinical grade cells, and physiologically relevant options to use hESCs in cell therapy. We report (1) a system for novel scalable defined culture conditions for propagating hESCs in serum-free and xeno-free culture conditions and (2) generation of a novel hESC reporter line that expresses GFP under the control of *Oct4* promoter. This reporter line will be an ideal tool for rapid screening of cultures for pluripotency and in drug discovery models.

Human embryonic stem cells (hESCs) exhibit high capacity for self-renewal, pluripotency, and proliferate indefinitely in culture and maintain epigenetic [1, 2] and karyotypic stability [3, 4]. This ability of hESCs, together with the increased efficiency of new cell-line generation and the overall functional similarity across various hESC lines, allows a virtual unlimited supply of cells for new

---

* Mohan.Vemuri@invitrogen.com

clinical trials, drug discovery, toxicology screening, and enables developing novel *ex vivo* gene therapy techniques. This article describes the status of:

- current hESCs culture systems, the options for feeder-free and serum-free culture of hESCs, development of a novel serum-free and feeder-free culture medium, StemPro® hESC-SFM, a humanized substrate CELLstart™, and an animal origin free cryopreservation medium, Synth-a-freeze, and

- the generation of a reporter hESCs line BG01v-Oct4-GFP that can indicate the stemness phenotype in an ongoing hESCs culture and thus identify the extent of differentiation, besides aid in tracking hESCs in pre-clinical transplantation studies.

CURRENT hESCs CULTURE SYSTEMS

The current model of using hESCs is built on the idea of a cell bank from which a qualified master bank is produced for a working bank of cells (Fig. 7.1). From this working bank of cells, a process for expansion and directed differentiation toward specific cell types is developed because undifferentiated cells cannot be used as such in most applications. Further, as it is difficult to direct differentiation into a single cell type, a process for selecting an appropriate differentiated phenotype needs to be established. Current protocols for differentiation are sub-optimal and do not permit differentiation at high efficiency [5]. Further, they require to use xenobiotic material, co-culture with fibroblasts, and the addition of several factors that trigger lineage-specific differentiation. In

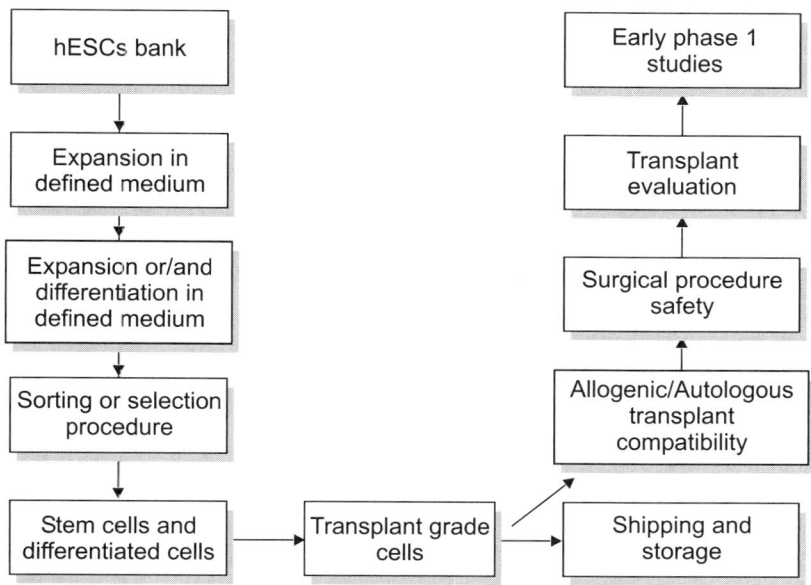

▶ **Fig. 7.1** *A general workflow process with hESCs.*

addition, most techniques used to derive differentiated populations, such as co-culture with differentiated cells and isolation based on difficult-to-quantify criteria such as colour of the embryoid bodies, represent inherently variable methodologies that are difficult to adapt to a large scale.

FEEDER-FREE AND SERUM-FREE DEFINED CULTURE SYSTEMS FOR hESCs

Several recent studies have contributed to the development of defined media using reagents that are better defined, qualified, and preferably derived from a non-animal source and promote the growth of hESCs in feeder-free, serum-free conditions while maintaining the undifferentiated stem state [6–13]. Different serum-free and feeder-free culture media for hESCs that are currently used to culture various hESCs lines by different investigators from published formulations and three from commercial sources are described in Table 7.1. The efficacy of these formulations is currently being evaluated by the International Stem Cell Initiative through ISCI project (www.stemcellforum.org.uk) to select preferred media that could work robustly across different hESC lines.

The choice of growth factors in these media plays a critical role in sustained stemness and pluripotency. Attempts to understand how hESCs maintain pluripotency in a feeder-free environment showed the critical role of growth factors such as fibroblast growth factor (FGF-2) [6], noggin [7], activin [8], and heregulin [9] (Allan Robin, personal communication), either in combination or alone in hESC media supplementation. In a recent study, nine xeno-free culture media comprising control hESC medium, human serum, Lipumin™ 10, Plasmanate SerEx 10, serum substitute supplement, SR3 medium, TeSR1, X-Vivo 10, and X-Vivo 20 were compared with the conventional KnockOut serum replacement (K-SR) media for the cultivation of hESC on human feeder cells [10]. The medium containing 20% human serum was found to sustain undifferentiated hESC proliferation to some extent, but was inferior to the conventional KnockOut-serum replacement-containing medium. None of the studied xeno-free media were able to maintain the undifferentiated growth of hESC in a manner that fulfills the criteria of pluripotency, differentiation into multilineage phenotypes, and maintenance of other hESC-specific criteria. Toward the development of a serum-free medium, one of the earliest products that was tested systematically and extensively is KNOCKOUT™ Serum Replacement (KSR). Different concentrations of KSR have been tested and several investigators found 20% to be the optimal concentration. Used with basal medium and bFGF, this cell culture serum replacement has supported the undifferentiated propagation of hESCs. When used with MatrigelÔ and KSR-conditioned medium, KSR can support serum-free, feeder-free growth of hESCs.

Even though several improvements in hESCs culture media have been made and new factors have been identified during the past few years, a completely defined xeno-free media still requires a significant improvement in order to provide a robust expansion that sustains pluripotency. To our knowledge, no study so far has developed a comprehensive hESCsculture system that includes serum-free, feeder-free culture medium, a substrate that is functionally equivalent to Matrigel™ but

Table 7.1 *Serum-free and Feeder-free Culture Media for the Propagation of hESCs*

Name of media	Substrate/matrix/GFS	Base media	Special GF additives	Passaging conditions	Reference
Geron	Matrigel (1:30) or hLaminin (2 μg/cm^2)	CM-DMEM/X-VIVO 10	KO-SR (20%) b-FGFSCF hFlt-3h-LIF	Collagenase IV	[35]
Bresagen/Invitrogen	Low-dose Matrigel (1:200)	DMEM/F12	Transferrin Ascorbicacid LR-IGF Activin Heregulin bFGF	Accutase	[11]Commercial name: StemPro âhESC SFM
Peter Donovan	Matrigel (1:30)	KO-DMEM	Wnt3a Some other factors CM BMP4 RA	Trypsin	[36]
Martin Pera	Matrigel	DMEM	KO-SR (15%) bFGF PDGF/SIP	Collagenase IV	[37]
Snyder lab Yale	Matrigel/Fibronectin (25 mg/mL)	DME/F12	KO-SR (20%) b-FGF ITS Wnt3a April or BAFF Cholesterol lipid-supplement	Collagenase IV	[38]
Scripps	Matrigel human serum albumin (0.5 mg/mL)	DME/F12	KO-SR or B27 N2 bFGF2	Collagenase IV	[39]
Technion	Fibronectin	DMEM	KO-SR (15%) bFGFTGFb1	Collagenase IV	[40]
Vallier	Gelatin (Porcine)	KO-DMEM/IMDM/	KO-SR/F12/NUTMIX Insulin, Transferrin Monothioglycerol	Collagenase IV	[41]
Liu et al.	Matrigel CM	DMEM/F12	B27 N2 bFGF2 Noggin	Collagenase IV	[42]
mTeSR1 Ludwig et al.	Matrigel Combinations of collagen IV Fibronectin Laminin Vitronectin	DMEM/F12	bFGF TGF beta 1 ITS Glutathione Pipeolic Acid GABA LiCl	Collagenase IV	[12]

is made of human compatible and user friendly ingredients, and a reporter hESC cell line that can indicate the stem state readily without going through extensive phenotypic marker evaluation.

Our group, in collaboration with colleagues at BresaGen, has developed a defined culture medium at GIBCO Invitrogen, termed StemPro® hESC-SFM, which allows expansion of hESCs in large numbers and was evaluated extensively on hESC growth, expansion, stemness of cells, retention of differentiation ability into three germinal lineages, and karyotypic stability [11].

StemPro® hESC SFM is a serum-free and feeder-free medium for the culture of hESCs, which supports hESCs' expansion and retains the undifferentiated state and long-term karyotypic stability over extended (>70) passages [11]. StemPro® hESC SFM also supports the ability of hESCs to differentiate into derivates of all three germ layers. Embryoid bodies generated from the BG01v line were harvested after 21 days, and the expression of various differentiation markers was determined by RT-PCR (Fig. 7.2).

More recently a second media termed mTeSR1| has been made available commercially. This media is based on the formulation described by Thomson and colleagues [12, 13] and appears to be

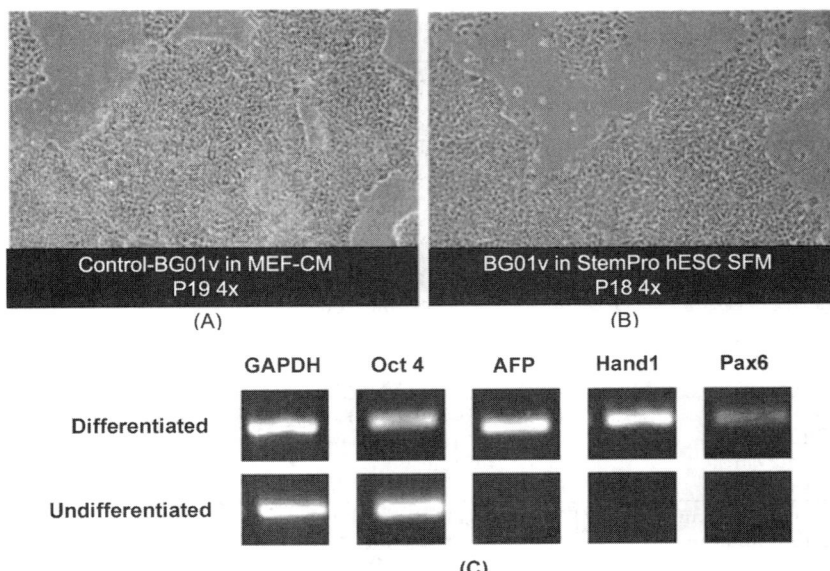

► Fig. 7.2 *hESCs grown in StemPro® hESC SFM retain normal morphology. BG01v cells grown (A) in MEF-CM (control) or (B) in StemPro® hESC SFM show normal and healthy morphology with pluripotency marker Oct4 expression by PCR (C) in BG01vs grown in StemPro® hESC SFM. The data also show that BG01v cells grown in StemPro® hESC SFM media exhibit a complete ability to differentiate into ectoderm (PAX6), mesoderm (HAND1) and endodermal (AFP) lineages. GAPDH is a house-keeping control.*

capable of maintaining ESCs for prolonged periods. The authors reported occasional karyotypic abnormalities and suggested the use of zebra fish FGF because the cost of recombinant FGF was prohibitive for large-scale propagation. Additional results on its use in multiple cell lines are awaited. Nevertheless this medium represents the first commercially available medium that is xeno-free and uses a defined substrate rather than feeders.

A third commercially available medium is from Millipore, termed HEScGRO. The formulation was released after testing on MEL-1, MEL-2 and H1. HEScGRO medium is serum-free, but still requires hESCs to be cultured on human feeders, and will not support hESC growth on murine feeders. Its performance in feeder-free conditions is yet to be examined. Other feeder-free and serum-free media have been described, but none are available commercially and many require growth on matrigel or feeders both of which contain undefined components (Table 7.1).

Matrix/substrate in hESCs Feeder-Free Cultures

Most of the feeder-free systems require the use of a substrate for hESCs to attach and grow while retaining their pluripotent character. Thus, the choice of substrate is critical for the optimal attachment and growth of these cells. A variety of substrates have been evaluated by researchers. These include Matrigel™, Geltrex™, fibronectin, laminin, collagen, human serum, VitroGro®, and nanofibrous materials using serum-free medium, or MEF-conditioned medium. Culture plates have been coated with extra-cellular matrices derived from MEFs or HFFs, or a combination of medium conditioned with human fibroblasts. Some of these matrices will allow cell growth, but the risk of potential xenogenic contamination remains due to the MEFs. Also, performance variability from HFF batches remains problematic. So far, defined substrates have not been used with defined medium, although defined substrate for hESCs is a critical requirement because Matrigel™ (the current standard) is neither defined nor xeno-free. CELLStart™ from Invitrogen and VitroGro® from Tissue Therapeutics Limited are two recent entries into this substrate arena; the characteristics and features of CELLstart are discussed further.

Humanised substrate CELLstart™

In a serum-free and feeder-free medium formulation, use of a substrate is critical to facilitate cell attachment and support continued growth of cells in the pluripotent state. In most of these studies, Matrigel| has been the preferred choice for investigators as a substrate for cell attachment. This is an extracellular matrix extracted from Engelbreth-Holm-Swarm tumour [14] that contains laminin, collagen, and several uncharacterized components and includes variable amounts of growth factors that bind to substrate and cannot be purified away from the matrix. Although the use of mouse embryonic fibroblast feeder conditioned medium together with Matrigel| as a substrate resulted in successful propagation of feeder-free hESC cultures, such cells may not be compatible for cell therapy applications. Studies carried out using laminin [15], fibronectin [16], human serum [17], 3D substrates such as alginate matrix [18], and a synthetic polyamide matrix (Ultra-Web) [19] suggest that each of them can support pluripotent growth of hESCs to a differing extent in xeno-free cultures with some advantages and limitations.

None of these factors, however, has been shown to support the propagation of hESC when used in combination with a serum-free, xeno-free medium. Our group has developed a humanized substrate, termed CELLstart™, which uses defined human or recombinant proteins and includes ECM and serum proteins. We have shown that this matrix supports the growth and large scale expansion of hESCs when used with defined as well as standard medium. Cultures grown on CELLstart™ replicated in a manner similar to those of cultures expanded on Matrigel™ (Fig. 7.3). Multiple lines could be maintained successfully over a long term on this substrate. Use of CELLstart™ allows removal of murine feeders, conditioned medium, and matrices such as Matrigel™. A combination of the defined medium described earlier along with a defined substrate such as CELLstart™ is close to an ideal culture system with no requirements for murine feeders or serum-dependent medium. Such culture systems that promote large-scale expansion of hESCs and require a retention of the stem cell phenotype are critical for drug discovery and therapeutic applications [20]. CELLstart™ is the first fully defined, completely animal origin free substrate in the marketplace for the attachment and expansion of embryonic, mesenchymal, and neural stem cells. It provides the serum-free and feeder-free scaffold upon which stem cells grow in both research and therapy applications. Being easy to use, CELLstart™ requires researchers to go through minimal or no changes in processes from current culture methods.

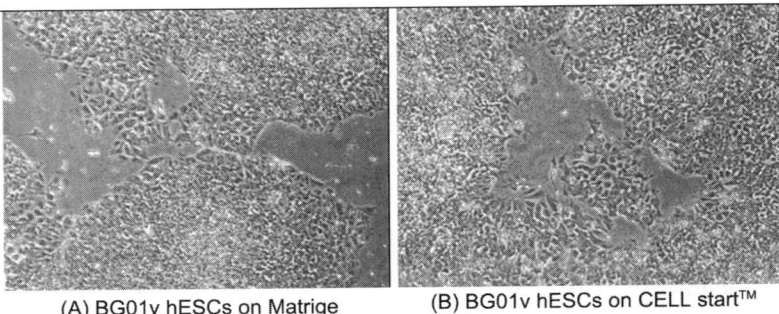

(A) BG01v hESCs on Matrige   (B) BG01v hESCs on CELL start™

▶ Fig. 7.3 *Show normal and healthy morphology, as that of (A) cells grown on Matrigel| and (B) BG01v hESCs grown in CELLstart®. Both photographs were taken under phase microscopy at 4↔.*

A Cryopreservation Medium, Synth-a-freeze

Note that not only one needs to develop a medium for culture and propagation of cells but one also needs a freezing medium for cells in a defined formulation using clinical grade material. With this freezing medium, one also needs a quick dependable process of shipping cells from the site of manufacture to the site of use with greater cell viability and yield. Equally important is the development of a monitoring system to allow one to assess cells on a routine basis.

    Current cryopreservation protocols for stem cells suspend hESC colonies in a growth medium containing fetal bovine serum (FBS) and dimethylsulphoxide (DMSO), followed by an automated or semi-automated slow freezing protocol overnight with subsequent transfer and storage in liquid nitrogen. Thawing is rapid. This protocol is effective for the preservation of human, murine, and porcine embryonic stem cells. Although this procedure works with hESCs, these cells frozen using this method suffer from low viability, and many cells fail to survive and differentiate upon thawing and expansion. The frequency of karyotypic abnormalities reported may also be increased due to the selection stress placed on cells going through multiple freeze-thaw protocols. The percentage of revived cells following a thaw is relatively low, and there is a great need to improve on the composition of freezing medium as well as the freezing protocol. Synth-a-freeze is an animal origin free (AOF) freezing medium that we have developed for use in this defined platform (www.cascadebio.com). This serum-free freezing formulation supports better revival and viability of hESCs following cryopreservation. Although Synth-a-freeze supports efficient cryopreservation of hESCs, it still requires the use of 10% DMSO. In our view, this represents only a partial solution. Considering the potential transplant and cell therapy applications of hESCs, particularly in the case of banking of hESCs one would ideally like to develop a DMSO-free freezing solution. Although a few alternatives have been proposed, such as the use of a combination of conventional cooling protocols [21] and vitrification protocols with different freezing medium components [22–24], we have so far been unsuccessful in identifying a robust cryopreservation medium without DMSO and thus leaves more scope for further development in this area.

Together, these novel reagents such as defined humanized substrate, serum-free and feeder-free culture medium to propagate hESCs, enzymatic dissociation agents that allow hESCs grow as single cells or in a colony format, animal origin free cryopreservation medium, and characterization kits to identify hESC markers, karyotypic abnormalities or a reporter line to assess the stem cell state, constitute a hESC platform that can enable therapeutical applications of hESCs (Fig. 7.4).

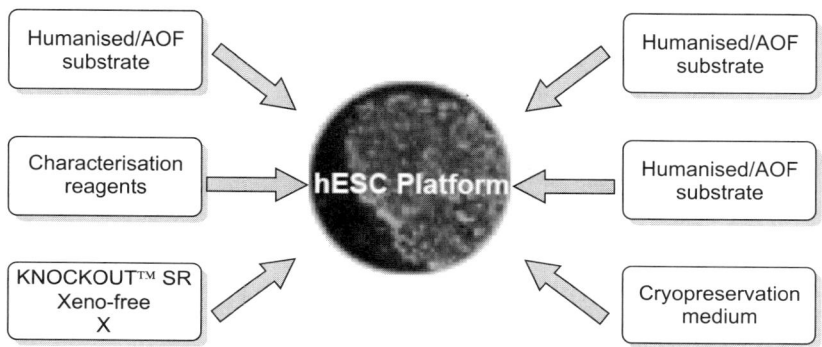

▶ **Fig. 7.4** *An ideal hESC platform with required tools and reagents to enable clinical transition of stem cells. (For colour figure see Plate 18).*

An alternative strategy pursued by some investigators is to take KNOCKOUT™ SR (KSR)-the preferred serum replacement for the growth of hESC—and develop a humanized or xeno—free clinical grade version of this product. Our laboratory has tested several versions of humanized KSR. Results with one xeno-free prototype are shown in Fig. 7.5. Additional evaluations will determine its utility with multiple hESC lines and other xeno-free systems.

Although there appears to be a satisfactory solution to growing and propagating cells in a defined, xeno-free condition, it is important to note that to our knowledge no one has derived a cell line from the ICM using these new reagents. Given the relative efficiency of hESCs and the multiplicity of lines supported, it is perhaps only a matter of time. Nevertheless, it remains to be formally proven.

Generation of a Novel Oct4-GFP hESC Reporter Line

Combined with the advances in developing a serum-free and feeder-free culture system for hESCs destined for cell therapy purposes, a quick screening tool that indicates stem cell state readily is essential. Such a reporter line will allow effective monitoring in the development of clinically relevant stem cell phenotypes. Current methods of hESC expansion are limited by morphological monitoring and manual processing of differentiated cell types from renewing cells. Developing reliable technical tools that can be readily used in routine screening is a considerable challenge [25]. We further describe the generation of an hESC line that expresses GFP under the control of *Oct4*

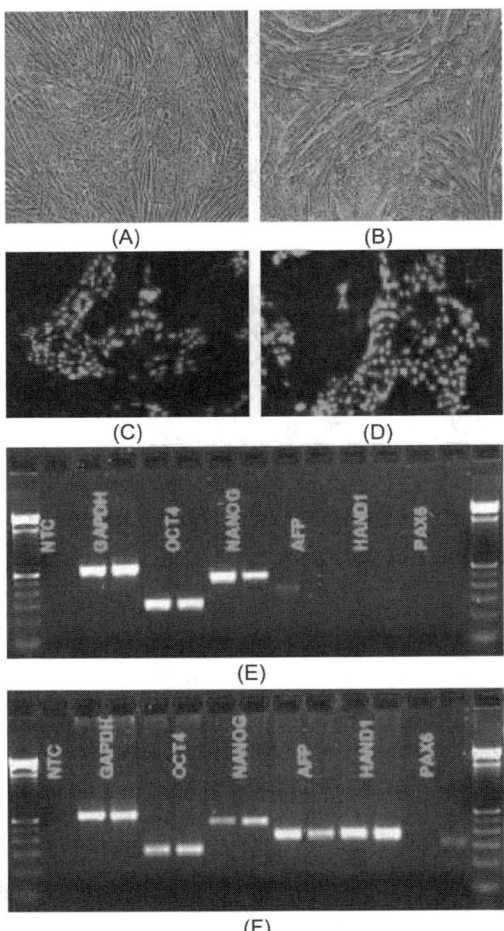

▶ **Fig. 7.5** *Performance of humanised KSR: BG01v cultured in either (A) KSR or (B) KSR Xeno-free on human foreskin fibroblasts (HFF) attached with CELLstart substrate (Passage 4). OCT-4 staining of BG01v cultured on HFF + CELLstart™ in (C) KSR or (D) KSR Xeno-free for five passages. (E) Following 10 passages in either KSR (left lane) or KSR Xeno-free (right lane) on HFF attached with CELLstart™ substrate, BG01v gene expression was examined. (F) Gene expression of embryoid bodies generated from the same P10 BG01v/HFF cultures . . . (For colour figure see Plate 1). (For colour figure see Plate 19).*

promoter. Such a reporter line will enable quick and continued monitoring of strictly controlled cultures toward pluripotency in a non-invasive fashion and may aid in hESC propagation in scalable high-throughput cell culture formats.

The BG01v-Oct4-GFP hESC reporter line was generated using the BG01v hESC line (BresaGen), a variant of parent BG01 line that was established and characterized by BresaGen, Inc.

(Athens, GA) and is listed on the National Institutes of Health (NIH) Stem Cell Registry (http://stemcells.nih.gov) [4]. This stable variant hESC line, designated BG01v, was derived from a cell line with a normal karyotype but possesses known stable chromosomal aberrations (XXY) [12, 17]. Although karyotypically abnormal, the cell line proliferates in culture, expresses markers of pluripotency, and retains the ability to differentiate into cell types representative of the three germ layers. Variant hESC lines, relative to their normal parental lines, are easier to manipulate *in vitro* and recover more rapidly after passaging and cryopreservation.

In order to generate the Oct4-GFP line, we used the phiC31 integrase system. This integrase is derived from the bacteriophage phiC31 and has been shown to function efficiently in mammalian cells [26]. The integrase can facilitate the integration of a plasmid containing an attB site into genomic locations that resemble the attP site, and termed pseudo attP sites [27]. These sites have been shown to exist in the genomes of all species tested so far, and recent studies have shown that there are a limited number of hotspots in the human genome that are targeted by the integrase [28]. These hotspots generally tend to occur in regions of open chromatin, thereby reducing the probability of the integrated transgene being shut down. In the three different human cell types studied previously [28], the integration hotspots were shared, even though the cells were from different tissues. Our hypothesis was that since the phiC31 integrase had been able to efficiently target pseudo sites in a variety of cell types derived from a number of species, it would be an effective method to integrate transgenes in a site-specific manner in human embryonic stem cells.

BG01v cells were transfected with a plasmid containing the Oct4-GFP expression cassette, a selectable marker, and a phiC31 *attB* site using Lipofectamine2000 (Invitrogen Corporation, Carlsbad, CA). The plasmid containing these elements was generated by using Multisite Gateway technology, which allows for assembly of multiple fragments in a single step. After selection for expression of the drug-resistance marker, individual clones were isolated, expanded, and analysed further. Clones selected from the transfections outlined earlier contained the *GFP* gene, and expression of this gene was driven by the human *Oct4* promoter. Since the *Oct4* promoter is only active in undifferentiated ESCs, expression of GFP was confined to undifferentiated cells. The site of integration of the transgene was determined by using a plasmid rescue protocol.

One of the reporter lines containing the Oct4–GFP expression marker was further characterized. The karyotype of the lines was found to be identical to the parent BG01v line, suggesting that the transfection of the cells, expression of phiC31 integrase, and subsequent manipulation of the cells had not led to any chromosomal abnormalities. Growing in the conditions favorable for maintenance of the undifferentiated phenotype, the cells maintained expression of GFP driven by the *Oct4* promoter (Fig. 7.6, Panel B). These data suggest that silencing of the transgene did not occur when integration was mediated by phiC31 integrase.

Upon formation of embryoid bodies, the expression of GFP was not detected (Fig. 7.6, Panel C), suggesting that the promoter fragment used in the generation of these lines contained all the information necessary for proper regulation. Embryoid bodies generated from the reporter line were harvested after 21 days, and the expression of various differentiation markers was determined by RT-PCR (Fig. 7.7). All three lineages were found in the embryoid bodies generated in this manner, indicating that the cells had maintained their pluripotency.

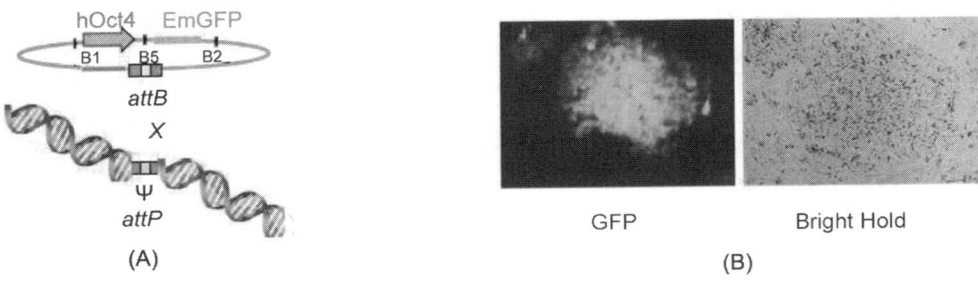

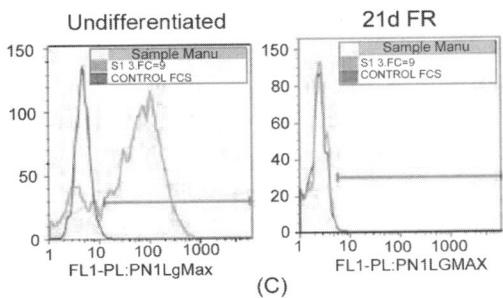

▶ **Fig. 7.6** *BGO1V/ human Oct4-GFP reporter line. (A) Vector configuration and phiC31-mediated genomic targeting. Clones resulting from transfection of GFP expression plasmid and phiC31 integrase were picked, expanded and their integration sites mapped. (B) Representative hOct4-GFP clone derived from BG01v was analysed by bright-field and fluorescent microscopy. (C) Clone analysis before and after 21 day embryoid body differentiation. GFP expression was analysed by using a Guava FACS analyser. The green curve represents GFP expression, and the red curve represents a control, unengineered cell line that did not express GFP. (For colour figure see Plate 20).*

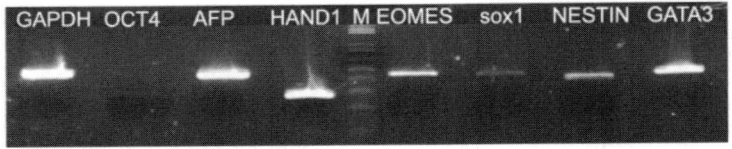

▶ **Fig. 7.7** *Expression of Candidate Markers of Differentiation measured by RT-PCR during differentiation of BG01v cells, grown in StemPro® hESC SFM, to embryoid bodies (EBs). The data show that BG01v cells grown in StemPro® hESC SFM media exhibit a complete ability to differentiate into ectoderm (SOX1, Nestin, Eomes), mesoderm (HAND1), and endodermal (AFP) and trophoblast (GATA3) lineages. GAPDH is a house keeping control and M denotes the marker. (For colour figure see Plate 1).*

As described previously, the reporter line was created using a phage integrase mediated system to target the Oct4–GFP construct to the hESC genome. While there have been other similar reporter hESC lines [29] constructed using methods such as viral delivery and homologous recombination [29, 30], we believe this represents the first example of a line created using this method. To our knowledge, this represents the first commercially available hESC reporter line generated using any of the available methods [31].

CONCLUSION

Current culture systems for human embryonic stem cells are sub-optimal due to the use of murine feeders, serum-containing medium, and substrates such as Matrigel™. There is a great need for inexpensive culture systems that can facilitate the propagation of hESCs in serum-free, feeder-free, and if possible in a substrate-free culture condition so that the cells produced in such culture conditions are suitable for cell therapy applications. At present, large-scale culturing of hESCs is complex, and efforts to mass produce hESCs for cell therapy applications through cell banking are at early stages and substantially differ from mouse ESC-based suspension bioreactor approaches [32], embryoid body [33], or cell factories and microcarriers [34]. Future attempts to generate hESCs in large scale with more humanized reagents, such as defined medium and defined substrate, can enable hESCs that are more clinically useful.

REFERENCES

1. Pannetier M., R. Feil (2007). Epigenetic stability of embryonic stem cells and developmental potential. Trends. *Biotechnol.* **25**(12):556–612 .
2. Rugg-Gunn P.J., A.C. Ferguson-Smith, R.A. Pedersen (2005). Epigenetic status of human embryonic stem cells. *Nat. Genet.* **37**(6): 585–587.
3. Hoffman L.M., M.K. Carpenter (2005). Human embryonic stem cell stability. Stem. *Cell. Rev.* **1**(2):139–144.
4. Brimble S.N., X. Zeng , D.A. Weiler, et al. (2004). Karyotypic stability, genotyping, differentiation, feeder-free maintenance, and gene expression sampling in three human embryonic stem cell lines derived prior to August 9, 2001. Stem. *Cells. Dev.* **13**(6): 585–597.
5. Ponsaerts P., .F. van Tendeloo V, P.G. Jorens, Z.N. Berneman, van D.R. Bockstaele (2004). Current challenges in human embryonic stem cell research: Directed differentiation and transplantation tolerance. J. Biol. Regul. Homeost. *Agents.* **18**(3-4): 347–351.
6. Xu R.H., R.M. Peck, D.S. Li, X. Feng, T. Ludwig, J.A. Thomson (2005). Basic FGF and suppression of BMP signaling sustain undifferentiated proliferation of human ES cells. *Nat. Methods.* **2**(3): 185–190.
7. Wang L., L. Li, P. Menendez, C. Cerdan, M. Bhatia (2005). Human embryonic stem cells maintained in the absence of mouse embryonic fibroblasts or conditioned media are capable of haematopoietic

development. *Blood.* **105**(12): 4598–4603.

8. Xiao L., X. Yuan, S.J. Sharkis (2006). Activin A maintains self-renewal and regulates fibroblast growth factor, Wnt, and bone morphogenic protein pathways in human embryonic stem cells. *Stem Cells.* **24**(6): 1476–1486.
9. Shin S., M. Mitalipova, S. Noggle, et al. (2006). Long-term proliferation of human embryonic stem cell-derived neuroepithelial cells using defined adherent culture conditions. *Stem Cells.* **24**(1): 125–138.
10. Rajala K., H. Hakala, S. Panula, et al. (2007). Testing of nine different xeno-free culture media for human embryonic stem cell cultures. *Hum. Reprod.* **22**(5): 1231–1238.
11. Wang L., T.C. Schulz, E.S. Sherrer, et al. (2007). Self-renewal of human embryonic stem cells requires insulin-like growth factor-1 receptor and ERBB2 receptor signaling. *Blood.* **110**(12): 4111–4119
12. Ludwig T.E., V. Bergendahl, M.E. Levenstein, J. Yu, M.D. Probasco, J.A. Thomson (2006). Feeder-independent culture of human embryonic stem cells. *Nat Methods.* **3**(8): 637–646.
13. Ludwig T.E., M.E. Levenstein, J.M. Jones, et al. (2006). Derivation of human embryonic stem cells in defined conditions. *Nat Biotechnol.* **24**(2): 185–187.
14. Kleinman H.K., M.L. McGarvey, L.A. Liotta, P.G. Robey, K. Tryggvason, G.R. Martin (1982). Isolation and characterization of type IV procollagen, laminin, and heparan sulfate proteoglycan from the EHS sarcoma. *Biochemistry.* **21**(24): 6188–6193.
15. Beattie G.M., A.D. Lopez, N. Bucay, et al. (2005). Activin A maintains pluripotency of human embryonic stem cells in the absence of feeder layers. *Stem Cells.* **23**(4): 489–495.
16. Amit M., C. Shariki, V. Margulets, J. Itskovitz-Eldor (2004). Feeder layer- and serum-free culture of human embryonic stem cells. *Biol. Reprod.* **70**(3): 837–845.
17. Stojkovic P., M. Lako, R. Stewart, et al. (2005). An autogeneic feeder cell system that efficiently supports growth of undifferentiated human embryonic stem cells. *Stem Cells.* **23**(3): 306–314.
18. Gerecht-Nir S., J. Itskovitz-Eldor (2004). The promise of human embryonic stem cells. *Best Pract Res Clin Obstet Gynaecol.* **18**(6): 843–852.
19. Nur E.K.A., I. Ahmed, J. Kamal, M. Schindler, S. Meiners (2006). Three-dimensional nanofibrillar surfaces promote self-renewal in mouse embryonic stem cells. *Stem Cells.* **24**(2): 426–433.
20. Skottman H., O. Hovatta (2006). Culture conditions for human embryonic stem cells. *Reproduction.* **132**(5): 691–698.
21. Heng B.C., K.J. Vinoth, H. Liu, M.P. Hande, T. Cao (2006). Low temperature tolerance of human embryonic stem cells. *Int J. Med. Sci.* **3**(4): 124–129.
22. Reubinoff B.E., P. Itsykson, T. Turetsky, et al. (2001). Neural progenitors from human embryonic stem cells. *Nat Biotechnol.* **19**(12): 1134–1140.
23. Richards M., C.Y. Fong, S. Tan, W.K. Chan, A. Bongso (2004). An efficient and safe xeno-free cryopreservation method for the storage of human embryonic stem cells. *Stem Cells.* **22**(5): 779–789.
24. Fleming K.K., A. Hubel (2006). Cryopreservation of haematopoietic and non-haematopoietic stem cells. *Transfus Apher. Sci.* **34**(3): 309–315.
25. Loring J.F., M.S. Rao (2006). Establishing standards for the characterization of human embryonic stem cell lines. *Stem Cells.* **24**(1): 145–150.
26. Groth A.C., E.C. Olivares, B. Thyagarajan, M.P. Calos (2000). A phage integrase directs efficient

site-specific integration in human cells. *Proc. Natl. Acad. Sci. USA.* **97**(11): 5995–6000.

27. Thyagarajan B., E.C. Olivares, R.P. Hollis, D.S. Ginsburg, M.P. Calos (2001). Site-specific genomic integration in mammalian cells mediated by phage phiC31 integrase. *Mol Cell Biol.* **21**(12): 3926–3934.
28. Chalberg T.W., J.L. Portlock, E.C. Olivares, et al. (2006). Integration specificity of phage phiC31 integrase in the human genome. *J. Mol. Biol.* **357**(1): 28–48.
29. Gerrard L., D. Zhao, A.J. Clark , W. Cui (2005). Stably transfected human embryonic stem cell clones express OCT4-specific green fluorescent protein and maintain self-renewal and pluripotency. *Stem Cells.* **23**(1): 124–133.
30. Rodda D.J., J.L. Chew, L.H. Lim, et al. (2005). Transcriptional regulation of nanog by OCT4 and SOX2. *J. Biol. Chem.* **280**(26): 24731–24737.
31. Thyagarajan B., Y. Liu, S. Shin, et al. (2008). Creation of Engineered Human Embryonic Stem Cell Lines Using phiC31 Integrase. *Stem Cells.* **26**(1): 119–126.
32. Cameron C.M., W.S. Hu, D.S. Kaufman (2006). Improved development of human embryonic stem cell-derived embryoid bodies by stirred vessel cultivation. *Biotechnol Bioeng.* **94**(5): 938–948.
33. Fok E.Y., P.W. Zandstra (2005). Shear-controlled single-step mouse embryonic stem cell expansion and embryoid body-based differentiation. *Stem Cells.* **23**(9): 1333–1342.
34. Abranches E., E. Bekman, D. Henrique, J.M. Cabral (2007). Expansion of mouse embryonic stem cells on microcarriers. *Biotechnol Bioeng.* **96**(6): 1211–1221.
35. Li Y., S. Powell, E. Brunette, J. Lebkowski, R. Mandalam (2005). Expansion of human embryonic stem cells in defined serum-free medium devoid of animal-derived products. *Biotechnol Bioeng.* **91**(6): 688–698.
36. Dravid G., H. Hammond, L. Cheng (2006). Culture of human embryonic stem cells on human and mouse feeder cells. *Methods Mol. Biol.* **331**: 91–104.
37. Pera M.F., A.O. Trounson (2004). Human embryonic stem cells: prospects for development. *Development.* **131**(22): 5515–5525.
38. Lu J., R. Hou, C.J. Booth, Yang S.H., Snyder M. (2006). Defined culture conditions of human embryonic stem cells. *Proc. Natl. Acad. Sci. USA.* **103**(15): 5688–5693.
39. Yao S., S. Chen, J. Clark, et al. (2006). Long-term self-renewal and directed differentiation of human embryonic stem cells in chemically defined conditions. *Proc. Natl. Acad. Sci. USA.* **103**(18): 6907–6912.
40. Amit M., J. Itskovitz-Eldor (2006). Sources, derivation, and culture of human embryonic stem cells. *Semin. Reprod. Med.* **24**(5): 298–303.
41. Vallier L., M. Alexander, R.A. Pedersen (2005). Activin/Nodal and FGF pathways cooperate to maintain pluripotency of human embryonic stem cells. *J. Cell. Sci.* **118**(Pt 19): 4495–4509.
42. Liu Y., Z. Song, Y. Zhao, et al. (2006). A novel chemical-defined medium with bFGF and N2B27 supplements supports undifferentiated growth in human embryonic stem cells. *Biochem Biophys Res. Commun.* **346**(1): 131–139.

8

EPIGENETICS AND EMBRYONIC STEM CELLS

*Jayant G Mehta**
Sub-Fertility Unit, Queens Hospital
Rom Valley way, Romford, Essex RM7 OAG, UK

INTRODUCTION

Genetics has become an integral part of medicine in the last 50 years since Watson and Crick first presented the DNA–double helix to the world. During this time, molecular biology progressed in leaps and bounds and provided science with stunning discoveries. Significant advances have been made in identifying and understanding specific mutations. The human Genome Project has furthered our opportunities to develop accurate diagnostic tools, thus paving way for targeted gene therapies.

Epigenetics is an emerging field, which also has an enormous impact on medicine. In 1942, Conrad Waddington coined the term 'epigenetic' for "the branch of biology which studies the causal interactions between genes and their products, which bring the phenotype into being" [1]. The word 'epigenetic' appears in the literature as far back as the mid 19th century, although it must be noted that its conceptual origins date back to Aristotle (384–322 BC).

Developmental biologists and cancer researchers define epigenetic as the 'meiotically and mitotically heritable changes in gene expression that are coded in the DNA sequence itself'. In other words, it broadly describes the additional information superimposed on the genome, all of which contributes to the heritable establishment and to the cellular identity [2–6]. In the nucleus, DNA, along with proteins, exists as a highly compressed structure—the chromatin.

---

* jayantgmehta@gmail.com

The epigenome can be considered as the sum of both the chromatin structure and the pattern of DNA methylation, which results from the interaction between genome and the environment. This interaction is achieved by post-translational modifications to the histones [7], DNA methylation of cytosine/guanine pairs (CpG) nucleotides [8, 9], ATP-dependent chromatin remodeling, exchange of histones and histone variants [10], and small RNA molecules [11] that may serve to guide epigenetic machinery to specific genomic loci.

It is, therefore, likely that a single genome may be modified to produce multiple epigenomes, thus allowing for cellular diversity—a necessary criterion for the mammalian development [6]. The altered patterns of gene expression can occur through development and is characterized by cycles of epigenetic reprogramming which occur in germ cells and early embryos. Epigenetic reprogramming, however, is associated with the erasure of imprints, genetic conflicts, and the return of embryonic genome to totipotency. Therefore, it needs to be determined how genome function is affected by mechanisms that regulate the way genes are processed.

Genomic imprinting describes the origin-specific, monoallelic expression of a subset of genes, which arises as a result of differential epigenetic/chromatin modifications during gametogenesis and preimplantation development [7, 8]. Post-fertilisation, the sperm and egg epigenomes are actively reprogrammed into a functional embryonic nucleus with major remodeling of chromatin, histone modifications, and DNA methylation [9–11].

Epigenetic disruption of imprinted genes, during assisted reproduction technologies (ART) has been described in the pre-implantation embryos of a range of species when ART have been applied [12–17]. It is, therefore, likely that at least some human embryonic stem cells (hESCs), derived from human embryos cultured *in vitro* to the blastocyst stage, may also stably inherit imprinting errors from the donor embryo. It has also been suggested that hESC lines derived from normal, non-ART embryos, inherit unstable imprints during *in vitro* stem cell culture [18]. Preliminary studies of a limited number of imprinted genes, in six ESC lines, have detected little variation in the allelic expression status of imprinted genes [19, 20]. However, a more recent study of 10 genes in 3–24 cell lines suggests that loss of imprinting can be more readily detected by increasing the number of lines analysed [21].

As different cell cycle, senescence, spontaneous transformation, and tumourigenic potential characteristics have been reported in mouse foetal fibroblasts with either two female (parthenogenetic) or two male (androgenetic) genomes [22], it is important to consider the possibility that disrupted maternal and paternal imprints could confer different phenotypes on hESCs. This aspect calls for careful investigation. Furthermore, imprinting can occur in a tissue-specific and developmental stage specific manner. However, for most genes imprinted in humans [8], the status of allelic expression status in the blastocyst and in the ESCs is not known.

Currently, efforts are directed toward elucidating the mammalian epigenetic mechanisms that regulate cell identity and fate, especially with reference to stem cells due to their emerging roles in development and disease [23–33].

ESCs are pluripotent. They can differentiate into any type of the specialized body cells, placenta being the exception. The ESCs can also undergo a process called 'self-renewal' to transform into new, undifferentiated stem cells. Understanding the genetic basis of self-renewal is integral to the long-term maintenance of viable ESC lines [34, 35]. ESCs-selective transcription factors such as octamer-binding transcription factor 4 (Oct4) [36], Nanog [37, 38], Sox2 [39], and extracellular signaling molecules such as leukemia inhibitory factor (LIF) [40], and bone morphogenic proteins [41, 42] are some of the factors responsible for maintaining pluripotency in ESCs. In addition, studies have revealed that histone modification, DNA methylation, and chromatin remodeling are vital for the ESCs to differentiate [43, 44]. This observation is suggestive of a possible link between epigenetic and pluripotency of ESC. This unique epigenetic profile may not only allow ESC to prevent transcription of genes associated with differentiation, but also to allow transcription, should the correct developmental signals be received.

A major challenge that exists for developmental biologists is to elucidate mechanisms whereby a structurally and functionally heterogeneous organism is built from genetically homogeneous cells. An interesting scenario has been to determine how pluripotent stem cells, within close proximity of one another within a developing blastocyst (and thus exposed to very similar if not identical environmental cues), ultimately form different cell lineages [45–47]. It has been documented that apparently 'identical' ESCs, cultured under carefully controlled cell culture conditions, can be induced to form multiple, different cell types, despite sharing very similar local environmental cues [48, 49].

Epigenetic research is, therefore, of key importance to stem cell technology. It asks what makes a stem cell a stem cell. Since stem cells can give rise to all cell types, it is thought, and indeed studies have indicated, that their DNA is more open to instruction and manipulation. In other words, their epigenetic instructions are missing [48–53].

By deciphering the nature of epigenetic modifications and by trying to understand how these modifications are established, maintained, and how they can be influenced by the cells' environment, developmental biologists would strive to improve culture techniques for converting stem cells into specific types. Ultimately, there is scope to manipulate the stem cells directly, in a patient by using carefully targeted drugs.

According to the studies in the 1980s, the loss of normal DNA methylation patterns is the best understood epigenetic cause of disease. These studies focused on X chromosome inactivation [54], genomic imprinting [55] and cancer [56]. DNA involves the addition of a methyl group to cistern within the CpG pairs [57, 58]). Unmethylated clusters, or CpG "islands," have been shown to be targets for proteins that initiate gene transcription. In contrast, methylated CpGs are generally associated with silent DNA, which block methylation-sensitive proteins and which are easily mutated. DNA methylation patterns are established and maintained by the enzymes DNA methyltransferases (DNMTs), which are essential for proper gene expression patterns [59]. In animal experiments, the removal of genes that encode DNMTs has been shown to be lethal; in humans, over-expression of these enzymes has been linked to a variety of cancers. These reversible modifications ensure that specific genes can be expressed or silenced depending on specific developmental or biochemical cues, such as changes in hormone levels, dietary components, or drug exposures [60, 61].

This chapter reviews the principles of epigenetic mechanisms and their role in maintaining the pluripotency of the ESCs. It also attempts to explain the molecular aspects of epigenetic pathways in novel therapeutic approaches.

MOLECULAR MECHANISMS IN EPIGENETIC: TAGGING OF HISTONES

The chromatin present in the nucleus contains the genomic DNA. The fundamental unit of the chromatin—nucleosome—is composed of four core histones, each in two copies. These histones—H2A, H2B, H3 and H4—are highly basic proteins surrounded by DNA, which are 147 base pairs long.

It has been reported that the nucleosomes are connected by linker DNA and the histone H1 [49, 62, 63]. Several post-translational sites, within the N-terminal tails of histone proteins, modulate the overall structure of chromatin during modification [64]. Different post-translational histone modifications have recently been shown to be the part of epigenetic mechanisms regulating selective gene expression patterns to maintain cell identity and/or cell lineage specifications [51, 52]. So far five such post-translational modifications have been characterized: acetylation [65], methylation, [66], ubiquitylation [67–71], phosphorylation [72–78], and ADP-ribosyalation, all of which can serve as epigenetic tags [22]. In addition, histone modifications are known to occur as a secondary consequence of DNA methylation. They can also be mediated by mechanisms that are independent of DNA methylation and controlled by intracellular signaling, directing the activity of numerous transcription factors, co-factors, and transcriptional machinery in general [24].

Post-translation acetylation of histone reduces adjacent nucleosome interaction, which in turn is responsible for the unfolding of compacted chromatin fiber and the exposure of the DNA template to transcription factors [79]. Further studies have shown that acetylated histones and H3 Lys-4 methylation bind the transcriptional activators to chromatin [80, 81].

Gene silencing associated with methylation of H3 Lys-9 and of H3 Lys-27 plays a different role. The repression mediated by H3 Lys-27 methylation is only related to facultative (developmentally regulated) heterochromatin formation, while H3 Lys-9 repression is associated with constitutive (permanent) heterochromatin in several circumstances [82–84]. It is, therefore, possible that histone modifications might be used during cellular differentiation to stably program different gene subsets for transcriptional activation or silencing. This, in turn, would give rise to different cell lineages and help in maintaining cell lineage "memory" [85–87]. For example, it has been shown that the promoter loci of genes specific to the smooth muscle cell lineage acquired high levels of H3 Lys-4 methylation and H4 acetylation during development of the smooth muscle cell lineage. However, the same gene loci were programmed with H3Lys-27 methylation in other cell lineages [88]. Based on the above observations, it is possible that the histone modifications not only are the markers reflecting the transcriptional activity status of the genes, but also actively participate in the regulation of gene transcription in differentiated somatic cells.

Histone Modifications in ESCs

Studies have shown that the regulation of gene expression patterns in ESCs is influenced by histone modifications. These studies also confirm that the promoter region of active genes in ESCs, such as *Oct4* and *Nanog*, is marked by acetylation of H3 and H4 and that these histone modifications are important for active gene transcription [89–91]. It is worth noting that ESCs employ similar epigenetic mechanisms for active gene transcription, as the differentiated somatic cells. However, it has been suggested that there may be some unique histone modification mechanisms in ESCs for silencing the lineage-control genes, which are not actively transcribed in ESCs but may be activated later during differentiation [91].

Using quantitative chromatin immunoprecipitation (ChIP) assays, Azuara et al. [43] have reported that a number of critical transcription factors for cell lineage determination, such as Sox1, Nkx2-2, Msx1, Irx3, and Pax3, were not expressed in mouse ESCs. However, they were shown to be associated with both activating (H3 Lys-9 acetylation and H3 Lys-4methylation) and repressive (H3 Lys-27 methylation) histone modifications within their promoter loci. These studies also demonstrated that expression of these lineage-control genes was aberrantly induced in during embryonic ectoderm development (EED) protein gene expression. They concluded, therefore, that the presence of H3 Lys-27 methylation is functionally important for preventing expression of lineage-control genes in ESCs.

Similarly studies of Bernstein et al. [44] reported 'bivalent' histone modifications. These studies cleverly demonstrated that the chromatin of key developmental genes, which included members of the *Sox*, *Hox*, *Pax*, and *Pou* gene family, displayed a unique histone modification patterns in mouse ESCs. They found the presence of large stretches of repressive H3 Lys-27 methylation within the genes studied, while simultaneously harboring activating H3 Lys-4 methylation around the transcriptional start sites. Extending their study and using the sequential ChIP assays for H3 Lys-27 methylation and H3 Lys-4 methylation, they conclusively demonstrated that these apparently conflicting histone modifications were indeed present on the same physical chromosomes and that bivalent modification patterns were not due to the presence of two distinct subpopulations of ESCs. Their studies also showed that bivalent histone modification patterns resolved during the differentiation from ESCs into a neuronal cell lineage. Only H3 Lys-4 methylation remained within neuron-specific gene promoters in neuronal precursors, whereas H3 Lys-27 methylation disappeared from these neuron-specific promoters. However, it was also observed that promoters of other genes that remained silent in the neurons lost H3 Lys-4 methylation but retained enrichment for H3 Lys-27 methylation. These studies have provided significant evidence that key lineage-control genes in ESCs are marked with a unique combination of activating and repressive histone modifications.

It is not surprising that it has been postulated that this unique ESC-specific histone modification pattern could contribute to the maintenance of pluripotency in ESCs, because of the bivalent histone modifications, which silenced the genes in ESCs due to the dominant effect of H3 Lys-27 methylation over H3 Lys-4 methylation. Simultaneously it preserved their potential to become activated upon initiation of ESC differentiation [43, 44, 82, 83, 92].

It is possible that permissive chromatin conformation at lineage-control gene loci in ESCs would allow committed ESCs to differentiate into a particular lineage. This would be achieved by removing the repressive histone modification from the required lineage-control gene loci. This activation would then allow initiation of transcription and ensure that these genes are accessible to chromatin remodeling complexes, and to transcription factors for subsequent transcriptional initiation in response to appropriate environmental cues.

Therefore, when ESCs undergo induction to transform into other cell lineages, it is likely that the activating histone modifications would be removed from the lineage-control gene loci, conserving the repressive modifications, resulting in the genes being stably silenced. The bivalent histone modifications, at the level of lineage-control gene loci, can be considered to provide the initial platform for either later activation or inactivation in ESCs. It is worth noting that such a transcriptional-permissive status of lineage-control genes may, therefore, play a key role in the maintenance of pluripotency in ESCs. However, the bivalent histone modifications were not detected at promoter loci of the multiple lineage-control genes, including Myf5 and Mash1 [43, 44, 93]. Myf5 is a member of MyoD transcription factor family and has been shown to be a critical regulator for muscle lineage determination [94], whereas Mash1 is a critical transcription factor for the production of neural precursor cells [93]. In addition, the studies by Bernstein et al. [44] have also shown that a number of key developmental genes in ESCs were marked only by H3 Lys-4 methylation (e.g. *Pbx3*) or did not exhibit either H3 Lys-4 methylation or H3Lys-27 methylation (e.g., *Foxp1*). It is more than likely that these genes may allow activation at later developmental stages; the mechanism is yet to be identified.

It is worth investigating the fact that it is only a subset of lineage-control genes that exhibit bivalent histone modification, which are then programmed into the selective lineage-control gene loci. To answer this fundamental question, it is important to examine the findings of a number of studies reported recently. These studies, conducted on subset of cell lineage-specific genes, revealed that transcription factors are not associated with activating histone modifications in mouse ESCs [46, 47, 85]. These studies have clearly demonstrated that some of the promoter regions such as pancreatic cell specific *insulin* gene, neuron-specific *synaptotagmin-4* gene, and pre-B cell specific *5-VpreB1* genes were marked by acetylation of H3 and methylation of H3 Lys-4 in ESCs [85, 95, 96]. In addition, RNA polymerase II transcription factor was shown to be associated with the *synaptotagmin-4* promoter and the *5-VpreB1* intergenic locus [95, 96]. Therefore, it is possible that the epigenetic priming mechanisms may not appear solely on the key lineage-control genes.

Studies of Boyer et al. [97] and Loh et al. [98] have reported that association exists between these ESC-selective transcription factors and a number of transcriptionally silent lineage-control genes such as *Hox* and *Pax*. These observations confirm the hypothesis that ESC-selective transcription factors are important mediators for the assembly of bivalent histone modifications in ESCs [42]. However, Oct4, Nanog, and Sox, which are transcriptional activators, have not been found to possess intrinsic enzymatic activities for histone modifications in their target genes [42]. These studies also suggest that ESC-selective transcription factors introduce activating histone modifications into the selective lineage control gene loci by recruiting histone-modifying enzymes.

Recently, it has also been shown that the polycomb protein complexes (PcG) required for methylation of H3 Lys-27 may be associated with a number of key lineage-control gene loci, both in human and in mouse ESCs. Further, their target genes have been shown to highly overlap with those for ESCs' selective transcription factors [99, 100]. This raises a very important question, whether ESC-selective transcription factors are capable of directly recruiting the PcG protein complexes and introduce repressive histone modifications at bivalently modified lineage-control gene loci in ESCs? Before we can attempt to answer this question, it is worth considering the fact that more than 100 loci have been identified to be associated with bivalent histone modifications in mouse ESCs, but only half of them were found to be associated with Oct4 and Nanog [44]. It is also important to note that although the promoter region of the Myf5 gene has been reported to be "unassociated" with both activating and repressive histone modifications, studies by Azuara et al. [43] and Loh et al. [98] have reported this promoter region to be still occupied by Oct4 in mouse ESCs. It can, therefore, be concluded that the bivalent histone modifications assembly can not be explained by these ESC-selective transcription factors.

Further, Bernstein et al. [44] have reported a strong correlation between the localisation of CpG islands and the presence of methylated H3 Lys-4 in mouse ESCs. Their finding suggests that specific *cis*-regulatory elements present within the promoter's lineage control genes may also be responsible for the existence of bivalent histone modification patterns observed in ESCs. However, Lee and Skalnik [101] have recently shown that CXXC finger protein 1—a binding factor of unmethylated CpG islands—is a component of the mammalian Set1 H3Lys-4 methyltransferase complex. Although, there is no evidence to suggest that CpG islands at the key lineage-control gene loci are selectively unmethylted in ESCs, it is more than likely that unmethylated CpG islands may be the direct target of this methyltransferase [9]. It is also possible that the subset of key lineage-control genes may be mediated by the recruitment of Set1 methyltransferase into CpG islands at these promoter regions in ESCs.

Although there is compelling evidence to suggest the formation of bivalent histone modifications at key lineage-control gene loci in ESCs, the specific mechanisms that give rise to these bivalent histone modifications and their precise role in the control of embryonic stem pluripotency and in the subsequent lineage determination still remain to be elucidated. Consistent with this evidence, it is important to consider the possible role of other epigenetic regulatory factors, such as polycomb group (PcG) and trithorax group (trxG) proteins, in maintaining the pluripotency of ESCs.

Polycomb Group Proteins

PcG and trxG proteins have emerged as key players in gene regulation and have been demonstrated to be the functional coordination of DNA-accessibility during the developmental stages of an organism [2, 102–104]. Unlike bivalent histone modifications, these epigenetic regulators act antagonistically, through regulation of specific amino acid modification in histones [103, 104]. Also, trxG promotes, while PcG represses transcription.

Several studies reported so far have identified the PcG implication in regulation of other cellular processes such as cell cycle control [105], actin polymerisation [106], cellular senescence [107]; X-inactivation [108], genomic imprinting [109] and haematopoiesis [110, 111] have been shown to be PcG coordinated. Further, chromatin disrupting processes, during DNA replication and transcription, is also maintained by PcG proteins [102, 112].

These proteins are known to comprise of two biochemically and functionally distinct multimeric polycomb repressive complexes, namely PRC1 and PRC2. The core components of PRC2—enhancer of zeste-2 (EZH2), suppressor of zeste-12 (SUZ12) and EED protein [102]—have been reported to maintain complex stability and help with the methyltransferase activity of the EZH2 [113]. Working in an orchestrated manner, PRC2 is responsible for initiating transcription repression with the help of PRC1. It has also been reported that Ezh2-mediated H3K27 methylation has been shown to help achieve PRC1 and Hox gene silencing [114]. It is known that functional mutations in the PRC components eliminate the ability of ESCs to be pluripotent [32,115], leading to embryonic lethality [116,117]. ESCs that fail to repress the expression of differentiation-specific transcription factors have been shown to be Eed-deficient [118]. This finding is supportive of the orchestral transcriptional repression by the functional collaboration between the components of PRC1 and PRC2.

It is, therefore, highly probable that the composition of PRC components may be histone-substrate specific, which in turn interact with chromatin and influence the eventual epigenetic outcome [119]. Additional, physical interactions between the PcG protein, EZH2, and DNA-DNMTs have been reported to be essential for transcriptional repression in somatic cells [120]. Although the precise signal that triggers PRC2-mediated recruitment of DNMTs is not known, it has been suggested that EZH2 provides recruitment platform for DNMTs. It is likely that some other unknown interaction partners may confer specificity to gene-specific promoter DNA for methylation [121, 122], ensuring the establishment and maintenance of cellular memory in somatic cells [120].

PcG AND STEM CELL MAINTENANCE

PcG proteins have been shown to be highly abundant in the inner cell mass (ICM). These proteins, however, begin to decline as the ICM in cultured cells start to differentiate [123]. By silencing key lineage-specific transcription factor genes and preventing differentiation, PcG proteins play a critical role in maintaining pluripotency in ESCs [115, 118]. It has been reported that EZH2, the PRC2 component, is also highly expressed in ESCs and is required for the derivation of ESCs [116]. Recent studies have shown the Suz12 levels to remain constant, while those of EZH2 and Eed decline during stem cell differentiation [124].

PcG-protein-mediated stem cell maintenance is a biologically dynamic process and not a consequence of permanent silencing. Studies have shown that PcG mutants lose their ability to maintain ESCs in the undifferentiated state [99,100]. These studies are suggestive of the fact that PcG proteins help to maintain the silencing of these genes in undifferentiated ESCs. It is worth noting the studies of Azuara et al. [125], who have shown that some of the PcG protein targets are associated with activating transcription factors but remain silent in expression (poised for activation) or active in stem cells but become repressed upon differentiation.

Further in ESCs, the active promoters of pluripotency genes such as (*OCT4*) and *NANOG* have been shown to be associated with transcriptionally active histone modifications [126]. To maintain the ESCs' identity, the key developmental regulators were shown to be silenced and reported to co-occupy a significant subset of PcG proteins [97–99].

In order to conclusively confirm a link between ESCs' pluritotency transcription factors and PcG proteins in maintaining identity of stem cells, more specific studies on transcriptional factors and other gene regulators are needed. However, studies reviewed here conclusively suggest a possible role for PcG proteins in regulating 'self-renewal' and the maintenance of stem cell identity. It appears that PcG-mediated gene repression may be a dynamic process, in both early embryos and in ESCs.

MOLECULAR TARGETS OF PcG PROTEIN

As discussed, PcG proteins play a major role in cellular identity and stem cell maintenance. Further, studies have been reported to identify and understand the molecular targets of PcG, and the epigenetic events associated with PcG [99, 100].

In mouse ESCs [104], it has been shown that components of PRC1 (Phc1, Rnf2) and PRC2 (Suz12, Eed) including the repressive histone modification H3K27me3 were simultaneously localized at promoter regions of hundreds of target genes. These studies demonstrated less target gene transcripts in mouse ESCs, when compared with differentiating embryoid bodies.

These observations may confirm the direct role of PcG proteins as gene repressors, which are associated with transition from cellular pluripotency to lineage commitment. In addition, genes containing PcG proteins also encoded for regulatory factors. These regulatory factors are known to play a role in developmental processes such as cell differentiation, embryonic development, and cell-fate commitment. It is interesting to note that these regulatory factors have also been identified as lineage-specific transcription factors [99, 100]. It can, therefore, be concluded that PcG proteins maintain pluripotency by targeting a large cohort of developmentally vital regulators in ESCs, whose expressions would, otherwise, lead to differentiation [99, 127].

In summary, it is highly probable that PcG and TrxG proteins fine-tune gene expression with the help of cell type specific accessory factors, which are responsible for changing the overall composition of PcG complexes during the development [128]. It has been observed that transcription factors such as Oct4, Sox2, and Nanog occupy almost one-third of PRC2 regions in human ESCs and direct repression of target genes through various interactions with PcG [99]. It is, therefore, conceivable that Oct4, Sox2, and Nanog proteins are capable of directing both Peg and Trig proteins to target sites in ESCs. Although, sufficient evidence exists, the molecular mechanisms responsible for both initiation and maintenance of these processes still remain to be established.

CONCLUSION

Chromatin is a dynamic structure that potentially integrates hundreds of signals from the cell surface, and affects a coordinated and appropriate transcriptional response. It is increasingly clear that epigenetic marking of chromatin, and the DNA itself, is an important component of the signal integration, which is performed by the genome as a whole. However, the questions remain as to how chromatin architecture remains stable over many cell generations by the propagation of its prior active and inactive states. Also, the mechanism by which these molecular states are changed upon lineage specification needs to be answered.

Several studies reviewed here have shown PcG proteins as epigenetic switches, which function by delineating both repressive and depressive gene signals to ensure that a particular cell identity is maintained or that the cell differentiates.

The characteristic properties of stem cells, such as self-renewal, maintaining pluripotency, or being lineage–specific, have been modulated by and attributed to the epigenetic processes. ESCs exhibit unique histone modification patterns for priming the lineage-control genes for later activation. The histone modification patterns of lineage-control genes in ESCs are classified into three groups: (a) genes marked by "bivalent" histone modifications consisting of activating H3 Lys-4 methylation and repressive H3 Lys-27 methylation; (b) genes not marked by any known or detectable histone modifications; and (c) genes marked only by activating H3 Lys-4 methylation. These genes are poised for activation with the appropriate environmental cues and signals. In differentiated somatic cells, the inactive lineage-control genes are stably silenced by repressive epigenetic modifications. These inactive lineage-control genes are normally resistant to transcriptional activation. The actively transcribed lineage-control genes are accompanied by an open chromatin structure and are marked by activating histone modifications, both in the ESCs and in the differentiated somatic cells.

Although, the pluripotent transcription factors exert both the positive and negative control on the expression factors related to stem cells, there is no evidence that PcG protein mediated histone modifications are responsible for cellular epigenetic state. In addition, it is still not clear whether PcG homeostasis can be maintained as a function of cell identity. Bernstein et al. have suggested that DNA sequence and associated PcG proteins contribute to the overall epigenetic landscape in stem cells [44].

Given that the balance between activating and repressive histone modifications is important for the maintenance of ESC pluripotency as well as for cell lineage determination, extensive studies on the large repertoire of histone modifications, in ESCs, is likely to provide further insights into this field. Similarly, mapping of the epigenetic mechanisms responsible for the regulation of genome function, in its natural environment, is essential to our understanding of development.

If cell transplantation and regenerative therapies for wide range of human diseases using ESCs is to be successful, a better understanding of epigenetic mechanisms that are responsible for controlling pluripotency and line determinations need to be achieved. With recent developments in the isolation and characterisation technology, it is more than likely that the epigenetic molecular signature of stem cells may soon be uncovered.

ACKNOWLEDGEMENTS

I am indebted to Dr. Reeja Tharu for her valuable time, effort, and suggestions in preparation of this manuscript.

REFERENCES

1. Waddington C.H. (1942). The epigenotype. *Endeavour.* **1**: 18–20.
2. Grimaud C., N. Negre, G. Cavalli (2006). From genetics to epigenetics: The tale of Polycomb group and trithorax group genes. *Chromosome. Res.* **14**: 363–375.
3. Richards E.J. (2006). Inherited epigenetic variation-revisiting soft inheritance. *Nat. Rev. Genet.* **7**: 395–401.
4. Orlando V. (2003). Polycomb, epigenomes, and control of cell identity. *Cell.* **112**: 599–606.
5. Ptashne M. (2007). On the use of the word 'epigenetic.' *Curr. Biol.* **17**: R233–236.
6. Bernstein B.E., A. Meissner, E.S. Lander (2007). The mammalian epigenome. *Cell.* **128**: 669–681.
7. Ferguson-smith A.C., M.A. Surani (2001). Imprinting and the epigenetic asymmetry between parental genomes. *Science.* **293**: 1086–1089.
8. Morison I.M., J.P. Ramsay, H.G. Spencer (2005). A census of mammalian imprinting. Trends. *Genet.* **21**: 457–465.
9. Meissner A., T.S. Mikkelsen, H.Gu, et.al. (2008). Genome-scale DNA methylation maps of pluripotent and differentiated cells. *Nature.* **454**: 766–770.
10. Jaenisch R., A. Bird (2003). Epigenetic regulation of gene expression: How the genome integrates intrinsic and environmental signals. *Nat. Genet.* **33**: 245–254.
11. Morgan H.D., F.Santos, K. Green, et al. (2005). Epigenetic reprogramming in mammals. *Hum. Mol. Genet.* **14**(Spec No 1): R47–58.
12. Mann M.R., Y.G. Chung, L.D.Nolen, et al. (2003). Disruption of imprinted gene methylation and expression in cloned preimplantation stage mouse embryos. *Biol. Reprod.* **69**: 902–914.
13. Mann M.R., S.S. Lee, A.S. Doherty, et al. (2004). Selective loss of imprinting in the placenta following preimplantation development in culture. *Development.* **131**: 3727–3735.
14. Maher E.R. (2005). Imprinting and assisted reproductive technology. *Hum. Mol. Genet.* **14**: R133–R138.
15. Sato A., E. Otsu, H. Negishi, et al. (2007). Aberrant DNA methylation of imprinted loci in superovulated oocytes. *Hum. Reprod.* **22**: 26–35.
16. Allegrucci C., L.E. Young (2007). Differences between human embryonic stem cell lines. Hum. Reprod. *Update.* **13**: 103–120.
17. Allegrucci C., C. Denning, H. Priddle, et al. (2004). Stem-cell consequences of embryo epigenetic defects. *Lancet.* **364**: 206–208.
18. Rugg-Gunn P.J., A.C. Ferguson-Smith, R.A. Pedersen (2005). Epigenetic status of human embryonic stem cells. *Nat. Genet.* **37**: 585–587.
19. Sun B.W., A.C. Yang, Y. Feng, et al. (2006). Temporal and parental-specific expression of imprinted genes in a newly derived Chinese human embryonic stem cell line and embryoid bodies. *Hum. Mol. Genet.* **15**: 65–75.
20. Allegrucci C., Y.Z. Wu, A. Thurston, et al. (2007). Restriction Landmark Genome Scanning identifies culture-induced DNA methylation instability in the human embryonic stem cell epigenome. *Hum. Mol. Genet.* **16**: 1253–1268.
21. Hernandez L., S. Kozlov, G. Piras (2003). Paternal and maternal genomes confer opposite effects on proliferation, cell-cycle length, senescence, and tumour formation. *Proc. Natl. Acad. Sci.* **100**: 13344–13349.

22. Jenuwein T., C.D. Allis (2001). Translating the histone code. *Science.* **293**: 1074–1080.
23. Bird A.P., A.P. Wolffe (1999). Methylation-induced repression-belts, braces, and chromatin. *Cell* **99**: 451–454.
24. Ahmad K., S. Henikoff (2002). Histone H3 variants specify modes of chromatin assembly. *Proc. Natl. Acad. Sci. USA.* **99**: 16477–16484.
25. Sheardown S.A., S.M. Duthie, C.M. Johnston, et al. (1997). Stabilisation of Xist RNA mediates initiation of X chromosome inactivation. *Cell.* **91**: 99–107.
26. Valk-Lingbeek M.E., S.W. Bruggeman, M. van Lohuizen (2004). Stem cells and cancer: The polycomb connection. *Cell.* **118**: 409–418.
27. Sparmann A., M. van Lohuizen (2006). Polycomb silencers control cell fate, development and cancer. *Nat. Rev. Cancer.* **6**: 846–856.
28. Lotem J., L. Sachs (2006). Epigenetics and the plasticity of differentiation in normal and cancer stem cells. *Oncogene.* **25**: 7663–7672.
29. Feinberg A.P., R. Ohlsson, S. Henikoff (2006). The epigenetic progenitor origin of human cancer. *Nat. Rev. Genet.* **7**: 21–33.
30. Ting A.H., K.M. McGarvey, S.B. Baylin (2006). The cancer epigenome-components and functional correlates. Genes. *Dev.* **20**: 3215–3231.
31. Jones P.A, S.B. Baylin (2007). The epigenomics of cancer. *Cell* **128**: 683–692.
32. Surani M.A., K. Hayashi, P. Hajkova (2007). Genetic and epigenetic regulators of pluripotency. *Cell.* **128**: 747–762.
33. Spivakov M., A.G. Fisher (2007). Epigenetic signatures of stem-cell identity. *Nat. Rev. Genet* **8**: 263–271.
34. Evans M.J., M.H. Kaufman (1981). Establishment in culture of pluripotential cells from mouse embryos. *Nature.* **292**: 154–156.
35. Bradley A., M. Evans, M.H. Kaufman, et al. (1984). Formation of germ-line chimaeras from embryo-derived teratocarcinoma cell lines. *Nature.* **309**: 255–256.
36. Nichols J., B. Zevnik, K. Anastassiadis, et al. (1998). Formation of pluripotent stem cells in the mammalian embryo depends on the POU transcription factor Oct4. *Cell.* **95**: 379–391.
37. Chambers I., D. Colby, M. Robertson, et al. (2003). Functional expression cloning of Nanog, a pluripotency sustaining factor in embryonic stem cells. *Cell.* **113**: 643–655.
38. Mitsui K., Y. Tokuzawa, H. Itoh, et al. (2003). The homeoprotein nanog is required for maintenance of pluripotency in mouse epiblast and ES cells. *Cell.* **113**: 631–642.
39. Botquin V., H. Hess, G. Fuhrmann, et al. (1998). New POU dimer configuration mediates antagonistic control of an osteopontin preimplantation enhancer by Oct-4 and Sox-2. Genes. *Dev.* **12**: 2073–2090.
40. Nichols J., I. Chambers, T. Taga, et al. (2001). Physiological rationale for responsiveness of mouse embryonic stem cells to gp130 cytokines. *Development.* **128**: 2333–2339.
41. Ying Q.L., J. Nichols, I. Chambers, et al. (2003). BMP induction of Id proteins suppresses differentiation and sustains embryonic stem cell self-renewal in collaboration with STAT3. *Cell.* **115**: 281–292.
42. Boiani M., H.R. Scholer (2005). Regulatory networks in embryo-derived pluripotent stem cells. *Nat. Rev. Mol. Cell. Biol.* **6**: 872–881.
43. Azuara V., P. Perry, S. Sauer, et al. (2006). Chromatin signatures of pluripotent cell lines. *Nat. Cell. Biol.* **8**: 532–538.

44. Bernstein B.E., T.S. Mikkelsen, X. Xie, et al. (2006). A bivalent chromatin structure marks key developmental genes in embryonic stem cells. *Cell.* **125**: 315–326.
45. Smith A.G. (2001). Embryo-derived stem cells: Of mice and men. *Annu. Rev. Cell. Dev. Biol.* **17**: 435–462.
46. Keller G.M. (1995). *In vitro* differentiation of embryonic stem cells. *Curr. Opin. Cell. Biol.* **7**: 862–869.
47. Tabata T., Y. Takei (2004). Morphogens, their identification and regulation. *Development.* **131**: 703–712.
48. Barrera L.O., B. Ren (2006). The transcriptional regulatory code of eukaryotic cells-insights from genome-wide analysis of chromatin organisation and transcription factor binding. Curr. Opin. *Cell. Biol.* **18**: 291–298.
49. Quina A.S., M. Buschbeck, L. Di Croce (2006). Chromatin structure and epigenetics. *Biochem. Pharmacol.* **72**: 1563–1569.
50. Narlikar G.J., H.Y. Fan, R.E. Kingston (2002). Cooperation between complexes that regulate chromatin structure and transcription. *Cell.* **108**: 475–487.
51. Carey B. Li, M., J.L. Workman J.L. (2007). The role of chromatin during transcription. *Cell.* **128**: 707–719.
52. Rea S., F. Eisenhaber, D. O'Carroll, et al. (2000). Regulation of chromatin structure by site-specific histone H3 methyltransferases. *Nature.* **406**: 593–599.
53. Klose R.J., A.P. Bird (2006). Genomic DNA methylation: The mark and its mediators. *Trends. Biochem. Sci.* **31**: 89–97.
54. Avner P., E. Heard (2001). X-chromosome inactivation: counting, choice and initiation. *Nat. Rev. Genet.* 259–267.
55. Verona R.I., M.R. Mann, M.S. Bartolomei (2003). Genomic imprinting: intricacies of epigenetic regulation in cluster. *Ann. Rev. Cell. Dev. Biol.* **19**: 237–259.
56. Feinberg A.P., B. Tycko (2004). The history of cancer epigenetics. *Nat. Rev. Cancer.* **4**: 143–153.
57. Ehrlich M., R.Y.Wang (1981). 5-Methylcytosine in eukaryotic DNA. *Science.* **212**: 1350–1357.
58. Laird P.W., R. Jaenisch (1994). DNA methylation and cancer. *Hum. Mol. Genet.* **3**: 1487–1495.
59. Robertson K. D. (2002). DNA methylation and chromatin-unraveling the tangled web. *Oncogene.* **21**: 5361–5379.
60. Espino P.S., B. Drobic, K.L. Dunn et al. (2005). Histone modifications as a platform for cancer therapy. *J. Cell. Biochem.* **94**: 1088–1102.
61. Elgin S.C., S. I. Grewal (2003). Heterochromatin: Silence is golden. *Curr. Biol.* **13**: R895–898.
62. Barrera L.O., B. Ren (2006). The transcriptional regulatory code of eukaryotic cells-insights from genome-wide analysis of chromatin organisation and transcription factor binding. *Curr. Opin. Cell. Biol.* **18**: 291–298.
63. Luger K., A.W. MadeR, R. K. Richmond, et al. (1997). Crystal structure of the nucleosome core particle at 2.8 Å resolution. *Nature.* **389**: 251–260.
64. Narlikar G.J., H.Y Fan, R.E. Kingston (2002). Cooperation between complexes that regulate chromatin structure and transcription. *Cell.* **108**: 475–487.
65. Tanner K.G., M.R. Langer, J. M. Denu (2000). Kinetic mechanism of human histone acetyltransferase P/CAF. *Biochemistry.* **39**: 11961–11969.

66. Murray K. (1964). The occurrence of e-N-methyl lysine in histones. *Biochemistry*. **13**: 10–15.
67. Pickart C.M. (2001). Mechanisms underlying ubiquitination. *Annu. Rev. Biochem*. **70**: 503–533.
68. Nickel B.E., J. R. Davie (1989). Structure of polyubiquitinated histone H2A. *Biochemistry*. **28**: 964–968.
69. West M.H., W. M. Bonner (1980). Histone 2B can be modified by the attachment of ubiquitin. Nucleic. *Acids. Res*. **8**: 4671–4680.
70. Chen H.Y., J.M. Sun, Y. Zhang, et al. (1998). Ubiquitination of histone H3 in elongating spermatids of rat testes. *J. Biol. Chem*. **273**: 13165–13169.
71. Pham A.D., F. Sauer (2000). Ubiquitin-activating/conjugating activity of TAFII250, a mediator of activation of gene expression in Drosophila. *Science*. **289**: 2357–2360.
72. Bradbury E.M., R.J Inglis, H.R. Matthews, et al. (1973). Phosphorylation of very-lysine-rich histone in Physarum polycephalum. Correlation with chromosome condensation. *Eur. J. Biochem*. **33**: 131–139.
73. Gurley L.R., R.A. Walters, R.A. Tobey (1974). Cell cycle-specific changes in histone phosphorylation associated with cell proliferation and chromosome condensation. *J. Cell Biol*. **60**: 356–364
74. Mahadevan L.C., A.C. Willis, M.J, Barratt (1991). Rapid histoneH3 phosphorylation in response to growth factors, phorbolesters, okadaic acid, and protein synthesis inhibitors. *Cell*. **65**: 775–783.
75. Agostino S. Di, P. Rossi, R. Geremia, et al. (2002). The MAPKpathway triggers activation of Nek2 during chromosome condensation in mouse spermatocytes. *Development*. **129**: 1715–1727.
76. Goto H., Y. Yasui, E. A. Nigg, Inagaki, M., et al. (2002). Aurora-B phosphorylates Histone H3 at serine28 with regard to the mitotic chromosome condensation. *Genes Cells*. **7**: 11–17.
77. Ajiro K., K. Yoda, K. Utsumi, et al. (1996). Alteration of cell cycle-dependent histone phosphorylations by okadaic acid. Induction of mitosis-specific H3 phosphorylation and chromatin condensation in mammalian interphase cells. *J. Biol. Chem*. **271**: 13197–13201.
78. Nowak S.J., C.Y. Pai, V.G. Corces (2003). Protein phosphatase 2A activity affects histone H3 phosphorylation and transcription in Drosophila melanogaster. *Mol. Cell. Biol*. **23**: 6129–6138.
79. Shogren-Knaak M., H. Ishii, J.M. Sun, et al. (2006). Histone H4-K16 acetylation controls chromatin structure and protein interactions. *Science*. **311**: 844–847.
80. Kingston R. E., G. J. Narlikar (1999). ATP-dependent remodeling and acetylation as regulators of chromatin fluidity. *Genes. Dev*. **13**: 2339–2352.
81. Pray-Grant M. G., J.A. Daniel, D. Schieltz et al. (2005). Chd1 chromodomain links histone H3 methylation with SAGA- and SLIK-dependent acetylation. *Nature*. **433**: 434–438.
82. Plath K., J. Fang, S. K. Mlynarczyk-Evans, et al. (2003). Role of histone H3 lysine 27 methylation in X inactivation. *Science*. **300**: 131–135.
83. Koyanagi M., A. Baguet, J. Martens, et al. (2005). EZH2 and histone 3 trimethyl lysine 27 associated with Il4 and Il13 gene silencing in TH1 cells. *J. Biol. Chem*. **280**: 31470–31477.
84. Nakayama J., J.C. Rice, B.D. Strahl, et al. (2001). Role of histone H3 lysine 9 methylation in epigenetic control of heterochromatin assembly. *Science*. **292**: 110–113.
85. Chakrabart S.K., J. Francis, S.M. Ziesmann, et al. (2003). Covalent histone modifications underlie the developmental regulation of insulin gene transcription in pancreatic beta cells. *J. Bio. Chem*. **278**: 23617–23623.

86. Fish J.E., C.C. Matouk, A. Rachlis, et al. (2005). The expression of endothelial nitric-oxide synthase is controlled by a cell-specific histone code. *J. Biol. Chem.* **280**: 24824–24838.
87. Edelstein L.C., A. Pan, T. Collins (2005). Chromatin modification and the endothelial-specific activation of the E-selectin gene. *J. Biol. Chem.* **280**: 11192–11202.
88. McDonald O.G., B.R. Wamhoff, M.H. Hoofnagle, et al. (2006). Control of SRF binding to CArG box chromatin regulates smooth muscle gene expression *in vivo*. *J. Clin. Invest.* **116**: 36–48.
89. Hattori N., K. Nishino, Y.G. Ko, et al. (2004). Epigenetic control of mouse Oct-4 gene expression in embryonic stem cells and trophoblast stem cells. *J. Biol. Chem.* **279**: 17063–17069.
90. Kimura H., M. Tada, N. Nakatsuji, et al. (2004). Histone code modifications on pluripotential nuclei of reprogrammed somatic cells. *Mol. Cell. Biol.* **24**: 5710–5720.
91. O'Neill L.P., M.D. VerMilyea, B.M. Turner (2006). Epigenetic characterisation of the early embryo with a chromatin immunoprecipitation protocol applicable to small cell populations. *Nat. Genet.* **38**: 835–841.
92. Jiang G., F. Yang, C. Sanchez, et al. (2004). Histone modification in constitutive heterochromatin versus unexpressed euchromatin in human cells. *J. Cell. Biochem.* **93**: 286–300.
93. Williams R.R.E., V. Azuara, P. Perry, et al. (2006). Neural induction promotes large-scale chromatin reorganisation of the Mash1 locus. *J. Cell. Sci.* **119**: 132–140.
94. Pownall M.E., M.K. Gustafsson, C.P. Emerson Jr. (2002). Myogenic regulatory factors and the specification of muscle progenitors in vertebrate embryos. *Annu. Rev. Cell. Dev. Biol.* **18**: 747–783.
95. Ballas N., C. Grunseich, D.D. Lu, et al. (2005). REST and its corepressors mediate plasticity of neuronal gene chromatin throughout neurogenesis. *Cell.* **121**: 645–657.
96. Szutorisz H., C. Canzonetta, A. Georgiou, et al. (2005). Formation of an active tissue-specific chromatin domain initiated by epigenetic marking at the embryonic stem cell stage. *Mol. Cell. Biol.* **25**: 1804–1820.
97. Boyer L.A., T.I. Lee, M.F. Cole, et al. (2005). Core transcriptional regulatory circuitry in human embryonic stem cells. *Cell.* **122**: 947–956.
98. Loh Y.H., Q. Wu, J.L. Chew, et al. (2006). The Oct4 and Nanog transcription network regulates pluripotency in mouse embryonic stem cells. *Nat. Genet.* **38**: 431–440.
99. Lee T.I., R.G. Jenner, L.A. Boyer, et al. (2006). Control of developmental regulators by Polycomb in human embryonic stem cells. *Cell.* **125**: 301–313.
100. Boyer L.A., K. Plath, J. Zeitlinger, et al. (2006). Polycomb complexes repress developmental regulators in murine embryonic stem cells. *Nature.* **441**: 349–353.
101. Lee J.H., D.G. Skalnik. (2005). CpG-binding protein (CXXC finger protein 1) is a component of the mammalian Set1 histone H3-Lys4 methyltransferase complex, the analogue of the yeast Set1/COMPASS complex. *J. Biol. Chem.* **280**: 41725–41731.
102. Ringrose L. (2006). Polycomb, trithorax and the decision to differentiate. Bioessays. 28: 330–334.
103. Schwartz Y.B., V. Pirrotta (2007). Polycomb silencing mechanisms and the management of genomic programmes. *Nat. Rev. Genet.* **8**: 9–22.
104. Schuettengruber B., D. Chourrout, M. Vervoort, et al. (2007). Genome regulation by polycomb and trithorax proteins. *Cell.* **128**: 735–745.
105. Martinez A.Z., G. Cavalli (2006). The role of polycomb group proteins in cell cycle regulation during development. *Cell. Cycle.* **5**: 1189–1197.

106. Su I.H., M.W. Dobenecker, E. Dickinson, et al. (2005). Polycomb group protein ezh2 controls actin polymerisation and cell signaling. *Cell.* **121**: 425–436.

107. Guney I., J.M. Sedivy (2006). Cellular senescence, epigenetic switches and c-Myc. Cell. *Cycle.* **5**: 2319–2323.

108. Heard E. (2005). Delving into the diversity of facultative heterochromatin: The epigenetics of the inactive X chromosome. *Curr. Opin. Genet. Dev.* **15**: 482–489.

109. Mager J., N.D. Montgomery, F.P. de Villena, et al. (2003). Genome imprinting regulated by the mouse Polycomb group protein Eed. *Nat. Genet.* **33**: 502–507.

110. Hosokawa H., M.Y. Kimura, R. Shinnakasu, et al. (2006). Regulation of Th2 cell development by Polycomb group gene bmi-1 through the stabilisation of GATA3. *J. Immunol.* **177**: 7656–7664.

111. Oguro H., A. Iwama, Y. Morita, et al. (2006). Differential impact of Ink4a and Arf on haematopoietic stem cells and their bone marrow microenvironment in Bmi1-deficient mice. *J. Exp. Med.* **203**: 2247–2253.

112. Groth A., W. Rocha, A. Verreault, et al. (2007). Chromatin challenges during DNA replication and repair. *Cell.* **128**: 721–733.

113. Pasini D., A.P. Bracken, M.R. Jensen, et al. (2004). Suz12 is essential for mouse development and for EZH2 histone methyltransferase activity. *EMBO. J.* **23**: 4061–4071.

114. Cao R., Y. Tsukada, Y. Zhang (2005). Role of Bmi-1 and Ring1A in H2A ubiquitylation and Hox gene silencing. Mol. *Cell.* **20**: 845–854.

115. Boyer L.A., D. Mathur, R. Jaenisch (2006). Molecular control of pluripotency. *Curr. Opin. Genet. Dev.* **16**: 455–462.

116. O'Carroll D., S. Erhardt, M. Pagani, et al. (2001). The polycomb-group gene Ezh2 is required for early mouse development. *Mol. Cell. Biol.* 21: 4330–4336.

117. Voncken J.W., B. A, Roelen, M. Roefs, et al. (2003). Rnf2 (Ring1b) deficiency causes gastrulation arrest and cell cycle inhibition. *Proc. Natl. Acad. Sci. USA.* **100**: 2468–2473.

118. Jorgensen H.F., S. Giadrossi, M. Casanova. et al. (2006). Stem cells primed for action: Polycomb repressive complexes restrain the expression of lineage-specific regulators in embryonic stem cells. *Cell. Cycle.* **5**: 1411–1414.

119. Kuzmichev A., R. Margueron, A. Vaquero, et al. (2005). Composition and histone substrates of polycomb repressive group complexes change during cellular differentiation. *Proc. Natl. Acad. Sci. USA.* **102**: 1859–1864.

120. Vire E., C. Brenner, R. Deplus, et al. (2006). The Polycomb group protein EZH2 directly controls DNA methylation. *Nature.* **439**: 871–874.

121. Taghavi P., M. van Lohuizen (2006). Developmental biology: Two paths to silence merge. *Nature.* **439**: 794–795.

122. Hernandez-Munoz I., P. Taghavi, C.Kuijl, et al. (2005). Association of BMI1 with polycomb bodies is dynamic and requires PRC2/EZH2 and the maintenance DNA methyltransferase DNMT1. *Mol. Cell. Biol.* **25**: 11047–11058.

123. Iwama A., H. Oguro, M. Negishi, et al. (2005). Epigenetic regulation of haematopoietic stem cell self-renewal by polycomb group genes. *Int. J. Haematol.* **81**: 294 –300.

124. de la Cruz C.C., J. Fang, K. Plath, et al. (2005). Developmental regulation of Suz 12 localisation. *Chromosoma.* **114**: 183–192.

125. Roh T.Y., S. Cuddapah, K. Cui, et al. (2006). The genomic landscape of histone modifications in human T cells. *Proc. Natl. Acad. Sci. USA.* **103**: 15782–15787.
126. Tada T. (2006) Toti-/pluripotential stem cells and epigenetic modifications. *Neurodegener Dis.* **3**: 32–37.
127. Bracken A.P., N. Dietrich, D. Pasini, et al. (2006). Genome-wide mapping of Polycomb target genes unravels their roles in cell fate transitions. *Genes. Dev.* **20**: 1123–1136.
128. Ringrose L., R. Paro (2004). Epigenetic regulation of cellular memory by the Polycomb and Trithorax group proteins. *Annu. Rev. Genet.* **38**: 413–443.

9

EMBRYONIC STEM CELLS AS A THERAPY FOR DIABETES

Jennifer C.Y. Wong, Steven Y. Gao, Bernard E. Tuch*

*Diabetes Transplant Unit, Prince of Wales Hospital and
The University of New South Wales, Australia*

INTRODUCTION

Diabetes mellitus, more commonly known as diabetes, is a chronic disorder that is characterised by the presence of excess glucose in the blood and tissues of the body. Two types of diabetes exist: type 1 diabetes, or insulin-dependent diabetes, resulting from the autoimmune destruction of insulin producing β-cells within the islets of Langerhans in the pancreas; and type 2 diabetes, or non-insulin dependent diabetes, resulting from a combination of the body's resistance to the action of insulin and/or deficient insulin synthesis by β-cells.

The adult pancreas is composed of two distinct tissues types that serve two main functions. The exocrine pancreas, which consists of acinar cells and duct cells, is responsible for the production and secretion of digestive enzymes that promote nutrient digestion and adsorption. The endocrine pancreas, which constitutes only 1% of the total pancreatic mass, consists of compartmentalized structures known as the islets of Langerhans, or simply referred to as islets (Fig. 9.1). These are scattered throughout the exocrine pancreas and function to maintain blood glucose homeostasis through the actions of the five cell types found within each islet—α, β, δ, PP and ε cells which produce the hormones glucagon, insulin, somatostatin, pancreatic polypeptide and ghrelin respectively. β-cells comprise approximately 70% of the cells within an islet and are the only cells in the body that can produce and secrete insulin—a vital hormone that acts in response to changing levels of glucose. Both type 1 and type 2 diabetes results from a defect in glucose homeostasis in a

---

* b.tuch@unaw.edu.au

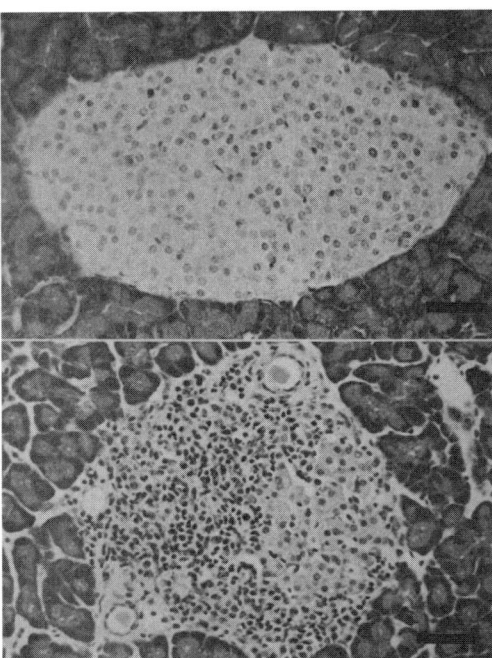

▶ **Fig. 9.1** *Haematoxlin and eosin stains of an islet from the pancreas of (top) a non-diabetic mouse and (bottom) a mouse with diabetes caused by the destruction of β-cells in the islet. Bar is 50 μm. Source: Ref. [1]. (For colour figure see Plate 21).*

way that the body cannot maintain blood glucose levels within the normal physiological range of 4–7 mmol/L. As a result, microvascular complications can arise if the disorder is left untreated, such as damage to the retina of the eye which may lead to retinopathy (blindness), damage to the nephrons of the kidney that may progress to nephropathy (renal failure) and damage to the nerves of the peripheral nervous system which may result in neuropathy (loss of feeling). Macrovascular complications as a result of atherosclerosis can also develop such as ischaemic heart disease, cerebrovascular disease and peripheral vascular disease.

Extensive research in the field of diabetes has provided a relatively in depth understanding of the causes and pathogenesis of the disease. The precise reasons as to how and why this disease occurs, however, remain to be elucidated. Currently, it is known that the cells responsible for the autoimmune destruction of β-cells include $CD8^+$ cytotoxic cells and $CD4^+$ helper T-cells. Specific genes, such as *HLA D3* and *D4* in the major histocompatibility complex HLA class II region, are known to be involved in the susceptibility to and protection from the development of the disorder.

Moreover, environmental factors such as the exposure to toxins and viral infections are believed to have a role in initiating the prediabetic phase in a type 1 diabetic patient. Type 2 diabetes on the other hand may be instigated by a mutation in the genes that are involved in the pathways which are associated with insulin, thus resulting in a defect. Genetic mutations that have been reported include those involving the insulin receptor and glycogen synthase. More commonly, however, the occurrence of type 2 diabetes develops after the age of 40 years and is associated with the health effects of obesity and an inactive lifestyle. Therefore, unlike type 1 diabetes, the onset of type 2 diabetes may be prevented or delayed by reducing risk factors such as lack of exercise, high blood pressure and high levels of cholesterol.

Type 1 diabetes currently affects at least 13 million people worldwide, and the incidence of this disorder is projected to increase at a rate of 3.2% per annum. It can occur at any age but is usually diagnosed in children with a peak incidence at 12 years of age, and young adults between the ages of 20 and 35 years. While a cure for this disease has yet to be found, it can be treated with regular injections of exogenous insulin in combination with an even intake of carbohydrates throughout the day. However, this treatment is not ideal as it remains difficult to maintain blood glucose levels within the physiological range because the meticulous control that is exercised by β-cells cannot be adequately mimicked. In addition, this treatment does not necessarily prevent the microvascular and macrovascular complications that are associated with the disease which were mentioned previously.

A more promising approach that is available at present, which aims to properly treat and potentially cure typehFi 1 diabetes, is the development of therapies that can replace the destroyed β-cells (Fig. 9.2). Two such therapies currently exist—whole pancreas transplantation and the transplantation of human islets. Both of these are obtained from human cadavers and have been able to achieve the reversal of diabetes for varying periods of time.

Whole pancreas transplantation. The transplantation of donor pancreas is now a well-established and successful long-term procedure with patient survival similar to that achieved with other solid

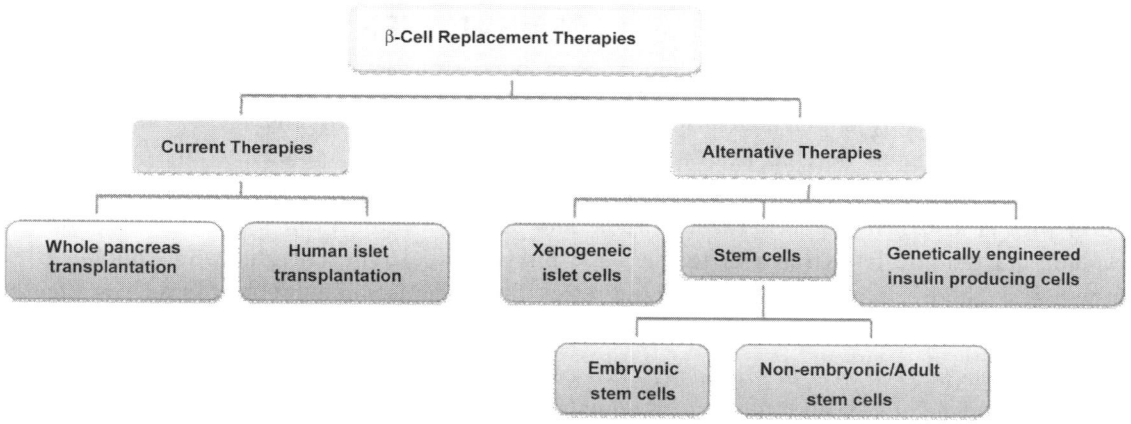

▶ **Fig. 9.2** *Current and alternative β-cell replacement therapies. Current β-cell replacement therapies include the transplantation of whole pancreas and human islets from donor human cadavers. Alternative cellular therapies that are currently under examination include xenogeneic islet cells such as porcine islets, stem cells, both embryonic and non-embryonic or adult, and genetically engineered insulin producing cells. (For colour figure see Plate 21).*

organ transplants. At 1-year post-transplantation, 85% of grafts remain functioning with the recipient not requiring injections of insulin while the patient survival rate is 91%. At the end of 5 years, 64% of grafts remain functioning with the patient survival rate at 80% [2]. Whole pancreas transplantation, however, is mostly performed in patients who are also in need of a kidney transplant due to the reported success of the combined procedure [3].

Human islet transplantation. The transplantation of human islets as a research tool has been performed since 1982, with the first patient being able to cease exogenous insulin in 1989. This approach has shown more promise since the development of the Edmonton Protocol [4] in 1999. The key feature of this protocol was the removal of steroids from the immunosuppressive regime which remarkably improved the success rate of the procedure. By 2007, 1049 people world-wide had received this cellular therapy, which was able to restore endogenous insulin production and glycaemic stability for up to 1 year. However by 5-years post-transplantation, almost all recipients had reverted back to treatment by insulin injection due to progressive islet failure [5].

Significant shortcomings that continue to persist with both whole pancreas and human islet transplantation include the need for life-long immunosuppression that greatly increases the risks of associated morbidity. The most critical limitation, however, is the lack of donor organs because their demand greatly exceeds their supply. In 2007 for example, there were at least 1.82 million Americans and 130,000 Australians with type 1 diabetes but only 1930 and 48 pancreases were donated in the USA and Australia, respectively. To address the issue of donor organ shortages, the development of alternative β-cell replacement therapies is urgently needed.

1.2 ALTERNATIVE β-CELL REPLACEMENT THERAPIES

Alternative sources of β-cell surrogates that are currently under examination include xenogeneic islets cells, genetically engineered insulin-producing cells and stem cells. *Xenogeneic islet cells* denote the use of islets derived from a different species. Of particular interest are porcine islets, which are advantageous because they provide a plentiful and readily available supply of β-cells. The use of pigs is also favourable because porcine and human insulin differ by only one amino acid residue, and porcine islets respond to the same physiological range of glucose as human islets. Nonetheless, there remains a small risk of pig retrovirus transmission to humans as well as a feeling of discomfort in using animal rather than human tissue.

Genetically engineered insulin-producing cells can be derived by inserting specific β-cell-related genes into non-β-cells in an attempt to modulate their function to resemble normal β-cells. This process, known as transdifferentiation, has been experimented on gut K-cells and liver cells, with the resulting genetically modified cells shown to be effective and glucose responsive in animal models [6,7]. Transdifferentiation, however, is a complex and poorly understood area of research as it requires explicit knowledge of the individual changes that occur at the cellular and molecular level. In addition, it remains difficult to ensure the long-term stability and the proper functioning of genetically modified cells.

Stem cells, both non-embryonic and embryonic, have been the subject of extensive investigations as an alternative source of β-cell surrogates. Non-embryonic stem cells, also colloquially known as adult stem cells, are undifferentiated cells that are found in tissues and organs within the body. They function to maintain, repair and enable the regeneration of the tissue in which they are found. Adult stem cells have a limited proliferation capacity and a restricted differentiation capability to the lineage of their tissue of origin. Despite this, they have been successfully used for the treatment of certain diseases, a common example of which is the use of haematopoietic stem cells derived from the bone marrow for the treatment of leukemia. To date, however, there has been no convincing evidence to show that adult stem cells can be successfully used to treat type 1 diabetes. Indeed, studies in mice have shown that murine β-cells during adult life appear to be sourced from pre-existing terminally differentiated β-cells that maintain a significant proliferative capacity *in vivo*, rather than an adult stem cell population [8–10]. While multipotent adult stem cells such as umbilical cord blood stem cells have demonstrated the ability to be converted into insulin-producing cells [11,12], the efficacy of these differentiation protocols require considerable improvement and the hurdles associated with transdifferentiation remain.

Alternatively, embryonic stem cells (ESCs), which are derived from the inner cell mass of blastocyst stage embryos, are particularly advantageous because they possess two distinct characteristics. First, ESCs are capable of unlimited self-renewal, in which the cells can be maintained and expanded for an extended period of time. This will be essential to obtain the abundant supply of β-cells for a treatment therapy for type I diabetes. Second, ESCs are pluripotent, with the unique ability to differentiate into all of the cell types representative of the three primary germ layers-ectoderm, mesoderm and endoderm. Pancreatic cells are derived from the endoderm lineage, where development is initiated from the definitive endoderm germ layer. A disadvantage of ESCs is their tendency to form teratomas upon transplantation if residual pluripotent cells remain present. In addition, a strong ethical concern overshadows the use of ESCs in scientific and medical research by some. Nonetheless, recent rapid progress in ESC research has shown them to be a frontrunner as a promising source of renewable, insulin-producing β-cell surrogates.

NON-ONTOGENY-BASED DIFFERENTIATION OF ESCs INTO INSULIN-PRODUCING β-CELLS

The potential of ESCs as a source of β-cell surrogates was originally validated by *in vitro* studies, which applied a variety of distinct approaches to manipulate ESC cultures into insulin-producing cells. This was initially shown to be possible with mouse embryonic stem cells (mESCs), and later with human embryonic stem cells (hESCs). The differentiation strategies that were used include the following: (1) genetic manipulation; (2) permitting the spontaneous differentiation of ESCs; and (3) an ectodermal approach based on the initial selection of nestin positive cells. These initial strategies were non-ontogeny-based approaches, in that they did not follow the normal *in vivo* process of pancreatic development.

Genetic manipulation. Soria et al. [13] generated an insulin-secreting clone from mESCs by the transfection of undifferentiated mESCs with a gene construct that bestowed neomycin resistance under the control of the human insulin (*Ins*) gene promoter. The resulting transfected cells were grown as spherical clusters, known as embryoid bodies (EBs), in both suspension and adherent culture conditions. This was followed by a final incubation at the physiological glucose concentration of 5 mM, which was hypothesized to be crucial in the production of higher levels of insulin by the differentiated cells. This is because studies have shown that β-cells exposed to sustained high glucose concentrations exhibit loss of differentiation capabilities [14], an altering in cellular metabolism [15] and impairment of insulin secretion [16].

After growth of the transfected cells in glucose-containing media, the insulin content of the cells progressively increased to a level that was equivalent to approximately 90% of that found in mouse islets. The pattern of glucose-induced insulin secretion also resembled that of normal mouse islets, which indicated the achievement of regulated insulin secretion *in vitro*. Furthermore, the transfected cells could normalize blood glucose levels in streptozotocin-induced diabetic mice within 1 week after transplantation, and the mice increased body weight within 4 weeks. Therefore, while the insulin-producing cells were not characterized as β-cells by gene and protein expression analysis, this study was nonetheless significant as it was the first to provide evidence of the potential of ES-derived insulin-producing cells for the treatment of diabetes.

Spontaneous differentiation of ESCs. The potential of human ESCs as a prospective β-cell replacement therapy was initially demonstrated by Assady et al. [17], where analyses were made of spontaneously differentiated hESCs in both adherent and suspension culture conditions. Immunocytochemistry analysis of the resulting EBs revealed that 60–70% of the differentiated cells stained positively for insulin, while only 1–3% of the cells stained at maximum density. An enzyme immunoassay suggested that these cells secreted insulin, as demonstrated by significantly higher concentrations of insulin in the surrounding medium of the differentiated cells in comparison to the control samples. Gene expression analysis confirmed the expression of β-cell-related genes such as the critical transcription factors Pancreatic and duodenal homeobox 1 (*Pdx-1*) and Neurogenin 3 (*Ngn3*), which are important in the regulation of endocrine cell differentiation. The cells also expressed genes such as Glucose transporter 2 (*Glut2*) and islet-specific Glucokinase (*GCK*), which are essential in the maintenance of β-cell function and glucose-stimulated insulin secretion. This study, therefore, demonstrated the possibility that the complex differentiation pattern of hESCs should include a proportion of cells that possess the characteristics of β-cells including insulin synthesis and the expression of markers specific for β-cells.

The ectodermal approach. Preliminary investigations found that nestin—an intermediate filament present in neural cells—was also expressed in small groups of pancreatic cells thought to be islet precursors [18,19]. It was thus hypothesised that neural cells and pancreatic cells could both be derived from the same nestin-expressing precursor cells. This resulted in the development of a five-stage protocol by Lumelsky et al. (2001) [20] that involved the generation of highly enriched populations of nestin-positive cells from mouse ESCs. These nestin-positive progenitor cells were selected and induced to differentiate into insulin-producing cells by culture with various growth factors and supplements. The resulting differentiated cells expressed a number of definitive endoderm markers as well as the murine pancreatic β-cell markers *insulin I* and *insulin II*. This study was most significant, however, because confocal microscopy indicated the presence of islet-like cell clusters which stained positively for the pancreatic endocrine hormones glucagon and somatostatin. This suggested that the differentiated cells self-assembled into multicellular structures topographically similar to pancreatic islets found *in vivo*, which was the first demonstration of the *in vitro* production of islet-like structures from ESCs. Moreover, the cells in some cases were glucose-responsive [21]. Nevertheless, the applicability of this protocol was limited by the inefficiency of differentiation, in that only 10–30% of the differentiated cells stained positively for insulin. Furthermore, the insulin content of the cells was much lower than that of normal pancreatic islet cells and upon transplantation, the cells failed to correct hyperglycaemia in streptozotocin-induced-diabetic mice.

LIMITATIONS IN NON-ONTOGENY BASED DIFFERENTIATION STRATEGIES

The crucial limitation in each of the differentiation approaches examined here is the failure to correlate the expression of insulin with an endodermal origin—the lineage from which β-cells are derived. This has particular implications for the study by Lumelsky et al. [20] since it was based on the selection of cells that express the neuroectodermal marker, nestin. A number of studies have since questioned the reliability and validity of the results provided by them as they have not been able to reproduce the data. It was discovered that insulin from the culture medium could be adsorbed to the surface of the cells and this may have given a false positive stain as determined by immunocytochemistry [22]. Another application of the protocol failed to detect the presence of storage granules that are characteristic of β-cells, and the expression of key β-cell gene markers [23]. Furthermore, insulin-positive cells often did not stain with an antibody for C-peptide, which is a by-product of de novo insulin synthesis and is secreted on an equimolar basis with insulin from β-cells [24]. Those cells that did stain for insulin and the pancreatic transcription factor Pdx-1 also stained for neural markers and nestin. This led to the conclusion that the approach by Lumelsky et al. actually resulted in the development of neural cells and not β-cells [25]. Therefore, while it was originally believed that nestin was expressed in β-cells, the role of nestin-expressing cells in the development of the pancreas is now strongly disputed and the ectodermal approach to differentiation has been abandoned [26, 27]. Indeed, it has been suggested that until proper characterisation is completed to prove otherwise, the insulin-producing cells derived by the spontaneous differentiation of hESCs or from genetic manipulation may also be of the neuroectodermal origin that can both produce and secrete low levels of insulin. The ability of neurons to express insulin has been observed in the developing nervous system in transgenic mice, where insulin expression was detected by the expression of the SV40 T-antigen under the control of the insulin promoter [28].

The failure of ESCs to produce true pancreatic β-cell surrogates from a non-ontogeny based differentiation strategy prompted the development of a new approach, which was to direct the differentiation of hESCs following the *in vivo* process of pancreatic development, also known as an ontogeny-based strategy. To successfully conform to an ontogeny-based strategy, a comprehensive understanding of foetal pancreatic development is required. The understanding of foetal pancreatic development was initially generated from studies of mouse organogenesis and has been complemented with data from the human foetal pancreas.

Foetal Pancreatic Development

The pancreas is initially derived from the definitive endoderm germ layer (Fig. 9.3), which consists of precursor cells that later give rise to the epithelial lining of the respiratory and digestive tracts, as

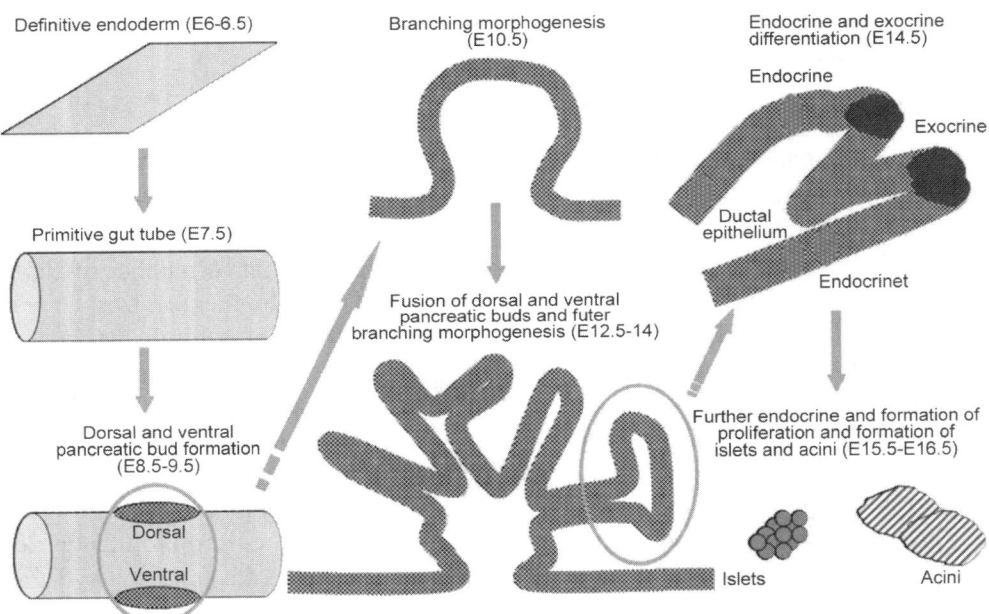

► **Fig. 9.3** *A basic schematic summary of fetal pancreatic development in the mouse. Development initiates from the definitive endoderm germ layer which appears as a flat sheet of cells on embryonic day 6–6.5 (E 6–6.5). This sheet of cells further develops into the primitive gut tube by E 7.5. Between E 8.5 and E 9.5, the dorsal and ventral pancreatic buds form from a population of Pdx-1 expressing epithelial cells of the gut tube. The pancreatic epithelium of the two buds then undergoes extensive proliferation and branching morphogenesis to form the basis of the complex ductal network of the pancreas (E 10.5). Fusion of the dorsal and ventral pancreatic buds forms the definitive pancreas, and after further branching morphogenesis (E 12.5–14.5), exocrine and endocrine differentiation begins, which results in the formation of the exocrine ducts and acini and the endocrine islets of Langerhans (E 15.5–16.5). (For colour figure see Plate 22).*

well as to the structural formation of gut derivatives such as the lungs, liver and gall bladder. It is formed during the gastrulation stage of embryogenesis, the process where pluripotent epiblast cells within the inner cell mass are allocated to the three primary germ layers. Epiblast cells are one of the two types of cells found within the inner cell mass that gives rise to foetal tissues. The other type of cell is the hypoblast, which gives rise to extraembryonic tissues such as visceral endoderm and parietal endoderm. These form the lining of the embryonic cavities, but are not known to contribute to the development of any foetal tissues.

The initiation of gastrulation is characterised by the appearance of the primitive streak in the developing embryo through which the epiblast cells ingress and undergo an epithelial-to-mesenchymal transition to emerge as either definitive endoderm or mesoderm. Studies have suggested that these cells develop from a common precursor, which is an intermediate cell types called mesendoderm. Pluripotent epiblast cells that do not pass through the primitive streak differentiate to form the embryonic ectoderm, which gives rise to neuronal tissues.

Fate-mapping studies of cultured mouse embryos has revealed that definitive endoderm begins to form at mouse embryonic day 6-6.5 (E 6-6.5) and is evidenced by a flat layer of cells that have an anterior–posterior (frontal and behind) pattern formation. At the end of gastrulation (E 7.5), this sheet of cells further develops to form the primitive gut tube, also known as the foregut in mammals, as a result of morphogenetic movements such as embryonic folding and midline fusion. The anterior-posterior axis along the early gut tube is patterned with specific domains that develop into various endoderm organ primordial such as the thyroid, liver and pancreas. The pancreas develops from the posterior foregut, where it emerges as buds (E 8.5–9.5) from a population of Pdx-1-expressing epithelial cells at the dorsal (back) and ventral (front) sides of the gut tube.

The pancreatic epithelium of the two buds then undergoes extensive proliferation and branching morphogenesis to form the basis of the complex ductal network of the pancreas (E 10.5). This results in the formation of two primordial pancreas organs (E 12.5) that are primarily composed of an undifferentiated ductal epithelium that represents the progenitor cells to the pancreatic exocrine and endocrine tissue. Between E 13 and E 14, continued growth, branching and fusion of the dorsal and ventral pancreatic buds produce a single organ, the definitive pancreas.

Exocrine and endocrine cell differentiation from the ductal epithelium begins from E 14.5 and is largely dependent on signals from the surrounding tissue called the mesenchyme. By E 15.5, acini, which are exocrine cell clusters, are clearly distinguishable from duct cells. From E 14, the endocrine cells, which are arrayed as single cells within the ductal epithelium, undergo extensive proliferation. The nascent endocrine cells then migrate from the branched epithelium into the surrounding mesenchyme to form the islets of Langerhans (E 16.5). The hormone-expressing endocrine cells within the islets undergo further differentiation to a mature functional state. For β-cells, this involves the ability for glucose-dependent insulin release. The islets are not fully formed until shortly before birth on E 18–19 and additional remodelling and maturation continues to take place 2–3 weeks after birth.

Development of the pancreas in humans is similar to that in the mouse, but understandably it takes considerably longer time [29]. The foetal pancreas develops from two diverticula, which appear as buds on the dorsal and ventral surfaces of the primitive gut at the fourth week of gestation. These buds meet and fuse during the seventh week of gestation, with endocrine tissue identified from the eighth week and exocrine tissue only after the twelfth week. Initially, β- and other endocrine cells are scattered throughout the pancreas as either single cells or in islet-like cell clusters, the dominant cell of which is the pancreatic precursor. By 16 weeks of gestation, islets can be identified in the pancreas in several forms. These are (a) mantle islets, with β-cells in the centre and the other endocrine cells forming a mantle around them; (b) bipolar islets, where β-cells are located at one pole and the other endocrine cells at the other pole; and (c) mature islets, where all endocrine cells are mixed.

Although foetal endocrine cells can initially express multiple endocrine hormones, in time they will only secrete one which characterizes their designated cell type. Similar to the adult β-cell, the foetal β-cell does possess the machinery required to synthesize, process, and store insulin. Unlike its adult counterpart, however, the foetal β-cell is unable to secrete insulin acutely in response to glucose. This is in contrast to its ability to secrete insulin in response to other β-cell stimuli such as agents that increase levels of cyclic AMP and Ca^{2++}, activate protein kinase C, amino acids including arginine and leucine, and cholinergic factors. Moreover, insulin and C-peptide stored in foetal secretory granules are morphologically different to that stored in adult cells. The reason for the inability to secrete insulin acutely in response to glucose relates to an immaturity of one or more enzymes in the Krebs citric acid cycle, including those involved in NADH shuttles. Closer to birth, the β-cell begins to mature with enhanced responsiveness observed in the first week of neonatal life. It is not until 6 months of life, however, that the insulinogenic response to glucose is mature. Most exocrine development also occurs post-natally.

Ontogeny-Based Directed Differentiation of hESCs into Insulin-Producing Cells

In order to follow an ontogeny-based strategy, ESCs must be guided through the *in vivo* process of pancreatic development *in vitro*. Therefore, such an approach is heavily reliant on developmental biology and the *in vitro* replication of the biochemical signalling pathways and transcription factors that are activated and inhibited in pancreatic organogenesis *in vivo*. This can be achieved by the supplementation of a cocktail of growth factors and chemical inhibitors in the culture environment. As previously described, an hESC ontogeny-based differentiation protocol involves the replication of five key stages in pancreatic development (Fig. 9.4): (1) the generation of definitive endoderm from pluripotent cells; (2) development of the primitive gut tube from definitive endoderm; (3) production of Pdx-1-expressing pancreatic progenitor cells; (4) inducing the differentiation of pancreatic progenitor cells to endocrine cells; and (5) endocrine cell specification to mature insulin-producing β-cell surrogates. By following such an approach, it is hoped that mature insulin-producing cells can be derived that closely resemble natural β-cells in terms of their gene and protein expression, metabolism and secretory function, and proliferative capacity.

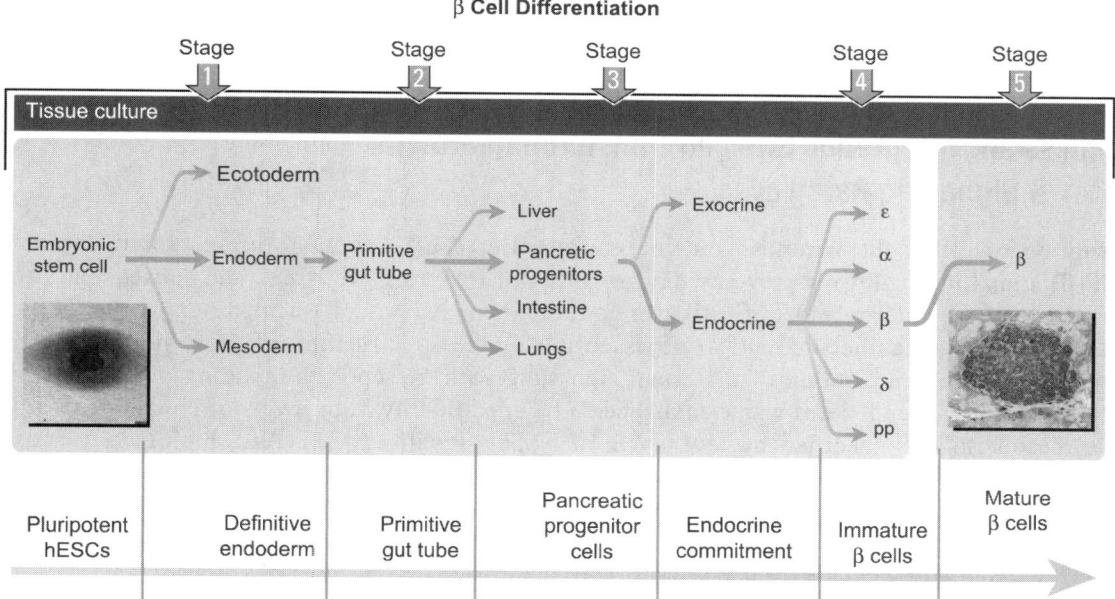

► **Fig. 9.4** *An ontogeny-based approach to differentiate hESCs into insulin-producing β-cell surrogates* **in vitro** *involves the replication of five key stages in pancreatic development* **in vivo**: *(1) the generation of definitive endoderm from pluripotent cells; (2) development of the primitive gut tube from definitive endoderm; (3) production of Pdx-1-expressing pancreatic progenitor cells; (4) inducing the differentiation of pancreatic progenitor cells to endocrine cells and (5) endocrine cell specification to mature insulin-producing β-cell surrogates. (For colour figure see Plate 23).*

THE GENERATION OF DEFINITIVE ENDODERM FROM PLURIPOTENT CELLS

Based on the knowledge elucidated from foetal pancreatic development, all insulin-producing cells are derived from the definitive endoderm germ layer. Therefore, the effective derivation of definitive endoderm is essential for the successful production of pancreatic β-cell surrogates from ESCs. While a specific marker for definitive endoderm has yet to be discovered, an accepted practice for the identification of definitive endoderm from ESCs is to examine for the presence of multiple markers of endoderm in general which include the transcription factors SRY box containing gene 17 (SOX17) and Forkhead box A2 (FOXA2). Since these transcription factors are also commonly expressed by extraembryonic tissues such as visceral endoderm, it is also necessary to investigate for the absence of extraembryonic endoderm markers such as alpha-fetoprotein (AFP), thrombomodulin (THBD) and SRY box containing gene 7 (SOX7). Genetic analyses in mice have disclosed several biochemical pathways that are involved in the generation of definitive endoderm, as discussed further.

Wingless and Integration Site and Transforming Growth Factor-β Signalling Pathways

Disruption of either the wingless and integration site (Wnt) or transforming growth factor-β (TGF-β) signalling pathways prevents the formation of the primitive streak and subsequently the derivation of mesoderm and definitive endoderm in mice. Nodal—a member of the TGF-β superfamily [30]—has been found to be essential for the specification of germ lineage during gastrulation through activation of the Smad signalling pathway upon the binding of Nodal to its receptor. Definitive endoderm was established to be specified by high levels of Nodal signalling, while mesoderm was specified by low levels of Nodal signalling [31]. As a biologically active source of the Nodal protein is not available however, activin A—another member of the TGF-β superfamily—can be used as a substitute because it is an analogue of Nodal and can bind to all but one of the same receptors as Nodal. Activin A can thus mimic Nodal activity by triggering similar intracellular signalling events [32].

Based on these analyses, the first ontogeny-based differentiation protocol involved the differentiation of mouse ESCs into definitive endoderm via the initial sorting of cells that express Brachyury [33]—an early primitive streak and mesendoderm marker. This was followed by the exposure of Brachyury-positive cells to activin A in serum-free conditions, which was able to effectively derive definitive endoderm as determined by the expression of the endoderm markers SOX17 and FOXA2. However the efficiency of definitive endoderm formation was not clear, because the principle marker that was used at the single cell level, FOXA2, is also expressed in axial mesoderm. Importantly though, this study proved that an ontogeny-based strategy was a plausible approach to derive lineage-specific cells.

These findings were translated to hESCs by the demonstration that the *in vitro* differentiation of definitive endoderm from hESCs could be efficiently achieved through the exposure of pluripotent hESCs to high concentrations of activin A in serum-depleted media [34]. This treatment resulted in a temporal sequence of gene expression that was similar to the transitions that occur during vertebrate gastrulation to derive definitive endoderm. That is, peak expression of primitive streak markers such as the gene Mix-1 homeobox-like 1 (*MIXL1*) early in the course of treatment, followed by mesendoderm markers such as *Brachyury* and high expression of the definitive endoderm markers *SOX17* and *FOXA2*. At least 80% of the differentiated cell population expressed SOX17 protein with no detectable immunoreactivity for the extraembryonic endoderm markers AFP and THBD. Low levels of expression of multiple markers of mesoderm, ectoderm and trophectoderm were detected, which suggested that a small proportion of cells differentiated along these lineages. Nonetheless, this study was significant due to the efficiency of hESC differentiation into definitive endoderm, which has been generally considered a 'successful approach to achieving gastrulation in a dish' [35]. The effective derivation of definitive endoderm as a result of this treatment also suggested that there are similarities in the signalling pathways between humans and mice. An improvement to this protocol was made by shortening the activin A treatment period and the addition of Wnt3a during the first day of activin exposure. These modifications increased the efficiency of mesendoderm specification and the synchrony of definitive endoderm formation [36], as the addition of Wnt3a aimed to promote the development of the primitive streak.

The Phosphatidylinositol-3 Kinase Signalling Pathway

It has since been shown that the requirement for serum depletion to induce the formation of definitive endoderm is in order to suppress the phosphatidylinositol 3-kinase (PI3-K) signalling pathway, the activation of which is important in the self-renewal of cells and in the maintenance of hESC pluripotency and survival. This was substantiated by an investigation which found that the addition of a chemical PI3-K inhibitor with lower concentrations of activin A was as effective at deriving definitive endoderm from hESCs as established protocols that require the addition of high concentrations of activin A in serum-depleted conditions [37]. While the mechanism by which PI3-K antagonizes hESC differentiation remains unknown, the agonists that are responsible for the activation of PI3-K signalling were identified to be insulin and insulin-like growth factor present in the supplements commonly used in hESC culture media.

A Potential Role of Sodium Butyrate and Bmp4

Both sodium butyrate and bone morphogenetic protein 4 (Bmp4) have been separately implicated as having a potential role in the formation of definitive endoderm. This was the result of findings that the treatment of hESCs with sodium butyrate and activin A [38], as well as Bmp4 and activin A [39], induced higher levels of endodermal gene expression and Pdx-1 expression in comparison to treatment with activin A alone. The role of Bmp signalling in early endoderm regionalisation and pancreas development remains to be elucidated. The means by which sodium butyrate influences endodermal differentiation also remains unknown. It has been suggested, however, that sodium butyrate may serve a function in hepatocyte rather than pancreatic differentiation, both of which require the initial formation of definitive endoderm [40]. This was based on research which has shown that sodium butyrate contributes to more homogeneous hepatocyte differentiation from ESCs [41, 42]. Indeed, further differentiation of definitive endoderm derived by treatment with activin A and sodium butyrate has successfully resulted in the efficient derivation of hepatocytes from hESCs [40].

These studies have established the absolute requirements for the *in vitro* differentiation of ESCs into definitive endoderm: high levels of Nodal signalling as mimicked by activin A and the inhibition of the PI3-K signalling pathway through the addition of chemical inhibitors or the culture of cells in serum-depleted conditions. Cellular differentiation may be further enhanced by the additional supplementation of exogenous growth factors such as sodium butyrate and Bmp4 (Fig. 9.5). The discovery of the mechanisms by which these additional factors influence differentiation will further enhance current knowledge of pancreatic development, and hence provide extra tools to refine the procedures used in stem cell differentiation. Nonetheless, the generation of definitive endoderm from ESCs-the first pivotal step in pancreatic differentiation-is now reasonably well established, with multiple studies able to successfully and effectively derive definitive endoderm based on these principles [36–39, 43–46].

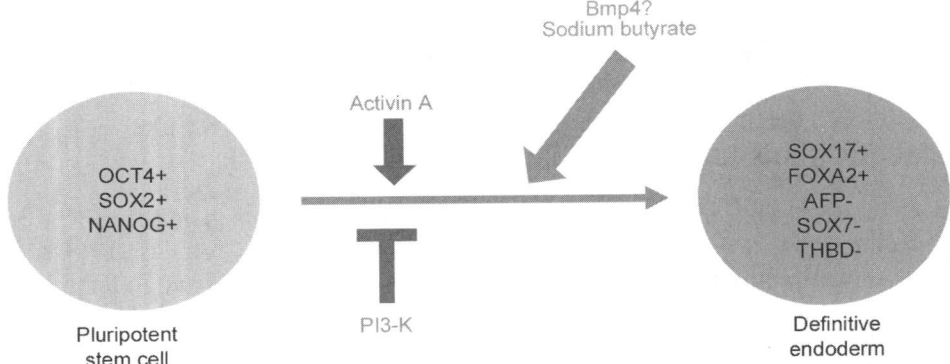

▶ **Fig. 9.5** *The absolute requirements for the* **in vitro** *differentiation of human embryonic stem cells into definitive endoderm are high levels of Nodal signalling as mimicked by activin A and the inhibition of the phosphatidylinositol 3-kinase (PI3-K) signalling pathway. Cellular differentiation may be further enhanced by the additional supplementation of exogenous growth factors such as sodium butyrate and Bmp4. The expression patterns of key genes, which are characteristic of pluripotent stem cells and definitive endoderm are noted. (For colour figure see Plate 23).*

DEVELOPMENT OF THE PRIMITIVE GUT TUBE FROM DEFINITIVE ENDODERM

The *in vitro* replication of the development of the primitive gut tube from definitive endoderm is a stage that has largely been ignored in current ESC-directed differentiation protocols. This is largely because little is known regarding how the primitive gut tube develops and how different domains along the gut tube become specialized to generate distinct organ primordial. In current hESC differentiation protocols, removal of activin A from the culture media following the development of definitive endoderm was found to allow the transition of definitive endoderm to a stage that resembled the primitive gut tube [36]. This was evidenced by the expression of the gut tube markers hepatocyte nuclear factor 1β (HNF1β) and (HNF4α) at both the messenger RNA (mRNA) and protein levels.

Studies of mouse pancreatic development, however, have suggested an important role of Notch signalling and Hedgehog signalling in guiding definitive endoderm through the primitive gut tube transition and towards a pancreatic cell fate. Specifically, the upregulation of Notch signalling in conjunction with an inhibition of Hedgehog signalling was found to be essential for pancreatic specification during this period in mice. This was further proved to be true during human development as the addition of fibroblast growth factor 10 (FGF10)—a Notch signalling promoter—and KAAD-cyclopamine (cyclopamine)—a Hedgehog signalling inhibitor—was sufficient to guide the hESC-derived definitive endoderm through the primitive gut tube transition [36]. The omission of this stage in most ontogeny-based differentiation protocols, however, has not appeared to affect the ability to subsequently obtain Pdx-1-expressing pancreatic progenitor cells from differentiating hESCs *in vitro*. A more detailed explanation of the role of Notch and Hedgehog signalling in pancreatic development is presented in the following section.

PRODUCTION OF PDX-1-EXPRESSING PANCREATIC PROGENITOR CELLS

One of the key steps in the successful production of β-cell surrogates from hESCs is the development of Pdx-1 expressing pancreatic progenitor cells. This is because Pdx-1, also known as Ipf-1, is a critical transcription factor that is recognised as the earliest specific marker of pancreatic endoderm. It is also considered to be a master regulator of pancreatic development. In mice, targeted disruption of the *Pdx-1* gene resulted in failure of the pancreas to develop (pancreatic agenesis) as it led to multiple deficiencies such as regression of the pancreatic buds and the inability of the pancreatic epithelium to respond to signals from the mesenchyme [47, 48]. Similarly, pancreatic agenesis in humans was found to be the result of a mutation that inactivates the *Pdx-1* gene [49].

The *in vitro* derivation of a high proportion of Pdx-1-expressing cells from hESCs is, therefore, a pre-requisite in further differentiation towards the pancreatic lineage. Activators of Pdx-1 include the transcription factors FOXA2 and HNF6. Homeobox HB9 (*HLXB9*) is also a key gene during this stage of development as it is a pivotal marker of the dorsal pancreatic bud. In foetal pancreatic development, the mechanism by which cells of the gut endoderm acquire a pancreatic fate involves signalling from the surrounding mesodermal and ectodermal cell populations such as the notochord, aorta and septum transversum. These biochemical signalling pathways, which include the retinoic acid (RA), Notch and Sonic hedgehog signalling pathways among others, can be mimicked *in vitro* to induce the differentiation of hESC-derived definitive endoderm through the primitive gut tube transition and into Pdx-1-expressing pancreatic progenitor cells (Fig. 9.6).

Retinoic Acid Signalling

Retinoic acid is an active metabolite of vitamin A derived by a modification of the vitamin by the retinaldehyde dehydrogenase (RALDH) enzymes. Of particular biological importance are two isomers of RA; *all-trans* RA (atRA) that can bind to both the RA receptors (RAR) and retinoid X receptors (RXR) to activate the RA signalling pathway; and 9-*cis* RA that specifically binds to RXR. Gain and loss of function studies in vertebrate species including zebrafish, Xenopus and mouse have suggested a role of RA in specifying the pancreatic domain along the anterior-posterior axis of the gut [50–52]. In addition, RA signalling is required for the generation of Pdx-1-expressing cells and in the development of the dorsal pancreas in mice as evidenced from transgenic mice with a targeted deletion of the RA synthesizing *Raldh2* gene [53]. Furthermore, due to the varying distribution and expression of RAR and RXR in the developing mouse pancreas [54], 9-*cis* and at RA are believed to exert different effects during pancreatic development. For example, the addition of atRA to the *in vitro* culturing of mouse embryonic pancreas resulted in the early expression of high levels of *Pdx-1*, the premature formation of endocrine cell clusters and suppression of branching morphogenesis and exocrine differentiation [55]. This suggested a specific role of atRA for endocrine differentiation in mice.

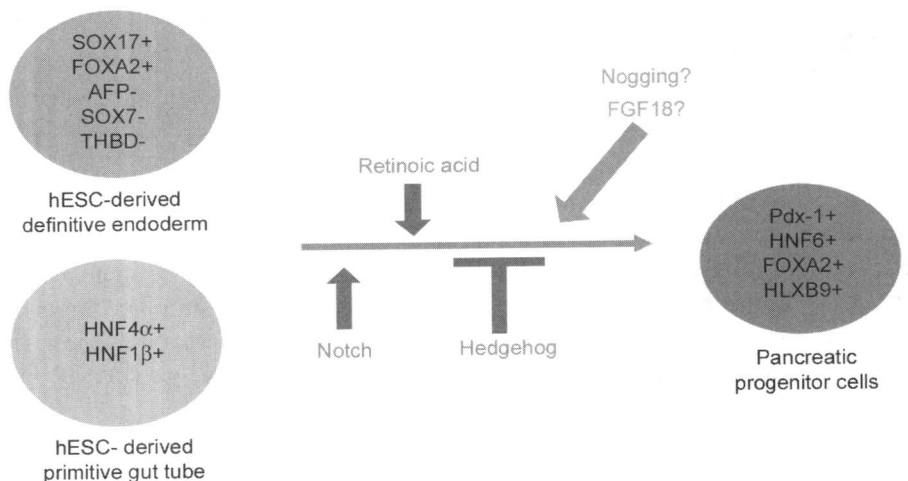

▶ **Fig 9.6** *Current approaches to derive pancreatic progenitor cells from either hESC-derived definitive endoderm or primitive gut tube revolve around the themes of activated Notch and retinoic acid signalling and inhibition of the Sonic hedgehog signalling pathway. Additional growth factors such as Noggin and FGF18 have also been supplemented at this stage of differentiation to generate pancreatic progenitor cells, but their role in differentiation is unknown. This suggests that an interaction between a variety of signalling pathways is required for the development of pancreatic progenitor cells from hESCs. The expression patterns of key genes which are the characteristic of hESC-derived definitive endoderm, primitive gut tube and pancreatic progenitor cells are noted. (For colour figure see Plate 24).*

On this basis, the addition of atRA by itself has been sufficient to induce the derivation of Pdx-1-expressing cells from mouse ESCs [56]. These cells also expressed the genes *FOXA2*, *HNF6*, *HNF4α* and *HNF1β*, which are characteristic of early pancreatic endoderm. However genes characteristics of later-stage pancreatic development, such as Ngn3, Neurogenic differentiation factor 1 (*NeuroD*), NK2 transcription factor related locus 2(*Nkx2.2*), and Ins, were not detected. This indicated that while early pancreatic endoderm cells could be produced by the addition of RA to mESCs, other signalling factors may be required to drive further pancreatic differentiation.

The results from mESC differentiation were translated to human ESCs with the finding that the addition of atRA to either hESC-derived definitive endoderm [46] or gut tube endoderm [36, 57] could induce the production of Pdx-1-expressing cells. These cells also expressed the dorsal pancreatic gene *HLXB9*, which provided proof of the necessity of RA signalling in early human pancreatic development. While hESCs have been found to express Pdx-1 in the absence of RA [36, 46], upon further differentiation these cells failed to express later-stage markers of pancreatic development [36]. Therefore, it is possible that in the absence of RA signalling, the ultimate fate of the Pdx-1-expressing cells derived may not be pancreatic, because while Pdx-1 is the earliest specific marker of pancreatic endoderm, its expression subsequently expands to include the posterior stomach, bile duct and duodenum.

Notch Signalling

The activation and inhibition of the Notch signalling pathway, which acts via the Hairy and enhancer of split 1 (*Hes1*) gene, requires strict and timely control during pancreatic development. This is because Notch signalling plays a critical regulatory role in the differentiation of pancreatic progenitor cells and in the subsequent determination of exocrine and endocrine cell fate. In the early stages of pancreatic organogenesis, studies in mice have shown that Notch activation is essential for two purposes. First, it is required for the proliferation and expansion of the pancreatic epithelium. This was determined by multiple studies which found that the addition of the Notch activator FGF10 to the *in vitro* culture of pancreas explants resulted in the maintenance and proliferation of Pdx-1-expressing pancreatic epithelial cells. In the absence of FGF10, these cells failed to proliferate [58–60]. Second, Notch activation is necessary to prevent the premature differentiation of pancreatic progenitor cells. This was observed in mice where the absence of *Hes1* activity resulted in a depletion of the pool of pancreatic precursor cells and the subsequent early development of the endocrine pancreas [61, 62].

The important role of Notch activation during the stage of Pdx-1 development has been acknowledged and incorporated in some of the current hESC differentiation protocols. Activators of Notch signalling which have been supplemented into hESC culture conditions include FGF10 [36] and keratinocyte growth factor (KGF) [57]. KGF—also known as FGF7—binds to the same receptor as FGF10 to activate the Notch signalling pathway and has also been reported to support the growth of the pancreatic epithelium while suppressing endocrine development [63–65].

Sonic Hedgehog Signalling

Extensive investigations have recognized the necessity to repress Sonic hedgehog (Shh) signalling during multiple stages of pancreatic development. For example, induction of Shh signalling in mouse pancreatic endoderm resulted in the loss of Pdx-1 expression [66] and severe alterations to pancreatic endocrine and exocrine differentiation [67]. This demonstrated the requirement for Shh inhibition in the early stages of pancreatic development. Moreover, investigations of transgenic mice in which the Shh gene was under the control of the Paired box 4 (*Pax4*) promoter showed a severe reduction in pancreatic mass with 80% loss of islet cells, 90% loss of acinar cells and the appearance of dilated ducts [68]. As *Pax4* is a gene characteristic of early endocrine progenitor cells, this study, therefore, also showed the requirement for Shh inhibition in the later stages of pancreatic development.

The repression of Shh signalling *in vivo* is provided by signals from the notochord following contact with the dorsal endoderm [69] as well as signalling from the endothelia of the dorsal aorta [70]. These signals can be mimicked *in vitro* by the combination of FGF2 and activin B [69]. Based on this, FGF2 has been added to the hESC culture of activin A induced definitive endoderm in an attempt to imitate notochord signalling *in vitro* for the development of Pdx-1-expressing cells [38]. From this approach, up to 25% of cells expressed Pdx-1 protein. Although it is important to note that while the only difference between activin A and activin B is in their dimer subunits, activin B has been identified to be the more potent inducer of *Pdx-1* expression in hESCs in comparison to activin A [71]. Another growth factor, cyclopamine, has been found to permit expansion of the area with Hedgehog inhibition leading to pancreatic differentiation in a larger portion of the foregut endoderm [72]. Other hESC differentiation protocols have thus utilized this factor to directly inhibit Shh signals during the stage of Pdx-1 development [36,57], which has resulted in the majority of cells expressing Pdx-1 protein [57].

Other Signalling Pathways

Current approaches to derive Pdx-1-expressing cells from hESCs revolve around the themes of RA signalling, Notch activation and Shh inhibition. Additional growth factors, however, have also been supplemented at this stage of hESC differentiation including other members of the FGF family such as FGF18 [39] and the TGF–β inhibitor and BMP antagonist, Noggin [38, 57]. Using a combination of growth factors to induce Pdx-1 expression suggests that an interaction between a variety of signalling pathways is required for the development of pancreatic progenitor cells from hESCs. Further research is necessary to define these interactions in order to maximize the efficiency of generating Pdx-1-expressing cells from hESCs.

DIFFERENTIATION OF PANCREATIC PROGENITOR CELLS TO ENDOCRINE CELLS

Following the production of Pdx-1-expressing pancreatic progenitor cells, the next step to differentiation is the induction of these cells to an endocrine cell fate (Fig. 9.7). This can be identified by the expression of Ngn3, the central intrinsic regulator and marker of endocrine specification [73, 74]. Ngn3 subsequently initiates the expression of a cascade of transcription factors such as Islet-1 (Isl1) and NeuroD that directs endocrine pancreas and expression of the genes, which encode all five endocrine cell types. The genes encoding β-cell specification include Nkx2.2, NK6 transcription factor related locus 1 (*Nkx6.1*) and the re-expression of *Pdx1* and *Pax4*.

However, in developmental biology there is little information regarding the mechanisms by which pancreatic progenitor cells assume an endocrine cell fate. Therefore, in current hESC differentiation strategies, the choice of growth factors to induce endocrine differentiation varies widely between protocols. Some of the growth factors, however, appear to have been supplemented based on their effect in the culture of pancreas explants. Exendin-4, for example, has been found to cause the growth and differentiation of human foetal β-cells from undifferentiated precursor cells *in vivo* [75] and porcine foetal β-cells *in vitro* [76]. It has also been found to accelerate the functional maturation of foetal β-cells as evidenced by their glucose stimulated insulin secretion [75]. Another factor called nicotinamide has been incorporated in a number of differentiation strategies [38, 46] as it has been identified as a potent inducer of endocrine

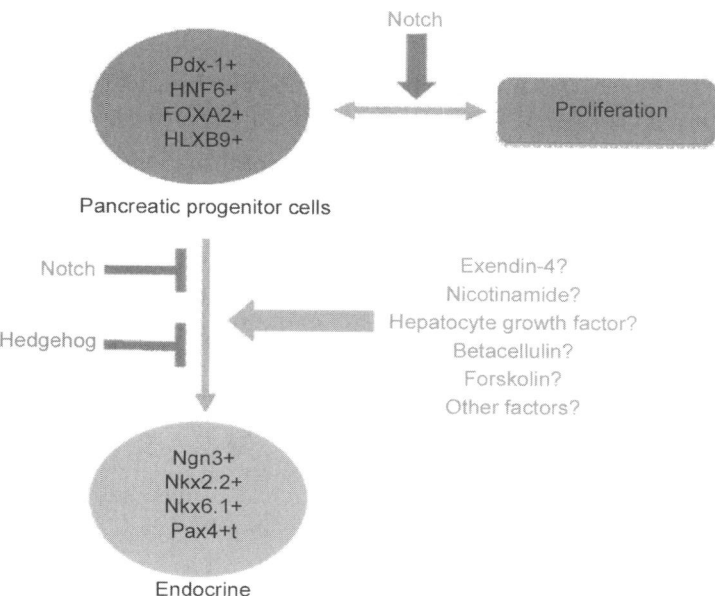

▶ **Fig. 9.7** *Activation of Notch signalling can permit the proliferation of pancreatic progenitor cells while inhibition of Notch and Sonic hedgehog signalling can induce endocrine differentiation. Aside from this, little is known in developmental biology regarding the mechanisms by which pancreatic progenitor cells assume an endocrine cell fate. In current hESC differentiation strategies, the choice of growth factors to stimulate endocrine differentiation varies widely between protocols. The expression patterns of key genes, which are the characteristic of pancreatic progenitor cells and differentiation into endocrine cells, are noted. (For colour figure see Plate 25).*

differentiation. In the culture of human foetal pancreatic islet cells, the addition of nicotinamide resulted in enhanced expression of the genes encoding all of the endocrine hormones, increased insulin content and insulin secretion in response to glucose [77]. In addition, the Notch signalling inhibitor DAPT has been used at the stage of endocrine differentiation of hESCs [36]. This was on the basis of studies that have found that Notch inhibition resulted in the rapid expression of *Ngn3* and is necessary to induce the differentiation of pancreatic progenitors into endocrine cells at the expense of exocrine [61, 78, 79]. A variety of other growth factors have been applied to current hESC strategies for the induction of endocrine differentiation. These include insulin-like growth factor I and II, betacellulin, hepatocyte growth factor and forskolin among others, which have resulted in obtaining some insulin-producing cells from hESCs.

In 2006, the first complete *in vitro* differentiation of hESCs into insulin-producing cells following an ontogeny-based protocol was described by D'Amour et al. [36]. The publication of this protocol, despite yielding only 7.3% insulin—positive cells, was an important milestone because it was the first to convert hESCs into endocrine cells that synthesised all of the pancreatic hormones-insulin, glucagon, somatostatin, PP and ghrelin. It was also the first approach to differentiate cells with an insulin content similar to foetal, but not adult β-cells. In addition, the detection of structures similar to secretory granules further suggested the formation of β-cell surrogates.

A number of other ontogeny-based hESC differentiation protocols have since been developed, which have shown a gene expression profile consistent with pancreatic development, and the development of insulin as well as other hormone-producing cells by immunocytochemistry [38,39,46]. However, a limitation that continues to persist is a low yield of insulin-producing cells. Furthermore, while C-peptide positive cells have been detected in differentiated cell populations, a key drawback to current protocols including the initial publication in 2006 is the inability of the differentiated cells to secrete insulin in response to an acute challenge with glucose. This is a function characterizing an adult β-cell and thus showing that the insulin-producing cells which have been derived from current protocols are immature β-cell surrogates. Further elucidation of the developmental biology behind endocrine differentiation and β-cell maturation will be required to refine the later stages of current differentiation strategies to derive mature insulin-producing β-cell surrogates from hESCs *in vitro*.

ENDOCRINE CELL SPECIFICATION TO MATURE INSULIN-PRODUCING β-CELLS

While the final stage in the development of β-cell surrogates from hESCs has yet to be achieved *in vitro*, it has been achieved *in vivo* [57]. In 2008, Kroon et al. demonstrated that β-cell maturation could occur *in vivo* by transplanting hESC-derived pancreatic progenitor cells which co-expressed PDX1, FOXA2, HNF6 and NKX6.1 into immunodeficient mice. While the process of maturation took several months to achieve, the human endocrine cells in the graft were able to respond acutely to a glucose stimulus. This approach of maturing β-cells *in vivo* has previously been used in the 1980s when immature foetal β-cells were transplanted into immunodeficient mice and developed glucose responsiveness [80]. Diabetic immunodeficient mice with the mature human β-cells derived from hESCs became normoglycaemic, with blood glucose levels in the human reference range as would be expected. Moreover, the resulting cells possessed secretory granules, which are the hallmark of a mature β-cell as they are required for the storage and rapid release of insulin upon exposure to glucose.

However, a negative effect of the *in vivo* maturation of the human pancreatic progenitor cells was the formation of teratomas in a small number of mice. This was likely due to the presence of residual pluripotent cells within the transplanted population. Further investigations will be required to determine how these cells can be sorted and removed. Nonetheless, this study was significant as it described the first complete differentiation of hESCs into not only insulin-producing cells, but glucose-responsive insulin-producing cells using the combination of an ontogeny-based *in vitro* strategy and an *in vivo* strategy for maturation. This has opened the door to the ultimate representation of the potential of hESCs in their ability to differentiate into insulin-producing β-cells and restore hypoglycaemia in primates.

OTHER APPROACHES TO EMBRYONIC STEM CELL DIFFERENTIATION INTO INSULIN-PRODUCING β-CELLS

While the ontogeny-based directed differentiation of ESCs into insulin-producing cells is currently the favoured approach to generate β-cell surrogates, a variety of alternative approaches are still under investigation. These include, but are not limited to, (1) the genetic modification of ESCs by transfection of plasmid vectors that contain pivotal pancreatic transcription factors; and (2) the co-culture of ESCs with pancreatic tissue for the development of insulin-producing cells.

Genetic Modification of ESCs

The introduction of essential pancreatic transcription factors into ESCs aims to alter the genetic profile of the cells to a β-cell phenotype. This technique was initially utilised by Soria et al. [13] for the creation of an insulin-expressing undifferentiated mESC clone as described in section *Non-Ontogeny-Based Differentiation of ESCs Into Insulin-Producing β-Cells* earlier. Genetic modification is also useful as it provides a method for the selection of specific populations of differentiating ESCs. As our understanding of critical pathways that dictate *in vivo* pancreatic organogenesis has broadened since the original application of this technique, transfection targets have since been refined to not only the insulin gene alone, but also to other factors that are pivotal in pancreatic development. Such strategies have resulted in the creation of transfected cells that display a phenotype that more closely resembles a normal pancreatic β-cell in comparison to the cells generated in the initial study.

In mouse ESCs, the introduction of the promoter sequences for *Ins* and *Nkx6.1*, and the cDNA for *Pax4* and *Pdx-1*, resulted in the development of some cells that produced insulin. The enhanced expression of key β-cell-related genes was also observed, with normalisation of blood glucose levels in some diabetic mice [81–83]. Encouraging results were also demonstrated by genetic modifications of human ESCs, where the over-expression of *Pax4* by stable transfection resulted in the expression of key β-cell-related markers and some *Ins* expression [84]. In all of these studies, however, the insulin-producing cells derived by genetic modifications differed from pancreatic β-cells by way of lower insulin content and a negligible or abnormal glucose dose response curve for insulin secretion. This indicates that the genetic interactions in the development and regulation of a β-cell are extremely complex and cannot be completely imitated by the alteration or introduction of a small number of genes. Nonetheless, these studies provide a proof of concept that genetic manipulation of ESCs can be used as a tool to assist in the generation of insulin-producing cells.

Co-culture of ESCs with Pancreatic Tissue

Another approach to ESC differentiation into insulin-producing cells is by the co-culture of ESCs either directly with foetal pancreatic extracts or indirectly with media conditioned by foetal pancreatic tissue. This is in an attempt to reproduce and extrapolate currently unknown intrinsic signals and interactions provided *in vivo* by associated tissues for the development of the pancreas and insulin-producing cells. Indeed, human ESCs co-transplanted with mouse pancreatic tissue have shown the ability to differentiate into insulin-producing cells [85]. Mouse ESCs, which were cultured in media conditioned by mouse embryonic pancreas, differentiated and expressed the pancreatic markers Pdx-1 and Ins. These cells also secreted C-peptide in response to stimuli and were capable of normalising blood glucose levels upon transplantation into diabetic mice [86]. Pdx-1 expression and the formation of insulin-producing cell clusters was also observed in mESCs co-cultured *in vitro* or co-transplanted *in vivo* with mouse foetal pancreatic tissue [87, 88]. These studies confirm that pancreatic tissues do release unknown soluble factors and extracellular signals that can stimulate the *in vitro* differentiation of ESCs into pancreatic endocrine cells. Further research on the co-culture of ESCs with pancreatic tissue or conditioned media may, therefore, lead to the identification of novel factors or signalling pathways that are involved in pancreatic development.

THE FUTURE

The ability of hESCs to differentiate into insulin-producing cells has been rapidly progressing over the past few years and has now reached a stage where insulin-producing cells can be derived *in vitro* with their potential to mature and respond to glucose shown *in vivo*. The possibility of using differentiated hESCs in a clinical context for the treatment of type 1 diabetes has become foreseeable, where such a period of time to achieve a clinical benefit on par with other new therapies such as the use of insulin-producing cells from organ donors. Before hESC-derived mature insulin-producing β-cells can be used in a clinical context, however, certain issues will need to be addressed. These include (1) elucidation of the biology to efficiently produce glucose responsive β-cells *in vitro*, (2) expansion of β-cell surrogates, (3) the risk of teratoma formation, (4) immunogenicity, (5) compliance with good manufacturing practices (GMP) and (6) regulatory and ethical approval.

Elucidation of the Biology to Efficiently Produce Glucose Responsive Insulin Producing β-cells *in vitro*

While the cues to produce definitive endoderm and Pdx-1-expressing pancreatic progenitor cells are increasingly becoming known, the biology to induce endocrine differentiation, β-cell specification and in particular, glucose responsiveness from insulin-producing cells remains to be fully determined. Further research is, therefore, required to elucidate the signalling pathways and interactions that specify pancreatic progenitor cells to mature β-cells, and this is the work currently being examined by a variety of research groups. For example, in early 2008, the gene expression profile of the developing human foetal pancreas at different stages of development was generated in an attempt to elucidate the critical biochemical signalling pathways currently unknown in pancreatic development [89]. In addition, experiments are being carried out with three-dimensional (3D) culturing techniques such as the use of 3D scaffolds to generate insulin-producing cells from hESCs that more closely mimic the *in vivo* cell-cell interactions *in vitro*. The ultimate goal is for hESC differentiation protocols to properly recapitulate the stepwise commitment of cells from definitive endoderm to a mature insulin-producing β-cell as seen during embryogenesis.

Expansion of β-cell Surrogates

Once the method to derive a mature insulin-producing β-cell *in vitro* is known, investigations will then need to focus on how this population of cells can be expanded on a large scale. This is because an abundant supply of surrogate β-cells will be required to be used as a cellular therapy for type 1 diabetes. While hESCs do provide an unlimited starting source of cells, their individual differentiation into insulin-producing β-cell surrogates may potentially create problems such as batch-to-batch variations. Therefore, the large-scale expansion of the insulin-producing β-cell surrogates may provide an ideal solution. Such a procedure may involve the culturing of cells in a bioreactor such as a perfusion bioreactor, though the method by which this could be achieved will need to be perfected.

The Risk of Teratoma Formation

A real concern with the use of hESCs in a clinical context is that any residual cells which are not terminally differentiated may develop into teratomas upon transplantation into a patient. Methods will, therefore, need to be developed that can effectively sort any remaining pluripotent stem cells from terminally differentiated insulin-producing β-cells prior to transplantation.

Immunogenicity

Upon transplantation of any hESC-derived cells into the human body, an issue with immunogenicity remains as the transplantation of any foreign cells may trigger an immune response from the host, which leads to the rejection of the graft. A variety of methods are currently being investigated to overcome the possible immune rejection of hESC-derived cellular products. These include microencapsulation, the encasing of cells in micro-sized capsules prior to transplantation [90], where restrictions on the pore size of the capsules permits essential nutrients to be transported in and out of the capsule but does not allow the passage of immune cells. Furthermore, advancements in somatic cell nuclear transfer, from which patient-specific hESC lines can be derived, may also overcome the issue with immunogenicity. Most recently, the discovery of induced pluripotent stem cells (iPS cells) has beneficial implications from both an immunological and ethical perspective. This is because iPS cells are skin cells which can be reprogrammed to an embryonic state with a phenotype that is essentially the same as hESCs. iPS cells may, therefore, also lead to the development of patient-specific cell lines. Further investigations, however, will still be needed from all of these methodologies to address the issues of long-term cell stability, differentiation efficiency, tumour formation and immunogenicity.

Compliance with Good Manufacturing Practices

For regulatory approval of a cellular therapy from hESCs, the means by which the β-cell surrogate product is derived will need to comply with current GMP. This includes the initial creation of the hESC lines and the steps taken throughout the differentiation process. A small number of GMP-hESC lines have been produced, but their use may be limited by the fear of the potential transmission of xenogeneic viruses to humans due to the utilisation of animal proteins in the derivation of the cell lines. The derivation of further GMP-hESC lines under xeno-free conditions is being examined.

Regulatory and Ethical Approval

Regulatory approval for the use of hESCs in medical research differs between different countries. In the United Kingdom, India, Australia, Israel and Singapore for example, hESCs can be derived and used in laboratory experimentation. In some countries such as Italy, Poland and the state of Nebraska in the USA however, neither the derivation nor the use of hESCs is permitted. A number of other countries have intermediate policies. In Germany for example, hESCs may not be derived, but can be used for non-commercial research. In those countries permitting the derivation of hESCs, most, but not all, advise that only embryos which are in excess of those required for *in vitro* fertilization may be used. Permission to use these embryos is required from those who provided the eggs and sperm.

The creation of hESCs by somatic cell nuclear transfer is permitted in some countries including the United Kingdom, Australia, Sweden, Israel, and China. As yet however, no cell lines have been created using this technology although embryos have. The creation of iPS cells is likely to be accepted in most countries as the pluripotent cells are derived from somatic cells alone. Whether the derivation of β-cells from either of these pluripotent cells will be possible, and if so, how they will compare to those derived from the more traditionally derived hESCs remains to be determined.

CONCLUSION

When the first human ESCs were derived in 1998, it was hoped that these cells might eventually be used for drug screening and in developmental biology to study human and disease development. In particular however, it was hoped that these cells might be able to differentiate into mature β-cells as a cellular replacement therapy for insulin-dependent diabetes (Fig. 9.8). Ten years later with the derivation of the first functional β-cell from an hESC, we are that much closer to achieving this goal. It would seem that the commencement of clinical trials can only be a matter of time, perhaps by 2012, if all the outstanding issues mentioned in section *The Future* earlier can be addressed. The time frame for commencing such trials will be accelerated if suitable resources are provided to those with expertise in this area and with a track record of translational research. Venture capitalists in particular can play a major role here. Therefore, it is quite reasonable to believe that over the next few decades, the use of exogenous insulin in the treatment of insulin-dependent diabetes will be replaced by surrogate β-cells derived from pluripotent stem cells.

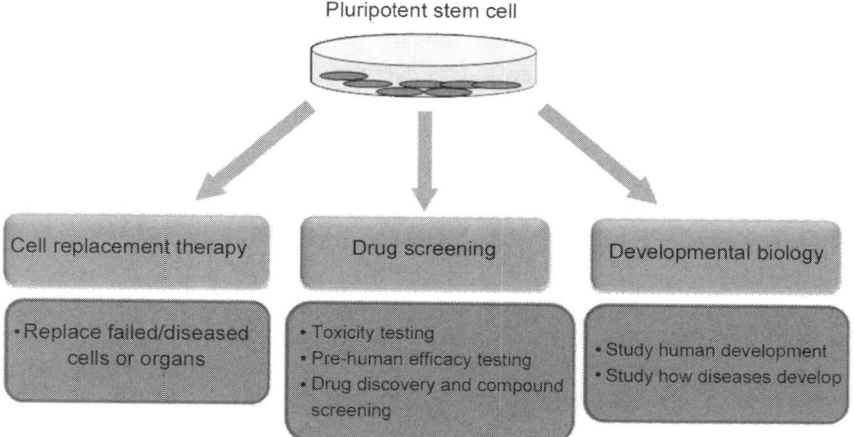

➤ **Fig 9.8** *Three potential uses of pluripotent stem cells in basic research and medicine: cellular replacement therapy to replace failed/diseased cells or organs; drug screening and developmental biology to study human development and disease development. (For colour figure see Plate 25).*

ACKNOWLEDGEMENTS

We are grateful for financial assistance from the following organizations in allowing us to pursue the goal of making mature β-cells from hESCs: Australian Foundation for Diabetes Research, Australian Centre for Stem Cell Research, Clive & Vera Ramaciotti Research Foundation, Diabetes Australia, Juvenile Diabetes Research Foundation, Rebecca L. Cooper Medical Research Foundation, and the Sydney Medical Research Foundation.

REFERENCES

1. Tuch B.E., M. Dunlop, and J. Proietto (2000). Diabetes research—a guide for postgraduates. Overseas Publishers Association. Singapore.
2. Robertson R.P., D.E. Sutherland, and K.J. Lanz (1999). Normoglycemia and preserved insulin secretory reserve in diabetic patients 10-18 years after pancreas transplantation. *Diabetes.* **48**: 1737–1740.
3. Smets Y.F., R.G. Westendorp, J.W. van der Pijl, F.T. de Charro, J. Ringers, J.W. de Fijter, and H.H. Lemkes (1999). Effect of simultaneous pancreas-kidney transplantation on mortality of patients with type-1 diabetes mellitus and end-stage renal failure. *Lancet.* **353**: 1915–1919.
4. Shapiro A.M., J.R. Lakey, E.A. Ryan, G.S. Korbutt, E. Toth, G.L. Warnock, N.M. Kneteman and R.V. Rajotte (2000). Islet transplantation in seven patients with type 1 diabetes mellitus using a glucocorticoid-free immunosuppressive regimen. *New Engl. J. Med.* **343**: 230–238.

5. Ryan E.A., B.W. Patyl, P.A. Senior, D. Bigam, E. Alfadhli, N.M. Kneteman, J.R.T. Lakey and A.M.J. Shapiro (2005). Five-year follow-up after clinical islet transplantation. *Diabetes.* **54**: 2060–2069.
6. Cheung A.T., B. Dayanandan, J.T. Lewis, G.S. Korbutt, R.V. Rajotte, M. Bryer-Ash, M.O. Boylan, M.M. Wolfe and T.J. Kieffer (2000). Glucose-dependent insulin release from genetically engineered K cells. *Science.* **290**: 1959–1962.
7. Lee H.C., S.J. Kim, K.S. Kim, H.C. Shin and J.W. Yoon (2000). Remission in models of type I diabetes by gene therapy using a single-chain insulin analogue. *Nature.* **408**: 483–488.
8. Dor Y., J. Brown, O.I. Martinez and D.A. Melton (2004). Adult pancreatic beta-cells are formed by self-duplication rather than stem-cell differentiation. *Nature.* **429**: 41–46.
9. Georgia S. and A. Bsushan (2004). Beta cell replication is the primary mechanism for maintaining postnatal beta cell mass. *J. Clin. Invest.* **114**: 963–968.
10. Kushner J.A., M.A. Ciemerych, E. Sicinska, L.M. Wartschow, M. Teta, S.Y. Long, P. Sicinski and M.F. White (2005). Cyclins D2 and D1 are essential for postnatal pancreatic beta-cell growth. *Mol. Cell Biol.* **25**: 3752–3762.
11. Oh S.H., Muzzonigro, T.M., Bae, S.H., LaPlante, J.M., Hatch, H.M. and B.E. Petersen (2004). Adult bone marrow-derived cells trans-differentiating into insulin-producing cells for the treatment of type I diabetes. *Lab. Invest.* **84**: 607–617.
12. Moriscot C., de Fraipont, F., Richard, M.J., Marchand, M., Savatier, P., Bosco, D., Favrot, M. and Benhamou, P.Y. (2005). Human bone marrow mesenchymal stem cells can express insulin and key transcription factors of the endocrine pancreas developmental pathway upon genetic and/or microenvironmental manipulation *in vitro*. *Stem Cells.* **23**: 594–603.
13. Soria B., E. Roche, G. Berna, T. Leon-Quinto, J.A. Reig and F. Martin (2000). Insulin-secreting cells derived from embryonic stem cells noramlize glycemia in streptozotocin-induced diabetic mice. *Diabetes.* **49**: 157–162.
14. Jonas J.C., A. Sharma, W. Hasenkamp, H. Ilkova, G. Patane, R. Laybutt, S. Bonner-Weir and G.C. Weir (1999). Chronic hyperglycemia triggers loss of pancreatic β-cell differentiation in an animal model of diabetes. *J. Biol. Chem.* **274**: 14112–14121.
15. Roche E., S. Farfari, L.A. Witters, F. Assimacopoulos-Jeannet, S. Thumelin, T. Brun, B.E. Corkey, A.K. Saha and M. Prentki (1998). Long-term exposure of beta-INS cells to high glucose concentrations increases anaplerosis, lipogenesis and lipogenic gene expression. *Diabetes.* **47**: 1086–1094.
16. Eizirik D.A., G.S. Korbutt and C. Hellerstrom (1992). Prolonged exposure of human pancreatic islets to high glucose concentrations *in vitro* impairs the β-cell function. *J. Clin. Invest.* **90**: 1263–1268.
17. Assady S., G. Manor, M. Amit, J. Itskovitz-Eldor, K.L. Skorecki and M. Tzukerman (2001). Insulin production by human embryonic stem cells. *Diabetes.* **50**: 1691–1697.
18. Hunziker E. and M. Stein (2000). Nestin-expression cells in the pancreatic islets of Langerhands. *Biochem. Biophys. Res. Comm.* **271**: 116–119.
19. Zulewski H., E.J. Abraham, M.J. Gerlach, P.B. Daniel, W. Moritz, V. Muller, M. Vallejo, M.K. Thomas and J.F. Habener (2001). Multipotential nestin-positive stem cells isolated from adult pancreatic islets differentiate ex vivo into pancreatic endocrine, exocrine and hepatic phenotypes. *Diabetes.* **50**: 523–533.
20. Lumelsky N., O. Blondel, P. Laeng, I. Velasco, R. Ravin and R. McKay (2001). Differentiation of embryonic stem cells to insulin-secreting structures similar to pancreatic islets. *Science.* **292**: 1389–1394.

21. Bai L., G. Meredith and B.E. Tuch (2005). Glucagon-like peptide-1 enhances production of insulin in insulin-producing cells derived from mouse embryonic stem cells. *J. Endocrinol.* **186**: 343–352.
22. Rajagopal J., W.J. Anderson, S. Kume, O.L. Martinez and D.A. Melton (2005). Insulin staining of ES cell progeny from insulin uptake. *Science.* **299**: 363.
23. Rajagopal J., W.J. Anderson, S. Kume, O.I. Martinez and D.A. Melton (2003). Insulin staining of ES cell progeny from insulin uptake. *Science.* **299**: 363.
24. Polonsky K.S., J. Licinio-Paixao, B.D. Given, W. Pugh, P. Rue, J. Galloway, T. Karrison and B. Frank (1986). Use of biosynthetic human C-peptide in the measurement of insulin secretion rates in normal volunteers and type I diabetic patients. *J. Clin. Invest.* **77**: 98–105.
25. Sipione S., A. Eshpeter, J.G. Lyon, G.S. Korbutt and R.C. Bleackley (2004). Insulin expressing cells from differentiated embryonic stem cells are not beta cells. *Diabetologia.* **47**: 499–508.
26. Humphrey R.K., N. Bucay, G.M. Beattie, A. Lopez, C.A. Messam, V. Cirulli and A. Hayek (2003). Characterization and isolation of promoter-defined nestin-positive cells from the human foetal pancreas. *Diabetes.* **52**: 2519–2525.
27. Selander L. and H. Edlund (2002). Nestin is expressed in mesenchymal and not epithelial cells of the developing mouse pancreas. *Mech. Dev.* **113**: 189–192.
28. Alpert S., D. Hanahan and G. Teltelman (1988). Hybrid insulin genes reveal a developmental lineage for pancreatic endocrine cells and imply a relationship with neurons. *Cell* 53
29. Tuch B.E. (1992) Clinical results of Transplanting foetal Pancreas, Cambridge University Press, Cambridge.
30. Kelly O.G., K.I. Pinson and W.C. Skarnes (2004). The Wnt co-receptors Lrp5 and Lrp6 are essential for gastrulation in mice. *Development.* **131**: 2803–2815.
31. Kaufman D.B. and W.L. Lowe, Jr. (2003). Clinical islet transplantation. *Curr Diab Rep* **3**: 344–350.
32. Vallier L., M. Alexander and R.A. Pedersen (2005). Activin/Nodal and FGF pathways cooperate to maintain pluripotency of human embryonic stem cells. *J. Cell. Sci.* **118**: 4495–4509.
33. Kubo A., K. Shinozaki, J.M. Shannon, V. Kouskoff, M. Kennedy, S. Woo, H.J. Fehling and G. Keller (2004). Development of definitive endoderm from embryonic stem cells in culture. *Development.* **131**: 1651–1662.
34. D'Amour K.A., A.D. Agulnick, S. Eliazer, O.G. Kelly, E. Kroon and E.E. Baetge (2005). Efficient differentiation of human embryonic stem cells to definitive endoderm. *Nat. Biotechnol.* **23**: 1534–1541. Epub 2005 Oct 1528.
35. Jensen J. (2007). Pathway decision-making strategies for generating pancreatic β-cells: systems biology or hit and miss- Curr. Opin. Endocrinol. *Diabetes Obesity.* **14**: 277–282.
36. D'Amour K.A., A.G. Bang, S. Eliazer, O.G. Kelly, A.D. Agulnick, N.G. Smart, M.A. Moorman, E. Kroon, M.K. Carpenter and E.E. Baetge (2006). Production of pancreatic hormone-expressing endocrine cells from human embryonic stem cells. *Nat. Biotechnol.* **24**: 1392–1401.
37. McLean A.B., K.A. D'Amour, K.L. Jones, M. Krishnamoorthy, M.J. Kulik, D.M. Reynolds, A.M. Sheppard, H. Liu, Y. Xu, E.E. Baetge and S. Dalton (2007). Activin A efficiently specifies definitive endoderm from human embryonic stem cells only when phosphatidylinositol 3-kinase signaling is suppressed. *Stem Cells.* **25**: 29–38.
38. Jiang J., M. Au, K. Lu, A. Eshpeter, G.S. Korbutt, G. Fisk and A.S. Majumdar (2007). Generation of insulin-producing islet-like clusters from human embryonic stem cells. *Stem Cells.* 1940–1953.

39. Phillips B.W., H. Hentze, W.L. Rust, Q.P. Chen, H. Chipperfield, E.K. Tan, S. Abraham, A. Sadasivam, P.L. Soong, S.T. Wang, R. Lim, W. Sun, A. Colman and N.R. Dunn (2007). Directed differentiation of human embryonic stem cells into the pancreatic endocrine lineage. Stem. *Cell. Dev.* **16**: 561–578.
40. Hay D.C., D. Zhao, J. Fletcher, Z.A. Hewitt, D. McLean, A. Urruticoechea-Uriguen, J.R. Black, C. Elcombe, J.A. Ross, R. Wolf and W. Cui (2008). Efficient differentiation of hepatocytes from human embryonic stem cells exhibiting markers recapitulating liver development *in vivo*. *Stem Cells.* **26**: 894–902.
41. Rambhatla L., C.-P. Chiu, P. Kundu, Y. Peng and M.K. Carpenter (2003). Generation of hepatocyte-like cells from human embryonic stem cells. *Cell Transplan.* **12**: 1–11.
42. Sharma N.S., R. Shikhanovich, R. Schloss and M.L. Yarmush (1994). Sodium-buyturate treated embryonic stem cells yield hepatocyte-like cells expressing a glycolytic phenotype. *Biotechnol Bioeng.* **94**: 1053–1063.
43. D'Amour K.A., A.D. Agulnick, S. Eliazer, O.G. Kelly, E. Kroon and E.E. Baetge (2005). Efficient differentiation of human embryonic stem cells to definitive endoderm. *Nat. Biotechnol.* **23**: 1534–1541.
44. Shim J.H., S.E. Kim, D.R. Woo, S.K. Kim, C.H. Oh, R. McKay and J.H. Kim (2007). Directed differentiation of human embryonic stem cells towards a pancreatic cell fate. *Diabetologia.* **50**: 1228–1238.
45. Shi Y., L.L. Hou, F. Tang, W. Jiang, W. Peigang, M. Ding and H. Deng (2005). Inducing embryonic stem cells to differentiate into pancreatic beta cells by a novel three step approach with activin A and all-trans retinoic acid. *Stem Cells.* **23**: 656–662.
46. Jiang W., Y. Shi, D. Zhao, S. Chen, J. Yong, J. Zhang, T. Qing, X. Sun, P. Zhang, M. Ding, D. Li and H. Deng (2007). *In vitro* derivation of functional insulin-producing cells from human embryonic stem cells. *Cell Research.* **17**: 333–344.
47. Offield M.F., J.L. Jetton, P.A. Labosky, M. Ray, R.W. Stein, M.A. Magnuson, B.L.M. Hogan and C.V.E. Wright (1996). PDX-1 is required for pancreatic outgrowth and differentiation of the rostral duodenum. *Development.* **122**: 983–955.
48. Holland A.M., M.A. Hale, H. Kagami, R.E. Hammer and R.J. MacDonald (2002). Experimental control of pancreatic development and maintenance. *Proc. Nat. Acad. Sci.* **99**: 12236–12241.
49. Stoffers D.A., N.T. Zinkin, V. Stanojevic, W.L. Clarke and J.F. Habener (1997). Pancreatic agenesis attributable to a single nucleotide deletion in the human IPF1 gene coding sequence. *Nat. Genet.* **15**: 106–110.
50. Chen Y., F.C. Pan, N. Brandes, S. Afelik, M. Solter and T. Pieler (2004). Retinoic acid signalling is essential for pancreas development and promotes endocrine at the expense of exocrine cell differentiation in Xenopus. *Develop. Biol.* **271**: 144–160.
51. Stafford D. and V.E. Prince (2002). Retinoic acid signalling is required for a critical early step in zebrafish pancreatic development. *Curr. Biol.* **12**: 1215–1220.
52. Molotkov A., N. Molotkova and G. Duester (2005). Retinoic acid generated by Raldh2 in mesoderm is required for mouse dorsal endodermal pancreas development. Develop. *Dynamics.* **232**: 950–957.
53. Martin M., J. Gallego-Llamas, V. Ribes, M. Kedinger, K. Neiderreither, P. Chambon, P. Dolle and G. Gradwohl (2005). Dorsal pancreas agenesis in retinoic acid-deficient Raldh2 mutant mice. *Develop. Biol.* **284**: 399–411.

54. Kadison A., J.H. Kim, T. Maldonado, C. Crisera, K. Prasadan, P. Manna, B. Preuett, M. Hembree, M. Longaker and G. Gittes (2001). Retinoid signalling directs secondary lineage selection in pancreatic organogenesis. *J. Ped. Surg.* **36**: 1150–1156.
55. Shen C.-N. , A. Marguerie, C.-Y. Chien, C. Dickson, J.M.W. Slack and D. Tosh (2007). All-trans retinoic acid suppresses exocrine differentiation and branching morphogenesis in the embryonic pancreas. *Differentiation.* **75**: 62–74.
56. Micallef S.J., M.E. Janes, K. Knezevic, R.P. David, A.G. Elefanty and E.G. Stanely (2005). Retinoic acid induces Pdx-1 positive endoderm in differentiating mouse embryonic stem cells. *Diabetes.* **54**: 301–305.
57. Kroon E., L.A. Martinson, K. Kadoya, A.G. Bang, O.G. Kelly, S. Eliazer, H. Young, M. Richardson, N.G. Smart, J. Cunningham, A.D. Agulnick, K.A. D'Amore, M.K. Carpenter and E.E. Baetge (2008). Pancreatic endoderm derived from human embryonic stem cells generates glucose-responsive insulin secreting cells *in vivo*. *Nat. Biotechnol.* **26**(4): 443–452.
58. Norgaard G.A., J.N. Jensen and J. Jensen (2003). FGF10 signaling maintains the pancreatic progenitor cell state revealing a novel role of Notch in organ development. *Develop. Biol.* **264**: 323–338.
59. Hart A., S. Papadopoulou and H. Edlund (2003). Fgf10 maintains notch activation, stimulates proliferation and blocks differentiation of pancreatic epithelial cells. Develop. *Dynamics.* **228**: 185–193.
60. Miralles F., P. Czernichow, K. Ozaki, K. Itoh and R. Scharfmann (1999). Signaling through fibroblast growth factor receptor 2b plays a key role in the development of the exocrine pancreas. *Proc. Nat. Acad. Sci.* **96**: 6267–6272.
61. Apelqvist A., H. Li, L. Sommer, P. Beatus, D.J. Anderson, T. Honjo, M. Hrabe de Angelis, U. Lendhal and H. Edlund (1999). Notch signalling controls pancreatic cell differentiation. *Nature.* **400**: 877–881.
62. Jensen J., E.E. Pedersen and P. Galante (2000). Control of endodermal endocrine development by Hes-1. *Nat. Genet.* **32**: 128–134.
63. Bottaro D.P., J.S. Rubin, D. Ron, P.W. Finch, C. Florio and S.A. Aaronson (1990). Characterisation of the receptor for keratinocyte growth factor: evidence for multiple fibroblast growth factor receptors. *J. Biol. Chem.* **265**: 12767–12770.
64. Elghazi L., C. Cras-Meneur, P. Czernichow and R. Scharfmann (2002). Role for FGFR2IIIb-mediated signals in controlling pancreatic endocrine progenitor cell proliferation. *Proc. Nat. Acad. Sci.* **99**: 3884–3889.
65. Ye F., B. Duvillie and R. Scharfmann (2005). Fibroblast growth factors 7 and 10 are expressed in the human embryonic pancreatic mesenchyme and promote the proliferation of embryonic pancreatic epithelial cells. *Diabetologia.* **48**: 277–281.
66. Hebrock M., S.K. Kim, B. St Jacques, A.P. MacMahon and D.A. Melton (2000). Regulation of pancreas development by hedgehog signalling. *Development.* **127**: 4905–4913.
67. Apelqvist A., U. Ahlgren and H. Edlund (1997). Sonic hedgehog directs specialised mesoderm differentiation in the intestine and pancreas. *Curr. Biol.* **7**: 801–804.
68. Kawahira H., D.W. Scheel, B.H. Smith, M.S. German and M. Hebrock (2005). Hedgehog signalling regulates expansion of pancreatic epithelial cells. *Develop. Biol.* **280**: 111–121.

69. Hebrok M., S.K. Kim and D.A. Melton (1998). Notochord repression of endodermal Sonic hedgehog permits pancreas development. Genes. *Develop.* **12**: 1705–1713.
70. Lammert E.C.O. and D.A. Melton (2002). Induction of pancreatic differentiation by signals from blood vessels. *Science.* **294**: 564–567.
71. Frandsen U., A. Dorte Porneki, C. Floridon, B.M. Abdallah and M. Kassem (2007). Activin B mediated induction of Pdx1 in human embryonic stem cell derived embryoid bodies. Biochem. Biophys. *Res. Comm.* **362**: 568–574.
72. Kim S.K. and D.A. Melton (1998). Pancreas development is promoted by cyclopamine, a Hedgehog signaling inhibitor. Proc. *Nat. Acad. Sci.* **95**: 13036–13041.
73. Gu G., J. Dubauskaite and D.A. Melton (2002). Direct evidence for the pancreaticl ineage: NGN3+ cells are islet progenitors and are distinct from duct progenitors. *Development.* **129**: 2247–2457.
74. Wilson M.E., D. Scheel and M.S. German (2003). Gene expression cascades in pancreatic development. *Mech. Dev.* **120**: 65–80.
75. Movassat J., G.M. Beattie, A.D. Lopez and A. Hayek (2002). Exendin 4 up-regulates expression of PDX1 and hastens differentiation and maturation of human foetal pancreatic cells. *J. Clin. Endocrinol. Metabolism.* **87**: 4775–4781.
76. Hardikar A.A., X.Y. Wang, L. Williams, J. Kwok, R. Wong, M. Yao and B.E. Tuch (2002). Functional maturation of foetal porcine β cells by glucagon-like peptide 1 and cholecystokin. *Endocrinology.* **143**: 3505–3514.
77. Otonkoski T., G.M. Beattie, M.I. Mally, C. Ricordi and A. Hayek (1993). Nicotinamide is a potent inducer of endocrine differentiation in cultured human foetal pancreatic cells. *J. Clin. Inves.* **92**: 1459–1466.
78. Rooman I., N. De Medts, L. Baeyens, J. Lardon, S. De Breuck, H. Heimberg and L. Bouwens (2006). Expression of the Notch signaling pathway and effect on exocrine cell proliferation in adult rat pancreas. *Am. J. Pathol.* **169**: 1206–1214.
79. Murtaugh L.C., B.Z. Stanger, K.M. Kwan and D.A. Melton (2003). Notch signaling controls multiple stels of pancreatic differentiation. *Proc. Nat. Acad. Sci.* **100**: 14920–14925.
80. Tuch B.E., A. Jones and J.R. Turtle (1985). Maturation of the response of human foetal pancreatic explants to glucose. *Diabetologia.* **28**: 28–31.
81. Blyszczuk P., J. Czyz, G. Kania, M. Wagner, U. Roll, L. St-Onge and A.M. Wobus (2003). Expression of Pax4 in embryonic stem cells promotes differentiation of nestin-positive progenitor and insulin-producing cells. *Proc. Nat. Acad. Sci.* **100**: 998–1003.
82. Leon-Quinto T., J.M. Jones, A. Skoudy, M. Burcin and B. Soria (2004). *In vitro* directed differentiation of mouse embryonic stem cells into insulin producing cells. *Diabetologia.* **47**: 1442–1451.
83. Miyazaki S., E. Yamato and J. Miyazaki (2004). Regulated expression of pdx-1 promotes *in vitro* differentiation of insulin producing cells from embryonic stem cells. *Diabetes.* **53**: 1030–1037.
84. Gee Liew C., N.N. Shah, S.J. Briston, R.M. Shepherd, C. Peen Khoo, M.J. Dunne, H.D. Moore, K.E. Cosgrove and P.W. Andrews (2008). PAX4 enhances beta-cell differentiation of human embryonic stem cells. *PLoS One.* **3**: e1783.
85. Brolen G.K., N. Heins, J. Edsbagge and H. Semb (2005). Signals from the embryonic mouse pancreas induce differentiation of human embryonic stem cells into insulin-producing β-cell-like clusters. *Diabetes.* **54**: 2867–2874.

86. Vaca P M.F., J.M. Vegara-Meseguer, J.M. Rovira, G. Berná, B. Soria (2006). Induction of differentiation of embryonic stem cells into insulin-secreting cells by foetal soluble factors. *Stem Cells*. **24**: 258–265.
87. Shiraki N., C. Lai, Y. Hishikari and S. Kume (2005). TGF-β signaling potentiates differentiation of embryonic stem cells to Pdx-1 expressing endodermal cells. *Genes. Cells*. **10**: 503–516.
88. Brolén G.K., N. Heins, J. Edsbagge and H. Semb (2005). Signals from the embryonic mouse pancreas induce differentiation of human embryonic stem cells into insulin-producing beta-cell-like cells. *Diabetes*. **54**: 2867–2874.
89. Sarkar S.A., S. Kobberup, R. Wong, A.D. Lopez, N. Quayum, T. Still, A. Kutchma, J.N. Jensen, R. Gianani, G.M. Beattie, J. Jensen, A. Hayek and J.C. Hutton (2008). Global gene expression profiling and histochemical analysis of the developing human foetal pancreas. *Diabetologia*. **51**: 285–297.
90. Dean S.K., Y. Yulyana, G. Williams, K.S. Sidhu and B.E. Tuch (2006). Differentiation of encapsulated embryonic stem cells after transplantation. *Transplantation*. **82**: 1175–1184.

10

MANIPULATION OF BIOMATERIAL INTERFACE BY BIOMIMETIC STRATEGIES: POSSIBILITIES OF EXPLORING STEM CELLS

K. Kaladhar, Chandra P Sharma*

Biosurface Technology Division, BMT Wing, Sree Chitira Tirunal Institute for Medical Science and Technology, Thiruvananthapuram, Kerala, India

INTRODUCTION

The interfacial modification of biomaterials using biomimetic strategies is an important area of research today. The biological interaction at the material-biology interface is initiated with ligand supplementation from the material surface. This is modulated with the adsorption of proteins and other biological agents present in the biological fluid; intern depends upon the surface chemical and topographical features. Thin solid films (TSF), mimicking biological cell membranes and proteins, are promising for the post synthetic surface modification of biomaterials or devices in the range of macro to nano level. In this chapter, we discuss about the exploration of thin film strategy with the help of stem cell technology for the interfacial manipulation of medical devices.

Regenerative therapeutic strategies are progressing with combination of drugs, devices, cells, scaffolds and other physical stimuli to restore the structural and functional proficiency of lost tissue due to degenerative diseases. The different methodologies are aimed at accelerated tissue in growth, and sequential functional restoration, through guided tissue regeneration. The immediate functional restoration is achieved through artificial implants, and subsequent guided tissue regeneration, is accepted as an effective methodology. However, often these strategies are failing due to immunological or inflammatory reactions at the implantation site resulting in lysis and

---

* drcpsharma@rediffmail.com

rejection of the implant. The reason for the failure varies with different clinical conditions like age, sex, method of implantation, sterility of the implant, disease state of the patient and depending upon the stay of the implant in the body. Stem cells by virtue of its stemness and low immune profile are considered as an ideal candidate for regenerative therapeutic applications.

In 2006, the Royal Melbourne Hospital [1] has performed the world's first implant of culture specialist stem cells (Adult stem cell technology, Mesoblast Ltd., Australia) into an orthopedic patient who suffered a broken femur 9 months ago which failed to heal, and reported that the approach is promising. This latest development of accelerated healing in a wound site with less immunological and inflammatory complications is an important development. This initiates a lot of expectations whether these "intelligent young cells" can be modulated for "interfacial modification of biomaterials and devices" for an "optimum" biological response. The first chapter of this book provides a brief introduction about the stem cells and the clinical opportunities of stem cell therapy. In this chapter particularly, we deal about the critical biological environment in an open wound and in a tissue degeneration/regeneration site. We attempted to review the following questions: (1) how the stem cells behave in this environment, (2) how the presence of an implant modulate the scenario, (3) and the problems to be addressed and its future prospectives.

STEM CELLS: ORIGIN AND EXPLORATION FOR TISSUE ENGINEERING AND REGENERATIVE MEDICINE

Briefly, stem cells can be divided into embryonic stem cells (ESCs) and adult stem cells (ASCs). Research using ESCs, derived from embryos, is in political controversy on moral and social grounds. The ASCs has been extensively exploited for the anticipated clinical potential of the stem cells. ASCs has been chiefly isolated from bone marrow (BM), and is available in niches of different connective tissue.

Friedenstein et al. [2] first reported that an adherent, fibroblast-like population of adult BM derived cells is osteogenic *in vitro*. These mesenchymal stem cells (MSC) were later isolated from adipose, cord blood [3], and foetal liver, blood, BM and lung [4]. It appears that MSCs reside within the connective tissue of various organs [5]. Since then, studies from different laboratories have demonstrated that this cell population is capable of differentiating into multiple connective tissue cell types like bone, cartilage, muscle, marrow stoma, tendon, fat and dermis at a clonal level [6], and demonstrated mesogenic process, as explained by Caplan [7].

Engraftment of a xenograft essentially needs the nieghbouring tissue regeneration. MSCs are proven candidates of connective tissue repair. Their co-existence with blood-forming stem cells and immune cell progenitors in the BM or in tissue stem cell niches is evident for its immune and inflammatory modulation potential. However, the implant site is catabolic and all the immune and inflammatory processes might have upregulated.

INFLAMMATION AT AN IMPLANT SITE

In the case of an implant, if the surface cues are bioactive, the implant will get integrated well into the nieghburing connective tissue. In the absence of surface cues for a preferred bioactive response, a wound-healing process with excessive fibroblastic and smooth muscle activity toward the formation of fibrous capsule, or rejection due to lysis, is observed in soft and hard tissues. The cellular changes with respect to time are explained in Fig. 10.1.

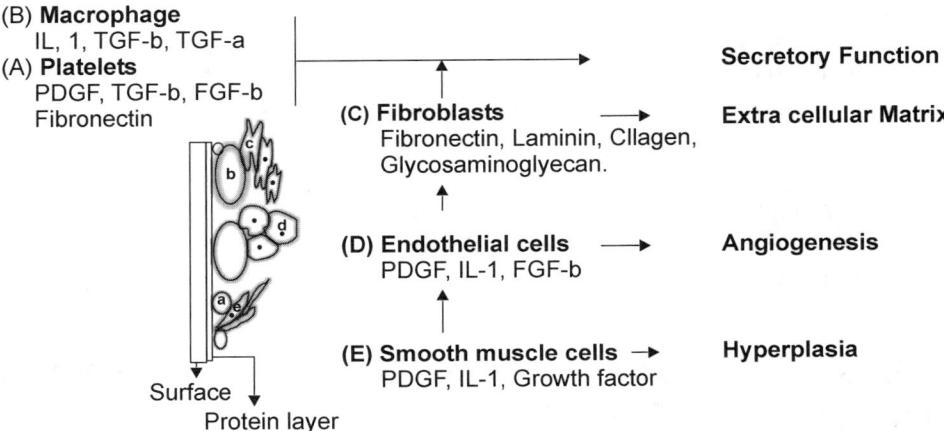

▶ **Fig. 10.1** *Time-based phenomena at an inflammatory site. Concept conceived from Ref. [8] and redrawn.*

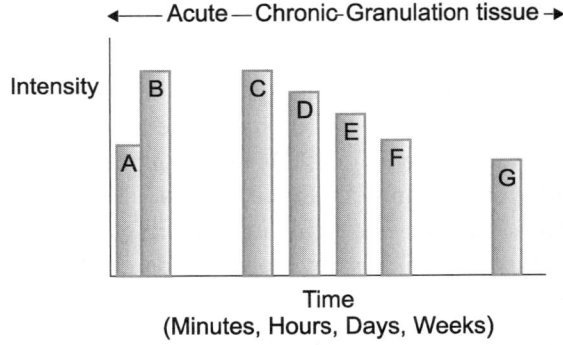

▶ **Fig. 10.2** *Surface activation of cellular responses.*

Generally these pathological events are surface-activated due to the adsorption and exchange of proteins to the implant surface. The integrin-mediated cellular response dominates the other soluble factor mediated responses initially, inducing the foreign body response in endogenous adherent pathological cells. After the initial acute inflammation, a wound-healing pattern is followed to form a temporary tissue, gradually regain its functional proficiency. In this process, the surface activation and cellular response is detailed as in Fig. 10.2. However, under chronic inflammatory conditions, more local to systemic signaling (Fig. 10.3) is made and chaemotaxis of inflammatory and immune cells happens. Eventually due to the chronic inflammatory and immune response, the implant is either lysed or rejected.

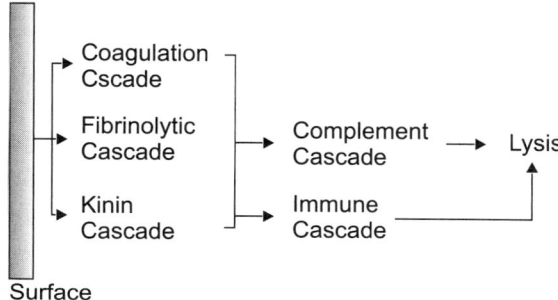

▶ **Fig. 10.3** *Different acute inflammatory cascades during enhanced vascular permeability.*

Therefore, an idea about the interaction of stem cells with this various kind of cells and environment is essential to regulate the inflammatory and immune response at the implantation site.

STEM CELL BEHAVIOUR IN RELATION TO INFLAMMATORY RESPONSE

Mechanisms of stem cell trafficking preciously to the regions of injury or degeneration is of particular importance in understanding the migratory response of stem cells with environment cues. Rolling of the extravasating cells on endothelium followed by cell adhesion and transmigration are the three sequential steps in the systemic-local immune or inflammatory cell response. Mueller et al. [9] demonstrated by *in vitro* that stem cell rolling and homing in a inflamed vascular milieu (under TNF-α, stimulation) are mediated by a subclass of integrins-$\alpha 2$, $\alpha 6$ and $\beta 1$, but not $\alpha 4$, $\alpha 5$ or the chemokine-mediated pathway, CXCR4-stromal cell-derived factor-1α, indicating that stem cell homing via the vasculature differs from that through parenchyma. TNF-α stimulation also upregulates VCAM-1 expression on the human neural stem cells (hNSCs) themselves and increases NSC-endothelial interactions.

Gronthos et al. [10] had demonstrated this integrin-mediated response in BM-derived stromal progenitors and demonstrated that growth on Col, Ln and Fn was mediated by multiple $\beta 1$ integrins. Later Van der Velde- Zimmermann-D et al. [11] demonstrated the role of $\alpha 5$ subunit in the Fn-mediated cellular interactions between BM stromal stem cells and tumour cells, using anti-$\alpha 5\beta 1$ and Fn antibodies. Thibault et al. [12] have demonstrated the chemotactic and haptotactic potential of ECM proteins like Fn, Ln and Col1 as motogenic factors on MSCs by a transmembrane assay and concluded that they can act as well like other soluble factors on cellular migration.

All theses evidences clearly indicate that supplementation of soluble and insoluble agents regulates the phenotypic behaviour of stem cells at the inflammatory milieu. In this process, stem cells express variety of surface proteins and receptors indicating its ability to interact with various other pathological cells (see Table 10.1) also.

Soluble factors regulate close control of most of these surface proteins. At the inflammation site, the concentration and the gradients as well as the ratio of specific soluble factor to the receptor regulate the stem cell behaviour.

Stem cells also express variety of soluble factors to up- or down- regulate the inflammatory processes depending upon the concentration and time. Kovach et al. [14] demonstrated that stem cell factor (SCF) produced a transient increase in adherence to cytokine-activated HUVEC cells and VCAM1 transfected CHO cells, with peak adherence at 30 min. However on prolonged incubation on CSF has significantly reduced integrin-mediated adherence (no change in integrin expression by FACS). This suggests that a subpopulation of expressed $\alpha 4\beta 1$ and $\alpha 5\beta 1$ integrins is disengaged by prolonged incubation with SCF. Recently Snyder et al. [15] have demonstrated that NSCs migrate to the inflamed tissue due to the production of the chemokine SDF-1 by the inflamed or injured neural tissue [15]. This indicates that number of stem cells at the site of implantation is very important in modulating the inflammatory response.

Table 10.1 *Cell Surface Antigens Expressed by Mesenchymal Stem Cells* [13]

Lineage markers	$CD45^-$, $CD34^-$, $CD31^-$, $CD11b^-$ (Mac1), $CD19^-$, Glycophorin A^-, Stro-1^+, SSEA-4^+
Chemokine receptors	$CXCR1^-$, $CXCR2^-$, $CXCR3^-$, $CXCR4^{+/-}$, $CXCR5^-$, $CXCR6^-$, $CCR1^-$, $CCR2^-$, $CCR3^-$, $CCR4^-$, $CCR5^-$, $CCR6^-$, $CCR7^-$, $CCR8^-$, $CCR9^-$, $CX3CR1^-$, $CX3CL1^-$
Growth factor receptors	TGF-bs-R^+, $CD105^+$ (endoglin), PDGF-R^+, EGF-R^+, IGF1-R^+, NGF-R^+, FGF-R^+, VEGF-R^+, Frizzled$^+$, LRP6$^+$
Cytokine receptors	IL-1R^+, IL-3R^+, IL-4R^+, IL-6R^+, IL-7R^+, IL-15R^+, IFN-gR^+, TNF-a 1R^+, TFN-α 2R^+
Adhesion proteins	$CD44^+$, VCAM-1^+, PECAM-1^-, NCAM-1^+, HCAM-1^+, ALCAM$^+$ (CD166), ICAM-1^+, ICAM-2^+, ICAM-3^+, MUC18$^+$, $CD90^+$, LFA3^{++}, α-4 integrin$^-$, α-5 integrin$^-$, $CD29^+$ (VLA-4^+, VLA-5^+), b-2 integrin$^-$, b-4 integrin$^+$, L-selectin$^-$, PSGL-1^-, $CD24^-$, Cadherin-5^-
Immune-related proteins	MHC I low^a, MHC II2^a, $CD80^-$, $CD86^-$, $CD40^-$, TLR-2^+ TLR-9^-, B7H1$^{+/-}$, Nectin-2^{+a}, PVR^{+a}, ULBP-3^+, ULBP-$1^{+/-}$, ULBP-$2^{+/-}$, ULBP-$4^{+/-}$, MICA$^+$ (30% of human donors), $CD48^-$, NTBA$^-$
Others	CD 73$^+$

Note: ALCAM, activated leukocyte cell adhesion molecule; EGF, epidermal growth factor; FGF, fibroblast growth factor; HCAM, homing-associated cell adhesion molecule; ICAM, intercellular adhesion molecule; IFG, insulin-like growth factor; IFN-g, interferon-g; IL, interleukin; LFA, lymphocyte function-associated antigen; LRP, low-density lipoproteinreceptor-related protein; MHC, major histocompatibility complex; MICA, MHC class-I-related chain A; NCAM, neural cell adhesion molecules; NGF, nerve growth factor; NTBA, NK-, T-, and b-cell antigen; PDGF, platelet-derived growth factor; PECAM, platelet endothelial cell adhesion molecule; PSGL, P-selectin glycoprotein ligand; PVR, poliovirus receptor; TGF, transforming growth factor; TLR, Toll-like receptor; TNF, tumor necrosis growth factor; VCAM, vascular cell adhesion molecule; VEGF, vascular endothelial growth factor; ULBP, UL16-binding protein. Upregulated by IFN-g. Exposure of hMSCs to IL-1a, TNF-a or IFN-g upmodulated ICAM-1 surface expression, whereas only IFN-g increased both HLA-class I and class II molecules on the cell surface.

STEM CELLS AND CHRONIC INFLAMMATION

Stem cells significantly alter the cellular phenotypic behaviour of immune and inflammatory cells. Lee et al. [16] demonstrated that early intravenous NSCs (from foetal brain tissue) injection displayed anti-inflammatory properties due to the suppression of splenic inflammatory activity and macrophages. However under chronic inflammation due to a xenogenic agent, it has also been demonstrated that they can also form teratoma. This is especially observed in the case of BMDCs. Houghton et al. [17] in mouse model demonstrated the relationship between chronic inflammation, haematopoietic stem-cell recruitment, mutation, and teratogenesis in the inflamed target tissue. Liu et al. [18] also have demonstrated that multiple tumour types could also be developed out of BMDCs. Therefore the maintenance of balance between differentiation and teratoma formation is important when considering stem cells as a therapeutic agent.

STEM CELLS AND CONTROLLED DIFFERENTIATION

The balance between proliferation, differentiation, and apoptosis is very important in tissue regeneration. Calof et al.[19] have observed that olfactory nerve population is closely controlled by a feedback mechanism of (a) different stages of stem and transit amplifying progenitor cell populations, and (b) differential expression of soluble mediators locally delivered. They have observed that FGF and TGF-β [20] have a role in neurogenesis, while BMPs [21] strongly inhibit neurogenesis. Buckland et al. [22] explained the role of neuro trophic factors like glial cell line derived neurotrophic factor (GDNF), brain-derived neuro trophic factor (BDNF), and ciliary neuro trophic factors (CNTF) and their differential expression in the rat neuro epithelium and olfactory bulb on postbulbectomy. Recently, looking at receptors in a different fashion based on activity in presence or absence of its ligand has emerged, and which has been explored to address the receptor differential mechanism to address apoptopsis [23]. They explained that certain receptors remain active even in the absence of the bound ligand while others remain active in the presence of ligand only. The so-called dependence receptors may exert an apoptopsis signal in the absence of the specific ligand. This has been proposed as the crucial mechanism regulating directed neuronal regeneration [24] as reviewed by Arakawa et al. Development of feedback strategies may help in controlled differentiation of the stem cells.

STEM CELLS AND ENGRAFTMENT

Stem cells found to enhance implant engraftment and subsequent tissue regeneration at a faster phase. Fan et al. [25] have demonstrated with various polymers like PLGA-gelatin/chondroitin/hyaluronate stem cells differentiated in vivo helps in implant engraftment better than that differentiated *in vitro*. Studies on allograft and xenograft engraftment using BM aspirate have shown better implantation [26] in human. This may be due to the presence of other chaemotactic and haptotactic factors at the site. Graziano et al. [27] demonstrated the role of surface geometry in stem cell attachment and growth. The phenotypic behaviour is better on concave surfaces as compared to smooth surface and is worst on convex surfaces. However we have to modify the surface based on the terminal use of the implant as it also controls the differentiation process.

STEM CELLS AND ANGIOGENESIS

Galiano et al [28] have demonstrated that locally upregulating growth factors like vascular endothelial growth factor (VEGF) could be explored for the recruitment of BM-derived progenitor cells for neovascularisation in diabetic wounds. Ziegelhoeffer et al. [29] demonstrated that BM-derived progenitor cells do not engraft at growing adult vasculature; they in fact act as a support for neovascularization through trophic factors. Zhang et al. [30] on the other hand demonstrated that BM-derived progenitor cells affect the substructure of the choroids plexus after ischemia and induce angiogenesis and vasculaogenesis. Ii et al. [31] have demonstrated that vascular endothelial progenitor cells (EPCs) produce cardioprotective cytokine nitric oxide synthase isoforms and contribute to the cardioprotective effect. All these evidences indicate the role of progenitor cells in the neovascularisation, which is important in the directed tissue regeneration around an implant.

STEM CELL BEHAVIOUR IN RELATION TO IMMUNE RESPONSE

Modulation of immune response is important to control the chronic inflammation around an implant. Recently many groups have demonstrated that MSCs can suppress both the innate and humoral response (Fig. 10.4) as reviewed by Le Blanc [32].

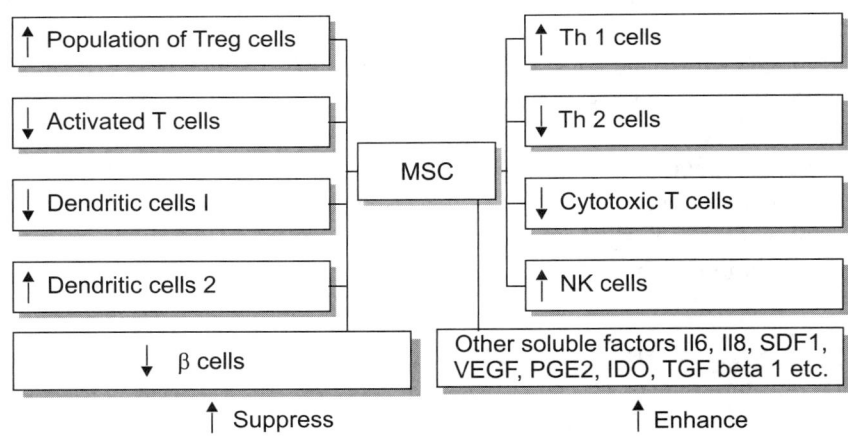

► **Fig. 10.4** *Immune modulation by stem cells.*

MSCs have been found to suppress T-cell proliferation and have been reported to induce split T-cell energy [33]. Corcione et al. [34] recently reported that MSCs co-cultured with β-cells significantly inhibited β-cell proliferation, in the presence of external stimuli. Spaggiarri et al. [35] showed that allogeneic MSCs could inhibit IL-2- and IL-15-induced proliferation of resting NK-cells. NK-cells exposed to IL-2 showed strong cytotoxicity against MSCs, related to expression of surface ligands involved in NK-cell activation like NKG2D ligands (MICA and ULBPs) and DNAM-1 ligands (PVR and nectin-2). IFN-g-exposed MSCs were protected from NK-cell mediated killing.

Jiang et al. [36] demonstrated that MSCs co-cultured with blood monocytes significantly inhibited the generation of differentiated DCs, and MSCs co-cultured with matured DCs caused a significant decrease in MHC class II molecules, CD80 and CD86 expression on DCs. It also reduced cytokine (TNF-a) production by DCs in response to LPS [37]. All these studies indicate that allogeneic MSCs can inhibit the generation and function of professional antigen presenting cells.

MODULATING IMMUNE RESPONSE: A LESSON FROM FOETAL ENGRAFTMENT [38]

The fertilized egg (embryos) seeks nutrients from mother by forming contact through the placenta. The reason for tolerance of the embryos by the maternal immune system despite the presence of paternal MHC histocompatibility antigens has invited a lot of research. The reason is that the local immune response at the site of contact is suppressed by the anti-immune cytokines produced locally at the site of contact. There is also clear evidence that the maternal immune system during pregnancy can enhance or inhibit the development of feto-placental unit. Recent data support that some cytokines produced by both T-cells and non-T-cells (IL-3, GM-CSF, TGF-β, IL-4, IL-10) favor foetal survival and growth. In contrast, other cytokines such as IFN-γ, TNF-β, and TNF-α can rather compromise pregnancy. The human lymphocyte CD4+T helper cells have been classified into two classes (T helper 1 and T helper 2) based upon these cytokine secretion profiles. The T helper 1 (Th1) cells produces IFN-γ and TNF-β, while the second type, T helper 2 (Th2), cells produces IL-4 and IL-5, and a third type (Th0) is also observed that produced both Th1 and Th2 cytokines. They concluded that the cytokine network maintaining the foetal survival mainly belongs to Th2 pathway, whereas the failure of pregnancy is associated with the dominance by Th1- type cytokines. *In vitro* studies by Joachim et al. [39] suggest that progesterone enhances the preferential development of Th2-like cells and enhances transient IL-4 production. While study using relaxin [40]—another corpus luteum derived hormone—by Piccinni et al. mainly promotes development of Th1-like cells. Further during the development of the umbilical cord, this principle is maintained and the cord blood contains more Th2 cytokines rather than the proimmune cytokines. At situations like urinary infection and associated abortion of the foetus, the shift in paradigm of increased production of Th1 cytokines has been observed. This clearly demonstrates the endocrine-immune relationship in maintaining the pregnancy.

Here is an equivalent situation when an implant—whether tissue or artificial—is placed in a wound (*implant site*). When the body identifies the implant as a xenobiotic, it first tries to destroy the implant, if not possible to isolate with the help of a fibrous capsule. If the inflammatory reaction persists due to variety of reasons like leachables, abration of the implant, infection at the implant site etc., this will lead to the immune response against the implant. This can be Type-I–IV depending upon the solubility, size, and surface chemistry of the xenobiotic formed from the implant. Most of the materials used for developing the implants are tested for each class of immune reaction preclinically, and qualify for the implantation [41]. However taking care of the difference in pathogenesis of various microorganisms, between the animals (using for preclinical studies) and human, certain clinical evaluation protocols are also to be followed to avoid any hypersensitivity cascades during implantation [42]. Both the humoral immunity due to the formation of antibodies against the soluble leachables and cell-mediated immune reactions against the particulate material affect the biomaterials. Cell-mediated immunity, when activated due to the surface identification of the biomaterials, invites chronic inflammatory responses, which complementarily activate the humoral immune system and further decides the fate of implant. The nature of the lymphocytes at the implant site and their cytokine responses regulate the cell-cell signaling during chronic inflammatory response. Here also the Th1 cell response is higher leading to excessive secretion of the IL-1, IFN-γ, TNF-β, and TNF-α at the implant site [43]. These lymphokines are proinflammatory in nature, as said earlier.

When comparison is made between the fetus and the implant, the implant exposes to a big area, and so more amount of proinflammatory agents are released, and with time the environment at the implant site decides the paradigm shifts toward tissue integration or immune rejection. The lymphocytes at the implant site through different cytokine profiles regulate the "paradigm shift", which is similar to the maintenance of feto-maternal system. Recently Zhou et al. [44] have demonstrated that stem cells could be explored for improving the tolerance of total artificial heart transplant by inducing a shift in the Th1/Th2 profile toward Th2, anti-inflammatory.

SURFACE THE CULPRIT

However, its interaction with the surface of various kinds of materials is extremely important to know for exploring MSCs for overcoming graft rejection. All the inflammatory or immune reactions related to the biomaterials and devices are originating from the biomaterial surface. Various kinds of strategies have been explored for the surface modification of these biomaterials either to reduce "fouling" or for a "bioactive response", as detailed in Fig. 10.5 [45].

STEM CELL INTERACTION WITH BIOMATERIALS

Interaction of stem cells to various kinds of surfaces is an under-studied area. The surface adhesion cues presented by these materials induce morphological and phenotypic changes onto the cells. Recently, Engler et al. [46] demonstrated that the differentiation of MSCs is greatly influenced by the elasticity of the extracellular matrix to which they are exposed—a mechanism possibly involved in MSC-mediated tissue repair. It was shown that soft matrices that mimic the brain microenvironment induced MSCs to express neurogenic markers, while stiffer matrices induced the expression of myogenic markers and hard matrices induced the expression of osteogenic markers. Therefore, it is mandatory to control the surface cues (controlled ligand

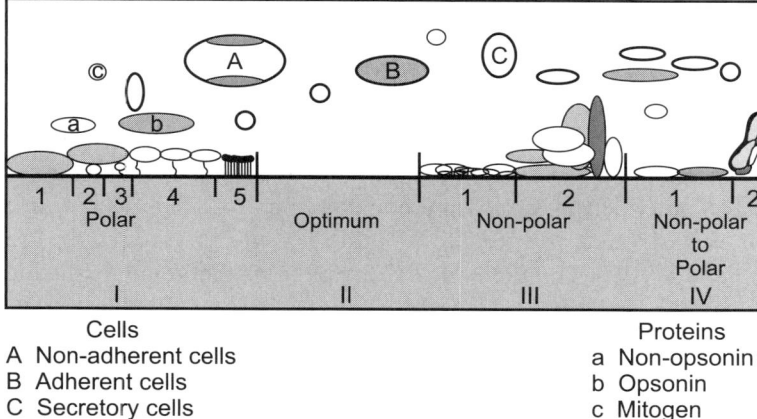

▶ Fig. 10.5 *Pictorial demonstration of the biological interaction at biomaterial surfaces. The section in grey color indicates the surface and in white indicates the biological fluids (blood, inflammatory exudates, tissue culture, tissue regeneration fluids, etc.). Section I: (1) polar; direct cell interaction; (2) growth factor adsorbed or immobilized on surface; (3) tethered growth factor; (4) mediated cell interaction, non-opsonin (albumin) binding renewable surfaces, where least cell interaction happens; (5) cell membrane mimetic surfaces; Section II: optimum; ideal surface. Section III; (1) nonpolar, preferential albumin adsorption; (2) Vroman effect, cell adhesion and activation, Section IV: (1) non-polar to polar (thermoresponsive), initial protein adsorption; (2) protein-mediated cell adhesion and adsorbed layer removal.*

supplementation (CLS)). Recently we have developed a cell mimetic strategy for the post-synthetic surface modification of biomaterials and devices for CLS based on biomimetic thin film strategy.

BIOMIMETIC APPROACHES IN SURFACE MODIFICATION USING DENSE THIN SOLID FILMS

Dense thin solid films (TSF) are self-assembled 2D architectures with the ability to interact with various 2D and 3D systems and form defined 2D or 3D supramolecular assemblies with desired functional properties. Development of such 2D systems using biological membrane mimicry helps in rapid screening of the essential components. The lipids form the major class of the structural elements, and are heterogeneous across the plasma membrane. The trans-membrane lipid heterogeneity is closely controlled in living systems, and is disrupted under pathological conditions. Surfaces modified with the heterogeneous lipid compositions and lipid protein as well as lipid-macromolecular interactions could be explored for tissue [47] and blood-compatible surface modification [4] for directed tissue regeneration.

OPTIMISATION OF HETEROGENEOUS BIOMIMETIC LIPID COMPOSITION

We have attempted to immobilize cell mimetic lipid monolayers over polymeric substrates. The cell membrane components of the endothelial cell membrane, phosphatidylcholine (PC) for phospholipid, galactocerebroside (GalC) for glycolipid, and cholesterol (Chol) based on the head group structure to represent the major lipid components of the endothelial luminal cell membrane. We have optimized an outer cell mimetic lipid composition (OCMC), (PC: Chol: GalC) (1: 0.35: 0.125), based on the air/ water interfacial studies, which is further incorporated with phosphatidylethanolamine (PE). The fundamental interactions that regulate the blood compatibility of these stabilized surfaces have been studied by *In vitro* blood cell adhesion from washed cells, protein adsorption, and calcification studies. Our studies indicated that the blood cell interaction to this lipid-modified surfaces was depending upon the orientation and packing density of the monolayers. Most interestingly, actin-mediated events in platelet activation processes were suppressed in lipid-modified surfaces [48]. We have also observed similar effects in fibroblast and macrophage adhesion studies (unpublished data). Therefore, we have further attempted to explore these surfaces for controlling various actin-mediated events. Graziano et al. [49] demonstrated that stem cells should hold its native phenotype for its optimum performance in the presence of xenografts.

STEM CELL ADHESION AND PROLIFERATION STUDIES

Our studies using ESCs (Fig. 10.6) show that cell spreading is minimal on lipid-modified surfaces and they form embryoid bodies (EBs) on lipid-modified surfaces [50].

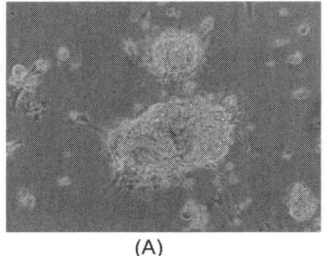
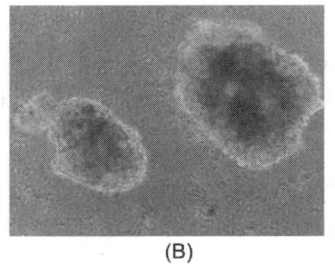
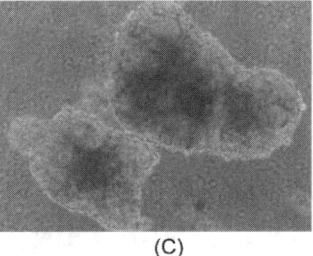

(A)   (B)   (C)

▶ **Fig. 10.6** *Embryonic stem cell behaviour on cell mimetic surfaces. (A) Bare polymer, (B) bare polymer modified with OCMC, (C) PCMC at 3-days.*

Large EBs have been formed on lipid-modified surfaces after 3 days. In the case of bare surface, many small EBs with irregular shape are formed. The cell density in the EBs is less on the bare as compared to the lipid-modified surfaces. On detailed examination at different time points, it is established that lipid-modified surfaces induce EB formation. In this process, enhanced cell–cell interaction happens at the PE incorporated surface. This clearly indicates that this lipid-modified surfaces could be explored for the surface modification of implants for the CLS; further addition of implants into the interface will help the stem cells to show a normal phenotype which is not affected by the implant surface.

CONCLUSION

A detailed understanding of stem cells in inflammatory and immune milieu is required for exploring it as an ideal candidate for regenerative therapies. Development of TSFs based on self-assembling of amphiphiles by biomimetic approaches has immense potential in controlling both the inflammatory and immune interactions. A combination of stem cells with thin film modified surfaces may effectively modulate the inflammatory and immune responses.

ACKNOWLEDGEMENTS

We express our sincere gratitude to the Director SCTIMST, and Head BMT Wing for providing the facilities and to the DST and CSIR for the financial support.

REFERENCES

1. http://www.mesoblast.com/news_mediareleases17.pdf.
2. Friedenstein A. J., K. V. Petrakova, A. I. Kurolesova, G. P. Frolova (1968). Heterotopic of bone marrow. Analysis of precursor cells for osteogenic and hematopoietic tissues. *Transplantation.* **6**: 230–247.
3. Erices A., P. Conget, J. J. Minguell (2000). Mesenchymal progenitor cells in human umbilical cord blood. *Br. J. Haematol.* **109**: 235–242.
4. Campagnoli C., I. A. Roberts, S. Kumar, P. R. Bennett, I. Bellantuono, N. M. Fisk (2001). Identification of mesenchymal stem/progenitor cells in human first- trimester foetal blood, liver, and bone marrow. *Blood.* **98**: 2396–2402.
5. Young H. E., M. L. Mancini, R. P. Wright, et al. (1995). Mesenchymal stem cells reside within the connective tissues of many organs. *Dev. Dyn.* **202**: 137–144.
6. Dennis J. E., A. Merriam, A. Awadallah et al. (1999). Quadri-potential mesenchymal progenitor cell isolated from the marrow of an adult mouse. *J. Bone. Miner. Res.* **14**: 700–709.
7. Caplan A. I. (1994). The mesengenic process. *Clin. Plast. Surg.* **21**: 429–435.
8. Anderson J. M. et al. (2004). Host reaction to biomaterials and their evaluation. In: An Introduction to Materials in Medicine (B. D. Ratner, A. S. Hoffman, F. J. Schoen, J. E. Lemons, eds.), 2nd edn, 237–291. Biomaterial Science, Elsevier Academic Press.
9. Mueller F. J., N. Serobyan, I. U. Schraufstatter, R. DiScipio, D. Wakeman, J. F. Loring, E. Y. Snyder, S. K. Khaldoyanidi (2006). Adhesive interactions between human neural stem cells and inflamed human vascular endothelium are mediated by integrins. *Stem Cells.* **24**(11): 2367–2372 .

10. Gronthos S., P. J. Simmons, S. E. Graves, P. G. Robey (2001). Integrin-mediated interactions between human bone marrow stromal precursor cells and the extracellular matrix. *Bone.* **28**(2): 174–181.
11. Van der Velde-Zimmermann D., M. A. Verdaasdonk, L. H. Rademakers, R. A. De Weger, J. G. Van den Tweel, P. Joling (1997). Fibronectin distribution in human bone marrow stroma: Matrix assembly and tumour cell adhesion via alpha5 beta1 integrin. *Exp Cell Res* 10. **230**(1): 111–120.
12. Thibault M. M., C. D. Hoemann, M. D. Buschmann (2007). Fibronectin, vitronectin, and collagen I induce chaemotaxis and haptotaxis of human and rabbit mesenchymal stem cells in a standardised transmembrane assay. *Stem Cells Dev.* **16**(3): 489–502.
13. Stagg J. (2007). Immune regulation by mesenchymal stem cells: Two sides to the coin. *Tissue Antigens.* **69**(1): 1–9.
14. Kovach N. L., N. Lin, T. Yednock, J. M. Harlan, V. C. Broudy (1995). Stem cell factor modulates avidity of alpha 4 beta 1 and alpha 5 beta 1 integrins expressed on hematopoietic cell lines. *Blood 1.* **85**(1): 159–167.
15. Imitola J., K. Raddassi, K. I. Park, F. J. Mueller, M. Nieto, Y. D. Teng, D. Frenkel, J. Li, R. L. Sidman, C. A. Walsh, E. Y. Snyder, S. J. Khoury (2004). Directed migration of neural stem cells to sites of CNS injury by the stromal cell-derived factor 1alpha/CXC chaemokine receptor 4 pathway. *Proc. Natl. Acad. Sci. USA.* 28, **101**(52): 18117–18122.
16. Lee S. T., K. Chu, K. H. Jung, S. J. Kim, D. H. Kim, K. M. Kang, N. H. Hong, J. H. Kim, J. J. Ban, H. K. Park, S. U. Kim, C. G. Park, S. K. Lee, M. Kim, J. K. Roh (2007). Anti-inflammatory mechanism of intravascular neural stem cell transplantation in haemorrhagic stroke. *Brain.* Dec. 20.
17. Houghton J., C. Stoicov, S. Nomura, A. B. Rogers, J. Carlson, H. Li, X. Cai, J. G. Fox, J. R. Goldenring, T. C. Wang (2004). Gastric cancer originating from bone marrow-derived cells. *Science.* 26, **306** (5701): 1568–1571.
18. Liu C., Z. Chen, Z. Chen, T. Zhang, Y. Lu (2006). Multiple tumour types may originate from bone marrow-derived cells. *Neoplasia.* **8**(9): 716–724.
19. Calof A. L., P. C. Rim, K. J. Askins, J. S. Mumm, M. K. Gordon, P. Iannuzzelli, J. Shou (1998). Factors regulating neurogenesis and programmed cell death in mouse olfactory epithelium. *Ann. N. Y. Acad. Sci.* 30. **855**: 226–229.
20. Kawauchi S., C. L. Beites, C. E. Crocker, H. H. Wu, A. Bonnin, R. Murray, A. L. Calof (2004). Molecular signals regulating proliferation of stem and progenitor cells in mouse olfactory epithelium. *Dev. Neurosci.* **26**(2): 166–180.
21. Shou J., P. C. Rim, A. L. Calof (1999). BMPs inhibit neurogenesis by a mechanism involving degradation of a transcription factor. *Nat Neurosci.* **2**(4): 339–345.
22. Buckland M. E., A. M. Cunningham (1998). Alterations in the neurotrophic factors BDNF, GDNF and CNTF in the regenerating olfactory system. *Ann. N.Y. Acad. Sci.* 30. **855**: 260–265.
23. Bernet A., P. Mehlen (2007). Dependence receptors: When apoptosis controls tumour progression. *Bull Cancer.* 1. **94**(4): E12–17.
24. Mehlen P., F. Llambi. (2005). Role of netrin-1 and netrin-1 dependence receptors in colorectal cancers. *Br. J. Cancer.* 11. **93**(1): 1–6.
25. Fan H., H. Liu, R. Zhu, X. Li, Y. Cui, Y. Hu, Y. Yan (2007). Comparison of chondral defects repair with *in vitro* and *in vivo* differentiated mesenchymal stem cells. *Cell Transplant.* **16**(8): 823–832.

26. Soltan M., D. Smiler, H. S. Prasad, M. D. Rohrer (2007). Bone Block Allograft Impregnated With Bone Marrow Aspirate. *Implant Dent.* **16**(4): 329–339.
27. Graziano A., R. d'Aquino, M. G. Cusella-De Angelis, F. De Francesco, A. Giordano, G. Laino, A. Piattelli, T. Traini, A. De Rosa, G. Papaccio (2008). Scaffold's surface geometry significantly affects human stem cell bone tissue engineering. *J. Cell. Physiol.* **214**(1): 166–172.
28. Galiano R. D., O. M. Tepper, C, R. Pelo, K. A. Bhatt, M. Callaghan, N. Bastidas, S. Bunting, H. G. Steinmetz, G. C. Gurtner (2004). Topical vascular endothelial growth factor accelerates diabetic wound healing through increased angiogenesis and by mobilizing and recruiting bone marrow-derived cells. *Am. J. Pathol.* **164**(6): 1935–1947.
29. Ziegelhoeffer T., B. Fernandez, S. Kostin, M. Heil, R. Voswinckel, A. Helisch, W. Schaper (2004). Bone marrow-derived cells do not incorporate into the adult growing vasculature. *Circ. Res. 6*, **94**(2): 230–238.
30. Zhang Z. G., L. Zhang, Q. Jiang, M. Chopp (2002). Bone marrow-derived endothelial progenitor cells participate in cerebral neovascularization after focal cerebral ischemia in the adult mouse. *Circ. Res. 22.* **90**(3): 284–288.
31. Ii M., H. Nishimura, A. Iwakura, A. Wecker, E. Eaton, T. Asahara, D. W. Losordo (2005). Endothelial progenitor cells are rapidly recruited to myocardium and mediate protective effect of ischemic preconditioning via "imported" nitric oxide synthase activity. *Circulation. 8.* **111**(9): 1114–1120.
32. Blanc Le, O. Ringden (2007). Immunomodulation by mesenchymal stem cells and clinical experience. *J. Intern. Med.* **262**, 509–525.
33. Glennie S., I. Soeiro, P. J. Dyson, E. W. Lam, F. Dazzi (2005). Bone marrow mesenchymal stem cells induce division arrest energy of activated T cells. *Blood.* **105**: 2821–2827.
34. Corcione A., F. Benvenuto, E. Ferretti et al. (2006). Human mesenchymal stem cells modulate B-cell functions. *Blood.* **107**: 367–372.
35. Spaggiari G. M., A. Capobianco, S. Becchetti, M. C. Mingari, L. Moretta (2006). Mesenchymal stem cell-natural killer cell interactions: Evidence that activated NK cells are capable of killing MSCs, whereas MSCs can inhibit IL-2-induced NK-cell proliferation. *Blood.* **107**: 1484-1490.
36. Jiang X. X., Y. Zhang, B. Liu et al. (2005). Human mesenchymal stem cells inhibit differentiation and function of monocyte-derived dendritic cells. *Blood.* **105**: 4120–4126.
37. Aggarwal S., M. F. Pittenger (2005). Human mesenchymal stem cells modulate allogeneic immune cell responses. *Blood.* **105**: 1815–1822.
38. Kaladhar K., C. P. Sharma (2006). Possibilities of using cord blood for improving the biocompatibility of implants. In: Frontiers of Cord Blood Science (N. Bhattacharya, P. Stubblefield, eds.). Radcliffe Publishing House, Oxford, UK .
39. Joachim R., A. C. Zenclussen, B. Polgar, A. J. Douglas, S. Fest, M. Knackstedt, B. F. Klapp, P. C. Arck (2003). The progesterone derivative dydrogesterone abrogates murine stress-triggered abortion by inducing a Th2 biased local immune response. *Steroids.* **68**(10): 931–940.
40. Piccinni M. P., D. Bani, L. Beloni, C. Manuelli, C. Mavilia, F. Vocioni, M. Bigazzi, T. B. Sacchi, S. Romagnani, E. Maggi (1999). Relaxin favors the development of activated human T cells into Th1-like effectors. *Eur. J. Immunol.* **29**(7): 2241–2247.
41. Park J. C., D. H. Lee, H. Suh (1999). Preclinical evaluation of prototype products. Yonsei. *Med. J.* **40**(6): 530–535.

42. Leuschner J., M. Rimpler (1990). Preclinical safety testing of plastic products intended for use in man. Biomed. *Tech. (Berl)* **35**(3): 44–47.
43. Mohanty M. (1996). Cellular basis for failure of joint prosthesis. Biomed. Mater Eng. 6(3): 165–172.
44. Zhou H., Z. Jin, J. Liu, S. Yu, Q. Cui, D. Yi (2007) Mesenchymal stem cells might be used to induce tolerance in heart transplantation. *Med. Hypotheses*. Oct 22.
45. Kaladhar K., C. P. Sharma (2007). Surface passivation and controlled ligand supplementation of cellular activation processes—Strategies for bottom up synthesis of bioactive surfaces. Trends. Biomater. Artif. *Organs*. **21**(1), 29–62.
46. Engler A. J., S. Sen, H. L. Sweeney, D. E. Discher (2006). Matrix elasticity directs stem cell lineage specification. *Cell*. **126**: 677–689.
47. Svedhem S., D. Dahlborg, J. Ekeroth, J. Kelly, F. Ho''o''k, J. Gold (2003). In Situ Peptide-Modified Supported Lipid Bilayers for Controlled Cell Attachment. *Langmuir*. **19**, 6730.
48. Kaladhar K., C. P. Sharma (2004). Supported cell mimetic monolayers and their interaction with blood. *Langmuir 7*. **20**(25): 11115–11122.
49. Graziano A., R. d'Aquino, M. G. Cusella-De Angelis, F. De Francesco, A. Giordano, G. Laino, A. Piattelli, T. Traini, A. De Rosa, G. Papaccio (2008). Scaffold's surface geometry significantly affects human stem cell bone tissue engineering. *J. Cell. Physiol*. **214**(1): 166–172.
50. Kaladhar K., K.D. Deb, C.P. Sharma (2008). Preliminary investigation on formation of embryoid bodies on cell mimetic surfaces and controlled ligand supplementation, *World Biomaterials Congress*, Netherlands.

11
IMMUNOMODULATORY ASPECTS OF HUMAN EMBRYONIC AND ADULT STEM CELLS

Pollyanna A. Tat, Karen L. Jones, Huseyin Sumer, Paul. Verma*
Program Leader, Stem Cell Biology, Monash Institute of Medical Research, Clayton, VIC 3168, Australia

INTRODUCTION

The ability to harness the pluripotent nature of human embryonic stem cells (hESCs) would provide a major breakthrough in modern medicine. Provided hESCs could be differentiated uniformly and in a controlled manner, treatments for a range of currently incurable conditions, such as neurodegenerative diseases, will be possible. However, given their origin, hESCs and their derivatives will be genetically incompatible with patients and thus likely to face immune rejection on transplantation to many sites in the body. Clearly, if stem cells could be derived from the patient themselves, this would largely overcome the risk of immune rejection. Patient-derived adult stem cells have, therefore, been identified as a candidate for cellular therapies. While overcoming the problem of immune rejection, patient-derived adult stem cells have limited potential as they are programmed to differentiate into progenitors and mature cells of only one lineage, typically that of their tissue of origin. There have been several reports of tissue-specific adult stem cells being able to differentiate across lineages in response to certain environmental cues [1]. However, these findings are still debated and further research is needed before adult stem cells can be used for broader applications [2–4]. While the use of hESCs is controversial, they are postulated to have the greatest potential in research and therapy due to their intrinsic ability to form cells of all three germ lineages. This chapter will focus on the issues associated with transplanting adult and ES cells and examine the current strategies for modulating the immune response.

---

* *paul.verma@med.monash.edu.au*

TRANSPLANTATION IMMUNOLOGY: REJECTION ISSUES

Transplantation has become a routine means to replace diseased or damaged organs, tissues or cells in modern medicine and can be categorised into four different types as follows:

1. Autologous (transfer between different sites on the same animal or person)
2. Syngeneic (transfer between genetically identical animals or people)
3. Allogeneic (transfer between genetically different animals or people of the same species)
4. Xenogeneic (transfer between different species; this will be discussed later)

Autologous and syngeneic transplants are usually successful, whereas allogeneic transplants are normally rejected. However, with a better understanding of immunology, rejection has largely been overcome due to tissue matching and administration of immunosuppressive drugs. Despite these advances, rejection remains an issue with approximately 7% of adult kidney transplants and 12% of adult heart transplants being rejected post-operatively in the US (from 1st July, 2004 to 31st December, 2007; source: www.ustransplant.org, last accessed on 20th February, 2008). The underlying immune mechanisms behind allogeneic rejection, tissue matching and immunosuppression are introduced here (for a more in depth review see Ref. [5]).

Table 11.1 *Organ, Tissue and Cell Transplants Routinely Performed in the Clinic*

Type	Transplantable
Organs	Kidneys, heart, lungs, pancreas, liver and intestines
Tissues	Cornea, middle ear, skin, heart valves, bone, veins, cartilage, tendons and ligaments
Cells	Bone marrow (haematopoietic) stem cells, peripheral blood stem cells, cord blood stem cells, whole blood (or separated into red blood cells, plasma, platelets)

Note: US source: www.organdonor.gov, last accessed on 20th February, 2008.

Tissue Matching and the Major Histocompatibility Complex

The major histocompatibility complex (MHC) is a series of highly polymorphic groups of genes located on mouse chromosome 17 and human chromosome 6 [6]. In humans, MHC is also known as human leukocyte antigen (HLA). Differences in HLA alleles between recipients and donors have been found to be important determinants of transplantation outcomes [7, 8]. Tissue typing refers to the matching of recipient and donor at HLA alleles (see Table 11.2 for most common HLA tissue matching alleles).

Table 11.2 *HLA Alleles Commonly Used in Tissue Matching*

MHC Class I	MHC Class II
HLA-A, HLA-B, HLA-C	HLA-DR, HLA-DQ, HLA-DP

Given the co-dominant expression of HLA genes, matches are found for each locus. Statistics from kidney transplant registries have shown that the greatest graft survival rates (from live donors) are obtained when 0 HLA alleles are mismatched, followed by two mismatches etc. [9]. While it is important to tissue match donors and recipients perfectly, the urgency status of the patient, the limited number of available donors and the extensive polymorphisms found in HLA alleles prevent this from being practiced.

Table 11.3 *The Percentage of Transplanted Allogeneic Kidneys Surviving after 5 years from Ref. [9]*

Number of HLA mismatches (HLA-A, -B, -DR)	Percentage of kidneys surviving after 5 years (n = 2281)
0	85
1	82
2	79
3	75
4	69
5	72
6	56

There are three different classes of MHC molecules. The most characterized are MHC class I (MHC-I) and class II (MHC-II) molecules, which are involved in the processing and presentation of antigens to T-cells. MHC class III genes are not as well characterized but are known to encode several complement and cytokine proteins. MHC-I molecules are expressed on the surface of all nucleated cells and bind stably to antigens derived from proteins synthesized and degraded in the cytosol. In contrast, MHC-II molecules are only present on T-cells, B-cells, dendritic cells and macrophages, and they bind stably to antigens derived from proteins degraded in endocytic vesicles. Antigens are normally derived from an individual's own proteins and are known as "self-antigens". If an individual's cells become infected, for instance with a virus, antigens may be derived from the viral proteins and are known as "foreign antigens".

The CD8 co-receptor and T-cell receptor (TCR) of cytotoxic T-cells recognize and bind only MHC-I-foreign-antigen complexes and they become activated to directly kill cells which display foreign antigens. The CD4 co-receptor and TCR of T helper cells recognize and bind only MHC-II-foreign-antigen complexes and become activated to release cytokines which recruit other immune cells such as B-cells and macrophages to act against the cells displaying foreign antigens. In order to prevent T-cells from reacting with MHC-self-antigen complexes and killing the host's own healthy cells, T-cells undergo a process known as clonal deletion in the thymus. Thymocytes (mature T-cell precursors) with TCRs exhibiting a high affinity for MHC-self-antigens are forced to undergo apoptosis. On the other hand, thymocytes with TCRs which exhibit low to moderate affinity for MHC-self-antigens are positively selected and released from the thymus into the periphery to mature into CD4+ or CD8+ T-cells. Clonal deletion leads to central tolerance by eliminating potential self-reactive T-cells, which cause autoimmune diseases. There are several other peripheral mechanisms, which also aid tolerance to self-antigens (for more comprehensive reading see Ref. [6]).

Antigens derived from or displayed on graft's cells are known as "alloantigens". Rejection is the result of an immune response mounted against alloantigens, which are ultimately recognized as being foreign antigens. Alloantigens can trigger an immune response by two distinct pathways: direct or indirect. For direct recognition, the donor cells themselves act as the antigen-presenting cells to the recipient's CD4+ and CD8+ T-cells. In contrast, indirect recognition involves the processing of donor alloantigens by the recipient's antigen-presenting cells which are then presented to the recipient's CD4+ T helper cells. If the recipient is immunosuppressed and tissue matched at HLA alleles with the graft, alloantigens are recognized as "self" and do not elicit an immune response.

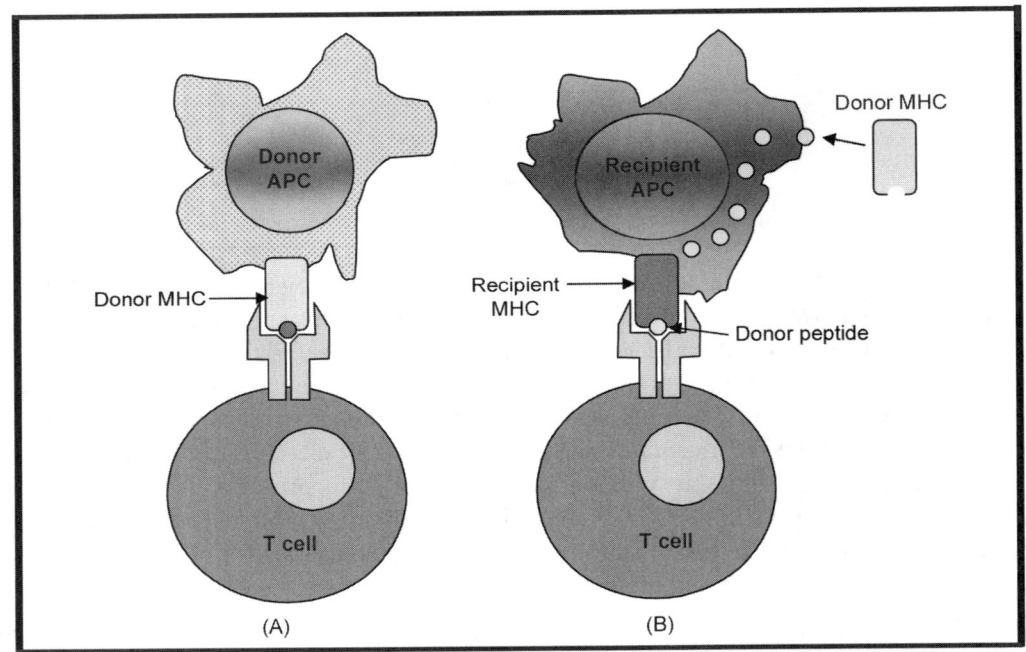

➤ **Fig. 11.1** *(A) Direct recognition, (B) Indirect recognition. APC, antigen presenting cell. (For colour figure see Plate 26).*

In spite of tissue matching, if a patient has a fully functional immune system, transplantation almost always leads to some form of rejection. To avoid or delay rejection, patients must be immunosuppressed, which commonly involves the administration of drugs which inhibit or lyse T-cells. Antibodies against T-cell surface structures and anti-inflammatory agents are also routinely used (see Table 11.4). Continued immunosuppression is necessary for enhancing the long term survival of grafts post transplantation. As immunosuppression is generalized and cannot specifically suppress immune attack on the graft, the recipient's immune defenses are compromised. Given that T-cells are the targets of immunosuppression, susceptibility to viral infections and virus-related malignancies such as B-cell lymphomas is increased. Other long-term effects include drug toxicity, hypertension and diabetes.

Table 11.4 *Clinical Methods of Immunosuppression and Potential Side Effects*

Method	Mechanism of action	Side effects
Cyclosporine/FK-506 (drug)	Blocks activation of T-cells and transcription of Interleukin-2 (IL-2) and other cytokine genes	Renal toxicity, hypertension, hyperglycaemia leading to diabetes and higher risk of neoplasia and opportunistic infections
Anti-CD25 (antibody)	Blocks IL-2 activation of T-cells and lyses already activated T-cells	Increased risk of neoplasia and opportunistic infections
Corticosteroids (anti-inflammatory)	Blocks cytokine production in antigen presenting cells and T-cells and interrupts macrophage activation	Hypertension, weight gain, osteoporosis, hyperglycaemia leading to diabetes, higher risk of neoplasia and opportunistic infections

Note: Adapted from Ref. [6]

Classes of Rejection: Hyperacute, Acute and Chronic

Rejection is classified as three types based on the histopathology and time course of rejection after transplantation. Hyperacute rejection occurs minutes to hours after the patient's and graft's blood vessels are connected. It is mediated by pre-existing antibodies in the patient's circulation, which bind to the graft's endothelial alloantigens. Binding of the antibody activates complement and together the two trigger rapid intravascular thrombosis and necrosis of the vessel walls. These pre-existing antibodies usually arise from exposure to alloantigens via prior blood transfusions, transplantation or multiple pregnancies. Hyperacute rejection is largely avoided by screening the recipient's antibodies with the potential graft.

In contrast to hyperacute rejection's antibody response, acute rejection occurs about one week after transplantation and is essentially a T-cell response to the direct recognition of alloantigens. CD8+ T-cells are the key players in acute rejection causing direct lysis of grafted cells. CD4+ T helper cells can also be implicated in acute rejection and act by secreting cytokines that recruit and activate inflammatory cells, which cause necrosis of the graft. Continued immunosuppression and obtaining minimal HLA mismatches between the graft and recipient is vital in avoiding hyperacute rejection.

Long-term transplant survivors may face chronic rejection despite immunosuppression and accurate tissue matching. Chronic rejection is characterized by fibrosis and vessel occlusion, ultimately leading to organ failure. Chronic rejection is not as well-characterized as hyperacute and acute rejection but is thought to be due to a delayed CD4+ T helper cell response through indirect recognition of alloantigens. There is currently no treatment strategy available for chronic rejection, and so it remains a major hurdle for transplantation.

Bone Marrow Transplantation, A Special Case: Graft Versus Host Disease

Another group of alloantigens known as the minor histocompatibility antigens can sometimes trigger rejection in recipients with HLA-matched grafts. An example is the male specific *HY* gene, which triggers rejection of male donor tissue in female recipients [10]. Differences in minor antigens are a significant problem for bone marrow transplants and may often lead to graft-versus-host disease (GvHD), which is essentially the converse of graft rejection. Bone marrow transplantation is usually used to treat leukaemia and lymphomas and involves the ablation of the recipient's bone marrow by aggressive chaemotherapy. This can result in the graft's transplanted T-cells recognizing the tissues of the recipient as foreign resulting in diarrhoea, jaundice, nausea, cramping and rashes. In some cases, a positive therapeutic outcome can be gained by GvHD known as the graft-versus-leukaemia effect. Graft-versus-leukaemia refers to the graft's T-cells recognizing tumour-specific antigens expressed by leukaemic cells and subsequently killing them. Elimination of mature T-cells in the donor's bone marrow prior to transplantation is one method of treating GvHD; however, this increases the risk of leukaemic relapse due to the loss of the graft-versus-leukaemia effect.

Xenotransplantation

Xenotransplantation has been suggested as an alternative method for organ and tissue transplantation due to the limited availability of human donors. Porcine species are seen as the primary candidates given their suitable organ sizes, ease of farming, large litter size, short gestation and limited public objection relative to non-human primates. The major medical impediment to xenotransplantation is hyperacute rejection due to the abundance of human antibodies, which recognize and react to the endothelial antigens of pigs. Since hyperacute rejection is characterized by the activation of complement, xenografts are more vulnerable since they lack human specific complement regulatory proteins which help slow down hyperacute rejection in allogeneic transplants. Pigs deficient for $\alpha 1,3$GalT, which is the main xenoantigen responsible for eliciting hyperacute rejection, have been generated to circumvent antibody and complement immune responses [11]. Despite this advance, recipients of xenografts must remain heavily immunosuppressed to avoid T-cell-mediated mechanisms of rejection, which may increase the risk of infection by opportunistic porcine pathogens that appear innocuous. The transmission of porcine-specific pathogens, such as the porcine endogenous retrovirus (PERV), to human cells has been shown *in vitro* [11]. This finding raises concern over the possibility of PERV mutating or recombining with human viruses to gain virulence and spreading among the general population. Such difficulties have, at present, rendered xenotransplantation of a clinical impracticality. For an in-depth review on xenotransplantation, see Refs. [11, 12].

Summary of Transplantation Immunology

In spite of current advances in transplantation immunology, there still remain many barriers to successful allogeneic transplants. Given that genes encoding MHC are highly polymorphic, even when donor and patient are related or tissue-matched, small genetic differences in some minor histocompatibility loci can still trigger rejection without warning. Furthermore, patients must remain immunosuppressed causing a plethora of side effects and increasing their susceptibility to T-cell-related infections and diseases. These current issues in transplantation highlight many immununological barriers faced when transplanting allogeneic stem cells and the need for modulating the immune response.

TRANSPLANTING ADULT STEM CELLS

Derivation of cells from the individual (autologous) requiring treatment ensures full acceptance of the graft. Evidence of adult stem or progenitor cells residing in various adult tissues is continually being updated. Adult stem cells have been found in tissues including, but not confined to, the bone marrow, intestinal crypts, brain, epidermis, heart, lung, liver, placenta, dental pulp, muscle, prostate, hair follicle and endometrium (for an in-depth review see Ref. [13]). The observation that adult stem cells can actively eliminate hoescht H33342 and other fluorescent dyes resulting in a relatively non-fluorescent "side population" of cells, which can be sorted using a fluorescent activated cell sorter (FACS) machine, has resulted in the identification of many new adult stem cell types [14]. Classically, adult stem cells are defined as being multipotent-capable of differentiating into cell types of only one lineage. However, there have been reports of adult stem cells being able to "trans-differentiate" across lineages [15]. Trans-differentiation describes the direct cellular transition of one cell type into another distinct cell type without being "de-differentiated" to a transient pluripotent or totipotent state first. These reports were contradicted by Ying et al. and Terada et al.'s 2002 studies, which show that spontaneous fusion with ES cells was responsible for the "trans-differentiation" seen with their cultured bone marrow and neurosphere cells [2, 3]. Therefore, the inherent plasticity of adult stem cells is still contentious.

Nevertheless, autologous adult stem cells have already been used successfully in therapy. Apart from bone marrow transplants that have been conducted since 1968, transplantation of keratinocytic stem cells for restoring epidermis or repairing damage to the cornea is widely accepted as one of the easiest successes in adult stem cell therapy [16]. The success of keratinocytic stem cell transplantations is due to its ease of collection; only a small biopsy from the epidermis and limbal region of the cornea is needed, and these can then be rapidly grown *in vitro* for grafting. The treatment of diseases and disorders using autologous adult stem cells is limited to certain types of genetic disorders and cannot be used where the adult stem cell population has been ablated in the patient as is the case in patients undergoing chaemotherapy. Thus allogeneic adult stem cell transplants are important for treating such cases. Allogeneic adult stem cells that are routinely transplanted include haematopoietic stem cells (HSCs), peripheral blood stem cells and umbilical cord blood (UCB) stem cells. UCB stem cells appear to be less prone to rejection than the earlier-mentioned stem cell types and are not as susceptible to GvHD (see Table 11.5 for clinical characteristics of stem cell sources). However, strict HLA matching is required since the risk of GvHD is still high. Mesenchymal stem cells (MSCs) have gained attention due to their immuno-modulatory and migratory properties, and they have been shown to have therapeutic potential in the treatment of GvHD.

MESENCHYMAL STEM CELLS

Within the bone marrow, two distinct stem cell populations with distinct progenies can be found: HSCs and MSCs. HSCs give rise to all cells of the myeloid and lymphoid lineages, and its renewal and differentiation are tightly regulated by the bone marrow microenvironment or niche. Cells of this microenvironment include osteoblasts, adipocytes, vascular pericytes and endothelial cells, which are all derived from the MSCs (with the exception of endothelial cells which can be derived

Table 11.5 *Clinical Characteristics of Various Sources of Stem Cells (www.emedicine.com/PED/topic2593.htm)*

Cellular characteristics	Source		
	Peripheral blood	Haematopoietic stem cells	UCB stem cells
HLA matching	Close matching	Close matching	Less restrictive
Engraftment	Fastest	Faster than UCB stem cells but slower than peripheral blood	Slowest
Risk of acute GvHD	Same as bone marrow	Same as peripheral blood	Lowest
Risk of chronic GvHD	Highest	Lower than peripheral blood	Lowest

from multiple stem cell sources). The "gold standard" definition of human MSCs was described by Pittenger et al. [17] as bone marrow derived fibroblasts which can be induced to differentiate into three lineages: adipocytic, chondrocytic or osteocytic. MSCs are an ideal adult stem cell source for therapies as they are multipotent and can be purified and expanded easily *in vitro*. Therapeutic use of human MSCs appears promising given the many pre-clinical and clinical trials currently underway for treatment of GvHD, Crohn's disease, osteogenesis imperfecta, osteoarthritis, cardiac regeneration, vaccine development and many others (for a comprehensive review on the therapeutic applications of MSCs, see Ref. [18]). Human MSCs have also been isolated from other sources including the placenta, adipose tissue, cord blood and liver [18]. However, a uniform set of markers to define human MSCs phenotypically is currently lacking, which hinders the interpretation of results across different lab groups. Generally, human MSCs are accepted to be CD34-, CD45-, CD14-, CD73+, CD105+, CD106+, CD166+, CD90+ and CD29+ [19].

MSCs express medium levels of MHC-I molecules and no extra-cellular MHC-II molecules unless stimulated with IFN-γ [19]. MSCs are a prime candidate for allogeneic transplants as they are postulated to possess immunosuppressive properties which are not yet fully understood, but several studies have implicated a mechanism of T-cell suppression. In pro-inflammatory conditions, MSCs have been shown to produce chaemokines and inducible nitric oxide synthase (iNOS). The chaemokines attract the T-cells to the MSCs, and the nitric oxide acts to suppress T-cell responsiveness [20]. Accordingly, MHC mismatched and haplo-identical (partially matched) MSCs are currently used for treatment of Crohn's disease and GvHD which have begun phase III clinical trials (www.osiristx.com). Another useful property of MSCs is their ability to preferentially "home" or migrate and engraft to the bone marrow when injected intravenously into a healthy animal. Interestingly, when inflammation is induced in the animal, the MSCs will preferentially home to the site of inflammation instead [21]. The molecular mechanisms underlying MSC homing is not fully understood but is thought to involve rolling and tethering akin to leukocyte migration. Such homing abilities are highly desirable in a therapeutic setting since MSCs can be administered intravenously and non-invasively. This homing property has been exploited in recent clinical trials.

Le Blanc et al. [22] were the first to publish data that suggested that MSCs may be used in the treatment of steroid-refractory GvHD. A follow-up study investigated this further by infusing eight patients with varying degrees of GvHD using allogeneic MSCs [23]. Symptoms of acute GvHD disappeared completely in six out of eight patients, and the other two patients died soon after MSC infusion with no obvious response. The MSCs appeared to have homed to the gut, liver and skin. Five patients were long-term survivors suggesting that allogeneic MSC infusion is well-tolerated by patients and has a clinical benefit in alleviating GvHD. Commercial MSC clinical trials are currently being undertaken by Osiris Therapeutics (www.osiristx.com). In phase II clinical trials, 67% of patients experienced complete resolution of all symptoms of acute gastrointestinal GvHD compared to 35–40% resolution with classical treatments.

Le Blanc and Ringden [19] have also transplanted allogeneic foetal MSCs (of male origin) into a female foetus presenting severe osteogenesis imperfecta in utero. At 9 months of age, co-culture experiments were performed *in vitro* between the recipient's lymphocytes and the injected foetal MSCs, and patient lymphocyte proliferation was not seen. By 2 years of age, the recipient showed normal development and growth and only three fractures were noted. This study indicates that allogeneic foetal MSCs can home, engraft and differentiate into bone in a human foetus even if the recipient is immuno-competent. These clinical studies thus far show a promising role for MSC transplantation.

Amnion Stem Cells

Another adult stem cell type with postulated immuno-modulatory properties is the human amniotic epithelial cell (hAEC). These cells are isolated from term-delivered foetal membranes and have been found to express hESC markers including POU domain class 5 transcription factor 1 (OCT3/4), SRY-box 2 (SOX2) and SSEA-4 [24]. hAECs are also clonogenic and can be directed to differentiate into cells of all three lineages akin to hESCs *in vitro*; however, they were unable to form teratomas. This inability to trigger teratoma formation in severe combined immune deficient (SCID) mice may be advantageous from a therapeutic point of view. Flow cytometric analysis also revealed that only a small proportion of hAECs express MHC-I and MHC-II molecules. Upon differentiation into the hepatic and pancreatic lineages, hAECs showed a significant increase in MHC-I but not MHC-II molecules, suggesting that hAECs may possess some immune privilege [24]. More studies are needed to assess this point further. hAECs may provide a useful source of multipotent stem cells for transplantation given their abundance and ease of isolation.

Summary of Transplanting Adult Stem Cells

Despite showing therapeutic promise, adult stem cells from different sources have varying ability to proliferate and are rare. While ES cells can be genetically manipulated easily, adult stem cells can only be manipulated through introduction of retroviral vectors which could cause insertional mutagenesis and activate cancer genes. Further research is required before adult stem cells can be widely used for therapy, but the promising results from clinical trials suggest that the first clinical applications will likely emerge through the use of adult stem cells.

TRANSPLANTING ALLOGENEIC EMBRYONIC STEM CELLS

Given their origin, hESCs are genetically incompatible with patients. Following immunological principles, grafted hESCs and/or their derivatives will result in graft rejection. However, it has been hypothesized that allogeneic hESCs can be transplanted into certain "immune privileged sites" without rejection. hESCs are also thought to have a special immunological status which may render them safe from immune destruction. Strategies which can circumvent immune rejection of allogeneic hESCs have been postulated, and these include genetic modification of hESCs, the development of hESC banks and tolerance induction.

Transplantation into Immune-Privileded Areas

Certain sites in the human body are regarded as being "immune privileged" because allogeneic grafts to these sites do not illicit immune responses. Immune-privileged sites include the brain, eye, testis and uterus. One of the most commonly accepted allografts is the developing foetus expressing paternal antigens that should be recognized by the mother's immune system as foreign. A special subset of CD4+ T helper cells—known as the T regulatory cells—has been implicated in the protection of the foetus from eliciting an allogeneic immune response [25]. Another example of a suppressed allograft response in an immune-privileged site is corneal transplantation. HLA tissue matching is not routinely performed, and immunosuppression is limited to the topical application of corticosteroids, yet only a low frequency of rejection is observed in allogeneic corneal transplants [26]. The immune privilege displayed by the foetus and in corneal allografts is thought to be due to three mechanisms as follows:

- The absence of lymphatic vessels to drain the immune-privileged tissue of antigenic extracellular fluid and the existence of a physical barrier such as the placenta [6].
- The secretion of anti-inflammatory cytokines such as TGF-β, which directs T-cells away from destructive inflammatory T helper cell class I responses [27].
- The expression of Fas ligand (FasL or CD95L) on the immune-privileged tissues which bind to the Fas death receptor (CD95) expressed on lymphocytes leading to the induction of apoptosis of the Fas receptor expressing lymphocytes [28].

Such immune protection has been exploited in recent studies which transplanted allogeneic stem cells into immune privileged sites. Several researchers have injected MHC-mismatched hESC derived neurons into patients with Parkinson's disease without administering immunosuppression and seen marked improvement in the disease without graft rejection (for an in-depth review see Ref. [29]). Thus, if non-autologous stem cells are transplanted into immune-privileged sites, the immune barriers confronting allogeneic graft rejection may not be applicable.

Immunological Status of hES Cells

Recent reports have shown that hESCs express low levels of MHC-I molecules and do not express MHC-II molecules at all [30]. Upon *in vitro* differentiation into embryoid bodies, a two- to four-fold increase in expression was seen; however relative to the somatic HeLa cells tested, expression levels of MHC-I molecules are still lower in hESC derivatives. These results have been supported by another study showing that hESCs injected into immune-competent mice failed to induce an immune response [31]. In addition, undifferentiated and differentiated hESCs failed to stimulate proliferation of alloreactive primary human T-cells and inhibited third-party allogeneic dendritic cell-mediated T-cell proliferation. Given the low expression of MHC-I, it is postulated that hESCs and their derivatives may be susceptible to natural killer (NK) cell destruction. NK cells kill all cells that do not express MHC-I molecules and, unlike T-cells, do not have special receptors to distinguish between self- or allogeneic cells. The main function of NK cells is to kill self cells that have lost expression of MHC-I molecules as a result of viral infections or that have become tumourigenic. Therefore to test if hESCs would be subject to NK cell mediated killing, Drukker et al. [32] devised several experiments. As a control, hESCs were injected into the kidney capsule of SCID mice which lead to massive teratoma formation. When injected into nude mice that were characterized by an absence of T-cells but had normal levels of NK- and B-cells, teratoma formation was also observed. In contrast, hESC injection into mice that were characterized by an absence of either NK cells (Lyst^{bg} mice) or of B-cells (Btk^{xid} mice) resulted in vigorous rejection as no teratoma growth resulted. These experiments clearly demonstrate that xenorejection of hESCs is a T-cell mediated process. To determine whether hESCs were "immune privileged", the authors determined the level of FasL expression by microarray and flow cytometric analyses. Both analyses indicated that hESCs and their derivatives expressed significantly lower levels of FasL than other somatic tissues, ruling out FasL protection from immune destruction. The co-stimulatory molecules, CD80 and CD86, play an important role in T-cell recognition of MHC–antigen complexes and hESCs were found to lack expression of both. Even upon addition of IFN-γ, a potent stimulator of the co-stimulatory molecules, CD80 and CD86 expression, could not be induced in hESCs. Furthermore, a DNA microarray analysis of 306 genes known to be involved in immune responses was investigated. It was found that hESCs and their derivatives clustered together and only half of the immune genes expressed in haematopoietic cells, lymphoid organs, and other tissues was expressed in hESCs and their derivatives. The authors hypothesized that this may be because hESCs have not reached their immunological maturity. Taken together, these studies suggest that allogeneic hESC derivatives may elicit a reduced immune response following transplantation. Long-term studies are required to examine this more extensively since it may be possible that hESC derivatives could mature further after transplantation and start to express higher levels of MHC antigens.

Immune-Matched Banks of hES Lines

A recent study by Taylor et al. [33] suggested that approximately 150 hESC lines are needed to generate a clinically relevant HLA-typed bank of hESC lines. Such a bank would be perfectly matched at the HLA-A, HLA-B, and HLA-DR loci for 20% of individuals in the general population, 25–50% of individuals would have a beneficial match at only one loci: HLA-A or HLA-B and HLA-DR will be matched for most individuals. However, the rarity of obtaining homozygous hESC lines from IVF clinics limits development of such a bank. Furthermore, certain ethnic backgrounds, such as Asians may be under-represented in this estimation since they are less likely to become donors than Caucasians (based on current organ donor registries in the UK). Increasing the size of the hESC bank 10-fold would not increase the number of matched individuals significantly but interestingly; retrospective selection of 10 individuals would provide a full match for 37.7% of individuals. Currently, the only way to achieve this would be to isolate hESCs from individuals by somatic cell nuclear transfer (SCNT) or cell reprogramming. A limitation to this study is that it assumes that all hESC lines are equal and have the same differentiation potential-an assumption that is clearly not true at present.

Tolerance Induction of Grafted Cells

One strategy devised to circumvent rejection of allogeneic stem cells is the use of the patient's own immune system to induce tolerance to alloantigens. Tolerance would lead to the patient being non-responsive to alloantigens. This is achieved by establishing thymic chimerism of both host- and donor-derived HSCs in the patient. Akin to self-tolerance, this would result in the deletion of any T-cells that would be reactive to the donor cells or tissues derived from hESCs. A limitation of this strategy is that a functional thymus is necessary to induce chimerism. Since the thymus degenerates with age, its rejuvenation is another hurdle that must be overcome before this strategy can be realized in the clinic. For an in-depth review, see Ref. [34].

STRATEGIES FOR GENERATING AUTOLOGOUS ES CELLS

Generating autologous hESCs would offer the ideal solution to the immune problems posed in allogeneic transplantation. However, the ability to achieve this has remained elusive for researchers until the birth of Dolly [35]. The first mammal produced from an adult cell, Dolly, proved that it is possible to wind back the developmental clock. The reversion of an adult somatic cell which has a restricted differentiation potential, back to a pluripotent state, is known as reprogramming. Four different methods of reprogramming have received attention-SCNT, cell fusion, treatment with cytoplasmic lysates or extracts and more recently, induced pluripotency by viral transduction. In addition to reprogramming, another strategy for generating autologous hESCs is known as parthenogenesis. Here we review parthenogenesis and the four reprogramming technologies and discuss their limitations with regard to generating autologous hESCs for clinical application.

Plate 17

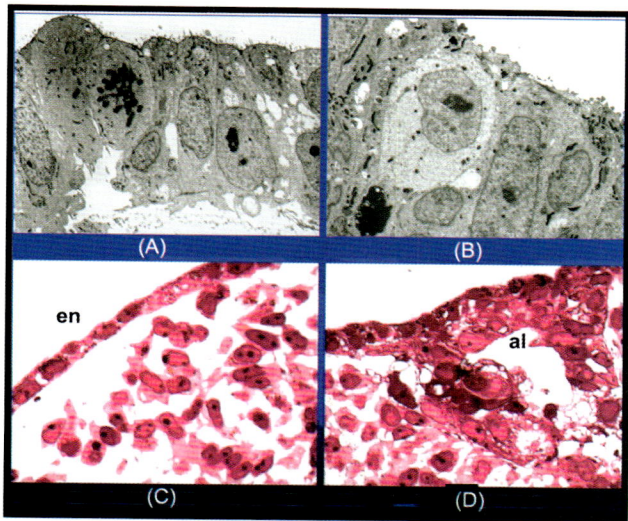

➤ *Fig. 6.14 Spontaneous differentiation of ESC (LM and TEM). (A) Epithelioid layer with dividing cell. (B) Epthelioid layer with clear germ cell; ✗3400. (C) Mesenchyme stem cells and endoderm (en). (D) Lung cells in alveolus (al); ✗1000 (from Ref. [25]).*

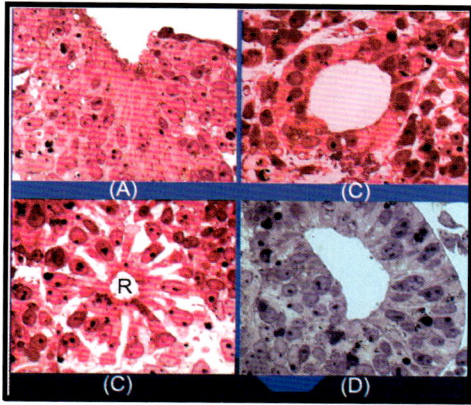

➤ *Fig. 6.15 Spontaneous cell differentiation in EBs (LM). (A) Surface invagination in outer ectoderm. (B) Neural rosette [R] with neural stem cells. (C) Tubular formation with cuboidal cells. (D) Tubular formation like neural tube. ✗1000 (from Ref. [25]).*

Plate 18

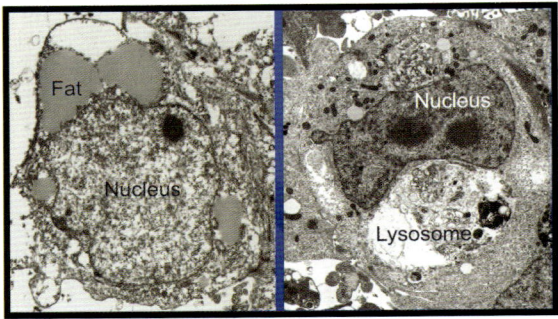

► *Fig. 6.16 ESC colony—lipid and phagocytic cell (P35). TEM The lipid cell has globules of fat and the phagocyte resembles an endoderm cell of blastocysts. Note that lysosome with cellular debris. ✗10,200 (from Ref. [25]).*

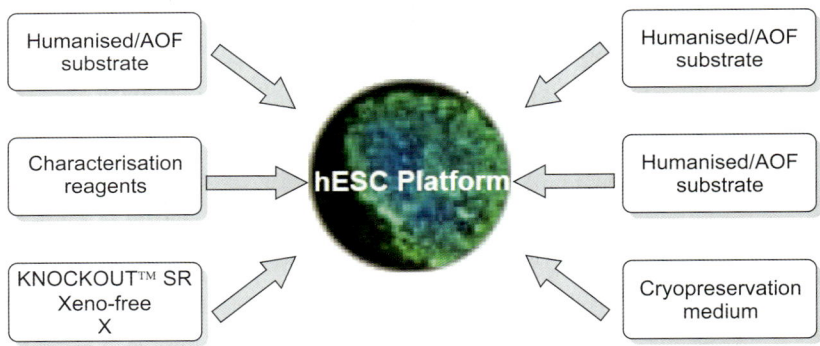

► *Fig. 7.4 An ideal hESC platform with required tools and reagents to enable clinical transition of stem cells.*

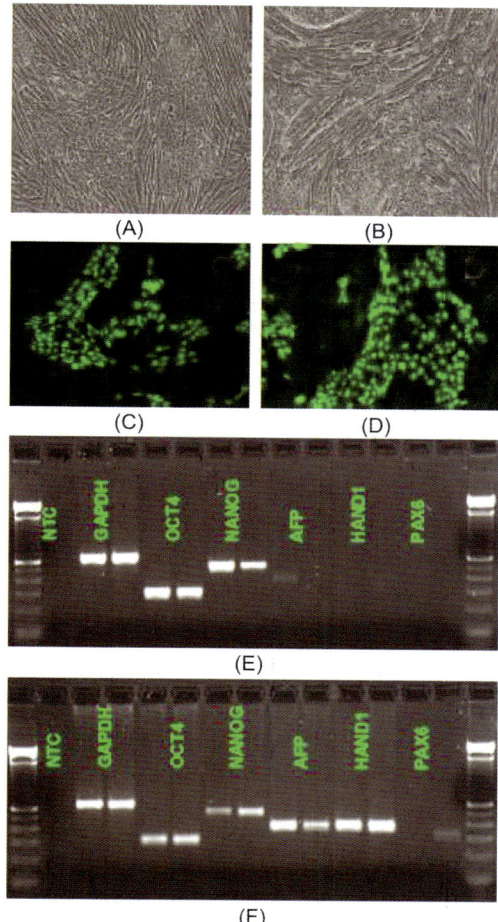

▶ *Fig. 7.5 Performance of humanised KSR: BG01v cultured in either (A) KSR or (B) KSR Xeno-free on human foreskin fibroblasts (HFF) attached with CELLstart substrate (Passage 4). OCT-4 staining of BG01v cultured on HFF + CELLstart™ in (C) KSR or (D) KSR Xeno-free for five passages. (E) Following 10 passages in either KSR (left lane) or KSR Xeno-free (right lane) on HFF attached with CELLstart™ substrate, BG01v gene expression was examined. (F) Gene expression of embryoid bodies generated from the same P10 BG01v/HFF cultures.*

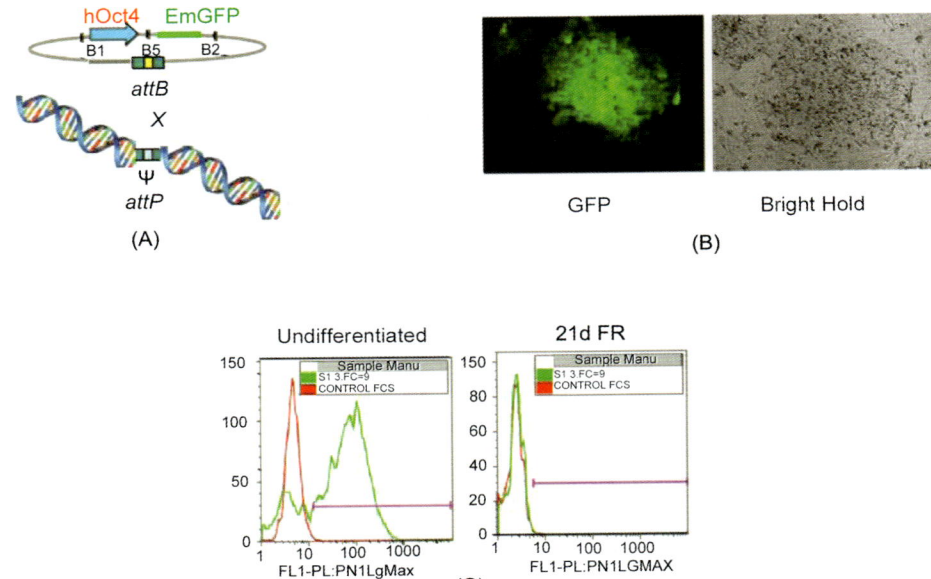

▶ *Fig. 7.6 BGO1V/ human Oct4-GFP reporter line. (A) Vector configuration and phiC31-mediated genomic targeting. Clones resulting from transfection of GFP expression plasmid and phiC31 integrase were picked, expanded and their integration sites mapped. (B) Representative hOct4-GFP clone derived from BG01v was analysed by bright-field and fluorescent microscopy. (C) Clone analysis before and after 21 day embryoid body differentiation. GFP expression was analysed by using a Guava FACS analyser. The green curve represents GFP expression, and the red curve represents a control, unengineered cell line that did not express GFP.*

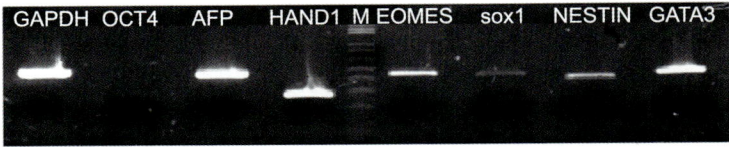

▶ *Fig. 7.7 Expression of Candidate Markers of Differentiation measured by RT-PCR during differentiation of BG01v cells, grown in StemPro® hESC SFM, to embryoid bodies (EBs). The data show that BG01v cells grown in StemPro® hESC SFM media exhibit a complete ability to differentiate into ectoderm (SOX1, Nestin, Eomes), mesoderm (HAND1), and endodermal (AFP) and trophoblast (GATA3) lineages. GAPDH is a house keeping control and M denotes the marker.*

Plate 21

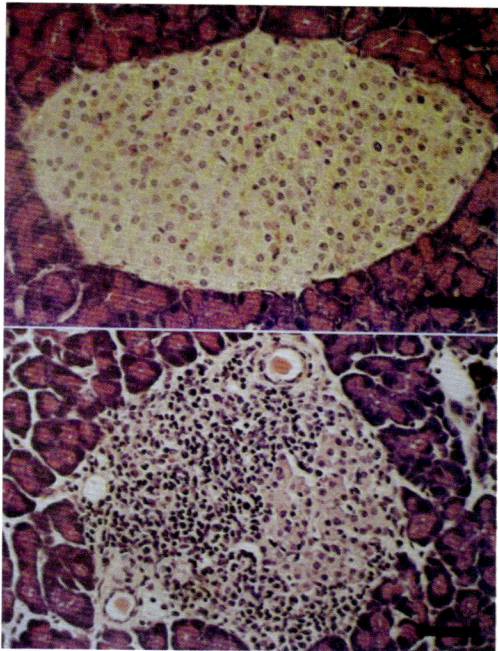

➤ *Fig. 9.1 Haematoxlin and eosin stains of an islet from the pancreas of (top) a non-diabetic mouse and (bottom) a mouse with diabetes caused by the destruction of β-cells in the islet. Bar is 50 μm. Source: Ref. [1].*

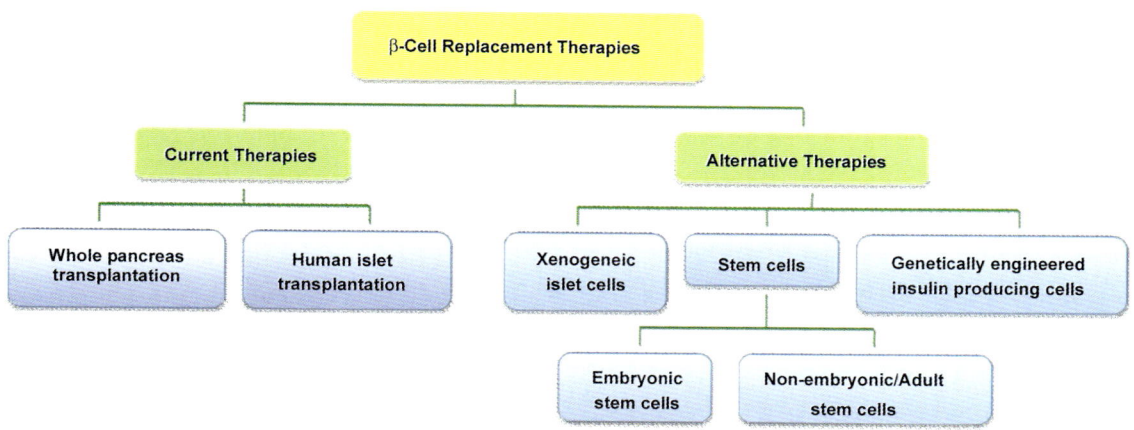

➤ *Fig. 9.2 Current and alternative β-cell replacement therapies. Current β-cell replacement therapies include the transplantation of whole pancreas and human islets from donor human cadavers. Alternative cellular therapies that are currently under examination include xenogeneic islet cells such as porcine islets, stem cells, both embryonic and non-embryonic or adult, and genetically engineered insulin producing cells.*

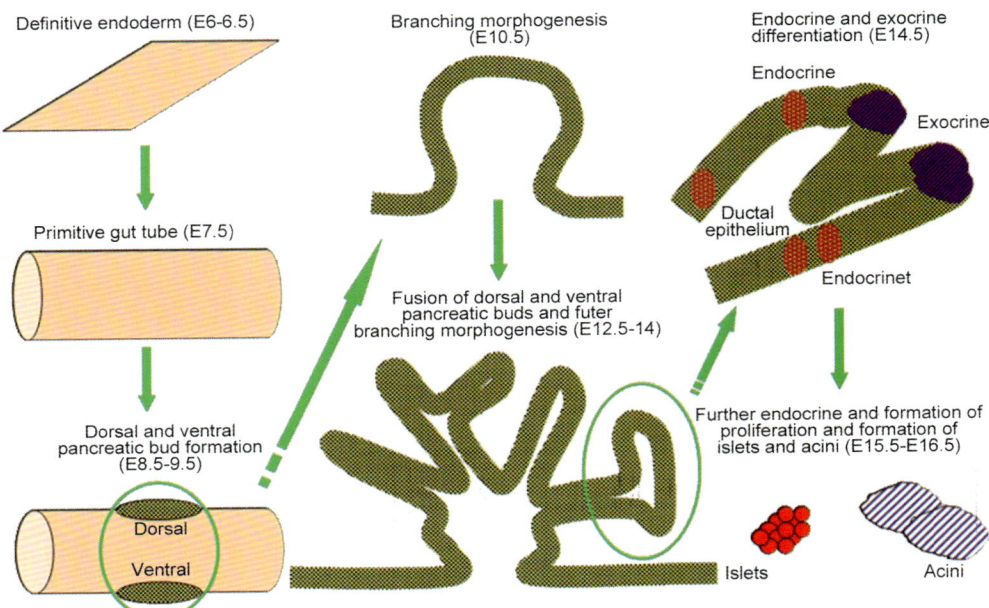

▶ *Fig. 9.3 A basic schematic summary of fetal pancreatic development in the mouse. Development initiates from the definitive endoderm germ layer which appears as a flat sheet of cells on embryonic day 6–6.5 (E 6–6.5). This sheet of cells further develops into the primitive gut tube by E 7.5. Between E 8.5 and E 9.5, the dorsal and ventral pancreatic buds form from a population of Pdx-1 expressing epithelial cells of the gut tube. The pancreatic epithelium of the two buds then undergoes extensive proliferation and branching morphogenesis to form the basis of the complex ductal network of the pancreas (E 10.5). Fusion of the dorsal and ventral pancreatic buds forms the definitive pancreas, and after further branching morphogenesis (E 12.5–14.5), exocrine and endocrine differentiation begins, which results in the formation of the exocrine ducts and acini and the endocrine islets of Langerhans (E 15.5–16.5).*

Plate 23

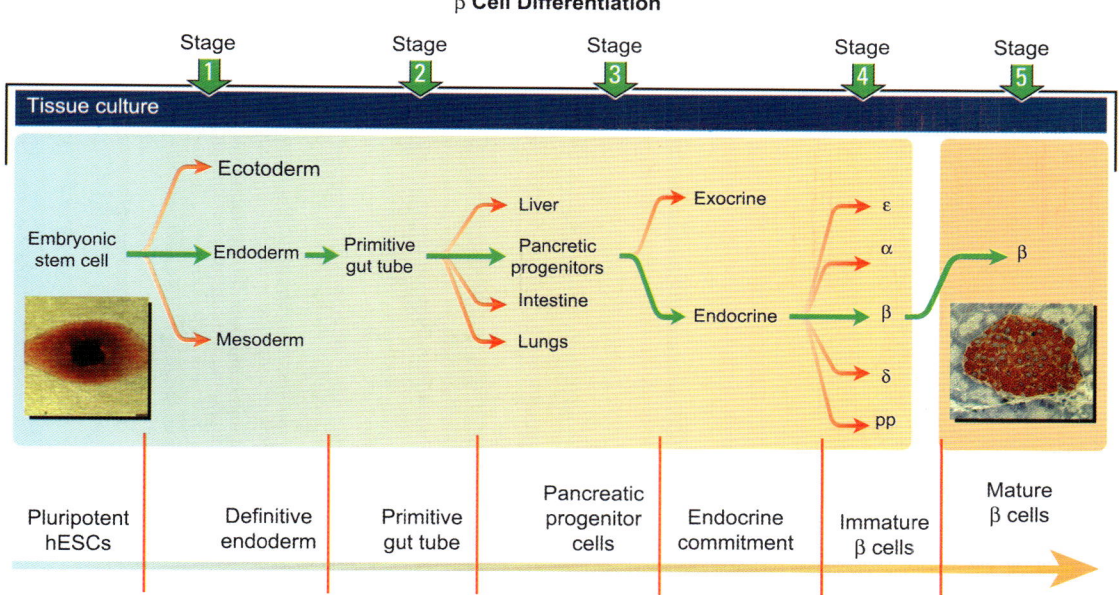

➤ *Fig. 9.4* **An ontogeny-based approach to differentiate hESCs into insulin-producing β-cell surrogates** in vitro *involves the replication of five key stages in pancreatic development* in vivo: *(1) the generation of definitive endoderm from pluripotent cells; (2) development of the primitive gut tube from definitive endoderm; (3) production of Pdx-1-expressing pancreatic progenitor cells; (4) inducing the differentiation of pancreatic progenitor cells to endocrine cells and (5) endocrine cell specification to mature insulin-producing β-cell surrogates.*

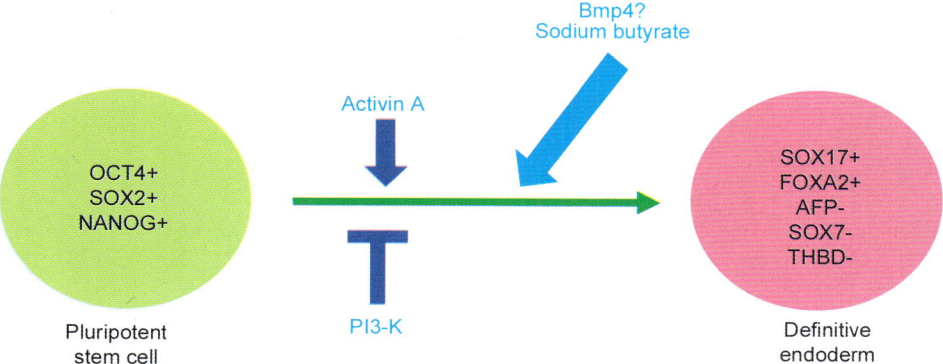

➤ *Fig. 9.5 The absolute requirements for the* **in vitro** *differentiation of human embryonic stem cells into definitive endoderm are high levels of Nodal signalling as mimicked by activin A and the inhibition of the phosphatidylinositol 3-kinase (PI3-K) signalling pathway. Cellular differentiation may be further enhanced by the additional supplementation of exogenous growth factors such as sodium butyrate and Bmp4. The expression patterns of key genes, which are characteristic of pluripotent stem cells and definitive endoderm are noted.*

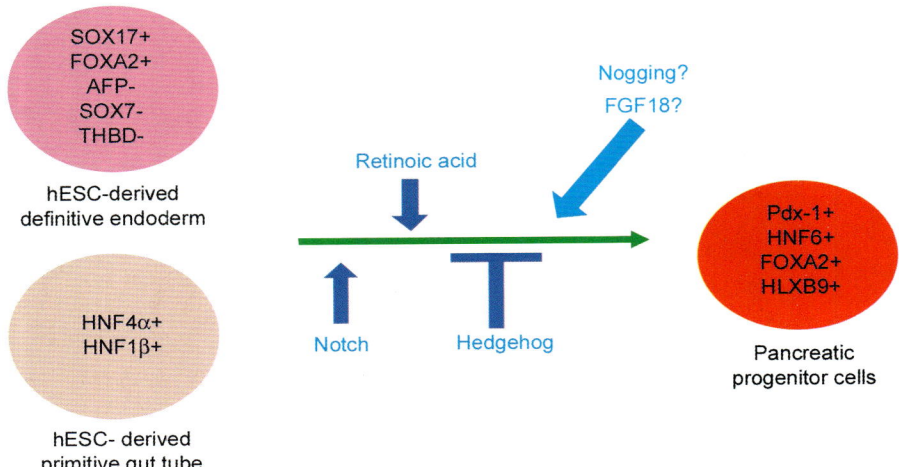

▶ *Fig 9.6 Current approaches to derive pancreatic progenitor cells from either hESC-derived definitive endoderm or primitive gut tube revolve around the themes of activated Notch and retinoic acid signalling and inhibition of the Sonic hedgehog signalling pathway. Additional growth factors such as Noggin and FGF18 have also been supplemented at this stage of differentiation to generate pancreatic progenitor cells, but their role in differentiation is unknown. This suggests that an interaction between a variety of signalling pathways is required for the development of pancreatic progenitor cells from hESCs. The expression patterns of key genes which are the characteristic of hESC-derived definitive endoderm, primitive gut tube and pancreatic progenitor cells are noted.*

Plate 25

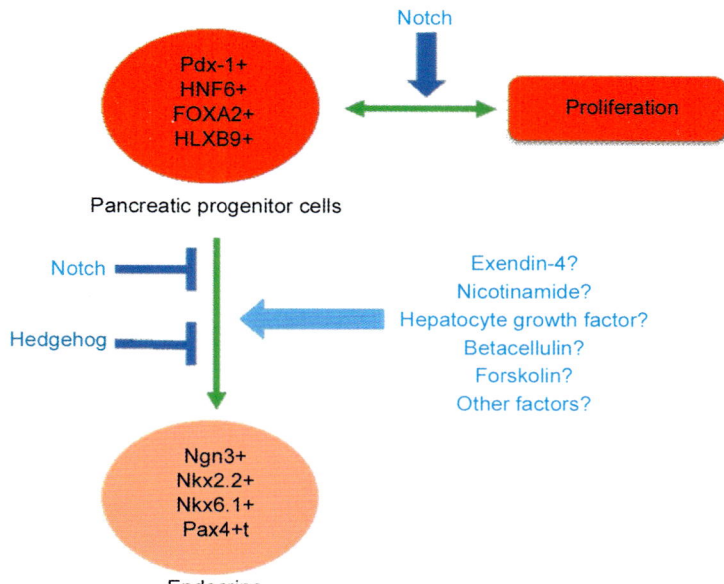

► *Fig. 9.7 Activation of Notch signalling can permit the proliferation of pancreatic progenitor cells while inhibition of Notch and Sonic hedgehog signalling can induce endocrine differentiation. Aside from this, little is known in developmental biology regarding the mechanisms by which pancreatic progenitor cells assume an endocrine cell fate. In current hESC differentiation strategies, the choice of growth factors to stimulate endocrine differentiation varies widely between protocols. The expression patterns of key genes, which are the characteristic of pancreatic progenitor cells and differentiation into endocrine cells, are noted.*

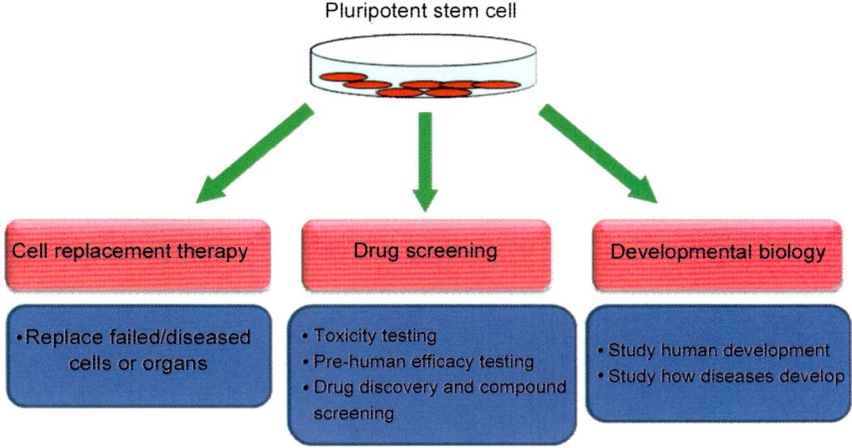

► *Fig 9.8 Three potential uses of pluripotent stem cells in basic research and medicine: cellular replacement therapy to replace failed/diseased cells or organs; drug screening and developmental biology to study human development and disease development.*

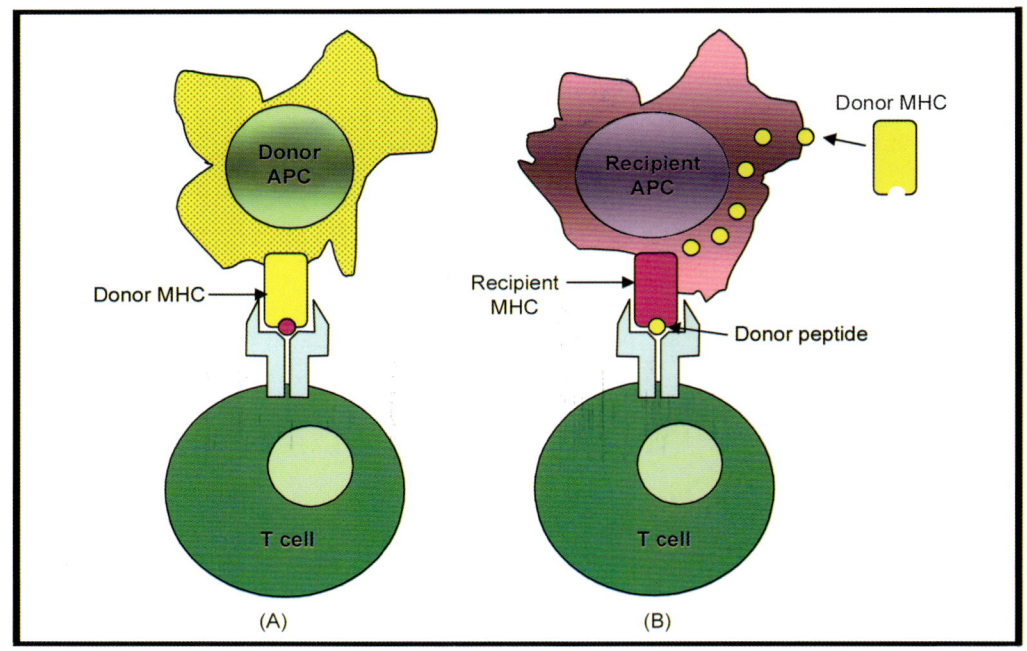

▶ *Fig. 11.1 (A) Direct recognition, (B) Indirect recognition. APC, antigen presenting cell.*

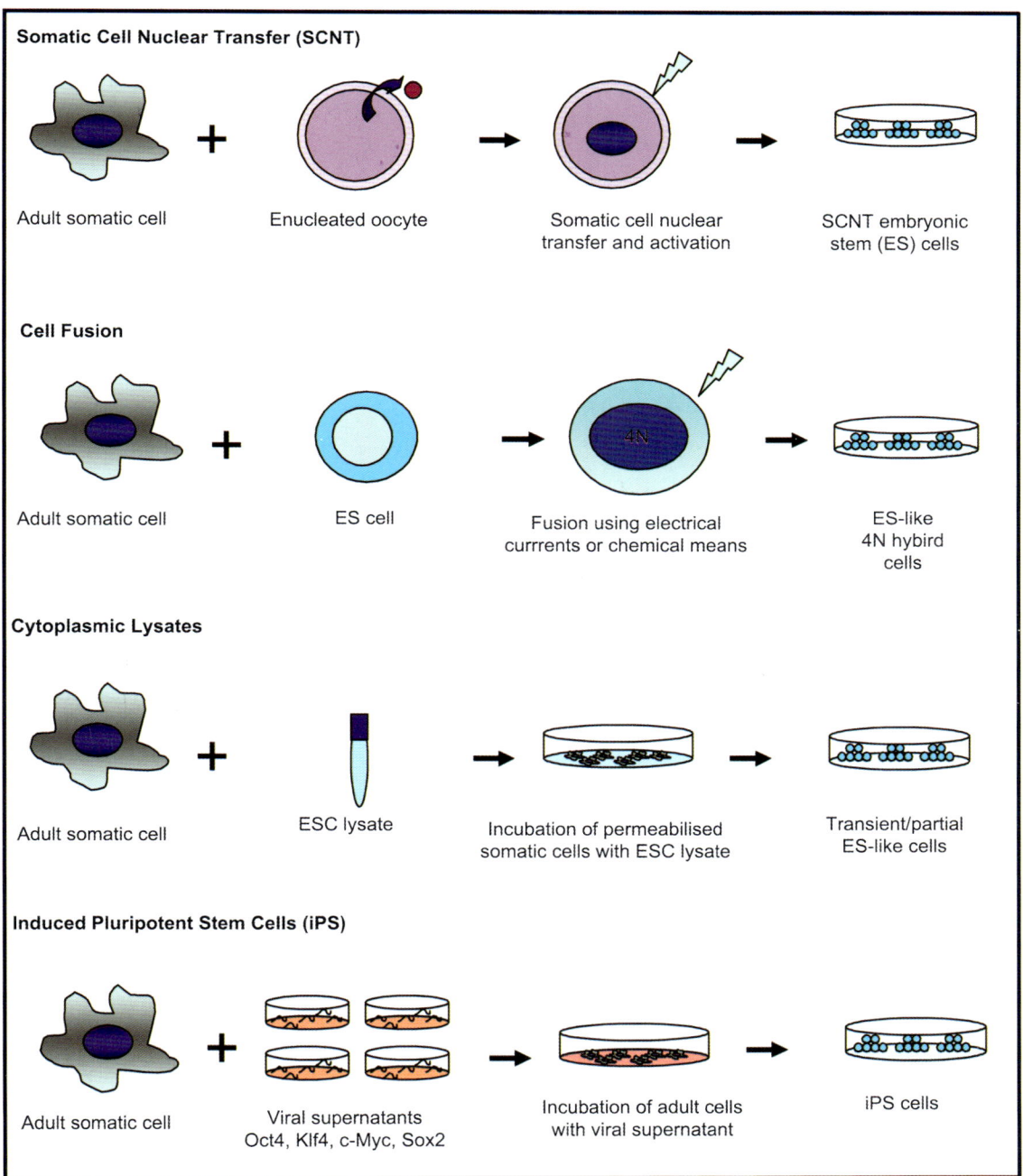

➤ *Fig. 11.2 Various reprogramming methods.*

Plate 28

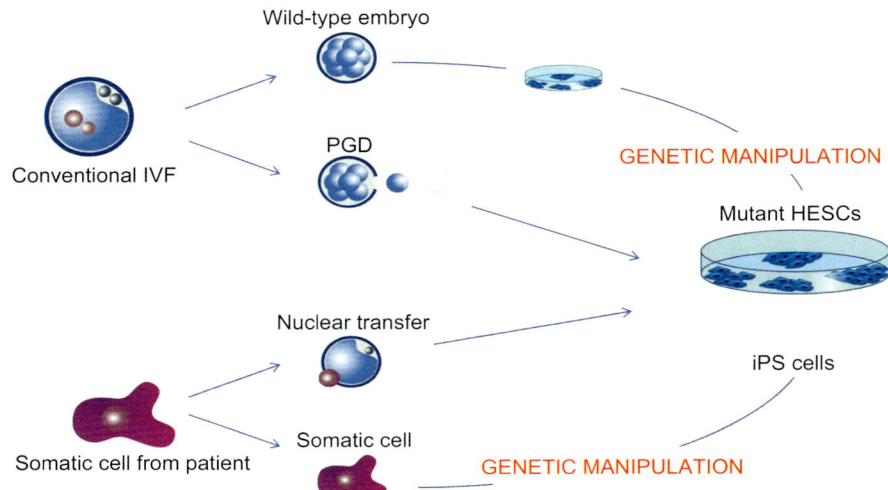

▶ *Fig. 12.1 Different approaches to obtain mutant pluripotent cells (from top to bottom): (A) artificially introduce specific mutations in a pre-existing cell line; (B) directly derive a hESC line from a genetically abnormal embryo; (C) reprogramming a nucleus from a somatic cell of a patient by nuclear transfer; (D) reprogramming a somatic cell by genetic manipulation to produce transcription factor iPS cells.*

Plate 29

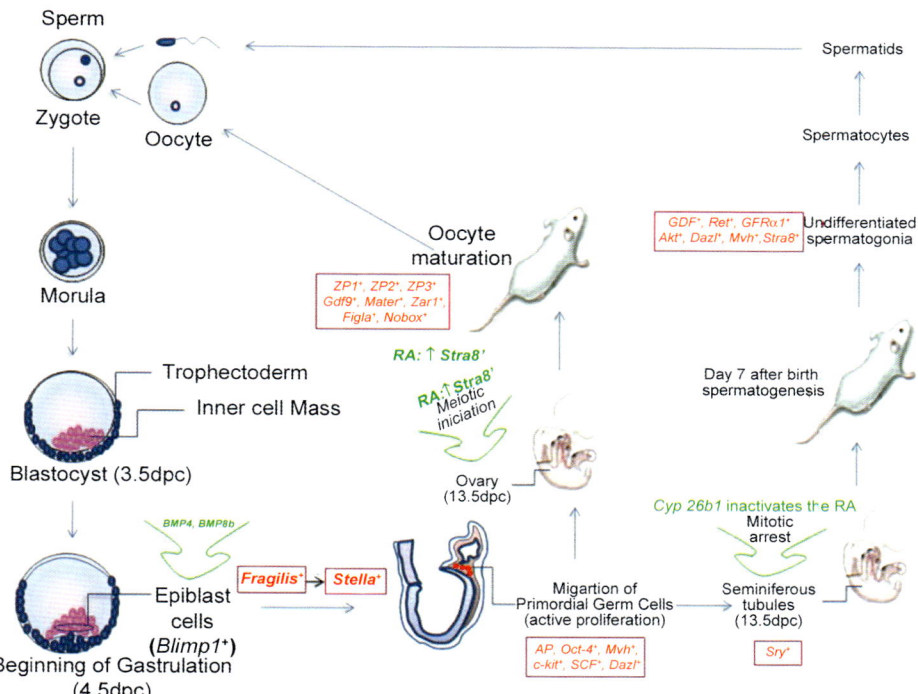

▶ *Fig. 13.1 Mouse germ cell development in vitro. After fertilization, the zygote initiates the embryonic development yielding stem cells, which are responsible for originating all embryo cell types. Cells from the ICM epiblast acquire the ability to contribute to the germ line. Around 7 dpc, primordial germ cells migrate to the genital ridge, where they proliferate and complete further differentiation into gametes. Stage-specific gene expression is represented within squares, while green arrows represent the influence of specific factors in GC differentiation.*

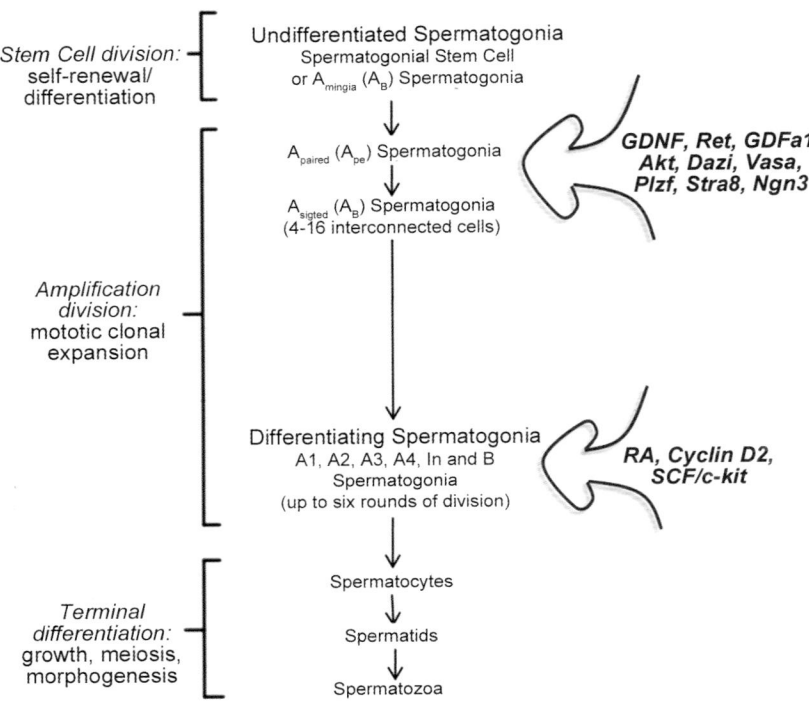

▶ *Fig. 13.2 Male germ line differentiation. The scheme begins with spermatogonial stem cells (A_s), which can either self-renew or give rise to A_{paired} (A_{pr}) spermatogonia, which are destined to proceed through spermatogenic lineage. After a series of 9–11 mitotic divisions, spermatocytes are formed. After that, terminal differentiation yielding spermatozoa occurs throughout meiosis and morphological changes. Stage-specific gene expression and influence of specific factors in spermatogenesis are represented within arrows (modified from Ref. [22]).*

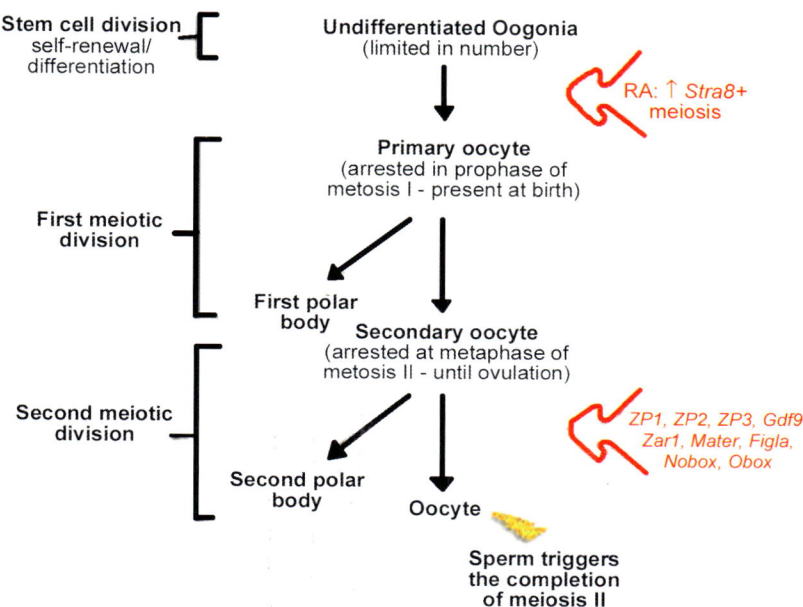

▶ Fig. 13.3 Female germ-line differentiation. The scheme begins with undifferentiated oogonia, which can either self-renew or give rise to the primary oocytes. At this stage, primary oocyte is arrested at prophase of meiosis I until birth. After that, the first meiotic division is accomplished, originating the first polar body as well as the secondary oocyte, which is now arrested at metaphase of meiosis II until ovulation. The second meiotic division is triggered by sperm fertilisation. Stage-specific gene expression and influence of specific factors in oogenesis are represented within arrows.

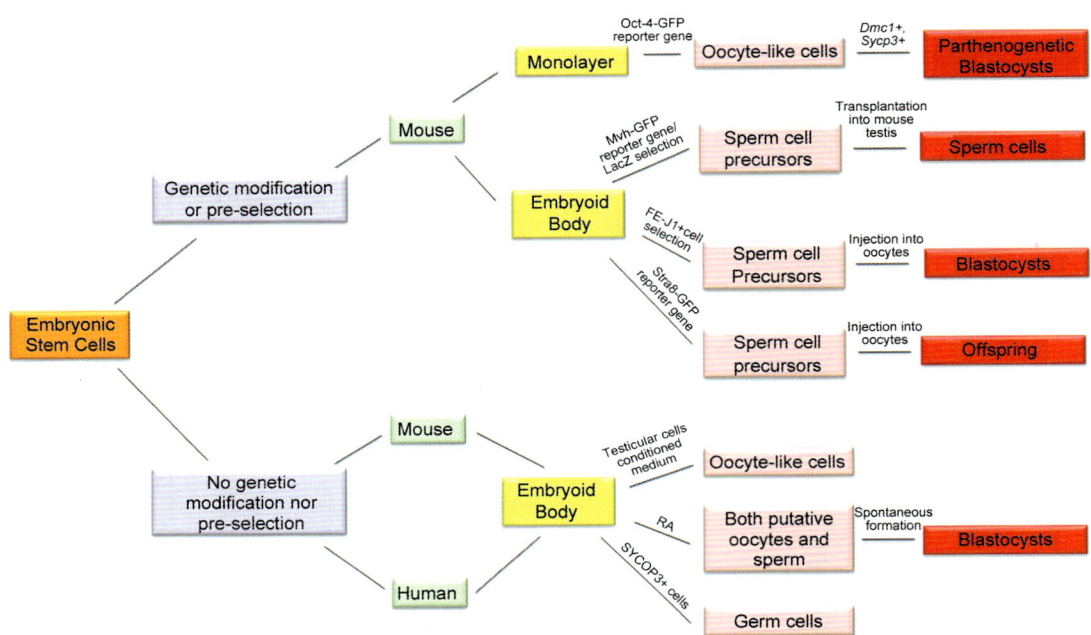

▶ *Fig. 13.4 Scheme summarising the strategies described at the recent publications about artificial gametes production ([48, 47, 49–53]; from top to bottom panels). FE-J1, cell surface binding anti-anterior acrosome monoclonal antibody; MVH, mouse vasa homologue; Oct4-GFP, Oct4-green fluorescent protein; RA, retinoic acid; Stra8–eGFP, Stra8-enhanced green fluorescent protein; Sycp3, synaptonemal complex protein 3; Dmc1, disruption of meiotic control1 (modified from Ref. [54]).*

Parthenogenesis

Parthenogenesis describes the artificial activation and subsequent development of unfertilized oocytes. This is achieved routinely by retention of the second polar body after activation using chemical treatments including cytochalasin B or 6-DMAP [36]. Blastocysts derived in this manner can be used to derive diploid hESC lines which should be homozygous for half of the maternal chromosomal complement, including the MHC loci. Depending on the approach used to generate diploid pESCs, MHC matching of pESCs with the oocyte donor of 70.9% or greater can be anticipated [37, 38]. pESC lines can, therefore, be transplanted into haplo-identical patients and may be subject to reduced immune rejection. Revazova et al. [36] have described the parthenogenetic activation of human oocytes and subsequent derivation of six pESC lines. These pESC lines were found to be identical to hESC lines morphologically, expressed many markers of pluripotency, and showed high telomerase activities akin to hESCs. pESC lines could also form derivatives of the three germ layers when differentiated *in vitro* and *in vivo*. All cell lines except one had a normal 46, XX karyotype, but all lines were MHC-matched to oocyte donors. This study shows that it is possible to generate pESC lines which can be differentiated to form functional cells. Whether these derivatives will be accepted fully by their oocyte donors or not remains to be determined. Since the generation of pESC lines requires oocytes, non-reproductive female and male patients that are not haplo-identical to oocyte donors would not be able to utilize this strategy.

Reprogramming Methods
Somatic cell nuclear transfer

In 1952, Briggs and King were the first to successfully demonstrate that the frog (*Rana pipiens*) oocyte could be enucleated and replaced with the nuclei of blastula cells without hindering embryo development [39]. This was followed by Gurdon's experiments which showed that fully differentiated adult intestinal cells could be reprogrammed by enucleated *Xenopus laevis* oocytes to form adult cloned frogs [40]. For an in-depth review of the history of nuclear transfer in non-mammalian animals, see Ref. [41]. Decades later, a breakthrough came with cloning of the first mammal, Dolly, from an adult mammary epithelial cell [35]. This has been followed closely by the cloning of other species including mice, cattle, pigs and many others [42–44]. Although these cloning experiments used adult somatic cells, it could not be proven unequivocally that they were derived from terminally differentiated cells that do not already possess stem-cell-like properties. This was resolved by Hochedlinger and Jaenisch [45], who reported the cloning of mice from mature lymphocytes, and this was confirmed by the presence of immune receptor rearrangements in the clones' cells. Immunoglobulin and TCR rearrangements only occur in terminally differentiated lymphocytes and, as such, act as a definitive in-built marker of these cells. In addition, cloned mice and SCNT ES cell lines have also been produced using genetically labelled olfactory neurons indicating that postmitotic cells can re-enter the cell cycle and regain totipotency [46, 47]. The possibility of stem cell contamination in the donor cells was eliminated in these studies and proves that it is possible to reprogram terminally differentiated cells.

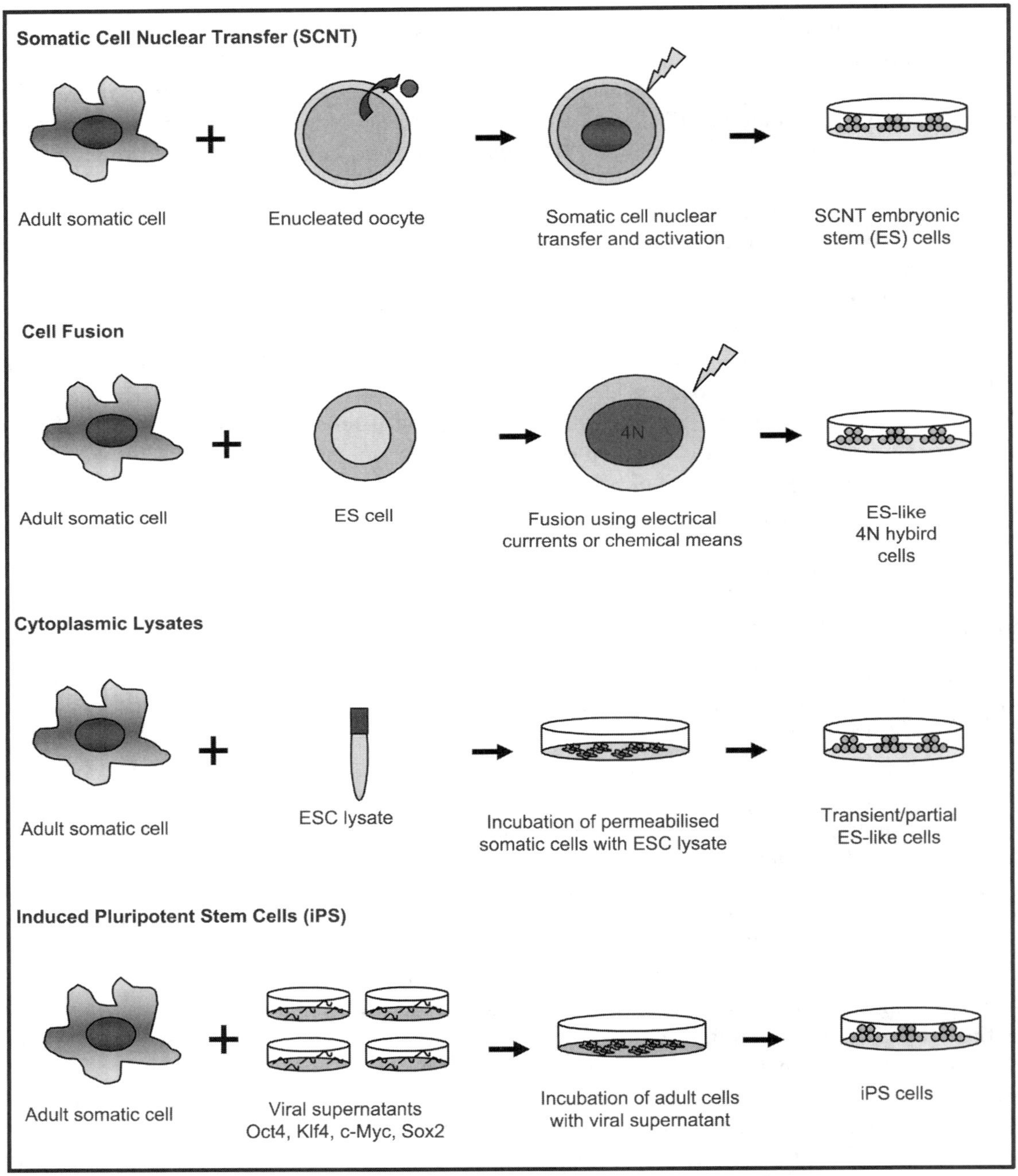

➤ **Fig. 11.2** *Various reprogramming methods. (For colour figure see Plate 27).*

The generation of clones is achieved using the SCNT technique. SCNT involves the removal of an oocyte's maternal chromosomes, followed by the injection and integration/fusion of the donor cell nucleus with the enucleated oocyte. This reconstructed oocyte is then triggered to begin development by either chemical or electrical stimulation. Once these develop into blastocysts, they can be transferred into a recipient animal for full-term development into a cloned animal. The production of cloned animals is known as "reproductive cloning", which contrasts with "therapeutic cloning" where the SCNT blastocysts are not transferred into recipients but used to derive ES cell lines *in vitro*. These SCNT ES cell lines would be autologous to the donor cells and could be theoretically transplanted into the donor without immune rejection. A proof of principle study was conducted by the Jaenisch group in 2002 [48]. Isogenic (from the same inbred strain) mouse ES cells were derived from SCNT blastocysts using donor cells from immune-deficient $Rag2^{-/-}$ mice. These $Rag2^{-/-}$ mice lack mature B- and T-lymphocytes in the lymphoid organs and do not have antibodies in their serum which is the typical phenotype observed in humans with SCID. After isolation of these SCNT ES cell lines, the Rag2 recombinase gene was corrected by targeted homologous recombination. These corrected SCNT ES cells ($Rag2^{+R/-}$) were used in two different ways:

- They were used in tetraploid complementation studies to generate clones derived entirely from $Rag2^{+R/-}$ ES cells. Cloned animals derived via this method had a restored immune system. In addition, these cloned $Rag2^{+R/-}$ mice were used as HSC donors for irradiated $Rag2^{-/-}$ mice recipients. The recipients showed restored immune function following transplantation. $Rag2^{+R/-}$ ES cells were differentiated *in vitro* to form embryoid bodies, and then retrovirally transduced with GFP and HoxB4—a gene that has been found to enable embryonic HSCs to stably engraft and chimerize the lymphoid and myeloid lineages of transplanted mice long term [49]. Next, these transduced Rag2+R/- ES cells were directed to differentiate into HSCs and subsequently transplanted into the $Rag2^{-/-}$ mice. The transduced $Rag2^{+R/-}$ ES cells engrafted and were able to reconstitute the haematopoietic system.

These results show that it is feasible to combine therapeutic cloning with gene therapy to treat a genetic disorder. However, it was observed that the HSCs derived from $Rag2^{+R/-}$ ES cells expressed low levels of MHC-I molecules, which would leave them vulnerable to NK-cell-mediated immune attack. Indeed, a low-level engraftment of HSCs was seen (1.5% chimerism) in $Rag2^{-/-}$ recipients, which have functional NK cells. When the $Rag2^{-/-}$ recipients had their NK cell function ablated, 95% chimerism was seen. This interesting result shows that even genetically matched cells and their derivatives derived from SCNT may be ineffective for the treatment of some disorders.

Reports of primate SCNT using adult donor cells are limited, possibly indicating a higher degree of difficulty in reprogramming these species. A study by Mitalipov et al. [50] showed that unlike other mammals, enucleated primate oocytes cannot induce essential nuclear remodelling required for reprogramming. This was determined by monitoring the protein, laminin A/C, which plays an integral role in maintaining the structure of the nuclear envelope and is removed by the maturation promoting factor (MPF) during nuclear envelope breakdown and premature chromosome condensation (PCC) at fertilisation. Since various re-organisations of the nuclear envelope take place during reprogramming, changes in laminin A/C allow the assessment of nuclear remodelling. It was found that there was a delayed and inefficient removal of laminin A/C due to lower levels of MPF. This was attributed to monkey oocytes being particularly vulnerable to MPF degradation during micromanipulations. Given the lack of nuclear remodelling, incomplete reprogramming was blamed for the current failures in monkey SCNT. Mitalipov and colleagues [51] proposed that this could be overcome by alterations to the conventional SCNT protocol and this proved successful when deriving Rhesus macaque monkey ES cell lines from blastocysts produced by SCNT. The authors attributed their success to the employment of the Oosight imaging system to visualize chromosomes using "birefringence" (polarized light) instead of using the conventional hoescht H33342 staining of chromosomes. It was postulated that the UV exposure might be damaging mitochondrial DNA, hindering success in non-human primate SCNT. Furthermore, calcium and magnesium was removed from the handling media to prevent MPF levels from dropping due to spontaneous oocyte activation. Using R. macaque monkey oocytes and adult fibroblasts as donor cells, 35 of 213 SCNT reconstructed oocytes showing cleavage developed into blastocysts (16%) using this modified protocol. Twenty of these blastocysts were used for ES cell derivation and two cell lines were created. Overall 304 monkey oocytes were used to generate these two SCNT cell lines, giving a low derivation efficiency of 0.7%. Independent genotyping using nuclear and mitochondrial DNA analyses verified that these ES cell lines were indeed derived from primate SCNT embryos and not from *in vitro* fertilized or parthenogenetic embryos [52].

While Byrne et al.'s study [51] is a landmark advance towards human therapeutic cloning, human SCNT studies are limited owing to the restricted availability of oocytes. Several studies have described production of early cleavage stage human SCNT embryos, but these did not proceed past the morula stage [53, 54]. Stojkovic et al. [55] were the first to report the development of a single human SCNT embryo to the blastocyst stage using an undifferentiated hESC as the donor cell and confirm their findings with DNA fingerprinting. Later, French et al. [56] showed development to the blastocyst stage of five SCNT embryos derived from differentiated adult fibroblast cells. In this study, comparatively high rates of blastocyst development were seen, 5 out of 21 SCNT embryos (23%), and the authors attributed this to the better quality of oocytes obtained from young women donors (20–24 year olds). One of these blastocysts was confirmed via nuclear and mitochondrial DNA analyses to be the result of SCNT. DNA fingerprinting of two other blastocysts indicated that they were generated by SCNT, but this could not be conclusively supported by mitochondrial DNA analysis. Unfortunately, none of these studies have been able to generate any hESC lines from SCNT-generated blastocysts.

Since it is possible to acquire only limited numbers of human oocytes, alternative sources have been suggested, including IVF aged or failed fertilisation oocytes. However, there are now reports that suggest that these oocytes are less capable of supporting human embryo development following SCNT, due to high levels of aneuploidy and aberrant spindles [57]. It would, however, be important to examine these alternative sources of oocytes using the modified SCNT protocols developed in non-human primates. In addition, French et al.'s work [56] suggests that using fresh oocytes from young fertile women is a critical factor for the development of human SCNT embryos. This further highlights the need to optimize human SCNT methodologies to increase the efficiencies of blastocyst formation and hESC derivation.

While the proof of principle mouse study has shown that SCNT-derived ES cells can be used in cellular therapy, the efficacy of this approach in circumventing the allogeneic immune response remains uncertain. The transfer of some donor cell cytoplasm containing mitochondrial DNA is common in the SCNT process and despite removal of maternal chromosomes from the oocyte, maternal mitochondrial DNA remains in enucleated oocytes. Therefore, reconstructed SCNT oocytes consist of a mixed pool of mitochondrial DNA which may be transmitted down to ES cell lines. It has not been determined if mitochondrial heteroplasmy will lead to chimeric expression of MHC molecules and whether this in turn leads to immune recognition and rejection. Furthermore, many animal clones have shown an array of problems from developmental abnormalities to aberrant gene expression patterns suggesting incomplete epigenetic reprogramming (animal clone abnormalities are reviewed by Li et al. [58]). It is unclear whether hESCs derived from SCNT-generated embryos will also have aberrations. It will first be necessary to generate hESC lines from human SCNT blastocysts as proof of principle before these concerns can be resolved.

There are many issues involved with the use of SCNT for the generation of autologous hESCs. It is a labour-intensive procedure involving the use of large numbers of human oocytes, making it impractical for widespread therapeutic use. Even if the technical and logistical aspects could be solved and efficiencies increased, the controversy surrounding destruction of SCNT blastocysts will remain. SCNT is a useful model for understanding the mechanisms behind reprogramming; however, its future use is envisaged more as a research tool and unlikely to translate into the clinical arena.

Cell fusion

Cell fusion describes the hybridisation of cells and can be induced experimentally using Sendai viruses, chemicals such as polyethylene glycol (PEG) and electrical currents [59]. In hybrids of pluripotent cells and more differentiated cell types, the phenotype of the pluripotent cell dominates. This was first shown in Miller and Ruddle's pioneering study [60], where fusion was performed between diploid murine embryonal carcinoma (EC) cells and diploid thymocytes. They found that the somatic cell hybrid took on the pluripotent characteristics of the EC cell, namely the hybrids, were able to produce subcutaneous tumours which contained derivatives of all three germ layers. However, it is unclear whether the chromosomes of the somatic cell partner had actually been reprogrammed or if they just remained silent.

Tada et al. [61] answered this question by demonstrating extensive demethylation of the somatic cell genome in embryonal germ cell (EGC)-thymocyte hybrids, a significant indicator of reprogramming since somatic cells are normally methylated. Furthermore, the silent imprinted maternal allele, Peg1/Mest, was reactivated in the thymocyte nucleus further supporting evidence of reprogramming to an EGC-like state. While ECs and EGCs appear to have powerful reprogramming abilities, their use in therapy is unlikely since ECs are derived from teratocarcinomas and EGCs cause the loss of parental imprints. Several studies now show that fusion of mouse and human ES cells to somatic cells can also produce hybrids with pluripotent properties.

Tada et al. [62] reported molecular reprogramming of the somatic cell genome in hybrids generated from murine male ES cells and female thymocytes carrying an Oct4-GFP transgene. GFP expression was seen 48 hours after fusion, suggesting rapid reprogramming of the somatic Oct4. Furthermore, it was found that the silent X chromosome derived from the somatic female nucleus adopted some characteristics of the active male X chromosome after fusion, specifically the replication and pattern of *Xist* (inactive X-specific transcript) expression found only in undifferentiated cells. The ability of ES cell-thymocyte hybrids to contribute to different tissues of chimeras provided further evidence of reprogramming. Interestingly, maternally imprinted genes *H19* and *Igf2r* remained methylated following fusion to ES cells which contrasted to Tada et al.'s earlier results [61] with EGCs. As there is no loss of parental imprints, this shows that ES cells are the best candidate for fusion-based reprogramming of somatic cells.

In similar studies conducted by Matveeva et al. [63, 64], mouse male hypoxanthine phosphoribosyl transferase (HPRT)-deficient ES cells were fused to female splenocytes. In the following selection on HAT medium, hybrids were found to have three synchronously replicating X chromosomes suggesting reprogramming of the somatic female X chromosome. Also, the ES hybrids were able to form embryoid bodies *in vitro* and differentiate into derivatives of all three germ layers. Notably, it was reported that three out of four hybrid clones obtained were near diploid. Microsatellite analysis was used to discriminate between the two different strains of mice representing the parental cells used. In the near-diploid hybrids, it was shown that most of the somatic partner chromosomes had been eliminated. This finding correlates with the extensive chromosome loss seen in hybridoma fusion studies, highlighting the karyotypic instability of hybrids.

Evidence that the reprogramming ability of pluripotent cells is conserved in humans was provided in a study by Cowan et al. [65]. In this study, hygromycin-resistant and constitutively expressing GFP hESCs were fused to puromycin-resistant human fibroblasts. After double antibiotic selection, resulting hybrids were found to have morphology and growth characteristics comparable to hESCs and also expressed human embryonic specific antigens such as SSEA-1 and TRA-1-61. In addition, the fibroblast nucleus showed expression of pluripotency genes, whereas fibroblast-specific genes were repressed. Lastly, the promoter region of Oct-4, which is methylated in fibroblasts, became unmethylated, emulating hESCs. Yu et al. [66] fused hESCs with hES-cell-derived myeloid precursors that possessed an Oct4-GFP transgene knocked in at the Oct4 locus [66]. The hybrids generated expressed GFP, indicating that Oct4 was re-activated while myeloid-specific antigens could not be detected. When hybrid cells were differentiated *in vitro* into embryoid bodies, genes characteristic of the three germ layers were upregulated. These studies provide the closest evidence yet of using hESCs in fusion-based reprogramming of human somatic cells to a pluripotent state.

Unfortunately, all hybrid cells generated are tetraploid and contain DNA from both cell types and so will have limited therapeutic applications since they are not autologous and likely to be rejected. This was highlighted in Tada et al.'s [67] study where they injected somatic cells differentiated from mouse ES cell-thymocyte hybrids subcutaneously into the thymocyte donor resulting in transplant rejection. This was supported by the evidence of co-expression of somatic cell and ES cell MHC-I molecules in the teratomas generated by the injection of hybrid cells in SCID mice. Therefore, for cell fusion to be a viable approach for generating autologous hESCs, the ES cell nucleus must be removed. An important question arising from this is where the cellular location of the factors responsible for reprogramming in ES cells resides. Do and Scholer [68] attempted to investigate this by separating the nucleus (karyoplast) and cytoplasm (cytoplast) of mouse ES cells and then fusing each to neomycin-resistant LacZ-positive neurosphere cells with an *Oct4-GFP* transgene [68]. GFP-positive cells were observed only in karyoplast fusions, prompting the authors to claim that nuclear factors play a vital role in reprogramming. These findings contrast with Strelchenko's study [69] showing that human somatic cells could be reprogrammed with hESC cytoplasts as evidenced by the expression of pluripotent markers, Oct4 and TRA-2-3. The somatic cells used were lymphocytes, fibroblasts and UCB cells with a male karyotype, while the hESC cytoplasts had a female karyotype. Interestingly, a majority of the cybrids (36/41) had a female karyotype and only five clones had a Y chromosome but in mixed and abnormal karyotypes. However, it should be noted that both these studies could not unequivocally demonstrate that fusion had occurred between cytoplasts and somatic cells. Development of molecular methods to track the cytoplast in fusions is essential so that true cybrids can be isolated.

It has been argued that enucleating ES cells leaves insufficient cytoplasm for reprogramming somatic nuclei due to the high nucleus-to-cytoplasm volume ratio of ES cells [70]. It has been suggested that larger polyploid ES cells could be generated first to create more ES cell cytoplasm for reprogramming [71, 72]. Pralong and colleagues [71] produced mouse tetraploid ES cells carrying two different antibiotic resistance cassettes that retained expression of pluripotent markers, were able to form teratomas containing derivatives of all three germ layers and could contribute to the inner cell mass of blastocysts. As proof of principle, the authors described a method known as "post-fusion enucleation" where they created heterokaryons by fusing tetraploid ES cells with diploid ES cell karyoplasts and then selectively removed the heavier tetraploid nucleus by differential centrifugation 48 hours later. Since formation of the heterokaryon is transient, it will be vital to determine whether there is sufficient time to reprogram a somatic nucleus especially in the absence of cell division or DNA replication. If post-fusion enucleation could be achieved between tetraploid ES cells and diploid somatic cells with evidence of reprogramming of the somatic cell genome, this would be of great therapeutic interest.

An alternative method of eliminating the ES cell genome was proposed by Matsumura and colleagues [73]. In this method, hybrids were generated between mouse thymocytes and ES cells containing a universal chromosome elimination cassette (CEC), which consisted of a GFP reporter and puromycin resistance flanked between oppositely oriented loxP sites. These CECs were generated at either chromosomes 11 (CEC11) and 12 (CEC12) for their selective elimination. It was hypothesized that cre-mediated sister-chromatid recombination in the late S and G2 phases of the cell cycle would create dicentric and nullicentric chromosomes, which would be naturally deleted from the cell during cell division. After cre-selection of CEC11 hybrids, only three copies of chromosome 11 were detected in 26.1% of GFP-negative hybrids and in CEC12 hybrids, only three copies of chromosome 12 were detected in 88.4% of GFP-negative hybrids. Despite differences in chromosome elimination frequency between the two integration sites, Matsumura and colleagues [73] have proven that it is possible to target elimination of any desired chromosome. As the MHC-I and II genes are clustered on mouse chromosome 17 and human chromosome 6, targeted elimination of these chromosomes from the ES cell genome in ES-somatic-cell hybrids may provide a source of MHC-matched hybrid cells. However, this potentially would result in karyotypic instability due to loss of chromosomes.

Cytoplasmic lysates or extracts treatment

As SCNT and cell fusion studies show that oocytes and pluripotent cells possess reprogramming factors, it has been suggested that treatment with extracts of these cells can induce reprogramming and/or transdifferentiation [74, 75]. Treatment with cytoplasmic lysates or extracts usually involves the reversible permeabilisation of differentiated cells to ensure uptake of the reprogramming factors present in the extracts. Reversible permeabilisation has been achieved using streptolysin O (SLO)—a pore-forming toxin that allows molecules of up to 100 kDa to pass into the cytosol and cells can reseal upon treatment with calcium calmodulin without compromising viability [76]. On the other hand, cell extracts are prepared by ultra sonication of large numbers of cells in the presence of a HEPES-buffered cell lysis solution [75].

The first studies to demonstrate transdifferentiation using mammalian differentiated somatic cell extracts were completed by Collas and colleagues [75, 77]. Their 2002 study demonstrated transdifferentiation of HEK293T (human embryonic kidney) cells and human 293T fibroblasts into lymphoid-like cells and T-cells. This was achieved by permeabilizing HEK293T cells with SLO and exposing them for a short period to stimulated T-cell (Jurkat cell line) extracts. Subsequent analysis revealed expression of T-cell-specific antigens and expression of the IL-2 gene which is normally repressed in fibroblasts. To further support epigenetic reprogramming in these 293T fibroblasts, histone H4 acetylation of the IL-2 proximal promoter region was observed which indicates active chromatin remodelling. Microarray analysis also revealed activation of haematopoietic genes and downregulation of 293 T-specific genes. Most of these modifications persisted over weeks of cell culture.

Using *Xenopus* oocyte extracts, Hansis et al. [78] demonstrated incomplete reprogramming of digitonin permeabilized HEK293T cells and human primary leukocytes. Only upregulation of Oct4 and germ cell alkaline phosphatase (GCAP) was seen in extract treated cells. Interestingly, the authors went on to screen for factors responsible for reprogramming and found that antibody depletion of the chromatin remodelling ATPase, BRG1 or the expression of dominant negative BRG1 abolished the reprogramming ability of *Xenopus* extracts.

Experiments using mammalian pluripotent cell extracts have now been described. Taranger et al. [79] treated human HEK293T cells and mouse NIH3T3 fibroblasts with human EC cell (NCCIT) extracts and with mouse ES cell extracts, respectively. The NCCIT-treated HEK293T cells formed colonies that were maintained for at least 23 passages and showed upregulation of ES-cell-specific markers and downregulation of HEK293T-specific genes. The *Oct4* promoter of these cells was also demethylated indicating an active state. NCCIT-treated cells could also be induced to differentiate towards the neurogenic, adipogenic, osteogenic and endothelial antigens.

Extract-treated NIH3T3 fibroblasts lead to the formation of ES-cell-like colonies after 4–9 days and these expressed Oct4 transcripts and alkaline phosphatase—key indicators of pluripotency. They also formed embryoid-like bodies and differentiated *in vitro* into derivatives of the three germ layers. Neri et al. [80] showed partial reprogramming of the immortalized mouse cell lines: STO and NIH3T3 cells when treated with mouse ES cell extracts. Approximately 0.003–0.04% of treated cells were positive for alkaline phosphatase and showed expression of Oct4 and Rex1 via RT-PCR analyses. However, cells were not positive for Oct4 by immunohistochemistry analyses suggesting that RT-PCR analyses may only be detecting transcripts already present in the extracts. Furthermore, there is a significant probability that ES cells could have contaminated the experiments since it was shown that they could survive extract preparation. The extent of reprogramming is unclear from these studies since the extract-treated mouse and human cells were not shown to differentiate *in vivo* by teratoma formation in SCID mice, nor were the treated mouse cells shown to contribute to chimeric embryos—the most strict tests of pluripotency. Also, cell extract studies have mainly concentrated on transformed cell lines. It remains to be seen whether this can be applied to primary cell lines. Very few studies have shown stable reprogramming and transdifferentiation using cytoplasmic cell extracts. Neri et al. [80] pointed out that their experiments were not as effective as those performed by Taranger et al. [79, 80]. This highlights a need for robust protocols to be developed. In comparison to SCNT and cell fusion, the potential for generating autologous hES-like cells without genetic modification is limited at present.

Induced pluripotency by viral transduction

In 2006, Yamanaka's work [81] describing the viral transduction of genes to induce pluripotency in mouse fibroblasts provided a major breakthrough in reprogramming studies. In this study, the authors selected 24 genes thought to be pivotal in maintaining the pluripotency of ES cells, and generated retroviral vectors encoding the cDNA of each of these 24 genes. These retroviral vectors were then introduced into mouse embryonic fibroblasts (mEFs) that possessed a knock-in of a bgeo cassette (a fusion of the β-galactosidase and neomycin resistance genes) at the *Fbx15* locus [82]. *Fbx15* is a novel target gene of Oct3/4 that is expressed specifically in mouse ES cells and early embryos but is considered dispensable for the maintenance of pluripotency and mouse development. As such, all ES cells homozygous for the *Fbx15*–βgeo construct (Fbx15$^{\beta geo/\beta geo}$) are resistant to high concentrations of G418 treatment, but Fbx15$^{\beta geo/\beta geo}$ mEFs are sensitive to normal concentrations of G418. When all 24 candidate genes were transfected into Fbx15$^{\beta geo/\beta geo}$ mEFs and put under G418 treatment for 2–3 weeks, surviving colonies displayed ES-cell-like morphology and proliferation properties. Furthermore, the colonies expressed many pluripotency markers and the endogenous *Fbx15* and *Nanog* promoters had become demethylated. Interestingly, the *Oct3/4* promoter remained methylated. These surviving colonies were subsequently named "induced pluripotent stem (iPS) cells". When the authors only transfected Fbx15$^{\beta geo/\beta geo}$ mEFs with one candidate gene, no G418-resistant colonies were seen suggesting that no single candidate gene was sufficient to activate the Fbx15 locus. In order to determine which genes or factors were most critical for inducing the expression of ES cell marker genes, the authors transfected Fbx15$^{\beta geo/\beta geo}$ mEFs and sequentially withdrew candidate genes. By repeating this approach, the authors limited to four pool of candidate genes that are essential for obtaining iPS colonies:

1. *Oct4* (a key pluripotency gene)
2. *Sox2* (a key pluripotency gene working in conjunction with *Oct4*)
3. *c-myc* (an oncogene that regulates cell proliferation)
4. *Klf-4* (supports the self-renewal of ES cells and augments the levels of *Oct4*).

The four factors were also transfected into mouse tail tip fibroblasts (TTFs) and yielded iPS cells. Using primers that amplify only transcripts of the endogenous genes, it was reported that iPS cells (from both mEFs and TTFs) expressed most markers of pluripotency. iPS cells could also be differentiated into derivatives of three germ layers *in vitro* following embryoid body formation and *in vivo* after teratoma formation in SCID mice. In addition, when injected into mouse blastocysts, iPS cells contributed to various tissues of the chimeric embryos. However, several observations suggest that iPS cells are not identical to ES cells. Namely, the *Oct3/4* and *Nanog* promoters of iPS cells were partially methylated; iPS cells did not contribute to any tissues of the chimeric animals after full-term development. They did not express the ES cell marker, ECAT1; and global gene expression profiles of iPS cells were closely clustered to ES cells but not identical. Taken together, these results show that it is possible to directly return adult somatic cells to a pluripotent-like state without the use of oocytes or ES cells, but it was uncertain whether the reprogramming in these iPS cells was complete. More complete reprogramming is seen in ES cells derived from SCNT and in hybrids from cell fusion; therefore, it has been suggested that iPS cells are either trapped in an intermediate state between somatic cells and ES cells or a completely different type of pluripotent cell such as EC cells [38].

Similar studies have now replicated Yamanaka's results by transfecting the four factors into the *Oct4* and *Nanog* promoters in mEFs and mouse TTFs instead of at the Fbx15 locus [83–86]. The Oct4 and Nanog-selected iPS cells expressed ECAT1, were able to form viable adult chimeric animals with iPS contribution to the germ line and could form live late-term embryos when injected into tetraploid blastocysts [83, 84]. Another study transfected female mEFs with the four factors into the Nanog locus and the iPS cells generated showed reactivation of the silent X chromosome and histone methylation patterns that were very similar to ES cells [85]. These reports suggest that Oct-4- and Nanog-selected iPS cells result in a different pluripotent cell type with a greater pluripotent potential than with Fbx15-selected iPS cells. Surprisingly, Nanog was found to be dispensable and not one of the crucial four factors, as discussed earlier. These Nanog-selected iPS cell studies in conjunction with the fact that Nanog is a target of Oct4 and Sox2 genes demonstrate that Nanog also plays a critical role in epigenetic reprogramming. This is supported by Silva et al.'s report [87] showing that when mouse ES cells overexpressed Nanog and were subsequently fused with somatic cells, it would lead to 200-fold more colonies than in fusions with normal ES cells.

The iPS strategy has now been translated into human systems by transfecting the four factors into hES-cell-derived foetal fibroblasts, MSCs, foetal, neonatal and adult primary dermal fibroblasts [88–90]. iPS cells derived from human somatic cells were found to be very similar to hESCs with regard to morphology, proliferation, feeder dependence, surface markers, gene expression, promoter demethylation patterns, telomerase activities and capacity for *in vitro* and *in vivo* differentiation into derivatives of the three germ layers. However, the human iPS cells generated were not identical to hESCs as shown by DNA microarray data [87]. Interestingly, the four introduced vectors were found to be silenced in human iPS cells suggesting that the somatic cells are not dependent on the continual expression of transgenes to maintain ES-like properties [89, 90].

Evidence that iPS cells can be used in therapy was given in a proof of principle study conducted by Hanna et al. [91]. The authors generated mouse iPS cells by transfecting the four factors into TTFs of mice with a humanized knock-in of sickle cell anaemia gene, α-*globin*. The iPS cells generated were corrected for sickle cell anaemia by targeted homologous recombination using the wild-type β-*globin* gene and then differentiated into haematopoietic progenitors *in vitro*. These iPS-cell-derived haematopoietic progenitors were transplanted into isogenic mice which showed multi-lineage reconstitution of the bone marrow and rescued the disease phenotype. These experiments show that it is possible to generate autologous iPS cells, correct a genetic defect, differentiate them into a desired cell type and transplant them in therapy without the use of immunosuppressive drugs. Despite this proof of principle study in mice, the human therapeutic use of iPS cells generated using current methods is doubtful given the use of retroviruses and *c-myc*. This was outlined by Okita et al. [84] who found that 20% of offspring derived from mouse iPS cells developed tumours which could be attributed to the reactivation of the *c-myc* transgene. This issue has been somewhat resolved in recent studies showing that both *Klf4* and *c-myc* are dispensable for the induction of iPS cells. Park et al. [89] showed that either *c-myc* or *Klf4* alone is necessary for iPS colony formation, albeit at a lower efficiency [89]. While Yu et al. [92] showed that *c-myc* and *Klf4* could be replaced by Nanog and LIN28. In addition, Nakagawa et al. [93] have shown that it is possible to generate both mouse and human iPS cells without *c-myc*. Fewer colonies were observed when Nanog-selected iPS cells were generated with only three factors but the iPS cells expressed ES cell marker genes at levels similar to ES cells and could contribute to tissues of the three germ layers in adult chimeras. Notably, none of the chimeric mice had developed tumours but longer term monitoring is required to determine whether chimeras will develop tumours later in life. The withdrawal of c-myc from the four factors increases the potentiality of iPS methodologies in therapy. However, the use of retroviral vectors must still be overcome since they could potentially introduce mutations into the insertion site upon integration. Use of adenoviral vectors and replacement of DNA factors with chemical compounds are currently being investigated [94–97]. Extensive research is still required to determine the safety and efficacy of the iPS approach for human therapy. Non-etheless, iPS technology is an invaluable tool as it is now easier to produce panels of many different cell lines for drug screening and research.

CONCLUSION

Cell therapy promises to alleviate a range of currently incurable diseases; however, there are inherent limitations to this approach whether using adult or embryonic stem cells. A key question is whether niches can be recapitulated *in vitro*. Cells normally act synergistically *in vivo* and being able to mimic this in culture conditions presents a huge challenge to researchers. For instance, how can a neuron be stimulated *in vitro*? Furthermore, it remains uncertain whether cell therapy would be beneficial for treatment of diseases, such as Parkinson's disease, which are also environmentally influenced. Other issues that need consideration include the long-term propagation of adult stem cells *in vitro* and the scale up of hESC production and differentiation, all of which needs to be done in the presence of defined media without animal components. Non-etheless, techniques described here are still valuable scientific and diagnostic tools with potential for use in research of complex diseases (with genetic components) and drug discovery. The earliest biomedical benefits of stem cell technologies are envisaged not from therapeutic use but in the understanding of diseases, early diagnosis of debilitating diseases and drug discovery.

REFERENCES

1. Orlic D., J. Kajstura, S. Chimenti, I. Jakoniuk, S.M. Anderson, B. Li, J. Pickel, R. Mckay, B. Nadal-Ginard, D.M. Bodine, A. Leri, P. Anversa (2001). 'Bone marrow cells regenerate infarcted myocardium'. *Nature*. **410**: 701–705 .
2. Ying Q., J. Nichols, E.P. Evans, A. Smith (2002). 'Changing potency by spontaneous fusion'. *Nature*. **416**: 545–548.
3. Terada N., T. Hamazaki, M. Oka, M. Hoki, D.M. Mastalerz, Y. Nakano, E.M. Meyer, L. Morel, B.E. Petersen, E.W. Scott (2002). 'Bone marrow cells adopt the phenotype of other cells by spontaneous cell fusion'. *Nature*. **416**: 542–545.
4. Pells S., A.I. Di Domenico, E.J. Gallagher and J. Mcwhir (2002). 'Multipotentiality of neuronal cells after spontaneous fusion with embryonic stem cells and nuclear reprogramming *in vitro'*. *Cloning. Stem Cells.* **4**: 331–338.
5. Bradley J.A., E.M. Bolton and R.A. Pedersen (2002). 'Stem cell medicine encounters the immune system'. *Nat Rev Immunol.* **2**: 859–871.
6. Janeway C.A., P. Travers, M. Walport and M. Shlomick. 'Immunobiology: The Immune System in Health and Disease'. (2001). Garland Publishing, New York.
7. Pescovitz M. D., J.R.J. Thistlethwaite, H.J. Auchincloss, S.T. Ildstad, T.G. Sharp, R. Terrill and D.H. Sachs (1984). 'Effect of class II antigen matching on renal allograft survival in miniature swine'. *J. Exp. Med.* **160**: 1495–1508.
8. Morita N., B. Munkhbat, B. Gansuvd, N. Kanai, M. Hagihara, J. Shimazaki, K. Tsubota and K. Tsuji (1998). 'Effect of HLA-A and -DPB1 matching in corneal transplantation. Transplantation'. *Proc.* **30**: 3491–3492.
9. Opelz G. (1997). 'Impact of HLA compatibility on survival of kidney transplants from unrelated live donors'. *Transplantation.* **64**: 1473–1475.
10. Simpson E., D. Scott and P. Chandler (1997). 'The male specific histocompatibility antigen, H-Y: A history of transplantation'. *Ann. Rev. Immunol.* **15**: 39–61.
11. Yang Y.G. and M. Sykes (2007). 'Xenotransplantation: Current status and a perspective on the future'. *Nat. Rev. Immunol.* **7**: 519–531.
12. Schuurman H.J. and R.N. Pierson (2008). 'Progress towards clinical xenotransplantation'. *Fron. Biosci.* **13**: 204–220.
13. Mimeault M. and S.K. Batra (2006). 'Concise review: Recent advances on the significance of stem cells in tissue regeneration and cancer therapies'. *Stem Cells.* **24**: 2319–2345.

14. Challen G.A. and M.H. Little (2006). 'A side order of stem cells: The SP phenotype'. *Stem Cells.* **24**: 3–12.
15. Verfaillie C.M. (2002). 'Adult stem cells: assessing the case for pluripotency'. *Trend. Cell. Biol.* **12**: 502–508.
16. Pelligrini G., C.E. Traverso, P. Paterna, A. Lambiase, S. Bonini, P. Rama and M. De Luca (1997). 'Long-term restoration of damaged corneal surfaces with autologous cultivated corneal epithelium'. *Lancet.* **349**: 990–993.
17. Pittenger M.F., A.M. Mackay, S.C. Beck, R.K. Jaiswal, R. Douglas, J.D. Mosca, M.A. Moorman, S.D.W., S. Craig and D.R. Marshak (1999). 'Multilineage potential of adult human mesenchymal stem cells'. *Science.* **284**: 143–147.
18. Brooke G., M. Cook, C. Blair, R. Han, C. Heazlewood, B. Jones, M.Kambouris, K. Kollar, S. Mctaggart, R. Pelekanos, A. Rice, T. Rossetti and K. Atkinson (2007). 'Therapeutic applications of mesenchymal stromal cells'. *Semin. Cell. Dev. Biol.* **18**: 846–858.
19. Le Blanc K. and O. Ringden (2005). 'Immunobiology of human mesenchymal stem cells and future use in haematopoietic stem cell transplantation'. *Biol. Blood. Marrow. Transplant.* **11**: 321–334.
20. Ren G., L. Zhang, X. Zhao, G. Xu, Y. Zhang, A.I. Roberts, R.C. Zhao and Y. Shi (2008). 'Mesenchymal Stem Cell-Mediated Immunosuppression Occurs via Concerted Action of Chaemokines and Nitric Oxide'. *Cell. Stem Cell.* **2**: 141–150.
21. Ortiz L. A., F. Gambelli, C. Mcbride, D. Gaupp, M. Baddoo, N. Kaminiski and D.G. Phinney (2003). 'Mesenchymal stem cell engraftment in lung is enhanced in response to bleomycin exposure and ameliorates its fibrotic effects'. *Proc. Nat. Acad. Sci. USA.* **100**: 8407–8411.
22. Blanc K. Le, I. Rasmusson, B. Sundburg, C. Gotherstrom, M. Hassan, M. Uzunel and O. Ringden (2004). 'Treatment of severe acute graft-versus-host disease with third party haploidentical mesenchymal stem cells'. *Lancet.* **363**: 1439–1441.
23. Ringden O., M. Uzunel, I. Rasmusson, M. Remberger, B. Sundburg, H. Lonnies, H.U. Marschall, A. Dlugosz, A. Szakos, Z. Hassan, B. Omazic, J. Aschan, L. Barkholt and K. Le Blanc (2006). 'Mesenchymal stem cells for treatment of therapy-resistant graft-versus-host disease'. *Transplantation.* **81**: 1390–1397.
24. Ilancheran S., A. Michalska, G. Peh, E.M. Wallace, M. Pera and U. Manuelpillai (2007). 'Stem cells derived from human foetal membranes display multilineage differentiation potential'. *Biol. Reprod.* **77**: 577–588.
25. Aluvihare V.R., M. Kallikourdis and A.G. Betz (2004). 'Regulatory T cells mediate maternal tolerance to the foetus'. *Nat. Immunol.* **5**: 266–271.
26. Niederkorn J.Y. (1999). 'The immune privilege of corneal allografts'. *Transplantation.* **67**: 1503–1508.
27. Wilbanks G.A. and J.W. Streilein (1992). 'Fluids from immun privileged sites endow macrophages with the capacity to induce antigen-specific immune deviation via a mechanism involving transforming growth factor-beta'. *Eur. J. Immunol.* **22**: 1031–1036.
28. Bellgrau D., D. Gold, H. Selawry, A. Moore, A. Franzusoff and R.C. Duke (1995). 'A role for CD95 ligand in preventing graft rejection'. *Nature.* **377**: 630–632.
29. Morizane A., J.Y. Li and P. Brundin (2008). 'From bench to bed: the potential of stem cells for the treatment of Parkinson's disease. Cell'. *Tissue. Res.* **331**: 323–336.
30. Drukker M., G. Katz, A. Urbach, M. Schulinder, G. Markel, J. Itskovitz-Eldor, B. Reubinoff, O. Mandelboim and N. Benvenisty (2002). 'Characterisation of the expression of MHC proteins in human embryonic stem cells. *Proc. Nat. Acad. Sci. USA.* **99**: 9864–9869.

31. Li L., M.L. Baroja, A. Majumdar, K. Chadwick, A. Rouleau, L. Gallacher, I. Ferber, J. Lebkowski, T. Martin, J. Madrenas and M. Bhatia (2004). 'Human embryonic stem cells possess immune-privileged properties'. *Stem Cells*. **22**: 448–456.
32. Drukker M., H. Katchman, G. Katz, S.E.-T. Friedman, E. Shezen, E. Hornstein, O. Mandelboim, Y. Reisner and N. Benvenisty (2006). 'Human embryonic stem cells and their differentiated derivatives are less susceptible to immune rejection than adult cells'. *Stem Cells*. **24**: 221–229.
33. Taylor C.J., E.M. Bolton, S. Pocock, L.D. Sharples, R.A. Pedersen and J.A. Bradley (2005). 'Banking on human embryonic stem cells: estimating the number of donor cell lines needed for HLA matching'. *Lancet*. **366**: 2019–2025.
34. Seach N., D. Layton, J. Lim, A. Chidgey and R. Boyd (2007). 'Thymic generation and regeneration: A new paradigm for establishing clinical tolerance of stem cell-based therapies'. *Curr. Opin. Biotechnol.* **18**: 441–447.
35. Wilmut I., A.E. Schnieke, J. Mcwhir, A.J. Kind and K.H.S. Campbell (1997). 'Viable offspring derived from foetal and adult mammalian cells'. *Nature*. **385**: 810–813.
36. Revazova E.S., N.A. Turovets, O.D. Kochetkova, L.B. Kindarova, L.N. Kuzmichev, J.D. Janus and M.V. Pryzhkova (2007). 'Patient-specific stem cell lines derived from human parthenogenetic blastocysts'. *Cloning. Stem. Cell.* **9**: 1–18.
37. Kim K., K. Ng, P.J. Rugg-Gunn, J.H. Shieh, O. Kirak, R. Jaenisch, T. Wakayama, M.A. Moore, R.A. Pedersen and G.Q. Daley (2007). 'Recombination Signatures Distinguish Embryonic Stem Cells Derived by Parthenogenesis and Somatic Cell Nuclear Transfer'. *Cell. Stem. Cell.* **1**: 346–352.
38. Rodolfa K.T. and K. Eggan (2006). 'A transcriptional logic for nuclear reprogramming'. *Cell*. **126**: 652–655.
39. Briggs R. and T.J. King (1952). 'Transplantation of living nuclei from blastula cells into enucleated frogs' eggs'. *Zoology*. **38**: 455–463.
40. Gurdon J.B (1962). 'The developmental capacity of nuclei taken from intestinal epithelial cells of feeding tadpoles'. *J. Embryol. Exp. Morphol.* **10**: 622–640.
41. Berardino M.A. Di (2006). 'Origin and progress of nuclear transfer in non-mammalian animals. In: Nuclear Transfer Protocols Cell Reprogramming and Transgenesis' (P.J. Verma, A.O. Trounson, eds.), 3–31. Humana Press Inc., Toronto.
42. Wakayama T., A.C.F. Perry, M. Zuccotti, K.R. Johnson and R. Yanagimachi (1998). 'Full-term development of mice from enucleated oocytes injected with cumulus cell nuclei'. *Nature*. **394**: 318–319.
43. Kato Y., T. Tani, Y. Sotomaru, K. Kurokawa, J. Kato, H. Doguchi, H. Yasue and Y. Tsunoda (1998). 'Eight calves cloned from somatic cells of a single adult'. *Science*. **282**: 2095–2098.
44. Onishi A., M. Iwamoto, T. Akita, S. Mikawa, K. Takeda, T. Awata, H. Hanada and A.C.F. Perry (2000). 'Pig cloning by microinjection of foetal fibroblast nuclei'. *Science*. **289**: 1188–1190.
45. Hochedlinger K. and R. Jaenisch (2002). 'Monoclonal mice generated by nuclear transfer from mature B and T donor cells'. *Nature*. **415**: 1035–1038.
46. Li J., T. Ishii, P. Feinstein and P. Mombaerts (2004). 'Odorant receptor gene choice is reset by nuclear transfer from mouse olfactory sensory neurons'. *Nature*. **428**: 393–399.
47. Eggan K., K. Baldwin, M. Tackett, J. Osborne, J. Gogos, A. Chess, R. Axel and R. Jaenisch (2004). 'Mice cloned from olfactory sensory neurons'. *Nature*. **428**: 44–49.
48. Rideout W.M. Iii, K. Hochedlinger, M. Kyba, G.Q. Daley and R. Jaenisch (2002). 'Correction of a genetic defect by nuclear transplantation and combined cell and gene therapy'. *Cell*. **109**: 17–27.

49. Kyba M., R.C.R. Perlingeiro and G.Q. Daley (2002). 'HoxB4 Confers Definitive Lymphoid-Myeloid Engraftment Potential on Embryonic Stem Cell and Yolk Sac Haemopoietic Progenitors'. *Cell.* **109**: 29–37.

50. Mitalipov S.M., Q. Zhou, J.A. Byrne, W.Z. Ji, R.B. Norgren and D.P. Wolf (2007). 'Reprogramming following somatic cell nuclear transfer in primates is dependent upon nuclear remodelling'. *Hum. Reprod.* **22**: 2232–2242.

51. Byrne J.A., D.A. Pedersen, L.L. Clepper, M. Nelson, W.G. Sanger, S. Gokhale, D.P. Wolf and S.M. Mitalipov (2007). 'Producing primate embryonic stem cells by somatic cell nuclear transfer'. *Nature.* **450**: 497–502.

52. Cram D.S., B. Song and A.O. Trounson (2007). 'Genotyping of Rhesus SCNT pluripotent stem cell lines'. *Nature.* **450**: E12–E14.

53. Heindryckx B., P. De Sutter, J. Gerris, M. Dhont and J. Van Der Elst (2007). 'Embryo development after successful somatic cell nuclear transfer to *in vitro* matured germinal vesicle oocytes'. *Hum. Reprod.* **22**: 1982–1990.

54. Cibelli J.B., A.A. Kiessling, K. Cuniff, C. Richards, R. Lanza and P. West (2001). 'Somatic cell nuclear transfer in humans: pronuclear and early embryonic development'. *J. Regen. Med.* **2**: 25–31.

55. Stojkovic M., P. Stojkovic, C. Leary, V.J. Hall, L. Armstrong, M. Herbert, M. Nesbitt, M. Lako and A. Murdoch (2005). 'Derivation of a human blastocyst after heterologous nuclear transfer to donated oocytes'. *Reprod. Biomed. Online.* **11**: 226–231.

56. French A.J., C.A. Adams, L.S. Anderson, J.R. Kitchen, M.R. Hughes and S.H. Wood (2008). 'Development of human cloned blastocysts following somatic cell nuclear transfer (SCNT) with adult fibroblasts'. *Stem. Cells.* **26**: 494–495.

57. Hall V.J., D. Compton, P. Stojkovic, M. Nesbitt, M. Herbert, A. Murdoch and M. Stojkovic (2007). 'Developmental competence of human *in vitro* aged oocytes as host cells for nuclear transfer'. *Hum. Reprod.* **22**: 52–62.

58. Li X., Z. Li, A. Jouneau, Q. Zhou and J.P. Renard (2003). 'Nuclear transfer: progress and quandaries'. *Reprod. Biol. Endocrinol.* **1**: 84–89.

59. Lucas J.L. and N. Terada (2003). 'Cell fusion and plasticity'. *Cytotechnology.* **41**: 103–109.

60. Miller R.A. and F.H. Ruddle (1976). 'Pluripotent teratocarcinoma-thymocyte somatic cell hybrids'. *Cell.* **9**: 45–55.

61. Tada M., T. Tada, L. Lefebvre, S.C. Barton and M.A. Surani (1997). 'Embryonic germ cells induce epigenetic reprogramming of somatic nucleus in hybrid cells'. *EMBO. J.* **16**: 6510–6520.

62. Tada M., Y. Takahama, K. Abe, N. Nakatsuji and T. Tada (2001). 'Nuclear reprogramming of somatic cells by *in vitro* hybridisation with ES cells'. *Curr. Biol.* **11**: 1553–1558.

63. Matveeva N.M., A.G. Shilov, E.M. Kaftanovskaya, L.P. Maximovsky, A.I. Zhelezova, A.N. Golubitsa, S.I. Bayborodin, M.M. Fokina and O.L. Serov (1998). '*In vitro* and *in vivo* study of pluripotency in intraspecific hybrid cells obtained by fusion of murine embryonic stem cells with splenocytes. Mol. Reprod'. *Dev.* **50**: 128–138.

64. Serov O.L., N.M. Matveeva, S. Kuznetsov and E.M. Kaftanovskaya (2001). 'Embryonal cell hybrids: New opportunities for studying pluripotency and reprogramming of the differentiated cell chromosomes'. *Biol. Bull. Acad. Sci. USSR* **28**: 601–605.

65. Cowan C.A., J. Atienza, D.A. Melton and K. Eggan (2005). 'Nuclear reprogramming of somatic cells after fusion with human embryonic stem cells'. *Science.* **309**: 1369–1373.

66. Yu J., M.A. Vodyanik, P. He, I.I. Slukvin and J.A. Thomson (2006). 'Human embryonic stem cells reprogram myeloid precursors following cell-cell fusion'. *Stem. Cells.* **24**: 168–176.
67. Tada M., A. Morizane, H. Kimura, H. Kawasaki, J.F.X. Ainscough, Y. Sasai, N. Nakatsuji, T. Tada (2003). *Pluripotency of reprogrammed somatic genomes in embryonic stem hybrid cells.* **227**(4): 504–510.
68. Do J.T. and H.R. Scholer (2004). 'Nuclei of embryonic stem cells reprogram somatic cells'. *Stem. Cells.* **22**: 941–949.
69. Strelchenko N., V. Kukharenko, A. Shkumatov, O. Verlinsky, A. Kuliev and Y. Verlinsky (2006). 'Reprogramming of human somatic cells by embryonic stem cell cytoplast. Reprod. Biomed'. *Online.* **12**: 107–111.
70. Pralong D., A.O. Trounson and P.J. Verma. 'Cell fusion for reprogramming pluripotency: Towards elimination of the pluripotent genome'. *Stem Cells Rev.* **2**: 331–340.
71. Pralong D., M.L. Lim, I. Vassiliev, K. Mrozik, N. Wijesundara, P. Rathjen and P.J. Verma (2005). 'Tetraploid embryonic stem cells contribute to the inner cell mass of mouse blastocysts. Cloning'. *Stem. Cells.* **7**: 272–278.
72. Pralong D., K. Mrozik, F. Occhiodoro, N. Wijesundara, H. Sumer, A.L. Van Boxtel, A. 'Trounson and P. J. Verma (2005). A novel method for somatic cell nuclear transfer to mouse embryonic stem cells. Cloning'. *Stem. Cells.* **7**: 265–271.
73. Matsumura H., M. Tada, T. Otsuji, K. Yasuchika, N. Nakatsuji, A. Surani and T. Tada (2007). 'Targeted chromosome elimination from ES-somatic hybrid cells'. *Nat. Methods.* **4**: 23–25.
74. Freberg C.T., J.A. Dahl, S. Timoskainen and P. Collas (2007). 'Epigenetic Reprogramming of OCT4 and NANOG Regulatory Regions by Embryonal Carcinoma Cell Extract'. *Mol. Biol. Cell.* **18**: 1543–1553.
75. Hakelien A., H.B. Landsverk, J.M. Robl, B.S. Skalhegg and P. Collas (2002). 'Reprogramming fibroblasts to express T-cell functions using cell extracts'. *Nat. Biotechnol.* **20**: 460–466.
76. Walev I., S.C. Bhakdi, F. Hofmann, N. Djonder, A. Valeva, K. Aktories and S. Bhakdi (2001). 'Delivery of proteins into living cells by reversible membrane permeabilisation with streptolysin-O'. *Proc. Natl. Acad. Sci. USA.* **98**: 3185–3190.
77. Landsverk H.B., A. Hakelien, T. Kuntziger, J.M. Robl, B.S. Skalhegg and P. Collas (2002). 'Reprogrammed gene expression in a somatic cell-free extract'. *EMBO. J.* **3**: 384–389.
78. Hansis C., G. Barreto, N. Maltry and C. Niehrs (2004). 'Nuclear reprogramming of human somatic cells by Xenopus egg extract requires BRG1'. *Curr. Biol.* **14**: 1475–1480.
79. Taranger C.K., A. Noer, A.L. Sorensen, A. Hakelien, A.C. Boquest and P. Collas (2005). 'Induction of de-differentiation, genomewide transcriptional programming, and epigenetic reprogramming by extracts of carcinoma and embryonic stem cells'. *Mol. Biol. Cell.* **16**: 5719–5735.
80. Neri T., M. Monti, P. Rebuzzini, V. Merico, S. Garagna, C.A. Redi and M. Zuccotti (2007). 'Mouse fibroblasts are reprogrammed to Oct-4 and Rex-1 gene expression and alkaline phosphatase activity by embryonic stem cell extracts'. *Cloning. Stem Cells.* **9**: 394–406.
81. Takahashi K. and S. Yamanaka (2006). 'Induction of pluripotent stem cells from mouse embryonic and adult fibroblast cultures by defined factors'. *Cell.* **126**: 1–14.
82. Tokuzawa Y., E. Kaiho, M. Maruyama, K. Takahashi, K. Mitsui, M. Maeda, H. Niwa and S. Yamanaka (2003). 'Fbx15 is a novel target of Oct3/4 but is dispensable for embryonic stem cell self-renewal and mouse development'. *Mol. Cell. Biol.* **23**: 2699–2708.

83. Wernig M., A. Meissner, R. Foreman, T. Brambink, M. Ku, K. Hochedlinger, B.E. Bernstein and R. Jaenisch (2007). 'In vitro reprogramming of fibroblasts into a pluripotent ES-cell-like state'. *Nature.* **448**: 318–324.
84. Okita K., T. Ichisaka and S. Yamanaka (2007). 'Generation of germline-competent induced pluripotent stem cells'. *Nature.* **448**: 313–317.
85. Maherali N., R. Sridharan, X. Wei, J. Utikal, S. Eminli, K. Arnold, M. Stadtfeld, R. Yachechko, J. Tchieu, R. Jaenisch, K. Plath and K. Hochedlinger (2007). 'Directly reprogrammed fibroblasts show global epigenetic remodeling and widespread tissue contribution'. *Cell. Stem. Cell.* **1**: 55–70.
86. Qin D., W. Li, J. Zhang and D. Pei (2007). 'Direct generation of ES-like cells from unmodified mouse embryonic fibroblasts by Oct4/Sox2/Myc/Klf4. Cell'. *Res.* **17**: 959–962.
87. Silva J., I. Chambers, S. Pollard and A. Smith (2006). 'Nanog promotes transfer of pluripotency after cell fusion'. *Nature.* **441**: 997–1001.
88. Takahashi K., K. Tanabe, M. Ohnuki, M. Narita, T. Ichisaka, K. Tomoda and S. Yamanaka (2007). 'Induction of pluripotent stem cells from adult human fibroblasts by defined factors'. *Cell.* **131**: 1–12.
89. Park I.H., R. Zhao, J.A. West, A. Yabuuchi, H. Huo, T.A. Ince, P.H. Lerou, M.W. Lensch and G.Q. 'Daley (2007). Reprogramming of human somatic cells to pluripotency with defined factors'. *Nature.* **451**: 141–147.
90. Lowry W.E., L. Richter, R. Yachecko, A.D. Pyle, J. Tchieu, R. Sridharan, A.T. Clark and K. Plath (2008). 'Generation of human induced pluripotent stem cells from dermal fibroblasts'. *Proc. Natl. Acad. Sci. USA.* **105**: 2883–2888.
91. Hanna J., M. Wernig, S. Markoulaki, C.W. Sun, A. Meissner, J.P. Cassady, C. Beard, T. Brambink, Wu L.-C., T. M. Townes and R. Jaenisch (2007). 'Treatment of sickle cell anemia mouse model with iPS cells generated from autologous skin'. *Science.* **318**: 1920–1923.
92. Wu J., M.A. Vodyanik, K. Smuga-Otto, J. Antosiewicz-Bourget, J.L. Frane, S. Tian, J. Nie, G.A. Jonsdottir, V. Ruotti, R. Stewart, I.I. Slukin and A.J. Thomson (2007). 'Induced pluripotent stem cell lines derived from human somatic cells'. *Science.* **318**: 1917–1920.
93. Nakagawa M., M. Koyanagi, K. Tanabe, K. Takahashi, T. Ichisaka, T. Aoi, K. Okita, Y. Mochiduki, N. Takizawa and S. Yamanaka (2008). 'Generation of induced pluripotent stem cells without Myc from mouse and human fibroblasts'. *Nat. Biotechnol.* **26**: 101–106.
94. Huangfu D., K. Osafune, R. Maehr, W. Guo, A. Eijkelenboom, S. Chen, W. Muhlestein and D.A. Melton (2008). 'Induction of pluripotent stem cells from primary human fibroblasts with only Oct4 and Sox2. Nat'. *Biotechnol.* **26**: 1269–1275.
95. Huangfu K., M. Nakagawa, H. Hyenjong, T. Ichisaka and S. Yamanaka (2008). 'Generation of Mouse Induced Pluripotent Stem Cells Without Viral Vectors'. *Science.* **322**: 949–953.
96. Stadtfeld M., M. Nagaya, J. Utikal, G. Weir and K. Hochedlinger (2008). 'Induced Pluripotent Stem Cells Generated Without Viral Integration'. *Science.* **322**: 945–949.
97. Shi Y., J. Tae Do, C. Desponts, H.S. Hahm, H.R. Schöler and S. Ding (2008). 'A Combined Chemical and Genetic Approach for the Generation of Induced Pluripotent Stem Cells'. *Cell. Stem Cell.* **2**: 525–528.

12

HUMAN EMBRYONIC STEM CELLS AS A MODEL SYSTEM TO STUDY HUMAN GENETICS

Eiges Rachel*

Stem Cell Research Laboratory, Medical Genetics Unit,
Shaare Zedek Medical Center, Jerusalem, Israel

INTRODUCTION

Human embryogenesis is one of the most exciting fields in biology and medicine. However, apart from the very early stages of preimplantation development, human embryos are inaccessible for research and their investigation is restricted to section studies of diseased foetuses. One approach to deal with this limitation is to utilize tissue culture based systems obtained from somatic cell biopsies. These include primary cell cultures and transformed cell lines. Primary cell cultures, however, are restricted in their availability and are often difficult to maintain *in vitro*. In addition, they undergo cell senescence, restricting their life span in culture. It is possible to overcome the problem of cell senescence by using transformed cell lines. Yet, these cell lines frequently display chromosomal irregularities and are characteristically malignant. Moreover, these cellular models allow the study of only a restricted time point during a continuous, developmentally regulated pathway. An alternative approach to model human embryo development is to make use of animal models, specifically mice, taking advantage of their well-defined genetic and reproductive characteristics. Using mice as a model for human development is justified by the strong conservation between the species in developmental processes and genes throughout evolution. Moreover, the relative ease by which their genome can be genetically manipulated and used for the introduction of specific mutations by targeted mutagenesis has made them extremely important for studying specific genes and pathways involved in embryonic development. Despite the similarities

---

* *rachela@szmc.org.il*

between mice and humans, there are still major differences between the species in size, growth, and anatomy that result from critical developmental events that vary between the two. All these are a product of genetic variations, leading to differences in biochemical pathways and phenotypes [1]. These crucial discrepancies may explain why in many instances the mouse model falls short in mimicking the biological phenomena as is manifested in humans [2], emphasizing the need for an improved research model to study human embryo development and genetic malformations. In this respect, human embryonic stem cells (hESCs) promise to be extremely useful, as they are undifferentiated cells capable of forming practically any cell type, including germ cells and extra-embryonic tissues [3].

HUMAN EMBRYONIC STEM CELLS

hESCs are derived from the ICM cells of balstocyst stage embryos (day 5–7 post-fertilization) [4, 5] and can be grown for an unlimited period in culture without losing their properties. In addition, hESCs are characterized by the expression of specific cell surface antigens, undifferentiated markers, and certain enzymes, including SSEA-4 and SSEA-3, Oct4, Nanog, alkaline phosphatase, and telomerase that are typical of pluripotent stem cells [6]. Not only do these cells remain undifferentiated and karyotypically stable during prolonged passages, they also readily differentiate *in vivo* and *in vitro* into cells representing the three germ layers. *In vivo*, hESCs can differentiate spontaneously by forming non-malignant tumours known as teratomas [4, 5]. Teratomas are generated by injecting undifferentiated hESCs into immune-compromised mice. They are composed of many different cell types and structures including cartilage, squamous epithelium, primitive neuroectoderm, ganglionic structures, muscle, bone, and glandular epithelium. Similarly to teratoma induction, hESCs can be induced to spontaneously differentiate *in vitro* by growing them in suspension culture. Under these conditions, they tend to form cell aggregates, which are known as embryoid bodies (EBs) [7]. Human EBs are dynamic structures which grow and proliferate extensively. They initiate by forming densely packed cell aggregates, which begin to cavitate as they grow and gradually accumulate fluid. While they expand *in vitro*, differentiation takes place spontaneously, resulting in the formation of many different cell types including nerve, skin, adrenal, blood, endothelial, kidney, heart, bone, muscle, and liver cells [7, 8]. As normal pluripotent cells, hESCs provide a powerful research model system to study fundamental questions related to early human development, including those that are associated with genetic disorders in humans.

DIFFERENTIATING hESCs AS A MODEL SYSTEM TO STUDY EMBRYOGENESIS

As hESCs undergo differentiation, they recapitulate molecular events and pathways as they occur in the embryo. This has been illustrated by profiling human EBs (hEBs) for gene expression at different stages during their differentiation using cDNA microarrays [9]. By comparing the expression profile between early, partially, and fully matured hEBs, several transiently expressed gene clusters can be identified. These correspond to the different stages of embryo development, from blastocyst to late organogenesis through gastrulation. The power of EB formation as a model system to study molecular events that take place during embryo development was further confirmed by examining the temporal expression of specific gene cascades that are known to be involved in a given developmentally regulated pathway and are active in succession. One such example is the globin gene switching that normally occurs during human haematopoiesis, from embryo to adulthood. By following the expression mode of globin genes during *in vitro* differentiation of hESCs, transient expression patterns are observed [10]. The switch in gene expression from embryonic to adult with time, in a chronological order, illustrates the usefulness of this system for studying the regulation of developmentally regulated pathways as they occur *in vivo*. X inactivation is a different biological phenomenon that is developmentally regulated and is tightly linked with cell differentiation. In this process, a single X chromosome undergoes transcriptional silencing in every XX cell in females during early development in mammals. It is initiated by the upregulation and *cis*-accumulation of a non-coding RNA molecule termed *Xist*. *Xist* accumulation recruits silencing complexes in a multistep process, which is acquired in a developmentally controlled fashion and results in the establishment of a heritable inactive state of the entire chromosome. Analyzing differentiating XX ES in mice and humans has allowed the definition of the developmental time window and order of events that take place during embryo development that lead to X inactivation and dosage compensation of X-linked genes in somatic cells of females [11–16].

The power of differentiating hESCs as an *in vitro* model system to study developmentally regulated processes is not only in its ability to follow dynamic changes along time but also by its ability to intervene and carry out functional studies involving genetic manipulation. This has been well practiced in mice, where the effect of specific gene disruptions (either germline transmitted or induced in culture) have been examined in ES cells following differentiation. The strength of this system is most pronounced in cases where the genes involved lead to early embryonic lethality, preventing their role from being elucidated in whole animals. A good example of this is the use of differentiating mESCs to study the role of *OCT-3/4* as a master regulator of pluripotency. *OCT-3/4* is a mammalian POU transcription factor expressed by early embryonic cells and germ cells. *OCT-3/4*-null embryos produced by homologous recombination develop to the blastocyst stage, but their ICM cells are not pluripotent. Instead, they die early during development at the peri-implantation stage, just before egg cylinder formation [17]. In order to assess the exact role of *OCT-3/4* in early developing embryos, mESCs have been studied following altered gene expression, either by inactivating both *OCT-3/4* alleles or by changing its precise expression levels [18]. These studies have demonstrated that *OCT-3/4* is essential not only for the initial formation of a pluripotent founder cell population in the mammalian embryo but also that critical amounts of it are required to sustain stem cell self-renewal and that up- and downregulation of *OCT-3/4* by only twofold triggers differentiation to endoderm/mesoderm or trophectoderm lineages, respectively. The same holds true for other developmentally regulated genes like the polycomb-group genes (*EZH2* and *SUZ12*) [19, 20], DNA methyl-transferases (*DNMTs*) [21], homeobox genes (*Msx1* and *Msx2*) [22], and many others, which when mutated lead to embryo lethality, limiting the potential of conventional knockouts. Under such circumstances, targeted ESCs have been highly instrumental for investigating the function of distinct genes. As human embryos are inaccessible for research purposes and cannot be experimentally applied to gene function studies, mutant hESCs may serve as a complementary model system to carry out these types of investigations. The use of mutant hESCs for studying the function of specific genes, including those related to genetic disorders in humans, may be extremely useful when no good animal or cellular models are available, and/or when the regulation of expression is developmentally controlled.

DERIVATION OF MUTANT hESC LINES

There are several approaches to obtain mutant hESCs (Fig. 12.1). One such approach is to artificially introduce specific mutations in a pre-existing cell line, while the others rely on the establishment of hESC lines from cells that naturally carry inherited mutations.

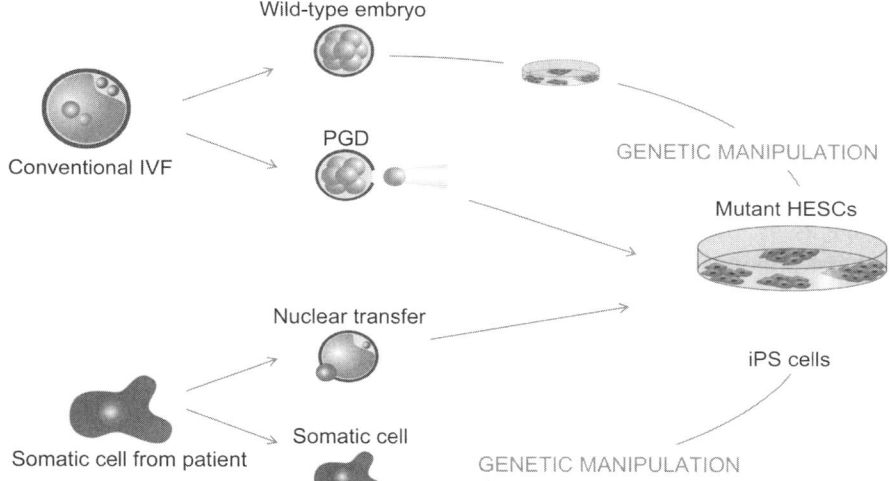

▶ **Fig. 12.1** *Different approaches to obtain mutant pluripotent cells (from top to bottom): (A) artificially introduce specific mutations in a pre-existing cell line; (B) directly derive a hESC line from a genetically abnormal embryo; (C) reprogramming a nucleus from a somatic cell of a patient by nuclear transfer; (D) reprogramming a somatic cell by genetic manipulation to produce transcription factor iPS cells. (For colour figure see Plate 28).*

Introduction of Genetic Alteration in Pre-Existing hESC Lines

The establishment of mutant hESC lines by genetic manipulation *in vitro* requires the availability of wild-type hESC lines. Such cell lines can be derived from unused embryos produced by *in vitro* fertilization (IVF) to infertile couples that have completed their family planning and wish to donate their spare embryos specifically for this purpose. One of the great advantages of ESCs over other cell types is their ability to undergo genetic manipulation. They can easily be genetically modified and still remain pluripotent. They can also be selectively propagated, allowing the clonal expansion of genetically altered cells in culture. Since the first isolation of ESCs in mice, many effective techniques have been developed for gene delivery and manipulation. These include transfection, electroporation, and infection protocols, as well as different approaches for inserting, deleting, or changing the expression of genes. These methods proved to be extremely useful in mESCs for monitoring and directing differentiation, discovering unknown genes and studying their function, and are now being applied in hESCs.

There are basically four types of strategies to genetically engineer hESCs—overexpression, knockout, knock-in and knock-down experiments. Overexpression of genes is usually based on random integration of an exogenous DNA sequence into the genome. It can be applied for constitutive or facultative expression of either cellular or foreign genes and may also be used for the introduction of reporter or selection genes, under the regulation of tissue-specific promoters. These procedures allow specific cell lineages to be labelled and tracked following induced differentiation of hESCs in culture [23]. Moreover, it can be employed for the isolation of pure populations of specific cell types by using selectable markers [23]. Overexpression experiments may also be employed for directing the cell fate of differentiating hESCs in culture. This can be carried out by introducing master genes that play a dominant role in cell commitment, forcing the cells to differentiate into specific lineages that otherwise are rarely obtained [24]. In addition, overexpression may be employed for the generation of cell-based delivery systems by producing therapeutic agents at the site of the damaged tissue. The use of ES-derived cells as therapeutic vectors has been previously shown to be possible in mice, where grafting of ES-derived insulin-secreting cells normalized glycemia in streptozotocin-induced diabetic mice [25].

Apart from tagging, selecting, and directing the differentiation of specific cell types, it is possible to inactivate endogenous genes to study their function. This can be achieved either by disrupting both copies of the gene or by downregulating its activity *in trans*. The most widely used technique for inactivating genes in ESCs is site-directed mutagenesis [26]. This procedure involves the replacement of a specific sequence in the genome by a mutated copy through homologous recombination with a targeting vector. The targeting vector that contains the desired mutation and a selectable marker, flanked by sequences that are interchangeable with the genomic target, pairs with the wild-type chromosomal sequence and replaces it through homologous recombination. By targeting both alleles, using distinct selection markers, it is possible to create "loss-of-function" or so-called knockout phenotypes in ESCs that can be used for functional studies of specific genes. This technology has been well practiced in mice for gene function studies, where genetically altered cells are introduced into wild-type embryos, resulting in the creation of germ-line transmitting chimeras. The genetically manipulated animals can be further mated to generate animals that are homozygous for the desired mutation. The creation of hESCs with a null genotype for specific genes may have great importance for modeling human diseases, and for the study of crucial developmental genes that, in their absence, are lethal at the embryonic stage. One example for generating an hESC-based disease model by homologous recombination is the targeting of the X-linked gene hypoxanthine phosphoribosyl transferase 1 (*HPRT1*), which when mutated is responsible for the development of Lesch-Nyhan syndrome [27, 28]. Lesch-Nyhan syndrome is a severe genetic disease that results in impaired purine metabolism and is characterized by hyperuricemia leading to fatal uric acid urinary stones and a gout-like phenotype. In addition, the disease is manifested by neurological symptoms including mental retardation and self-destructive biting of fingers and lips. It is yet unclear how the deficiency in *HPRT1* activity leads to the neurological symptoms, although damage of dopaminergic neurons is involved. In rodents, the biochemical features leading to overproduction of uric acid as a result of *HPRT1* deficiency do not exist, and nor do the neurobehavioural defects. By introducing large deletions at the *HPRT1* locus through homologous recombination, hESCs that recapitulate the major biochemical defect characterizing Lesch-Nyhan syndrome were produced. These novel *HPRT1*-deficient cell lines accumulate uric acid, confirming the molecular indications of the disease in patients [27]. Thus, *HPRT1* pluripotent cell lines may serve as an improved model system for investigating the disease in humans, in particular for examining the effect of *HPRT1* on dopaminergic neurons. This will enable the exploration of new drugs and gene therapy based treatments in the future.

Similarly to the knockout strategy, it is possible to generate clones of hESCs in which the gene of interest is deleted by inserting a promoterless reporter gene through homologous recombination. This method—termed knock-in—allows the positioning of a reporter gene under the transcription regulation of a native gene [29]. Therefore, it can be applied to monitor the expression of a target gene in situ during ESC differentiation. In this way, Zwaka and Thomson have created human knock-in hESC clones that express either green fluorescent protein (GFP) or a neomycin resistance gene, under the regulation of the endogenous *OCT-3/4* promoter [28]. The *OCT-3/4* gene encodes for a transcription factor that is specifically expressed by pluripotent stem cells. Thus, by replacing *OCT-3/4* with such reporters, the authors were able to monitor and select for undifferentiated hESCs in culture. The same can be applied to any gene that is related to a certain disorder, allowing its mode of expression and effect on cellular or molecular events that take place as a result of differentiation to be followed.

Downregulation of particular genes can also be achieved by overexpressing specific RNA molecules that promote degradation of the mRNA transcripts through the production of short interfering RNA molecules (siRNAs). Because siRNAs operate *in trans* and are not involved in the modification of the targeted gene, it is relatively simple to achieve transient or conditional gene silencing using this method. The use of RNA interference (RNAi) was demonstrated to be feasible in mESCs to inactivate genes and was shown to be equally as effective as the knockout models in generating null mutant embryos [30]. Downregulation by RNAi in hESCs was demonstrated for *OCT-3/4*, β2-microglobulin [31] and *FANCD2* (unpublished data), which is a key player in the evolvement of Fanconi anemia (FA) disease.

Derivation of Mutant hESCs from Genetically Affected Embryos

A different approach to obtain mutant hESC lines is to derive them directly from diseased embryos. Such embryos are usually obtained as a product of preimplantation genetic diagnosis (PGD). PGD is performed in couples at high risk of transmitting a genetic defect who would like to guarantee the birth of a healthy baby. It involves IVF treatment, which helps in performing a biopsy and molecularly analyzing the embryos while they grow in culture. The embryos are tested for the genetic defect carried by the parent(s) and only disease-free embryos are transferred to the womb. The remaining embryos, which were found to be affected and are usually discarded, can be used to establish novel hESC lines that carry naturally inherited mutations. The great advantage of deriving hESC from genetically diseased embryos is that it does not require gene manipulation in order to target a specific gene. In addition, it allows generating cell lines that harbor genetic modifications which are otherwise unattainable, such as triplet repeat expansions and chromosomal translocations. To date, derivation of hESC lines from genetically abnormal embryos has been reported by only a small number of groups, including ours, generating cellular models for CF (cystic fibrosis) [32], MD (myotonic dystrophy) [33], FRAX (Fragile X) [34], and several others [35]. The value of these and similar cell lines in unraveling unresolved molecular pathways that are involved with specific genetic disorders is well illustrated by a Fragile X hESC line we have established [34]. This cell line was used for investigating the mechanism by which inappropriate gene silencing takes place in affected foetuses. Fragile X syndrome (FRAXA) is the most common form of inherited mental retardation. It is an X-linked disease and is caused by the inactivation of the *FMR1* gene. In most patients, *FMR1* is inactivated due to a CGG tri-nucleotide repeat expansion in the regulatory region of the gene. Yet, the timing and mechanism by which *FMR1* is inactivated in Fragile-X-affected embryos are unknown due to the lack of appropriate animal or cell models. The fragile-X-affected cell line that we derived displays all characteristics typical to hESCs. In addition, it has a full CGG repeat expansion at the *FMR1* gene that characterizes FRAXA patients. Studying *FMR1* inactivation using this novel cell line demonstrated that the CGG repeat expansion leads to gene inactivation by a multistep process in a developmentally regulated mode. This unique cellular system highlights the power of genetically aberrant hESCs as a model system to study molecular mechanisms related to human genetic disorders and are in action during early human embryogenesis.

A different source of embryos available to establish mutant hESC lines is genetically abnormal embryos that have been produced through preimplantation genetic screening (PGS) programs [36, 37]. PGS is tailor-made assisted reproduction technique, especially designed for infertile couples experiencing recurrent miscarriages or IVF failure involving aneuploidy screening by fluorescent *in situ* hybridization (FISH) analysis of single blastomeres biopsied from IVF-derived embryos. As such, it allows the selective transfer of embryos with a normal karyotype and occasionally results in the disposal of embryos with numerical chromosomal abnormalities.

A fine example that highlights the potential of aberrant hESCs as a tool to study genetic syndromes that result from chromosomal abnormalities in humans is trisomy for chromosome 21—also known as Down syndrome (DS). DS is the most common viable chromosomal anomaly, having an incidence of about 1 : 800 live births. It usually results in mental retardation and is accompanied by many other diverse clinical symptoms. A most typical disturbance manifested by patients is haematological anomalies. Among those, the most serious one is the development of leukemia. In general, children with DS have a 10- to 40-fold increase in the incidence of leukemia [38]. However, the association between the presence of an extra copy of chromosome 21 and the etiology of the disease, including the link to leukemia, remains poorly understood. Although a murine model for DS has been created, it fails to faithfully reproduce important aspects of the DS phenotype, including those which are associated with the haematological anomalies [39]. The availability of hESCs with trisomy 21 will contribute to the elucidation of the effect of an additional chromosome 21 on the commitment and proliferation of cells and also on the preferential occurrence of leukemia by providing a source of primitive, unspecialized blood cells, which may be forced to further differentiate in culture.

Reprogramming Somatic Cells from Patients

A different approach to establish hESC lines with naturally inherited mutations is to derive them directly from somatic cells of patients. This can be achieved by reprogramming somatic nuclei or cells through embryo cloning or genetic manipulation, respectively. Both the techniques have been shown to be feasible in mice and primates.

NUCLEAR-TRANSFER-DERIVED EMBRYOS

It may be possible to obtain ESC lines from blastocysts, which have been obtained by somatic cell nuclear transfer (SCNT). In this method, a nucleus from a somatic cell of a patient is introduced into an enucleated oocyte, resulting in a cloned embryo. The SCNT-derived blastocyst can be used for the establishment of an ESC line perfectly matched to the donor of the nucleus, which can serve as an unlimited source of cells for autologous transplantation. In addition, this approach is especially advantageous for the derivation of affected hESC lines for a wide range of multifactorial disorders, such as heart disorders, diabetes, and cancer, which result from mutations in multiple genes. The complexity of these diseases makes them especially difficult to model by genetic manipulation, to study, and to treat. Although sophisticated procedures, termed therapeutic cloning, are not unreasonable as it has been previously demonstrated to be feasible in mice and recently in non-human primates [40–42]. By injecting skin cells from a 9-year old male rhesus macaque into enucleated eggs collected from 14 females, researchers were able to establish two ESC lines which show all the requisite pluripotent stem cell markers. These NT-derived ESC lines are able to generate heart and brain neurons in the lab dish and are capable of forming teratomas that are composed of many different cell types when injected into immune-competent mice [40]. However, the problem of egg supply must be resolved before therapeutic cloning can be considered practical for clinical application. In addition, the low success rate (0.7% in non-human primates) requires more research before women can be asked to donate their oocytes for this purpose. Above all, safety concerns should be carefully addressed and it should be determined how well a somatic nucleus can be reprogrammed without being transmitted through the germ line [43].

TRANSCRIPTION-FACTOR-INDUCED PLURIPOTENT STEM CELLS

If an efficient method for de-differentiating somatic cells directly in culture was available, the technical and ethical complications associated with egg donation and embryo destruction for research purposes would be resolved. Indeed, studies in mice and humans show that terminally differentiated cells can be induced to reprogram to an ES-like pluripotent state by over-expression of only a few defined transcription factors (*Oct4*, *Sox2*, *Nanog and c-Myc, Klf4* or *Lin-28*) [44–50]. The transcriptionally induced pluripotent stem (iPS) cells display all characteristics typical of ESCs. They are morphologically similar to ES cells. Their self-renewing ability is comparable with ES cells as determined by their doubling time in culture. They express cell surface markers that are typical of hESCs such as SSEA-3, SSEA-4, TRA-1-60 and TRA-1-81. They also express genes which are specific to undifferentiated pluripotent cells like *OCT-3/4*, *Sox2*, Nanog, *GDF3*, *REX1* and *FGF4*. In addition, human iPS cells show high alkaline phosphatase and telomerase activity, which sustain self-renewal and high proliferation rates. The iPS cells have a wide developmental potential and are capable of differentiating *in vivo* and *in vitro* into teratomas and EBs, similarly to hESCs. They can generate many different cell types, derivatives of all three germ layers—endoderm, ectoderm, and mesoderm—and can even form germ-line transmitting chimeras when injected into blastocysts (mouse only). In terms of their epigenetic status, it seems that the genes that are known to be associated with pluripotency undergo proper DNA demethylation and histone modifications, suggesting that reprogramming has indeed occurred. If these cells will prove to be indistinguishable from hESCs in terms of their biological potential and epigenetic state, and safety issues related to the use of viral vectors to transmit and express the transcription factors for reprogramming will be resolved, this approach would bypass the technical and ethical difficulties that are involved with hESC derivation that impede their use in therapeutic applications. Recently, new iPS cell lines were generated by the reprogramming of somatic cells from six different diseases, including Huntigton's disease, DS, and a type of muscular dystrophy [51]. These cell lines can now be used to study how these diseases affect fundamental events during embryo development and can be applied for drug screening and improvement, accelerating the development of new therapies for these devastating diseases.

To conclude, it remains to be determined how well the epigenetic status of the DNA can be re-established to an embryonic state following somatic cell/nuclei reprogramming. Moreover, these experimental artefacts need to be further improved before they can be considered for application. Reprogramming by nuclear transfer, for example, is technically difficult, labor intensive, extremely inefficient, and may be ethically unacceptable. Somatic cell reprogramming, on the other hand, although relatively easy to perform, has not yet been fully characterized and involves the introduction of cancer promoting transcription factors.

POTENTIAL APPLICATIONS AND FUTURE GOALS

Mass Production of Impaired Cells in Culture

As hESCs are pluripotent cells that can grow to large numbers continuously, they are a powerful tool for basic and applied research. In addition to providing an unlimited cell source for tissue regeneration, they are valuable for investigating human pathologies by offering sufficient supply of impaired cells in culture. Once efficient protocols for induced differentiation are established, it will be possible to produce many cells that manifest the disease from mutant hESCs, increasing the accessibility of affected tissues for investigation. In this respect, differentiated cell derivatives of hESCs are favorable over currently available human-based cellular models, as they do not require establishing cultures from biopsies of patients, are unrestricted in their life span, and allow the generation of all sorts of cell types upon demand, bypassing the problem of the limited range of tissues that can be obtained from patients. The availability of these cells will facilitate basic research related to the mechanisms that bring about specific disorders, at the molecular and cellular levels. Moreover, they may be utilized as a practical tool for cell-based therapy, drug screening and development.

One good example demonstrating the potential of mutant hESCs in producing impaired cells in culture are the *FANCD2*-mutant hESCs that are employed to study FA. This is a rare autosomal recessive disease characterized by progressive bone marrow failure, developmental anomalies, and cancer predisposition [52]. It is caused by chromosomal instability as a result of defects in the cellular DNA repair machinery [53]. One of the hallmarks of FA patients is the failure to produce all types of haematopoietic stem cells. However, the study of human FA stem cell biology has been restricted due to the few CD34+ cells that can be isolated from FA patients. Moreover, studies using animal models have been uninformative as they do not develop haematological abnormalities that are typical to FA patients (bone marrow failure and leukemia), though they do manifest both a DNA repair defect and hypersensitivity to apoptosis-inducing agents [54]. By targeting FANCD2 in hESCs through the expression of specific siRNA molecules and inducing the manipulated cells to differentiate into CD+34 cells in culture, it was possible to successfully generate a proper experimental model for investigating FA phenotype. The hESC-*FANCD2* cells, which hardly express FANCD2 protein, are extremely sensitive to DNA cross-linking agents and, as a result, display the cytogenetic aberrations that are typical to FA patients. Moreover, differentiation *in vitro* allows the production of a small, but discrete, number of CD+34/c-kit+ cells that are able to form haematopoietic colonies, providing a source of FA multipotent blood progenitors for investigation (unpublished data).

Modeling Genetic Pathologies in Culture
Examining the effect of specific mutations

Mutant hESCs may be specifically instrumental for studying particular aspects in embryo development as they can reproduce early stages of embryogenesis. Furthermore, by allowing the consequences of distinct mutations to be examined, they provide a unique research tool for investigating genetically associated pathologies in humans. The use of aberrant hESC lines to elucidate the function of specific mutations can be illustrated by the following examples. There are several hereditary syndromes in human that are associated with premature ageing, chromosomal instability, and predisposition to various types of cancer [55]. One group of disorders are those that result from mutations in the DNA repair proteins and include Nijmegen breakage syndrome (Nbs1), Ataxia telangiectasia-like (*Mre11*), Werner syndrome (*WRN*), Bloom syndrome (BLM), Ataxia telangiectasia (ATM), and FA (*FANC* genes) [56]. The gradual loss of telomere sequences as a result of impaired function of the DNA repair machinery in the cells of these patients triggers the

DNA damage response to induce the senescence or programmed cell death. Alternatively, the loss of telomere protection permits inappropriate joining of chromosome ends to yield fused chromosomes. Fused chromosomes are highly vulnerable to breakage, resulting in genomic instability. Both, activation of the DNA damage response and genomic instability, contribute to premature aging, impaired stem cell function, and tumour induction [57]. However, how diseased embryos survive to term and overcome DNA damage and genomic instability is unclear. It is hypothesized that embryonic cells are distinct from somatic cells in their ability to maintain genome integrity. Accordingly, they are predicted to have robust mechanisms to resist and/or repair DNA damage. Understanding the mechanisms by which hESCs, representing embryonic cells, respond to impaired DNA repair function or to telomerase deficiency may shed light on the molecular pathways that are critical to premature aging as well as to impaired stem cell function.

A different example in which the function of genes is elucidated by the employment of mutant hESCs is to use them as a cell source for various types of nerve cells which are regularly inaccessible for research. These types of cells may be beneficial for investigating genetic neurological disorders such as Parkinson's disease, Alzheimer, Huntington's disease (HD), and many others [58]. In the case of HD, the production of specific nerve cells that display the common mutation that is typical of all patients may allow uncovering the pathogenetic mechanism by which neurons are lost in the brain, leading to the manifestation of the disease. HD is an autosomal dominant disorder which displays a "gain-of-function" phenotype as a result of a tri-nucleotide repeat expansion (CAG) in the coding region of the huntington gene [59]. The expansion results in the production of an altered form of the wild-type protein, containing an amplified number of polyglutamine amino acids in tandem [60]. The aberrant protein has a toxic effect that leads to the degeneration of specific nerve cells (dopaminergic neurons) in the brain. Although aggregates of aberrant proteins have been detected in some brain samples of patients and diseased animals, the mechanism by which the polyglutamine-dependent function is involved in triggering the neuropathology condition in affected individuals is poorly understood. The availability of hESC-derived nerve cells that harbor repeat expansion in the huntingin gene may provide an appropriate model system to examine the effect of polyglutamine tracts on the differentiation and survival of specific nerve cells in the brain. These types of studies will allow new insights, not only into the mechanism that underlies HD but also on those that relate to the etiology of other polyglutamine tri-nucleotide neurodegenerative disorders such as Machado-Joseph, Kennedy's disease (SBMA), and several spinocerebellar ataxias (SCAs) [61].

Investigating genetically controlled evolutionary process

Mutant hESCs can also be used to follow genetically controlled evolutionary processes resulting from multiple alterations over time, an example of which is cancer development. Leukemias, for instance, are considered to arise by malignant transformation of a critical stem cell population and a key step in this transformation is the gain of unlimited capacity for self-renewal or immortality. Nevertheless, leukemic stem cells are difficult to obtain and maintain in culture. Consequently, many studies on leukemia are based on the analysis of fully matured malignant cells, which have multiple genetic alterations. Such cells are inadequate for studying early events in tumour formation and vital events in tumour progression, since they represent an "end stage" of these processes, where primary, causative events cannot be distinguished from secondary ones. In this regard, hESCs can be instrumental for studying cancer initiation and progression, as they can be used to follow, in a controlled fashion, the effect and order of accumulating heritable changes along the tumourigenic process.

Studying developmentally regulated disorders

Modeling human genetic diseases through the use of genetically aberrant hESCs is especially useful under circumstances where the driving mechanism that leads to the manifestation of the disease is developmentally regulated. A fine example of this is FRAXA. It is caused by the inactivation of the X-linked *FMR1* gene. In most patients, this gene is inactivated due to a CGG tri-nucleotide repeat expansion in the regulatory region of the gene. CGG repeat expansion in affected individuals is accompanied by DNA hypermethylation of the repeat and its upstream flanking region, a CpG-island type promoter [62]. It also coincides with other epigenetic modifications characteristic to heterochromatin including histone deacetylation and histone H3-K9 methylation [63–66]. Nevertheless, the molecular mechanisms by which the repeat expansion alters the FMR1 locus and leads to the earlier mentioned epigenetic modifications and transcriptional silencing are not yet known due to the lack of appropriate animal and cellular models. We have established a XY hESC line from a PGD-derived fragile X affected embryo [34]. This hESC line contains a full CGG repeat expansion at the *FMR1* gene that is typical of FRAXA patients. Yet, despite the presence of the full mutation, the FRAXA-hESCs express *FMR1* RNA and protein at normal levels and are DNA unmethylated at the promoter region of the gene when in an undifferentiated state. Epigenetic silencing of *FMR1*, as determined by downregulation of RNA transcription, specific histone modifications, and DNA hypermethylation, occurs only as the hESCs are induced to differentiate. Moreover, the order in which these changes are acquired can be determined, demonstrating downregulation of RNA transcription and histone modifications prior to DNA methylation. Study of *FMR1* inactivation in fragile X affected hESCs indicates that the CGG repeat expansion is required, but not sufficient, for *FMR1* inactivation and that other, developmentally regulated epigenetic modifications are needed in order to initiate the inactivation process. In addition, it demonstrates that DNA methylation of *FMR1* plays a role in the maintenance, rather than in the establishment, of the inactive state of the gene. *FMR1* silencing using fragile X affected hESC lines illustrates the power of this system to study abnormal events during early human embryogenesis, events which are otherwise difficult to explore, and to follow dynamic changes which are acquired along with differentiation and are developmentally controlled.

Discovering the function of embryonic lethal genes

The creation of hESCs with irregularity in specific genes will aid in studying crucial developmental genes whose absence are lethal to embryos. Of particular interest are hESCs which carry lethal nullisomies and trisomies like Turner syndrome. This is one of the most common chromosomal abnormalities in females and results from the loss of a single X chromosome [67]. It is typically manifested by short stature, gonadal dysgenesis, and webbed neck, and affects one in every 2500 girls. Yet, only 1% of diseased foetuses survive to term while the rest are spontaneously lost during gestation at first trimester [67, 68]. In addition, over 50% of patients are mosaic for the missing chromosome, suggesting that the survival frequency of 45,XO foetuses is even less than 1% [67]. Although several mechanisms and candidate genes have been proposed to explain embryo death as a result of X chromosome loss in females, none have been confirmed. This is mainly due to the fact that existing patients are actually the exception, representing rare cases which escape from early embryonic lethality. Thus, no conclusion regarding the mechanism by which this syndrome results can be reached based on these individuals. Furthermore and in contrast to humans, XO mice are viable, anatomically normal, and fertile [69]. Hence, XO mice are inadequate for modeling Turner syndrome and for examining the lethal effect attributed to X monosomy. The availability of hESCs with a single X (e.g. 45,XO) are of great advantage as they may be employed to study the relatively common cause of early miscarriages that result from a loss of a single X chromosome in females.

Gene Therapy

Owing to the great potential of hESCs to undergo genetic manipulation, the possibility of repairing genetic defects in patients by replacing aberrant genes with intact sequences arises. This can be accomplished by coupling embryo cloning or iPS cells with gene therapy so that genetic manipulation will be carried out on isogenic hESC lines obtained from patients. Using this technique, it is possible to cure the disease by providing the patients with genetically engineered autologous grafts, overcoming the difficulties in graft rejection as a result of the immune response. This approach has been previously shown to be feasible in mice, where immune-defficient *Rag2–/–* mutant mice were used as nuclear donors for generating blastocysts and isolating isogenic ESC lines [41]. The mutant ESCs were corrected for *Rag2* gene activity by homologous recombination and used as a source for haematopoietic committed cells. The genetically engineered blood cells were then grafted into the *Rag–/–* mutant mice, resulting in the rescue of the immune-deficient phenotype. Likewise, a recent report by Hanna et al. shows that symptoms of sickle cell anemia can be ameliorated with iPS cells in a humanized knock-in mouse model [70]. Adult somatic cells taken from the tail of a mouse bearing a mutated version of the human β-globin gene were converted into iPS cells. The iPS cells were genetically corrected by replacing the mutant with a wild-type allele through homologous recombination, and then induced to differentiate *in vitro* into blood progenitors. *In vitro* generated haematopoietic progenitors were then transplanted into irradiated genetically identical adult recipients and rescued them by multi-lineage blood reconstitution. This therapeutic approach, which involves reprogramming of mutant donor fibroblasts, *ex-vivo* repair through gene-specific targeting, commitment to primitive blood cells, and rescue of diseased mice by autologous cell transplantation, provides a proof of principle for using transcription-factor-induced reprogramming, combined with gene and cell therapy, for curing genetic diseases, at least in an animal model.

Drug Screening and Toxicology

Much effort is invested in the development of new drugs by trying to understand the pathology of a disease and its underlying treatment. This approach, which is disease-oriented, requires model systems based on human cells rather than animals, which in many cases are inadequate. Obviously, the best model system would be specific cell types of human that manifest the clinical phenotype. However, the currently available cellular models in human are limited in their potential by the lack of relevant and validated cell types. Most are suboptimal since they are based on the use of abnormal cancer cell lines or on primary cell cultures. hESCs, however, may be exceptionally useful for drug screening and development, as they have the capacity to self-renew in culture and can potentially differentiate into all cell types in the body. Provided that efficient protocols for induced differentiation of hESCs will be established, it will be possible to generate specific cell types in large numbers so that the targeted tissues will be accessible for large-scale drug screening and development. Moreover, hESCs can be genetically modified to allow the introduction of reporter genes, under appropriate regulation, that will facilitate the analysis of the tested compounds. In addition, establishing mutant cell lines to generate impaired cells for research may lead to the identification of new drug targets, improving the currently available treatments. Finally, since differentiating hESCs can imitate, to some extent, early human embryogenesis, they may have great value in assessing the potential toxicity of new drug candidates and their teratogenic effects.

CONCLUSION

Mutant hESCs provide a powerful system for basic and applied research by making impaired cells accessible in culture. They can be used to study unresolved questions related to the etiology and/or biology of particular disorders in human, specifically those for which no good animal or cellular models are available. The outcome of these studies may lead to exploration and development of new therapeutic protocols, including gene-therapy-based treatments and disease-oriented drug screenings. The potential implications of mutant hESC-based research, therefore, are far-reaching and may have great importance not only in delineating the mechanisms that are associated with specific diseases but also in improving the quality of life of patients.

REFERENCES

1. Dvash T., N. Benvenisty (2004). Human embryonic stem cells as a model for early human development. Best. Pract. Res. Clin. Obstet. *Gynaecol.* **18**: 929–940.
2. Elsea S.H., R.E. Lucas (2002). The mousetrap: What we can learn when the mouse model does not mimic the human disease. *ILAR. J.* **43**: 66–79.
3. Dvash T., D. Ben-Yosef, R. Eiges (2006). Human embryonic stem cells as a powerful tool for studying human embryogenesis. *Pediatr Res.* **60**: 111–117.
4. Reubinoff B.E., M.F. Pera, C.Y. Fong, A. Trounson, A. Bongso (2000). Embryonic stem cell lines from human blastocysts: Somatic differentiation *in vitro*. *Nat Biotechnol.* **18**: 399–404.

5. Thomson J.A., J. Itskovitz-Eldor, S.S. Shapiro, M.A. Waknitz, J.J. Swiergiel, et al. (1998). Embryonic stem cell lines derived from human blastocysts. *Science.* **282**: 1145–1147.
6. Eiges R., N. Benvenisty (2002). A molecular view on pluripotent stem cells. *FEBS Lett* **529**: 135–141.
7. Itskovitz-Eldor J., M. Schuldiner, D. Karsenti, A. Eden, O. Yanuka, et al. (2000). Differentiation of human embryonic stem cells into embryoid bodies compromising the three embryonic germ layers. *Mol. Med.* **6**: 88–95.
8. Schuldiner M., O. Yanuka, J. Itskovitz-Eldor, D.A. Melton, N. Benvenisty (2000). Effects of eight growth factors on the differentiation of cells derived from human embryonic stem cells. *Proc. Natl. Acad. Sci USA.* **97**: 11307–11312.
9. Dvash T., Y. Mayshar, H. Darr, M. McElhaney, D. Barker, et al. (2004). Temporal gene expression during differentiation of human embryonic stem cells and embryoid bodies. *Hum. Reprod.* **19**: 2875–2883.
10. Qiu C., E. Hanson, E. Olivier, M. Inada, D.S. Kaufman, et al. (2005). Differentiation of human embryonic stem cells into haematopoietic cells by coculture with human foetal liver cells recapitulates the globin switch that occurs early in development. *Exp Hematol.* **33**: 1450–1458.
11. Brockdorff N. (2002). X-chromosome inactivation: closing in on proteins that bind Xist RNA. *Trends Genet.* **18**: 352–358.
12. Dhara S.K., N. Benvenisty (2004). Gene trap as a tool for genome annotation and analysis of X chromosome inactivation in human embryonic stem cells. *Nucleic Acids Res* **32**: 3995–4002.
13. Heard E., C. Rougeulle, D. Arnaud, P. Avner, C.D. Allis, et al. (2001). Methylation of histone H3 at Lys-9 is an early mark on the X chromosome during X inactivation. *Cell.* **107**: 727–738.
14. Keohane A.M., L.P. O'Neill, N.D. Belyaev, J.S. Lavender, B.M. Turner (1996). X-Inactivation and histone H4 acetylation in embryonic stem cells. *Dev. Biol.* **180**: 618–630.
15. Sheardown S.A., S.M. Duthie, C.M. Johnston, A.E. Newall, E.J. Formstone, et al. (1997). Stabilization of Xist RNA mediates initiation of X chromosome inactivation. *Cell.* **91**: 99–107.
16. Wutz A., R. Jaenisch (2000). A shift from reversible to irreversible X inactivation is triggered during ES cell differentiation. *Mol. Cell.* **5**: 695–705.
17. Nichols J., B. Zevnik, K. Anastassiadis, H. Niwa, D. Klewe-Nebenius, et al. (1998). Formation of pluripotent stem cells in the mammalian embryo depends on the POU transcription factor Oct4. *Cell.* **95**: 379–391.
18. Niwa H., J. Miyazaki, A.G. Smith (2000). Quantitative expression of Oct-3/4 defines differentiation, dedifferentiation or self-renewal of ES cells. *Nat. Genet.* **24**: 372–376.
19. O'Carroll D., S. Erhardt, M. Pagani, S.C. Barton, M.A. Surani, et al. (2001). The polycomb-group gene Ezh2 is required for early mouse development. *Mol. Cell. Biol.* **21**: 4330–4336.
20. Pasini D., A.P. Bracken, J.B. Hansen, M. Capillo, K. Helin (2007). The polycomb group protein Suz12 is required for embryonic stem cell differentiation. *Mol. Cell. Biol.* **27**: 3769–3779.
21. Jackson M., A. Krassowska, N. Gilbert, T. Chevassut, L., Forrester et al. (2004). Severe global DNA hypomethylation blocks differentiation and induces histone hyperacetylation in embryonic stem cells. *Mol. Cell. Biol.* **24**: 8862–8871.
22. Fu H., M. Ishii, Y. Gu, R. Maxson (2007). Conditional alleles of Msx1 and Msx2. *Genesis.* **45**: 477–481.
23. Eiges R., M. Schuldiner, M. Drukker, O. Yanuka, J. Itskovitz-Eldor, et al. (2001). Establishment of human embryonic stem cell-transfected clones carrying a marker for undifferentiated cells. *Curr. Biol.* **11**: 514–518.

24. Lavon N., O. Yanuka, N. Benvenisty (2006). The effect of overexpression of Pdx1 and Foxa2 on the differentiation of human embryonic stem cells into pancreatic cells. *Stem Cells.* **24**: 1923–1930.
25. Soria B., E. Roche, G. Berna, T. Leon-Quinto, J.A. Reig, et al. (2000). Insulin-secreting cells derived from embryonic stem cells normalize glycemia in streptozotocin-induced diabetic mice. *Diabetes.* **49**: 157–162.
26. Thomas K.R., M.R. Capecchi (1987). Site-directed mutagenesis by gene targeting in mouse embryo-derived stem cells. *Cell.* **51**: 503–512.
27. Urbach A., M. Schuldiner, N. Benvenisty (2004). Modeling for Lesch-Nyhan disease by gene targeting in human embryonic stem cells. *Stem Cells.* **22**: 635–641.
28. Zwaka T.P., J.A. Thomson (2003). Homologous recombination in human embryonic stem cells. Nat *Biotechnol.* **21**: 319–321.
29. Irion S., H. Luche, P. Gadue, H.J. Fehling, M. Kennedy, et al. (2007). Identification and targeting of the ROSA26 locus in human embryonic stem cells. *Nat. Biotechnol.* **25**: 1477–1482.
30. Kunath T., G. Gish, H. Lickert, N. Jones, T. Pawson, et al. (2003). Transgenic RNA interference in ES cell-derived embryos recapitulates a genetic null phenotype. *Nat. Biotechnol.* **21**: 559–561.
31. Matin M.M., J.R. Walsh, P.J. Gokhale, J.S. Draper, A.R. Bahrami, et al. (2004). Specific knockdown of Oct4 and beta2-microglobulin expression by RNA interference in human embryonic stem cells and embryonic carcinoma cells. *Stem Cells.* **22**: 659–668.
32. Pickering S.J., S.L. Minger, M. Patel, H. Taylor, C. Black, et al. (2005). Generation of a human embryonic stem cell line encoding the cystic fibrosis mutation deltaF508, using preimplantation genetic diagnosis. *Reprod Biomed Online.* **10**: 390–397.
33. Mateizel I., N. De Temmerman, U. Ullmann, G. Cauffman, K. Sermon, et al. (2006). Derivation of human embryonic stem cell lines from embryos obtained after IVF and after PGD for monogenic disorders. *Hum. Reprod.* **21**: 503–511.
34. Eiges R., A. Urbach, M. Malcov, T. Frumkin, T. Schwartz, et al. (2007). Developmental Study of Fragile X Syndrome Using Human Embryonic Stem Cells Derived from Preimplantation Genetically Diagnosed Embryos. *Cell Stem Cell.* **1**: 568–577.
35. Verlinsky Y., N. Strelchenko, V. Kukharenko, S. Rechitsky, O. Verlinsky, et al. (2005). Human embryonic stem cell lines with genetic disorders. *Reprod Biomed Online.* **10**: 105–110.
36. Lavon N., K. Narwani, T. Golan-Lev, N. Buehler, D. Hill, et al. (2008). Derivation of Euploid Human Embryonic Stem Cells from Aneuploid Embryos. *Stem Cells.* **26**(7): 1874–1882.
37. Peura T.T., A. Bosman, T. Stojanov (2007). Derivation of human embryonic stem cell lines. *Theriogenology.* **67**: 32–42.
38. Webb D., I. Roberts, P. Vyas (2007). Haematology of Down syndrome. *Arch Dis Child Fetal Neonatal Ed.* **92**: F503–507.
39. Olson L.E., J.T. Richtsmeier, J. Leszl, R.H. Reeves (2004). A chromosome 21 critical region does not cause specific Down syndrome phenotypes. *Science.* **306**: 687–690.
40. Byrne J.A., D.A. Pedersen, L.L. Clepper, M. Nelson, W.G. Sanger, et al. (2007). Producing primate embryonic stem cells by somatic cell nuclear transfer. *Nature.* **50**: 497–502.
41. Rideout W.M., K. Hochedlinger, M. Kyba, G.Q. Daley, R Jaenisch (2002). Correction of a genetic defect by nuclear transplantation and combined cell and gene therapy. *Cell.* **109**: 17–27.
42. Wakayama T., V. Tabar, I. Rodriguez, A.C. Perry, L. Studer, et al. (2001). Differentiation of embryonic stem cell lines generated from adult somatic cells by nuclear transfer. *Science.* **292**: 740–743.

43. Hochedlinger K., R. Jaenisch (2003). Nuclear transplantation, embryonic stem cells, and the potential for cell therapy. *N. Engl. J. Med.* **349**: 275–286.
44. Meissner A., M. Wernig, R. Jaenisch (2007). Direct reprogramming of genetically unmodified fibroblasts into pluripotent stem cells. *Nat. Biotechnol.* **25**: 1177–1181.
45. Nakagawa M., M. Koyanagi, K. Tanabe, K. Takahashi, T. Ichisaka, et al. (2008). Generation of induced pluripotent stem cells without Myc from mouse and human fibroblasts. *Nat. Biotechnol.* **26**: 101–106.
46. Okita K., T. Ichisaka, S. Yamanaka (2007). Generation of germline-competent induced pluripotent stem cells. *Nature.* **448**: 313–317.
47. Park I.H., R. Zhao, J.A. West, A. Yabuuchi, H. Huo, et al. (2008). Reprogramming of human somatic cells to pluripotency with defined factors. *Nature.* **451**: 141–146.
48. Takahashi K., K. Okita, M. Nakagawa, S. Yamanaka (2007). Induction of pluripotent stem cells from fibroblast cultures. *Nat Protoc.* **2**: 3081–3089.
49. Takahashi K., S. Yamanaka (2006). Induction of pluripotent stem cells from mouse embryonic and adult fibroblast cultures by defined factors. *Cell.* **126**: 663–676.
50. Wernig M., A. Meissner, R. Foreman, T. Brambrink, M. Ku, et al. (2007). *In vitro* reprogramming of fibroblasts into a pluripotent ES-cell-like state. *Nature.* **448**: 318–324.
51. Lensch M.W., G.Q. Daley (2006). Scientific and clinical opportunities for modeling blood disorders with embryonic stem cells. *Blood.* **107**: 2605–2612.
52. Taniguchi T., A.D. D'Andrea (2006). Molecular pathogenesis of Fanconi anemia: recent progress. *Blood.* **107**: 4223–4233.
53. Joenje H., K.J. Patel (2001). The emerging genetic and molecular basis of Fanconi anaemia. *Nat. Rev. Genet.* **2**: 446–457.
54. Haneline L.S., T.A. Gobbett, R. Ramani, M. Carreau, M. Buchwald, et al. (1999). Loss of FancC function results in decreased haematopoietic stem cell repopulating ability. *Blood.* **94**: 1–8.
55. Artandi S.E. (2006). Telomeres, telomerase, and human disease. *N Engl J Med* **355**: 1195–1197.
56. Callen E., J. Surralles (2004). Telomere dysfunction in genome instability syndromes. *Mutat Res.* **567**: 85–104.
57. Blasco M.A. (2007). Telomere length, stem cells and aging. *Nat. Chem. Biol.* **3**: 640–649.
58. Rice C.M., C.A Halfpenny, N.J. Scolding (2003). Stem cells for the treatment of neurological disease. *Transfus Med.* **13**: 351–361.
59. Imarisio S., J. Carmichael, V. Korolchuk, C.W. Chen, S. Saiki, et al. (2008). Huntington's disease: From pathology and genetics to potential therapies. *Biochem. J.* **412**: 191–209.
60. The Huntington's Disease Collaborative Research Group (1993). A novel gene containing a trinucleotide repeat that is expanded and unstable on Huntington's disease chromosomes. *Cell.* **72**: 971–983.
61. Truant R., L.A. Raymond, J. Xia, D. Pinchev, A. Burtnik, et al. (2006). Canadian Association of Neurosciences Review: polyglutamine expansion neurodegenerative diseases. *Can. J. Neurol. Sci.* **33**: 278–291.
62. Oberle I., F. Rousseau, D. Heitz, C. Kretz, D. Devys, et al. (1991). Instability of a 550-base pair DNA segment and abnormal methylation in fragile X syndrome. *Science.* **252**: 1097–1102.
63. Coffee B., F. Zhang, S. Ceman, S.T. Warren, D. Reines (2002). Histone modifications depict an aberrantly heterochromatinized FMR1 gene in fragile x syndrome. *Am. J. Hum. Genet.* **71**: 923–932.

64. Coffee B., F. Zhang, S.T. Warren, D. Reines (1999). Acetylated histones are associated with FMR1 in normal but not fragile X-syndrome cells. *Nat. Genet.* **22**: 98–101.
65. Pietrobono R., E. Tabolacci, F. Zalfa, I. Zito, A. Terracciano, et al. (2005). Molecular dissection of the events leading to inactivation of the FMR1 gene. *Hum. Mol. Genet.* **14**: 267–277.
66. Tabolacci E., R. Pietrobono, U. Moscato, B.A. Oostra, P. Chiurazzi, et al. (2005). Differential epigenetic modifications in the FMR1 gene of the fragile X syndrome after reactivating pharmacological treatments. *Eur J. Hum. Genet.* **13**: 641–648.
67. Saenger P. (1996). Turner's syndrome. *N Engl. J. Med.* **335**: 1749–1754.
68. Cockwell A., M. MacKenzie, S. Youings, P. (Jacobs 1991). A cytogenetic and molecular study of a series of 45,X foetuses and their parents. *J. Med. Genet.* **28**: 151–155.
69. Ashworth A., S. Rastan, R. Lovell-Badge, G. Kay (1991). X-chromosome inactivation may explain the difference in viability of XO humans and mice. *Nature.* **351**: 406–408.
70. Hanna J., M. Wernig, S. Markoulaki, C.W. Sun, A. Meissner, et al. (2007). Treatment of sickle cell anemia mouse model with iPS cells generated from autologous skin. *Science.* **318**: 1920–1923.

13

GERM CELLS AND ARTIFICIAL GAMETES PRODUCTION FROM EMBRYONIC STEM CELLS

Thais M.C. Lavagnolli[1,2,3], *Irina Kerki**[2,3]

[1]*Department of Genetics and Morphology, Federal University of São Paulo, São Paulo, Brazil*
[2]*Laboratory of Genetics, Butantan Institute, Sao Paulo, Brazil*
[3]*Roger Abdelmassih Human Reproduction Clinic and Research Centre, Sao Paulo, Brazil*

INTRODUCTION

Germ cells (GCs) represent a highly specialised cell population, which is indispensable for the continuation and evolution of the species. The separation between germ line and soma occurs early in the development ensuring that genetic or regulatory modifications occurring within somatic cells during development have no effect on gamete formation, and in this way, are not passed on to the next generation. Over the last decade, several research groups have shown that these unique cells can be produced *in vitro* from pluripotent stem cells. Although there are still a number of questions left unanswered, these researches suggest numerous new possibilities for stem cell research and assisted reproductive technology (ART).

Germ Cell Development *in vivo*

GCs are the precursors of the mature gametes. After fertilisation, a totipotent zygote initiates the whole program of embryonic development, leading to the formation of the stem cells of all adult tissues as well as the next generation of GCs.

The embryo proceeds through mitotic divisions known as embryonic cleavage, which originate a morula (8–16 cells). Further cell divisions lead to the formation of a blastocyst, which is made up

---

* *ikerkis@butantan.gov.br*

of two distinct layers—the outer epithelial cell layer, known as the trophoectoderm, and a cluster of cells attached to one side of the inside surface of the trophoectoderm, known as the inner cell mass (ICM). The epiblast cells, which arise from the ICM of blastocysts, are the precursors of all cells that constitute an embryo. Until approximately 6 days postcoitum (dpc) in mice, all epiblast cells are apparently equivalent in their developmental potential. Regional specialisation occurs in the epiblast shortly before gastrulation begins, and only the cells in the proximal epiblast, which directly faces extraembryonic ectoderm, acquire the ability to contribute to the germ line [1].

Recent studies indicate that as early as 6.25 dpc, germ-line competence can be identified in a founder population of perhaps as few as six epiblast cells that express the protein *Blimp1* (B-lymphocyte-induced maturation protein 1) [2–4]. Blimp1 was first identified as a transcriptional repressor that enables the further differentiation of immunoglobulin-secreting plasma cells by inhibiting the expression of genes involved in alternative B-cell development [5].

GC competence is induced in the murine proximal epiblast in response to signals emanating from the extra-embryonic ectoderm including the synergistic action of the growth factors, particularly bone morphogenic proteins (BMP) 4 and 8b; both members of the transforming growth factor-β (TGF-β) superfamily of secreted proteins [6–8]. However, BMP4 or BMP8 alone are unable to induce GCs from cultured epiblast, while they can when combined, which suggests that signaling for various BMPs may occur through separate receptor complexes.

BMP4 stimulates the expression of the *Fragilis* gene in the proximal epiblast, marking the first step to the germ-line commitment (Fig. 13.1). Fragilis is a transmembrane protein and part of a larger interferon-inducible family of genes that is evolutionarily conserved and has human homologues. Interferon-inducible proteins, such as Fragilis, have an anti-proliferative function and may serve to increase the length of the cell cycle in GCs. As GC fate is induced, there is only transient expression of *Fragilis*, but this gene is also expressed in embryonic stem cells (ESCs) and embryonic germ cells (EGCs), suggesting a potential role in pluripotency status [8].

Soon after gastrulation begins, about 7 dpc in mice, presumptive GCs lose their responsiveness to BMP4, and a subset of *Fragilis*$^+$ cells, that migrated to the extra-embryonic mesoderm, start expressing the *Stella* gene. They represent the first population of foetal GC (primordial germ cells or PGC). Stella may have a function during the development of pluripotency and is associated with chromatin remodeling or RNA processing. *Stella*$^+$ nascent GCs exhibit repression of homeobox genes, which may explain their escape from a somatic cell fate and the retention of pluripotency [8]. The term PGCs is strictly applied to diploid GC precursors that transiently exist in the embryo before they enter into close association with somatic cells of the gonad and become irreversibly committed as GC.

Over many years, the activity of tissue non-specific alkaline phosphatase (TNAP) has been used to mark PGC and monitor their transition from the base of allantois, through the hindgut to the dorsal body wall, where they enter into the genital ridges of gonadal anlagen [9]. During this process, the expression of the POU domain transcription factor *Oct4* has been shown to have a role in PGC survival, while it becomes repressed in somatic cell lineages [10]. It has been known for some time that this transcription factor is crucial for maintaining pluripotency in the ICM of the blastocyst and in ESCs. In human foetal tissue, *Oct4* is strongly expressed in migrating PGC as well

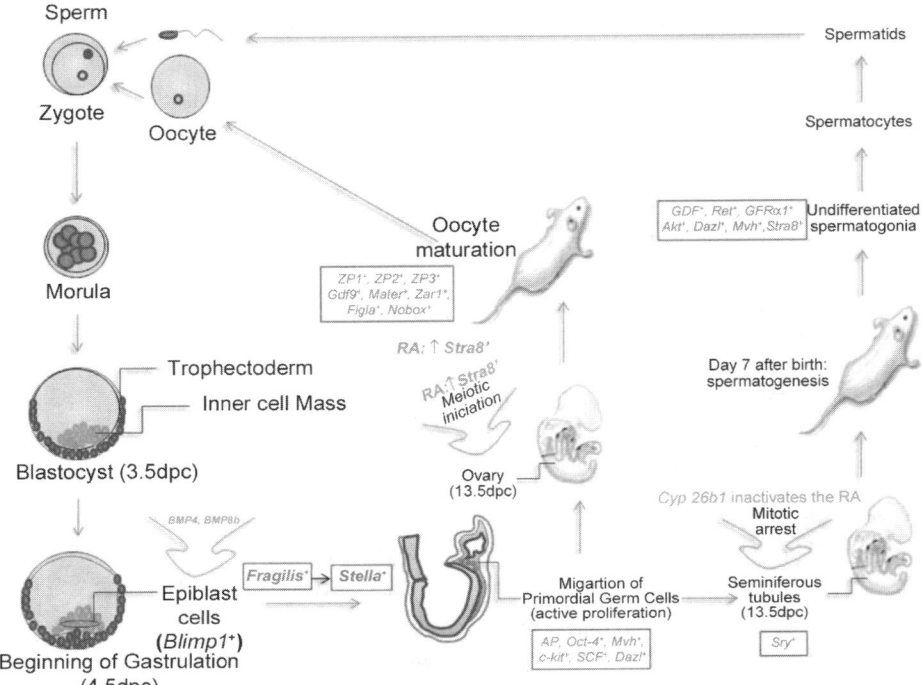

▶ **Fig. 13.1** *Mouse germ cell development in vitro. After fertilization, the zygote initiates the embryonic development yielding stem cells, which are responsible for originating all embryo cell types. Cells from the ICM epiblast acquire the ability to contribute to the germ line. Around 7 dpc, primordial germ cells migrate to the genital ridge, where they proliferate and complete further differentiation into gametes. Stage-specific gene expression is represented within squares, while green arrows represent the influence of specific factors in GC differentiation. (For colour figure see Plate 29).*

as in human GC tumours and EGCs. *Oct4* expression is down-regulated rapidly in the human female gonad and silenced as oocytes enter the first meiotic prophase. In contrast, the same process occurs much more gradually in the male with *Oct4* expression often persisting in some gonocytes and spermatogonia [5].

The migration of PGC coincides with their active proliferation (Fig. 13.1). Upon completion of migration by 11.5 dpc, PGCs start expressing a marker gene for postmigratory GCs, *Mvh* (mouse vasa homologue) [11]. The tyrosine-kinase receptor c-kit and its ligand, stem cell factor (*SCF*), are also essential for the maintenance of PGC in both sexes. In the adult testis, the c-kit receptor is re-expressed in differentiating spermatogonia, but not in spermatogonial stem cells, whereas SCF is expressed by Sertoli cells under follicle-stimulating hormone (FSH) stimulation [12]. Another set of genes involved in GC development is *DAZ* (deleted in azoospermia) genes. Men with deletions encompassing the Y-chromosome *DAZ* genes have few or no GC, indicating that they are defective in the formation or maintenance of GC. A *DAZ* homolog, *DAZL* (*DAZ*-like), found in diverse organisms including humans is required for GC development in males and females.

By 13.5 dpc, the ovary and testis become morphologically distinguishable, and PGCs initiate sex-specific development (Fig 13.1) [13]. Female PGCs enter meiosis and become arrested in meiotic prophase, followed by periodic recruitment to folliculogenesis after birth. On the other hand, male PGCs enter mitotic arrest following migration to the genital ridge, and after birth, male GCs are reactivated to initiate spermatogenesis. A uniqueness of the male GC is that it remains self-renewing stem cells throughout life.

Recent studies [14,15] indicate that instead of an intrinsic program to enter meiosis, GCs respond to the external signal of retinoic acid (RA) and its metabolism, while determining whether GCs will develop as oocytes or as prospermatogonia. The identification of two genes with particular gonadal expression patterns pointed toward the involvement of RA signaling in the control of the initiation of GC meiosis during embryogenesis. The RA-responsive gene *Stra8* (*stimulated by retinoic acid 8*) was first identified in a screen for genes induced by RA in embryonal carcinoma cells [16]. Analysis of its expression pattern in the developing gonad showed that it was specific to GC of female embryos at a stage that preceded the onset of meiosis in the ovary by a day [17]. This suggested that *Stra8* was an early marker of meiotic initiation. At 13.5 dpc, the gene *Cyp26b1*, which encodes an enzyme that metabolizes RA to an inactive form, was found to be expressed exclusively in the somatic cells of the male gonad. These expression patterns suggested that RA might be responsible for inducing meiosis in the female embryonic gonad and that in the male this signal was inhibited by the action of CYP26B1. It should also be noted that *in viro* RA can stimulate mitotic proliferation of PGC up to 13.5 dpc [18]. Therefore, while migratory and postmigratory PGCs appear to respond differently to RA, the gonadal somatic environment also has an important role in regulating sex-specific GC development.

These new data tell us a lot about sex determination in GC. The decision is postponed until somatic sex is determined in the gonad through the action of the testis-determining gene, *Sry*. The action of *Sry* leads to the upregulation of *Cyp26b1* expression in the XY gonad, while it is being downregulated in the XX gonad. Consistent with this, *Cyp26b1*-deficient mice show premature expression of meiotic markers in the XX gonad [19]. Previous studies suggested that the testis may also use RA signaling to induce meiosis postnatally, as spermatogenesis was found to be blocked in male rats fed a diet deficient in vitamin A—a precursor of RA [20]. Consistent with this, Koubova et al., [15] injected RA into testis of these males and observed an increase in *Stra8* expression. This indicates that the ovary and testis use the same signaling system to induce GC meiosis, although they do so at different times.

Once within the gonad, the PGC differentiate in a sex-specific manner, including a distinct program of proliferation.

Male Germ Cells

In males, PGCs are enclosed in seminiferous cords to form the gonads. The somatic environment in the seminiferous tubules triggers the male-specific differentiation of PGCs into gonocytes, which originate male germ-line stem cells that further initiate the first round of spermatogenesis that will produce sperm at the onset of reproductive age (Fig. 13.1).

Male GC differentiation is a highly regulated, complex process that involves a number of unique stages, including meiosis, haploid gene expression, formation of the acrosome and the flagellum, removal of histones from chromatin and their replacement with protamines, and nuclear condensation [21].

Spermatogenesis occurs in successive mitotic, meiotic, and post-meiotic phases (Fig. 13.2). During the mitotic phase, spermatogonial stem cells (As) can either self-renew or give rise to A_{paired} (A_{pr}) spermatogonia, which are destined to proceed through spermatogenic lineage [22]. Beginning before puberty and continuing in the adult animal, the spermatogonial stem cells undergo continuous replication, thereby maintaining their number. Spermatogonial stem cells are thus essentially the only self-renewing cell type in the adult capable of providing a genetic contribution to the next generation. In addition, a fraction of the proliferative spermatogonia population undergoes the differentiation process resulting in the production of spermatozoa (Fig. 13.2). The final division of these spermatogonia produces preleptotene spermatocytes, which begin the meiotic phase and undergo the last cell cycle S phase of spermatogenesis. During the meiotic phase (leptotene, zygotene, pachytene, diplotene, and diakinesis stages), chromosomes condense, synaptonemal complexes form, and homologous chromosomes synapse and recombine to exchange genetic materials. This is followed by two meiotic divisions that occur in rapid succession without DNA replication to produce spermatids, the post-meiotic phase cells. Spermatids are then remodeled into spermatozoa by the processes of acrosome formation, nuclear condensation, flagellar development, and loss of the majority of the cytoplasm (Fig. 13.2) [21, 22].

The patterns of gene expression in spermatogenic cells are likely to be the result of multiple processes. Owing to the complexity and uniqueness of the process, it is not surprising that a large number of chauvinist genes (expressed in developmentally regulated patterns and are transcribed only in, or produce mRNA unique to, spermatogenic cells) are expressed during spermatogenesis [21]. These genes often encode proteins for unique structural components of spermatogenic cells, such as the synaptonemal complex, acrosome, and flagellum, and for unique functional processes, such as meiotic recombination and transcriptional regulation.

When translation activity of spermatids are studied, only few proteins are derived from the stored mRNA transcribed early in the nucleus of spermatocytes, demonstrating that spermatids are transcriptionally and translationally very active cells [21]. This finding suggests that at least the gene products of sex chromosomes and most likely haploid-cell-specific genes are actively transported between spermatids ensuring the synchronous development of the GCs. The function of these haploid-cell-specific genes are only elusively known, but most likely they are involved in the formation of haploid-cell-specific organelles, such as sperm tail, chromatoid body, and acrosomic system [21, 22].

Somatic cells of the testis such as Sertoli, peritubular myoid, and Leydig cells are responsible for the hormonal regulation of the spermatogenesis. Gonadotrophins—luteinizing hormone (LH) and FSH—are secreted from the cells in the anterior pituitary. LH-stimulation of Leyidg cells causes the

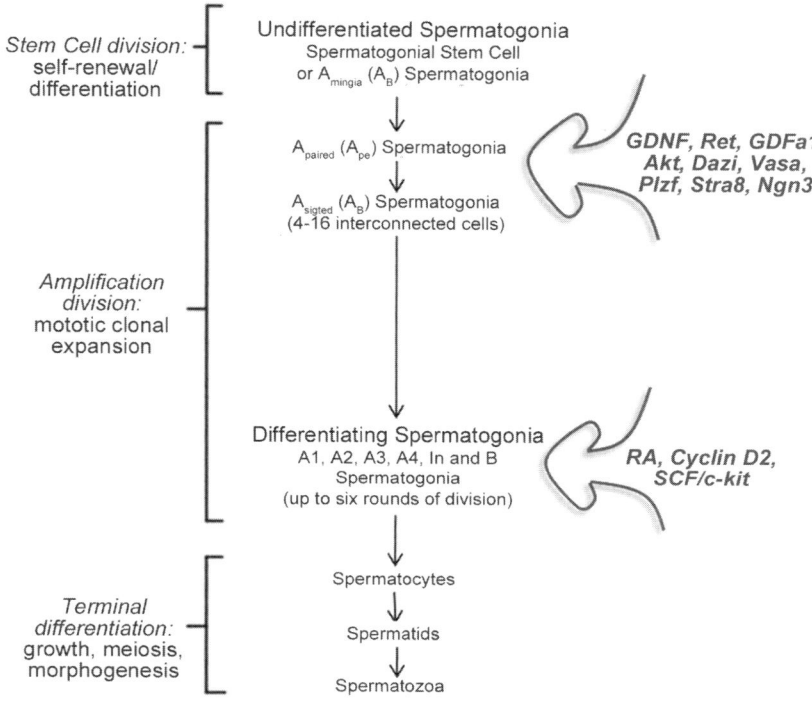

▶ **Fig. 13.2** *Male germ line differentiation. The scheme begins with spermatogonial stem cells (A_s), which can either self-renew or give rise to A_{paired} (A_{pr}) spermatogonia, which are destined to proceed through spermatogenic lineage. After a series of 9–11 mitotic divisions, spermatocytes are formed. After that, terminal differentiation yielding spermatozoa occurs throughout meiosis and morphological changes. Stage-specific gene expression and influence of specific factors in spermatogenesis are represented arrows (modified from Ref. [22]). (For colour figure see Plate 30).*

production of testosterone, which is the major hormone supporting adult spermatogenesis. The action of testosterone is mediated through nuclear androgen receptors, which are localized in Sertoli, peritubular, and Leydig cells [23]. In immature testis, FSH is essential for initiation and maintenance of spermatogenesis. In adult testis, FSH supports the function of Sertoli cells, which in turn support many aspects of sperm cell maturation.

Lastly, male germ line offers a powerful system to study central question in stem cell biology, as cell production in the seminiferous epithelium is a very efficiently organised process. An even density of cells is produced constantly as a result of several emergency and fine-tuning mechanisms to enable the testis to cope with local or overall problems with cell production.

Female Germ Cells

During ovarian development, oogonia and granulose cells characteristically arrange in cords and sheets without specific organisation. The population of oogonia increases by undergoing multiple divisions, after which they become oocytes, stop proliferating, and enter the first steps of meiosis (Fig. 13.3).

In mice, the vast majority of oocytes have entered meiosis during embryonic life, and at birth some oocytes are in the transitory stages of prophase (pachytene and early diplotene), while others have entered late diplotene and dictyate in which they apparently remain until meiosis resumes shortly before ovulation. Primordial follicles in the adult ovary are located in the periphery of the ovary underneath the epithelial surface. The number of GC clusters declines rapidly after birth with few clusters remaining beyond postnatal day 7 [24]. By postnatal day 3, some primordial follicles become primary follicles as granulose cells undergo a transition and oocytes grow beyond 20 µm. The transcription of numerous oocyte-specific genes is initiated during these early stages of the primordial to primary follicle transition.

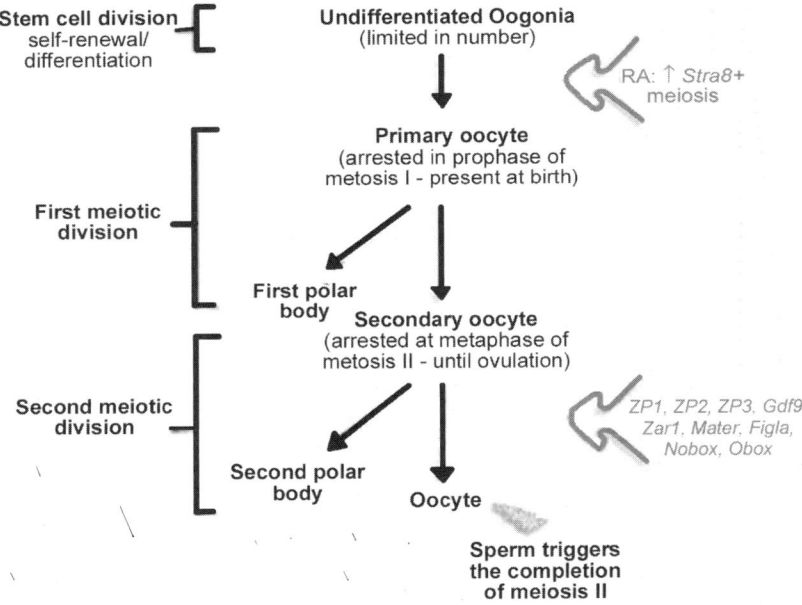

▶ **Fig. 13.3** *Female germ-line differentiation. The scheme begins with undifferentiated oogonia, which can either self-renew or give rise to the primary oocytes. At this stage, primary oocyte is arrested at prophase of meiosis I until birth. After that, the first meiotic division is accomplished originating the first polar body as well as the secondary oocyte, which is now arrested at metaphase of meiosis II until ovulation. The second meiotic division is triggered by sperm fertilisation. Stage-specific gene expression and influence of specific factors in oogenesis are represented within arrows. (For colour figure see Plate 31).*

Oocyte-specific transcriptional factors are likely to control reproductive life span, success in fertilisation, early embryo development, and formation of ovarian tumours. Oocyte transcriptional control must orchestrate expression of oocyte-specific genes necessary for oocyte growth and early embryonic development. Oocyte-specific genes are some of the most abundant transcripts in the ooplasm and include zona pellucid genes (*ZP1*, *Zp2*, and *ZP3*) [25], growth factor 9 (*Gdf9*) [26], maternal antigen that embryo require (Mater) [27], and zygote arrest 1 (Zar 1) [28]. Multiple approaches have been undertaken to identify genes preferentially expressed in the oocyte and many new genes, including transcriptional factors, such as factor in the germ line (*Figla*), Gpbox, newborn ovary homeobox gene (*Nobox*), and *Obox* have been discovered due to these efforts [29, 30].

Figla is a basic helix-loop-helix transcription factor that is expressed exclusively in GCs and regulates transcription of zona pellucids (*ZP*) genes. Figla is, therefore, a critical transcription factor for early steps in folliculogenesis. Human *FIGLA* also binds an E-box in the human *ZP2* promoter, suggesting a similar conserved function of the human and mouse FIGLA proteins [31].

Nobox mRNA is preferentially expressed in GCs, as early as 15 dpc, and throughout folliculogenesis, including GC cysts, primordial, growing, and antral follicles. Ovaries lacking *Nobox* formed what appeared to be histologically primordial follicles, but the oocytes rarely grow beyond 20 μm and the number of somatic cells surrounding oocytes rarely exceeds seven [32]. Newborn ovaries that lack Nobox present a reduced expression of transcripts oocyte-specific, such as *Gdf9*, *Bmp15*, *Rfp14*, *Zar1*, and *Mos*. Future studies are necessary to determine whether Nobox affects the amount of oocyte-specific transcripts by directly or indirectly regulating the transcription of such genes.

During follicular growth, a remarkable increase in the size of the murine oocyte occurs—from approximately 20 μm in the primordial follicle to greater than 70 μm in the antral follicle—with accompanying proliferation of somatic granulose and differentiation of thecal cells. Oocyte-expressed transcriptional factors—*Nr6a1*, *Figla*, and *Nobox*—are expressed throughout folliculogenesis and most likely play a critical function in accumulation of transcripts necessary for follicular growth (such as *Gdf9* and *Bmp15*) as well as transition from fertilisation to embryonic development (*Zar1* and *Mater*).

Identification and functional characterisation of transcription factors that regulate genesis of GC as well as growth and development of oocyte during folliculogenesis will be essential not only to understand genetic pathways that give rise to oocytes but also to understand mechanistically the origin of pluripotency. Oocytes, however, are limited in number in adult mammals. In future, efforts to derive eggs from somatic cells will benefit from understanding oocyte-specific transcription factors and their downstream target genes. Ultimately, the better we understand transcriptional control in oogenesis, the more rational our approach will be to modulate human fertility.

Embryonic Germ Cells

When compared to ESCs and embryonal carcinoma cells, EGCs are a lesser-known type of pluripotent stem cell, which can be found in early embryos. EGCs are derived from PGCs [33].

Mouse EGCs share many characteristics with mouse ESCs such as high level of alkaline phosphatase (AP) activity, the presence of certain embryonic cell surface antigens, and growth as tightly adherent multicellular colonies. EGCs could be isolated based on the expression of β_1-*integrins*, α_6-*integrins*, *EB-cam*, *Hsp90a*, *Thy* (*CD90*), and *CD9* markers [34–36]. They can be continuously cultured while retaining a normal karyotype, but unlike mouse ES cells, stable XX EGC lines can be derived and propagated [37–40].

Cells from mouse EGCs lines can participate in embryogenesis when introduced into a blastocyst and can contribute to all tissues including the germ line [41]. However, imprinting patterns are erased during GC development—fact that can compromise the developmental potential of EG cultures if they are established from late stage PGCs [42]. Detailed examination of the methylation status of imprinting in the insulin-like growth factor 2 receptor (*Igf2r*) gene in several mouse ES and EG lines demonstrated that although the methylation state of most EGC lines is different from ESC and somatic cell lines, there was no correlation between the methylation pattern and the ability to contribute to the germ line of chimeric mice. It is not clear whether the methylation differences observed between EG lines and ES and somatic lines were due to differences inherent to PGC or to their response to EG derivation and culture [43].

AP-positive human PGCs are observed in the yolk sac and migrate through the embryo to the developing gonads. This information, and well-developed protocols for the derivation of mouse EGCs, led to the derivation of human XX and XY EG cultures from 5 to 11 week post-fertilisation gonadal tissue [44]. Undifferentiated human EGCs are *Oct4* positive, express *SSEA-1*, *SSEA-3*, and *SSEA-4* antigens, are immunoreactive for *TRA-1-60* and *TRA-1-81*, and also have elevated levels of telomerase. These markers are rapidly lost during the differentiation that accompanies routine culture.

GAMETES DERIVED FROM EMBRYONIC STEM CELLS

ESCs are derived from ICM of blastocysts and can be maintained and expanded indefinitely in culture, while retaining their ability to produce all cell types in the body. Such a robust differentiation potential of ESCs provides a unique tool to generate various cell types, which could be beneficial in regenerative medicine.

A significant difficulty in studies of *in vitro* gamete generation from ESCs arises from the fact that many PGC markers are identical to ESC markers; thus, it is a challenging task to distinguish early embryonic germ-line cells from ESCs [45].

In spite of that difficulty, a number of recent papers have described the successful derivation of egg and sperm precursor cells from mouse ESCs—so-called "artificial" gametes. Hence, this section is dedicated to provide a current overview of *in vitro* gamete formation from ESCs.

Even though the increasing interest in gametogenesis from ESCs was motivated by a need for an endless supply for oocytes [46], target cell for somatic nuclear cell transfer, less attention has been given to the GC research area, when compared to other cell lineages derived from ESCs. Only throughout the first decade of 2000 had scientist successfully derived GC from mouse ESCs *in viro* [47] and *in vitro* [48–52] and human ESCs [53] (Fig. 13.4 and Table 13.1).

Table 13.1 *ESC Types and Culture Conditions Described to Achieve Germ Cell Differentiation in the Recent Works.*

Works	ESC type	Differentiation medium	GC formation
Hübner et al. [48]	XX, genetically modified (Oct4-GFP) and selection of cells positive for c-kit-GFP	MEM supplemented with 3 mg/ml BSA, 0.23 mM pyruvic acid, 5 µg/ml transferrin, 5 ng/ml selenium and 10 µg/ml insulin, 1 ng/ml EGF and 1 U/ml of each gonadotropin	Female
Toyooka et al. [47]	XY, genetically modified (Mvh-GFP)	DMEM supplemented with 10% FCS for EB formation	Male
Geijsen et al. [48]	XY, genetically modified (SSEA1-GFP) and selection of male GC positive for FE-J1	EB medium (IMDM supplemented with 15% IFS, 200 mg/ml iron saturated transferrin, 4.5 mM monothiolglycerol, 50mg/ml ascorbic acid and 2mM glutamine	Male
Nayernia et al. [50]	XY, genetically modified (Stra8-eGFP and Prm1-dsRED)	DMEM supplemented with 15% FCS, 2mM L-glutamine, 50 mM β-mercaptoethanol and 10 µM RA	Male (production of spermatogonial stem cells)
Clark et al. [53]	Human XX, XY	Knockout DMEM- high glucose, 20% foetal calf serum, 1mM glutamine and 0,1 mM β-mercaptoethanol.	Male
Lacham-Kaplan et al. [51]	XY	DMEM supplemented with 10% FCS or in testicular cell (TC) conditioned DMEM to a concentration of 100,000 cells per ml.	Female
Kerkis et al. [52]	XY	Neurobasal, B27, 1% Peniciline/Streptomicine, retinoic acid (1µM)	Concurrently male and female gametes

Note: GC, germ cells; EB, embryoid body; GFP, green fluorescent protein; FCS, foetal calf serum; RA, retinoic acid.

The first study of *in vitro* female gamete production was reported by Hübner et al., [48]. The authors cultured ESCs carrying an *Oct4*-reporter gene without feeder cells or any growth factors addition. The formation of ovarian follicle-like structures was observed, which included oocyte-like cells expressing Oct4 and Mvh. These cells also expressed meiotic marker genes (*Dmc1* and *Sycp3*), and were released from follicles after 3–4 weeks of culture. Although the GC potential of oocyte-like cells was not tested by fertilisation, structures that closely resemble blastocysts emerged in approximately 6-weeks, suggesting spontaneous parthenogenetic activation. Thus this study suggested that ESC-derived cells have a potential to undergo gametogenesis *in vitro*.

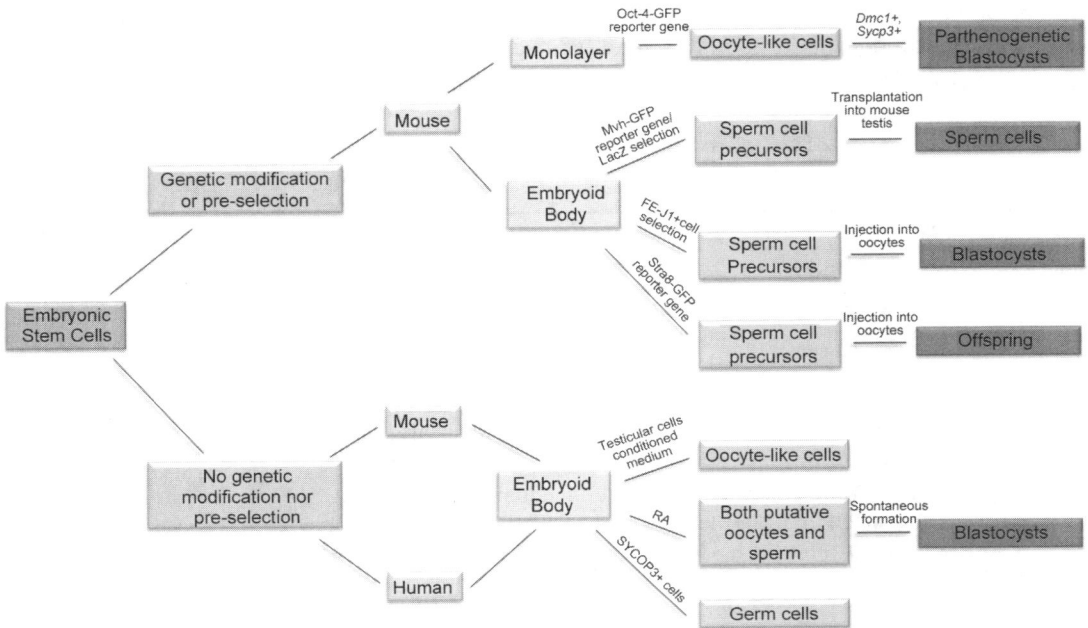

▶ **Fig. 13.4** *Scheme summarising the strategies described at the recent publications about artificial gametes production ([48, 47, 49–53]; from top to bottom panels). FE-J1, cell surface binding anti-anterior acrosome monoclonal antibody; MVH, mouse vasa homologue; Oct4-GFP, Oct4-green fluorescent protein; RA, retinoic acid; Stra8–eGFP, Stra8-enhanced green fluorescent protein; Sycp3, synaptonemal complex protein 3; Dmc1, disruption of meiotic control1 (modified from Ref. [54]). (For colour figure see Plate 32).*

Toyooka et al. [47] reported the derivation of male GC from mouse ESCs, using a different approach: ESCs carrying a *Mvh*-reporter construct were induced to form embryoid bodies (EB), structures obtained *in vitro* resembling a preimplantation embryo, in which differentiation occurs randomly. Then, Mvh^+ cells were purified from the EB and aggregated within male genital ridge cells of wild-type embryos. Following implantation into adult mouse testis, the cell aggregates composed by Mvh^+ positive cells undergo further maturation, formed seminiferous tubule-like structures that have a capacity to support complete spermatogenesis and to produce mature putative sperm cells. Interestingly, the authors also found that the exposure of EB to BMP4 led to the emergence of Mvh^+ cells within 24-h. Although the ability of spermatozoa to activate eggs was not examined, this study suggested that the germ-line specification and the emergence of postmigratory PGC can occur spontaneously or be induced in EB, which can be completed in testis environment.

Geijsen et al., [49], throughout EB induction, also showed a spontaneous emergence of male PGC from mouse ESCs *in vitro* and even without transplantation into testis environment.

Furthermore, the authors detected and isolated haploid cells from EB and showed that the injection of these cells into eggs resulted in the formation of blastocyst-like structures. Although the isolated cells did not produce cells resembling spermatozoa and analyses of further embryonic development were not complete, this study suggested that male PGC arise from ESCs and spontaneously become postmeiotic cells that are capable of activating eggs.

Recently, Nayernia et al. [50] not only reported the induction of male gametes from ESCs but also production of offspring. First they created an ESC line to select for spermatogonial stem cells by introducing a promoter (*Stra8*) active in early male GC linked to a marker gene encoding enhanced green fluorescent protein (eGFP). The selected cell population already had the characteristics of male GC ready to enter the initial stages of meiosis. Using these enriched cells, they performed another round of selection by introducing the promoter of a gene expressed in more mature haploid male GC (*Prm1*) linked to another fluorescence marker gene, *dsRED*. After inducing differentiation with RA treatment, some of them appeared mobile. The appearance of red fluorescent cells suggested that the emerging haploid cells had undergone the final stages of spermatogenesis. The shape of the resulting sperm was, however, abnormal. Following repeated RA treatments and selection of *Stra8$^+$* cells, they isolated *Prm1$^+$* cells and injected them into eggs. When the resulting 65 embryos were transferred to surrogate mothers, seven live pups carrying the *Prm1*-reporter gene were derived, which apparently had growth abnormalities and died short time after birth. Although the production of progeny needs to be confirmed by other laboratories, this study presented the potential of haploid male GC derived from mouse ESC to activate the egg and lead to live births of progeny.

The ability of human ESCs to enter the germ line was examined by Clark et al., [53]. The authors generated EB from female and male human ESCs and found by using PCR and immunohistochemistry that some cells in EB express marker genes specific to different stages of germ-line development. Furthermore, cells expressing a meiotic marker gene, *SYCP3*, were identified in human EB. Thus, this study suggested the possibility that human ESC of both sexes may spontaneously enter the germ line and undergo meiosis.

Lacham-Kaplan et al. [51] reported putative oocytes formation within EB culture obtained from normal, without any genetic modification of ESCs, in a testicular cells conditioned medium, once testis of newborn males contain most growth factors required for the transformation of germ stem cells into differentiated gametes. The oocytes observed were surrounded by one to two layers of flattened cells and did not have visible zona pellucida. However, oocyte-specific markers, such as Figla and *ZP3*, were found expressed by the putative ovarian structures.

Recently, our group published a pioneer work [52], which demonstrated the production of both types of gametes using RA to induce differentiation from male ESCs without any genetic modification or preselection. We described that gamete-like cell formation occurred in the correct manner based on the expression of early and late GC-specific genes, such as *Oct-4, Mvh, Stella, Dazl, Piwil 2, Pdrd 1, Rex 14, Rnf 17, Bmp8b, Acrosin, Stra-8, Haprin*, LH-R, *Gdf9, Zp3, Zp2, Sycp1*, and *Sycp3*. Our immunofluorescence analysis of morphologically well-formed GC and presumptive gametes showed positive labeling with SSEA-1, *Oct-4, EMA-1, FE-J1, Dazl, Fragilis, Mvh, Acrosin*, and *acetylated* α-tubulin. Thus, we showed the presence of specific genes that are expressed within later stages of spermatogenesis and oogenesis as well as that these cells underwent chromosome reduction. Moreover, our electron microscopy analysis revealed that the sperm-like cells we obtained *in vitro* were morphologically similar to normal ones (Fig. 13.5).

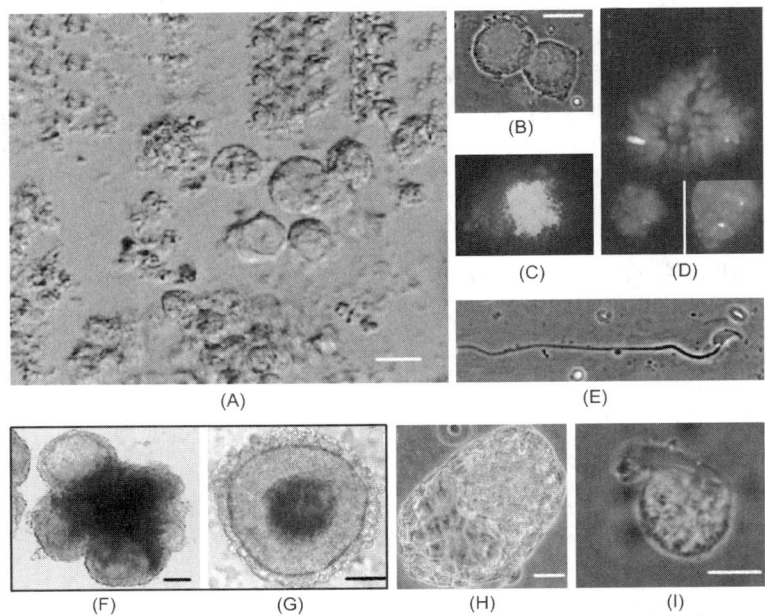

➤ **Fig. 13.5** *Mouse embryonic stem cell differentiation into both presumptive gametes* **in vitro** *[52]. (A) General view of the beginning of GC differentiation. (B) Round spermatids stained with DAPI. Observe the nuclei in a central position (blue). (C) CG positive for FE-J1 antibody (green). Nucleus stained with DAPI. (D) Reduction of chromosome number observed by Fish analysis. Cells showing signals for whole chromosome paintings X (green) and Y (red). Nuclei stained with DAPI. (E) sperm-like cell observed* **in vitro**. *(F) EB aggregate resembling multiple early ovarian follicles (phase contrast image). (G) Oocyte-like cell observed* **in vitro**. *(H) Presumptive blastocyst (phase contrast image). (I) Round spermatid showing positive staining for anti-α-tubulin (red) and anti-acrosin (green) antibodies. Nucleus stained with DAPI. Scale bars: (A),(B),(I) 5 μm; (F)100 μm; (G) and (H) 40 μm. (For colour figure see Plate 33).*

An intriguing aspect of the generation of ESC-derived GC *in vitro* is how the normal checkpoints of the development process, which would naturally span a timeline from early foetal development to puberty and lead to arrest of cells at different stages, seem to be overcome and then apparently compressed into a relatively much short culture period. Curiously, all the groups mentioned earlier noticed that the gene-expression pattern of the PGC showed highly accelerated development. Developmental timers are normally precisely regulated, and so it will be important to know why the timing went shortened in these cases. It has been suggested that in the absence of environmental cues, GC may develop according to an intrinsic clock.

Another aspect essential to analyse involves the fact that developing sperm and eggs must acquire their characteristic identity tags, or 'imprints', which regulate their complementary functions when embryonic development begins after fertilisation. However, it is not yet known whether the eggs and sperm now generated *in vitro* have the appropriate marks. So, although Geijsen et al., [49] did obtain blastocysts when they injected their sperm-like cells into unfertilized eggs, we cannot make any predictions about the long-term development of these early embryos. It is also interesting that the eggs generated by Hübner et al. [48] developed spontaneously to the blastocyst stage once released from the surrounding cells—even though mature eggs should remain "arrested" until they are fertilized or artificially activated. Encouragingly, these blastocysts exhibited appropriate gene-expression pattern. However their full development potential is unknown.

In conclusion, exactly how ESC cultures mimic the somatic environment that encapsulates either developing oogonia (the follicle) or the sperm stem cell (seminiferous epithelium) is unclear. The appropriate growth factor and hormonal microenvironment required to support and sustain these complex niches probably depends to some extent on the type of culture system adopted for the differentiation process. The reports described earlier adopted ESC cultures usually in two basic ways: monolayer adherent cultures of ESCs can be allowed to differentiate directly to form an appropriate niche, or ESC can be induced to aggregate to form EB that forms a more three-dimensional micro-environment. Therefore, generally, the culture conditions attempt to create an environment conductive for germ-cell proliferation and differentiation.

ARE ARTIFICIAL GAMETES FUNCTIONAL?

Of the many differentiated cell types that have been derived *in vitro* from mouse ESCs, certainly the most intriguing are the ones resembling male and female gametes. Such ESC-derived GCs have been shown to undergo meiosis to form haploid gametes, which could support early development, but their capacity to support postnatal development remained conflicting.

The high rate of abnormal development observed in these manipulated gametes is linked to be epigenetic, rather than genetic. Such epigenetic changes may involve DNA modifications (such as methylation) and/or chromatin modifications that in turn regulate gene expression, which is essential for normal development. In this context, imprinted genes-a subset of mammalian genes that are methylated either in the male or the female germ line and hence expressed only from one of the parental chromosomes in the offspring—are of particular interest. These genes tend to affect foetal growth, and their parent-specific DNA methylation marks need to be erased in PGC, and re-established according to the sex of the gamete at later stages of gametogenesis.

An important question is whether the epigenetic reprogramming occurs normally as ESCss undergo GC development *in vitro*; conversely if reprogramming never happens, the resulting sperm or egg would have abnormal patterns of imprinted methylation. Further experiments that rigorously investigate the methylation and expression status of a wider panel of imprinted genes in the ESC-derived GCs during the differentiation process will be important to help clarify the extent to which reprogramming occurs normally in such engineered GCs.

BENEFITS OF ESC-DERIVED GAMETES

Human ESC-derived gametes would allow studies of human GCs, about of which very little is known. Use of such cells might also illuminate the causes of infertility and GC tumours. It might even be possible to use artificial sperm to treat male infertility. And, with improvements, the culture systems could be used to examine many complex processes, including the roles of key genes and the mechanisms underlying imprinting and the halving of chromosome numbers.

Human eggs derived in culture could also have an even more exciting use. By following the same procedure, it might be possible to use these eggs to generate ESCs that produce diseased tissues—the adult nuclei for the process being taken from patients with complex diseases such as diabetes. This might, in turn, lead to new treatments.

And simply being able to study human GCs in culture might allow more thorough investigations into the origin and properties of these remarkable cells. This could give us a grip on our destiny in more ways than we can image.

Clinically, *in vitro* GC induction systems have the potential of developing into a novel ART. A clinical impact of such a technology would be more valuable for female GCs than for their male counterparts, as the supply of oocytes is a major limiting factor in ART. Considering that only one study has been reported on oocyte derivation from ESC, more intensive studies are essential to efficiently produce healthy oocytes *in vitro*.

CONCLUSION

It remains to determine if the *in vitro* process faithfully recapitulates normal GC development. In fact, as described earlier, the study of *in vitro* GC induction is still at a stage where we apply knowledge obtained in studies *in viro* to situations *in vitro*, but not vice versa. Hence, we need to assess if the three critical steps in GC development (specification, migration/proliferation, and sex-specific pre- and postnatal development) indeed take place in the *in vitro* systems.

In addition, it will be crucial to develop means of controlling ESC differentiation into GCs more tightly, once current methods rely on spontaneous and stochastic events, which makes it difficult to analyse each of the complex steps leading to the production of sperm and eggs. Systematic studies of GC-specification and properties will benefit from greater control over these steps *in vitro*. Moreover, the great majority of studies have used genetically manipulated mouse ESCs or pre-selection to achieve GC differentiation, which is a huge limitation to their use in cell therapy. The establishment of efficient protocols that allow gametes production which do not need such molecular techniques will be of great benefit for future application in reproductive medicine [54].

Hence, what can be done with the 'artificial' eggs and sperm? At present, we are largely in the ground of fantastical thought experiments—can we, for instance, generate viable embryos from artificial GCs? This could find applications in animal breeding, although researchers have yet to make ESCs from most mammalian species.

Given the continuing advances in the proliferation and maturation of GC *in vitro*, the production of *in vitro* gametes may be a practical proposition in the not too distant future. However, there remains major uncertainty about the genetic/epigenetic processing of GC *in vitro* and very careful consideration would need to be given to safety aspects of such cells if ever they are to be used for clinical applications.

ACKNOWLEDGEMENTS

We are grateful to Drs. Simone S.A. Fonseca, Virgínia S. Pereira and Alexandre Kerkis for their careful reading of this manuscript and their suggestions. We are in debt to those studies which could not be included, or were cited only partially or indirectly, due to the space constraints. Our research group is supported by grants from Roger Abdelmassih Human Reproduction Clinic and Research Centre and FAPESP.

REFERENCES

1. Saiti D., O. Lacham-Kaplan (2007). 'Mouse germ cell development *in-vivo* and *in-vitro*'. *Biom. Ins.* **2**: 241–252.
2. McLaren A., K.A. Lawson (2005). 'How is the mouse germ-cell lineage established?' *Differentiation.* **73**: 435–437.
3. Ohinata Y., B. Payer, D. O'Carroll, K. Ancelin, et al. (2005). 'Blimp1 is a critical determinant of the germ cell lineage in mice'. *Nature.* **436**: 207–213.
4. Vincent S.D., N.R. Dunn, R. Sciammas, et al. (2005). 'The zinc finger transcriptional repressor Blimp1/Prdm1 is dispensable for early axis formation but is required for specification of primordial germ cells in the mouse'. *Development.* **132**: 1315–1325.
5. Aflatoonian B. and H. Moore (2006). 'Germ cells from mouse and human embryonic stem cells'. *Reproduction.* **132**: 699–707.
6. Shimasaki S., R.K. Moore, F. Otsuka and G.F. Erickson (2004). 'The bone morphogenetic protein system in mammalian reproduction'. *Endocr. Rev.* **25**: 72–101.
7. Surani M.A., K. Ancelin, P. Hajkova, et al. (2004). 'Mechanism of mouse germ cell specification: a genetic program regulating epigenetic reprogramming'. *Cold Spring Harbor Symp Quant Biol.* **69**: 1–9.
8. Saitou M., S.C. Barton and M.A. Surani (2002). 'A molecular programme for the specification of germ cell fate in mice'. *Nature.* **418**: 293–300.
9. McLaren A. (2003). 'Primordial germ cells in the mouse'. *Dev. Biol* **262**: 1–15.
10. Boiani M. and H.R. Scholer (2005). 'Regulatory networks in embryo-derived pluripotent stem cells'. *Nat Rev Mol Cell Biol.* **6**: 872–884.
11. Fujiwara Y., T. Komiya, H. Kawabata, et al. (1994). 'Isolation of a DEAD-family protein gene that encodes a murine homolog of Drosophila vasa and its specific expression in germ cell lineage'. *Proc. Natl. Acad. Sci. USA.* **25**: 12258–12262.
12. Rossi P., C. Sette, S. Dolci and R. Geremia (2000). 'Role of c-kit in mammalian spermatogenesis. J. Endocrinol'. *Invest.* **23**: 609–615.

13. Brennan, J. and B. Capel, (2004). 'Two fates: molecular genetic events that underlie testis versus ovary development'. *Nat. Rev. Genet.* **5**: 509–521.
14. Bowles J., D. Knight, C. Smith, et al. (2006). 'Retinoid signaling determines germ cell fate in mice'. *Science.* **312**: 596–600.
15. Koubova, J., D.B. Menke, Q. Zhou, et al. (2006). 'Retinoic acid regulates sex-specific timing of meiotic initiation in mice'. *Proc. Natl. Acad. Sci. USA* **103**: 2474–2479.
16. Oulad-Abdelghani M., P. Bouillet, D. Décimo, et al. (1996). 'Characterisation of a premeiotic germ cell-specific cytoplasmic protein encoded by Stra8, a novel retinoic acid-responsive gene'. *J Cell Biol.* **135**: 469–477.
17. Menke D.B., J. Koubova, D.C. Page (2003). 'Sexual differentiation of germ cells in XX mouse gonads occurs in an anterior-to-posterior wave'. *Dev. Biol.* **262**: 303–312.
18. Koshimizu U., M. Watanabe and N. Nakatsuji (1995). 'Retinoic acid is a potent growth activator of mouse primordial germ cells *in vitro*'. *Dev. Biol.* **168**: 683–685.
19. Swain A. (2006). 'Sex determination: time for meiosis? The gonad decides'. *Curr Biol* **16**: R507–R509.
20. Thompson J.N., J.M. Howell and G.A. Pitt (1964). 'Vitamin A and reproduction in rats'. *Proc. R. Soc. Lond. B. Biol. Sci.* **159**: 510–535.
21. Eddy E.M. (1998). 'Regulation of gene expression during spermatogenesis'. *Cell and Dev. Biol.* **9**: 451–457.
22. de Rooij D.G. (2001). 'Proliferation and differentiation os spermatogonial stem cells'. *Reproduction.* **121**: 347–354.
23. Bremner W.J., M.R. Millar, R.M. Sharpe and P.T. Saunders (1994). 'Immunohistochemical localisation of androgen receptors in the rat testis: Evidence for stage-dependent expression and regulation by androgens'. *Endocrinology.* **135**: 1227–1234.
24. Pepling M.E. and A.C. Spradling (2001). 'Mouse ovarian germ cell cysts undergo programmed breakdown to form primordial follicles'. *Dev. Biol.* **234**: 339–351.
25. Rankin T.L., M. O'Brien, E. Lee, K. Wigglesworth, J. Eppig and J. Dean (2001). 'Defective zonae pellucidae in Zp-2null mice disrupt folliculogenesis, fertility and development'. *Development.* **128**: 1119–1126.
26. McGrath S.A., A.F. Esquela and S.J. Lee (1995). 'Oocyte-specific expression of growth/differentiation factor-9'. *Mol Endocrinol.* **9**: 131–136.
27. Tong Z.B., L.M. Nelson, and J. Dean (2000). 'Mater encodes a maternal protein in mice with a leucine-rich repeat domain homologous to porcine ribonuclease inhibitor'. *Mamm. Genome.* **11**: 281–287.
28. Wu M.H., A. Rajkovic, K.H. Burns, et al. (2003). 'Sequence and expression of testis-expressed gene 14 (Tex14): A gene encoding a protein kinase preferentially expressed during spermatogenesis'. *Gene. Expr. Patterns.* **3**: 231–236.
29. Rajkovic A., W. Yan, M. Klysik and M.M. Matzuk (2002). 'Obox, a family of homeobox genes preferentially expressed in germ cells'. *Genomics.* **79**: 711–717.
30. Agoulnik A.I., B. Lu, Q. Zhu, et al. (2002). 'A novel gene, Pog, is necessary for primordial germ cell proliferation in the mouse and underlies the germ cell deficient mutation, gdc.' *Hum Mol Genet.* **11**: 3047–3053.

31. Bayne R.A., S.J. Martins da Silva and R.A. Anderson (2004). 'Increased expression of the FIGLA transcription factor is associated with primordial follicle formation in the human foetal ovary'. *Mol. Hum. Reprod.* **10**: 373–381.
32. Rajkovic A., S.A. Pangas D. Ballow, N. Suzumori and M.M. Matzuk(2004). 'NOBOX deficiency disrupts early folliculogenesis and oocyte-specific gene expression'. *Science.* **305**: 1157–1159.
33. Resnick J.L., L.S. Bixter, L. Cheng and P.J. Donovan (1992). 'Long-term proliferation of mouse primordial germ cells in culture'. *Nature.* **359**: 550–551.
34. Hamra K. F., K. M. Chapman, D. M. Nguyen, et al. (2005). 'Self renewal, expansion, and transfection of rat spermatogonial stem cells in culture'. PNAS. **102**(48): 17430–17435.
35. Tokuda M., Y. Kadokawa, H. Kurahashi, and T. Marunouchi (2007). 'CDH1 Is a Specific Marker for Undifferentiated Spermatogonia in Mouse Testes'. *Biol. Reprod.* **76**: 130–141.
36. Brinster R. L. (2007). 'Male Germline Stem Cells: From Mice to Men;. *Science.* **315**: 404–405.
37. Guan K., K. Nayernia, L. S. Maier, et al. (2006). 'Pluripotency of spermatogonial stem cells from adult mouse testis'. *Nature.* **440**: 1199–1203.
38. Hamra F. K., N. Schultz, K. M. Chapman, et al. (2004). 'Defining the spermatogonial stem cell'. *Dev. Biol.* **269**: 393–410.
39. Nagano M., M.A. Avarbock, E.B. Leonida, C.J. Brinster and R.L. Brinster (1998). 'Culture of mouse spermatogonial stem cells'. *Tissue Cell.* **30**(4): 389–397.
40. Seandel M., D. James, S.V. Shmelkov, et al. (2007). 'Generation of functional multipotent adult stem cells from GPR125+ germline progenitors'. *Nature.* **449**: 346–350.
41. Stewart C., I. Gadi and H. Bhatt (1994). 'Stem cells from primordial germ cells can reenter the germ line'. *Dev. Biol.* **161**: 626–628.
42. Tada T., M. Tada, K. Hilton, S.C. Barton, et al. (1998). 'Epigenotype switching of imprintable loci in embryonic germ cells'. *Dev. Genes. Evol.* **207**: 551–561.
43. Labosky P., D. Barlow, B. Hogan (1994). 'Mouse embryonic germ (EG) cell lines: transmission through the germline and differences in the methylation imprint of insulin-like growth factor 2 receptor (igf2r) gene compared with embryonic stem (ES) cell lines'. *Development.* **120**: 3197–3204.
44. Shamblott M.J., J. Axelman, S. Wang et al (1998). 'Derivation of pluripotent stem cells from cultured human primordial germ cells'. *Proc. Natl. Acad. Sci. USA.* **95**: 13726–13731.
45. Nagano M.C. (2007). '*In vitro* gamete derivation from pluripotent stem cells: progress and perspective'. *Biol. Reprod.* **76**: 546–551.
46. Daley, G.Q. (2007). 'Gametes from embryonic stem cells: a cup half empty or half full?' *Science.* **316**: 409–410.
47. Toyooka Y., N. Tsunekawa, R. Akasu, et al. (2003). 'Embryonic stem cells can form germ cells *in vitro*'. *Proc. Natl. Acad. Sci. USA.* **100**: 11457–11462.
48. Hübner K., G. Fuhrmann, L.K. Christenson, et al. (2003). 'Derivation of oocytes from mouse embryonic stem cells'. *Science.* **300**: 1251–1256.
49. Geijsen, N., M. Horoschak, K. Kim, et al. (2004). 'Derivation of embryonic germ cells and male gametes from embryonic stem cells'. *Nature.* **427**: 148–154.
50. Nayernia K., J. Nolte, H.W. Michelmann, et al. (2006). '*In vitro*-differentiated embryonic stem cells give rise to male gametes that can generate offspring mice'. *Dev. Cell.* **11**: 125–132.

51. Lacham-Kaplan O. and A. Trounson (2006). 'Testicular cell conditioned medium supports differentiation of embryonic stem cells into ovarian structure containing oocytes'. *Stem Cells.* **24**: 266–273.
52. Kerkis I., S.A.S Fonseca, R.C. Serafim et al. (2007). '*In vitro* differentiation of male mouse embryonic stem cells into both presumptive sperm cells and oocytes'. *Clon. Stem Cells.* **9**(4): 535–48.
53. Clark A.T., M.S. Bodnar, M. Fox, et al. (2004). 'Spontaneous differentiation of germ cells from human embryonic stem cells *in vitro*'. *Hum. Mol. Genet.* **13**: 727–739.
54. Nagy Z.P., I. Kerkis and C. Chang (2008). 'Development of artificial gametes. RBM Online Symposium: Genetic and Epigenetic Aspects of Assisted Reproduction'. **16**(4): 539–544.

14

USING EMBRYONIC STEM CELLS TO INTRODUCE MUTATIONS INTO THE MOUSE GERM LINE: ANIMAL MODELS AND FUTURE APPLICATIONS

Mariane Secco*, Mayana Zatz

Human Genome Research Centre, Department of Genetic and Evolutive Biology, University of São Paulo, São Paulo, SP, Brazil

INTRODUCTION

Extensive investigation of mouse embryonic stem cells (ESCs) for the last two decades resulted in better understanding of the molecular basis involved during the differentiation process of ESCs to the adult mature phenotype. Under specific culture conditions, ESCs can proliferate indefinitely and are able to differentiate into all specific cell lineages when induced by the appropriate signals. By using different methods, protocols have been established to differentiate ESCs into cardiogenic, myogenic, neurogenic, adipogenic, vascular, and haematopoietic cells; consequently, more complex processes of cell differentiation, as angiogenesis, neuromuscular interaction and synaptogenesis, can be analysed in culture [1–3].

In addition, the potential of mouse ESCs can be evaluated by means of *in vivo* tests: when they are replaced inside a carrier embryo, they resume normal development and contribute to all the tissues of the live-born chimeric animal [4–6].

The ability of ESCs to differentiate into any tissue represents an enormous therapeutic potential. Induced to differentiate *in vitro* into specific cell types, they may be an unlimited source of tissues for transplant in the treatment of several diseases [3,7]. Furthermore, the aforementioned

---

* *marianesecco@usp.br*

properties of ESCs have led to their extensive use in developmental biology, genetics, and biotechnology as an *in vitro* model for early embryogenesis. Currently, ESCs are used to study the mechanisms of early embryonic cell differentiation, the process and mechanisms of X chromosome inactivation, and the effects of biologically active and toxic substances *in vitro* [8–10].

Over the past few years, ESCs have been extensively used for the production of transgenic mice as genetic models of heritable human diseases [11]. The introduction of site-specific mutations into ESCs by homologous recombination, for example, permits the creation of specific genetic alterations in the mammalian genome and conditional or tissue-specific expression of gene products in transgenic mice [12,13]. Genetically modified ESCs, when introduced into blastocysts and transferred to a foster mother, can colonize the germ-line of the resulting chimeras. Transmission of the genetic alteration by breeding leads to the production of mutant animals, the so-called knockout mice powerful tools for the study of gene function *in vivo*. Moreover, gene transfer allows us to test the possible therapeutic use of gene products for gene therapy [12,14].

The purpose of this chapter is to introduce the reader to the system of the generation of transgenic mice, which utilizes ESC lines, and the application of these methods to solve problems in biology. While the technology promises to provide enormous insight into mammalian development genetics, we hope that this review will stimulate the application of these strategies in ESCs to unravel problems in complex regulatory pathways.

ESCs CAN BE GENETICALLY MANIPULATED

The development of techniques for introducing genes (or mutations) stably into the germ-line of experimental mammals, referred to as "transgenic technology," has provided unique insights into complex biologic phenomena.

Although simplistic, this technology can now be broadly defined into two experimental categories: "gain-of-function" mutations, typically created by microinjection of the gene (transgene) of interest directly into the zygote stage of development (e.g. to define the controlling elements for a muscle-specific gene) ; and "loss-of-function" mutations, which employ ESCs.

The technique of DNA microinjection results in random integration of the transgene, giving rise to the founder animal(s). The founder animals are then bred individually to establish independent lines of transgenic animals that can undergo further characterisation (e.g. pattern(s), levels, and consequences of expression of the transgene). However, many of these lines exhibit variable expression of the gene of interest since the transgene may integrate in a manner that alters its expression (e.g. by integration near controlling elements that affect the pattern and level of expression), and this may confound the interpretation of results [15].

In contrast, the use of ESCs as a vehicle has both extended and refined the types of genetic modifications that can be transferred into the germ-line. Currently, two specific applications of this systems are available. The first of these is the use of a random insertional mutagenesis strategy (random transgenesis) to allow the identification of novel genes involved in embryonic development processes. The second is the use of homologous recombination vectors to introduce defined mutations into specific genetic loci, typically used to create the null genotype (*"knockout"* mice). ESCs carrying genes disruption can be identified, cultured in their undifferentiated stem cell state, and introduced back into a blastocyst by microinjection, where they will associate with the inner cell mass cells of the embryo and chimerize the resulting animal. Thus, it is now possible to directly test the role of a given gene in the context of the whole animal. This system affords unique possibilities for studying fundamental aspects of mammalian developmental biology and human diseases.

Historically, the first remarkable property of ESCs described was that they retain the potential to reform an embryo: when they are replaced inside a carrier embryo, they resume normal development and contribute to all the tissues of the live-born chimeric animal. Previous studies have demonstrated the efficient and widespread incorporation of XY cells into XX host embryos, resulting in phenotypic sex conversion to give male animals to transmit only ESC-derived sperm in the germ-line [16]. Further, other experiments showed that ESCs, like other cell types, are amenable to a variety of experimental manipulations in tissue culture, thus opening new avenues to genetic analysis of developmental biology and disease *in vitro*. For example, ESCs have been used to understand the expression of human type II collagen gene in the developing embryo. Other studies have employed the "knockout" mice technology to analyse disease models for humans with inherited cancer predisposition syndromes. Recently, the generation of mice with germ-line tumor suppressor gene mutations through gene targeting in ESCs has provided another dimension by allowing experimental studies of tumor suppressor function in an organismal context [17]. In fact, it is now possible to create mice with several genotypes, and to assess the consequences of these mutations in the context of a developing and intact mammal.

These technologies and others will be discussed further.

USING ESCs TO STUDY EMBRYONIC DEVELOPMENT VIA RANDOM TRANSGENESIS

The random transgenesis or insertional mutagenesis technology are performed through transfection of multiple proviral sequences into the germ-line following infection of cells in tissue culture with a replication deficient viral vector. Some of these proviral sequences may have inserted into genes that play an important role in the development.

Previous studies have highlighted this type of approach. Currently, the retroviral vectors have been used for the delivery of the genetic material into the cells. The advantage of a retroviral system is that genetic sequences can easily, efficiently, and permanently be introduced into target cells. Furthermore, the use of sequences that confer antibiotic resistance (e.g., neomycin, puromycin, hygromycin, and others) for clonal selection or of reporter genes (e.g., green fluorescent protein (GFP/EGFP, LacZ)) to identify specific lineages has facilitated the rapid genomic cloning of the integration site. Additional constructs have been designed to overexpress transcription factors (GATA-4, Twist), signaling molecules (insulin-like growth factor II (IGF-II), Cripto), or cellular proteins in differentiated phenotypes of myogenic, erythroid, pancreatic, and cardiomyocytic cell lines.

Despite the advantages aforementioned, retroviral mediated mutagenesis is clearly limited by a low rate at which germ-line mutations are generated. Most recently, a number of laboratories have been testing new strategies that will enable the prescreening and identification of ESCs harboring mutations *in vitro*, prior to their use for the generation of chimeric mice.

The technique known as "gene trapping" is the most commonly employed insertional mutagenesis strategy, that it has been extensively reviewed by others. This approach involves the use of reporter gene constructs to detect *cis*-acting regulatory elements of genes expressed in specific patterns in the developing embryo and is based on very elegant experiments carried out in Drosophila. To accomplish this, "gene trap" vectors have been designed to recover ESC clones carrying integration events that have occurred specifically in actively transcribed regions of the genome.

The three essential features of these trapping vectors are the inclusion of a reporter gene sequence (e.g. a bacterial *lacZ* reporter gene) between a splice acceptor (SA) and a polyadenylation signal (pA), and a selectable marker such as the bacterial neo gene, to allow the recovery of ESCs colonies that have integrated vector DNA (Fig. 14.1). When inserted in a functional gene, the

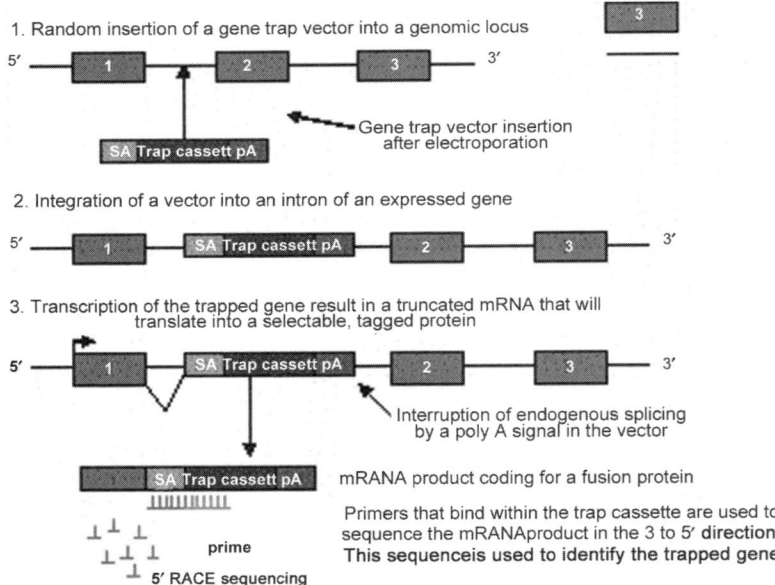

▶ **Fig. 14.1** *The conventional trapping process. The schematic gene trap vector is represented for SA (splice acceptor sequence); trap cassette (selection/reporter construct, e.g., βgeo); pA (polyadenylation signal). (For colour figure see Plate 34).*

endogenous splice donor (SD) and gene trap slice acceptor are processed to form a fusion transcript to activate the reporter gene contained in the gene trap construct. Because integration of the copies of the vector occurs at random in the genome, this event will be rare. However, expression of the gene trap is assayed for reporter gene expression (β-galactosidase activity), and the staining is indicative of a insertion event. The transgene is only activated when it integrates correctly within an active transcriptional unit. Some translation fusions (frame shifts) may inactivate the reporter activity or may target the translated proteins into subcellular locations where reporter activity is not easily detectable. Gene trapping, therefore, selects for integration events in functional genes, and it is especially useful for the analysis of mammalian cells that have complex genomic organisations that consist of promoters and exons that are separated by introns. The main disadvantage to using gene-trap vectors is that because the insertion occurs in an intron, alternative splicing can sometimes take place, leading to lower levels of wild-type transcripts and often resulting in hypomorphic alleles. In addition, these constructs require gene to be expressed in undifferentiated ESCs [18].

Other trap vectors were simultaneously being developed for ESC mutagenesis. PolyA gene trap vectors, for example, employ a different strategy, shown in Fig. 14.2. These vectors contain a promoter signal and a transcriptional start site, allowing genes to be trapped that are not normally expressed or expressed at very low levels under experimental conditions. A SD sequence is present at the end of the gene trap cassette, causing the mRNA product of the vector construct to be fused with any downstream exons. Since these vectors do not have their own polyA sequences to signal the end of translation, only cell lines in which the vector inserts upstream of a terminal exon will produce the selectable reporter tag. It is important to note that the sequence used to identify a gene

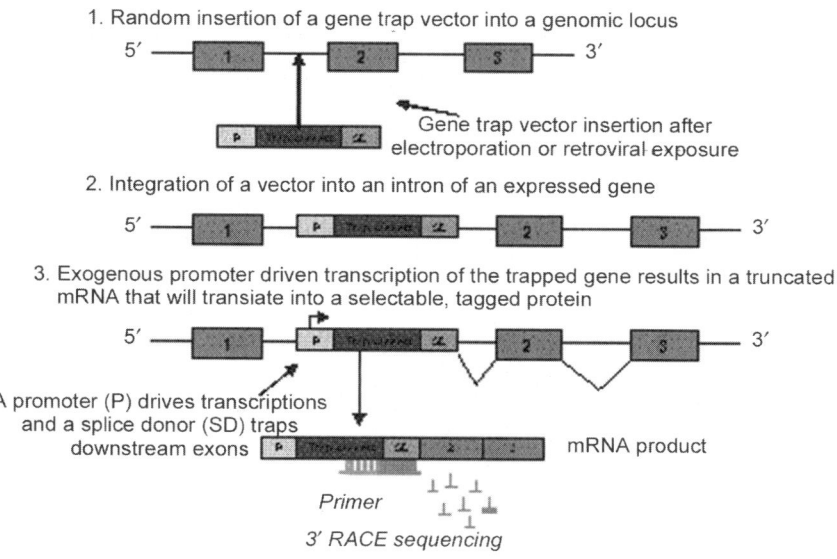

▶ **Fig. 14.2** *PolyA gene trap vectors. (For colour figure see Plate 34).*

inactivated by a polyA gene trap vector is taken from exons 3' of the vector insertion site. This differs from typical gene trap sequence which is taken from the exons 5' of the vector insertion site [18].

In vivo gene trap screens in mice have permitted the identification of many developmentally regulated genes that are expressed within specific tissues in a spatiotemporal pattern, including novel RA responsive, neuronal, glial, chondrocytic, myocytic, and haematopoietic genes. Moreover, novel genes identified by trapping can be immediately analysed phenotypically and can often point to shared molecular pathways. In accordance of this, an increasing number of success experiments are emerging showing that gene trapping is a fruitful approach. For example, homozygous embryos that contain an insertion in *Lrp6*—a novel low-density lipoprotein receptor family gene—had a truncated body axis, loss of distal limb structures, microophthalmia, and cranial facial defects that resembled phenotypes caused by mutations in the Wnt signaling pathway. Biochemical analysis confirmed that *Lrp6* signals in the Wnt pathway, shedding new light on an already well-studied pathway. In other cases, sequence can indicate molecular pathways, as in the case of the gene-trap insertion into a novel mouse gene called *shroom*, because the neural folds mushroom outward and do not converge at the dorsal midline in homozygous mutant embryos. Sequence analysis showed shroom to be a PDZ-domain-containing, actin-binding protein, and biochemical analysis showed that shroom was required for proper subcellular localisation of F-actin in the neural tube [18].

Homologous Recombination: Introduction of Precise Mutations into Resident Genes

It is expected that the earlier mentioned approaches will reveal new genes involved in development. An equally important goal is the generation of mutations in previously characterized genes to determine whether they have a developmental function. For example, the expression of many growth factors, receptors, and proto-oncogenes appears to be temporally and spatially regulated in the developing embryo. However, these studies are purely descriptive. In order to determine whether a given gene has a controlling influence on development, it will be important to test the consequences of altering the normal pattern of expression.

During the past years, methods have been described to generate site-specific targeted mutations in the mammalian genome using homologous recombinant between introduced, cloned DNA sequences and corresponding chromosomal locus.

In an elegant series of papers using mammalian cell lines, based primarily on prior work in *Saccharomyces cerevisiae* [19], it became clear that cloned DNA could be precisely altered *in vitro*, and when introduced into cells via a number of methods (infection, transfection) would homologously recombine with the resident gene and introduce the desired mutation at that site in the genome.

Several transfection experiments in higher eukaryotes have demonstrated that in contrast to yeast, the introduced DNA sequences integrated largely at random into the genome. Experiments using the X-linked *HPRT* gene as a model genetic locus have shown that the frequency of homologous integration exceeds that of homologous recombination by a factor of 1000-fold or more. Thus, much effort has gone into devising suitable strategies, applicable for both transcriptionally active and inactive genetic loci in ESCs, to enable the isolation of clones carrying the desired homologous recombination event.

A number of procedures have been described including the use of large-scale screening, screening pools of cells using PCR, and so-termed "positive–negative" protocol. In this last technique, the use of sequences that confer antibiotic resistance (e.g. neomycin) for clonal selection has greatly facilitated this approach both *in vitro* and *in vivo*. The procedure consists of the transfer of the vectors to ESCs via electroporation, and identifies those that contain the targeting construct by their survival in the presence of a drug such as G418 (an aminoglycoside related to neomycin). Vectors are usually transferred to ESCs via electroporation, and cells that contain the targeting construct are identified by their survival in the presence of a drug such as G418 (an aminoglycoside related to neomycin). This is referred to as "positive selection". The fact that most cells surviving the S1 selection will have integrated the targeting vector randomly by non-homologous recombination into the genome, rather than at the target locus by homologous recombination, led to the development of a negative selection scheme. Inclusion of another selectable marker in the targeting vector just outside one of the regions of homology allows for concomitant negative selection. The herpes simplex thymidine kinase (*TK*) gene in conjunction with the drug ganciclovir has been very successful in this context, as was originally described. Most of the cells taking up the targeting vector randomly into their genomes retain and express the *TK* gene; they are consequently killed by ganciclovir. In the much smaller proportion of cells that integrate the vector by homologous recombination, the *TK* gene is lost in the homologous crossover event, allowing the targeted cells to survive in the presence of ganciclovir. A DNA-screening method is usually required to identify correctly targeted cells. When the targeting frequency is high, the screening can be done by Southern blot analysis. In less favorable cases, a PCR-based recombinant fragment assay is more efficient. This combined strategy selection routinely increases the targeting frequency by an order of magnitude or more.

Once these mutated ESCs are isolated as a pure clone, they can be introduced into the blastocele cavity of a normal embryo, where they participate in the development of all tissues and result in the production of chimeras. In subsequent matings of these chimeric mice, if the ESCs have contributed to formation of germ cells, the mutant gene is transmitted to their progeny. By mating heterozygotes, each harboring a mutated copy of the gene of interest (detected by analysis of the isolated DNA), one can derive embryos and mice which are homozygous for the mutation. These techniques are illustrated schematically in Fig. 14.3.

The general strategy for generating gene-targeted mice so far includes the use of 129 derived ESC lines and the back-crossing of the induced mutation onto other inbred strain backgrounds such as C57BL6 or BALBc. The mutation can then be back-crossed onto different inbred strain backgrounds to generate congenic strains and to identify modifier loci. There may also be benefits to studying mixed strain backgrounds. Mixed genetic background knockout mice often have a wider range of phenotypes (Fig. 14.3).

The ability to produce mice that carry altered genomic DNA has greatly facilitated the study of many biological processes; however, not all biological processes can be studied by gene inactivation. Gene-targeting that have little effect on the health and viability of the mutant animal or

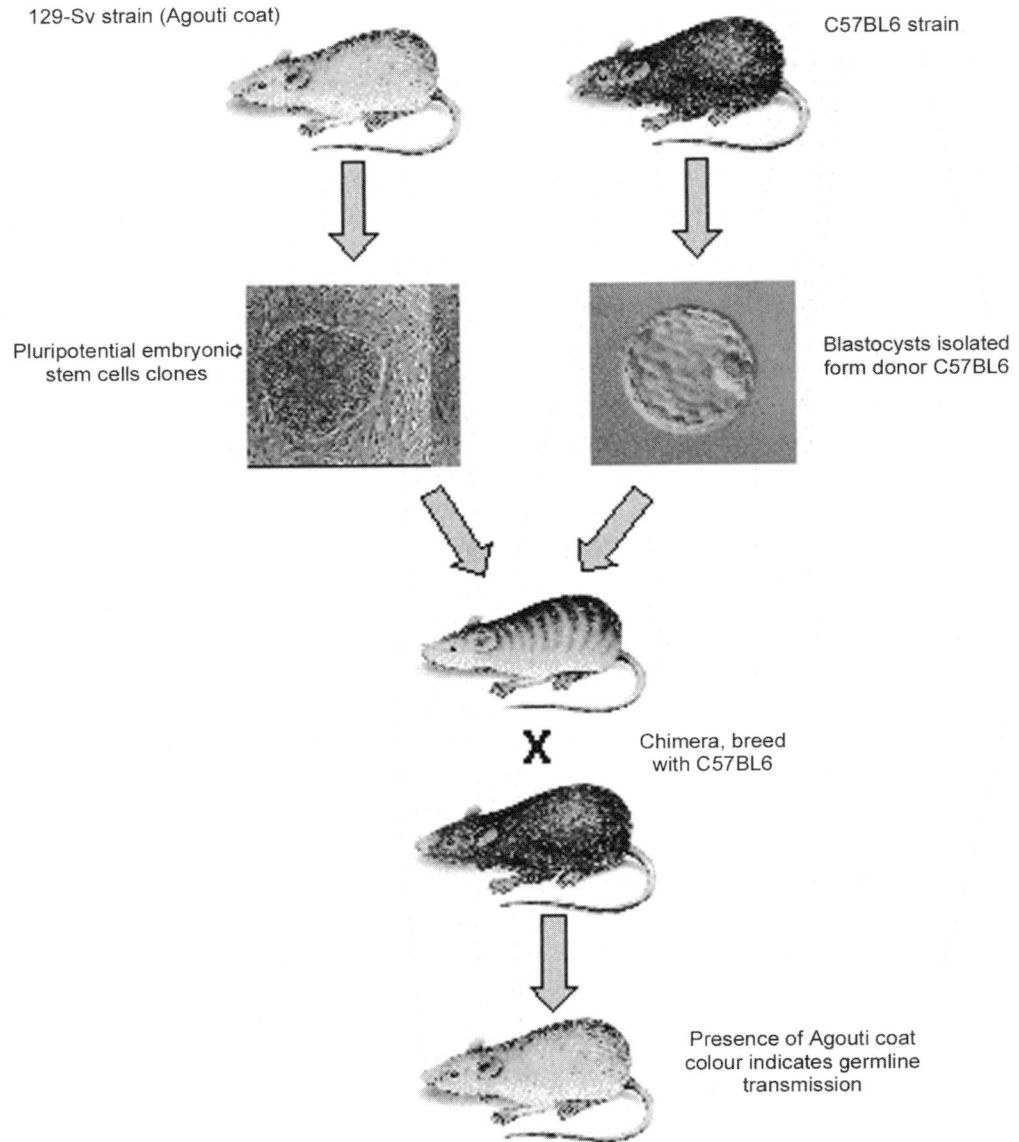

▶ Fig. 14.3 *General strategy for creating knockout mouse. (For colour figure see Plate 35).*

those that result in developmental arrest or embryonic lethality *in vivo* reflect the earliest non-redundant role of a gene and precludes analysis of function at later stages. An example of the former class is the mice resulting from inactivation of the gene coding for α_2-microglobulin—a protein subunit of the major histocompatibility class I molecules [20]. These mice have only very low levels of class I proteins on cell surfaces and a virtual absence of CD8+ T-cells, which are dependent on class I proteins for their development in the thymus. Yet the animals are born normal and show no obvious phenotype. Despite the easily detectable alterations in the immune system, only when the animals were studied more extensively did it become apparent that the mice have increased susceptibility to a small subset of infectious agents. In contrast, disruption of the gene coding for GATA-1—an erythroid-specific transcription factor—proved to be lethal in early foetal development [21]. In such cases of lethality, ESCs that are homozygous or hemizygous for a given targeted mutation can be injected into blastocysts and their fate can be determined. In this example, in which the gene for GATA-1 is X-linked and the ESCs were male, the hemizygous mutant ESCs contributed to most tissues, but only cells of the wild-type host blastocyst were detected in the erythroid compartment of the resulting chimeric embryos. Thus, lack of the *GATA-1* gene product precludes normal erythroid development, although the mutant cells can contribute normally to many other tissues.

Additionally, some genes have functions during embryogenesis that may differ from those in the adult (e.g. *LIF* and *vimentin*). Inactivation of these genes may lead to adaptations that preclude their functional analysis at later stages. To address these problems, a number of modifications to the original gene-targeting strategies have been developed. Embryonic lethality can be overcome by generating conditional knockout or knockin ESCs and mice, which can be used to activate or inactivate a gene both spatially and temporally. Typically, a conditionally targeted allele is made by inserting loxP or frt sites into two introns or at the opposite ends of a gene. Expression of P1 bacteriophage-derived Cre or yeast-derived Flp recombinases in mice carrying the conditional allele catalyzes recombination (insertions, deletions, inversions, duplications) between the loxP/frt sites, respectively, to inactivate (or activate) the gene. By expressing Cre recombinase from an endogenous or tissue-specific promoter, the conditional allele can be recombined in a restricted lineage or cell type (Fig. 14.4). The timing of recombinase expression can also be controlled using inducible expression systems or viral delivery systems such as adenovirus or lentivirus, which makes it possible to inactivate a gene in a temporal-specific fashion. This technique has been widely used in the analysis of mice, and its use in adult mice overcomes a major limitation associated with standard transgenics, for example, the developmental consequences of inactivated genes [22].

This system has also been adapted for ESC lines, both for *in vitro* studies and the generation of new mouse models (e.g. allele replacement by double loxP recombination. The use of site-specific recombination events (insertions, deletions, inversions, or duplications) can also be extended to the engineering of long-range modifications in the ESC genome.

The first description of gene targeting via homologous recombination in murine ESCs was published in 1987. To date, several hundred novel mouse mutants—"knockout" mice—have been created, with dozens being reported monthly. A synopsis of these mutations is obviously beyond the scope of this chapter, but it is readily apparent from a perusal of these mutations that the technology has become one of the most powerful methods in the repertoire of approaches to gain insight into the functions of genes. To cite only a few applications, these include developmental biology, behaviour and cognition, pharmaceutical research, and generation of models of human disease, such as cystic fibrosis and familial hypercholesterolemia [22].

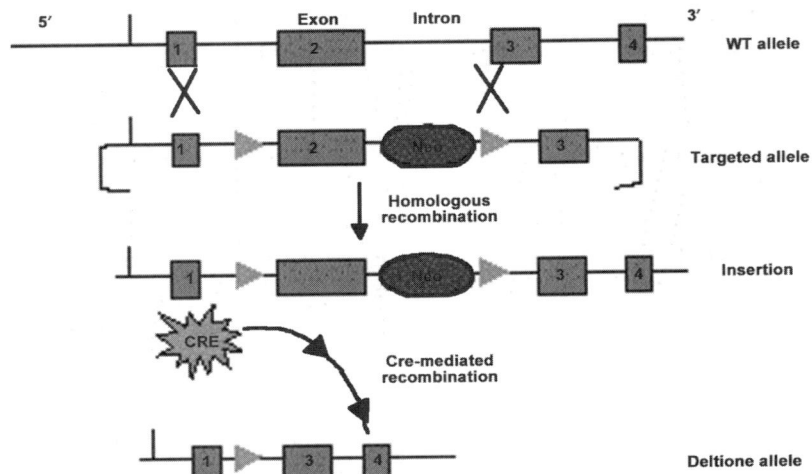

▶ **Fig. 14.4** *Gene targeting and conditional expression events in ESCs mediated by Cre recombinase-loxP recombination. In this example, a gene locus in ESCs has been targeted by homologous recombination to insert a Neo cassette flanked by two loxP sites. Cre recombinase is an enzyme that works like scissors to cut out a gene that is in between two target sequences called loxP. Because this enzyme is expressed only in certain cell types, the targeted gene will be knocked out of only those cells and only when the researcher wants them to be. As demonstrated in the figure, when Cre recombinase is expressed in the cell, the loxP site will be cut and joined together, removing the piece of DNA between the two sites. (For colour figure see Plate 36).*

ESCs AS CELLULAR MODELS IN DEVELOPMENTAL BIOLOGY AND PATHOLOGY

Genetic approaches involving transgenic mice have also greatly advanced our knowledge of development and disease. This has been accomplished primarily through (1) the isolation and cultivation of ESCs, which retain the ability to colonize all tissues of a host embryo including its germ line; (2) the resiliency of mammalian embryos/blastocysts to tolerate the addition or loss of embryonic cells; and (3) gene inactivation by homologous recombination or overexpression of transgenes to assess gene function and genetic labeling of precursor cells to determine cell lineages.

The earliest use of targeted animal models for gene therapy emphasized mouse models that simulated inherited disease, but these often proved disappointing. Subsequent studies have identified many useful mouse models for the study of human disease; however, the utility of these transgenic models frequently depends on the impact of environment and genetic background. A

The earliest use of targeted animal models for gene therapy emphasized mouse models that simulated inherited disease, but these often proved disappointing. Subsequent studies have identified many useful mouse models for the study of human disease; however, the utility of these transgenic models frequently depends on the impact of environment and genetic background. A good example is seen with mouse models of cystic fibrosis (CF), where the CF transmembrane conductance regulator (*CTFR*) gene was interrupted or mutated. The initial *CFTR*-deficient mice did not develop pulmonary pathologies before death; however, subsequent genetic and environmental modifications have increased its usefulness to model CF [22].

Currently over 1200 papers with transgenic mice can be found on-line (PubMed), and some 7000 mutant mice have been described. While not all of these models have proven useful, some have been critical for determining promoter and gene function, functional gene redundancy, spatial distributions of expression, and lineage tracing. Numerous papers have also documented the use of gene targeting for mouse models of development and disease. Some mouse and ESC studies have shed light on factors indispensable for haematopoiesis, while a number of knockout mouse models have been employed to reveal the critical roles for transcription factors (Ets family members) in guiding haematopoiesis, vasculo/angiogenesis, and other cellular differentiation processes. Others studies have used this technology to begin to unravel problems in complex regulatory pathways, specifically intermediary metabolism and physiology. In addition, many reviews have already been published showing how gene targeting has been employed to study cardiovascular, pancreatic, or renal systems, while still others have employed Cre/loxP systems for conditional regulation [22].

Animal models of human diseases are critical to the early development and evaluation of gene- and cell-based therapies; therefore, studies with mouse ESCs in the context of transgenic models form the foundation for current and future work with human ESCs and their derivatives for studies in human. Experimentally, it remains unclear whether human ESCs will be as versatile as mouse ESCs with respect to self-renewal, genetic manipulation, or developmental capacity, but the ability to test these cells in disease models, transgenic or otherwise, will be critical to this evaluation [22].

CONCLUSION

Manipulation of the mouse germ-line through the genetic modification of ESCs is clearly a very useful and extremely versatile tool for analyzing mammalian development and physiology and for creating models of human diseases. In addition to the widely utilized method of gene disruption, homologous recombination can mediate many other types of genetic changes, including both qualitative and quantitative modifications, making it practical to analyse the effects of many types of genetic changes on complex traits. The versatility of genetically modified ESCs is facilitating the rapid expansion of a very informative pool of induced mutations in mice and is allowing important questions to be addressed *in vivo* regarding gene expression and protein function and their impact on normal and disease states.

REFERENCES

1. Rippon H.J., A.E. Bishop (2004). 'Embryonic stem cells'. *Cell Prolif.* **37**(1): 23–34.
2. Doss M. X., C.I. Koehler, C. Gissel, J. Hescheler, A. Sachinidis (2004). 'Embryonic stem cells: A promising tool for cell replacement therapy;. *J. Cell. Mol. Med.* **8**(4): 465–473.
3. Murry C.E., G. Keller (2008). 'Differentiation of embryonic stem cells to clinically relevant populations: Lessons from embryonic development'. *Cell.* **132**(4): 661–680.
4. Bradley A., M. Evans, M. H. Kaufman, E. Robertson (1984). 'Formation of germ-line chimeras from embryo-derived teratocarcinoma cell lines'. *Nature.* **309**(5965): 255–256
5. Nagy A., E. Gocza, E.M. Diaz, V.R. Prideaux, E. Ivanyi, M. Marrkula, J. Rossant (1990). 'Embryonic stem cells alone are able to support foetal development in the mouse'. *Development.* **110**(3): 815–821.
6. Wang Z.Q., F. Kiefer, P. Urbanek, E.F. Wagner (1997). 'Generation of completely embryonic stem cell-derived mutant mice using tetraploid blastocyst injection'. *Mech. Dev.* **62**: 137–145.
7. Deb K.D., K. Sarda (2008). 'Human embryonic stem cells: Preclinical perspectives'. *J. Transl. Med.* **29**(6):7.
8. Keller G.M. (1995). '*In vitro* differentiation of embryonic stem cells. Curr. Opin'. *Cell. Biol.* **7**: 862–869.
9. Heard E., F. Mongelard, D. Arnaud, C. Chureau, C. Vourc'h, P. Avner (1999). 'Human XIST yeast artificial chromosome transgenes show partial X inactivation centre function in mouse embryonic stem cells'. *Proc. Natl. Acad. Sci. USA.* **96**: 6841–6846.
10. Thomson J.A., J. Itskovitz-Eldor, S.S. Shapiro, M.A. Waknitz, J.J. Swiegiel, V.S. Marshall, J.M. Jones (1998). 'Embryonic stem cell lines derived from human blastocysts'. *Science.* **282**: 1145–1147.
11. Clarke A.R. (1994). 'Murine genetic models of human disease'. *Curr. Opin. Genet. Dev.* **4**: 453–460.
12. Capecchi M.R. (1989). 'The new mouse genetics: Altering the genome by gene targeting'. *Trends Genet.* **5**: 70–76.
13. Robertson E. J. (1991). 'Using embryonic stem cells to introduce mutations into the mouse germ line'. *Biol. Reprod.* **44**: 238–245.
14. Gossler A., T. Doetschman, R. Korn, E. Serfling, R. Kemler (1986). 'Transgenesis by means of blastocyst-derived embryonic stem cell lines'. *Proc. Natl. Acad. Sci. USA.* **83**: 9065–9069.
15. Gordon J.W., G.A. Scangos, D.J. Plotkin, J.A Barbosa, F.H. Ruddle (1980). 'Genetic transformation of mouse embryos by microinjection of purified DNA'. *Proc. Natl. Acad. Sci .USA.* **77**: 7380–7384.
16. Robertson E.J. (1986). 'Pluripotential stem cell lines as a route into the mouse germ line'. *Trends Genet.* **2**: 9–14.
17. Ghebranious N., L.A. Donehower (1998). 'Mouse models in tumor suppression'. *Oncogene.* **17**: 3385–3400.
18. Stanford W.L., J. B. iCoh, S.P. Cordes (2001). 'Gene-trap mutagenesis: past, present and beyond'. *Nat. Rev. Genet.* **2**(10): 756–768.
19. Lin F., K. Sperle, N. Sternberg (1985). 'Recombination in mouse L cells between DNA introduced into cells and homologous chromosomal sequences'. *Proc. Natl. Acad. Sci. USA.* **82**: 1391–1395.
20. Koller B.H., P. Marrack, J.W. Kappler, O. Smithies (1990). 'Normal development of mice deficient in beta 2M, MHC class I proteins, and CD8+ T cells'. *Science.* **248**(4960): 1227–1230.
21. Pevny L., M.C. Simon, E. Robertson, W.H. Klein, S. F. Tsai, V.D. Agati, S. H. Orkin, F. Costantini (1991). 'Erythroid differentiation in chimaeric mice blocked by a targeted mutation in the gene for transcription factor GATA-1. *Nature.* **349**: 257–260.
22. Wobus A.M., K.R. Boheler (2005). 'Embryonic Stem Cells: Prospects for Developmental Biology and Cell Therapy'. *Physiol. Rev.* **85**: 635–678.

15

APPLICATION OF HUMAN EMBRYONIC STEM CELLS IN DRUG DISCOVERY

Rajarshi Pal[1], Satish Totey[2], Kaushik Deb*

[1]*Manipal Institute of Regenerative Medicine, Manipal University Branch Campus, Domlur Layout, Bangalore 560 071, INDIA Advanced Neuroscience Allies(ANSA) Pvt. Ltd . Indiranagar , Bangalore 560038, INDIA*

INTRODUCTION

The successful establishment of human embryonic stem cell (hESC) in 1998 ushered a revolution in biomedical research. These pluripotent stem cells possess unique properties that make them exceptionally useful in a range of applications. Therapeutic applications of ESCs are still in an "embryonic stage" and is certainly at least a decade away before it is transformed from the bench to bedside. However, hESCs hold tremendous potentials beyond their use in regenerative medicine with prospective applications that are likely to be more appropriate and immediate. Two such areas in the drug development process are drug discovery, with screening as one key activity, and toxicity testing. The hESC technology is anticipated to become particularly important for development of improved research tools for more cost-effective and efficient, rapid, robust and better precision, eventually leading to increased patient safety and reduce the number of laboratory animals needed for toxicological and pharmacological testing.

In the current drug discovery paradigm, validated recombinant targets constitute the backbone of *in vitro* high-throughput screening (HTS) assays. Isolated proteins cannot, however, be regarded as representative of complex biological systems. Therefore, cell-based systems can support *in vitro* data, providing greater confidence. The scarcity of foetal or human samples and the lack of proliferative capacity of primary cell cultures, combined with problems associated with

---

* *kaushik.deb@stempentics.com*

the use of somatic stem cells, make immortalized cell lines the most commonly used source of cells for HTS. While such cell lines have improved proliferative capacity, they often exhibit aberrant genetic and functional characteristics. Consequently, interest has focused on creating new cell lines using defined molecular strategies that overcome senescence signals and telomere shortening. These strategies include expression of viral oncogenes and the catalytic subunit of telomerase (TERT) to generate appropriate immortalized cells.

Human embryonic stem cells (hESCs) have been isolated from chromosomally euploid, aneuploid, and mutant human embryos that are available from *in vitro* fertilisation (IVF) clinics treating patients for infertility or preimplantation genetic diagnosis [1–6]. These hESC lines are valuable resources for functional genomics, drug screening, and, perhaps eventually, cell replacement therapy. The methods for deriving hESCs are well established and reproducible, and are relatively successful from 4- to 8-day-old morula and blastocysts and from isolated inner cell mass (ICM) cell clusters of human blastocysts. The hESCs can be produced and maintained on mouse or human somatic cells in humanized serum-free culture conditions and for several passages in cell-free culture systems. Their gene expression profiles and immunological characteristics have been described [6–12]. They may be grown indefinitely *in vitro* while retaining their original karyotype and epigenetic status. The hESCs can also be transfected with DNA constructs [13]. Furthermore, hESCs spontaneously differentiate in the absence of the appropriate cell feeder layer in the absence of fibroblast growth factor (FGF) 2. All three major embryonic lineages are produced in differentiating static cultures and embryoid bodies (EBs) in suspension cultures [14–16]. Cell progenitors/precursors of interest can be identified by markers, expression of reporter genes, and characteristic morphology. The cells thereafter can be enriched for further culture into mature cell types. Directed differentiation systems are well developed for ectodermal pathways that result in neural and glial cells and the mesodermal pathway for cardiac muscle cells, and many other cell types including haematopoietic progenitors and endothelial cells. Nonetheless, differentiation induction into definitive endodermal phenotype such as hepatocytes and pancreatic beta-islet cells has proved to be difficult.

In the last decade, ESC technology has been fancied in obtaining high-value knowledge on safety liability of novel drugs [17, 18]. Indeed, the hESCs appear to combine advantages of their unique characteristics for *in vitro* HTS techniques. Thus, if appropriately evaluated, it can offer undeniable advantages in drug discovery for identification of target and off-target effects. In fact, this *in vitro* platform can be employed in order to identify impacts that need investigation and further elucidation early in the drug discovery process which would eventually lead to reduction in the current, high degree of attrition in drug development. To be precise, many of the shortcomings in the classical drug development process could theoretically be improved by exploitation of hESC technology, optimally promoting the discovery of new drugs and treatments.

In vitro screening tests for reproductive toxicology are required by the 7th amendment of the directive 67/548 EEC and the OECD-program on interesting chemicals. Unfortunately, suitable methods for testing developmental toxicity or impairment of fertility are not available. In fact, the field of developmental toxicology is a milieu of embryology, toxicology, and pharmacology with a primary aim to identify and evaluate substances which perturb normal homeostasis in the developing organism. Moreover, pattern formation and growth in the embryo are controlled by highly complex temporal and spatial signaling. This exchange of information governs the delicately poised cascade of cell division, cell migration, cell death, gene transcription, and protein production.

In this chapter, we review the basic science behind this exciting field by describing the maintenance and characterisation of pluripotent hESCs along with their directed differentiation. We further discuss the current and future prospects for hESCs in some upcoming and improved stem cell based applications for drug discovery and toxicity testing. The focus throughout this chapter is on hESC-derived EBs, cardiomyocytes, and hepatocytes, since these two cell types are key to the drug development process.

ISOLATION, PROPAGATION AND CHARACTERISATION OF hESCs

An embryo at the blastocyst stage contains putative stem cells, which disappear after the seventh day to form the three embryonic germ layers [19, 20]. Nevertheless, ESCs isolated from the ICM during the blastocyst stage cultured *ex vivo* under appropriate conditions proliferate indefinitely and still maintain the developmental potential to form advanced derivatives of all three germ layers. It is at this stage of embryogenesis, toward the end of the first week of development, that ESCs can be isolated from the ICM. However, the basic approach for the derivation of hESC lines remained the same as of mouse and primate ESCs [1, 21–23]. In summary, the method involves proteolytic digestion (pronase) of the zona pellucida of an expanded blastocyst followed by immunosurgery to segregate the trophectoderm by an antibody/complement reaction. The isolated ICM cells are subsequently plated on a layer of growth-arrested mouse or human embryonic fibroblasts (MEF or HEF) feeder cells in tissue culture dishes. Approximately after 1–2 weeks, the initial outgrowth from the ICM is dissected manually and transferred to new culture dishes with fresh feeders. Successful propagation of the ICM is implicated with the appearance of colonies of cells with undifferentiated hESC morphology [24]. Moreover, alternative methods such as whole embryo culture, partial embryo culture, or laser ablation have also been employed successfully (US Patent No: WO03018783) [25, 26]. Overall, more than 400 hESC lines have been reported till date, though the extent of characterisation of these lines varies considerably [27]. These hESC lines can be maintained indefinitely in culture and exhibit a stable developmental potential to differentiate into all the cells of human body including trophectoderm cells [28]. However, the success rate of hESC derivation is somewhat associated with the quality of the blastocyst, in particular the integrity of the ICM cells.

hESCs are commonly maintained in an undifferentiated state in the presence of a feeder layer or feeder-free condition on an appropriate extra cellular matrix (ECM) supplemented with serum or conditioned medium [1, 7, 12]. In the absence of MEF, the hESCs tend to rapidly differentiate or cease to survive [29]. Unlike murine ESCs, the presence of exogenously added LIF does not prevent differentiation of hESCs. In light of the fact that secretion factors and direct cell-to-cell interactions control *in vitro* survival, proliferation, and differentiation of the stem cells, an ideal microenvironment should consist of healthy feeder cells with normal microstructures and functions. However, the most widely used method to propagate hESCs has been manual micro-dissection. In this delicate process, flame-pulled glass micropipettes are used for dissection of individual hESC colonies into smaller clumps. These clusters of hESCs are subsequently transferred to fresh culture dishes. The greatest advantage of the mechanical transfer method lies in the success to perform a positive selection at every passage by isolating undifferentiated hESCs from differentiated cells. But, this method is labor-intensive and time-consuming, posing significant challenge to process cells at a larger scale applicable to clinical scenario. This has evoked the development of alternative methods for hESC expansion. The use of enzymes for cell dissociation during passage is considerably faster and simpler than micro-dissection, and different enzymes such as collagenase IV [7], trypsin [3], dispase [8] and Tryple™ Select [30] have been used for the expansion of hESCs. More interestingly, in a recent study, enzymatic passaging was combined with a synthetic ROCK-inhibitor resulting in increased hESC survival during propagation of the cells [31]. A major disadvantage with the use of enzymes for hESC passaging is the increased susceptibility of incorporating genomic aberrations during long-term propagation *in vitro* [32]. hESCs have also been shown to grow in the absence of feeder cells on a suitable growth substrate such as Matrigel™ using feeder cell conditioned medium [7]. Other reports indicate that culture additives which activate the canonical Wnt pathway [33], a combination of growth factors such as LIF, transforming growth factor (TGF)-β1, and basic fibroblast growth factor (bFGF) [11], a combination of noggin and bFGF [34] or high levels of bFGF alone [35, 36], may be adequate to sustain undifferentiated hESCs in the absence of supporting feeders. Although, development of chemically defined media appears to be best suited [37, 38], continuous evaluation of methods for derivation and propagation of undifferentiated hESCs on feeder-free matrices to obtain xeno-free cultures of hESCs has been on the rise [39–41].

Undifferentiated hESCs are primarily characterized by demonstrating the presence of a set of molecular markers, largely consisting of markers previously used to distinguish human embryonic carcinoma cells and mouse ESCs [42]. The key markers include important cell surface antigens SSEA-3, SSEA-4, TRA-1-60, and TRA-1-81. In addition, expression of the transcription factors such as POU5F1/Oct3/4 and Nanog are very closely associated with the undifferentiated state of the cells [43, 44]. In nutshell, putative hESCs can be characterized by their distinct morphology, evaluation of surface antigens by immunocytochemistry or flow cytometry, gene expression analysis, karyotyping, DNA fingerprinting, epigenetic profiling, estimation of telomerase enzyme activity, and assessment of *in vitro* and *in vivo* differentiation potential [1,12, 14–16, 42, 45–47]. Nevertheless, the most convincing way to demonstrate the pluripotency of hESCs is by xenografting undifferentiated hESCs to immunodeficient mice where the human cells give rise to teratomas [1, 23]. The teratomas contain various types of tissues representing all three embryonic germ layers. Striated muscle, cartilage, bone, gut epithelium, and neural rosettes are commonly observed in the teratomas by histological analysis [1].

SPONTANEOUS DIFFERENTIATION OF hESCs

In contrast to human embryonal carcinoma cells, undifferentiated hESCs have tremendous potential to spontaneously differentiate into various cell types *in vitro*. However, too little is known about the intricate mechanisms which govern the spontaneous differentiation of pluripotent hESCs into a heterogeneous population of differentiated cell types. Undesired spontaneous differentiation of hESCs is also observed to a varying degree in most routine cultures and still presents one of the major challenges in hESC cultivation. The unique capacity of spontaneous differentiation in hESCs can, on the other hand, be harnessed to recapitulate certain aspects of early human embryonic development *in vitro* and to easily generate a variety of hESC progenies. Differentiation of hESCs can be easily initiated experimentally by withdrawal of MEF-conditioned medium or FGF-2. More complex differentiation can be induced by cultivation of the hESCs as three-dimensional aggregates, known as EBs [48]. To initiate the formation of EBs, the hESC colonies are selectively isolated manually or enzymatically and allowed to aggregate into spheres. The spheres are usually maintained in suspension culture and soon give rise to differentiated cell phenotypes of all three germ layers in the absence of FGF-2. Thereafter, further specialized differentiation can be obtained by subsequently plating the whole EBs onto specific ECM components with appropriate cocktail of growth factors mimicking *in vivo* embryogenesis.

DRAWBACKS WITH hESCs IN REGENERATIVE MEDICINE

Though functional hESC derivatives are considered as the most promising adjunct for degenerative diseases in humans, still there are safety, ethical, and technical concerns that need to be addressed. Strategies designed to develop refined protocols for differentiation, isolation, expansion, purification, and delivery of hESC derivatives are crucial to produce a sufficient number of pure neurons or cardiomyocytes that home to the right location. As of today, no protocol that yields a completely pure progeny population has been demonstrated. The general assumption that hESC-derived somatic cells are identical to the corresponding adult cell type generated through normal development remains to be scientifically established. In most cases, the hESC-derived cells have failed to demonstrate appropriate functional efficacy as they are ontogenetically closer to their foetal counterparts, which lack certain features. Next, the likelihood of teratoma formation and graft rejection are two burning issues in case of hESC transplantation. Furthermore, the fate of transplanted hESC derivatives has to be examined for longer time points in terms of toxicity and efficacy. In fact, there should be literature suggesting that caution should be raised when considering hESC derivatives for therapeutic interventions. Therefore, although, hESCs appear to be extremely promising as a potential new therapeutic strategy, several obstacles need to be overcome before their clinical application becomes a reality.

hESCs IN DRUG DISCOVERY AND DEVELOPMENT

Much of the attention focused on stem cells relates to their use in cell replacement therapy; however, stem cells may also transform the way in which therapeutics are discovered and validated. Drug discovery, drug development, clinical trials and regulatory approval are the imperatives preceding marketing of new pharmaceuticals. The critical phase of drug discovery covers the synthesis of a chemical library and the identification of candidate drugs with therapeutic potential. Owing to the advancements made in the understanding of molecular and physiological mechanisms underlying various diseases, scientists today aim to identify compounds to a biological target of interest and to evaluate the expected effects in animal models. Therefore, drug development includes pre-clinical studies, which address the parameters such as drug safety, toxicokinetics, toxicogenomics, chronic toxicity, reproductive toxicity, carcinogenicity, and immunotoxicity. More importantly, non-clinical studies are even continued once the drug enters into human clinical trials for long-term monitoring of the side effects, if any.

Roughly 250 drug candidates are identified for every 5000–10,000 compounds tested in drug discovery (PhRMA Industry profile 2006). Only 5 of these 250 candidate molecules will progress into human testing, and only one among these five is finally approved for marketing by regulatory authorities. The drug development costs for 2005 have been calculated to be at least in the order of US$40 billion. In 2004, 36 new drugs were approved by the US FDA, while only 20 new drugs were approved during 2005 [49]. Understandably, the processes preceding marketing of new pharmaceuticals are extremely costly and drawn out. In fact, estimates predict that from discovering one new drug to reaching the pharmacist's shelf typically takes 10–15 years and costs around US$900 million [50]. Obviously, more efficient and accurate predictions would eventually lead to lower attrition rates and safe new drugs and hence yield greater financial benefits. Furthermore, such improvements could simplify ethical considerations, by substantially reducing the number of animal and human subjects used in drug discovery.

ESC lines hold huge potential to improve the efficacy of drug discovery and development. Few scientific breakthroughs have raised so high expectations and enthusiasm from all quarters including industry, research community, and the general public as the successful establishment of hESCs in 1998 [1]. Following this discovery, a large number of basic research projects across the globe applying hESC models have substantially complemented knowledge on early human embryonic development. Therefore, hESCs can be employed for target validation and safety assessments in the pharmaceutical sector. Humanized tests based on hESCs not only provide information on interspecies variations but also may be rapid and cost-effective. However, extensive research is still looked for to standardize and to prove the scientific relevance of hESC *in vitro* assay.

Understandably, the use of animal-derived cells or tissues to develop selective drugs has its drawbacks, since activity in animals does not essentially translate into efficacy for humans. Although cell culture systems are well-established as cellular screening models in toxicology, in many cases, the *in vitro* models, such as primary cultures or established cell lines, do not represent the functional properties of specialized somatic cells. Long-term cultivation often leads to cessation of proliferation capacity, viability, chromosomal instability, and also tissue-specific properties [51, 52]. More importantly, ESCs, during *in vitro* differentiation, recapitulate cellular developmental processes and gene expression patterns of early embryogenesis and can give rise to functionally competent mature cell types [53, 54]. Therefore, hESCs present an excellent cell-based model which promises to be central for evaluating novel targets and compounds in a close to physiological environment.

DEVELOPMENTAL TOXICITY

Major birth defects are a critical social and healthcare hazard owing to their occurrence in 2–8% of human infants and foetuses (Eurocat Annual Report to WHO 2004–05) [55]. Less than 1% of birth defects are implicated to drugs [56]. The alarm of high risk of drugs is entrenched to the thalidomide disaster. Thalidomide, which demonstrated no effects in mice, was given to pregnant women in mid 19th century as a harmless sedative led to severe malformations in human embryos and more than 7000 children were born with birth defects. This tragedy triggered the imposition of stringent regulatory requirements for developmental toxicity testing [57]. In general, since long, physicians dealing with uncertainty during prescribing drugs for pregnant women are another major reason behind this upcoming field of research.

Developmental toxicity is a core subject of the ICH M3(R1) guideline and is addressed in detail in the ICH S5(R2) guideline on "Detection of Toxicity to Reproduction of Medical Products and Toxicity to Male Fertility". The field of developmental toxicology essentially combines embryology and toxicology to identify and evaluate substances that upset normal homeostasis in the developing organism. Pattern formation and growth in the embryo are regulated by highly complex temporal and spatial signaling along with physical interactions. This cross-talk orchestrates the delicately concerted cascade of cell division, cell migration, cell death, gene transcription, and protein production.

The evaluation of chemical-induced developmental toxicity requires extensive testing which essentially covers the period of organogenesis and effects from pre-implantation, through the entire gestation period. Non-clinical safety assessment of drug-induced developmental toxicity offers several challenges to overcome. Further, the currently used *in vivo* tests, being time-consuming, laborious, and expensive, require a high number of laboratory animals [58]. It has also been demonstrated that high interspecies variation is a hurdle for assessing developmental toxicity *in vivo* [59]. The model organisms differ from humans in many aspects, such as in size, appearance, longevity, physiology, and performance despite originating from a common ancestor during evolution. Developmental toxicity testing is commonly carried out in animal experiments using rodents or in embryo culture experiments. For obvious ethical reasons, *in vitro* or *in vivo* developmental toxicity testing using human embryos is not an option. Considerable advances toward an *in vitro* test have been achieved using mouse ESCs for an embryonic stem cell test (EST) [60]. The EST evaluates the inhibition of growth and differentiation of murine ESCs to determine the embryotoxic potential of chemicals and has been validated by the European center for the validation of alternative methods (ECVAM) in an international study [61]. Moreover, the original test has undergone several levels of improvements throughout the past decade [62]. The EST clearly delineates the potential of ESCs for developmental toxicity testing. But use of murine ESCs may raise cynicism about the validity of the predictions for humans. The potential for a "humanized EST" for developmental toxicity lies in pluripotency and self-renewal of hESCs and their ability to spontaneously differentiate and develop tissues from all three germ layers *in vitro*. Assays based on hESCs would probably have the utmost impact on reproductive toxicity testing for pharmaceuticals. Nevertheless, it should be pointed out that only a concerted effort will provide sufficient insight to support embryotoxic hazard identification of drugs due to complex mechanisms underlying the development of unborn child.

MEASUREMENT CRITERIA

We know that differentiation can be promoted in pluripotent stem cells by inducing EB formation in suspension cultures, eventually giving rise to many different cell types (Fig. 15.1 A–F). This ES/EB model presents an excellent stage to evaluate the effects of compounds on differentiation by simultaneously measuring the expression of key gene markers characteristic of different cell types belonging to three important germ layers. Likewise, the extent of cytotoxicity could also be determined in the EBs at different time points. This approach has commonly been employed by many groups including ECVAM [63–65].

DIFFERENTIAL GENE EXPRESSION DURING EB DEVELOPMENT

The list of genes selected for measuring the impact of compounds on the differentiation of hESCs needs careful thought. It is crucial that early and late stage markers from all three germ layers namely ectoderm, endoderm, and mesoderm are chosen in order to ensure a better insight of the ectopic gene expression during EB development. These genes are measured by real-time quantitative PCR using the Taqman Low Density Arrays (TLDA). A large-scale screening analysis using cDNA microarray technology is also interesting, owing to the volume of results that is obtained, despite being qualitative. The data generated thereafter would facilitate in establishing a prediction model. The basic model was constructed around the premises that (a) any change in gene expression (positive or negative) would represent an adverse effect on the embryo; (b) the greater the change in expression, the stronger the adverse impact on the embryo; (c) more toxic compounds would show greater change in gene expression with increasing concentration; and (d) smaller changes in the expression of many genes, not significant among themselves, could be summed to show the collective effect of a treatment [66]. Except some contradictions, compounds were successfully classified by this model, which opens up some scope of improvement.

CYTOTOXICITY

Another important step toward improving the overall performance of embryotoxicity assay is the incorporation of cytotoxicity testing as measured by the number of cells at progressive days of EB development in comparison to human fibroblasts. Fluorimetry-based methods like CYQUANT™ (Invitrogen) kit and DNA analysis by flow cytometry of fixed cells could be extremely useful approaches to examine the proliferative status of hESCs upon drug abuse. Furthermore, broadscale gene chip method, to check the dysregulation of various apoptotic and necrotic genes, pro-inflammatory cytokines, and death domain receptors could actually save time in achieving this objective.

ROLE OF GENETICALLY MODIFIED hESCs IN TOXICITY SCREENING

To keep pace with the promise of hESCs as a research tool and its diverse applications, it is important to be able to precisely and efficiently modify the hESC genome. The most widely used strategy is to engineer the mammalian genome by random integration, in which there is no control of the sites that foreign DNA is introduced and can, therefore, produce variable gene expression, silencing, and mutagenesis [67]. However, over the past two decades, some methods have been developed to induce controlled genetic modifications in the mammalian genome. These include homologous recombination, which allows for modification of particular genes of interest [68, 69], site-specific recombination of DNA into a predetermined natural or artificial site in the genome by integrases [70, 71], and site-directed insertion of genes by modified transposons [72, 73].

Owing to the versatile applications of genetically manipulated hESCs, a number of reports have been published establishing the role of baculovirus and Eppstein-barr-virus-based episomal vectors in transient and stable transgene expression in hESCs [74,75]. Recently, Liew et al. have compared the efficiency of four promoters (pUbiC, pR26, pCMV, and pCAGG) to mediate transient and stable transfection in hES and human embryonic carcinoma cells [76]. Furthermore, Oct-4/GFP reporter hESC lines have been established and their subsequent silencing employing siRNA technology has been demonstrated [77, 78].

Coercing the power of targeted gene manipulation with the unique ability of ESCs to grow and differentiate indefinitely allows for novel uses of ESCs in addition to bolstering their potential therapeutic application. Although any genetic manipulation to stem cells is associated with regulatory concerns due to the involvement of animal vectors/viruses, the technologies in use demonstrate that it is feasible to target a specific site in the genome rather than random integration. Moreover, the site itself can be evaluated in vivid detail prior to the development of a therapeutic product. In fact, genetic manipulation in hESCs offers the potential to develop universal stem cells by deletion of the immune complex and to cure genetic defects via homologous recombination or insertion of the deleted gene into a known locus. It can also be possible to knock out the gene/s responsible for teratoma formation, which is a key obstacle in ESC-mediated allogenic transplantation models. However, it is not only the direct applications of gene targeting which makes it a promising area of research. Facilitating discovery and HTS methodologies by using genetically modified stem cells (e.g., pluripotent/lineage-specific promoter reporter cells) offer the potential of discovering novel signaling pathways, small molecule mimics of growth factors, and other reagents. For example, a beta myosin heavy chain GFP hESC line could be used for screening molecules that are toxic and/or protective to cardiomyocytes, or a nanog/Sox-2-GFP hESC line could be used to monitor the hESCs at an undifferentiated state. Therefore, highly controlled genetic modification makes it possible to create knockout and knockin hESC lines to investigate gene function and lineage specification in human cells, to track and purify lineage-specific cells for biomarker identification and drug screening, and to develop hESC mutants for disease models [13].

APPLICATION FOR CARDIOMYOCYTES DERIVED FROM hESCs

Pro-arrhythmia (development of cardiac arrhythmias as a pharmacological side-effect) has emerged as the single most common cause of the withdrawal or constraint of the existing drugs in market. Therefore, drug-induced modulation of ion channel activities leading to anti- or pro-arrhythmic effects in cardiac tissue is critical for safety in pharmacology. However, the development of new medications, free from these side-effects, is challenged by the lack of an *in vitro* assay for human cardiac tissue. However, hESCs can produce spontaneously beating cardiomyocytes. During *in vitro* differentiation, cardiac-specific gene coding for proteins, e.g., receptors and ion channels (among them, sarcolemmal L- and T-type Ca2+ channels, Na+ and K+ channels) are expressed in a developmental continuum that closely mimics the developmental, morphological, and biochemical pattern from early (cardiac precursor cells) to terminally differentiated cells *in vivo* (e.g., atrial-like, ventricular-like, sinus nodal-like, and Purkinje-like cells). Further, the developmental succession of electrophysiological properties of EBs matches the sequence of electrophysiological changes described for the embryonic heart. Hence, beating cardiomyocyte clusters within EBs could serve as a potential model for testing cardio-active drugs.

CARDIOMYOCYTES: DERIVATION AND CHARACTERISTICS

Due to the scarcity of donor material coupled with a complicated method of cell isolation, human primary cardiomyocytes are not easily obtainable for preclinical drug discovery projects. Thus, there is a lot of optimism based on the use of pluripotent hESCs to derive functional cardiomyocytes for *in vitro* applications in drug development.

Following the first isolation of hESCs, several groups have demonstrated that they can readily be differentiated into contracting cardiomyocytes (Fig. 15.1D), having structural and functional characteristics comparable to their primary counterparts including cardiac ion channels, long action potential duration, and responsiveness to beta-adrenergic and muscarinic pharmacological stimulation [79–82]. In most reports, characterisation has been based on the expression of markers for cardiomyocytes, structural architecture, and functionality. At a molecular level, several markers expressed by adult cardiomyocytes are also expressed by hESC-derived cardiomyocytes, including transcription factors (Tbx5, Nkx2.5, GATA-4, MEF-2C), structural proteins (alpha/beta MHC, cardiac troponins I/T, alpha-actinin, titin, tropomyosin, MLC-2A/2V), hormones (ANP, SERCA, NCX, RyR), and tight junction proteins (connexin 43/45) [83–87]. However, one of the major obstacles, today, for the utilisation of hESC-derived cardiomyocytes is the insufficient number of cells achieved by the currently described differentiation protocols; in fact, the cardiomyocyte population that has so far been isolated is contaminated with other cell types. In addition, transplanted hESC-derived cardiomyocytes can mend the hearts of swine, sheep and rodent models with myocardial infarction leading to complete atrioventricular block, providing strong evidence for their *in vivo* functionality [88–93]. These findings have emerged as the main support for the development of cardiotoxicity assays at such a swift pace using hESC-derived cardiomyocytes as an *in vitro* system.

CARDIOMYOCYTES: SAFETY PHARMACOLOGY AND DISCOVERY OF NOVEL DRUG TARGETS

The use of hESC-derived cardiomyocytes in drug development can be either in drug discovery or in cardiac safety assessment of novel drug candidates. According to recommendations from all regulatory agencies, every compound has to undergo cardiac safety screening during development. Furthermore, cardiotoxicity is the core subject of the ICH M3(R1) guideline and is described in detail in the ICH S7B guideline on non-clinical evaluation of the potential of a drug to induce delayed ventricular repolarisation (http://www.emea.europa.eu/pdfs/human/ich/042302en.pdf). Delayed ventricular repolarisation, observed as QT interval prolongation in the electrocardiogram (ECG) is a significant risk indicator for the life-threatening form of ventricular tachycardia, "torsades-de-pointes".

Hence, there is a pressing need for strategies aimed at identifying the risk of drug-induced QT prolongation during early preclinical and clinical phases of drug development. The present preclinical models, in which a large library of substances can be screened rapidly, barely represent any synonymy with human physiology. The *in vitro* studies can address effects on the molecular, cellular, tissue, as well as organ level. So, various *in vitro* models are concurrently being employed including cell lines heterologously expressing human cardiac ion channels, primary cardiomyocyte cell lines, isolated tissue preparations, and perfused animal hearts. Owing to the restricted availability of human cardiomyocytes, researchers are bound to use animal-derived cells or tissues. Therefore, many of these experimental models have poor correlation for human *in vivo* response. Electrophysiological properties of the hESC-derived ventricular cells can be investigated in single cell preparations using standard patch-clamp techniques [94–96]. hESC-derived cardiomyocyte clusters can be applied in other platforms such as Micro Electrode Arrays (MEA) in which rhythm, route, and origin of excitation, repolarisation, and conduction can be analysed [96]. Further, heterogeneously expressed human ion channels (hERG potassium channels) are used to address impacts on specific proteins. In addition, automated platforms for high-throughput electrophysiological recordings in combination with relevant hESC-derived cardiomyocytes and novel *in silico* modeling approaches will provide cost-effective ways for investigating potential pro-arrhythmic risk of novel compounds [97]. Further, the field of cardiac pharmacological safety assessment is of considerable economic interest as evident by the presence of biotech companies such as Cellartis, Pfizer, Johnson and Johnson, Cellular dynamic international in this space. This may also be attributed to cardiotoxicity being among the leading reasons for drug attrition.

hESC-derived cardiomyocytes can also be employed for target identification, validation, and evaluation studies in the cardiac drug discovery process. To evaluate a new target in a clinically relevant environment is surely beneficial over the use of animal models or transformed cell lines. Moreover, genetically modified mouse models still offer substantial risk due to species difference. The successful insertion and deletion of genes in the human pluripotent stem cells has opened up possibilities for the development of *in vitro* cell-based assays toward initial target studies. In the case of complex polygenetic disorders, new hESC lines could potentially be developed using patient-specific embryos or somatic cell nuclear transfer (SCNT), and the isolated stem cells subsequently differentiated to cardiomyocytes. The phenotype of these cells could provide important insight on etiology and pathogenesis of diseases, and screens using siRNAs and compound libraries could potentially reveal new targets. In addition, there are several relevant cellular responses for different cardiac disorders that could be successfully evaluated on hESC-derived cardiomyocytes including contractile function, cardiac arrhythmia, and response to oxidative stress, resistance to apoptosis, and protection from ischemia.

APPLICATION FOR HEPATOCYTES DERIVED FROM hESCs

Primary cultured hepatocytes are a valuable *in vitro* model for drug metabolism studies. However, their widespread use is mired by the scarcity of suitable human liver samples. Moreover, the well-known *in vitro* phenotypic instability of hepatocytes, the irregular availability of fresh human liver for cell harvesting purposes, and the high batch-to-batch functional variability of hepatocyte preparations obtained from different human liver donors seriously complicate their use in routine testing. To overcome these limitations, different cell-line models have been proposed for drug metabolism screening. Human liver derived cell lines would be ideal models for this purpose given their availability, unlimited life-span, stable phenotype, and the fact that they are easy to handle. However, the human hepatocarcinoma cells currently used (i.e., HepG2, Mz-Hep-1) show negligible levels of drug-metabolizing and do not offer a real alternative to primary hepatocytes. Further, newer strategies have been proposed to generate metabolically competent immortalized hepatocytes (transformation of human hepatocytes with plasmids encoding immortalizing genes, hepatocyte-like cells derived from stem cells, cell lines generated from transgenic animals, hepatocyte/hepatoma hybrid cells). Moreover, recombinant models heterologously expressing P450 enzymes in different host cells have been developed and successfully used in drug-metabolism testing. In addition, latest strategies have recently been used to upregulate the expression of drug-metabolizing enzymes in cell lines of a human origin (i.e., transfection with expression vectors encoding key hepatic transcription factors). Among metabolic-based drug–drug interactions, P450 inhibition seems to be the most important. Hence, a major application of hESC-derived hepatocytes might be in the screening of potential enzyme inhibitors to study drug metabolism and drug-induced hepatotoxicity.

HEPATOCYTES: DERIVATION AND CHARACTERISTICS

The genetic switches that prompt human liver development and the molecular machinery downstream still remain hard to pin down. In fact, the different molecular mechanisms which are active during early liver development are an area which still evokes considerable debate among the developmental biologists. Stem cell differentiation into hepatocytes is of utmost interest, since access to adequate numbers of these cells would enable their use in the development of new drug discovery strategies and provide possibilities to perform *in vitro* metabolism studies and toxicity assessment.

Earlier studies have shown the *in vitro* ability of ESCs to differentiate into endoderm [98, 99]. Subsequently, *in vitro* differentiation of mESCs to hepatic lineage in the presence of cells from other lineage has been reported [100–102]. Further, the functionality of the ESC-derived hepatocytes has been demonstrated *in vivo* [103, 104] and *in vitro* [105]. However, relatively fewer reports are available, demonstrating hESC-derived cells with hepatocyte-like morphology and functions [106–110]. These observations were a result of directed differentiation of hESCs via EB formation (Fig. 15.1E) in modified culture conditions closely mimicking *in vivo* liver development. The cells obtained displayed appropriate polygonal-shaped morphology with multiple nuclei and expressed hepatocyte-associated markers, such as albumin, α-1-antitrypsin, hePAR1, and cytokeratin 19. In addition, hepatic transcription factors, such as FoxA2, HNF-1/4/6, and GATA-4 were also expressed by the hepatocyte-like cells. Functional analysis of the cells indicated glycogen accumulation, inducible cytochrome P450 activity, production of urea and albumin, and uptake of indocyanine green. Interestingly though, the first few studies on actual drug-metabolizing effects in hESC- and mESC-derived hepatic cells were recently published [111–113]. Besides the expression of several mature liver markers, these cells express functional glutathione transferase (GST) activity at levels comparable to primary human hepatocytes. Furthermore, recent studies have shed some light on the derivation of definitive endoderm from hESCs [114], followed by derivation of hepatocyte-like cells from definitive endoderm [115]. More recently, ECMs have been shown to play an essential role in hepatic differentiation from hESCs [116]. From the standpoint of the industrial use of stem cell derived hepatocytes, the presence of specific bio-transforming enzymes in the cells are of paramount importance. For broader drug discovery and toxicological applications, it is speculated that inducing hepatocyte derivation via definitive endoderm route may be a smart approach due to the larger number of differentiated cells obtained. However, the ability of these cells to perform critical liver functions like metabolic competence, biotransformation capacity, and transportation of exogenous compounds requires further investigation.

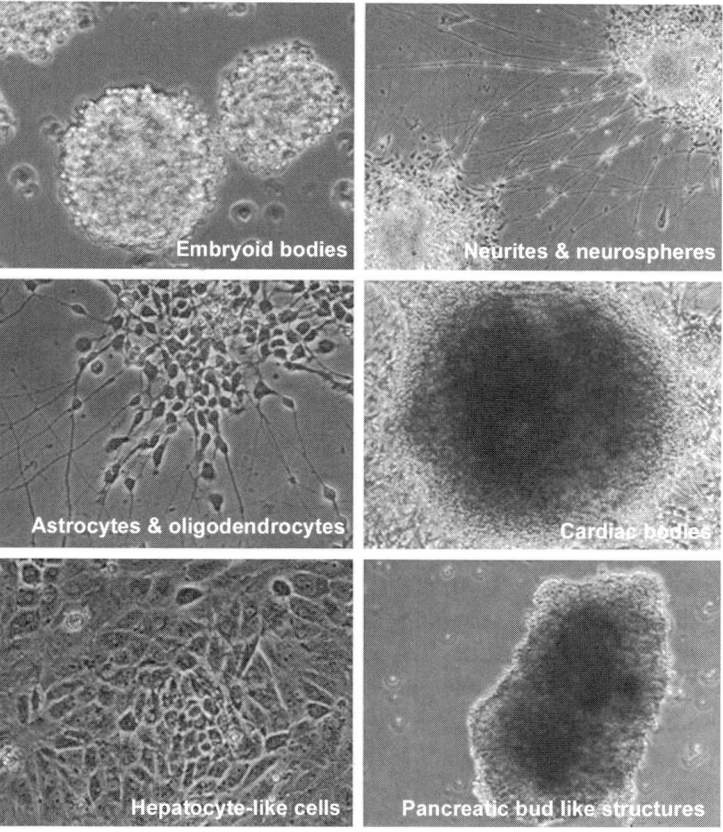

➤ Fig. 15.1 *Differentiation of hESCs into various lineages.*

HEPATOCYTES: STUDIES OF DRUG METABOLISM AND DRUG-INDUCED HEPATOTOXICITY

Hepatocytes happen to be a key component in the complex and lengthy drug discovery process. They are utilized for investigations of novel targets in metabolic diseases. These cells are also employed in studies of liver metabolism and pharmacokinetic properties of new chemical entities (NCE), as well as in hepatotoxicity testing. Notably, abnormal metabolism is primarily associated with the withdrawal of potential new drugs. Furthermore, liver toxicity and troubled liver function are the most common instances for toxicology among drug molecules. Hence, improved *in vitro* models to study human hepatocytes are required. Pluripotent stem cells may represent an extremely valuable source of functional hepatocytes. However, additional data are required to establish the state of maturation and usefulness of the hESC-derived hepatocytes.

Studies of metabolism and pharmacokinetic properties have emerged as a critical module of the drug discovery screening programs. This is primarily because 40% of NCE were recognized to fail late in the clinical phases because of pharmacokinetic problems. Hence, *in vitro* tools to predict pharmacokinetics of new compounds early in drug discovery will facilitate the selection of high-quality compounds that are safe and easy to administer. This is why the pharmaceutical companies have shown no hesitation in committing large investments to screen metabolic properties early in the drug discovery process [117].

Among the various human drug metabolizing enzymes, CYPs have been associated with the metabolism of most therapeutics whose elimination is facilitated by metabolism. CYPs are monooxygenases and are the major enzymes in phase 1 metabolism of xenobiotics. The major site of CYP expression is the liver, and CYP3A4 is the most abundant CYP isozyme in human adult liver. The most important drug metabolizing enzymes belong to the families one to three, responsible for about 70–80% of all phase 1, dependent metabolisms of clinically used drugs [118, 119]. Moreover, CYPs can be induced several-fold or inhibited by specific drugs, resulting in variability of metabolic activity [120]. Other metabolizing enzymes, which also may be of importance, are conjugating enzymes like UDP glucurunosyltransferases and flavin monooxygenases. Overall, early metabolism studies are related to metabolic degradation of the compound, mechanisms of the metabolism, and induction or inhibition of any drug-metabolizing enzymes. Furthermore, hepatic transporters may also play an important role in the excretion of drugs and their metabolites from the hepatocyte into bile [121].

In many cases, hepatotoxicity of unknown chemical compounds is detected late in the drug discovery process. The species difference may be considered as the biggest contributor to this anomaly since most of these studies are still performed on small animal models. Therefore, there has been a great interest in order to establish predictive human liver systems that could be assayed *in vitro*. The alternatives available today are human liver cancer cell lines or human primary hepatocytes isolated from liver biopsies. Transformed cell lines, such as the HepG2 human hepatocarcinoma cell line, are easily accessible for hepatotoxicity testing, but these cells have limited potential in terms of functional activity. Human primary hepatocytes obtained from different donor undermine the reproducibility of any assay performed with these cells. Further, available hepatic cell lines produce very low levels of metabolizing enzymes. Moreover, the toxic effects of xenobiotics are often dependent on their biotransformation into toxic metabolites and, therefore, the presence and distribution of biotransforming systems in potential test systems must be systematically evaluated.

In addition to the phase 1 class of liver proteins, GSTs belong to the important class of phase 2 metabolizing enzymes, which catalyze the conjugation of xenobiotics. Cytosolic GSTs detoxify electrophilic xenobiotics, such as chemical carcinogens, environmental pollutants, and anti-tumour agents. They also inactivate endogenous α- and β-unsaturated aldehydes, quinines, and epoxides as well as hydroperoxides, formed as secondary metabolites during oxidative stress [122]. Further, the conjugates formed by GSTs are transported by multidrug resistance protein 1 and 2, which are part of the phase 3 biotransformation system [123]. The highest levels of human GSTs are present in liver and testis [124]. In a recent study from Cellartis, hESC-derived hepatocyte-like cells were found to express a pattern of GST protein similar to that of human adult liver cells [112].

Recently, a proof-of-principle study was reported demonstrating the establishment of hepatotoxicity model using mouse ESCs differentiated into functional hepatocyte-like cells [105]. These results are encouraging toward the goal of developing models based on their human counterparts. However, there remains a vital issue to upscale the number of hepatocytes generated from hESCs, thereby maintaining their fidelity in the test system. Finally, the human liver being the largest gland constituting various cell types besides the hepatocyte such as kuppffer cells, stellate cells, and cholangiocytes, adds on to its complex architecture. Therefore, to be able to fully understand, and thereby predict, positive and negative effects of new pharmaceutical compounds *in vitro*, more complex models need to be developed. This may be possible by seeding hESC-derived EBs on tissue-engineered scaffolds that allow the cells to grow three-dimensionally. On the basis of the earlier discussion, there is a tremendous optimism that hESC research will help to mirror simple liver tissue, leading to dramatic improvement in accuracy and safety in prediction of *in vitro* human toxicity. It is also likely that these cells will find a variety of applications in target identification and validation studies.

CONCLUSION

Despite magnificent advancements made in the last decade, research on hESCs is still a young discipline. We are in very early stages of both understanding what can be accomplished with the great potential of hESCs as well as putting into practice the new research opportunity they provide. However, it is evident that the field of hESCs has a genuine potential to revolutionize the area of medical biotechnology and the understanding of human embryogenesis.

In the context of this chapter, we have emphasized on some interesting novel opportunities of hESCs and their differentiated derivatives in drug discovery and toxicology. This is motivated by the escalating need to move away from animal sources of experimentation toward humanized assays, in particular cell-based HTS with both good success rate and high clinical relevance.

Further, hESCs offer an exclusive source for generating large number of different specialized cell types. Although hESCs have been shown to differentiate efficiently, aiding adoption of hESC-derived functional cells as tools in drug screening needs to be carefully validated in terms of purity and reproducibility, and scaling-up production needs to be developed. In parallel, it is also important to develop robust and effective characterisation schemes to obtain better readouts in order to authenticate the quality of the cells being manufactured. Other key areas are the establishment of competent protocols to genetically manipulate hESCs and their progenies, which is becoming increasingly useful in maneuvering propensity of hESCs and monitor differentiation induction over time. The stem cell community anticipates that with the right resources, in concert with good regulatory practices and guidelines, the field of hESCs has a true prospect to reform many aspects of human biomedicine and the entire perspective of normal and abnormal human development. There is optimism that the biotechnology companies that successfully amalgamate stem cell based technologies will be benefited in the long run.

REFERENCES

1. Thomson J.A., J. Itskovitz-Eldor, S.S. Shapiro, M.A. Waknitz, J.J. Swiergiel, V.S Marshall, J.M. Jones (1998). 'Embryonic stem cell lines derived from human blastocysts'. *Science.* **282**: 1145–1147.
2. Mitalipova M., J. Calhoun, S. Shin, D. Wininger, T. Schulz, S. Noggle, A. Venable, I. Lyons, A. Robins, S. Stice (2003). 'Human embryonic stem cell lines derived from discarded embryos'. *Stem Cells.* **21**(5): 521–526.
3. Cowan C.A., I. Klimanskaya, J. McMahon, J. Atienza, J. Witmyer, J.P. Zucker, S. Wang, C.C. Morton, A.P. McMahon, D. Powers, D.A. Melton (2004). 'Derivation of embryonic stem-cell lines from human blastocysts'. *N. Engl. J. Med.* **350**: 13536.
4. Klimanskaya I., Y. Chung, L. Meisner, J. Johnson, M.D. West, R. Lanza (2005). 'Human embryonic stem cells derived without feeder cells'. *Lancet.* **365**(9471): 1636–1641.
5. Klimanskaya I., Y. Chung, S. Becker, S.J. Lu, R. Lanza (2006). 'Human embryonic stem cell lines derived from single blastomeres'. *Nature.* **444**(7118): 481–485.
6. Cheng L., H. Hammond, Z. Ye, X. Zhan, G. Dravid (2003). 'Human adult marrow cells support prolonged expansion of human embryonic stem cells in culture. *Stem Cells.* **21**(2): 131–142.
7. Xu C., M.S. Inokuma, J. Denham, K. Golds, P. Kundu, J.D. Gold, M.K. Carpenter (2001). 'Feeder-free growth of undifferentiated human embryonic stem cells'. *Nat. Biotechnol.* **19**: 971–974.
8. Richards M., C.Y. Fong, W.K. Chan, P.C. Wong, A. Bongso (2002). 'Human feeders support prolonged undifferentiated growth of human inner cell masses and embryonic stem cells'. *Nat Biotechnol.* **20**: 933–936.
9. Richards M., C.Y. Fong, S. Tan, W.K. Chan, A. Bongso (2004). 'An efficient and safe xeno-free cryopreservation method for the storage of human embryonic stem cells'. *Stem Cells.* **22**: 779–789.
10. Amit M., J. Itskovitz-Eldor (2002). 'Derivation and spontaneous differentiation of human embryonic stem cells. *J. Anat.* **200**: 225–232.
11. Amit M., V. Margulets, H. Segev, K. Shariki, I. Laevsky, R. Coleman, J. Itskovitz-Eldor (2003). 'Human feeder layers for human embryonic stem cells. *Biol. Reprod.* **68**: 2150–2156.

12. Amit M., C. Shariki, V. Margulets, J. Itskovitz-Eldor (2004). 'Feeder layer and serum-free culture of human embryonic stem cells'. *Biol. Reprod.* **70**: 837–845.
13. Zeng X., M.S. Rao (2008). 'Controlled genetic modification of stem cells for developing drug discovery tools and novel therapeutic applications'. *Curr. Opin. Mol. Ther.* **10**(3): 207–213.
14. Carpenter M.K., E.S. Rosler, G.J. Fisk, R. Brandenberger, X. Ares, T. Miura, M. Lucero, M.S. Rao (2004). 'Properties of four human embryonic stem cell lines maintained in a feeder-free culture system'. *Dev. Dyn.* **229**: 243–258.
15. Bhattacharya B., J. Cai, Y. Luo, T. Miura, J. Mejido, S.N. Brimble, X. Zeng, T.C. Schulz, M.S. Rao, R.K. Puri (2005). 'Comparison of the gene expression profile of undifferentiated human embryonic stem cell lines and differentiating embryoid bodies'. *BMC. Dev. Biol.* **5**: 22.
16. Brimble S.N., X. Zeng, D.A. Weiler, Y. Luo, Y. Liu, I.G. Lyons, W.J. Freed, A.J. Robins, M.S. Rao, T.C. Schulz (2004). 'Karyotypic stability, genotyping, differentiation, feeder-free maintenance, and gene expression sampling in three human embryonic stem cell lines derived prior to August 9, 2001'. *Stem Cells Dev.* **13**: 585–597.
17. Davila J.C., G.G. Cezar, M. Thiede, S. Strom, T. Miki, J. Trosko (2004). 'Use and application of stem cells in toxicology'. *Toxicol Sci.* **79**(2): 214–223.
18. Seiler A., A. Visan, R. Buesen, E. Genschow, H. Spielmann (2004). 'Improvement of an *in vitro* stem cell assay for developmental toxicity: The use of molecular endpoints in the embryonic stem cell test'. *Reprod Toxicol.* **18**(2): 231–240.
19. Andrews P.W., G. Banting, I. Damjanov, D. Arnaud, P. Avner (1984). 'Three monoclonal antibodies defining distinct differentiation antigens associated with different high molecular weight polypeptides on the surface of human embryonal carcinoma cells'. *Hybridoma.* **3**: 347–361.
20. Shamblott M.J., J. Axelman, S. Wang, E.M. Bugg, J.W. Littlefield, P.J. Donovan, P.D. Blumenthal, Huggins G.R., J.D. Gearhart (1998). 'Derivation of pluripotent stem cells from cultured human primordial germ cells'. *Proc Natl Acad Sci. USA.* **95**(23): 13726–13731.
21. Evans M.J. and M.H. Kaufman (1981). 'Establishment in culture of pluripotential cells from mouse embryos'. *Nature.* **292**: 154–156.
22. Thomson J.A., J. Kalishman, T.G. Golos, M. Durning, C.P. Harris, R.A. Becker, J.P. Hearn (1995). 'Isolation of a primate embryonic stem cell line'. *Proc. Natl. Acad. Sci. USA.* **92**: 7844–7848.
23. Reubinoff B.E., M.F. Pera, C.Y. Fong, A. Trounson, A. Bongso A (2000). 'Embryonic stem cell lines from human blastocysts: somatic differentiation *in vitro*'. *Nat. Biotechnol.* **18**: 399–404.
24. Amit M. and J. Itskovitz-Eldor (2002). 'Derivation and spontaneous differentiation of human embryonic stem cells'. *J. Anat.* **200**(Pt 3): 225-232.
25. Baharvand H., S.K. Ashtiani, M.R. Valojerdi, A. Shahverdi, A. Taee, D. Sabour (2004). 'Establishment and *in vitro* differentiation of a new embryonic stem cell line from human blastocyst'. *Differentiation.* **72**: 224–229.
26. Kim H.S., S.K. Oh, Y.B. Park, H.J. Ahn, K.C Sung, M.J. Kang, L.A. Lee, C.S. Suh, S.H. Kim, D.W. Kim, Moon S.Y. (2005). 'Methods for derivation of human embryonic stem cells'. *Stem Cells.* **23**: 1228–1233.
27. Guhr A., A. Kurtz, K. Friedgen, P. Löser (2006). 'Current state of human embryonic stem cell research: An overview of cell lines and their use in experimental work'. *Stem Cells.* **24**: 2187–2191.

28. Gerami-Naini B., O.V. Dovzhenko, M. Durning, F.H. Wegner, J.A. Thomson, T.G. Golos (2004). 'Trophoblast differentiation in embryoid bodies derived from human embryonic stem cells'. *Endocrinology.* **145**(4): 1517–1524.
29. Watt F.M., B.L. Hogan (2000). 'Out of Eden: stem cells and their niches'. *Science.* **287**(5457): 1427–1430.
30. Ellerstrom C., R. Strehl, K. Noaksson, J. Hyllner, H. Semb (2007). 'Facilitated expansion of human embryonic stem cells by single cell enzymatic dissociation'. *Stem Cells.* **25**(7): 1690–1996.
31. Watanabe K., M. Ueno, D. Kamiya, A. Nishiyama, M. Matsumura, T. Wataya, J.B. Takahashi, S. Nishikawa, S. Nishikawa, K. Muguruma, Y. Sasai (2007). 'A ROCK inhibitor permits survival of dissociated human embryonic stem cells'. *Nat. Biotechnol.* **25**: 681–686.
32. Mitalipova M.M., R.R. Rao, D.M. Hoyer, J.A. Johnson, L.F. Meisner, K.L. Jones, S. Dalton, S.L. Stice (2005). 'Preserving the genetic integrity of human embryonic stem cells'. *Nat. Biotechnol.* **23**: 19–20.
33. Sato N., L. Meijer, L. Skaltsounis, P. Greengard, A.H. Brivanlou (2004). 'Maintenance of pluripotency in human and mouse embryonic stem cells through activation of Wnt signaling by a pharmacological GSK-3-specific inhibitor;. *Nat. Med.* **10**: 55–63.
34. Wang G., H. Zhang, Y. Zhao, J. Li, J. Cai, P. Wang, S. Meng, J. Feng, C. Miao, M. Ding, D. Li, H. Deng (2005). 'Noggin and bFGF cooperate to maintain the pluripotency of human embryonic stem cells in the absence of feeder layers'. *Biochem. Biophys. Res. Commun.* **330**: 934–942.
35. Levenstein M.E., T.E. Ludwig, R.H. Xu, R.A. Llanas, K. VanDenHeuvel-Kramer, D. Manning, J.A. Thomson (2006). 'Basic fibroblast growth factor support of human embryonic stem cell self-renewal'. *Stem Cells.* **24**: 568–574.
36. Pal R., A. Mandal, H.S. Rao, M.S. Rao, A. Khanna (2007). 'A panel of tests to standardize the characterisation of human embryonic stem cells'. *Regen. Med.* **2**(2): 179–192.
37. Yao S., S. Chen, J. Clark, E. Hao, G.M. Beattie, A. Hayek, S. Ding (2006). 'Long-term self-renewal and directed differentiation of human embryonic stem cells in chemically defined conditions'. *Proc. Natl. Acad. Sci. USA.* **103**: 6907–6912.
38. Ludwig T.E., Levenstein ME, Jones JM, W.T. Berggren, E.R. Mitchen, J.L. Frane, L.J. Crandall, C.A. Daigh, K.R. Conard, M.S. Piekarczyk, R.A. Llanas, J.A. Thomson (2006). 'Derivation of human embryonic stem cells in defined conditions'. *Nat. Biotechnol.* **24**: 185–187.
39. Lu J., R. Hou, C.J. Booth, S.H. Yang, M. Snyder (2006). 'Defined culture conditions of human embryonic stem cells'. *Proc. Natl. Acad. Sci. USA.* **103**(15): 5688–5693.
40. Crook J.M., T.T. Peura, L. Kravets, A.G. Bosman, J.J. Buzzard, R. Horne, H. Hentze, N.R. Dunn, R. Zweigerdt, F. Chua, A. Upshall, A. Colman (2007). 'The generation of six clinical-grade human embryonic stem cell lines. Cell'. *Stem. Cell.* **1**(5): 490–494.
41. Bigdeli N., M. Andersson, R. Strehl, K. Emanuelsson, E. Kilmare, J. Hyllner, A. Lindahl (2008). 'Adaptation of human embryonic stem cells to feeder-free and matrix-free culture conditions directly on plastic surfaces'. *J. Biotechnol.* **133**(1): 146–153.
42. Draper J.S., K. Smith, P. Gokhale, H.D. Moore, E. Maltby, J. Johnson, L. Meisner, T.P. Zwaka, J.A. Thomson, P.W. Andrews (2004). 'Recurrent gain of chromosomes 17q and 12 in cultured human embryonic stem cells'. *Nat. Biotechnol.* **22**: 53–54.
43. Pesce M. and H.R. Schöler (2001). '*Oct-4*: Gatekeeper in the beginnings of mammalian development'. *Stem Cells.* **19**(4): 271–278.

44. Chambers I., D. Colby, M. Robertson, J. Nichols, S. Lee, S. Tweedie, A. Smith (2003). 'Functional expression cloning of Nanog, a pluripotency sustaining factor in embryonic stem cells'. *Cell.* **113**: 643–655.
45. Ginis I., Y. Luo, T. Miura, S. Thies, R. Brandenberger, S. Gerecht-Nir, M. Amit, A. Hoke, M.K. Carpenter, J. Itskovitz-Eldor, M.S. Rao (2004). 'Differences between human and mouse embryonic stem cells'. *Dev Biol.* **269**(2): 360–380.
46. Zeng X., T. Miura, Y. Luo, B. Bhattacharya, B. Condie, J. Chen, I. Ginis, I. Lyons, J. Mejido, R.K. Puri, M.S. Rao, W.J. Freed (2004). 'Properties of pluripotent human embryonic stem cells BG01 and BG02'. *Stem Cells.* **22**(3): 292–312.
47. Bhattacharya B., T. Miura, R. Brandenberger, J. Mejido, Y. Luo, A.X. Yang, B.H. Joshi, I. Ginis, R.S. Thies, M. Amit, I. Lyons, B.G. Condie, J. Itskovitz-Eldor, M.S. Rao, R.K. Puri (2004). 'Gene expression in human embryonic stem cell lines: unique molecular signature'. *Blood.* **103**(8): 2956–2964.
48. Itskovitz-Eldor J., M. Schuldiner, D. Karsenti, A. Eden, O. Yanuka, M. Amit, H. Soreq, N. Benvenisty (2000). 'Differentiation of human embryonic stem cells into embryoid bodies compromising the three embryonic germ layers'. *Mol Med.* **6**: 88–95.
49. Sartipy P., P. Björquist, R. Strehl, J. Hyllner (2007). 'The application of human embryonic stem cell technologies to drug discovery. Drug. Discov'. *Today.* **12**(17–18): 688–699.
50. Kola I. and J. Landis (2004). 'Can the pharmaceutical industry reduce attrition rates?' *Nat. Rev. Drug. Discov.* **3**: 711–715.
51. Wobus A.M., T. Kleppisch, V. Maltsev, J. Hescheler (1994). 'Cardiomyocyte-like cells differentiated *in vitro* from embryonic carcinoma cells P19 are characterized by functional expression of adrenoceptors and Ca^{2+} channels. *In Vitro Cell. Dev'. Biol. Anim.* **30A**(7): 425–434.
52. Gottlieb D. (2002). 'Signposts on the neural path'. *Nat. Biotechnol.* **20**(12): 1208–1210.
53. Guan K., H. Chang, A. Rolletschek, A.M. Wobus (2001). 'Embryonic stem cell-derived neurogenesis. Retinoic acid induction and lineage selection of neuronal cells'. *Cell. Tissue. Res.* **305**(2): 171–176.
54. Rohwedel J., K. Guan, C. Hegert, A.M. Wobus (2001). 'Embryonic stem cells as an *in vitro* model for mutagenicity, cytotoxicity and embryotoxicity studies: present state and future prospects'. *Toxicol. in vitro.* **15**(6): 741–753.
55. Queisser-Luft A., G. Stolz, A. Wiesel, K. Schlaefer, J. Spranger (2002). 'Malformations in newborn: Results based on 30,940 infants and foetuses from the Mainz congenital birth defect monitoring system (1990–1998)'. *Arch. Gynecol. Obstet.* **266**(3): 163–167.
56. Webster W.S., J.A. Freeman (2001). 'Is this drug safe in pregnancy?' *Reprod. Toxicol* **15**(6): 619–629.
57. Collins T.F. (2006). 'History and evolution of reproductive and developmental toxicology guidelines'. *Curr. Pharm. Des.* **12**(12): 1449–1465.
58. Höfer T., I. Gerner, U. Gundert-Remy, M. Liebsch, A. Schulte, H. Spielmann, R. Vogel, K. Wettig (2004). 'Animal testing and alternative approaches for the human health risk assessment under the proposed new European chemicals regulation'. *Arch. Toxicol.* **78**(10): 549–564.
59. Hurtt M.E., G.D. Cappon, A. Browning (2003). 'Proposal for a tiered approach to developmental toxicity testing for veterinary pharmaceutical products for food-producing animals. Food'. *Chem. Toxicol.* **41**(5): 611–619.

60. Genschow E., G. Scholz, N.A. Brown, A.H. Piersma, M. Brady, N. Clemann, H. Huuskonen, F. Paillard, S. Bremer, H. Spielmann (1999). 'Development of prediction models for three *in vitro* embryotoxicity tests which are evaluated in an ECVAM validation study'. *ALTEX*. **16**(2): 73–83.
61. Genschow E., H. Spielmann, G. Scholz, I. Pohl, A. Seiler, N. Clemann, S. Bremer, K. Becker. (2004). 'Validation of the embryonic stem cell test in the international ECVAM validation study on three *in vitro* embryotoxicity tests'. *Altern. Lab. Anim.* **32**(3): 209–244.
62. Buesen R., A. Visan, E. Genschow, B. Slawik, H. Spielmann, A. Seiler (2004). 'Trends in improving the embryonic stem cell test (EST): An overview'. *ALTEX*. **21**(1): 15–22.
63. Pellizzer C., S. Adler, R. Corvi, T. Hartung, S. Bremer (2004). 'Monitoring of teratogenic effects *in vitro* by analysing a selected gene expression pattern'. *Toxicol in vitro*. **18**(3): 325–335.
64. Pellizzer C., S. Bremer, T. Hartung (2005). 'Developmental toxicity testing from animal towards embryonic stem cells'. *ALTEX* **22**(2): 47–57.
65. Bremer S., T. Hartung (2004). 'The use of embryonic stem cells for regulatory developmental toxicity testing *in vitro*—the current status of test development'. *Curr. Pharm. Des.* **10**(22): 2733–2747.
66. Chapin R., D. Stedman, J. Paquette, P. Streck, S. Kumpf, S. Deng (2007). 'Struggles for equivalence: *In vitro* developmental toxicity model evolution in pharmaceuticals in 2006'. *Toxicol. In vitro*. **21**(8): 1545–1551.
67. Heaney J.D., S.K. Bronson (2006). 'Artificial chromosome-based transgenes in the study of genome function'. *Mamm. Genome.* **17**(8): 791–807.
68. Zwaka T.P. and J.A. Thomson (2003). 'Homologous recombination in human embryonic stem cells'. *Nat. Biotechnol.* **21**(3): 319–321.
69. Yang Y., B. Seed (2003). 'Site-specific gene targeting in mouse embryonic stem cells with intact bacterial artificial chromosomes'. *Nat. Biotechnol.* **21**(4): 447–451.
70. Belteki G., M. Gertsenstein, D.W. Ow, A. Nagy (2003). 'Site-specific cassette exchange and germline transmission with mouse ES cells expressing öC31 integrase'. *Nat. Biotechnol.* **21**(3): 321–324.
71. Belteki L., F. Edenhofer, S. Haupt, P. Koch, F.T. Wunderlich, H. Siemen, O. Brüstle (2006). 'Site-specific recombination in human embryonic stem cells induced by cell-permeant Cre recombinase'. *Nat. Methods.* **3**(6): 461–467.
72. Hackett P.B., S.C. Ekker, D.A. Largaespada, R.S. McIvor (2005). 'Sleeping beauty transposon-mediated gene therapy for prolonged expression'. *Adv. Genet.* **54**: 189–232.
73. Ivics Z., P.B. Hackett, R.H. Plasterk, Z. Izsvák (1997). 'Molecular reconstruction of Sleeping Beauty, a Tc1-like transposon from fish, and its transposition in human cells'. *Cell.* **91**(4): 501–510.
74. Ren C., M. Zhao, X. Yang, D. Li, X. Jiang, L. Wang, W. Shan, H. Yang, L. Zhou, W. Zhou, H. Zhang (2006). 'Establishment and applications of epstein-barr virus-based episomal vectors in human embryonic stem cells'. *Stem Cells.* **24**(5): 1338–1347.
75. Zeng J., J. Du, Y. Zhao, N. Palanisamy, S. Wang (2007) Baculoviral vector-mediated transient and stable transgene expression in human embryonic stem cells'. *Stem Cells.* **25**(4): 1055–1061.
76. Liew C.G., J.S. Draper, J. Walsh, H. Moore, P.W. Andrews (2007). 'Transient and stable transgene expression in human embryonic stem cells'. *Stem Cells.* **25**(6): 1521–1528.
77. Matin M.M., J.R. Walsh, P.J. Gokhale, J.S. Draper, A.R. Bahrami, I. Morton, H.D. Moore, P.W. Andrews (2004). 'Specific knockdown of Oct4 and beta2-microglobulin expression by RNA

interference in human embryonic stem cells and embryonic carcinoma cells'. *Stem Cells.* **22**(5): 659–668.

78. Gerrard L., D. Zhao, A.J. Clark, W. Cui (2005). 'stably transfected human embryonic stem cell clones express OCT4-specific green fluorescent protein and maintain self-renewal and pluripotency'. *Stem Cells.* **23**(1): 124–133.

79. Gerrard I., D. Kenyagin-Karsenti, M. Snir, H. Segev, M. Amit, L. Gepstein, E. Livne, O. Binah, J. Itskovitz-Eldor, L. Gepstein (2001). 'Human embryonic stem cells can differentiate into myocytes with structural and functional properties of cardiomyocytes'. *J. Clin. Invest.* **108**(3): 407–414.

80. Gerrard M., C. Boettinger, J. Hescheler (2004). 'Beta-adrenergic and muscarinic modulation of human embryonic stem cell-derived cardiomyocytes. Cell. Physiol'. *Biochem* **14**(4-6): 187–196.

81. Norström A., K. Akesson, T. Hardarson, L. Hamberger, P. Björquist, P. Sartipy (2006). 'Molecular and pharmacological properties of human embryonic stem cell-derived cardiomyocytes'. *Exp. Biol. Med* (Maywood). **231**(11): 1753–1762.

82. Huber I., I. Itzhaki, O. Caspi, G. Arbel, M. Tzukerman, A. Gepstein, M. Habib, L. Yankelson, I. Kehat, L. Gepstein (2007). 'Identification and selection of cardiomyocytes during human embryonic stem cell differentiation'. *FASEB. J.* **21**(10): 2551–2563.

83. Snir M., I. Kehat, A. Gepstein, R. Coleman, J. Itskovitz-Eldor, E. Livne, L. Gepstein (2003). 'Assessment of the ultrastructural and proliferative properties of human embryonic stem cell-derived cardiomyocytes'. *Am. J. Physiol. Heart. Circ. Physiol.* **285**(6): H2355–2363.

84. Snir R., D.W. Oostwaard, J. Snapper, J. Kloots, R.J. Hassink, E. Kuijk, B. Roelen, A.B. de la Riviere, C. Mummery (2005). 'Increased cardiomyocyte differentiation from human embryonic stem cells in serum-free cultures'. *Stem Cells.* **23**(6): 772–780.

85. Xu C., S. Police, N. Rao, M.K. Carpenter (2002). 'Characterisation and enrichment of cardiomyocytes derived from human embryonic stem cells'. *Circ Res.* **91**(6): 501–508.

86. Synnergren J., K. Akesson, K. Dahlenborg, H. Vidarsson, C. Améen, D. Steel, A. Lindahl, B. Olsson, P. Sartipy (2008). 'Molecular signature of cardiomyocyte clusters derived from human embryonic stem cells'. *Stem Cells.* **26**(7): 1831–1840.

87. Pal R. and A. Khanna (2007). 'Similar pattern in cardiac differentiation of human embryonic stem cell lines, BG01V and ReliCellhES1, under low serum concentration supplemented with bone morphogenetic protein-2'. *Differentiation.* **75**(2): 112–122.

88. Kehat I., L. Khimovich, O. Caspi, A. Gepstein, R. Shofti, G. Arbel, I. Huber, J. Satin, J. Itskovitz-Eldor, L. Gepstein (2004). 'Electromechanical integration of cardiomyocytes derived from human embryonic stem cells'. *Nat Biotechnol.* **22**(10): 1282–1289.

89. Ménard C., A.A. Hagège, O. Agbulut, M. Barro, M.C. Morichetti, C. Brasselet, A. Bel, E. Messas, A. Bissery, P. Bruneval, M. Desnos, M. Pucéat, P. Menasché (2005). 'Transplantation of cardiac-committed mouse embryonic stem cells to infarcted sheep myocardium: a preclinical study'. *Lancet.* **366**(9490): 1005–1012.

90. Laflamme M., J. Gold, C. Xu, M. Hassanipour, E. Rosler, S. Police, V. Muskheli, C.E. Murry (2005). Formation of human myocardium in the rat heart from human embryonic stem cells'. *Am. J. Pathol.* **167**(3): 663–671.

91. Laflamme M.A., K.Y. Chen, A.V. Naumova, V. Muskheli, J.A. Fugate, S.K. Dupras, H. Reinecke, C. Xu, M. Hassanipour, S. , C. O'Sullivan, L. Collins, Y. Chen, E. Minami, E.A. Gill, S. Ueno, C. Yuan,

J. Gold, C.E. Murry (2007). 'Cardiomyocytes derived from human embryonic stem cells in pro-survival factors enhance function of infarcted rat hearts'. *Nat. Biotechnol.* **25**(9): 1015–1024.

92. van Laake L.W., R. Passier, J. Monshouwer-Kloots, M.G. Nederhoff, D. Ward-van Oostwaard, L.J. Field, C.J. van Echteld, P.A. Doevendans, C.L. Mummery (2007). 'Monitoring of cell therapy and assessment of cardiac function using magnetic resonance imaging in a mouse model of myocardial infarction'. *Nat. Protoc.* **2**(10): 2551–2567.

93. Dai W., L.J. Field, M. Rubart, S. Reuter, S.L. Hale, R. Zweigerdt, R.E. Graichen, G.L. Kay, A.J. Jyrala, A. Colman, B.P. Davidson, M. Pera, R.A. Kloner (2007). 'Survival and maturation of human embryonic stem cell-derived cardiomyocytes in rat hearts'. *J. Mol. Cell. Cardiol.* **43**(4): 504–516.

94. Caspi O., I. Huber, I. Kehat, M. Habib, G. Arbel, A. Gepstein, L. Yankelson, D. Aronson, R. Beyar, L. Gepstein (2007). 'Transplantation of human embryonic stem cell-derived cardiomyocytes improves myocardial performance in infarcted rat hearts'. *J. Am. Coll. Cardiol.* **50**(19): 1884–1893.

95. Mummery C., D. Ward-van Oostwaard, P. Doevendans, R. Spijker, S. van den Brink, R. Hassink, M. van der Heyden, T. Opthof, M. Pera, A.B. de la Riviere, R. Passier, L. Tertoolen (2003). 'Differentiation of human embryonic stem cells to cardiomyocytes: role of coculture with visceral endoderm-like cells'. *Circulation.* **107**(21): 2733–2740.

96. Reppel M., F. Pillekamp, K. Brockmeier, M. Matzkies, A. Bekcioglu, T. Lipke, F. Nguemo, H. Bonnemeier, J. Hescheler (2005). 'The electrocardiogram of human embryonic stem cell-derived cardiomyocytes'. *J. Electrocardiol.* **38**(4 Suppl): 166–170.

97. Muzikant A.L., E.W. Hsu, P.D. Wolf, C.S. Henriquez (2002). 'Region specific modeling of cardiac muscle: comparison of simulated and experimental potentials'. *Ann. Biomed. Eng.* **30**(7): 867–883.

98. Abe K., H. Niwa, K. Iwase, M. Takiguchi, M. Mori, S.I. Abé, K. Abe, K.I. Yamamura (1996). 'Endoderm-specific gene expression in embryonic stem cells differentiated to embryoid bodies'. *Exp. Cell. Res.* **229**(1): 27–34.

99. Morrisey E.E., Z. Tang, K. Sigrist, M.M. Lu, F. Jiang, H.S. Ip, M.S. Parmacek (1998). 'GATA6 regulates HNF4 and is required for differentiation of visceral endoderm in the mouse embryo'. *Genes. Dev.* **12**(22): 3579–3590.

100. Hamazaki T., Y. Iiboshi, M. Oka, P.J. Papst, A.M. Meacham, L.I. Zon, N. Terada (2001). 'Hepatic maturation in differentiating embryonic stem cells *in vitro*'. *FEBS. Lett.* **18**: 497(1): 15–19.

101. Jones E.A., D. Tosh, D.I. Wilson, S. Lindsay, L.M. Forrester (2002). 'Hepatic differentiation of murine embryonic stem cells'. *Exp. Cell Res.* **272**(1): 15–22.

102. Pal R. and A. Khanna (2005). 'Role of hepatocyte-like cells in the differentiation of cardiomyocytes from mouse embryonic stem cells'. *Stem Cells. Dev.* **14**(2): 153–161.

103. Chinzei R., Y. Tanaka, K. Shimizu-Saito, Y. Hara, S. Kakinuma, M. Watanabe, K. Teramoto, S. Arii, K. Takase, C. Sato, N. Terada, H. Teraoka (2002). 'Embryoid-body cells derived from a mouse embryonic stem cell line show differentiation into functional hepatocytes'. *Hepatology.* **36**(1): 22–29.

104. Choi D., H.J. Oh, U.J. Chang, S.K. Koo, J.X. Jiang, S.Y. Hwang, J.D. Lee, G.C. Yeoh, H.S. Shin, J.S. Lee, B. Oh (2002). '*In vivo* differentiation of mouse embryonic stem cells into hepatocytes'. *Cell Transplant.* **11**(4): 359–368.

105. Kulkarni J.S. and A. Khanna (2006). 'Functional hepatocyte-like cells derived from mouse embryonic stem cells: a novel *in vitro* hepatotoxicity model for drug screening'. *Toxicol. in vitro.* **20**(6): 1014–1022.

106. Rambhatla L., C.P. Chiu, P. Kundu, Y. Peng, M.K. Carpenter (2003). 'Generation of hepatocyte-like cells from human embryonic stem cells'. *Cell Transplant.* **12**(1): 1–11.
107. Lavon N., O. Yanuka, N. Benvenisty (2004). 'Differentiation and isolation of hepatic-like cells from human embryonic stem cells'. *Differentiation.* **72**(5): 230–238.
108. Schwartz R.E., J.L. Linehan, M.S. Painschab, W.S. Hu, C.M. Verfaillie, D.S. Kaufman (2005). 'Defined conditions for development of functional hepatic cells from human embryonic stem cells'. *Stem. Cells Dev.* **14**(6): 643–655.
109. Baharvand H., S.M. Hashemi, S. Kazemi Ashtiani, A. Farrokhi (2006). 'Differentiation of human embryonic stem cells into hepatocytes in 2D and 3D culture systems *in vitro*'. *Int. J. Dev. Biol.* **50**(7): 645–652.
110. Shiraki N., K. Umeda, N. Sakashita, M. Takeya, K. Kume, S. Kume (2008). 'Differentiation of mouse and human embryonic stem cells into hepatic lineages'. *Genes. Cells.* **13**(7): 731–746.
111. Ek M., T. Söderdahl, B. Küppers-Munther, J. Edsbagge, T.B. Andersson, P. Björquist, I. Cotgreave, B. Jernström, M. Ingelman-Sundberg, I. Johansson (2007). 'Expression of drug metabolizing enzymes in hepatocyte-like cells derived from human embryonic stem cells'. *Biochem. Pharmacol.* **74**(3): 496–503.
112. Söderdahl T., B. Küppers-Munther, N. Heins, J. Edsbagge, P. Björquist, I. Cotgreave, B. Jernström (2007). 'Glutathione transferases in hepatocyte-like cells derived from human embryonic stem cells'. *Toxicol. In vitro.* **21**(5): 929–937.
113. Maezawa K., K. Miyazato, T. Matsunaga, Y. Momose, T. Imamura, K. Johkura, K. Sasaki, S. Ohmori (2008). 'Expression of cytochrome P450 and transcription factors during *in vitro* differentiation of mouse embryonic stem cells into hepatocytes'. *Drug Metab Pharmacokinet.* **23**(3): 188–195.
114. D'Amour K.A., A.D. Agulnick, S. Eliazer, O.G. Kelly, E. Kroon, E.E. Baetge (2005). 'Efficient differentiation of human embryonic stem cells to definitive endoderm'. *Nat. Biotechnol.* **23**(12): 1534–1541.
115. Cai J., Y. Zhao, Y. Liu, F. Ye, Z. Song, H. Qin, S. Meng, Y. Chen, R. Zhou, X. Song, Y. Guo, M. Ding, H. Deng (2007). 'Directed differentiation of human embryonic stem cells into functional hepatic cells'. *Hepatology.* **45**(5): 1229–1239.
116. Ishii T., K. Fukumitsu, K. Yasuchika, K. Adachi, E. Kawase, H. Suemori, N. Nakatsuji, I. Ikai, S. Uemoto (2008). 'Effects of extracellular matrixes and growth factors on the hepatic differentiation of human embryonic stem cells. Am. J. Physiol. Gastrointest'. *Liver. Physiol.* **295**(2): G313–321.
117. Masimirembwa C.M., R. Thompson, T.B. Andersson (2001) *In vitro* high throughput screening of compounds for favorable metabolic properties in drug discovery. Comb. Chem. High'. *Throughput. Screen.* **4**(3): 245–263.
118. Bertz R.J., G.R. Granneman (1997). 'Use of *in vitro* and *in vivo* data to estimate the likelihood of metabolic pharmacokinetic interactions'. *Clin. Pharmacokinet.* **32**(3): 210–258.
119. Ingelman-Sundberg M. (2004). 'Pharmacogenetics of cytochrome P450 and its applications in drug therapy: The past, present and future. Trends'. *Pharmacol. Sci.* **25**(4): 193–200.
120. Pelkonen O., J. Mäenpää, P. Taavitsainen, A. Rautio, H. Raunio (1998). 'Inhibition and induction of human cytochrome P450 (CYP) enzymes'. *Xenobiotica.* **28**(12): 1203–1253.
121. Chandra P., K.L. Brouwer (2004). 'The complexities of hepatic drug transport: current knowledge and emerging concepts'. *Pharm. Res.* **21**(5): 719–735.

122. Hayes J.D., J.U. Flanagan, I.R. Jowsey (2005). 'Glutathione transferases'. *Ann. Rev. Pharmacol. Toxicol.* **45**: 51–88.
123. Haimeur A., G. Conseil, R.G. Deeley, S.P. Cole (2004). 'Mutations of charged amino acids in or near the transmembrane helices of the second membrane spanning domain differentially affect the substrate specificity and transport activity of the multidrug resistance protein MRP1 (ABCC1)'. *Mol. Pharmacol'.* **65**(6): 1375–1385.
124. Rowe J.D., E. Nieves, I. Listowsky (1997). 'Subunit diversity and tissue distribution of human glutathione S-transferases: Interpretations based on electrospray ionisation-MS and peptide sequence-specific antisera'. *Biochem. J.* **325**(Pt 2): 481–486.

Part 2

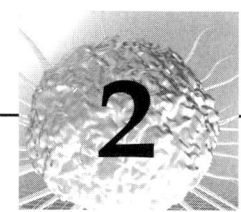

Adult Stem Cells

16 *Recent Advancements on Tissue-Resident Adult Stem/Progenitor Cell Biology and their Therapeutic Applications in Regenerative Medicine and Cancer Therapies*

17 *Mesenchymal Stem Cells and their Therapeutic Potential in the Nervous System*

18 *Mesenchymal Stem Cells: Biology, Culture Optimisation, and Potential Clinical Use*

19 *Mesenchymal Stem Cells: Culture, Characterisation, and Therapeutic Applications in Neurodegenerative Diseases*

20 *Dental Pulp Stem Cells and Perspectives of Future Application in Cell Therapy*

21 *Hematopoietic Stem Cells and their Dynamics*

22 *Spermatogonial Stem Cells*

23 *Epigenetics as a Cause for the Eloquent Silence of Stem Cells*

24 *Cancer Stem Cells: The Enemy within Cancers*

25 *Exploring the Potential of Adult Stem Cells: Why Umbilical Cord Blood?*

26 *Cell Therapy and Tissue Engineering Strategies in Regenerative Medicine*

27 *Biomaterials for Enhancing Corneas and Spinal Cord Regeneration*

16

RECENT ADVANCEMENTS IN TISSUE-RESIDENT ADULT STEM/PROGENITOR CELL BIOLOGY AND THEIR THERAPEUTIC APPLICATIONS IN REGENERATIVE MEDICINE AND CANCER THERAPIES

Murielle Mimeault*, Surinder K. Batra

Department of Biochemistry and Molecular Biology, Eppley Institute for Research in Cancer and Allied Diseases, University of Nebraska Medical Centre, Omaha, NE 68198-5870, USA

INTRODUCTION

Recent advances in the adult stem/progenitor cell research have inspired great interest, because these immature cells from your own body may act as the potential cell sources for tissue regeneration and gene therapy. The tissue-resident stem/progenitor cells, which generally possess a self-renewal ability and multilineage differentiation potential are able to regenerate all the mature cells in the tissue from their origin along the lifespan of an individual. The presence of a rare population of adult stem/progenitor cells in most tissues and organs offers the possibility of stimulating their *in vivo* differentiation, or using their ex vivo expanded progenies for cell-replacement therapies with multiple applications in humans. Among the diseases that could be treated by the adult stem cell-based therapies, there are haematopoietic and immune disorders, multiple degenerative disorders such as Parkinson's and Alzheimer's diseases, types 1 and 2 diabetes mellitus as well as skin, eye, liver, lung, and cardiovascular disorders. In addition, the molecular therapeutic targeting of

---

* *mmimeault@unmc.edu*

their malignant counterpart, cancer stem/progenitor cells also represents a promising strategy to treat the aggressive, metastatic, and recurrent cancers. Particularly, the genetically modified adult stem/progenitor cells could be used as a delivery system for expressing the therapeutic molecules in specific damaged areas including malignant neoplasms in humans.

Recent progress in stem/progenitor cell biology has allowed researchers to identify distinct adult stem/progenitor cells in most mammalian tissues and organs [1–5]. Among the tissues and organs harboring a very small subpopulation of specific multipotent and undifferentiated adult stem/progenitor cells, there are bone marrow (BM), vascular walls, heart, brain, skeletal muscles, adipose tissues as well as the epithelium of the skin, eye, lung, liver, digestive tract, pancreas, breast, ovary, uterus, prostate, and testis (Fig. 16.1) [1, 3–24]. Numerous studies have approved the unique features of each tissue-resident adult stem/progenitor cells and their specialized local microenvironment designated as niche [3–6, 8, 10, 25–29]. The longevity of these immature cells has been associated with their high telomerase activity, detoxification capacity and active DNA repair mechanisms that permit them to resist adverse effects caused by cytotoxic agents and environmental insults [5, 21, 30–50]. The tissue-resident adult stem/progenitor cells and their early progenies endowed with a self-renewal and multilineage differentiation potential generally provide critical physiological functions in the regenerative process for tissue homeostatic maintenance, and repair after intense injuries such as chronic inflammatory atrophies and fibrosis [3–5, 51, 52]. Multipotent adult stem/progenitor cells are able to give rise to different differentiated cell lineages in tissues from which they originate in physiological conditions, and thereby regenerate the tissues and organs throughout the lifespan of an individual (Fig. 16.2A). Importantly, certain adult stem cells including BM-derived stem cells may also be attracted at distant extramedullary peripheral sites after intense injuries and participate in the tissue repair through remodeling and regeneration process of damaged areas [3–5, 18, 20, 53–57].

Of clinical interest, the small pools of endogenous adult stem/progenitor cells represent the attractive sources of immature cells with important applications for cell replacement-based therapies in regenerative medicine in human [3–6, 8, 10, 11, 15, 16, 18–22, 25, 29, 58]. Particularly, the *in vivo* stimulation of normal tissue-resident adult stem/progenitor cells or the replacement of unfunctional or lost adult stem/progenitor cells by new *ex vivo* expanded immature cells or their differentiated progenies has been recognized as promising therapeutic strategies [3–5,10, 25, 26, 55, 58, 59]. These cell-replacement therapies could transform current palliative care toward curative treatments for many chronic and lethal degenerative disorders and life-threatening diseases associated with a poor quality of life. Among the human diseases that could be treated by stem cell-based therapies, there are haematopoietic and immune disorders, type 1 or 2 diabetes mellitus, cardiovascular, neurodegenerative, and musculoskeletal diseases and skin, eye, liver, lung, and gastrointestinal disorders (Fig. 16.1) [2–5, 8, 9, 11, 13, 18, 29, 54–57, 60–86].

Recent accumulating lines of experimental evidence have also revealed that the genetic and/or epigenetic alterations occurring in adult stem/progenitor cells may lead to their malignant transformation into cancer stem/progenitor cells also designated as cancer-initiating cells [2–5, 51, 52, 87–92]. The cancer stem/progenitor cell concept is notably supported by the identification and isolation of a very small subpopulation of malignant immature cells with stem cell-

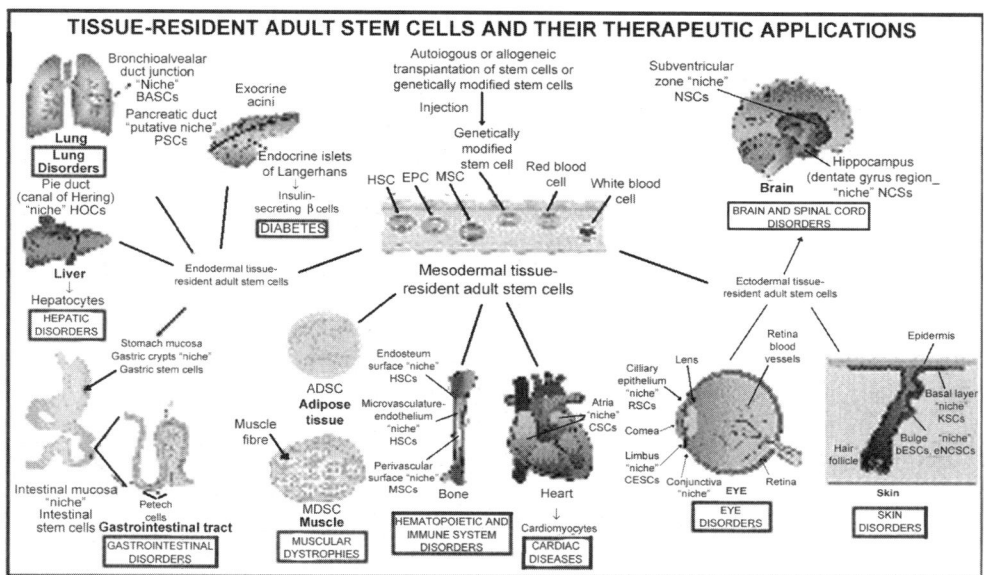

▶ **Fig. 16.1** *Scheme showing the anatomic localisations of tissue-resident adult stem/progenitor cells and their niches and stem cell-based therapies against diverse human disorders. The clinical treatments consisting of an injection of autologous or allogeneic stem cell transplant including bone marrow (BM)-derived stem cells (HSC, EPC and MSC), peripheral blood (PB) or genetically modified stem cells into peripheral circulation or diseased areas in the same patient or host patient is illustrated. The tissue-specific degenerating disorders and diseases, which might be treated by the autologous or allogeneic transplantation of adult stem/progenitor cells, are also indicated. (For colour figure see Plate 37).*

like markers such as telomerase, aldehyde dehydrogenase, CD133, CD44, Oct-3/4, c-KIT, and/or ATP-binding cassette (ABC) multidrug transporters, in most of cancers including leukemias, lymphomas, sarcomas, melanomas, and epithelial cancers [2–5, 51, 52, 87–101]. It has also been shown that the cancer stem/progenitor cells endowed with aberrant proliferation and/or differentiation potential are able to give rise to the total bulk mass of further differentiated cancer cells *in vitro* and *in vivo* [92, 93, 95–97, 102,103]. Particularly, the sustained activation of diverse developmental cascades such as hedgehog, epidermal growth factor receptor (EGFR), Wnt/β-catenin, and/or stromal cell-derived factor (SDF-1)/CXC chaemokine receptor-4 (CXCR4) signaling pathways in tumourigenic cancer stem/progenitor cells has been associated with their acquisition of a more malignant behaviour including a migratory phenotype during cancer progression [2–5, 51, 52, 88–91, 104]. Thus, the targeting of the deregulated signaling pathways in the cancer stem/progenitor cells, which may contribute to their sustained growth, survival, invasion, metastasis, and/or treatment resistance, is also of immense therapeutic interest for eradicating these cancer-initiating cells [2–5, 88–91, 105, 106]. In addition, gene therapy by using adult stem/progenitor cells as vehicles is also an attractive approach for a specific delivery and long-term expression of therapeutic gene products to damaged areas [106]. These stem cell-based gene therapies could be used to treat diverse inherited disorders and aggressive cancers. In regard with this, we describe here the recent progress in basic and clinical research in the adult stem/progenitor cell field. The emphasis is on the characterisation of the phenotypic and functional properties of tissue-resident adult stem/progenitor cells and their niches in terms of their implication in tissue regeneration in the physiological and pathological conditions. We also discuss the diseases that may be associated with a dysfunctional behaviour of adult stem/progenitor cells including the implication of cancer stem/progenitor cells in cancer initiation and progression. The provided information should help to develop novel therapeutic strategies that could be translated in effective and curative treatments for the patients diagnosed with diverse degenerative disorders and lethal diseases including the metastatic and recurrent cancers.

PHENOTYPIC AND FUNCTIONAL PROPERTIES OF ADULT STEM/PROGENITOR CELLS

The tissue-resident adult stem/progenitor cells possessing a high self-renewal and differentiation capacity, and which have a generally small size relative to the terminally differentiated cells, have been identified in the specific anatomic localisations in most tissues and organs in human (Fig. 16.1) [1, 3–24]. In general, the poorly differentiated adult stem/progenitor cells, are localised within a specialized microenvironment designated as niches consisting of the neighboring cells such as fibroblasts, endothelial cells, and/or stromal components that tightly regulate their functions through the direct interactions and release of specific soluble factors [3–6, 10, 25–29, 88, 90, 107]. Many efforts have allowed researchers to define the unique features of each tissue-resident adult stem/progenitor cell type as well as the molecular signaling pathways and niche interactions regulating their behaviour in the normal and pathophysiological conditions. A very small subpopulation of adult stem cells corresponding to about 0.1–3% from the total cell mass has notably been identified and isolated from a variety of mammalian tissues (Fig. 16.1) [1,3–24]. Among the method frequently used for adult stem cell enrichment and isolation, there are fluorescence-activated cell sorting (FACS) with the antibodies directed against the specific stem cell markers such as CD133, CD44, stem cell factor (SCF) receptor (KIT), ABC transporters, and stem cell antigen-1 (Sca-1 for the mice). Moreover, certain cell subpopulations with the stem cell-like properties have also been identified by the techniques based on their high aldehyde dehydrogenase activity and/or Hoechst dye efflux ability (Table 16.1) [42, 43, 47, 49, 50, 101, 108–111]. In addition, since primitive adult stem cells generally exist under a quiescent state and they replicate infrequently in homeostatic conditions, they also display the capacity to retain the DNA nucleotide label (5-bromo-2-deoxyuridine, BrdU) for a longer time than the mature and dividing cells [112]. Based on this property, the specific anatomic localisations of slow-cycling and label retaining cells (LRCs) possessing the stem cell-like properties have also been identified within

Plate 33

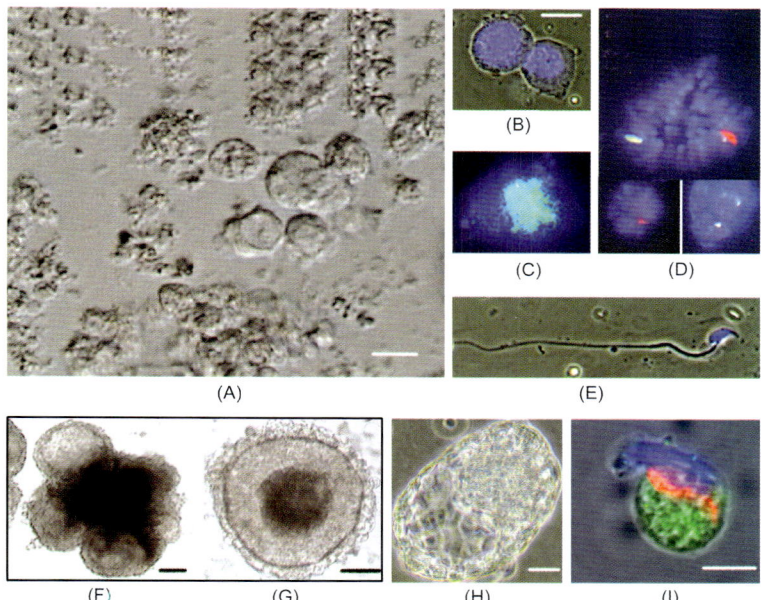

➤ *Fig. 13.5 Mouse embryonic stem cell differentiation into both presumptive gametes* in vitro *[52]. (A) General view of the beginning of GC differentiation. (B) Round spermatids stained with DAPI. Observe the nuclei in a central position (blue). (C) CG positive for FE-J1 antibody (green). Nucleus stained with DAPI. (D) Reduction of chromosome number observed by Fish analysis. Cells showing signals for whole chromosome paintings X (green) and Y (red). Nuclei stained with DAPI. (E) sperm-like cell observed* in vitro. *(F) EB aggregate resembling multiple early ovarian follicles (phase contrast image). (G) Oocyte-like cell observed* in vitro. *(H) Presumptive blastocyst (phase contrast image). (I) Round spermatid showing positive staining for anti-α–tubulin (red) and anti-acrosin (green) antibodies. Nucleus stained with DAPI. Scale bars: (A),(B),(I) 5 μm; (F)100 μm; (G) and (H) 40 μm.*

Plate 34

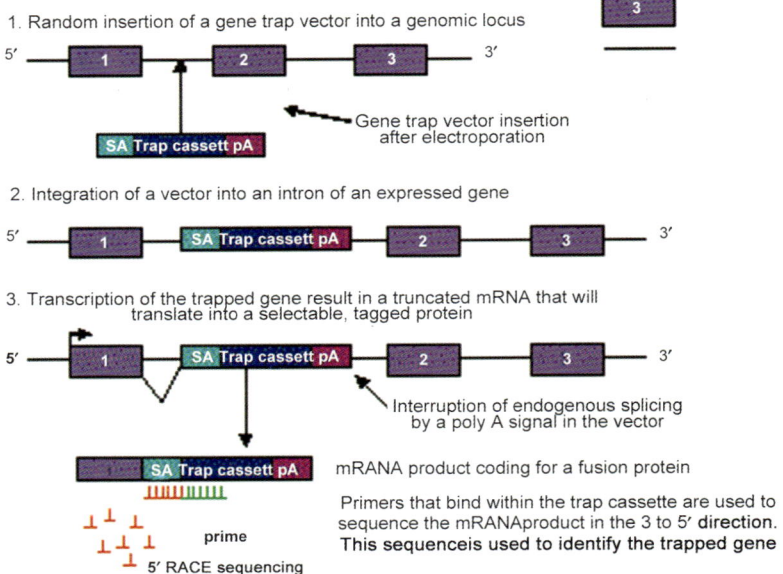

► Fig. 14.1 The conventional trapping process. The schematic gene trap vector is represented for SA (splice acceptor sequence); trap cassette (selection/reporter construct, e.g. βgeo); pA (polyadenylation signal).

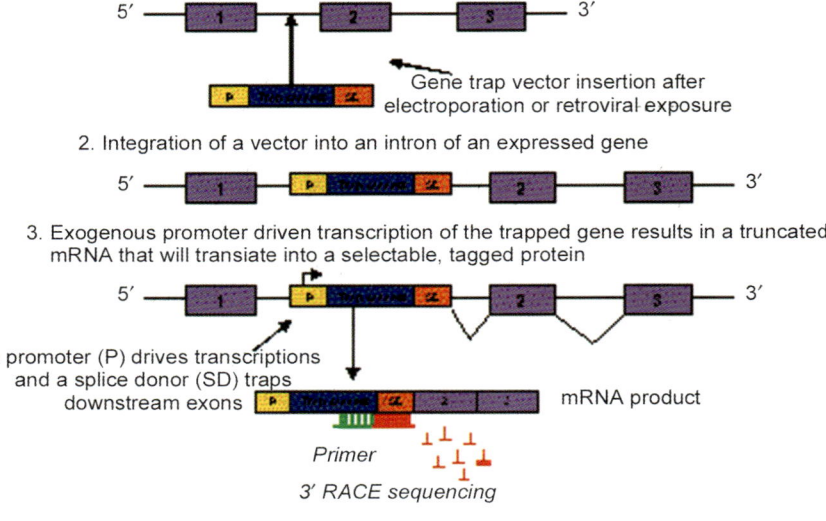

► Fig. 14.2 PolyA gene trap vectors.

Plate 35

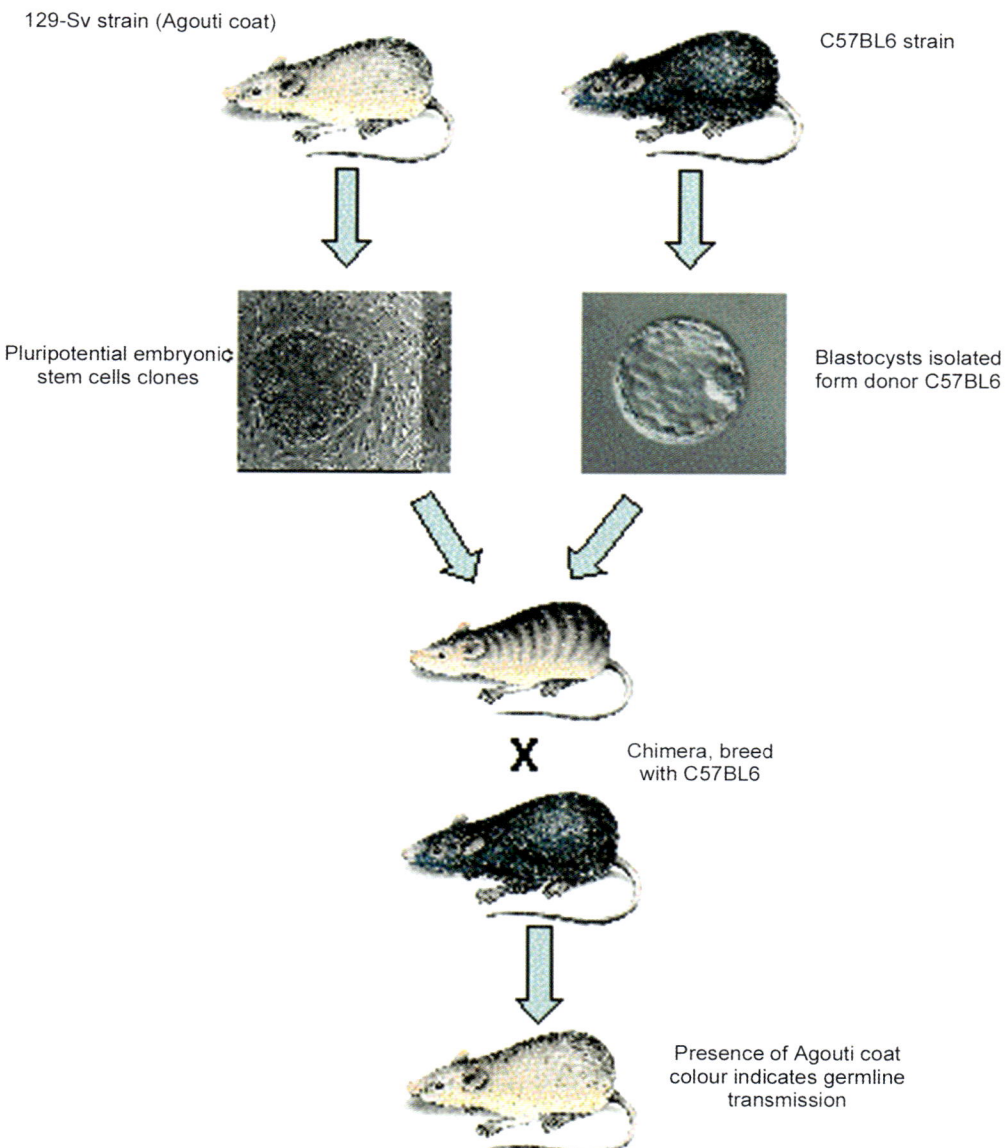

▶ *Fig. 14.3 General strategy for creating knockout mouse.*

Plate 36

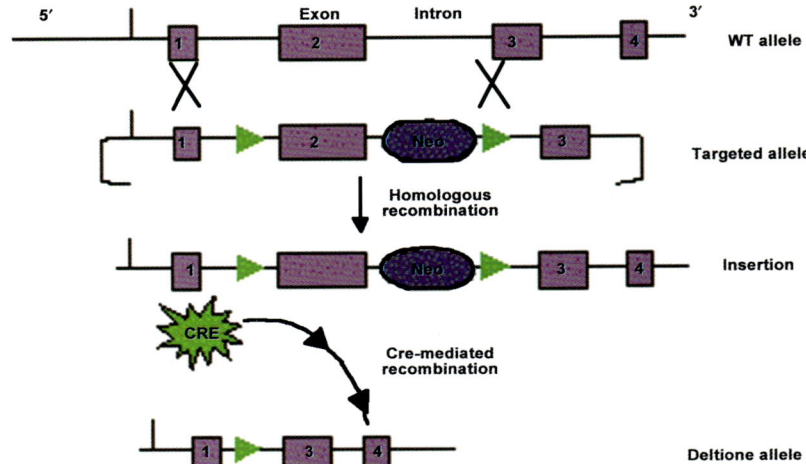

➤ *Fig. 14.4 Gene targeting and conditional expression events in ESCs mediated by Cre recombinase-loxP recombination. In this example, a gene locus in ESCs has been targeted by homologous recombination to insert a Neo cassette flanked by two loxP sites. Cre recombinase is an enzyme that works like scissors to cut out a gene that is in between two target sequences called loxP. Because this enzyme is expressed only in certain cell types, the targeted gene will be knocked out of only those cells and only when the researcher wants them to be. As demonstrated in the figure, when Cre recombinase is expressed in the cell, the loxP site will be cut and joined together, removing the piece of DNA between the two sites.*

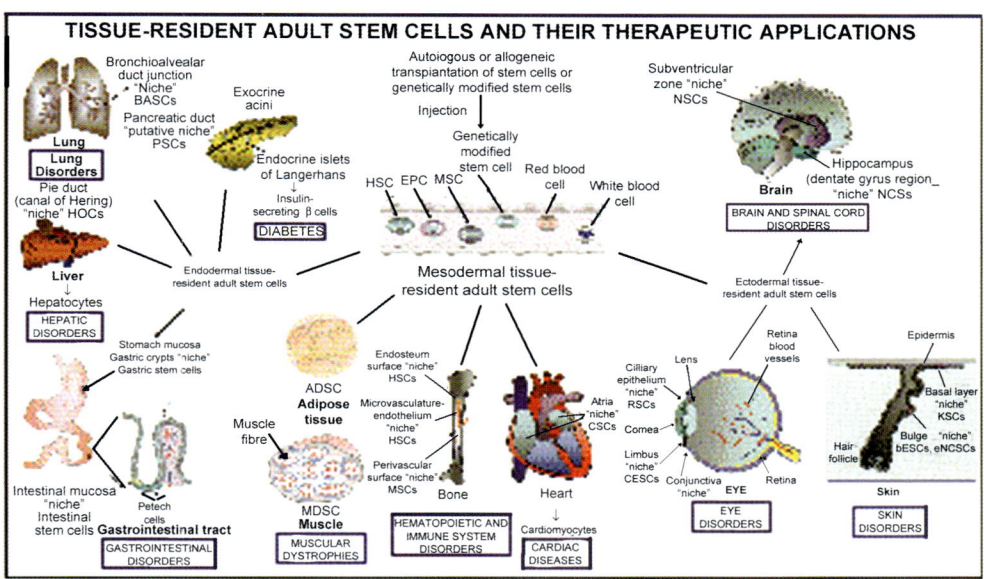

▶ *Fig. 16.1 Scheme showing the anatomic localisations of tissue-resident adult stem/progenitor cells and their niches and stem cell-based therapies against diverse human disorders. The clinical treatments consisting of an injection of autologous or allogeneic stem cell transplant including bone marrow (BM)-derived stem cells (HSC, EPC and MSC), peripheral blood (PB) or genetically modified stem cells into peripheral circulation or diseased areas in the same patient or host patient is illustrated. The tissue-specific degenerating disorders and diseases, which might be treated by the autologous or allogeneic transplantation of adult stem/progenitor cells, are also indicated.*

Plate 38

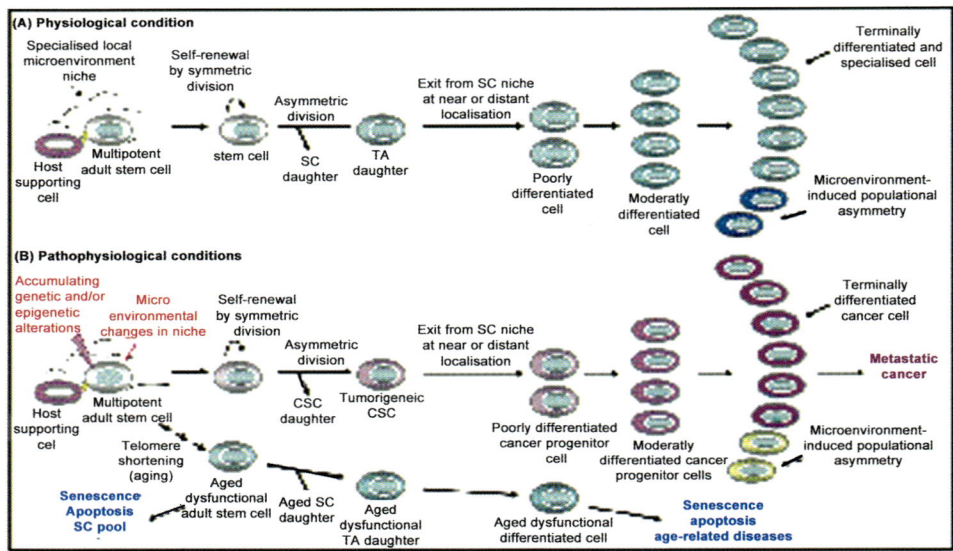

► *Fig. 16.2 Proposed model of the molecular events associated with tissue regeneration via adult stem/ progenitor cells in physiological conditions, aging and cancer initiation and progression through their malignant transformation. This scheme shows (A) the symmetric or asymmetric division of normal adult stem cells (SC) into transit-amplifying (TA)/intermediate cells that in turn may regenerate the bulk mass of further differentiated cells in the tissue from the origin in homeostatic conditions or after tissue injury. Moreover, this scheme shows (B) the aging process initiated through telomere shortening and malignant transformation of adult stem/progenitor cells into tumourigenic cancer stem/progenitor cells, which may be induced through the genetic and/or epigenetic alterations, leading to the sustained activation of distinct tumourigenic cascades during cancer progression to the invasive and metastatic disease stages.*

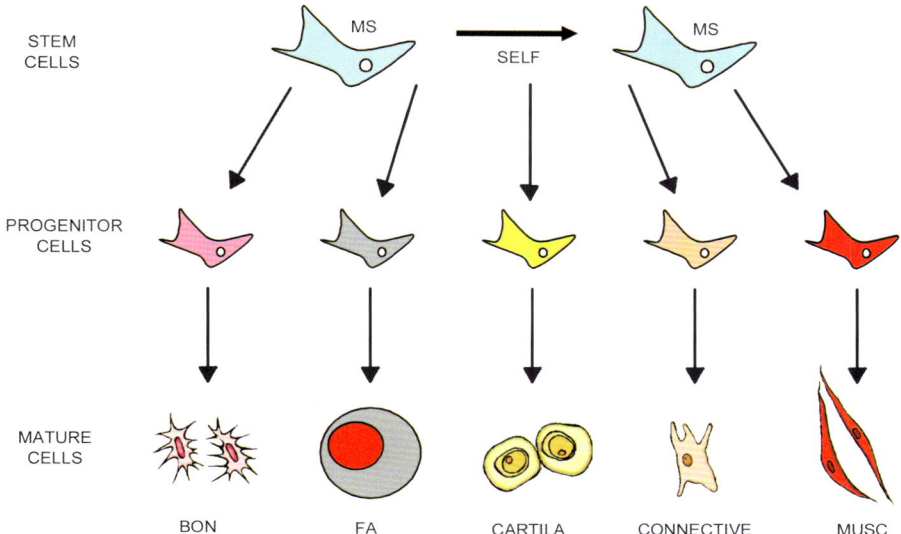

▶ *Fig. 17.1 Diagram describing the developmental potential of MSCs. Following cell division, MSCs may form one of two different progeny: they may produce a daughter stem cell, in a process known as self-renewal, or they may give rise to a progenitor cell that divides more rapidly but is more restricted in terms of its overall proliferative capacity and developmental potential. Progenitor cells derived from MSCs follow a defined path of differentiation. MSCs have the capability to differentiate into a wide range of mesodermal cell types, including bone, fat, cartilage, connective tissue and muscle.*

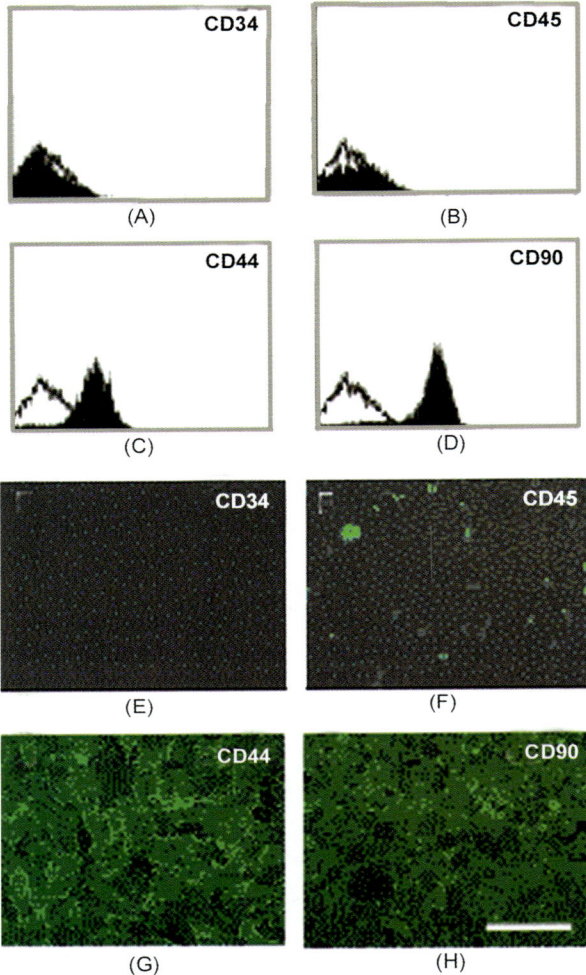

► *Fig. 17.3 Analysis of cell marker expression in undifferentiated MSCs. Marker expression as determined by (A–D) flow cytometry and (E–H) immunocytochemistry for (A and E) CD34, (B and F) CD45, (C and G) CD44 and (D and H) CD90. MSCs are largely negative for the haematopoietic markers CD34 and CD45, indicating that levels of haematopoietic contamination is low. MSCs are positive for CD44 and CD90 expression, which is indicative of MSCs. Scale bar: 30 μm.*

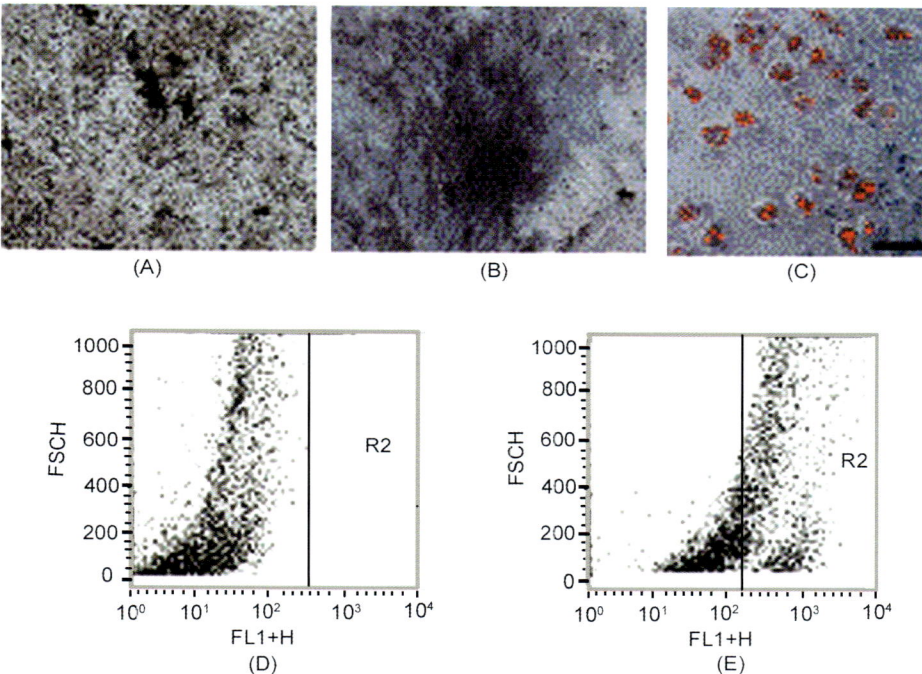

▶ *Fig. 17.4 Developmental potential of MSCs grown under conditions to induce either bone or fat differentiation. Osteogenic differentiation was assessed using (A) von Kossa and (B) Masson's trichrome staining to visualise extra cellular calcium deposition and collagen formation, respectively. (A) Bone nodule formation is the evidence as black deposits, and (B) the formation of collagen matrix as blue-green staining. Adipogenic differentiation was assessed using (C) Oil Red O and (D, E) Nile Red staining. Oil Red O staining showed the accumulation of intracellular lipid droplets within the cells (C, red droplets). MSCs stained with Nile Red were subject to flow cytometric analysis, and representative traces are shown for undifferentiated (D) and differentiated (E) MSCs. A population shift to the right is indicative of adipogenic differentiation. Approximately one third of the total MSC population are positive for Nile Red. Scale bar: 50 μm.*

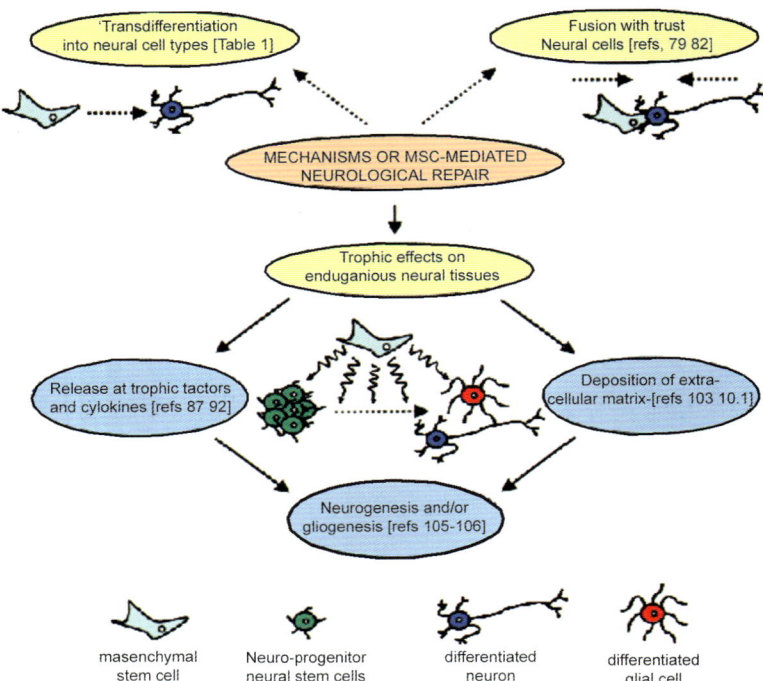

▶ *Fig. 17.5 Diagrammatic representation summarising the mechanisms by which MSCs may elicit functional neural recovery following their transplantation into models of neurological disorders. See text for further details.*

Plate 43

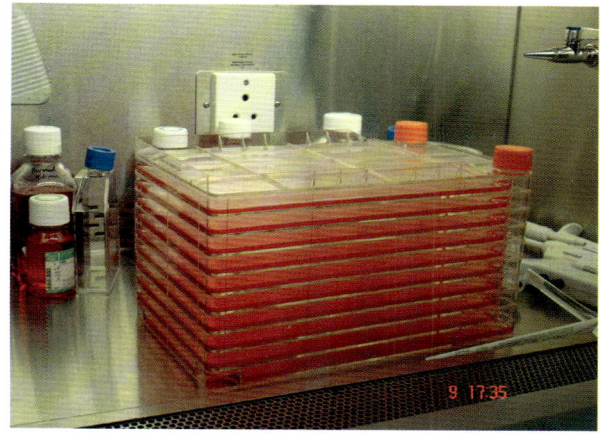

Courtesy: Stempeutics Research Private Limited, Bangalore, India

▶ *Fig. 18.1 Up-scaling of bone marrow derived mesenchymal stem cells in 10-cell stack.*

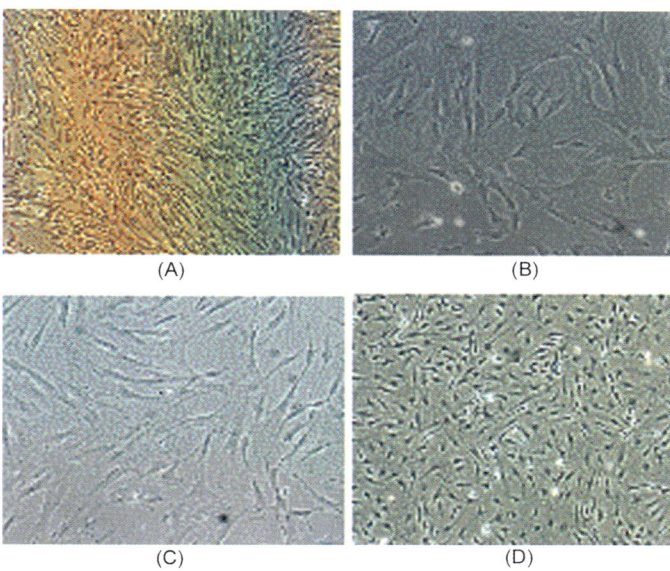

▶ *Fig. 18.2 Mesenchymal stem cells can be isolated from various tissues including bone marrow, adipose tissue, umbilical cord, synovial fluid and dental pulp. They have similar properties such as spindle-shaped morphology and plastic adherence. This figure shows morphology of MSCs derived from various tissues. (A) bone marrow; (B) umbilical cord; (C) adipose tissue and (D) synovial fluid.*

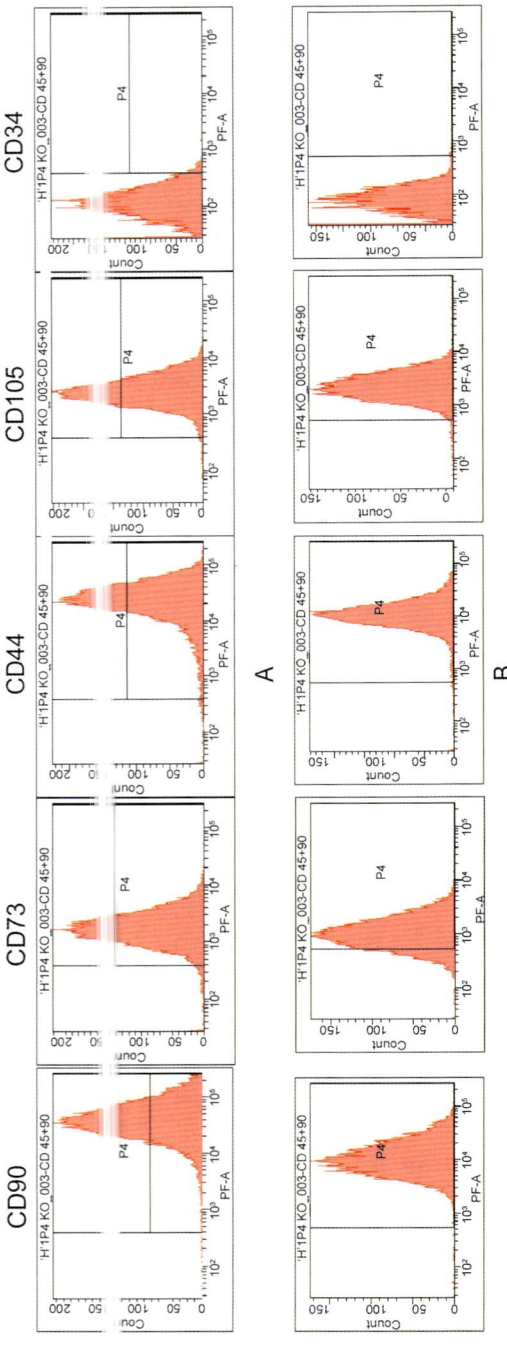

▶ *Fig. 18.3 Mesenchymal stem cell specific cell surface marker expression on bone marrow derived MSC at: (A) early passage (Passage 3) and (B) late passage (Passage 20).*

Plate 45

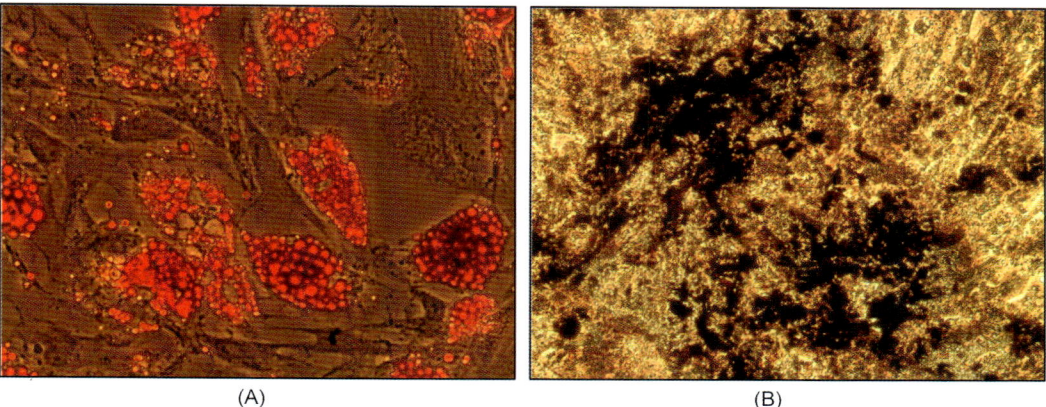

➤ *Fig. 18.4 Differentiation of bone marrow derived mesenchymal stem cells into (A) adipogenic and (B) osteogenic cells.*

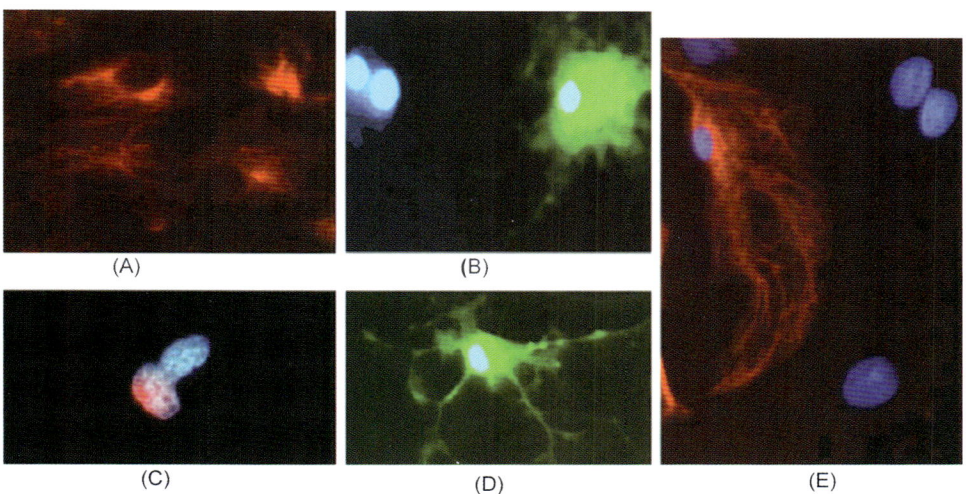

➤ *Fig. 19.1 Neuronal differentiation of MSCs in vitro. After RA treatment, NFH-/- MSCs can be induced to differentiate into cells expressing (A) Nestin marker (for neural precursor cells, red), (B) MAP-2 (B) (for neurons, green), (C) NeuN (for neurons, red), (D) O4 (for oligodendrocytes, green) or (E) GFAP (for astrocytes, red). In MAP-2, NeuN, GFAP and O4 staining, the nuclei are counterstained with DAPI (blue). A NeuN-positive cell with its nucleus stained in red (C) is observed to locate closely with a NeuN-negative cell. Final magnification, ✗400 in (A–E).*

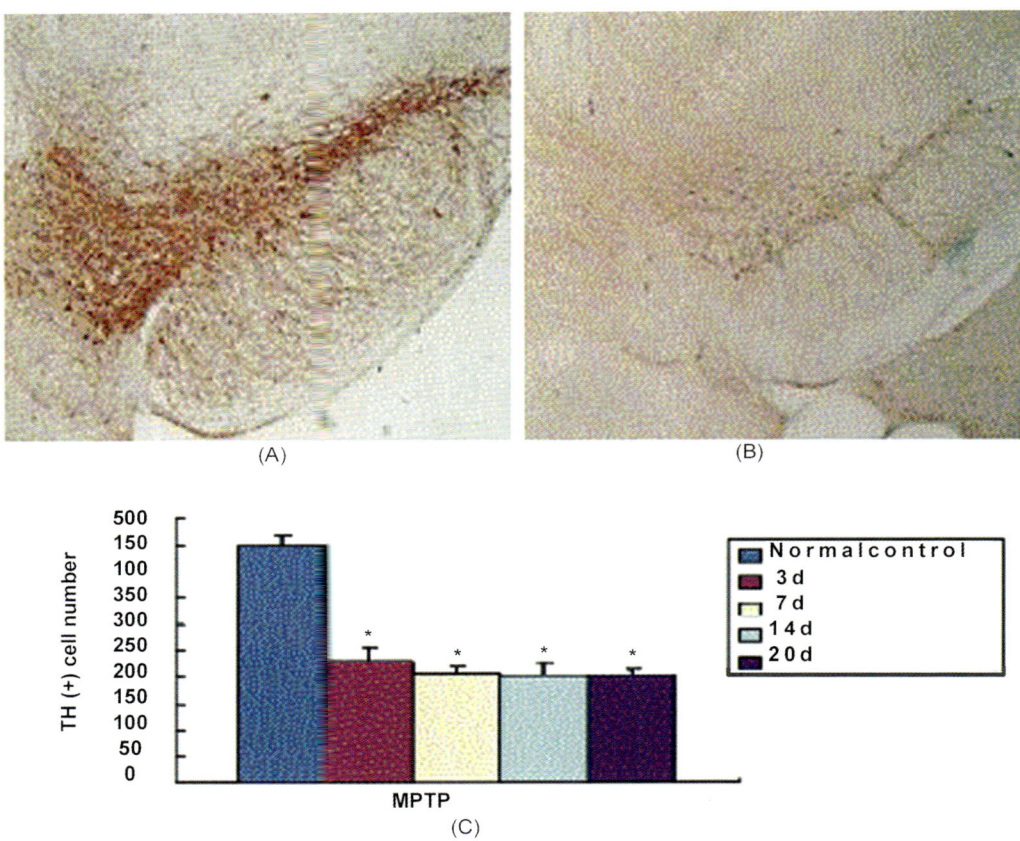

▶ *Fig. 19.2 Cell number counting. Tyrosine hydroxylase (TH) positive dopaminergic neurons in the substantia nigra area compacta (SNc) of (A) normal controls and (B) MPTP-treated mice. (C) Numbers of TH-positive cells at different time points post-MPTP injection (C, *p < 0.01). Final magnification, ✗200 in (A and B).*

Plate 47

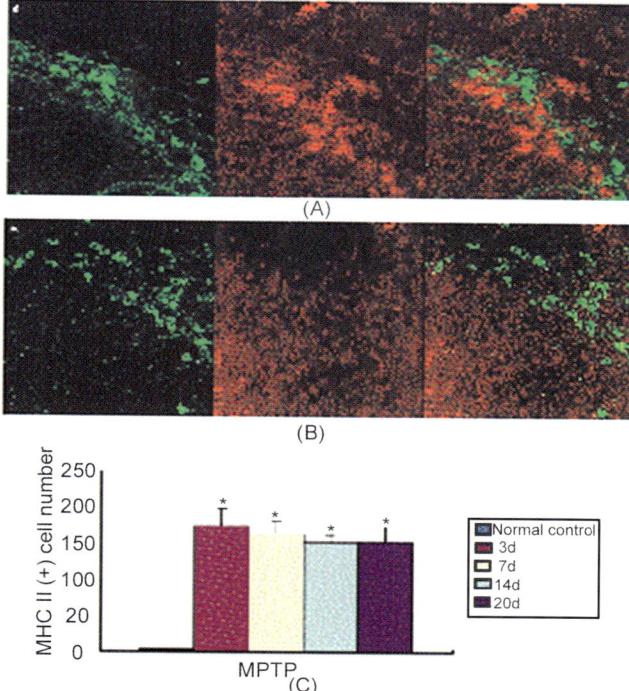

➤ *Fig. 19.3 Microglial activation in the MPTP-treated mice. MHC II staining in the substantia nigra area compacta (SNc) of (A) MPTP-treated mice became amoeboid compared with the (B) ramified normal controls and (C) the positive cell number increased *$p < 0.05$). Final magnification, ×400 in (A and B).*

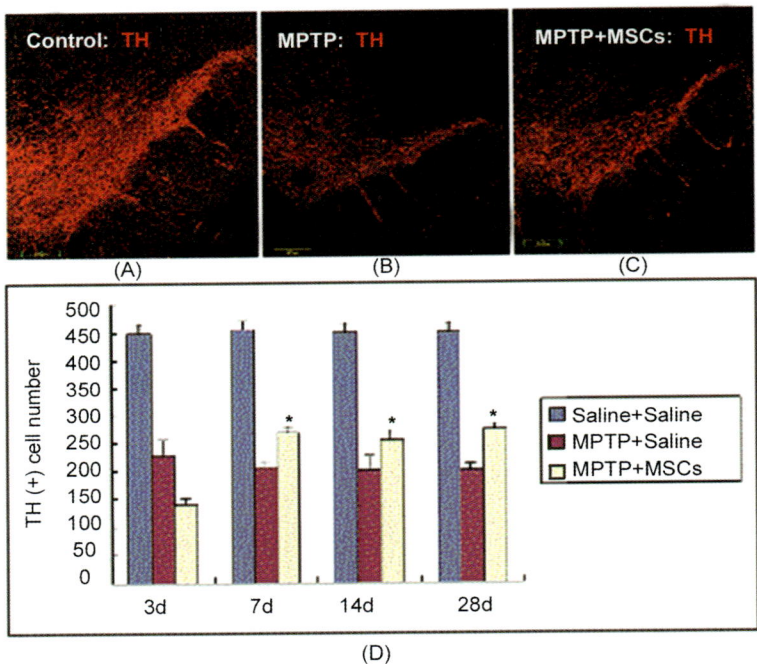

► Fig. 19.4 *Dopaminergic neuron rescue. TH staining of dopaminergic neurons in the (A) SNc of normal controls, (B) MPTP-treated mice and (C) MPTP-treated plus MSCs transplanted mice. (D) TH-positive cell number counting in different treatment groups at different time points post-MPTP injection shows that the difference is statistically significant (*$p < 0.01$). Final magnification, ×200 in (A)–(C).*

diverse mammalian tissues including skin, intestine, bone, prostate, pancreas, bladder, uterus, and eye [16, 40, 113–115]. It has been observed that the tissue-resident adult stem/progenitor cells generally possess several characteristics common with embryonic stem cells (ESCs), foetal and umbilical stem cells, but display a more limited self-renewal ability and restricted differentiation potential.

Intrinsic Properties of Adult Stem/Progenitor Cells

The persistence of tissue-resident adult stem/progenitor cell pools along the lifespan has been associated with several intrinsic properties of these immature cells. Among the molecular mechanisms that may contribute to extend the longevity of adult stem/progenitor cells, there are their expression levels of telomerase, slow division and high resistance to internal and external damages [4, 5, 21, 30–43, 51, 116]. Particularly, adult stem/progenitor cells of self-renewing organs and the germinal stem cells in ovaries and testis generally display a telomerase activity, while most somatic cells express low to undetectable levels of telomerase [30–38]. Telomerase activity of adult stem/progenitor cells may also be enhanced in response to different growth factor stimulus. Telomerase is an enzyme complex consisting of an RNA component designated as TER, which acts as a template for a reverse transcriptase component telomerase reverse transcriptase (TERT), that catalyses the addition of telomere repeats $(TTAGGG)_n$ at the chromosome end [117]. In the absence of a sufficient telomerase activity, the telomere length generally shortens at each cell division in normal somatic cells, and this molecular event may result in the genetic instability, cellular dysfunctions, and ultimately to a state of permanent growth arrest called replicative senescence and/or programmed cell death "apoptosis" (Fig. 16.2B). Therefore, the active telomerase detected in human adult stem/progenitor cells may help maintain their telomere length, chromosomal integrity, self-renewal, and viability during aging [31, 117–120]. In addition to its function in telomere maintenance, it has been reported that the telomerase catalytic subunit, TERT may directly promote the entrance of quiescent adult stem cells including haematopoietic, skin, and gastrointestinal stem/progenitor cells into the cell cycle through non-canonical pathways [121–124].

On the other hand, the adult stem/progenitor cells are usually characterized by the high expression levels of anti-apoptotic factors such as Bcl-2 and ABC multidrug transporters including ABCG2/brain cancer resistance protein (BCRP), multidrug resistance-1 (MDR-1) encoding P-glycoprotein gene product, and/or multidrug resistance proteins (MRPs) [4, 5, 43–51]. The ABC efflux pumps may protect the adult stem/progenitor cells against the adverse effects induced by xenobiotic substances such as toxins and chemical drugs by excluding intracellular cytotoxic compounds outside the cells. The high capacity of adult stem/progenitor cells to efflux Hoechst 33342 dye due to their high expression of ABC membrane transporters is notably at the basis of the Hoechst dye exclusion method which permits to isolate a very small cell fraction designated side population "SP" from the large cell mass [43, 46, 47, 49, 50, 111, 125, 126]. In fact, the small SP cell subpopulation generally shows the stem cell-like properties including its high expression levels of multidrug efflux pumps relative to the non-SP cell fraction. Moreover, other adult stem/progenitor cell attributes responsible for their resistance to metabolic cytotoxins and external insults include their high ability to actively repair DNA damages due to their elevated expression levels and the activity of enzymes involved in DNA repair mechanisms [21, 39, 40]. Hence, all these

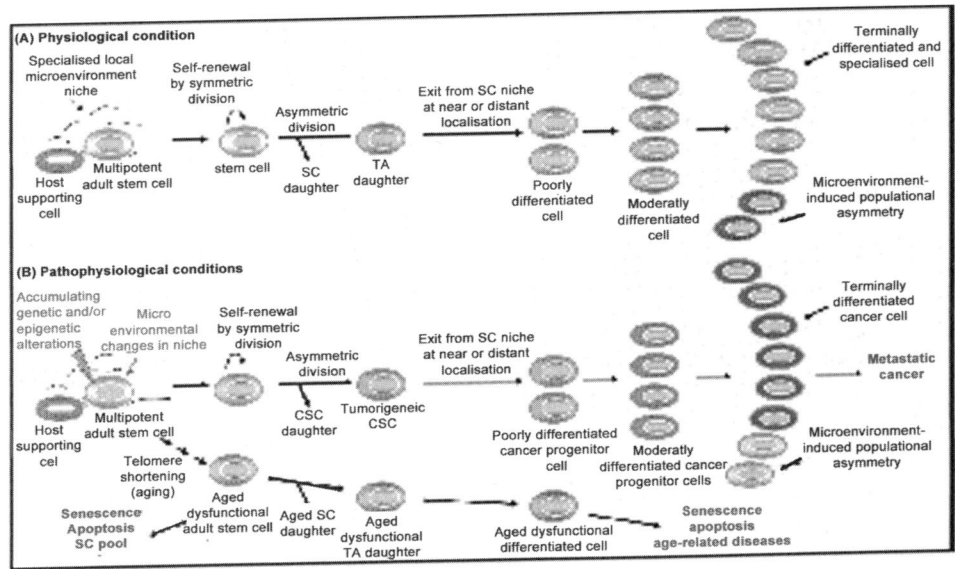

▶ **Fig. 16.2** *Proposed model of the molecular events associated with tissue regeneration via adult stem/progenitor cells in physiological conditions, aging and cancer initiation and progression through their malignant transformation. This scheme shows (A) the symmetric or asymmetric division of normal adult stem cells (SC) into transit-amplifying (TA)/intermediate cells that in turn may regenerate the bulk mass of further differentiated cells in the tissue from the origin in homeostatic conditions or after tissue injury. Moreover, this scheme shows (B) the aging process initiated through telomere shortening and malignant transformation of adult stem/progenitor cells into tumourigenic cancer stem/progenitor cells, which may be induced through the genetic and/or epigenetic alterations, leading to the sustained activation of distinct tumourigenic cascades during cancer progression to the invasive and metastatic disease stages. (For colour figure see Plate 38).*

aforementioned intrinsic properties common to a large variety of tissue-resident adult stem/progenitor cells may provide them survival advantages as compared to their differentiated progenies.

Functions of Adult Stem/Progenitor Cells in Tissue Regeneration

All of the multipotent or bi-potent adult stem/progenitor cell types display a long-term self-renewing capacity and can give rise to the mature and specialized cell types of distinct lineages in the tissues/organs from which they originate during tissue regeneration in homeostatic conditions and after intense injuries (Fig. 16.2A) [3–6, 8,10,11,15,16,18–20, 22, 25, 29]. In this context, certain adult stem/progenitor cells found in the BM and liver, skin, and gastrointestinal tract epitheliums usually show a continuous and rapid turnover to replenish the mature cells with a short live time loss along the lifespan [3, 4, 20, 21, 27]. At the opposite end, other tissue-resident adult stem/progenitor cell types usually remain under a quiescent state and rarely divide in normal conditions, and undergo only a sustained proliferation after intense tissue injuries. In regard with this, we review here the concept on the hierarchical organisation of tissue-resident adult stem/progenitor cells as well as the molecular mechanisms that may regulate their behaviours.

Hierarchical Organisation of Adult Stem/Progenitor Cells

Although normal tissue-resident adult stem cells are able to regenerate the tissues and organs throughout the lifespan of an individual, a growing body of scientific evidence supports the concept that these immature cells and their progenies, and more particularly those found in haematological system and epithelial tissues, may be organized as a cellular hierarchy. In fact, the hierarchical organisation may provide the primitive adult stem cells, early, intermediate and late transit-amplifying (TA) with high proliferative potential but more restricted differentiation ability and terminally differentiated cells (Fig. 16.2A) [4, 5, 51]. In regard with this, in the prevailing model of tissue homeostasis, the expansion of the adult stem/progenitor cell pool within a niche is generally accomplished though a symmetric cell division that gives rise to two identical stem cell daughters (Fig. 16.2A) [4, 5, 51]. The generation of differentiated cell lineages rather generally implicates an asymmetric division of a stem cell that gives rise to one stem cell daughter and one cell termed as early TA cells [3–5, 14, 20, 26, 27, 51, 88, 90]. The early TA cells, which possess a high proliferative potential and migratory ability, may exit the stem cell niche and give rise to intermediate and late TA cells. The late TA cells in turn can undergo a limited number of cell divisions before initiating terminal differentiation and are responsible for the immediate replenishment of mature cells that are lost to the environment along the lifespan. The changes in the local environmental of early, intermediate and late TA cells during the amplification process and their migration at distant sites from the niche may also influence the phenotype of their further terminally differentiated progenies, and thereby contribute to the populational asymmetry and cellular diversity characterizing each tissue and organ (Fig. 16.2A) [3–5, 51]. In addition, BM-derived stem/progenitor cells and/or vascular resident stem/progenitor cells may also be recruited to injured tissues or organs such as the skin, eye, gastrointestinal tract, kidney, and the lung where they can participate in the tissue repair by promoting vasculogenesis and/or trans-differentiating into epithelial cells [3–5, 18, 20, 51, 54–57,127].

Molecular Mechanisms Regulating Adult Stem/Progenitor Cell Functions

A complex network of several developmental signaling pathways appear to be involved in the maintenance of adult stem/progenitor cells in a quiescent state in homeostatic conditions as well as in the regulation of their rate of proliferation and differentiation after their activation [3–5, 51]. For instance, several transcription factors involved in the regulation of embryonic and foetal development and the determination of cell fate during tissue patterning are also expressed in certain adult stem cells in postnatal tissues. Moreover, the tissue regeneration mediated via adult stem/

progenitor cells is usually accompanied by environmental changes in the niche and orchestrated by several growth factors, cytokines and integrins [3–5, 51, 90]. More specifically, the reciprocal interactions of adult stem/progenitor cells with neighboring cells via the formation of adherens junctions and the secretion of diverse soluble factors might contribute to their restricted mobility and the adoption of a quiescent state within niches (Fig. 16.2A). The stringent regulation of the balance between a quiescent and dividing state of adult stem cells is also mediated via the activation of a complex network of diverse developmental signaling pathways. Among them, there are fibroblast growth factor (FGF)-FGFR, epidermal growth factor (EGF)-EGFR, sonic hedgehog (SHH)-PTCH, Wnt/â-catenin, notch and/or bone morphogenic protein (BMP)-initiated cascades [3–6, 27, 51, 128]. The activation of these signaling cascades may lead to an up-regulation of the self-renewal and/or differentiation of adult stem cells under specific physiological and pathological conditions. Furthermore, the soluble factors may be released by tissue-resident activated stem/progenitor cells and stromal cells including the myofibroblasts, endothelial cells and immune cells such as macrophages. In certain pathological conditions including intense wound and chronic inflammatory atrophies, certain adult stem/progenitor cells, and more particularly, BM derived-stem/progenitor cells, may also be attracted to distant injured tissues [3–5, 18, 51, 54–57, 88, 90]. BM-derived stem/progenitor cells may thereby contribute to the tissue repair by the release of diverse soluble factors and/or to trans-differentiate into specific cell types within the host tissue.

IMPLICATIONS OF ADULT STEM/PROGENITOR CELL DYSFUNCTIONS IN DISEASE DEVELOPMENT

Each human cell including adult stem/progenitor cell has needed to repair a large number of different DNA damages to survive in response to environmental insults including physiological stress [21, 39, 40, 129–133]. Along the long lifespan of most adult stem/progenitor cells, numerous lines of experimental evidence have revealed that an accumulation of DNA damages due to a genomic instability and/or abnormalities in DNA repair mechanisms may lead to an adult stem cell loss during aging and disease development [3, 4, 6, 8, 10, 20, 21, 26–28, 39–41, 107, 131, 132, 134–136]. In fact, the genetic and/or epigenetic alterations in these immature cells as well as their deregulated interactions with the niche components resulting in their acquisition of a dysfunctional behaviour have been associated with the occurrence of particular human disorders and cancers (Fig. 16.2B) [3, 4, 6, 8, 10, 20, 26–28, 41, 107, 131, 132, 134, 135]. A decline in the functions of tissue-resident adult stem/progenitor cells concomitant with the changes in their aged microenvironment appear notably to occur during physiological aging or in disorders associated with a premature or accelerated pathological aging such as dyskeratosis congenital, aplastic anemia, idiopathic pulmonary fibrosis, atherosclerosis, vascular dementia, and premature myocardial infraction [31, 33, 124, 132, 134, 135, 137–147]. For instance, a cumulative loss of telomeric DNA during aging of adult stem/progenitor cells and their differentiated progenies may trigger DNA repair and cell cycle checkpoint mechanisms initiated through tumour suppressor gene products such as p53 transcription factor, p16INK4a, and retinoblastoma protein (RB), and thereby inhibit stem cell functions [31, 33, 41,120, 124,132,135,138,139,148,149]. These molecular events may culminate in a genomic instability, chromosomal fusions, cell cycle arrest and replicative senescence, and/or apoptosis and development of age-associated diseases (Fig. 16.2B) [31, 41, 120, 124, 132, 134, 135, 138, 139, 148, 150]. The adoption of a senescent state by adult stem/progenitor cells and their differentiated progenies may also involve other molecular mechanisms such as acute and chronic stress signals [120,151]. The subsequent occurrence of certain molecular events, such as a down-regulation of cell cycle checkpoint pathways including p53 and RB and/or an aberrant up-regulation of telomerase activity, may also allow adult stem cells and their progenies to bypass or overcome the senescence process [119, 120, 124, 138, 139, 151, 152]. This may lead to a re-activation of the cell division program. In addition, the accumulation of additional genomic and epigenetic alterations in senescent adult stem/progenitor cells may ultimately culminate in their malignant transformation into cancer stem/progenitor cells showing an uncontrolled cell growth, and cancer formation [2–5, 41, 51, 52, 87–97, 99, 100, 102, 119, 120, 138, 139, 150, 152]. In support with this model of age-related carcinogenesis, it has been observed that the cancer cells in precancerous lesions such as ductal breast carcinoma *in situ* and prostatic, pancreatic and cervical intraepithelial neoplasias (PINs), are characterized by shortened telomeres [120, 153]. This suggests then that the telomere loss during aging may pre-dispose cells to the development of certain malignancies. Moreover, it has also been observed that an up-regulation or re-activation of telomerase activity occurs in numerous human cancers [119, 120, 139, 154]. Hence, it appears that an accumulation of telomere dysfunctions during aging may promote the chromosomal instability, which in turn may trigger the early stage of pre-cancerous lesions, while a subsequent up-regulation of telomerase activity plus other genetic/epigenic alterations could induce cancer initiation and progression in aged individuals.

In addition, the expression of ABC multidrug transporters and sustained activation of distinct developmental signaling pathways, such as hedgehog, EGFR, Wnt/β-catenin and/or SDF-1/CXCR4 cascades that are involved in the stringent regulation of normal adult stem/progenitor cell behaviour, may also confer a more malignant phenotype to cancer stem/progenitor cells [2–5, 28, 51,52,88–91,105,155,156]. Therefore, the molecular targeting of these deregulated oncogenic signaling elements contributing to cancer progression and ABC transporters is of great therapeutic interest to eradicate the total cancer cell mass including highly leukemic or tumourigenic cancer stem/progenitor cells. These targeting strategies should improve the efficacy of current clinical therapies, and thereby prevent treatment resistance and disease relapse. In regard with this, we describe here the specific functions provided by different tissue-resident adult stem cell types and their niches as well as their potential therapeutic applications in cell-replacement and molecular targeting-based therapies for treating diverse degenerative disorders, diseases and aggressive cancers in humans.

TISSUE-RESIDENT ADULT STEM/PROGENITOR CELL TYPES AND THEIR THERAPEUTIC APPLICATIONS

The tissue-resident adult stem/progenitor cells are unique compared with other embryonic, umbilical cord, foetal and placenta-derived stem cell sources, in being enriched in an anatomic location that may be easy to access (Fig. 16.1) [4, 5, 51]. Thus, in that way, the adult stem cells may be stimulated *in vivo* in their respective environment by the exogenous application of specific growth factors and cytokines in the damaged areas that restores the endogenous tissue regeneration program. Moreover, the use of the *ex vivo* expanded adult stem/progenitor cells and their differentiated progenies also offer great promise for cell-replacement therapies with multiple applications in humans [4, 5, 51]. As a matter of fact, we are reporting in a more detailed manner, the specific functional properties of the adult stem/progenitor cells, with an emphasis on the immature cells localized in the BM, heart, brain, skin, and pancreas as well as their potential therapeutic applications.

Bone-Marrow-Derived Stem Cells

Haematopoietic stem/progenitor cells and their clinical applications

The BM-derived haematopoietic stem cells (HSCs) provide a critical role by generating all of the new mature and differentiated white and red blood cell lineages along the lifespan of an individual [3, 4, 26, 55, 157]. The most immature and quiescent multipotent HSCs, which are characterized by the expression of specific biomarkers including telomerase, high levels of aldehyde dehydrogenase and $CD34^-$ or $CD34^+/CD38^{-/low}/Thy1^+/CD90^+/C\text{-}kit^{-/lo}/Lin^-/CD133^+$/vascular endothelial growth factor receptor 2 ($VEGFR2^+$)/ABCG2 are co-localized with the osteoblasts in a specialized niche within a BM region designated as the endosteum (Fig. 16.1) [3, 4, 26, 27, 30, 32, 48, 158]. Moreover, another subpopulation of HSCs, which is found in a BM microvasculature-sinusoidal endothelium niche, appears to represent the stem cells that may rapidly supply new mature blood cell lineages cells which have a short life into peripheral circulation (Fig. 16.1) [4, 26, 107, 158]. In regard with this, the results from a recent study have also revealed the presence of postnatal $CD34^+/Lin^-/CD10^+/CD24^-$ progenitors co-expressing recombination activating gene 1, terminal deoxynucleotide transferase, PAX5, interleukin 7 receptor-α, and CD3ϵ in the BM and peripheral blood [159]. These haematopoietic progenitors, which possessed a very low myeloid potential, were able to migrate from BM to the thymus and generate B-, T-, and natural killer (NK)-lymphocytes [159]. Additionally, the primitive KIT^+ haematopoetic progenitor cells with a long-term self-renewal capacity have also been identified in the adult spleen in humans [111].

In clinical practice, autologous or allogeneic HSC transplantation is currently used to treat the patients with diverse haematopoietic disorders to reconstitute the haematopoietic cell lineages and immune system defense [3, 4, 25, 160]. BM-derived HSCs may be collected from BM aspirate or by aphaeresis after their mobilisation in peripheral blood (PB) by using diverse mobilizing agents such as granulocyte-stimulating factor (G-CSF), granulocyte colony-stimulating factor (GM-CSF) and/or synthetic chemical compounds like bicyclam derivative, AMD 3100 (Plerixafor) [3, 4, 161]. Hence, BM or mobilized PB HSC-containing samples or isolated HSC preparations may be retransplanted into the same patients (autografts) or different patients (allografts) by injection into the bloodstream (Fig. 16.1). Transplanted HSCs spontaneously migrate and engraft at the BM compartment where they establish their novel homing, and thereby contribute to replenish all the mature blood cell types and restore immune system functions [162]. Moreover, treatment of patients with a myeloablative conditioning regimen consisting of high-dose chaemotherapy or radiotherapy is generally made prior to allogeneic HSC transplantation to reduce the immune response and risk of graft rejection and improve the anti-tumoural efficacy. Particularly, HSC transplants, alone or in combination therapies, may be used to treat HSC aging related-intrinsic functional defects, inherited immunodeficient and autoimmune diseases such as multiple sclerosis, refractory and severe aplastic anemias, congenital thrombocytopenia, osteoporosis, cardiovascular disorders, chronic inflammatory Bowel disorders (IBDs), including Crohn's disease and ulcerative colitis and diabetes mellitus [3, 4, 125, 134]. Particularly, HSC transplant may improve the immune response of patients, and thereby help both to repair damaged tissues at distant sites in diverse pathological conditions and prevent infectious diseases after the transplantation of tissue or organ grafts [3, 4, 25]. Moreover, high-dose or intermittent systemic chaemotherapy or ionizing radiation therapy plus HSC transplantation represents a potential therapeutic option to treat and even cure the high risk patients with advanced and/or relapsed/refractory cancers. Among them, there are leukemias, multiple myelanomas, Hodgkin's and non-Hodgkin's lymphomas, melanomas and aggressive and metastatic solid tumours such as sarcomas, retinoblastoma, kidney, brain, lung, pancreatic, prostate, breast and ovarian cancers [2–5, 20, 81–83, 88, 106, 160, 163, 164]. In fact, HSC transplant may restore the haematopoietic and immune system after myeloablative effects induced by high-dose irradiation or chaemotherapy following the treatment of cancer patients [160].

Although the important advances in HSC transplantation procedures, the graft-versus-host diseases (GVHDs), toxicity of cytoreductive conditioning regimens, the presence of residual malignant cells in allograft as well as the lack of appropriate donors for some patients yet represent the major limiting factors for their clinical applications in safe conditions [165]. Particularly, several secondary effects may be prevalent after HSC transplantation in certain patients and contribute to a poor quality of life [166]. Among them, acute or GVHD is a common late complication of allogeneic transplantation characterized by specific clinical and pathologic signs related with the fact that immunocompetent donor cells may attack fast proliferating recipient tissues such as skin, liver and gastrointestinal tract [167]. GVHD may be associated with the occurrence of severe vascular and fibrotic lesions. In order to reduce the toxicity of myeloablative regimens, non-myeloablative or reduced-intensity myeloablative conditioning regimens may be used in certain cases, and more particularly, for old patients or patients with comorbidities that are unable to tolerate this immunosuppressor treatment [4, 5, 106, 168]. Moreover, an autograft after purging of malignant cells may constitute another alternative treatment in patients with high-risk leukemic relapse when no stem cell donor is available [165, 169]. For instance, the autologous transplantation of $CD133^+$ selected HSCs may be used for pediatric patients with relapsed $CD34^+/CD133^-$ leukemia [83]. The results from a recent investigation revealed that the homing and engraftment of $BCR-ABL^+$ leukemic stem cells (LSCs) in the BM of patients with chronic myelogenous leukemias (CMLs) is highly dependent on CD44 adhesion molecule expression in respect to normal HSCs[170]. Therefore, the targeting of CD44 LSC using an anti-CD44 antibody may constitute another approach to improve the efficacy of HSC transplantation in CML patients [170]. In addition, bank stored-umbilical cord (UC) cells including umbilical cord blood (UCB) and placenta cells and foetal tissue-derived HSC transplants, which generally induce a less intense detrimental alloreactive response, may also constitute other HSC sources for autograft or allograft in certain clinical or experimental settings [3, 4, 171–173].

Mesenchymal stem cells and endothelial progenitor cells and their therapeutic applications

The BM stroma, PB as well as the walls of large and small blood vessels in most tissues/organs including brain, spleen, liver, kidney, lung, muscle, adipose tissues, thymus, uterus and pancreas also contain the multipotent mesenchymal stem cells (MSCs) and endothelial progenitor cells (EPCs) [3–5, 34, 47, 55–57, 174–177]. Much of the work conducted on adult stem/progenitor cells has focused on MSCs found within the BM stroma. More particularly, the MSCs expressing CD49a and CD133 markers are localized in a perivascular niche in BM, and may give rise to the osteoblasts that are co-localized with HSCs, and which may support the haematopoiesis by producing the growth factors and cytokines that promote the expansion and/or differentiation of HSCs (Fig. 16.1) [3–5,55,178,179]. Recently, it has also been reported that lung-resident mice subpopulations of $CD45^-$ SP containing MSCs and expressing high telomerase level and mesenchymal markers (CD44, CD90, CD105, CD106, CD73, and Sca-I) can differentiate into chondrocytes, adipocytes, and osteocytes [180]. The BM-derived or tissue-resident MSCs can generate diverse mesodermal cell lineages involved in osteogenesis, adipogenesis, cartilage, and muscle formation including the osteoblasts, osteocytes, adipocytes, chondrocytes, myoblasts, and myocytes under appropriate culturing conditions *ex vivo* and *in vivo*. Moreover, MSCs may also be induced to differentiate into fibroblasts, neuronal cells, pulmonary cells, pancreatic islet β-cells, corneal epithelial cells and cardiomyocytes *ex vivo* and/or *in vivo* using specific growth factors and cytokines [3–5, 14, 34, 178,181–187]. In the case of EPCs, which are derived like HSCs from the embryonic hemangioblasts, they may be distinguished by the expression of different biomarkers, including $CD34^+$ or $CD34^-$, CD133, $VEGFR2^+$ also designated as Flk-1 (foetal liver kinase-1), KIT and CXCR4[5,88]. EPCs may contribute in a significant manner to give rise to mature endothelial cells that form new vascular walls of vessels after intense injury and vascular diseases as well as the new vessel formation in tumours [3–5, 35, 55–57, 106, 176, 188–190]. The critical role of circulating EPCs in endothelial cell maintenance after tissue injury is notably supported by the observation that their number and function is inversely associated with the progression of atherosclerosis and an enhanced risk of cardiovascular diseases. For instance, it has been proposed that the number of circulating EPCs may be increased by down-regulating EPC senescence through plasma high-density lipoprotein (HDL)-induced nitride oxide (NO) production and telomerase activity via PI3K/Akt signaling pathway [35]. These molecular events may promote the angiogenic process, and thereby decrease the incidence of atherosclerosis-related ischemic diseases [35].

All of the aforementioned functional properties of MSCs and EPCs made them promising sources of immature cells for treating numerous degenerative and vascular disorders in humans [112, 191]. The autologous or allogeneic transplantation of BM or PB samples can lead to the homing and engraftment of functional HSCs, MSCs, and EPCs and/or their differentiated progenies in BM and distant damaged tissues. Thus, this supports the feasibility of this strategy for improving the tissue remodeling and healing processes after severe injury as well as in the treatment of diverse human disorders including osteogenesis imperfecta, atherosclerotic lesions, ischemic cardiovascular and muscular diseases (Table 16.1) [3–5, 112, 125, 176, 182, 183, 186, 191]. It has been reported that MSCs, EPCs and their progenies can contribute to the vasculogenesis and regenerative process of several tissues including bone, cartilage, tendon, muscle, adipose, brain, heart, lung, skin, pancreas, kidney and eye [5,112,191–193]. Importantly, adult BM-derived and tissue-resident MSCs are little immunogenic and display immunomodulatory and anti-inflammatory effects in host *in vivo* [160,193,194]. Therefore, these therapeutic properties of MSCs also support their possible clinical applications to prevent tissue/organ allograft rejection and severe acute and chronic GVHDs as well as to treat the autoimmune disorders such as inflammatory bowel diseases and inflammation of the heart muscle walls associated with autoimmune myocarditis in which immunomodulation and tissue repair are required [160,193]. Indeed, MSCs can prolong skin allograft survival and reverse severe acute GVHDs *in vivo* supporting their use in treating skin diseases as well as in the maxillofacial surgery [112,195]. In counterpart, the migration and proliferation of vascular smooth-muscle cells (SMCs) derived BM cells including HSCs and MSCs in the vascular injured area leading to an excessive cell accumulation, may contribute to the development of vascular pathologies such as intimal hyperplasia and atherosclerotic lesions [5,196]. Therefore, future investigations are necessary to optimize the BM-derived cell transplantation strategies and establish the specific mechanism(s) of action and the physiological effects of HSCs, MSCs, and EPCs in the long term. This should allow for the improvement their therapeutic and curative benefits and prevent their detrimental clinical effects in treated patients.

Cardiac stem/progenitor cells and their therapeutic applications

The heart is a muscular organ that provides a vital function by pumping blood through the circulatory system through its repeated and rhythmic contractions. The heart muscle, which is constituted by the cardiac involuntary striated muscle cells also called cardiomyocytes or cardiac myocytes, has the capacity to repair itself through the lifespan of an individual. The myocardial regeneration and cardiac function recovery in physiological conditions and after severe injuries may occur via the activation of a small subpopulation of interstitial cardiac stem/progenitor cells (CSCs) found within the specialized niches localized at the apex and atria of heart (Fig. 16.1) [3–5, 7–9, 145, 146, 197–201]. Mammalian heart-resident adult CSCs or their early progenies expressing different stem cell-like markers including telomerase, nestin, KIT (also designated CD117), MDR-1, ABCG2 multidrug transporters, Islet1 transcription factor and/or Sca-1 (in mice), are endowed with a self-renewal capacity and multilineage differentiation potential [3–5, 7–9, 36, 37, 146, 197–202]. These immature cells are able to give rise to three major cell types constituting the myocardium including cardiomyocytes, smooth muscles and vascular endothelial cells in homeostatic conditions and after myocardial injuries [3, 4, 7–9, 146, 197–200]. Therefore, the *in vivo* stimulation of CSCs and early progenies by diverse agents such as IGF-1 and HGF or the intravascular, intramyocardial or catheter-based delivery of *ex vivo* expanded CSCs or their differentiated progenies may represent the potential therapeutic strategies for the cardiac cell replacement-based therapies to treat and even cure diverse heart diseases (Table 16.1) [3, 4, 7–9, 145, 200, 203–206]. Particularly, the CSC-based therapies could be used to replace the aged, dysfunctional or loosed CSCs by new functional cardiomyocytes and regenerate coronary vessels after cardiac injury. The transplantation of genetically modified adult stem cells also offers great promise by allowing the delivery of a specific therapeutic gene product such as an angiogenic agent or cardioprotective factor in the ischemic or non-ischemic heart disease areas [4, 5, 207]. These therapeutic strategies could be used to improve the survival of heart-resident CSCs and/or their progenies including cardiomyocytes, and thereby the myocardial regeneration process. These treatment types combined with the conventional clinical therapies by using pharmacological agents could also lead to a long-term outcome of patients diagnosed with heart failures resulting from ischemic heart disease, hypertension and myocardial infarction [3–5, 7–9, 60, 200, 203–205, 208].

In addition, the use of other stem/progenitor cell types including ESCs, UCB-derived stem cells ($CD133^+$ cells, HSCs or MSCs), AECs, BMSCs ($CD133^+$ cells, HSCs, MSCs or EPCs), ADSCs, MDSCs, PSCs, and adult testicular stem cells or their progenies, which can differentiate into functional and contractile cardiomyocytes and/or vascular endothelial cells *in vitro* and/or *in vivo* also represent potent therapeutic stem cell sources [3–5, 8, 9, 11, 34, 60, 125, 182, 183, 197, 204, 209–212]. In support with this, the results from numerous investigations carried out on animal injury models *in vivo* have revealed the potential benefit of using these stem cell types or their further differentiated progenies with the cardiomyogenic properties to repair the damaged myocardium and improving the coronary revascularisation and cardiac function [4, 7, 11, 60, 125, 182, 183, 200, 203, 204, 209–211]. For instance, it has been observed that the transplantation of *ex vivo* differentiated cardiomyocytes derived from hESCs resulted in a stable cardiomyocyte engraftment and improvement of myocardial performance in rats with extensive myocardial infraction [213]. The data from small clinical trials consisting of the transplantation of human BMSC, mobilized PB cells or purified $CD133^+$ BMSCs into patients with advanced ischemic heart diseases have also indicated that this treatment generally improves the vascularisation process and/or myocardial function [60, 205, 214]. More specifically, the BM-derived cells (HSCs, MSCs, and EPCs), ADSCs and MDSCs may contribute to the repair of the injured cardiovascular system via multiple molecular mechanisms. Among them, there are the transdifferentiation of these adult stem cells into new cardiomyocytes, smooth muscle cells and/or endothelial cells as well as their release of diverse paracrine factors such as HGF, IGF-I, and VEGF that may in turn stimulate the angiogenesis and endogenous CSCs and inhibit their apoptotic/necrotic cell death [125, 188, 215, 216]. It has notably been reported that BM-side population (SP) cells endowed with a haematopoietic stem cell activity were able to transdifferentiate into cardiomyocyte-like cells when co-cultured on neonatal cardiomyocytes fixed to inhibit cell fusion [125].

These recent advancements generated by basic and clinical research in cardiology have revealed the potential of using diverse stem/progenitor cell sources for cardiac regeneration therapies in humans, and more particularly for treating late-stage heart failure patients that have little hope of survival without an opportunity of heart transplantation due to a massive loss of functional cardiomyocyte mass. Nevertheless, future investigations appear necessary to more precisely establish the specific biomarkers and anatomic localisation, niche of endogenous CSCs and their early committed progenies within the heart as well as the extrinsic factors that regulate their self-renewal, differentiation, and/or migration capacities in homeostatic and pathological conditions. Additional studies are also required to determine the functional properties of transplanted stem/

progenitor cells and their progenies for the long-term as well as the molecular mechanisms responsible for their beneficial versus detrimental effects in the animal models *in vivo* and in the clinical setting. An optimisation of cell delivery methods and combination therapies consisting of cardiac cell-replacement therapies plus conventional pharmacotherapies currently used in the clinics also merit further investigations. These future works should permit a better definition of the potential clinical risks associated with cell replacement therapy to treat the ischemic and non-ischemic heart before their possible applications in the safe conditions in humans.

Neural stem/progenitor cells and their therapeutic applications

Adult neurogenesis in central and peripheral nervous tissues may occur through the activation of self-renewal and multipotent adult neural stem/progenitor cells (NSCs) [5, 85]. More specifically, NSCs found in the adult human brain are localized within two specific neurogenic regions, niches designated as the subventricular zone of the lateral ventricle in the forebrain and dentate gyrus in hippocampus (Fig. 16.1) [3–5, 10, 29, 217]. Multipotent $CD133^+$/nestin NSCs with an astroglia-like cell phenotype are endowed with a self-renewal potential and capable of giving rise to the progenitors that can proliferate and migrate at distant damaged areas of the brain where they can generate further differentiated and functional progenies [3, 4, 10, 29, 217, 218]. NSCs localize in the subraventricular zone, and can give rise to three principal neural cell lineages, including mature neurons and glial cells, astrocytes, and oligodentrocytes, while those found in the subgranular cell layer of hippocampus may generate the granule cell projection neurons [3, 4, 10, 29, 217]. Hence, NSCs and their progenitors can participate to regenerate and repair the injured tissues after neurological damage and trauma in humans. NSCs may notably give rise to diverse neural and glial cell lineages in appropriate conditions *ex vivo* and *in vivo* [3–5, 61, 217–220]. Numerous developmental signaling cascades [EGF-EGFR, sonic hedgehog SHH-PTCH-GLI, Wnt/β-catenin, Notch, basic fibroblast growth factor (bFGF), nerve growth factor (NGF), neuregulins, BMPs, platelet-derived growth factors (PDGFs), ciliary neutrophic factor, vascular endothelial growth factor (VEGF), thyroid hormone T3, dopamine, TGF-β, integrins, Ephrins/Ephs, leukemia inhibitory factor (LIF), and/or RNA-binding proteins, Musashi (Msi-1 and Msi-2)] may contribute to the stringent regulation of the proliferation and cell fate decision of NSCs and astroglial progenitor cells in the developing and adult central nervous system (CNS) [4, 5, 10]. The genetic alterations in NSCs leading to a sustained activation of these mitotic cascades including EGF-EGFR and sonic hedgehog pathways may also result in their malignant transformation into brain tumour stem cells (BTSCs) and cancer formation [2, 4, 5, 88, 90, 91]. The changes in the local microenvironment, niche components of NSCs, including neighboring endothelial cells co-localized with NSCs in the subraventricular zone, may also influence their behaviour in homeostatic and neuropathologic conditions [3, 4, 10, 29, 217]. Additionally, the adult stem/progenitor cells derived from neural crest-derived stem cells have also been identified in the peripheral nervous system within a germinal centre designated carotid body (CB) [221]. These multipotent CB-resident adult stem cells, which represent the glia-like sustentacular cells expressing the glial markers can give rise to the dopaminergic glomus cells that produce the glial cell line-derived neurotrophic factor [221].

The *in vivo* stimulation of NSCs or replacement by new functional cells offer great promise and an ultimate hope for treating diverse incurable CNS degenerative disorders and incurable diseases including Parkinson's, Alzheimer's, Lou Gehrig's and Huntington's diseases, temporal lobe epilepsy, stroke, multiple sclerosis, and amyotrophic lateral sclerosis (Table 16.1) [3–5, 29, 61–64, 84–86, 217, 222–227]. Accumulating lines of experimental evidence revealed that *ex vivo* expanded NSCs or their progenies may be transplanted in damaged areas of the brain where they can proliferate, survive, migrate, and differentiate into functional neural and glial cells *in vivo* [3–5, 85, 86, 224–227]. It has been reported that the transplantation of adult neural precursor cells (aNPCs) from the brain of adult transgenic mice into the spinal cords of adult Shiverer (shi/shi) mice with congenitally dysmyelinated adult CNS axons, give rise to the cells expressing the oligodendrocyte markers and resulted in the formation of nodes of Ranvier and improved axonal conduction [224]. Interestingly, it has also been shown that the neural progenitor cells from the olfactory organ of patients with Parkinson's disease were able to generate dopaminergic cells *in vitro* and reduce the behavioural asymmetry resulting from the dopaminergic neuron loss in the rat model of Parkinson's disease [84]. In regard with this, the intrastriatal transplantation of CB-stem/progenitor cells or their progenies also constitute a potential cell source for anti-parkinsonian therapy [221]. Moreover, the combination of stem cell-based treatment with the administration of β-amyloid precursor protein (APP), which can promote the differentiation of NSCs and neurogenesis, also represents a potential therapeutic strategy for treating Down syndrome or Alzheimer's disease [86]. Additionally, ESCs, foetal stem/progenitor cells, UC-derived stem cells, AECs, BMSCs including MSCs, ADSCs, and pluripotent epidermal neural crest stem cells (eNCSCs) found in bulge areas within the hair follicle of the skin may also be induced to differentiate or trans-differentiate into functional neurons (tubulin-β and Tuj1), astrocytes (glial fibrillary acidic protein, GFAP), or oligodendrocytes (O4) *in vitro* and/or *in vivo* [3–5, 11, 61, 228, 229]. For instance, it has been observed that the transplanted dopaminergic neurons derived from human and mouse ES cells survived for more than one year and displayed the functional properties of an animal model of Parkinson's disease [226, 227]. Hence, on the basis of these observations, it is now possible to envision using these immature cells or their further differentiated progenies in the near future for treating diverse incurable neurodegenerative diseases. Additional investigations are required before their use in clinical settings in humans in order to more precisely establish the intrinsic and extrinsic factors that control their behaviour within the niches *in vivo* and migration at distant sites within the brain as well as their therapeutic effect in the long term after treatment initiation.

Skin stem/progenitor cells and their therapeutic applications

Human skin epidermis and its appendages exhibit important different site-specific morphologies and functions including the protection against environmental injury, infection and excessive dehydration [230]. Normal skin integrity and functions are maintained by different adult stem/progenitor cell subpopulations localized within the interfollicular epidermis (IFE) and hair follicle bulge (Fig. 16.1) [230, 231]. In particular, the epidermal barrier is affected by a continuous loss of terminally differentiated keratinocytes at the skin surface, the stratum corneum which are shipped out during the desquamation process [230]. The small clusters of multipotent keratinocyte stem cells (KSCs) expressing cytokeratins (CK5/14/15), p63, $\alpha_6\beta_4$- and $\alpha_3\beta_1$-integrins, and ABC transporters, which are localised in the basal layer near the basement membrane of the epithelial compartment, may then provide critical roles in participating in the cell replenishment of mature keratinocytes in homeostasis conditions and after skin injuries [50, 231–235]. The maintenance of the stratified epidermis is notably accomplished by the division of KSCs that give arise to early TA cells. Early TA cells can leave the basal compartment into the suprabasal layer and give rise after subsequently division to more committed suprabasal TA cells forming multiple epithelial layers and terminally differentiated keratinocytes constituting the outside surface of skin. Multiple signaling networks initiated by different growth factor signaling including Notch, EGF, Myc, and TGF-β, appear to be involved in the regulation of the proliferation, migration and/or differentiation of KSCs [230–232, 236–238].

In addition, under specific physiological and pathological conditions, the adult stem/progenitor cells resident in the hair follicle bulge may also serve as a source of immature cells to regenerate the skin appendages or damaged epithelial tissue in response to intense injuries. More specifically, the multipotent epithelial stem cells (bESCs) within the bulge area, which express CD34, CK5/14, p63, and $\alpha_6\beta_4$- and $\alpha_3\beta_1$-integrins, are able to proliferate and give rise to the follicular epithelium as well as new cells constituting IFE and sebaceous glands after severe injury [6, 230, 231, 239–241]. In fact, bESC progenitors can emigrate along the outer root sheath (ORS) forming the outermost layer toward the germinal matrix and dermal papilla during the hair cycle. It has been reported that the induction of telomerase or its catalytic subunit TERT in mouse skin epithelium caused a rapid transition from the resting phase of the hair follicle cycle (telogen) into the active phase (anagen) by activating hair follicle stem cell proliferation [123, 242, 243]. In this matter, several growth factors including sonic hedgehog, Wnt/β-catenin, Notch/RBP-J and BMP, also appear to contribute to the maintenance and/or regeneration of the hair follicles [123, 230, 240, 241, 244, 245]. Particularly, a recent investigation revealed that the quiescent state of stem cells in the bulge niche may be maintained via BMP signaling that activates nuclear factor activated T cells c1 (NFATc1 also called NFAT2), which in turn represses the CDK4 protein [246]. In contrast, a NFATc1 down-regulation may relieve its inhibitory effect during new hair follicle growth [246]. The transient stimulation of sonic hedgehog and Wnt/β-catenin cascades also appears to provide an essential function for hair follicle growth by inducing the transition from a resting phase (telogen) to an active phase (anagen) [247]. Moreover, the activation of Wnt/β-catenin and Notch pathways also seems to be necessary for the fate specification of bESCs into hair follicular keratinocytes than IFE keratinocytes during the hair growth cycle [248, 249]. In addition, bESCs and a small population of progenitor cells expressing CK5/14 and Blimp1 residing near the basement membrane surrounding the sebaceous gland, may also contribute to the regeneration of the gland including the terminally differentiated sebatocytes [230, 231, 250, 251].

The bulge area in the adult mammalian hair follicle also contains pluripotent eNCSCs that show several properties similar to embryonic neural-crest stem cells (Fig. 16.1) [252]. In fact, it has been proposed that eNCSCs, as bESCs, might emigrate from the bulge region during the hair cycle and migrate along the entire length of the hair follicle in the inner layers of ORS toward the base of hair follicles. More specifically, the dermal papilla is also enriched in eNCSC progenitors that might give rise to the progeny constituting the germinal zone matrix whose cells, in turn, can proliferate and differentiate into mature cells and form a new hair follicle [252]. The pluripotent eNCSCs in the bulge area are also able to self-renew and give rise to multiple cell lineages *in vivo* including melanocytes, neurons, Schwann cells, smooth muscle cells and chondrocytes [252]. Moreover, eNCSCs in the bulge area may also differentiate to other neural crest derivatives, the Merkel cells that are characterized by specific marker K8, and whose cells remain localized around the bulge zone [252]. In addition, the multipotent adult skin-derived precursors (SKPs), which reside within the dermal papillae of hair follicles, also appear to exhibit properties similar to those of eNCSCs [253, 254]. SKPs can be expanded in the presence of EGF, bFGF, and TGF-â and give rise to neurons, glia, smooth muscle cells and adipocytes in culture *in vitro*.

Skin disorders and stem-cell-based therapies

The physical and chemical perturbations of skin may cause severe damage to the stratified epidermis and its appendages and lead to skin pathologies including chronic non-healing wounds and ulcers, ectodermal dysplasia congenital disorders and cancers. The use of skin adult stem/progenitor cells and their progenies for cell-replacement therapy and tissue engineering represent then a cell source of great hope and major interest in view of their applications for replacing unfunctional or lost skin cells or obtaining physiologic tissue-engineered human skin equivalents for skin grafting. Several new therapeutic strategies have been investigated for skin reconstructive surgery after intractable and severe wounds such as burns, deep skin injuries, infections and decubitus ulcers including cell grafting with diverse biocompatible artificial materials and/or skin fragments [112,191]. For instance, the results from a recent study have revealed that the use of cultured autologous BM-derived mesenchymal cells plus artificial dermis composite graft may improve mixed chimerism and vascularized skin graft survival, and thereby promote the recovery of diverse types of intractable dermatopathies and prevent graft rejection [112]. Similarly, the incorporation of *ex vivo* expanded adult circulating EPCs from an autologous source onto tissue-engineered human skin substitutes also resulted in the formation of functional microvessels that improved the perfusion and survival of bioengineered tissues [191]. Even if these techniques offer great promise, the regeneration stability and functions of skin tissue grafts in the long-term must be established on a greater number of patients with different skin disorders before they can be safely incorporated in clinical settings in safe conditions.

In addition, the dysregulation of gene products involved in stringent regulation of cell cycle and differentiation of KSCs and bulge stem cells may lead to epidermal hyperproliferative disorders, skin epithelial cancers and melanomas [255, 256]. Particularly, the sustained activation of TERT, sonic hedgehog and EGFR cascades in skin stem cells may contribute to the maintenance of the proliferative basal and/or suprabasal cell populations in the benign hyperproliferative disorders such as wound healing as well as in basal cell carcinomas (BCCs) and/or squamous cell carcinomas (SCCs) [123, 230, 241]. Additional investigations are however necessary to more precisely determine the specific implication of KSCs residing near the basement membrane in epithelium versus bulge ESCs during the different stages leading to the epidermis regeneration after trauma and carcinogenesis. In regard with this, the results from a recent study have notably revealed that certain epithelial cancer types could derive from the malignant transformation of primitive bulge stem cells in mice models of carcinogenesis [255]. In addition, the data from a recent investigation have indicated that certain melanoma types may derive from the malignant transformation of skin stem cells like eNCSCs or their progenitors. It has been reported that the *ex vivo* culture of a tumourigenic cell subpopulation from metastatic melanomas corresponding to an enriched $CD20^+$ fraction of melanoma cells in the growth medium, which is used for human ESCs, may result in their propagation under the form of non-adherent spheres [97]. In fact, it has been observed that each individual multipotent cell from melanoma spheres was able to differentiate like eNCSCs into multiple cell types including melanocytes, adipocytes, osteocytes, and chondrocytes under well-defied conditions [97].

Pancreatic stem/progenitor cells and their therapeutic applications

The pancreas is a glandular organ in the gastrointestinal tract that is constituted by an exocrine compartment, comprising the ductal epithelium and acinar cells, that secrete the enzymes directly into the small intestine which aid in the digestion of food (Fig. 16.1) [3–5, 257, 258]. The pancreas is also composed of an endocrine compartment designated as islets of Langerhans containing four different types of cells: insulin-producing β-cells, glucagon-releasing α-cells, somatostatin-producing δ-cells and pancreatic polypeptide-containing cells. More specifically, islet β-cells provide critical functions by releasing the insulin into the bloodstream, and which in turn controls the level of blood glucose in peripheral circulation [3, 4, 257–259]. Importantly, an accumulating body of experimental evidence has indicated that the human and rodent mature insulin-producing islet β-cells could arise in part from adult pancreatic stem/progenitor cells (PSCs) resident in the adult pancreas. Putative PSCs expressing ductal epithelial cell markers, cytokeratin 19 ($CK19^{high}$), neural (nestin) and endocrine nuclear pancreatic and duodenal homeobox factor-1 (PDX-1) markers and/or more committed $nestin^+/PDX-1^+/CD19^{low}$ islet precursors appear to be localized in the ductal regions and/or within the islet compartment (Fig. 16.1) [3–5, 58, 258, 260–263]. Therefore, these putative adult PSCs or their early progenies within the adult pancreas represent a potential renewal source of immature cells for pancreatic cell replacement or gene therapy for treating chronic pancreatitis and life-threatening diseases such as type 1 or 2 diabetes mellitus [264–266]. More particularly, the stimulation of PSCs *in vivo* or transplantation of *ex vivo* expanded pancreatic β-cells in the host diseased recipient may constitute a therapeutic strategy for restoring the β-cell mass lost over time in diabetic patients [3–5, 11, 58, 71–74, 266, 267]. In support with this, it has also been reported that the transplantation of purified pancreatic duct cells from islet-depleted human pancreatic tissue plus stromal cell preparation generated the insulin-producing cells in the normoglycemic NOD/SCID mice model *in vivo* [268].

In addition, the differentiation of other stem cell sources such as embryonic, foetal and UCB stem/progenitor cells, hAECs, PDMSCs, and adult stem cells including HSCs, MSCs, HOCs, NSCs, hAECs, and ADSCs into pancreatic insulin-producing β-cell like progenitors has also been performed *in vitro* by using specific growth factors (Table 16.1) [3–5, 11, 58, 73, 74, 187, 219]. Moreover, the results from numerous *in vitro* pre-clinical investigations and β-cell-based transplantation studies in diabetic animal models *in vivo* have revealed the potential therapeutic beneficial to use these stem/progenitor cell types for generating insulin-producing β-cells for the treatment of type 1 or 2 diabetes [3–5, 219, 266, 269]. Importantly, the gene therapy by using insulin-producing cells such as hADMSC with unfractionated cultured BM has also been observed to be effective for treating insulinopenic patients with type 1 diabetes mellitus [270]. In spite of these significant advancements, additional *in vivo* studies appear to be necessary to establish the beneficial effects of using these stem cell types and their further differentiated progenies for restoring blood sugar levels after long-term treatment in diabetic patients.

On the other hand, it will also be important to establish the potential implication of putative pancreas-resident adult PSCs in the development of other pancreatic disorders such as autoimmune or chronic pancreatitis and pancreatic cancers. These additional studies should provide important information for the development of novel molecular targeting therapies for eradicating pancreatic cancer stem/progenitor cells which may play a major role in the formation of ductal pancreatic adenocarcinomas, treatment resistance and disease relapse. Hence, these advancements could offer an ultimate hope for treating and even curing the aggressive, metastatic and recurrent ductal pancreatic adenocarcinomas which remain incurable with the conventional treatment by surgical resection, radiation and/or chaemotherapies [5, 88, 92, 271, 272].

Other stem cell types and their therapeutic implications

Among the other tissues/organs harboring an adult stem/progenitor cell population, there are lung (bronchioalveolar stem cells "BASCs") [14, 15, 17], liver (hepatic oval cells "HOCs") [5, 18, 19, 68, 273–275], intestinal crypts and gastic glands [5, 20, 21, 70, 275–278], adipose tissues (ADSCs adipose tissue-derived stem cells "ADSCs"; muscles (muscle-derived stem cells "MDSCs") and eye (corneal epithelial stem cells "CESCs', retinal stem cells "RSCs" and conjunctival stem cells) (Fig. 16.1) [5, 16, 80, 279–281]. Hence, the *in vivo* stimulation of these adult stem/progenitor cells and/or the replacement of their dysfunctional counterparts and/or their further differentiated progenies, also constitute the potential stem cell-based strategies for the treatment of numerous pathological disorders in humans (Fig. 16.1; Table 16.1) [3, 5, 6, 55, 59]. This could result in the restoration of the regeneration program, and thereby prevent the progressive loss of functions of these adult stem cells with aging and lead to the treatment of diverse human disorders. Amongst them, there are lung disorders (interstitial lung diseases, cystic fibrosis, asthma, chronic bronchitis, and emphysema) [14, 17, 65–67], chronic liver injuries (hepatitis and liver cirrhosis) [5, 18, 19, 54, 68, 69, 274], gastrointestinal disorders (chronic inflammatory bowel diseases and ulcers) [5, 20, 21, 70, 282, 283], bone, cartilage (osteoporosis, osteogenesis imperfecta), musculoskeletal disorders (Duchenne and Becker dystrophies and amyotrophic lateral sclerosis) [5, 11, 13, 76–78, 284–286] and eye diseases (partial or total limbal and/or conjunctival stem cell deficiency, bullous keratopathy, glaucoma and retinal damages) [5, 16, 79, 80, 280]. In addition, it has been reported that the human endometrial gland-derived mesenchymal cells (EMCs) and menstrual blood-derived mesenchymal cells (MMCs) expressing CD29 and CD105 were more proliferative than MSCs from umbilical cords [177, 287]. These pluripotent immature cells were able to differentiate into cardiomyocytic, respiratory epithelial, neurocytic, myocytic, endothelial, pancreatic, hepatic, adipocytic, and osteogenic cells *in vitro* and transdifferentiate into the cardiac tissue-layer *in vivo* [177, 287]. This suggest then EMCs and MMCs may constitute new potential pluripotent stem cell sources, easily accessible for cell replacement-based therapy [177]. On the other hand, the targeting of their malignant counterparts, cancer stem/progenitor cells and their local microenvironment involved in cancer development also offers great promises for the development of new therapeutic approaches for treating the aggressive and metastatic cancers derived from these different tissue-resident adult stem cells.

CONCLUSION

Together these recent advancements in the basic and clinical research on tissue-resident adult stem/progenitor cells have provided important information on the unique features and functions of these immature cells with a long lifespan and multilineage differentiation potential for the tissue regeneration in homeostatic and pathological conditions in human. Hence, multipotent or pluripotent tissue-resident adult stem/progenitor cells represent promising sources for cell-replacement and gene therapies with multiple potential applications in regenerative medicine and cancer treatment. Adult stem/progenitor cells could be used for treating numerous devastating diseases including haematopoietic, skeletomuscular, cardiovascular, neurodegenerative, skin and eye disorders and diabetes mellitus as well as aggressive and metastatic cancers. In counterpart, further research is necessary to more precisely establish the specific biomarkers of adult stem/progenitor cells versus their differentiated progenies as well as their functional properties in the long-term for tissue repair. The identification of the specific intrinsic factors that govern the decision between the self-renewal versus the differentiation of tissue-resident adult stem cells as well as the influence of the extracellular signals from their local microenvironment "niche" on their behaviour is notably of immense interest for the design of new therapeutic strategies. These future studies should aid to optimize the methods for their *ex vitro* and *in vivo* expansion and differentiation into desired cell lineages as well as the most appropriate administration mode for their delivery in the specific damaged tissue areas *in vivo*. These additional investigations should permit to specify the therapeutic benefit of each tissue-resident adult stem cell/progenitor cell type and minimize the potential risks for their successful use in stem cell-based approaches. These new novel therapeutic strategies could be translated for treating and even curing diverse human diseases which remain incurable in the clinics with the current conventional therapies.

ABBREVIATIONS

ABC, ATP-binding cassette; ADSCs, adipose tissue-derived stem cells; ATP, adenosine triphosphate; BM, bone marrow; BMP, bone morphogenic protein; bESCs, bulge epithelial stem cells; CB, carotid body; CESCs, corneal epithelial stem cells; CNS, central nervous system; CXCR4, CXC chaemokine-receptor-4; EGF, epidermal growth factor; EGFR, epidermal growth factor receptor; eNCSCs, epidermal neural crest stem cells, EPCs, endothelial progenitor cells; ESCs, embryonic stem cells; hAECs, human amniotic epithelial cells; hESCs, human embryonic stem cells; HGF, hepatocyte growth factor; hMMCs, Human menstrual blood-derived mesenchymal cells; HSCs, haematopoietic stem cells; IGF-1, insulin-like growth factor-1; KIT, SCF receptor tyrosine kinase; KSCs, keratinocyte stem cells; MDSCs, muscle-derived stem cells; MSCs, mesenchymal stem cells; NCSCs, neural crest stem cells; NSCs, neural stem cells; Oct-3/4, octamer-binding protein; PSCs, pancreatic stem cells; PDMSCs, placenta-derived multipotent stem/progenitor cells; RB, retinoblastoma protein; RSCs, retinal stem cells; SC, stem cell; Sca-1, stem cell antigen-1; SCF, stem cell factor; SDF-1, stromal cell-derived factor-1; SHH, sonic hedgehog ligand; SP, side population; TA, transit-amplifying; TERT, telomerase reverse transcriptase; UCB, umbilical cord blood; VEGF, vascular endothelial growth factor; VEGFR2, vascular endothelial growth factor receptor 2; Wnt, Wingless ligand.

ACKNOWLEDGEMENTS

This work was supported by grants from the U.S. Department of Defense (PC04502, OC04110) and the National Institutes of Health (CA78590, CA111294). We thank Ms. Kristi L. Berger for editing the manuscript.

REFERENCES

1. de Paiva C.S., S.C. Pflugfelder and D.Q. Li (2006). 'Cell size correlates with phenotype and proliferative capacity in human corneal epithelial cells'. *Stem Cells.* **24**: 368–375.
2. Mimeault M., R. Hauke and S.K. Batra (2007). 'Recent advances on the molecular mechanisms involved in drug-resistance of cancer cells and novel targeting therapies'. *Clin. Pharmacol. Ther.* **83**: 673–691.
3. Mimeault M. and S.K. Batra (2006). 'Recent advances on the significance of stem cells in tissue regeneration and cancer therapies'. *Stem Cells.* **24**: 2319–2345.
4. Mimeault M., R. Hauke and S.K. Batra (2007). 'Stem cells—a revolution in therapeutics—recent advances on the stem cell biology and their therapeutic applications in regenerative medicine and cancer therapies'. *Clin. Pharmacol. Ther.* **82**: 252–264.
5. Mimeault M. and S.K. Batra (2008). 'Recent progress on tissue-resident adult stem cell biology and their therapeutic implications'. *Stem Cell Rev.* **4**(1): 27–49.
6. Li L. and T. Xie (2005). 'Stem cell niche: structure and function. *Ann. Rev. Cell Dev. Biol.* **21**: 605–631.6.
7. Beltrami A.P., L. Barlucchi, D. Torella, M. Baker, F. Limana, S. Chimenti, H. Kasahara, M. Rota, E. Musso, K. Urbanek et al. (2003). 'Adult cardiac stem cells are multipotent and support myocardial regeneration'. *Cell.* **114**: 763–776.

8. Leri A., J. Kajstura and P. Anversa (2005). 'Cardiac stem cells and mechanisms of myocardial regeneration'. *Physiol. Rev.* **85**: 1373–1416.

9. van Vliet P., J.P. Sluijter, P.A. Doevendans and M.J. Goumans (2007). 'Isolation and expansion of resident cardiac progenitor cells'. *Expert. Rev. Cardiovasc. Ther.* **5**: 33–43.

10. Watts C., H. McConkey, L. Anderson and M. Caldwell (2005). 'Anatomical perspectives on adult neural stem cells'. *J. Anat.* **207**: 197–208.

11. Schaffler A. and C. Buchler (2007). 'Concise review: adipose tissue-derived stromal cells—basic and clinical implications for novel cell-based therapies'. *Stem Cells.* **25**: 818–827.

12. Dhawan J. and T.A. Rando (2005). 'Stem cells in postnatal myogenesis: molecular mechanisms of satellite cell quiescence, activation and replenishment'. *Trends Cell Biol.* **15**: 666–673.

13. Peault B., M. Rudnicki, Y. Torrente, G. Cossu, J.P. Tremblay, T. Partridge, E. Gussoni, L.M. Kunkel and J. Huard (2007). 'Stem and progenitor cells in skeletal muscle development, maintenance and therapy'. *Mol. Ther.* **15**: 867–877.

14. Griffiths M.J., D. Bonnet and S.M. Janes (2005). 'Stem cells of the alveolar epithelium'. *Lancet* **366**: 249–260.

15. Kim C.F., E.L. Jackson, A.E. Woolfenden, S. Lawrence, I. Babar, S. Vogel, D. Crowley, R.T. Bronson and T. Jacks (2005). 'Identification of bronchioalveolar stem cells in normal lung and lung cancer'. *Cell.* **121**: 823–835.

16. Lavker R.M., S.C. Tseng and T.T. Sun (2004). 'Corneal epithelial stem cells at the limbus: looking at some old problems from a new angle'. *Exp. Eye Res.* **78**: 433–446.

17. Liu X., R.R. Driskell and J.F. Engelhardt (2004). 'Airway glandular development and stem cells'. *Curr. Top. Dev. Biol.* **64**: 33–56.

18. Guettier C. (2005). 'Which stem cells for adult liver?' *Ann. Pathol.* **25**: 33–44.

19. Herrera M.B., S. Bruno, S. Buttiglieri, C. Tetta, S. Gatti, M.C. Deregibus, B. Bussolati and G. Camussi (2006). 'Isolation and characterisation of a stem cell population from adult human liver'. *Stem Cells.* **24**: 2840–2850.

20. Brittan M. and N.A. Wright (2002). 'Gastrointestinal stem cells. *J. Pathol*'. **197**: 492–509.

21. Potten C.S. and J.R. Ellis (2006). 'Adult small intestinal stem cells: identification, location, characteristics and clinical applications'. *Ernst. Schering. Res. Found.* Workshop. **60**: 81–98.

22. Koblas T., K. Zacharovova, Z. Berkova, M. Mindlova, P. Girman, E. Dovolilova, L. Karasova and F. Saudek (2007). 'Isolation and characterisation of human CXCR4-positive pancreatic cells'. *Folia Biol.* (Praha) **53**: 13–22.

23. Gotte M., M. Wolf, A. Staebler, O. Buchweitz, R. Kelsch, A. Schuring and L. Kiesel (2008). 'Increased expression of the adult stem cell marker Musashi-1 in endometriosis and endometrial carcinoma'. *J. Pathol.* **215**: 317–329.

24. Aponte P.M., van M.P. Bragt, de D.G. Rooij and A.M. van Pelt (2005). 'Spermatogonial stem cells: characteristics and experimental possibilities'. *APMIS.* **113**: 727–742.

25. Bryder D., D.J. Rossi and I.L. Weissman (2006). 'Haematopoietic stem cells: the paradigmatic tissue-specific stem cell'. *Am. J. Pathol.* **169**: 338–346.

26. Wilson A. and A. Trumpp (2006). 'Bone-marrow haematopoietic-stem-cell niches'. *Nat. Rev. Immunol.* **6**: 93–106.

27. Moore K.A. and I.R. Lemischka (2006). 'Stem cells and their niches'. *Science.* **311**: 1880–1885.
28. Rizo A., E. Vellenga, de G. Haan and J.J. Schuringa (2006). 'Signaling pathways in self-renewing haematopoietic and leukemic stem cells: do all stem cells need a niche?' *Hum. Mol Genet.* **15**(2): R210–R219.
29. Lim D.A., Y.C. Huang and A. Alvarez-Buylla (2007). 'The adult neural stem cell niche: lessons for future neural cell replacement strategies'. *Neurosurg. Clin. N. Am.* **18**: 81–92.
30. Yui J., C.P. Chiu and P.M. Lansdorp (1998). 'Telomerase activity in candidate stem cells from foetal liver and adult bone marrow'. *Blood.* **91**: 3255–3262.
31. Allsopp R.C., G.B. Morin, R. ePinho, C.B. Harley and I.L. Weissman (2003). 'Telomerase is required to slow telomere shortening and extend replicative lifespan of HSCs during serial transplantation'. *Blood.* **102**: 517–520.
32. Chen J. (2005). 'Senescence of haematopoietic stem cells and bone marrow failure'. *Int. J. Haematol* **82**: 190–195.
33. Gonzalez A., M. Rota, D. Nurzynska, Y. Misao, J. Tillmanns, C. Ojaimi, M.E. Padin-Iruegas, P. Muller, G. Esposito, C. Bearzi et al. (2008). 'Activation of cardiac progenitor cells reverses the failing heart senescent phenotype and prolongs lifespan'. *Circ. Res.* **102**: 597–606.
34. Madonna R., J.T. Willerson and Y.J. Geng (2008). 'Myocardin a enhances telomerase activities in adipose tissue mesenchymal cells and embryonic stem cells undergoing cardiovascular myogenic differentiation'. *Stem Cells.* **26**: 202–211.
35. Pu D.R. and L. Liu (2008). 'HDL slowing down endothelial progenitor cells senescence: a novel anti-atherogenic property of HDL'. *Med. Hypotheses.* **70**: 338–342.
36. Tateishi K., E. Ashihara, S. Honsho, N. Takehara, T. Nomura, T. Takahashi, T. Ueyama, M. Yamagishi, H. Yaku, H. Matsubara et al. (2007). 'Human cardiac stem cells exhibit mesenchymal features and are maintained through Akt/GSK-3beta signaling. Biochem. Biophys'. *Res. Commun.* **352**: 635–641.
37. Schneider M.D. (2006). 'Dual roles of telomerase in cardiac protection and repair. Novartis'. *Found. Symp* **274**: 260–267.
38. Urbanek K., D. Torella, F. Sheikh, A.A. De, D. Nurzynska, F. Silvestri, C.A. Beltrami, R. Bussani, A.P. Beltrami, F. Quaini et al. (2005). 'Myocardial regeneration by activation of multipotent cardiac stem cells in ischemic heart failure'. *Proc. Natl. Acad. Sci. USA.* **102**: 8692–8697.
39. Kenyon J. and S.L. Gerson (2007). 'The role of DNA damage repair in aging of adult stem cells'. *Nucleic Acids Res.* **35**: 7557–7565.
40. Potten C.S. (2004). 'Keratinocyte stem cells, label-retaining cells and possible genome protection mechanisms'. *J. Investig. Dermatol. Symp. Proc.* **9**: 183–195.
41. Vaish M. (2007). 'Mismatch repair deficiencies transforming stem cells into cancer stem cells and therapeutic implications'. *Mol. Cancer.* **6**: 26.
42. Martin C.M., A. Ferdous, T. Gallardo, C. Humphries, H. Sadek, A. Caprioli, J.A. Garcia, L.I. Szweda, M.G. Garry and D.J. Garry (2008). 'Hypoxia-inducible factor-2alpha transactivates Abcg2 and promotes cytoprotection in cardiac side population cells'. *Circ. Res.* **102**: 1075–1081.
43. Bunting K.D. (2002). 'ABC transporters as phenotypic markers and functional regulators of stem cells'. *Stem Cells.* **20**: 11–20.
44. Gottesman M.M., T. Fojo and S.E. Bates (2002). 'Multidrug resistance in cancer: role of ATP-dependent transporters'. *Nat. Rev. Cancer.* **2**: 48–58.

45. Krishnamurthy P. and J.D. Schuetz (2006). 'Role of ABCG2/BCRP in biology and medicine'. *Ann. Rev. Pharmacol. Toxicol.* **46**: 381–410.
46. Johnnidis J.B. and F.D. Camargo (2008). 'Isolation and functional characterisation of side population stem cells'. *Methods Mol. Biol.* **430**: 183–193.
47. Conrad C., E. Zeindl-Eberhart, S. Moosmann, P.J. Nelson, C.J. Bruns and R. Huss (2008). 'Alkaline phosphatase, glutathione-S-transferase-P and cofilin-1 distinguish multipotent mesenchymal stromal cell lines derived from the bone marrow versus peripheral blood'. *Stem Cells Dev.* **17**: 23–27.
48. Ahmed F., N. Arseni, H. Glimm, W. Hiddemann, C. Buske and M. Feuring-Buske (2008). 'Constitutive expression of the ATP-binding cassette transporter ABCG2 enhances the growth potential of early human haematopoietic progenitors'. *Stem Cells.* **26**: 810–818.
49. Challen G.A. and M.H. Little (2006). 'A side order of stem cells: the SP phenotype'. *Stem Cells.* **24**: 3–12.
50. Larderet G., N.O. Fortunel, P. Vaigot, M. Cegalerba, P. Maltere, O. Zobiri, X. Gidrol, G. Waksman and M.T. Martin (2006). 'Human side population keratinocytes exhibit long-term proliferative potential and a specific gene expression profile and can form a pluristratified epidermis'. *Stem Cells.* **24**: 965–974.
51. Mimeault M. and S.K. Batra (2008). 'Stem cell applications in disease research: recent advances on stem cell and cancer stem cell biology and their therapeutic implications. In Progress in Stem Cell Applications,' eds. Allen V. Faraday and Jonathon T. Dyer, NOVA Publisher. Hauppauge, New York. 57–98.
52. Mimeault M., P.P. Mehta, R. Hauke and S.K. Batra (2007). 'Functions of normal and malignant prostatic stem/progenitor cells in tissue regeneration and cancer progression and novel targeting therapies against advanced prostate cancers'. *Endocr. Rev.* **29**: 234–252.
53. Herzog E.L. and D.S. Krause (2006). 'Engraftment of marrow-derived epithelial cells: the role of fusion'. *Proc. Am. Thorac. Soc.* **3**: 691–695.
54. Furst G., E.J. Schulte am, L.W. Poll, S.B. Hosch, L.B. Fritz, M. Klein, E. Godehardt, A. Krieg, B. Wecker, V. Stoldt et al. (2007). 'Portal vein embolisation and autologous CD133+ bone marrow stem cells for liver regeneration: initial experience'. *Radiology.* **243**: 171–179.
55. Asahara T. and A. Kawamoto (2004). 'Endothelial progenitor cells for postnatal vasculogenesis'. *Am. J. Physiol. Cell. Physiol.* **287**: C572–C579.
56. Friedrich E.B., K. Walenta, J. Scharlau, G. Nickenig and N. Werner (2006). 'CD34-/CD133+/VEGFR-2+ endothelial progenitor cell subpopulation with potent vasoregenerative capacities'. *Circ. Res.* **98**: E20–E25.
57. Schatteman G.C., M. Dunnwald and C. Jiao (2007). 'Biology of bone marrow-derived endothelial cell precursors'. *Am. J. Physiol. Heart Circ. Physiol.* **292**: H1–18.
58. Bonner-Weir S. and G.C. G.C. Weir (2005). 'New sources of pancreatic beta-cells. *Nat. Biotechnol.* **23**: 857–861.
59. Pessina A. and L. Gribaldo (2006). 'The key role of adult stem cells: therapeutic perspectives'. *Curr. Med. Res. Opin.* **22**: 2287–2300.
60. McMullen N.M. and K.B. Pasumarthi (2007). 'Donor cell transplantation for myocardial disease: does it complement current pharmacological therapies?' *Can. J. Physiol. Pharmacol.* **85**: 1–15.
61. Garbuzova-Davis S., A.E. Willing, S. Saporta, P.C. Bickford, C. Gemma, N. Chen, C.D. Sanberg, S.K. Klasko, C.V. Borlongan and P.R. Sanberg (2006). 'Novel cell therapy approaches for brain repair'. *Prog. Brain Res.* **157**: 207–222.

62. Chang Y.C., W.C. Shyu, S.Z. Lin and H. Li (2007). 'Regenerative therapy for stroke'. *Cell Transplant.* **16**: 171–181.
63. Geraerts M., O. Krylyshkina, Z. Debyser and V. Baekelandt (2007). 'Concise review: therapeutic strategies for Parkinson disease based on the modulation of adult neurogenesis'. *Stem Cells.* **25**: 263–270.
64. Trzaska K.A. and P. Rameshwar (2007). 'Current advances in the treatment of Parkinson's disease with stem cells'. *Curr. Neurovasc. Res.* **4**: 99–109.
65. Gharaee-Kermani M., M.R. Gyetko, B. Hu and S.H. Phan (2007). 'New insights into the pathogenesis and treatment of idiopathic pulmonary fibrosis: a potential role for stem cells in the lung parenchyma and implications for therapy'. *Pharm. Res.* **24**: 819–841.
66. Sueblinvong V., B.T. Suratt and D.J. Weiss (2007). 'Novel therapies for the treatment of cystic fibrosis: new developments in gene and stem cell therapy'. *Clin. Chest Med.* **28**: 361–379.
67. Andrade C.F., A.P. Wong, T.K. Waddell, S. Keshavjee and M. Liu (2007). 'Cell-based tissue engineering for lung regeneration'. *Am. J. Physiol. Lung Cell Mol. Physiol* **292**: L510–L518.
68. Sharma A.D., T. Cantz, M.P. Manns and M. Ott (2006). 'The role of stem cells in physiology, pathophysiology and therapy of the liver'. *Stem Cell Rev.* **2**: 51–58.
69. Fiegel H.C., C. Lange, U. Kneser, W. Lambrecht, A.R. Zander, X. Rogiers and D. Kluth (2006). 'Foetal and adult liver stem cells for liver regeneration and tissue engineering'. *J. Cell. Mol. Med.* **10**: 577–587.
70. Fox J.G. and T.C. Wang (2007). 'Inflammation, atrophy, and gastric cancer'. *J. Clin. Invest.* **117**: 60–69.
71. Fellous T.G., N.J. Guppy, M. Brittan and M.R. Alison (2007). 'Cellular pathways to beta-cell replacement'. *Diabetes Metab. Res. Rev.* **23**: 87–99.
72. Santana A., R. Ensenat-Waser, M.I. Arribas, J.A. Reig and E. Roche (2006). 'Insulin-producing cells derived from stem cells: recent progress and future directions'. *J. Cell Mol. Med.* **10**: 866–883.
73. Gangaram-Panday S.T., M.M. Faas and P. de Vos (2007). 'Towards stem-cell therapy in the endocrine pancreas'. *Trends Mol. Med.* **13**: 164–173.
74. Lu P., F. Liu, L. Yan, T. Peng, T. Liu, Z. Yao and C.Y. Wang (2007). 'Stem cells therapy for type 1 diabetes'. *Diabetes Res. Clin. Pract.* **78**: 1–7.
75. Kuroda R., A. Usas, S. Kubo, K. Corsi, H. Peng, T. Rose, J. Cummins, F.H. Fu and J. Huard (2006). 'Cartilage repair using bone morphogenetic protein 4 and muscle-derived stem cells'. *Arthritis Rheum.* **54**: 433–442.
76. Gimble J.M., A.J. Katz and B.A. Bunnell (2007). 'Adipose-derived stem cells for regenerative medicine'. *Circ. Res.* **100**: 1249–1260.
77. Dragoo J.L., G. Carlson, F. McCormick, H. Khan-Farooqi, M. Zhu, P.A. Zuk and P. Benhaim (2007). 'Healing full-thickness cartilage defects using adipose-derived stem cells'. *Tissue Eng.* **13**: 1615–1621.
78. Helder M.N., M. Knippenberg, J. Klein-Nulend and P.I. Wuisman (2007). 'Stem cells from adipose tissue allow challenging new concepts for regenerative medicine'. *Tissue Eng.* **13**: 1799–1808.
79. Limb G.A., J.T. Daniels, A.D. Cambrey, G.A. Secker, A.J. Shortt, J.M. Lawrence and P.T. Khaw (2006). 'Current prospects for adult stem cell-based therapies in ocular repair and regeneration'. *Curr. Eye Res.* **31**: 381–390.

80. MacLaren R.E., R.A. Pearson, A. MacNeil, R.H. Douglas, T.E. Salt, M. Akimoto, A. Swaroop, J.C. Sowden and R.R. Ali (2006). 'Retinal repair by transplantation of photoreceptor precursors'. *Nature.* **444**: 203–207.
81. Rossi D.J., D. Bryder and I.L. Weissman (2007). 'Haematopoietic stem cell aging: mechanism and consequence'. *Exp. Gerontol.* **42**: 385–390.
82. Ju Z., Jiang H., M. Jaworski, C. Rathinam, A. Gompf, C. Klein, A. Trumpp and K.L. Rudolph (2007). 'Telomere dysfunction induces environmental alterations limiting haematopoietic stem cell function and engraftment'. *Nat. Med.* **13**: 742–747.
83. Barfield R.C., G.A. Hale, K. Burnette, F.G. Behm, K. Knapp, P. Eldridge and R. Handgretinger (2007). 'Autologous transplantation of CD133 selected haematopoietic progenitor cells for treatment of relapsed acute lymphoblastic leukemia. Pediatr'. *Blood Cancer.* **48**: 349–353.
84. Murrell W., A. Wetzig, M. Donnellan, F. Feron, T. Burne, A. Meedeniya, J. Kesby, J. Bianco, C. Perry, P. Silburn et al. (2008). 'Olfactory mucosa is a potential source for autologous stem cell therapy for Parkinson's disease'. *Stem Cells.* **26**(8): 2183–2192.
85. Zhao C., W. Deng and F.H. Gage (2008). 'Mechanisms and functional implications of adult neurogenesis'. *Cell.* **132**: 645–660.
86. Sugaya K., Y.D. Kwak, O. Ohmitsu, A. Marutle, N.H. Greig and E. Choumrina (2007). 'Practical issues in stem cell therapy for Alzheimer's disease'. *Curr. Alzheimer Res.* **4**: 370–377.
87. Al-Hajj M. and M.F. Clarke (2004). 'Self-renewal and solid tumour stem cells'. *Oncogene.* **23**: 7274–7282.
88. Mimeault M., R. Hauke, P.P. Mehta and S.K. Batra(2007). 'Recent advances on cancer stem/progenitor cell research: therapeutic implications for overcoming resistance to the most aggressive cancers'. *J. Mol. Cell. Med.* **11**: 981–1011.
89. Mimeault M. and S.K. Batra (2006). 'Recent advances on multiple tumourigenic cascades involved in prostatic cancer progression and targeting therapies'. *Carcinogenesis.* **27**: 1–22.
90. Mimeault M. and S.K. Batra (2007). 'Interplay of distinct growth factors during epithelial-mesenchymal transition of cancer progenitor cells and molecular targeting as novel cancer therapies'. *Ann. Oncol.* **18**: 1605–1619.
91. Mimeault M. and S.K. Batra (2007). 'Functions of tumourigenic and migrating cancer progenitor cells in cancer progression and metastasis and their therapeutic implications'. *Cancer Metastasis Rev.* **26**: 203–214.
92. Hermann P.C., S.L. Huber, T. Herrler, A. Aicher, J.W. Ellwart, M. Guba, C.J. Bruns and C. Heeschen (2007). 'Distinct populations of cancer stem cells determine tumour growth and metastatic activity in human pancreatic cancer'. *Cell Stem Cell.* **13**: 313–323.
93. Hope K.J., L. Jin and J.E. Dick (2004). 'Acute myeloid leukemia originates from a hierarchy of leukemic stem cell classes that differ in self-renewal capacity'. *Nat. Immunol.* **5**: 738–743.
94. Matsui W., C.A. Huff, Q. Wang, M.T. Malehorn, J. Barber, Y. Tanhehco, B.D. Smith, C.I. Civin and R.J. Jones (2004). 'Characterisation of clonogenic multiple myeloma cells'. *Blood.* **103**: 2332–2336.
95. Singh S.K., C. Hawkins, I.D. Clarke, J.A. Squire, J. Bayani, T. Hide, R.M. Henkelman, M.D. Cusimano and P.B. Dirks (2004). 'Identification of human brain tumour initiating cells'. *Nature.* **432**: 396–401.
96. Ponti D., A. Costa, N. Zaffaroni, G. Pratesi, G. Petrangolini, D. Coradini, S. Pilotti, M.A. Pierotti and M.G. Daidone (2005). 'Isolation and *in vitro* propagation of tumourigenic breast cancer cells with stem/progenitor cell properties'. *Cancer Res.* **65**: 5506–5511.

97. Fang D., T.K. Nguyen, K. Leishear, R. Finko, A.N. Kulp, S. Hotz, P.A. Van Belle, X. Xu, D.E. Elder and M. Herlyn (2005). 'A tumourigenic subpopulation with stem cell properties in melanomas'. *Cancer Res.* **65**: 9328–9337.
98. Wright M.H., A.M. Calcagno, C.D. Salcido, M.D. Carlson, S.V. Ambudkar and L. Varticovski (2008). 'Brca1 breast tumours contain distinct CD44+/CD24- and CD133+ cells with cancer stem cell characteristics'. *Breast Cancer Res.* **10**: R10.
99. Eramo A., F. Lotti, G. Sette, E. Pilozzi, M. Biffoni, V.A. Di, C. Conticello, L. Ruco, C. Peschle and R. De Maria (2008). 'Identification and expansion of the tumourigenic lung cancer stem cell population'. *Cell Death. Differ.* **15**: 504–514.
100. Yang Z.F., D.W. Ho, M.N. Ng, C.K. Lau, W.C. Yu, P. Ngai, P.W. Chu, C.T. Lam, R.T. Poon and S.T. Fan (2008). 'Significance of CD90(+) cancer stem cells in human liver cancer'. *Cancer Cell.* **13**: 153–166.
101. Ginestier C., M.H. Hur, E. Charafe-Jauffret, F. Monville, J. Dutcher, M. Brown, J. Jacquemier, P. Viens, C.G. Kleer, S. Liu et al. (2007). 'ALDH1 Is a marker of normal and malignant human mammary stem cells and a predictor of poor clinical outcome'. *Cell Stem Cell.* **1**: 555–567.
102. Ricci-Vitiani L., D.G. Lombardi, E. Pilozzi, M. Biffoni, M. Todaro, C. Peschle and M.R. De (2007). 'Identification and expansion of human colon-cancer-initiating cells'. *Nature.* **445**: 111–115.
103. Bhatia M., D. Bonnet, B. Murdoch, O.I. Gan and J.E. Dick (1998). 'A newly discovered class of human haematopoietic cells with SCID-repopulating activity'. *Nat. Med.* **4**: 1038–1045.
104. Katoh M. (2007). 'Networking of WNT, FGF, Notch, BMP and Hedgehog signaling pathways during carcinogenesis'. *Stem Cell Rev.* **3**: 30–38.
105. Dean M., T. Fojo and S. Bates (2005). 'Tumour stem cells and drug resistance'. *Nat. Rev. Cancer.* **5**: 275–284.
106. Mimeault M. and S.K. Batra (2008). 'Targeting of cancer stem/progenitor cells plus stem cell-based therapies: an ultimate hope for treating and curing the aggressive and recurrent cancers'. *Panminerva Medica.* **50**: 3–18.
107. Arai F. and T. Suda (2007). 'Maintenance of quiescent haematopoietic stem cells in the osteoblastic niche'. *Ann. N. Y. Acad. Sci.* **1106**: 41–53.
108. Christ O., K. Lucke, S. Imren, K. Leung, M. Hamilton, A. Eaves, C. Smith and C. Eaves (2007). 'Improved purification of haematopoietic stem cells based on their elevated aldehyde dehydrogenase activity'. *Haematologica.* **92**: 1165–1172.
109. Pearce D.J. and D. Bonnet (2007). 'The combined use of Hoechst efflux ability and aldehyde dehydrogenase activity to identify murine and human haematopoietic stem cells'. *Exp. Haematol.* **35**: 1437–1446.
110. Gentry T., S. Foster, L. Winstead, E. Deibert, M. Fiordalisi and A. Balber (2007). 'Simultaneous isolation of human BM haematopoietic, endothelial and mesenchymal progenitor cells by flow sorting based on aldehyde dehydrogenase activity: implications for cell therapy'. *Cytotherapy.* **9**: 259–274.
111. Dor F.J., M.L. Ramirez, K. Parmar, E.L. Altman, C.A. Huang, J.D. Down and D.K. Cooper (2006). 'Primitive haematopoietic cell populations reside in the spleen: studies in the pig, baboon and human'. *Exp. Haematol.* **34**: 1573–1582.
112. Yoshikawa T., H. Mitsuno, I. Nonaka, Y. Sen, K. Kawanishi, Y. Inada, Y. Takakura, K. Okuchi and A. Nonomura (2008). 'Wound therapy by marrow mesenchymal cell transplantation. Plast'. *Reconstr. Surg.* **121**: 860–877.

113. Fuchs E., T. Tumbar and G. Guasch (2004). 'Socializing with the neighbors: stem cells and their niche'. *Cell.* **116**: 769–778.
114. Kurzrock E.A., D.K. Lieu, L.A. Degraffenried, C.W. Chan and R.R. Isseroff (2008). 'Label-retaining cells of the bladder: candidate urothelial stem cells'. *Am. J. Physiol. Renal Physiol.* **294**: F1415–F1421.
115. Teng C., Y. Guo, H. Zhang, H. Zhang, M.Ding and H. Deng (2007). 'Identification and characterisation of label-retaining cells in mouse pancreas'. *Differentiation.* **75**: 702–712.
116. Urbanek K., F. Quaini, G. Tasca, D. Torella, C. Castaldo, B. Nadal-Ginard, A. Leri, J. Kajstura, E. Quaini and P. Anversa (2003). 'Intense myocyte formation from cardiac stem cells in human cardiac hypertrophy'. *Proc. Natl. Acad. Sci. USA.* **100**: 10440–10445.
117. Blasco M.A. (2005). 'Telomeres and human disease: ageing, cancer and beyond'. *Nat. Rev. Genet.* **6**: 611–622.
118. Lansdorp P.M. (2005). 'Major cutbacks at chromosome ends'. *Trends Biochem. Sci.* **30**: 388–395.
119. Deng Y. and S. Chang (2007). 'Role of telomeres and telomerase in genomic instability, senescence and cancer'. *Lab. Invest.* **87**: 1071–1076.
120. Shay J.W. and W.E. Wright (2007). 'Hallmarks of telomeres in ageing research'. *J. Pathol.* **211**: 114–123.
121. Sarin K.Y., P. Cheung, D. Gilison, E. Lee, R.I. Tennen, E. Wang, M.K. Artandi, A.E. Oro and S.E. Artandi (2005). 'Conditional telomerase induction causes proliferation of hair follicle stem cells'. *Nature.* **436**: 1048–1052.
122. Flores I., M.L. Cayuela and M.A. Blasco (2005). 'Effects of telomerase and telomere length on epidermal stem cell behaviour'. *Science.* **309**: 1253–1256.
123. Choi J., L.K. Southworth, K.Y. Sarin, A.S. Venteicher, W. Ma, W. Chang, P. Cheung, S. Jun, Artandi M.K., N. Shah et al. (2008). 'TERT promotes epithelial proliferation through transcriptional control of a Myc- and Wnt-related developmental program'. *PLoS. Genet.* **4**: e10.
124. Rajaraman S., J. Choi, P. Cheung, V. Beaudry, H. Moore and S.E. Artandi (2007). 'Telomere uncapping in progenitor cells with critical telomere shortening is coupled to S-phase progression in vivo'. *Proc. Natl. Acad. Sci. USA.* **104**: 17747–17752.
125. Yoon J., S.C. Choi, C.Y. Park, J.H. Choi, Y.I. Kim, W.J. Shim and D.S. Lim (2008). 'Bone marrow-derived side population cells are capable of functional cardiomyogenic differentiation'. *Mol. Cells.* **25**: 216–223.
126. Mimeault M. and S. Batra (2008). 'Characterisation of non-malignant and malignant prostatic stem/progenitor cells by Hoecsht side population method'. *Methods Mol. Biol.* (in press)
127. Krause D.S. (2008). 'Bone marrow-derived cells and stem cells in lung repair'. *Proc. Am. Thorac. Soc.* **5**: 323–327.
128. Heissig B., Y. Ohki, Y. Sato, S. Rafii, Z. Werb and K. Hattori (2005). 'A role for niches in haematopoietic cell development'. *Haematology.* **10**: 247–253.
129. Barnes D.E. and T. Lindahl (2004). 'Repair and genetic consequences of endogenous DNA base damage in mammalian cells'. *Ann. Rev. Genet.* **38**: 445–476.
130. Lindahl T. and D.E. Barnes (2000). 'Repair of endogenous DNA damage. Cold Spring Harb'. *Symp. Quant. Biol.* **65**: 127–133.
131. O'Driscoll M. and P.A. Jeggo (2008). 'The role of the DNA damage response pathways in brain development and microcephaly: insight from human disorders. *DNA Repair (Amst).* **7**: 1039–1050.
132. Hinkal G. and L.A. Donehower (2008). 'How does suppression of IGF-1 signaling by DNA damage affect aging and longevity?' *Mech. Ageing Dev.* **129**: 243–253.

133. Mattson M.P., S.L. Chan and W. Duan (2002). 'Modification of brain aging and neurodegenerative disorders by genes, diet and behaviour'. *Physiol. Rev.* **82**: 637–672.
134. Calado R.T. and N.S. Young (2008). 'Telomere maintenance and human bone marrow failure'. *Blood.* **111**: 4446–4455.
135. Young N.S., P. Scheinberg and R.T. Calado (2008). 'Aplastic anemia. Curr. Opin'. *Haematol.* **15**: 162–168.
136. Nijnik A., L. Woodbine, C. Marchetti, S. Dawson, T. Lambe, C. Liu, N.P. Rodrigues, T.L. Crockford, E. Cabuy, A. Vindigni et al. (2007). 'DNA repair is limiting for haematopoietic stem cells during ageing'. *Nature.* **447**: 686–690.
137. Pekcec A., W. Baumgartner, J.P. Bankstahl, V.M. Stein and H. Potschka (2008). 'Effect of aging on neurogenesis in the canine brain'. *Aging Cell.* **7**: 368–374.
138. Lansdorp P.M. (2008). 'Telomeres, stem cells and haematology'. *Blood.* **111**: 1759–1766.
139. Shin J.S., A. Hong, M.J. Solomon and C.S. Lee (2006). 'The role of telomeres and telomerase in the pathology of human cancer and aging'. *Pathology.* **38**: 103–113.
140. Gopinath S.D. and T.A. Rando (2008). 'Aging of the skeletal muscle stem cell niche'. *Aging Cell.* **7**(4): 590–598.
141. Zhou Z., A. Flesken-Nikitin and A.Y. Nikitin (2007). 'Prostate cancer associated with p53 and Rb deficiency arises from the stem/progenitor cell-enriched proximal region of prostatic ducts'. *Cancer Res.* **67**: 5683–5690.
142. Samani N.J., R. Boultby, R. Butler, J.R. Thompson and A.H. Goodall (2001). 'Telomere shortening in atherosclerosis'. *Lancet.* **358**: 472–473.
143. Tsakiri K.D., J.T. Cronkhite, P.J. Kuan, C. Xing, G. Raghu, J.C. Weissler, R.L. Rosenblatt, J.W. Shay and C.K. Garcia (2007). 'Adult-onset pulmonary fibrosis caused by mutations in telomerase'. *Proc. Natl. Acad. Sci. USA.* **104**: 7552–7557.
144. Armanios M.Y., J.J. Chen, J.D. Cogan, J.K. Alder, R.G. Ingersoll, C. Markin, W.E. Lawson, M. Xie, I. Vulto, J.A. PhillipsIII, et al. (2007). 'Telomerase mutations in families with idiopathic pulmonary fibrosis'. *N. Engl. J. Med.* **356**: 1317–1326.
145. Dimmeler S. and A. Leri (2008). 'Aging and disease as modifiers of efficacy of cell therapy'. *Circ. Res.* **102**: 1319–1330.
146. Ballard V.L. and J.M. Edelberg (2008). 'Stem cells for cardiovascular repair-The challenges of the aging heart'. *J. Mol. Cell Cardiol.* **45**(4): 582–92.
147. Cheng A., K. Shin-ya, R. Wan, S.C. Tang, T. Miura, H. Tang, R. Khatri, M. Gleichman, X. Ouyang, D. Liu et al. (2007). 'Telomere protection mechanisms change during neurogenesis and neuronal maturation: newly generated neurons are hypersensitive to telomere and DNA damage'. *J. Neurosci.* **27**: 3722–3733.
148. Janzen V., R. Forkert, H.E. Fleming, Y. Saito, M.T. Waring, D.M. Dombkowski, T. Cheng, R.A. DePinho, N.E. Sharpless and D.T. Scadden (2006). 'Stem-cell ageing modified by the cyclin-dependent kinase inhibitor p16INK4a'. *Nature.* **443**: 421–426.
149. Sherr C.J. and F. McCormick (2002). 'The RB and p53 pathways in cancer. *Cancer Cell* **2**: 103–112.
150. Shiras A., S.T. Chettiar, V. Shepal, G. Rajendran, G.R. Prasad and P. Shastry (2007). 'Spontaneous transformation of human adult non-tumourigenic stem cells to cancer stem cells is driven by genomic instability in a human model of glioblastoma'. *Stem Cells.* **25**: 1478–1489.

151. Campisi J. (2005). 'Senescent cells, tumour suppression and organismal aging: good citizens, bad neighbors'. *Cell.* **120**: 513–522.

152. Shay J.W. and W.E. Wright (2002). 'Telomerase: a target for cancer therapeutics'. *Cancer Cell.* **2**: 257–265.

153. Meeker A.K., J.L. Hicks, C.A. Iacobuzio-Donahue, E.A. Montgomery, W.H. Westra, T.Y. Chan, B.M. Ronnett and A.M. De Marzo (2004). 'Telomere length abnormalities occur early in the initiation of epithelial carcinogenesis'. *Clin. Cancer Res.* **10**: 3317–3326.

154. Rudolph K.L., M. Millard, M.W. Bosenberg and R.A. DePinho (2001). 'Telomere dysfunction and evolution of intestinal carcinoma in mice and humans'. *Nat. Genet.* **28**: 155–159.

155. de Jonge-Peeters S.D., F. Kuipers, E.G. de Vries and E. Vellenga (2007). 'ABC transporter expression in haematopoietic stem cells and the role in AML drug resistance'. *Crit. Rev. Oncol. Haematol.* **62**: 214–226.

156. Milas L., U. Raju, Z. Liao and J. Ajani (2005). 'Targeting molecular determinants of tumour chaemo-radioresistance'. *Semin. Oncol.* **32**: S78–S81.

157. Neiva K., Y.X. Sun and R.S. Taichman (2005). 'The role of osteoblasts in regulating haematopoietic stem cell activity and tumour metastasis'. *Braz. J. Med. Biol. Res.* **38**: 1449–1454.

158. Kiel M.J., O.H. Yilmaz, T. Iwashita, O.H. Yilmaz, C. Terhorst and S.J. Morrison (2005). 'SLAM family receptors distinguish haematopoietic stem and progenitor cells and reveal endothelial niches for stem cells'. *Cell.* **121**: 1109–1121.

159. Six E.M., D. Bonhomme, M. Monteiro, K. Beldjord, M. Jurkowska, C. Cordier-Garcia, A. Garrigue, C.L. Dal, B. Rocha, A. Fischer et al. (2007). 'A human postnatal lymphoid progenitor capable of circulating and seeding the thymus'. *J. Exp. Med.* **204**: 3085–3093.

160. Ringden O. (2007). 'Immunotherapy by allogeneic stem cell transplantation'. *Adv. Cancer Res.* **97C**: 25–60.

161. De Clercq E. (2005). 'Potential clinical applications of the CXCR4 antagonist bicyclam AMD3100'. *Mini. Rev. Med. Chem.* **5**: 805–824.

162. Burger J.A., A. Spoo, A. Dwenger, M. Burger and D. Behringer (2003). 'CXCR4 chaemokine receptors (CD184) and alpha4beta1 integrins mediate spontaneous migration of human CD34+ progenitors and acute myeloid leukaemia cells beneath marrow stromal cells (pseudoemperipolesis)'. *Br. J. Haematol.* **122**: 579–589.

163. Hale G.A. (2005). 'Autologous haematopoietic stem cell transplantation for pediatric solid tumours'. *Expert Rev. Anticancer Ther.* **5**: 835–846.

164. Bregni M., M. Bernardi, F. Ciceri and J. Peccatori (2004). 'Allogeneic stem cell transplantation for the treatment of advanced solid tumours'. *Springer Semin. Immunopathol.* **26**: 95–108.

165. Vogel W., H.G. Kopp, L. Kanz and H. Einsele (2005). 'Myeloma cell contamination of peripheral blood stem-cell grafts can predict the outcome in multiple myeloma patients after high-dose chaemotherapy and autologous stem-cell transplantation'. *J. Cancer Res. Clin. Oncol.* **131**: 214–218.

166. Petersen S.L. (2007). 'Alloreactivity as therapeutic principle in the treatment of haematologic malignancies. Studies of clinical and immunologic aspects of allogeneic haematopoietic cell transplantation with non-myeloablative conditioning'. *Dan. Med. Bull.* **54**: 112–139.

167. Hausermann P., R.B. Walter, J. Halter, Biedermann B.C., Tichelli A., P. Itin and A. Gratwohl (2008). 'Cutaneous graft-versus-host disease: a guide for the dermatologist'. *Dermatology.* **216**: 287–304.

168. Aoudjhane M., M. Labopin, N.C. Gorin, A. Shimoni, T. Ruutu, H.J. Kolb, F. Frassoni, J.M. Boiron, J.L. Yin, J. Finke et al. (2005). 'Comparative outcome of reduced intensity and myeloablative conditioning regimen in HLA identical sibling allogeneic haematopoietic stem cell transplantation for patients older than 50 years of age with acute myeloblastic leukaemia: a retrospective survey from the Acute Leukemia Working Party (ALWP) of the European group for Blood and Marrow Transplantation (EBMT)'. *Leukemia.* **19**: 2304–2312.
169. Oyan B., Y. Koc and E. Kansu (2005). 'Successful salvage with high-dose sequential chaemotherapy coupled with *in vivo* purging and autologous stem cell transplantation in 2 patients with primary refractory mantle cell lymphoma presenting in the leukemic phase'. *Int. J. Haematol.* **81**: 155–158.
170. Krause D.S., K. Lazarides, von U.H. Andrian and R.A. Van Etten (2006). 'Requirement for CD44 in homing and engraftment of BCR-ABL-expressing leukemic stem cells'. *Nat. Med.* **12**: 1175–1180.
171. Bonanno G., A. Perillo, S. Rutella, D.G. De Ritis, A. Mariotti, M. Marone, F. Meoni, G. Scambia, G. Leone, S. Mancuso et al. (2004). 'Clinical isolation and functional characterisation of cord blood CD133+ haematopoietic progenitor cells'. *Transfusion.* **44**: 1087–1097.
172. Rollini P., S. Kaiser, H.E. Faes-van't, U. Kapp and S. Leyvraz (2004). 'Long-term expansion of transplantable human foetal liver haematopoietic stem cells'. *Blood.* **103**: 1166–1170.
173. Weiss M.L. and D.L. Troyer (2006). 'Stem cells in the umbilical cord'. *Stem Cell Rev* **2**: 155–162.
174. Delorme B., S. Chateauvieux and P. Charbord (2006). 'The concept of mesenchymal stem cells'. *Regen. Med.* **1**: 497–509.
175. da Silva M.L., P.C. Chagastelles and N.B. Nardi (2006). 'Mesenchymal stem cells reside in virtually all post-natal organs and tissues'. *J. Cell Sci.* **119**: 2204–2213.
176. Zengin E., F. Chalajour, U.M. Gehling, W.D. Ito, H. Treede, H. Lauke, J. Weil, H. Reichenspurner, N. Kilic and S. Ergun (2006). 'Vascular wall resident progenitor cells: a source for postnatal vasculogenesis'. *Development.* **133**: 1543–1551.
177. Hida N., N. Nishiyama, S. Miyoshi, S. Kira, K. Segawa, T. Uyama, T. Mori, K. Miyado, Y. Ikegami, C. Cui et al. (2008). 'Novel cardiac precursor-like cells from human menstrual blood-derived mesenchymal cells'. *Stem Cells.* **26**(7): 1695–704.
178. Bernardo M.E., J.A. Emons, M. Karperien, A.J. Nauta, R. Willemze, H. Roelofs, S. Romeo, A. Marchini, G.A. Rappold, S. Vukicevic et al. (2007). 'Human mesenchymal stem cells derived from bone marrow display a better chondrogenic differentiation compared with other sources. Connect'. *Tissue Res.* **48**: 132–140.
179. Gindraux F., Z. Selmani, L. Obert, S. Davani, P. Tiberghien, P. Herve and F. Deschaseaux (2007). ' Human and rodent bone marrow mesenchymal stem cells that express primitive stem cell markers can be directly enriched by using the CD49a molecule'. *Cell Tissue Res.* **327**: 471–483.
180. Martin J., K. Helm, P. Ruegg, M. Varella-Garcia, E. Burnham and S. Majka (2008). 'Adult lung side population cells have mesenchymal stem cell potential'. *Cytotherapy.* **10**: 140–151.
181. Chang Y.J., D.T. Shih, C.P. Tseng, T.B. Hsieh, D.C. Lee and S.M. Hwang (2006). 'Disparate mesenchyme-lineage tendencies in mesenchymal stem cells from human bone marrow and umbilical cord blood'. *Stem Cells.* **24**: 679–685.
182. Hattan N., H. Kawaguchi, K. Ando, E. Kuwabara, J. Fujita, M. Murata, M. Suematsu, H. Mori and K. Fukuda (2005). 'Purified cardiomyocytes from bone marrow mesenchymal stem cells produce stable intracardiac grafts in mice'. *Cardiovasc. Res.* **65**: 334–344.

183. Kajstura J., M. Rota, B. Whang, S. Cascapera, T. Hosoda, C. Bearzi, D. Nurzynska, H. Kasahara, E. Zias, M. Bonafe et al. (2005). 'Bone marrow cells differentiate in cardiac cell lineages after infarction independently of cell fusion'. *Circ. Res.* **96**: 127–137.

184. Ma Y., Y. Xu, Z. Xiao, W. Yang, C. Zhang, E. Song, Y. Du and L. Li (2006). 'Reconstruction of chemically burned rat corneal surface by bone marrow-derived human mesenchymal stem cells'. *Stem Cells.* **24**: 315–321.

185. Ookura N., Y. Fujimori, K. Nishioka, S. Kai, H. Hara and H. Ogawa (2007). 'Adipocyte differentiation of human marrow mesenchymal stem cells reduces the supporting capacity for haematopoietic progenitors but not for severe combined immunodeficiency repopulating cells'. *Int. J. Mol. Med.* **19**: 387–392.

186. Bobis S., D. Jarocha and M. Majka (2006). 'Mesenchymal stem cells: characteristics and clinical applications. Folia Histochem'. *Cytobiol.* **44**: 215–230.

187. Sun Y., L. Chen, X.G. Hou, W.K. Hou, J.J. Dong, L. Sun, K.X. Tang, B. Wang, J. Song, H. Li et al. (2007). 'Differentiation of bone marrow-derived mesenchymal stem cells from diabetic patients into insulin-producing cells *in vitro*'. *Chin Med. J. (Engl.).* **120**: 771–776.

188. Balbarini A., M.C. Barsotti, S.R. Di, A. Leone and T. Santoni (2007). 'Circulating endothelial progenitor cells characterisation, function and relationship with cardiovascular risk factors'. *Curr. Pharm. Des.* **13**: 1699–1713.

189. Rafii S., D. Lyden, R. Benezra, K. Hattori and B. Heissig (2002). 'Vascular and haematopoietic stem cells: novel targets for anti-angiogenesis therapy?' *Nat. Rev. Cancer.* **2**: 826–835.

190. Peters B.A., L.A. Diaz, K. Polyak, L. Meszler, K. Romans, E.C. Guinan, J.H. Antin, D. Myerson, S.R. Hamilton, B. Vogelstein et al. (2005). 'Contribution of bone marrow-derived endothelial cells to human tumour vasculature'. *Nat. Med.* **11**: 261–262.

191. Kung E.F., F. Wang and J.S. Schechner (2008). '*In vivo* perfusion of human skin substitutes with microvessels formed by adult circulating endothelial progenitor cells'. *Dermatol. Surg.* **34**: 137–146.

192. Wagner W., F. Wein, A. Seckinger, M. Frankhauser, U. Wirkner, U. Krause, J. Blake, C. Schwager, V. Eckstein, W. Ansorge et al. (2005). 'Comparative characteristics of mesenchymal stem cells from human bone marrow, adipose tissue and umbilical cord blood'. *Exp. Haematol.* **33**: 1402–1416.

193. Le Blanc K. and O. Ringden (2007). 'Immunomodulation by mesenchymal stem cells and clinical experience'. *J. Intern. Med.* **262**: 509–525.

194. Hoogduijn M.J., M.J. Crop, A.M. Peeters, G.J. Van Osch, A.H. Balk, J.N. Ijzermans, W. Weimar and C.C. Baan (2007). 'Human heart, spleen and perirenal fat-derived mesenchymal stem cells have immunomodulatory capacities'. *Stem Cells Dev.* **16**: 597–604.

195. Shanti R.M., W.J. Li, L.J. Nesti, X. Wang and R.S. Tuan (2007). 'Adult mesenchymal stem cells: biological properties, characteristics and applications in maxillofacial surgery'. *J. Oral Maxillofac. Surg.* **65**: 1640–1647.

196. Wang C.H., W.J. Cherng, N.I. Yang, L.T. Kuo, C.M. Hsu, H.I. Yeh, Y.J. Lan, C.H. Yeh and W.L. Stanford (2007). 'Late-outgrowth endothelial cells attenuate ntimal hyperplasia contributed by mesenchymal stem cells after vascular injury. Arterioscler'. *Thromb. Vasc. Biol.* **28**: 54–60.

197. Anton R., M. Kuhl and P. Pandur (2007). 'A molecular signature for the "master" heart cell'. *Bioessays.* **29**: 422–426.

198. Bearzi C., M. Rota, T. Hosoda, J. Tillmanns, A. Nascimbene, A.A. De, S. Yasuzawa-Amano, I. Trofimova, R.W. Siggins, N. Lecapitaine et al. (2007). 'Human cardiac stem cells'. *Proc. Natl. Acad. Sci. USA.* **104**: 14068–14073.

199. Scobioala S., R. Klocke, M. Kuhlmann, W. Tian, L. Hasib, H. Milting, S. Koenig, M. Stelljes, A. El-Banayosy, G. Tenderich et al. (2007). 'Up-regulation of nestin in the infarcted myocardium potentially indicates differentiation of resident cardiac stem cells into various lineages including cardiomyocytes'. *FASEB J.* **22**: 1021–1031
200. Urbanek K., M. Rota, S. Cascapera, C. Bearzi, A. Nascimbene, A.A. De, T. Hosoda, S. Chimenti, M. Baker, F. Limana et al. (2005). 'Cardiac stem cells possess growth factor-receptor systems that after activation regenerate the infarcted myocardium, improving ventricular function and long-term survival'. *Circ. Res.* **97**: 663–673.
201. Parmacek M.S. (2006). 'Cardiac stem cells and progenitors: developmental biology and therapeutic challenges'. *Trans. Am. Clin Climatol. Assoc.* **117**: 239–256.
202. Moretti A., L. Caron, A. Nakano, J.T. Lam, A. Bernshausen, Y. Chen, Y. Qyang, L. Bu, M. Sasaki, S. Martin-Puig et al. (2006). 'Multipotent embryonic isl1+ progenitor cells lead to cardiac, smooth muscle and endothelial cell diversification'. *Cell.* **127**: 1151–1165.
203. Dawn B., A.B. Stein, K. Urbanek, M. Rota, B. Whang, R. Rastaldo, D. Torella, X.L. Tang, A. Rezazadeh, J. Kajstura et al. (2005). 'Cardiac stem cells delivered intravascularly traverse the vessel barrier, regenerate infarcted myocardium and improve cardiac function'. *Proc. Natl. Acad. Sci. USA.* **102**: 3766–3771.
204. Christoforou N. and J.D. Gearhart (2007). 'Stem cells and their potential in cell-based cardiac therapies. Prog. Cardiovasc'. *Dis.* **49**: 396–413.
205. Sohn R.L., M. Jain and R. Liao (2007). 'Adult stem cells and heart regeneration. Expert. Rev Cardiovasc'. *Ther.* **5**: 507–517.
206. Roccio M., M.J. Goumans, J.P. Sluijter and P.A. Doevendans (2008). 'Stem cell sources for cardiac regeneration'. *Panminerva Med.* **50**: 19–30.
207. Haider H.K. and M. Ashraf (2008). 'Strategies to promote donor cell survival: combining preconditioning approach with stem cell transplantation'. *J. Mol. Cell Cardiol.* **45**(4): 554–566
208. Haider H.K., I. Elmadbouh, M. Jean-Baptiste and M. Ashraf (2008). 'Non-viral vector gene modification of stem cells for myocardial repair'. *Mol. Med.* **14**: 79–86.
209. Bonanno G., A. Mariotti, A. Procoli, M. Corallo, S. Rutella, G. Pessina, G. Scambia, S. Mancuso and L. Pierelli (2007). 'Human cord blood CD133+ cells immunoselected by a clinical-grade apparatus differentiate *in vitro* into endothelial- and cardiomyocyte-like cells'. *Transfusion.* **47**: 280–289.
210. Behfar A., C. Perez-Terzic, R.S. Faustino, D.K. Arrell, D.M. Hodgson, S. Yamada, M. Puceat, N. Niederlander, A.E. Alekseev, L.V. Zingman et al. (2007). 'Cardiopoietic programming of embryonic stem cells for tumour-free heart repair'. *J. Exp. Med.* **204**: 405–420.
211. Yamada Y., S. Yokoyama, X.D. Wang, N. Fukuda and N. Takakura (2007). 'Cardiac stem cells in brown adipose tissue express CD133 and induce bone marrow non-haematopoietic cells to differentiate into cardiomyocytes'. *Stem Cells.* **25**: 1326–1333.
212. van de Ven C., D. Collins, M.B. Bradley, E. Morris and M.S. Cairo (2007). 'The potential of umbilical cord blood multipotent stem cells for non-haematopoietic tissue and cell regeneration'. *Exp. Haematol.* **35**: 1753–1765.
213. Caspi O., I. Huber, I. Kehat, M. Habib, G. Arbel, A. Gepstein, L. Yankelson, D. Aronson, R. Beyar and L. Gepstein (2007). 'Transplantation of human embryonic stem cell-derived cardiomyocytes improves myocardial performance in infarcted rat hearts'. *J. Am. Coll. Cardiol.* **50**: 1884–1893.

214. Ahmadi H., H. Baharvand, S.K. Ashtiani, M. Soleimani, H. Sadeghian, J.M. Ardekani, N.Z. Mehrjerdi, A. Kouhkan, M. Namiri, M. Madani-Civi et al. (2007). 'Safety analysis and improved cardiac function following local autologous transplantation of CD133(+) enriched bone marrow cells after myocardial infarction'. *Curr. Neurovasc. Res.* **4**: 153–160.

215. Braga L.M., K. Rosa, B. Rodrigues, C. Malfitano, M. Camassola, P. Chagastelles, S. Lacchini, P. Fiorino, A.K. De, S.B. D'Agord et al. (2008). 'Systemic delivery of adult stem cells improves cardiac function in spontaneously hypertensive rats'. *Clin. Exp. Pharmacol. Physiol.* **35**(2): 113–119.

216. Sadat S., S. Gehmert, Y.H. Song, Y. Yen, X. Bai, S. Gaiser, H. Klein and E. Alt (2007). 'The cardioprotective effect of mesenchymal stem cells is mediated by IGF-I and VEGF'. *Biochem. Biophys. Res. Commun.* **363**: 674–679.

217. Lindvall O., Z. Kokaia and A. Martinez-Serrano (2004). 'Stem cell therapy for human neurodegenerative disorders-how to make it work'. *Nat. Med.* **10**: S42–S50.

218. Ourednik J., V. Ourednik, W.P. Lynch, M. Schachner and E.Y. Snyder (2002). 'Neural stem cells display an inherent mechanism for rescuing dysfunctional neurons'. *Nat. Biotechnol.* **20**: 1103–1110.

219. Trounson A. (2006). 'The production and directed differentiation of human embryonic stem cells'. *Endocr. Rev.* **27**: 208–219.

220. Walton N.M., B.M. Sutter, H.X. Chen, L.J. Chang, S.N. Roper, B. Scheffler and D.A. Steindler (2006). 'Derivation and large-scale expansion of multipotent astroglial neural progenitors from adult human brain'. *Development.* **133**: 3671–3681.

221. Pardal R., P. Ortega-Saenz, R. Duran and J. Lopez-Barneo (2007). 'Glia-like stem cells sustain physiologic neurogenesis in the adult mammalian carotid body'. *Cell.* **131**: 364–377.

222. Ramaswamy S., K.M. Shannon and J.H. Kordower (2007). 'Huntington's disease: pathological mechanisms and therapeutic strategies'. *Cell Transplant.* **16**: 301–312.

223. Shetty A.K. and B. Hattiangady (2007). 'Concise review: prospects of stem cell therapy for temporal lobe epilepsy'. *Stem Cells.* **25**: 2396–2407.

224. Eftekharpour E., S. Karimi-Abdolrezaee, J. Wang, B.H. El, C. Morshead and M.G. Fehlings (2007). 'Myelination of congenitally dysmyelinated spinal cord axons by adult neural precursor cells results in formation of nodes of Ranvier and improved axonal conduction'. *J. Neurosci.* **27**: 3416–3428.

225. Roy N.S., C. Cleren, S.K. Singh, L. Yang, M.F. Beal and S.A. Goldman (2006). 'Functional engraftment of human ES cell-derived dopaminergic neurons enriched by coculture with telomerase-immortalized midbrain astrocytes'. *Nat. Med.* **12**: 1259–1268.

226. Rodriguez-Gomez J.A., J.Q. Lu, I. Velasco, S. Rivera, S.S. Zoghbi, J.S. Liow, J.L. Musachio, F.T. Chin, H. Toyama, J. Seidel et al. (2007). 'Persistent dopamine functions of neurons derived from embryonic stem cells in a rodent model of Parkinson disease'. *Stem Cells.* **25**: 918–928.

227. Geeta R., R.L. Ramnath, H.S. Rao and V. Chandra (2008). 'One year survival and significant reversal of motor deficits in parkinsonian rats transplanted with hESC derived dopaminergic neurons. Biochem. Biophys'. *Res. Commun.* **373**(2): 258–64.

228. Lee M.W., Y.J. Moon, M.S. Yang, S.K. Kim, I.K. Jang, Y.W. Eom, J.S. Park, H.C. Kim, K.Y. Song, S.C. Park et al. (2007). 'Neural differentiation of novel multipotent progenitor cells from cryopreserved human umbilical cord blood'. *Biochem. Biophys. Res. Commun.* **358**: 637–643.

229. Ning H., G. Lin, T.F. Lue and C.S. Lin (2006). 'Neuron-like differentiation of adipose tissue-derived stromal cells and vascular smooth muscle cells'. *Differentiation.* **74**: 510–518.

230. Mimeault M., D. Bonenfant and S.K. Batra (2004). 'New advances on the functions of epidermal growth factor receptor and ceramides in skin cell differentiation, disorders and cancers. Skin Pharmacol'. *Physiol.* **17**: 153–166.
231. Fuchs E. and V. Horsley (2008). 'More than one way to skin'. *Genes Dev.* **22**: 976–985.
232. Watt F.M. (2002). 'Role of integrins in regulating epidermal adhesion, growth and differentiation'. *EMBO J.* **21**: 3919–3926.
233. Le Douarin N.M., G.W. Calloni and E. Dupin (2008). 'The stem cells of the neural crest'. *Cell Cycle.* **7**.
234. Jones P.H., B.D. Simons and F.M. Watt (2007). 'Sic transit gloria: farewell to the epidermal transit amplifying cell?' *Cell Stem Cell.* **1**: 371–381.
235. Senoo M., F. Pinto, C.P. Crum and F. McKeon (2007). 'p63 Is essential for the proliferative potential of stem cells in stratified epithelia'. *Cell.* **129**: 523–536.
236. Pasonen-Seppanen S., S. Karvinen, K. Torronen, J.M. Hyttinen, T. Jokela, M.J. Lammi, M.I. Tammi and R. Tammi (2003). 'EGF upregulates, whereas TGF-beta downregulates, the hyaluronan synthases Has2 and Has3 in organotypic keratinocyte cultures: correlations with epidermal proliferation and differentiation. J. Invest'. *Dermatol.* **120**: 1038–1044.
237. Jensen K.B. and F.M. Watt (2006). 'Single-cell expression profiling of human epidermal stem and transit-amplifying cells: Lrig1 is a regulator of stem cell quiescence'. *Proc. Natl. Acad. Sci. USA.* **103**: 11958–11963.
238. Nguyen B.C., K. Lefort, A. Mandinova, D. Antonini, V. Devgan, G.G. Della, M.I. Koster, Z. Zhang, J. Wang, d.V. Tommasi, et al. (2006). 'Cross-regulation between Notch and p63 in keratinocyte commitment to differentiation'. *Genes Dev.* **20**: 1028–1042.
239. Tumbar T., G. Guasch, V. Greco, C. Blanpain, W.E. Lowry, M. Rendl and E. Fuchs (2004). 'Defining the epithelial stem cell niche in skin'. *Science.* **303**: 359–363.
240. Millar S.E. (2005). 'An ideal society? Neighbors of diverse origins interact to create and maintain complex mini-organs in the skin'. *PLoS. Biol.* **3**: 1873–1877.
241. Levy V., C. Lindon, B.D. Harfe and B.A. Morgan (2005). 'Distinct stem cell populations regenerate the follicle and interfollicular epidermis'. *Dev. Cell.* **9**: 855–861.
242. Flores I., M.L. Cayuela and M.A. Blasco (2005). 'Effects of telomerase and telomere length on epidermal stem cell behaviour'. *Science.* **309**: 1253–1256.
243. Sarin K.Y., P. Cheung, D. Gilison, E. Lee, R.I. Tennen, E. Wang, M.K. Artandi, A.E. Oro and S.E. Artandi (2005). 'Conditional telomerase induction causes proliferation of hair follicle stem cells'. *Nature.* **436**: 1048–1052.
244. Blanpain C. and E. Fuchs (2006). 'Epidermal stem cells of the skin. Annu'. *Rev Cell Dev. Biol.* **22**: 339–373.
245. Plikus M.V., J.A. Mayer, C.D. de la, R.E. Baker, P.K. Maini, R. Maxson and C.M. Chuong (2008). 'Cyclic dermal BMP signalling regulates stem cell activation during hair regeneration'. *Nature.* **451**: 340–344.
246. Horsley V., A.O. Aliprantis, L. Polak, L.H. Glimcher and E. Fuchs (2008). 'NFATc1 balances quiescence and proliferation of skin stem cells'. *Cell.* **132**: 299–310.
247. Vidal V.P., M.C. Chaboissier, S. Lutzkendorf, G. Cotsarelis, P. Mill, C.C. Hui, N. Ortonne, J.P. Ortonne and A. Schedl (2005). 'Sox9 is essential for outer root sheath differentiation and the formation of the hair stem cell compartment'. *Curr. Biol.* **15**: 1340–1351.

248. Huelsken J., R. Vogel, B. Erdmann, G. Cotsarelis and W. Birchmeier (2001). 'beta-Catenin controls hair follicle morphogenesis and stem cell differentiation in the skin'. *Cell.* **105**: 533–545.

249. Yamamoto N., K. Tanigaki, H. Han, H. Hiai and T. Honjo (2003). 'Notch/RBP-J signaling regulates epidermis/hair fate determination of hair follicular stem cells'. *Curr. Biol.* **13**: 333–338.

250. Ghazizadeh S. and L.B. Taichman (2001). 'Multiple classes of stem cells in cutaneous epithelium: a lineage analysis of adult mouse skin'. *EMBO J.* **20**: 1215–1222.

251. Horsley V., D. O'Carroll, R. Tooze, Y. Ohinata, M. Saitou, T. Obukhanych, Nussenzweig M., A. Tarakhovsky and E. Fuchs (2006). 'Blimp1 defines a progenitor population that governs cellular input to the sebaceous gland'. *Cell.* **126**: 597–609.

252. Sieber-Blum M. and M. Grim (2004). 'The adult hair follicle: cradle for pluripotent neural crest stem cells'. *Birth Defects Res. C. Embryo Today.* **72**: 162–172.

253. Kawase Y., Y. Yanagi, T. Takato, M. Fujimoto and H. Okochi (2004). 'Characterisation of multipotent adult stem cells from the skin: transforming growth factor-beta (TGF-beta) facilitates cell growth. Exp'. *Cell Res.* **295**: 194–203.

254. Fernandes K.J., I.A. McKenzie, P. Mill, K.M. Smith, M. Akhavan, F. Barnabe-Heider, J. Biernaskie, A. Junek, N.R. Kobayashi, J.G. Toma et al. (2004). 'A dermal niche for multipotent adult skin-derived precursor cells'. *Nat. Cell Biol.* **6**: 1082–1093.

255. Malanchi I., H. Peinado, D. Kassen, T. Hussenet, D. Metzger, P. Chambon, M. Huber, D. Hohl, A. Cano, W. Birchmeier et al. (2008). 'Cutaneous cancer stem cell maintenance is dependent on beta-catenin signalling'. *Nature.* **452**: 650–653.

256. Bachmann I.M., H.E. Puntervoll, A.P. Otte and L.A. Akslen (2008). 'Loss of BMI-1 expression is associated with clinical progress of malignant melanoma'. *Mod. Pathol.* **21**: 583–590.

257. Trucco M. (2005). 'Regeneration of the pancreatic beta cell'. *J. Clin Invest.* **115**: 5–12.

258. Bouwens L. and I. Rooman (2005). 'Regulation of pancreatic beta-cell mass'. *Physiol. Rev.* **85**: 1255–1270.

259. Nikolova G., N. Jabs, I. Konstantinova, A. Domogatskaya, K. Tryggvason, L. Sorokin, R. Fassler, G. Gu, H.P. Gerber, N. Ferrara et al. (2006). 'The vascular basement membrane: a niche for insulin gene expression and Beta cell proliferation'. *Dev. Cell.* **10**: 397–405.

260. Zulewski H., E.J. Abraham, M.J. Gerlach, P.B. Daniel, W. Moritz, B. Muller, M. Vallejo, M.K. Thomas and J.F. Habener (2001). 'Multipotential nestin-positive stem cells isolated from adult pancreatic islets differentiate *ex vitro* into pancreatic endocrine, exocrine and hepatic phenotypes'. *Diabetes.* **50**: 521–533.

261. D'Alessandro J.S., K. Lu, B.P. Fung, A. Colman and D.L. Clarke (2007). 'Rapid and efficient *in vitro* generation of pancreatic islet progenitor cells from non-endocrine epithelial cells in the adult human pancreas. *Stem Cells Dev.* **16**: 75–89.

262. Lin H.T., S.H. Chiou, C.L. Kao, Y.M. Shyr, C.J. Hsu, Y.W. Tarng, L.L. Ho, C.F. Kwok and H.H. Ku (2006). 'Characterisation of pancreatic stem cells derived from adult human pancreas ducts by fluorescence activated cell sorting'. *World J. Gastroenterol.* **12**: 4529–4535.

263. Xu X., J. D'Hoker, G. Stange, S. Bonne, L.N.De, X. Xiao, C.M. Van De, G. Mellitzer, Z. Ling, D. Pipeleers et al. (2008). 'Beta cells can be generated from endogenous progenitors in injured adult mouse pancreas'. *Cell.* **132**: 197–207.

264. Halban P.A. (2004). 'Cellular sources of new pancreatic beta cells and therapeutic implications for regenerative medicine'. *Nat. Cell Biol.* **6**: 1021–1025.

265. Suen P.M. and P.S. Leung (2005). 'Pancreatic stem cells: a glimmer of hope for diabetes?' JOP. **6**: 422–424.
266. Lock L.T. and E.S. Tzanakakis (2007). 'Stem/Progenitor cell sources of insulin-producing cells for the treatment of diabetes'. *Tissue Eng.* **13**: 1399–1412.
267. Banerjee M., M Kanitkar . and R.R. Bhonde (2005). 'Approaches towards endogenous pancreatic regeneration. Rev Diabet'. *Stud.* **2**: 165–176.
268. Wu D.C., A.S. Byod and K.J. Wood (2007). 'Embryonic stem cell transplantation: potential applicability in cell replacement therapy and regenerative medicine'. *Front Biosci.* **12**: 4525–4535.
269. Ezquer F.E., M.E. Ezquer, D.B. Parrau, D. Carpio, A.J. Yanez and P.A. Conget (2008). 'Systemic administration of multipotent mesenchymal stromal cells reverts hyperglycemia and prevents nephropathy in type 1 diabetic mice'. *Biol. Blood Marrow Transplant.* **14**: 631–640.
270. Trivedi H.L., A.V. Vanikar, U. Thakker, A. Firoze, S.D. Dave, C.N. Patel, J.V. Patel, A.B. Bhargava and V. Shankar (2008). 'Human adipose tissue-derived mesenchymal stem cells combined with haematopoietic stem cell transplantation synthesize insulin'. *Transplant. Proc.* **40**: 1135–1139.
271. Omary M.B., A. Lugea, A.W. Lowe and S.J. Pandol (2007). 'The pancreatic stellate cell: a star on the rise in pancreatic diseases. *J. Clin Invest.* **117**: 50–59.
272. Mimeault M., R.E. Brand, A.A. Sasson and S.K. Batra (2005). 'Recent advances on the molecular mechanisms involved in pancreatic cancer progression and therapies'. *Pancreas.* **31**: 301–316.
273. Alison M.R., P. Vig, F. Russo, B.W. Bigger, E. Amofah, M. Themis and S. Forbes (2004). 'Hepatic stem cells: from inside and outside the liver?' *Cell Prolif.* **37**: 1–21.
274. Cantz T., M.P. Manns and M. Ott (2008). 'Stem cells in liver regeneration and therapy'. *Cell Tissue Res.* **331**(1): 271–282.
275. Burke Z.D., S. Thowfeequ, M. Peran and D. Tosh (2007). 'Stem cells in the adult pancreas and liver'. *Biochem. J.* **404**: 169–178.
276. Modlin I.M., M. Kidd, K.D. Lye and N.A. Wright (2003). 'Gastric stem cells: an update'. *Keio J. Med.* **52**: 134–137.
277. Barker N., van J.H. Es, J. Kuipers, P. Kujala, B.M. van den, M. Cozijnsen, A. Haegebarth, J. Korving, H. Begthel, P.J. Peters et al. (2007). 'Identification of stem cells in small intestine and colon by marker gene Lgr5'. *Nature.* **449**: 1003–1007.
278. Yen T.H. and N.A. Wright (2006). 'The gastrointestinal tract stem cell niche'. *Stem Cell Rev.* **2**: 203–212.
279. Chen Z., C.S. dePaiva, L. Luo, F.L. Kretzer, S.C. Pflugfelder and D.Q. Li (2004). 'Characterisation of putative stem cell phenotype in human limbal epithelia'. *Stem Cells.* **22**: 355–366.
280. Das A.V., J. James, J. Rahnenfuhrer, W.B. Thoreson, S. Bhattacharya, X. Zhao and I. Ahmad (2005). 'Retinal properties and potential of the adult mammalian ciliary epithelium stem cells'. *Vision Res.* **45**: 1653–1666.
281. Moshiri A., J. Close and T.A. Reh (2004). 'Retinal stem cells and regeneration'. *Int. J. Dev. Biol.* **48**: 1003–1014.
282. Khalil P.N., V. Weiler, P.J. Nelson, M.N. Khalil, S. Moosmann, W.E. Mutschler, M. Siebeck and R. Huss (2007). 'Non-myeloablative stem cell therapy enhances microcirculation and tissue regeneration in murine inflammatory bowel disease'. *Gastroenterology.* **132**: 944–954.
283. Latella G., C. Fiocchi and R. Caprilli (2007). 'Late-breaking news from the "4th International Meeting on Inflammatory Bowel Diseases" Capri, 2006. Inflamm'. *Bowel. Dis.* **13**: 1031–1050.

284. Lin Y., L. Liu, Z. Li, J. Qiao, L. Wu, W. Tang, X. Zheng, X. Chen, Z. Yan and W. Tian (2006). 'Pluripotency potential of human adipose-derived stem cells marked with exogenous green fluorescent protein'. *Mol. Cell Biochem.* **291**: 1–10.
285. Niemela S.M., S. Miettinen, Y. Konttinen, T. Waris, M. Kellomaki, N.A. Ashammakhi and T. Ylikomi (2007). 'Fat tissue: views on reconstruction and exploitation'. *J. Craniofac. Surg.* **18**: 325–335.
286. Usas A. and J. Huard (2007). 'Muscle-derived stem cells for tissue engineering and regenerative therapy'. *Biomaterials.* **28**: 5401–5406.
287. Meng X., T.E. Ichim, J. Zhong, A. Rogers, Z. Yin, J. Jackson, H. Wang, W. Ge, V. Bogin, K.W. Chan et al. (2007). 'Endometrial regenerative cells: a novel stem cell population'. *J. Transl. Med.* **5**: 57.

17
MESENCHYMAL STEM CELLS AND THEIR THERAPEUTIC POTENTIAL IN THE NERVOUS SYSTEM

Steven A. Hardy[1,2], *Daniel J. Maltman*[1,2], *Stefan A. Pryzborski**[1,2]

[1]*School of Biological and Biomedical Sciences, Durham University, South Road, Durham DH1 3LE, United Kingdom*
[2]*ReInnervate Limited, Old Shire Hall, Old Elvet, Durham DH1 3HP, United Kingdom*

INTRODUCTION

Mesenchymal stem cells (MSCs) are promising candidates for use in cellular therapies and tissue engineering strategies. They are an adult stem cell population and derive their name from an intrinsic ability to differentiate into multiple cell lineages of mesodermal origin such as bone, fat and cartilage [1] (Fig. 17.1). Given the developmental potential of these cells, MSCs can be described as *multipotent*, meaning they are able to differentiate into closely related cell types of the same germ layer. The majority of adult stem cells fall into this classification. Stem cells displaying a greater differentiative potential than multipotent stem cells can be described as *pluripotent*, meaning that they demonstrate the ability to differentiate into multiple cell types derived from any of the three germ layers; however, it cannot contribute to extra-embryonic tissue and therefore cannot give rise to a new organism. *Unipotent* stem cells are considerably more restricted in their differentiative potential, in that they can only give rise to one cell type. *Totipotency* is a term used exclusively to describe the differentiative potential of embryonic stem cells, meaning that embryonic stem cells have the potential to differentiate into all cell types of all three germ layers, and can also contribute to extra-embryonic tissue such as placenta. By definition, totipotent embryonic stem cells are the

---

* *stefan.przyborski@durham.ac.uk*

only stem cell type capable of giving rise to a new organism. The difference between totipotent embryonic stem cells and more restricted adult stem cells reflects their different functions. While embryonic stem cells have the capacity to differentiate into all tissue types and form a new organism, adult stem cells have a role in tissue maintenance and general cell turnover.

Mesenchymal stem cells were first identified in 1976 in a fundamental work by Friedenstein [2]. At the time of this work, considerable research had been done and emphasis placed on haematopoietic cells present within the bone marrow. Friedenstein identified the presence of a second cell population within the bone marrow, distinct from the haematopoietic lineage. These cells displayed two key properties: (1) colony-forming potential and (2) ability to differentiate into bone and cartilage. Both the ability to form colonies (self-renew) and differentiate into progeny displaying a more mature phenotype indicated that Friedenstein had identified another type of stem cell in the bone marrow. Morphologically, these cells appeared fusiform and fibroblastic, and as such were initially described as *colony-forming unit-fibroblasts* (*CFU-F*). This original terminology proposed by Friedenstein [2] has generally been replaced and nowadays these cells are typically referred to as mesenchymal stem cells (MSCs) or marrow stromal cells.

Mesenchymal stem cells can be found ubiquitously throughout the body and are routinely isolated from sources such as bone marrow, adipose tissue, adult blood, umbilical cord blood,

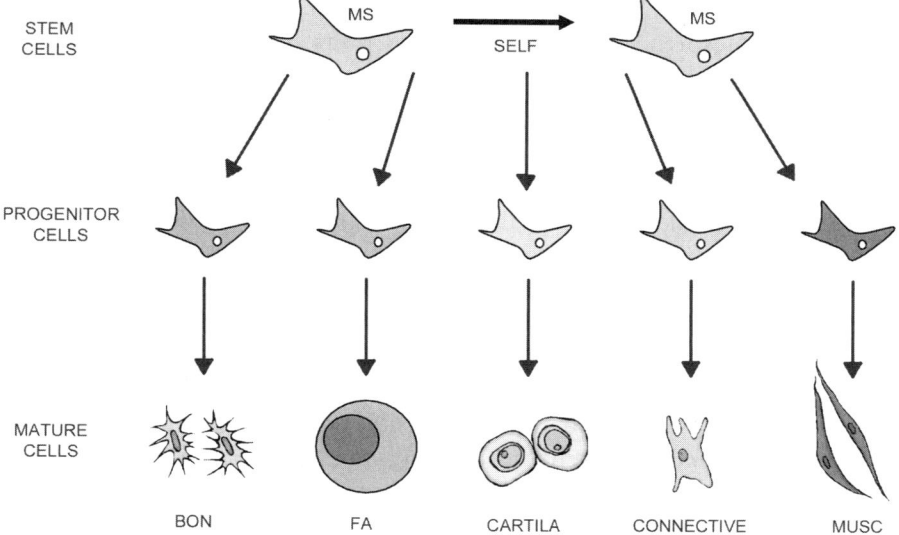

▶ Fig. 17.1 *Diagram describing the developmental potential of MSCs. Following cell division, MSCs may form one of two different progeny: they may produce a daughter stem cell, in a process known as self-renewal, or they may give rise to a progenitor cell that divides more rapidly but is more restricted in terms of its overall proliferative capacity and developmental potential. Progenitor cells derived from MSCs follow a defined path of differentiation. MSCs have the capability to differentiate into a wide range of mesodermal cell types, including bone, fat, cartilage, connective tissue, and muscle. (For colour figure see Plate 39).*

amniotic fluid, tendon and ligaments, chorionic villi of the placenta, synovial membranes, deciduous teeth and foetal liver, lung and spleen [3–12]. In fact, the distribution of MSCs throughout the body may be even more widespread, as alluded to in a recently published study by da Silva Meirelles et al., entitled 'Mesenchymal stem cells reside in virtually all post-natal organs and tissues' [13]. In this study, it has been found that MSCs could be isolated and subsequently cultured *in vitro* from all of the organs and tissues examined in a murine system, including brain, spleen, liver, kidney, lung, muscle, thymus, pancreas and, of course, bone marrow. In doing so, authors convincingly show that MSCs can be isolated from tissues arising from all three germ layers. Morphologically, MSCs isolated from the different tissues were very similar and marker expression profiles were indicative of an MSC population although there were some subtle differences in the expression of markers noticed between MSCs isolated from different sources. Furthermore, even though all cells examined had capabilities to undergo osteogenic and adipogenic differentiation (and hence may be correctly termed MSC), there were some differences in their capacity to form such derivatives which related to their site of origin. Such differences aside, cells isolated from these various sources were similar enough in biological terms to suggest a close relationship between them. Da Silva Meirelles et al., [13] went on to examine the isolation and culture of MSCs from vascular tissue, both complex (aorta and vena cava) and simple (capillaries of the kidney glomerulus). The isolation of an MSC population from such sources is in keeping with previous reports in the literature that claim MSCs may arise from perivascular cells [14, 15]. Da Silva Meirelles, therefore, concluded that the widespread distribution of MSCs throughout the body was a result of its perivascular location [13].

Despite the wide distribution of MSCs throughout the body, the bone marrow remains the most commonly used source for MSC isolation to date, even though MSCs represent only a tiny fraction of the total marrow population, around 0.01% [1]. Commonly used sources of bone marrow for the isolation of MSCs include the iliac crest of the pelvis, femur, tibia and thoracic and lumbar spine [1, 16–18]. The iliac crest of the pelvis is the most commonly used human source of MSCs, whereas the most predominant source in rodents are the longer bones (femurs and tibiae). A typical protocol for the isolation of MSCs from rodent femurs and tibiae follows. Firstly, the rodent must be euthanised and their femurs and tibiae extracted, cleaned of all connective tissue and placed in complete culture medium (CCM) on ice. CCM typically consists of Dulbecco's Modified Eagles Medium (DMEM) supplemented with 10% foetal calf serum (FCS), non-essential amino acids, L-glutamine, penicillin and streptomycin, although the exact components of CCM can vary somewhat between different laboratories. The ends of the bones must then be cut in order to expose the bone marrow, and the marrow aspirated with CCM by inserting a 21-gauge needle into the shaft of the bone and flushing with CCM. Bone marrow aspirates must then be plated into a tissue culture flask with more CCM and incubated, typically at 37°C and 5% CO_2 in a humidified incubator. Aspirates must be left untouched for 48 h to allow the stromal component of the marrow (MSCs) to adhere to the tissue culture plastic. The second cell population present within bone marrow—the haematopoietic population—should remain in suspension and not attach to the plastic substrate. Adherence to tissue culture plastic is the major criterion to discriminate between and separate the haematopoietic and stromal components of the bone marrow. The resulting cultures of MSCs are therefore highly heterogenous, consisting of a mixture of cells at various stages of development. Alternative isolation methods exploiting marker expression are currently being sought in order to isolate a more homogenous starting population of MSCs; however, adherence to tissue culture plastic remains the most highly used method to date. Phase contrast micrographs of MSCs at various stages during the cell adherence isolation process are shown in Fig. 17.2. After 48 h, non-adherent haematopoietic cells are removed from cultures and the remaining adherent stromal (MSC) cells maintained in CCM (Fig. 17.2(A)). These cells are designated as *passage 0 (P0)*. While haematopoietic cells do not adhere to tissue culture plastic, a small degree of contamination is inevitable, especially during early passages. Contaminating haematopoietic cells, mainly erythrocytes, are still apparent after several washes and remain as phase bright, highly refractile cells in suspension (Fig. 17.2(A)). Within 2–3 days, the small spindle-shaped adherent cells begin to divide and flatten out (Fig. 17.2(B)). The adherent MSCs proliferate and appear spherical in morphology as they undergo cytokinesis. Following division, cells begin to notably increase in size and spread out, acquiring a more stromal, fibroblastic-like morphology. Cells continue to divide, initially forming colonies, and eventually form a monolayer culture when colonies merge (Fig. 17.2(C)).

▶ **Fig. 17.2** *Phase contrast images showing cultured MSCs at various stages during the cell isolation process from adult rat bone marrow. (A) At 48 h post-isolation, MSCs adhere to the tissue culture plastic, whereas haematopoietic contaminants remain in suspension as highly refractile, phase-bright cells (arrows). (B) Adherent MSCs begin to divide and flatten out. Arrows indicate proliferating cells. (C) Confluent monolayer of MSCs at passage 0. Scale bar: 30 μm.*

Although the morphology of cells shown in Fig. 17.2 is characteristic of the MSC phenotype, such characterisation alone is insufficient to confirm the isolation of a true MSC population. Consequently, there is a requirement for cell-type specific markers (both cell surface and intracellular) that allow more accurate identification of different cell types. It is uncommon to find a protein whose expression is completely specific to one particular cell type and it is therefore usually necessary to build an expression 'profile' which considers a range of markers known to be expressed by a given cell type. Cell marker analysis is particularly important with regards the identification of stem cells and their derivatives. At present, there is no single definitive marker to define individual populations of MSCs, largely due to the heterogeneous nature of MSC cultures. To confirm the MSC phenotype, flow cytometry is often used to examine the expression of a range of positive and negative markers. Positive cell surface markers include CD29, CD44, CD54, CD55, CD73 (SH3 and SH4), CD90 (Thy-1), CD105 (SH2, or endoglin), CD106 (vascular cell adhesion molecule, VCAM-1), CD117 and CD166. Negative cell surface markers are usually markers of haematopoietic cells, and include CD11b, CD14, CD31, CD33, CD34 and CD45 [1, 19]. In addition to cell surface antigens, markers of MSCs also include the intracellular proteins fibronectin, vimentin and α-smooth muscle actin (α-SMA) [1]. Considerable attention is currently focussed on the utilisation of cell surface marker expression as a means of producing more homogenous MSC cultures by fluorescence-activated cell sorting (FACS) [20, 21]. Figure 17.3 gives an example of typical flow cytometric and immunocytochemical analysis of cell marker expression in undifferentiated MSCs.

Analysis of marker expression is not the only factor to consider in confirming the isolation of a true MSC population. In addition, functional characterisation together with analysis of marker expression provides more compelling evidence of a true MSC population. For purposes of this chapter, we will consider the ability of MSCs to undergo osteogenic and adipogenic differentiation to demonstrate bone and fat phenotypes, respectively. However, it should be recognized that MSCs are capable of differentiating into a range of other mesodermal cell types including cartilage and muscle (Fig. 17.1). Differentiation of MSCs to form osteogenic and adipogenic lineages can be easily achieved *in vitro* by the addition of several specific inductive factors to the culture medium. For osteogenic differentiation, MSCs can be treated with a cocktail of dexamethasone, ascorbic acid 2-phosphate and α-glycerophosphate. Adipogenesis, on the other hand, is slightly more complex, requiring a cyclic treatment with dexamethasone, indomethacin, insulin and 3-isobutyl-1-methylxanthine (IBMX) (induction medium) for 3 days, followed by treatment with insulin alone (maintenance medium) for 2 days. Although slight variations in the differentiation protocols are inevitable between different laboratories, e.g. concentrations of differentiation factors, frequency of media changes etc., the inducing factors used are highly standardised.

To assess the phenotype of MSCs induced to undergo osteogenic and adipogenic differentiation, a range of techniques must be employed. These techniques are largely histological and require cells to be stained with solutions that allow identification of various structures associated with either lineage. In the example given herein, we show how MSCs cultured under osteogenic conditions for a period of three weeks subsequently form boned nodules (Fig. 17.4). Demonstration of a mature osteoblastic phenotype can be achieved using standard procedures involving staining for deposition of mineralised matrix using von Kossa staining to visualize calcium and collagen deposition using Masson's trichrome. In both instances, cells are counterstained with haematoxylin, a basic dye that stains cell nuclei with a blue–purple colour. Bone nodule formation is detectable by von Kossa staining as early as one week post-induction of differentiation, although such nodules are small in size and very rare in occurrence (data not shown). After three weeks, however, large bone nodules become evident and the majority of the surface had been calcified. Another indication of osteogenic differentiation is the deposition of collagen in the extracellular matrix. Masson's trichrome staining is the most commonly used histological method to demonstrate the presence of collagen in the extracellular matrix, staining collagen with a blue–green colour. When performing histology, it is of the utmost importance to include undifferentiated negative controls, to allow distinction between specific and non-specific staining.

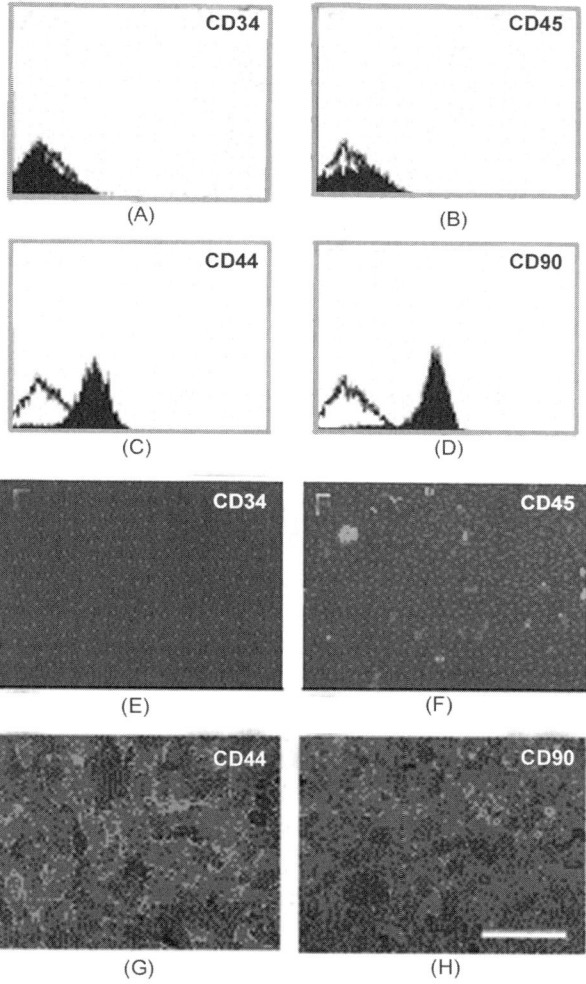

▶ **Fig. 17.3** *Analysis of cell marker expression in undifferentiated MSCs. Marker expression as determined by (A–D) flow cytometry and (E–H) immunocytochemistry for (A and E) CD34, (B and F) CD45, (C and G) CD44 and (D and H) CD90. MSCs are largely negative for the haematopoietic markers CD34 and CD45, indicating that levels of haematopoietic contamination is low. MSCs are positive for CD44 and CD90 expression, which is indicative of MSCs. Scale bar: 30 μm. (For colour figure see Plate 40).*

To demonstrate the formation of a mature adipogenic phenotype, MSCs cultured under adipogenic conditions for a period of three weeks form lipid droplets as determined by Oil Red O staining or Nile Red staining followed by flow cytometric analysis (Fig. 17.4). Using Oil Red O staining lipid droplets are detectable as early as one week post-induction of differentiation, although

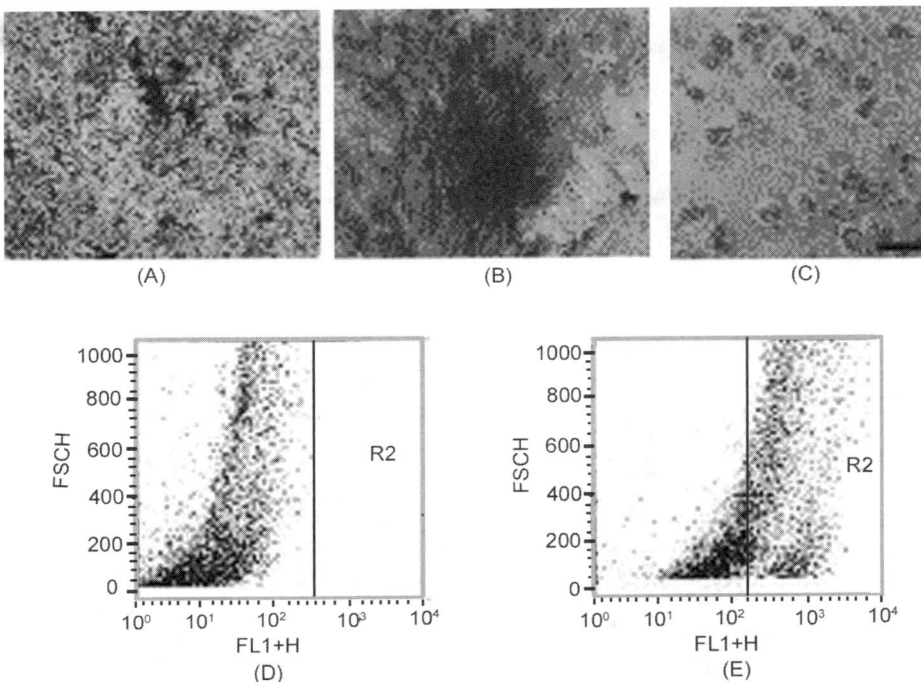

▶ **Fig. 17.4** *Developmental potential of MSCs grown under conditions to induce either bone or fat differentiation. Osteogenic differentiation was assessed using (A) von Kossa and (B) Masson's trichrome staining to visualise extra cellular calcium deposition and collagen formation, respectively. (A) Bone nodule formation is the evidence as black deposits, and (B) the formation of collagen matrix as blue-green staining. Adipogenic differentiation was assessed using (C) Oil Red O and (D, E) Nile Red staining. Oil Red O staining showed the accumulation of intracellular lipid droplets within the cells (C, red droplets). MSCs stained with Nile Red were subject to flow cytometric analysis, and representative traces are shown for undifferentiated (D) and differentiated (E) MSCs. A population shift to the right is indicative of adipogenic differentiation. Approximately one third of the total MSC population are positive for Nile Red. Scale bar: 50 µm. (For colour figure see Plate 41).*

these appear infrequently and are small in size. After three weeks, however, large lipid droplets are readily detectable in the cell cytoplasm, indicating that MSCs are adopting the phenotype of an adipocyte. An alternative (and much less commonly used) method to demonstrate the presence of lipid within the cell cytoplasm is using the lipophilic Nile red fluorescent dye, followed by flow cytometric analysis. Representative flow cytometric traces are shown in Fig. 17.4 for undifferentiated MSCs and cells exposed to inductive conditions to form fat. The representative trace for undifferentiated MSCs is important to allow background levels of Nile red fluorescence to be determined, and therefore distinction between cells that are negative (undifferentiated) and positive (differentiated) for Nile red staining. The representative trace for MSCs induced to undergo adipogenic differentiation shows a marked population shift to the right compared to the control. This population shift confirms the presence of Nile red-positive cells within the differentiated population and supports the Oil Red O staining data, further confirming the accumulation of lipid within differentiated cells. In this example, the number of Nile red-positive cells is around one-third of the total cell population.

Stem cells can be classified according to their potency and their ability to differentiate into other cell types. By showing that MSCs have the potential to differentiate into multiple cell lineages that are closely related to each other (bone and fat both arise from the mesoderm), MSCs can be described as a multipotent stem cell population. It is generally accepted in stem cell biology that adult stem cells are significantly more restricted in their potential to differentiate than embryonic stem cells. Multipotent adult stem cells typically form lineages specific to their germ layer of origin, whereas embryonic stem cells can form all cell types from any of the three germ layers as well as extra-embryonic tissues such as placenta. Consequently, the potency of MSCs was considered to be such that they could only form tissues of mesodermal origin, as has been demonstrated above. However, in recent years this philosophy has been challenged, with numerous reports suggesting that certain stem cells display the ability to 'cross' the germ layer boundary and form tissues of a different lineage. This process has been termed '*trans*'-differentiation. The concept that adult stem cells might possess a far wider potential for differentiation than previously anticipated has led to considerable excitement with respect to their potential therapeutic application, since the use of adult stem cells is not met with the same criticism or ethical issues associated with the use of embryonic stem cells. There have been several reports claiming that MSCs possess a greater potential for differentiation than what was previously anticipated, including the ability of MSCs to form tissues of both endodermal and ectodermal origins [22–27]. This chapter will focus particularly on the formation of neural-like cells from MSCs and address the issue of whether or not they represent an alternative cell source to neural stem cells for neurological therapies.

POTENTIAL THERAPEUTIC APPLICATIONS OF MSCs

Stem cells are recognised as promising candidates for clinical use in cellular therapies and tissue engineering strategies. Cellular therapies are medical processes that involve the administration of healthy, fully functioning cells to replace dysfunctional cells that have been damaged or destroyed in disease. Although adult stem cells can be isolated from various sources throughout the body, MSCs are recognised as one of the most promising for use as potential therapeutic agents. There are a number of reasons as to why this is the case. MSCs can be found and isolated from a wide range of autologous sources, as has been previously discussed [3–12]. Many of these sources, particularly bone marrow and adipose tissue, are clinically accessible, and MSCs can be easily isolated using robust, well-established techniques [1, 7, 13]. Since MSCs represent such a low proportion of the total cells isolated (around 0.01% for bone marrow), it is necessary to expand cells to generate suitable numbers for clinical application.

MSCs have a high proliferative potential and can be readily and rapidly expanded *ex vivo*, while crucially maintaining their multi-potentiality to differentiate into clinically significant derivatives such as bone, fat and cartilage. Perhaps most encouraging is the very real possibility that MSCs are suitable for use in allogeneic transplantation as well as autologous transplantation [28]. Allogeneic transplantation can be clinically problematic, as the recipient's body recognizes the allogeneic material as foreign, resulting in an immune response and possible rejection of the transplanted material. As such, allogeneic transplantation must be accompanied by rigorous immune suppression. MSCs appear to be immunologically unique in that they express intermediate levels of MHC Class I antigens and negligible levels of MHC Class II antigens [29]. It has also been reported that MSCs do not express co-stimulatory molecules [30]. Crucially, it is not just undifferentiated MSCs that lack expression of MHC Class II antigens, but their differentiated derivatives also. Le Blanc et al. [29] examined expression of MHC Class I and II antigens in undifferentiated MSCs and MSCs that had been differentiated into their gold-standard derivatives bone, fat and cartilage. They reported that both undifferentiated MSCs and their differentiated counterparts lacked expression of MHC Class II antigens, strongly suggesting that MSCs are not immunogenic and would not evoke an immune response upon transplantation [29]. In support of these observations, Aggarwal et. al., reported that, following allogeneic transplantation, MSCs evaded immune recognition and were readily detectable in recipients at extended time points [28]. A common complication of allogeneic transplantation is graft-versus-host disease (GVHD), in which functional immune cells carried over from the donor in the transplanted material recognize the recipient as foreign and launch an immune response. Transplantation of MSCs has been reportedly associated with reduced incidence of GVHD, and there have been even reports of MSC transplantation being used in the treatment of GVHD [31]. The potential of MSCs for use in allogeneic transplantation as well as autologous transplantation has major clinical significance, since material could be transplanted between mismatched individuals.

MSC-BASED THERAPY IN THE TREATMENT OF OSSEOUS AND ADIPOSE TISSUE DEFECTS

As MSCs are defined functionally owing to their ability to differentiate into mesodermal derivatives such as bone and fat as considered herein, the most obvious clinical application of MSCs at present is in the treatment of osseous and adipose tissue defects, hence exploiting their intrinsic developmental potential for the replacement of diseased bone and fat tissues [1, 32, 33]. Adipose tissue defects, commonly referred to as soft tissue defects, can be defined as substantial voids within the subcutaneous fat layer. Such defects can be a consequence of numerous conditions, such as traumatic injury e.g. burns, hereditary and congenital conditions e.g. Poland syndrome and Romberg's disease, tumour resection e.g. breast mastectomies, facial reconstructions and lipodystrophy associated with type II diabetes [34–39]. Currently, treatment options for adipose tissue defects consist mainly of plastic and reconstructive surgical procedures. Logic might suggest that autologous adipose tissue from alternative sources on the patient's body would provide the best and most obvious source of material for such surgeries, as adipose tissue is usually present in excess, providing a readily available yet readily expendable source of material [40]. However, although the implementation of autologous fat transplantation has been trialled in many cases, the outcome of these procedures has unfortunately been unpredictable and somewhat inconsistent [34, 41]. There are several reasons for this. Due to inadequate revascularisation of transplanted adipose tissue, graft volume can be reduced by up to 60%, while the viable tissue that remains is unable to function appropriately [41]. Furthermore, adipocytes are extremely vulnerable to mechanical/pressure damage during aspiration procedures due to the high lipid content of their cytoplasm (up to 90% cytoplasm volume is lipid)—up to 90% of adipocytes are damaged during aspiration [42]. This can result in only a small number of viable cells being isolated, and *ex vivo* expansion to generate sufficient numbers for clinical use is not an option since mature adipocytes are terminally differentiated, and therefore non-proliferative. Because of such limitations, the use of autologous adipose tissue in cellular therapies for the treatment of soft tissue defects is not ideal and, as such, there is a search for alternative cell sources. MSCs in particular present themselves as promising candidates for use in the cellular treatment of soft tissue defects, as they can be easily isolated and rapidly expanded in the tissue culture laboratory, and under appropriate culture conditions can be manipulated to undergo adipogenic differentiation to form pre-adipocytes and adipocytes.

One of the most obvious and realistically achievable clinical uses of MSCs at present is in bone tissue engineering as a means of replacing damaged bone and osseous defects. New cellular therapies and tissue engineering strategies in the treatment of osseous defects are a major clinical goal and this impetus is set to significantly increase in the future as improved quality of life leads to increased life expectancy and consequently a higher incidence of age-related disorders such as osteoporosis. A number of factors can contribute towards the development of osseous defects and bone trauma, such as fractures, tumour invasion or genetic disorders, e.g. osteogenesis imperfecta [43, 44]. Current treatment strategies consist predominantly of autologous bone grafts, and to a lesser extent allogeneic bone grafts, metallic implants and bioceramics [45, 46]. Unfortunately, as is the case for adipose tissue defects, the success of these current treatment methods is somewhat inadequate, not to mention distressful for the patient and extremely painful due to their invasive nature. Reasons for limited success include limited autograft supply/volume, infection due to pathogen transfer/immunological rejection following allografts, loss of function/donor site morbidity and, in some cases, even nerve damage [47]. It is therefore hardly surprising that alternative strategies in the treatment of osseous defects are currently sought, with a particular clinical goal being the development of bone tissue engineering technologies to replace tissue damaged as a result of disease. MSCs are therefore highly appealing candidates for use in the cellular treatment of osseous defects due to their intrinsic osteogenic ability upon appropriate stimulation.

MSC-BASED THERAPY IN THE TREATMENT OF NEUROLOGICAL DEFICITS

Besides the more generally accepted therapeutic applications of MSCs as mentioned above, MSCs are also attracting considerable interest as a potential resource for the treatment of neurological and neurodegenerative disorders. The remainder of this chapter will focus on the recent developments in this area, firstly by considering the use of MSC transplantation as a means of treating such deficits.

Neurological disorders result from disruption in the structure and/or function of nervous tissue, either in the central nervous system (CNS), peripheral nervous system (PNS), or both.

Neurodegenerative disorders are a further sub-classification of neurological disorders, and this term is used to describe disorders that result from a loss and/or degeneration of nervous tissue. Examples of neurodegenerative disorders include Huntington's disease, Parkinson's disease, multiple sclerosis and Alzheimer's, among many others. Current therapies for the treatment of neurological and neurodegenerative disorders are largely ineffective and, as such, inadequate. The development of effective therapies in this area is further complicated owing to the significantly varying pathologies between different disorders and even between individuals suffering from the same disorder. Consequently, the development of a standardised therapy in the treatment of such conditions has not been achieved, as patients require treatments that are tailored specifically to their unique clinical symptoms. With the major developments in stem cell research and regenerative medicine over recent years, there is the potential of cellular therapy involving the administration of stem cells and/or their derivatives to replace damaged and/or degenerated nervous tissue. With evidence to suggest that MSCs display a greater plasticity than previously anticipated, concomitant with their unique immunological properties which hold them with such promise for both autologous and allogeneic transplantation, MSCs might present themselves as promising candidates in the treatment of neurological and neurodegenerative deficits. The following will consider some of the key studies in this area.

There are a number of factors that need to be taken into consideration in order to ascertain whether MSCs are suitable candidates in the cellular treatment of neurological and neurodegenerative disorders. The first of these considerations is whether or not MSCs remain viable and functional upon transplantation into the nervous system, as the microenvironment of nervous tissue is far-reached from the native niche of MSCs in the bone marrow. It is becoming increasingly more apparent that the stem cell niche plays a crucial role in the maintenance and function of stem cells, providing a delicate balance between cell–cell interactions, extracellular matrix contact, oxygen tension, pH and exposure to growth factors, etc. [48]. These are just a handful of factors which interplay within the niche to contribute towards stem cell maintenance and function. As such, it is necessary to investigate whether the microenvironment of the nervous system is permissive to the survival and engraftment of transplanted MSCs and to the formation of terminally differentiated neural cells.

A number of studies have demonstrated that, upon transplantation into the brain, MSCs survive and undergo successful engraftment and migration. A report published in 1998 by Azizi et. al., [49] demonstrates this concept. In this study, Azizi et al. examined the transplantation of both rat and human MSCs, administering them directly into the corpus striatum of rat brains. The corpus striatum consists predominantly of the basal ganglia and its primary function is in the coordination of movement. Periodically up to 72 days post-transplantation, brain sections were examined in order to ascertain whether transplanted MSCs had (a) successfully engrafted and (b) undergone migration. Azizi et al. also injected rat astrocytes in the corpus striatum, which served as a suitable control, as astrocytes have been previously shown to successfully engraft and migrate throughout the CNS following transplantation, in an analogous manner to neural stem cells [50, 51]. Two weeks post-transplantation, an engraftment success rate of 20% for human MSCs, and a comparable 30% for rat MSCs, was observed. This observation is of particular interest, with the group reporting that the success rate of human MSC transplantation into rat brain (xeno-transplantation) was only 10% less than that observed for rat MSC transplantation (autologous transplantation). If such observations hold true, then this would suggest that the unique immunological properties of MSCs, as previously discussed, would render them suitable for xeno-transplantation as well as allogeneic and, of course, autologous transplantation. Azizi et. al., [49] also examined whether engrafted MSCs were capable of undergoing migration throughout the brain. Recruitment of MSCs to neurologically damaged areas will be a crucial event if they are to elicit any kind of beneficial effect that would ultimately lead to functional recovery. There is some evidence that MSCs are capable of undergoing migration following engraftment by following a known cell migration pathway. MSCs were found in several successive layers of the brain up to 72 days post-transplantation, and this was mirrored in the pattern of astrocyte migration [49]. The work by Azizi et. al., [49] provides a logical and progressive example that clearly highlights effective MSC transplantation, survival and migration in the nervous system. While we have only considered this single study, there have been many other reports of similar observations from other systems including mouse and higher order primates [52–56]. Such observations served to intensify the promise surrounding MSCs as potential candidates in cellular therapies of neurological and neurodegenerative disorders.

Besides engraftment and migration, it is important to address whether transplantation of MSCs results in any degree of functional recovery in models of neurological and neurodegenerative disorders. There is evidence to suggest that MSC transplantation does elicit functional recovery to various degrees. This has been reported in many models of neurological and neurodegenerative disorders including, but by no means limited to, Huntington's disease, Parkinson's disease, multiple sclerosis, amyotrophic lateral sclerosis, stroke, traumatic brain injury and also damage to the spinal cord and peripheral nerves [57–63]. Restoration of functional activity can be assessed in many ways and these can range from simple behavioural tests to more complex cell function assays. Many of these proof of principle studies suggest that MSC transplantation may be a promising therapy for neurological and neurodegenerative disorders; however, the large majority of work in this area has been performed using animal models [64–67]. Before this can be significantly trialled in humans, a number of technical issues must be addressed. For example, in terms of the cellular material to be transplanted, should whole bone marrow be transplanted or should attempts be made to culture a more homogeneous starting population of MSCs? Once this has been addressed, how should cells be transplanted? Should they be administered directly into the damaged nervous tissue or should they be administered systemically? If MSCs were administered systematically, would they home to the neurologically damaged areas in sufficient quantity to elicit a beneficial effect? In terms of timescale, when should MSCs be administered? Does administration need to be immediate following injury, or can a longer timescale be employed for the treatment of more chronic disorders? What numbers of cells should be transplanted? Is it possible to predict the success rate of engraftment? If MSCs were to become serious contenders as a cellular therapy to treat neural deficits then these questions represent just a handful of the technical issues that must be considered prior to any human trial.

MECHANISMS BY WHICH TRANSPLANTED MSCs ELICIT FUNCTIONAL RECOVERY IN THE DAMAGED NERVOUS SYSTEM

Before MSC-based cellular therapies become serious propositions for use in the clinic, a significant advance in our knowledge and understanding of the complex mechanism(s) via which MSCs commit to specific lineages and undergo differentiation to demonstrate mature and functioning phenotypes is required. A thorough understanding of these biological phenomena will not only benefit from a clinical perspective, but will also be of major significance in a research setting, as MSCs provide an ideal cellular system in which to study the complex events of cell differentiation. In terms of MSC transplantation in the therapy of neurological and neurodegenerative disorders, proof of principle studies in animal models demonstrate that MSC transplantation can result in significant functional improvements. However, the mechanism(s) by which MSCs elicit neurological recovery remain largely unknown. At present, opinion is divided between three main hypotheses:

(a) MSCs undergo '*trans*'-differentiation to produce neural cell types, thereby directly replacing neural tissue that has been damaged as a result of disease;

(b) MSCs undergo spontaneous fusion events with host neural cells, and in doing so acquire the phenotypic properties of the host neural cells;

(c) MSCs migrate to neurologically damaged areas but, rather than differentiate into neural cells, they release a complement of trophic factors and cytokines that promote endogenous repair mechanisms, predominantly by the instructive, neurogenic effects of these factors on resident neural stem and progenitor populations.

Each of these three possible mechanisms will be considered in the following sections and are summarised schematically in Fig. 17.5.

IN VITRO 'TRANS'-DIFFERENTIATION OF MSCs INTO NEURAL CELL TYPES

Reports of MSCs undergoing '*trans*'-differentiation *in vitro* to form neural cell types have been numerous within the current literature and, as such, it is not possible to provide a complete overall discussion of these studies here. Therefore, for purposes of this chapter, we will draw the reader's attention to the two seminal studies that demonstrated the concept of *in vitro* '*trans*'-differentiation of MSCs. These studies are those of Sanchez-Ramos and Woodbury [22, 25, 26].

Sanchez-Ramos et al. [26] reported that MSCs had the potential to demonstrate neural differentiation when cultured in 'differentiation medium', consisting of DMEM supplemented with a cocktail of factors that had been previously reported to exert a neurogenic effect, including all-*trans*-retinoic acid (ATRA) and brain-derived neurotrophic factor (BDNF). Following treatment of MSCs with this cocktail of neurogenic factors, Sanchez-Ramos et. al., examined cellular morphology and expression of cell-specific markers. When cultured under standard conditions,

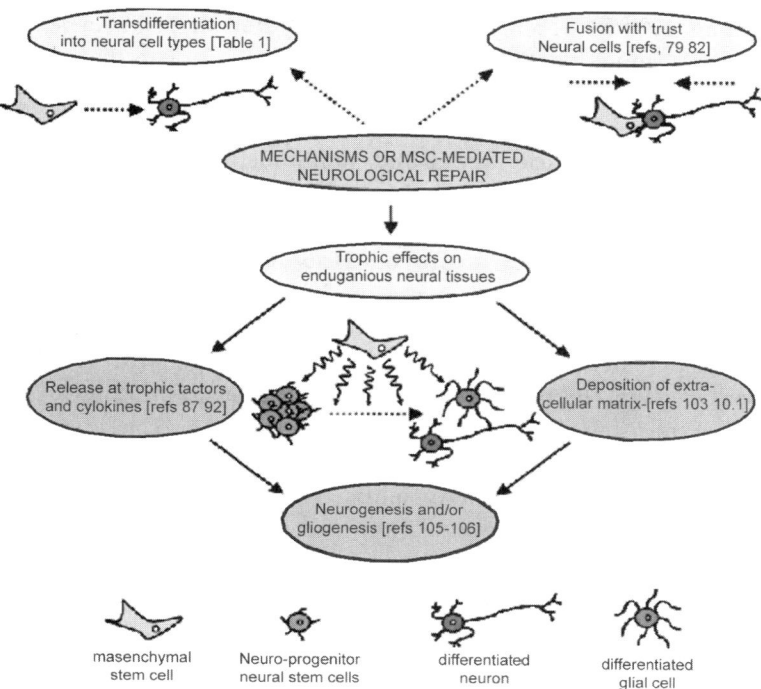

▶ **Fig. 17.5** *Diagrammatic representation summarising the mechanisms by which MSCs may elicit functional neural recovery following their transplantation into models of neurological disorders. See text for further details. (For colour figure see Plate 42).*

MSCs display a very characteristic, stromal-like morphology, with cultures consisting predominantly of large, flat, fusiform cells. However, MSCs cultured in 'differentiation medium' underwent significant radical changes and the distinctive stromal morphology of MSCs was replaced by a structure which resembled that of neural cell types whereby cells became spindle-shaped and possessed several long processes. Marker analyses revealed that MSCs cultured in 'differentiation medium' expressed the neural progenitor marker, nestin, the neuronal marker, NeuN and the glial marker, glial fibrillary acidic protein (GFAP). These data support the hypothesis that MSCs may be able to cross the germ layer boundary to form non-mesodermal cell types such as neural cells, which are ectodermal in origin. Adding more strength to the argument that MSCs were undergoing '*trans*'-differentiation to form neural cell types was the observation that the up-regulation of these neural antigens was concomitant with a down-regulation of fibronectin expression, a marker of MSCs. However, despite the observed down-regulation in fibronectin expression, the percentage of cells expressing neural markers was proportionately lower than the percentage that retained fibronectin expression (<1% compared with 30% fibronectin).

Woodbury et al. [25] adopted a slightly different approach to that of Sanchez-Ramos [26], by culturing MSCs with a cocktail of chemicals as opposed to well-characterised neurogenic factors. In their chemical induction protocol, Woodbury et. al., removed MSCs from their standard culture conditions, containing 10% FCS, and subsequently cultured them under serum-free conditions followed by exposure to antioxidants. However, immediately prior to the removal of serum from their culture medium, MSCs were first 'pre-induced' for 24 h by the addition of β-mercaptoethanol (BME) to their standard media. 'Pre-induction' was followed by 'induction', which involved removal of serum from the culture media and supplementation with antioxidants butylated hydroxyanisole (BHA) and dimethylsulphoxide (DMSO). Similarly to observations of Sanchez-Ramos, MSCs underwent radical changes in both cell morphology and marker expression in response to the chemical induction protocol. However, these changes occurred over an extraordinarily short timescale. Within a matter of hours following induction, MSCs lost their typical stromal morphologies and exhibited tight, multipolar soma with many processes and cytoplasmic extensions emanating from the cell body, as is typical of certain neural cell types. During the very early stages of the experiment, the cytoplasmic processes were very simple, although by the end of the 5-h time course their morphology had become somewhat more complex, with processes demonstrating secondary branching. These morphological observations suggested that MSCs were acquiring neuronal phenotype, and subsequent marker analysis supported these morphological observations with an up-regulation in expression of the neuronal marker, neuron-specific enolase (NSE).

There have been a substantial number of reports demonstrating the ability of MSCs to 'trans'-differentiate and adopt neuronal and/or glial fates in vitro. While these studies are too numerous to discuss in detail here, a summary of some of these additional studies is provided in Table 17.1. Of interest are the widely varying culture conditions employed by different groups in order to attempt to induce neural differentiation by MSCs. The various differentiation protocols have been divided into three main categories: (a) chemical induction protocols, which involve the supplementation of MSC culture medium with cocktails of various small molecules or chemicals; (b) treatment of MSCs with known agents known to mediate neurogenesis or transfection of MSCs with neurogenic protein-encoding genes; (c) co-culture of MSCs with neural cell types, in which MSCs are induced to adopt neural phenotypes as a result of direct cell–cell interactions and/or the interplay of soluble factors produced by the co-cultured neural cell type (Table 17.1).

IN VITRO '*TRANS*'-DIFFERENTIATION OF MSCs INTO NEURAL CELL TYPES—AN ARTEFACT OF CELL CULTURE?

It is not surprising that the concept of MSCs displaying a greater plasticity than previously anticipated such that they undergo 'trans'-differentiation into neural cell types is somewhat controversial. MSCs have long been held as a multipotent stem cell population which, by definition, means that they are restricted in their plasticity and form cell types specific to their germ layer of origin. Since MSCs are mesodermal in origin, it was readily accepted that MSCs should possess an intrinsic ability to differentiate into mesodermal cell types, such as bone, fat and cartilage. As discussed above, however, there are some data which suggest that MSCs are able to cross the germ layer boundary and differentiate to acquire a neural phenotype. While this holds major promise for therapeutic applications, it also challenges the central dogma of adult stem cell biology.

Table 17.1 *Summary of Investigations Reporting the Potential of Cultured MSCs to form Neural Derivatives (see Text for Further Details)*

Treatment chemical	Neural phenotype	Reference
Isobutylmethylxanthine(IBMX) + dibutyryl cAMP	Neuronal	[110, 111]
Â-mercaptoethanol (BME), then butylated hydroxyanisole (BHA) and dimethylsulphoxide (DMSO)	Neuronal	[22, 25, 78, 111, 112]
5-azacytidine	Neuronal	[113]
BME + retinoic acid (RA)	Neuronal	[114]
BHA + dibutyryl cAMP + IBMX + RA + ascorbic acid	Neuronal	[115, 116]
bFGF for 24 h, then BHA + KCl + valproic acid + forskolin	Neuronal/glial	[117]
Neurotrophic/glial factors or gene transfection		
GDNF + PACAP + dibutyryl cAMP	Neuronal	[118]
Generation of neurospheres, then induction with RA and Shh	Neuronal	[119]
Transfection with Notch intracellular domain (NICD), then treatment with neurotrophic factors	Neuronal	[62, 120]
Transfection with *noggin*	Neuronal	[113]
Glial growth factor (GGF)	Glial (astrocytic)	[121]
Culture conditions as for neural stem cells (NSCs)	Neurosphere formation	[122]
bFGF, then BME+NT-3, then BHA + valproic acid + forskolin + hydrocortisone + insulin + NT-3 + NGF + BDNF	Neuronal	[123]
bFGF for 24 h, then Shh + FGF-8 + RA, then BHA + valproic acid + forskolin + hydrocortisone + insulin + NT-3 + NGF + BDNF + GDNF	Neuronal (dopaminergic)	[123]
Shh + FGF-8 + bFGF	Neuronal (dopaminergic)	[124]
EGF + FGF for 72 h, then cAMP + IBMX for 72 h, then GDNF + TGFâ3 + RA	Neuronal (dopaminergic)	[125]
BDNF + EGF + NGF	Neuronal	[126]
Co-culture with neural cell types		
Co-culture with fetal midbrain cultures	Neuronal	[26]
Co-culture with astrocytes	Neuronal	[127, 128]
Co-culture with NSCs	Glial (astrocytic)	[107]
Co-culture with cerebellar granule neurons	Neuronal/ glial (astrocytic)	[129]
Co-culture with Schwann cells	Glial (Schwann cells)	[117]

Recent research raises considerable doubt over the authenticity of the neural derivatives that result from the various differentiation protocols used in the earlier studies which report the *'trans'*-differentiation of cultured MSCs into neural lineages. Issues such as the remarkably short timescale over which the differentiation of neural derivatives occurred are a concern which in some instances amounted to only a few hours [25]. Neural differentiation is a highly complex and multifaceted process. *In vivo* neurogenesis requires spatial and temporal regulation of the expression of a wide array of genes. The concept that MSCs display the ability to undergo the complex molecular events associated with neural differentiation within a matter of hours to form mature, functional neural derivatives, is highly questionable and improbable. Indeed, for MSCs to differentiate and form mesenchymal derivatives such as bone or fat occurs over a timescale of several days and weeks, with the earliest phenotypic indications of differentiation appearing only after 5 days post-induction of differentiation.

It was generally thought to be highly unlikely that true neural derivatives could be formed in this short timescale. However, rather than just dismiss these observations, several groups set about trying to explain the mechanisms by which MSCs were changing under these alternative culture conditions. A number of studies have been published between 2004 and 2006 that provided convincing and highly compelling evidence that *'trans'*-differentiation of MSCs into neural cell types did not represent a pathway of true neural differentiation as originally thought, but was in fact an artificial response by the MSCs to the growth conditions employed. Most notably, the acquirement of a neural morphology by MSCs was attributed to cell-stress in response to these culture conditions [68–71]. The radical changes in MSC morphology, from a large unbranched stromal cell to a cell possessing a highly distinctive neural morphology were considered. A principal event that occurs during neuronal differentiation is that of neurite outgrowth which is a highly dynamic process involving microtubule re-modelling and the outgrowth of neurites from a central cell soma. The transformation of MSCs into cells possessing a neural-like structure did not follow this traditional process of neurite extension. Rather, the MSC cytoplasm retracted from its periphery towards the centre of the cell, forming a tight, spherical cell body around the nucleus, reminiscent of a neural soma. What were originally perceived to be neurites were in fact cytoplasmic extensions that remained following cytoplasmic retraction, a result of strong focal contacts made between cells and the underlying substrata. Such extensions of the cytoplasm resembled neurites but did not equate to actual neurogenesis. Experiments confirmed that retraction of the cytoplasm was the consequence of cytoskeletal collapse. This affect could be readily mimicked by treating MSCs with pharmacological agents, such as cytochalasin B and D, which disrupt the actin cytoskeleton resulting in cell morphologies that resemble neural phenotypes [69–71].

Although such work proved that culturing MSCs with chemical induction protocols did not induce the formation of a true neural phenotype, they did not discount the possibility of MSCs possessing an intrinsic neurogenic potential that may have contributed in part to the acquirement of a neural phenotype. To address this issue, several independent groups applied the chemical induction protocol to a range of other cell types in addition to MSCs. When treated with the chemical induction protocol, primary fibroblasts, HEK293 cells (a human embryonic kidney cell line) and PC-12 cells (a human pheochromocytoma cell line) all underwent morphological changes that were indistinguishable from the morphological changes in control MSCs [68–71]. These data indicated that the morphological changes observed in MSCs following chemical induction were most likely to be attributable to an aberrant response by the cell to the artificial culture conditions rather than any intrinsic neurogenic potential possessed by MSCs.

It appears therefore that under certain culture conditions, MSCs adopt a neural-like morphology as a result of cytoskeletal collapse. However, in addition to structural changes, Woodbury et. al., [25] also reported regulated expression of known neural markers. While such data may partly support the concept of neural differentiation by MSCs, it must be recognised that analysis of marker expression is insufficient in itself to conclusively prove true differentiation and cell identity. Functional data provides the strongest evidence that a cell has differentiated into a mature, functional derivative and, in terms of assessing neural differentiation, this classically involves measurement of electrical activity. MSCs also have been shown to spontaneously express neural proteins under their standard culture conditions, that is, even before induction to undergo neural differentiation [72–74]. Furthermore, in a study published by Lu et al., MSCs were able to acquire a neuronal morphology even when protein synthesis was totally blocked, which casts further doubt that neural morphology of MSCs relates to true neural differentiation [68]. It is a highly unlikely scenario that a cell would be able to undergo differentiation without modifying its protein complement, let alone make the protein changes necessary to differentiate across the germ layer boundary into a cell type representative of a different germ layer. An extension of this argument has been provided in a study by Bertani et al. who, using a microarray to examine the transcriptome of induced and non-induced MSCs, reported that chemical induction of MSCs did not significantly alter the transcriptome compared to non-induced controls [70]. In this study, the expression of neural markers did not significantly differ between induced and non-induced MSCs.

Analysis of differentiation in cell biology and discrimination between stem cells and their differentiated counterparts can sometimes be more complicated than originally anticipated. Morphological observations, although useful, are insufficient on their own to conclude the identification of a particular cell type in culture. By far the most common method of discriminating between different cell types involves analysis of cell-specific marker expression. While data provided from such cell marker analyses is usually informative, it should nonetheless be treated with caution, as has been highlighted above and careful validation of results is required before drawing a solid conclusion. If possible, gene and protein expression data are best complemented by functional analysis of the differentiated cell phenotype to provide the best indication of cell type. Only when all such information is brought together can a firm view on cell identity be determined.

IN VIVO 'TRANS'-DIFFERENTIATION OF MSCs INTO NEURAL CELL TYPES

It is generally accepted that the growth of cells outside the body in the culture environment is artificial and cells might respond abnormally to such conditions. For example, culturing cells on the two-dimensional (2D) environment of tissue culture plastic places them under immense stress by causing cells to adopt a highly unnatural conformation which in turn influences their behaviour and function. It is therefore also appropriate to study MSC behaviour in an *in vivo* setting to gain a more physiologically relevant understanding of whether MSCs can '*trans*'-differentiate into neural cell types. Using the work of Azizi et al. as an example, MSCs are not only capable of surviving transplantation into the unfamiliar environment of the nervous system, but are also capable of undergoing successful engraftment and migration [49]. Numerous reports have stated that MSCs demonstrate marker expression profiles indicative of neural differentiation following transplantation into the nervous system. Kopen et al. proposed MSCs as potential alternatives to neural stem cells in the cellular therapy of neurological and neurodegenerative disorders [53]. In this study, MSCs proliferated and underwent migration throughout the forebrain and cerebellum, in a manner pertinent to that of neural stem and progenitor cells following administration into the lateral ventricle of neonatal mice. By labelling MSCs, it was possible to distinguish between transplanted cells and host neural cells. Transplanted MSCs demonstrated strong expression of GFAP, indicative of astrocytic differentiation, and to a lesser degree expression of neurofilament, indicative of neuronal differentiation [53]. Since these observations were based on the transplantation of undifferentiated MSCs that had not been manipulated *in vitro* prior to transplantation, Kopen et. al., proposed that the acquirement of a neural phenotype by MSCs was a consequence of exposure to the brain microenvironment alone [53]. If such findings hold true, this would be of major interest from a clinical perspective, as it would permit MSCs to be administered to patients in their undifferentiated state, without any requirement for prior *in vitro* manipulation to produce the desired neural phenotype prior to transplantation.

In another example, a study by Li et al., reported that MSCs had the potential to undergo '*trans*'-differentiation *in vivo* to produce dopaminergic neurons, following transplantation into a mouse model of Parkinson's disease [65]. The acquirement of a dopaminergic phenotype was supported by the expression of tyrosine hydroxylase (TH), a marker of dopaminergic neurons, in MSCs post-transplantation. Of course, it cannot be overstated that if its true, such observations hold much promise for the treatment of neurological diseases such as Parkinson's. The main causative mechanism of Parkinson's disease is insufficient action and/or efficacy of dopamine, which is a neurotransmitter produced predominantly by dopaminergic neurons. Whether transplanted MSCs definitively adopt a bona fide, dopaminergic phenotype following transplantation was not fully established. If so, expression of functional TH, an enzyme required for the conversion of tyrosine to dihydroxyphenylalanine (DOPA, a precursor of dopamine), would be expected. The same group reported in a later study that MSCs were able to demonstrate expression of two mature neuronal markers, NeuN and MAP2, and the mature astrocytic marker, GFAP following transplantation [75]. The number of MSCs induced to express markers of neural differentiation, as percentage of the total number of MSCs that successfully engraft following transplantation is very low [74]. Only 1% express neuronal markers and only 5% of these express glial markers. While it is not possible to discount the possibility that cells expressing markers of neural differentiation are functional, what is recognised is that such a rate of '*trans*'-differentiation would be unable to fully account for the neurological improvements reported following administration of MSCs in models of neurological and neurodegenerative disorders. There have been several other studies reporting the '*trans*'-differentiation of MSCs into neural cell types *in vivo* [73, 76–78]; however, the exact mechanism(s) by which such observations occur remains to be fully elucidated.

SPONTANEOUS CELL FUSION OF MSCs WITH HOST NEURAL CELLS

It appears that *'trans'*-differentiation of MSCs into neural cell types *in vivo* occurs at too low a frequency to explain functional recovery in models of neurological disease. Alternative mechanisms to explain apparent improvements in neural function have been considered. Spontaneous fusion between transplanted MSCs and host neural cells has been previously proposed. Two independent studies by Ying et al. [79] and Terada et al. [80] examined such spontaneous cell fusion events. They proposed that the ability of different cell types to demonstrate alternative phenotypes, originally perceived as *'trans'*-differentiation, could be attributable to an alternative mechanism of spontaneous cell fusion. Specifically, the acquirement of alternative phenotypes was found not necessarily to be the result of *'trans'*-differentiation, since similar phenomena were observed following spontaneous fusion of two different cell types. The resultant hybrid cells displayed the phenotypic traits of the recipient cell. Detailed cytogenetic analysis would therefore be required to distinguish between cells that had truly undergone *'trans'*-differentiation and cells that had undergone spontaneous fusion. It is known that that bone marrow cells can spontaneously fuse with other cell types in co-culture systems and, in doing so, acquire phenotypic traits of these cells [79, 80] and the acquirement of these phenotypic properties can extend to plasticity. There are studies for and against such observations, making it difficult to determine exactly how cells are behaving. For example, cell fusion events may not completely discount the phenomenon of *'trans'*-differentiation as Crain et al. demonstrated that MSCs can *'trans'*-differentiate into neural cell types by mechanisms independent of cell fusion [81]. In a study by Alvazrez-Dolado et. al.,, it has been shown that MSCs could spontaneously fuse with neural cell types [82].

Controversy aside, by applying principles introduced by Ying et al. [79] and Terada et al. [80], spontaneous fusion of MSCs with host neural cells following transplantation into the nervous system could cause MSCs to undergo genetic reprogramming and subsequently adopt phenotypic traits of the recipient neural cell. As such, spontaneous fusion events might provide a potential mechanism by which MSC transplantation contributes towards functional recovery in neurological and neurodegenerative systems. However, as is the case for *'trans'*-differentiation of MSCs, spontaneous cell fusion occurs at an extremely low frequency. In the study by Terada and co-workers, a fusion rate of 2–11 clones per million cells has been reported [80]. Together, the low frequency of both *'trans'*-differentiation and spontaneous cell fusion suggest that both of these phenomena are unlikely to fully account for the observed neurological improvements.

MSCs AS TROPHIC MEDIATORS OF NEURAL DIFFERENTIATION

The low frequency at which *'trans'*-differentiation and cell fusion events occur means that they are unlikely to account for the neurological recovery observed following MSC transplantation in models of neurological and neurodegenerative disorders. However, that is not to discount these mechanisms completely. It is possible that both might contribute in part towards observed functional recovery; however, the low frequency with which they occur suggests there may also be other mechanisms at play.

By definition, stem cells (a) display the capacity for self-renewal and (b) possess potential to differentiate into more mature cell types. This functional definition continues to evolve and there is considerable evidence to suggest that stem cells have an important role within their niche including the release and/or uptake of trophic factors and cytokines. This intercellular communication occurs between cells in close proximity to one another and may be *paracrine* or *autocrine*. The term 'paracrine' is used to describe signalling that occurs between two different cell types i.e. a cell-derived soluble factor acting upon a different cell type to that which secreted it. Conversely, the term 'autocrine' describes signalling that occurs between two cells of the same type i.e. a cell-derived soluble factor acting upon a cell of the same type as that which secreted it. The concept of stem cells having a trophic role in addition to their capacity to differentiate has been described for both embryonic and adult stem cell populations, including haematopoietic and neural stem cells, among others [83–85]. In regard to MSCs, there are several studies demonstrating trophic roles *in vivo*, and perhaps the best characterised of these is the trophic role MSCs perform in the regulation of haematopoiesis [86]. For the remainder of the chapter, we consider the third and likely predominant mechanism by which MSCs are understood to promote functional recovery from neurological deficits which involves the trophic influence of MSCs on neighbouring cells. More specifically, a cocktail of 'factors' secreted by MSCs has been proposed to ameliorate neural damage by a paracrine mechanism, exerting a neurogenic effect on resident neural stem and progenitor cells to promote endogenous repair processes within the host nervous system [87, 88]. Such factors released by MSCs can be loosely categorised into two groups: *neurotrophic factors* and *non-neurotrophic factors*.

SECRETION OF NEUROTROPHIC FACTORS BY MSCs

There have been numerous reports suggesting that MSCs secrete a wide array of neurotrophins, growth factors, cytokines and other soluble factors. Work from both *in vitro* and *in vivo* settings indicate that such factors from MSCs promote cell proliferation, survival and differentiation. Several of the soluble factors secreted by MSCs are known to exert neurogenic effects on neural stem and progenitor cells and include molecules such as nerve growth factor (NGF), brain-derived neurotrophic factor (BDNF), glial cell line-derived neurotrophic factor (GDNF), neurotrophin-3 (NT-3), basic fibroblast growth factor (bFGF), vascular endothelial growth factor (VEGF), hepatocyte growth factor (HGF) and ciliary neurotrophic factor (CNTF) [87–89]. In addition, the 'secretome' of MSCs also consists of a wide variety of interleukin (IL) molecules, including IL-6, -7, -8, -11, -12, -14 and 15, and this is in keeping with their housekeeping role as trophic mediators of haematopoiesis. It has long been established that a complement of IL molecules play key roles throughout the process of haematopoiesis, although it is now becoming increasingly apparent that IL molecules may also function during neural development. This possibly indicates crosstalk between haematopoietic and neural differentiation pathways, further supporting MSCs as trophic mediators in both capacities [90, 91].

Proteomic analysis of the MSC secretome suggests that it is likely to be a combination of soluble factors and cytokines, rather than the activity of any single factor that is responsible for neurological recovery following MSC transplantation [92]. The beneficial effects of MSC-derived factors in mediating neurological recovery are highly documented in the literature, and effects of these factors can be broadly classified into five main categories: *angiogenic effects* (involving the formation of new blood vessels), *neurogenic effects* (involving the formation of new neural tissue), *neuroprotective effects* (protection of neural tissue from degeneration and apoptosis), *synaptogenic effects* (involving the formation of synapses and synaptic contacts) and *inhibition of scarring* (prevention of scarring that would prevent reconstruction of neural circuitry following injury or damage) [93–99]. Here we will focus on a few key studies demonstrating MSCs as trophic mediators of neurological recovery.

In 2002, Chen et al., demonstrated that the major mechanism behind MSC-mediated repair of neural tissue in an *in vitro* model of stroke was a result of cytokine and soluble factor secretion by MSCs [89]. Even more intriguing was the convincing argument that the MSC secretome was significantly influenced by the microenvironment in which MSCs were cultured, suggesting that MSCs are capable of modifying their secretory profile in response to various environmental cues. Demonstration of this latter point involved an *in vitro* model system in which Chen and co-workers collected supernatant media that had been exposed either to normal rat brain tissue or to rat brain tissue that had been subjected to middle cerebral artery occlusion (MCAo), a model of stroke in rat. Human MSCs were cultured in these supernatants and allowed to condition their media, which was subsequently collected and analysed using a human-specific ELISA to determine levels of the neurotrophic factors BDNF, NGF, bFGF, VEGF and HGF. However, authors did point out that, such was the nature of ELISA technique used in their study, a degree of cross-reactivity with rat-derived cytokines could not be discounted (potentially carried over in supernatants). Following exposure to ischaemic brain supernatant compared to normal brain supernatant, MSCs produced and secreted elevated quantities of the neurotrophic factors BDNF, NGF, VEGF and HGF [89]. These data suggest that perhaps the main mechanism *in vivo* by which MSCs ameliorate neural damage is the result of production and secretion of neurotrophic factors and cytokines that stimulate endogenous recovery. This study has been successful in demonstrating that not only do MSCs secrete neurotrophic factors and cytokines but, perhaps more importantly in terms of cellular therapy of neurological deficits, that MSCs are able to alter their secretory profile in response to specific micro-environmental cues i.e. increased secretion of neurotrophic factors in response to neural damage. Of additional interest from a clinical perspective, the secretome of MSCs not only varied in response to different environmental cues, but also varied depending on the time point at which the brain tissue was extracted following MCAo. This is a crucial factor requiring further investigation if MSC-based cellular therapies are to be used successfully in the treatment of neurological and neurodegenerative disorders. Delayed administration of MSCs post-injury might not be as efficacious as when administered immediately.

In a more recent study by the same group, Chen et al. used sciatic nerve injury in rat as a model to ascertain whether MSC transplantation could be used as a means of treating peripheral nerve damage in addition to CNS damage [87]. The sciatic nerve is one of the most commonly used in such studies, and is perhaps the most useful as it is one of the longest nerves in the body and is found in the lower limbs. Functional recovery following MSC transplantation was assessed using the sciatic function index (SFI). While a detailed description of SFI is superfluous herein, it is useful to point out the scale on which SFI is measured, which runs from 0 to −100, where a value of 0 is indicative of normal nerve function and a value of −100 is indicative of complete sciatic nerve dysfunction. Sciatic nerve injury was induced, followed by MSC administration, to ascertain whether MSC transplantation could mediate functional recovery following damage to peripheral nerves. To assess functional recovery, SFI was determined periodically for up to 10 weeks post-transplantation. Significantly higher SFI values have been reported in animals subjected to MSC transplantation compared to controls from the fifth week after treatment onwards. It appears therefore that transplantation of MSCs also results in clear beneficial effects following peripheral nerve damage. In terms of the potential mechanism(s) by which MSC transplantation mediated functional repair, an increase in levels of NGF, BDNF, GDNF and CNTF, and similarly increased expression of the extracellular matrix proteins, collagens I and IV, fibronectin and laminin, was observed. This suggested that not only were MSCs secreting soluble neurotrophic factors and cytokines that potentially stimulated endogenous repair mechanisms via their effects on immature neural stem and progenitor cells, but they also produced molecules that modified the microenvironment within the neurologically damaged area. This phenomenon may also have been key in mediating functional recovery by encouraging beneficial events, such as neurite outgrowth, which would contribute toward the reconstruction of neural circuitry.

These data support the notion that transplanted MSCs release trophic factors and cytokines that mediate functional recovery following neurological damage in the CNS and to peripheral nerves. However, there is little evidence in these studies to demonstrate that these MSC-derived factors act directly on neural stem and progenitor cells. To investigate this issue, Munoz et al., administered MSCs directly into the dentate gyrus of the hippocampus, an area known to be rich in neural stem and progenitor cells [100]. A substantial increase in the proliferation and migration of host neural stem cells was noted, at the expense of MSC proliferation and migration. The increase in proliferation and migration of resident neural stem and progenitor cells was attributed to increased levels of NGF, VEGF, CNTF and bFGF in the hippocampus following MSC transplantation, strongly suggesting that MSCs were at least partly responsible for their secretion. This study also claimed that MSC-derived soluble factors not only influence neural stem and progenitor cells to undergo proliferation and differentiation events, but also function by 'activating' resident astrocytes within the niche. A contribution by 'activated' astrocytes towards the increased neurotrophin levels observed in the hippocampus following MSC transplantation cannot be discounted, nor can the possibility of 'activated' astrocytes directly influencing neurogenesis of resident neural stem and progenitor cells. An astrocytic contribution to neural recovery concomitant with the direct effects of MSC-derived factors is not unlikely, as there have been numerous reports that astrocytes have a trophic role in neurogenesis [101, 102].

SECRETION OF NON-NEUROTROPHIC FACTORS BY MSCs

As MSCs are essentially a population of stromal cells it comes as no surprise that in addition to the array of soluble factors and cytokines they also secrete and deposit a wide range of extracellular matrix proteins, including collagen, laminin and fibronectin that may also contribute to modify the niche of neural stem cells [87]. Protein modification of the niche in which stem cells reside, might influence the behaviour of these stem cells and that of their neighbours, thereby promoting events that would facilitate neurological recovery, for example, neurite extension.

By analysing the transcriptome of a clonal population of MSCs using a serial analysis of gene expression (SAGE) approach, Crigler and co-workers demonstrated (and validated) that MSCs express a range of extracellular matrix and adhesion molecules that might contribute to MSC-mediated recovery of neural injury *in vivo* [103, 104]. Examples of molecules expressed by MSCs include netrin-4 and reticulon-4 (previously reported to be involved in axonal guidance), ninjurin-1/2 and astrotactin (previously reported to be involved in neural cell adhesion), and prosaposin and pleiotrophin (previously reported to be neurite-inducing factors), among others. Hence, expression and secretion of extracellular matrix and/or adhesion molecules may also contribute to the mechanism of MSC-mediated recovery of neural damage. In reality, MSC-mediated amelioration of neural damage is likely to occur by a combination of neurotrophic factors acting directly on resident neural stem and progenitor cells, and also supporting extracellular proteins that exert a chaemotactic effect on neurite outgrowth facilitating neural recovery.

ANALYSIS OF MSC-DERIVED SOLUBLE FACTORS *IN VITRO*: A CONDITIONED MEDIA APPROACH

One of the primary methods that has been used to analyse the secretome of various cell types involves the production and subsequent analysis of conditioned media (CM) produced by such cells under certain growth conditions. CM is a term used to describe media in which cells have already been cultured, and thus might contain a mixture of trophic factors and cytokines that have been actively secreted by cells. A cocktail of cell-derived soluble factors may be useful in the subsequent culture of other cell types whether this be for routine growth or to induce other events such as cellular differentiation. Accordingly, the study of the effect(s) of MSC-conditioned media (MSC-CM) on cell fate decisions by neural stem and progenitor cells coupled with the appropriate analysis of MSC-CM media might lead to the identification of key combinations of factors responsible for such activity.

In 2006, Rivera et al. demonstrated that MSC-derived soluble factors and cytokines induced neural stem cells to acquire a predominately oligodendrocytic phenotype at the expense of an astrocytic cell fate [105]. In this study, rat MSCs were cultured under standard culture conditions in the presence of 10% FCS and were allowed to condition the medium for 3 days. The presence of serum in any CM can be problematic in subsequent analysis, since serum itself is a complex mixture of unknown nutrients, hormones and growth factors that are often necessary for routine cell culture. It can be extremely difficult to distinguish between proteins that have been secreted by cells in culture and extraneous protein from the serum, unless sophisticated proteomic techniques are employed to distinguish between species-specific proteins. Additionally, high levels of proteins such as albumin and transferrin in serum might dominate subsequent analysis, masking cell-specific factors that have been secreted and which are likely to be present in considerably lower concentrations. After allowing MSCs to condition the medium for 3 days, Rivera et al., used the MSC-CM to culture adult rat neural stem cells for a period of 7 days. To assess the effect of MSC-CM, and by extension MSC-derived soluble factors on the differentiation of adult rat neural stem cells, analysis of a range of neural-specific markers was employed by the group in order to (a) determine whether the neural stem cells had been induced to differentiate and, if so, (b) determine whether differentiation resulted in the formation of neurons, astrocytes or oligodendrocytes. To serve as an appropriate control, neural stem cells cultured in unconditioned medium were also analysed. Neural stem cells cultured in MSC-CM for 7 days demonstrated a significant increase in expression of the oligodendrocytic markers galactocerebroside (GalC) and myelin basic protein (MBP), and this was concomitant with a significant decrease in the expression of the astrocytic marker, GFAP. These findings suggest that the cocktail of soluble factors secreted by MSCs induced the adult neural stem cells to preferentially form oligodendrocytes at the expense of astrocytes.

It is important, however, to determine whether the MSC-derived factors were exerting an instructive or selective effect. If operating by an instructive mechanism, then the soluble factors present in MSC-CM could direct immature, non-committed neural stem cells to undergo oligodendrocyte differentiation. Conversely, if operating via a selective mechanism, then these factors may be providing conditions that select for the enhanced growth of precursors already committed toward an oligodendrocytic fate and/or providing sub-optimal growth conditions for precursors committed to astrocytic and neuronal fates. To address this issue, Rivera and co-workers [105] evaluated the effect of MSC-CM on rates of cell proliferation and cell death within different subpopulations of the neuro-progenitor cells that differentially expressed cell type specific markers of glia and neurons. These experiments concluded that the acquirement of an oligodendrocyte phenotype produced by the neural stem cells occurred independently of rates of cell proliferation and cell death in the different progenitor subpopulations, suggesting that MSC-derived soluble factors were functioning by an instructive rather than selective mechanism.

In a separate study conducted in 2004, Wislet-Gendebien et al., investigated effects of CM derived from rat MSCs on the differentiation of mouse neuro-progenitor cells from the embryonic (E16) mouse [106]. Specifically, the Wislet-Gendebien group produced and analysed media that had been conditioned by nestin-positive rat MSCs (npMSCs). Under standard growth conditions, MSCs are largely negative for expression of neural antigens such as nestin. However, appropriate manipulation of the culture conditions can induce MSCs to express nestin, such as the removal of serum from the culture medium and continued culture [107]. As MSCs have been reported in numerous studies to demonstrate neural antigen expression following transplantation *in vivo*, it may be considered more relevant to examine media conditioned by neural antigen-positive MSCs rather than neural antigen-negative MSCs. Mouse embryonic neural progenitors were grown in the presence of CM from npMSCs for a period of 5 days. During this time, expression of the astrocytic marker, GFAP, increased while there was a concomitant decrease in the percentage of cells expressing neuronal (Tuj1) and oligodendrocytic (O4) markers [106]. Similarly, rates of cell proliferation and cell death were examined in the appropriate subpopulations of the mouse embryonic progenitor cultures following culture with npMSC-CM. Although no changes in cell proliferation were noted, a significant decrease in the rate of cell death of the GFAP-positive subpopulation was recorded after the 48-h treatment with npMSC-CM. It is possible that a component of the npMSC-CM inhibited the death of GFAP-positive cells. In an attempt to identify the factor(s) responsible for influencing the differentiation of astrocytes, npMSCs were screened for a number of candidate factors. It was found that npMSCs showed an up-regulation of the biologically active form of bone morphogenetic protein 4 (BMP4) at both the mRNA and protein level. MSCs that were negative for nestin expression did not express active BMP4. Accordingly, it was concluded that the effect of npMSC-CM on astrocyte differentiation by embryonic neuro-progenitor cells was attributable, at least in part, to the secretion of active BMP4 [106].

It is now becoming increasingly accepted that the likely predominant mechanism by which MSC transplantation results in functional recovery following neural injury is via the release of trophic factors and cytokines that act on resident neural stem and progenitor cells, thereby promoting endogenous repair processes. While MSCs have been reported to secrete a wide range of neurotrophic factors and cytokines, our knowledge in this area remains at an early stage. The two studies highlighted herein demonstrate that the effect of MSC-derived soluble factors on the cell fate decisions of neural stem and progenitor cells can vary significantly, depending on a range of factors, including the developmental state of the MSC population, the source of neural stem cells (embryonic versus adult), etc. However, despite the obvious gaps in our understanding, there have already been several efforts to exploit the natural beneficial properties of MSCs, by genetically engineering cells to over-express the neurotrophic factors and cytokines that promote functional recovery from neurological deficits. There have recently been two studies using herpes simplex virus (HSV) as a vector to deliver target genes that enhance the expression and secretion of the growth factors HGF and FGF [108, 109]. Results achieved by this research have been promising, with high rates of transfection efficiency reported, and the efficacy of genetically engineered MSCs in mediating functional recovery from neural injury being enhanced above that of control MSCs. As a consequence, MSCs might provide useful vehicles for the delivery of highly efficacious levels of neurotrophic factors and cytokines to areas of neurological and neurodegenerative damage.

REFERENCES

1. Pittenger M.F., A.M. Mackay, S.C. Beck, et al. (1999). 'Multilineage potential of adult human mesenchymal stem cells'. *Science.* **284**(5411): 143–147.
2. Friedenstein A.J., J.F. Gorskaja, N.N. Kulagina (1976). 'Fibroblast precursors in normal and irradiated mouse haematopoietic organs'. *Exp. Haematol.* **4**(5): 267–274.
3. Kuznetsov S.A., M.H. Mankani, S. Gronthos, K. Satomura, P. Bianco, P.G. Robey (2001). 'Circulating skeletal stem cells'. *J. Cell. Biol.* **153**(5): 1133–1140.
4. Rosada C., J. Justesen, D. Melsvik, P. Ebbesen, M. Kassem (2003). 'The human umbilical cord blood: A potential source for osteoblast progenitor cells'. *Calcif. Tissue. Int.* **72**(2): 135–142.
5. F. Vandenabeele, C. De Bari, M. Moreels, et al. (2003). 'Morphological and immunocytochemical characterisation of cultured fibroblast-like cells derived from adult human synovial membrane'. *Arch. Histol. Cytol.* **66**(2): 145–153.

6. Miura M., S. Gronthos, M. Zhao, et al. (2003). 'SHEDx Stem cells from human exfoliated deciduous teeth. Proc. Natl'. *Acad. Sci. USA.* **100**(10): 5807–5812.
7. Gronthos S., D.M. Franklin, H.A. Leddy, P.G. Robey, R.W. Storms, J.M. Gimble (2001). 'Surface protein characterisation of human adipose tissue-derived stromal cells'. *J. Cell. Physiol.* **189**(1): 54–63.
8. Igura K., X. Zhang, K. Takahashi, A. Mitsuru, S. Yamaguchi, T.A. Takashi (2004). 'Isolation and characterisation of mesenchymal progenitor cells from chorionic villi of human placenta'. *Cytotherapy.* **6**(6): 543–553.
9. Tsai M.S., J.L. Lee, Y.J. Chang, S.M. Hwang (2004). 'Isolation of human multipotent mesenchymal stem cells from second-trimester amniotic fluid using a novel two-stage culture protocol'. *Hum. Reprod.* **19**(6): 1450–1456.
10. Anker P.S. in't, W.A. Noort, S.A. Scherjon, et al. (2003). 'Mesenchymal stem cells in human second-trimester bone marrow, liver, lung, and spleen exhibit a similar immunophenotype but a heterogeneous multilineage differentiation potential'. *Haematologica.* **88**(8): 845–852.
11. Salingcarnboriboon R. H. Yoshitake, K. Tsuji, et al. (2003). 'Establishment of tendon-derived cell lines exhibiting pluripotent mesenchymal stem cell-like property'. *Exp Cell Res.* **287**(2):289–300.
12. Seo BM, M. Miura, S. Gronthos, et al. (2004). 'Investigation of multipotent postnatal stem cells from human periodontal ligament'. *Lancet.* **364**(9429): 149–155.
13. da Silva Meirelles L, P.C. Chagastelles, N.B. Nardi (2006). 'Mesenchymal stem cells reside in virtually all post-natal organs and tissues'. *J Cell Sci.* **119**(Pt 11):2204–2213.
14. Farrington-Rock C., N.J. Crofts, M.J. Doherty, B.A. Ashton, C. Griffin-Jones, AE. Canfield (2004). 'Chondrogenic and adipogenic potential of microvascular pericytes'. *Circulation* H.H. **110**(15): 2226–2232.
15. Shi S, S. Gronthos (2003). 'Perivascular niche of postnatal mesenchymal stem cells in human bone marrow and dental pulp'. *J Bone Miner Res.* **18**(4): 696–704.
16. Digirolamo CM, D. Stokes, D. Colter, D.G. Phinney, R. Class, D.J. Prockop (1999). 'Propagation and senescence of human marrow stromal cells in culture: A simple colony-forming assay identifies samples with the greatest potential to propagate and differentiate'. *Br J Haematol.* **107**(2): 275–281.
17. Murphy J.M., K. Dixon, S. Beck, D. Fabian, A. Feldman, F. Barry (2002). 'Reduced chondrogenic and adipogenic activity of mesenchymal stem cells from patients with advanced osteoarthritis'. *Arthritis Rheum.* **46**(3): 704–713.
18. D'Ippolito G., P.C. Schiller, C. Perez-stable, W. Balkan, B.A. Roos, G.A. Howard. (2002). 'Cooperative actions of hepatocyte growth factor and 1, 25-dihydroxyvitamin D3 in osteoblastic differentiation of human vertebral bone marrow stromal cells'. *Bone.* **31**(2): 269–275.
19. Shyu K.G., B.W. Wang, H.F. Hung, C.C. Chang, D.T. Shih (2006). 'Mesenchymal stem cells are superior to angiogenic growth factor genes for improving myocardial performance in the mouse model of acute myocardial infarction'. *J Biomed Sci.* **13**(1): 47–58.
20. Alsalameh S., R. Amin, T. Gemba, M. Lotz (2004). 'Identification of mesenchymal progenitor cells in normal and osteoarthritic human articular cartilage'. *Arthritis Rheum.* **50**(5): 1522–1532.
21. Dennis J.E., J.P. Carbillet, A.I. Caplan, P. Charbord (2002). 'The STRO-1+ marrow cell population is multipotential'. *Cells Tissues Organs.* **170**(2–3): 73–82.
22. Woodbury D., Reynolds K., I.B. Black (2002). 'Adult bone marrow stromal stem cells express germline, ectodermal, endodermal, and mesodermal genes prior to neurogenesis'. *J Neurosci. Res.* **69**(6): 908–917.

23. Krause D.S., N.D. Theise, M.I. Collector, et al. (2001). 'Multi-organ, multi-lineage engraftment by a single bone marrow-derived stem cell'. *Cell.* **105**(3): 369–377.
24. Petersen B.E., W.C. Bowen, K.D. Patrene, et al. (1999). 'Bone marrow as a potential source of hepatic oval cells'. *Science.* **284**(5417): 1168–1170.
25. Woodbury D., E.J. Schwarz, D.J. Prockop, I.B. Black. (2000). 'Adult rat and human bone marrow stromal cells differentiate into neurons'. *J Neurosci Res.* **61**(4): 364–370.
26. Sanchez-Ramos J., S. Song, F. Cardozo-Pelaez, et al. (2000). 'Adult bone marrow stromal cells differentiate into neural cells *in vitro*'. *Exp Neurol.* **164**(2): 247–256.
27. Sato Y., H. Araki, J. Kato, et al. (2005). 'Human mesenchymal stem cells xenografted directly to rat liver are differentiated into human hepatocytes without fusion'. *Blood.* **106**(2): 756–763.
28. Aggarwal S., M.F. Pittenger (2005). 'Human mesenchymal stem cells modulate allogeneic immune cell responses'. *Blood.* **105**(4): 1815–1822.
29. Le Blanc K., C. Tammik, K. Rosendahl, E. Zetterberg, O. Ringden (2003). 'HLA expression and immunologic properties of differentiated and undifferentiated mesenchymal stem cells'. *Exp Haematol.* **31**(10): 890–896.
30. Majumdar M.K., M. Keane-Moore, D. Buyaner, et al. (2003). 'Characterisation and functionality of cell surface molecules on human mesenchymal stem cells'. *J Biomed Sci.* **10**(2): 228–241.
31. Le Blanc K., I. Rasmusson, B. Sundberg, et al. (2004). 'Treatment of severe acute graft-versus-host disease with third party haploidentical mesenchymal stem cells'. *Lancet.* **363**(9419): 1439–1441.
32. Barry F.P., J.M. Murphy (2004). 'Mesenchymal stem cells: Clinical applications and biological characterisation'. *Int J Biochem Cell Biol.* **36**(4): 568–584.
33. Prockop D.J (1997). 'Marrow stromal cells as stem cells for nonhaematopoietic tissues'. *Science.* **276**(5309): 71–74.
34. Gomillion C.T., K.J. Burg (2006). 'Stem cells and adipose tissue engineering'. *Biomaterials.* **27**(36): 6052–6063.
35. Patrick C.W., Jr. (2001). 'Tissue engineering strategies for adipose tissue repair'. *Anat Rec.* **263**(4): 361–366.
36. Langstein H.N., G.L. Robb (1999). 'Reconstructive approaches in soft tissue sarcoma'. *Semin Surg Oncol.* **17**(1): 52–65.
37. Katz A.J., R. Llull, M.H. Hedrick, J.W. Futrell (1999). 'Emerging approaches to the tissue engineering of fat'. *Clin Plast Surg.* **26**(4): 587–603, viii.
38. Tzikas T.L. (2004). 'Lipografting: Autologous fat grafting for total facial rejuvenation'. *Facial Plast Surg.* **20**(2): 135–143.
39. Rooney D.P., M.F. Ryan (2006). 'Diabetes with partial lipodystrophy following sclerodermatous chronic graft vs. host disease'. *Diabet Med.* **23**(4): 436–440.
40. Mauney J.R., V. Volloch, D.L. Kaplan (2005). 'Matrix-mediated retention of adipogenic differentiation potential by human adult bone marrow-derived mesenchymal stem cells during *ex vivo* expansion'. *Biomaterials.* **26**(31): 6167–6175.
41. Patrick C.W., Jr. Adipose tissue engineering: (2000). 'The future of breast and soft tissue reconstruction following tumor resection'. *Semin Surg Oncol.* **19**(3): 302–311.
42. Mandrup S., M.D. Lane (1997). 'Regulating adipogenesis'. *J Biol Chem.* **272**(9): 5367–5370.
43. Hsu T.L., F.Y. Chiu, C.M. Chen, T.H. Chen (2005). 'Treatment of nonunion of humeral shaft fracture with dynamic compression plate and cancellous bone graft'. *J Chin Med Assoc.* **68**(2): 73–76.

44. Zeitlin L., Fassier F., F.H. Glorieux (2003). 'Modern approach to children with osteogenesis imperfecta'. *J Pediatr Orthop B.* **12**(2): 77–87.
45. Hatano H., A. Ogose, T. Hotta, N. Endo, H. Umezu, T. Morita (2005). 'Extracorporeal irradiated autogenous osteochondral graft: A histological study'. *J Bone Joint Surg Br.* **87**(7): 1006–1011.
46. Salgado A.J., O.P. Coutinho, R.L. Reis (2004). 'Bone tissue engineering: State of the art and future trends'. *Macromol Biosci.* **4**(8): 743–765.
47. Damien C.J., J.R. Parsons (1991). 'Bone graft and bone graft substitutes: A review of current technology and applications'. *J Appl Biomater.* **2**(3): 187–208.
48. Fuchs E., T. Tumbar, G. Guasch (2004). 'Socializing with the neighbors: Stem cells and their niche'. *Cell* **116**(6): 769–778.
49. Azizi S.A., D. Stokes, B.J. Augelli, C. Di Girolamo, D.J. Prockop (1998). 'Engraftment and migration of human bone marrow stromal cells implanted in the brains of albino rats—similarities to astrocyte grafts'. *Proc Natl Acad Sci USA.* **95**(7): 3908–3913.
50. Andersson C., M. Tytell, J. Brunso-Bechtold (1993). 'Transplantation of cultured type 1 astrocyte cell suspensions into young, adult and aged rat cortex: Cell migration and survival'. *Int J Dev Neurosci.* **11**(5): 555–568.
51. Zhou H.F., R.D. Lund (1992). 'Migration of astrocytes transplanted to the midbrain of neonatal rats'. *J Comp Neurol.* **317**(2): 145–155.
52. Deng Y.B., X.G. Liu, Z.G. Liu, X.L. Liu, Y. Liu, G.Q. Zhou (2006). 'Implantation of BM mesenchymal stem cells into injured spinal cord elicits de novo neurogenesis and functional recovery: Evidence from a study in rhesus monkeys'. *Cytotherapy.* **8**(3): 210–214.
53. Kopen G.C., D.J. Prockop, D.G. Phinney (1999). 'Marrow stromal cells migrate throughout forebrain and cerebellum, and they differentiate into astrocytes after injection into neonatal mouse brains'. *Proc Natl Acad Sci USA.* **96**(19): 10711–10716.
54. Isakova I.A., K. Baker, J. Dufour, D. Gaupp, D.G. Phinney (2006). 'Preclinical evaluation of adult stem cell engraftment and toxicity in the CNS of rhesus macaques'. *Mol Ther.* **13**(6): 1173–1184.
55. Phinney D.G., M. Baddoo, M. Dutreil, D. Gaupp, W.T. Lai, I.A. Isakova (2006). 'Murine mesenchymal stem cells transplanted to the central nervous system of neonatal versus adult mice exhibit distinct engraftment kinetics and express receptors that guide neuronal cell migration'. *Stem Cells Dev.* **15**(3): 437–447.
56. Ankeny D.P., D.M. McTigue, L.B. Jakeman (2004). 'Bone marrow transplants provide tissue protection and directional guidance for axons after contusive spinal cord injury in rats'. *Exp Neurol.* **190**(1): 17–31.
57. Chen J., Y. Li, M. Katakowski, et al. (2003). 'Intravenous bone marrow stromal cell therapy reduces apoptosis and promotes endogenous cell proliferation after stroke in female rat'. *J Neurosci Res.* **73**(6): 778–786.
58. Li Y., J. Chen, M. Chopp (2001). 'Adult bone marrow transplantation after stroke in adult rats'. *Cell Transplant.* **10**(1): 31–40.
59. Mahmood A., D. Lu, L. Yi, J.L. Chen, M. Chopp 'Intracranial bone marrow transplantation after traumatic brain injury improving functional outcome in adult rats'. (2001). *J Neurosurg.* **94**(4): 589–595.
60. Lescaudron L., D. Unni, G.L. Dunbar (2003). 'Autologous adult bone marrow stem cell transplantation in an animal model of huntington's disease: Behavioural and morphological outcomes'. *Int J Neurosci.* **113**(7): 945–956.

61. Mazzini L., K. Mareschi, I., Ferrero et al. (2007). 'Stem cell treatment in Amyotrophic Lateral Sclerosis'. *J Neurol Sci.*
62. Dezawa M., H. Kanno, M. Hoshino, et al. (2004). 'Specific induction of neuronal cells from bone marrow stromal cells and application for autologous transplantation'. *J Clin Invest.* **113**(12): 1701–1710.
63. Chopp M., X.H. Zhang, Y. Li, et al. (2000). 'Spinal cord injury in rat: Treatment with bone marrow stromal cell transplantation'. *Neuroreport.* **11**(13): 3001–3005.
64. Cuevas P., F. Carceller, M. Dujovny, et al. (2002). 'Peripheral nerve regeneration by bone marrow stromal cells'. *Neurol Res.* **24**(7): 634–638.
65. Li Y., J. Chen, L. Wang, L. Zhang, M. Lu, M. Chopp (2002). 'Intracerebral transplantation of bone marrow stromal cells in a 1-methyl-4-phenyl-1,2,3,6-tetrahydropyridine mouse model of Parkinson's disease'. *Neurosci Lett.* **316**(2): 67–70.
66. Mandalfino P., G. Rice, A. Smith, J.L. Klein, L. Rystedt, G.C. Ebers (2000). 'Bone marrow transplantation in multiple sclerosis'. *J Neurol.* **247**(9): 691–695.
67. Li Y., J. Chen, C.L. Zhang, et al. (2005). 'Gliosis and brain remodeling after treatment of stroke in rats with marrow stromal cells'. *Glia.* **49**(3): 407–417.
68. Lu P., A. Blesch, M.H. Tuszynski (2004). 'Induction of bone marrow stromal cells to neurons: Differentiation, transdifferentiation, or artefact'? *J Neurosci Res.* **77**(2): 174–191.
69. Neuhuber B., G. Gallo, L. Howard, L. Kostura, A. Mackay, I. Fischer (2004). 'Reevaluation of *in vitro* differentiation protocols for bone marrow stromal cells: Disruption of actin cytoskeleton induces rapid morphological changes and mimics neuronal phenotype'. *J Neurosci Res.* **77**(2): 192–204.
70. Bertani N., P. Malatesta, G. Volpi, P. Sonego, R. Perris (2005). 'Neurogenic potential of human mesenchymal stem cells revisited: Analysis by immunostaining, time-lapse video and microarray'. *J Cell Sci.* **118**(Pt 17): 3925–3936.
71. Croft A.P., S.A. Przyborski (2006). 'Formation of neurons by non-neural adult stem cells: Potential mechanism implicates an artefact of growth in culture'. *Stem Cells.* **24**(8): 1841–1851.
72. Lamoury F.M., J. Croitoru-Lamoury, B.J. Brew (2006). 'Undifferentiated mouse mesenchymal stem cells spontaneously express neural and stem cell markers Oct-4 and Rex-1'. *Cytotherapy.* **8**(3): 228–242.
73. Deng J., B.E. Petersen, D.A. Steindler, M.L. Jorgensen, E.D. Laywell (2006). 'Mesenchymal stem cells spontaneously express neural proteins in culture and are neurogenic after transplantation'. *Stem Cells.* **24**(4): 1054–1064.
74. Tondreau T., L. Lagneaux, M. Dejeneffe, et al.(2004). 'Bone marrow-derived mesenchymal stem cells already express specific neural proteins before any differentiation'. *Differentiation.* **72**(7): 319–326.
75. Li Y, Chen J., X.G. Chen, et al. (2002). 'Human marrow stromal cell therapy for stroke in rat: Neurotrophins and functional recovery'. *Neurology.* **59**(4): 514–523.
76. Mezey E., K.J. Chandross (2000). 'Bone marrow: A possible alternative source of cells in the adult nervous system'. *Eur J Pharmacol.* **405**(1–3): 297–302.
77. Lu J., S. Moochhala, X.L. Moore, et al. (2006). 'Adult bone marrow cells differentiate into neural phenotypes and improve functional recovery in rats following traumatic brain injury'. *Neurosci Lett.* **398**(1–2): 12–17.
78. Munoz-Elias G., A.J. Marcus, T.M. Coyne, D. Woodbury, I.B. Black 'Adult bone marrow stromal cells in the embryonic brain: Engraftment, migration, differentiation, and long-term survival'. (2004). *J Neurosci.* **24**(19): 4585–4595.

79. Ying Q.L., J. Nichols, E.P. Evans, A.G. Smith 'Changing potency by spontaneous fusion'. (2002). *Nature.* **416**(6880): 545–548.
80. Terada N., T. Hamazaki, M. Oka, (2002). et al. 'Bone marrow cells adopt the phenotype of other cells by spontaneous cell fusion'. *Nature.* **416**(6880): 542–545.
81. Crain B.J., S.D. Tran, E. Mezey (2005). 'Transplanted human bone marrow cells generate new brain cells'. *J Neurol Sci.* **233**(1–2): 121–123.
82. Alvarez-Dolado M., R. Pardal, J.M. Garcia-Verdugo, (2003). et al. 'Fusion of bone-marrow-derived cells with Purkinje neurons, cardiomyocytes and hepatocytes'. *Nature.* **425**(6961): 968–973.
83. Lu P., L.L. Jones, E.Y. Snyder, M.H. Tuszynski (2003). 'Neural stem cells constitutively secrete neurotrophic factors and promote extensive host axonal growth after spinal cord injury'. *Exp Neurol.* **181**(2): 115–129.
84. Guo Y., B. Graham-Evans, H.E. Broxmeyer (2006). 'Murine embryonic stem cells secrete cytokines/growth modulators that enhance cell survival/anti-apoptosis and stimulate colony formation of murine haematopoietic progenitor cells'. *Stem Cells.* **24**(4): 850–856.
85. Cabanes C., S. Bonilla, L. Tabares, S. Martinez (2007). 'Neuroprotective effect of adult haematopoietic stem cells in a mouse model of motoneuron degeneration'. *Neurobiol Dis.* **26**(2): 408–418.
86. Haynesworth S.E., M.A. Baber, A.I. Caplan (1996). 'Cytokine expression by human marrow-derived mesenchymal progenitor cells *in vitro*: Effects of dexamethasone and IL-1 alpha'. *J Cell Physiol.* **166**(3): 585–592.
87. Chen C.J., Y.C. Ou, S.L. Liao, (2007). et al. 'Transplantation of bone marrow stromal cells for peripheral nerve repair'. *Exp Neurol.* **204**(1): 443–453.
88. Chen Q., Y. Long, X. Yuan, et al. (2005). 'Protective effects of bone marrow stromal cell transplantation in injured rodent brain: Synthesis of neurotrophic factors'. *J Neurosci Res.* **80**(5): 611–619.
89. Chen X., Y. Li, L. Wang, et al. (2002). 'Ischemic rat brain extracts induce human marrow stromal cell growth factor production'. *Neuropathology.* **22**(4): 275–279.
90. Majumdar M.K., M.A. Thiede, J.D. Mosca, M. Moorman, S.L. Gerson (1998). 'Phenotypic and functional comparison of cultures of marrow-derived mesenchymal stem cells (MSCs) and stromal cells'. *J Cell Physiol.* **176**(1): 57–66.
91. Mehler M.F., R. Rozental, M. Dougherty, D.C. Spray, J.A. Kessler (1993). 'Cytokine regulation of neuronal differentiation of hippocampal progenitor cells'. *Nature.* **362**(6415): 62–65.
92. Schinkothe T., W. Bloch, A. Schmidt (2008). '*In vitro* secreting profile of human mesenchymal stem cells'. *Stem Cells Dev.* **17**(1): 199–206.
93. Fiore M., J. Korf, A. Antonelli, L. Talamini, L. Aloe (2002). 'Long-lasting effects of prenatal MAM treatment on water maze performance in rats: Associations with altered brain development and neurotrophin levels'. *Neurotoxicol Teratol.* **24**(2): 179–191.
94. Schanzer A., F.P. Wachs, D. Wilhelm, et al. (2004). 'Direct stimulation of adult neural stem cells *in vitro* and neurogenesis *in vivo* by vascular endothelial growth factor'. *Brain Pathol* **14**(3): 237–248.
95. Palmer T.D., E.A. Markakis, A.R. Willhoite, F. Safar, F.H. Gage (1999). 'Fibroblast growth factor-2 activates a latent neurogenic program in neural stem cells from diverse regions of the adult CNS'. *J Neurosci.* **19**(19): 8487–8497.

96. Hsu Y.C., D.C. Lee, I.M. Chiu (2007). 'Neural stem cells, neural progenitors, and neurotrophic factors'. *Cell Transplant.* **16**(2): 133–150.
97. Hagg T. (2005). 'Molecular regulation of adult CNS neurogenesis: An integrated view'. *Trends Neurosci.* **28**(11): 589–595.
98. Chopp M., Y. Li (2002). 'Treatment of neural injury with marrow stromal cells'. *Lancet Neurol.* **1**(2): 92–100.
99. Chen J., M. Chopp (2006). 'Neurorestorative treatment of stroke: Cell and pharmacological approaches'. *NeuroRx.* **3**(4): 466–473.
100. Munoz J.R., B.R. Stoutenger, A.P. Robinson, J.L. Spees, D.J. Prockop (2005). 'Human stem/progenitor cells from bone marrow promote neurogenesis of endogenous neural stem cells in the hippocampus of mice'. *Proc Natl Acad Sci USA.* **102**(50): 18171–18176.
101. Rudge J.S., R.F. Alderson, E. Pasnikowski, J. McClain, N.Y. Ip, R.M. Lindsay (1992). 'Expression of Ciliary Neurotrophic Factor and the Neurotrophins-Nerve Growth Factor, Brain-Derived Neurotrophic Factor and Neurotrophin 3-in Cultured Rat Hippocampal Astrocytes'. *Eur J Neurosci.* **4**(6): 459–471.
102. Nakayama T., T. Momoki-Soga, N. Inoue (2003). 'Astrocyte-derived factors instruct differentiation of embryonic stem cells into neurons'. *Neurosci Res.* **46**(2): 241–249.
103. Crigler L., R.C. Robey, A. Asawachaicharn, D. Gaupp, D.G. Phinney (2006). 'Human mesenchymal stem cell subpopulations express a variety of neuro-regulatory molecules and promote neuronal cell survival and neuritogenesis'. *Exp Neurol.* **198**(1): 54–64.
104. Tremain N., J. Korkko, D. Ibberson, G.C. Kopen, C. DiGirolamo, D.G. Phinney (2001). 'MicroSAGE analysis of 2,353 expressed genes in a single cell-derived colony of undifferentiated human mesenchymal stem cells reveals mRNAs of multiple cell lineages'. *Stem Cells.* **19**(5): 408–418.
105. Rivera F.J., S. Couillard-Despres, X. Pedre, et al. (2006). 'Mesenchymal stem cells instruct oligodendrogenic fate decision on adult neural stem cells'. *Stem Cells.* **24**(10): 2209–2219.
106. Wislet-Gendebien S., F. Bruyere, G. Hans, P. Leprince, G. Moonen, B. Rogister (2004). 'Nestin-positive mesenchymal stem cells favour the astroglial lineage in neural progenitors and stem cells by releasing active BMP4'. *BMC Neurosci.* **5**: 33.
107. Wislet-Gendebien S., P. Leprince, G. Moonen, B. Rogister (2003). 'Regulation of neural markers nestin and GFAP expression by cultivated bone marrow stromal cells'. *J Cell Sci.* **116**(Pt 16): 3295–3302.
108. Zhao M.Z., N. Nonoguchi, N. Ikeda, et al. (2006). 'Novel therapeutic strategy for stroke in rats by bone marrow stromal cells and *ex vivo* HGF gene transfer with HSV-1 vector'. *J Cereb Blood Flow Metab.* **26**(9): 1176–1188.
109. Ikeda N., N. Nonoguchi, M.Z. Zhao, et al. (2005). 'Bone marrow stromal cells that enhanced fibroblast growth factor-2 secretion by herpes simplex virus vector improve neurological outcome after transient focal cerebral ischemia in rats'. *Stroke.* **36**(12): 2725–2730.
110. Deng W., M. Obrocka, I. Fischer, D.J. Prockop (2001). '*In vitro* differentiation of human marrow stromal cells into early progenitors of neural cells by conditions that increase intracellular cyclic AMP'. *Biochem Biophys Res Commun.* **282**(1): 148–152.
111. Rismanchi N., C.L. Floyd, R.F. Berman, B.G. Lyeth (2003). 'Cell death and long-term maintenance of neuron-like state after differentiation of rat bone marrow stromal cells: A comparison of protocols'. *Brain Res.* **991**(1–2):46–55.

112. Jori F.P., M.A. Napolitano, M.A. Melone, et al. (2005). 'Molecular pathways involved in neural *in vitro* differentiation of marrow stromal stem cells'. *J Cell Biochem.* **94**(4): 645–655.
113. Kohyama J., H. Abe, T. Shimazaki, et al. (2001). 'Brain from bone: Efficient "meta-differentiation" of marrow stroma-derived mature osteoblasts to neurons with Noggin or a demethylating agent'. *Differentiation.* **68**(4–5):235–244.
114. Hung S.C., H. Cheng, C.Y. Pan, M.J. Tsai, L.S. Kao, H.L. Ma (2002). '*In vitro* differentiation of size-sieved stem cells into electrically active neural cells'. *Stem Cells.* **20**(6): 522–529.
115. Levy Y.S., D. Merims, H. Panet, Y. Barhum, E. Melamed, D. Offen (2003). 'Induction of neuron-specific enolase promoter and neuronal markers in differentiated mouse bone marrow stromal cells'. *J Mol Neurosci.* **21**(2): 121–132.
116. Hellmann M.A., H. Panet, Y. Barhum, E. Melamed, D. Offen (2006). 'Increased survival and migration of engrafted mesenchymal bone marrow stem cells in 6-hydroxydopamine-lesioned rodents'. *Neurosci Lett.* **395**(2): 124–128.
117. Krampera M., S. Marconi, A. Pasini, et al.(2007). 'Induction of neural-like differentiation in human mesenchymal stem cells derived from bone marrow, fat, spleen and thymus'. *Bone.* **40**(2): 382–390.
118. Tzeng S.F., M.J. Tsai, S.C. Hung, H. Cheng (2004). 'Neuronal morphological change of size-sieved stem cells induced by neurotrophic stimuli'. *Neurosci Lett.* **367**(1): 23–28.
119. Locatelli F., S. Corti, C. Donadoni, et al. (2003). 'Neuronal differentiation of murine bone marrow Thy-1- and Sca-1-positive cells'. *J Haematother Stem Cell Res.* **12**(6): 727–734.
120. Kamada T., M. Koda, M. Dezawa, et al. (2005). 'Transplantation of bone marrow stromal cell-derived Schwann cells promotes axonal regeneration and functional recovery after complete transection of adult rat spinal cord'. *J Neuropathol Exp Neurol.* **64**(1): 37–45.
121. Tohill M., C. Mantovani, M. Wiberg, G. Terenghi (2004). 'Rat bone marrow mesenchymal stem cells express glial markers and stimulate nerve regeneration'. *Neurosci Lett.* **362**(3): 200–203.
122. Hermann A., R. Gastl, S. Liebau, et al. (2004). 'Efficient generation of neural stem cell-like cells from adult human bone marrow stromal cells'. *J Cell Sci.* **117**(Pt 19): 4411–4422.
123. Tatard V.M., G. D'Ippolito, S. Diabira, et al. (2007). 'Neurotrophin-directed differentiation of human adult marrow stromal cells to dopaminergic-like neurons'. *Bone.* **40**(2): 360–373.
124. Trzaska K.A., E.V. Kuzhikandathil, P. Rameshwar (2007). 'Specification of a dopaminergic phenotype from adult human mesenchymal stem cells'. *Stem Cells.* **25**(11): 2797–2808.
125. Barzilay R., I. Kan, T. Ben-Zur, S. Bulvik, E. Melamed, Offen D. (2008). 'Induction of human mesenchymal stem cells into dopamine-producing cells with different differentiation protocols'. *Stem Cells Dev.* **17**(3): 547–554.
126. Choong P.F., P.L. Mok, S.K. Cheong, C.F. Leong, K.Y. Then (2007). 'Generating neuron-like cells from BM-derived mesenchymal stromal cells *in vitro*'. *Cytotherapy.* **9**(2): 170–183.
127. Jiang Y., B.N. Jahagirdar, R.L. Reinhardt, et al. (2002). 'Pluripotency of mesenchymal stem cells derived from adult marrow'. *Nature.* **418**(6893): 41–49.
128. Jiang Y., D. Henderson, M. Blackstad, A. Chen, R.F. Miller, C.M. Verfaillie (2003). 'Neuroectodermal differentiation from mouse multipotent adult progenitor cells'. *Proc Natl Acad Sci USA* **100** Suppl. 1:11854–11860.
129. Wislet-Gendebien S., G. Hans, P. Leprince, J.M. Rigo, G. Moonen, B. Rogister (2005). 'Plasticity of cultured mesenchymal stem cells: Switch from nestin-positive to excitable neuron-like phenotype'. *Stem Cells.* **23**(3): 392–402.

18

MESENCHYMAL STEM CELLS: BIOLOGY, CULTURE OPTIMISATION, AND POTENTIAL CLINICAL USE

Satish Totey[1]*, Rakhi Pal[3], Chuah Chong Boon[2], K.G. Vijayendran[2], Rajarshi Pal[2], Swapnil Totey[2]

[1]*Advanced NeuroScience Allies (ANSA) Pvt. Ltd. Indiranagar, Bangalore 560038, INDIA*
[2]*Stempeutics Research Malaysia Sdn Bhd, Kuala Lumpur, Malaysia;*
[3]*Stempeutics Research Private Limited, Bangalore, India*

INTRODUCTION

Stem cells have emerged as one of the most fascinating areas of biology. It provides an excellent model to acquire in-depth knowledge about how an organism develop from a single cell and how healthy cell can replace a damaged cell in adult organisms. This opens up a potential avenue of investigating possibilities of cell-based therapies to treat debilitating diseases for which we do not have any cure today.

Stem cells have two major important characteristics. Firstly, stem cells are unspecialised cells that can renew themselves throughout their life and secondly they have a unique ability to differentiate into multiple tissues. Post-natal tissues have reservoir of tissue-specific stem cells called tissue-specific precursor cells that possess the capacity to differentiate into the committed cells of the homing tissue and contribute to its maintenance and regeneration. For example, epithelial stem cell can give rise to epidermal lineages [1], neural stem cell into central nervous system [2], oval cells into liver cells [3] muscle stem cell into muscle [4]. Bone marrow is a repository for stem cells and progenitor cells [5–7] that is instrumental in constant remodeling of the entire skeleton in order to keep bone tissue and their biochemical properties unchanged [8].

---

* *satish.totey@stempeutics.com*

Bone marrow is composed of two major types of stem cells. One is haematopoietic stem cells (HSC) and other is non-HSCs also called mesenchymal stem cells (MSCs) or marrow stromal cells [9]. MSCs are pluripotent stem cells that are capable to self-renew throughout the life and give rise to skeletal tissue such as cartilage, bone tendon, ligament, adipose tissue and marrow stroma [10]. Given that MSC is such a huge and ever expanding field, this chapter will focus particularly on the MSC and their major biological features, recent progress in our basic understanding and their clinical application.

HISTORICAL BACKGROUND

German scientist Cohnheim was first to suggest that non-HSC are present in the bone marrow almost a century ago. His work raised the possibility that bone marrow may be the source of fibroblastic cells that deposit collagen fibers as a part of the normal process of wound repair [11, 12]. However, in 1960s Alexander Friedenstein was the first to observe progenitor cells for bone marrow connective tissue [13], when they seeded rabbit and rat bone marrow cells at low density in plastic culture dish in a medium containing serum. They observed the discrete colonies of heterogeneous adherent cells which are of fibroblastic in appearance. They further observed that these cells have the capability to differentiate into bone or cartilage [13–16]. Thereafter, this remarkable experiment provided two major hypotheses. First, progenitor cells exist in bone marrow, which is capable of differentiating into bone and cartilage tissue and second, these fibroblastic cells provide stem cell niche and adequate microenvironment for HSCs [16, 17]. MSCs help to produce mature and specialized cells that interact with HSC and are directly involved with regulation of the haematopoietic process [18]. Although, several attempts were made to define, isolate and characterize these cells, Friedenstein's procedure is still considered as a standard protocol to isolate bone marrow MSC till date [19]. Caplan (1991) defined these cells that give rise to bone, marrow stroma, cartilage, tendon and muscle [6] whereas Prockop (1997) described them as MSC or marrow stromal cells [11]. Another group has further identified a subpopulation of bone marrow-derived cells termed as multipotent adult progenitor cells (MAPC) or mesodermal progenitor cells that posses longer proliferative capacity and broader differentiation potential than MSC [20, 21]. Nevertheless, isolation and phenotypic characterisation of MSC have been demonstrated in several vertebrate species including murine [22], canine [23], ovine [24], avian [25], bovine [26, 27], equine [28], porcine [29] and primates [30].

MSC ARE PRESENT IN ALL THE TISSUES

Although MSC is considered to be ubiquitous, location, origin and their distribution in different tissues is always a matter of argument and subject of intense study. Mesenchymal stem cells have been isolated conventionally from bone marrow [9]. However, there is enough good evidence that MSC exist not only in bone marrow, but virtually in all organs of body. These include adipose tissue [31], periosteum [32], tendon [33], periodontal ligament [34], muscle [35], synovial membrane [36], skin [37], lungs [38], amniotic fluid [39], human placenta [40, 41], umbilical cord [42], umbilical cord blood [43], dental pulp [44], limbal tissue [45] and menstrual blood [46, 47]. MSCs were also derived and cultured from brain, spleen, liver, kidney, lung, thymus and pancreas [48].

MSCs isolated from various tissues could be grown in long-term cultures and exhibited the capacity of prolonged self-renewal and differentiation along with mesoderm cells. MSC population derived from all tissues showed similar morphology and some extent surface markers expression profile [41, 43, 49]. On the contrary, differentiation capabilities of the MSC derived from various tissues or organs showed variability in term of frequency and degree of differentiation. It is therefore speculated that this variation might be due to effect of local environment from where they were originated [48].

This widespread distribution of MSC in several organs raised the question of relationship among the population in different tissues. Thus, three hypotheses were considered: first, MSCs are tissue resident cells and can be collected from individual tissue; second, MSCs are resident cells in tissue and circulate in the peripheral blood; third, MSCs are derived from peripheral blood and lodge in the various tissues for physiological regeneration [48]. However, for a long time it was considered that MSCs are present only in bone marrow from where they were released into the circulation and migrated at the site of injury to replenish the cell population that undergo apoptosis. However, MSC could not be recovered in peripheral blood collections obtained from either normal donors or patients who underwent peripheral blood mononuclear cell collections after mobilization therapy but could be obtained routinely from bone marrow samples [50, 51]. Although the existence of circulating MSC has been proven in foetal and neonatal blood [52, 53], their isolation in adult peripheral blood depends on investigators, with reported successes [54] and failures [50, 51], which may be related to the low frequency of such cells at steady state. In one of the reports, it was shown that MSCs are regularly observed in the circulating blood of rats and that the circulating MSC pool is consistently and dramatically increased by almost 15-fold when animals are exposed to chronic hypoxia [55]. However, we are still unclear about the origin of MSC in peripheral blood. Interestingly, Bianco et al. [56] suggested that the MSC present in bone marrow might originate from microvascular pericytes, because many of the other tissues from which MSC have been isolated are also rich in microvessels. Therefore the question is that 'do the MSC derived from pericytes or MSC located in various tissues represent pericytes'? In support of this hypothesis, Farrington-Rock et al. [57] have demonstrated that pericytes can differentiate into osteoblast, chondrocytes and adipocytes and hence they are multipotent cells and might contribute to growth, wound healing repair and development [58–60]. This hypothesis was further confirmed that MSC from bone marrow and dental pulp reside in microvasculature of their tissue of origin [61]. MSC population within adult adipose tissue also appears to be associated with perivascular cells surrounding blood vessels [62].

MSC NICHE

Stem cells in human and animal tissues are often located and controlled by special localised environment known as niche. This special microenvironment helps to nurture stem cells and enable them to maintain tissue homeostasis. Niche cells have mainly three functions: First, there is a continuous cross-talk between stem cells and niche cells in order to fulfill lifelong demands of differentiated cells, which is required during physiological regeneration or during injury. Second, niche cells provide a protective environment and safeguards against unwanted differentiation, apoptosis and any other stimuli that challenges stem cell reserve. Third, stem cell niche control the number of stem cells in the body and protect it from excessive stem cell proliferation that might lead to cancer. Thus, the dynamic balance between maintenance and regulation of normally quiescent stem cells population is tightly controlled and regulated by the local microenvironment according to the requirement of the host tissue [63–66].

Stem cell niche is identified in number of adult tissues including bone marrow, dental pulp, brain, pancreas, intestine and skin and generally in highly vascular site [61, 67]. Evidence based on several studies also suggests that MSC are associated with the microvasculature of their respective tissues and they are situated throughout the body as pericytes [60, 68]. Cell morphology of MSC and pericytes are similar both under light and transmission electron microscopy. The phenotypes of MSC and pericytes defined by the several markers are also similar and are negative for CD34, CD33, CD45, CD14, HLA-DR, KDR and CD31 and positive for CD73, CD90, CD29, CD44 and HLA-I. Differentiation potential into osteocytes and adipocytes for both MSC and pericytes is also similar. Cluster analysis of semi-quantitative expression of the 39 selected genes confirmed that all the MSC lines and pericytes share a common expression profile distinct from other normal cells [69].

Recently, several groups have identified microvasculature cells and pericytes as an important osteo-progenitor [70–72]. Pericytes express early feature of the vascular smooth muscle cell (VSMC) lineages including a-smooth muscle cell (SMC) and species-specific gangliosides 3G5 [57, 58, 73]. Anatomically, pericytes are intimately associated with endothelial cells of blood vessels and capillary network. In the marrow environment, MSC exhibits the histoanatomic characteristics of the pericytes [60]. Thus, from this perspective the bone marrow MSC can be viewed as a tissue-specific pericytes.

In many of the regenerating tissues, STRO-1$^+$ cells were found in a perivascular and vascular endothelial location that represent the population of mesenchymal progenitor cells [74]. Evidence showed that there is a link between vascular damage and formation of myofibroblasts from pericytes [75]. These myofibroblasts are derived from mesenchymal precursors. Under the influence of various growth factors they become either endothelial cells or pericytes/smooth muscle cells. These populations are identified as STRO-1$^+$ [76]. Mesenchymal stem cells isolated from bone marrow and dental pulp also showed high expression of STRO-1 and further recognizes an antigen on perivascular cells in bone marrow. STRO-1 was also shown to be localized on large blood vessels in different tissues such as brain, gut, heart, kidney, liver, lung, muscle and thymus [56]. Therefore, it appears that STRO-1 could be an early candidate marker of different MSC populations and infers a possible perivascular niche for these stem cell populations. However, recently focus was given more on CD34$^+$ cells and showed that these cells are widely distributed among adipocytes and predominantly associated with vascular structures and showed expression of pericytes markers. This result suggests that adipocytes-derived stem cells are CD34$^+$ pericytes [77]. However, these results were contradicted by other group [78] suggesting that pericytes has specificity for smooth muscle marker α-SMA. Therefore anti-α-SMA positive cells, which are implicated as pericytes and are invariably CD 34⁻, strongly suggest that pericytes in the adipose vasculature do not express CD34 CD34 has long been regarded as a reliable marker for HSCs. However, recent studies have demonstrated the existence of CD34 negative HSCs and that two populations—$CD34^+$ and $CD34^-$—can differentiate into one another [79]. Several papers including our data clearly show that CD34 is highly expressed in freshly isolated adipose-derived stem cells; however, its expression is quickly lost in culture when it grows beyond two passages [80]. Therefore, it is suggested that primitive MSC or pericytes can be a source of several tissue including cartilage and bone. This adequately explains why MSC can be isolated from all tissues.

OPTIMAL CULTURE CONDITIONS AND UP-SCALING

One of the most important criteria for application of MSC for clinical therapy is that they should be up-scaled according to the current manufacturing practice guideline and should be able to maintain for prolonged passages (Fig. 18.1). It is believed that MSC enter senescence from the moment of *in vitro* culturing. They tend to lose their stem cell characteristics and differentiation potential and hence prolonged culture of MSC was believed to be difficult [81]. Considering this fact, MSC can only be used for clinical purposes at early stage of *in vitro* culture. Limited passages of MSC for clinical purposes not only require continuous supply of bone marrow donors but also significantly increase the cost of the culture and therapy. There have been continuous attempts in optimizing culture conditions for MSC that could be beneficial for clinical and therapeutic applications. Hence, identifying appropriate media, serum supplement or replacement, growth factors and tissue culture wares are most crucial in order to achieving this challenge.

Courtesy: Stempeutics Research Private Limited, Bangalore, India

➤ **Fig. 18.1** *Up-scaling of bone marrow derived mesenchymal stem cells in 10-cell stack. (For colour figure see Plate 43).*

In order to identify optimum culture condition, several different basal media were tried. Most commonly used media are Dulbecco's modified Eagles's medium low glucose (DMEM-LG) or Dulbecco's modified Eagles's medium high glucose (DMEM-HG), Iscove's modified Dulbecco's medium (IMDM), Minimum essential medium-alpha (MEM-alpha), DMEM-F12, DMEM- knock out (DMEM-KO). DMEM-LG, MEM-alpha, DMEM-F12 and DMEM-KO showed to be most consistent in isolation and expansion of MSC and support long-term growth.

In one of our studies, we compared different media compositions for MSC isolation and expansion so as to maintain them for longer passages without compromising on quality or losing their characteristics and functionality [82].

In our studies five different basal media were used: DMEM-KO, DMEM-F12, DMEM-LG, MesenCult and DMEM-high glucose (DMEM-HG). Our results demonstrated that DMEM-KO and DMEM-F12 supplemented with 10% fetal bovine serum (FBS) provides optimum and suitable culture conditions for MSC isolation and *ex vivo* expansion and maintain morphology, population doubling time and immunophenotype for more than 30 passages (Fig. 18.2), whereas, DMEM-LG support the expansion till 10 passages and become senescence. On the contrary, DMEM-HG and commercially available MesenCult failed to support human MSC growth in our culture system beyond 5 passages.

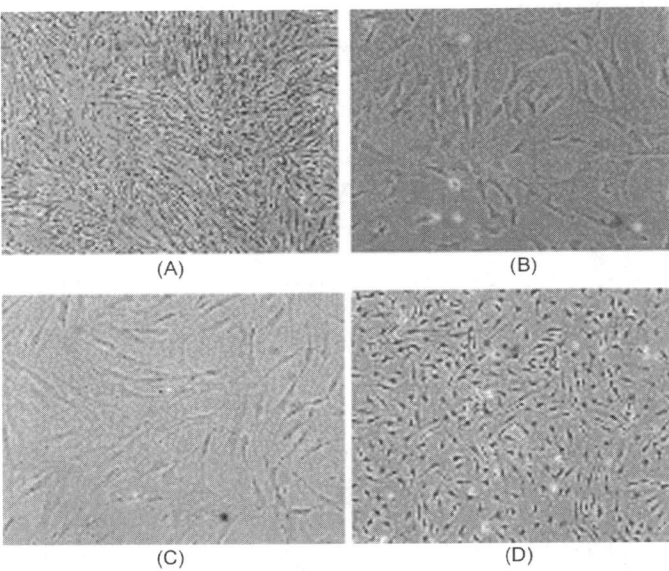

▶ Fig. 18.2 *Mesenchymal stem cells can be isolated from various tissues including bone marrow, adipose tissue, umbilical cord, synovial fluid and dental pulp. They have similar properties such as spindle-shaped morphology and plastic adherence. This figure shows morphology of MSCs derived from various tissues. (A) bone marrow; (B) umbilical cord; (C) adipose tissue and (D) synovial fluid. (For colour figure see Plate 43).*

Mesenchymal stem cells cultured in DMEM-KO and DMEM-F12 were identically negative for haematopoietic surface markers CD34 and CD45 and positive for CD44, CD73, CD90, CD105 and CD166 up to 30 passages. Differentiation capacity of the MSC into osteogenic and adipogenic lineages cultured in DMEM-KO and DMEM-F12 at passage 30 was also comparable to passage 5 (Fig. 18.3).

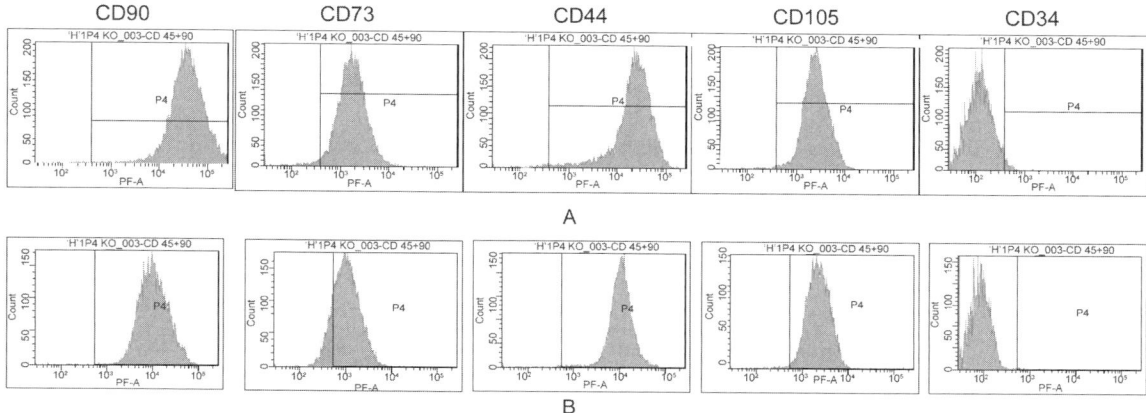

▶ **Fig. 18.3** *Mesenchymal stem cell specific cell surface marker expression on bone marrow derived MSC at: (A) early passage (Passage 3) and (B) late passage (Passage 20). (For colour figure see Plate 44).*

To our knowledge this is the first report using DMEM-KO or DMEM-F12 for culturing MSC for more than 30 passages. An earlier report by Sotiropoulou et al. [83] using various culture conditions—media, serum, glucose concentrations, stable glutamine, plating density and plastic wares—revealed that those based on αMEM are more suitable for both isolation and expansion of multipotent MSC. Low glucose concentration in DMEM-based media and Glutamax instead of L-Glutamine in all different basal media used consistently supported MSC growth. However, the proliferation was checked up to 3 passages. Therefore, it is difficult to conclude that αMEM can be used for commercial MSC production. MSCs were also cultivated, expanded and differentiated into GMP-accepted medium (EMEA medium) in the early passages. However, there is no data on prolonged passaging.

One of the limitations of the existing culture systems is the use of animal-derived FBS. Media supplemented with 10% FBS yields better than no serum. However, several clinical trials using MSC supplemented with FBS were approved by the US Food and Drug Administrations (US-FDA) [84] and the risk of transmission of prion diseases and zoonoses from the use of FBS is considered to be small. Perhaps the greater risk associated with using FBS is the immunogenicity of the xenogenic proteins [85]. However, the complete clinical impact of this observation remains to be investigated. In our laboratory, in an attempt to minimize traces of FBS in the cell suspension, MSCs are washed with injectable saline several times. However, this does not eliminate the risk factor of having xenogenic protein completely. Therefore, we have attempted to establish serum-free culture conditions for the up-scaling of MSC for clinical applications. Preliminary results indicate that MSC may be propagated *in vitro* with reduced serum concentration or with human plasma supplemented with certain combination of cytokines. However, their effectiveness on long passages is not same as 10% FBS. Plasma alone does not support isolation and derivation of BM-MSC [82]. We also used several serum-free media that were recently available commercially; however, these media are suitable for maintaining on-going culture or already isolated BM-MSC with other methods using FBS. Thus, serum-free media is not optimum for isolation and derivation of BM-MSC from fresh bone marrow [82].

The use of fetal calf serum (FCS) for the culture of cells to be used in clinical trials raises potential hazards that cannot be neglected, but this is a regulatory issue. However, as specifically regards the isolation and expansion of human MSCs, unfortunately serum-free media have not yet been defined except few. The alternative of using autologous serum or platelet lysate is feasible only for the minority of clinical protocols involving low numbers of MSCs. Therefore logistically it is important to develop serum-free media for expansion of MSC. There are few commercially available serum-free media such as StemPro from InVitrogen and Medicult's SSR media from Medicult. Serum-free media cannot promote human MSC growth without the addition of cytokines and or growth factors. In both bioreactor and static cultures, daily feeding with dilution of cells improve the expansion of MSC as well as total cell counts compared with feeding every other day with total cell retention. In composition of a serum-free media for adult bone marrow-derived MSCs, addition of stem cell factor (SCF), interleukin (IL)-3 and IL-6 and freshly dissolved 10^{-6} M hydrocortisone may be effective. This media composition with a rotary bioreactor reported a rapid expansion of MSCs. Static culture reported similar results although not superior to supplementation with FBS. FBS can be substituted with a recombinant human serum albumin. A new product, recombinant human serum albumin, is available under the commercial name Cellastim™ (InVitria). The use of Cellastim™ in the culturing of cell lines and bio-processing would allow for animal-free culture conditions. Studies have shown that Cellastim™ enhances cell growth in hybridoma cells and antibody productivity in CHO cells better than or equivalent to plasma-derived human serum albumin. The optimal concentration of Cellastim™ varies with the cell line and media composition. Many classical media formulations include albumin at concentration between 0.5 and 1 g/L.

Seeding density is another critical factor to ensure maximum yield. It is recommended that initial plating of mononuclear cells (MNC) should be high and range from $150 \infty 10^3$ to $200 \infty 10^3$ per cm^{-2}. Plating density from the subsequent passages should be very low. Sotiropoulou et al. [83] demonstrated that low density MSC yield higher proliferation rate. Increasing the plating density from 10 cells per cm^{-2} to $1 \infty 10^3$ cells per cm^{-2} resulted into decrease in the amount of expansion by several fold. It is difficult to achieve such a low density in large scale production and hence optimum seeding density should be calculated for large scale production. However, plating density of 10 000 cells cm^{-2} is reasonable for high yield.

PHENOTYPIC PROPERTIES OF MSC

Although there is considerable progress made towards phenotypic characterisation of MSC, unfortunately none of them are specific to MSC. However, Mesenchymal and Tissue Stem Cell

Committee of International Society for Cellular Therapy has proposed minimal criteria to define human MSC [86]. First, MSC must be plastic-adherent when maintained in standard culture conditions. Second, MSC must express CD73, CD90 and CD105 and lack expression of CD11b, CD14, CD19, CD34, CD45, CD79α and HLA-DR surface molecules. Third, MSC must differentiate into osteoblast, adipocytes and chondroblast *in vitro* (Fig. 18.4). It has been mentioned earlier in this chapter that MSC are located throughout the body as pericytes and one of the early markers for pericytes is STRO-1. Although STRO-1 is a best known marker for MSC, it is not exclusive to the MSC and its expression is lost over prolonged cultures. It is generally known that MSC are non-haematopoietic cells and do not express CD11, CD14, CD34 and CD45. They also do not express co-stimulatory molecules such as CD11A, CD11B, CD80, CD86, CTLA4, and adhesion molecules such as CD18, CD31 and CD56 but express CD29, CD 44, CD71, CD73, CD90, CD105, CD166 and ICAM-1 [9, 87] (see Table 18.1 for detail).

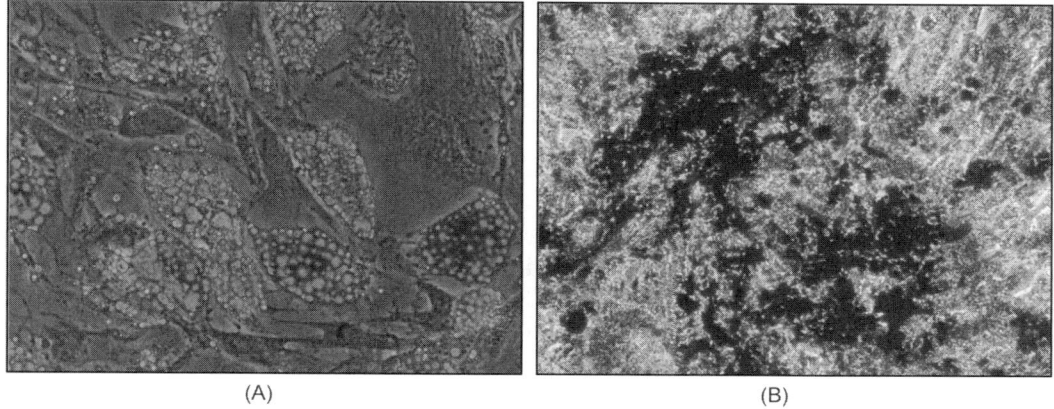

(A)   (B)

➤ **Fig. 18.4** *Differentiation of bone marrow derived mesenchymal stem cells into (A) adipogenic and (B) osteogenic cells. (For colour figure see Plate 44).*

Table 18.1 *Surface Marker Expression Profile of MSC*

Surface marker	Expression
CD4	Negative
CD9	Positive
CD10	Positive
CD11a	Negative
CD11b	Negative
CD13	Positive
CD14	Negative
CD15	Negative

(Contd.)

Table 18.1 (Contd.)

	CD18	Negative
	CD25	Negative
	CD29	Positive
	CD31	Negative
	CD34	Negative
	CD44	Positive
	CD45	Negative
	CD49a	Positive
	CD49b	Positive
	CD49c	Positive
	CD49e	Positive
	CD50	Negative
	CD51	Positive
	CD54	Positive
	CD58	Positive
	CD61	Positive

It is well known that MSC are present in almost all the tissue of the body and they reside in a perivascular niche. MSCs derived from various tissues of human and other mammals are found to display phenotypic consistency regardless of their diverse ontogeny and developmental potential. However, relationship between MSC and pericytes has not been established although they share many functional properties. For example, adipose-derived mesenchymal stem cells (AD-MSC) express both phenotypic characteristics for MSC and perivascular pericytes such as STRO-1, CD146 and 3G5. Immuno-selection of each population revealed that each fraction selected from either STRO-1, CD146 or 3G5 fraction showed expression of CD44, CD90, CD105, CD106, CD146, CD166 and alkaline phosphatase that is usually present on bone marrow derived MSC. Cultures established from either STRO-1, CD146 or 3G5 populations are capable of forming ectopic bone when transplanted into NOD/SCID mice [61]. However, relationship of all the three different population and their *in vivo* functions is still unknown.

IMMUNOGENICITY OF MSC

One of the most surprising features of MSC to all immunologists is that they do not obey the normal rules of allogenic rejection. Evidence suggests that MSC does not provoke a proliferative T-cell response in allogenic mixed lymphocyte reaction (MLR) that suggests immunosuppressive role of MSC even in inflammatory situation after INFγ stimulation. MSCs, therefore, appear to invade allogenic rejection because of being hypo-immunogenic, by modulating T-cell phenotypes and by creating immunosuppressive local milieu. MSCs do not express MHC-II and also do not express co-stimulatory molecules such as CD40, CD40L, CD80 and CD86 [87, 88]. *In vivo* studies demonstrated that MSC avoid normal allo-responses. These characteristics support the possibility of exploiting universal donor MSC for therapeutic application.

MSCs derived from other tissues or organ also demonstrate similar immunophenotypes as seen in bone marrow. MSC derived from placenta also demonstrated that these cell do not express MHC-II molecules and suppress umbilical cord blood lymphocytes proliferation induced by cellular or non-specific mitogenic stimuli. This strongly implies that MSC derived from placenta have potential application in allograft transplantation.

IMMUNOMODULATORY PROPERTIES OF MSC

Immunomodulatory nature of MSC was first described in human [89], primate [90] and mouse [91] and suggested that MSC can suppress T lymphocyte activation and proliferation *in vitro*. This suppression affects the proliferation of T lymphocytes by CD3/CD28, mitogens and alloantigens [87, 91, 92]. Suppression of T lymphocyte proliferation has not been associated with induction of apoptosis, but to suppression of cell division. The proliferation of antigen-primed cytotoxic T lymphocytes (CTLs) is also inhibited by MSC. However, supernatant from MSC cultures do not show any inhibitory effect unless MSC have been co-cultured with T lymphocytes. This suggests that in order to have suppressive effect, direct cross-talk between MSC and T lymphocytes is required [93]. Although, mechanisms responsible for these effects are not fully understood, reports showed that soluble factors such as prostaglandin E2 (PGE2), interleukin-10 (IL10), transforming growth factor-β (TGF-β), nitric oxide and indolamine dioxygenase (IDO)-mediated tryptophan metabolites secreted by MSC have all been found to suppress T-cell mediated antigen response *in vitro*.

Immunomodulatory potential of human amniotic MSC, human amniotic epithelial cells and adipose-derived MSC demonstrated a dose-dependent inhibition of peripheral blood mononuclear cell (PBMC) immune responses in mixed lymphocyte reaction. Sub-cultivation does not show any alteration in immunomodulation but cryopreservation significantly reduces immunomodulatory properties of MSC [94].

MSC FACILITATE TISSUE REPAIR

Mesenchymal stem cells are naturally found as perivascular cells and normally referred to as pericytes. During the process of injury they are released at sites of injury, where they secrete large quantities of bioactive factors that are both immunomodulatory and trophic. The trophic activity inhibits ischemia-caused apoptosis and scarring while stimulating angiogenesis and the mitosis of tissue intrinsic progenitor cells. The immunomodulation inhibits lymphocyte surveillance of the injured tissue, thus preventing autoimmunity and allows allogenic MSCs to be used in various clinical situations. Thus, a new, enlightened era of experimentation and clinical trials has been initiated with xenogenic and allogenic MSCs. Close relationship of MSC with pericytes as operationally defined by culture methods are similar cells located in the wall of the vasculature at the basement membrane where they function as a source of precursor cells for repair and tissue maintenance and maintain homeostasis. During the event of tissue injury, MSC loses contact with basement membrane and endothelial cells. MSCs start intense proliferation that leads to migration to adjacent damaged area. MSCs divide and secrete several bioactive factors that function to protect and repair or regenerate the damaged tissue. Bioactive factor of MSC that secrete at the site of injury strongly repress immune surveillance and inhibit T- and B-cell mediated destruction of injured site. Trophic factors secreted by the MSC also exert immunosuppressive effect on the surrounding tissue that lead to minimize inflammatory response, reduce apoptosis, inhibit scarring, stimulate angiogenesis and vasculogenesis and induce chemoattractant effects on other cells. This also leads to stimulation of tissue—intrinsic progenitor to regenerate the damaged tissue and help in remodeling tissue [95].

Several hypotheses have been postulated by different investigator regarding the mechanism through which MSC help in repair and regenerate the tissue. First, MSC secrete cytokines, trophic factors and growth factors that stimulate recovery in paracrine manner. Secondly, MSC modulate the stem cell niche and stimulate recruitment of endogenous stem cells to the injured site and promote differentiation. Third, MSC also release antioxidant chemicals, free radical scavengers, chaperon and heat shock protein at the injured site. This results in removal of toxic substances from the injured site and thereby promotes recovery of the surviving cells [96]. Recently it was seen that MSC can deliver new mitochondria to damaged cells thereby rescuing aerobic metabolism [97].

Furthermore, MSC not only help to regenerate injured tissue but also play a significant role in normal tissue regeneration throughout the life term of mammals. This supports the view that the annual antler regeneration in deer is a stem cell based process and the regenerating antler is build up by the progeny of MSC located in the cambial layer of pedicle periosteum. These MSC are $STRO^{1+}$ cells and are present in a perivascular and vascular endothelial location both in the subcutaneous tissue of the pedicle and in the cartilaginous zone. It is suggested that these $STRO\text{-}1^+$ cells exists in stem cell niche, which is located in the cambial layer of the periosteum and help in regeneration of antler by periodic activation of these stem cells [74].

Similar mechanism and role of MSC have been witnessed in teeth eruption and remodeling dental pulp after injury. MSCs present in the dental niche proliferate and migrate to the wounded site, where in cooperation with local cells, participate in tooth repair [98, 99]. Stem cells isolated from the dental pulp are capable of forming osteodentin *in vitro* and *in vivo* and they have been suggested to originate from pericytes [60]. Recent findings have demonstrated that Notch-3 and Rgs5 is activated in response to dental injury and they are strongly expressed in pericytes and vascular wall of both developing and injured teeth. Activation of Notch-3 expression during tooth repair might be important for the regulation of the fate of pericytes-derived stem cells and may be playing central role during tooth repair and neovascularisation [99].

MSC also secrete a wide array of arteriogenic cytokines and contribute to collateral remodeling in ischemic limb via paracrine mechanism. It is demonstrated that MSCs express gene encoding as a broad spectrum of arteriogenic and angiogenic cytokines including vascular endothelial growth factor (VEGF), fibroblast growth factor (FGF), angiopoentin-1 (Ang-1), matrix metalloproteinase (MMPs) and transforming growth factor-β (TGF-β) [100], and stimulates the proliferation of endothelial cells and promote neovascularisation. Thus transplanted MSC might initiate angiogenesis in diabetic ischemic limb or injured pancreas by producing angiogenic factors [101].

MSCs FOR CLINICAL APPLICATION

Application of stem cell in regenerative medicine should meet certain criteria. First stem cell should be found in abundance. Second, they should be able to harvest with minimally invasive procedure. Third, they should able to differentiate into various lineages. Fourth, they should be transplantable to either an autologous or allogenic host and finally, they can be scaled up in concordance with the current manufacturing practice guideline. MSCs indeed abide to all these criteria and hence can be an excellent candidate in clinical application for the treatment of a wide range of medical conditions.

Although the mechanistic underpinnings of stem cell therapy are still intensely debated, the concept of cell therapy has already been introduced into the clinical setting, where a flurry of small, mostly uncontrolled trials indicate that stem cell therapy may be feasible in patients. The overall clinical experience also suggests that stem cell therapy can be safely performed, if the right cell type is used in the right clinical setting.

MSC are widely being used for clinical application and therapy. Preliminary efficacy data indicate that stem cells have the potential to enhance myocardial perfusion and/or contractile performance in patients with acute myocardial infarction, advanced coronary artery disease and chronic heart failure [102]. The field now is rapidly moving toward intermediate-size, double-blinded trials to gather more safety and efficacy data. Ultimately, large outcome trials will have to be conducted.

MSC have been successfully used for amyotrophic lateral sclerosis (ALS) [103].

As per US National Institute of Health website on clinical trials, there are totally 55 clinical trials currently ongoing using allogenic and autologous MSC for various disease indications (Table 18.2). Many of them completed phase-II trials and few of them have already released products in the market. This indicates that MSC has tremendous potential for clinical application. However, our knowledge about MSC biology is still limited and there is a need to increase our understanding of MSC biology, although many challenges lie ahead before we realize full therapeutic potential of MSC.

Table 18.2 *Current Clinical Trials Using Mesenchymal Stem Cells as on 2008*

Disease	Phase	Place
Graft Versus Host Disease	II	University Hospital of Liege, Belgium
	I and II	University of Salamanca, Spain
	I and II	Christian Medical College, Vellore, India
	I	National Cancer Institute, USA
	I and II	Hadassah Medical Organisation, Israel
Graft Rejection	II	University Hospital of Liege, Belgium
Chronic allograft Nephropathy	I and II	Fuzhou General Hospital, China
Sub-clinical Rejection	I and II	Leids University Medical Centre, The Netherlands
Crohn's Disease	II	Osiris Therapeutics, USA

(Contd.)

Table 18.2 (Contd.)

Type-I Diabetes	I and II	Fuzhou General Hospital, China
	II	Osiris Therapeutics, USA
Cirrhosis	II	University of Tehran, Iran
Lupus Nephritis	I and II	Organ Transplant Institute, China
End-stage liver disease	I and II	Shaheed Beheshti Medical University, Iran
Multiple Sclerosis	I and II	University of Cambridge, UK
Myocardial Infarction	I and II	National heart, Lung and Blood Institute, John Hopkins University, USA
	II	Angioblast Systems, USA
	I	Osiris Therapeutics, USA
	II	Helsinki University, Finland
	I and II	Rigshospitalet, Denmark
Renal Transplant	I and II	Mario Negri Institute of Pharmacological Research, Italy
	I and II	Fuzhou General Hospital, China
Systemic Lupus Erythaematous	I and II	Nanjing Medical University, China
Tibia Fracture	I and II	Hadassah Medical Organisation, Israel
	I	University Hospital, Tours, France
Burn Injury	I	Ohio State University, USA
Bone Fracture	I and II	University Hospital, Clermont-Ferrand, France
Cartilage Injury	II	Osiris Therapeutics, USA
Chronic Obstructive Pulmonary Disease	II	Osiris Therapeutics, USA
Osteogenesis Imperfecta	I	St. Jude Children's Research Hospital, USA
Periodontitis	I and II	Translational Research Informatics Centre, Kobe Hyogo, Japan
Osteodysplasia	I	St. Jude Children's Research Hospital Drexel University Wayne State University
Dilated Cardiomyopathy	II	University Medical centre, Ljubljana, Republic of Slovenia
	II	Hospital Universitario Reina Sofia, Spain
Posterolateral Lumbar Fusion	I and II	Mesoblast, USA

Source: www.clinicaltrial.gov

CONCLUSION

Our understanding of MSC biology is gradually increasing although many challenges lie ahead before we can realize the full potential of these cells for clinical and therapeutic applications. This chapter has highlighted some of the important research, investigations and key areas for future investigation which include the identification of specific markers to enable the isolation of pure populations of MSC, as well as a better understanding of the regulatory pathways that control MSC behaviour. Advances in these areas increase our basic knowledge of MSC as well as facilitate the development of MSC-based therapies.

REFERENCES

1. Slack J.M. (2000). 'Stem cells in epithelial tissues'. *Science.* **287**: 1431–1433.
2. McKay R. (1997). 'Stem cells in the central nervous system'. *Science.* **276**: 66–71.
3. Yoon B.I., Y.K. Choi, D.Y. Kim (2004). 'Differentiation process of oval cells into hepatocytes: Proposals based on morphological and phenotypical traits in carcinogen treated hamster liver'. *J. Comp. Pathology.* **131**: 1–9.
4. Charge S.B., M.A. Rudnicki (2004). 'Cellular and molecular regulation of muscle regeneration'. *Physiol. Rev.* 84: 209–238.
5. Beresford J.N., J.H. Bennett, C. Devlin, P.S. Leboy, M.E. Owen (1992). 'Evidence for an inverse relationship between the differentiation of adipocytic and osteogenic cells in rat marrow stromal cell cultures'. *J. Cell. Sci.* **102**: 341–351
6. Caplan A.I. (1991). 'Mesenchymal stem cells'. *J. Orthop. Res.* **9**: 641–650.
7. Owen M. (1988). 'Marrow stromal stem cells'. *J. Cell. Sci. Suppl.* **10**: 63–76.
8. Parfitt A.M. (1984). 'The cellular basis of bone remodeling: the quantum concept reexamined in light of recent advances in the cell biology of bone'. *Calcif. Tissue. Int.* **36**: 37–45.
9. Pittenger M.F., A.M. Mackay, S.C. Beck, R.K. Jaaiswal, R. Douglas, J.D. Mosca, M.A. Moorman, D.W. Simonetti, S. Craig, D.R. Marshak (1999). 'Multilineage potential of adult human mesenchymal stem cells'. *Science.* **284**: 143–147.
10. Muraglia A., R. Cancedda, R. Quarto (2000). 'Clonal mesenchymal progenitors from human bone marrow differentiate *in vitro* according to hierarchical model'. *J. Cell. Sci.* **113**: 1161–1166.
11. Prockop D.J. (1997). 'Marrow stromal cells as stem cells for non-haematopoietic tissues'. *Science.* **276**: 71–74.
12. Chamberlain G., J. Fox, B. Ashton, J. Middleton (2007). 'Concise review: Mesenchymal stem cells: Their phenotype, differentiation capacity, immunological featuresand potential for homing'. *Stem Cells.* **25**: 2739–2749.
13. Friedenstein A.J., R.K. Chailakhyan, K.S. Lalykina (1970). 'The development of fibroblast colonies in monolayer cultures of guinea-pig bone marrow and spleen cells'. *Cell Tissue Kinet.* **3**: 393–403.
14. Friedenstein A.J. (1976). 'Precursor cells of mechanocytes'. *Int. Rev. Cytol.* **47**: 327–359.
15. Friedenstein A.J., N.W. Latzinik, A.G. Grosheva, U.F. Gorskaya (1982). 'Marrow microenvironment transfer by heterotopic transplantation of freshly isolated and cultured cells in porous sponges'. *Exp. Haematol.* **10**: 217–227.
16. Friedenstein A.J., R.K. Chailakhyan, U.V. Gerasimov (1987). 'Bone marrow osteogenic stem cells: in vitro cultivation and transplantation in diffusion chambers'. *Cell Tissue. Kinet.* **20**: 263–272.
17. Delorme B., S. Chateauvieux, P. Charbord (2006). 'The concept of mesenchymal stem cells'. *Regen Med.* **1**: 497–509.
18. Schofield R. (1978). 'The relationship between the spleen colony-forming cell and the haemopoietic stem cell'. *Blood. Cells.* **4**: 7–25.
19. Alhadlaq A., J.J. Mao (2004). 'Mesenchymal stem cells: isolation and therapeutics'. *Stem Cells. Dev.* **13**: 436–448.
20. Reyes M., T. Lund, T. Lenvik, D. Aguiar, L. Koodie, C.M. Verfaillie (2001). 'Purificcation and ex vivo expansion of postnatal human marrow mesodermal progenitor cells'. *Blood.* **98**: 2615–2625.

21. Jiang Y., B. Vaessen, T. Lenvik, M. Blackstad, M. Reyes, C.M. Verfaillie (2002). 'Multipotent progenitor cells can be isolated from postnatal murine bone marrow, muscle and brain'. *Exp. Haematol.* **30**: 896–904.
22. Meirelles L., S. da, N.B. Nardi (2003). 'Murine marrow derived mesenchymal stem cell: isolation, in vitro expansion, and characterisation'. *Br. J. Haematol.* **123**: 702–711.
23. Kadiyala S., R.G. Young, M.A. Thiede, S.P. Bruder (1997). 'Culture expanded canine mesenchymal stem cells posses osteochondrogenic potential *in vivo* and *in vitro*'. *Cell. Transplant.* **6**: 125–134.
24. Mrugala D., C. Bony, N. Neves, L. Caillot, S. Fabre, D. Moukoko, C. Jorgensen, D. Noel (2008). 'Phenotypic and functional characterisation of ovine mesenchymal stem cells: application to a cartilage defect model'. *Ann. Rheum. Dis.* **67**: 288–295.
25. Lennon, D.P., S.P. Bruder, S.E. Haynesworth, N. Jaiswal and A.I. Caplan, (1996). 'Human and Animal Mesenchymal Progenitor Cells From Bone Marrow: Selection of Serum for Optimal Proliferation'. *in vitro Cell and Developmental Biology*. **32**: 602–611.
26. Niku M., L. Ilmonen, T. Pessa-Morikawa, A. Iivanainen (2004). 'Limited contribution of circulating cells to the development and maintenance of non-haematopoietic bovine tissues'. *Stem Cells.* **22**: 12–20.
27. Bosnakovski D., M. Mizuno, G. Kim, S. Takagi, M. Okumura, T. Fujinaga (2005). 'Isolation and multilineage differentiation of bovine bone marrow mesenchymal stem cells'. *Cell Tissue Res.* **319**: 243–253.
28. Worster A.A., A.J. Nixon, B.D. Brower-Toland, J. Williams (2000). 'Effect of transforming growth factor beta 1 on chondrogenic differentiation of cultured equine mesenchymal stem cells'. *Am J. Vet. Res.* **61**: 1003–1010.
29. Ringe J., C. Kaps, B. Schmitt, K. Buscher, J. Bartel, H. Smolian, O. Schultz, G.R. Burmester, T. Haupl, M. Sittinger (2002). 'Porcine mesenchymal stem cells. Induction of distinct mesenchymal stem cell lineages'. *Cell Tissue Res.* **307**: 321–317.
30. Bartholomew A., S. Patil, A. Mackay, M. Nelson, D. Buyaner, W. Hardy, J. Mosca, C. Sturgeon, M. Siatskas, N. Mahmud, K. Ferrer, R. Deans, A. Moseley, R. Hoffman, S.M. Devine (2001). 'Baboon mesenchymal stem cells can be genetically modified to secret human errytropoietin *in vivo*'. *Hum. Gen. Ther.* **12**: 1527–1541.
31. Zuk P.A., M. Zhu, P. Ashjian, De D.A. Ugarte, J.I. Huang, H. Mizuno, Z.C. Alfonso, J.K. Fraser, P. Benhaim, M.H. Hedrick (2002). 'Human adipose tissue is a source of multipotent stem cells'. *Mol. Bio. Cell.* **13**: 4279–4295.
32. Nakahara H., J.E. Dennis, S.P. Bruder, S.E. Haynesworth, D.P. Lennon and A.I. Caplan (1991). '*in vitro* differentiation of bone nad hypertrophic cartilage from periosteal derived cells'. *Exp. Cell Res.* **195**: 492–503.
33. Salingcarnboriboon R., H. Yoshitake, K. Tsuji, M. Obinata, T. Amagasa, A. Nifuji, M. Noda (2003). 'Establishment of tendon-derived cell lines exhibiting pluripotent mesenchymal stem cell-like property'. *Exp. Cell. Res.* **287**: 289–300.
34. Seo M., S. Miura, P. Gronthos, S. Mark Bartold, J. Batouli, M. Brahim, P. Young, C. Gehron Robey, S. Wang, B. Shi (2004). 'Investigation of multipotent postnatal stem cells from human periodontal ligament'. *The Lancet.* **364**: 149–155
35. Kuroda R., A. Usas, S. Kubo, K. Corsi, H. Peng, T. Rose, J. Cummins, F.H. Fu, J. Huard (2006). 'Cartilage repair using bone morphogenetic protein 4 and muscle-derived stem cells'. *Arthritis Rheum.* **54**: 433–442.

36. De Bari C., F. Dell'Accio, P. Tylzanowski, F.P. Luyten (2001). 'Multipotent mesenchymal stem cells from adult human synovial membrane'. *Arthritis Rheum.* **44**: 1928–1942.
37. Toma J.G., M. Akhavan, K.J. Fernandes, F. Barnabe-Heider, A. Sadikot, D.R. Kaplan, F.D. Miller (2001). 'Isolation of multipotent adult stem cells from the dermis of mammalian skin'. *Nat. Cell Biol.* **3**: 778–784.
38. Sabatini F., L. Petecchia, M. Tavian, Jodon de V. Villeroche, G.A. Rossi, D. Brouty-Boye (2005). 'Human bronchial fibroblasts exhibit a mesenchymal stem cell phenotype and multilineage differentiating potentialities'. *Lab. Invest.* **85**: 962–971.
39. Tsai M.S., J.L. Lee, Y.J. Chang, S.M. Hwang (2004). 'Isolation of human multipotent mesenchymal stem cells from second-trimester amniotic fluid using a novel two stage culture protocol'. *Hum. Reprod.* **19**: 1450–1456.
40. Zhang W., W. Ge, C. Li, S. You, L. Liao, Q. Han, W. Deng, R.C. Zhao (2004). 'Effects of mesenchymal stem cells on differentiation, maturation and function of human monocytes-derived dendritic cells'. *Stem cells. Dev.* **13**: 263–271.
41. Li C.D., W.Y. Zhang, H.L. Li, X.X. Jiang, Y. Zhang, P.H. Tang, N. Mao (2005). 'Mesenchymal stem cells derived from human placenta suppress allogenic umbilical cord blood lymphocyte proliferation'. *Cell Res.* **15**: 539–547.
42. Fu Y.S., Y.C.Cheng, M.Y. Lin, H. Chen, P.M. Chu, S.C. Chou, Y.H. Shih, K.O. M.H., M.S. Sung (2005). 'Conversion of human umbilical cord mesenchymal stem cells in Wharton's Jelly to dopaminergic neurons in vitro: Potential therapeutic application for Parkinson's Disease'. *Stem Cell.* **24**: 115–124.
43. Lee O.K., T.K. Kuo, E.M. Chen, K.D. Lee, S.L. Hsieh, T.H. Chen (2004). 'Isolation of multipotent mesenchymal stem cells from umbilical cord blood'. *Blood.* **103**: 1669–1675.
44. Lee R.H., B. Kim, I. Choi, H. Kim, H.S. Choi, K. Suh, Y.C. Bae, J.S. Jung (2004). 'Characterisation and expression analysis of mesenchymal stem cells from bone marrow and adipose tissue'. *Cell Physiol Biochem.* **14**: 311–324.
45. Huang A.H.C., Y.K. Chen, M.L. Lin, T.Y. Shieh, A.W.S. Chan (2008). 'Isolation and characterisation of dental pulp stem cells from supernumerary tooth'. *J. Oral. Path. Med.* **37**: 571–574.
46. Polisetty N., A. Fatima, S.L. Madhira, V.S. Sangwan, G.K. Vemuganti (2008). 'Mesenchymal cells from limbal stroma of human eye'. *Mol Vis.* **14**: 431–442.
47. Meng X., T.E. Ichim, J. Zhong, A. Rogers, Z. Yin, J. Jackson, H. Wang, W. Ge, V. Bogin, K.W. Chan, B. Thebaud, N.H. Riordan (2007). 'Endometrial regenerative cells: a novel stem cell population'. *J. Trans. Med.* **5**: 57.
48. Schwab K.E., C.E. Gargett (2007). 'Co-expression of two perivascular cell markers isolates mesenchymal stem like cells from human endometrium'. *Human Reprod.* **22**: 2903–2911.
49. Meirelles L. Da S., P.C. Chagastelles, N.B. Nardi (2006). 'Mesenchymal stem cells reside in virtually all postnatal organ and tissues'. *J. Cell Sci.* **119**: 2204–2213.
50. Lazarus H.M., S.E. Haynesworth, S.L. Gerson, A.I. Caplan (1997). 'Human bone marrow-derived mesenchymal progenitor cells (MPCs) cannot be recovered from peripheral blood progenitor cell collections'. *J. Haematother.* **6**: 447–455.
51. Wexler S.A., C. Donaldson, P. Denning-Kendall, C. Rice, B. Bradley, J.M. Hows (2003). 'Adult bone marrow is a rich source of human mesenchymal stem cells but umbilical cord and mobilized adult blood are not'. *Br J. Haematol.* **121**: 368–374.

52. Gutierrez-Rodriguez M., E. Reyes-Maldonado, H. Mayani (2000). 'Characterisation of the adherent cells developed in dexter-type long term culture from human umbilical cord'. *Stem Cells.* **18**: 46–52.
53. Naruse K., K. Urabe, T. Mukaida, T. Ueno, F. Migishima, A. Oikawa, Y. Mikuni-Takagaki, M. Itoman (2004). 'Spontaneous differentiation of mesenchymal stem cells obtained from foetal rat circulation'. *Bone.* **35**: 850–858.
54. Zvaifler N.J., L. Marinova-Mutafchieva, G. Adams, C.J. Edwards, J. Moss, J.A. Burger, R.N. Miani (2000). 'Mesenchymal precursor cells in the blood of normal individuals'. *Arthritis Res.* **2**: 477–488.
55. Rochefort G.Y., B. Delorme, A. Lopez, O. Herault, P. Bonnet, P. Charbord, V. Eder, J. Domenech (2006). 'Multipotential mesenchymal stem cells are mobilized into peripheral blood by hypoxia'. *Stem Cells.* **24**: 2202–2208.
56. Bianco P., M. Riminucci, S. Gronthos, P.G. Robey (2001). 'Bone marrow stromal stem cells: nature, biology and potential applications'. *Stem Cells.* **19**: 180–192.
57. Farrington-Rock C., N.J. Croft, M.J. Doherty, B.A. Ashton, C. Griffin-Jones, A.E. Canfield (2004). 'Chondrogenic and adipogenic potential of microvascular pericytes'. *Circulation.* **110**: 2226–2232.
58. Schor A.M., T.D. Allen, A.E. Canfield, P. Sloan, S.L. Schor (1990). 'Pericytes derived from the retinal microvasculature undergo calcification *in vitro*'. *J. Cell Sci.* **97**: 449–461.
59. Brighton C.T., D.G.Lorich, R. Kupcha, T.M. Reilly, A.R. Jones, R.A. Woodbury 2nd (1992). 'The pericytes as a possible osteoblast progenitor cell'. *Clin. Orthop. Relat. Res.* 287–299.
60. Doherty M.J., B.A. Ashton, S. Walsh, J.N. Beresford, M.E. Grant, A.E. Canfield (1998). 'Vascular pericytes express osteogenic potential *in vitro* and *in vivo*'. *J. Bone Miner Res.* **13**: 828–838.
61. Shi S., S. Gronthos (2003). 'Perivascular niche of postnatal mesenchymal stem cells in human bone marrow and dental pulp'. *J. Bone. Miner. Res.* **18**: 696–704.
62. Zannettino A.C., P.J. Psaltis, S. Gronthos (2008). 'Home is where the heart is: via the frount'. *Cell Stem Cell.* **2**: 513–514.
63. Fuchs E., J.A. Segre (2000). 'Stem cells: a new lease on life'. *Cell.* **100**: 143–155.
64. Ohlstein B., T. Kai, E. Decotto, A. Spradling (2004). 'The stem cell niche: theme and variations'. *Curr. Opin. Cell Biol.* **16**: 693–699.
65. Li L., T. Xie (2005). 'Stem cell niche: structure and function'. *Ann. Rev. Cell. Dev. Biol.* **21**: 605–631.
66. Moore K.A., I.R. Lemischka (2006). 'Stem cell and their niches'. *Science.* **311**: 1880–1885.
67. Spradling A., D. Drummond-Barbosa, T. Kai (2001). 'Stem cells find their niche'. *Nature.* **414**: 98-104.
68. Ozerdem U., K. Alitalo, P. Salven, A. Li (2005). 'Contribution of bone marrow-derived pericytes precursor cells to corneal vasculogenesis'. *Invest Ophthalmol. Vis Sci.* **46**: 3502–3506.
69. Covas D., R.A. Panepucci, A.M. Fontes, W.A. Silva-Jr, M.D. Orellana, L. Neder, A.R.D. Santos, L.C. Perez, M.C. Jamur, M.A. Zago (2007). 'The similarities of human mesenchymal stem cells and pericytes'. *Haematologica.* **92**: 118.
70. Tintut Y., Z. Alfonso, T. Saini, K. Radcliff, K. Watson, K. Bostrom, L.L. Demer, (2003). 'Multilineage potential of cells from the artery wall'. *Circulation.* **108**: 2505–2510.
71. Cheng S.L., J.S. Shao, N. Charlton-Kachigian, A.P. Loewy, D.A. Towler (2003). 'MSX2 promotes osteogenesis and suppresses adipogenic differentiation of multipotent mesenchymal progenitors'. *J. Biol Chem.* **278**: 45969–45977.
72. Kalajzic I., Z. Kalajzic, L. Wang, X. Jiang, K. Lamothe, S.M. San Miguel, H.L. Aguila, D.W. Rowe (2006). 'Pericyte/myofibroblast phenotype of osteoprogenitor cell'. *Musculoskelet neuronal Interact.* **7**: 320–322.

73. Towler D.A. (2007). 'Vascular biology and bone formation: hints from HIF'. *J. Clin Invest.* **117**: 1477–1480.
74. Rolf H.J., Kierdorf U., H. Kierdorf, J. Schulz, N. Seymour, H. Schliephake, J. Napp, S. Niebert, H. Wolfel, K.G. Wiese (2008). 'Localisation and characterisation of STRO^{-1+} cells in the deer pedicle and regenerating antler'. *PLOS. one.* **3**: 2064–2075.
75. Rajkumar V.S., K. Howell, K. Csiszar, C.P. Denton, C.M. Black, D.J. Abraham (2005). 'Shared expression of phenotypic markers in systemic sclerosis indicates a convergence of pericytes and fibroblasts to a myofibroblast lineage in fibrosis'. *Arthritis. Res. Ther.* **7**: 1113–1123.
76. Zvaifler N.J. (2006). 'Relevance of the stroma and epithelial-mesenchymal transition (EMT) for the rheumatic diseases'. *Arthritis Research and Therapy.* **8**: 210.
77. Traktuev D.O., J. S. Merfeld-Clauss Li, M. Kolonin, W. Arap, R. Pasqualini, B.H. Johnstone, K.L. March (2008) 'A population of multipotent Cd34 positive adipose stromal cells share pericytes and mesenchymal surface markers, reside in a periendothelial location and stabilize endothelial networks'. *Circ. Res.* **102**: 77–85
78. Lin G., M. Garcia, H. Ning, L. Banie, Y.L. Guo, T.F. Lue, C.S. Lin (2008) 'Defining stem and progenitor cells with adipose tissue'. *Stem Cells and Development.* **17**: 1053–1064.
79. Gangenahalli G.U., V.K. Singh, Y.K. Verma, P. Gupta, R.K. Sharma, R. Chandra, P.M. Luthra (2006) 'Haematopoietic stem cell antigen Cd34: role in adhesion or homing'. *Stem Cells Dev.* **15**: 305–313.
80. Bonab M.M., K. Alimoghaddam, F. Talebian, S.H. Ghaffari, A. Ghavamzadeh, B. Nikbin (2006). 'Aging of mesenchymal stem cell *in vitro*'. *BMC Cell Biol.* **10**: 7–14.
81. Pal R., M. Hanwate, S.M. Totey (2008). 'Effect of holding time, temperature, and different parenteral solutions on viability and functionality of adult bone marrow-derived mesenchymal stem cells before transplantation'. *J. Tissue Eng. Regen Med.* **2**: 436–444.
82. Pal R., M. Hanwate, M. Jan, S. Totey (2008). 'Phenotypic and Functional comparison of optimum culture condition for upscaling of Bone marrow derived Mesenchymal stem cells'. *J. Tissue Eng. Regen Med.* (in press)
83. Sotiropoulou P.A., S.A. Perez, M. Salagianni, C.N. Baxevanis, M. Papamichail (2006). 'Characterisation of the optimal culture conditions for clinical scale production of human mesenchymal stem cells'. *Stem Cells.* **24**: 462–471.
84. Horwitz E.M. (2006). 'MSC: a coming of age in regenerative medicine'. *Cytotherapy.* **8**: 194–195.
85. Shahdadfar A., K. Fronsdal, T. Haug, F.P. Reinholt, J.E. Brinchmann (2005). '*in vitro* expansion of human mesenchymal stem cells: choice of serum is a determinant of cell proliferation, differentiation, gene expression and transcriptome stability'. *Stem Cells.* **23**: 1357–1366.
86. Dominici M., K. Le Black, I. Mueller, I. Slaper-Cortenbach, F. Marini, D. Krause, R. Deans, A. Keating, D.J. Prockop, E. Horwitz (2005). 'Minimal criteria for defining multipotent mesenchymal stromal cells. The International Society for Cellular Therapy Position Statement'. *Cytotherapy.* **8**: 315–317.
87. Le Blanc K., L. Tammik, B. Sundberg, S.E. Haynesworth, O. Ringden (2003). 'Mesenchymal stem cells inhibit and stimulate mixed lymphocyte cultures and mitogenic responses independently of the major histocompatibility complex'. *Scand J Immunol.* **57**: 11–20.
88. Majumdar M.K., M. Keane-Moore, D. Buyaner, W.B. Hardy, M.A. Moorman, K.R. McIntosh, J.D. Mosca (2003). 'Characterisation and functionality of cell surface molecules on human mesenchymal stem cells'. *J Biomed Sci.* **10**: 228–241.

89. De Nicola M., C. Carlo-Stella, M. Magni, M. Milanesi, P.D. Longoni, P. Matteucci, S. Grisanti, A.M. Gianni (2002). 'Human bone marrow stromal cells suppress T-lymphocyte proliferation induced by cellular or non-specific mitogenic stimuli'. *Blood.* **99**: 3838–3843.

90. Bertholomew A., C. Sturgeon, M. Siatskas, K. Ferrer, S. Patil, W. Hardy, S. Devine, D. Ucker, R. Deans, A. Moseley, R. Hoffman (2002). 'Mesenchymal stem cells suppress lymphocyte proliferation *in vitro* and prolong skin graft survival *in vivo*'. *Exp. Haematol.* **30**: 42–48.

91. Krampera M., S. Glennie, J. Dyson, D. Scott, R. Laylor, E. Simpson, F. Dazzi (2003). 'Bone marrow mesenchymal stem cells inhibit the response of naïve and memory antigen specific T cells to their cognate peptide'. *Blood.* **101**: 3722–3729.

92. Tse W.T., J.D. Pendleton, W.M. Beyer, M.C. Egalka, E.C. Guinan (2003). 'Suppression of allogeneic T-cell proliferation by human marrow stromal cells: implications in transplantation'. *Transplantation.* **75**: 389–397.

93. Potian J.A., H. Aviv, N.M. Ponzio, J.S. Harrison, P. Rameshwar (2003). 'Veto-like activity of mesenchymal stem cells: Functional discrimination between cellular responses to alloantigens and recall antigens'. *J. Immunol.* **171**: 3426–3434.

94. Wolbank S., A. Peterbauer, M. Fahrner, S. Hennerbichler, M. van Griensven, G. Stadler, H. Red, C. Gabriel (2007). 'Dose dependent immunomodulatory effect of human stem cells from amniotic membrane: a comparison with human mesenchymal stem cells from adipose tissue'. *Tissue Eng.* **13**: 1173–1183.

95. Meirelles L., S. Da, A.I. Caplan, N.B. Nardi (2008). 'In search of the *in vivo* identity of mesenchymal stem cells'. *Stem Cells.* **26**: 2287–2299.

96. Gimble J.M., A.J. Katz, B.A. Bunnell (2007). 'Adipose-derived stem cells for regenerative medicine'. *Circ. Res.* **100**: 1249–1260.

97. Spees J.L., S.D. Olson, M.J. Whitney, D.J. Prockop (2006). 'Mitochondrial transfer between cells can rescue aerobic respiration'. *Proc. Natl. Acad. Sci. USA.* **103**: 1283–1288.

98. Sonoyama W., Y. Liu, D. Fang, T. Yamaza, B. Moo-Seo, C. Zhang, H. Liu, S. Gronthos, C.Y. Wang, S. Shi, S. Wang (2006). 'Mesenchymal stem cell-mediated functional tooth regeneration in swine'. *PLOS. One.* **1**: e79.

99. Lovschall H., T. Mitsiadis, K. Poulsen, K.H. Jensen, A.L. Kjeldsen (2007). 'Coexpression of Notch3 and Rgs5 in the pericytes-vascular smooth muscle cell axis in response to pulp injury'. *Int J. Dev. Biol.* **51**: 715–721.

100. Kinnaird T., E. Stabile, M.S. Burnett, C.W. Lee, S. Barr, S. Fuchs, S.E. Epstein (2004). 'Marrow-derived stromal cells express genes encodidng a broad spectrum of arteriogenic cytokines and promote *in vitro* and *in vivo* arteriogenesis through paracrine mechanisms'. *Circ. Res.* **94**: 678–685.

101. Liu M., Z.C. Han (2008). 'Mesenchymal stem cells: biology and clinical potential in type 1 diabetes therapy'. *J. Cell Mol. Med.* **12**: 1155–1168.

102. Wollert K.C., Drexler H. (2005). 'Clinical application of stem cells for the heart'. *Circulation Research.* **96**: 151–163.

103. Letizia M., M. Katia, F. Ivana, V. Elena, O. Giuseppe, B. Riccardo, T. Lucia, L. Sergio, F. Franca (2006). 'Autologous mesenchymal stem cells: Clinical application in amyotrophic lateral sclerosis'. *Neurological Research.* **28**: 523–526.

19

Mesenchymal Stem Cells: Culture, Characterisation, and Therapeutic Applications in Neurodegenerative Diseases

*Samuel SW Tay**

*Department of Anatomy, Yong Loo Lin School of Medicine,
National University of Singapore, Singapore 117597*

INTRODUCTION

Bone marrow contains precursors for haematopoietic and stem-like cells, for a variety of non-haematopoietic tissues. The precursors for non-haematopoietic tissues were initially referred to as plastic-adherent cells or colony-forming-units fibroblasts because of their ability to adhere to culture dishes and form fibroblast-like colonies [1, 2]. Mesenchymal stem cells (MSCs) in the bone marrow have been established as a mixed population of cells that generate bone, cartilage, fat and fibrous connective tissues [3, 4]. Scientists in laboratories are trying to grow MSCs *in vitro* and manipulate them to transform into specific cell types for treating injuries or diseases. Three methods have been employed to identify and test stem cells:

(1) Labelling the cells in a living tissue with molecular markers to determine the specialized cell types they generate.
(2) Extracting the cells from a living animal, labelling them in cell culture and transplanting them back into another animal to determine whether the cells repopulate their tissue of origin.

---

* *anttaysw@nus.edu.sg*

(3) Isolating the cells, propagating them in culture and manipulating them, often by adding growth factors or incorporating new genes to determine what differentiated cell type they can be transformed into. It is well-established that adult stem cells occur in many tissues and they are engaged in normal differentiation pathways to form the specialized cell types of the tissue they reside in. Adult stem cells also exhibit the ability to form specialized cell types of other tissues—a process known as transdifferentiation.

One of the main objectives of contemporary research in neuroscience is to find a way to overcome the debilitating effects of brain diseases (especially neurodegenerative diseases). Cell therapy has been advocated as an attractive approach to treat such diseases [5]. To date, the realization of gene therapy is still off target.

The animal body contains a variety of stem cells capable of both repeated renewal and production of specialized, differentiated progeny. Crucial to an understanding of stem cells is a thorough knowledge of which external signals in the microenvironment provide the one to control their fate decision in terms of cell proliferation or differentiation into the desired specific phenotype [6]. These signals must be incorporated into a research strategy or therapeutic approach to achieve success. MSCs derived from bone marrow consist of distinct cell populations that have very similar differentiation abilities but differ in their expressions of several stem cell markers [7]; hence these cells appear to have all the crucial components to complete specified differentiation programs necessary for the acquisition of a fully functional phenotype. Although most problems associated with cellular isolation and culture have been solved, the low frequency of these cells and the lack of special identification markers make their isolation and search difficult (including post-transplantation). The technique of transferring new genetic material into stem cell and the expression of the gene product in the daughter cells is an exciting approach to the treatment of congenital and acquired diseases [8].

It has been well-established that transplantation of MSCs into the adult brain reduces the functional deficits associated with stroke and traumatic brain injury. However, there is no evidence that the transplanted cells replaced the host brain cells. It appears that stem cell transplantation through cell replacement or as vector of gene delivery is a potential strategy for the treatment of neurodegenerative diseases [9]. Recent findings of the cell fusion phenomenon have raised some doubts about stem cell plasticity. The improvement observed does not seem to be dependent on the replacement of the lost neurons, but may be due to increased expression levels of certain neurotrophic factors, thereby suggesting a beneficial effect of somatic cells regardless of transdifferentiation [9,10].

ISOLATION AND CULTURE OF RAT MSCs

Whole bone marrow plugs were obtained by flushing the bone marrow cavity of the femurs and tibiae of adult male Wistar rats (200–250 g) with a 21-gauge syringe filled with alpha minimal essential medium (Invitrogen Life Technologies, Carlsbad, CA, USA), supplemented with 15% foetal calf serum (FCS), 2 mmM L-glutamine and 100 μ/mL penicillin, 100 μg/mL streptomycin, and 0.25 mg/mL amphotericin B (Sigma, St Louis, USA). Cells were passed through a 70-μm cell strainer

(BD Bioscience Discovery Labware, Bedford MA, USA) to remove the remaining clumps of tissue and placed in a 75-cm^2 flask and incubated at 37° C in a humidified chamber with 95% air and 5% CO_2. The first medium change was done in 24 h to remove the non-adherent haematopoietic lineage cells and the remaining adherent cells were defined as passage zero (P zero) cells. For further expanding and processing, the P-zero cells at 90% confluency were detached by incubating with 0.25% trypsin (Sigma, St Louis, USA) and 1mM EDTA (Sigma, St Louis, USA) for 5 min at 37° C, plated as P1 cells in 75 cm^2 flasks at a density of 5000 cells/cm^2 and then grown to 90% confluency.

CHARACTERISATION OF MSCs

Differentiation Assays

For osteogenic differentiation, detached MSCs were plated at 1000 cells/cm^2 and allowed to grow to 70% confluency in 5 days. They were incubated in an osteogenic medium containing $10^{-2}\mu M$ dexamethasone, 0.2 mM ascorbic acid and 10 mM β-glycerol phosphate (Sigma Chemical Company, St Louis, USA). The medium was replaced every 3–4 days for 21 days. Cultured cells were washed with PBS, fixed in a solution of ice-cold 70% ethanol for 1 h and stained for 10 min with 1 mL of 40 mM Alizarin red (at pH 4.1).

For adipogenic differentiation, 70% confluent cultures were incubated in complete medium supplemented with 0.5 μM hydrocortisone, 0.5mM isobutyl-methylxanthine and 60 μM indomethacin (Sigma, St Louis, USA). The medium was replaced every 3–4 days for 21 days and then washed in PBS, fixed in 10% formalin for 10 min and stained for 15 min with fresh oil red-0 solution (Fisher Scientific, Pittsburgh, PA, USA).

For neurogenic differentiation, 70% confluent cultures were incubated in complete medium supplemented with EGF and/or FGF2 and containing 10% FCS, 0.1 mM non-essential amino acid (Invitrogen Life Technologies, Carlsbad, CA, USA), 100 μ/mL penicillin, 100 μg/mL streptomycin and 0.25 μg/mL amphotericin (Sigma, St Louis, USA). Medium was changed every 3 days for 21 days. Cells were stained for neuronal markers such as neuronal marker (NeuN), protein gene product 9.5 (PGP 9.5), neuronal cell adhesion molecule (NCAM) to confirm their identity as neuronal cells (Unpublished data, see Fig. 19.1).

However, although MSCs have been broadly investigated, currently there is no specific marker or combination of markers identified to specifically define MSCs. However, phenotypically, *in vitro* expanded MSCs express a number of non-specific markers, including CD 105 (SH2 or endoglin), CD73 (SH3 or SH4), CD90, CD166, CD44, and CD29 (Pittenger, 2008), while MSCs are negative for most haemtopoietic and endothelial cell markers (CD45, CD34, CD11a, CD235a, HLA-DR, CD144 [11,12].

TRANSPLANTATION OF MSCs FOR CNS REPAIRS

With the discovery of stem cell plasticity, recent reports have raised hopes that stem cell transplantation could be within reach in the near future. Recent reports have demonstrated that stem

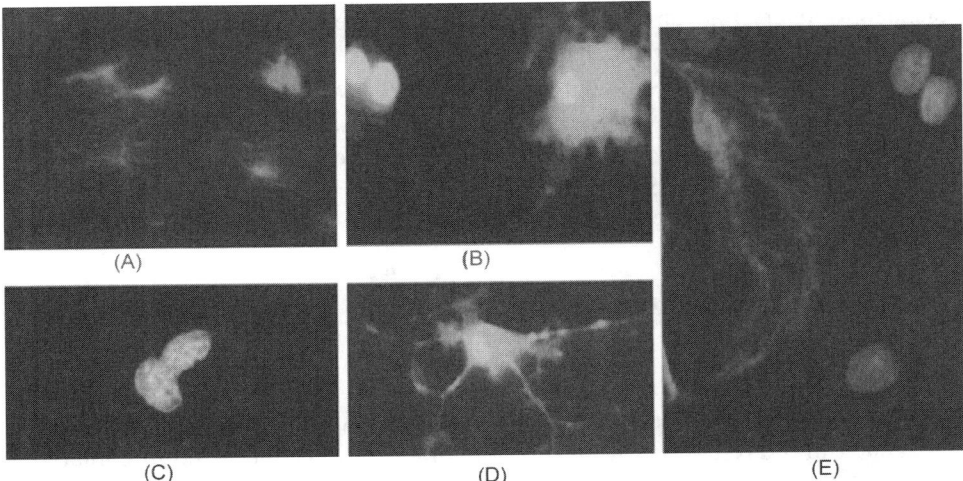

▶ **Fig. 19.1** *Neuronal differentiation of MSCs* **in vitro.** *After RA treatment, NFH-/- MSCs can be induced to differentiate into cells expressing (A) Nestin marker (for neural precursor cells, red), (B) MAP-2 (B) (for neurons, green), (C) NeuN (for neurons, red), (D) O4 (for oligodendrocytes, green) or (E) GFAP (for astrocytes, red). In MAP-2, NeuN, GFAP and O4 staining, the nuclei are counterstained with DAPI (blue). A NeuN-positive cell with its nucleus stained in red (C) is observed to locate closely with a NeuN-negative cell. Final magnification, ↔400 in (A–E). (For colour figure see Plate 45).*

cells in cultures can retain their stem cell potentials and can differentiate into the desired cell type in an appropriate environment [13,14]. The mechanism of transdifferentiation of a committed cell directly into another cell type in response to the environmental cues is still unknown. Direct transdifferentiation is limited by the number of cells that can be introduced into an organ without affecting/removing some of the resident cells. If bone marrow stem cells could give rise to stem cells of another tissue, then they could, in theory, repopulate whole organs with a few starting cells, and it appears that this model of dedifferentiation is consistent with results obtained in animal models.

Another mechanism postulated recently for stem cells after transplantation is that of fusion of the host and donor cells, which can in turn produce mature tissue cells without trans- or dedifferentiation. It has been well-established that glial activation plays an important role in the pathogenesis of neurodegenerative disorders and that direct transplantation of bone marrow derived MSCs results in the alleviation of inflammatory response associated with the cerebellum of Niemann-Pick disease Type C (NP-C) mice model. Using GFAP and F4/80 antibodies, it has been shown that transplanted MSCs reduced significantly both astrocytic and microglial activation, respectively [15], thereby indicating the potential of MSCs to treat NP-C and other neurodegenerative diseases, such as Parkinson's disease (PD) in animal models (see Figs. 19.2 and 19.3).

Recent newer techniques involve the use of MSCs as a vehicle to deliver therapeutic genes to the brain to achieve biologically significant functional recovery. Recently, Lu and others (2005)

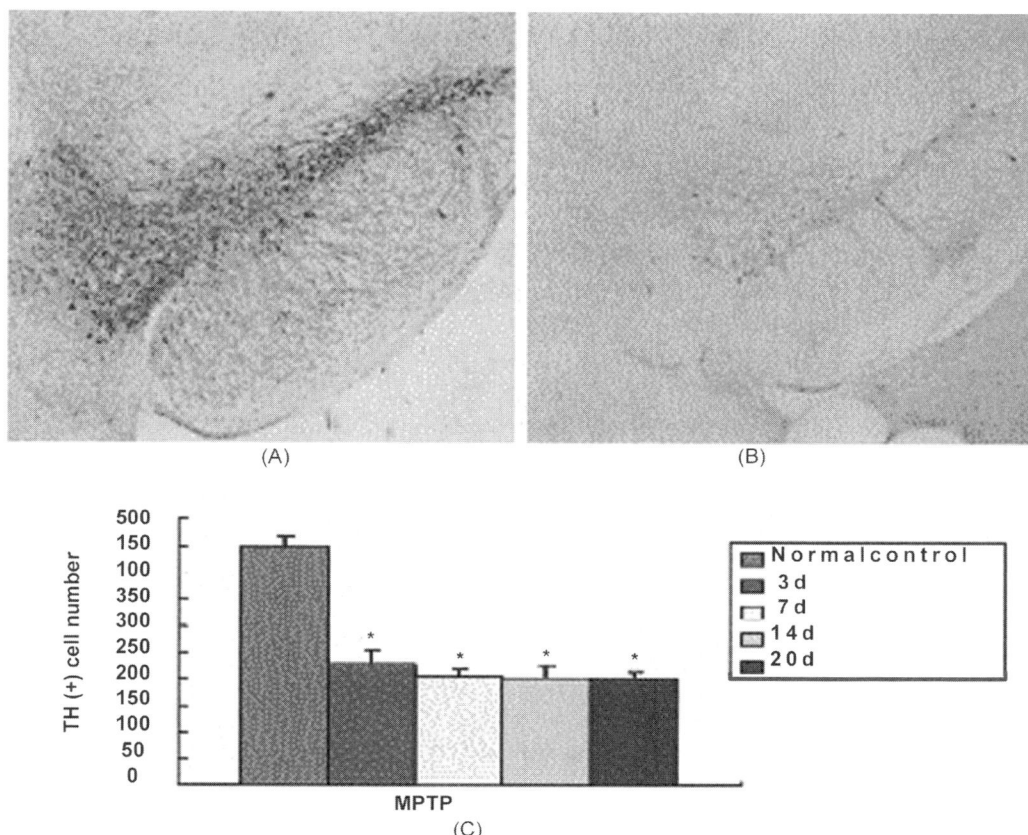

▶ Fig. 19.2 *Cell number counting. Tyrosine hydroxylase (TH) positive dopaminergic neurons in the substantia nigra area compacta (SNc) of (A) normal controls and (B) MPTP-treated mice. (C) Numbers of TH-positive cells at different time points post-MPTP injection (C, *$p < 0.01$). Final magnification, ↔200 in (A and B). (For colour figure see Plate 46).*

transfected the tyroxine hydroxylase gene into MSCs with an adeno-associated virus vector [16]. The transfected MSCs were transplanted into the striatum of PD rats, and they reported that the gene expression efficiency was about 75%, with most rats showing decreased rounds of asymmetric rotation after apomorphine administration.

Phinney and Isakova (2005) reviewed the evidence of differentiation of MSCs into neural lineages and evaluated the utility of MSCs as cellular vectors for treating neurological disorders and spinal cord injury [17]. They theorized that the varied functions of MSCs and their progeny are potential therapeutic agents for the treatment of neurological diseases.

There is growing evidence that reservoirs of stem cells may reside in several types of adult tissues, including adipose tissues [18]. Adult adipose tissue is a rich source of MSCs, providing an abundant and accessible source of adult stem cells, which can transdifferentiate into neural cells.

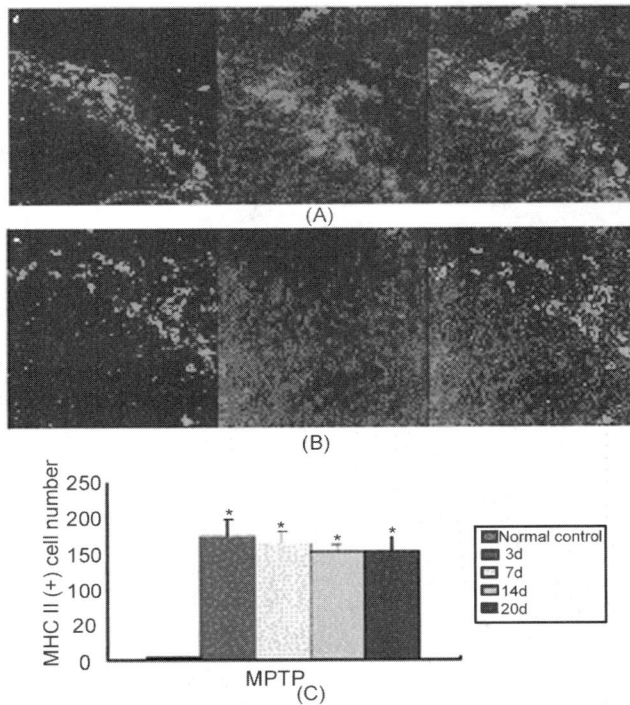

▶ **Fig. 19.3** *Microglial activation in the MPTP-treated mice. MHC II staining in the substantia nigra area compacta (SNc) of (A) MPTP-treated mice became amoeboid compared with the (B) ramified normal controls and (C) the positive cell number increased* $*p < 0.05$). *Final magnification, ↔400 in (A and B). (For colour figure see Plate 47).*

Hence, adipocytes can be a great source of mesenchymal cells for the treatment of various neurological disorders [18].

Ischemia is known to induce the mobilisation of MSCs in both animal models and humans. The tissue injury is "sensed" by the MSCs and they migrate to the site of cellular damage where they undergo differentiation [19]. The plasticity, differentiation and migration of MSCs are dependent on the specific signals present in the local microenvironment of the damaged tissue. Thus, the ischemic microenvironment has the critical patho-biological functions essential for the seeding, expansion, survival, renewal, growth and differentiation of MSCs in brain remodelling [20]. Specific molecular markers such as SDF-1 and CXCR4 have been identified to be essential for the interactions between MSCs and the damaged brain tissue to promote migration of MSCs [19]. It is well-established that MSCs can produce various cell types such as non-neuronal cells [21].

Deng and others (2006) clearly demonstrated that MSCs spontaneously express neural proteins in culture and are neurogenic after transplantation into the lateral ventricles of neonatal mouse brain, and they migrate extensively and differentiate into olfactory bulb granule cells and periventricular astrocytes, with no evidence of cell fusion [22]. Moviglia and others (2006) reported the successful transplantation of MSCs (transdifferentiated into neural stem cells) into the spinal cord of a man and a woman, both of whom had partial recovery of their senses at the affected spinal cord levels [23].

Stem cells have been well-recognized as the potential resource for restorative approaches for degenerative diseases and traumatic injuries [24]. In the central nervous system (CNS), stem cell based strategies have been proposed to replace lost neurons in degenerative diseases, Huntington's disease, amyotrophic lateral sclerosis (ALS) or to replace lost oligodendrocytes as in multiple sclerosis. Stem cells have also been used to repair the injured/damaged spinal cord [24]. It has been emphasized that in the replacement therapy, *in vitro* techniques must be developed to generate the desired cell type for the regenerative repairs. Dezawa (2006) described the discovery of special induction systems that control the MSC-derived cells for "auto-cell transplantation therapy" in neuro- and muscular degenerative diseases [25]. MSCs were isolated from bone marrow and grown in culture medium; the plastic-adherent precursor/stem cells (pMSCs) were grown to confluency in appropriate culture conditions as fibroblast-like cells for transplantation into the host [26]. They further demonstrated that pMSCs and MSCs cultured and analysed by immunohistochemistry and RT-PCR were successfully induced to become nestin-positive neurospheres in the presence of EGF and bFGF.

With the withdrawal of the mitogens, the cells could differentiate into neurofilament-positive glia cells, thereby suggesting the potential use of MSCs for cell therapy in the CNS [27]. Stem cells strategies were also employed for Alzheimer's disease therapy by various research groups [28]. Magnetic resonance imaging (MRI) has been incorporated into the study of the fate of transplanted cells *in vivo* [29]. MSCs labelled with iron oxide nanoparticles (endorem) and CD 34+ cells labelled with magnetic microbeads (miltenyi) were examined in rats with cortical or spinal cord lesion. Results indicate that MRI of grafted adult as well as embryonic stem cells labelled with iron oxide nanoparticles is a useful method for evaluating their migration and fate in the CNS [30].

Phinney and collaborators (2006) compared the engraftment kinetics and anatomical distribution of murine MSCs injected intracranially into neonatal versus adult mice [31]. They reported low-level engraftment in both neonatal and adult recipients at 12 days post-injection. However, at 60 and 150 dpi, the engraftment level was higher in the neonates than the adults. Similarly, Isakova and collaborators (2007) have reported that MSCs transplanted into the CNS of adult macaques engrafted at low levels without adversely affecting the animal health, behaviour or motor function [32].

Autologous transplantation of MSCs entirely circumvents the problem of immune rejection, does not promote the formation of teratomas and raises few ethical concerns. Although recent studies showed that the transplantation of MSCs have resulted in clinical improvements, the exact mechanisms responsible for the beneficial outcome have not been elucidated. Some of the possible rationalizations for the recovery processes include the following:

(1) *Cell replacement.* This theory hinges on the idea that replacement of degenerated neural cells with alternate functioning cells would induce long-lasting clinical improvement. It has been concluded that the transplanted cells survive, integrate into the endogeneous neural network and lead to some forms of functional improvement/recovery.

(2) *Trophic factor delivery.* This is credited to the transplanted MSCs secreting neurotrophic factors that support neuronal cell survival, which in turn induce endogenous cell proliferation and promote nerve fibre regeneration at sites of injury.

(3) *Immunomodulation.* Neuroimmunomodulation refers to the mechanism whereby neuroinflammation plays an important role in the pathogenesis of neurodegenerative disorders and that the inhibition of chronic inflammatory stress may explain the beneficial effects that accompany MSCs transplantation *in vivo* (Unpublished data, see Fig. 19.4) [33].

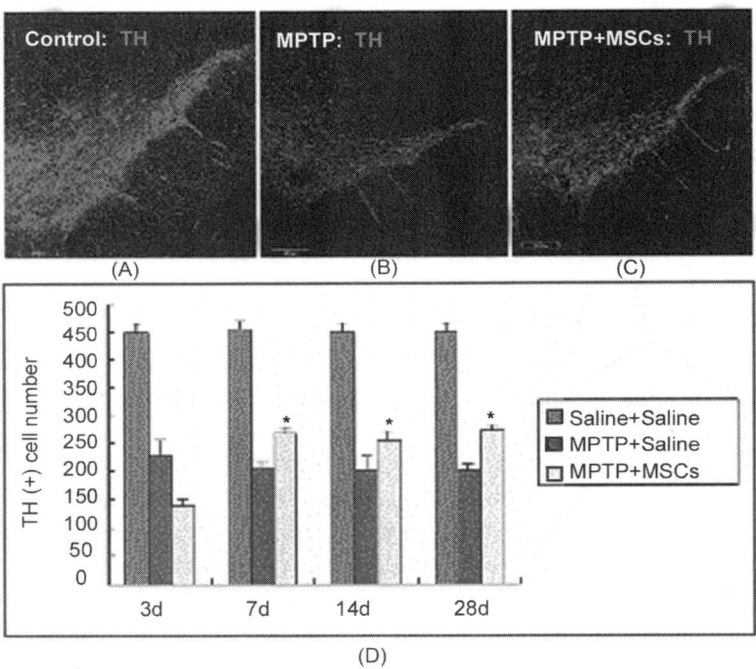

▶ **Fig. 19.4** *Dopaminergic neuron rescue. TH staining of dopaminergic neurons in the (A) SNc of normal controls, (B) MPTP-treated mice and (C) MPTP-treated plus MSCs transplanted mice. (D) TH-positive cell number counting in different treatment groups at different time points post-MPTP injection shows that the difference is statistically significant (*p < 0.01). Final magnification, ↔200 in (A)–(C). (For colour figure see Plate 48).*

EFFECTS OF MSCs TRANSPLANTATION IN THE CENTRAL NERVOUS SYSTEM

Three mechanisms have been postulated for the effects of MSCs after transplantation into the brain of animal models:

(1) transdifferentiation of the grafted cells, with partial or total replacement of the degenerating neural cells,

(2) cell fusion, and

(3) neuroprotection of the dying neurons (Unpublished data, see Figs. 19.5 and 19.6) [34].

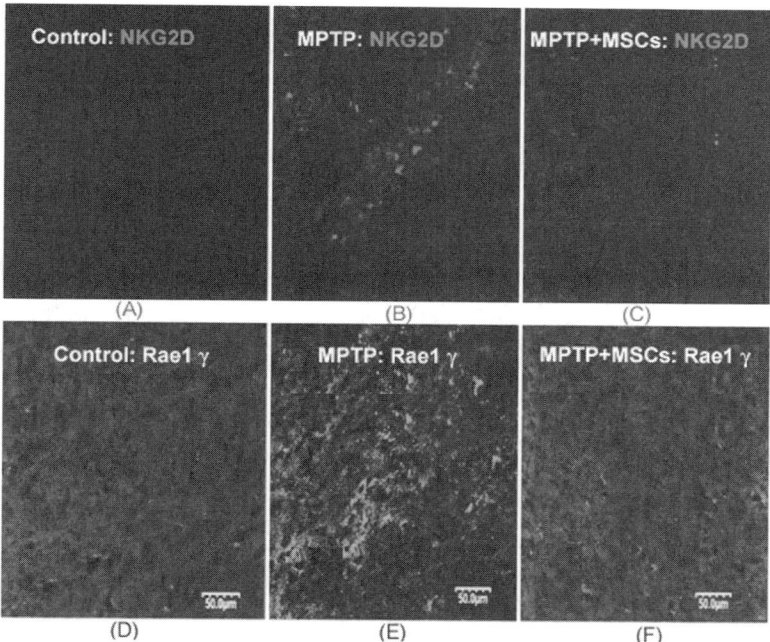

▶ **Fig. 19.5** *Suppression of natural killer (NK) cell activation by MSC transplantation. Immunofluorescence revealed (B) NKG2D and (E) Rae-1γ positive staining increased in the SNc of MPTP-treated mice compared with normal controls (A) and (D), respectively, while MSC transplantation suppressed NKG2D expression (C) and Rae-1ã expression (F). Final magnification, ↔200 in (A)–(F). (For colour figure see Plate 49).*

Results from recent experiments have shown that adult MSCs have a limited capacity to differentiate into neurons, whereas a substantial number of them transdifferentiate into astroglial cells [34]. Recently, it was demonstrated that MSCs were capable of transdifferentiating into neuron-like cells and carry with them genetical and pathological deficiency, thereby indicating the importance of selecting the genetic quality of donor cells before transplantation [10]. Kim and others (2006b) transplanted cultured human MSCs and basic fibroblast growth factor (bFGF) infusion into the CSF space after weight-driven spinal cord injury in rats [27]. In this study, based on improved functional outcome, they concluded that glial differentiation of MSCs was not uncommon but neural differentiation was doubtful (Unpublished observation, see Figs. 19.3 and 19.4). Boucher and others (2008) demonstrated partial recovery of dopaminergic pathway after

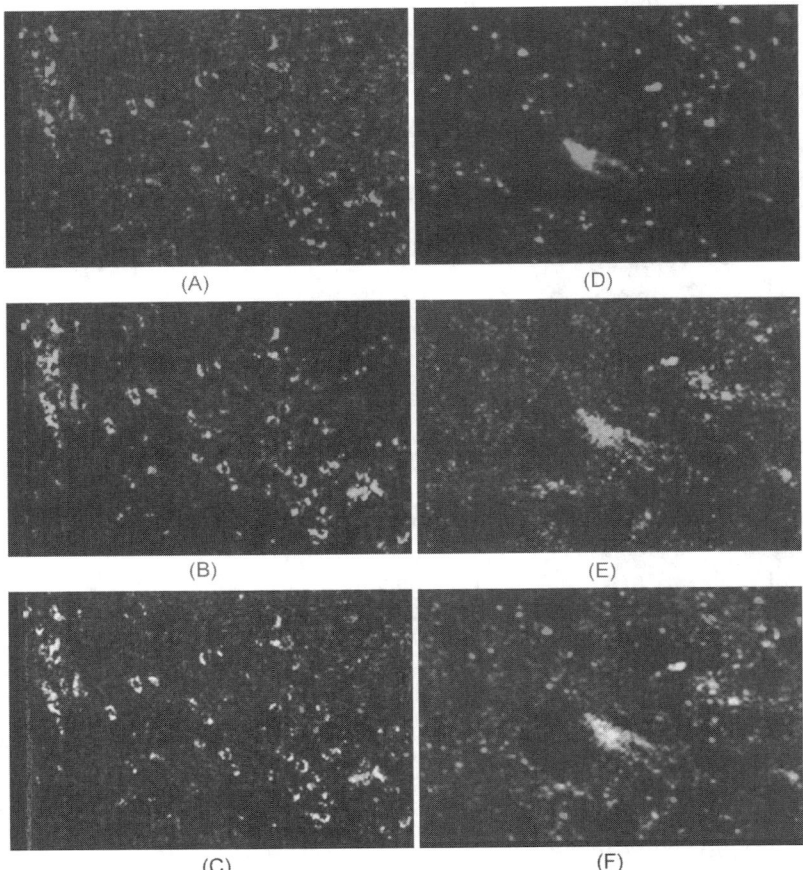

▶ Fig. 19.6 *TGF-β1 and IL-6 release in MSCs* **in vivo**. *Double-labelling of (A) TGF-β1 or (D) IL-6 (D) with CFDA SE pre-labelled MSCs (B) and (D), respectively, showed co-localisation of (C) TGF-β1 and (IL-6) with MSCs in the SNc of transplanted mice. Final magnification, ↔200 in (A)–(C) ↔400 in (D)–(F). (For colour figure see Plate 50).*

graft of adult MSCs in a rat model of PD [35]. Kim and others (2008) demonstrated that MSCs transplanted into middle cerebral artery occlusion rats showed the migration of the transplanted cells to the infarcted area exclusively in both ipsilateral and contralateral injections, thereby exhibiting pathotropism of the cells as seen by using MRI [36]. They speculated that the extensive migratory ability of MSCs represents a good therapeutic potential for developing efficient cell transplantation strategies for stroke.

Greco and Rameshwan (2007) reported the enhancing effect of IL-1α on the neurogenesis from transplanted adult human MSCs and reiterated the importance of inflammatory factors in the regenerative processes [37]. Mazzini and others (2008) have reported that MSCs represent a good chance for stem cell based therapy for ALS in patient [38]. They reported no side effects in patients injected intraspinally using autologous MSCs at the thoracic level over a 4-year period. This has been confirmed by Satake and collaborators (2004) that transplanted MSCs injected into the subarachnoid space of the lumbar spine can migrate to the injured thoracic spinal cord tissue [39]. Huang and others (2004) showed that brain-derived neurotrophic factor (BDNF) can induce rat bone marrow stromal cells to differentiate into neurons *in vitro* and that such cells have a longer life-span to better serve the purpose of transplantation and gene therapy for the nervous system in the near future [40].

DISCUSSION

In the classical review of developmental biology, organisms are shaped by the cellular differentiation that is unidirectional, with cells never switching lineages. This classical concept has been challenged over the past few years by reports suggesting that stem cells from various tissues might be more plastic than previously anticipated, with the ability to generate cells of lineages thought of as unrelated [41, 42]. This suggests that cells from one lineage can be reprogrammed (transdifferentiate) to change into a completely distinct phenotype of cells from another lineage.

Owing to the simple harvesting, culturing and expansion procedures, MSCs represent one of the most accessible sources of multipotent stem cells for therapeutic use in the treatment of CNS diseases. Transplantation of MSCs has been addressed in a number of animal models of CNS diseases and injuries including stroke [43, 44], traumatic brain injury [45, 46], PD [43, 47–49] and demyelinating diseases [50, 51]. Transplanted MSCs seem to promote the functions of the impaired brains in these animal models. Accumulated evidence seems to suggest their therapeutic potential to regenerate or protect the neural cells in the injured or diseased brain [52–54].

Despite the systemic injection of MSCs into mice with brain injury has positive effects on the injury repair, incompatible genetic background and the low transdifferentiation rate may limit its employment in therapeutic intervention for cell replacement [10]. It is hypothesized that the therapeutic and beneficial effects *in vivo* of cultured MSCs is most likely due to the immune modulatory effect on immune cells associated with a pro-survival trophic effect on the injured cells.

In line with the advancement of medical technologies, it is now possible to isolate MSCs from bone marrow, expand them in culture and genetically modify them to serve as cell carriers for local or systematic therapy [55]. Until today, despite their known ability to differentiate into cytoblasts, chondrocytes, adipocytes, myocytes and neuronal cells under appropriate stimuli, the distinct molecular signals that guide the migration of MSCs to their specific targets are still largely unknown.

CONCLUSION

The immunomodulatory effects of MSCs in the CNS appears to affect the inflammatory processes in the brain and spinal cord, at least in some circumstances whereby the MSC-mediated tissue repair is associated with a reduction of the inflammatory components. One challenge would be to inject antibiotics into the brain (via CSF or the blood circulation) prior to the transplantation of

MSCs. This treatment regime should result in higher proliferation of the MSCs, reduced apoptosis and better functional recovery of the damaged (diseased) site in the CNS. Another important area of interest is the ischemic brain or spinal cord, commonly known as stroke. MSCs would serve as an important therapeutic agent to ameliorate the ischemic tissues in the CNS by suppressing the associated immune responses resulting from neuronal and non-neuronal tissue damage. MSCs can be introduced either intracerebrally, intraventrically or intravenously, expecting that they will migrate to the site of ischemia where they could then facilitate the reparatory processes, acting as an immunomodulatory agent. In CNS repairs involving nerve grafts, transplantation of MSCs and their final migration to the site of repair should enhance the regeneration of the grafted nerve tissue, stimulated mainly by the inhibition of intrinsic immune response. Last, MSCs can also serve as an efficacious vehicle for gene transfer in the treatment of neurodegenerative diseases in the CNS. More research work should be targeted to optimize the therapeutic strategies for possible cures in the near future.

ACKNOWLEDGEMENTS

The author would like to thank Dr. Chao Yinxia for the bench work and the provision of the photographs. Special thanks to Associate Professor ST Dheen and Dr. SD Kumar for their critical comments on the manuscript, and to the National University of Singapore for the Academic Research Fund.

REFERENCES

1. Friedenstein A.J., U.F. Deriglasova, N.N. Kulagina, A.F. Panasuk, S.F. Rudakowa, E.A. Luria, I.A. Ruadkow (1974). 'Precursors for fibroblasts in different populations of haematopoietic cells as detected by the *in vitro* colony assay method'. *Exp. Haematol.* **2**: 83–92.
2. Piersma A.H., K.G. Brockband, R.E. Ploemacher, E.van Vliet, K.M. Brakel-van Peer, P.J. Visser (1985). 'Characterisation of fibroblastic stromal cells from murine bone marrow'. *Exp. Haematol.* **13**: 237–243.
3. Caplan A.I. (1991). 'Mesenchymal stem cells'. *J. Orthop. Res.* **9**: 641–650.
4. Caplan A.I., S.P. Bruder (2001). 'Mesenchymal stem cells: Building blocks for molecular medicine in the 21st century'. *Trends. Mol. Med.* **7**(5): 259–264.
5. Feron F. (2007). 'Current cell therapy strategies for repairing the central nervous system'. *Rev. Neurol.* (Paris) **163**(1): 3S23–3S30.
6. Alsberg E., H.A. von Recum, M.J. Mahoney (2006). 'Environmental cues to guide stem cell fate decision for tissue engineering applications'. *Expert Opin. Biol.Ther.* **6**(9): 847–866.
7. Bedada F.B., S. Günther, T. Kubin, T. Braun (2006). 'Differentiation versus plasticity: Fixing the fate of undetermined adult stem cells'. *Cell Cycle.* **5**(3): 223–236.
8. Galamb O., B. Molnár, F. Sipos, Z. Tulassay (2003). 'Possibilities of investigation and clinical application of adult human stem cells'. *Orv Hetil.* **144** (46): 2263–2270.
9. Corti S., F. Locatelli, S. Strazzer, M. Guglieri, G.P. Comi (2003). 'Neuronal generation from somatic stem cells: Current knowledge and perspectives on the treatment of acquired and degenerative central nervous system disorders'. *Curr Gen Ther.* **3**(3): 247–272.

10. Chao Y.X., B.P. He, Q. Cao, S.S.W. Tay (2007). 'Protein aggregate-containing neuron-like cells are differentiated from bone marrow mesenchymal stem cells from mice with neurofilament light subunit gene deficiency'. *Neurosci Lett.* **417**(3): 240–245.

11. Dvorakova J., A. Hruba, V. Velebny, I. Kubala (2008). 'Isolation and characterization of mesenchymal stem cell population entrapped in bone marrow collection sets'. *Cell Biol. Int.* **32**(9): 1116–1125.

12. Pittenger M.F. (2008). 'Mesenchymal stem cells from adult bone marrow'. *Methods Mol. Biol.* **449**: 27–44.

13. Karussis D., I. Kassis, B.G. Kurkalli, S. Slavin (2008). 'Immunomodulation and neuroprotection with mesenchymal bone marrow stem cells (MSCs): A proposed treatment from multiple sclerosis and other neuroimmunological/neurodegenerative diseases'. *Neurol Sci.* **265**(1–2): 131–135.

14. Kashofer K., D. Bonnet (2005). 'Gene therapy progress and prospects: Stem cell plasticity'. *Gene. Ther.* **12**(16): 1229–1234.

15. Bae J.S., S. Furuya, S.J. Ahn, S.J. Yi, Y. Hirabayashi, H.K. Jin (2005). 'Neuroglial activation in Niemann-Pick Type C mice is suppressed by intracerebral transplantation of bone marrow-derived mesenchymal stem cells'. *Neurosci. Lett.* **381**(3): 234–236.

16. Lu L., C. Zhao, Y. Liu, X. Sun, C. Duan, M. Ji, H. Zhao, Q. Xu, H. Yang (2005). 'Therapeutic benefit of TH-engineered mesenchymal stem cells for Parkinson's disease'. *Brain Res Brain Res Protoc.* **15**(1): 46–51.

17. Phinney D.G., I. Isakova (2005). 'Plasticity and therapeutic potential of mesenchymal stem cells in the nervous system'. *Curr. Pharm Des.* **11**(10): 1255–1265.

18. Safford K.M., H.E. Rice (2005). 'Stem cell therapy for neurologic disorders: Therapeutic potential of adipose-derived stem cells'. *Curr. Drug Targets.* **6**(1): 57–62.

19. Ji J.F., B.P. He, S.T. Dheen, S.S.W. Tay (2004). 'Interactions of chaemokines and chemokine receptors mediate the migration of mesenchymal stem cells to the impaired site in the brain after hypoglossal nerve injury'. *Stem Cells.* **22**: 415–427.

20. Shyu W.C., Y.J. Lee, D.D. Liu, S.Z. Lin, H. Li (2006). 'Homing genes, cell therapy and stroke'. *Front Biosci.* **11**: 899–907.

21. Ji J.F., S.T. Dheen, D.K. Srinivasan, B.P. He, S.S.W. Tay (2005). 'Expressions of cytokines and chemokines in the dorsal motor nucleus of the rat after right vagotomy'. *Brain Res.* **142**: 47–57.

22. Deng J., B.E. Petersen, D.A. Steindler, M.L. Jorgensen, E.D. Laywell (2006). 'Mesenchymal stem cells spontaneously express neural proteins in culture and are neurogenic after transplantation'. *Stem Cells.* **24**(4): 1054–1064.

23. Moviglia G.A., R. Fernandez Viña, J.A. Brizuela, J. Saslavsky, F. Vrsalovic, G. Varela, F. Bastos, G. Etchegaray, M. Barbieri, G. Martinez, F. Picasso, Y. Schmidt, P. Brizuela, C.A. Gaeta, H. Costanzo, M.T. Moviglia Brandolino, S. Merino, M.E. Pes, M.J. Veloso, C. Rugilo, I. Tamer, G.S. Shuster (2006). 'Combined protocol of cell therapy for chronic spinal cord injury. Report on the electrical and functional recovery of two patients'. *Cytotherapy.* **8**(3): 202–209.

24. Nandoe R.D. Tewarie, A. Hurtado, A.D. Levi, J.A. Grotenhuis, M. Oudega (2006). 'Bone marrow stromal cells for repair of the spinal cord: Towards clinical application'. *Cell Transplant.* **25**(7): 563–577.

25. Dezawa M. (2006). 'Insights into autotransplantation: The unexpected discovery of specific induction systems in bone marrow stromal cells'. *Cell Mol Life Sci.* **63**(23): 2764–2772.

26. Kim S., O. Honmou, K. Kato, T. Nonaka, K. Houkin, H. Hamada, J.D. Kocsis (2006a). 'Neural differentiation potential of peripheral blood- and bone-marrow-derived precursor cells'. *Brain Res.* **1123**(1): 27–33.
27. Kim K.N., S.H. Oh, K.H. Lee, D.H. Yoon (2006b). 'Effect of human mesenchymal stem cell transplantation combined with growth factor infusion in the repair of injured spinal cord'. *Acta Neurochir Suppl.* **99**: 133–136.
28. Sugaya K., A. Alvarez, A. Marutle, Y.D. Kwak, E. Choumkina (2006). Stem cell strategies for Alzheimer's disease therapy. *Panminerva Med.* **48**(2): 87–96.
29. Syková E., P. Jendelová (2006). 'Magnetic resonance tracking of transplanted stem cells in rat brain and spinal cord'. *Neurodegener. Dis.* **3**(1–2): 62–67.
30. Syková E., P. Jendelová (2007). 'Migration, fate and *in vivo* imaging of adult stem cells in the CNS'. *Cell Death Diff.* **14**(7): 1336–1342.
31. Phinney D.G., M. Baddoo, M. Dutreil, D. Gaupp, W.T. Lai, I.A. Isakova (2006). 'Murine mesenchymal stem cells transplanted to the central nervous system of neonatal versus adult mice exhibit distinct engraftment kinetics and express receptors that guide neuronal cell migration'. *Stem Cells Dev.* **15**(3): 437–447.
32. Isakova I.A., K. Baker, M. DuTreil, J. Dufour, D. Gaupp, D.G. Phinney (2007). 'Age- and dose-related effects on MSC engraftment levels and anatomical distribution in the central nervous systems of non-human primates: Identification of novel MSC subpopulations that respond to guidance cues in brain'. *Stem Cells.* **25**(12): 3261–3270.
33. Kan I., E. Melamed, D. Offen (2007). 'Autotransplantation of bone marrow-derived stem cells as a therapy for neurodegenerative diseases'. *Handb. Exp. Pharmacol.* (180): 219–242.
34. Pisati F., P. Bossolasco, M. Meregalli, L. Cova, M. Belicchi, M. Gavina, C. Marchesi, C. Calzarossa, D. Soligo, G. Lambertenghi-Deliliers, N. Bresolin, V. Silani, Y. Torrente, E. Polli (2007). 'Induction of neurotrophin expression via human adult mesenchymal stem cells: Implication of cell therapy in neurodegenerative diseases'. *Cell Transplant.* **16**(1): 41–55.
35. Bouchez G., L. Sensebé, P. Vourc'h, L. Garreau, S. Bodard, A. Rico, D. Guilloteau, P. Charbord, J.C. Besnard, S. Chalon (2008). 'Partial recovery of dopaminergic pathway after graft of adult mesenchymal stem cells in a rat model of Parkinson's disease'. *Neurochem Int.* **52**(7): 1332–1342.
36. Kim D., B.G. Chun, Y.K. Kim, Y.H. Lee, C.S. Park, I. Jeon, C. Cheong, T.S. Hwang, H. Chung, B.J. Gwag, K.S. Hong, J. Song (2008). '*In vivo* tracking of human mesenchymal stem cells in experimental stroke'. *Cell Transplant.* **16**(10): 1007–1012.
37. Greco S.J., P. Rameshwar (2007). 'Enhancing effect of Il-1alpha on neurogenesis from adult human mesenchymal stem cells: Implication for inflammatory mediators in regenerative medicine'. *J. Immunol.* Sep 1. **179**(5): 3342–3350.
38. Mazzini L., K. Mareschi, I. Ferrero, E. Vassallo, G. Oliveri, N. Nasuelli, G.D. Oggioni, L. Testa, F. Fagioli (2008). 'Stem cell treatment in Amyotrophic Lateral Sclerosis'. *J. Neurol Sci.* **265**(1–2): 78–83.
39. Satake K., J. Lou, L.G. Lenke (2004). 'Migration of mesenchymal stem cells through cerebrospinal fluid into injured spinal cord tissue'. *Spine.* **29**(18): 1971–1979.
40. Huang W., C. Zhang, S.L. Chen, W.X. Zhang, X.L. Yao, Y. Zeng, H. Huang, S.W. Feng, T.Y. Liu (2004). 'Brain-derived neutrophic factor induces rat bone marrow stromal cells to differentiate into neuron-like cells *in vitro*'. *Di Yi Jun Yi Da Xue Xue Bao.* **24**(8): 854–858.

41. Clarke D., J. Frisen (2001). 'Differentiation potential of adult stem cells'. *Curr. Opin. Genet. Dev.* **11**: 575–580.
42. Anderson S.A., D.D. Eisenstat, L. Shi, J. L. Rubenstein (1997). 'Interneuron migration from basal forebrain to neocortex: Dependence on *Dlx* genes'. *Science*. **278**: 474–476.
43. Li Y., J. Chen, L. Wang, L. Zhang, M. Lu, M. Chopp (2001). 'Intracerebral transplantation of bone marrow stromal cells in a 1-methyl-4-phenyl-1,2,3,6-tetrahydropyridine mouse model of Parkinson's disease'. *Neurosci. Lett.* **316**: 67–70.
44. Li Y., J. Chen, X.G. Chen, L. Wang, S.C. Gautam, V.X. Xu, M. Katakowski, L.J. Zhang, M. Lu, N. Janakiraman, M. Chopp (2002). 'Human marrow stromal cell therapy for stroke in rat: Neurotrophins and functional recovery'. *Neurology.* **59**: 514–523.
45. Mahmood A., D. Lu, L. Wang, M. Chopp (2002). 'Intracerebral transplantation of marrow stromal cells cultured with neurotrophic factors promotes functional recovery in adult rats subject to traumatic brain injury'. *J. Neurotrauma.* **19**: 1609–1617.
46. Mahmood A., D. Lu, M. Lu, M. Chopp (2003). 'Treatment of traumatic brain injury in adult rats with intravenous administration of human bone marrow stromal cells'. *Neurosurgery.* **53**: 697–703.
47. Storch A., J. Schwarz (2002). 'Neural stem cells and Parkinson's disease'. *J. Neurol.* **249**(Suppl 3): III/30–2.
48. Dezawa M., M. Hoshino, Y. Nabeshima. C. Ide (2005). 'Marrow stromal cells: Implications in health and disease in the nervous system'. *Curr. Mol. Med.* **5**: 723–732.
49. Tropel P., N. Platel, JC Platel. D. Noel, M. Albrieux, A.L. Benabid, F. Berger (2006). 'Functional neuronal differentiation of bone marrow-derived mesenchymal stem cells'. *Stem Cells.* **24**: 2868–2876.
50. Aikiyama Y., C. Radka, J. Kocsis (2002a). 'Remyelination of the rat spinal cord by transplantation of identified bone marrow stromal cells'. *J. Neurosci.* **22**: 6623–6630.
51. Akiyama Y., C. Radha, J. Kocsis (2002b). 'Remyelination of the spinal cord following intravenous delivery of bone marrow cells'. *GLIA.* **39**: 229–238.
52. Prockop D.J. (1997). 'Marrow stromal cells as stem cells for non-haematopoietic tissues'. *Science.* **276**: 71–74.
53. Prockop D.J., CA Gregory, JL Spees (2003). 'One strategy for cell and gene therapy: Harnessing the power of adult stem cells to repair tissues'. *Proc Natl Acad Sci.* **100**: 11917–11923.
54. Shintani A., N. Nakao, K. Kakishita, T. Itakura (2007). 'Protection of dopaminergic neurons by bone marrow stromal cells'. *Brain Res.* **1186**: 48–55.
55. Kumar. S., D.Chandra, S. Ponnazhagan (2008). 'Therapeutic potential of genetically modified mesenchymal stem cells'. *Gene Ther.* **15**(10): 711–715.

20

DENTAL PULP STEM CELLS AND PERSPECTIVES OF FUTURE APPLICATION IN CELL THERAPY

Irina Kerkis[1], *Sonja E. Lobo,*[1] *and, Alexandre Kerkis*[2]

[1] *Laboratory of Genetics, Butantan Institute,* São Paulo, Brazil*
[2] *Applied Genetics, Veterinary Activities Ltd. (Genética Aplicada),* São Paulo, Brazil*

INTRODUCTION

During life, adult organisms present high regeneration ability in response to different types of injuries or diseases. This occurs due to the presence of special types of cells located in each organ, in their niche, the so-called adult stem cells (ASCs). Within their niche, ASCs maintain a quiescent state when they are not dividing. In case of organ damage, tissues and/or any type of disease these cells received signals from surrounding environment, which lead to activation of their division and migration into injured local. Due to their role in repairing processes, ASCs can be defined as self-renewing cells, which can differentiate into several functional cell types *in vitro* and after their re-introduction into an injured organ they are able to functionally restore tissues environment throughout their differentiation into definitive functional cell type. Technological advances, which provide a better understanding of potential therapeutic uses of ASC and their capacity to regenerate tissues in animal models, are widely highlighted [1–7]. Studies concerning ASC niche, isolation, manipulation, differentiation and mechanisms of their action could contribute significantly and unbeatable for our knowledge about basic process of human development as well as the practical application of these cells in regenerative medicine. It is noteworthy that although ASCs are not as versatile as embryonic stem cells (ESCs), they did not present ethical problems, since embryo destruction is not required for their isolation.

For the first time, stromal ASCs—which present fibroblast-like morphology and potential for osteogenic differentiation—were isolated from bone marrow by Alexandre Friedenstein [8, 9].

---

* ikerkis@butantan.gov.br

Later, it has been shown that these cells were also capable of producing other derivates of mesoderm, such as cartilage and adipocytes. Thus they were denominated mesenchymal progenitor cells and/or mesenchymal stem cells (MSC) by Arnold Caplan [3]. At the same time, potential therapeutic use of MSCs, particularly for bone repair and skeletal regeneration therapy, has been proposed [10–12]. Development of techniques for ASC isolation led to obtainment and characterization of homogeneous population of human MSCs, which can grow from a single clonal cell and differentiate into several derivates of mesoderm [13]. Furthermore, it described a subpopulation of MSC, known as multipotent adult progenitor cells (MAPC), which express several markers of ESCs and present a narrower differentiation potential, since they can mainly produce cells with visceral mesoderm, neuroectoderm and endoderm characteristics *in vitro* [14]. Currently, it has been shown that ASCs—similar to bone marrow-derived MSCs (BM-MSC) and MAPC—can be isolated from different tissues, such as nervous tissue, skeletal muscles, skin and dental pulp [15–21]. This chapter will focus on the isolation, culture, characterization of the stem cells, which can be isolated from dental pulp, as well as on the development of therapeutic approaches using these cells.

ISOLATION OF STEM CELLS FROM DENTAL PULP

Process of Dental Pulp Formation

The tooth formation is completely dependent on the constant interactions between the stomodium-derived epithelium and the cranial neural crest-derived ectomesenchyme [22–26]. Accompanying the arch of the maxilla and the mandible which are being formed and starting from the epithelium that is recovering the primary stomodium, there is the development of a thickened U-shaped epithelial band, called dental lamina, which invaginates into the subjacent ectomesenchymal tissue [23–25].

Along the dental lamina, at the sites correspondent to the teeth that will be formed, there is the formation of the epithelial placodes, which are small thickenings of the epithelium. This process resembles the development of all organs that form as appendages of the ectoderm. It is followed by the formation of an epithelial bud which leads to the condensation of the ectomesenchymal cells around it [23, 24]. The influence of sonic Hedgehog (*Shh*) gene expression has been described in this phase of tooth morphogenesis [26].

This complex undergoes a specific folding and, in the centre of the bud, there is the formation of the enamel knot, characterized by the presence of a transient epithelial cluster [23, 24]. This stage is termed as the cap stage. The enamel knot, where the enamel organ epithelium begins to be formed, marks the onset of the tooth crown development and expresses more than ten signal molecules belonging mainly to the TGFβ (including BMPs and activins), FGF, sonic hedgehog (Shh) and Wnt families beyond the Msx1, Pax9 and Runx2 transcription factors [23]. At that level, the differentiation of the ameloblasts (that will give rise to the dental enamel) and the odontoblasts (which will synthesize the dentin) will take place. The ameloblasts are differentiated from the epithelium cells and will form the enamel organ epithelium, while the odontoblasts are formed from the ectomesenchymal cells that are underlaying the epithelium and forming a tissue called dental papilla [22].

The cap stage is followed by the bell stage, characterized by the presence of the enamel organ, dental papilla and dental follicle [25]. The enamel organ is composed of the inner enamel epithelium, outer enamel epithelium, stellate reticulum and stratum intermedium [25]. The dental papilla will give rise to the dental pulp. The dental follicle—a transient structure that surrounds the dental germ—is the origin of cementoblasts, osteoblasts and fibroblasts, taking place not only in the formation of the periodontium but also in the formation of the bone tissue by the alveolar process, through the differentiation of osteoblasts from the ectomesenchymal cells [22, 25].

It is known that the Runx2 expression plays a crucial role in all these phases. This gene, also named *Cbfa1*, *PEBP2A1* and *AML3*, is necessary for the differentiation of MSCs to osteoblasts and controls the maturation of osteoblasts and odontoblasts [22].

As the enamel organ epithelium proliferates apically from the cervical dental crown, it will form the epithelium root sheath—the Hertwig's epithelial root sheath—that will determine the root morphology, maintaining the interactions with the underlying ectomesenchymal cells. Depending on the pattern of invagination of this sheath into the dental pulp ectomesenchymal cells, the tooth will present different root morphology. When it does not happen, the tooth will present only one root [27].

As the ectomesenchymal cells from dental pulp keep in contact with the epithelium cells, they differentiate into odontoblasts and begin the deposition of dentin. On the other hand, when the epithelium sheath interacts with dentin, their cells differentiate into cementoblasts, which will deposit cementum. Several transcription and growth factors have been shown to be expressed by the cells involved in this process among which are Shh, Dlx2, Patched1, Patched2, Nfic, Gli1, Smoothened, amelogenin and MMPs [27].

The cementum allows the attachment of collagen fibres that anchores the tooth to the alveolar bone [27], while the dental pulp presents the blood vessels that will assure the survival of all tissues that form the dental organ. Fibroblasts, nerves, lymphatic ducts and odontoblasts, which will be adjacent to the inner enamel epithelium, also take part in the composition of the dental pulp tissue [25].

At the meantime the root is being formed, the dental organ will be directed to the oral cavity until its eruption occurs [25]. The apical region of the root will remain opened allowing the entrance of vascular and nervous tissues that will be present as much as at the crown as at the root regions of the dental pulp tissue.

Stem Cells Isolated from Dental Pulp

The presence of undifferentiated cells in rat dental pulp capable of differentiating into odontoblasts, which subsequently produce a dentin matrix, was first reported by Yamamura [28]. Extending these findings, Arnold Caplan's group demonstrated that cultured rat dental pulp cells presented osteogenic and chondrogenic potential and could differentiate into dentin *in vivo* [29]. Over the next years, several stem cells populations have been isolated from human dental pulp of permanent, deciduous and supernumerary teeth [19–21, 30–35] (Table 20.1).

For the first time human postnatal dental pulp stem cells (DPSCs) were isolated from normal molar teeth of adult patients [19]. These cells were isolated using enzymatic digestion and colony-forming-unit-fibroblast assay (CFU-F), when each DPSC colony was originated from a single

Table 20.1 *Summarised Methods and Characterisation Described for Several Dental Pulp Stem Cell Populations*

Works	Cell types	Method of isolation	Culture	Markers conditions	In vitro differentiation	in vivo differentiation
Gronthos et al., 2000 [19]	DPSC Third molars adults 19–29 years old	Enzymatic digestion CFU-F	α-MEM 20% (FBS) 100 mM L-ascorbic acid 2-phosphate 2 mM glutamine	CD44, integrin β1, VCAM-1, MUC-18 (CD146), collagen-I and II, osteocalcin, alkaline phosphatase (ALP), FGF-2	Osteogenic	Dentin after transplantation into immunocompromised mice
Miura et al., 2003 [20]	SHED 7–8 years old children	Enzymatic digestion CFU-F	Same as Gronthos et al., 2000 [19]	STRO-1, MUC-18 (CD146), ALP, extracellular phosphaglycoprotein (MEPE), bFGF, endostatin	Functional odontoblasts like cells neural adipogenic	Odontoblasts neural cells after transplantation into immuno-compromised mice
Laino et al., 2005 [31]	SBP-DPCS Molars, adult 30–45 years old	Enzymatic digestion FACSorting	Same as Gronthos et al., 2000 [19]	c-kit (CD117), CD34	Osteoblast precursors, fibrous bone tissue	Lamellar bone, osteocytes after transplantation into immuno-compromised rats
Pierdomenico et al., 2005 [32]	DP-MSC with immunosuppressive activity Molars, adults 40 years old, different sexes	Tissue explants	DMEM 10% FBS	SH2, SH3, SH4, CD29, CD166	Osteogenic Adipogenic	Not analyzed
Kerkis et al., 2006 [21]	IDPSC, exfoliated deciduous teeth 5–7 years old children	Tissue explants CFU-F	DMEM Ham's F12 15% FBS (Hyclone) 2 mM glutamine 2mM non-essential amino acid	SH2, SH3, SH4, CD13, CD31, Oct4, Nanog, SSEA-3, SSEA-4, TRA-1-60, TRA-1-81	Osteogenic Chondrogenic Neural Skeletal muscle Smooth muscle	Engraftment After transplantation into immuno-mised mice in lung, brain, liver, kidney

(Contd.)

Table 20-1 (Contd.)

Jo et al., 2007 [33]	DPSC Molars, adults 18–35 years old	Enzymatic digestion CFU-F	Same as Gronthos et al., 2000 [19]	STRO-1 CD29 CD44	Osteogenic Adipogenic	Not analyzed
Huang et al., 2008 [35]	DPSC, supernumerary tooth 20-year-old-male, Deciduous tooth (left lower canine)	Enzymatic digestion CFU-F	α-MEM 20% FBS (hyclone) 100 mM L-ascorbic acid 2-phosphate 2 mM glutamine	Oct4, nanog, rex 1, nestin, osteonectin	Osteogenic Adipogenic	Not analyzed
Mendonça et al., 2008 [34]	DPSC, Exfoliated deciduous teeth 5–7 years old children	Tissue explants	DMEM Ham's F12 15% FBS (hyclone) 2 mM glutamine	SH2, SH3, SH4, CD29	Osteogenic Chondrogenic Adipogenic Skeletal muscle	Bone production in critical size defect in non-immuno- suppressed rats

Note: Abbreviations: DPSC, dental pulp stem cells; SHED, stem cells from exfoliated deciduous teeth; SBP-DPCS, stromal bone producing DPSC; DP-MSC, dental pulp mesenchymal stem cells; IDPSC, immature dental pulp stem cells; CFU-F, colony-forming-unit-fibroblast.

progenitor cell. The cells within each colony presented a typical fibroblast-like morphology resembling those of BM-MSC. They showed higher colony-forming efficiency and cell proliferation *in vitro* when compared to BM-MSC. Studies of the DPSC immunophenotype demonstrated that they failed to react with the haematopoietic markers, such as CD14 (monocyte/macrophage), CD45 (common pan-leukocyte antigen), CD34 (haematopoietic stem/progenitor/endothelium), while expressing variety of markers associated with endothelium (vascular cell adhesion molecule 1 (VCAM-1)), smooth muscle (α-SM actin), bone (collagen type I, osteonectin) and some other markers, such as CD29 (integrin β1), protein involved in cell adhesion and recognition in a variety of processes including embryogenesis, haemostasis, tissue repair, immune response and metatastatic diffusion of tumour cells [19, 30, 33].

The same group isolated stem cells from human exfoliated deciduous teeth (SHED) from young patients, demonstrating similar characteristics to DPSC, while their proliferative and clonogenic potential was even higher than those of DPSC. SHED presented approximately 140 population doublings (PD), while DPSC presented ~100 PD. For SHED characterization, STRO-1 antigen, which potentially defines MSC progenitor subset, and CD146 (MUC18), a marker of early MSC, were used. However, after SHED isolation and *ex vivo* expansion, only 9% of the cells presented positive reaction with anti-STRO-1 antibody [20].

Another DPSC population, which was denominated stromal bone producing DPSC (SBP-DPSC), was isolated using enzymatic digestion and fluorescence-activated cell sorting (FACSorting). This cell population was also highly clonogenic, and it was able to self-maintain for a long period in culture (6 months), while expressing markers such as, CD34 and c-kit (CD117), or marker of neural crest derived precursors [36]. The authors suggested that they isolated a population of stromal cells with neural crest origin, from dental pulp, which seems to be different from DPSC isolated by Gronthos's group [31].

Using the tissue explants method, a population of dental pulp MSC (DP-MSC), with immunosuppressive activity, was obtained [32]. After expansion *ex vivo*, 94 ± 4% of DP-MSC expressed traditional MSC antigens, such as CD105 (SH2, endoglin), and CD73 (SH3, SH4), indicating that this isolated population was highly homogeneous regarding these markers. Additionally, DP-MSCs were positive for CD29 and CD166 (activated leukocyte cell adhesion molecule (ALCAM), which was shown to be expressed in human blastocysts and endometrial epithelial cells [37]). Analysis of their proliferation activity demonstrated that it was significantly higher in DP-MSC, when compared to those from bone marrow. Furthermore, similarly to BM-MSC, these cells inhibited the proliferation of PHA (phytohaemagglutinin) stimulated T-cells, presenting even stronger affect than in BM-MSC [32, 38].

Two works reported the isolation of stem cells from dental pulp (Fig. 20.1A–C), which expressed markers of human embryonic stem cells (hESCs), such as *Oct4* (Pou transcription factor Octamer-4), *Nanog* (transcription factor, key factor of pluripotency), *Rex 1* (ZFP42, zinc finger protein 42 homolog (mouse), SSEA-3 and 4 (stage-specific embryonic antigen), TRA-1-60, TRA -1-81(tumour recognition antigen), and being negative for haematopoietic markers. The cells expressing these markers were isolated from human exfoliated deciduous teeth and supernumerary tooth, were denominated human immature dental pulp cells (hIDPSC) [21] and DPSC [35], respectively. Both works used similar cultivation conditions for their isolation as those used to hESCs and CFU-F assay (Fig. 20.1D–F). These cell populations present rapid proliferation, great expansion *in vitro* high clonogenic capacity and self-maintain for a long period in culture. Moreover, hIDPSC co-express markers of MSC, such as SH2, SH3 and SH4, which were observed in approximately 97% of hIDPSC, as well as CD31 and CD13. The expression pattern of hESCs and MSC markers were maintained also in different colonies of hIDPSC obtained after CFU-F [21, 34].

Normal karyotype is a pre-requisite for the potential use of stem cells in different types of therapies. Analysis of karyotype of hIDPSC at advanced passages (P10) was performed and showed that they did not present structural chromosomal rearrangements, although a low rate of tetraploid cells was observed possibly due to cell fusion [21]. Unfortunately, such important karyotype studies are lacking for other cell populations isolated from dental pulp.

DPSCs, which present similar properties to human DPSCs, were isolated from several species, such as bovine, canine, porcine and rat [39–41].

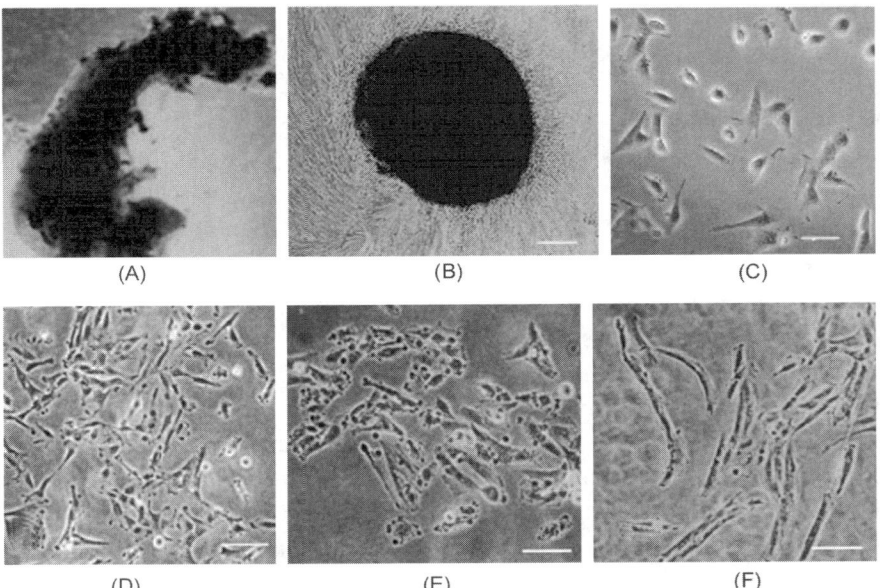

▶ Fig. 20.1 *Morphology of hIDPSC (according to Kerkis et al., 2006 [21]). (A) Image of dental pulp after isolation from tooth. (B) Tissue explant of dental pulp with outgrowing hIDPSC. (C) Adherent fibroblast-like morphology of rapidly expanding hIDPSC culture. (D)–(F). Morphological differences observed between some single-cell-derived colonies of hIDPSC (CFU-F assay). Phase contrast. A, 50 X. Scale bars: B, 50 µm, C–D, 20 µm.*

TISSUE PROCESSING, EXPANSION IN CULTURE AND DENTAL PULP CELL BANKING

Dental Pulp Extraction and Tissue Processing

Dental pulp can be reached through the crown as well from the root of the dental organ.

Whether the dental pulp will be obtained from a deciduous teeth, which is following through the natural exfoliation process, or from the one that presents a incomplete root formation (deciduous or permanent tooth), the access to the cells can be done using the root via, once the dental pulp will be already exposed.

In other cases, the access to the dental pulp can be established by using the dental crown. A drill under low or high rotation and abundant irrigation with 0.9% physiological saline can eliminate the enamel and dentin that covers the crown of the dental pulp (its number and anatomy depend on the tooth used).

In all cases, an endodontic file can be introduced into the dental pulp, involving its crown and root regions, and after gentle rotation movements, it is pulled out of the dental organ. The dental pulp tissue that will be attached to endodontic device can then be submitted to the process for the isolation of the stem cells.

The cleaning of the tooth before its manipulation and the maintenance of an aseptic chain during all the procedure are crucial cares that must be respected to avoid the contamination of the cells.

Expansion in Culture

Two different methods were used for DPSC isolation: enzymatic digestion and tissues explants (Table 20.1). Following enzymatic digestion, pulp tissue is incubated for 1 h at 37°C in a solution composed by 3 mg/mL collagenase type 1 and 4 mg/mL dispase. After that, single-cell suspension is obtained by passing the cells thought a 70-μm strainer and they are seeded into 6-well plates at a density of 100 to 1 $\leftrightarrow$ 10^5 cells/well in alpha-modification of Eagle's medium (α-MEM) supplemented with 20% of foetal calf serum (FCS), 100μM L-ascorbic acid 2-phosphate, 2 mM glutamine and antibiotics. Under such or slightly modified conditions, SHED and different populations of DPSC were obtained [19, 20, 30, 31, 33]. DP-MSCs were obtained using combined approach: tissue explants and culture medium used for DPSC cultivation [32]. Stem cells isolated from dental pulp, which express markers of hESCs, were isolated using both enzymatic digestion and tissues explants, under culture conditions routinely used for hESCs cultivation, such as Dulbeco's Modified Eagle's Medium (DMEM)/Ham's F12; 15–20% foetal bovine serum (FBS, Hyclone), 2 mM glutamine, 2 mM non-essential amino acid and antibiotics [21]. Further, independently on the isolation method, the primary culture was reseeded and expanded following regular trypsinization (0.25% trypsin) usually, when they achieved semi-confluence in order to avoid spontaneous differentiation. Passage 1 (P1) was denoted after the first primary culture reseeding. Culture medium was replaced each 3 days and the cells were cultivated in humid atmosphere at 37°C in 5% CO_2 chamber.

Dental Pulp Stem Cells Banking

From the moment DPSC and SHED were isolated, it has been suggested that dental pulp is a source of easily accessible stem cells that could be harvested and cryopreserved for a long period of time aiming at a possibility for their further manipulation to repair damage teeth, bone lesions and, perhaps, to treat neural injury or diseases. Due to this reason, it became relevant to evaluate their potential for long-term storage. Independently, two groups tested the proliferating activity and multilineage differentiation potential of DPSC and SBP-DPSC after cryopreservation [42, 43]. Two years after storage, SBP-DPSCs proliferate rapidly, continuously express all their respective surface antigens and produce woven bone, which after *in vivo* transplantation were remodelled into lamellar bone [43]. Different strains of DPSCs showed STRO-1 positive staining, adipogenic, neurogenic, odontogenic/osteogenic, chondrogenic and myogenic differentiation pathway when cultivated in appropriate culture medium even after cryopreservation [42]. These data taken together with the recent publications indicate that dental pulp tissue is a suitable source of multipotent stem cells for tissue engineering and cell-based therapies. Currently, the first tooth cell bank "BiOEDEN" was opened, which isolate and preserve stem cells from child's baby tooth from families throughout the UK, Europe and Middle East (http://bioeden.co.uk/).

IN VITRO DIFFERENTIATION ASSAYS

Odontogenic/osteogenic, adipogenic, chondorogenic, myogenic and neural differentiation of human stem cells isolated from dental pulp from adult and young patients was reported, mainly using protocols previously developed for human BM-MSCs and tested in MSCs obtained from other sources (Table 20.1, Figs. 20.2 and 20.3). These cells present similar timing to osteogenic and adipogenic differentiation with BM-MSCs, as well as similar morphological and hystochemical characteristics [39, 44]. The capacity of DPSCs to undergo chondorogenic differentiation has been shown in pellet culture, aggregates formed by DPSC suspension [42] and by differentiation of hIDSC in monolayer culture [21] (Fig. 20.2, A–C). Differentiation of DPSC and hIDPSC into skeletal smooth muscle cells, which express myogenic specific markers, such as, titin, a-actinin (sarcomeric), myosin, MyoD1, smooth muscle actin have also been demonstrated [21, 34, 42] (Fig. 20.2 D, E).

Neurogenic potential of cultivated SHED was evidenced by the expression of neural markers, such as nestin, β3 tubulin, neuronal nuclei (NeuN), neurofilament M (NFM), glial fibrillary acidic protein (GFAP) and 2′,3′-cyclic nucleotide-3′-phosphodiesterase (CNPase) [20]. For the first time, neurosphere formation was demonstrated for hIDPSC, which showed positive immunostaining for intermediate filament nestin, a marker for neural progenitors, suggesting that the cells within neurospheres present similar characteristics of neural stem and progenitor cells from central nervous system [45]. Furthermore, when adherent, they were capable to form nestin- and â3 tubulin-positive rosettes, which were later transformed into a uniform layer of well-distinguished glial cells overlaid with neurons (Fig. 20.3) [21]. More recently, confirming our study, the other group generates neurospheres from dental pulp of adult rat incisor [41].

IN VIVO DIFFERENTIATION ASSAYS

Gronthos et al., [19] were the first to show that clonogenic and *ex vivo* expanded human DPSCs were capable of generating a dentin/pulp-like complex after their *in vivo* transplantation into immunocompromised mice. This complex was characterized by the formation of a well-defined layer of aligned odontoblast-like cells, which express dentin-specific protein DSPP (dentin sialophosphoprotein), and associated pulp tissue. Furthermore, they re-established and characterized single-colony-derived DPSC strains from aforementioned primary DPSC transplants confirming their "stemness". Some of these strains following their new *in vivo* transplantation were able to produce abundant ectopic dentin, while others presented only limited amount of it, indicating that these strains differed each other regarding their rate of odontogenesis *in vivo* [30, 46, 47].

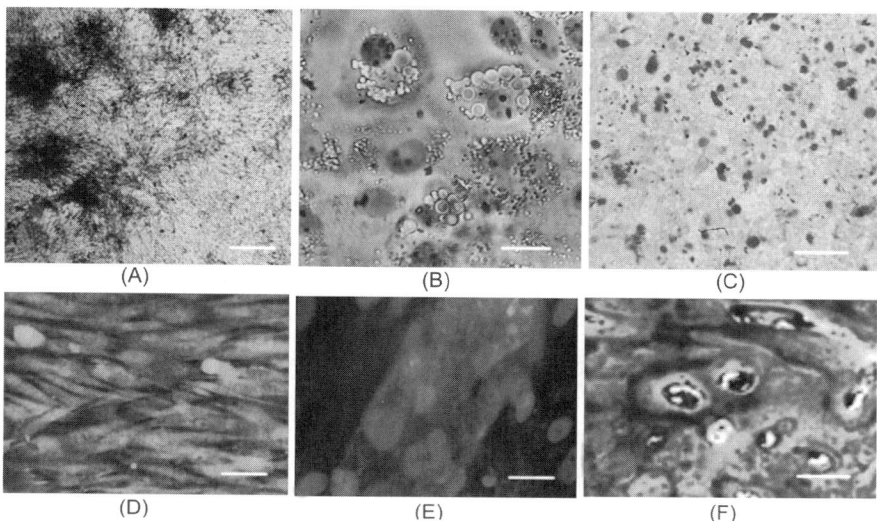

► **Fig. 20.2** *in vitro and in vivo differentiation of hIDPSC (according to Kerkis et al., 2006 [21]; Mendonça et al., 2008 [34]). (A) Osteogenic differentiation. Von Kossa staining revealed calcified extracellular matrix (arrows) in hIDPSC cultured in osteogenic medium. (B) Oil Red O staining showed accumulation of lipid-filled droplets (red) within the differentiated hIDPSC cultured in adipogenic medium. (C) Chondrogenic differentiation. Positive Toluidine blue staining observed in pellet culture of hIDPSC. (D) Smooth muscle differentiation, immunopositive staining for smooth muscle anti-actin antibody (green). (E) Skeletal muscle differentiation, immunopositive staining for anti-sarcomeric á-actinin antibody (red). (F) In vivo bone formation from hIDPSC after their transplantation into critical size cranial defect in non-immunosuppressed rat (Haematoxylin and Eosin staining). Nuclei stained with DAPI (Blue). A–C, F: light microscopy. D, E epi-fluorescence. Scale bars: A, C, D = 50 µm; B, E, F = 10 µm. (for colour figure see Plate 51).*

In vivo osteogenic and neurogenic differentiation potential of human SHED was demonstrated after their transplantation into immunocompromised mice. In this work, the SDED showed contribution at the initial site of bone formation: differentiating into odontoblasts or residing in the connective tissue and an abundant bone formation after transplantation using hydroxyapatite/triticalcium phosphate ceramic powder, as evidenced by the use of human-specific *Alu* sequence *in situ* hybridization. Neurogenic potential of SHED has been demonstrated after their injection into the dentate gyrus of the hippocampus of mice. The cells survived after application of more than 10 days while expressing neural markers such as NFM [20].

Stromal cells from human dental pulp (SBP-DPSC) also demonstrated osteogenic differentiation *in vivo*, as well as, the expression of a set of human genes associated with this type of differentiation [31, 43, 44]. After SBP-DPSC transplantation in immunosuppressed rats, either with

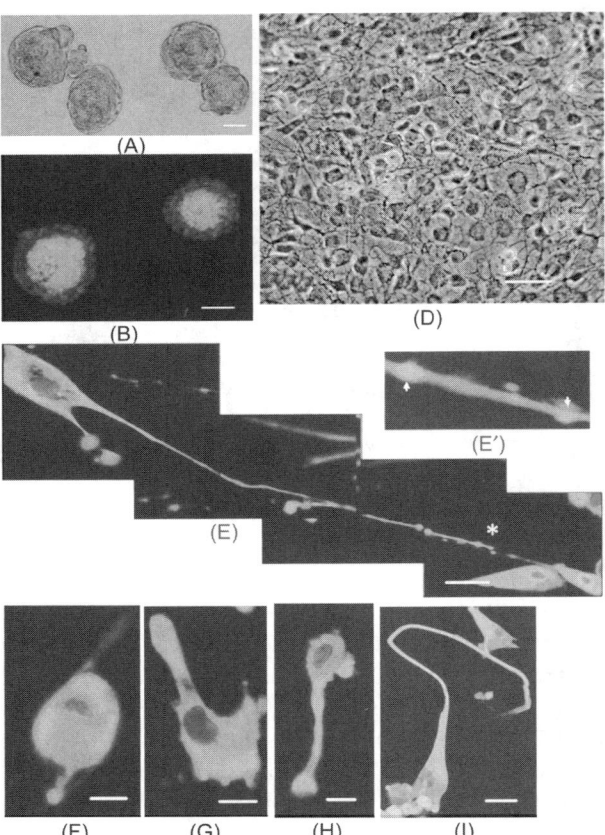

▶ **Fig. 20.3** *in vitro neuronal differentiation of hIDPSC (according to Kerkis et al., 2006 [21]). (A) Neurosphere formation from hIDPSC. (B) In vitro growing neurospheres from hIDPSC presenting positive immunostaining with anti-proliferating cell nuclear antigen (green) antibody. (D) Neuron/glial sandwich resembling the primary culture obtained from tissues of central neurvous system. (D). (E) Anti-β-tubulin III antibody positive neurons from hIDPSC showing the synapse, point of connection between two neurons (asterisk). (E') Higher magnification: synapse contains a small gap separating neurons (arrows). (F–I) Different steps of in vitro neuron formation from hIDPSC, axon formation. Nuclei stained with DAPI (Blue). A,D: phase contrast. B, E–I: epi-fluorescence. Scale bars: A,B = 300 μm; G = 50 μm, E–I= 10 μm. (for colour figure see Plate 52).*

woven bone tissue samples obtained *in vitro* after 60 days of culture (chips) or with the cells cultured 30 days on polymer scaffolds, bone tissue formation was observed around the location of transplantation. Importantly, within bone tissues as well as on its borders, formation of HLA-1-positive vessels has also been observed, indicating their human origin. This finding suggested complete integration of vessels within the bone chips, which leads to the formation of vascularized bone tissue [44].

Finally, hIDPSC—which is constituted by homogeneous population positive for MSCs, such as CD105, CD73 and CD29—was shown to be capable of contributing into reconstruction of large cranial defects produced in non-immunosuppressed rats after their transplantation onto collagen membrane without presenting any graft rejection [34] (Fig. 20.2F).

Human DPSC, which presented similar expression pattern of MSC markers with hIDPSC, can survive and engraft in ischemic environments in non-immunosuppressed rats with acute myocardial infarction. Although they did not show any differentiation into cardiac or smooth muscle as well as into endothelial cells, they contribute to the improvement of left ventricular function, induced angiogenesis and reduced infarct size. The benefits observed after DPSC transplantation was supposed to be due to secretion of paracrine factors by these cells [48] (Table 20.1).

CONCLUSION

According to current knowledge, different stem cell populations of ectomesenchymal origin can be isolated from dental pulp from adult and young patients. Recent study demonstrated that DPSC can differentiate into non-mesenchymal melanocytes and present label-retaining and sphere-forming abilities—traits attributed to multipotent stem cells, providing evidence that these might be multipotent neural crest stem cells [49]. It seems that characteristics of DPSC populations depend on the method, as well as on the culture medium used for their isolation and *ex vivo* expansion. DPSC colonies obtained from single cell also showed similar characteristics; however, their differential potential varies between the colonies. Comparison with BM-MSC showed that DPSCs apparently present higher proliferative potential and self-renewing characteristics. It is important to mention that BM-MSCs and DPSCs harbour in perivascular niches, while presenting phenotype consistent with different perivascular cell populations due to their diverse ontogeny and developmental potential [50].

Both cell types showed similar differentiating potential, which may be regulated by distinct mechanism at least in the case of osteogenesis and dentinogenesis [46]. The differentiation potential of DPSC represents important implications for clinical uses in the future. Despite the differences between animal models and human beings, the first ones have provided important data related to tissue responses that guide the scientific community to the establishment of new clinical therapies. DPSCs represent an optimal option for skeletal reconstructions, considering the similarity of its biology, when compared to the bone tissue formation. Treatment of craniofacial and orthopaedic defects is being analyzed not only when the cells are used alone but also when they have been associated with biomaterials. In these situations, tissue-engineering techniques are applied in different animal models such as mice, rabbits, dogs and sheep. In all these cases, histological techniques, *in situ* hybridisation and imunohistochemistry are some of the methods used to identify the presence and viability of the stem cells applied and its real contribution to the regeneration of the tissue that has been evaluated.

The knowledge of all normal process and steps involving cell differentiation and tissue formation, beyond the challenge of applying such knowledge to solve clinical situations, are goals to be reached with cell therapy. The advances that have been observed with DPSCs have provided evidences of their great potential and applicability in cell therapy, as well as in the enhancement of healing processes through the production of cytokines. Moreover the reproducibility, easily handled characteristic, the facility of being obtained in a relatively easy way, not only in adults but also in children, and the advantage of the dental organ to be stored in banks are remarkable points that have to be taken into account concerning the tissue-engineering science.

GLOSSARY

Supernumerary is an extra tooth. The most common is called mesiodens and is located in the anterior superior region, between the central superior incisors. Predominantly it occurs in the male gender.

Exfoliation is the process through which the root of the primary tooth resorbs as the permanent tooth erupts from beneath.

Molar teeth are the ones localized at the back of the mouse. Adults have 12 molars in four groups each having three, while children have four groups of two. The third molar in adults is called a wisdom tooth. The first permanent molar begins its eruption around 6 years.

ACKNOWLEDGEMENTS

We would like to pay a tribute to Dr. Alexander Friedenstein of the Gamaleya Institute in Moscow, Russia, and Dr. Arnold Caplan of the Skeletal Research Centre, Case Reverse University, Cleveland, Ohio, USA. Their publications have generated a lot of scientific excitement and inspired a great deal of experiments with ASC by other researches. Their pioneer contribution in this area represents important gear lever for the stem cell biology and biomedical communities. The authors of this chapter want to thank Humberto Cerruti Filho, well-known Brazilian dentistry surgeon from Sao Paulo of Clinica CERA, for his valuable advices to the development of our research and clinical vision for the potential application of DPSC. Our group is supported by grants from Clinic CERA and FAPESP.

REFERENCES

1. Friedenstein A.J., S. Piatetzky, II, K.V. Petrakova (1966). 'Osteogenesis in transplants of bone marrow cells'. *J. Embryol. Exp. Morphol.* **16**: 381–390.
2. Friedenstein A.J., R.K. Chailakhyan, U.V. Gerasimov (1987). 'Bone marrow osteogenic stem cells: *In vitro* cultivation and transplantation in diffusion chambers'. *Cell. Tissue. Kinet.* **20**: 263–272.
3. Caplan A.I. (1991). 'Mesenchymal stem cells'. *J. Orthop. Res.* **9**: 641–650.

4. Caplan A.I. (2000). 'Tissue engineering designs for the future: New logics, old molecules'. *Tissue. Eng.* **6**: 1–8.
5. Caplan A.I. (2005). 'Review: mesenchymal stem cells: cell-based reconstructive therapy in orthopedics'. *Tissue Eng.* **11**: 1198–1211.
6. Caplan A.I., S.P. Bruder (2001). 'Mesenchymal stem cells: building blocks for molecular medicine in the 21st century'. *Trends Mol Med.* **7**: 259–264.
7. Pittenger M.F. (2008). 'Mesenchymal stem cells from adult bone marrow'. *Methods Mol Biol.* **449**: 27–44.
8. Owen M.E. and A.J. Friendenstein (1988). 'Stromal stem cells: marrow derived osteogenic precursors'. *Ciba Found.Symp.* **136**: 42–60.
9. Friendenstein A.J. (1995). 'Marrow Stromal Fibroblasts'. *Calsif. Tissue Int* **56** Suppl. 1: S17.
10. Bruder SP, Fink DJ, A.I. Caplan (1994). 'Mesenchymal stem cells in bone development, bone repair and skeletal regeneration therapy'. *J Cell Biochem.* **56**: 283–294.
11. Lazarus H.M., S.E. Hayneworth, L. Gerson, N.S. Rosenthal, and A.I. Caplan (1995). '*Ex vivo* expantion and subsequent infusion of human bone marrow-derived stromal progenitor cells (mesenchymal progenitor cells): Implications for therapeutic use'. *Bone Marrow Transpl.* **16**: 557–564.
12. Prockop D.J. (1998). 'Marrow stromal cells as stem cells for non-haematopoietic tissues'. *Science.* **276**: 71–74.
13. Pittenger M.F., A.M. Mackay, S.C. Beck et. al. (1999). 'Multilineage potential of adult human mesenchymal stem cells'. *Science.* **284**: 143–147.
14. Jiang Y., B.N. Jahagirdar, R.L. Reinhardt, et. al. (2002). 'Pluripotency of mesenchymal stem cells derived from adult marrow'. *Nature.* **418**: 41–49.
15. Harada H., P. Kettunen, H.S. Jung, et. al. (1999). 'Localization of putative stem cells in dental epithelium and their association with Notch and FGF signaling'. *J Cell Biol.* **147**: 105–120.
16. Fuchs E, J.A. Segre (2000). 'Stem cells: a new lease on life'. *Cell.* **100**: 143–155.
17. Bianco P, M. Riminucci, S. Gronthos, P.G. Robey (2001). 'Bone marrow stromal stem cells: Nature, biology, and potential applications'. *Stem Cell.* **19**: 180–192.
18. Blau H.M., T.R. Brazelton, J.M. Weimann (2001). 'The evolving concept of a stem cell: entity or function?' *Cell.* **105**: 829–841.
19. Gronthos S., M. Mankani, J. Brahim, P.G. Robey, and S. Shi, (2000). 'Postnatal human dental pulp stem cells (DPSCs) *in vitro* and *in vivo*'. *Proc. Natl. Acad. Sci. USA.* **97**: 13625–13630.
20. Miura M., S. Gronthos, M. Zhao, B. Lu, L.W. Fisher, P.G. Robey, and S. Shi, (2003). 'SHED: Stem cells from human exfoliated deciduous teeth'. *Proc. Natl. Acad. Sci. USA.* **100**: 5807–5812.
21. Kerkis I., D. Dozortsev, G.C. Stukart-Parsons et. al. (2006). 'Isolation and characterization of sub-population of dental pulp stem cells expressing OCT-4 and other key embryonic stem cells markers'. *Cells Tissues and Organs.* **184**: 105–116.
22. Camilleri S. and F. McDonald (2006). 'Runx2 and dental development'. *Eur J Oral Sci.* 114: 361–373.
23. Thesleff I. (2006). 'The genetic basis of tooth development and dental defects'. *Am J Med Genet A.* **140**: 2530–2535.
24. Lin D., Y. Huang, F. He, et. al. (2007). 'Expression survey of genes critical for tooth development in the human embryonic tooth germ'. *Den Dyn.* **236**: 1307–1312.

25. Yen A.H., P.T. Sharpe (2008). 'Stem cells and tooth tissue engineering'. *Cell Tissue Res.* **331**: 359–372.
26. Miyado M., H. Ogi, G. Yamada, et. al. (2007). 'Sonic hedgehog expression during early tooth development in Suncus murinus'. *Biochem Biophys Res Commun.* **363**: 269–275.
27. Wright, T. (2007). 'The molecular control of and clinical variations in root formation'. *Cells Tissues Organs.* **186**: 86–93.
28. Yamamura T. (1985). 'Differentiation of pulpal cells and inductive influences of various matrices with reference to pulpal wound healing'. *J Dent Res.* **64** Spec: 530–540.
29. L.M. Mann, D.P. Lennon, A.I. Caplan (1996). 'Cultured rat pulp cells have the potential to form bone, cartilage, and dentin *in vivo*'. In: *Biological Mechanisms of Tooth Movement and Craniofacial Adaptation.* (Z. Davidovitch, L.A. Norton, eds.), 7–10. Harvard Society of the Advancement of Orthodontics: Boston.
30. Gronthos S., J. Brahim, W. Li, et. al. (2002). 'Stem cell properties of human dental pulp stem cells'. *J Dent Res.* **81**: 531–535.
31. Laino G, R. d'Aquino, A. Graziano, et. al. (2005). 'A new population of human adult dental pulp stem cells: a useful source of living autologous fibrous bone tissue' (LAB). *J Bone Miner Res.* **20**: 1394–1402.
32. Pierdomenico L., L. Bonsi, M. Calvitti, et. al. (2005). 'Multipotent mesenchymal stem cells with immunosuppressive activity can be easily isolated from dental pulp'. *Transplantation.* **80**: 836–842.
33. Jo Y.Y., H.J. Lee, S.Y. Kook et. al. (2007). 'Isolation and characterization of postnatal stem cells from human dental tissues'. *Tissue Eng.* **13**: 767–773.
34. Mendonça C. A., D.F. Bueno, M.T. Martins, I. Kerkis, A. Kerkis et. al. (2008). 'Reconstruction of large cranial defects in non-immunosuppressed rats'. 'Experimental design with human dental pulp stem cells'. *J Craniofac Surg.* **19**: 204–210.
35. Huang A.H-C., Y. K. Chen, L. M. Lin et. al. (2008). 'Isolation and characterization of dental pulp stem cells from supernumerary tooth'. *J Oral Pathol Med.* **37**: 571–574.
36. Wehrle-Haller B. (2003). 'The role of Kit-ligand in melanocyte development and epidermal homeostasis'. *Pigment Cell Res.* **16**: 287–296.
37. Fujiwara H, K. Tatsumi, K. Kosaka, *et. al.* (2003). 'Human blastocysts and endometrial epithelial cells express activated leukocyte cell adhesion molecule (ALCAM/CD166).' *J. Clin. Endocrinol. Metab.* **88**: 3437–3443.
38. Di Nicola, C. Carlo-Stella, M. Magni, et. al.. (2002). 'Human bone marrow stromal cells suppress T-Lymphocyte proliferation induced by cellular or non-specific mitogenic stimuli'. *Blood.* **99**: 3838–3843.
39. Iohara K, L. Zheng, M. Ito, A. Tomokiyo, K. Matsushita and M. Nakashima (2006). 'Side population Cells isolated from porcine dental pulp tissue with self-renewal and multipotency for dentinogenesis, chodrogenesis, adipogenesis, and neurogenesis'. *Stem Cells.* **24**: 2493–2503.
40. Yu J., U. Wang, Z. Deng et. al. (2007). 'Odontogenic capacity: bone marrow stromal stem cells versus dental pulp stem cells'. *Biol. Cell.* **99**: 465–474.
41. Sasaki R., S. Aoki, M. Yamato, H. Uchiyama, K. Wada, T. Okano and H. Ogiuchi (2008). 'Neurosphere generation from dental pulp of adult rat incisor'. *Europ. J. Neurosci.* **27**: 538–548.
42. Zhang W., X.F. Walboomers, H. Shi, et. al. (2006). 'Multilineage differentiation of stem cells derived from human dental pulp after cryopreservation'. *Tissue Eng.* **12**: 2813–2823.

43. Papaccio G., A. Graziano, R. d'Aquino, et. al. (2006). 'Long-term cryopreservation of dental pulp stem cells (SBP-DPSCs) and their differentiated osteoblasts: a cell source for tissue repair'. *J Cell Physiol.* **208**: 319–325.

44. d'Aquino R., A. Graziano, M. Sampaolesi et. al. (2007). 'Human postnatal dental pulp cells co-differentiate into osteoblasts and endotheliocytes: a pivotal surgery leading to adult bone tissue formation'. *Cell Death and Dif.* **14**: 1162–1171.

45. Lendahl U., L.B. Zimmerman & R.D. McKay (1990). 'CNS stem cells express a new class of intermediate filament protein'. *Cell.* **60**: 585–595.

46. Batouli S., M. Miura, J. Brahim et al (2003). 'Comparison of item-cell-mediate osteogenesis and dentinogenesis. *J Dent Res.* **82**: 976–981.

47. Goldberg M., A.J. Smith (2004). 'Cells and extracellular matrices of dentin and pulp: a biological basis for repair and tissue engineering'. *Crit Rev Oral Biol Med.* **15**: 13–27.

48. Gandia C., A. Armiñan, J.M. Garcia-Verdugo, E. Lledó, et. al. (2007). 'Human dental pulp stem cells improve left ventricular function, induce angiogenesis and reduce infarct size in rats with acute myocardial infarction'. *Stem Cells.* **26**(3): 638–645.

49. Stevens A., T. Zuliani, C. Olejnik (2008). 'Human dental pulp stem cells differentiate into neural crest- derived melanocytes and have label-retaining and sphere-forming abilities'. *Stem Cells Dev.* **17**(6): 1175–1184.

50. Shi H, S. Gronthos (2003). 'Perivascular niche of postnatal mesenchymal stem cells in human bone marrow and dental pulp'. *J Bone Mineral Res.* **18**: 696–704.

21

HEMATOPOIETIC STEM CELLS AND THEIR DYNAMICS

David Dingli[1,*], Arne Traulsen[2] and, Jorge M. Pacheco[3]

[1] Division of Hematology, Mayo Clinic College of Medicine, Rochester, MN 55905;
[2] Max Planck Institute for Evolutionary Biology, 24306 Ploen, Germany;
[3] ATP-Group, CFTC and Departamento de Física da Faculdade de Ciências, P-1649-003 Lisboa Codex, Portugal

INTRODUCTION

Hematopoiesis is an amazingly complex system. Through the course of this chapter, we review arguments supporting the idea that, similar to all other mammals, humans have hematopoietic stem cell (HSC) at the root of hematopoiesis. We concentrate on consequences of this view. Let us take humans as an example, at any time, roughly 400 HSC actively contribute to hematopoiesis, each cell replicating, on average, once a year. At the other extreme of the hematopoietic tree, $3.5 \leftrightarrow 10^{11}$ cells are replaced every single day. The amazing scaling of sheer cell number and replication rates poses a challenging problem for anyone who envisions a mathematical description of this multi-scale system of cells which functions as a complex organ. In the following we summarise the salient biology of HSC and attempt to show how some simple scaling and dynamical considerations might shed light on dynamics of hematopoiesis.

EVIDENCE FOR HEMATOPOIETIC STEM CELLS

Haematopoiesis couples together in a complex hierarchical organisation, many types of blood cells. The various types of circulating blood cells have a finite lifespan and are continuously being replaced by new cells produced in the bone marrow. Under equilibrium conditions, cellular output from the bone marrow is of the order of $3.5 \leftrightarrow 10^{11}$ cells per day [1] in humans. Indeed, approximately 1% of the erythrocytes are replaced daily while the average time that neutrophils

*Dingli.David@mayo.edu

remain in the circulation is of the order of hours [2–4]. Bone marrow output increases in response to higher demands (e.g. bleeding or infection) or as a result of neoplastic transformation of haematopoiesis as in chronic myeloid leukemia (CML) [5] or polycythemia vera (PV) [6].

It is well known that the different types of blood cells originate at different stages of haematopoiesis, exhibiting a high level of interdependence, with a tree-like organization. At the root of blood formation are the HSC [7, 8]. Nowadays, the existence of HSC is taken for granted since they make bone marrow transplantation possible; a procedure that is performed almost routinely in many major medical centers and has provided curative therapy for a variety of otherwise lethal genetic/metabolic or neoplastic disorders [9].

However, despite this unanimous recognition, isolation or direct observation of HSC has not been achieved to date. Indeed, the existence of HSCs is often inferred indirectly and almost always as an *a posteriori* event. For many years, the presence of these cells was demonstrated by marking with retroviral or lentiviral vectors [10–12] or studied from informative gene marking experiments such as X-linked genes (e.g. glucose-6-phosphate dehydrogenase (G6PD) [13] or chronic granulomatous disease (CGD)) [14]. The presence of HSCs was initially inferred from the observation that lethally irradiated mice died due to haematopoietic failure and this could be averted by transplantation of syngeneic bone marrow [7, 15]. Infusion of different limiting numbers of bone marrow cells led to the formation of spleen colony forming units in mice, with the number of colonies being proportional to that of the injected cells [15]. Subsequent experiments showed that each spleen colony arose from a single cell and the progeny of each colony included cells from both myeloid as well as lymphoid lineages [16, 17].

Although some investigators believe that HSCs can be isolated, others are of the view that a single HSC may not be isolated in pure form for various reasons. However, with the advent of fast, multicolor flow cytometry, populations of long-term reconstituting cells have been isolated and shown to rescue irradiated mice and can be used for serial transplantation [18–20].

The cell-surface expression profile of HSC is becoming increasingly well defined, although controversy here also exists. Essentially all investigators agree that human HSC are lineage negative (Lin$^-$, i.e. negative for CD4, 8, 10, 14, 15, 16, 19, 20), Thy1^{lo}, c-kit$^+$ and CD38$^-$, but there is no agreement on their CD34 expression, some believe that they are positive [21], while others have shown that CD34$^-$ cells isolated from the bone marrow can reconstitute lethally irradiated mice and give rise to CD34$^+$ cells (Fig. 21.1) [22–25]. Murine HSCs are also Lin$^-$ (CD4, 8, B220, Gr-1, Mac-1), Thy1^{lo}, c-kit$^+$ and Sca-1$^+$. Transplantation of a single HSC (murine or human) in sublethally irradiated severe combined immune deficient (SCID) mice allows haematopoietic recovery and maintenance of haematopoiesis for the lifetime of the animal, confirming the long-term self-renewal and differentiation ability of these cells [21].

PROPERTIES OF HAEMATOPOIETIC STEM CELLS

A stem cell is operationally defined as a cell that has long-term self-renewal capability and at the same time is able to give rise to progeny cells that differentiate along various possible lineages.

Progenitor cells differ from HSC in that they have less self-renewal potential and a more restricted differentiation profile (Fig. 21.1). In the case of HSCs, the progeny are erythrocytes, platelets, various types of granulocytes (neutrophils, eosinophils and basophils), monocytes/macrophages, lymphocytes (T and B), natural killer (NK) cells and mast cells. The two daughter cells that arise from mitosis of a single HSC can have completely different fates, suggesting that 'decisions' along the path of differentiation can occur within one replication event and widely interpreted as being due to a stochastic process [15, 26, 27]. Limiting dilution studies and competitive repopulation experiments in murine models suggest that perhaps a single stem cell may be enough to maintain haematopoiesis in the mouse [28], illustrating the long-term self-renewal and differentiation ability of these cells. Moreover, the progeny of these cells can be used for serial transplantation up to four five times, confirming the long-term self-renewal properties of the cells [29]. Studies in larger mammals such as cats and non-human primate show that haematopoiesis is a polyclonal process, in other words different stem cells are contributing to blood formation at any time [30, 31]. HSC can be operationally divided into an active and a reserve pool. Cells in the active pool contribute to haematopoiesis and are associated with the endosteal surface of bone [32]. There is evidence that cells selected to contribute to haematopoiesis, can do so for a very long time, perhaps the lifetime of the individual [33].

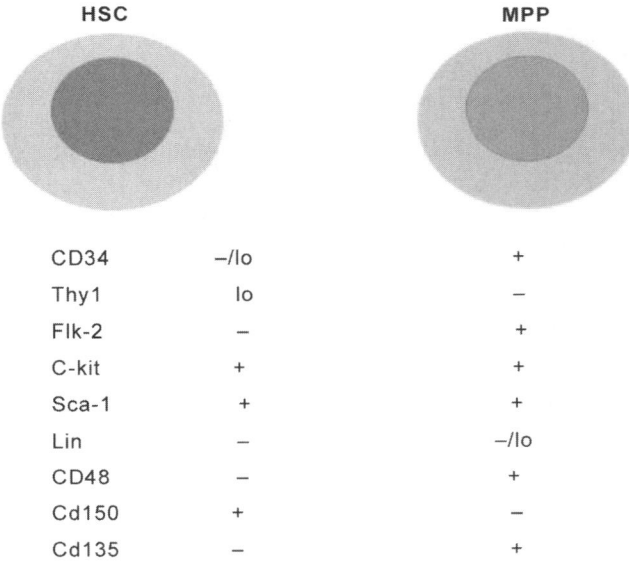

	HSC	MPP
CD34	–/lo	+
Thy1	lo	–
Flk-2	–	+
C-kit	+	+
Sca-1	+	+
Lin	–	–/lo
CD48	–	+
Cd150	+	–
Cd135	–	+

▶ **Fig. 21.1** *Hematopoietic stem cells and multipotent progenitor (MPP) cells. The immunophenotype of HSC and MPP is becoming increasingly defined. The two terms are often used interchangeably but they refer to distinct cell types. Only the HSC have the ability for long-term reconstitution of hematopoiesis. There is increasing evidence that HSC are CD34 negative. Both cell types appear to be uncommitted to any lineage, hence the designation Lin⁻. This requires that cells are negative for CD10, 14, 15, 16, 19 and 20. (for colour figure see Plate 553).*

SOURCES OF HAEMATOPOIETIC STEM CELLS

Harvesting

For many years, the main source of HSC has been the bone marrow itself. Harvesting of the bone marrow in humans requires a surgical procedure with general anesthesia. For safe stem cell transplantation in humans, a minimum of $2 \leftrightarrow 10^8$ mononuclear cells per kilogram is required. The cells are harvested by multiple aspirations from both posterior iliac crests. Many of the cells collected are not stem cells but more committed progenitors that engraft rapidly and start producing mature cells.

Mobilisation

With the recognition that both haematopoietic stem and progenitor cells circulate in the blood at low levels and that their numbers increase after chaemotherapy and/or cytokine therapy (e.g., cyclophosphamide and granulocyte colony-stimulating factor (G-CSF), respectively), most centres now collect these cells by apheresis after therapy with G-CSF (a process known as mobilisation). HSC mobilisation with G-CSF is through an indirect mechanism, the cytokine stimulates cells of the myelomonocytic lineage to release proteolytic enzymes such as elastase, cathepsin G, proteinase 3 and gelatinase (MMP-9) that disrupt cell adhesion molecule interactions such as VCAM-1-$\alpha_4\beta_1$, *c-kit* receptor with surface expressed *c-kit* [34] and perhaps most importantly alters the gradient of stromal cell derived factor 1 (SDF-1, also known as CXCL12) between the bone marrow and the blood (higher in the latter) inducing cells to emigrate to the vascular compartment [35]. The current view is that SDF-1 is the main anchor of HSC to their endosteal niche (discussed further). There are additional cytokines that mobilize HSC into the circulation including GM-CSF, IL-3 and stem cell factor (SCF); however, their exact mechanisms of action are not so clear [36]. The optimal time for HSC collection can be determined by serial monitoring of the circulating $CD34^+$ cell count. The main advantage of transplantation with peripheral blood stem/progenitor cells is faster haematopoietic reconstitution and therefore a shorter interval of neutropenia and thrombocytopenia, minimizing the risk of fatal infections or hemorrhage. In addition, the donor (especially in the case of allogeneic transplantation) avoids a surgical procedure with its associated risk and the pain due to the multiple aspiration sites. However, it must be pointed out that for specific diseases, bone marrow is still the preferred source of stem cells compared to apheresis collections of mobilised progenitors.

The last source of HSC is umbilical cord blood that is harvested soon after delivery of the neonate. Cord blood is enriched with haematopoietic stem and progenitor cells and can be used for transplantation in the pediatric population, especially if the recipient mass is less than 25 kg [37]. As experience with the procedure has expanded, multiple cord blood donors have been pooled and used to reconstitute haematopoiesis in adult recipients [38]. In general, bone marrow reconstitution and haematopoietic recovery are slower after cord blood transplantation compared to autologous or allogeneic peripheral blood/bone marrow transplantation [39, 40].

THE STEM CELL NICHE

It has been next to impossible to expand HSCs in culture and demonstrate persistence of both self-renewal and multilineage commitment. This observation suggests that an HSC can only function properly when it resides in an appropriate microenvironment, now known as the stem cell niche, a concept initially proposed by Schofield in 1978 [41]. Consequently, the definition of a stem cell becomes a functional one. One can even venture further and state that a cell is not a stem cell anymore when it is outside the niche since the latter seems to determine the fate of the HSC, cell cycle status, replication rate and symmetry of division/differentiation [42–44]. At present, the exact nature of the growth factors that regulate HSC behavior is not clear and it is possible that these requirements change as the host animal ages. However, insightful experiments from transgenic mice are starting to shed important light on the nature of the stem cell niche.

The intimate association between haematopoiesis and bone led to the discovery of the role of osteoblasts in HSC niche formation [32, 45]. Mice with a conditional knockout of bone morphogenic protein (BMP) receptor 1A have ectopic trabecular-like bone along their long bones and concomitantly have an increase in the number of HSC [46]. Transgenic mice with osteoblasts that constitutively express parathyroid hormone and parathyroid hormone related receptors experience an increase in osteoblast numbers as well as in HSC [47]. The high calcium concentration present in the osteoblast niche is essential for homing and retention of HSC to this niche and for this purpose, HSCs express the calcium-sensing receptor that enables them to respond to the calcium signal [48]. Osteopontin, produced by osteoblasts in response to various stimuli and secreted into the extracellular matrix is another important regulator of the number of HSC. Low levels of osteopontin lead to an increase in the number of HSC, suggesting that it is a negative regulator of stem cell number [49].

More recently, a second niche, in relation to endothelial cells, has been identified and called the vascular niche [50, 51]. HSCs interact with endothelial cells via signaling lymphocyte activation molecule (SLAM) receptors on sinusoidal cells both in the bone marrow as well as the spleen [52]. The current view is that perhaps the osteoblastic niche provides a quiescent environment for HSC maintenance while the vascular niche provides the environment required for differentiation and mobilization. From an ontological perspective, the intimate relationship between HSC, osteoblasts, and vascular endothelial cells has been rationalized on the basis of the close embryological origins of these cells. HSC and vascular endothelial cells are both derived from hemangioblasts [52], while the haematopoietic system is derived from the mesoderm around the aorta while osteoblasts originate from the surrounding mesenchyme [53].

The microenvironment in the stem cell niche is composed of cell-to-cell interactions via cell-surface molecule expression, chemokines and growth factors. This orchestra determines the fate of the cell(s) within the niche. The list of molecular mediators that regulate the HSC is increasing and includes angiopoietin-1/Tie-2, SDF-1 (CXCL12)/CXCR4, very late antigen 4 (VLA-4), leukocyte function antigen 1 (LFA-1), FGF-4, VEGF, SCF/c-kit, Wnt, *N*-cadherin and indirectly G-CSF, GM-CSF, FLT-ligand and possibly many others [54]. The latest discovery has been the role of the sympathetic nervous system on stem cell function. Reticular cells, present in the bone marrow stroma, appear to respond to noradrenaline by expressing the β_3 adrenergic receptor [55].

Adrenergic stimulation of reticular cells leads to dephosphorylation of the transcription factor Sp-1 that leads to destabilization of the factor and a reduction in CXCL12 production. Lower concentrations of CXCL12 allow HSC to enter the circulation. As a result of this interaction of the sympathetic nervous system with HSC, the number of circulating HSCs varies with the circadian rhythm, with the peak in HSC seen during maximal lighting, while their number falls during the dark part of the cycle [55].

HOW MANY STEM CELLS?

Although most of biology is a descriptive science, quantification is essential, since ultimately, number and dynamics can tell us what is possible. Indeed, for many situations, an understanding of the phenotype requires a dynamic description. The desire to understand details of haematopoiesis and how it responds to varying demands and the origin and evolution of various haematopoietic disorders make the determination of the number of HSCs important. A few years ago, it was proposed that the total number of HSCs is conserved across mammals and may be between 11 000 and 22 000 cells [56]. Perhaps this suggestion might appear unusual; however, there is experimental evidence from mice, rats, and cats that can accommodate such a proposition [30, 57–59]. Theoretical arguments based on this hypothesis have been used to address the largest land mammals—elephants [59]. If we assume that these observations/inferences are true, they beg several questions including (i) why is the number of HSC conserved, (ii) are mammals born with the full complement of HSC, (iii) what are the reasons and evolutionary mechanisms that have selected or imposed such a constraint on this pool of cells, or (iv) are there similar constraints on other tissue-specific stem cells. There is no doubt that demands on haematopoiesis are different across mammals [30]. One can show that the total marrow output during the lifetime of a mouse is roughly the same as an adult human produces in a day [30, 60, 61]. Therefore, we need to tackle the issue of how many stem cells actively contribute to haematopoiesis in a given adult mammal. Until now, no unambiguous answer has been provided in the laboratory. It has been proposed that haematopoiesis is maintained by a subset of 'active' HSC and supported by a quiescent 'reserve' that may be called upon to contribute depending on circumstances [62]. With such a hypothesis, perhaps the number of 'active' stem cells might differ across mammals without requiring the total number of cells to change.

One approach that might help us to understand this potential conundrum is to resort to the field of allometry [63] combined with the earlier definitions of active and reserve HSC pools. In biological systems, many observables (Y) that are related to nutrient transport, such as metabolic rate and tree trunk thickness, scale with mass as $Y \sim Y_0 M^b$ where b is typically a multiple of 1/4. There are competing explanations for the origin of these exponents and this issue is by no means resolved [64–66]. To apply these principles with respect to HSC and haematopoiesis, some general observations are in order: (i) haematopoiesis has appeared only once during evolution (as seen by similarities between the process across mammals, enabling detailed studies in the murine model and applied to other species), (ii) while haematopoiesis is distributed across various bones, it is functionally coupled by the circulation and so HSC effectively function collectively as a single haematopoietic organ, and (iii) from the definition of the HSC, every active HSC is equally represented in the circulation of a given mammal. With these premises, we were able to provide an allometric estimate of the size of the active stem cell pool in mammals (N_{SC}) as a function of their mass [67]. The simplest marker of cellular marrow output is the total number of circulating reticulocytes (R_T) that can be estimated in many mammals from knowledge of their circulating blood volume, the red cell concentration and the percentage of reticulocytes present in circulation as a function of erythrocytes. Taking into account the variable maturation rate of reticulocytes across species, we could determine the R_T scales with the mass of the adult mammal as $R_T \sim M^{3/4}$ [67]. From (iii) above, we conclude that $N_{SC} \sim M^{3/4}$. Therefore, if N_{SC} and the mass of *any* adult mammal are known, N_{SC} for *any other* adult species can be determined based on premise (i) earlier. Using this relationship and the estimated number of active HSC in Safari cats, it allowed us to calculate that in humans $N_{SC} \cup 400$ in agreement with experimental observations derived from patients with CGD [14]. The same relationship suggests that after bone marrow transplantation, a typical adult human has $N_{SC} \cup 116$, which is again in excellent agreement with experimental data [68]. This model also predicts that a number of the order of 1 HSC can maintain haematopoiesis in a mouse for its lifetime [67], a prediction that is supported by experimental observations [28]. Consequently, the smallest mammal, a shrew with a mass of 3 g would also require one or very few HSC to maintain haematopoiesis. On the other hand, while there is no experimental validation of the active SC pool in elephants, our scaling predicts that an Asian Elephant (*Elephas maximus*, approximately 4500 kg weight) has an active stem pool comprising almost 9600 cells. Therefore, the size of the HSC pool based on allometry falls within the limit proposed by Abkowitz et al. [14]. However, our estimates are based on the number of cells that are *active* in blood formation and hence represented in the circulation. The remaining cells are inactive and we consider them to constitute a reserve pool of cells. If the number of HSC is conserved, then one should consider the sum of these two pools, compatible with the hypothesis proposed by Kay [62]. Interestingly, the species-specific HSC replication rate (B) naturally also follows an allometric relationship with adult mass in any given species; HSCs replicate at a rate which decreases with increasing mass ($B \sim M^{-1/4}$) [67].

EXPANSION OF THE HSC DURING HUMAN ONTOGENY

Several hereditary/congenital HSC and non-HSC disorders are amenable to gene therapy approaches. Although individually these disorders are rare, they make the HSC compartment an important therapeutic target (e.g., X-SCID and retroviral vector therapy) [69]. Thus, it is essential to determine the size of the active HSC pool in newborns and how the active pool increases over time to reach adult levels. A small active HSC pool requires highly efficient cell transduction if the genetic/metabolic defect is to be corrected successfully. In order to gain some insight on the size of this pool during human ontogeny, we have applied the previously discussed allometric scaling that relates the reticulocyte count with mass to data derived during normal human growth. We found that the active HSC pool scales linearly with mass during human ontogeny. Perhaps the most relevant observation was that the number of active HSCs in newborn babies is quite small, in the range of ~20 [70], making them a difficult target for gene therapy. It is at present unclear whether babies are born already equipped with the full pool of HSC or whether their population expands in time to reach the estimated adult size.

STEM CELL AGING/CLONAL SUCCESSION

Aging is a natural process that is experienced by every living organism and associated with a slow but continuous modification in phenotype and physiology. The process is poorly understood and sometimes conflicting hypotheses have been proposed in an attempt to explain it. Given the importance of stem cells in tissue maintenance and their long-term contribution to tissue homeostasis and repair, it is understandable that considerable effort has been expended to understand how stem cell number and function change as a function of age. For ethical reasons and experimental convenience, most of the work has been done in murine models. Intriguingly, in most common strains of mice, the number of HSC seems to increase with age [71]. It has been reported that in older mice, there are higher numbers of circulating HSC and progenitors and mobilisation is easier in older species [72]. However, the data on HSC mobilisation as a function of age in humans show that the number of HSC mobilised and obtained in collections decreases as a function of age when healthy, allogenic donors are considered [73].

Competitive repopulation assays show that HSCs taken from older mice have a reduced potential for haematopoietic reconstitution, and in time, cells experience a progressive bias in their ability to give rise to progeny cells, HSCs taken from young mice preferentially produce cells of the lymphoid lineage, while HSCs derived from older mice seem to be skewed toward the myeloid lineage. These behaviors appear to be inherited and related to epigenetic changes in cells that accumulate with age. However, changes in number and phenotype are not solely cell autonomous. There is increasing evidence that the stem cell niche function also changes with age and exhibit differences in signaling patterns that may have an impact on HSC behavior. This has been best described in the case of *Drosophila* germline stem cells. Moreover, calorie restriction, that is known to increase longevity in animal models, also 'improves' HSC function [74]. Thus, HSC harvested from calorie restricted aged mice, exhibit improved function even when transplanted in non-calorie-restricted mice.

The various mechanisms postulated to influence the aging process and its speed of development include mutations in genomic DNA via reactive oxygen species, ultraviolet irradiation and the intrinsic error rate of the DNA replication machinery [75]. Although HSCs have DNA repair mechanisms as well as checkpoint controls, these are not perfect and mutations accumulate that could contribute both to the process of aging as well as carcinogenesis. The DNA replication machinery is unable to replicate ends of chromosomes and as a result, ~50–70 bases of DNA are lost with each replication cycle from ends of each double helix. Chromosomes have a specialized complex of nucleoproteins called telomeres that protect ends of DNA and serve as an internal clock for the number of divisions a given cell has undergone. If telomeres lose a critical part of their length, the cell activates pathways that lead to cell death. HSCs have some ability to maintain their telomeres by expression of telomerase, the enzyme complex that utilizes an RNA template to extend the nucleotide repeats present in telomeres. However, it is well known that even HSCs experience serial shortening of their telomeres [1] leading to functional decline. The importance of telomere maintenance in HSCs is further supported by the syndrome of dyskeratosis congenita where the main cause of death is bone marrow failure due to loss of HSC function [76]. These diseases are characterized by defects in telomere maintenance mechanisms. The most recent studies suggest that the serial accumulation of DNA damage is the main determinant of HSC senescence [77, 78].

Despite HSC aging, serial transplantation experiments show that HSC can function for longer than the lifetime of the animal. Indeed, HSC have been transplanted serially up to five times without loss of repopulating ability in murine species [29]. One potential explanation for this observation is the unusual long telomeres possessed by murine chromosomes. Yet, calculations in humans would suggest that with replication rates of approximately 1 per year, HSCs can contribute to haematopoiesis for many years if not the lifetime of the mammal if only telomeres attrition was the only determinant of the duration of stem cell function [67]. There is evidence that once HSCs start to contribute to haematopoiesis, they tend to do so for a long time [33]. The question then arises, whether there is clonal succession in the active HSC pool. In other words, is there a continuous shift in the use of stem cells that are active in blood formation or are the same cells utilized till exhaustion (including telomere attrition) and then stochastically replaced from HSCs present within the reserve pool (clonal succession)? The answer to this question is not yet fully known; however, there is some evidence to support clonal succession in a feline model [13]. However, the strongest evidence for clonal succession in this model required observations of animals after stem cell transplantation when fluctuations in cells contributing to haematopoiesis are expected. Indeed, even for cats under steady-state conditions, it has been estimated that HSC clones can contribute to haematopoiesis for more than 300 weeks, which is equivalent to at least half the lifetime of this mammal [13].

STOCHASTIC DYNAMICS WITHIN THE HSC POOL

The small number and slow replication rate of HSC in the active pool can have important consequences on the evolutionary dynamics of mutations that arise within this cellular compartment. As discussed earlier, HSCs tend to contribute to haematopoiesis for a long time (years in humans) and, therefore, mutations in these cells can be retained leading to a clonal population that might expand or decay over time. Serial accumulation of mutations makes these cells a dangerous target for the development of cancer. In the absence of immortal DNA strand co-segregation [79], these cells have various mechanisms to reduce the appearance and rapid expansion of mutant cells including (i) the slow replication rate of these cells [67, 80], (ii) checkpoint control from various tumor suppressor genes, and (iii) expression of plasma membrane carriers that pump out of the cells a broad spectrum of genotoxic agents (e.g., P-glycoprotein) [81]. However, mutations in HSCs do occur and are responsible for a number of neoplastic and nonneoplastic disorders such as CML [82] and paroxysmal nocturnal hemoglobinuria (PNH) [83], respectively. It would seem that the number of active HSCs is determined by the availability of active niches that can house them, that in turn, must be correlated with demands for haematopoietic cell output. HSCs are under the influence of various signals from

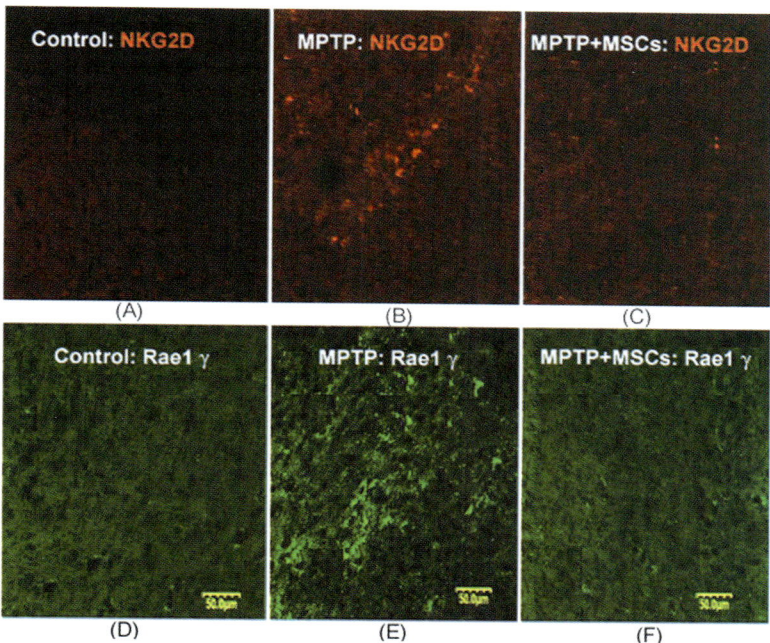

➤ *Fig. 19.5 Suppression of natural killer (NK) cell activation by MSC transplantation. Immunofluorescence revealed (B) NKG2D and (E) Rae-1λ positive staining increased in the SNc of MPTP-treated mice compared with normal controls (A) and (D), respectively, while MSC transplantation suppressed NKG2D expression (C) and Rae-1ã expression (F). Final magnification, ✗200 in (A)–(F).*

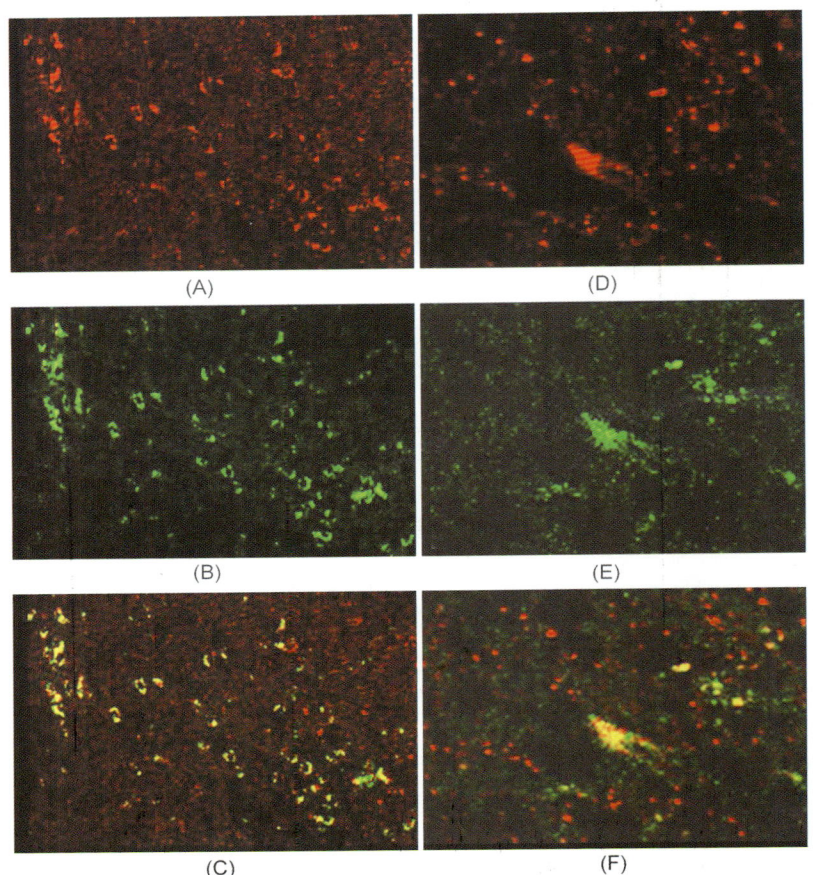

➤ *Fig. 19.6 TGF-β1 and IL-6 release in MSCs in vivo. Double-labelling of (A) TGF-β1 or (D) IL-6 (D) with CFDA SE pre-labelled MSCs (B) and (D), respectively, showed co-localisation of (C) TGF-β1 and (IL-6) with MSCs in the SNc of transplanted mice. Final magnification, ✗200 in (A)–(C) ✗400 in (D)–(F).*

Plate 51

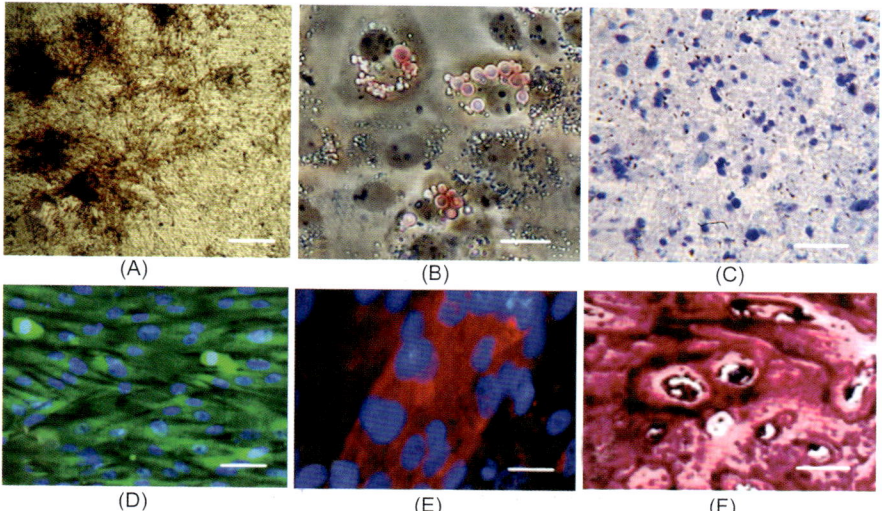

➤ *Fig. 20.2 in vitro and in vivo differentiation of hIDPSC (according to Kerkis et al., 2006 [21]; Mendonça et al., 2008 [34]). (A) Osteogenic differentiation. Von Kossa staining revealed calcified extracellular matrix (arrows) in hIDPSC cultured in osteogenic medium. (B) Oil Red O staining showed accumulation of lipid-filled droplets (red) within the differentiated hIDPSC cultured in adipogenic medium. (C) Chondrogenic differentiation. Positive Toluidine blue staining observed in pellet culture of hIDPSC. (D) Smooth muscle differentiation, immunopositive staining for smooth muscle anti-actin antibody (green). (E) Skeletal muscle differentiation, immunopositive staining for anti-sarcomeric á-actinin antibody (red). (F) In vivo bone formation from hIDPSC after their transplantation into critical size cranial defect in non-immunosuppressed rat (Haematoxylin and Eosin staining). Nuclei stained with DAPI (Blue). A–C, F: light microscopy. D, E epi-fluorescence. Scale bars: A, C, D = 50 μm; B, E, F = 10 μm.*

Plate 52

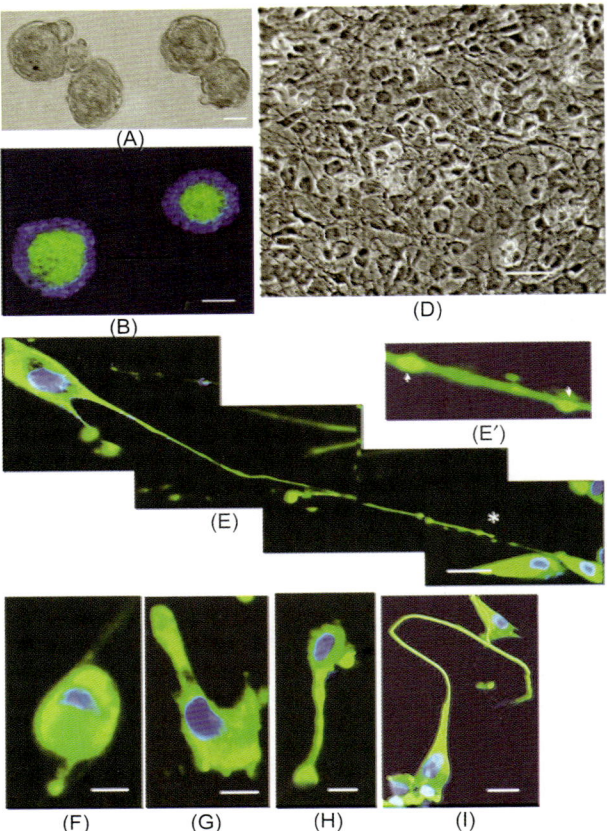

▶ Fig. 20.3 in vitro neuronal differentiation of hIDPSC (according to Kerkis et al., 2006 [21]). (A) Neurosphere formation from hIDPSC. (B) In vitro growing neurospheres from hIDPSC presenting positive immunostaining with anti- proliferating cell nuclear antigen (green) antibody. (D) Neuron/glial sandwich resembling the primary culture obtained from tissues of central nervous system. (D). (E) Anti-β-tubulin III antibody positive neurons from hIDPSC showing the synapse, point of connection between two neurons (asterisk). (E') Higher magnification: synapse contains a small gap separating neurons (arrows). (F–I) Different steps of in vitro neuron formation from hIDPSC, axon formation. Nuclei stained with DAPI (Blue). A,D: phase contrast. B, E–I: epi-fluorescence. Scale bars: A,B = 300 μm; G = 50 μm, E–I = 10 μm.

Plate 53

	HSC	MPP
CD34	–/lo	+
Thy1	lo	–
Flk-2	–	+
C-kit	+	+
Sca-1	+	+
Lin	–	–/lo
CD48	–	+
Cd150	+	–
Cd135	–	+

▶ *Fig. 21.1 Hematopoietic stem cells and multipotent progenitor (MPP) cells. The immunophenotype of HSC and MPP is becoming increasingly defined. The two terms are often used interchangeably but they refer to distinct cell types. Only the HSC have the ability for long-term reconstitution of hematopoiesis. There is increasing evidence that HSC are CD34 negative. Both cell types appear to be uncommitted to any lineage, hence the designation Lin⁻. This requires that cells are negative for CD10, 14, 15, 16, 19 and 20.*

Plate 54

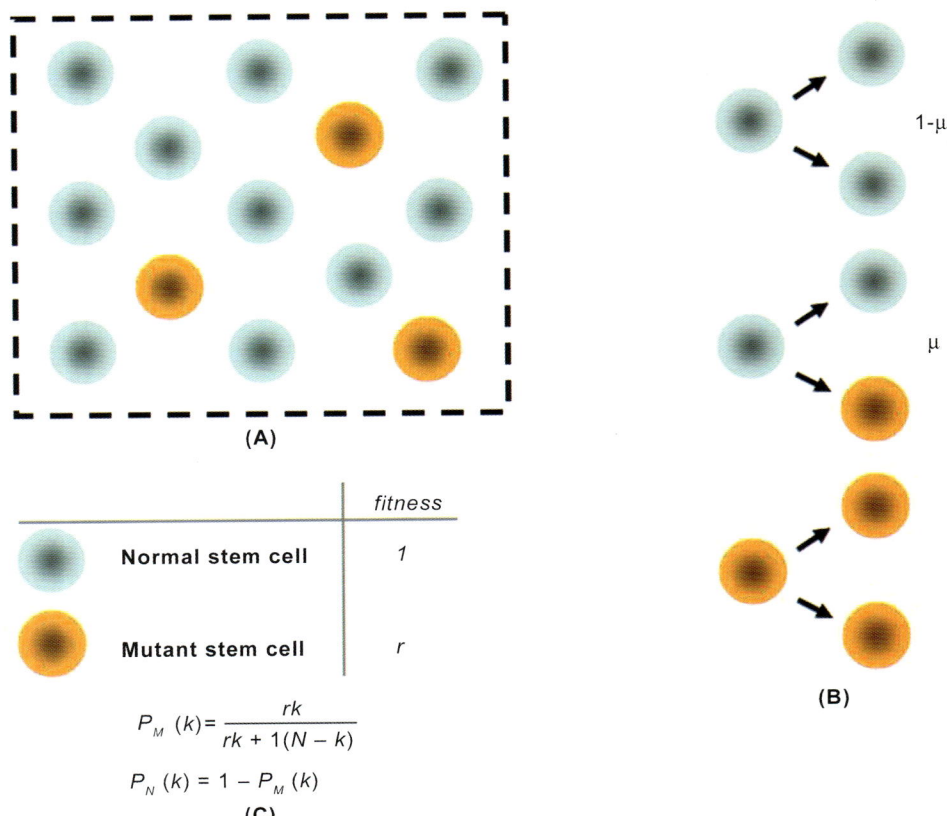

▶ *Fig. 21.2 Haematopoietic stem cell dynamics. (A) The number of active HSCs is considered to be fairly constant in time in adults. (B) Whenever an HSC divides, it can mutate to produce one normal cell and a mutated cell. Mutated cells cannot revert to normal state (theoretically, this is a very low probability event). (C) Stem cells are chosen for reproduction at random but based on their fitness, r. Mutant cells with a higher fitness (r > 1) are chosen more often for reproduction.*

Plate 55

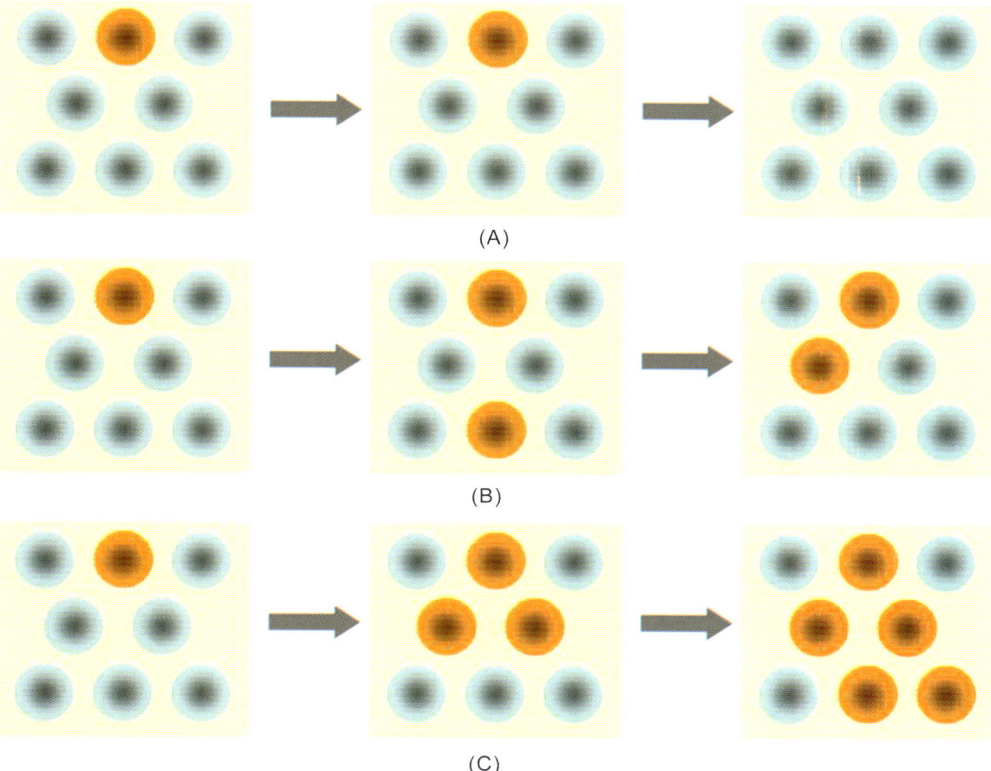

➤ *Figure 21.3 Stochastic dynamics within the stem cell pool. In a Moran process, the total number of cells is kept constant by first selecting a cell for reproduction, with selection being proportional to fitness and then selecting a cell at random for elimination (differentiation). Such a process can lead to three outcomes due to the finite lifespan of mammals: (A) the mutant clone can be eliminated, (B) it can remain latent and not cause disease, or (C) it can expand to generate a burden high enough to cause disease. Note that full replacement of the HSC with mutant cells is not required to cause disease.*

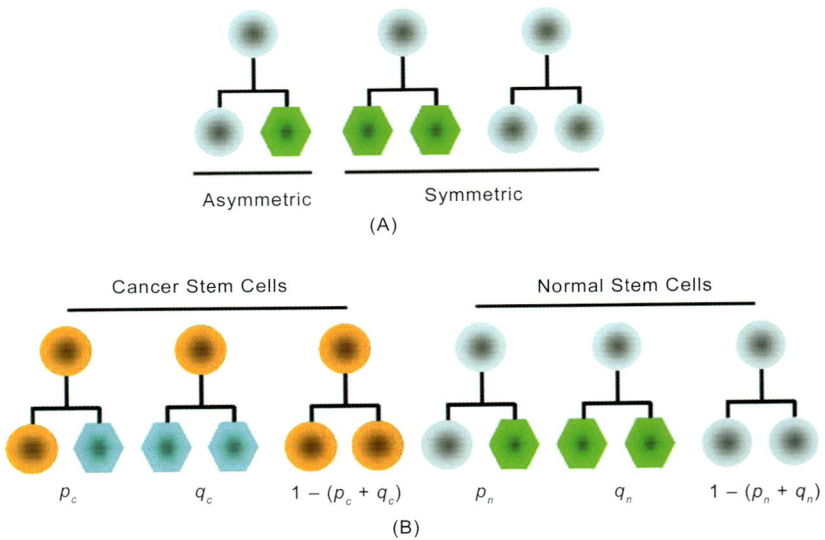

► *Fig. 21.4 Hamatopoietic stem cell reproduction. The fate of the daughter cells that arise from HSC replication define the symmetry of the reproductive event. Whenever the daughter cells have the same fate, the division is considered symmetric while when the two daughter cells have a different fate, it is an asymmetric division. These fates can be defined by probabilities, and mutations can alter these probabilities and the outcome of mutant HSC dynamics.*

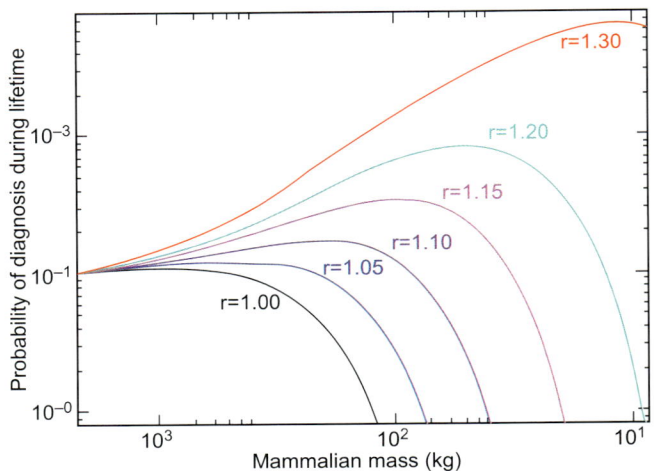

► *Fig. 21.5 Stem cell reproduction, longevity and the risk of cancer. The number of active HSC increases with mammalian mass, while the rate at which these cells divide decreases with mass. However, larger mammals live longer than smaller ones. This aspects are conflicting but results of simulations suggest that larger mammals are at higher risk of acquired HSC disorders compared to smaller mammals. The population of cells at risk and the longer lifespan of larger mammals eliminate the beneficial effect of slower replication. It appears that the onset of the first mutation is the rate limiting step in carcinogenesis.*

Plate 57

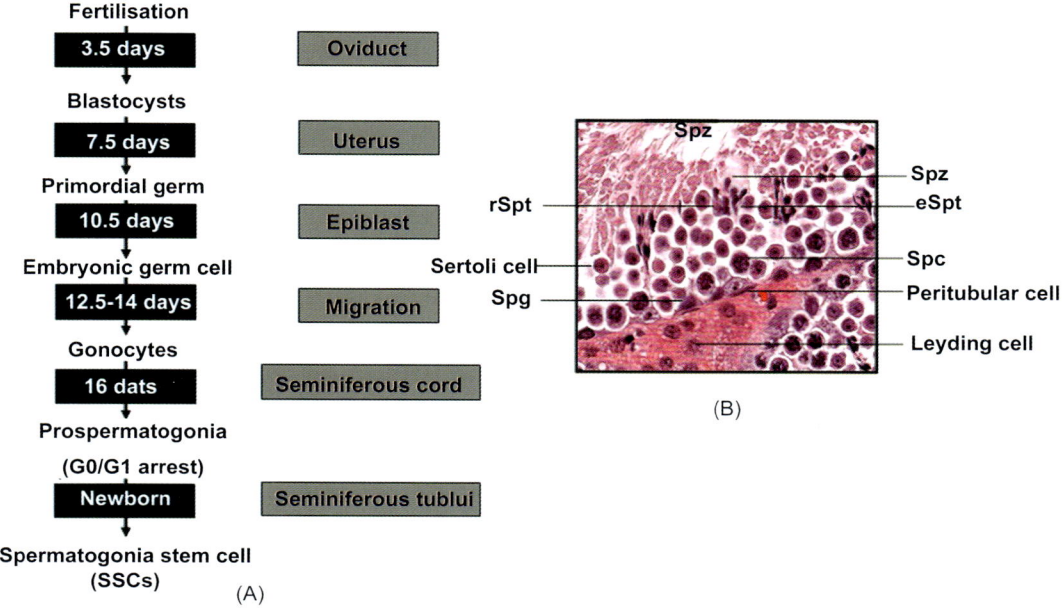

▶ Fig. 22.1 *Development stages and localisation of mouse spermatogonial stem cells. (A) After fertilisation and blastocyst development, primordial germ cells are formed and continue to develop toward SSCs in the newborn offspring. (B) The development stages of SSCs in the seminiferous tubule are depicted. Leydig cells are located in the interstitial tissue of the testis. SCs and SCCs (Spg; attached to the basement membrane), spermatocytes (Spc), round spermatids (rSpt), elongated spermatids (eSpt) and spermatozoa (Spz) are located in the seminiferous tubules (B).*

GENOMIC IMPRINTING

Functional differences between paternal and maternal genomes

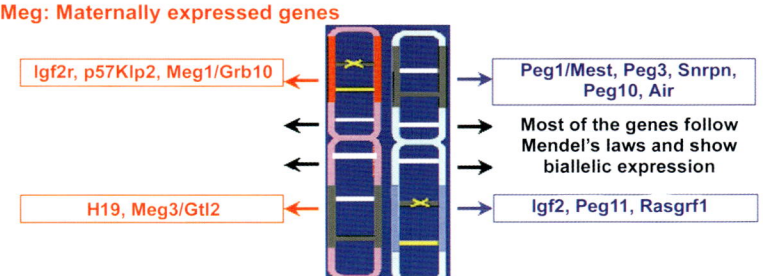

▶ Fig. 23.2 *Effect of genomic imprinting on gene expression. Conventionally, imprinting implies repression. Human chromosome 11 depicted: in pink is the maternal chromosome and blue is the paternal chromosome. Note genes that are expressed either only from maternal chromosome or from paternal chromosome; e.g. H19 gene is expressed from maternal chromosome while it is repressed on the paternal complement, similarly, Igf2 is expressed from paternal chromosome and silenced on the maternal.*

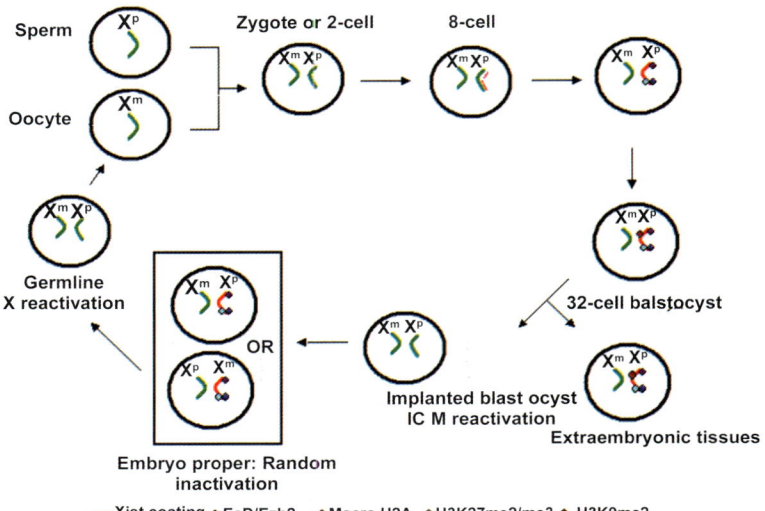

▶ *Fig. 23.3 X chromosome inactivation in mammalian females. Sperm containing an inactive X from the father (Xp) fertilises an oocyte containing an active X from the mother (Xm) to form a female zygote with two X chromosomes. Partial Xist RNA coating of the Xp during early development (pink lines) correlates with incomplete silencing of genes on the Xp. Accumulation of the EED–EZH2 protein complex (purple circle), macroH2A variant (blue circle) and H3K27 di/tri-methylation (light blue circle) on the Xp are first seen at the 16-cell morula stage. Di-methylation of H3K9 (maroon circle) on the Xp is detected at the 32-cell blastocyst stage. Imprinted Xp inactivation stabilizes in extraembryonic tissues. After the blastocyst implants, the inner cell mass (ICM) undergoes reactivation of the Xp, as characterized by the loss of Xist coating, of EED–EZH2 and macroH2A accumulation, and of H3K27 and H3K9 methylation. Cells of the embryo proper then undergo random inactivation in which either the Xm or the Xp are inactivated (red). In the germline of the female embryo, the inactive X is reactivated to insure that all of the oocytes inherit an active X to pass on to the next generation.*

Plate 59

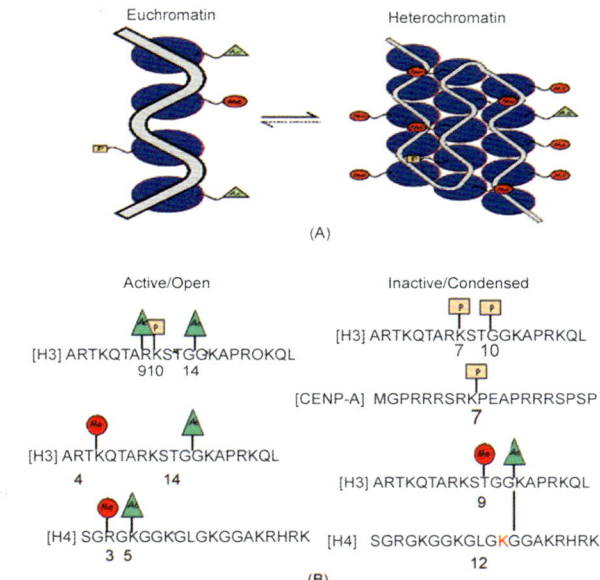

➤ *Fig. 23.4 A diagrammatic representation of epigenetic marking on histones. (A) Schematic representation of euchromatin and heterochromatin as accessible or condensed nucleosome containing acetylated (Ac), phosphorylated (P) and methylated (Me) histone NH2-termini (B) Examples of combinatorial modifications in histone NH2-termini that are associated with active (green flag) or inactive (red circles) chromatin. Single-letter abbreviations for amino acid residues: A, Ala; E, Glu; G, Gly; H, His; K, Lys; L, Leu; M, Met; P, Pro; Q, Gln; R, Arg; S, Ser; and T, Thr.*

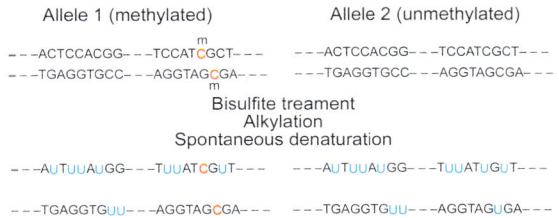

➤ *Fig. 23.5 Outline of Bisulfite sequencing.*

Plate 60

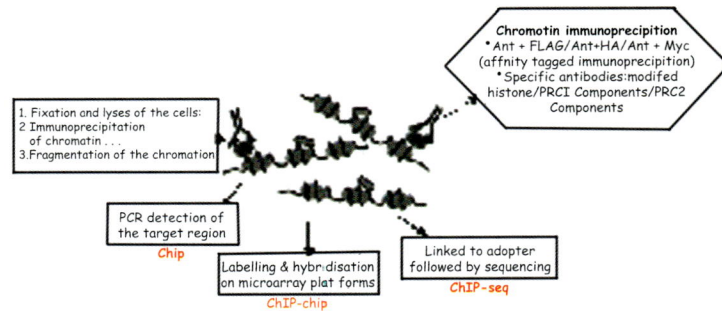

Fig. 23.6   Outline of ChIp.

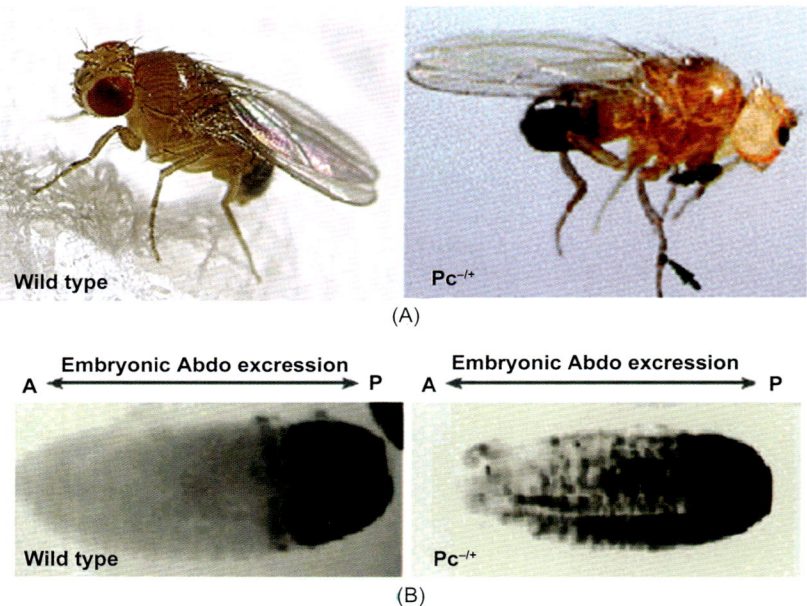

➤ *Fig. 23.8   Mutations in Polycomb genes mimic homeotic mutations. (A) In wild type flies, the sex comb is restricted to the first pair of legs, while in the heterozygous state (Pc–/+), mutant adult flies manifest extra sex combs on second and the third pairs of legs, therefore, transforming posterior segments to anterior like. The arrow in wild type points to the first pair of legs while that in Pc–/+ points to the second pair. The first pair is not visible from this angle. (B) Effect of the mutation at early embryonic stages traced intragenic flies carrying β-galactosidase as reported under homeotic gene promoter. Mutations in Polycomb genes cause a de-repression of specific Hox members, which leads to extended zone of expression and hence homeotic transformations of one body segment into the identity of another.*

Plate 61

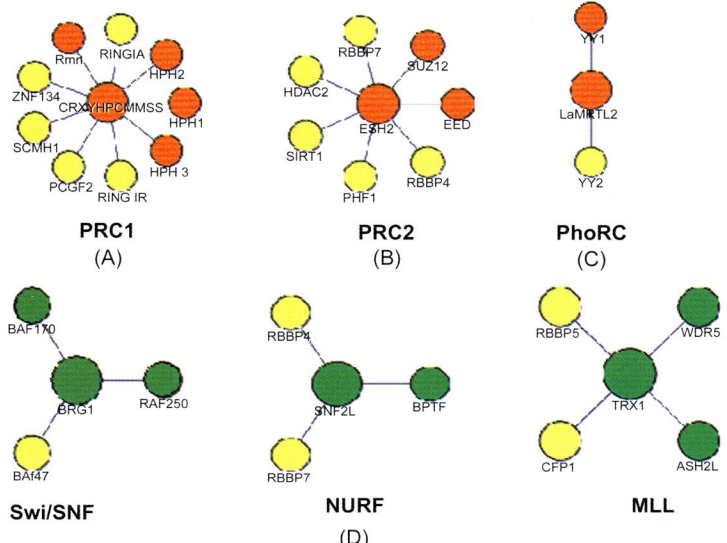

► *Fig. 23.9 Networks of components of human Polycomb (PcG) and Trithorax (TrxG) proteins. The core enzymatic subunits of PcG (red) and TrxG (green) complexes are shown in interaction with other complex members (yellow). Note the two components (RBBP4 and RBBP7) shared between PRC2 and NURF that are known to have antagonistic activity, an example of cros-stalk and combinatorial use of parts to make two distinct complexes. The networks were generated using Cytoscape software from the public domain.*

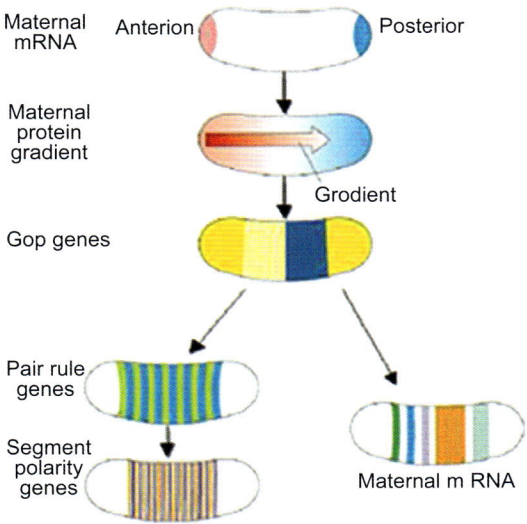

► *Fig. 23.10 Cascade of gene activity in early development.*

Plate 62

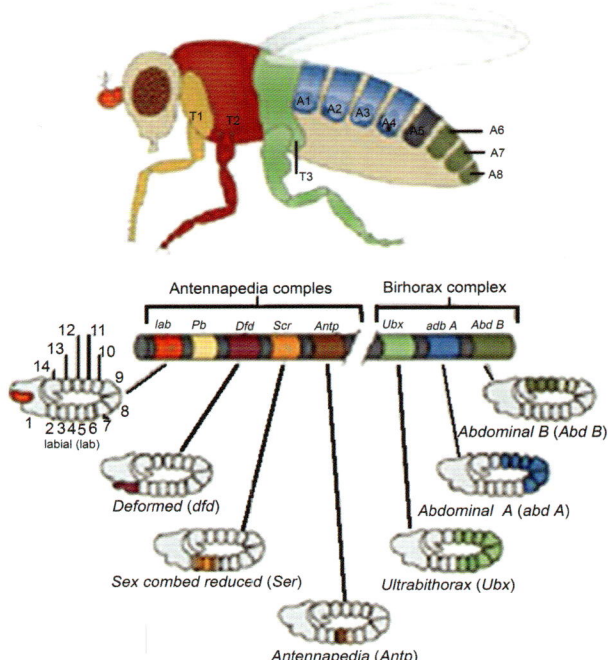

▶ *Fig. 23.11 Homeotic gene expression in Drosophila. In the centre are genes of antennapedia and bithorax complex and their functional domains. Below and above the gene map, the region of the homeotic gene expression (both mRNA and protein) in the blastoderm of Drosophila embryo and regions that form from them in the adult fly are shown.*

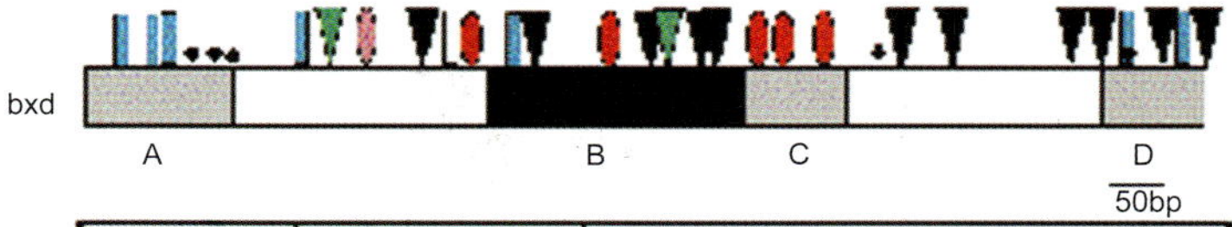

SYMBOL	MOTIF	CONSENSUS	Ref.
	Zeste	YGAGYG (Y – C/T)	129
	GAF/Pipsqueak	GAGAG	122
	PHO	GCCATHWY (H=A/C/T: W=A/T)	124
	PHO core	GCCAT	127
	TGC repeats	TGCTGC	127
	A rich	AAAAA	127
	GT	GTGTG	127
	TRE motif	AACAA	128
	Minimal PRE		

▶ *Fig. 23.12 Diagrammatic representation of a minimal Polycomb responsive element. The PRE from the bi-thorax complex is shown with motif positions. The minimal fragment that has PRE activity is shown as a black box. The boxes A, C and D indicate regions outside the minimal PRE that have additional functions. DNA motifs along with symbols used in the line diagram are listed in the table.*

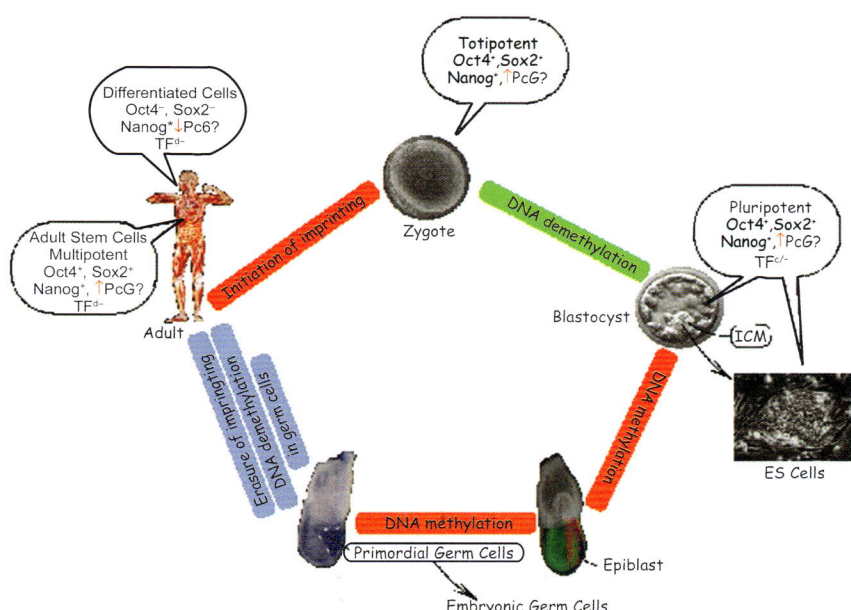

▶ *Fig. 23.13 A summary of epigenetic regulatory cycle of pluripotency. The alteration in epigenetic modifiers (PcG), DNA methylation and transcription factors that maintain stemness, such as OCT4, NANOG and SOX2 are shown. ICM-Inner cell mass, TF^d, differentiation-specific transcription factors, + is for presence, − for absence, upward/downward arrow indicates elevated and decreased levels, respectively. The totipotent zygote contains maternally inherited epigenetic modifiers and transcription factors, including OCT4, SOX2 and NANOG. These, together with the embryonic transcripts, regulate development to the blastocyst stage, where the pluripotent cells are established in the inner cell mass (ICM), from which ES cells are derived both of which share several common epigenetic imprints. In the post-implantation embryo, pluripotent epiblast cells are controlled by diverse repressive mechanisms during their differentiation into somatic and germ-cell lineages. The early germ cells exhibit epigenetic and transcriptional states that are associated with pluripotency, and can give rise to embryonic germ cells. Epigenetic reprogramming, including genomic imprinting within this lineage re-generates totipotency.*

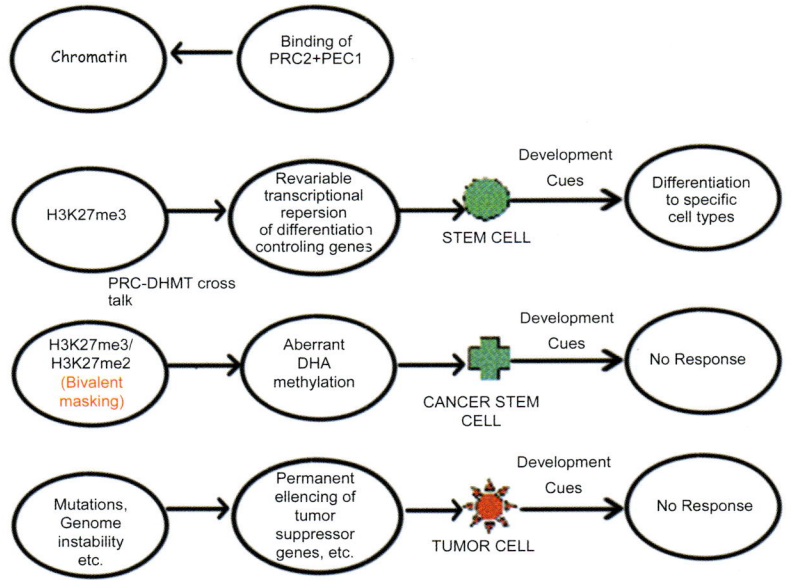

▶ *Fig. 23.14 A comparison of epigenetic regulation in stem cells and cancer cells.*

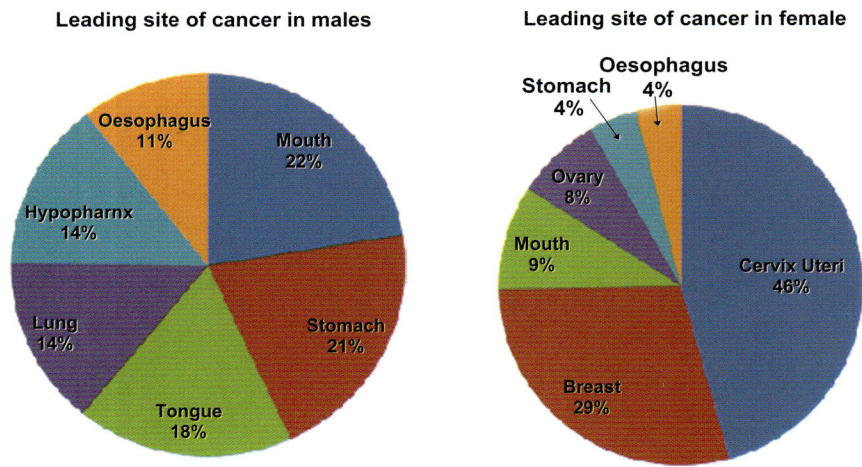

Courtesy: Adopted from Indian Council of Medical Research

▶ *Fig. 24.1 Leading site of cancer in males and females in India. The National Cancer Registry Programme by the Indian Council of Medical Research (ICMR) conducted before 2005 came up with a detailed statistics with respect to the various cancers prevalent among males and females in India. While mouth cancer came up as the most prevalent in males, cervical cancer came up as the most common cancer in females. The 2005 cancer atlas by ICMR revealed that incidence of breast cancer had surpassed that of cervical cancer in females.*

cells forming the niche, the extracellular stroma, cytokines, and growth factors (discussed earlier). The overall input from these various sources together with the transcriptome of the cell presumably define the behavior of the HSCs and determining whether a given cell replicates, the symmetry of replication, and the commitment to differentiation of the progeny cells along specific pathway(s). Although the detailed molecular pathways that regulate stem cell behavior are incompletely understood, the HSC dynamics can be approximated by a Markov process. Such a coarse-grained approach is also in line with the experimental evidence that haematopoiesis may be a stochastic process [26, 27]. Despite this stochastic behavior at the individual cell level, the number of terminally differentiated blood cells appears to be fairly constant. This observation is due to the sheer number of cells present at the end of the haematopoietic tree that ultimately masks the stochastic behavior of individual cells at all levels of haematopoiesis. As expected, the impact of this stochastic behavior is strongest closer to the root of the process, i.e. at the stem and progenitor cell levels [60, 84] since these are the smaller cell populations.

Under steady-state conditions, it is thought that the number of HSC is constant and the population behaves in a homogenous way since cells are chemically coupled despite being separated in space [85]. The evolutionary dynamics of mutated cells in such a population can be modeled by a simple stochastic birth–death process known as a Moran process [86] (Fig. 21.2A). In each cycle of this process, one cell is chosen for reproduction and another is chosen for elimination so that the total population remains constant. The rate at which events take place is dictated by the replication rate of HSC and deduced from the allometric relations specific for each adult mammalian species. A normal cell that is chosen for reproduction can mutate with probability μ Experimental observations estimate that the normal mutation rate per gene is of the order of $\mu \cup 10^{-7}$ per replication [87]. Thus, when a normal stem cell is selected to reproduce, with probability $1 - \mu$, it produces two normal daughter cells, while with probability μ, the mitotic event produces one normal cell and one mutated cell (Fig. 21.2B). When a mutated HSC reproduces, it generates more mutated cells since the probability of a mutant cell to revert to normal (wild-type) is essentially negligible. A mitotic event increases the number of HSCs by one and, therefore, one cell from the whole pool is chosen at random for elimination. More precisely, the cell chosen for elimination is exported, and it initiates its journey of increasing specialisation down the haematopoietic cascade. The probability that a cell is chosen for reproduction is proportional to its fitnesse r. Normal cells have fitness $r = 1$, cells with mutations that enhance their fitness have $r > 1$, while mutations that confer a fitness disadvantage have $r < 1$ Figure 21.2C illustrates the process of stem cell replication. Since normal HSCs replicate approximately 1 per year [67, 80], then in one year approximately 400 HSC divisions take place; this provides the time scale for our *in silico* evolutionary dynamics. By repeating the Moran process many times, probability distribution functions are obtained that illustrate the evolutionary histories of mutations in a population of virtual individuals [84].

The Moran process has two absorbing states (of course, if we wait an arbitrarily long time, ultimately all cells will be mutated), either the clone expands to take over the entire population or it is stochastically eliminated and goes extinct. During the finite lifetime of each individual organism, clonal invasion (i.e. complete replacement of the HSCs by a mutant clone) is rarely achieved, and what is normally observed is an intermediate state. Moreover, it is often the case that the appearance of disease does not require that all the HSC are mutated (discussed further). A small

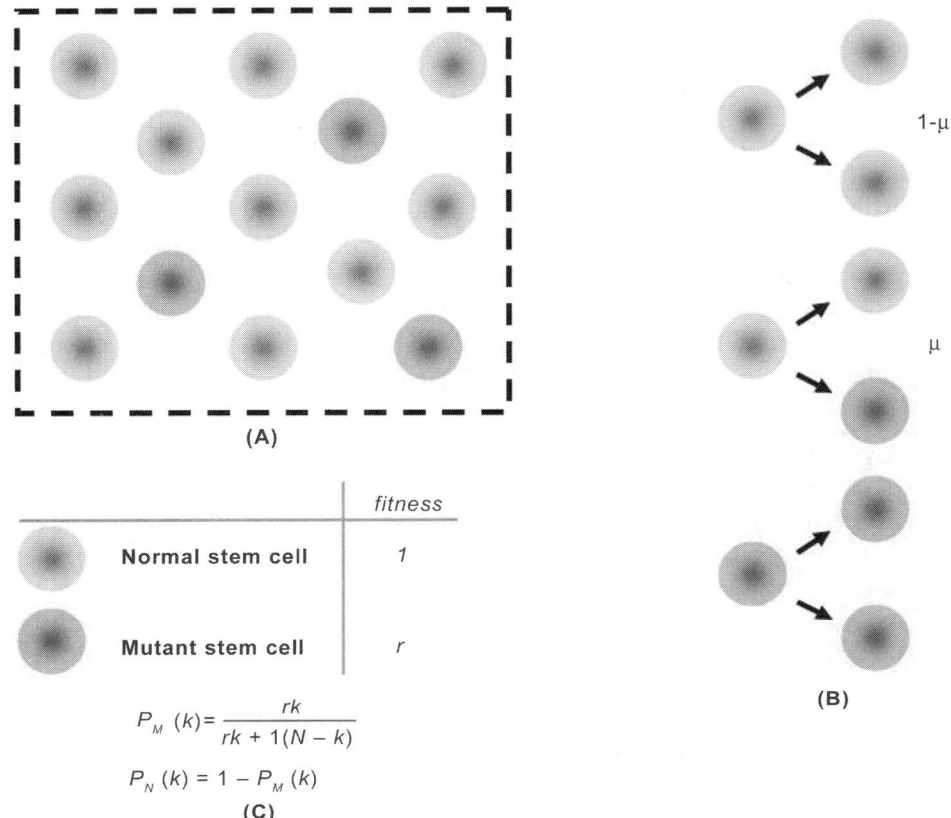

▶ **Fig. 21.2** *Haematopoietic stem cell dynamics. (A) The number of active HSCs is considered to be fairly constant in time in adults. (B) Whenever an HSC divides, it can mutate to produce one normal cell and a mutated cell. Mutated cells cannot revert to normal state (theoretically, this is a very low probability event). (C) Stem cells are chosen for reproduction at random but based on their fitness, r. Mutant cells with a higher fitness (r > 1) are chosen more often for reproduction. (for colour figure see Plate 54).*

population of mutated HSCs (e.g., Cancer Stem Cells (CSC)) may be enough to induce disease, if not a lethal burden [84, 88]. Interestingly, such modeling predicts that a mutant HSC clone can invade the whole population (rarely), go extinct or persist at relatively 'stable' levels that may or may not cause disease (Fig. 21.3). A major determinant of the outcome is the fitness advantage of the mutant cells compared to their normal counterparts. *In silico* studies of such a population of cells show that starting from one mutated HSC, the probability of invasion increases with higher fitness, while the chance of extinction falls down. However, even for a fitness advantage of 2, which is a very high fitness advantage [89, 90]—such a cell has a 50% probability of clonal extinction (for large populations). While it is clear that the fitness conferred by the mutation is an important parameter when describing the phenotype of cancer cells, it is easier to define 'r' than to provide experimental estimates of its value, even for well-defined mutations such as *bcr-abl*.

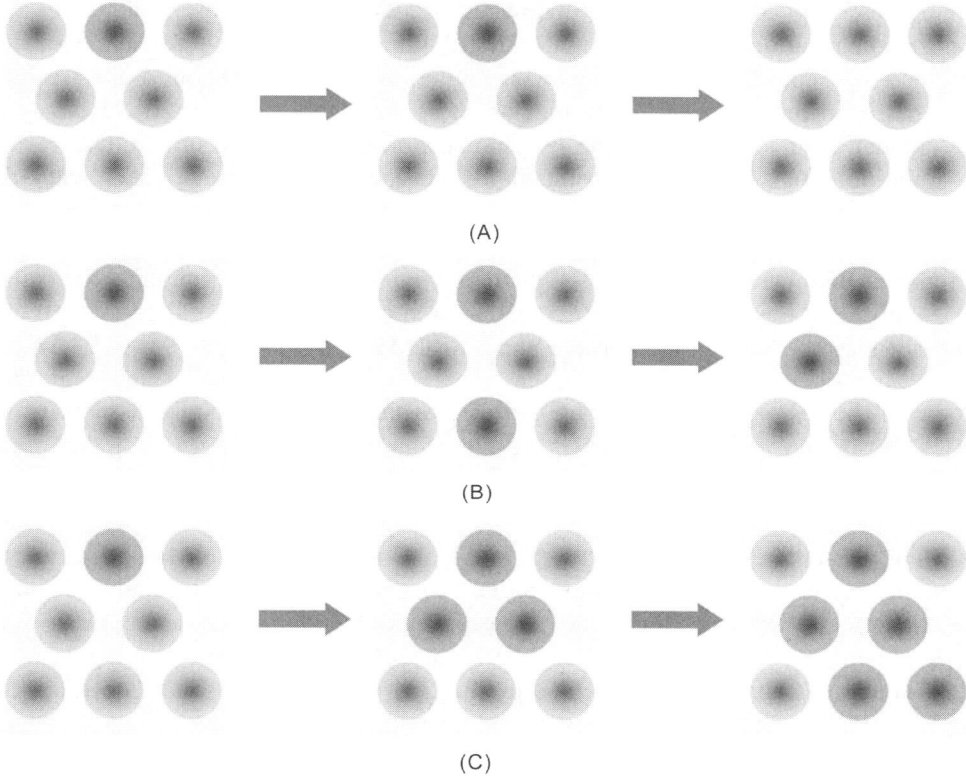

▶ **Figure 21.3** *Stochastic dynamics within the stem cell pool. In a Moran process, the total number of cells is kept constant by first selecting a cell for reproduction, with selection being proportional to fitness and then selecting a cell at random for elimination (differentiation). Such a process can lead to three outcomes due to the finite lifespan of mammals: (A) the mutant clone can be eliminated, (B) it can remain latent and not cause disease, or (C) it can expand to generate a burden high enough to cause disease. Note that full replacement of the HSC with mutant cells is not required to cause disease. (for colour figure see Plate 55).*

The diagnosis of various haematopoietic tumors requires that a minimum fraction of cells in the bone marrow must be abnormal for the tumor to be clinically detectable or defined. These definitions are somewhat arbitrary and subject to revision. For example, at least 10% plasma cells have to be present for multiple myeloma to be diagnosed and 20% bone marrow blasts define acute leukemia [88]. If we take these thresholds into consideration, it can be shown that when the mutant clone expands and approaches say 20% of N_{SC} stochastic clonal extinction becomes an increasingly rare event, although it is still possible [84]. The evolutionary trajectories of the mutant cells are described by probability distribution functions that are always 'one humped' functions of time [84, 88]. The fitness advantage of the mutant cells has a major impact on the time required for the clone to either completely invade or reach the diagnostic threshold. As the fitness advantage increases, the peak of the distributions moves to the left (shorter time to reach threshold). The variance of the distributions also depends on the fitness associated with the mutation. For a small fitness advantage, distributions are very wide but, as the fitness advantage increases, they become narrower (smaller variance), because the process becomes more deterministic. Our results show that for a mutation with a low fitness advantage, the 'average' time to reach a diagnostic threshold or invasion (the mean of the corresponding distribution) can be of the order of the variance of the distribution, and as such lacks a precise meaning. In other words, these results suggest that each patient's tumor is unique and has different dynamics even when arising from the same cell type. Of course, these observations relate to disorders where a single mutation could be enough to explain some aspect of the disease (e.g. the chronic phase of CML due to *bcr-abl* [91–93]). An important corollary of this model is that for a single gene mutation, knowledge of the time required for the development of the disease can provide an estimate of the fitness advantage associated with that mutation. For example, if retinoblastoma were due to mutations in the *Rb* gene only, the fitness advantage of the mutation must be >1.7 to be compatible with the time frame in which the disease appears [84].

The concept of clonal expansion and tumor progression are well-established features in cancer, but perhaps the idea of stochastic clonal extinction or latency are less obvious and one might think that these are artificial results of the model. However, there is clinical evidence for 'spontaneous' elimination of malignant clones, even though it is rare. Perhaps the most striking case is that of transient leukemia (TL) that often develops in children with Down's syndrome [94, 95]. Some studies suggest that perhaps up to 10% of children with Down's syndrome might develop TL, usually within 5 days of birth [96]. This potentially lethal disease clearly affects an early progenitor cell in haematopoiesis (if not the HSC) [94, 96]; yet, in up to 85% of cases, the disease resolves with minimal or only supportive therapy [95]. Our model of stochastic dynamics accommodates this behavior. There are also reports of 'spontaneous' resolution of myelodysplastic syndromes as well as acute myeloid leukemia in patients [97] who did not receive any disease-modifying therapy. Finally, the model suggests that malignant clones can experience latency and stability; in other words they do not change in size appreciably over long periods of time (including years). The best evidence for this is provided by a group of patients with essential thrombocythemia that were untreated. In this cohort, the size of the $JAK2^{V617F}$ clone was determined serially and shown to be stable over several years [98], again compatible with our model of stochastic HSC dynamics.

SYMMETRY OF STEM CELL REPLICATION AND FITNESS

One of the *sine qua* nonproperties of stem cells is self-renewal. The fate of the daughter cells defines the symmetry of division; 'symmetric' division gives rise to two cells that have the same fate (both cells either differentiate or retain stem cell properties). However, whenever the daughter cells have different fates, the division is considered to be 'asymmetric' (Fig. 21.4A) [99]. In principle, asymmetric division should be enough to maintain tissues by simultaneously keeping the HSC population constant and feeding downstream compartments. However, this would not allow HSC expansion that is necessary during ontogeny of the haematopoietic (and other) systems or in response to injury or bone marrow transplantation. There is now good experimental evidence from DNA methylation patterns that all three types of SC division as depicted in Fig. 21.4A occur, depending on demands imposed on the SC pool [33, 100] by the host.

The determinants of the (a) symmetry of HSC replication are not completely understood, although interactions of the HSC with the stem cell niche and signals via the *Notch*, *Hedgehog* and *Wnt* pathways (including β-catenins) are known to be quite important [101–103]. Recent work in model organisms such as *Drosophila* is starting to decipher the potential impact of mutations on the (a)symmetry of stem cell replication. Some of the genes that have been recently shown to determine cell division fate include *partner of inscuteable (PINS)* [104], *lethal giant larvae (LGL)* [105], *HUGL-1* and *adenomatous polyposis coli (APC)*. Mutations in *PINS, LGL* and *HUGL-1* result in the formation of tumor-like tissue in these animals.

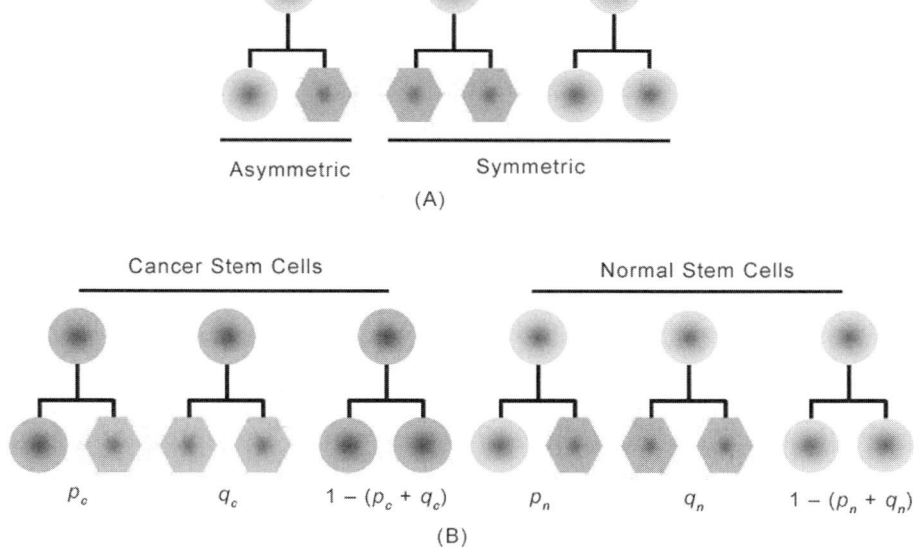

▶ **Fig. 21.4** *Hamatopoietic stem cell reproduction. The fate of the daughter cells that arise from HSC replication define the symmetry of the reproductive event. Whenever the daughter cells have the same fate, the division is considered symmetric while when the two daughter cells have a different fate, it is an asymmetric division. These fates can be defined by probabilities, and mutations can alter these probabilities and the outcome of mutant HSC dynamics. (for colour figure see Plate 56).*

In order to understand the impact of mutations that alter the symmetry of HSC replication, we have developed the model illustrated in Fig. 21.4 [106]. The fate of HSC daughter cells after a single HSC replication is determined by parameters p and q that define the probability of (a)symmetric cell replication. A normal HSC divides asymmetrically with probability p_n and with probability q_n the cell divides symmetrically to give rise to two differentiated cells. Finally, with probability $p_n - q_n$ the cell divides symmetrically to give rise to two HSC (self-renewal). Mutations confer to the cell a relative fitness r and cells are chosen for reproduction according to their fitness (Fig. 21.2C). The dynamics resemble a Moran process such that we impose the condition that the total cell population must remain constant. To meet this condition, cells may be forced to divide symmetrically to replenish the pool if a cell is lost [106]. Under such a model scenario, what would be the impact of a mutation that increases the probability of self-renewal compared to that of a normal HSC (i.e. $p_c + q_c < p_n + q_n$)? One can show that such mutated cells will take over the compartment as they possess a selective advantage (even for $r = 1$). Such mutations enable the mutant to invade the population and impart an effective 'reproductive fitness' advantage to cells. Naturally, if the mutation also gives a selective advantage whereby the mutant cells are chosen more often for reproduction, the average time needed for fixation will be significantly reduced [106]. Therefore, mutations that increase the self-renewal capability of cells may be an early event in cancer. Indeed, these mutations favor the mutated population to outgrow the normal cell population number, thereby increasing the population of cells at risk, with the possibility to accumulate additional mutations and progression. This is perhaps one of the reasons why there is so much interest in the transformation of normal stem cells to CSC.

Due to their location at the root of the haematopoietic tree, HSC would be expected to be the only ones exhibiting long-term self-renewal capability, as progenitor cells can self-renew but for a limited time [60]. However, it has been shown that acute myeloid leukemia can arise in progenitor cells such as CFU-GEMM that re-acquire stem cell-like properties and can drive the tumor. Indeed, there is a good evidence for this both in the case of some *de novo* acute myeloid leukemia as well as the blast crisis that often accompanies CML [107, 108]. Our modeling suggests that for a tumor to arise due to mutations in a progenitor cells, such a mutant cell must acquire the potential for long-term self-renewal at an early step in its path to cancer, otherwise it will be eliminated relatively rapidly down the haematopoietic cascade [109, 110].

STEM CELL POOL, LONGEVITY AND CANCER

There is increasing evidence that like normal blood cells, haematopoietic tumors are also driven by CSC [111]. Presumably CSCs arise from HSCs by mutations, although not all CSCs necessarily result from the malignant transformation of normal stem cells. Hence, HSCs are potentially a double-edged sword; a small and slowly replicating pool minimizes the risk of mutations, but at the same time facilitates the expansion of neutral or disadvantageous mutations. Mutations in HSCs can lead to cancer, giving rise to tumors such as the myeloproliferative disorders (e.g. CML and PV) [6, 82]. Normally, cells are at highest risk of acquiring mutations during DNA replication. As already discussed, the rate of replication of the HSC *decreases* with the mass of a given species as $B_c = B_0 M^{-1/4}$ [67]. This implies that in larger mammals, the HSCs divide more slowly and should acquire mutations due to replicative errors at a slower rate. On the other hand, the size of the active stem cell pool *increases* with mass ($N_{SC} \sim M^{3/4}$) and so the population of cells at risk is higher. It is also clear that a tumor clone is relevant and can cause disease only if it expands during the lifespan of the mammal. Interestingly, the expected lifetime (L_E) of a mammal also scales allometrically (*increases*) with mass ($L_E = L_0 M^{1/4}$, $L_0 = 8.6$) [112, 113]. Finally, the probability of reaching a given fraction of mutated cells *decreases* as the size of the active stem cell pool expands. To explore the impact of these potentially conflicting variables on the development of CSC-derived disorders, we performed simulations of HSC dynamics taking these contrasting variables into consideration. For the purpose of our analysis, any species is defined by the average adult mass characteristic of that species and we estimated N_{SC}, L_E and B_c and with the above definitions. The size of the species-specific active stem cell pool was assumed to be constant during the lifetime of a given mammal. We also considered that the mutation rate, ($10^{-7} \le \mu \le 10^{-6}$), is constant across all mammals [75, 87] given the similarity of the DNA replication machinery across eukaryotes and back mutations do not occur (very low probability event). A contamination threshold of 20% was assumed to be necessary to define disease in all species.

We illustrate some of the results in Fig. 21.5, where we plot the probability of disease during the mammal's lifetime as a function of both mass and fitness advantage of the mutant cell. The data suggest that larger mammals are better suited at resisting clonal expansion of mutations that are either neutral or confer only a marginal advantage. However, for mutations with a significant fitness advantage, the larger species are worse off compared to smaller mammals such as rodents. The reason behind these dynamics is that once such a mutation occurs, the probability of diagnosis does not necessarily decrease for larger mammals [113]. On the other hand, for any species, the expected number of mutations (n_μ) within the HSC pool (i.e. the population at risk) is given by, $n_\mu = \mu\, N_{SC}\, B_c\, L_E \cup M^{3/4}$ which favors smaller mammals [113]. The conclusion from all this is that the rate limiting or decisive step is the occurrence of the first mutation. As a consequence, larger mammals, with their larger active stem cell pool, are protected from the expansion of *neutral* mutations but not from advantageous mutations. The results of these simulations are robust and remain true with respect to variations of the mutation rate and of the fraction of mutated cells necessary to cause disease within the HSC pool [113].

These *in silico* results are supported by clinical observations on humans and other animals. For example, a search of the mouse tumor database (http://www.nih.gov/science/models/mouse/resources/mtbdb.html) shows that the spontaneous development of a myeloproliferative disorder (CMPD) has not been reported in this mammal. The incidence of CMPD is uncommon in dogs that share our environment and presumably are exposed to the same levels of background radiation as humans [114] (the only known external risk factor for the development of these diseases). Moreover, CMPD are quite uncommon in humans younger than 20 years of age (<10% of cases of CML occur in such people), again compatible with our model [88]. An unavoidable conclusion of our modeling is that the human species is a victim of its own success, while the average human lifespan provides the largest deviation from the predicted allometric lifespan (considerably longer than predicted); this comes at the price of higher risk of CMPD. Our modeling also naturally explains why the incidence of cancer tends to increase with longevity.

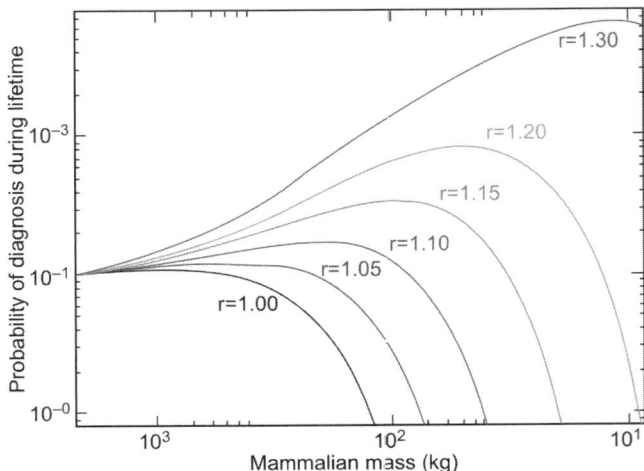

▶ **Fig. 21.5** *Stem cell reproduction, longevity and the risk of cancer. The number of active HSC increases with mammalian mass, while the rate at which these cells divide decreases with mass. However, larger mammals live longer than smaller ones. This aspects are conflicting but results of simulations suggest that larger mammals are at higher risk of acquired HSC disorders compared to smaller mammals. The population of cells at risk and the longer lifespan of larger mammals eliminate the beneficial effect of slower replication. It appears that the onset of the first mutation is the rate limiting step in carcinogenesis.*

INHERITED AND ACQUIRED HAEMATOPOIETIC STEM CELL DISORDERS

Given the importance of HSCs, it is understandable that both genetic and acquired disorders that affect them are rare. However, inherited HSC disorders have provided considerable insights into the fundamental biology of various basic cellular events (e.g., telomere maintenance and ribosomal biogenesis in the various forms of dyskeratosis congenita, and adenosine deaminase (ADA) or the common γ chain (γc) deficiency, characteristic of the most common subtypes of severe combined immune deficiency [115]. These inherited disorders generally manifest themselves in one of two ways, a marrow failure syndrome (single or multiple lineages) that may or may not be associated with an elevated risk of neoplasia or a profound immune deficiency. In general, these disorders can be cured by bone marrow transplantation if a suitable donor is available. Prototypic examples include dyskeratosis congenita (DC) which presents as an inherited syndrome with bone marrow failure as the most important feature that threatens the life of the individual, while ADA or gc deficiency exhibit profound immunological defects and high susceptibility to infections. Acquired HSC disorders can also lead to bone marrow failure (aplastic anemia), neoplastic or 'benign' proliferation of mutant cells (e.g., CML and PNH). A partial list of inherited and acquired HSC disorders is presented in Table 21.1.

Table 21.1 *Inherited and Acquired Haematopoietic Stem Cell Disorders*

Inherited	Acquired
Marrow failure	*Marrow failure*
Fanconi anemia	Idiopathic aplastic anemia
Dyskeratosis congenita	Post infection
Shwachman-Diamond	*Drug induced*
Diamond-Blackfan	Radiation
Kostmann	PNH
Seckel	*Autoimmune*
CAT	
TAR	
Immune deficiency	*Myeloproliferation*
Chronic granulomatous disease	CML
SCID	Polycythemia vera
X-linked (gc deficiency)	Essential thrombocythemia
Adenosine deaminase	*Idiopathic myelofibrosis*
JAK3 deficiency	Myelodysplastic syndromes
RAG-1, RAG-2 deficiency	
Artemis	
ZAP70 deficiency	
Bloom	
Ataxia telangiectasia	
Wiskott-Aldrich	
Nijmegen breakage syndrome	
Hemoglobin-related disorders	
Thalassemia syndromes	
Sickle cell disease	
*Inborn errors of metabolism**	
Osteopetrosis	
Glycogen storage diseases	
Mucopolysaccharidoses	
Leukodystrophies	
Glycoprotein disorders	

Note: PNH, paroxysmal nocturnal hemoglobinuria; CAT, congenital amegakaryocytic thrombocytopenia; TAR, thrombocytopenia with absent radii; CML, chronic myeloid leukemia; SCID, severe combined immune deficiency.

* These disorders may or may not be amenable to therapy with stem cell transplantation.

ACQUIRED HAEMATOPOIETIC STEM CELLS DISORDERS WITH CML AND PNH AS PROTOTYPES

CML has been a paradigm setting neoplastic disorder for many reasons. It was the first tumor associated with a specific chromosomal abnormality [the Philadelphia chromosome (Ph), t(9,22)(p34;q11)] that subsequently was shown to bring the abl kinase proto-oncogene into the breakpoint cluster region (*bcr*) on chromosome 22, with the formation of a fusion product (*bcr-abl*) that leads to the uncontrolled activity of the abl kinase and therefore an oncoprotein. Expression of *bcr-abl* in HSC is associated with a phenotype compatible with the chronic phase of CML. Thus, it is thought that this single mutation may be enough to explain the chronic phase of the disease. However, in time, the mutant clone acquires additional mutations that induce progression to the accelerated phase and finally to the terminal blast crisis. CML is a true HSC disorder since the *bcr-abl* fusion gene is present in all types of blood cells including a small fraction of T- and NK-cells [116].

The identification of the *abl* kinase enabled the discovery of relatively specific inhibitors for this enzyme, initially with imatinib and now with more novel agents such as dasatinib and nilotinib that all bind reversibly with their target. This was another first in cancer therapy—the availability of a small molecule, inhibitor of a crucial pathway within cancer cells. Therapy with imatinib has changed the way CML is now treated and the majority of patients achieve excellent responses that can last many years. It is unusual for the blood counts not to return to normal (haematologic response) and many patients lose the signal of the Ph chromosome on marrow cytogenetics. A significant fraction of patients become negative for *bcr-abl* by quantitative real-time polymerase chain reaction (Q-RT-PCR) assay, implying a minimum tumor reduction of 4 orders of magnitude. The first studies were performed in a mixed population of patients, those with recently diagnosed disease and others who had been treated with prior agents such as interferon and hydroxyurea. In these initial cohorts, imatinib failures were observed in time mainly due to the accumulation of mutations that led to drug resistance. Interestingly, the more recently diagnosed and treated patients have experienced fewer problems with resistance, perhaps in part because they were diagnosed at an earlier stage with a smaller tumor burden and so the population of cells at risk of acquired resistance and/or progression is reduced. Imatinib therapy is administered continuously but only a fraction of CML cells are responding to the drug (~5%). However, the response kinetics are quite fast (usually within a couple of months). This is due to the mechanism of action of imatinib, while *bcr-abl* expression gives the CML cells (progenitors and precursors) a fitness advantage due to a higher probability of self-renewal; imatinib therapy reverses this fitness advantage to levels that are inferior to those of normal progenitor cells [117]. The result is that normal progenitors take over haematopoiesis quickly, despite the persistence of CML cells [118]. If therapy is stopped, the persistent progenitors ensure that relapse is rapid, usually within a matter of months [118, 119].

One of the major outstanding questions in CML is whether therapy with these agents can cure the disease. There is no clear cut answer to this and tentative answers to this question have been based on a combination of disease biology and mathematical models. Some have suggested that imatinib can cure the disease [120] while others believe that it is unlikely that imatinib or any of the second generation tyrosine kinase inhibitors can eradicate the disease [118, 119]. The reason behind the different predictions lies in details of the models. While Roeder et al. [120] stated that imatinib can kill CML stem cells depending on their cell cycle status, the other models do not make this assumption [118, 119]. It has been shown experimentally that the majority of CML stem cells are quiescent and it is very difficult to stimulate them to enter the cell cycle [5]. There is experimental evidence that CML stem cells are independent of *bcr-abl* for their existence and therefore not sensitive to the drug (Holyoake). Moreover, CML stem cells express membrane pumps that belong to the ATP-binding cassette family that actively pump out imatinib and similar TKIs, providing further protection of the cells from these agents. In addition, the number of CML stem cells was recently estimated to be quite small in many patients [118]. This in itself presents a very steep hurdle for therapy since it is similar to the proverbial needle in a haystack problem. The answer to the question will only be answered in a definite way after extended follow-up of patients currently on the drug, although preliminary results seem to support models which do not require imatinib to kill stem cells [121, 122].

PNH is another acquired genetic disorder of HSC [83]. In this disease, red cells are unusually sensitive to complement attack due to complete or partial deficiency in membrane proteins that protect them from the membrane attack complex of complement. These proteins include CD55, CD59 and CD46. A mutation in a single gene known as *PIG-A*, (phosphatidylinositol glycan anchor biosynthesis, class A) a subunit of an enzyme required for GPI biosynthesis, leading to a block of N-acetylglucosaminyl phosphatidylinositol (GPI) biosynthesis that is required to anchor all these proteins to the plasma membrane of cells. Many mutations in *PIG-A* have been described that alter its transcription and/or activity [83, 123, 124]. The most severe cases are associated with a complete deficiency of *PIG-A* with no expression of these membrane proteins. Many patients with PNH have more than one distinct clone of GPI-anchored proteins which means that at least two cells *independently* acquired different mutations in PIG-A. The cellular origin of these mutations is important, mutations in the HSC are expected to persist for a long time while mutations in progenitor cells would be expected to be transient [60]. Analysis of this problem has to take into consideration the fact that one clone must initially appear and expand to induce the disease phenotype (normally 20% of the neutrophils must be deficient in GPI-linked proteins for the disease to be clinically significant). While the first clone is expanding, a second independent mutation has to appear to produce the second clone. A mathematical analysis of such a scenario shows that it is very unlikely that both mutations occur at the level of the HSCs [109]. It is far more likely that one of the mutations occurs in the progenitor pool, resulting in a smaller clone that also remains detectable for a shorter time interval. This is supported by clinical observation.

The second major issue with PNH is the identification of the mechanism for clonal expansion of *PIG-A* mutated cells. The consensus that has pervaded the longest has been that the mutant cells must have a fitness advantage that allows them to take over haematopoiesis. This concept was provided with some additional support by the observation that every healthy adult human has small numbers of cells with mutations in *PIG-A*, perhaps implying that mutations in *PIG-A* are not enough to explain the disease [125]. Two hypothesis have been proposed that could lead to this scenario: (i) either the cells have an intrinsic advantage perhaps due to a mutation in a second gene or (ii) an immune attack against normal HSCs gives mutant cells an indirect (cell extrinsic) proliferative advantage. Recently, some evidence in support of proposal (i) has been provided by the description of two patients with mutations in the *HMGA2* gene that is often mutated in benign tumors [126]. It was proposed that aberrant expression of this gene provides the necessary proliferative advantage to the mutant cells. However, dynamic considerations suggest that this is unlikely to explain clonal expansion in the majority of patients with this disease since this scenario requires that two independent mutations occur in the same cell, one to mutate *PIG-A* and the second to mutate *HMGA2*. Given that the mutation rate in PNH cells is normal (and therefore very low), it would be very difficult to observe this phenomenon, even more so if more than one clone of mutant cells is present [109, 110]. It can be shown that the *PIG-A* mutated cells present in healthy adults have their origin in progenitor cells that are present in much higher numbers compared to the active HSC pool. This also explains why such clones tend to be transient [125]. There is indirect evidence in support of an immune attack in PNH, presumably directed against the normal HSC [127–129]. However, the disease is not treatable with immunosuppressive therapy, implying that the immune attack hypothesis may not explain clonal expansion either. Hence, to date, a good explanation for clonal expansion in PNH is not available. However, we hope that with these two examples, we have illustrated to the reader the importance of understanding the dynamic behavior of cells in order to make sense of data and generate plausible hypotheses based on mathematical models. Only a quantitative description of the dynamics of haematopoietic cells allows us to determine what is possible or not in haematopoiesis.

FUTURE DEVELOPMENTS
Haematopoietic Stem Cell Plasticity

The relative ease with which HSCs can be harvested, fueled by the expectation of *in vitro* HSC expansion have raised hopes that HSCs may be able to generate cells that belong to other tissues which may or may not be derived from other germ layers (e.g., epithelial cells, neurons, cardiomyocytes, hepatocytes, etc.). This process has been described as cell 'plasticity' or 'transdifferentiation' and considered a violation of one of the central dogmas of developmental biology [130]. There is evidence in nature that such a phenomenon can occur. The example that is most often cited in support of transdifferentiation is the formation of Barrett's esophagus where in response to chronic acid exposure—the lower esophageal epithelium changes from squamous cells to columnar cells—similar to those that line the stomach. However, we hasten to note that this is an example of an epithelial cell changing to a different type of epithelial cell and there is no crossing of germ layers. Some investigators have reported that HSC can give rise to neural cells, hepatocytes, cardiomyocytes and others although skepticism remains. Some leaders in the field of HSC biology are quite pessimistic that HSC can behave in this fashion in a meaningful way, namely that significant number of HSC can differentiate in a specific tissue to regenerate it after injury [131, 132]. It is possible that some of the confusion in the field could be due to semantics and in an attempt to rectify this problem, clear definitions have been proposed [133]. One of the major issues is how should a transdifferentiated cell be defined. Is it enough for the cell to have the shape of surrounding cells, express one or a few tissue specific markers or perform some function(s) typical of the target tissue? Another important question has been whether the number of cells that exhibit plasticity should be taken into consideration; is one cell among millions enough to prove the point? Proponents of HSC plasticity argue that in such a case, it is the exception that proves the rule, while opponents claim that such a rare event may not have any biological relevance. However, recent rigorous experiments suggest that the promise of HSC transdifferentiation is limited and most of the (rare) instances where this was observed were due to fusion of the HSC with the surrounding cells of the injured organ [134, 135]. Perhaps it is the exchange of DNA that leads to change in cell fate rather that intrinsic plasticity of the cells.

Stem Cell Engineering

HSCs are the most accessible stem cell pool available. Hence they are attractive targets for gene therapy approaches to potentially cure many inherited disorders (see Table 21.1). Recent advances in vector design, especially the introduction of retroviral and lentiviral vectors that can integrate in the genome of the target cell seem to hold great promise. One of the success stories in stem cell engineering has been the expression of the gc gene in HSC from patients with X-SCID. This vector transduced ~40% of the haematopoietic cells (with activation) and corrected the defect in a small fraction of cells that expanded in time [69, 136]. Importantly, therapy was associated with an improvement in the lymphocyte profile in 9 out of 10 treated patients with sustained gene expression for over 3 years in some of the children. Unfortunately, 2 of the 10 children treated have developed T cell leukemia due to aberrant expression of the *LMO2* gene [137], that is known to be associated with *de novo* acute leukemia in childhood. Subsequent work showed that the vector had integrated upstream of exon 1 of the gene in one patient and within the first intron in the second patient but in either case the gene was aberrantly expressed. This highlights the potential risk of insertional mutagenesis and the need to better understand the 'selection' of insertion sites by these viruses and perhaps more importantly improved vector design to limit the possibility of such events. In the vector used (MFG), gene expression was under the control of the viral long terminal repeat that must have an enhancer element which led to the high-level expression of the proto-oncogene [137]. Perhaps the use of transcriptionally targeted vectors serves as a further check against the problem of insertional mutagenesis although this problem cannot be eliminated completely. This approach requires though the identification of critical gene regulatory elements where tissue-specific transcription factors bind to enable the recruitment of the transcriptional machinery. To date, very few such regulatory elements have been identified, including those for the globin genes and immunoglobulins. In addition, for even higher safety, gene silencers may be included to restrict the promoter activity residing in the vector to that transgene of interest. Again the best example of a tissue-specific gene silencer is that for the globin cluster. Recent work has shown the feasibility of using critical elements of these sequences to achieve tissue-specific and regulated (balanced) gene expression for the correction of thalassemia syndromes in animal models [138, 139]. The incorporation of such elements in vectors is limited by the size restriction of the genomes that can be efficiently packaged into the viral capsid.

Ex vivo Stem Cell Expansion

Expansion of HSCs *in vitro* has been one of the holy grails of stem cell transplantation since it would open many possibilities for the manipulation of these cells (such as gene correction) and also expand the utility of cord blood derived HSC. However, to date, expansion of HSCs has been next to impossible. The task at hand is not simple since what is required are the necessary conditions that allow the HSCs to expand and retain both self-renewal and differentiation ability. One of the reasons behind the reported experimental failures has been the choice of cytokines and growth factors used to achieve this aim [140] but perhaps even more important are the provision of a more realistic stem cell niche environment. Recently, the generation of pluripotent stem cells from adult human fibroblasts by expression of *Oct3/4*, *Sox2*, *Klf4* and *c-Myc* has been described [141]. We believe that it is only a question of time before the detailed molecular mechanisms behind HSC self-renewal and expansion are defined, allowing efficient expansion of this important pool of cells *in vitro*.

In conclusion, the biology of HSCs and their dynamic behaviour are being deciphered at a fast pace. A complete understanding of HSC behavior requires both a detailed molecular description, as well as, a consideration of the dynamic behavior of these cells. Conceptually, HSC behavior can be rationalized using allometric principles together with stochastic dynamics. The small size of the active HSC pool requires a probabilistic description of these cells and stochastic considerations can have a major impact on dynamics of this cellular compartment. Dynamics also help us understand clinically relevant observations pertaining to various neoplastic and acquired genetic disorders of HSCs. In principle, these considerations should apply to other (non-haematopoietic) stem cell compartments as well.

ACKNOWLEDGEMENTS

Financial support from Mayo Foundation (DD) and FCT-Portugal (JMP) is gratefully acknowledged.

REFERENCES

1. Vaziri H., W. Dragowska, R.C. Allsopp, T.E. Thomas, C.B. Harley, et. al. (1994). 'Evidence for a mitotic clock in human haematopoietic stem cells: Loss of telomeric DNA with age'. *Proc. Natl. Acad. Sci. USA*. **91**: 9857–9860.
2. Cronkite E.P. , T.M. Fliedner (1964). 'Granulocytopoiesis'. *N. Engl. J. Med.* **270**: 1347–1352.
3. Donohue D.M., R.H. Reiff, M.L. Hanson, Y. Betson, C.A. Finch (1958). 'Quantitative measurement of the erythrocytic and granulocytic cells of the marrow and blood'. *J. Clin. Invest.* **37**: 1571–1576.
4. Finch C.A., L.A. Harker, J.D. Cook (1977). 'Kinetics of the formed elements of human blood'. *Blood*. **50**: 699–707.
5. Holyoake T.L., X. Jiang, M.W. Drummond, A.C. Eaves, C.J. Eaves (2002). 'Elucidating critical mechanisms of deregulated stem cell turnover in the chronic phase of chronic myeloid leukemia'. *Leukemia*. **16**: 549–558.
6. Tefferi A. (2003). 'Polycythemia vera: A comprehensive review and clinical recommendations'. *Mayo. Clin. Proc.* **78**: 174–194.
7. McCulloch E.A., J.E. Till (1964). 'Proliferation of hemopoietic colony-forming cells transplanted into irradiated mice. *Radiat. Res.* **22**: 383–397.

8. McCulloch E.A., J.E. Till (2005). 'Perspectives on the properties of stem cells'. *Nat. Med.* **11**: 1026–1028.
9. Appelbaum F.R. (2007). 'Haematopoietic-cell transplantation at 50'. *N. Engl. J. Med.* **357**: 1472–1475.
10. Williams D.A., I.R. Lemischka, D.G. Nathan, R.C. Mulligan (1984). 'Introduction of new genetic material into pluripotent haematopoietic stem cells of the mouse'. *Nature.* **310**: 476–480.
11. Lemischka I.R., D.H. Raulet, R.C. Mulligan (1986). 'Developmental potential and dynamic behavior of haematopoietic stem cells'. *Cell.* **45**: 917–927.
12. Dunbar C.E., M. Cottler-Fox, J.A. O'Shaughnessy, S. Doren, C. Carter, et. al. (1995). 'Retrovirally marked CD34-enriched peripheral blood and bone marrow cells contribute to long-term engraftment after autologous transplantation'. *Blood.* **85**: 3048–3057.
13. Abkowitz J.L., M.L. Linenberger, M.A. Newton, G.H. Shelton, R.L. Ott, et. al. (1990). 'Evidence for the maintenance of haematopoiesis in a large animal by the sequential activation of stem-cell clones'. *Proc. Natl. Acad. Sci. USA.* **87**: 9062–9066.
14. Buescher E.S., D.W. Alling, J.I. Gallin (1985). 'Use of an X-linked human neutrophil marker to estimate timing of lyonisation and size of the dividing stem cell pool'. *J. Clin. Invest.* **76**: 1581–1584.
15. Till J.E., E.A. McCulloch, L. Siminovitch (1964). 'A Stochastic Model of Stem Cell Proliferation, Based on the Growth of Spleen Colony-Forming Cells'. *Proc. Natl. Acad. Sci. USA.* **51**: 29–36.
16. Becker A.J., C.E. Mc, J.E. Till (1963). 'Cytological demonstration of the clonal nature of spleen colonies derived from transplanted mouse marrow cells'. *Nature.* **197**: 452–454.
17. Bramson S.A., R.G. Miller, R.A. Phillips (1977). 'The identification in adult bone marrow of pluripotent and restricted stem cells of the myeloid and lymphoid systems'. *J. Exp. Med.* **145**: 1567–1579.
18. Spangrude G.J., S. Heimfeld, I.L. Weissman (1988). 'Purification and characterisation of mouse haematopoietic stem cells'. *Science.* **241**: 58–62.
19. Uchida N., I.L. Weissman (1992). 'Searching for haematopoietic stem cells: Evidence that Thy-1.1lo Lin- Sca$^{-/+}$ cells are the only stem cells in C57BL/Ka-Thy-1.1 bone marrow'. *J. Exp. Med.* **175**: 175–184.
20. Ikuta K., N. Uchida, J. Friedman, I.L. Weissman (1992). 'Lymphocyte development from stem cells'. *Ann. Rev. Immunol.* **10**: 759–783.
21. Baum C.M., I.L. Weissman, A.S. Tsukamoto, A.M. Buckle, B. Peault (1992). 'Isolation of a candidate human haematopoietic stem-cell population'. *Proc Natl Acad Sci. USA.* **89**: 2804–2808.
22. Osawa M., K. Hanada, H. Hamada, H. Nakauchi (1996). 'Long-term lymphohaematopoietic reconstitution by a single CD34-low/negative haematopoietic stem cell'. *Science.* **273**: 242–245.
23. Bhatia M., D. Bonnet, B. Murdoch, O.I. Gan, J.E. Dick (1998). 'A newly discovered class of human haematopoietic cells with SCID-repopulating activity'. *Nat. Med.* **4**: 1038–1045.
24. Goodell M.A., M. Rosenzweig, H. Kim, D.F. Marks, M. DeMaria, et. al. (1997). 'Dye efflux studies suggest that haematopoietic stem cells expressing low or undetectable levels of CD34 antigen exist in multiple species'. *Nat. Med.* **3**: 1337–1345.
25. Zanjani E.D., G. Almeida-Porada, A.G. Livingston, A.W. Flake, M. Ogawa (1998). 'Human bone marrow CD34- cells engraft *in vivo* and undergo multilineage expression that includes giving rise to CD34+ cells'. *Exp. Haematol.* **26**: 353–360.

26. Gordon M.Y., N.M. Blackett (1994). 'Routes to repopulation—a unification of the stochastic model and separation of stem-cell subpopulations'. *Leukemia.* **8**: 1068–1072; *discussion.* 1072–1063.
27. Abkowitz J.L., S.N. Catlin, P. Guttorp (1996). 'Evidence that haematopoiesis may be a stochastic process in vivo'. *Nat. Med.* **2**: 190–197.
28. Spangrude G.J., L. Smith, N. Uchida, K. Ikuta, S. Heimfeld, et. al. (1991). 'Mouse haematopoietic stem cells'. *Blood.* **78**: 1395–1402.
29. Allsopp R.C., S. Cheshier, I.L. Weissman (2001). 'Telomere shortening accompanies increased cell cycle activity during serial transplantation of haematopoietic stem cells'. *J Exp. Med.* **193**: 917–924.
30. Abkowitz J.L., M.T. Persik, G.H. Shelton, R.L. Ott, J.V. Kiklevich, et. al. (1995). 'Behavior of haematopoietic stem cells in a large animal'. *Proc. Natl. Acad. Sci. USA.* **92**: 2031–2035.
31. Shepherd B.E., H.P. Kiem, P.M. Lansdorp, C.E. Dunbar, G. Aubert, et. al. (2007). 'Haematopoietic stem cell behavior in non-human primates'. *Blood.* **110**: 1806–1813.
32. Moore K.A., I.R. Lemischka, (2006). 'Stem cells and their niches'. *Science.* **311**: 1880–1885.
33. McKenzie J.L., O.I. Gan, M. Doedens, J.C. Wang, J.E. Dick (2006). 'Individual stem cells with highly variable proliferation and self-renewal properties comprise the human haematopoietic stem cell compartment'. *Nat. Immunol.* **7**: 1225–1233.
34. Heissig B., K. Hattori, S. Dias, M. Friedrich, B. Ferris, et. al. (2002). 'Recruitment of stem and progenitor cells from the bone marrow niche requires MMP-9 mediated release of kit-ligand'. *Cell.* **109**: 625–637.
35. Petit I., M. Szyper-Kravitz, A. Nagler, M. Lahav, A. Peled, et. al. (2002). 'G-CSF induces stem cell mobilisation by decreasing bone marrow SDF-1 and up-regulating CXCR4'. *Nat. Immunol.* **3**: 687–694.
36. Mohle R., L. Kanz (2007). 'Haematopoietic growth factors for haematopoietic stem cell mobilisation and expansion'. *Semin. Haematol.* **44**: 193–202.
37. Gluckman E., A. Devergie, H. Bourdeau-Esperou, D. Thierry, R. Traineau, et. al. (1990). 'Transplantation of umbilical cord blood in Fanconi's anemia'. *Nouv. Rev. Fr. Haematol.* **32**: 423–425.
38. Haspel R.L., K.K. Ballen (2006). 'Double cord blood transplants: Filling a niche'? *Stem Cell. Rev.* **2**: 81–86.
39. Thomson B.G., K.A. Robertson, D. Gowan, D. Heilman, H.E. Broxmeyer, et. al. (2000). 'Analysis of engraftment, graft-versus-host disease, and immune recovery following unrelated donor cord blood transplantation'. *Blood.* **96**: 2703–2711.
40. Eapen M., P. Rubinstein, M.J. Zhang, C. Stevens, J. Kurtzberg, et. al. (2007). 'Outcomes of transplantation of unrelated donor umbilical cord blood and bone marrow in children with acute leukaemia: A comparison study'. *Lancet.* **369**: 1947–1954.
41. Schofield R. (1978). 'The relationship between the spleen colony-forming cell and the haemopoietic stem cell'. *Blood Cells.* **4**: 7–25.
42. Spradling A., D. Drummond-Barbosa, T. Kai (2001). 'Stem cells find their niche'. *Nature.* **414**: 98–104.
43. Fuchs E., T. Tumbar, G. Guasch (2004). 'Socializing with the neighbors: stem cells and their niche'. *Cell.* **116**: 769–778.
44. Li L., T. Xie (2005). 'Stem cell niche: Structure and function'. *Ann. Rev Cell Dev. Biol.* **21**: 605–631.
45. Scadden D.T. (2006). 'The stem-cell niche as an entity of action'. *Nature.* **441**: 1075–1079.

46. Zhang J., C. Niu, L. Ye, H. Huang, X. He, et. al. (2003). 'Identification of the haematopoietic stem cell niche and control of the niche size'. *Nature*. **425**: 836–841.
47. Calvi L.M, G.B. Adams, K.W. Weibrecht, J.M. Weber, D.P. Olson, et. al. (2003). 'Osteoblastic cells regulate the haematopoietic stem cell niche'. *Nature*. **425**: 841–846.
48. Adams G.B., D.T. Scadden (2006). 'The haematopoietic stem cell in its place'. *Nat Immunol.* **7**: 333–337.
49. Stier S., Y. Ko, R. Forkert, C. Lutz, T. Neuhaus, et. al. (2005). 'Osteopontin is a haematopoietic stem cell niche component that negatively regulates stem cell pool size'. *J. Exp. Med.* **201**: 1781–1791.
50. Avecilla S.T., K. Hattori, B. Heissig, R. Tejada, F. Liao, et. al. (2004). 'Chemokine-mediated interaction of haematopoietic progenitors with the bone marrow vascular niche is required for thrombopoiesis'. *Nat. Med.* **10**: 64–71.
51. Sipkins D.A., X. Wei, J.W. Wu, J.M. Runnels, D. Cote, et. al. (2005). 'In vivo imaging of specialized bone marrow endothelial microdomains for tumour engraftment'. *Nature*. **435**: 969–973.
52. Kopp H.G., S.T. Avecilla, A.T. Hooper, S. Rafii (2005). 'The bone marrow vascular niche: Home of HSC differentiation and mobilisation'. *Physiology. (Bethesda).* **20**: 349–356.
53. Li F., S.J. Lu, G.R. Honig (2006). 'Haematopoietic cells from primate embryonic stem cells'. *Methods Enzymol.* **418**: 243–251.
54. Arai F., A. Hirao, M. Ohmura, H. Sato, S. Matsuoka, et. al. (2004). 'Tie2/angiopoietin-1 signaling regulates haematopoietic stem cell quiescence in the bone marrow niche'. *Cell.* **118**: 149–161.
55. Mendez-Ferrer S., D. Lucas, M. Battista, P.S. Frenette (2008). 'Haematopoietic stem cell release is regulated by circadian oscillations'. *Nature*. **452**: 442–447.
56. Abkowitz J.L., S.N. Catlin, M.T. McCallie, P. Guttorp (2002). 'Evidence that the number of haematopoietic stem cells per animal is conserved in mammals'. *Blood.* **100**: 2665–2667.
57. McCarthy K.F. (1997). 'Population size and radiosensitivity of murine haematopoietic endogenous long-term repopulating cells'. *Blood.* **89**: 834–841.
58. McCarthy K.F. (2003) 'Marrow frequency of rat long-term repopulating cells: Evidence that marrow haematopoietic stem cell concentration may be inversely proportional to species body weight'. *Blood.* **101**: 3431–3435.
59. Gordon M.Y., J.L. Lewis, S.B. Marley (2002). 'Of mice and men…and elephants'. *Blood.* 100: 4679–4680.
60. Dingli D., A. Traulsen, J.M. Pacheco (2007). 'Compartmental architecture and dynamics of haematopoiesis'. *PLoS ONE.* **2**: e345.
61. Dingli D., A. Traulsen, J.M. Pacheco (2008). 'Haematopoietic stem cell behavior across mammals'. *Proc. R. Soc.* B 22. **275**(1649): 2389–2392.
62. Kay M.M., T. Makinodan (1976). 'Immunobiology of aging: Evaluation of current status'. *Clin Immunol. Immunopathol.* **6**: 394–413.
63. Huxley J.S. (1932). 'Problems of relative growth'. *Dial Press, New York, USA.*
64. West G.B., J.H. Brown, B.J. Enquist (1999). 'The fourth dimension of life: Fractal geometry and allometric scaling of organisms'. *Science.* **284**: 1677–1679.
65. West G.B., J.H. Brown (2005). 'The origin of allometric scaling laws in biology from genomes to ecosystems: Towards a quantitative unifying theory of biological structure and organisation'. *J. Exp. Biol.* **208**: 1575–1592.

66. Banavar J.R., A. Maritan, A. Rinaldo (1999). 'Size and form in efficient transportation networks'. *Nature.* **399**: 130–132.
67. Dingli D., J.M. Pacheco (2006). 'Allometric scaling of the active haematopoietic stem cell pool across mammals'. *PLoS ONE.* **1**: e2.
68. Nash R., R. Storb, P. Neiman (1988). 'Polyclonal reconstitution of human marrow after allogeneic bone marrow transplantation'. *Blood.* **72**: 2031–2037.
69. Calvo-M Cavazzana, S. Hacein-Bey, G. de Saint Basile, F. Gross, E. Yvon, et. al. (2000). 'Gene therapy of human severe combined immunodeficiency (SCID)-X1 disease'. *Science.* **288**: 669–672.
70. Dingli D., J.M. Pacheco (2007). 'Ontogenic growth of the haemopoietic stem cell pool in humans'. *Proc. Biol Sci.* **274**: 2497–2501.
71. Morrison S.J., A.M. Wandycz, K. Akashi, A. Globerson, I.L. Weissman (1996). 'The aging of haematopoietic stem cells'. *Nat. Med.* **2**: 1011–1016.
72. Xing Z., M.A. Ryan, D. Daria, K.J. Nattamai, G. Van Zant, et. al. (2006). 'Increased haematopoietic stem cell mobilisation in aged mice'. *Blood.* **108**: 2190–2197.
73. Suzuya H., T. Watanabe, R. Nakagawa, H. Watanabe, Y. Okamoto, et. al. (2005). 'Factors associated with granulocyte colony-stimulating factor-induced peripheral blood stem cell yield in healthy donors'. *Vox. Sang.* **89**: 229–235.
74. Rossi D.J., C.H. Jamieson, I.L. Weissman (2008). 'Stems cells and the pathways to aging and cancer'. *Cell.* **132**: 681–696.
75. Kunkel T.A., K. Bebenek (2000). 'DNA replication fidelity'. *Ann. Rev. Biochem.* **69**: 497–529.
76. Vulliamy T.J., I. Dokal (2008). 'Dyskeratosis congenita: The diverse clinical presentation of mutations in the telomerase complex'. *Biochimie.* **90**: 122–130.
77. Rossi D.J., D. Bryder, J. Seita, A. Nussenzweig, J. Hoeijmakers, et. al. (2007). 'Deficiencies in DNA damage repair limit the function of haematopoietic stem cells with age'. *Nature.* **447**: 725–729.
78. Nijnik A., L. Woodbine, C. Marchetti, S. Dawson, T. Lambe, et. al. (2007). 'DNA repair is limiting for haematopoietic stem cells during ageing'. *Nature.* **447**: 686–690.
79. Kiel M.J., S. He, R. Ashkenazi, S.N. Gentry, M. Teta, et. al. (2007). 'Haematopoietic stem cells do not asymmetrically segregate chromosomes or retain BrdU'. *Nature.* **449**: 238–242.
80. Rufer N., T.H. Brummendorf, S. Kolvraa, C. Bischoff, K. Christensen, et. al. (1999). 'Telomere fluorescence measurements in granulocytes and T lymphocyte subsets point to a high turnover of haematopoietic stem cells and memory T cells in early childhood'. *J. Exp. Med.* **190**: 157–167.
81. Zhou S., J.D. Schuetz, K.D. Bunting, A.M. Colapietro, J. Sampath, et. al. (2001). 'The ABC transporter Bcrp1/ABCG2 is expressed in a wide variety of stem cells and is a molecular determinant of the side-population phenotype'. *Nat. Med.* **7**: 1028–1034.
82. Goldman J.M., (2004). 'Chronic myeloid leukemia-still a few questions'. *Exp Haematol.* 32: 2–10.
83. Luzzatto L., M. Bessler, B. Rotoli (1997). 'Somatic mutations in paroxysmal nocturnal hemoglobinuria: A blessing in disguise'? *Cell.* **88**: 1–4.
84. Dingli D., A. Traulsen, J.M. Pacheco (2007). 'Stochastic dynamics of haematopoietic tumor stem cells'. *Cell. Cycle.* **6**: 441–446.
85. Lemischka I.R. (1997). 'Microenvironmental regulation of haematopoietic stem cells'. *Stem Cells.* **15**(Suppl 1): 63–68.
86. Moran P. (1962). 'The statistical processes of evolutionary theory'. *Clarendon Press.* Oxford.

87. Araten D.J., D.W. Golde, R.H. Zhang, H.T. Thaler, L. Gargiulo, et. al. (2005). 'A quantitative measurement of the human somatic mutation rate'. *Cancer Res.* **65**: 8111–8117.
88. Jaffe E.S., N.L. Harris, H. Stein, J.W. Vardiman, (2001). 'Tumours of haematopoietic and lymphoid tissues'. *World Health Organisation*: 77–80.
89. Tomlinson I., P. Sasieni, W. Bodmer (2002). 'How many mutations in a cancer?' *Am. J. Pathol.* **160**: 755–758.
90. Beerenwinkel N., T. Antal, D. Dingli, A. Traulsen, K.W. Kinzle, V.E. Velculescu, B. Vogelstein, M.A. Nowak, (2007). 'Genetic progression and the waiting time to cancer'. *PLoS Computational Biology.* **3**: e225.
91. Pear W.S., J.P. Miller, L. Xu, J.C. Pui, B. Soffer, et. al. (1998). 'Efficient and rapid induction of a chronic myelogenous leukemia-like myeloproliferative disease in mice receiving P210 bcr/abl-transduced bone marrow'. *Blood.* **92**: 3780–3792.
92. Koschmieder S., B. Gottgens, P. Zhang, J. Iwasaki-Arai, K. Akashi, et. al. (2005). 'Inducible chronic phase of myeloid leukemia with expansion of haematopoietic stem cells in a transgenic model of BCR-ABL leukemogenesis'. *Blood.* **105**: 324–334.
93. Michor F., Y. Iwasa, M.A. Nowak (2006). 'The age incidence of chronic myeloid leukemia can be explained by a one-mutation model'. *Proc. Natl. Acad. Sci. USA.* **103**: 14931–14934.
94. Lange B. (2000). 'The management of neoplastic disorders of haematopoiesis in children with Down's syndrome'. *Br. J. Haematol.* **110**: 512–524.
95. Massey G.V., A. Zipursky, M.N. Chang, J.J. Doyle, S. Nasim, et. al. (2006). 'A prospective study of the natural history of transient leukemia (TL) in neonates with Down syndrome (DS): Children's Oncology Group (COG) study POG-9481'. *Blood.* **107**: 4606–4613.
96. Zipursky A., E.J. Brown, H. Christensen, J. Doyle (1999). 'Transient myeloproliferative disorder (transient leukemia) and haematologic manifestations of Down syndrome'. *Clin. Lab. Med.* **19**: 157–167, vii.
97. Tricot G., C. Mecucci, H. Van den Berghe (1986). 'Evolution of the myelodysplastic syndromes'. *Br. J. Haematol.* **63**: 609–614.
98. Gale R.E., A.J. Allen, M.J. Nash, D.C. Linch (2006). 'Long-term serial analysis of X-chromosome inactivation patterns and JAK2 V617F mutant levels in patients with essential thrombocythemia show that minor mutant-positive clones can remain stable for many years'. *Blood.* **109**: 1241–1243.
99. Morrison S.J., J. Kimble (2006). 'Asymmetric and symmetric stem-cell divisions in development and cancer'. *Nature.* **441**: 1068–1074.
100. Yatabe Y., S. Tavare, D. Shibata (2001). 'Investigating stem cells in human colon by using methylation patterns'. *Proc. Nat. Acad. Sci. USA.* **98**: 10839–10844.
101. Reya T., A.W. Duncan, L. Ailles, J. Domen, D.C. Scherer, et. al. (2003). 'A role for Wnt signalling in self-renewal of haematopoietic stem cells'. *Nature.* **423**: 409–414.
102. Reya T. (2003). 'Regulation of haematopoietic stem cell self-renewal'. *Recent. Prog. Horm. Res.* **58**: 283–295.
103. Reya T., H. Clevers (2005). 'Wnt signalling in stem cells and cancer'. *Nature.* **434**: 843–850.
104. Albertson R., C.Q. Doe (2003). 'Dlg, Scrib and Lgl regulate neuroblast cell size and mitotic spindle asymmetry'. *Nat. Cell. Biol.* **5**: 166–170.
105. Lee C.Y., K.J. Robinson, C.Q. Doe (2006). 'Lgl, Pins and aPKC regulate neuroblast self-renewal versus differentiation'. *Nature.* **439**: 594–598.

106. Dingli D., A. Traulsen, F. Michor (2007). '(A)symmetric stem cell replication and cancer'. *PLoS Comput. Biol.* **3**: e53.
107. Jamieson C.H., L.E. Ailles, S.J. Dylla, M. Muijtjens, C. Jones, et. al. (2004). 'Granulocyte-macrophage progenitors as candidate leukemic stem cells in blast-crisis CML'. *N. Engl. J. Med.* **351**: 657–667.
108. Krivtsov A.V., D. Twomey, Z. Feng, M.C. Stubbs, Y. Wang, et. al. (2006). 'Transformation from committed progenitor to leukaemia stem cell initiated by MLL-AF9'. *Nature.* **442**: 818–822.
109. Traulsen A., J.M. Pacheco, D. Dingli (2007). 'On the Origin of Multiple Mutant Clones in Paroxysmal Nocturnal Hemoglobinuria'. *Stem. Cells.* **25**(12): 3081–3084.
110. Dingli D., J.M. Pacheco, A. Traulsen (2008) 'Multiple mutant clones in blood rarely coexist'. *Phys. Rev. E. Stat. Nonlin. Soft. Matter. Phys.* **77**(2 Pt 1): 021915.
111. Reya T., S.J. Morrison, M.F. Clarke, Weissman IL (2001). 'Stem cells, cancer, and cancer stem cells'. *Nature.* **414**: 105–111.
112. Schmidt-Nielsen K. (1984). 'Why is animal size so important'? Cambridge University Press, New York, *USA*.
113. Lopes J.V., J.M. Pacheco, D. Dingli (2007). 'Acquired haematopoietic stem cell disorders and mammalian size'. *Blood.* **110**: 4137–4139.
114. Squire R.A. (1969). 'Spontaneous haematopoietic tumors in dogs'. *National Cancer Institute Monographs.* **32**: 97–116.
115. Otsu M., F. Candotti (2002). 'Gene therapy in infants with severe combined immunodeficiency'. *BioDrugs.* **16**: 229–239.
116. Primo D., J. Flores, S. Quijano, M.L. Sanchez, M.E. Sarasquete, et. al. (2006). 'Impact of BCR/ABL gene expression on the proliferative rate of different subpopulations of haematopoietic cells in chronic myeloid leukaemia'. *Br. J. Haematol.* **135**: 43–51.
117. Marley S.B., M.W. Deininger, R.J. Davidson, J.M. Goldman, M.Y. Gordon (2000). 'The tyrosine kinase inhibitor STI571, like interferon-alpha, preferentially reduces the capacity for amplification of granulocyte-macrophage progenitors from patients with chronic myeloid leukemia'. *Exp. Haematol.* **28**: 551–557.
118. Dingli D., A. Traulsen, J.M. Pacheco (2008). 'Chronic myeloid leukemia: Origin, development, response to therapy and relapse'. *Clin. Leukemia.* **2**: 133–139.
119. Michor F., T.P. Hughes, Y. Iwasa, S. Branford, N.P. Shah, et. al. (2005). 'Dynamics of chronic myeloid leukaemia'. *Nature.* **435**: 1267–1270.
120. Roeder I., M. Horn, I. Glauche, A. Hochhaus, M.C. Mueller, et. al. (2006). 'Dynamic modeling of imatinib-treated chronic myeloid leukemia: Functional insights and clinical implications'. *Nat. Med.* **12**: 1181–1184.
121. Savona M., M. Talpaz (2008). 'Getting to the stem of chronic myeloid leukaemia'. *Nat. Rev. Cancer.* **8**: 341–350.
122. Heaney N., M. Drummond, J. Kaeda, F. Nicolini, R. Clark, et. al. (2007). 'A phase 3 study of continuous imatinib versus pulsed imatinib with or without G-CSF in patients with chronic phase CML who have achieved a complete cytogenetic response to imatinib'. *Blood.* **110**: 313A (Abstract 1033).
123. Bessler M., P. Mason, P. Hillmen, L. Luzzatto (1994). 'Somatic mutations and cellular selection in paroxysmal nocturnal haemoglobinuria'. *Lancet.* **343**: 951–953.

124. Bessler M., P.J. Mason, P. Hillmen, L. Luzzatto (1994). 'Mutations in the PIG-A gene causing partial deficiency of GPI-linked surface proteins (PNH II) in patients with paroxysmal nocturnal haemoglobinuria'. *Br. J. Haematol.* **87**: 863–866.

125. Araten D.J., K. Nafa, K. Pakdeesuwan, L. Luzzatto (1999). 'Clonal populations of haematopoietic cells with paroxysmal nocturnal hemoglobinuria genotype and phenotype are present in normal individuals'. *Proc. Natl. Acad. Sci. USA.* **96**: 5209–5214.

126. Inoue N., T. Izui-Sarumaru, Y. Murakami, Y. Endo, J.I. Nishimura, et. al. (2006). 'Molecular basis of clonal expansion of haematopoiesis in two patients with paroxysmal nocturnal hemoglobinuria (PNH)'. *Blood.* **108**(13): 4232–4236.

127. Karadimitris A., K. Li, R. Notaro, D.J. Araten, K. Nafa, et. al. (2001). 'Association of clonal T-cell large granular lymphocyte disease and paroxysmal nocturnal haemoglobinuria (PNH): Further evidence for a pathogenetic link between T cells, aplastic anaemia and PNH'. *Br. J. Haematol.* **115**: 1010–1014.

128. Karadimitris A., J.S. Manavalan, H.T. Thaler, R. Notaro, D.J. Araten, et. al. (2000). 'Abnormal T-cell repertoire is consistent with immune process underlying the pathogenesis of paroxysmal nocturnal hemoglobinuria'. *Blood.* **96**: 2613–2620.

129. Karadimitris A., R. Notaro, G. Koehne, I.A. Roberts, L. Luzzatto (2000). 'PNH cells are as sensitive to T-cell-mediated lysis as their normal counterparts: Implications for the pathogenesis of paroxysmal nocturnal haemoglobinuria'. *Br. J. Haematol.* **111**: 1158–1163.

130. Gilbert S.F., S. Sarkar (2000). 'Embracing complexity: Organicism for the 21st century'. *Dev. Dyn.* **219**: 1–9.

131. Wagers A.J., R.I. Sherwood, J.L. Christensen, I.L. Weissman (2002). 'Little evidence for developmental plasticity of adult haematopoietic stem cells'. *Science.* **297**: 2256–2259.

132. Lemischka I. (2002). 'A few thoughts about the plasticity of stem cells'. *Exp Haematol.* **30**: 848–852.

133. Moore B.E., P.J. Quesenberry (2003) 'The adult hemopoietic stem cell plasticity debate: Idols vs new paradigms'. *Leukemia.* **17**: 1205–1210.

134. Nygren J.M., S. Jovinge, M. Breitbach, P. Sawen, W. Roll, et. al. (2004). 'Bone marrow-derived haematopoietic cells generate cardiomyocytes at a low frequency through cell fusion, but not transdifferentiation'. *Nat. Med.* **10**: 494–501.

135. Vieyra D.S., K.A. Jackson, M.A. Goodell (2005). 'Plasticity and tissue regenerative potential of bone marrow-derived cells'. *Stem Cell Rev.* **1**: 65–69.

136. Hacein-Bey-Abina S., F. Le Deist, F. Carlier, C. Bouneaud, C. Hue, et. al. (2002). 'Sustained correction of X-linked severe combined immunodeficiency by ex vivo gene therapy'. *N. Engl. J. Med.* **346**: 1185–1193.

137. Hacein-Bey-Abina S., C. Von Kalle, M. Schmidt, M.P. McCormack, N. Wulffraat, et. al. (2003). 'LMO2-associated clonal T cell proliferation in two patients after gene therapy for SCID-X1'. *Science.* **302**: 415–419.

138. May C., S. Rivella, J. Callegari, G. Heller, K.M. Gaensler, et. al. (2000). 'Therapeutic haemoglobin synthesis in beta-thalassaemic mice expressing lentivirus-encoded human beta-globin'. *Nature.* **406**: 82–86.

139. Lisowski L., M. Sadelain (2007). 'Locus control region elements HS1 and HS4 enhance the therapeutic efficacy of globin gene transfer in beta-thalassemic mice'. *Blood.* **110**: 4175–4178.

140. Zubler R.H. (2006). 'Ex vivo expansion of haematopoietic stem cells and gene therapy development'. *Swiss. Med. Wkly.* **136**: 795–799.
141. Takahashi K., K. Tanabe, M. Ohnuki, M. Narita, T. Ichisaka, et. al. (2007). 'Induction of pluripotent stem cells from adult human fibroblasts by defined factors'. *Cell.* **131**: 861–872.

22
SPERMATOGONIAL STEM CELLS

Mahmoud Huleihel[*,1,2], *Mahmoud AbuElhija*[1,2], *Eitan Lunenfeld*[2,3]

[1]*The Shraga Segal Department of Microbiology and Immunology, Beer-Sheva, Israel;*
[2]*Faculty of Health Sciences, Ben-Gurion University of the Negev. Beer-Sheva, Israel;*
[3]*Department of Obstetrics and Gynecology, Soroka University Medical Centre, Beer-Sheva, Israel*

INTRODUCTION

Spermatogenesis is a complicated process regulated mainly by endocrine factors, as well as by testicular paracrine/autocrine growth factors. These factors are produced by SCs, differentiated and undifferentiated spermatogonial stem cells (SSCs), peritubular cells and interstitial cells, mainly Leydig cells and macrophages. In order to culture SSCs *in vitro*, researchers attempted to overcome a number of obstacles—such as the low number of stem cells in the testis, absence of specific markers to identify SSCs—in addition to difficulties in keeping the SSCs alive in culture. Recently, some growth factors which are important for the proliferation and differentiation of SSCs have been identified, such as glial cell line derived neurotrophic factor (GDNF), stem cell factor (SCF) and leukemia inhibitory factor (LIF); also, markers for SSCs at different stages have been reported. Therefore, some groups have succeeded in culturing SSCs (under limitations) or more differentiated SSCs, and have even been able to produce in vitro germ cells from embryonic stem cells.

Thus, we propose that success in culturing SSCs is dependent on understanding the molecular mechanisms behind self-renewal and differentiation. Culture of SSCs should be a good tool for discovering new therapeutic avenues for some infertile men or for patients undergoing chemotherapy/radiotherapy (pre-puberty or post-puberty).

---

* huleihel@bgu.ac.il

BASIC PRINCIPLES AND JARGONS RELATED TO SPERMATOGONIAL STEM CELLS

Origin and Location of Spermatogonial Stem Cells in the Testis

Spermatogenesis is a highly organised process, which contains complicated steps of cell proliferation and differentiation [1, 2]. In male mammals, spermatogenesis is active throughout the reproductive lifetime of the animal. It begins with the asymmetric division within the spermatogonial stem cells (SSCs) that produces cell differentiation; primary spermatocytes, secondary spermatocytes, spermatids, round spermatids and spermatozoa [3, 4].

Before delivery, the germ-line passes through several developmental stages ending with SSCs [5, 6] (Fig. 22.1). After fertilisation, the blastocyst develops primordial germ cells (PGCs). Progenitors in mice have been recognised in the proximal epiblast [7]. These constitute the majority of cells within the testis; PGCs start proliferation in the original site and migrate to the future position of the gonad, wherein, they start the interaction with the somatic gonadal cells to give the form of the gonad [7] (Fig. 22.1A).

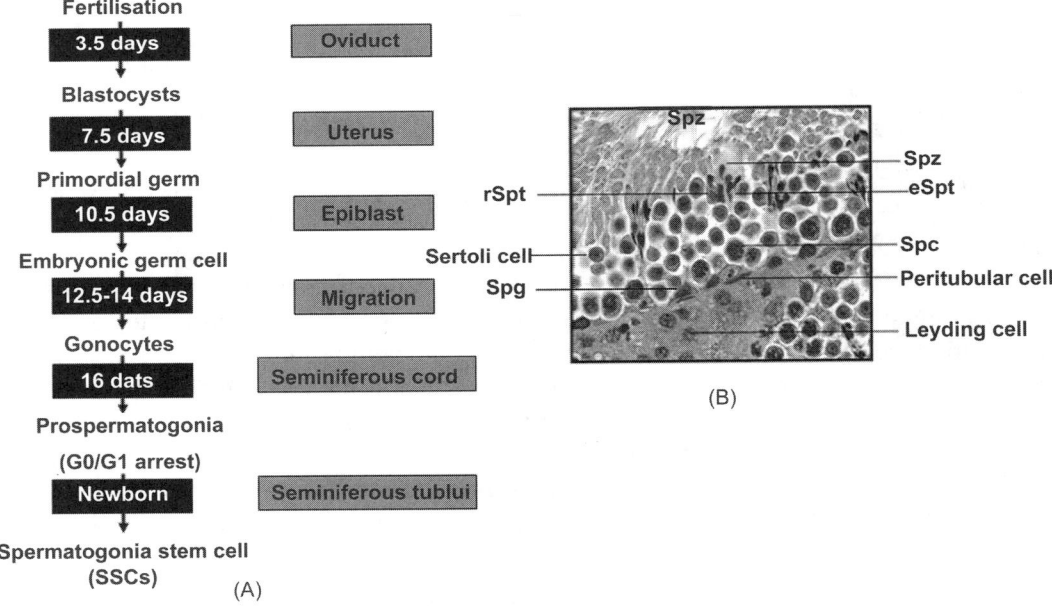

▶ **Fig. 22.1** *Development stages and localisation of mouse spermatogonial stem cells. (A) After fertilisation and blastocyst development, primordial germ cells are formed and continue to develop toward SSCs in the newborn offspring. (B) The development stages of SSCs in the seminiferous tubule are depicted. Leydig cells are located in the interstitial tissue of the testis. SCs and SCCs (Spg; attached to the basement membrane), spermatocytes (Spc), round spermatids (rSpt), elongated spermatids (eSpt) and spermatozoa (Spz) are located in the seminiferous tubules (B). (for colour figure see Plate57).*

In the gonad, PGCs become associated with the somatic cells that support the progress of PGCs' differentiation. These somatic cells are Sertoli cells (SCs). Together with PGCs, SCs construct the seminiferous cords. Later, these cords develop and become the seminiferous tubules. When the PGCs are enclosed in seminiferous cords, they change morphologically and are called gonocytes. After this stage, the cells rest in the G0/G1 phase of the cell cycle. In rodents (rats and mice), this process continues after birth and the gonocytes develop into spermatogonia stem cell [8] (Fig. 22.1A).

The processes of spermatogenesis take place in the seminiferous tubules (Fig 22.1B). This process continues in a spatial gradient within the testis allowing SSCs activity to be easily visualized as a continuous process of differentiating SSCs. The SSCs are located in the seminiferous tubules attached to the basement membrane (Fig. 22.1B). The process of spermatogenesis starts with SSCs that proliferate and differentiate in successive waves along the length of the tubules [9] (Fig. 22.1B). Spermatogonia stem cells differentiate until sperm formation and then are released into the central lumen of the seminiferous tubules.

Spermatogonial multiplication and stem cell renewal can best be studied in mounts of whole seminiferous tubules, because in this way, the topographical arrangement of the cells is preserved [10, 11].

TESTICULAR COMPARTMENT AND CELLULAR INTERACTION

The seminiferous tubules contain germ cells and SCs and peritubular myoid cells surround the tubules and are in contact with the basal surface of germ cells and SCs. Peritubular cells are smooth muscle cells that support the process of spermatogenesis by pushing the differentiated cell into the central lumen. Between the seminiferous tubules, in the interstitium, Leydig cells (LCs) and macrophages are located [12].

Cell–cell interactions are occurring between these cells and support the essential regulation for the process of spermatogenesis [12, 13].

It has been suggested that macrophages, in addition to their immune function, have a critical role in the cell–cell interaction. The products of macrophages (cytokines) after stimulation by infection or inflammation are increased and could down regulate spermatogenesis [12, 13].

LCs are located adjacent to the seminiferous tubules in the interstitium. The major functions of LCs are to secrete testosterone. LCs release a class of hormones called androgens (19-carbon steroids). They secrete testosterone following stimulation by the pituitary hormone, luteinizing hormone (LH) [14]. Follicle-stimulating hormone (FSH) increases the response of LCs to LH by increasing the number of LH receptors expressed on LCs [14].

SCs play an essential role in the spermatogenesis process. They provide a structural support and nutrition to developing germ cell. In addition, SCs have the phagocytosis ability of degenerating germ cells and residual bodies [15–24].

SCs are involved in the spermiation and production of host proteins that regulate the secretion of hormones that influence mitotic activity of spermatogonia [15–24].

The structure of SCs supports the development of germ cells which are located between two SCs. The structures of tight junction between the adjacent SCs are very special arrangement that create an immunologic barrier; the blood–testis barrier. This barrier is very critical because the immune system identifies differentiated germ cells, such as antigens, and stimulates the autoimmunity response [25, 26].

One of the mechanisms to regulate the number of SSCs is the relationship between the number of SCs and SSCs. The number of SCs is in relation to the magnitude of spermatogenesis. In humans, the ratio between SCs and daily sperm production per testis is 0.39; the number of SCs could be correlated also with the testicular weight (in horse the ratio is 0.68) [27–29]. Many studies have shown that the total SC number per testis is very important in determining the total daily sperm production per testis [28, 30, 31].

The interaction between SCs and germ cells play an essential role in the successful development of germ cells [1]. SCs regulate germ cells by limiting the expansion of germ cells population, and it is assumed that each SC supports a defined number of SSCs and germ cells [32].

PROLIFERATION AND DIFFERENTIATION OF SSCs

Spermatogonia are considered to be stem cells of spermatogenesis, and they are the only stem cells in the body that can be recognized and studied at the cellular level, during proliferation, differentiation and regulation of these two activities [33, 34]. The developments of SSCs are different in species of mammals; spermatogonia are classified into several subtypes according to the differentiation status. In non-primate mammals, the SSCs are A-single (As) [35–38]. As spermatogonia are single cells, they can self-renew into two new stem cells or differentiate to another type of spermatogonia. A-paired (Apr) spermatogonia are produced from the As and still connect with the As by an intercellular bridge. From this stage, the Apr spermatogonia are predestined to develop further to divide into chains of four A-aligned (Aal) spermatogonia. Aal spermatogonia can be divided further into chains of 8, 16 and 32 cells (Fig. 22.2).

The Apr and Aal spermatogonia are clonally expanded colonies which are not synchronized with the seminiferous epithelial; however, when Aal divides further into A1–A4 spermatogonia, the expansions of these cells will be synchronized with the seminiferous epithelial cycle. A spermatogonia (A1–A4) further develop into B spermatogonia. Finally type A intermediate and B spermatogonia differentiate into primary spermatocyte [36, 39–45]. The subtypes of SSCs could be differentiated based primarily on mottling of heterochromatin; in As, Apr and Aal spermatogonia the mottling of heterochromatin is throughout the nucleus in the absence of heterochromatin lining the nuclear envelope. The A1 spermatogonia display finely granular chromatin throughout the nucleus and have virtually no flakes of heterochromatin along the nuclear membrane. The A2–A4 spermatogonia contain more heterochromatin rimming the nucleus. Intermediate type of spermatogonia display flaky heterochromatin that completely rims the nucleus. Type B spermatogonia show rounded heterochromatin periodically along the nuclear envelope [46].

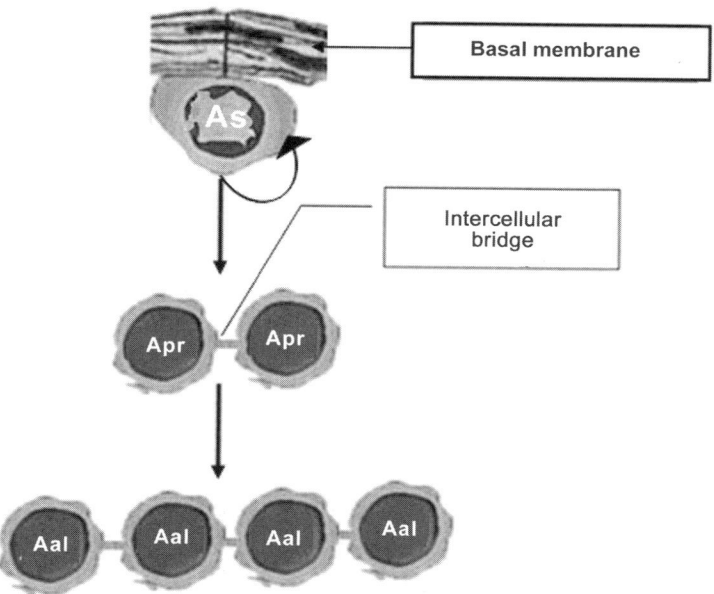

▶ **Fig. 22.2** *Intercellular bridges connecting A spermatogonia in the seminiferous tubule. A-single spermatogonia are single cells that can self-renew into two new stem cells or differentiate to another type of spermatogonia. A-paired (Apr) spermatogonia produce from the As and still connect with the As by an intercellular bridges. From this stage the Apr spermatogonia are predestined to develop further to divide into chains of four A-aligned (Aal) spermatogonia.*

All differentiating germ cells are derived clonally and directed from a single testicular stem cell. However, the large number of subsequent mitotic steps supplies rodents with a highly efficient germ cell generating system [44, 47, 48].

There are two differentiation steps that seem to take place in the developmental path of spermatogonia. First, there is step from the As spermatogonia to the Apr spermatogonia. After this step, i.e. the Apr spermatogonia onwards, all divisions are such that the daughter cells remain connected by intracellular bridges (Fig. 22.2). The second differentiation step is that from Aal to A1 spermatogonia, and this step brings about marked changes in cell behaviour. When A1–B spermatogonia are unable to divide at the appropriate time, they enter apoptosis [10].

In macaques and men, two morphologically distinguishable type of spermatogonia exist, the A (dark) and the A (pale) spermatogonia. Although both are commonly referred as SSCs, their biological functions are very different and the A (dark) shows characteristics indicating that it acts as testicular stem cells [41–45]. A (pale) shows typical characteristics of a progenitor and A (dark) shows high-proliferative activity during pre-pubertal testicular development when the pool of both types A spermatogonia expands [49].

A (dark) have been shown to be the SSCs and have the ability for self-renewal, and also produces differentiated forms of these cells, ending in spermatozoa in sexually mature males [44].

SSCs and their progeny are contained in the germinal epithelium of the seminiferous tubules in close association with the somatic SCs [12] (Fig. 22.3). Regulatory mechanisms mediated by growth factors induce or inhibit the proliferation and differentiation of SSCs [13, 46].

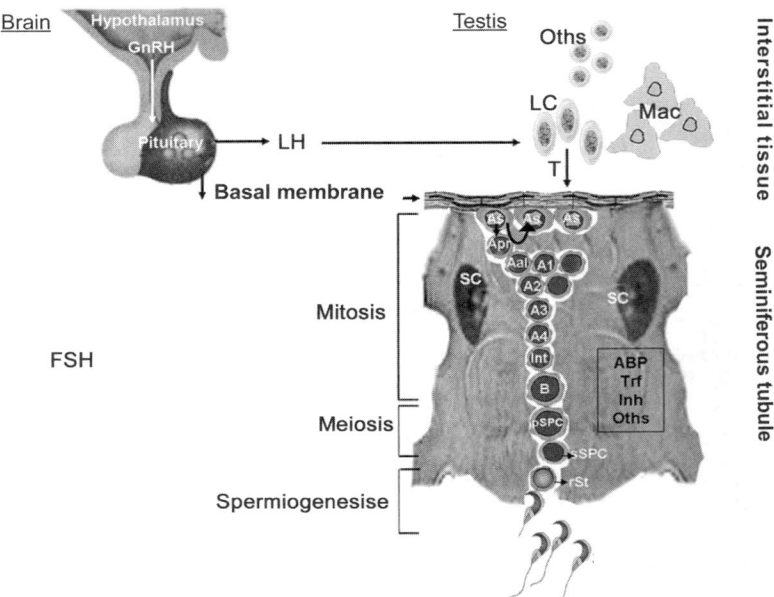

▶ **Fig. 22.3** *Endocrine and paracrine regulation of spermatogenesis. In the brain, the hypothalamus secretes gonadotropin-releasing hormone (GnRH) that induces the pituitary to secrete the gonadotropins leutinizing hormone (LH) and follicle stimulating hormone (FSH). These hormones affect different cells in the testis. LH affects LC in the interstitial tissue and FSH affects SCs in the seminiferous tubule. LH induces LC to secrete testosterone (T), while FSH induces SC to secrete androgen-binding protein (ABP), transferrin (Trf), inhibin (Inh) and other molecules and factors that are involved in spermatogenesis. Developed cells of spermatogenesis [including developed A, intermediate (Int) B, primary spermatocytes (pSPC), secondary spermatocytes (sSPC) round spermatids (rSt) and elongated spermatid (eSt)] are located between two adjacent SC. At the end of the spermatogenesis, the spermatozoa (SPZ) shed to the lumen of the seminiferous tubule. The interstitial tissue contains macrophages (Mac) and other cells (Oths) such as fibroblasts, dendritic cells, and cells of the connective tissue in addition to LC.*

REGULATION OF SSCs DEVELOPMENT
Endocrine Regulation in SSCs Development

The endocrine stimulation of spermatogenesis involves both FSH and LH (Fig. 22.3); the latter acting through the testosterone (T), produced by LCs in the testis. Several studies have demonstrated that T is critical for spermatogenesis [50–54]. Testosterone is essential in the development of round spermatids due to the premature detachment of round spermatids from the epithelium. Androgens regulate adhesion between SCs and round spermatids, either via effects on the cell adhesion molecules (CAMs) located between the two cell types, or on the intracellular junction apparatus located in SC. Testosterone also impacts on normal spermatogonial mitosis and the successful completion of meiosis [3, 54].

In rats, FSH might modulate the number of germ cells proceeding successfully through the mitotic and meiotic phase of spermatogenesis. It has been suggested that FSH might play a role in stimulating mitotic and meiotic DNA synthesis in type B spermatogonia and preleptotene spermatocytes, as well in preventing apoptosis of pachytene spermatocytes and round spermatids [3, 54]. The contribution of FSH to round spermatids adhesion is expected to be permissive rather than regulatory; it is well known that the intracellular organization of various SC cytoskeleton proteins is dependent on FSH [55].

It has been shown that high concentrations of T and FSH *in vitro* could be associated with rapid trans- and post-meiotic differentiation in biopsies from men with germ cell maturation arrest [56].

The role of FSH in both the initiation and the maintenance of spermatogenesis is controversial and exhibits important species differences [57–61]. In studies where male mice were rendered FSH deficient by a targeted disruption of the FSH subunit [57] or FSH receptor gene [58], fertility is preserved despite reduced testes size and partial spermatogenic failure. In humans, there is discordance between the phenotype of men with mutations in the FSH receptor gene who are oligospermic [59] and the phenotype of men with mutations of the FSH subunit who are azoospermic [60, 61]. This discrepancy in the degree of spermatogenic failure suggests that ligand deficiency may well have greater effect or expression than that suggested by its role as FSH receptor, and raises questions regarding the completeness of the receptor knockout models. The data indicate that FSH is required for quantitatively normal spermatogenesis; delineating the feedback regulation of FSH is a key to defining the pathophysiology of male infertility, as well as determining the feasibility of hormonal approaches to male contraception.

It has long been known that regulation of gonadotropin secretion in the human male involves a complex interplay between stimulation by GnRH secretion and inhibitory feedback by sex steroids found in T from LCs. However, in the last decade it has also been increasingly appreciated that for FSH, there is an additional level of complexity mediated by non-steroidal factors. This non-steroidal regulation of FSH comprises an endocrine negative feedback loop mediated by inhibin B secretion from SCs [62–68] in addition to autocrine/paracrine modulation within the pituitary mediated by activin and follistatin [69].

Thus, an integrated approach to the study of FSH regulation requires utilization of models that permit isolation of the effects of sex steroids from those of non-steroidal factors.

Follicle-stimulating hormone and testosterone regulate the spermatogenic process individually; in combination they show a synergistic effect [70]. Activin A, follistatin and FSH were suggested as crucial factors in germ cell maturation [71]. Activin was demonstrated to increase the ratio of germ cell/SCs, while FSH and follistatin increased the proliferation of spermatogonia [72].

Many reports suggest that FSH and testosterone act co-operatively and indicate that a lower dose of either is equally effective when the other is present [73–76]. This has led to speculation that FSH and testosterone may have common post-receptor pathways of action [24]. There are several points at which this co-operative relationship between FSH and testosterone may exist. The first relates to the SC cytoskeleton and associated Sertoli–germ cell junctions. Structural proteins might help determine the capacity of SCs to bind spermatids under the influence of testosterone. *In vitro* evidence supports the synergistic role of FSH on testosterone-mediated spermatid binding [77]. Similar cytoskeletal effects of FSH may be crucial in other Sertoli–germ cell interactions. Secondly, at the receptor level, FSH may synergize with testosterone by stimulating the synthesis of the androgen receptor (AR), based on *in vitro* studies in which FSH resulted in a two- to three-fold increase in AR number [78]. FSH treatment of hypophysectomized rats promoted the effect of testosterone on spermiogenesis and was associated with increased testicular content of androgen-binding protein (ABP) in the testis and epididymis [79] and it was suggested therefore that through this mechanism, FSH has a role in facilitating the transport and localization of T within SCs. In the absence of identified second messengers for T and FSH, one can only speculate as to the roles of FSH and T (individual or synergistic) in the production of SC growth factors involved in spermatogenesis.

Paracrine/Autocrine Regulation

There must be regulatory mechanisms controlling the ratio between self-renewal and differentiation of SSCs. In the normal case, the ratio between self-renewal and differentiation of SSCs should be about 1.0 [80]. More self-renewal than differentiation would reduce the seminiferous epithelium to only stem cells and a tumour might form. If differentiation prevailed, the stem cells would deplete themselves, leading to seminiferous tubules with only the supporting SCs [80]. Recent studies have shown that a specific regulatory mechanism controls the ratio between self-renewal and differentiation of SSCs [80].

The seminiferous tubule cells and the interstitial compartment, regulated by endocrine factors and autocrine/paracrine factors, are important to efficient growth and differentiation of SSCs [13, 54, 79] (Fig. 22.3).

SCs support the germ-cell development by releasing essential factors for proliferation and differentiation of SSCs to spermatozoa [13, 54, 81]. Spermatogonia stem cells and SCs are connected to the basement membrane (Figs. 22.1 and 22.3). Spermatogonia stem cells develop between two SCs (Fig. 22.3). Many morphological and functional changes in SSCs begin during their passing in the tight junction that is controlled by SCs [32].

The interaction between SCs and germ cells is regulated by autocrine/paracrine factors that are secreted from both cells [1, 12, 13, 54]. These factors affect the producer cells (autocrine) and the target cells (paracrine).

The progress development of germ cells depend on various factors that produce from SCs, such as lactate, transferrin (that transport the iron to germ cells) [1, 12, 13, 54], ABP and plasminogen activator [82].

The ABP is secreted by SCs following stimulation by FSH and is essential in the development of round spermatids during the spermatogenesis [13, 54, 83]. ABP has a high affinity for androgen; thus, it generates high concentrations of available androgen for germ cells during spermatogenesis [83].

On the other hand, the interactions between seminiferous tubule cells can be regulated by endocrine factors like FSH. It has been shown that FSH regulates the secretion of lactate, insulin and insulin-like factor-I (IGF-I) [84].

Studies have shown that paracrine/autocrine regulations are involved in this mechanism of renewal and division of SSCs [1, 54]. Most of these factors have been studied by transgenic loss-of-function and over-expression models for the regulation of spermatogonial mitosis and, more specifically, the control of balance between differentiation and proliferation to allow renewal of SSCs [85].

The paracrine/autocrine control of the seminiferous tubules are mediated by cytokines and growth factors, such as Interleukin (IL)-1, IL-6, tumour necrosis factor (TNF), glial cell line derived neurotrophic factor (GDNF), stem cell factor (SCF) and leukemia inhibitory factor (LIF). These factors and others are essential for maintenance of spermatozoa production throughout life [23, 54, 86–90].

LIF is secreted as multi-functional cytokine of the IL-6 family, sharing the common gp130 receptor subunit together with IL-6. LIF has a range of biological actions in various tissues [13, 54, 91, 92]. Although LIF is named for its ability to inhibit proliferation of myeloid leukemic cell line by inducing differentiation [92–95], it also regulates the growth and differentiation of embryonic stem cells [96, 97], PGCs [98], peripheral neurons [99], osteoblast [100], adipocytes [101] and endothelial cells [102], and improves mouse blastocyst development *in vitro* [103]. LIF is present in human follicular fluid and its level regulated according to the stage of follicle development [91–93].

It has been shown that LIF knock-out mice does not affect survival or male fertility [104]; on other hand, LIF receptor (LIF-R) knock-out mice leads to death shortly after birth of offspring, suggesting its involvement in many crucial physiological processes [105, 106].

In the testes, LIF promotes the proliferation and survival of murine PGCs, by preventing their apoptosis [107]. LIF is produced in the testicular tissue by peritubular cells and the interstitial compartments (Leydig and macrophages) [108]. This could suggest that LIF may be a paracrine/autocrine regulator of both testicular compartments [108]. Also LIF enhances the survival of SCs and gonocytes *in vitro* system [108–110].

Another crucial factor in the regulation of SSCs is called SCF. SCF is a haematopoietic cytokine that triggers its target by binding to c-kit receptor for different biological effects. It is involved in mast cell activation and chaemotaxis. SCF is produced in various cells in the tissues. SCF and c-kit are expressed in mast cells, melanocytes, testis and in bone marrow in the progenitor (CFU-c) compartment, as well as in non-hematopoietic cells, such as vascular endothelial cells, astrocytes, neurons and epithelial cells [13, 111–117].

SCF and c-kit play an important role in spermatogonial development; the crosstalk between the Sertoli and germ cells involves many paracrine/autocrine factors. SCF is one of the paracrine factors produced by SCs [13, 88]. Mutations in the gene encoding, either SCF or c-kit, results in infertility owing to defective migration, proliferation and survival of primordial spermatogonia stem cell [13, 118]. All differentiating types of spermatogonia were shown to express c-kit as compared to undifferentiated spermatogonia which do not express it [119, 120].

SCF stimulates proliferation and differentiation of type A spermatogonia and appears involved in regulating latter stages of spermatogenesis [121–123]. On the other hand, SCF decreases the germ cells apoptosis [124]. Thus, the expression levels of SCF might indicate the status of spermatogenesis.

GDNF, necessary for proliferation and differentiation of SSCs in mammalian [13, 125], is a neurotrophic polypeptide distantly related to transforming growth factor-β (TGF-β), originally identified as a survival factor for midbrain dopaminergic neurons [13, 126]. It supports the survival and regulates the differentiation of many neuron cells [126]. GDNF has another function in different tissues, in the regulation of kidney morphogenesis and spermatogenesis [127–129]. GDNF signals through specific receptors a (GFR-α), including four forms of GFR-α from 1–4 and RET receptor tyrosine kinase (RTK) [13, 130]. The interaction with the receptor leads to activated downstream signal pathway that is important for cell survival, proliferation and differentiation.

In the testis, GDNF is secreted from SC and is an essential growth factor for maintenance of SSCs [13, 86, 131]. Over-expression of GDNF in mouse testes leads to stimulating self-renew of SSCs and block the differentiation of SSCs [125, 132]. In mice with one GDNF null allele, undifferentiating SSCs disappeared and resulted in SC only like syndrome in the seminiferous tubules [54, 125, 132]. However, GDNF knockout mice died after birth [87]. These clearly established GDNF as one SC factor, which is necessary for SSC maintenance. These *in vivo* studies suggest that GDNF has a critical role in spermatogenesis by acting in a paracrine manner. *In vitro* experiments were shown that the development of SSCs failed and lost SSCs when GDNF was removed from the culture [54, 133].

It has been shown that SSCs express the GFR-α1 and c-Ret receptor tyrosine kinases [13, 54, 125, 132] located on the surface of SSCs. These receptors are very important and critical for the self-renewal and maintenance of SSCs [13, 54, 87]. However, like the GDNF knockout mice, the GFR-a1 knockout mice were also dying one day of birth [87].

Cytokines were also involved in testicular paracrine/autocrine regulation. Many cells of the testicular tissue produce cytokines, such as Interleukin (IL)-1 family, Tumour necrosis factor alpha (TNF-α)?, IL6, IL-10, IL-12, IL-18, IL-18 binding protein and others [13, 134–142].

It has been shown that IL-1β is secreted from LCs and testicular macrophages. In addition, IL-1α could be detected in seminiferous tubular and germ cells [135, 142, 143–146]. We have shown that SCs and germ cells could produce IL-1 receptor antagonist (IL-ra) under *in vitro* conditions [142, 145].

Recently, we have shown that IL-6, IL-18 and IL-18 binding protein are expressed in testicular germ cells (spermatogonia, spermatocytes and spermatozoa), interstitial cells (Leydig and macrophages) and peritubular cells [141, 147–149].

In other studies and unpublished data, it has been indicated that testicular cells produce under physiological conditions other cytokines, such as IL-10 and IL-12. Therefore, we suggest that these cytokines might play an important role in the regulation of testicular cell functions, tissue development, growth and differentiation.

A continuous balance is maintained between germ cell generation and death, which is regulated by apoptosis regulatory factors that are secreted from the testicular compartments, mainly by SCs, differentiated germ cells and also by interstitial cells [13, 54]. Fas/FasL is one of these mechanisms which could be involved in the regulation of germ cell growth and proliferation; germ cells express the FAS and SCs express the FasL [134, 150–160]. These factors involve activating different intracellular signaling pathways, including Bcl-2 family that are expressed by germ cells in immature testis and spermatocytes/spermatids in mature testis and are induced by the endocrine system (testosterone) and by the paracrine/autocrine system (Fas/FasL) [134, 150–160]. TNF-a-related apoptosis-inducing ligand (TRAIL), which are expressed in different human testicular germ cell types, LCs and peritubular and SCs, could also participate in the regulation of testicular cell functions and germ cell apoptosis [134, 150–152, 155–157, 159–161].

CURRENT STATUS OF THE RESEARCH
Isolation and Enrichment of SSCs

As mentioned before, SSCs include subtypes of spermatogonia. To understand the behavior and the properties of SSCs, we need to use pure SSCs that have the ability of self-renewal and differentiation [47, 162, 163].

The isolation of SSCs is difficult. Some of the limitations of SSCs include: (1) Very limited information regarding the kinetics of stem cell division; (2) The morphology of undifferentiating spermatogonia, which is very difficult for characterization after isolation; (3) The SSCs and their differentiation progeny which could be characterized according to location in the seminiferous tubules in the testis, but which disappears after destroying this structure in order to obtain SSC suspension; (4) Until now no specific marker for each stage in the spermatogenesis process, and the many types of germ cells which express the same marker through the mitotic divisions and differentiation steps; (5) The low number of SSCs in the mature testis (only 2–3 stem cells exist in 10^4 testicular cells in adult mouse); therefore, testes from immature animals have been used to isolate a large number of SSCs [48, 54, 162–165].

It was shown that up to 35% of spermatogonia degenerate, and apoptosis was implicated as a principal process by which these cells are removed from the epithelium [3]. The apoptosis and differences in the first step of SSC development between primates and non-primates make it difficult to culture *in vitro* SSCs, and the difference in the number of stages between human and non-human system make the conditions for *in vitro* culture very difficult.

The first method for isolation was based on cell size and shape, using elutriation or velocity sedimentation in BSA gradient at unit gravity [165], and Percoll density gradient centrifugation [166, 167]. The new information on specific markers on the membrane on SSCs, such as c-kit and GFR-α-1, help in developing new methods using these markers for sorting by magnetic cell sorting and systems of immunomagnetic beads [168, 169]. Other markers were used for sorting SSCs, such as Thy-1, α-6 and β-1-integrin antibodies [169, 186]. The specific markers can be also used for sorting by fluorescence-activated cell sorting (FACS) [167].

There are studies that used some of these markers and succeeded in increasing the number of isolated SSCs and to culture them *in vitro* [168, 169]. The limitation of these methods is the use of markers that could be expressed in low levels in SSCs or expressed in another subtypes of SSCs, which lead to a reduction in the yield of cells that we aimed to isolate and culture.

All the methods above try to isolate highly pure population of SSCs from primary cultures of testicular cells. A new method for such isolation was described recently, based on the principle that testicular somatic cells bind tightly to plastic and collagen matrices when cultured in serum-containing medium. But the spermatogonia and spermatocytes do not bind to plastic or collagen when cultured in serum-containing medium [141, 144, 146].

These methods described above used the collagen-non-binding testis cells and can get 97% pure SSCs. The SSCs can be isolated after short incubation in culture on laminin matrix. The spermatogenic cells that bind to laminin are more than 90% undifferentiated SSCs; the cells can develop into functional spermatozoa. This method does not require flow cytometry and can also be applied to obtain enriched cultures of mouse SSCs. These cells can be enriched for *in vitro* culture studies [141, 144, 146].

MARKERS EXPRESS ON SSCs DURING THEIR DEVELOPMENT

Some of the well-known pre-meiotic, meiotic and post-meiotic markers are presented [54] in Table 22.1.

Table 22.1 *Cellular Markers on Developed Spermatogonial Stem Cells during Spermatogenesis*

Stage of differentiation	Cellular markers
Pre-meiotic	Oct-4, C-kit, GFR-a-1, 6-a-integrin, CD9, Notch-1, Thy-1
Meiotic	Prohibitin, SRF-1, Boule
Post-meiotic	LDH, Protamine, Crem-1, SP-10, Acrosin

Oct-4 is a germ cell specific transcriptional factor; it is expressed in totipotent embryonic cells. In male mice, the expression of Oct-4 is maintained until the beginning of spermatogenesis and is confined to the type A spermatogonia [54, 170].

Several surface proteins markers, such as α-6-integrin and CD9 (tetraspanin transmembrane protein), are commonly expressed in undifferentiated spermatogonial cells [54, 171–173].

The second group of markers is Prohibitin, SRF [174] and boule [170], which express on meiotic spermatocytes. These markers are primarily expressed in leptotene spermatocytes.

Prohibitin is a mitochondrial protein and involved in DNA damage repair. They are expressed constitutively in adult LCs and SCs at all stages and at a very low level in preleptotene spermatocytes [175]. Prohibitin protein is expressed at a very high level in leptotene spermatocytes and at a very low level in zygotene spermatocytes. In pachytene spermatocytes, Prohibitin is expressed at a very high level in stages VII–XI and minimal during stages XII and XIV of the spermatogonia wave [175].

Spermatogenesis-related factor-1 (SRF-1) is a marker expressed only in testicular tissue. The SRF-1 gene product is involved in the meiotic process in spermatogonia or spermatocytes. The expression of the protein was detected mainly in the cytoplasm of leptotene or pachytene spermatocytes, with a higher level of expression observed at 7 weeks in rats [176].

Boule is a member of the DAZ family; a gene family that encodes proteins that are distinct from other RNA-binding proteins. Boule is expressed in the cytoplasm of pachytene spermatocytes, persists through meiosis and decreases in early spermatids [177]. Boule is detectable in secondary spermatocytes and early spermatids (round spermatids), then decreases until it is in undetectable level in spermatids. Boule is not expressed in spermatogonia or primary spermatocytes [177].

Post-meiotic markers that are expressed on haploid cells include Crem-1, LDH, Protamine 1, acrosin and SP10 [178, 179].

Lactate Dehydrogenase (LDH) participates in anaerobic glycolysis, a nearly universal pathway that converts glucose into pyruvate. LDH-C4 is a testis-specific lactate dehydrogenase. It is uniquely suited for satisfying the metabolic requirements of differentiating germ cells and functional spermatozoa. LDH-C4 gene is expressed exclusively during meiosis and spermiogenesis, beginning in leptotene/zygotene spermatocytes and continuing through to the elongated spermatids [180].

Protamine-1 is involved in histone synthesis during spermiogenesis. The histones are replaced by a set of transition proteins (TPs), which are subsequently replaced by protamines. Protamine gene is transcribed post-meiotically in the round spermatid stage of spermatogenesis; the Protamine-1 expression is a molecular marker that is useful to predict the presence of testicular spermatozoa [181].

Acrosin is considered a post-meiotic marker; the former associated with the acrosome and the latter involved in the compaction of the chromosomal matrix.

Sperm protein-10 (SP-10) is a mouse sperm acrosomal protein. It is associated with the outer and inner acrosomal membranes of the anterior acrosome and is also present within the equatorial segment. SP-10 is transcribed post-meiotically in the round spermatid and spermatozoa stages of development germ cell [182].

The cAMP-responsive element modulator (CREM) is involved in regulating gene expression in haploid spermatids. CREM transcriptional activity is controlled through interaction with testicular compartment cells. The CREM gene is highly and specifically expressed in post-meiotic male cells where it regulates a number of genes involved in the process of spermatogenesis [183].

In Vitro Culture of SSCs

Many methods for *in vitro* differentiation of SSCs have been developed, and success in differentiating SSCs into spermatids is potentially an extremely attractive treatment for infertility, particularly the one caused by spermatogenic arrest [54, 184–187].

The recent advances show the possibility of differentiating germ cells *in vitro* from embryonic stem cells [12, 13, 32, 54, 81]; all the methods have been focused on testicular culture that are directed at studies on germ cell differentiation. The aims were to achieve meiotic or post-meiotic differentiation of cultured male germ cells by using tissue culture, organ culture and co-culture systems [56, 185, 188–194].

The isolation of SSCs is a very critical obstacle in the *in vitro* culture. It has been shown that germ cells in culture dishes are not all stem cells, a lot of cells lost the stem cell activity and other cells were differentiated. In addition, some of the germ cell colonies in culture expressed the c-kit marker which indicates on differentiated spermatogonia [195]. It was also determined that only less than 1% of germ cells in the culture after isolation acted like stem cells in mice [195].

Therefore, according to the limitations mentioned above, the researchers should consider at least three points:

  i. *In vitro* culture and transplantations should be established on species other than mice.
 ii. The genetic manipulation of germ cells must be feasible.
iii. The entire process of spermatogenesis of germ cells *in vitro* should be outside the testis.

Meiosis *In Vitro*

Induction of meiosis is still the main obstacle for *in vitro* culture of SSCs. It appears that both the initiation of spermatogonial differentiation from stem cells and the entry of differentiating spermatogonia into meiosis are blocked under *in vitro* conditions. The development of meiosis, *in vivo*, depends on the microenvironment present in the seminiferous tubules, which provide crucial factors and signals needed in advancing to SSCs to the final stage of differentiation. Thus far these factors and conditions are unknown. As yet, organ and tubule cultures are the only tools for studying the factors and mechanisms involved in full pre-meiotic germ cell development up to meiotic progression of cells [107, 196]. Induction of rat primary rat spermatocytes to round spermatids with acrosomes has been reported in seminiferous segments cultured *in vitro* [197]. Also, co-culture of animal spermatocytes with SCs showed differentiation up to round spermatids [229, 198, 199], and even produced normal and fertile offspring following ooplasmic nuclear injection [200]. Addition of Vero cell conditioned media to human spermatogonial cells, co-cultured with SCs (in the presence of FSH and testosterone), induced their development to late spermatid stage [190]. This might suggest a fundamental role of SCs in the regulation of SSC growth and differentiation under *in vitro* conditions. On the other hand, the importance of growth factors produced by Vero cells in the induction of meiosis and even spermiogenesis has been demonstrated. Culturing of isolated primary spermatocytes from testicular biopsies of non-obstructive azoospermia on Vero cell monolayer induced differentiation up to round spermatid stage [190]. In addition, a culture of human round spermatids on Vero cell monolayer were induced to elongated spermatids and even to mature spermatozoa with normal morphology, which in turn were used even to fertilize human oocyte and to develop embryos [187, 196, 201, 154, 191].

Recently, mouse male gametes and embryonic germ cells were derived from embryonic stem cells (ES) [202]. In addition, it has been demonstrated that mouse ES can produce functional germ cells *in vitro* [203, 179]. Also, spontaneous differentiation of germ cells from human ES *in vitro* was published [204]. Recently, the importance of retinoic acid levels in the mechanism controlling meiosis has been suggested [205, 206].

A lot of methods have been successful in culturing germ cells *in vitro* using different growth factors—LIF [94, 95], GDNF [86] or SCF [38, 207]—which have been described as essential factors in the spermatogenesis process. In recent studies it was shown that LIF is useful in the initiation of mouse germline stem-cell culture and suggest that LIF is involved in the maturation of gonocytes into spermatogonia [208, 209]. GDNF with FCS can induce SSCs maintenance *in vitro* [210]; the using of GDNF in the media increased the gene expression of undifferentiated SSCs [211]. In the other hand, SCF did not increase the proliferation of SSCs, but did increase the differentiation SSCs [212]. Specific condition media, such as KSOM medium [159, 213, 214], foetal calf serum (FCS) in mice or bovine systems were also used for *in vitro* culture of SSCs [210, 214, 215].

Other methods to culture germ cell used different conditioned media for SCs, LCs [213, 216] and other conditioned media from Vero cells or mouse embryo fibroblasts (MEF) [86].

Moreover, for long-term culture, feeder layers from SCs or MEF have been used for culture SSCs to enhance the concentrations of growth factors in the media [217]. In order to enhance the differentiation of SSCs in the system with SCs layer, researchers have tried to culture the SSCs in the presence of FSH and testosterone in human system [155, 191]. It has been shown that the using of rFSH and testosterone *in vitro* with human testicular tissue can be helpful in the differentiation of primary spermatocytes into elongated spermatids [56, 152, 153, 194]. Moreover, the germ cell culture was induced with pathological conditions by vitamin A, 2.5-hexanedione, X-irradiation, cryptorchid conditions and S1 infertile mutant mice in order to reduce or eliminate the differentiation of SSCs [54].

There is no doubt that these systems, growth factors and conditions enhance the differentiation of germ cells in the culture as examined by different assays and markers [160].

Until now, there has been no full successful sequence for *in vitro* culturing of SSCs and differentiation to full spermatogenesis; this could be that more information regarding specific factors or conditions in the various stages of spermatogenesis is required. For example, one of the special structures in the testis is the microenvironment that controls the self-renewal of SSCs and the progeny production *in vivo* are known as niches [218]. Subtypes of spermatogonia (As, Apr and Aal) are located in these niches in mice and rats [219]. The positions of these niches are related to vasculature pattern [220] and have an important function in the assuring balance ratio between self-renewal and differentiation of SSCs. In order to imitate these niches many *in vitro* culture systems were performed using the 2D cultures for mammalian SSCs in dishes and flasks [212, 214, 216, 221, 222]; however, the 3D cell culture system has been suggested as representative of the condition *in vivo* [223].

Recently, the importance of a 3D structure for the differentiation of human and mouse testicular cells and the support of *in vitro* spermatogenesis were evaluated [223, 224]. The 3D cell culture has been attempted with reconstructed testes tissues to gain more insights into germ–somatic cell interaction or germ cell extracellular matrix (ECM) interactions during *in vitro* spermatogenesis [225–227].

In the recent years, several studies have attempted to define a new 3D culture system in an attempt to devise a long-term culture that might help improve *in vitro* spermatogenesis of human germ cells. The importance of somatic cells in stimulating germ cell progression during culture was demonstrated.

It has been reported that reaggregates of male germ cells and SCs on a filter lacking a collagen matrix flatten and no spermatocyte is formed, indicating that a 3D structure of the cell aggregates is required for further differentiation *in vitro* [227]. The possibility that the collagen matrix or other materials retain growth factors or other hormonal factors that are secreted by SCs in close proximity of germ cells for spermatogonia proliferation was considered [227, 228].

In our recent study we have shown the isolation and culture of SSCs *in vitro*. For enriching the isolation of SSCs, we used GFRα-1 as a specific surface marker and magnetic-activated cell sorting (MACS) as a separation approach. We used a novel 3D soft agar culture system (SACS). This system provides 3D structural support and multiple options for manipulations through the addition of endocrine factors and testicular cells [223]. In conclusion we suggest that our enrichment and culture approach is highly effective in research of SSC expansion and we have found indications that the system supports differentiation up to the level of post-meiotic germ cells [223].

Establishment of SSCs Lines

In the testes, all the stages of development take place at the same time and, therefore, it is very difficult to study the germ cells. *In vitro* experiments need isolated germ cells in each step. However, only a low number of germ cell can be isolated by the available methods and it is difficult to distinguish SSCs from more differentiated A spermatogonia *in vitro*, due to the limited information regarding specific markers for SSCs [213, 229–231].

One of the possible solutions for this problem is to establish SSCs lines. Several germ cell lines of spermatocytes have been developed [232–234].

One of the SSC lines was the GC-1 spg cells that are characteristic for the B spermatogonia cells and early spermatocytes [232]. The GC-2 spd was developed for spermatid cells, the GC-3 spc is a spermatocyte cell line [233] and GC-4 spc is an early pachytene spermatocyte cell line [234].

These germ cell lines help the scientist understand the biomolecular mechanisms involved in the proliferation and differentiation of SSCs during the developmental stages. However, these cell lines are limited in that none of them is able to differentiate in response to known growth factors. Given this limitation we conclude that the factors and conditions of the process of SSCs differentiation and meiosis are not yet clear enough to solve this problem.

Transplantation of SSCs

The transplantation technique started in 1994, using a mixed cellular solution of SSCs from immature mice; and these donor cells were transplanted into mouse seminiferous tubules. This experiment demonstrated that donor SSCs could interact with new environment, migrate and pass through the tight junction to enter the basal compartment [34, 235, 236].

The technique of the transplantation was performed by microinjection of SSCs into mice pretreated with busulfan to eliminate native spermatogenesis. Donor cells that express the LacZ were injected to the seminiferous tubules. These cells, when differentiated to the round spermatids stage, were stained blue and could characterize the cells [34].

In the transplantation experiments using transgenic SSCs, it has been found that donor-derived spermatogonia were responsible for generating offspring [33]. The successfully transplanted SSCs, localized at the basement membrane, began to show evidence of division by the first week after the transplantation. After one month, the donor SSCs migrated to the original origin in the seminiferous tubules and differentiated to spermatozoa [236–238].

The use of transplantation technique is a powerful functional method to analyse and to characterize SSCs activity in donor testes cell population, to evaluate the quality of SSCs niches in recipient testes and to assess the effect of *in vitro* manipulations on SSCs function. The purpose for using transplantation in the clinical setting is to restore fertility in patients.

FUTURE APPLICATIONS AND CAUTION

Investigations of spermatogonia transplantation have provided greater insight into the process of spermatogenesis in relationship to various diseases, and the successful development of transplantation in non-human primates provides new insights in an animal system that could have relevance for human physiology; the results of research in spermatogonia transplantation will be instructive for future clinical trials.

In the last few years, research on SSCs has developed new methods and the transplantation of germ cells in mouse testes has opened new vistas for basic information on SSCs development [33, 34].

Potential human clinical applications in the future could include treatment for salvaging testes in spermatogenic arrest or during risk of infertility (for boys or men) due to chaemo- or radiotherapy, and providing needed information concerning gene therapy for male infertility.

Today, one potential clinical application is transplanting SSCs into seminiferous tubules of infertile men to restore fertility. Results of these clinical trials should be available in next few years. Another potential clinical application using human SSCs is germline gene therapy [239]; these applications depend on the possibility of gene therapy to correct the non-functional, hypofunctional or hyperfunctional gene that causes these endocrine or metabolic abnormalities [240].

Numbers of different risks are associated with SSCs *in vitro* culture and should be considered. These include contamination of SSCs with animal viruses or other infectious agents, chromosomal abnormalities, imprinting patterns, degenerative cellular changes and lose of the ability to develop multipotent cells during long–term cell culture [54, 240].

CONCLUSION

Full differentiation of male germ cells starting at the level of SSCs and passing meiosis to mature gametes has not been achieved reliably *in vitro*, since the biomolecular factors and the specific microenvironment crucial for the development of each stage of spermatogenesis, *in vitro*, are not yet clear.

In vitro culture of SSCs could be used as a powerful tool in studying the biomolecular factors involved in the regulation of SSCs growth/proliferation and differentiation. In addition, it enables us to identify specific cell markers for differentiation stages and to increase the number of these cells. The development of *in vitro* culture of SSCs might lead to a new clinically relevant approach in treating male infertility in patients with germ cell arrest or patients (boys or men) who are at risk of infertility due to chaemo- or radiotherapy. On the other hand, ethic problems and risk factors should be considered while using these cultures.

REFERENCES

1. Russell L.D., R.A. Ettlin, A.P. Sinha Hikim, et. al. (1990). 'Mammalian spermatogenesis, in Histological and Histopathological Evaluation of the Testis'. Cache River Press. *Clearwater: FL*, pp. 1–40.
2. Nayernia K., M. Li, W. Engel (2004). 'Spermatogonial stem cells. methods in molecular biology'. vol. 253: *Germ. Cell. Protocols*. **1**: 105–120.
3. De Kretser D.M., K.L. de Loveland, A. Meinhardt, D. Simorangkir, N. Wreford (1998). 'Spermatogenesis'. *Hum. Reprod.* **13**: 1–8.
4. Ogawa T. (2001). 'Spermatogonial transplantation: the principle and possible applications'. *J. Mol. Med.* **79**: 368–374.
5. Saffan E.E., P. Lasko (1999). 'Germline development in vertebrates and invertebrates'. *Cell. Mol. Life Sci.* **55**: 1141–1163.
6. Wylie C. (1999). 'Germ cells'. *Cell.* **100**: 157–168.
7. Matsui, Y. (1998). 'Developmental fates of the mouse germ cell line'. *Int. J. Dev. Biol.* **42**: 1037–1042.
8. De Rooij D.G., F.M.F. Van Diessel-Emiliani (1997). 'Regulation of proliferation and differentiation of stem cells in the male germ line'. *Stem Cells (Potten, C.S., ed.). Academic: London.* 283–313.
9. Oakberg E.F. (1956). 'A description of spermiogenesis in the mouse and its use in analysis of the cycle of the seminiferous epithelium and germ cell renewal'. *Am. J. Anat.* **99**: 391–414.

10. De Rooij D.G. (2001). 'Proliferation and differentiation of spermatogonial stem cells'. *Reproduction*. **121**: 347–354.
11. Clermont Y., E. Bustos-Obregon (1968). 'Re-examination of spermatogonial renewal in the rat by means of seminiferous tubules mounted "in toto"'. *Am. J. Anat.* **122**: 237–247.
12. Skinner M.K. (1991). 'Cell–cell interactions in the testis. *Endocr. Rev.* **12**: 45–77.
13. Huleihel M., E. Lunenfeld (2004). 'Regulation of spermatogenesis by paracrine/autocrine testicular factors'. *Asian. J. Androl.* **6**: 259–268.
14. Al-Agha O., C. Axiotis (2007). 'An in-depth look at Leydig cell tumour of the testis. Arch. Pathol'. *Lab. Med.* **131**(2): 311–7.
15. Amann R.P. (1986). 'Detection of alterations in testicular and epididymal function in laboratory animals'. *Environ. Health Perspect.* **70**: 149–158.
16. Dym M., H.G.M. Raj (1977). 'Response of adult rat Sertoli cells and Leydig cells to depletion of luteinizing hormone and testosterone'. *Biol. Reprod.* **17**: 676–696.
17. Feig L.A., A.R. Bellve, N.H. Erickson, M. Klagsbrun (1980). 'Sertoli cells contain a mitogenic polypeptide'. *Proc. Natl. Acad. Sci. USA*. **77**: 4774–4778.
18. Jutte N.H., R. Jansen, J.A. Grootegoed, F.F. Rommerts, O.P. Clausen, H.J. van der Molen (1982). 'Regulation of survival of rat pachytene spermatocytes by lactate supply from Sertoli cells'. *J. Reprod. Fertil.* **65**: 431–438.
19. Jutte N.H., R. Jansen, J.A. Grootegoed, F.F. Rommerts, H.J. van der Molen (1983). 'FSH stimulation of the production of pyrovate and lactate by rat Sertoli cells may be involved in hormonal regulation of spermatogenesis'. *J. Reprod. Fertil.* **68**: 219–226.
20. Tres L.L., E.P. Smith, J.J. VanWyk, A.L. Kierszenbaum (1986). 'Immunoreactive sites and accumulation of somatomedin- C in rat Sertoli-spermatogenic cell co-cultures'. *Exp. Cell Res.* **162**: 33–50.
21. Buch J.P., D.J. Lamb, L.I. Lipshultz, R.G. Smith (1988). 'Partial characterization of a unique growth factor secreted by human Sertoli cells'. *Fertil. Steril.* **49**: 658–665.
22. Bellve A.R., W. Zheng (1989). 'Growth factors as autocrine and paracrine modulators of male gonadal functions'. *J. Reprod. Fertil.* **85**: 771–793.
23. Johnson L. (1991b). 'Spermatogenesis. In: P.T. Cupps, (Ed.). Reproduction in Domestic Animals, fourth ed'. *Academic Press Inc. San Diego*. CA. pp. 173–219.
24. Russell L.D., M.D. Griswold (1993). 'The Sertoli Cell'. *Cache River Press, Clearwater*. FL.
25. Wong C.H., C.Y. Cheng (2005). 'The blood-testis barrier: its biology, regulation, and physiological role in spermatogenesis'. *Curr. Top. Dev. Biol.* **71**: 263–296.
26. M.W. Li, W. Xia, D.D. Mruk, C.Q. Wang, H.H. Yan, M.K. Siu, W.Y. Lui, W.M. Lee, C.Y. Cheng (2006). 'Tumour necrosis factor "alpha" reversibly disrupts the blood-testis barrier and impairs Sertoli-germ cell adhesion in the seminiferous epithelium of adult rat testes'. *J. Endocrinol.* **190**: 313–329.
27. Johnson L., Jr. D.L. Thompson (1983). 'Age-related and seasonal variation in the Sertoli cell population, daily sperm production and serum concentrations of follicle-stimulating hormone, luteinizing hormone and testosterone in stallions'. *Biol. Reprod.* **29**: 777–789.
28. Johnson L., R.S. Zane, C.S. Petty, W.B. Neavesn (1984c). 'Quantification of the human Sertoli cell population: its distribution, relation to germ cell numbers, and age-related decline'. *Biol. Reprod.* **31**: 785–795.

29. Johnson L., H.B. Nguyen (1986). 'Annual cycle of the Sertoli cell population in adult stallions'. *J. Reprod. Fertil.* **76**: 311–316.
30. Johnson L., D.D. Varner, Jr. D.L. Thompson (1991b). 'Effect of age and season on the establishment of spermatogenesis in the horse'. *J. Reprod. Fertil. Suppl.* **44**: 87–97.
31. Johnson L., N.H. Ing, Jr. T.H. Welsh, D.D. Varner, W.L. Scrutchfield, M.T. Martin (2000). 'Efficiency of spermatogenesis in animals and humans'. *J. Anim. Sci.* **77**: 1–47.
32. Griswold M.D. (1995). 'Interaction between germ cells and Sertoli cells in the testis'. *Biol Reprod.* **52**: 211–216.
33. Brinster R.L., M.R. Avarbock (1994). 'Germline transmission of donor haplotype following spermatogonial transplantation'. *PNAS.* **91**: 11303–11307.
34. Brinster R.L., J.W. Zimmermann (1994). 'Spermatogenesis following male germ-cell transplantation'. *Proceedings National Academy of Sciences. USA.* **91**: 11298–11302.
35. Huckins C. (1971a). 'The spermatogonial stem cell population in adult rats I. Their morphology, proliferation and maturation'. *Anat. Record.* **169**: 533–557.
36. Oakberg E.F. (1971). 'Spermatogonial stem-cell renewal in the mouse'. *Anatomical Record.* **169**: 515–531.
37. Lok D., D. Weenk, D.G. de Rooij (1982). 'Morphology, proliferation, and differentiation of undifferentiated spermatogonia in the Chinese hamster and the ram'. *Anatomical Record.* **203**: 83–99.
38. De Rooij DG., J.A. Grootegoed (1998). 'Spermatogonial stem cells'. *Current Opinion in Cell Biology.* **10**: 694–701.
39. Chiarini-Garcia H., L.D. Russel (2001). 'High resolution light microscopic characterization of mouse spermatogonia'. *Biol. Reprod.* **85**: 1170–1178.
40. Huckins C., E.F. Oakberg (1978). 'Morphological and quantitative analysis of spermatogonia in mouse testis using whole mounted seminiferous tubules. I'. *The normal testis. Anat. Rec.* **192**: 519–528.
41. Clermont Y. (1996). 'Renewal spermatogenesis in man'. *Am. J. Anat.* **118**: 509–524.
42. Ramaswamy S., G.R. Marshall, A.S. McNeilly, T.M. Plant (2000). 'Dynamics of the follicle-stimulating hormone (FSH)-inhibin B feedback loop and its role in regulating spermatogenesis in the adult male rhesus monkey (Macaca mulatta) as revealed by unilateral orchidectomy'. *Endocrinology.* **141**: 18–27.
43. De Rooij D.G., H.J. van de Kant, R. Dol, G. Wagemaker, P.P. van Buul, A. van Duijn-Goedhat, F.H. de Jong, J.J. Broerse (2002). 'Long-term effects of irradiation before adulthood on reproductive function in the male rhesus monkey'. *Biol. Reprod.* **66**: 486–494.
44. Ehmcke J., J. Wistuba, S. Schlatt (2006). 'Spermatogonial stem cells. Questions, models and perspectives'. *Hum Reprod Update.* **12**: 275–282.
45. Ehmcke J.C.M., C.M. Luetjens, S. Schlatt (2005). 'Clonal organization of proliferating spermatogonial stem cells in adult males of two species of non-human primates, Macaca mulatta and Callithrix jacchus'. *Biol Reprod.* **72**: 293–300.
46. Jegou B. (1993). 'The Sertoli-germ cell communication network in mammals'. *Int. Rev. Cytol.* **147**: 25–96.
47. De Rooij D.G., L.D. Russell (2000). 'All you wanted to know about spermatogonia but were afraid to ask'. *J. Androl.* **21**: 776–798.

48. Aponte P.M., M.P.A. van Bragt, D.G. De Rooij, A.M.M. van Pelts (2005). 'Spermatogonial stem cells: Characteristics and experimental possibilities'. *APMIS.* **113**: 727–742.

49. Simorangkir D.R., D.M. de Krester, N.G. Wreford (1995). 'Increased numbers of Sertoli and germ cells in adult rat testis induced by synergistic action of transient neonatal hypothyroidism andneonatal hemicastration'. *J. Reprod. Fertil.* **104**: 207–213.

50. Awoniyi C.A., R. Santulli, R.L. Sprando, L.L. Ewing, BR. Zirkin (1989b). 'Restoration of advanced spermatogenic cells in the experimentally regressed rat testis: quantitative relationship to testosterone concentration within the testis'. *Endocrinology.* **124**: 1217–1223.

51. Sun YT., D.C. Irby, D.M. Robertson, D.M. de Kretser (1989). 'The effects of xogenously administered testosterone on spermatogenesis in intact and hypophysectomized rats'. *Endocrinology.* **125**: 1000–1010.

52. McLachlan RI., N.G. Wreford, S.J. Meachem, D.M. De Kretser, D.M. Robertson (1994a). 'Effects of testosterone on spermatogenic cell populations in the adult rat'. *Biol Reprod.* **51**: 945–955.

53. O'Donnell L., R.I. McLachlan, N.G. Wreford, D.M. Robertson (1994). 'Testosterone promotes the conversion of round spermatids between stages VII and VIII of the rat spermatogenic cycle'. *Endocrinology.* **135**: 2608–2614.

54. Huleihel M., M. abuelhija, E. Lunenfeld (2007). 'In vitro culture of testicular germ cells: Regulatory factorsand limitations'. *Growth Factors.* **25**(4): 236–252.

55. Muffly K.E., S.J. Nazian, D.F. Cameron (1994). 'Effects of follicle-stimulating hormone on the junction-related Sertoli cell cytoskeleton and daily sperm production in testosterone-treated hypophysectomized rats'. *Biol. Reprod.* **51**: 158–166.

56. Tesarik J., M. Guido, C. Mendoza, E. Greco (1998b). 'Human spermatogenesis in vitro: Respective effects of follicle-stimulating hormone and testosterone onmeiosis, spermiogenesis, and Sertoli cell apoptosis'. *J. Clin. Endocrinol. Metabol.* **83**: 4467–4473

57. Kumar T.R., Y. Wang, N. Lu, M.M. Matzuk (1997). 'Follicle stimulating hormone is required for ovarian follicle maturation but not male fertility'. *Nat. Genet.* **15**: 201–204.

58. Dierich A., M.R. Sairam, L. Monaco, G.M. Fimia, A. Gansmuller, M. LeMeur, P. Sassone-Corsi (1998). 'Impaired follicle-stimulating hormone (FSH) signaling *in vivo*: targeted disruption of the FSH receptor leads to aberrant ametogenesis and hormonal imbalance'. *Proc. Natl. Acad. Sci.* **95**: 13612–13617.

59. Tapanainen J.S., K. Aittomaki, J. Min, T. Vaskivuo, I.T. Huhtaniemi (1997). 'Men homozygous for an inactivating mutation of the follicle-stimulating hormone (FSH) receptor gene present variable suppression of spermatogenesis and fertility'. *Nat. Genet.* **15**: 205–206.

60. Phillip M., J.E. Arbelle, Y. Segev, R. Parvari (1998). 'Male hypogonadism due to a mutation in the gene for the ?-subunit of follicle-stimulating hormone'. *N. Engl. J. Med.* **338**: 1729–1732.

61. Lindstedt G., E. Nystro, C. Matthews, I. Ernst, P.O. Janson, K. Chaterjee (1998). 'Follitropin (FSH) deficiency in an infertile male due to FSH mutation. A syndrome of normal puberty and virilization but underdeveloped testicles with azoospermia, low FSH but high lutropin and normal serum testosterone concentrations'. *Clin. Chem. Lab. Med.* **36**: 663–665.

62. Dubey A.K., A.J. Zeleznik, T.M. Plant (1987). 'In the rhesus monkey (Macaca mulatta), the negative feedback regulation of follicle-stimulating hormone secretion by an action of testicular hormone directly at the level of the anterior pituitary gland cannot be accounted for by either testosterone or estradiol'. *Endocrinology.* **121**: 2229–223.

63. Illingworth P.J., N.P. Groome, W. Byrd, W.E. Rainey, A.S. McNeilly, J.P. Mather, W.J. Bremner (1996). 'Inhibin B: a likely candidate for the physiologically important form of inhibin in men. *J. Clin. Endocrinol. Metab.* **81**: 1321–1325.
64. Anawalt B.D., R.A. Bebb, A.M. Matsumoto, N.P. Groome, P.J. Illingworth, A.S. Mc-Neilly, W.J. Bremner (1996). 'Serum inhibin B levels reflect Sertoli cell function in normal men and men with testicular dysfunction'. *J. Clin. Endocrinol. Metab.* **81**: 3341–3345.
65. Nachtigall L.B., P.A. Boepple, S.B. Seminara, R.H. Khoury, P.M. Sluss, A.E. Lecain, W.F. Crowley Jr (1996). 'Inhibin B secretion in males with gonadotropin-releasing hormone (GnRH) deficiency before and after long-term GnRH replacement: relationship to spontaneous puberty, testicular volume, and prior treatment–a clinical research centre study'. *J. Clin. Endocrinol. Metab.* **81**: 3520–3525.
66. Seminara S.B., P.A. Boepple, L.B. Nachtigall, F.P. Pralong, R.H. Khoury, P.M. Sluss, A.E. Lecain, WF. Crowley (1996). 'Inhibin B in males with gonadotropinreleasing hormone (GnRH) deficiency: changes in serum concentrations after short term physiologic GnRH replacement–a clinical research centre study'. *J. Clin. Endocrinol. Metab.* **81**: 3692–3696.
67. Plant T.M., V. Padmanabhan, S. Ramaswamy, D.S. McConnell, S.J. Winters, N. Groome, A.R. Midgley Jr (1997). 'Circulating concentrations of dimeric inhibin A and B in the male rhesus monkey (Macaca mulatta)'. *J. Clin. Endocrinol. Metab.* **82**: 2617–2621.
68. Hayes F.J., J.E. Hall, P.A. Boepple, W.F. Crowley Jr. (1998). 'Differential regulation of gonadotropin secretion in the human: endocrine role of inhibin'. *J. Clin. Endocrinol. Metab.* **83**: 1835–184.
69. Ying S.Y. (1988). 'Inhibins, activins, and follistatins: gonadal proteins modulating the secretion of follicle-stimulating hormone'. *Endocr. Rev.* **9**: 267–293.
70. Marshall G.R., D.S. Zorub, T.M. Plant (1995). 'Follicle-stimulating hormone amplifies the population of differentiated spermatogonia in the hypophysectomized testosterone-replaced adult Rhesus monkey (Macaca mulatta)'. *Endocrinology.* **136**: 3504–3511.
71. Meehan T., S. Schlatt, M.K. O'Bryan, D.M. de Krester, K.L. Loveland (2000). 'Regulation of germ cell and Sertoli cell development by activin, folistatin, and FSH'. *Dev. Biol.* **220**: 225–237.
72. Toebosch A.M., D.M. Robertson, J. Trapman, P. Klaasen, R.A. de Paus, de Jong, F.H.J.A. Grootgoed, (1988). 'Effect of FSH and IGF-I on immature rat Sertoli cells: Inhibin-alpha- and beta-subunit mRNA levels and inhibin secretion'. *Mol. Cell. Endocrinol.* **55**: 101–105.
73. Bartlett J.M.S., G.F. Weibauer, E. Nieschlag (1989). 'Differential effects of FSH and testosterone on the maintenance of spermatogenesis in the adult hypophysectomized rat'. *J. Endocrinol.* **121**: 49–58.
74. Sun Y.T., D.C. Irby, D.M. Robertson, D.M. de Kretser (1989). 'Cloning and characterization of an extracellular Ca2+ sensing receptor from bovine parathyroid'. *Endocrinology.* **125**: 1000–1010.
75. Sun Y.T., N.G. Wreford, D.M. Robertson, D.M. de Kretser (1990). 'Quantitative cytological studies of spermatogenesis in intact and hypophysectomized rats: identification of androgen-dependent stages'. *Endocrinology.* **127**: 1215–1223.
76. Sinha A.P. Hikim, R.S. Swerdloff (1994). 'Suppression of Spermatogenesis in Man Induced by Nal-Glu Gonadotropin Releasing Hormone Antagonist and Testosterone Enanthate (TE) is Maintained by TE Alone'. *Endocrinology* **134**: 1627–1634.
77. Cameron D.F., K.E. Muffly (1991). 'Hormonal regulation of spermatid binding'. *J. Cell Science* **100**: 623–633.

78. Verhoeven G., J. Cailleau (1988). 'Interleukin-1 stimulates steroidogenesis in cultured rat Leydig cells'. *Endocrinology* **122**: 1541–1550.
79. Huang HFS., L.M. Pogach, E. Nathan, W. Giglio, J.J. Seebode (1991). 'Proposed mechanism for sperm chromatin condensation/decondensation in the male rat'. *Endocrinology.* **128**: 3152–3161.
80. Van Beek M.E.A.B., M.L. Meistrich, D.G. De Rooij (1990). 'Probability of self renewing divisions of spermatogonial stem cells in colonies, formed after fission neutron irradiation'. *Cell and Tissue Kinetics.* **23**: 1–16.
81. Sofikitis N., K. Ono, Y. Yamamoto, H. Papadopoulos, I. Miyagawa (1999). 'Influence of the male reproductive tract on the reproductive potential of round spermatids abnormally released from the seminiferous epithelium'. *Hum. Reprod.* **14**: 1998–2006.
82. Kangasniemi M., A. Kaipia, P. Mali, J. Toppari, I. Huhtaniemi, M. Parvinen (1990). 'Modulation of basal and FSH-stimulated cyclic AMP production in rat seminiferous tubules by an improved transillumination technique'. *Anat. Rec.* **227**: 62–76.
83. Martin du Pan R.C., A. Campana (1993). 'Physiopathology of spermatogenic arrest'. *Fertil. Steril.* **60**: 937–946.
84. Esposito G., M. Keramidas, C. Mauduti, J.J. Feige, A.M. Morera, M. Benahmed (1991). 'Direct regulating effect of transforming growth factor b1 on lactate production in cultured porcine Sertoli cells'. *Endocrinology.* **128**: 1441–1449.
85. Hiroshi K., R. Mary, Avarbock, L. Ralph, Brinster (2004). 'Culture Conditions and Single Growth Factors Affect Fate Determination of Mouse Spermatogonial Stem Cells'. *Biology of Reproduction.* **71**: 722–731.
86. Kubota H., MR. Avarbock, RL. Brinster (2004). 'Growth factors essential for self-renewal and expansion of mouse spermatogonial stem cells'. *PNAS* **101**: 16489–16494.
87. Naughton C.K., S. Jain, A.M. Strickland, A. Gupta, J. Milbrandt (2006). 'Glial cell-line derived neurotrophic factor-mediated RET signaling regulates spermatogonial stem cell fate'. *Biol. Reprod.* **74**: 314–321.
88. Mruk D.D., C.Y. Cheng (2004). 'Sertoli-Sertoli and Sertoli-germ cell interactions and their significance in germ cell movement in the seminiferous epithelium during spermatogenesis'. *Endocr. Rev.* **25**: 747–806.
89. Diemer T., D.B. Hales, W. Weinder (2003). 'Immune-endocrine interactions and Leydig cell function: the role of cytokines'. *Andrology.* **35**: 55–63.
90. Bornstein S.R., H. Rutkowski, I. Vrezas (2004). 'Cytokines and steroidogenesis'. *Molecular and Cellular Endocrinology.* **215**: 135–141.
91. Senturk L.M., A. Arici (1998). 'Leukemia inhibitory factor in human reproduction'. *Am. J. Reprod. Immunol* **39**: 144–151.
92. Arici A., E. Oral, O. Bahtiyar, O. Engin, E. Seli, E. Jones (1997). 'Leukaemia inhibitory factor expression in human follicular fluid and ovarian cells'. *Hum. Reprod.* **12**: 1233–1239.
93. Nilsson E., P. Kezele, M.K. Skinner (2002). 'Leukemia inhibitory factor (LIF) promotes the primordial to primary follicle transition in rat ovaries'. *Mol. Cell. Endocrinol* **188**: 65–73.
94. Moreau J., D. Donaldson, F. Bennett, J. Witek-Giannotti, S. Clark, G. Wong (1988). 'Leukaemia inhibitory factor is identical to the myeloid growth factor human interleukin for DA cells'. *Nature.* **366**: 690–692.

95. Gearing D., N. Gough, J. King, D. Hilton, N. Nicola, R. Simpson, E. Nice, A. Kelso, D. Metcalf (1987). 'Molecular cloning and expression of cDNA encoding a murine myeloid leukaemia inhibitory factor (LIF)'. *EMBO. J* **6**: 3995–4002.
96. Moassagh Pour A.A., M. Salehnia, A.A. Pourfatollah, M. Soleimani (2004). 'CFU-GM like colonies derived from embryonic stem cells cultured on the bone marrow stromal cells'. *Iran. Biomed. J.* **8**: 1–5.
97. Smith A.G., J. Nichols, M. Robertson, P.D. Rathjen (1992). 'Differentiation inhibiting activity (DIA/LIF) and mouse development'. *Dev. Biol.* **151**: 339–351.
98. Matsui Y., D. Toksoz, S. Nishikawa, S.I. Nishikawa, D. Williams, K. Zsebo, B.L.M. Hogan (1991). 'Effect of Steel factor and leukaemia inhibitory factor on mutine primordial germ cells in culture'. *Nature.* **353**: 750–752.
99. Yamamori T., K. Fukada, R. Aebersold, S. Korsching, M.J. Fann, P.H. Patterson (1989). 'The cholinergic neuronal differentiation factor from heart cells is identical to leukemia inhibitory factor'. *Science.* **246**: 1412–1416.
100. Reid L.R., C. Lowe, J. Cornish, S.J. Skinner, D.J. Hilton, T.A. Willson, D.P. Gearing, T.J. Martin (1990). 'Leukemia inhibitory factor: a novel boneactive cytokine'. *Endocrinology.* **126**: 1416–1420.
101. Mori M., K. Yamaguchi, K. Abe (1989). 'Purification of a lipoprotein lipase-inhibiting protein produced by a melanoma cell line associated with cancer cachexia'. *Biochem. Biophys. Res. Comm* **160**: 1085–1092.
102. K. Arai, J. Nishida, K. Hayashida, K. Hatake, T. Kitamura, A. Miyajima, N. Arai, T. Yokota (1990). 'Coordinate regulation of immune and inflammatory responses by cytokines (Japanese)'. *Rinsho. Byori. Jpn. J. Clin. Pathol.* **38**: 347–353.
103. Kauma S.W., D. W. Matt (1995). 'Co-culture cells that express leukemia inhibitory factor (LIF) enhance mouse blastocyst development in vitro'. *J. Assist. Reprod. Genet.* **12**: 153–156.
104. Stewart C.L., P. Kaspar, L.J. Brunet, H. Bhatt, I. Gadi, F. Kontgen, S.J. Abbondanzo (1992). 'Blastocyst implantation depends on maternal expression of leukemia inhibitory factor'. *Nature* **359**: 76–79.
105. Ware C.B., M.C. Horowitz, B.R. Renshaw, J.S. Hunt, D. Liggitt, S.A. Koblar, B.C. Gliniak, H.J. McKenna, T. Papayannopoulou, B. Thoma, L. Cheng, P. Dovovan, J. Peschon, P. Bartlett, C. Willis, B. Wright, M. Carpenter, B. Davison, D. Gearing (1995). 'Targeted disruption of the low-affinity leukemia inhibitory factor receptor gene causes placental, skeletal, neural and metabolic defects and results in perinatal death'. *Development.* **121**: 1283–1299.
106. Li M., M. Sendtner, A. Smith (1995). 'Essential function of LIF receptor in motor neurons'. *Nature.* **378**: 724–727.
107. Piquet-Pellorce C., I. Dorval-Coiffec, M. D. Pham, B. Jegou (2000). 'Leukemia Inhibitory Factor Expression and Regulation within the Testis'. *Endocrine Society.* Vol. 141. No. 3: 1136–1141.
108. De Miguel M.P., M. De Boer-Brouwer, M. Paniagua, R. van den Hurk, D.G. De Rooij, F.M. Van Dissel-Emiliani (1996). 'Leukemia inhibitory factor and ciliary neurotropic factor promote the survival of Sertoli cells and gonocytes in coculture system'. *Endocrinology.* **137**: 1885–1893.
109. Jenab S., P.L. Morris (1998). 'Testicular leukemia inhibitory factor (LIF) and LIF receptor mediate phosphorylation of signal transducers and activators of transcription(STAT)-3 and STAT-1 and induce c-fos transcription and activator protein-1 activation in rat Sertoli but not germ cells'. *Endocrinology.* **139**: 1883–1890.

110. Morris P.L., W.W. Vale, S. Cappel, C.W. Bardin (1988). 'Inhibin production by primary Sertoli cell-enriched cultures: regulation by follicle-stimulating hormone, androgens, and epidermal growth factor'. *Endocrinology* **122**: 717–725.

111. Zsebo K.M., D.A. Williams, E.N. Geissler, V.C. Broudy, F.H. Martin, H.L. Atkins, R.Y. Hsu, N.C. Birkett, K.H. Okino, D.C. Murdock (1990). 'Stem cell factor is encoded at the Sl locus of the mouse and is the ligand for the c-kit tyrosine kinase receptor'. *Cell.* **63**: 213–224.

112. Wershil B.K., M. Tsai, E.N. Geissler, K.M. Zsebo, S.J. Galli (1992). 'The rat c-kit ligand, stem cell factor, induces c-kit receptor-dependent mouse mast cell activation *in vivo*: evidence that signaling through the c-kit receptor can induce expression of cellular function'. *J. Exp. Med.* **175**: 245–255.

113. Valent P.E., E. Spanblochl, W.R. Sperr, C. Sillaber, K.M. Zsebo, H. Agis, H. Strobl, K. Geissler, P. Betterlheim, K. Lechner (1992). 'Induction of differentiation of human mast cells from bone marrow and peripheral blood mononuclear cells by recombinant human stem cell factor-kit-ligand in long-term culture'. *Blood.* **80**: 2237–2245.

114. Coleman J.W., M.R. Holiday, I. Kimber, K.M. Zsebo, S.J. Galli (1993). 'Regulation of mouse peritoneal mast cell secretory function by stem cell factor, IL-3, or IL-4'. *J. Immunol* **150**: 556–562.

115. Nilsson G., J.H. Butterfield, K. Nilsson, A. Siegbahn (1994). 'Stem cell factor is achaemotactic factor for human mast cells'. *J. Immunol.* **153**: 3717–3723.

116. Fujio K., R.P. Evarts, A. Hu, E.R. Marsen (1994). 'Expression of stem cell factor and its receptor, c-kit, during liver regeneration from putative stem cells in adult rat'. *Lab. Invest.* **70**: 511–516.

117. Fujio K., Z. Hu, R.P. Evarts, E.R. Marsden, C.H. Niu, S.S. Thorgeirsson (1996). 'Co-expression of stem cell factor and c-kit in embryonic and adult liver'. *Exp. Cell. Res.* **224**: 243–250.

118. Sette C., S. Dolci, R. Geremia, P. Rossi (2000). 'The role of stem cell factor and of alternative c-kit gene products in the establishment, maintenance and function of germ cells.' *Int. J. Dev. Biol* **44**(6): 599–608.

119. Ashman L.K. (1999). 'The biology of stem cell factor and its receptor c-kit'. *Int. J. Biochem. Cell Biol* **31**: 1037–1051.

120. Meachem S., V. von Schonfeldt, S. Schlatt (2001). 'Spermatogonia stem cell with a great perspective'. *Reproduction* **121**: 825–834.

121. Rossi P., C. Albanesi, P. Grimaldi, R. Geremia (1991). 'Expression of the mRNA for the ligand of c-kit in mouse Sertoli cells. Biochem'. *Biophys. Res. Commun* **176**: 910–914.

122. Tajima Y., H. Onoue, Y. Kitamura, Y. Nishimune (1991). 'Biologically active kit ligand growth factor is produced by mouse Sertoli cells and is defective in Sld mutant mice'. *Development.* **113**: 1031–1035.

123. Vincent S., D. Segretain, S. Nishikawa, SI. Nishikawa, J. Sage, F. Cuzin, M. Rassoulzadegan (1998). 'Stage-specific expression of the Kit receptor and its ligand (KL) during male gametogenesis in the mouse: a Kit-KL interaction critical for meiosis'. *Development.* **125**: 4585–4593.

124. Yan W., M. Samson, B. Jegou, J. Toppari (2000). 'Bcl-w forms complexes with Bax and Bak, and elevated ratios of Bax/Bcl-w and Bak/Bcl-w correspond to spermatogonial and spermatocyte apoptosis in the testis'. *Mol. Endocrinol.* **14**(5): 682–699.

125. Meng X., M. Lindahl, M.E. Hyvö̈nen, M. Parvinen, D.G. de Rooij, M.W. Hess, A. Raatikainen-Ahokas, K. Sainio, H. Rauvala, M. Lakso, J. G. Pichel, H. Westphal, M. Saarma, H. Sariola (2000). 'Regulation of cell fate decision of undifferentiated spermatogonia by GDNF. *Science.* **287**: 1489–1493.

126. L.F. Lin, D.H. Doherty, J.D. Lile, S. Bektesh, F. Collins (1993). 'GDNF: a glial cell line-derived neurotrophic factor for midbrain dopaminergic neurons'. *Science.* **260**: 1130–1132.
127. M.S. Airaksinen, M. Saarma (2000). 'The GDNF family: signaling, biological functions and therapeutic value.' *Nat. Rev. Neurosci.* **3**: 383–394.
128. Airaksinen M., A. Titievsky, M. Saarma (1999). 'GDNF family neurotrophic factor signaling: four masters, one servant'? *Mol. Cell. Neurosci.* **13**: 313–325.
129. Manié S., M. Santoro, A. Fusco, M. Billaud (2001). 'The RET receptor: function in development and dysfunction in congenital malformation'. *Trends Genet.* **17**: 580–589.
130. Baloh R.H., H. Enomoto, E.M. Johnson, J. Milbrandt (2000). 'The GDNF family ligands and receptors–implications for neural development'. *Curr. Opin. Neurobiol.* **10**: 103–110.
131. Viglietto G., S. Dolci, P. Bruni, G. Baldassarre, L. Chiariotti, RM. Melillo, G. Salvatore, G. Chiappetta, F. Sferratore, A. Fusco, M. Santoro (2000). 'Glial cell line-derived neutrotrophic factor and neurturin can act as paracrine growth factors stimulating DNA synthesis of Ret-expressing spermatogonia.' *Int. J. Oncol.* **16**: 689–694.
132. K. Yomogida, Y. Yagura, Y. Tadokoro, Y. Nishimune (2003). 'Developmental stage- and spermatogenic cycle-specific expression of transcription factor GATA-1 in mouse Sertoli cells'. *Biol. Reprod.* **69**: 1303–1307.
133. Kubota H., M.R. Avarbock, R.L. Brinster (2004). 'Spermatogonial stem cells share some, but not all, phenotypic and functional characteristics with other stem cells'. *Natl. Acad. Sci. USA.* **101**: 16489–16494.
134. Schlatt S., A. Meinhardt, E. Nieschlag (1997). 'Paracrine regulation of cellular interactions in the testis: an integrated system with hormones and local environment'. *Eur. J. Endocrinol.* **137**: 107.
135. Cudicini C., H. Lejeune, E. Gomez, et. al. (1997). 'Human Leydig Cells and Sertoli Cells are Producers of Interleukin-1 and -6. J'. *Clin. Endocr. Met.* **82**: 1426.
136. De SK., H.L. Chen, J.L. Pace, et. al. (1993). 'Expression of tumour necrosis factor-alpha in mouse spermatogenic cells'. *Endocrinology.* **133**: 389.
137. Huleihel M., E. Lunenfeld, S. Horowitz, et. al. (2000). 'Involvement of serum and LPS in the production of interleukin-1- and interleukin-6-like molecules by human sperm cells'. *Am. J. Reprod. Immunol.* **43**: 41.
138. Huleihel M., E. Lunenfeld, S. Horowitz, et. al. (2000). 'Production of IL-1- like molecules by human sperm cells'. *Fertil. Steril.* **73**: 1132.
139. Huleihel M., E. Lunenfeld, A. Blindman, et. al. (2003). 'Overexpression of interleukin-1?, interleukin-1? and interleukin-1 receptor antagonist in testicular tissues from sexually mature mice as compared to adult mice'. *Eur. Cytokine. Net.* **14**: 27–33.
140. Khan S.A., K. Schmidt, P. Hallin, et. al. (1988). 'Human testis cytosol and ovarian follicular fluid contain high amounts of interleukin-1-like factors'. *Mol. Cell. Endocrinol.* **58**: 221.
141. Potashnik H., M.Abu Elhija, E. Lunenfeld, et. al. (2005). 'IL-6 expression during normal maturation of the mouse testis'. *Eur. Cytokine. Netw.* **16**: 161.
142. Zeyse D., E. Lunenfeld, M. Beck, et. al. (2000). 'Interleukin-1 receptor antagonist is produced by Sertoli cells in vitro'. *Endocrinology.* **141**: 1521.
143. Lunenfeld E., D. Zeyse, M. Huleihel (1998). 'Cytokines in the testis, human sperm cells, and semen'. *Assist. Reprod* **9**: 157.

144. Elhija M.A., H. Potashnik, E. Lunenfeld, G. Potashnik, S. Schlatt, E. Nieschlag, M. Huleihel (2005). 'Testicular interleukin-6 response to systemic inflammation'. *Eur. Cytokine. Netw.* **16**(2): 167–172.
145. Huleihel M., E. Lunenfeld, D. Zeyse, M. Beck, I. Prinsloo, G. Potashnik, M. Mazor (2001). 'Immunohistochemical staining of IL-1 alpha and IL-1 receptor antagonist but not IL-1 beta in cultures of Sertoli cells'. *Am. J. Reprod. Immunol.* **45**: 135–141.
146. Wang J.E., G.M. Josefsen, V. Hansson, T.B. Haugen (1998). 'Residual bodies and IL-1alpha stimulate expression of mRNA for IL-1alpha and IL-1 receptor type I in cultured rat Sertoli cells'. *Mol. Cell. Endocrinol.* **137**: 139–144.
147. Kent Hamra F., M. Karen, W.U. Chapman Zhuoru, L. David (2008). 'Isolating Highly Pure Rat Spermatogonial Stem Cells in Culture'. *Germline Stem Cells.* 163–179
148. M.Abu Elhija, E. Lunenfeld, L. Persky, M. Huleihel (2008). 'Constitutive expression of IL-18 binding protein in murine testicular tissues and cells'. *Eur. Cyt. Netw.* **19**: 25–30.
149. M. Abu Elhija, E. Lunenfeld, T. Eldar-Geva, M. Huleihel (2008) Over-expression of IL-18, ICE and IL-18 R in testicular tissue from sexually immature as compared to mature mice. Eur. Cyt. Netw. **19**: 15–24.
150. O'Donnell L., K.M. Robertson, M.E. Jones, E.R. Simpson (2001). 'Estrogen and spermatogenesis'. *Endocr. Rev* **22**: 289–318.
151. Gotaroli R., D. Vindrieux, J. Selva, C. Felsenheld, A. Ruffion, M. Decaussin, M. Benahmed (2004). 'Characterization of tumour necrosis factor-alpha-related apoptosis-inducing ligand and its receptors in the adult human testis'. *Mol. Hum. Reprod* **10**: 123–128.
152. Tesarik J., C. Mendoza, R. Anniballo, E. Greco (2000a). 'In vitro differentiation of germ cells from frozen testicular biopsy specimens'. *Hum. Reprod.* **15**: 1713–1716.
153. Tesarik J., C. Mendoza, E. Greco (2000b). 'The effect of FSH on male germ cell survival and differentiation in vitro in mimicked by pentoxifylline but not insulin'. *Mol. Hum. Reprod.* **6**: 877–881.
154. Tesarik J., N. Cruz-Navaro, E. Moreno, M.T. Canete, C. Mendoza (2000c). 'Birth of healthy twins after fertilization with in vitro cultured spermatids from patients with massive *in vivo* apoptosis of postmeiotic germ cells'. *Fertil. Steril.* **74**: 1044–1046.
155. Tesarik J., E. Greco, C. Mendoza (2001). 'Assisted reproduction with in vitro cultured testicular spermatozoa in cases of severe germ cell apoptosis: A pilot study'. *Hum. Reprod.* **16**: 2640–2645.
156. Erkkila K., K. Henriksen, V. Hirvonen, S. Rannikko, J. Salo, M. Parvinen, L. Dunkel (1997). 'Testosterone regulates apoptosis in adult hum an seminiferous tubules in vitro'. *J. Clin. Endocrinol. Metabol.* **82**: 2314–2321.
157. Lee J., J.H. Richburg, S.C. Younkin, K. Boekelheide (1997). 'The Fas system is a key regulator of germ cell apoptosis in the testis'. *Endocrinology.* **138**: 2081–2088.
158. Creemers L.B., X. Meng, K. den Quden, A.M.M. Van Pelt, F. Izadyar, M. Santoro, H. Sariola, D.G. De Rooij (2002a). 'Transplantation of germ cells from glial cell line-derived neurotrophic factor over expressing mice to host testes depleted of endogenous spermatogenesis by fractionated irradiation'. *Biol. Reprod.* **66**: 1579–1584.
159. Creemers L.B., K. den Ouden, A.M.M. Van Pelt, D.G. De Rooij (2002b). 'Maintenance of adult mouse type A spermatogonia in vitro: Influence of serum and growth factors and comparison with prepubertal spermatogonial cell culture'. *Reproduction.* **124**: 791–799.
160. Sofikitis N., E. Pappas, A. Kawatani, D. Baltogiannis, D. Loutradis, N. Kanakas, D. Giannakis, F. Dimitriadis, K. Tsoukanelis, I. Georgiou, G. Makrydimas, Y. Mio, V. Tarlatzis, M. Melekos,

I. Miyagawa (2005). 'Efforts to create an artificial testis: Culture systems of male germ cells under biochemical conditions resembling the seminiferous tubular biochemical environment'. *Hum. Reprod. Update.* **11**: 229–259.

161. Gnessi L., A. Fabbri, G. Spera (1997). 'Gonadal peptides as mediators of development and functional control of the testis: An integrated system with hormones and local environment'. *Endocr. Rev.* **18**: 541–609.

162. Meistrich M.L., M.E. A.B. Van Beek (1993). 'Spermatogonial stem cells'. *Cell and Molecular Biology of the Testis. New York.* 266–295.

163. De Rooij D.G. (1998). 'Stem cells in the testis'. *Int. J. Exp. Pathol.* **79**: 67–80.

164. Bucci L.R., W.A. Brock, T.S. Johnson, M.L. Meistrich (1986). 'Isolation and biochemical studies of enriched populations of spermatogonia and early spermatocytes from rat testes'. *Biol. Reprod.* **34**: 195–206.

165. Bellve A.R. J.C. Cavicchia, D.A. Millette O'Brien, Y.M. Bhatnagar, M. Dym (1977). 'Spermatogenic cell of the prepuberal mouse: isolation and morphological characterization'. *J. Cell. Biol.* **74**: 68–85.

166. Marie-Claude H., S. Laura Braydich, D. Luis, J. Eric, D. Martyn (2005). 'Immortalization of Mouse Germ Line Stem Cells'. Stem Cells. **23**: 200–210.

167. Buom-Yong R., E. Kyle, H. Orwig, R. Mary, L. Avarbock. Ralph, Brinster, (2004). 'Phenotypic and functional characteristics of spermatogonial stem cells in rats'. *Developmental Biology.* **274**(1): 158–170.

168. Von Schonfeldt V., H. Krishnamurthy, L. Foppiani, S. Schlatt (1999). 'Magnetic cell sorting is a fast and effective method of enriching viable spermatogonia from Djungarian hamster, mouse, and marmoset monkey testes'. *Biol. Reprod.* **61**: 582–589.

169. Van der KS., E.W. Johnson, G. Dirami, M. Dym, M.C. Hofmann (2001). 'Immunomagnetic isolation and long term culture of mouse type A spermatogonia'. *J. Androl.* **22**: 696–704.

170. Amander T., S. Clark Megan, Bodnar, Mark, T. Fox Ryan, J. Rodriquez Michael, T. Abeyta. Meri, A. Firpo. Renee, P. Reijo (2004). 'Spontaneous differentiation of germ cells from human embryonic stem cells in vitro'. *Human Molecular Genetics.* **13**(7): 727–739.

171. Klassen H., M.R. Schwartz, A.H. Bailey, M.J. Young (2001). 'Surface marker expressed 440 d by multipotent human and mouse neural progenitor cells include tetraspanins and non-portein apitopes'. *Neurosci. Lett.* **312**: 180–182.

172. M. Oka, K. Tagoku, T.L. Russell, Y. Nakano, T. Hamazaki, E.M. Meyer, T. Yokota, N. Terada (2002). 'CD9 is associated with leukemia inhibitory factor-mediated maintenance of 445 embryonic stem cells'. *Mol. Biol. Cell.* **13**: 1274–1281.

173. Webb A., A. Li, P. Kaur (2004). 'Location and phenotype of human adult keratinocyte stem cells of the skin'. *Differentiation.* **72**: 387–395.

174. Yun J., et. al. (2004). 'Murine male germ cell apoptosis induced by bsulfan treatment correlates with loss of c-kit expression in a Fas/FasL and p53 independent manner'. *FEBS.* **575**: 41–51.

175. Choongkittaworn N.M., K.H. Kim, D.B. Danner, M.D. Griswold (1993). 'Expression of prohibitin in rat seminiferous epithelium'. *Biol. Reprod.* **49**(2): 300–310.

176. Yoshiaki Y., O. Kenji, S. Tomoaki, O. Masanori, S. Akinori, H. Yuri, S. Miyako, M. Isao (2001). 'A Novel Spermatogenesis-Related Factor-1 Gene Expressed in Maturing Rat Testis'. *Biochemical and Biophysical Research Communications.* **289**(4): 888–893.

177. Yujun. Xu. Eugene, L. Frederick, Moore, A. Renee, P. Pera (2001). 'A gene family required for human germ cell development evolved from an ancient meiotic gene conserved in metazoans'. *PNAS.* **98**(13): 7414–7419.
178. Wolkowicz, et. al. (1996). 'Refinement of the differentiated phenotype of the spermatogenic cell line GC-2spd(ts)'. *Bio. Of. Rep.* **55**: 923–932.
179. Nayernia K., J. Nolte, H.W. Michelmann, J.H. Lee, K. Rathsack, N. Drusenheimer A. Dev, G. Wulf, I. E. Ehrmann D.J. Elliott, V. Okpanyi, U. Zechner, T. Haaf, A. Meinhardt, W. Engel (2006a). 'In Vitro-differentiated embryonic stem cells give rise to male gametes that can generate offspring mice'. *Dev. Cell.* **11**: 125–132.
180. Siming LI., Z. Wentong, D. Lynn, G. Erwin (1998). 'Transgenic Mice Demonstrate a Testis-specific Promoter for Lactate Dehydrogenase', LDHC. *J. Biol. Chem.* **273**(47): 31191–31194.
181. Gyun Jee S., L. Hyun-Song, P. Yong-Suck, L. Ho-Joon, L. You-Sick, S. Joo Tae, K. Inn Soo (2000). 'Expression pattern of germ cell-specific genes in the testis of patients with non-obstructive azoospermia: usefulness as a molecular marker to predict the presence of testicular sperm'. *Fertility and Sterility.* **73**(6): 1104–1108.
182. Klaus S., B. Rüdiger, K. Ingrid, F. Gerhard, M. Weinbauer (2004). 'Expression of activator of CREM in the testis (ACT) during normal and impaired spermatogenesis: correlation with CREM expression'. *Molecular Human Reproduction.* **10**(2): 129–135.
183. Nicholas S., B. Foulkes, B. Nrico, S. Paolo (1992). 'Developmental switch of CREM function during spermatogenesis: from antagonist to activator'. *Nature.* **355**: 80–84.
184. Tesarik J., M. Bahceci, C. Ozcan, E. Greco, C. Mendoza (1999). 'Restoration of fertility by in-vitro spermatogenesis'. *Lancet.* 353–555.
185. Staub C.A. (2001). 'century of research on mammalian male germ cell meiotic differentiation in vitro'. *J. Androl.* **22**: 911–926.
186. Parks J.E., D.R. Lee, S. Huang, M.T. Kaproth (2003). 'Prospects for spermatogenesis in vitro'. *Theriogenology.* **59**: 73–86.
187. Cremades N., R. Bernabeu, A. Barros, M. Sousa (1999). 'In-vitro maturation of round spermatids using co-culture on Vero cells'. *Hum. Reprod.* **14**: 1287–1293.
188. Hue D., C. Staub, M.H. Perrard-Sapori, M. Weiss, J.C. Nicolle, M. Vigier et. al. (1998). 'Meiotic differentiation of germinal cells in three-week cultures of whole cell population from rat seminiferous tubules'. *Biol. Reprod.* **59**: 379–387.
189. Staub C., D. Hue, J.C. Nicolle, M.H. Perrard-Sapori, D. Segretain, P. Durand (2000). 'The whole meiotic process can occur in vitro in untransformed rat spermatogenic cells'. *Exp. Cell Res.* **260**: 85–95.
190. Sousa M., N. Cremades, C. Alves, J. Silva, A. Barros (2002). 'Developmental potential of human spermatogenic cells co-cultured with Sertoli cells'. *Hum. Reprod.* **17**: 161–172.
191. Tanaka A., M. Nagayoshi, S. Awata, Y. Mawatari, I. Tanaka, H. Kusunoki (2003). 'Completion of meiosis in human primary spermatocytes through in vitro coculture with Vero cells'. *Fertil. Steril.* **79**: 795–801.
192. Tesarik J., E. Greco, L. Rienzi, F. Ubaldi, M. Guido, P. Cohen-Bacrie et. al. (1998). 'Differentiation of spermatogenic cells during in-vitro culture of testicular biopsy samples from patients with obstructive azoospermia:effect of recombinant follicle stimulating hormone'. *Hum. Reprod.* **10**: 2772–2781.

193. Tesarik J. (2004). 'Overcoming maturation arrest by in vitro spermatogenesis: search for the optimal culture system'. *Fertil. Steril.* **81**: 1417–1419.
194. Kanatsu-Shinohara M., N. Ogonuki, K. Inoue, H. Miki, A. Ogura, S. Toyokuni, T. Shinohara (2003). 'Long-term proliferation in culture and germline transmission of mouse male germline stem cells'. *Biol. Reprod.* **69**: 612–616.
195. Kanatsu-shinohara M., H. Miki, K. Inoue, et. al. (2005). 'Long term culture of mouse male germ line stem cells under serum or feeder free conditions'. *Biol. Reprod.* **72**: 985–991.
196. Cremades N., M. Sousam, R. Bernabeu, A. Barros (2001). 'Developmental potential of elongating and elongated spermatids obtained after in-vitro maturation of isolated round spermatids'. *Hum. Reprod.* **16**: 1938–1944.
197. Hikim A.P.S., R.S. Swerdloff, Temporal (1995). 'stage-specific effects of recombinant follicle-stimulating hormone on the maintenance of spermatogenesis in gonadotropin-releasing hormone antagonist-treated rat'. *Endocrinology.* **136**: 253–261.
198. Tres L.L., A.L. Kierszenbaum (1983). 'Viability of rat spermatogenic cells in vitro is facilitated by their coculture with Sertoli cells in serum-free hormone-supplemented medium'. *PNAS.* **80**: 3377–3381.
199. Hadley M.A., S.W. Byers, C.A. Suarez-Quian, H.A. Kleinman, M. Dym (1985). 'Extracellular matrix regulates Sertoli cell differentiation, testicular cord formation, and germ cell development in vitro'. *J. Cell Biol.* **101**: 1511–1522.
200. Le Magueresse-Battistoni B., N. Gerard, B. Jegou (1991). 'Pachytene spermatocytes can achieve meiotic process in vitro. Biochemical Biophysical'. *Res. Commun.* **179**: 1115–1121.
201. Balaban B., B. Urman, A. Sertac, C. Alatas, S. Aksoy, R. Mercan, A. Nuhoglu (1999). 'In vitro culture of spermatozoa induced motility and increases implantation and pregnancy rates after testicular sperm extraction and intracytoplasmic sperm injection'. *Hum. Reprod.* **14**: 2808–2811.
202. Geijsen N., M. Horoschak, K. Kim, J. Gribnau, K. Eggan, G.Q. Daley (2004). 'Derivation of embryonic stem cells and male gametes from embryonic stem cells'. *Nature.* **427**: 148–154.
203. Toyooka Y., N. Tsunekawa, R. Akasu, T. Noce (2003). 'Embryonic stem cells can form germ cells in vitro'. *Proc. Nat. Acad. Sci.* **100**: 11457–11462.
204. Clark A.T., M.S. Bodnar, M. Fox, R.T. Rodriquez, M.J. Abeyta, M.T. Firpo, R.A. Reijo Pera (2004). 'Spontaneous differentiation of germ cells from human embryonic stem cells in vitro'. *Hum. Molec. Gene.* **13**: 727–739.
205. Bowles J., D. Knight, C. Smith, D. Wilhelm, J. Richman, S. Mamiya, K. Yashiro, K. Chawengsaksophak, MJ. Wilston, J. Rossant, H. Hamada, P. Koopman (2006). 'Retinoid signaling determines germ cell fate in mice'. *Science.* **312**: 596–600.
206. Koubova J., D.B. Menke, Q. Zhou, B. Capel, D. Michael, D.C. Page (2006) Retinoic acid regulates sex-specific timing of meiotic initiation in mice. *Proc. Natl. Acad. Sci.* **103**: 2474–2479.
207. Blanchard K.T., J. Lee, K. Boekelheide, Leuprolide (1998). 'A gonadotropin-releasing hormone agonist, reestablishes spermatogenesis after 2,5-hexanedione-induced irreversible testicular injury in the rat, resulting in normalized stem cell factor expression'. *Endocrinology.* **139**: 236–244.
208. Kanatsu-Shinohara M., K. Inoue, N. Ogonuki, H. Miki, S. Yoshida, S. Toyokuni et. al. (2007). 'Leukemia inhibitory factor enhances formation of germ cell colonies in neonatal mouse testis culture'. *Biol. Reprod.* **76**: 55–62.

209. Hasthorpe S., S. Barbic, P.J. Farmer, J.M. Hutson (2000). 'Growth factor and somatic cell regulation of mouse gonocytes-derived colony formation in vitro. *J. Reprod. Fertil.* **119**: 85–91.
210. Hasthorpe J.M., M.R. Avarbock, A.I. Telaranta, D.T. Fearon, R.L. Brinster (2006). 'Identifying genes important for spermatogonial stem cell self-renewal and survival. *Proc. Natl. Acad. Sci.* **103**: 9524–9529.
211. Hasthorpe S. (2003). 'Clonogenic culture of normal spermatogonia: In vitro regulation of postnatal germ cell proliferation'. *Biol. Reprod.* **68**: 1354–1360.
212. Dirami G., N. Ravindranath, V. Pursel, M. Dym (1999). 'Effects of stem cell factor and granulocyte macrophage-colony stimulating factor on survival of porcine type A spermatogonia cultured in KSOM'. *Biol. Reprod.* **61**: 225–230.
213. Izadyar F., K. Den Ouden, L.B. Creemers, G. Posthuma, M. Parvinen, D.G. De Rooij (2003). 'Proliferation and differentiation of bovine type A spermatogonia during long-term culture'. *Biol. Reprod.* **68**: 272–281.
214. Hasthorpe S., S. Barbic, P.J. Farmer, J.M. Hutson (1999). 'Neonatal mouse gonocytes proliferation assayed by an in vitro clonogenic method'. *J. Reprod. Fertil.* **116**: 335–344.
215. Feng L. X., Y. Chen, L. Dettin, R. A. Pera, J. C. Herr, E. Goldberg, M. Dym (2002). 'Generation and in vitro differentiation of a spermatogonial stem cell line'. *Science.* **297**: 392–95.
216. Nagano M., M.R. Avarbock, E.B. Leonida C.J. Brinster, R.L. Brinster (1998). 'Culture of mouse spermatogonial stem cells'. *Tissue and Cell.* **30**: 389–397.
217. Spradling A., D. Drummond-Barbosa, T. Kai (2001). 'Stem cells finds their niche. *Nature.* **414**: 98–104.
218. Chiarini-Garcia H., A.M. Raymer, L.D. Russell (2003). 'Non-random distribution of spermatogonia in rats: evidence of niches in the seminiferous tubules'. *Reproduction.* **126**: 669–680.
219. Yoshida S., M. Sukeno, Y. Nabeshima (2007). 'A vasculature-associated niche for undifferentiated spermatogonia in the mouse testis'. *Science.* **317**: 1722–1726.
220. Kanatsu-Shinohara M., K. Inoue, J. Lee, M. Yoshimoto, N. Ogonoki, H. Miki, S. Baba, T. Kato, Y. Kazuki, S. Toyokuni, M. Toyoshima, N. Ohtsura, M. Oshimura, T. Heike, T. Nakahata, F. Ishino, A. Ogura, T. Shinohara (2004a). 'Generation of pluripotent stem cells from neonatal mouse testis'. *Cell.* **119**: 1001–1012.
221. Kanatsu-Shinohara S., Toyokuni, T. Shinohara (2004b). 'CD-9 is a surface marker on mouse and male germline stem cells'. *Biol. Reprod.* **70**: 70–75.
222. Lee J.H. (2006). 'In vitro spermatogenesis by three-dimensional culture of rat testicular cells in collagen gel matrix'. *Biomaterials.* **27**: 2845–2853.
223. Stukenborg J.B., J. Wistuba, C.M. Luetjens, M. Abu Elhija, M. Huleihel, E. Lunenfeld, J Gromoll, E. Nieschlag, S. Schlatt (2007). 'Co-culture of spermatogonia with somatic cells in a novel three-dimensional Soft-Agar-Culture-System. *J. Androl.* [Epub ahead of print].
224. CukiermanE., R. Pankov, D.R. Stevens, K.M. Yamada (2001). 'Taking cell–matrix adhesion to the third dimension'. *Science.* **294**: 1708–1712.
225. Ito R., S.I. Abe (1999). 'FSH-initiated differentiation of newt spermatogonia to primary spermatocytes in germ-somatic cell reaggregates cultured within a collagen matrix'. *Int. J. Dev. Biol.* **43**: 111–116.
226. Rosso F., A. Giordano, M. Barbarisi, A. Barbarisi (2004). 'From cell–ECM interactions to tissue engineering'. *J. Cell Physiol.* **199**: 174–180.

227. Jae-Ho L., C. Myung, B. Gye, C. Kyoo Wan, J. Yup Hong, L. Yong Bok, P. Dong-Wook, L. Yeung Jae, M. Churl Min (2007). 'In vitro differentiation of germ cells from non-obstructive azoospermic patients using three-dimensional culture in a collagen gel matrix Fertility and Sterility'. **87**(4): 824–832.

228. Foresta C., A. Bettella, A. Ferlin, A. Garolla, M. Rossato (1998) Evidence for a stimulatory role of follicle-stimulating hormone on the spermatogonial population in adult male'. *Fertil. Steril.* **69**: 636–642.

229. Morena A.R., C. Boitani, M. Pesce, M. De Felici, M. Stefanini (1996) Isolation of highly purified type A spermatogonia from prepubertal rat testis'. *J Andro.* **117**: 708–717.

230. Li H., V. Papadopoulos, B. Vidic, M. Dym, M. Culty (1997). 'Regulation of rat testis tification of signaling mechanisms involved'. *Endocrinology.* **138**: 1289–1298

231. Schrans-Stassen B.H.G.J., H.J.G. Van de Kant, D.G. De Rooij, A.M.M. Van Pelt (1999). 'Differential expression of c-kit in mouse undifferentiated and differentiating type A spermatogonia'. *Endocrinology.* **140**: 5894–5900.

232. Hofmann M.C., S. Narisawa, R.A. Hess, J.L. Millan (1992) Immortalization of germ cells and somatic testicular cells using the SV40 large T antigen. Exp'. *Cell Res.* **201**: 417–435.

233. Hofmann M.C., R.A. Hess, E. Goldberg, J.L. Millan (1994) Immortalized germ cells undergo meiosis in vitro. *Proc. Natl. Acad. Sci.* **91**: 5533–5537.

234. Tascou S., K. Nayernia, A. Samani, J. Schmidtke, T. Vogel, W. Engel, P. Burfeind (2000). 'Immortalization of murine male germ cells at a discrete stage of differentiation by a novel directed promoter-based selection strategy'. *Biol. Reprod.* **63**: 1555–1561.

235. Griswold M. (2000). 'What can spermatogonial transplants teach us about male reproductive biology?' *Endocrinology.* **141**: 857–858.

236. Parreira G., T. Ogawa, M. Avarbock, L. Franca, R. Brinster, L. Russell (1998). 'Development of germ cell transplants in mice'. *Biol. Reprod.* **59**: 1360–1370.

237. Nagano M., M. Avarbock, R. Brinster (1999). 'Pattern and kinetics of mouse donor spermatogonial stem cell colonization in recipient testes'. *Biol. Reprod.* **60**: 1429–1436.

238. Kyle E., S. Orwig. S. Stefan (2005). 'Cryopresevation and transplantation of spermatogonia and testicular tissue for Preservation of male fertility. Journal of the national cancer institute monographs'. **34**: 51–56.

239. Barzon L. et. al. (2000). 'New perspectives for gene therapy in endocrinology'. *Eur. J. Endocrinol* **143**: 447–466.

240. Rubin H. (2002). 'The disparity between human cell senescence in vitro and lifelong replication *in vivo*'. *Nat. Biotechnol.* **20**: 675–681.

23

EPIGENETICS AS A CAUSE FOR THE ELOQUENT SILENCE OF STEM CELLS

*Shipra Bhatia and Vani Brahmachari**
Dr. B.R. Ambedkar Centre for Biomedical Research, University of Delhi, Delhi –110007

INTRODUCTION

Epigenetics is a fascinating phenomenon in biology. Its fundamental role in development and disease is appreciated only recently. The discussion in the following pages is an attempt to introduce the various biological processes where epigenetics has its footprints, but in no way it is exhaustive. The epigenetics of stem cells is a recent but rapidly growing field. Therefore, the information is derived largely from current research; hence resources are predominantly research papers published in journals. The phase of the field is similar to that of whole-genome sequencing efforts, when there is a considerable amount of data, entering the next phase of mechanistic understanding of the phenomena. The long list of references is an attempt to help the interested students and not meant to deter the reader.

WHAT IS EPIGENETICS?

The term 'Epigenetics' was coined by Conrad Hal Waddington in 1942 to define the phenotypic execution of developmental programs [1]. The discovery of DNA as the genetic material followed by deciphering of its structure led to its central place in biology. In 1980s, with the studies on chromatin structure, epigenetics came to the spotlight again. Today, epigenomics is recognized as a signature for the functional status of primary DNA sequences and has become one of the major foci

---

* vbrahmachari@acbr.du.ac.in

in the post-genome sequencing era. Epigenetic modifications refer to meiotically and mitotically heritable changes in gene expression that are not coded in the DNA sequence itself [2]. In mammals, epigenetic mechanisms including DNA-methylation [3], histone-modifications [4], chromatin-remodeling [5] and non-coding RNA-mediated regulatory processes [6] have profound regulatory roles in gene expression. Recent studies have shown that epigenetic regulators are key players in stem cell biology and their dysfunction can result in human diseases, such as cancer and neuro-developmental disorders [7].

Stem cells refer to cells that have two fundamental properties that define their 'stemness' characteristics: self-renewal and multipotency. During development, stem cells and the resulting progenitor cells are accountable for generating the functionally diverse cells and tissues of an organism. At adult stage, stem cells exist in many tissues throughout life and might play critical roles in tissue regeneration and repair. Mammalian embryonic stem cells (ESCs) derived from the inner mass of the blastocyst are pluripotent and not totipotent like the zygote, suggesting that they have lost the ability to differentiate into some of the cell types that are found in the organism during development. The zygote, which is said to be totipotent, has the potential to develop into a complete organism, as well as extraembryonic tissues. ESCs are experimentally derived from inner cell mass of blastocysts *in vitro* under specific culture conditions and have the ability to generate stem cells and subsequent differentiated cells of all the three germ layers, ectoderm, mesoderm and endoderm on transplantation into developing embryos as well as in culture dishes given an optimal *in vitro* condition [8]. ESCs are ideal stem cells for both research and cell-based therapies since they can be cultured almost indefinitely without altering their stem cell properties and that they have the potential to regenerate all types of cells and organs in an adult animal. The closest resemblance, but not identical counter part, of ESCs *in vivo* is epiblasts that appear at E6-6.5 (embryonic day) in mice. Unlike the zygote, ESCs cannot differentiate into extra-embryonic tissues such as yolk sac. The ESCs are the most accessible experimental system used to understand epigenetics of stem cells, as we will discuss in this chapter.

To manipulate the stem cells for therapeutics and as tools to understand development, it is important to understand all attributes of stem cells that lead either to the maintenance of 'stemness' and pluripotency or to differentiation. It is important to appreciate that the primary sequence of the genome in stem cells and differentiated cells are nearly identical, as we understand today. Though copy number variations and results of personal genome sequence have started throwing some surprises [9, 10]. Therefore the molecular signals on the genome for transcriptional silencing or activation should be over and above the primary sequence. This would basically imply that these would be the epigenetic (over the genetic) signals. It is also useful to remember that genes silenced in stem cells to maintain pluripotency would be activated during differentiation and similarly what is activated in stem cells need to be silenced to facilitate differentiation. Moreover, the completely differentiated adult once again produces germ cells that contribute new stem cells in the subsequent generation. Therefore we increase our demands on the epigenetic mechanisms that they have to be reversible, easily erasable and remade. It is also known that stem cells can be maintained as stem cells almost indefinitely in suitable culture medium. This demands a 'memory' component built into the epigenetic mechanisms that is stable through mitotic division.

There is a large effort towards understanding the epigenetic regulation resulting in the maintenance, loss and acquisition of pluripotency. In this chapter, we review epigenetic mechanisms in mammals and their role in determining stem cell signature and also in regulating stem cell functions.

STARK EXAMPLES OF EPIGENETIC PHENOMENA IN BIOLOGY

Historically, the existence of epigenetic regulation came to light through some of the interesting phenomena we encounter in genome regulation that have very little to do with genes alone. Though genome carries the blueprint of life in the DNA sequence which has been at the centre stage of biology for decades, it acts under the influence of an assortment of proteins and sometimes, RNAs, which pull strings, telling genes when and where to turn on or off. These regulatory networks constitute epigenetic factors. What is implied in the term 'epigenetic regulation' is more than gene activation and silencing in a tissue-specific manner. There is an element of 'memory' implied in the term as defined by Waddington. In most contexts, mitotic memory is implied; however, there are several epigenetic phenomena that could bear meiotic memory also. Thus, epigenetic regulation brings about phenotypic variation over and above that brought about by DNA sequence alterations due to mutation and recombination. A major part of the discussion in this chapter is focused on mitotically transmitted epigenetic regulation. However, there are several phenomena bearing meiotic memory that have puzzled biologists for over 50 years.

Genomic Imprinting

The work on genomic imprinting done till date is the outgrowth of one of those odd genetic phenomena observed by mule breeders 3000 years ago (Fig. 23.1). The observation that a mare crossed with a donkey yields a mule, whereas a stallion crossed with a donkey produces a hinny, which has shorter ears, a thicker mane and tail, and stronger legs than the mule, made researchers aware that there could be parent-specific effects in the offspring (Fig. 23.1). As you would note, even with an identical genetic constitution of one haploid set from donkey and the other from the horse, the two zygotes end up as different animals. This is a deviation from one of the well-known Mendelian tenants namely the equivalence of parental contribution. In other words, the non-equivalence of parental contribution to the zygote implies that genes remember their parental origin. The term genomic imprinting was coined by Helen Crouse to describe this phenomenon in relation to insects [11, 12].

The discovery that bi-parental contribution is necessary for successful development in mammals came from the elegant nuclear transfer experiments by two groups in 1984 [13, 14]. They discovered that embryos carrying either two maternal (egg) nuclei or paternal (sperm) nuclei could not complete development beyond early implantation stages. The basis of this is genomic imprinting, which leads to differences in transcription from the maternal and the paternal genomes; some genes being expressed exclusively either from the paternal genome (paternally expressed genes; *Peg*) or from the maternal genome (maternally expressed genes; *Meg*) (Fig. 23.2)

▶ **Fig. 23.1** *Epigenetics at work. The mule (foreground) is distinctly different from hinny (background), though both of them have similar diploid genome, one haploid set inherited from horse and the other from donkey, only the parental origin of the haploid genome in them is different and hence the set of imprinted genes are different.*

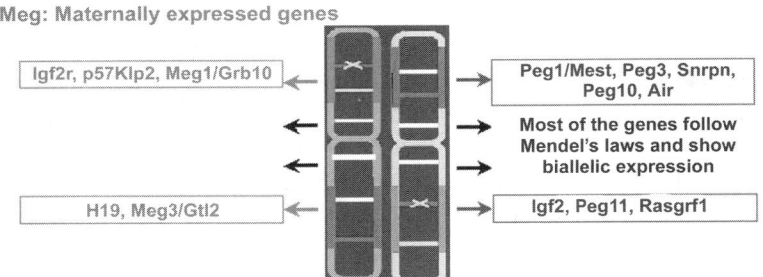

▶ **Fig. 23.2** *Effect of genomic imprinting on gene expression. Conventionally, imprinting implies repression. Human chromosome 11 depicted in pink is the maternal chromosome and in blue is the paternal chromosome. Note: genes that are expressed either only from maternal chromosome or from paternal chromosome, e.g., H19 gene is expressed from maternal chromosome while it is repressed on the paternal complement. Similarly, Igf2 is expressed from paternal chromosome and silenced on the maternal one. (for colour figure see Plate 57).*

About 80 genes are now known to be imprinted in mice and humans, and imprinting is conserved in protozoa, ruminants and angiosperms as well [15–18]. Imprinted genes play crucial role in embryonic and extraembryonic (placental) development and are often clustered in large domains [15]. Transgenic studies have identified that the allelic repression of imprinted genes is regulated by 'imprinting control regions' (ICRs). Imprinting control regions are regulated by epigenetic modifications, and have the following characteristic features [16, 19]: (i) they extend to several kilobases in length (ii) are rich in CpG dinucleotides (many correspond to CpG-islands) (iii) have DNA-methylation on one of the two parental alleles, the allelic methylation arising in the egg more often than in the sperm and (iv) allelic methylation marks are maintained throughout development after fertilization. There are several excellent reviews on genomic imprinting for interested readers [20, 21 www.geneimprint.com]. Since imprinted genes are involved in many aspects of development, including foetal and placental growth, cell proliferation and adult behaviour, alteration of normal imprinting patterns gives rise to numerous human genetic diseases, such as Prader-Willi and Angelman syndrome, childhood tumours like Wilms tumour (Table 23.1). The post-replication modification of DNA through cytosine methylation and post-translational modification of proteins interacting with DNA are part of the mechanisms of genomic imprinting, which are shared mechanisms with mitotically maintained epigenetic states and will be discussed in a later section.

Table 23.1 *Human Genetic Diseases Resulting from Aberrant Genomic Imprinting Patterns (http://www.geneimprint.com)*

Chromosome				
Human	Mouse	Imprinted allele	Relevant genes	Disease
4p16	–	Paternal	*RAF2* oncogene	Huntington disease
6pter–p12	17A–D	Paternal	Tumour necrosis factor (alpha and beta), *PIMI* oncogene	Insulin-dependent diabetes mellitus, spinocerebellar ataxia, juvenile myoclonic epilepsy
7q22–qter	6A–C	Maternal	*MET* oncogene	Cystic fibrosis
11pter-p15	–	Maternal	rhabdomyosarcoma	Weidemann–Beckwith syndrome
11p13–p15.5	7F	Maternal	Wilms, *WAGR*, *RAS* oncogene	Maturity-onset diabetes
15q11–.2	7	Maternal	PWCR1 non-coding RNA	Prader-Willi syndrome
15q11–q13	7	Paternal	UBE3A	Angelman syndrome
19cen–q13.32	7A-B	Maternal	$TGF-\beta$, *MEL* oncogene	Myotonic dystrophy

X Chromosome Inactivation

The mammalian answer to the dilemma of dosage compensation for X chromosome coded genes between males and females is X chromosome inactivation (XCI). This phenomenon has given us the insight into several fundamental mechanisms of epigenetic regulation and is also extensively reviewed [22, 23]. Out of the two X chromosomes in the female mammals, one of the X chromosomes gets inactivated around the 200-cell stage in development. The random inactivation

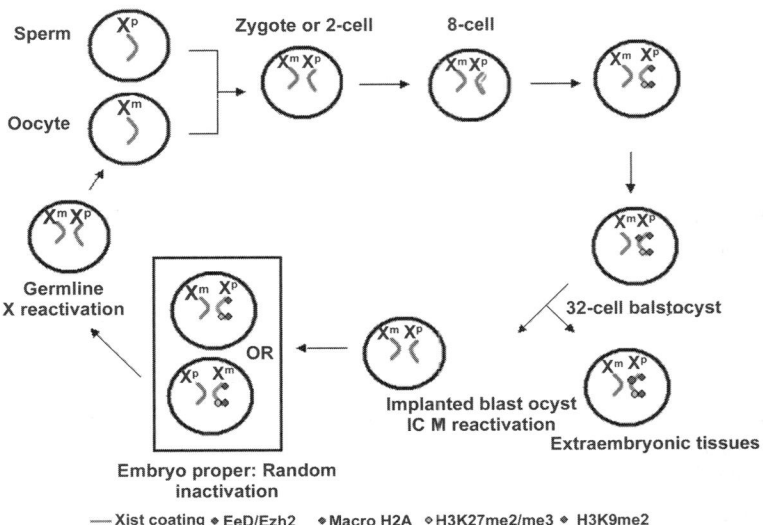

► **Fig. 23.3** *X chromosome inactivation in mammalian females. Sperm containing an inactive X from the father (Xp) fertilises an oocyte containing an active X from the mother (Xm) to form a female zygote with two X chromosomes. Partial Xist RNA coating of the Xp during early development (pink lines) correlates with incomplete silencing of genes on the Xp. Accumulation of the EED–EZH2 protein complex (purple circle), macroH2A variant (blue circle) and H3K27 di/tri-methylation (light blue circle) on the Xp are first seen at the 16-cell morula stage. Di-methylation of H3K9 (maroon circle) on the Xp is detected at the 32-cell blastocyst stage. Imprinted Xp inactivation stabilises in extraembryonic tissues. After the blastocyst implants, the inner cell mass (ICM) undergoes reactivation of the Xp, as characterised by the loss of Xist coating, of EED–EZH2 and macroH2A accumulation, and of H3K27 and H3K9 methylation. Cells of the embryo proper then undergo random inactivation in which either the Xm or the Xp are inactivated (red). In the germline of the female embryo, the inactive X is reactivated to ensure that all of the oocytes inherit an active X to pass on to the next generation. (for colour figure see Plate 58).*

of either the paternal or the maternal X chromosome in somatic tissues, disqualifies X inactivation as a genomic imprinting phenomenon [23]. However, once an X chromosome is inactivated in a cell in the developing embryo, the same chromosome is inactivated in all cells arising out of this cell, i.e. in the mitotic lineage, and therefore mitotic memory is maintained in X chromosome inactivation in mammals [22]. However, genomic imprinting operates in marsupials and the extraembryonic tissue of mouse embryos, where in only the paternal X chromosome is inactivated [22]. The analysis of X chromosome inactivation over several decades has revealed a plethora of interesting epigenetic mechanisms including the role of non-coding RNA molecules. A gene called *XIST* (Inactive X specific transcript coding gene) codes for a 17 kilobase RNA molecule that is not translated [22]. By binding to one of the two copies of the X chromosome, *XIST* sets the stage for a series of post-translational modifications of chromatin proteins and also DNA methylation resulting in the formation of the highly compacted X chromosome, which is cytologically described as heterochromatin and the Barr body [22] (Fig. 23.3). The mechanisms of this differential regulation of homologous chromosomes in the same nucleus has once again several overlaps with those of other epigenetic regulation in stem cells that are currently under investigation.

MARKERS OF EPIGENETIC IMPRINT

In mammals, epigenetic processes mainly include DNA-methylation [3], histone-modification [4] and non-coding RNA-mediated processes [6] (Table 23.2).

Table 23.2 *Epigenetic Modifications on the Genome and their Function in Gene Expression*

Epigenetic modification	Position or target	Effect
DNA methylation	CpG islands Non-CpG methylation (CpA, CpT)	Repression
Histone modification		
Methylation	H3 (K4,K36,K79)	Activation
	H3 (K9,K27), H4(K20)	Repression
	H3 (R17,R23), H4(R3)	Activation
Acetylation	H3 (K9,K14,K18,K56),	
	H4 (K5,K8,K13,K16)	Activatio
microRNA	mRNA	Repression

DNA Methylation

DNA methylation is a post-replication covalent modification of bases in DNA at cytosine at position C5 most commonly at CpG dinucleotides but not restricted to them. In mammals, over 70% of CpG dinucleotides are methylated and nearly all DNA methylation occurs on CpG dinucleotides that are underrepresented in the genome with the exception of CpG islands (CpG clusters). In fact, recently non-CpG methylation has been detected in stem cells at dinucleotides CpA and CpT [24]. The CpG islands of promoters and the first exon of active genes are generally unmethylated [25]. DNA methylation is catalyzed by several DNA methyltransferase (DNMTs). DNMT3a and 3b establish *de novo* DNA methylation, whereas the maintenance of DNA methylation depends on DNMT1 that specifically recognizes hemi-methylated DNA and methylates the unmethylated strand [26, 27]. DNMT2 does not have methyltransferase activity and its function is not clear [28]. DNMT3L (DNMT3-like) also lacks enzymatic activity, but could act as a cofactor for DNMT3a and 3b and modulate their enzymatic activity. Compared to DNA methylation, our knowledge of DNA demethylation is considerably more limited. It has been reported that MBD2 has DNA demethylase activity [29]; however, it remains controversial whether DNA demethylation is reversible and dynamically regulated in the absence of DNA replication.

DNA methylation represses gene transcription either by directly blocking the access of transcription factors to their binding sites or through indirectly recruiting methyl-CpG binding proteins (MBDs). The genes coding for MBDs family proteins include *MBD1, MBD2, MBD3, MBD4, MECP2,* and *KEISO* [30]. Mice lacking *Mbd1* protein (*Mbd1*-/- mice) have reduced neurogenesis and learning deficits [31], while *Mbd2*-/- mice have a reduced intestinal tumour incidence, defect in cytokine production and impaired maternal care behavior [32, 33]. $Mbd3^{-/-}$ is embryonic lethal [34], whereas mutation of *MBD4*, a DNA repair protein, leads to increased tumour incidence on susceptible genetic background [35]. Mutations in *MECP2* lead to neurodevelopmental deficits in both humans (Rett Syndrome) and in rodents [36, 37].

Mammalian DNA methylation has been implicated in a diverse range of cellular functions, including tissue-specific gene expression, cell differentiation, cell fate determination, genomic imprinting and X chromosome inactivation [38]. *Dnmt3a*-/- mice display normal phenotype at birth, but die at about 4 weeks after birth. *Dnmt3b*-/- mice exhibit developmental defects including growth impairment and rostral neural tube defects with variable severity at later stages [39]. In humans, *DNMT3b* mutation results in immunodeficiency, centromere instability and facial abnormality (ICF) syndrome with severe mental retardation [40, 41]. Thus DNA methylation is an essential epigenetic modification of the genome for normal development. Several cancers are characterized by hypermethylation resulting in an inappropriate gene silencing.

Histone Modifications

In the nuclei of all eukaryotic cells, genomic DNA is highly folded, constrained and compacted by histone and non-histone proteins generating a dynamic polymer called chromatin. For example, chromosomal regions that remain transcriptionally inert are highly condensed in the interphase nucleus and remain cytologically visible as heterochromatic foci or as the 'Barr body,' in case of inactive X chromosome in female mammalian cells [42]. The distinct levels of chromatin organization are dependent on the dynamic higher order structuring of nucleosomes, which represent the basic repeating unit of chromatin. In each nucleosome, roughly two superhelical turns of DNA wrap around an octamer of core histone proteins formed by four histone partners: one H3–H4 tetramer and two H2A–H2B dimers [43]. Histones are small basic proteins consisting of a globular domain and a more flexible and charged NH2-terminus (histone 'tail') that protrudes from the nucleosome. It remains unclear how nucleosomal arrays containing linker histone (H1) then twist and fold this chromatin fiber into increasingly more compacted filaments leading to defined higher order structures.

Central to our current discussion is that chromatin structure plays an important regulatory role and that multiple signaling pathways converge on histones [44]. Although histone proteins themselves come in generic or specialized forms [45], exquisite variation is provided by covalent modifications, such as acetylation, phosphorylation, methylation and ubiquitination of the histone tail domains (Fig. 23.4), which allow for variable affinity of histones and regulatable contacts with the underlying DNA. The enzymes transducing these histone tail modifications are highly specific for specific amino acid positions [46, 47], thereby extending the information content of the genome past the genetic/DNA code. This hypothesis, popularly known as the 'histone code hypothesis' predicts that:

(i) distinct modifications of the histone tails would induce alteration of interaction affinities for chromatin-associated proteins;

(ii) modifications on the same or different histone tails may be interdependent and generate various combinations on any one nucleosome;

(iii) distinct qualities of higher order chromatin, such as euchromatic or heterochromatic domains [47], are largely dependent on the local concentration and combination of differentially modified nucleosomes. It is envisioned that this 'nucleosome code' then permits the assembly of different epigenetic states [47], leading to distinct 'readouts' of the genetic information, such as gene activation versus gene silencing or, more inclusively, cell proliferation versus cell differentiation.

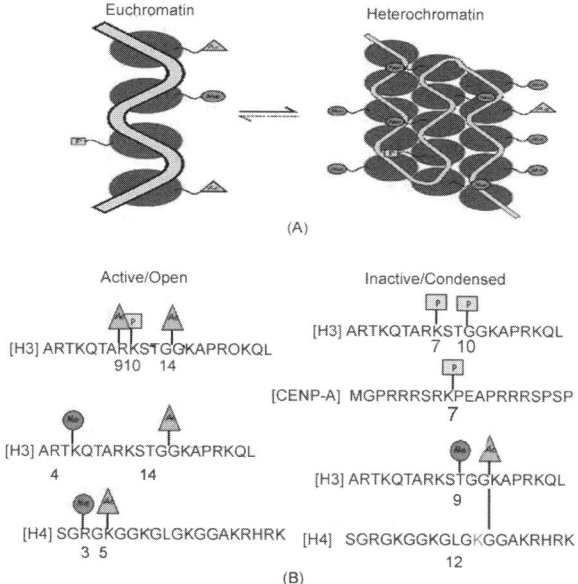

▶ **Fig. 23.4** *A diagrammatic representation of epigenetic marking on histones. (A) Schematic representation of euchromatin and heterochromatin as accessible or condensed nucleosome containing acetylated (Ac), phosphorylated (P) and methylated (Me) histone NH2-termini (B) Examples of combinatorial modifications in histone NH2-termini that are associated with active (green flag) or inactive (red circles) chromatin. Single-letter abbreviations for amino acid residues: A, Ala; E, Glu; G, Gly; H, His; K, Lys; L, Leu; M, Met; P, Pro; Q, Gln; R, Arg; S, Ser; and T, Thr. (for colour figure see Plate 59).*

Compared with DNA methylation, the post-translational modifications of histone display high levels of diversity and complexity. The core histones H2A, H2B, H3 and H4 are subject to dozens of different modifications, including acetylation, methylation and phosphorylation, [48]. Among these modifications, lysine (K) acetylation and methylation are the best-understood histone modifications. Lysine acetylation is generally associated with gene activation, whereas effects of lysine methylation depend on the specific lysine residue methylated in the histone. For example, methylation of H3K4 [49], H3K36 [50] or H3K79 [51] correlates with the active gene transcription; however, methylation at H3K9, H3K27 or H4K20 are usually linked to gene repression. In addition, mono-, di- and trimethylation at the same lysine residues leads to different levels of gene activation or repression and are involved in distinct cellular pathways. The term 'histone code' highlights the importance of histone modification in gene expression regulation.

The initial histone modification studies were focused largely on histone acetylation that is catalyzed by two opposing enzymes, histone acetyltransferase (HAT) and histone deacetylase (HDAC). To date, at least 6 HATs and HAT complexes and 18 HDACs have been identified in mammals [52, 53]. Many active transcription factors either recruit HATs or utilize their own internal HAT domains (e.g. CREB binding protein) to catalyze H3 and H4 acetylation and lead to accessible chromatin structure and gene activation. On the other hand, HDAC is frequently involved in gene repression. For example, neuron-specific genes are repressed by NRSF (Neuronal restricted silencing factor), which binds to a conserved DNA motif RE-1 (repressor element 1) and recruits HDAC1/2 and SIN3A to form a repressing complex [54–57]. NRSF is expressed in ESCs and the differentiation of ESCs into neural progenitors and neurons requires its degradation [58]. Treatment of adult hippocampal neural progenitors by Valproic acid (VPA), a HDACs inhibitor, leads to increased neuronal differentiation [59].

Similarly, histone methylation and demethylation enzymes are also involved in cellular functions. One of the best-studied histone methylations is the methylation of H3K9 by SUV39h, SETDB (mouse ESET) and G9a. These enzymes catalyze methylation of the histones at methylated DNA and are most probably recruited by methyl-CpG binding proteins (MBDs). This is followed by the binding of HP1 to form heterochromatin structure [60, 61]. In ESCs, proper H3K9 methylation is critical for regulated expression of imprinted genes [62]. The discovery of histone demethylases is highly significant as they provide the basis for reversibility of histone methylation, which is a hallmark of epigenetic mechanisms [63].

Non-coding RNA and Translational Regulation

Transcription factors are essential players in regulating stem cell functions [64, 65]. Recently, post-transcriptional gene regulation is emerging as another essential regulator of stem cell development [66]. It is likely that the coordinated regulation of gene clusters relies on transcriptional regulation for initial expression and post-transcriptional control for refinement. In 1960s, messenger RNAs (mRNAs) were demonstrated to carry the genetic information, while ribosomal RNAs (rRNAs) and transfer RNAs (tRNAs) that did not code for proteins but facilitated the protein synthesis were termed as non-coding RNAs. Twenty years later, small nuclear RNAs (snRNAs) involved in pre-mRNA splicing were identified. Today, non-coding RNA world has changed dramatically and some new members, such as microRNAs (miRNAs), small nucleolar RNAs (snoRNAs), short interfering RNAs (siRNAs), repeat-associated small interfering (rasi) RNA and PIWI interacting RNA (piRNA) have been discovered [67]. The variety of biological functions of these newly discovered non-coding RNAs are presently being revealed.

The biogenesis and functions of miRNAs are the best studied among these newly discovered non-coding RNAs. The gene encoding miRNA is transcribed into primary miRNA by RNA polymerase II, and RNase III DROSHA processes the primary miRNA into 70–100 nucleotide precursor miRNA. Precursor miRNA is transported into cytoplasm by exportin-5 and further processed by RNase III DICER to form mature miRNA [67]. Mature miRNAs are single stranded and composed of 17–25 nucleotides. Nearly 500 human miRNA sequences have been identified. miRNAs incorporate into a miRNA-induced silencing complex (miRISCs), and miRISC is directed to the target mRNAs based on sequence homology and leads to either degradation of mRNA or translational suppression by pairing with sequences in the 3' untranslated region of target mRNA [67–69]. miRNAs are especially attractive candidates for regulating stem cell functions as their interaction is not dependent on complete identity with the target mRNA; they can repress gene expression based on imperfect match and therefore regulate many targets simultaneously. This facilitates regulation of many genes with limited number of miRNA and also allows for coordinated control of many genes during stem cell differentiation. The importance of port-transcriptional regulation in development is illustrated by the lethality of *Dicer* mutations in mouse embryos as early as E7.5 [70]. Dicer-deficient ESCs fail to differentiate both *in vitro* and *in vivo* and exhibit defects in DNA methylation as well as histone modification patterns [71]. Argonaut proteins (AGO1-4) are key components of RISC complexes for miRNA pathway and are required for maintaining germline stem cells in *Drosophila, C. elegans* and mice [72].

Crosstalk among Epigenetic Mechanisms

The crosstalk among epigenetic pathways was initially demonstrated in invertebrates and plants first, and later confirmed in mammals. In *Arabidopsis thaliana*, the mutation of *KRYPTONITE*, a gene coding for H3K9 methyltransferase, leads to the loss of Cytosine DNA methylation, which resembles the phenotype produced by the mutation of DNMT CMT3 [73]. Similarly is heterochromatin which is repressive for gene transcription and is characterized by DNA methylation and enriched H3K9 trimethylation and heterochromatin protein 1 (HP1). The formation and maintenance of heterochromatin requires the coordination of DNA methylation, H3K9 methylation as well as the RNAi machinery [74]. The first evidence for the crosstalk between DNA methylation and histone modification in mammals is provided by studying *Suv39h* mutant mice. In ESCs with *Suv39h1* (H3K9 methylase) mutation, there was a decrease in DNA methylation because DNMT3b failed to localize to pericentric heterochromatic in absence of H3K9 methylation [75]. Although one study reported a contradictory result [76], MECP2 has been shown to recruit the SIN3-HDAC complex to methylated DNA and this effect could be reversed by HDAC inhibitor, trichostatin A [77, 78].

Functional interaction between small non-coding RNAs and other epigenetic mechanisms in gene expression regulation have recently been demonstrated in plants and invertebrates. Double-stranded RNA can induce both DNA and histone methylation in plants and yeast [79]. Small RNAs can, therefore, lead to mitotically heritable transcriptional silencing by the formation of heterochromatin in yeast [80–82]. However, the crosstalk between small non-coding RNA pathway and other epigenetic mechanisms in mammals have only been shown in a few examples, including X chromosome inactivation. Clustered miRNAs are localized in imprinted regions of human and mouse genome, suggesting potential role of DNA methylation in regulating the expression of these miRNAs [83]. Saito et al. found that treatment by DNA methylation inhibitor 5-aza-2'-deoxycytidine and histone deacetylase (HDAC) inhibitor 4-phenylbutyric acid leads to increased expression of certain miRNAs in human cancer cells [84]. *Mecp2* deficient neurons have altered expression levels of a subset of miRNAs [85]. On the other hand, DNMT3a, 3b and DNMT1 have been proposed as potential regulatory targets for miRNAs [86]. These data, along with the heterochromatin defects in *Dicer* mutant mice [87, 88] suggest that functional interaction between small RNA and epigenetic modulations is a critical mechanism regulating expression status of genes.

CELLULAR MEMORY MODULES

The development of an organism from a zygote to an adult involves a complex interplay of genes and gene products resulting in activation and silencing of a selection of genes depending upon the cell/tissue type. Many of the decisions are taken early in development as commitment which will be translated to a perceivable phenotype at a time point considerably removed from the primary commitment in a temporal scale. This requires that genes determined to follow a transcriptionally active course and the others chosen to be silenced should remember this determination through cell division so that the desired developmental end-point is achieved. Therefore, the ability to partition the genome into sets of active and quiescent genes, and to subsequently maintain this partitioning through cell divisions, underlies the process of cellular differentiation. In order to differentiate and maintain a specialized state, the gene expression profile of the progenitor cell has to be transmitted to daughter cells along lineages through mitosis. Experiments conducted in the 1960s and 1970s by Hadorn et al. [89] demonstrated that cells have an inherent 'cellular memory', also referred to as 'mitotic memory' allowing them to maintain developmental programs determined early in embryogenesis for the rest of the development process.

What is the molecular mechanism that maintains determined states? This question, reduced to the molecular level, inquires how the differential gene-expression patterns defining the specific cell type are maintained during DNA replication and at mitosis. This kind of functional segregation is also essential in stem cells to maintain their stemness or to differentiate. *Drosophila melanogaster*, the Cinderella of genetics, has also been developed as an ideal model to study the fundamental aspects of developmental biology as many of the genes involved in defining developmental states have been characterized painstakingly by ingenious genetic and molecular analysis that were recognized by Nobel Prizes in physiology or medicine to Edward B. Lewis, Christiane Nüsslein-Volhard and Eric F. Wieschaus in 1995. The extensive genetic analyses uncovered, for example, the homeotic (*HOX*) genes, the highly conserved class of regulators defining positions of structures and appendages along the anterior–posterior axis. Mutations in *HOX* genes transform one body segment into the identity of another segment (referred to as 'homeotic transformations'). It has been elucidated by experiments of Nüsslein–Volhard and Eric F. Wieschaus that the initial activation of the *HOX* genes is by what are referred to as segmentation genes classified into gap, pair-rule and segment polarity genes, which include transcription regulators and signaling proteins and have defined spatial and temporal sequence of expression during development of the Drosophila embryo (Box 23.2, Fig. 23.11).

Box 23.1 Techniques commonly used in the area of chromatin research

Bisulfite sequencing for mapping methylated cytosine residues (Fig. 23.5)

Bisulfite sequencing is based on the fact that treatment of DNA with *bisulfite* converts *cytosine* residues to *uracil*, but leaves *5-methylcytosine* residues unaffected. Bisulfite treatment thus introduces specific changes in the *DNA sequence* that depend on the *methylation* status of individual cytosine residues yielding very high-resolution DNA methylation profile.

Various analyses can be performed on the altered sequence to retrieve this information e.g. primer extension, Polymerase Chain Reaction (PCR), methylation sensitive restriction enzyme digestion, etc. The restriction enzyme digestion of PCR products of bisulfite-reacted DNA allows rapid analysis of patterns of regional methylation or demethylation of genomic DNA where an analysis of the methylation status of every CpG in the sequence is not required.

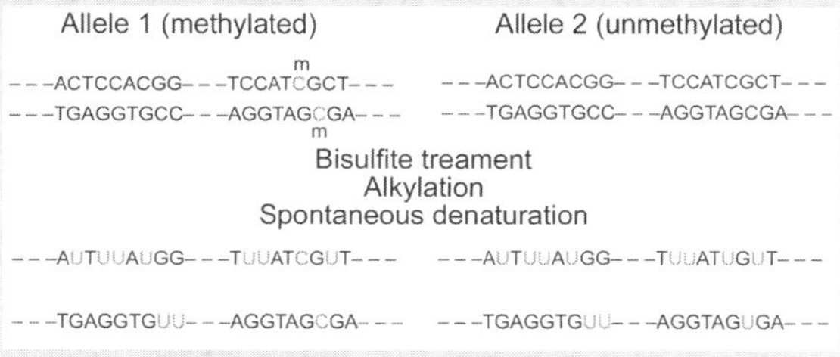

```
       Allele 1 (methylated)           Allele 2 (unmethylated)
                m
---ACTCCACGG---TCCATCGCT---      ---ACTCCACGG---TCCATCGCT---
---TGAGGTGCC---AGGTAGCGA---      ---TGAGGTGCC---AGGTAGCGA---
                         m
                    Bisulfite treament
                        Alkylation
                 Spontaneous denaturation

---AUTUUAUGG---TUUATCGUT---      ---AUTUUAUGG---TUUATUGUT---

---TGAGGTGUU---AGGTAGCGA---      ---TGAGGTGUU---AGGTAGUGA---
```

▶ **Fig. 23.5** *Outline of Bisulfite sequencing.*

Chromatin Immunoprecipitation (ChIP): *in vivo* **isolation of protein or protein complexes interacting with DNA (Fig. 23.6).**

The principle underpinning this assay is that *DNA*-bound proteins (including transcription factors) in living cells can be *cross-linked* to the *chromatin* where they are situated. This is usually accomplished by a gentle *formaldehyde* fixation, although it is sometimes advantageous to use the reversible crosslinker DTBP instead. Following fixation, cells are *lysed* and the DNA is sheared into fragments of 0.2–1 kb in length by *sonication*. Once proteins are immobilized on the chromatin and the chromatin is fragmented, whole protein-DNA complexes can be *immunoprecipitated* using an *antibody* specific for the protein in question. The DNA from the isolated protein/DNA fraction can then be purified. The identity of the DNA fragments in complex with the protein of interest can then be determined by *PCR* using primers specific for the DNA regions that the protein in question is hypothesized to bind. Alternatively, to map protein-binding sites across the whole *genome*, the DNA can be dissociated from the precipitated complex and hybridized on a *DNA microarray* (*ChIP-on-chip or ChIP-chip*) allowing for the characterization of the *cistrome*. As well, *ChIP-Sequencing* has recently emerged as a new technology that can localize protein-binding sites in a high-throughput, cost-effective fashion. Instead of hybridizing to DNA chip the DNA fragments are sequenced.

(*Contd.*)

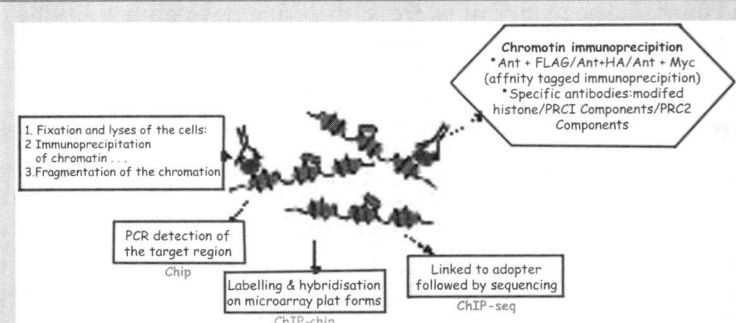

Fig. 23.6 *Outline of ChIp.*

The major disadvantage is the requirement for highly specific antibodies for each protein to be tested. This can be overcome by the construction of proteins fused to either *epitopes* (like HA, c-myc, FLAG) recognized by high affinity antibodies commercially available. ChIP is also widely applied for studying compositions of protein complexes. After de-crosslinking from the bound DNA, the immunoprecipitated proteins are identified by mass spectroscopy-based protein sequencing methods (MALDI, MUD-PIT, etc.).

ChIP-on-chip (also known as ChIP–chip) is a technique that combines chromatin immunoprecipitation ('ChIP') with microarray technology ('chip'). It allows the identification of sites of interaction of DNA-binding proteins efficiently and is also scalable. Recent developments include even whole-genome analysis to determine the location of such binding sites for any suitable protein of interest. The most prominent representatives of this class are transcription factors, such as p300, histones, their variants and modifications. The goal of ChIP-on-chip is to localize protein-binding sites, which might help in identifying functional elements in the genome. For example, in the case of a transcription factor as a protein of interest, one can determine its transcription factor binding sites throughout the genome. Other proteins allow the identification of promoter regions, enhancers, repressors and silencing elements, insulators, boundary elements, etc.

Starting with a biological question, a ChIP-on-chip experiment can be divided into three major steps: The first is to design the experiment by selecting system for analysis, the appropriate array and probe type. Second, the actual experiment (ChIP followed by hybridization on microarray) is performed in the wet-lab. Last, during the dry-lab portion of the cycle, gathered data are analysed to examine regions of interaction in the genome, genes and the non-coding DNA within these using bioinformatics tools.

ChIP-Sequencing, also known as **ChIP-Seq**, is the next frontier of technology used to analyze *protein* interactions with *DNA*. ChIP-Seq combines *chromatin immunoprecipitation* (ChIP) with massively parallel *DNA sequencing* to identify binding sites of DNA-associated proteins. It can be used to precisely and cost-effectively map global-binding sites for any protein of interest. After completion of an immunoprecipitation exercise, *oligonucleotide* adapters are then added to the small stretches of DNA that were bound to the protein of interest to enable massively parallel sequencing. After size selection, all the resulting ChIP-DNA fragments are sequenced simultaneously using a SOLEXA or a 454 platform, the new generation DNA sequencers. A single sequencing run can scan for genome-wide associations with high resolution, as opposed to large sets of *tiling arrays* required for lower resolution ChIP–Chip.

DamID: mapping of *in vivo* protein–genome interactions using tethered DNA adenine methyltransferase (Fig. 23.7).

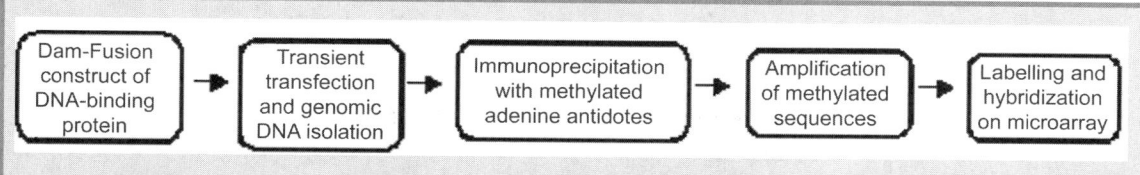

Fig. 23.7 *Outline of Dam ID*

A large variety of proteins bind to specific parts of the genome to regulate gene expression, DNA replication and chromatin structure. DamID is a powerful method used to map the genomic interaction sites of these proteins *in vivo*. It is based on fusing a protein of interest to DNA adenine methyltransferase (dam) of *Escherichia coli*. Expression of this fusion protein *in vivo* leads to preferential methylation of adenines in DNA surrounding the native binding sites of the dam fusion partner. Since adenine methylation does not occur endogenously in most eukaryotes, it provides a unique tag to mark protein interaction sites. The adenine-methylated DNA fragments are immunoprecipitated with antibodies against methylated adenine and then sequences are amplified by PCR and identified by microarray hybridization.

Independently, certain mutants discovered in Drosophila showed homeotic transformations similar to that seen in *HOX* gene mutants, but were not mutations in *HOX* genes [90]. These genes apparently were regulators of *HOX* genes and did not fall into any of the segmentation gene classes mentioned above. But still when they were mutated it was as if some of the *HOX* genes themselves were mutated. This suggested that these genes might be regulating homeotic genes and hence they mimic homeotic mutations. These were referred to as Polycomb (Pc) mutations referring to its phenotype; in heterozygous state, mutants have extra sex combs on the metathoracic segment, which is normally present on the mesothorax (Fig. 23.8). The homozygous mutations in Pc are late embryonic lethals [90]. Further genetic and molecular studies over the years demonstrated that Pc and other genes with similar functions could be classified into two antagonizing groups, Polycomb (*PcG*) and trithorax (*TrxG*) that were required to maintain gene-expression patterns of important developmental regulators like the *HOX* genes during cellular proliferation. Thus, the PcG and TrxG proteins along with the cognate-binding sites Polycomb/Trithorax Responsive Elements (PRE/TRE) appear to form the molecular basis of the cellular memory. Several mammalian, including human homologs of PcG and TrxG genes have been discovered across phyla [www.igh.cnrs.fr/equip/cavalli/]. Recently, their role in stem cell maintenance is also coming to light.

The Polycomb and Trithorax Complexes

Genetic and molecular analyses of cellular memory in Drosophila led to the basic understanding of the underlying mechanism, although many specific aspects are still unclear or controversial. In *D. melanogaster*, the role of PcG proteins in the control of the homeotic gene expression begins in the 3-h-old embryo, shortly after the homeotic genes have been turned on and their characteristic domains of expression have been defined by transient expression of segmentation gene products, which work as activators or repressors of homeotic genes. The action of PcG proteins first becomes detectable at gastrulation when the early regulators begin to disappear and they take over

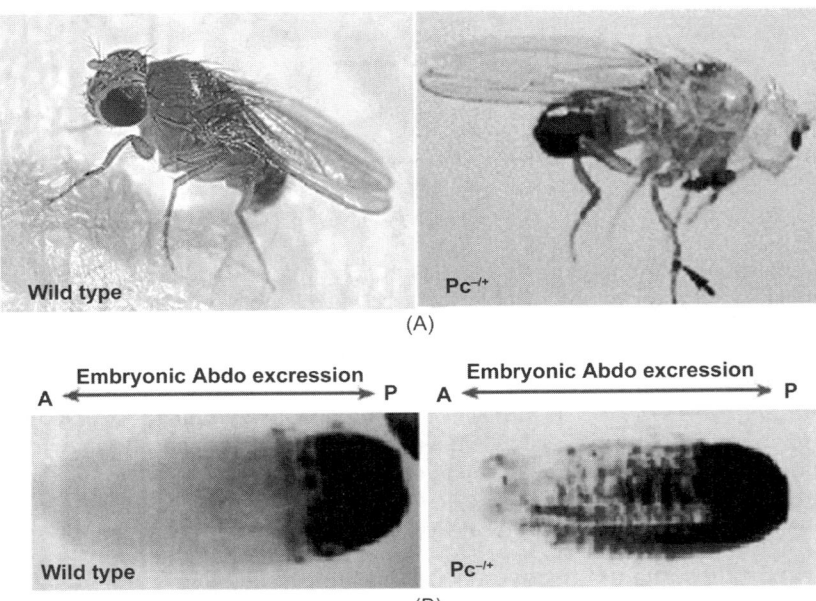

▶ **Fig. 23.8** *Mutations in Polycomb genes mimic homeotic mutations. (A) In wild type flies, the sex comb is restricted to the first pair of legs, while in the heterozygous state (Pc–/+), mutant adult flies manifest extra sex combs on second and the third pairs of legs, therefore, transforming posterior segments to anterior like. The arrow in wild type points to the first pair of legs while that in Pc–/+ points to the second pair. The first pair is not visible from this angle. (B) Effect of the mutation at early embryonic stages traced intragenic flies carrying β-galactosidase as reported under homeotic gene promoter. Mutations in Polycomb genes cause a de-repression of specific Hox members, which leads to extended zone of expression and hence homeotic transformations of one body segment into the identity of another. (for colour figure see Plate 60).*

the function to prevent gene reactivation. The experiments of Lewis suggested that silencing is as important as activation in gene regulation and more so in development. The maintenance of the repressed state of target homeotic genes which are chosen to be silenced early in development is the responsibility of PcG proteins (Fig. 23.8). As a consequence, a target gene is competent to be activated at later stages only in the progeny of cells in which it was active in the early embryo. This dependence on the history of the gene implies that cellular memory marks a previously silenced or the active state of the gene, so that it continues to maintain this status after every cell cycle. Genetic analysis has shown that a system antagonistic to PcG exists, which involves the Trithorax (TRX) proteins. These protein complexes can set a mark for the active state of a PcG target genes [91, 92] (Table 23.3). In the absence of TRX, a homeotic gene can become repressed by the PcG-mediated mechanism even in cells in which it was active in the early embryonic stages. A similar antagonistic relationship is thought to be involved in mammalian PcG mechanisms but considerably less direct evidence is available. The shared DNA interacting sites between these PcG and TrxG complexes

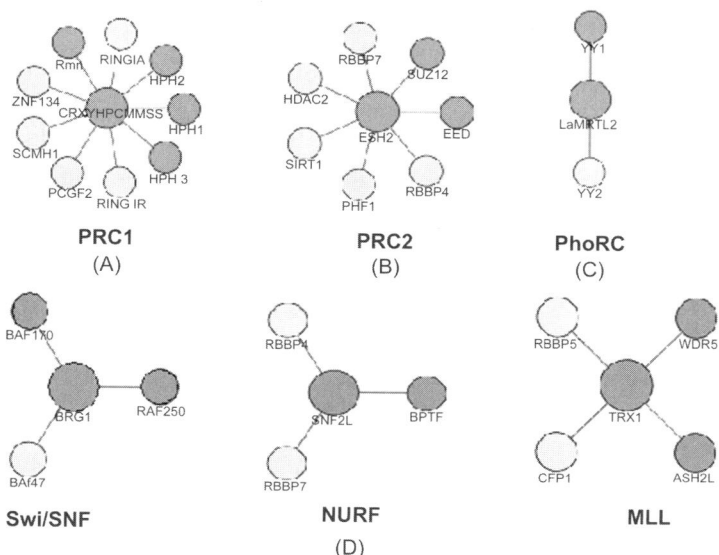

▶ **Fig. 23.9** *Networks of components of human Polycomb (PcG) and Trithorax (TrxG) proteins. The core enzymatic subunits of PcG (red) and TrxG (green) complexes are shown in interaction with other complex members (yellow). Note the two components (RBBP4 and RBBP7) shared between PRC2 and NURF that are known to have antagonistic activity, an example of cros-stalk and combinatorial use of parts to make two distinct complexes. The networks were generated using Cytoscape software from the public domain. (for colour figure see Plate 61).*

further reinforce the antagonistic effect of these complexes. These sites are referred to as Polycomb/Trithorax responsive elements (PRE/TRE). For more discussion of this aspect the reader can refer to excellent review articles [93].

The classification of a protein either as a PcG or Trx member is related to its interaction with Polycomb protein. In the classical examples any gene coding for a function which modulates the phenotype caused by a mutation in Polycomb gene is classified either as a PcG member or a Trx member depending on the effect it has on the phenotype of the mutant. For instance if the mutation enhances the PcG phenotype, it is classified as PcG member while the one which attenuates the Polycomb mutant phenotype is classified as a TrxG member [92]. These are generally assessed in Drosophila by genetic interaction between mutants. However, in the mammalian system it is only an extension of the criteria used in Drosophila because the classification is based on the homology with Drosophila proteins. To date there is no experimental system in mammals which facilitates phenotypic assessment similar to that in Drosophila. However, both systems share the common theme of chromatin modification as the molecular basis of CMM function. In this section of the chapter we present a brief overview of the known PcG, TrxG complexes and the molecular basis of their function as cellular memory modules.

Polycomb-Group (PcG) Proteins

PcG silencing involves at least three kinds of multiprotein complexes that work together. They are referred to as PRC1, PRC2 and PhoRC complexes (Fig. 23.9). All complexes accomplish the modification of histones and therefore the chromatin to achieve transcriptional repression. Every complex has a set of core components (four or five proteins) and additional proteins might associate with them depending on the site of interaction and the cell type. Each complex normally consists of proteins that aid in different steps to ultimately modify the chromatin components, most often histones. Typically, this would include DNA binding, opening up of the chromatin, histone methylation, deacetylation and recruiters and enhancers of these catalytic activities. Each complex has its own set of proteins, some being common between complexes and others unique to the complex. This is more like modules of the Lego set where different combinations of the pieces can be used to create unique structures. You would appreciate the clever strategy that nature has taken through this approach so that it can work with a limited number of basic components and use it in different combinations to generate a large repertoire of unique regulatory complexes.

Box 23.2 Homeotic genes

Path-breaking research endeavours of *Christiane Nüsslein-Volhard* and *Eric F. Wieschaus* identified and classified 15 genes of key importance in determining the body plan and the formation of body segments of the fruit fly *Drosophila melanogaster*. These set of genes, referred to as segmentation genes, classified into gap, pair-rule and segment polarity genes, which include transcription regulators and signaling proteins and have defined spatial and temporal sequence of expression during development of the Drosophila embryo (Fig. 23.10) *Edward B. Lewis* studied the downstream targets of segmentation genes – the homeotic genes that govern the development of a larval segment into a specific body segment.

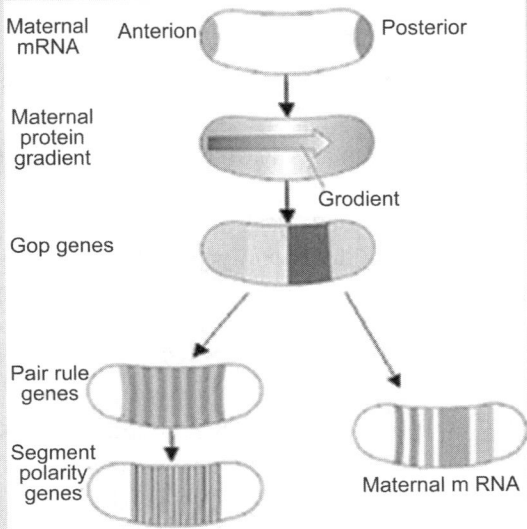

▶ **Fig. 23.10** *Cascade of gene activity in early development.*

Homeotic transformation in development means change of one body part to the likeness of another. Lewis found collinearity in time and space between the order of the genes in Drosophila homeotic complex, their expression in segments. These genes set up the basic regional layout of an organism, so that eyes form on the head and not on the abdomen, and limbs form at sides and not on the head. Even a single mutation in these genes can have drastic effects for the organism and so these genes have changed relatively less over time.

Homeotic genes are defined by a DNA sequence known as the homeobox, which codes for a 60 amino acid helix-turn-helix protein known as the homeodomain. The homeodomain endows a DNA-binding activity on products of Hox genes, enabling them to act as an 'on/off' switch for gene transcription by binding to specific sequences enhancers of a gene, which

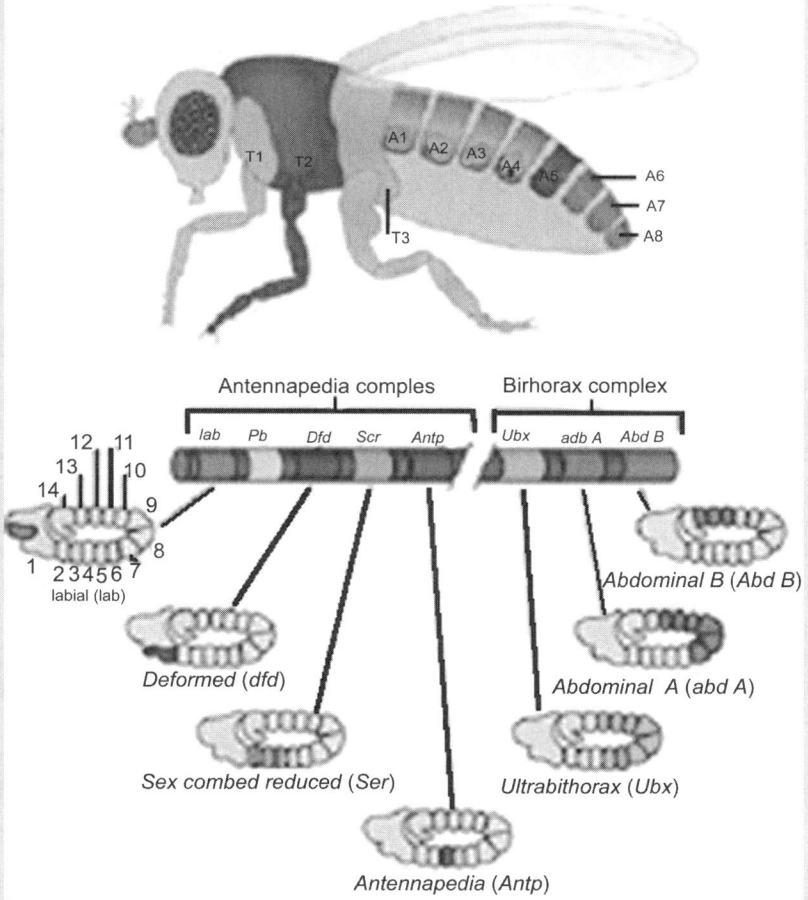

▶ **Fig. 23.11** *Homeotic gene expression in Drosophila. In the centre are genes of antennapedia and bithorax complex and their functional domains. Below and above the gene map, the region of the homeotic gene expression (both mRNA and protein) in the blastoderm of Drosophila embryo and regions that form from them in the adult fly are shown.*

(Contd.)

> either activates or represses the gene. In Drosophila, the protein product of the homeotic gene Antennapedia activates genes that specify structures of the second thoracic segment, which contains a leg and a wing, and represses genes involved in eye and antenna formation. Thus, legs and wings, but not eyes and antennae, will form wherever the Antennapedia protein is located. The genes regulated by homeobox proteins are called realisator genes, and it is the protein products of realisator genes that make tissues, organs and structures (legs, eyes, wings, etc).(Fig. 23.11) Apart from the pair-rule and gap genes, MicroRNA originating in HOX clusters have been shown to inhibit more anterior HOX genes, possibly to refine its expression pattern. Non-coding RNA (ncRNA) has been shown to be abundant in Hox clusters. In humans, 231 ncRNA may be present. One of these, HOTAIR, silences in transcription (it is transcribed from the HOXC cluster and inhibits late HOXD genes) by binding to Polycomb-group proteins (PRC2).

PRC1 Complex

This complex was first purified from Drosophila [94, 95]. The core of the complex contains a quartet of PcG proteins: Polycomb (PC), Posterior sex combs (PSC), polyhomeotic (PH) and RING (Table 23.3). Many additional proteins were also co-purified with these core components, including ZESTE, TBP (TATA-box binding protein)-associated factors TAFII250, TAFII110, TAFII185, and TAFII62, and elements of other multiprotein complexes, such as MI2, SIN3A, SMRTER (Table 23.3). The mammalian PRC1 complex (Fig. 23.9A) has been isolated from HeLa cells exploiting the interaction between proteins, using one protein as ligand to isolate other components of the complex. To facilitate this, the protein used as the ligand is cloned and tagged with a known peptide sequence that can be used to interact with complementarily tagged affinity matrix (Box 23.1). The core components of PRC1 from HeLa cells, are similar to those of the *D. melanogaster* PRC1 complex, although no transcription activating factors (TAF) have been detected in association with it.

The number of homologs of PcG proteins in mice and humans is higher relative to that in *D. melanogaster*. This presumably helps to generate complexes with tissue specific or gene specificity through combinatorial interactions. The purified complex(es) include HPC1, 2 and 3, HPH1, 2 and 3, RING1A and B, BMI1 and, potentially, its homolog MEL18 (Table 23.3). These are, correspondingly, homologous to PC, PH, RING and PSC proteins of Drosophila. It has been reported that in different immunoprecipitation experiments using the same cell line, for example HeLa cells, with the same antibody targeted against a component of PcG, one might pull out complexes with slightly different compositions [96]. This might indicate the dynamic nature of the complex to match tissue variability. However, one also needs to keep in mind variations in the affinity of the antibodies utilised.

The functional domains that are associated with the PRC1 complex are the chromodomain of Polycomb which bind specifically to trimethylated lysine27 of histone H3 (H3K27) and the RING domain of RING (Drosophila) and RING1A and B in mammals that function as E3 ubiquitin ligases that mono-ubiquitylate lysine 119 of histone H2A. Presence of BMI1 and the homolog of PSC in flies enhance this activity.

Table 23.3 *Major Components of Polycomb/Trithorax Maintenance System*

Drosophila	Complex	Human protein	Protein domains	Biochemical activity
Protein	PRC1		Chromodomain	Binding to H3K27Me3
	PRC1	HPC1/2/3	SAM	?
Polycomb-group	PRC1	HPH1/2/3	RING	Cofactor for SCE
PC	PRC1	BMI1	RING	E3 ubiquitin ligase specific to H2AK119
	PRC1?	RING1B	SAM, MBT, Zn-finger	H2AK119
PH				
PSC				
SCE (RING)				
SCM				
E(Z)	PRC2	EZH2	SET	Methylation of H3K9, H3K27
ESC	PRC2	EED	WD40	
ESCL	PRC2	—	WD40	Cofactor for E(Z)
SU(Z)12	PRC2	SUZ12	Zn-finger	Cofactor for E(Z)
PCL	PRC2	—	PHD, Tudor	?
				?
PHO	PhoRC	YY1	Zn-finger	DNA binding
PHOL	?	YY2	Zn-finger	DNA binding
CG16975 (SFMBT)	PhoRC	L3MBTL2	MBT, SAM	Binding to mono- and dimethyl H3K9,
SU(Z)2	?		RING,	H4K20
SXC	?		?	?
ASX	?		PHD	?
MXC	?		LA, RRM	?
E(PC)	?		?	?
	TAC1	TRX1/MLL	PHD, SET,	Methylation of H3K4
Trithorax	?	–	SET, PHD, BAH	Methylation of H3K4,
group	?	–	PHD, SPRY	H3K9, H4K20
TRX	SWI/SNF	BRG1	SNF2, HELICc,	?
ASH1	SWI/SNF	BAF170	Bromo	ATP-dependent nucleosome sliding
ASH2	SWI/SNF?	BAF250	SWIRM, SAINT	Cofactor for BRM
BRM	?		BRIGHT	?
MOR				
OSA				

PRC2 Complex

The core of the PRC2 complex purified from flies consists of four proteins: Enhancer of Zeste E(z), Extra Sex Comb, ESC or its homolog ESCL, P55 and Supressor of Zeste Su(Z)12 (Fig. 23.9B) (Table 23.3). All these proteins contribute to the methyltransferase activity of the complex in trimethylation of H3K27 and H3K9. Several isoforms of PRC2 have been detected in flies as well as in mammalian system. In Drosophila embryos, the PRC2 600 kDa complex is the most abundant and contains histone deacetylase RPD3 in addition to the core components. Another larger 1MDa embryonic complex includes both RPD3 and PCL (Polycomb-like). These two proteins seem to interact with each other but how they bind to the rest of the complex has not been described. In Drosophila larvae, another variant of PRC2 was found to contain the NAD^+-dependent histone deacetylase SIR2. In mammals, the story is more complicated. PRC2, PRC3 and PRC4 complexes have been biochemically characterized, and they differ by the presence of different isoforms of EED, the homolog of the fly ESC [97, 98]. In the presence of histone H1, PRC2, but not PRC3, preferentially methylates H1K26, which has an amino acid context similar to that of H3K27, although it is present in only one of the mammalian H1 variants and absent in H1 of *D. melanogaster*. PRC4 accumulates when EZH2, the mammalian E(Z), is over-expressed in cultured cells. It contains an EED isoform that is expressed only in undifferentiated ES cells, and SIRT1, a mammalian SIR2 homolog. In mammals, the knockout of *SUZ12* decreases the level of EZH2 protein and results in the loss of di- and trimethyl K27, but not monomethylation of H3K27 thus indicating on one hand that the formation of the complex stabilizes EZH2 and on the other that a different EED-containing complex monomethylates H3K27. The PRC2 complex and its components seem to be the functional core and the most ancient part of the PcG mechanism [99].

PhoRC Complex

PHO and its closely related homolog, PHOL, are the only PcG proteins that are known to bind directly to DNA (Fig. 23.9C) (Table 23.3) [100–102]. These two proteins are *D. melanogaster* homologs of the mammalian factor Yin–Yang 1 (YY1), so named because it has both activating and repressive functions [103]. Although it has been reported to interact with both PRC1 and PRC2 complexes in flies and mammals, PHO is not an important constituent of either of these two purified complexes. Instead, PHO has been found in two other kinds of complex. In one of these complexes it is associated with the chromatin remodeling machine, INO80, which is not known to be involved in PcG mechanisms [104, 105]. The second PHO-containing complex in *D. melanogaster*, PhoRC, is involved in homeotic gene silencing and includes an MBT (malignant Brain Tumour)-domain protein, SFMBT. The MBT domain is found in many mammalian homologs, one of which is the mouse SFMBT (*Scm*-related gene containing four MBT domains). Although not previously known in *D. melanogaster*, the fly SFMBT homolog functions as a bona fide PcG protein and is required for PcG silencing. Its MBT repeats bind specifically to mono- and dimethylated H3K9 and H4K20.

Trithorax Group (TrxG) Proteins

The trithorax group of proteins work antagonistically to Polycomb group of proteins. All PcG target genes are known to be positively regulated by the Trithorax group of proteins in regions where they are expressed. Whereas the PcG proteins seem to be dedicated to their target genes, many of the TrxG proteins form complexes that are involved in general transcriptional processes; thus, their function is not limited to epigenetic maintenance of gene activity (Fig. 23.9D) [106, 107]. Exceptions are the TRX and ASH1 proteins, which are involved more specifically in regulation at Polycomb Response Elements (PREs) [108, 109]. In *TRX* or *ASH1* mutants of Drosophila, expression of target genes is repressed in a way that is dependent on PcG regulation. If both PcG silencing and *Trx* are impaired, expression returns to near normal. A careful analysis shows that the loss of *TRX* causes PcG target genes to be silenced, even in regions in which they should remain derepressed. Therefore, it seems that ASH1 and TRX function as anti-repressors rather than typical activators. The molecular basis for this antagonistic activity is not obvious. Most or all known or presumptive PREs also bind TRX constitutively, whether or not the target gene is active or silenced [110–112], and thereby function as Trithorax response elements (TREs). The same DNA sequence can therefore behave as a PRE or a TRE, depending on the early events that set the epigenetic state of the gene. The mammalian TRX homolog, MLL (or ALL) was first discovered because translocations that fuse MLL to various transcriptional regulators are involved in certain human acute lymphoblastic leukemia. In mice, as in *D. melanogaster*, loss of MLL function is lethal. Heterozygosity for MLL causes homeotic transformations by decreasing the expression of Hox genes and reducing the anterior boundary of Hox gene expression in Drosophila. Other phenotypes are growth retardation and haematopoietic abnormalities [113].

Four complexes that contain TrxG proteins have been purified from *Drosophila* embryos, all with different chromatin-modifying properties. These complexes are reviewed in detail elsewhere [114]; here we give an overview and an update on recently identified enzymatic activities. The 2-MDa BRM complex (Fig. 23.9D) contains the TrxG proteins Brahma (BRM), Moira (MOR) and OSA, and at least four other accessory proteins [106, 115]. The BRM complex is highly related to yeast SWI/SNF nucleosome-remodeling complex, and the BRM protein (SWI2/SNF2 in yeast) functions as the ATPase subunit of this complex, using the energy of ATP hydrolysis to alter the position of histones. Multiple-related complexes are found in humans [115 and references therein]. Recently, two distinct versions of the BRM complex have been identified in *Drosophila* [116].

Two other *Drosophila* complexes of 2 MDa and 500 kDa, respectively, contain the ASH1 and ASH2 TrxG proteins [115]. The fourth known *Drosophila* TrxG complex, TAC1 [117], of 1 MDa contains the TrxG protein TRX, the histone acetyltransferase CBP (CREB binding protein) and the antiphosphatase Sbf1. The TRX protein contains a SET domain, which is highly similar to that of the yeast protein, SET1. The *S. cerevisiae* SET1 protein was the first histone methyltransferase shown to methylate lysine 4 on histone H3 (H3K4) [118]. This, together with the demonstration that SET1 is in a complex with a yeast homolog of ASH2 gave the first indication that TrxG function in higher eukaryotes might involve H3K4-specific methyltransferases [118]. Indeed, the SET domain containing proteins ASH1 and TRX were both subsequently shown to be histone methyltransferases, primarily targeting H3K4 [119, 120]. H3K4 methylation is typically associated with active chromatin [121].

In summary, all the TrxG complexes thus far identified contain enzymatic activities that help to activate transcription by modifying chromatin properties.

Polycomb and Trithorax—response elements (PREs/TREs)

Functional analysis of the regulatory regions of several PcG target genes allowed the identification of specific elements that are necessary and sufficient for PcG-mediated repression in *D. melanogaster*. All three PcG complexes, as well as TRX, bind to such sequences, called Polycomb response elements (PREs). Although several PREs have been identified in *D. melanogaster*, much of the information regarding their structure and function comes from work on a few specific PREs: the *engrailed* PRE, the *Fab7* PRE and, pre-eminently, the *bxd* PRE. These are compound elements of several hundred base-pairs, different parts of which, individually also have weaker PRE-like activity than the whole and contribute to the overall function [122] (Fig. 23.12).

Polycomb response elements (PREs) are not defined by a conserved sequence. Instead, like many complex enhancers, PREs include many conserved short motifs, several of which are recognized by known DNA-binding proteins. *Drosophila melanogaster* PREs often contain clusters of GAGAG motifs, which bind GAGA factor (GAF) and Pipsqueak (PSQ), both BTP/POZ proteins that reportedly associate with PcG complexes [122, 123]. PREs often contain binding sites for PHO and PHOL. SP1/KLF protein binding is important for the function of the *engrailed* PRE [124]. Binding of the high-mobility group (HMG)-like dorsal switch protein 1 (DSP1) is also involved in the function of at least some PREs [125]. A more comprehensive discussion of these factors can be found in the review by Müller and Kassis [126].

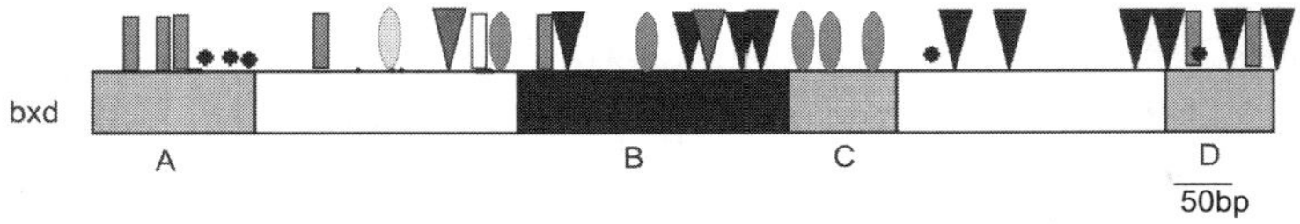

SYMBOL	MOTIF	CONSENSUS	Ref.
▼	Zeste	YGAGYG (Y – C/T)	129
▼	GAF/Pipsqueak	GAGAG	122
●	PHO	GCCATHWY (H=A/C/T: W=A/T)	124
○	PHO core	GCCAT	127
■	TGC repeats	TGCTGC	127
▮	A rich	AAAAA	127
▯	GT	GTGTG	127
•	TRE motif	AACAA	128
	Minimal PRE		

▶ **Fig. 23.12** *Diagrammatic representation of a minimal Polycomb responsive element. The PRE from the bi-thorax complex is shown with motif positions. The minimal fragment that has PRE activity is shown as a black box. The boxes A, C and D indicate regions outside the minimal PRE that have additional functions. DNA motifs along with symbols used in the line diagram are listed in the table. (for colour figure see Plate 62).*

No mammalian PRE has been identified to date, although it is clear that mammalian homeotic genes, and many other genes, are under PcG control, bind PcG proteins and bear methylated H3K27 [127–129]. So far, it is not known whether mammalian PREs exist but are more extensive or diffuse, and therefore harder to identify, or whether, in mammals, PcG proteins are recruited by a different mechanism (as discussed further).

Mechanism of PcG proteins mediated mitotic memory maintenance

PcG proteins are evolutionarily conserved and known to maintain specific repressive states of homeotic (Hox) gene expression patterns within body segments from flies to humans. Mechanisms by which this repression takes place at the level of transcription are not known. A general consensus reflected in the current models is that PRC2 initiates transcriptional repression by inhibiting transcription initiation, whereas the PRC1 maintains the repressive conditions. Initial interaction of certain DNA-binding factors facilitates the recruitment of PRC2. Enhancer of zeste-2 (EZH2), a component of PRC2, forms a trimethylation mark on H3K27 (H3K27me3) that serves as a flag for PRC1 recruitment. PRC1 is thereby recruited via additional cooperative effects with the DNA, DNA-binding factors and PRC2. Mono-ubiquitylation of histone H2A is accomplished by the PRC1 component RING. Different EZH2-containing PRCs might exhibit H1K26 methyltransferase activities. Cumulatively, PRC2 and PRC1 might induce chromatin compaction, as the former initiates transcriptional repression and the latter maintains the repression.

EPIGENETIC REGULATORS OF STEM CELL PLURIPOTENCY

Epigenetic differences in chromatin have been detected as early as the four-cell stage of the mouse preimplantation embryo (Fig. 23.13) [130]. These cells, referred to as blastomeres, are transiently totipotent. Subsequently the blastocyst develops, it differentiates into trophectoderm and inner cell mass, the latter of which is used in generating ES cells in the laboratory. On the other hand adult stem cells are found in differentiated tissues of the adult organs that maintain organ tissue homeostasis [131, 132]. ES cells can differentiate into all cell types of an adult organism (pluripotency); however, adult stem cells can form only cell types of their parental organ (multipotency). The genetic endowment of both these cell types being similar, differences in the developmental potential is closely associated with epigenetic markings on the genome and epigenetic factors within the microenvironment of the cell.

PcG proteins are highly abundant in ICM that forms the embryo proper and also in both embryonic and adult stem cells. These high levels of PcG proteins decline as stem cells begin to differentiate [133]. Several recent studies have highlighted the role for PcG proteins in the maintenance of pluripotency in both ES cells and adult stem cells in silencing the key lineage-specific transcription factor coding genes, thereby preventing differentiation [134, 135]. It is also interesting to know what signatures on the genome recruit PcG complexes to specific regions associated with maintenance of pluripotency in ESCs. Once again as we will discuss further the epigenetic modification of DNA and histones play an important part in this recognition (Fig. 23.13).

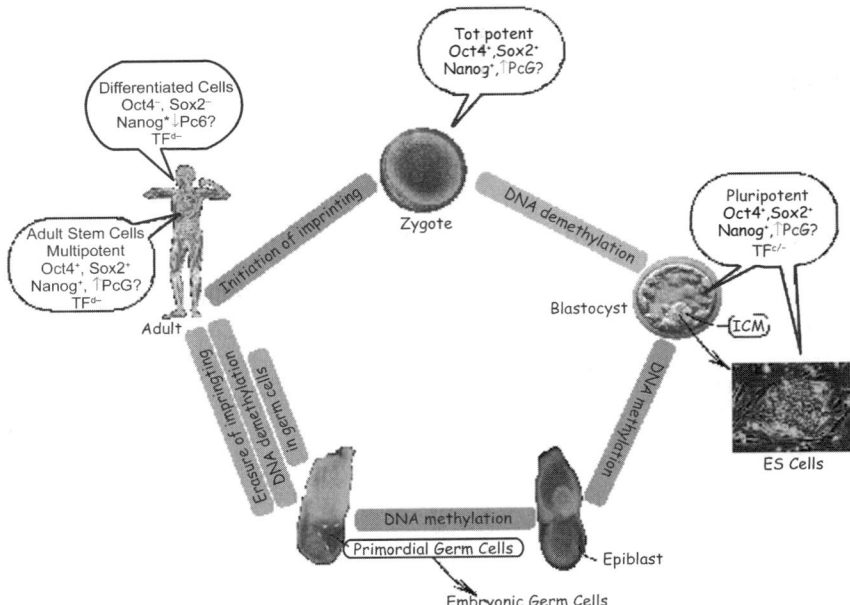

▶ **Fig. 23.13** *A summary of epigenetic regulatory cycle of pluripotency. The alteration in epigenetic modifiers (PcG), DNA methylation and transcription factors that maintain stemness, such as OCT4, NANOG and SOX2 are shown. ICM-Inner cell mass, TF^d – differentiation-specific transcription factors, + is for presence, – for absence, upward/downward arrow indicates elevated and decreased levels, respectively. The totipotent zygote contains maternally inherited epigenetic modifiers and transcription factors, including OCT4, SOX2 and NANOG. These, together with the embryonic transcripts, regulate development to the blastocyst stage, where the pluripotent cells are established in the inner cell mass (ICM), from which ES cells are derived both of which share several common epigenetic imprints. In the post-implantation embryo, pluripotent epiblast cells are controlled by diverse repressive mechanisms during their differentiation into somatic and germ-cell lineages. The early germ cells exhibit epigenetic and transcriptional states that are associated with pluripotency, and can give rise to embryonic germ cells. Epigenetic reprogramming, including genomic imprinting within this lineage re-generates totipotency. (for colour figure see Plate 63).*

Role of Histone Modification in Stem Cell Pluripotency

The initial step in deciphering the role of histone modification or the PcG complexes in maintenance of stemness is to locate regions where such alterations occur. Data for such an analysis is derived from methods referred to as chromatin immunoprecipitation (ChIP) using antibodies directed against either a specific protein component of the complex or a modified histone to pull down regions of the chromatin selectively (Box 23.1). The DNA sequences which are part of the pulled-down chromatin are sequenced or hybridized to an array of oligonucleotides representing the genome (DNA chip), to answer the question which DNA sequence is associated with which epigenetic factor. Based on such experiments, it is known that histone modifications associate with euchromatin indicating potentially active genic regions are abundant in ES cells, whereas as differentiation proceeds, silencing histone marks that signify heterochromatin accumulates [136]. It is now known that the regulation of the genes in ES cells most often occurs at the transcriptional level rather than translational level.

Using ChIP experiments on ES cells, PcG protein has been localized to promoters of a large number of genes associated with differentiation suggesting that PcG proteins help to maintain the repressed state of these genes in undifferentiated ES cells. Consistent with this, mutation in PcG gene results in the loss of ability to maintain ES cells in the undifferentiated state [137, 138]. However, the process of transition from repressed to activated state has to be accomplished as differentiation sets in. The identification of what is referred to as 'bivalent domains' provides some insight on the reversibility of the two states. To drive the point it is inevitable to introduce the highly conserved non-coding elements (HCNE), called so because of their extreme conservation across phyla or through evolution [139]. Interestingly, these HCNEs cluster near genes that encode developmentally important transcription factors, which are also reported to be targets of PcG-mediated transcriptional repression. Recently, Bernstein et al., analysed the epigenetic profile of mouse ES cells by ChIP experiments and hybridized the pulled-down DNA with oligonucleotide tiling arrays containing representatives of mouse loci enriched for highly conserved non-coding elements [140]. These authors identified a novel chromatin modification pattern consisting of both repressive H3K27me3 and activating H3K4me3 histone modifications called 'bivalent domains.' These bivalent domains resolved into either transcriptionally repressive H3K27me3 or transcriptionally permissive H3K4me3 in the differentiated cell types. Many of the bivalent domains in the HCNE regions overlap with transcription start sites, and it has been hypothesized that the bivalent domains signal repression while maintaining the adjacent genes in a state poised for activation during differentiation. These authors also raised the possibility that the HCNEs might promote chromosome conformations such that epigenetic switching is facilitated [140], thus suggesting that these can act like flags for recruitment of PcG complexes.

An independent study showed a similar coexistence of histone marks for both the active and the repressive chromatin in ES cells that maintain pluripotency [141]. This study also reported that genes associated with dual histone marks exhibited characteristics of euchromatin such as early DNA replication during S-phase, indicating that these silent genes are maintained in a transcriptionally permissive state. ES cells appear to be enriched with bivalent chromatin domains. Some of the genes coding for crucial transcription factors involved in lineage specification, such as *MSX1, NKX2–2, PAX3* and *SOX1*, are associated with both active (H3K4me3) and repressive (H3K27me3) histone modifications but are not expressed in ES cells. On the other hand, the active promoters of pluripotency genes such as octamer-binding transcription factor 4 (OCT4) and NANOG are associated with transcriptionally active histone modifications in ES cells [142]. It is apparently intriguing that decrease in OCT4 levels also reduced PcG protein binding at the target genes [143]. Several authors have observed that a significant subset of key developmental regulators that need to be in the repressed state to maintain ES cell identity are co-occupied by both the pluripotency regulators and PcG proteins [137, 144, 145]. Together, these results suggest a possible functional interaction between transcription factors for pluripotency and the PcG proteins in maintaining stem cell identity exists, although their direct interaction has not been firmly established.

In addition to their role in early embryonic development, PcG proteins have also been implicated as essential regulators of adult stem cell maintenance, including stem cell aging [145–148]. For example, in haematopoietic stem cell (HSC), proliferation appears to be regulated by a balance between Eed and Bmi1 levels, whereas overexpression of Ezh2 preserves the repopulating potential of the HSCs over serial transplantations [133, 148–151]. On the other hand, Bmi1 is also important for neural stem cell renewal [152–154], EZH2 is necessary for maintaining undifferentiated state of muscle cell precursors called myoblasts, which are like adult stem cells [155]. Together, these studies suggest a direct role for PcG proteins in regulating self-renewal and the maintenance of stem cell identity. The results of various studies on localization of epigenetic interactions in ES cells is summarized in Box 23.3.

Role of DNA Methylation in Stem Cell Pluripotency

DNA methylation profiles of various human ES cells are distinct from other human cell types including normal tissues, embryonal carcinoma cell lines, adult stem cell populations, lymphoblastoid cell lines and cancer cell lines [156]. However, this epigenetic signature is dispensable for ES cell propagation, as reported by several groups [157–161]. Extremely low levels of DNA methylation detected in the preimplantation embryo and its increase dramatically upon implantation and gastrulation, suggest that DNA methylation may not be a major mechanism for maintenance of silenced state of differentiation-specific genes in stem cells [136]. Thus, it appears that PcG-mediated gene repression is a dynamic process in early embryos and in ES cells that accommodates the global changes in gene expression accompanying differentiation. Genome-wide demethylation of DNA occurring in preimplantation embryos are considered essential for pluripotency development in epiblast cells, whereas demethylation in primordial germ cells (PGCs) ensures that aberrant epigenetic modifications are not transmitted to the progeny so that equivalent epigenetic state in the germline is maintained in male and female embryos [162].

There is a strict developmental regulation in the sequential conversion of one entire X chromosome from a euchromatin state to a heterochromatin state, which we discussed in an earlier section. This is triggered by the expression and cis-localization of a large non-coding RNA (*XIST* RNA—the chromosome X inactive specific transcript, encoded by the *XIST* gene) as well as mediated by its recruitment of PRC2 [163–165]. For example, SUZ12 levels are elevated during early stages of X chromosome inactivation (Xi) in a *XIST* RNA-dependent manner [166]. Although EED is dispensable for initiation of random X inactivation [167], it protects the inactive X-chromosome from differentiation-induced reactivation [168]. It has also been shown that the selective inactivation of the paternally inherited X chromosome in trophectoderm and primitive endoderm lineages of mouse embryos is *XIST* RNA-dependent and is mediated by EED [163]. The erasure and reprogramming of the epigenetic modification during development in primordial germ cells is very similar to the transition between stemness and differentiated state in ES cells. The mechanistic similarity between paternal X inactivation in trophectoderm and the random X inactivation in ICM is indicated by the common epigenetic modifications and the interacting PRC complexes shared by them. Interested reader can consult the elegant work reported by Surani and McLaren on embryonic germ cells [169]. It remains unclear how PcG proteins maintain the silent state of Xi once it is established and how they facilitate the Xi inheritance over subsequent cell generations throughout the life of an organism.

PcG Protein Targets in ESCs

Given the important role of PcG proteins in cellular identity and stem cell maintenance, it is essential to identify their molecular targets. Several groups have utilized ChIP-seq (Box 23.1) on mouse [138] and human ES cells [137] to identify the targets of PcG proteins and the epigenetic events associated with these cells by genome-wide analysis. The findings from various studies on PcG targets in ES cells can be summarized as follows:

- the components of PRC1 (Phc1, Rnf2) and PRC2 (Suz12, Eed) as well as the repressive histone modification H3K27me3 co-localize at promoter regions of hundreds of target genes in mouse ES cells [138].

- nearly all of the PcG-protein-bound regions are detected proximal to transcriptional start sites, and most of the target gene transcripts are lower in abundance in mouse ES cells compared with differentiating embryoid bodies.

- the genes occupied by PcG proteins are found to encode regulatory factors with known roles in developmental processes such as organogenesis, morphogenesis, pattern specification, neurogenesis, cell differentiation, embryonic development and cell fate commitment. Many of the developmental regulators could be identified as lineage-specific transcription factors.

- OCT4, SOX2 and NANOG localize to approximately one-third of PRC2-occupied regions in human ES cells suggesting that the transcription factors might direct the repression of target genes through interaction with PcG proteins [137]. On the other hand, these same transcription factors are also associated with RNA polymerase II-occupied regions that are transcriptionally active in human ES cells [137]. Thus, the OCT4, SOX2 and NANOG proteins might play a role in directing both PcG and TrxG proteins to target sites in ES cells; however, it is likely that there are other factors presently unknown.

We are at the first step of cataloguing all factors associated with maintenance or loss of stemness and perhaps mechanisms behind the initiation of these processes and how they are maintained still remain to be determined [170–176].

EPIGENETICS: A TRAIL FROM STEM CELLS TO CANCER

The components of epigenetic regulatory mechanism are functionally promiscuous as far as their interaction sites are concerned, interacting with many sites on the genome. Therefore, mutations in these genes are likely to have serious consequences and may be incompatible with development itself. However, in many disease conditions, they are altered either as a consequence of the disease or as a cause of the disease, though in most cases the two possibilities have not been distinguished. But what is well appreciated is that it is potentially a good indicator or a biomarker.

In the background of the discussion in the preceding sections, we can assume that epigenetic regulation is one of the most important player in the maintenance of stemness, which is characterised by the proliferative ability without undergoing differentiation. At the same time, a differentiated cell that acquires this ability has a dangerous prognosis. In this section, we briefly discuss the shared components of the epigenetic regulatory toolkit of cancer and stem cells.

Oncogenesis is an aberrant differentiation in cells that involves the inappropriate regulation of developmental genes and cellular signaling pathways. The known genetic and epigenetic changes that occur in cancer include: the deletion of tumour suppressor genes, activation/amplification of oncogenes, epigenetic lesions such as altered DNA methylation, misregulation of chromatin remodeling by histone modifications and aberrant expression of PcG proteins [143, 177–182]. In parallel, as we have seen, strict functional equilibrium between oncogenes, such as BMI and tumour suppressor genes, like RB1, regulates stem cell function [137, 138, 183–185]. Any dysregulation in such networks, such as those that normally occur with aging [186, 187] or inflammation [188, 189] might result in cancers. Thus, the epigenetic aberrations are thought to play a central role in early neoplastic transformation based on the contention that stem cells in tissues constitute a substrate for these early epigenetic alterations [190–192].

The adult stem cells in tissues maintain the homeostasis, similarly, several tumours are known to contain a minor population of cells called tumour-initiating cells, or stem-like cancer cells, or cancer stem cells [193, 194]. The cancer stem cells are thought to originate from mutations in adult stem cells that have sustained specific genetic alterations or by dedifferentiation of mature somatic cells to reacquire stem cell characteristics [195, 196]. The opposing effects of different PcG proteins, such as CBX2, MEL18 and EED against BMI1 might disrupt the delicate balance between self-renewal and differentiation, Therefore, if adult stem cells are absent, chances of getting cancer would be reduced, but this would compromise on the tissue renewal. Thus tipping the delicate balance between self-renewal and differentiation would result in cancer development [151, 197, 198]. The inappropriate expression of plueripotency maintenance factors, such as OCT4 and NANOG in tumours suggests that adult stem cells could be reservoirs for cancer cells [199–201].

The PcG proteins and histone modifications have a fundamental role in stem cell maintenance; it is suggested that the oncogenic potential of PcG genes such as BMI1 and EZH2 may also facilitate the maintenance of cancer stem cells [186, 193]. Moreover, several PcG proteins such as EZH2, EED and SUZ12 are over-expressed in cancer cells compared to differentiated cells. All these observations converge to suggest that silencing of differentiation-specific genes is fundamental requirement to progress towards pleuripotency and hence cancer. This goal can also be achieved by DNA hypermethylation of CpG islands upstream of differentiation-specific genes. In support of this, hypermethylation of PRC target genes has been reported in several cancer cells. However, there is still a distinct identity for ESCs and cancer cells. This is reflected in the fact that the PRC-bound hypermethylated genes in adult human cancers are not methylated in both the normal ES cells and the embryonal carcinoma (EC) cells [196]. What is common between ES and EC cells is that they can respond to normal developmental cues by activating genes. Thus, it appears that there exists a fine control of epigenetic gene silencing mechanism in adult tumours as in stem cells (Fig. 23.14) [202].

The stem cell microenvironment in adult tissues is known to form a physiological niche for their maintenance [131, 203, 204]. Similarly, the tumour microenvironment is required for the maintenance of cancer stem cells [205, 206]. This raises an important question with regard to the functional role of such niches in discriminating normal adult stem cells from cancer stem cells. Preliminary evidence has been reported for cell type-specific epigenetic changes at the level of DNA methylation in the stromal cells, such as epithelial and stromal fibroblasts from normal breast and breast carcinomas [207]. The molecular mechanisms and biological significance of such epigenetic changes in stromal cells of the tumour remain unknown.

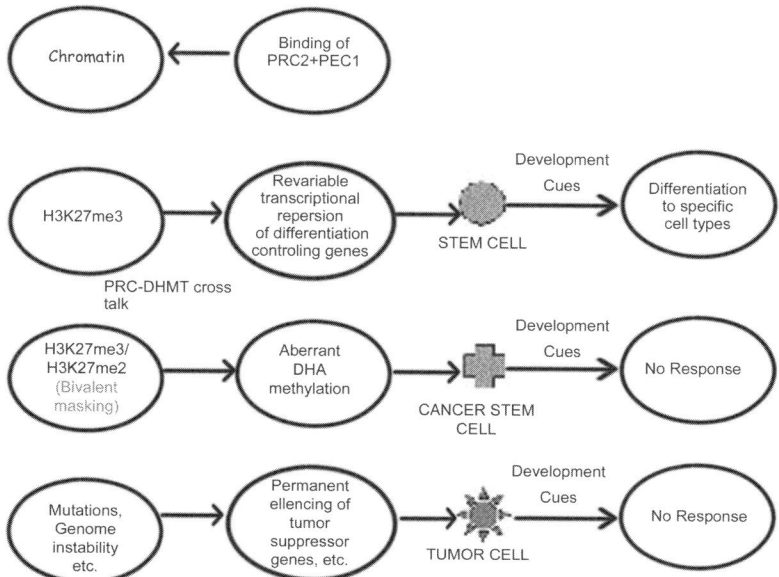

▶ **Fig. 23.14** *A comparison of epigenetic regulation in stem cells and cancer cells. (for colour figure see Plate 64).*

Box 23.3. Summary of localization of epigenetic interactions in ES cells

The names of the proteins used as primary bait in the ChIP experiment are indicated in the first column. (+)/(−) indicates presence/absence respectively of a particualr protein ont he promoters of expressed/repressed genes, number sin square brackets are the reference to of studies establishing the data.

Conclusions:

- the genes occupied by PcG proteins are found to encode regualatory factors with known roles in developmental processes such as organogenesis, morphogenesis, pattern specification, neurogenesis, cell differentation, embryonic development, and cell fate commitment.
- the components of PRC1 (*PHC1, RNF2*) and PRC2 (*SUZ12, FED*) as well as the repressive histone modification H3K27me3 colocalize at promoter regions of hundreds of target genes in mouse and human ES cells.
- *OCT4, SOX2*, and NANOG localize to approximately one-third of PRC2-occupied regions in human ES cells ssuggesting that these transcription factors may direct the repression of target genes through interaction with PcG proteins.

Looking Ahead

The progress in various technologies to address biological questions and consequently the design of experiments to address these aspects has yielded interesting leads that inspire confidence that we will be able to understand the dynamics of epigenetic regulation [208–214]. Some of the recent developments that are not discussed in this chapter include, the isolation and characterization of adult stem cells [215], determination of the 3D structure of derepressed and repressed chromatin [216], the progress in quantitative proteomic analysis of histone modifications [217], by differential fluorescence tags on isolated proteins, high-resolution mapping of chromatin modifications in the human genome [218, 219], genome-wide mapping of *in vivo* target loci of chromatin proteins and transcription factors (novel 'DamID' technique) [220], carrier ChIP protocol [221]. One of the consequences of the increased understanding of the regulatory network is the ability to induce stem-like cells from adult fibroblasts [222–225]. The use of stem cells holds a great promise in several disease treatment as elaborated in other chapters of this book.

The success of these attempts largely depends upon obtaining a desired and anticipated outcome of the interaction between the static information content of the genome (in the given cells/organism) with the dynamic epigenetic modulations. It is shown that parental imprints are maintained faithfully in human stem cells; however, there have been also reports of relatively higher methylation of certain genes in stem cells and in neuronal lineage induced from human stem cells [226–228].

Even as we understand nuts and bolts of 'stemness', it is essential to appreciate implications of dedifferentiation of diploid cells vis-à-vis the dynamic epigenetic alterations for effective utilization of stem cells in therapy. While it is important to understand epigenetic regulation of target genes, it is vital to study the regulation of epigenetic modifiers themselves and the crosstalk between these regulators. The detection and cataloging of post-replicational and post-translational modifications of DNA, histones and chromatin-interacting proteins have advanced remarkably and the time is ripe now to address mechanistic questions regarding the function of these modifications to progress from correlations to cause and effect.

ACKNOWLEDGEMENTS

The authors thank Dr. Debasis Dash, Ms. Meenakshi and Sourab for help with the artwork.

REFERENCES

1. Waddington C.H. (1942). 'The epigenotype'. *Endeavour*. 1: 18–20.
2. Levenson J.M., J.D. Sweatt (2005). 'Epigenetic mechanisms in memory formation'. *Nat. Rev. Neurosci.* **6**(2): 108–118.
3. Reik W., W. Dean, J. Walter (2001). 'Epigenetic reprogramming in mammalian development'. *Science*. **293**(5532): 1089–1093.
4. Jenuwein T., C.D. Allis (2001). 'Translating the histone code'. *Science*. 293(5532): 1074–1080.
5. Bouazoune K., A. Brehm (2006). 'ATP-dependent chromatin remodeling complexes in Drosophila'. *Chromosome Research*. **14**: 433–449.
6. Bernstein E., C.D. Allis (2005). 'RNA meets chromatin'. *Genes. Dev.* 19(14): 1635–1655.
7. Feinberg A.P., R. Ohlsson, S. Henikoff (2006). 'The epigenetic progenitor origin of human cancer'. *Nat. Rev. Genet.* **7**: 21–33.
8. Keller G. (2005). 'Embryonic stem cell differentiation: emergence of a new era in biology and medicine'. *Genes. Dev.* **19**(10): 1129–1155.
9. Venkatesh, B. E.F. Kirkness, Y.H. Loh, A.L. Halpern, A.P. Lee, J. Johnson, N. Dandona, L.D. Viswanathan, A. Tay, J.C. Venter, R.L. Strausberg, S. Brenner (2006). 'Ancient non-coding elements conserved in the human genome'. *Science*. **22**: 314(5807): 1892.
10. Venter J.C., et. al. (2001). *Science*. **16**: 291(5507): 1304–1351.
11. Brown S.W. (1966). Heterochromatin. *Science*. **151**: 417–425.
12. Khosla S., G. Mendiratta, V. Brahmachari (2006) 'Genomic imprinting in the mealybugs'. *Cytogenet Genome. Res.* **113**(1–4): 41–52.
13. Mc Grath J., D. Solter (1983). 'Nuclear transplantation in mouse embryos'. *J Exp Zool. Nov*D. **228**(2): 355–362
14. Surani M.A., S.C. Barton, M.L. Norris (1984). 'Development of reconstituted mouse eggs suggests imprinting of the genome during gametogenesis'. (5959). *Nature*. Apr. 5–11 **308**: 548–550.
15. Beechey C.V., B.M.C. Cattanach, R.L Selley: 'Mouse imprinting data and references'. URL: *http://www.mgu.har.mrc.ac.uk/imprinting/imptables.html*.
16. Reik W., J. Walter (2001). 'Genomic imprinting: parental influence on the genome'. *Nat. Rev. Genet.* **2**: 21–32.

17. Wilkins J.F. D. Haig (2003). 'What good is genomic imprinting: the function of parent-specific gene expression'. *Nat. Rev. Genet.* **4**: 359–368.
18. Young L.E. A.E. Schnieke, .J.K Mc Creath, S. Wieckowski, G. Konfortova, K. Fernandes, G. Ptak, A.J. Kind, I. Wilmut (2003). 'Pasqualino L et. al.: Conservation of IGF2-H19 and IGF2R imprinting in sheep: effects of somatic cell nuclear transfer'. *Mech. Dev.* **120**: 1433–1442.
19. Tilghman S.M. (1999). 'The sins of the fathers and the mothers: genomic imprinting in mammalian development'. *Cell.* **96**: 185–193.
20. Bartolomei M.S, S.M. Tilghman (1997) 'Genomic imprinting in mammals'. *Ann. Rev. Genet.* **31**: 493–525.
21. Delaval K., R. Feil (2004). 'Epigenetic regulation of mammalian genomic imprinting'. *Curr. Opin. Genet. Dev.* Apr;**14**(2): 188–195.
22. Avner P., E. Heard (2001). 'X-chromosome inactivation: counting, choice and initiation'. *Nat. Rev. Genet.* Jan; **2**(1): 59–67.
23. Reik W., A. Lewis (2005). 'Co-evolution of X-chromosome inactivation and imprinting in mammals'. *Nat. Rev. Genet.* May **6**(5): 403–410.
24. Ramsahoye B.H, D. Biniszkiewicz, F. Lyko, V. Clark, A.P. Bird, R. Jaenisch (2000). 'Non-CpG methylation is prevalent in embryonic stem cells and may be mediated by DNA methyltransferase' 3a. *Proc Natl. Acad. Sci. USA.* May 9; **97**(10): 5237–5242.
25. Jones, P.A. and D. Takai (5532). 'The role of DNA methylation in mammalian epigenetics'. *Science.* 2001. **293**: 1068–1070.
26. Bestor, T.H. (2000). 'The DNA methyltransferases of mammals'. *Hum. Mol. Genet.* **9**(16): 2395–2402.
27. Jaenisch, R. and A. Bird (2003). 'Epigenetic regulation of gene expression: how the genome integrates intrinsic and environmental signals'. *Nat. Genet.* **33** Suppl. 245–254.
28. Okano, M., S. Xie, and E. Li (1998). 'Dnmt2 is not required for de novo and maintenance methylation of viral DNA in embryonic stem cells'. *Nucleic. Acids. Res.* **26**(11): 2536–2540.
29. Bhattacharya, S.K., et. al. (1999). 'A mammalian protein with specific demethylase activity for mCpG DNA'. *Nature.* **397**(6720): 579–583.
30. Klose, R.J. and A.P. Bird (2006). 'Genomic DNA methylation: the mark and its mediators. *Trends Biochem. Sci.* **31**(2): 89–97.
31. Zhao, X., et. al. (2003). 'Mice lacking methyl-CpG binding protein 1 have deficits in adult neurogenesis and hippocampal function'. *Proc. Natl. Acad. Sci. USA.* **100**(11): 6777–6782.
32. Hutchins, A.S., et. al. (2002). 'Gene silencing quantitatively controls the function of a developmental trans-activator'. *Mol. Cell.* **10**(1): 81–91.
33. Sansom, O.J., et. al. (2003). 'Deficiency of Mbd2 suppresses intestinal tumourigenesis'. *Nat. Genet.* **34**(2): 145–147.
34. Hendrich, B., et. al. (2001). 'Closely related proteins MBD2 and MBD3 play distinctive but interacting roles in mouse development'. *Genes. Dev.* **15**(6): 710–723.
35. Millar, C.B., et. al. (2002). 'Enhanced CpG mutability and tumourigenesis in MBD4-deficient mice'. *Science.* **297**(5580): 403–405.
36. Amir, R.E., et. al. (1999). 'Rett syndrome is caused by mutations in X-linked MECP2 encoding methyl-CpG-binding protein 2'. *Nat. Genet.* **23**(2): 185–188.
37. Moretti, P. and H.Y. Zoghbi (2006). 'MeCP2 dysfunction in Rett syndrome and related disorders'. *Curr. Opin. Genet. Dev.* **16**(3): 276–281.

38. Bird, A. (2002). 'DNA methylation patterns and epigenetic memory'. *Genes. Dev.* **16**(1): 6–21.
39. Okano, M., et. al. (1999). 'DNA methyltransferases Dnmt3a and Dnmt3b are essential for de novo methylation and mammalian development'. *Cell.* **99**(3): 247–257.
40. Ueda, Y., et. al. (2006). 'Roles for Dnmt3b in mammalian development: a mouse model for the ICF syndrome'. *Development.* **133**(6): 1183–1192.
41. Hansen, R.S., et. al. (1999). 'The DNMT3B DNA methyltransferase gene is mutated in the ICF immunodeficiency syndrome'. *Proc. Natl. Acad. Sci. USA.* **96**(25): 14412–14417.
42. Turksen, K. and T.C. Troy (2006). 'Human embryonic stem cells: isolation, maintenance, and differentiation'. *Methods. Mol. Biol.* **331**: 1–12.
43. Gage, F.H. (2000). 'Mammalian neural stem cells'. *Science.* 287(5457): 1433–1438.
44. Ming, G.L. and H. Song (2005). 'Adult neurogenesis in the mammalian central nervous system'. *Ann.. Rev. Neurosci.* **28**: 223–250.
45. Kornack, D.R. and P. Rakic (2001). 'The generation, migration, and differentiation of olfactory neurons in the adult primate brain'. *Proc Natl. Acad. Sci. USA.* 98(8): 4752–4757.
46. Pencea, V., et. al. (2001). 'Neurogenesis in the subventricular zone and rostral migratory stream of the neonatal and adult primate forebrain'. *Exp. Neurol.* **172**(1): 1–16.
47. Curtis, M.A., et. al. (2007). 'Human neuroblasts migrate to the olfactory bulb via a lateral ventricular extension'. *Science.* **315**(5816): 1243–1249.
48. Lachner, M., R.J. O'Sullivan, and T. Jenuwein (2003). 'An epigenetic road map for histone lysine methylation'. *J. Cell. Sci.* **116**(Pt 11): 2117–2124.
49. Santos-Rosa, H., et. al. (2002). 'Active genes are tri-methylated at K4 of histone H3'. *Nature* **419**(6905): 407–411.
50. Krogan, N.J., et. al. (2003). 'Methylation of histone H3 by Set2 in Saccharomyces cerevisiae is linked to transcriptional elongation by RNA polymerase II'. *Mol. Cell. Biol.* **23**(12): 4207–4218.
51. Schubeler, D., et. al. (2004). 'The histone modification pattern of active genes revealed through genome-wide chromatin analysis of a higher eukaryote'. *Genes. Dev.* **18**(11): 1263–1271.
52. Lee, K.K. and J.L. Workman (2007). 'Histone acetyltransferase complexes: one size doesn't fit all'. *Nat. Rev. Mol. Cell. Biol.* **8**(4): 284–295.
53. Xu, W.S., R.B. Parmigiani, and P.A. Marks (2007). 'Histone deacetylase inhibitors: molecular mechanisms of action'. *Oncogene.* **26**(37): 5541–5552.
54. Ballas, N., et. al. (2005). 'REST and its corepressors mediate plasticity of neuronal gene chromatin throughout neurogenesis'. *Cell.* **121**(4): 645–657.
55. Lunyak, V.V., et. al. (2002). 'Corepressor-dependent silencing of chromosomal regions encoding neuronal genes'. *Science.* **298**(5599): 1747–1752.
56. Lunyak, V.V. and M.G. Rosenfeld (2005). 'No rest for REST: REST/NRSF regulation of neurogenesis'. *Cell.* **121**(4): 499–501.
57. Rice, J.C. and C.D. (2001). 'Allis, Histone methylation versus histone acetylation: new insights into epigenetic regulation'. *Curr. Opin. Cell. Biol.* **13**(3): 263–273.
58. Hsieh, J., et. al. (2004). 'Histone deacetylase inhibition-mediated neuronal differentiation of multipotent adult neural progenitor cells'. *Proc. Natl. Acad. Sci. USA.* **101**(47): 16659–16664.
59. Jessberger, S., et. al. (2007). 'Epigenetic modulation of seizure-induced neurogenesis and cognitive decline'. *J. Neurosci.* **27**(22): 5967–5975.

60. Agarwal, N., et. al. (2007). 'MeCP2 interacts with HP1 and modulates its heterochromatin association during myogenic differentiation'. *Nucleic Acids. Res.* **35**(16): 5402–5408.
61. Fujita, N., et. al. (2003). 'Methyl-CpG binding domain 1 (MBD1) interacts with the Suv39h1-HP1 heterochromatic complex for DNA methylation-based transcriptional repression'. *J. Biol. Chem.*1 **278**(26): 24132–24138.
62. Mikkelsen, T.S., et. al. (2007). 'Genome-wide maps of chromatin state in pluripotent and lineage committed cells'. *Nature.* **448**(7153): 553–560.
63. Klose, R.J. and Y. Zhang (2007). 'Regulation of histone methylation by demethylimination and demethylation'. *Nat. Rev. Mol. Cell Biol.* **8**(4): 307–318.
64. Howe, M.L., et. al. (2006). 'Transcription Factor IIA tau is associated with undifferentiated cells and its gene expression is repressed in primary neurons at the chromatin level *in vivo*'. *Stem. Cells. Dev.* **15**(2): 175–190.
65. Weissman, I.L., D.J. Anderson, and F. Gage (2001). 'Stem and progenitor cells: origins, phenotypes, lineage commitments, and transdifferentiations'. *Ann.. Rev. Cell. Dev. Biol.* **17**: 387–403.
66. Cheng, L.C., M. Tavazoie, and F. Doetsch (2005). 'Stem cells: from epigenetics to microRNAs'. *Neuron.* **46**(3): 363–367.
67. Bartel, D.P. (2004). 'MicroRNAs: genomics, biogenesis, mechanism, and function'. *Cell.* **116**(2): 281–297.
68. Kosik, K.S. (2006). 'The neuronal microRNA system'. *Nat. Rev. Neurosci.* **7**(12): 911–920.
69. Eulalio, A., E. Huntzinger, and E. Izaurralde. (2008). 'Getting to the Root of miRNA-Mediated Gene Silencing'. *Cell.* **132**(1): 9–14.
70. Bernstein, E., et. al. (2003). 'Dicer is essential for mouse development'. *Nat. Genet.* **35**(3): 215–217.
71. Kanellopoulou, C., et. al. (2005). 'Dicer-deficient mouse embryonic stem cells are defective in differentiation and centromeric silencing'. *Genes. Dev.* **19**(4): 489–501.
72. Carmell, M.A., et. al. (2002). 'The Argonaute family: tentacles that reach into RNAi, developmental control, stem cell maintenance, and tumourigenesis'. *Genes. Dev.* **16**(21): 2733–2742.
73. Jackson, J.P., et. al. (2002). 'Control of CpNpG DNA methylation by the KRYPTONITE histone H3 methyltransferase'. *Nature.* **416**(6880): 556–560.
74. Matzke, M.A. and J.A. Birchler (2005). 'RNAi-mediated pathways in the nucleus'. *Nat. Rev. Genet.* **6**(1): 24–35.
75. Lehnertz, B., et. al. (2003). 'Suv39h-mediated histone H3 lysine 9 methylation directs DNA methylation to major satellite repeats at pericentric heterochromatin'. *Curr. Biol.* **13**(14): 1192–1200.
76. Hu, K., et. al. (2006). 'Testing for association between MeCP2 and the brahma-associated SWI/SNF chromatin-remodeling complex'. *Nat Genet* **38**(9): 962–4; *author. reply.* 964–967.
77. Nan, X., et. al. (1998). 'Transcriptional repression by the methyl-CpG-binding protein MeCP2 involves a histone deacetylase complex'. *Nature.* **393**(6683): 386–389.
78. Jones, P.L., et. al. (1998). 'Methylated DNA and MeCP2 recruit histone deacetylase to repress transcription'. *Nat Genet.* **19**(2): 187–191.
79. Morris, K.V., et. al. (2004). 'Small interfering RNA-induced transcriptional gene silencing in human cells'. *Science.* **305**(5688): 1289–1292.
80. Hall, I.M., K. Noma, and S.I. Grewal (2003). 'RNA interference machinery regulates chromosome dynamics during mitosis and meiosis in fission yeast'. *Proc. Natl. Acad. Sci. USA.* **100**(1): 193–198.

81. Hall, I.M., et. al. (2002). 'Establishment and maintenance of a heterochromatin domain. Science'. **297**(5590): 2232–2237.
82. Volpe, T.A., et. al. (2002). 'Regulation of heterochromatic silencing and histone H3 lysine-9 methylation by RNAi'. *Science.* **297**(5588): 1833–1837.
83. Seitz, H., et. al. (2004). 'A large imprinted microRNA gene cluster at the mouse Dlk1-Gtl2 domain'. *Genome. Res.* **14**(9): 1741–1748.
84. Saito, Y., et. al. (2006). 'Specific activation of microRNA-127 with downregulation of the protooncogene BCL6 by chromatin-modifying drugs in human cancer cells'. *Cancer Cell.* **9**(6): 435–443.
85. Wu, H., et. al. (2007). 'Integrative genomic and functional analyses reveal neuronal subtype differentiation bias in human embryonic stem cell lines'. *Proc. Natl. Acad. Sci. USA.* **104**(34): 13821–13826.
86. Rajewsky, N. (2006). 'microRNA target predictions in animals'. *Nat. Genet.* **38** Suppl: S8–13.
87. Kanellopoulou, C., et. al. (2005). 'Dicer-deficient mouse embryonic stem cells are defective in differentiation and centromeric silencing'. *Genes. Dev.* **19**(4): 489–501.
88. Fukagawa, T., et. al. (2004). 'Dicer is essential for formation of the heterochromatin structure in vertebrate cells'. *Nat. Cell. Biol.* **6**(8): 784–791.
89. Hadorn E. (1978). 'Transdetermination in cells'. *Sci. Am.* **219**: 110–114
90. Lewis, E.B. (1978) 'A gene complex controlling segmentation in Drosophila'. *Nature.* **276**: 565–570.
91. Poux, S., Horard, B., Sigrist, C.J. and Pirrotta, V. (2002).'The Drosophila Trithorax protein is a coactivator required to prevent re-establishment of Polycomb silencing'. *Development.* **129**: 2483–2493.
92. Klymenko, T. and Muller, J. (2004).'The histone methyltransferases Trithorax and Ash1 prevent transcriptional silencing by Polycomb group proteins'. *EMBO. Re***5**, 373–377.
93. Ringrose L, Paro R. (2004). 'Epigenetic regulation of cellular memory by the Polycomb and Trithorax group proteins'. *Ann.. Rev. Genet.* **38**: 413–443.
94. Shao, Z. et. al. (1999). 'Stabilization of chromatin structure by PRC1, a Polycomb complex'. *Cell.* **98**, 37–46.
95. Saurin, A.J., Shao, Z., Erdjument-Bromage, H., P. Tempst, and Kingston, R.E.A (2001).'Drosophila Polycomb group complex includes Zeste and dTAFII proteins'. *Nature.* **412**: 655–660.
96. Levine, S.S. et. al. (2002). 'The core of the Polycomb repressive complex is compositionally and functionally conserved in flies and humans'. *Mol. Cell. Biol.* **22**: 6070–6078.
97. Kuzmichev, A., T. Jenuwein, Tempst, and D. Reinberg (2004). 'Different EZH2-containing complexes target methylation of histone H1 or nucleosomal histone H3'. *Mol. Cell.* **14**: 183–193.
98. Kuzmichev, A. et. al. (2005). 'Composition and histone substrates of Polycomb repressive group complexes change during cellular differentiation'. *Proc. Natl Acad. Sci. USA.* **102**: 1859–1864.
99. Kuzmichev A., K. Nishioka, H. Erdjument-Bromage, P. Tempst, D. Reinberg (2002). 'Histone methyltransferase activity associated with a human multiprotein complex containing the Enhancer of Zeste protein'. *Genes. Dev.* Nov. 15: **16**(22): 2893–2905.
100. Brown, J.L., D. Mucci, M. Whiteley, M.L. Dirksen and J.A. Kassis (1998). 'The Drosophila Polycomb group gene pleiohomeotic encodes a DNA binding protein with homology to the transcription factor YY1'. *Mol. Cell.* **1**: 1057–1064.

101. Fritsch, C., J.L. Brown, J.A. Kassis, and J.Muller, (1999). 'The DNA-binding Polycomb group protein pleiohomeotic mediates silencing of a Drosophila homeotic gene'. *Development.* **126**, 3905–3913.
102. Brown, J.L., Fritsch, C., J. Mueller and J. Kassis, (2003). 'A.The Drosophila pho-like gene encodes a YY1-related DNA binding protein that is redundant with pleiohomeotic in homeotic gene silencing'.*Development.* **130**: 285–294.
103. Wilkinson F.H.,0 K. Park, M.L. Atchison (2006). 'Polycomb recruitment to DNA *in vivo* by the YY1 REPO domain'. *Proc Natl Acad Sci USA.* Dec 19. **103**(51): 19296–19301. Epub 2006 Dec 8.
104. Klymenko T., B. Papp, W. Fischle, T. Köcher, M. Schelder, C. Fritsch, B. Wild, M. Wilm, J. Müller (2006). 'A Polycomb group protein complex with sequence-specific DNA-binding and selective methyl-lysine-binding activities'. *Genes. Dev.* May. 1: **20**(9): 1110–1122. Epub 2006 Apr 17.
105. Bakshi R., A.K. Mehta, R. Sharma, S. Maiti, S. Pasha, V. Brahmachari (2006). 'Characterization of a human SWI2/SNF2 like protein hINO80: demonstration of catalytic and DNA binding activity'. *Biochem. Biophys. Res. Commun.* Jan 6: **339**(1): 313–320. Epub 2005 Nov 15.
106. Collins R.T, J.E. Treisman (2000). 'Osa containing Brahma chromatin remodeling complexes are required for the repression of wingless target genes'. *Genes Dev* **14**: 3140–3152
107. Smith S.T., S. Petruk, Y. Sedkov, E. Cho, S. Tillib, et. al. (2004). 'Modulation of heat shock gene expression by the TAC1 chromatin-modifying complex'. *Nat. Cell. Biol.* **6**: 162–167.
108. Chinwalla V., E.P. Jane, P.J. Harte (1995).'The Drosophila Trithorax protein binds to specific chromosomal sites and is colocalized with Polycomb at many sites'. *EMBO. J.* **14**: 2056–2065
109. Rozovskaia T., S. Tillib, S. Smith, Y. Sedkov, O. Rozenblatt-Rosen, et. al. (1999). 'Trithorax and ASH1 interact directly and associate with the trithorax groupresponsive bxd region of the Ultrabithorax promoter'. *Mol. Cell. Biol.* **19**: 6441–6447
110. Papp, B. and J. Muller, (2006). 'Histone trimethylation and the maintenance of transcriptional ON and OFF states by TrxG and PcG proteins'. *Genes. Dev.* **20**: 2041–2054.
111. Kahn, T.G., Y.B. Schwartz, G.I. Dellino, and V. Pirrotta (2006). 'Polycomb complexes and the propagation of the methylation mark at the Drosophila Ubx gene'. *J. Biol.Chem.* **281**: 29064–29075.
112. Chinwalla, V., E. Jane, and P.J. Harte, (1995). 'The Drosophila Trithorax protein binds to specific chromosomal sites and is co-localized with Polycomb at many sites'. *EMBO. J.* **14**: 2056–2065.
113. Yu, B.D., J.L. Hess, S.E. Horning, G.A.J. Brown, and S.J. Korsmeyer, (1995). 'Altered Hox expression and segmental identity in Mll-mutant mice'. *Nature* **378**: 505–508.
114. Simon J.A., J.W. Tamkun (2002). 'Programming off and on states in chromatin: mechanisms of Polycomb and Trithorax group complexes'. *Curr. Opin. Genet. Dev.* **12**: 210–218
115. Papoulas O., S.J. Beek, S.L. Moseley, C.M. McCallum, M. Sarte, et. al. (1998). 'The Drosophila trithorax group proteins BRM, ASH1 and ASH2 are subunits of distinct protein complexes'. *Development.* **125**: 3955–3966
116. Mohrmann L., K. Langenberg, J. Krijgsveld, A.J. Kal, Heck A.J., C. Verrijzer (2004). 'Differential targeting of two distinct SWI/SNF-related Drosophila chromatin-remodeling complexes'. *Mol. Cell. Biol.* **24**: 3077–3088
117. Petruk S., Y. Sedkov, S. Smith, S. Tillib, V. Kraevski, et. al. (2001). 'Trithorax and dCBP acting in a complex to maintain expression of a homeotic gene'. *Science.* **294**: 1331–1334

118. Roguev A., D. Schaft, A. Shevchenko, W.W.M.P. Pijnappel, M. Wilm, R. Aasland, A.F. Stewart (2001). 'The Saccharomyces cerevisiae Set1 complex includes an Ash2 homolgue and methylates histone H3 lysine 4'. *EMBO. J.* **20**: 7137–7148

119. Smith S.T., S. Petruk, Y. Sedkov, E. Cho, S. Tillib, et. al. (2004). 'Modulation of heat shock gene expression by the TAC1 chromatin-modifying complex'. *Nat. Cell. Biol.* **6**: 162–167

120. Beisel C., A.Imhof, J. Greene, E. Kremmer, F. Sauer (2002). 'Histone methylation by the Drosophila epigenetic transcriptional regulator Ash1'. *Nature.* **419**: 857–862

121. Lachner M., R.J. O'Sullivan, T. Jenuwein (2003). ,An epigenetic road map for histone lysine methylation'. *J. Cell. Sci.* **116**: 2117–2124

122. Horard, B., C. Tatout, S. Poux, and V. Pirrotta, (2000). 'Structure of a Polycomb response element and *in vitro* binding of Polycomb group complexes containing GAGA factor'. *Mol. Cell. Biol.* **20**: 3187–3197.

123. Hodgson, J.W., B. Argiropoulos and H.W. Brock, (2001). 'Site-specific recognition of a 70-base-pair element containing d(GA)(n) repeats mediates bithoraxoid Polycomb group response element-dependent silencing'. *Mol. Cell. Biol.* **21**: 4528–4543.

124. Brown, J.L., D.J. Grau, S.K. DeVido and J.A. Kassis, (2005). 'An Sp1/KLF binding site is important for the activity of a Polycomb group response element from the Drosophila engrailed gene'. *Nucleic. Acids. Res.* **33**: 5181–5189.

125. Dejardin, J. et. al. (2005). 'Recruitment of Drosophila Polycomb group proteins to chromatin by DSP1'. *Nature.* **434**: 533–538.

126. Müller, J. and J.A. Kassis, (2006). 'Polycomb response elements and targeting of Polycomb group proteins in Drosophila'. *Curr. Opin. Genet. Dev* **16**: 476–484

127. Ringrose L., M. Rehmsmeier, J.M. Dura, R. Paro (2003). 'Genome-wide prediction of Polycomb/Trithorax response elements in Drosophila melanogaster'. *Dev. Cell.* **5**: 759–771.

128. Tillib S., S. Petruk, Y. Sedkov, A. Kuzin, M. Fujioka, et. al. (1999). 'Trithorax and Polycomb-group response elements within an Ultrabithorax transcription maintenance unit consist of closely situated but separable sequences'. *Mol. Cell. Biol.* **19**: 5189–5202.

129. Mulholland N.M., I.F. King, R.E. Kingston (2003). 'Regulation of Polycomb group complexes by the sequence-specific DNA binding proteins Zeste and GAGA'. *Genes. Dev.* Nov. 15, **17**(22): 2741–2746.

130. Torres-Padilla M.E., D.E. Parfitt, T. Kouzarides et. al. (2007). 'Histone arginine methylation regulates pluripotency in the early mouse embryo'. *Nature.* **445**: 214–218.

131. Rajasekhar V.K, M.C. Vemuri (2005). 'Molecular insights into the function, fate, and prospects of stem cells'. *STEM CELLS.* **23**: 1212–1220.

132. Rando T.A. (2006). 'Stem cells, ageing and the quest for immortality'. *Nature.* 441: 1080–1086.

133. Iwama A., H. Oguro, M. Negishi, et. al. (2005). 'Epigenetic regulation of haematopoietic stem cell self-renewal by polycomb group genes'. *Int. J. Haematol.* **81**: 294–300.

134. Boyer L.A., D. Mathur, R. Jaenisch (2006). 'Molecular control of pluripotency'. *Curr. Opin. Genet. Dev.* **16**: 455–462.

135. Jorgensen H.F., S. Giadrossi, M. Casanova, et. al. (2006). 'Stem cells primed for action: Polycomb repressive complexes restrain the expression of lineage-specific regulators in embryonic stem cells'. *Cell Cycle.* **5**: 1411–1414

136. Meshorer E., T. Misteli (2006). 'Chromatin in pluripotent embryonic stem cells and differentiation'. *Nat Rev Mol Cell Biol.* **7**: 540–546.
137. Lee T.I., R.G. Jenner, L.A. Boyer, et. al. (2006). 'Control of developmental regulators by Polycomb in human embryonic stem cells'. *Cell.* **125**: 301–313.
138. Boyer L.A., K. Plath, J. Zeitlinger, et. al. (2006). 'Polycomb complexes repress developmental regulators in murine embryonic stem cells'. *Nature.* **441**: 349–353.
139. Sun H., G. Skogerbø, R. Chen (2006). 'Conserved distances between vertebrate highly conserved elements'. *Hum Mol Genet.* Oct 1. **15**(19): 2911–22. Epub 2006 Aug 21
140. Bernstein B.E., T.S. Mikkelsen, X. Xie, et. al., (2006). 'A bivalent chromatin structure marks key developmental genes in embryonic stem cells'. *Cell* **125**: 315–326.
141. Azuara V., P. Perry, S. Sauer, et. al. (2006). 'Chromatin signatures of pluripotent cell lines'. *Nat. Cell. Biol.* **8**: 532–538.
142. Tada T. (2006). 'Toti-/pluripotential stem cells and epigenetic modifications'. *Neurodegener Dis.* **3**: 32–37.
143. Squazzo S.L., H. O'Geen, V.M. Komashko, et. al. (2006). 'Suz12 binds to silenced regions of the genome in a cell-type-specific manner'. *Genome. Res.* **16**: 890–900.
144. Boyer L.A., T.I. Lee, M.F. Cole, et. al. (2005). 'Core transcriptional regulatory circuitry in human embryonic stem cells'. *Cell.* **122**: 947–956.
145. Loh Y.H., Q. Wu, J.L. Chew, et. al. (2006). 'The Oct4 and Nanog transcription network regulates pluripotency in mouse embryonic stem cells'. *Nat Genet* **38**: 431–440.
146. de Haan G. (2007). 'Epigenetic control of haematopoietic stem cell aging, the case of Ezh2'. *Ann N Y Acad Sci.* [Epub ahead of print].
147. Yeoh J.S., G. de Haan (2007). 'Fibroblast growth factors as regulators of stem cell self-renewal and aging'. *Mech. Ageing. Dev.* **128**: 17–24.
148. Kamminga L.M., G. de Haan (2006). 'Cellular memory and haematopoietic stem cell aging'. *STEM CELLS.* **24**: 1143–1149.
149. Kamminga L.M., L.V. Bystrykh, A. de Boer, et. al. (2006). 'The Polycomb group gene Ezh2 prevents haematopoietic stem cell exhaustion'. *Blood* **107**: 2170–2179.
150. Iwama A., H. Oguro, M. Negishi, et. al. (2004). 'Enhanced self-renewal of haematopoietic stem cells mediated by the polycomb gene product Bmi-1'. *Immunity.* **21**: 843–851.
151. Lessard J., A. Schumacher, U. Thorsteinsdottir, et. al. (1999). 'Functional antagonism of the Polycomb-Group genes eed and Bmi1 in haemopoietic cell proliferation'. *Genes Dev* **13**: 2691–2703.
152. Leung C., M. Lingbeek, O. Shakhova, et. al. (2004). 'Bmi1 is essential for cerebellarc development and is overexpressed in human medulloblastomas'. *Nature.* **428**: 337–341.
153. Molofsky A.V., S. He, M. Bydon, et. al. (2005). 'Bmi-1 promotes neural stem cell self-renewal and neural development but not mouse growth and survival by repressing the p16Ink4a and p19Arf senescence pathways'. *Genes Dev.* **19**: 1432–1437.
154. Molofsky A.V., R. Pardal, T. Iwashita, et. al. (2003). 'Bmi-1 dependence distinguishes neural stem cell self-renewal from progenitor proliferation'.*Nature* **425**: 962–967.
155. Caretti G., M. Di Padova, B. Micales, et. al. (2004). 'The Polycomb Ezh2 methyltransferase regulates muscle gene expression and skeletal muscle differentiation'. *Genes Dev.* **18**: 2627–2638.
156. Bibikova M., E. Chudin, B. Wu, et. al. (2006). 'Human embryonic stem cells have a unique epigenetic signature'. *Genome Res.* **16**: 1075–1083.

157. Zvetkova I., A. Apedaile, B. Ramsahoye, et. al. (2005). 'Global hypomethylation of the genome in XX embryonic stem cells'. *Nat Genet.* **37**: 1274–1279.
158. Sado T., M. Okano, E. Li, et. al. (2004). 'De novo DNA methylation is dispensable for the initiation and propagation of X chromosome inactivation'. *Development.* **131**: 975–982.
159. Lehnertz B., Y. Ueda, A..A. Derijck, et. al. (2003). 'Suv39h-mediated histone H3 lysine 9 methylation directs DNA methylation to major satellite repeats at pericentric heterochromatin'. *Curr Biol* **13**: 1192–1200.
160. Mermoud J.E., B. Popova, A.H. Peters, et. al. (2002). 'Histone H3 lysine 9 methylation occurs rapidly at the onset of random X chromosome inactivation'. *Curr Biol* **12**: 247–251.
161. Maitra A., D.E. Arking, N. Shivapurkar, et. al. 'Genomic alterations in cultured human embryonic stem cells'. *Nat Genet* 2005; **37**: 1099–1103.
162. Tada M., T. Tada, L. Lefebvre, et. al., (1997). 'Embryonic germ cells induce epigenetic reprogramming of somatic nucleus in hybrid cells'. *EMBO J* **16**: 6510–6520.
163. Wang J., J. Mager, Y. Chen, et. al. (2001). 'Imprinted X inactivation maintained by a mouse Polycomb group gene'. *Nat. Genet.* **28**: 371–375.
164. Takagi N., M. Sasaki (1975). 'Preferential inactivation of the paternally derived X chromosome in the extraembryonic membranes of the mouse'. *Nature.* **256**: 640–642.
165. Muyrers-Chen I., I. Hernandez-Munoz, A.H. Lund, et. al. (2004). 'Emerging roles of Polycomb silencing in X-inactivation and stem cell maintenance'. *Cold. Spring. Harb. Symp. Quant. Biol.* **69**: 319–326.
166. de la Cruz C.C., J. Fang, K. Plath, et. al. (2005). 'Developmental regulation of Suz12 localization'. *Chromosoma.* **114**: 183–192.
167. Kalantry S., T. Magnuson (2006). 'The Polycomb group protein EED is dispensable for the initiation of random X-chromosome inactivation'. *PloS. Genet.* **2**: e66.
168. Kalantry S., K.C. Mills, D. Yee, et. al. (2006). 'The Polycomb group protein Eed protects the inactive X-chromosome from differentiation-induced reactivation'. *Nat. Cell. Biol.* **8**: 195–202.
169. Shovlin T.C., G. Durcova-Hills, A. Surani, A. McLaren (2008). 'Heterogeneity in imprinted methylation patterns of pluripotent embryonic germ cells derived from pre-migratory mouse germ cells'. *Dev. Biol.* Jan 15. **313**(2): 674–681. Epub. 2007 Nov 17.
170. Park, I.K. et. al. (2003). 'Bmi-1 is required for maintenance of adult self-renewing haematopoietic stem cells'. *Nature.* **423**: 302–305.
171. Willert, K. et. al. (2003). 'Wnt proteins are lipid-modified and can act as stem cell growth factors'. *Nature.* **423**: 448–452.
172. Korinek, V. et. al. (1998). 'Depletion of epithelial stem cell compartments in the small intestine of mice lacking Tcf-4'. *Nature. Genet.* **19**: 379–383.
173. St-Jacques, B. et. al. (1998). 'Sonic hedgehog signaling is essential for hair development'. *Curr. Biol.* **8**: 1058–1068.
174. Lai, K., B.K. Kaspar, F.H. Gage and D.V. Schaffer, (2003). 'Sonic hedgehog regulates adult neural progenitor proliferation in vitro and *in vivo*'. *Nature. Neurosci.* **6**: 21–27.
175. Varnum-Finney, B. et. al. (2000). 'Pluripotent, cytokine-dependent, haematopoietic stem cells are immortalized by constitutive Notch1 signaling'. *Nature. Med.* **6**: 1278–1281.
176. Groszer, M. et. al. (2001). 'Negative regulation of neural stem/progenitor cell proliferation by the pten tumour suppressor gene *in vivo*'. *Science.* 1, 1.

177. Hanahan D, R.A. Weinberg (2000). 'The hallmarks of cancer'. *Cell.* **100**: 57–70.
178. Enver T, S. Soneji, C. Joshi, et. al. (2005). 'Cellular differentiation hierarchies in normal and culture-adapted human embryonic stem cells'. *Hum. Mol. Genet.* **14**: 3129–3140.
179. Jones P.A., S.B. Baylin (2002). 'The fundamental role of epigenetic events in cancer'. *Nat Rev Genet.* **3**: 415–428.
180. Frigola J., J. Song, C. Stirzaker, et. al. (2006). 'Epigenetic remodeling in colorectal cancer results in coordinate gene suppression across an entire chromosome band'. *Nat. Genet.* **38**: 540–549.
181. Cadieux B., T.T. Ching, S.R. Vandenberg, et. al., (2006). 'Genome-wide hypomethylation in human glioblastomas associated with specific copy number alteration, methylenetetrahydrofolate reductase allele status, and increased proliferation'. *Cancer. Res.* **66**: 8469–8476.
182. Esteller M. (2007). 'Cancer epigenomics: DNA methylomes and histone-modification maps'. *Nat Rev. Genet.* **8**: 286–298
183. Bracken A.P., N. Dietrich, D. Pasini, et. al. (2006). 'Genome-wide mapping of Polycomb target genes unravels their roles in cell fate transitions'. *Genes. Dev.* **20**: 1123–1136.
184. Pardal R., M.F. Clarke, S.J. Morrison (2003). 'Applying the principles of stem-cell biology to cancer'. *Nat. Rev. Cancer.* **3**: 895–902
185. Parrish J.Z., K. Emoto, L.Y. Jan, et. al. (2007). 'Polycomb genes interact with the tumour suppressor genes hippo and warts in the maintenance of Drosophila sensory neuron dendrites'. *Genes. Dev.* **21**: 956–972.
186. Pardal R., A.V. Molofsky, S. He, et. al. (2005). 'Stem cell self-renewal and cancer cell proliferation are regulated by common networks that balance the activation of proto-oncogenes and tumour suppressors'. *Cold Spring Harb .Symp. Quant. Biol.* **70**: 177–185.
187. Sharpless N.E., R.A. DePinho (2005). 'Cancer: Crime and punishment'. *Nature.* **436**: 636–637.
188. Lu H, Ouyang W., C. Huang (2006). 'Inflammation, a key event in cancer development'. *Mol. Cancer Res.* **4**: 221–233.
189. Coussens L.M., Z. Werb (2002). 'Inflammation and cancer'. *Nature.* **420**: 860–867.
190. Feinberg A.P., R. Ohlsson, S. Henikoff (2006). 'The epigenetic progenitor origin of human cancer'. *Nat. Rev. Genet.* **7**: 21–33.
191. Jones P.A., S.B. Baylin (2007). 'The epigenomics of cancer'. *Cell.* **128**: 683–692.
192. Baylin S.B., J.E. Ohm (2006). 'Epigenetic gene silencing in cancer—a mechanism for early oncogenic pathway addiction'? *Nat. Rev. Cancer.* **6**: 107–116.
193. Reya T., S.J. Morrison, M.F. Clarke, et. al. (2001). 'Stem cells, cancer, and cancer stem cells'. *Nature.* **414**: 105–111.
194. Wicha M.S., S. Liu, G. Dontu (2006). 'Cancer stem cells: An old idea—a paradigm shift'. *Cancer Res.* **66**: 1883–1890.
195. Clarke M.F., M. Fuller (2006). 'Stem cells and cancer: Two faces of eve'. *Cell.* **124**: 1111–1115.
196. Ohm J.E., K.M.McGarvey, X. Yu, et. al. (2007). 'A stem cell-like chromatin pattern may predispose tumour suppressor genes to DNA hypermethylation and heritable silencing'. *Nat. Genet.* **39**: 237–242.
197. Gil J., D. Bernard, G. Peters (2005). 'Role of polycomb group proteins in stem cell self-renewal and cancer'. *DNA Cell Biol.* **24**: 117–125.
198. Dahiya A., S. Wong, S. Gonzalo, et. al. 'Linking the Rb and polycomb pathways'. *Mol. Cell.* 2001; 8: 557–569.

199. Monk M., C. Holding (2001). 'Human embryonic genes re-expressed in cancer cells'. *Oncogene.* **20**: 8085–8091.
200. Ezeh U.I., P.J. Turek, R.A. Reijo, et. al. (2005). 'Human embryonic stem cell genes OCT4, NANOG, STELLAR, and GDF3 are expressed in both seminoma and breast carcinoma'. *Cancer.* **104**: 2255–2265.
201. Gidekel S., G. Pizov, Y. Bergman et. al. (2003). 'Oct-3/4 is a dose-dependent oncogenic fate determinant'. *Cancer Cell.* **4**: 361–370.
202. Ohm J.E., S.B. Baylin (2007). 'Stem cell chromatin patterns: An instructive mechanism for DNA hypermethylation'? *Cell. Cycle.* **6**: 1040–1043.
203. Baglole C.J., D.M. Ray, S.H. Bernstein, et. al. (2006). 'More than structural cells, fibroblasts create and orchestrate the tumour microenvironment'. *Immunol. Invest.* **35**: 297–325.
204. Moore K.A., I.R. Lemischka (2006). 'Stem cells and their niches'. *Science.* **311**: 1880–1885.
205. Bissell M.J., M.A. Labarge (2005). 'Context, tissue plasticity, and cancer: Are tumour stem cells also regulated by the microenvironment'? *Cancer. Cell.* **7**: 17–23.
206. Li L., W.B. Neaves (2006). 'Normal stem cells and cancer stem cells: The niche matters'. *Cancer Res.* **66**: 4553–4557.
207. Hu M, J. Yao, L. Cai, et. al. (2005). 'Distinct epigenetic changes in the stromal cells of breast cancers'. *Nat. Genet.* **37**: 899–905.
208. Pritsker M., N.R. Ford, H.T. Jenq, et. al. (2006). 'Genomewide gain-of-function genetic screen identifies functionally active genes in mouse embryonic stem cells'. *Proc. Natl Acad. Sci. USA.* **103**: 6946–6951.
209. Cavalli G. (2006). 'Chromatin and epigenetics in development: Blending cellular memory with cell fate plasticity'. *Development.* **133**: 2089–2094.
210. Orkin S.H. (2005). 'Chipping away at the embryonic stem cell network'. *Cell.* **122**: 828–830.
211. Lee H., R. Habas, C. Abate-Shen (2004). 'MSX1 cooperates with histone H1b for inhibition of transcription and myogenesis'. *Science.* **304**: 1675–1678.
212. Lee H., J.C. Quinn, K.V. Prasanth, et. al. (2006). 'PIAS1 confers DNA-binding specificity on the Msx1 homeoprotein'. *Genes. Dev.* **20**: 784–794.
213. Chagraoui J., S.L. Niessen, J. Lessard, et. al. (2006). 'E4F1: A novel candidate factor for mediating BMI1 function in primitive haematopoietic cells'. *Genes. Dev.* **20**: 2110–2120.
214. Mohd-Sarip A., J.A. van der Knaap, C. Wyman, et. al. (2006). 'Architecture of a polycomb nucleoprotein complex'. *Mol. Cell.* **24**: 91–100.
215. Rajasekhar V.K. (2007). 'Analytical methods for cancer stem cells'. *In: Vemuri MC, ed. Stem Cell Assays*. Totowa, N.J: Humana Press, 83–95.
216. Wegel E., Shaw (2005). 'Gene activation and deactivation related changes in the three-dimensional structure of chromatin'. *Chromosoma.* **114**: 331–337.
217. Beck H.C., E.C. Nielsen, R. Matthiesen, et. al. (2006). 'Quantitative proteomic analysis of post-translational modifications of human histones'. *Mol Cell. Proteomics.* **5**: 1314–1325.
218. Heintzman N.D., R.K. Stuart, G. Hon, et. al. (2007). 'Distinct and predictive chromatin signatures of transcriptional promoters and enhancers in the human genome'. *Nat. Genet.* **39**: 311–318.
219. Barski A., S. Cuddapah, K. Cui, et. al. (2007). 'High-resolution profiling of histone methylations in the human genome'. *Cell.* **129**: 823–837.

220. de Wit E., F. Greil, B. van Steensel (2007). 'High-resolution mapping reveals links of HP1 with active and inactive chromatin components'. *PloS. Genet.* **3**: e38.
221. O'Neill L.P., M.D. VerMilyea, B.M. Turner (2006). 'Epigenetic characterization of the early embryo with a chromatin immunoprecipitation protocol applicable to small cell populations'. *Nat. Genet.* **38**: 835–841.
222. Takahashi K., S. Yamanaka (2006). 'Induction of pluripotent stem cells from mouse embryonic and adult fibroblast cultures by defined factors'. *Cell.* **126**: 663–676.
223. Wernig M., A. Meissner, R. Foreman, et. al. (2007). '*In vitro* reprogramming of fibroblasts into a pluripotent ES-cell-like state'. *Nature.* [Epub ahead of print].
224. Okita K., T. Ichisaka, S. Yamanaka (2007). 'Generation of germline-competent induced pluripotent stem cells'. *Nature.* [Epub ahead of print].
225. Maherali N., R. Sridharan, W. Xie, et. al. (2007). 'Directly reprogrammed fibroblasts show global epigenetic remodeling and wide spread tissue contribution'. *Cell Stem Cell.* **1**: 55–70.
226. Rugg-Gunn, P.J., A.C. Ferguson-Smith and R.A. Pedersen (2005). 'Epigenetic status of human embryonic stem cells'. *Nat. Genet.* **37**: 585–587.
227. Maitra, A., D.E. Arking, N. Shivapurkar, M. Ikeda, V. Stastny, K. Kassauei, G. Sui, D.J. Cutler, Y. Liu, S.N. Brimble, et. al. (2005). 'Genomic alterations in cultured human embryonic stem cells'. *Nat. Genet* **37**: 1099–1103.
228. Shen Y., J. Chow, Z. Wang, G. Fan (2006). 'Abnormal CpG island methylation occurs during *in vitro* differentiation of human embryonic stem cells'. *Hum. Mol. Genet.* 2006 Sep 1). **15**(17): 2623–3265. Epub Jul 26.
229. van Kemenade F.J., F.M. Raaphorst, T. Blokzijl, et. al. (2001). 'Coexpression of BMI-1 and EZH2 polycomb-group proteins is associated with cycling cells and degree of malignancy in B-cell non-Hodgkin lymphoma'. *Blood.* **97**: 3896–3901.
230. Arisan S., E.D. Buyuktuncer, N. Palavan-Unsal. et. al. (2005). 'Increased expression of EZH2, a polycomb group protein, in bladder carcinoma'. *Urol. Int.* **75**: 252–257.
231. Arisan S., E.D. Buyuktuncer, N. Palavan-Unsal, et. al. (2005). 'Increased expression of EZH2, a polycomb group protein, in bladder carcinoma'. *Urol. Int.* **75**: 252–257.
232. Raman J.D., N.P. Mongan, S.K. Tickoo, et. al. (2005). 'Increased expression of the polycomb group gene, EZH2, in transitional cell carcinoma of the bladder'. *Clin. Cancer. Res.* **11**: 8570–8576.
233. Ding L., C. Erdmann, A.M. Chinnaiyan, et. al. (2006). 'Identification of EZH2 as a molecular marker for a precancerous state in morphologically normal breast tissues'. *Cancer. Res.* **66**: 4095–4099.
234. Collett K., G.E. Eide, J. Arnes et. al. (2006). 'Expression of enhancer of zeste homologue 2 is significantly associated with increased tumour cell proliferation and is a marker of aggressive breast cancer'. *Clin. Cancer. Res.* **12**: 1168–1174.
235. Bachmann I.M., O.J. Halvorsen, K. Collett, et. al. (2006). 'EZH2 expression is associated with high proliferation rate and aggressive tumour subgroups in cutaneous melanoma and cancers of the endometrium, prostate, and breast'. *J Clin Oncol.* **24**: 268–273.
236. Kleer C.G., Q. Cao, S. Varambally, et. al. (2003). 'EZH2 is a marker of aggressive breast cancer and promotes neoplastic transformation of breast epithelial cells'. *Proc Natl Acad Sci USA* **100**: 11606–11611.
237. Mimori K., K. Ogawa, M. Okamoto, et. al. (2005). 'Clinical significance of enhancer of zeste homolog 2 expression in colorectal cancer cases'. *Eur. J. Surg. Oncol.* **31**: 376–380.

238. Raaphorst F.M., van F.J. Kemenade, T. Blokzijl, et. al. (2000). 'Coexpression of BMI-1 and EZH2 polycomb group genes in Reed-Sternberg cells of Hodgkin's disease'. *Am. J. Pathol.* **157**: 709–715.
239. Sudo T., T. Utsunomiya, K. Mimori, et. al., (2005). 'Clinicopathological significance of EZH2 mRNA expression in patients with hepatocellular carcinoma'. *Br. J. Cancer.* **92**: 1754–1758.
240. Visser H.P., M.J. Gunster, H.C. Kluin-Nelemans, et. al. (2001). 'The Polycomb group protein EZH2 is upregulated in proliferating, cultured human mantle cell lymphoma'. *Br. J. Haematol.* **112**: 950–958.
241. Varambally S., S.M. Dhanasekaran, M. Zhou, et. al. (2002). 'The polycomb group protein EZH2 is involved in progression of prostate cancer'. *Nature.* **419**: 624–629.
242. Berezovska O.P., A.B. Glinskii, Z. Yang, et. al. (2006). 'Essential role for activation of the Polycomb group (PcG) protein chromatin silencing pathway in metastatic prostate cancer'. *Cell. Cycle.* **5**: 1886–1901.
243. van Leenders G.J., D. Dukers, D. Hessels, et. al. (2007). 'Polycomb-group oncogenes EZH2, BMI1, and RING1 are overexpressed in prostate cancer with adverse pathologic and clinical features'. *Eur. Urol.* **52**: 455–463.
244. Beke L., M. Nuytten, A. Van Eynde, et. al. (2007). 'The gene encoding the prostatic tumour suppressor PSP94 is a target for repression by the Polycomb group protein EZH2'. *Oncogene.* **26**: 4590–4595.
245. Chen H., S.W. Tu, J.T. Hsieh (2005). 'Down-regulation of human DAB2IP gene expression mediated by polycomb Ezh2 complex and histone deacetylase in prostate cancer'. *J. Biol. Chem.* **280**: 22437–22444.
246. Matsukawa Y., S. Semba, H. Kato, et. al. (2006). 'Expression of the enhancer of zeste homolog 2 is correlated with poor prognosis in human gastric cancer'. *Cancer. Sci.* **97**: 484–491.
247. Kirmizis A., S.M. Bartley, P.J. Farnham (2003). 'Identification of the polycomb group protein SU(Z)12 as a potential molecular target for human cancer therapy'. *Mol. Cancer. Ther.* **2**: 113–121.
248. Saramaki O.R., T.L. Tammela, P.M. Martikainen, et. al. (2006). 'The gene for polycomb group protein enhancer of zeste homolog 2 (EZH2) is amplified in late-stage prostate cancer'. *Genes Chromosomes. Cancer.* **45**: 639–645.
249. Sanchez-Beato M., E. Sanchez, J. Gonzalez-Carrero, et. al. (2006). 'Variability in the expression of polycomb proteins in different normal and tumoural tissues'. 'A pilot study using tissue microarrays'. *Mod. Pathol.* **19**: 684–694.
250. Breuer R.H., P.J. Snijders, E.F. Smit, et. al. (2004). 'Increased expression of the EZH2 polycomb group gene in BMI-1-positive neoplastic cells during bronchial carcinogenesis'. *Neoplasia.* **6**: 736–743.
251. Tateishi K., M. Ohta, F. Kanai, et. al. (2006). 'Dysregulated expression of stem cell factor Bmi1 in precancerous lesions of the gastrointestinal tract'. *Clin Cancer Res.* **12**: 6960–6966.
252. Prince M.E., R. Sivanandan, A. Kaczorowski, et. al. (2007). 'Identification of a subpopulation of cells with cancer stem cell properties in head and neck squamous cell carcinoma'. *Proc. Natl. Acad. Sci. USA.* **104**: 973–978.
253. Dukers D.F., van J.C. Galen, C. Giroth, et. al. (2004). 'Unique polycomb gene expression pattern in Hodgkin's lymphoma and Hodgkin's lymphomaderived cell lines'. *Am. J. Pathol.* **164**: 873–881.
254. Sawa M., K. Yamamoto, T. Yokozawa, et. al. (2005). 'BMI-1 is highly expressed in M0-subtype acute myeloid leukemia'. *Int. J. Haematol.* **82**: 42–47.

255. Bea S., F. Tort, M. Pinyol, et. al. (2001). 'BMI-1 gene amplification and overexpression in haematological malignancies occur mainly in mantle cell lymphomas'. *Cancer Res.* 61: 2409–2412.
256. Song L.B., M.S. Zeng, W.T. Liao, et. al. (2006). 'Bmi-1 is a novel molecular marker of nasopharyngeal carcinoma progression and immortalizes primary human nasopharyngeal epithelial cells'. *Cancer. Res.* 66: 6225–6232.
257. Nowak K, K. Kerl., D. Fehr, et. al. (2006). 'BMI1 is a target gene of E2F-1 and is strongly expressed in primary neuroblastomas'. *Nucleic. Acids. Res.* 34: 1745–1754.
258. Vonlanthen S., J. Heighway, H.J. Altermatt, et. al. (2001). 'The bmi-1 oncoprotein is differentially expressed in non-small cell lung cancer and correlates with INK4A-ARF locus expression'. *Br. J. Cancer.* 84: 1372–1376.
259. Breuer R.H., P.J. Snijders, G.T. Sutedja, et. al. (2005). 'Expression of the p16(INK4a) gene product, methylation of the p16(INK4a) promoter region and expression of the polycomb-group gene BMI-1 in squamous cell lung carcinoma and premalignant endobronchial lesions'. *Lung. Cancer.* 48: 299–306.
260. Kang M.K., R.H. Kim, S.J. Kim, et. al. (2007). 'Elevated Bmi-1 expression is associated with dysplastic cell transformation during oral carcinogenesis and is required for cancer cell replication and survival'. *Br. J. Cancer.* 96: 126–133.
261. Berezovska O.P., A.B. Glinskii, Z. Yang, et. al. (2006). 'Essential role for activation of the Polycomb group (PcG) protein chromatin silencing pathway in metastatic prostate cancer'. *Cell. Cycle.* 5: 1886–1901.
262. Seligson D.B., S. Horvath, T. Shi, et. al. (2005). 'Global histone modification patterns predict risk of prostate cancer recurrence'. *Nature.* 435: 1262–1266.
263. Gordon S., G. Akopyan, H. Garban et. al. (2006). 'Transcription factor YY1: Structure, function, and therapeutic implications in cancer biology'. *Oncogene.* 25: 1125–1142.
264. Wang S., F. He, W. Xiong, et. al. (2007). 'Polycomblike-2-deficient mice exhibit normal left-right asymmetry'. *Dev Dyn.* 236: 853–861.
265. Tokimasa S., H. Ohta, A. Sawada, et. al. (2001). 'Lack of the Polycomb-group gene rae28 causes maturation arrest at the early B-cell developmental stage'. *Exp. Haematol.* 29: 93–103.

24

CANCER STEM CELLS: THE ENEMY WITHIN CANCERS

Devaveena Dey, Naseer M., Meera Saxena, Anurag N. Paranjape*, Annapoorni Rangarajan

Indian Institute of Science, Bangalore 560 012, India

INTRODUCTION TO CANCER

"Regulation" or control of any living system is critical for its survival, growth and maintenance. From a unicellular bacterium to the highly complex multicellular forms of life, there is evidence of "control checkpoints" for all the crucial cellular processes. One of the most important of such processes is proliferation. What happens when this cellular process goes out of control due to alterations in the checkpoints is best represented by "cancer": a fatal disease of uncontrolled proliferation of genetically altered cells.

The term cancer is derived from the Greek word "Karkinos" which means crab and this name comes from the appearance of the cut surface of a solid malignant tumour, with the veins stretched on all sides. In all known cases, cancer cells are derived from the repeated divisions of a mutant cell. Some of these mutations may be due to the effects of carcinogens, such as tobacco smoke, radiation, chemicals, or infectious agents. A few other cancer-promoting mutations may be acquired through errors in DNA replication.

Genetic alterations that render a normal cell cancerous usually arise in two classes of genes termed the oncogenes and tumours suppressors (Box 24.1). Such genetic abnormalities found in cancer cells typically activate the cancer promoting oncogenes and/or inactivate the tumour suppressor genes. Yet another class of genes involved in carcinogenesis are those which "sense" and repair DNA damage and also ensure proper chromosome segregation during mitosis. These

---

* anu@mrdg.iisc.ernet.in

genes are also seen to be inactivated in tumours, thereby rendering the cells highly "mutable", enabling them to acquire multiple mutations [1]. Along with genetic abnormalities, changes at the epigenetic level are also well known to operate in cancers. There are changes in promoter methylation–demethylation patterns in the genome of cancer cells, which is a typical phenomenon-utilized to repress or activate gene expression respectively [2]. One of the common examples is the promoter hypermethylation of the tumour suppressor, p16, which is found in several cancers and renders this gene inactive. There are also seen chromatin modifications, in the form of acetylation–deacetylation of lysine residues of histones, which operate in cancer cells by turning "on" proliferation and prosurvival genes while turning "off" the anti-proliferation (checkpoint) and pro-apoptosis genes [3].

Overall, cancer can be viewed at three levels (a) a disease of altered tissue behaviour which recognizes that a tumour is not just a collection of cancer cells, but a complex organ whose properties reflect the interactions between genetically altered cancer cells and the host (b) a disease of altered cellular behaviour, and (c) a disease of genomic instability.

The process of conversion of a normal cell into a cancerous cell, termed cellular transformation, involves the accumulation of multiple genetic alterations that confer proliferation and survival benefits. For example, while normal cells fail to divide after a finite number of divisions, termed the Hayflick limit [4], cancer cells show extensive proliferation potential. While normal cells respond to cell–cell contacts by negatively regulating proliferation, cancer cells lack such contact inhibition [5]. Also, normal cells need exogenous growth factors to proliferate, but most cancer cells are self-sufficient in their requirement for mitogenic signals. While most normal cells need a substratum for attachment and growth, cancer cells are anchorage independent and can survive in the absence of any substratum [5].

An abnormal growth of tissue resulting from uncontrolled, progressive proliferation generates a mass of cells which is called a tumour. Sometimes, such a mass can be self-limiting, which does not invade neighboring tissue and hence pose no danger to life. These are the benign tumours. At other times, a tumour continues to grow at its original site, breaches the basement membrane and gets into the circulatory system, whereby these cancer cells migrate to new sites in the body and initiate secondary tumours. This process is called metastasis, and such tumours are called malignant [6]. Thus cancer always refers to malignant neoplasms, whereas tumours can be either benign or malignant.

Cancer of almost all tissues is known today. The cancers of the epithelium are called carcinomas. Those originating in the cells of connective tissues, like bones, cartilages, blood vessels or muscle are called sarcomas. Cancers that start in blood forming tissues such as bone marrow are called leukemia, and those that begin in cells of the immune system are termed as lymphoma and myeloma. Cancers of the central nervous system that arise in the tissues of the brain and spinal cord include gliomas and astrocytomas [1].

According to the World Health Organisation survey of 2008, lung cancer is the most common cancer worldwide, followed by cancer of the breast, and in order colorectal, stomach, liver, cervical, esophageal, etc. The three leading cancer killers in the world are different from the three most common forms, with lung cancer being responsible for 17.8% of all cancer deaths; stomach,

10.4% and liver, 8.8%. Breast cancer is among the leading chronic conditions affecting adult women. In India, oral cancer is the most common, followed by throat and lung cancer. In fact, India has the highest number of oral and throat cancer cases in the world (Fig. 24.1). In the 1990s, cervical cancer was the number one cause of cancer deaths among women in India, with breast cancer being the second highest. India's first cancer atlas, produced by the Indian Council of Medical Research in 2005, however, confirmed that breast cancer has replaced cervical cancer as the leading site of cancer among women in Indian cities. Breast cancer accounts for 19–34% of all cancer cases among women. This increase has been attributed to changes in lifestyle and food habits.

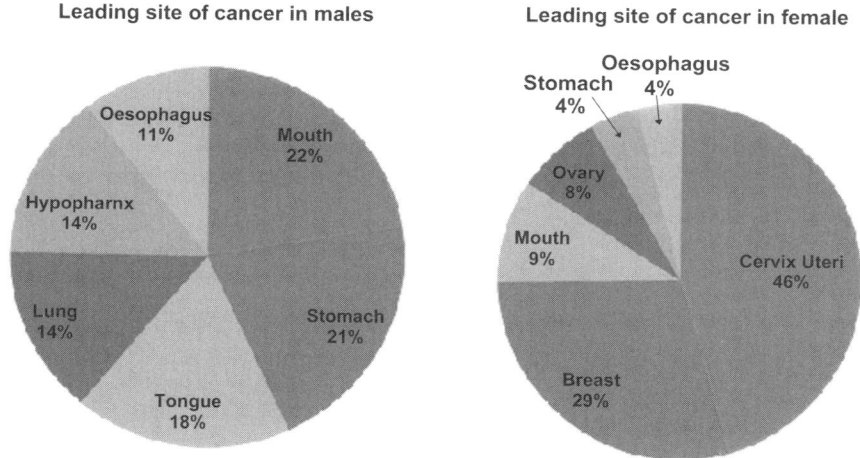

Courtesy: Adopted from Indian Council of Medical Research

▶ **Fig. 24.1** *Leading site of cancer in males and females in India. The National Cancer Registry Programme by the Indian Council of Medical Research (ICMR) conducted before 2005 came up with a detailed statistics with respect to the various cancers prevalent among males and females in India. While mouth cancer came up as the most prevalent in males, cervical cancer came up as the most common cancer in females. The 2005 cancer atlas by ICMR revealed that incidence of breast cancer had surpassed that of cervical cancer in females. (for colour figure see Plate 64).*

Chaemotherapy, surgery and radiotherapy have remained the main attempts of cancer treatment. Chemotherapy includes use of cytotoxic drugs to attack the cancer cells, most often exploiting the property of rapid proliferation of these cells. With succeeding generations of tumour cells, growth becomes less regulated, and tumours become less responsive to most chemotherapeutic agents [7]. Near the centre of some solid tumours, cell division effectively ceases, making them insensitive to chemotherapy. Another problem with solid tumours is the fact that the chemotherapeutic agent often does not reach the core of the tumour. Solutions to this problem include surgery and radiation therapy. Surgical cases include those where either the tumour size is very distinct or large. The goal of the surgery can be either the removal of only the tumour, or the entire organ. Radiotherapy refers to the medical use of ionising radiation for malignant tumours. The effects of radiation therapy are localised to the region being treated. This therapy injures or destroys cells in the area being treated (the "target tissue") by damaging their genetic material, making it impossible for these cells to continue to grow and divide. Usually, a combination of all the three treatment regimes is used against any given cancer for maximum benefit [8, 9].

CLONALITY IN CANCER

Studies as long back as 1930s indicated that cancer arises by mutations in a single cell, such that this one cell subsequently undergoes uncontrolled proliferation. This step is referred to as the "initiation" step. This is an essentially irreversible and a heritable process which increases the lifespan of the target cell and its resistance to toxins and apoptotic signals. This step is followed by extensive proliferation in a focal region, initiated in the target cell [10]. This step, known as "promotion" is reversible and may be the precursor to the carcinogenic process. As cells in this

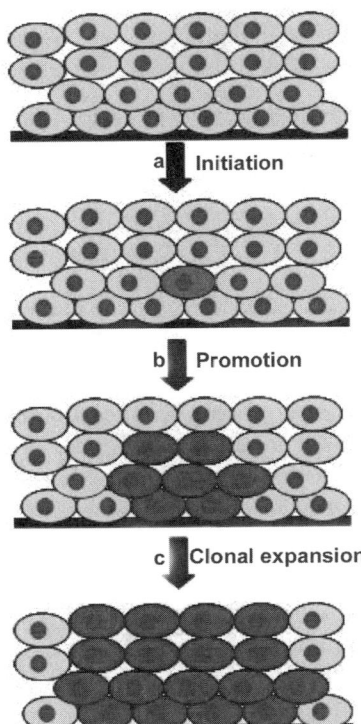

▶ **Fig. 24.2** *Clonal theory. Cancer arises from the clonal expansion of a single cell, in which a genetic change (mutation) had "initiated" tumorigenesis. Over a period, accumulation of multiple genetic and epigenetic changes results in "promotion" of tumorigenesis. (for colour figure see Plate 65).*

focal region continue to proliferate and become resistant to apoptosis, it marks the step of "progression" (Fig. 24.2). At every step of carcinogenic progression, the cancer cells acquire multiple mutations for their enhanced survival, to power their uncontrolled proliferation and overpower the normal cells [10]. The study of Chronic Myelogenous Leukemia (CML) provided the first evidence of a specific genetic change associated with a human cancer and of the clonal nature of these disorders [11]. The discovery of a consistent chromosomal abnormality in essentially every dividing haematopoietic cell of patients with CML suggested that the leukemia had arisen from the clonal expansion of a single cell in which this genetic change had occurred and that this change was crucial to the pathogenesis of the disease [12]. Thus, the clonal theory of cancer predicts that all cancer cells are descendants of a single parental cell.

CANCER HETEROGENEITY

Given that tumours have a clonal origin, does it mean that every cell within a tumour is identical? A closer look at the tumour cells would show that not all cells that form the tumour are identical. There exists variability within these cells with respect to their size, shape, granularity, size of nucleus, nucleo-cytoplasmic ratio, etc. When a biopsy specimen is stained with routine pathological stains, like eosin and haematoxylin, and observed under a regular microscope, one can observe the phenotypic differences, like the shapes and sizes of different cells (cellular morphology). While some cells are large and distinct, some tumour cells are small. Also while some cells stain intensely with the regular stains, some show very faint staining (Fig. 24.3). Marked differences between the proliferation status of tumour cells within a single tumour are common [13]. While some cells are dead, some are quiescent and other cells are in the cycle, but at different stages. Several cellular processes, like antigen expression, membrane characteristics etc. have been seen to vary between different cells within the same tumour. Another difference is that of karyotype between different cancer cells isolated from the same tumour [13]. Such distinct "subpopulations" of tumour cells varying in multiple ways from each other have been isolated from cancers of multiple tissues; from cancers induced by physical, chemical or viral agents, and from well-established cancer cell lines. One of the earlier experiments done to understand this heterogeneity in glioma separated out cells of different karyotypes from a fresh tumour specimen as distinct subpopulations. These populations differed in several parameters, like sensitivity to drugs and cell growth [14]. Hence, cellular differences within a tumour seem to be ubiquitous.

In vivo, a tumour does not exist in isolation. It is present in the context of complex vasculature, the stroma and infiltrating immune cells (Fig. 24.4). All the other cells, apart from the tumour cells are the ancillary or "supporting" cells that are recruited by the tumour cells to maintain and nourish the tumour. The most important "supporting" cells are the fibroblasts, which form a part of the tumour stroma [15]. These supporting cells, immune cells and blood vessels are present within an organized tissue framework in a normal tissue/organ, in association with the basement membrane. Through this basement membrane, the essential nutrients from the blood vessels are transported to the cells and via growth factors and various signaling molecules, the normal cells crosstalk with its microenviroenment. In case of a tumour, the hyperproliferative tumour cells increase in number

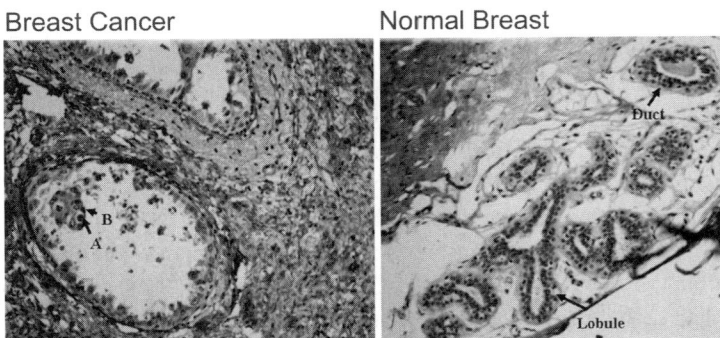

Courtesy: Suruchi Mittal.,
Indian Institute of Science (unpublished data)

➤ **Fig. 24.3** *Phenotypic heterogeneity within a tumour. Representative images of biopsy sections of breast cancer and normal breast, when observed through a simple microscope after staining with a routine dye like haematoxylin, indicates a distinct difference between the morphology of breast cancer tissue vs. its normal counterpart. It indicates that while the normal tissue shows a well-organised architecture of ducts and lobules, along with the surrounding stroma, this organisation is lost in the cancer counterpart. Within the cancer, there are cells of different morphologies. In this case, there are large (A) and small (B) tumour cells.*

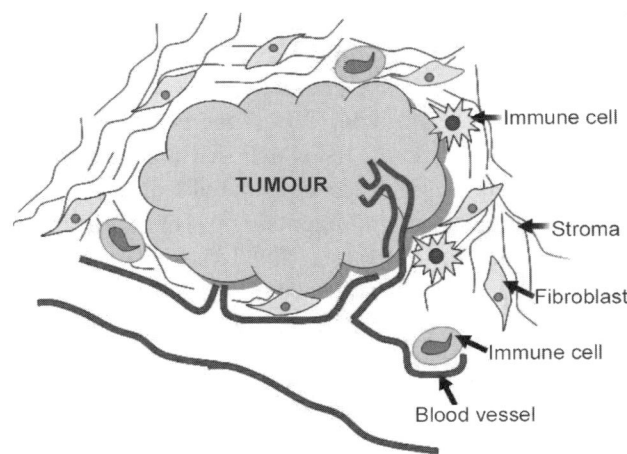

➤ **Fig. 24.4** *An illustration to represent a tumour in vivo. Within the body, a tumour doesn't exist in isolation. It is present in association with other components, like blood vessels, which nourish the tumour by supplying nutrients and oxygen. Stroma, which forms the extracellular matrix, supports the tumour and stromal cells such as fibroblasts are closely associated with the tumour. Cells of the immune system such as neutrophils, dendritic cells and WBCs are also known to be present in the vicinity of a tumour. (for colour figure see Plate 65)*

and move away from the basement membrane. This change has to be compensated by recruiting new blood vessels and supporting cells around the tumour. Hence a change in the crosstalk between a tumour and the stroma results in a change in the characteristics of the stromal cells, like fibroblasts, which are now referred to as "Cancer Associated Fibroblasts" (CAFs), such that now these very cells promote the growth of the tumour mass.

At any given point of time, most, but not all tumour cells are seen to be dividing [13]. An important issue that often surfaces in cancer biology involves the functionality of cancer cells. The "function" of a tumour cell is assessed by its ability to seed a new tumour and recapitulate the original tumour (termed as tumourigenicity). Two techniques routinely used to assess if a given population of cells is tumourigenic are as follows: The first is an *in vitro* assay, which scores for colony formation by single cells, when seeded in a 3D matrix, such as "soft agar" while the other technique is an *in vivo* assay, which scores for tumour formation in mice when these cells are injected (Fig. 24.5).

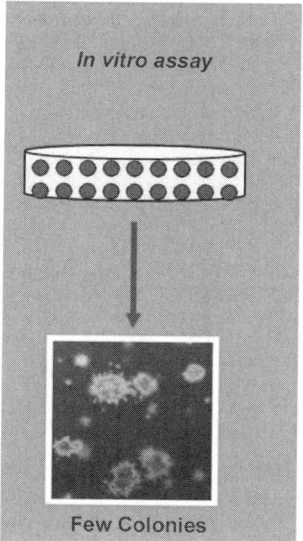
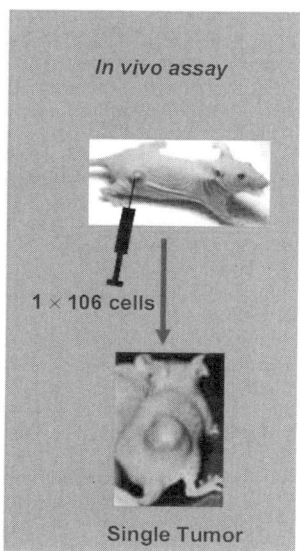

▶ **Fig. 24.5** *Functional heterogeneity within a tumour. The 'function' of a tumour cell is assessed by its ability to give rise to a tumour. Two techniques are routinely used to assess the tumourigenic potential of cells. First, an* **in** *vitro method, the 'soft agar' assay, which scores for formation of colonies by single cells seeded in a 3D matrix like agar. The second is an* **in** *vivo assay which scores for tumour formation when cells are injected in nude mice. It was observed that not every cell which was seeded in an in vitro matrix could give rise to a colony and cell numbers as high as 1×10^6 tumour cells had to be injected for the formation of a tumour, indicating a functional heterogeneity within tumours. (for colour figure see Plate 66).*

An interesting observation made decades ago was that when one seeded several thousands of cancer cells in soft agar one observed colony formation at a low frequency of 0.1–1%. That is, only one in every hundred or thousand cancer cells could actually behave like a cancer cell and give rise to colonies, while the rest failed to do so. This was true for several haematological malignancies, as well as solid tumours such as those arising in breast, lung, and skin. Both, primary tumours that have been resected from a patient, as well as cancer cell lines that have been in culture for several years behaved in the same way. Thus, it appeared that despite their clonal origin, all cells within a cancer were not identical with respect to their function at a given point of time [16].

Another interesting observation made involved the *in vivo* assay. In this assay, injection of at least one million cells was often required in order to initiate tumour formation in mice, whereas injection of lesser numbers failed to do so. Again, if all the cells within a tumour had the potential to initiate a new tumour, then injection of fewer numbers of these would suffice to give rise to tumours. Yet again it appeared that not all cells within a cancer are similar with respect to their functions. There appeared to be the existence of a "functional heterogeneity" within a tumour, whereby not every tumour cell had the potential to either generate a colony in soft agar, or seed a new tumour. These interesting observations led scientists to hunt for those cells within a tumour which had the potential to form a colony *in vitro* and to seed a new tumour *in vivo*. It was foreseen that the discovery of the identity of these within a cancer would enable one to get to the root of cancer and understand the cellular origin of a tumour.

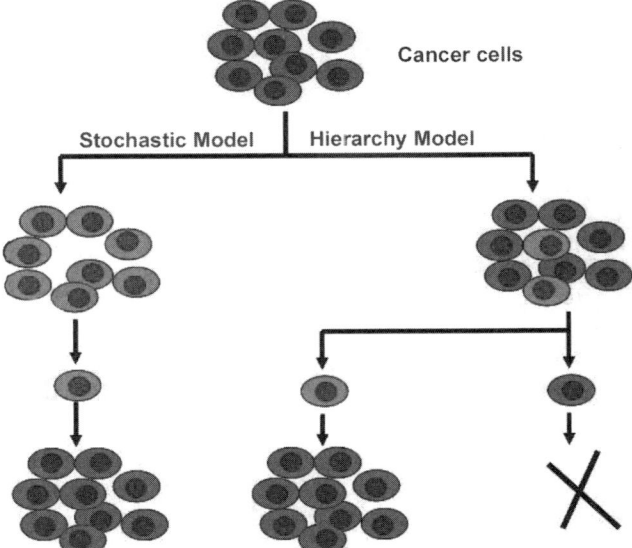

▶ **Fig. 24.6** *Stochastic and Hierarchy Models. Two hypotheses had been suggested to explain the functional heterogeneity as seen in tumours. First, that every cell within the tumour has the ability to give rise to a tumour (all red cells) and it is chance that determines which cell becomes tumourigenic (Stochastic Model). In the second hypothesis only few cells within the tumour have an inherent potential to give rise to the tumour (the red cells) while most of the cells cannot give rise to a tumour (Hierarchy Model). (for colour figure see Plate 66).*

STOCHASTIC AND HIERARCHY MODELS

Two hypotheses have been suggested to explain the functional heterogeneity within cancers as described above. One which assumes that every tumour cell is identical and has the potential to initiate another tumour, but it is only chance that determines which cell gives rise to the tumour [17]. This is called the "Stochastic Model". Hence, according to the Stochastic model, any cell can become a "cancer-initiating cell". The other hypothesis, assumes that to begin with there exists a "hierarchy" within the tumour, such that only some, very few cells, have the potential to initiate a new tumour, while the rest do not have this property. This hypothesis is known as the "Hierarchy Model" (Fig. 24.6). In order to prove that indeed "functionally" different types of cancer cells existed within a tumour, one needed to isolate these cells. In the absence of experimental evidence both hypothesis remained unsupported for several decades.

STEM CELLS AND HIERARCHY

Research in the field of stem cell biology is picking up pace. Such a functional hierarchy, as envisaged for cancers, began to unfold within normal tissue. It is now more than five decades from the time of bone marrow reconstitution experiments in lethally irradiated mice, which for the first time indicated the presence of the Haematopoietic Stem Cell (HSC), a cell whose existence was postulated way back in 1917. This finding was to pioneer the discovery of stem cells in multiple adult tissues [18]. The ease of availability of blood has made it possible to understand several cellular and molecular aspects of stem cell biology [17]. Simultaneously, with technical advancements attained with flow cytometry and modified culture conditions to maintain stem cells *in vitro*, stem cells could be isolated from a heterogeneous tissue followed by enrichment and characterisation.

Stem cells are the group of cells which can repopulate the entire tissue when required, such as, during disease or injury (Box 24.2). Two properties of stem cells that sets them aside from the other specialized cells of a tissue are (a) their ability to give rise to a cell identical to itself—a property known as "self renewal", and (b) their ability to give rise to all the cells comprising the tissue—a property known as "multilineage differentiation" (Fig. 24.7). By its ability to "self-renew", a stem cell can maintain a population of its own kind for the entire lifetime of an organism and by its ability to undergo multilineage differentiation; it ensures the maintenance of tissue homeostasis in normal and disease conditions [19].

Today, the molecular pathways underlying the unique property of self-renewal are known for multiple tissues. They appear to be the same pathways which are critical during embryonic development, like the Wnt, Notch, and Sonic Hedgehog pathways [17]. Activation of the Notch and Hedgehog pathways in HSCs in culture has been consistently shown to increase the number of primitive progenitors. The same was seen by activation of the Notch pathway in human mammary epithelial stem cells [20]. On the other hand, inhibition of this pathway resulted in a depletion of the stem cell population. Wnt pathway has been shown to be critical in the maintenance of stem cell pool in skin, intestine, and colon among many other organs [21]. Interestingly, the same pathways have also been shown to be critical for carcinogenesis and in most of the cancers, these pathways are dysregulated. Many players of these three pathways have been shown to be mutated in multiple cancers, for example the Notch pathway is seen to be dysregulated in most of the leukemias and epithelial cancers, mutations in members of the Wnt pathway is universal in cancers of the epidermis and colon cancers while the Hedgehog pathway is dysregulated in meduloblastomas and basal cell carcinomas [17]. In most of the cancers, these dysregulated pathways are seen to be constitutively upregulated.

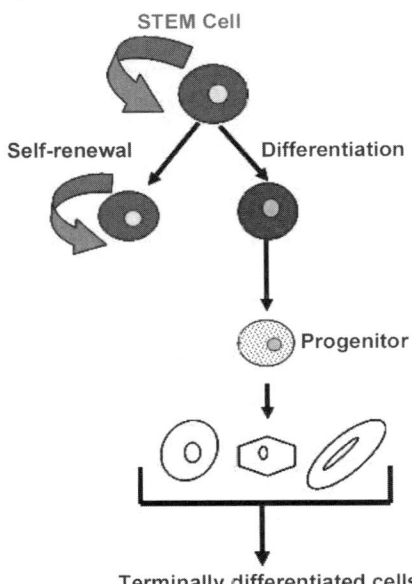

▶ **Fig. 24.7** *Stem cells and hierarchy. In a normal body, there exists a functional hierarchy within every tissue/organ and the cells present at the top of the hierarchy comprises of a small population of cells, known as the stem cells. These cells have two characteristics, which set them aside from all other somatic cells. First, their potential for "self-renewal" by which they are capable of giving rise to a cell of their own kind and second, multilineage differentiation potential, whereby a stem cell can give rise to all the other cell types of the particular tissue/organ via a pool of its progeny, the "progenitors." (for colour figure see Plate 67).*

In most adult tissues, it has been shown that there exists a "hierarchy", with stem cells at the top of the hierarchy, possessing the exclusive properties of indefinite self-renewal and multilineage differentiation. Lineage commitment starts with the stem cell giving rise to its progeny, the "progenitors", which have a limited lifespan. In case of the haematopoietic system, the HSC undergoes self-renewal divisions to maintain the HSC pool and it also divides to initiate the process of multilineage differentiation [17]. Through the latter, the HSC gives rise to two classes of progenitors, the lymphoid cells and the myeloid cells. The former gives rise to the NK cells and lymphocytes via a pool of progenitors, which have undergone further lineage commitment. The myeloid cells give rise to all other immune cells, platelets and RBCs via the Granulocyte Macrophage Precursor (GMP) and the

Megakaryocyte Erythrocyte Precursor (MEP). In the haematopoeitic system, this hierarchy has been thoroughly worked out, and cells at each stage of commitment have been characterized. A similar hierarchy exists in other adult tissues as well. Thus, a single stem cell from an adult tissue has the ability to give rise to all the cell types of that tissue (Fig. 24.8).

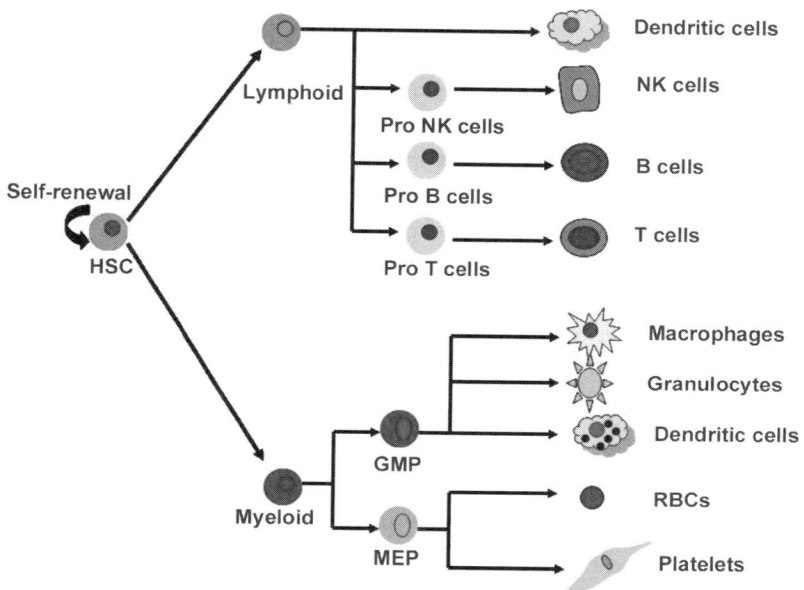

Courtesy: Adapted from Reya et al. Nature 2001, Vol. 414

▶ **Fig. 24.8** *Hierarchy in the haematopoeitic system. One of the best examples of stem cell hierarchy can be explained through the haematopoeitic system, where cells at every stage of the hierarchy have been well characterised. As the Haematopoetic Stem Cell (HSC) renew itself, it also gives rise to all the cell types of the blood, via two early progenitors, the lymphoid progenitor, which gives rise to the B and T lymphcytes, dendritic and NK cells, and the myeloid progenitor, which gives rise to the macrophages and granulocytes via the Granulocyte Macrophage Precursor (GMP) and the RBCs and platelets via the Megakaryocte Erythroid Precursor (MEP). A single HSC can, thus, give rise to all the cell types of the blood. (for colour figure see Plate 67).*

CANCER STEM CELLS

With advances in stem cell biology, cancer researchers began looking at cancer from a different perspective. Does the functional hierarchy found within cancers mirror the hierarchy seen in adult tissues and organs? Like in normal stem cells, clues to address this question also emerged from the haematological tissue, from studies of leukemia [18].

In spite of the decades of research on leukemia, and the observation of functional heterogeneity within tumours being prevalent from as early as 1960s, it was only in 1994 that the first experimental proof about the identity of the "tumour-initiating cell" emerged [22]. This was made possible because of technical advances in long term *in vitro* and *in vivo* experiments with tumour cells and development in flow cytometry based cell sorting [23, 24]. Leukemic cells isolated from seven patients of Acute Myelogenous Leukemia (AML) were transplanted into SCID mice, followed by analysis of the tumours so formed. To test for the clonogenicity of these AML cells, single cell suspension of the tumour was seeded in a 3D scaffold, methylcellulose to test whether "Colony Forming Units" (CFUs) were present in the bone marrow of the engrafted mice. AML-CFUs were seen to be present in all the transplanted mice. This was followed by the limiting dilution assay to calculate the frequency of leukemia-initiating cells in the AML patients. Using this technique, the frequency of leukemia-initiating cells was found to be 1 in $1 \infty 10^5$ cells. This clearly indicated that only a few cells within the tumour could initiate the tumour. In order to study the identity of these few tumour-initiating cells, tumour cells derived from the AML patients were fractionated based on a cell surface marker profile. Two markers were used; CD34, which is known to be present on progenitors and HSCs and CD38, a marker present on differentiated cells, such that a CD34+CD38− phenotype would mark a set of primitive, undifferentiated cells. It was found that the leukemia- or colony-initiating cells were CD34+CD38−. This was the first strong poof of principle for the Hierarchy Model. It was later shown that the marker profile exhibited by these cells was identical to normal HSCs [25]. Thus, the small number of leukemia-initiating cells showed both phenotypic (expressed the same markers) and functional (could initiate the parent tissue from which they were derived) resemblance to normal stem cells (Fig. 24.9).

Leukemia was followed by solid tumours, like breast cancer and brain tumour where the existence of a small, distinct population of cells was demonstrated within the tumours that represent the brain tumour or breast tumour-initiating population respectively (see Section *Identification and Isolation of Cancer Stem Cells from Solid Tumours*' for details). Such small populations of cancer-initiating cells have since then been discovered in multiple cancers by now [26] and today the Hierarchy Model is a well-established theory to explain the functional heterogeneity within tumours (Fig. 24.10). As few as 100 cells of this small population could initiate a tumour while as many as thousands of cells of the remaining (bulk) population could not do so. In addition, these tumour-initiating cells could repopulate themselves, as well as give rise to the ``bulk'' non-tumour-initiating cells. As stated in the Hierarchy Model, the small population of tumour-initiating cells within multiple cancers has the intrinsic potential to give rise to all the cell types of a tumour and also to cells of its own kind, which maintains and sustains the tumour over prolonged periods [27]. These properties exactly reflect the properties of multilineage differentiation and self-renewal of normal stem cells. Such striking similarities between these two cell types makes it apt to call the "tumour-initiating cells" within the tumour as "cancer stem cells" (CSCs) [28]. With the discovery of such cells in several cancers, the existence of CSCs is no longer a hypothesis, but a well-established fact. Thus, the leukemia-initiating cells, or the breast cancer-initiating cells are all CSCs within the respective tumour type (see Box 24.3).

TECHNIQUES USED TO ISOLATE CSCs

Since the proposition of the hierarchical/CSC model of carcinogenesis almost 40 years back [16], researchers all over the globe have doubled their efforts in developing techniques to identify and isolate these rare, evasive cells from a heterogeneous population of tumour cells. These include (i) Fluorescence Activated Cell Sorting (FACS) of cells based on expression of specific cell surface markers; (ii) isolation on the basis of a Side Population (SP) phenotype; (iii) isolation on the basis of cell size and (iv) isolation on the basis of activity of an enzyme, the Aldehyde dehydrogenase (ALDH). Almost all of these strategies to isolate CSCs were, in fact, initially developed for normal stem cells. Thus as many other concepts in normal stem cell biology, even the theory behind isolation of CSCs finds its roots in the studies of isolation and identification of normal stem cells. Let us in this part discuss these techniques in more detail.

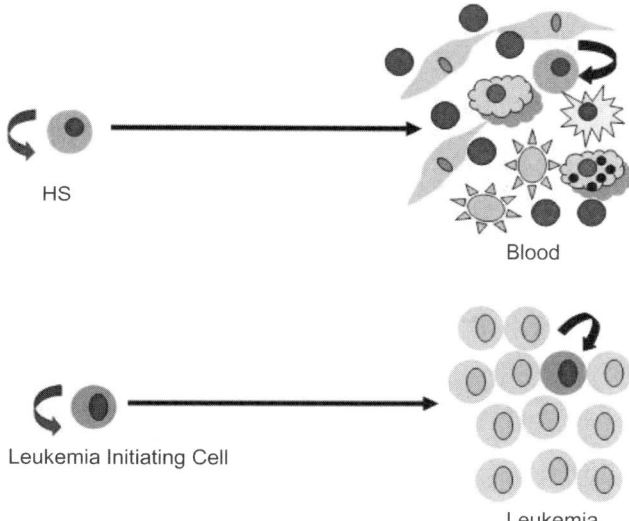

▶ **Fig. 24.9** *Schematic representation of the similarity in hierarchy between normal HSC and leukemia-initiating cell. A normal HSC sustains the haematopoeitic system, by renewing itself on the one hand and giving rise to all the cell types of the blood on the other hand, via progressively restricted progenitors. On similar lines, the Leukemia-Initiating Cell (LIC) sustains leukemia, by giving rise to all the cell types comprising leukemia and renewing itself at the same time to maintain the pool of LICs. (for colour figure see Plate 68).*

Cell Surface Marker Based Isolation of CSCs

All cells bear on their surface specific molecules (proteins or carbohydrates), which are essential for cell–cell communication, cell–cell adhesion and cell–extracellular matrix (ECM) interaction. These molecules provide a bridge for communication between the intracellular and the extracellular environment of a cell. Cells within each tissue of our body express on their cell surface certain molecules that are specific/characteristic only of that tissue. They are thus also often referred to as "cell surface markers". For example, cells of liver express molecules specific only to the liver and not to kidney or some other tissue. Many such cell surface markers are proteins belonging to the class of "CD markers", an acronym for "Cluster of Differentiation". There are almost 250 known CD markers in different tissues. Being cell surface molecules, it is technically much easier to detect them than intracellular molecules. This technical ease has been extensively exploited by several researchers and biotechnology companies to develop antibodies against known CD markers and other cell surface proteins which can be further used to identify and isolate cells of interest from any tissue.

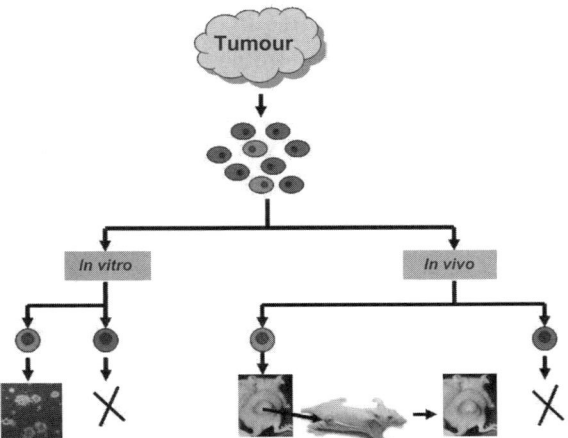

► **Fig. 24.10** *Proof of the principle of Hierarchy Model: the cancer stem cells (CSCs). Small populations of tumour-initiating cells have been discovered in multiple cancers by now so that as few as 100 of these cells (red cells) can give rise to colonies in a 3D matrix, like soft agar in vitro and a tumour in vivo. Furthermore, cells isolated from the tumour formed by these 100 cells could further seed a fresh tumour when transplanted in another animal, demonstrating the ability of self-renewal of this small population. Given the distinct similarities between the tumour-initiating population and the normal stem cells, this small population of cells in multiple cancers was termed as "Cancer Stem Cells". (for colour figure see Plate 68).*

While cell surface markers provide a tool for "identifying" the cells of interest, the technique of Fluorescence Activated Cell Sorting (FACS) provides the instrumentation to successfully "isolate" these same cells which can then be used for further analysis (Fig. 24.11). Cells to be isolated are incubated/ stained with antibodies against a specific cell surface marker. The antibodies are typically supplied conjugated to a fluorophore of choice (e.g. FITC, PE, APC), such that when this fluorophore is subjected to excitation by a laser source, it emits fluorescence in a specific wavelength. This fluorescence is then detected by special emission filters in a flow cytometer, and finally represented as a typical profile (FACS plot) (Fig. 24.12). On analysis of such plots which represent the marker profile of a population of cells, one can easily gate the population of interest and sort out only those cells for further experiments. Also it is possible to analyse more than one marker in the same experiment by using antibodies conjugated to different fluorophores against different cell surface antigens such that each of them have a different emission and thus can be distinguished from each other. With rapid technical advancements in the FACS technology, high speed FACS sorting machines and fluorophore development, it has become possible to carry out extremely intricate marker based experiments to study the detailed biology of stem cells. For example, using this technology, a pure population of murine mammary stem cells was isolated based on more than 5 different markers ($CD45^-$, $Ter119^-$, $CD31^-$, $CD140a^-$, $CD24^{medium}$, $CD49f^{high}$) and it was subsequently shown that a single cell having this marker profile could generate a complete functional mammary gland when injected in mice [29].

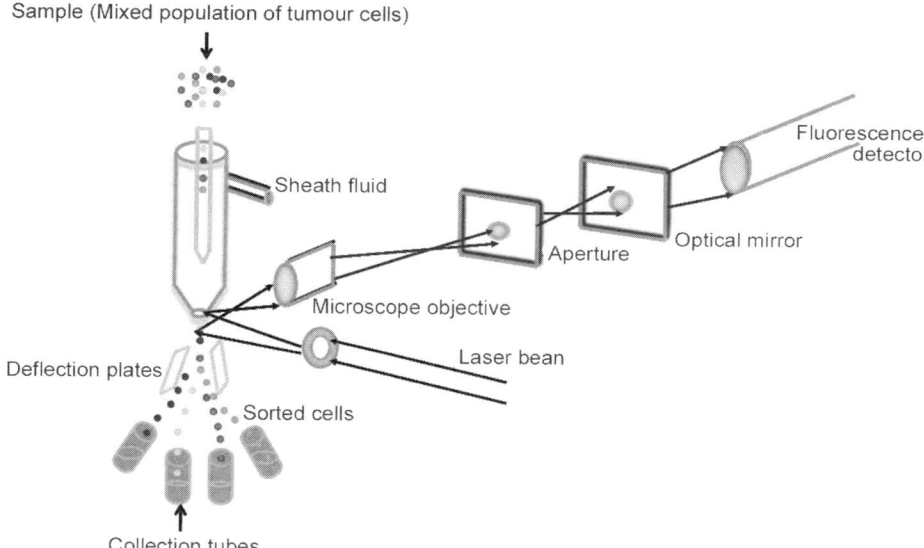

▶ **Fig. 24.11** *Principle of flow cytometry to analyse and sort cells. The suspension of heterogenous cell populations is introduced in the centre of a narrow, rapidly flowing stream of sheath fluid, such that these cells move in a single file. A vibrating mechanism causes the stream of cells to break into individual droplets at a nozzle, each droplet carrying with it a single cell. Laser of a particular wavelength is focused onto each droplet. When a cell labelled with a fluorescent antibody is probed with the laser, it gets excited and the fluorescent moiety emits light at a longer wavelength. The "unlabelled" cells are not detected. They pass down the stream into a waste collector. The light emitted by the "fluorescent" cells is detected by a number of detectors. Based on the fluorescence signal, a positive or a negative electric charge is assigned to each drop, which then gets deflected into respective tubes. The desired cells are thus sorted into different collection tubes by an electrostatic deflection system. (for colour figure see Plate 69).*

Besides specific detection, another advantage of this technique is that the emitted fluorescence can be quantified thereby giving accurate number of cells which are positive or negative for a given set of markers. For example, in a particular study, researchers were interested in identification of breast CSCs based on their $CD44^+$, $CD44^{-/low}$ phenotype [30]. In this FACS based study they first did a "negative sorting" to eliminate contaminating cells of the haematopoietic lineage. Such cells were eliminated by sorting out for Lineage$^+$ (Lin+) cells expressing CD2, CD3, CD10, CD16, CD18, CD31, and CD64, markers typical of leukocytes, endothelial cells, mesothelial cells and the fibroblasts. Next they performed a "positive sort" by staining for the CD44 and CD24 antigens in the Lin– cells. It was found that 15% of the primary breast cancer cells were CD44$^+$CD24$^-$ while 4% of the cells were positive for both the markers. They found that out of all the marker combinations (CD24$^+$CD44$^+$, CD24$^-$, CD44$^-$, CD24$^+$, CD44$^-$ and CD24$^-$, CD44$^+$), only the CD44$^+$, CD24$^-$/low fraction of cells demonstrated the properties of breast CSCs. Further, there was a 10-fold enrichment in the CSC-like cells on doing a further positive sorting of CD44$^+$CD24$^{-/low}$Lin$^-$ cells based on the expression of the Epithelial Specific Antigen (ESA). Thus, such quantitative studies have also enabled accurate enumeration of various subpopulations of cells, solely based on their marker expression.

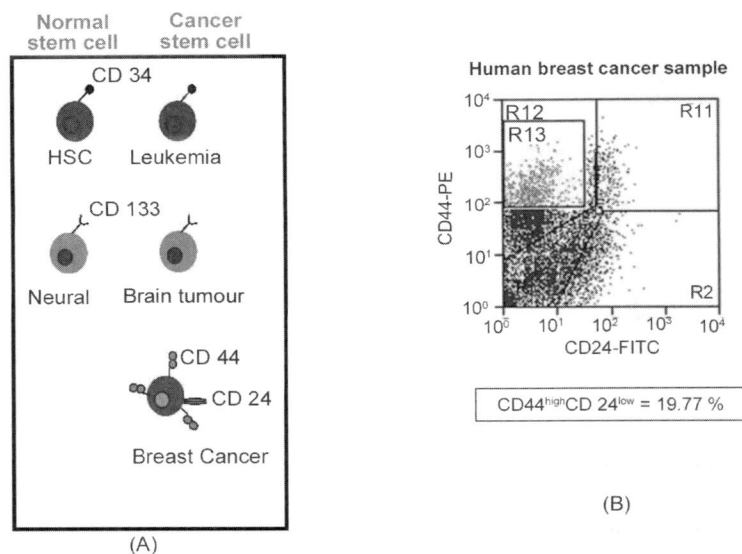

Courtesy: Nidhi Lal, IISc, Bangalore (unpublished data)

▶ **Fig. 24.12** *Cell surface marker based isolation of cancer stem cells.*
(A) Normal and cancer stem cells are isolated on the basis of cell surface markers. In Leukemia, leukemic stem cells were isolated on the basis of CD34+CD38– expression, which is also the marker of the HSC. Similarly, CD133, expression of which marks stem cells of the brain is also used for isolation of brain cancer stem cells. In breast cancer CD44^{high} CD24^{low} population has been found to be a marker of breast cancer stem cells. (B) Identification of breast cancer stem cells from human breast cancer samples. Primary human breast tumour cells were stained with CD44-PE and CD24-FITC and analysed by flow cytometry. Along the X axis of the FACS plot is represented the CD24 profile, while the Y axis represents CD44 profile. R2, R11, R12 represent various quadrants of the two marker combinations. In this case, CD44^{high} CD24^{low} population (Quadrant R13) was found to comprise 19.77% of the total population of cells. (for colour figure see Plate 70).

Doing a FACS based sorting of CSCs expressing specific markers has become a routine technique now. For example, using FACS, researchers have been able to isolate CSCs from leukemia as $CD34^+CD38^-$ cells, [22] from breast cancer as $CD44^+CD24^-$/low cells, [30] from brain cancer as $CD133^+$ cells [31] from pancreatic cancers as $CD44^+/\alpha2\beta1hi/CD133^+$ [32] and prostate cancer as $CD44^+ CD24^+$ cells [33].

Side Population Based Isolation of CSCs

Most major discoveries in science are often serendipitous. More than a decade ago, Margaret Goodell working in the lab of Richard Mulligan came across a phenotype that would revolutionize the identification of normal as well CSCs for the future generations. In an attempt to resolve the issue of haematopoietic stem cells (HSC) losing their capacity of self-renewal and multilineage reconstitution in serial transplantation experiments, these researchers decided to undertake a cell cycle analysis of dividing and quiescent HSCs. For this they used a fluorescent dye called Hoechst 33342 which can bind DNA of live cells and hence mark their DNA content. However, on exciting these cells using UV laser (355nm) what they obtained was an un-interpretable and complex DNA staining profile. When Hoechst fluorescence was viewed simultaneously at two emission wavelengths, (Hoechst blue at 450nm and Hoechst red at 670 nm) a small but distinct subset of cells constituting only 0.03–0.07% of whole bone marrow cells appeared. These rare cells appeared less fluorescent than the rest of the bone marrow cells and stood out at the side of a bivariate FACS plot as a distinct tail and hence the name the "Side Population" (SP) [34] (Fig. 24.13). On further analysis it was found that this side population had a Sca-1+/Lin– phenotype predominantly, a phenotype earlier shown to be characteristic of murine HSC. On doing competitive repopulation assays in mice the SP cells were found to be 1000 fold enriched in their HSC activity.

Attempts at explaining the SP phenotype showed that if bone marrow cells were concomitantly incubated with Hoechst and an inhibitor of ABCB1/ Multidrug Resistance-1 (MDR1) transporter (a protein belonging to the ATP Binding Cassette (ABC) of transporters; also called as P-glycoprotein), verapamil, the SP tail vanished [34] (Fig. 24.14) This finding thus suggested that MDR-1 could be responsible for extruding the dye from the cells that make up the SP, causing them to appear as less stained as compared the rest of the population. This result corroborated with an earlier landmark finding wherein the researchers demonstrated that primitive CD34+ HSCs when stained with Rhodamine 123 have low fluorescence and this phenotype could be abrogated with MDR-1 blockers verapamil and reserpine [35]. Other studies have demonstrated that though mice with ablated MDR1 homologs are not able to efflux the fluorescent dye, rhodamine 123, they are still capable of effluxing Hoechst and generate an SP [36]. This suggested the presence of other ABC transporters and indeed another protein, ABCG2 was shown to be able to efflux Hoechst and confer a SP phenotype on cells. These SP cells were shown to bear stem cell like properties like self-renewal and multilineage reconstitution. Thus stem cells can also be isolated as an SP, a phenotype conferred on them by elevated levels of certain ABC transporters. Discovery of SP in bone marrow was soon followed by uncovering SP in several other tissues like the mammary gland [37], lung [38], muscle [39], liver [40], and skin [41].

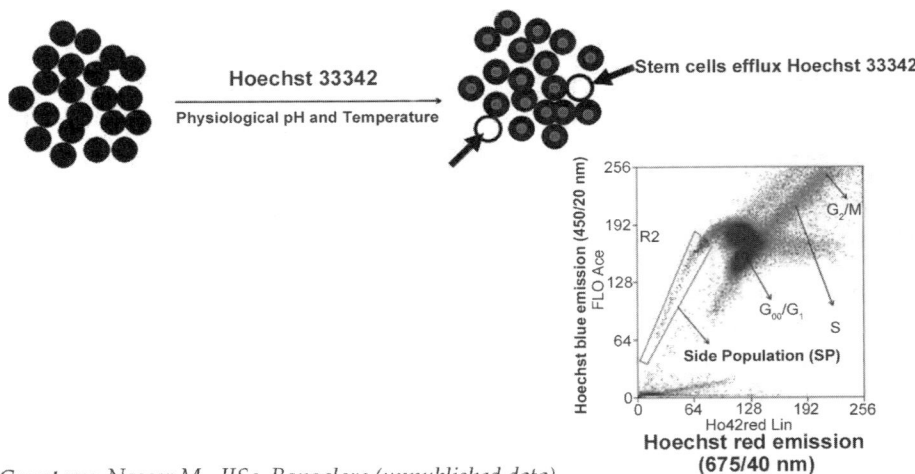

Courtesy: Naseer M., IISc, Bangalore (unpublished data)

▶ **Fig. 24.13** *Side population based isolation of cancer stem cells. Cancer stem cells have the ability to efflux, the lipohilic DNA binding dye, Hoechst 33342. Typically, Hoechst 33342 is excited in a flow cytometer equipped with a UV laser, when excited at a wavelength of 352nm. This dye emits in two wavelengths, Hoechst blue (450/20 nm) and Hoechst red (670/40 nm). SP stands out as a distinct and small population of cells as compared to the rest of the cells and has low Hoechst emission characteristics, which indicate low levels of the dye within these cells. It has been identified that ATP transporters like P-glycoprotein or ABCG2 which are preferentially expressed in cancer stem cells causes the Hoechst to efflux mediating the phenotype of Side population.*

From a SP profile, one can also get an idea of the cell cycle profile of the cells being analyzed. The distinct regions of the SP profile mark different phases of the cell cycle, as shown in the profile (G_0/G_1, S, G_2/M). (for colour figure see Plate 71).

The technique of isolating stem cells from normal compartment has also been extended to isolate stem cells from the cancer compartment. SP has been identified in cancer cell lines which have been in culture for long (Fig. 24.15). The very first experiments were done on the C6 glioma cell line. Researchers showed that the C6 SP cells, but not non-SP cells, were able to generate both SP and non-SP cells in culture and were responsible for the *in vivo* malignancy of this cell line. They also showed that these SP cells were capable of generating the differentiated cells of the brain, i.e. neurons and glial cells both *in vitro* and *in vivo* [42].

Soon thereafter, SP was also reported in primary neuroblastoma samples and cell lines from various sources like melanoma, ovarian, hepatocellular and glioma cancers and was shown to behave like CSCs [43]. SP cells were demonstrated to be capable of sustained expansion *ex vivo*, asymmetric division and generating heterogenous population of cells made up of both SP and non-SP progeny. Interestingly, a correlation between tumour grade and the percentage of SP cells has also been shown in primary mesenchymal tumours [44]. Data from several independent laboratories have demonstrated that when compared to the non-SP population, SP cells isolated

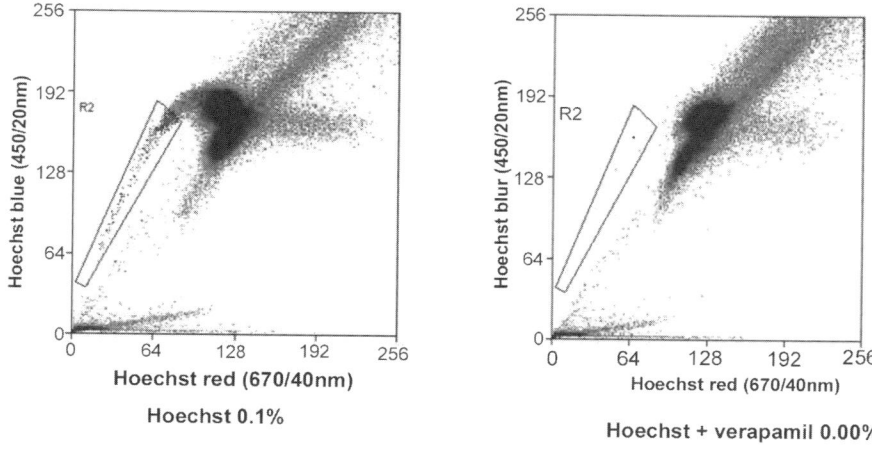

Courtesy: Naseer M., IISc, Bangalore (unpublished data)

▶ **Fig. 24.14** *Identification of side population in mouse bone marrow. Mouse bone marrow cells were isolated from C57/BL6 stain of mouse and stained with Hoechst 33342 for identification of SP cells by flow cytometry. 0.1% of the total population of bone marrow cells was found to be the "SP" (the boxed blue population). When in a control set, the cells were incubated with verapamil, a p-glycoprotein inhibitor, the efflux was completely blocked. As a consequence, the SP vanished completely (0.00%). (for colour figure see Plate 71).*

from hepatocellular, [45] lung [46] gastric [47], thyroid [48] and nasopharyngeal carcinoma cell lines [49] can also initiate tumour formation when xenografted into NOD/SCID mice. SP cells have also been shown to have increased expression of genes believed to be involved in the regulation of CSC function. These include the ABC transporter ABCG2 which confers a survival advantage on CSCs in face of chemotherapeutic challenge, [36,50] Wnt/ β-catenin signaling pathway which plays an important role in self-renewal of stem cells [47] and genes involved in cell cycle regulation like EXT1, INHBA and CCNT2 [51].

There have also been reports that have contested that SP cannot be used as a trustworthy universal assay for isolation of stem cells. For example, an SP has not been found in cell lines derived from Wilm's tumour, rhabdomyosarcoma or osteosarcoma and liver cancer (i.e. SK-NEP-1, A 204, and SaO^2, HepG2 respectively) [43, 47]. Another report demonstrated that both HSC and mammary gland repopulating cells are present within the SP as well as the non-SP compartments, suggesting that stem cells may well lie outside the SP population [52, 53]. Hence, presence of "SP" might not be a universal feature of all cancers.

Cell Size Based Isolation of CSCs

Another assay that many scientists have started exploring in search and identification of stem cells is based on the size of cells. A study dated as back as 1985 elegantly showed that the primary

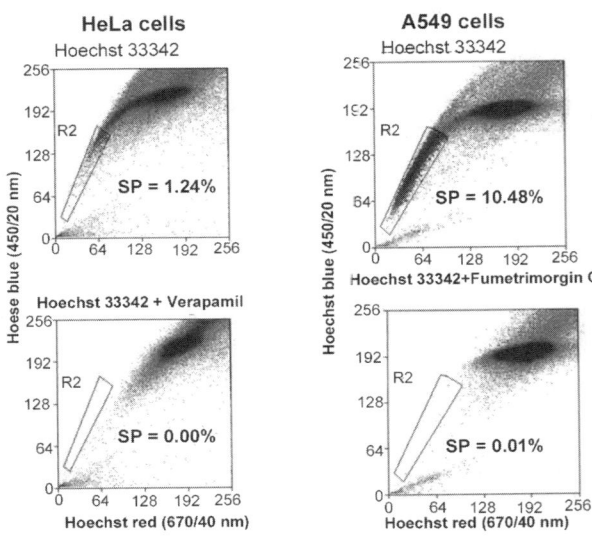

Courtesy: Naseer M., IISc, Bangalore (unpublished data)

▶ **Fig. 24.15** *Identification of side population in cancer cell lines (HeLa and A549 cells).*

Single cells of HeLa, a cervical cancer cell line and A549, a lung cancer cell line were stained with Hoechst 33342 for identification of SP cells. In HeLa and A549 cells, SP was found to be 1.24% and 10.48%, respectively. When cells were incubated with verapamil, SP vanished completely (0.00%). By a similar approach, when A549 cells were incubated with Fumetrimorgin C which is a ABCG2 specific inhibitor, SP was seen to vanish completely (0.01%). (for colour figure see Plate 72).

human neonatal foreskin keratinocytes as small as 11–20 µm were the most clonogenic cells and had the greatest colony forming ability [54]. Other researchers then showed that these smaller skin cells showed characteristics of stem/ progenitor cells both *in vivo* and *in vitro* [55–57] [58]. Interestingly it was noted with oral keratinocytes that as the size increased, their proliferative potential decreased, a sign of cellular differentiation and/or senescence [59]. This again suggested that the smaller cultured human oral keratinocytes may harbor the progenitor/stem cell population. Though the role and mechanism of cell size in stem cell regulation is not completely understood, it has been reported that cell size indeed correlates with cell phenotype, state of differentiation, and its proliferative capacity [60, 54, 61]. Thus, small size of cells seems to represent one of the characteristic features of adult stem cells.

To isolate cells on the basis of size, once again the FACS technique comes in handy. On doing a FACS experiment, one of the first important profiles observed is the forward scatter/side scatter (FSC/SSC) profile. Here the cells are distributed on the plot according to their size (FSC) and granularity (SSC) relative to each other (Fig. 24.16). The measure of these two parameters is based on bright light scattering; not on fluorescence. While the machine is able to judge the presence of different sized cells, it cannot determine the exact dimensions of these cells. However in this experiment if beads of known sizes are run through the sorter, one gets an easy "standard" to compare the cell size with. Cells of different sizes can thus be easily sorted by comparing their forward scatter profile with that of the beads.

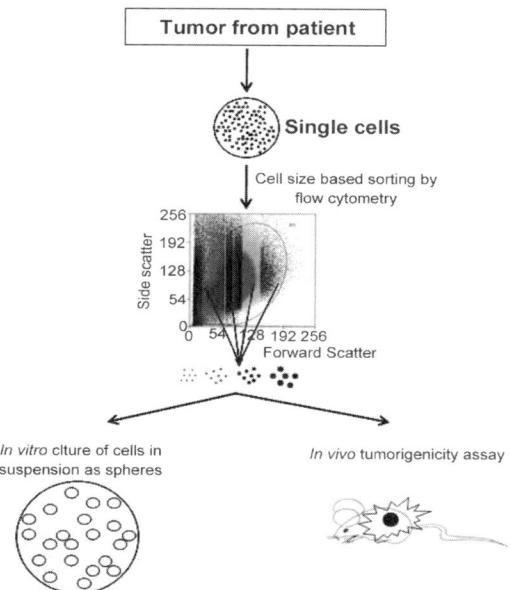

Courtesy: Nidhi Lal, IISc Bangalore (*unpublished data*)

► **Fig. 24.16** *Cell size based isolation of cancer stem cells.*

Cancer tissue is mechanically and enzymatically digested to yield single cells. Single cells are subsequently analysed by flow cytometry to compare the relative cell sizes based on forward scatter of light. The X axis represents the forward scatter and the Y axis represents the "side scatter", which is a measure of the granularity of cells. In this particular experiment, low cell size has been indicated by the blue population; medium cells are represented as the red and green populations while the grey population marks the large cells. These populations are sorted out on a flow cytometer, and subsequently, their clonogenic ability is studied **in vitro** *and* **in vivo**. *(for colour figure see Plate 73).*

Using the same technique, De Paiva et al in a study with corneal epithelial cells isolated different sized cell populations. They then showed that the population with the smallest size contained the greatest number of BrdU label-retaining slow-cycling cells, displayed the highest percentage of cells positive for p63 and ABCG2 (markers for stem cells) and negative for the differentiation markers, K3 and involucrin. These cells also possessed the highest colony forming efficiency and growth capacity. Alternatively, the larger sized cells were found to be terminally differentiated. The intermediately sized cells were shown to behave as transit amplifying cells and were capable of only limited proliferation [62].

As a parallel in cancer, one recent report on melanoma indeed demonstrated that the cells with the side population phenotype were also small in size. These cells were capable of giving rise to the heterogeneous population of cells and exhibited a low proliferative rate which is characteristic of stem cells [63]. Thus, cell size can be used as yet another parameter to isolate CSCs. However, this is yet to be proven for several other cancers.

Isolation of CSCs Based on Aldehyde Dehydrogenase (ALDH) Activity

Aldehyde dehydrogenases (ALDHs) are cytosolic enzymes that are able to oxidize intracellular aldehydes. The ALDH enzyme is thought to play an important role in oxidation of alcohol and vitamin A and also in cyclophosphamide chemoresistance [64]. Using western blotting and immunocytochemistry, an increased expression of ALDH has previously been observed in murine and human normal haematopoietic progenitors as compared to the other cells of the haematopoietic lineage [65].

Recently an elegant assay has been developed combining the activity of ALDH enzyme and the technique of FACS for sorting out ALDH+ stem cells. In this technique, cells to be characterized are incubated with a fluorescent substrate of ALDH like BAAA-DA (BODIPY Amino Acetaldehyde Diethyl Acetal). Being hydrophobic, these substrates can diffuse into the cell and subsequently be oxidized by active ALDH enzyme. This generates BODIPY-AminoAcetate (BAA) product which cannot cross the plasma membrane at physiological pH due to the presence of a charged carboxylate group. Pretreatment of cells with DEAB (Diethyl Amino Benzaldehyde) a specific inhibitor of cytosolic ALDH [66] decreases the BODIPY fluorescence thereby confirming the presence of ALDH activity within a cell. Therefore, only the cells with ALDH activity will fluoresce when incubated with its substrate and can be easily sorted using FACS (Fig. 24.17).

This technique is now being routinely used to isolate stem cells from different types of normal tissues. CD34+ haematopoietic progenitors have been shown to express high levels of cytosolic ALDH. SSC^{low} $ALDH^{bright}$ cell population from the human umbilical cord was also shown to harbor an enriched population of CD34+ (74%) haematopoietic stem/progenitor cells. $ALDH^{bright}$ Lin^- cells from the human cord blood have been shown to harbor cells with capability of repopulating the haematopoietic system of irradiated NOD/SCID mice [67]. These cells also co-express the markers of primitive HSC, i.e. CD34+CD38–. SSC^{low} $ALDH^{bright}$ cells have also been identified from the brain tissue which is capable of self-renewal and generating neurospheres in *in vitro* suspension cultures. Even the intestinal crypt stem cells display high levels of cytosolic ALDH activity [68].

Likewise, ALDH based technique has been recently used to characterize CSCs also. For example, in breast cancer cell lines $ALDH^{bright}$CD44+CD24– (MDA-MB-231) and $ALDH^{bright}$CD44+CD133+ (MDA-MB-468) cells demonstrated increased growth, colony formation, adhesion, migration, and invasion properties. Furthermore, following tail vein or mammary fat pad injection of NOD/SCID/IL2γ receptor null mice, $ALDH^{bright}$CD44+CD24– and $ALDH^{bright}$ CD44+CD133+ cells showed

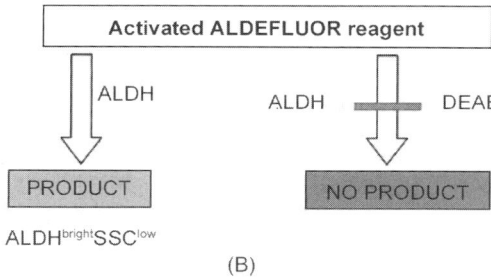

Courtesy: Robert W. Storms et al., PNAS 96, 1999.

▶ **Fig. 24.17** *ALDH based isolation of cancer stem cells assay.*

(A) A fluorescent substrate for ALDH (BAAA-DA) was synthesized by binding of the fluorescent compound BODIPY with Amino Acetaldehyde Diethyl Acetal (AAA-DA) to yield the stable compound BAAA-DA. Immediately before adding to the cell, this compound is converted into a more labile compound, BAAA (BIODIPY-amino acetaldehyde) in the presence of acid. This is the substrate for .ALDH. This substrate can freely diffuse into cells and then be converted by cytoplasmic ALDH into a carboxylate ion (BODIPY amino acetate, BAA), which is trapped intracellularly due to its charge. Consequently, cells expressing high levels of ALDH can be identified based on their bright fluorescence intensity.

(B) Utilising flow cytometry, cancer stem cells expressing high levels of ALDH can be identified by the use of the commercially available Aldefluor (Stem cell technologies), which is the activated substrate BAAA (BIODIPY-amino acetaldehyde). These cells are seen to be low on granularity, and hence identified as $ALDH^{bright}/SSC^{low}$ population. When cells are incubated with an ALDH inhibitor, Diethyl aminobenzaldehyde (DEAB), ALDH activity is inhibited leading to absence of the $ALDH^{bright}/SSC^{low}$ cells.

enhanced tumourigenicity and metastasis relative to ALDH^{low}CD44$^{low/-}$ cells [69]. Similarly ALDH activity has also been used to discriminate the CSC population in liver cancer and leukemia [70, 71]

Identifying and isolating stem cells on the basis of ALDH activity has recently gained much popularity because of many advantages like cost effectiveness of the requisite reagents and no requirement for specific antibodies. This method is also less toxic to the cells as compared to the side population analysis using toxic Hoechst dye. Thus, stem cells isolated by this method are healthier and can be propagated for extended times in culture. In addition, emission spectrum of BAAA does not overlap significantly with other fluorochromes so that BAAA staining can be readily combined with additional markers to isolate a further enriched population of stem cells.

TECHNIQUES USED TO ASSESS CSCs

It has been repeatedly emphasized in this text that the two hallmark and fundamental properties of both normal and CSCs are self-renewal and differentiation. Once putative stem cells have been identified and isolated, they need to be tested for their functionality. Many such functional assays have been developed and usually cells need to pass more than one of these tests before they can be conclusively and assertively classified as the authentic stem cells. The gold standard and routine techniques used for assaying the functionality of stem cells include (i) *in vivo* transplantation assays in mice and (ii) *in vitro* clonogenicity or colony forming assays (Fig. 24.18). Again, most advances in developing functional assays for CSCs have been made by getting a lead from the field of normal stem cell biology.

In vivo Transplantation Assays

One of the gold standard assays for testing stem cell functionality is based on its transplantation in an *in vivo* mouse model. Since a stem cell can undergo unlimited self-renewal and multilineage differentiation, a normal stem cell on transplantation should be able to reconstitute or repopulate the particular tissue from which it was originally isolated. Likewise a CSC on transplantation should be able to generate a tumour recapitulating the heterogeneity of the parent tumour. One of the first kinds of these experiments was done with normal haematopoietic stem cells wherein these cells were transplanted into myeloablated mice and then assessed for reconstitution of blood cells. This measured both the self-renewal and the multilineage differentiation capacities of HSCs. However such an assay comes along with other problems. For example, there are different hierarchies of cells present in the haematopoietic lineage with different proliferative and differentiation potentials. Researchers observed that on engrafting these cells, there was short term repopulation of the haematopoietic system which could prevent early death of the organism but over time these cells get depleted. Also these cells could not repopulate the bone marrow of lethally irradiated mice on serial transplantation questioning the self-renewal aspect of stem cells. This strongly suggested that the cells transplanted in the first place could belong to more mature subsets and could actually be progenitors and transit amplifying cells with limited self-renewal and proliferative potential. One of

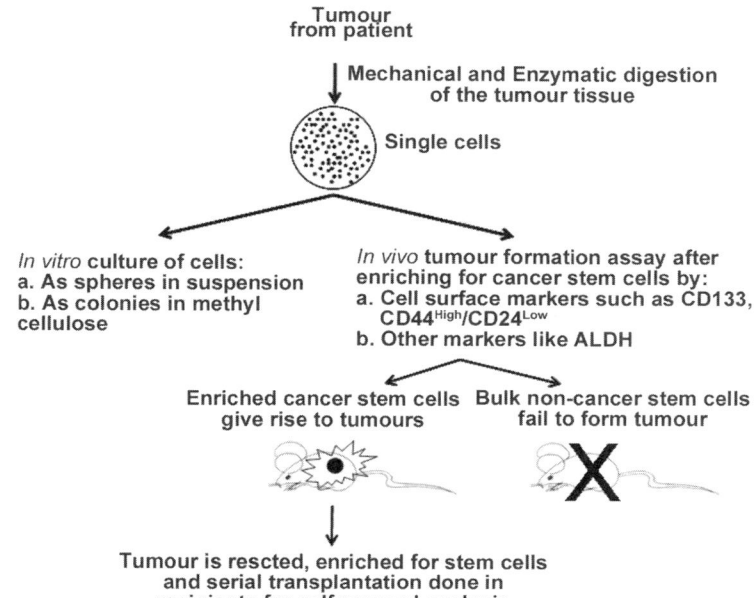

▶ Fig. 24.18 *Techniques used for isolation of cancer stem cells.*

Cancer tissue is mechanically and enzymatically digested to yield single cells. One way to enrich for CSCs is to seed the single cells subsequently in non-adherent dishes, in serum free medium containing defined growth factors. Formation of spheres under these conditions assesses the property of clonogenicity of cells. Subsequent passaging of the spheres so formed is used to study self-renewal property **in vitro**. *Similarly, cells can also be seeded in methyl cellulose to evaluate the colony formation ability in vitro.*

Isolated single cells from cancer tissue can directly be enriched for cancer stem cells based on cell surface markers, side population analysis and ALDH assay by flow cytometry. CSCs, thus isolated are injected to NOD/SCID mice to assess the tumor-initiating properties. As controls, the bulk cells are also injected in another set of mice. Enriched breast cancer stem cells form tumors in mice while the bulk cells do not. Serial transplantation of the tumor from one recipient to another is done to study self-renewal properties of cancer stem cells.

the other *in vivo* engraftment assays routinely used in testing for stem cells is called the **"spleen colony forming assay"** [72]. This assay utilises the transplantation of haematopoietic cells followed by assessment of number of haematopoietic colonies (colony forming units –CFUs) generated in the spleen over a time course of 8-10 days in lethally irradiated mice. Such assays helped elucidate that CFUs generated at different time points post transplantation represent cells of differing proliferative potential. While the CFUs developing at earlier time points represented more mature cells like progenitors, the CFUs developing at a later stage consisted of cells capable of long term reconstitution. The disadvantages of this assay however are that (a), it is a short term assay and hence can falsely classify a progenitor also as a stem cell and (b) it measures cells grown only in the spleen and not the entire haematopoietic system and thus could be an inefficient way to predict the self-renewal ability of HSCs. A much more refined assay for stem cells termed the "**competitive repopulation assay**" was then designed by Harrisson et al in 1980 [73]. Here, the stem cells from the marrow of a donor of a particular genotype are mixed with a constant number of marrow cells from a second donor with a distinguishable phenotypic marker and transplanted into lethally irradiated mice. The transplanted marrow cells compete with each other in repopulating the recipients and in producing cells of the haematopoietic lineage. This procedure can be used to compare repopulating and proliferative abilities of stem cells for as long as several years. The relative stem cell activity of these different marrow cells is measured by determining the percentage of peripheral blood cells derived from the each of the input populations in the transplanted recipients. These recipients are typically analyzed for a long period (4–6 months) after transplant when the majority of peripheral blood cells are derived from long-term reconstituting stem cells in the transplant instead of committed progenitors. Though this assay was first developed for haematopoietic cells, its principle has been extended to other systems also like the epidermal stem cells and the breast stem cells.

Similar to normal stem cells, assaying for CSC activity evaluates their potential for self-renewal and their ability to reconstitute the cellular heterogeneity present within the parent tumour. Therefore on transplanting putative CSCs into immunocompromized mice, one should observe the generation of a tumour. Also, CSCs now derived from this new tumour should be able to generate a tumour when serially transplanted into another mouse and should mimic the heterogeneity in cell types as of the parent tumour. While serial transplantation confirms the self-renewal attribute of CSCs, assessing for differentiation markers by techniques like immunocyto/histochemistry illustrates the multilineage reconstitution capacity of CSCs.

There are various experimental mouse models that have been generated over the last many years and have proved to be indispensable in stem cell research. These will be briefly discussed below. ***Nude mice.*** A nude mouse gets its name due to its "nude" or no hair phenotype. These mice are homozygous for the recessive mutant gene designated *Foxn1*. These mice do not have a thymus gland and thus have a greatly reduced number of T lymphocytes. This inhibits the immune system of the mouse to a great extent because of which it is not able to mount many of the immune responses. Such mice do not reject allografts and xenografts and thus become a valuable tool in cancer research.

SCID mice

Severe Combined Immunodeficient (SCID) mice are unable to make T or B lymphocytes due to a defect in the catalytic sub-unit of a DNA-dependent protein kinase "*Prkdc*" gene. Due to a compromised immune system these mice also do not exhibit transplant or tumour rejection and they have been extensively used as hosts for normal and malignant tissue transplants. Many variants of

SCID mice have been generated like the Natural Killer cell depleted C57BL/6J-SCID, DBA/2J-SCID without hemolytic complement and macrophage activation defective C3H/HeJ-SCID mice which show improved engraftment levels and niche-cell interactions. The best mouse model however generated thus far has been the NOD/SCID mice.

NOD/SCID mice

NOD/SCIDs have low levels of Natural killer (NK) cell activity, defects in macrophage activity, and low hemolytic complement activity with enhanced ability to be engrafted. These features together make NOD/SCID mice the most attractive mouse model for carrying out *in vivo* experiments with stem cells. It has been confirmed by various studies that human stem cell engraftment is significantly increased in NOD-SCID mice [74].

There can also be different modes of performing these *in vivo* transplantation experiments which in turn can to some level affect the experimental outcome. The two most common modes of performing transplantation are by introducing cells either through a subcutaneous injection or by an orthotropic transplant as described below:

Subcutaneous injections. In a transplant by a subcutaneous injection, cells are injected just below the skin. The cells thus enter the blood stream of the host and have the potential to travel systemically. Expressions of homing proteins thus guide these cells to the appropriate tissue. Most human tumour xenograft experiments employ conventional subcutaneous injection procedures. Despite the simplicity of this procedure, it poses some serious potential drawbacks. It sometimes becomes difficult to assess the induction of tumourigenicity or metastatic properties of cancer cells when they are administered by subcutaneous injections.

Orthotopic injections. In orthotopic injection, the cells of interest of a particular tissue are injected at the site of that tissue in the mouse. For example, injection of putative breast CSCs isolated from humans into the mammary fat pad of a mouse (site of mammary gland in mouse) is referred to as an orthotopic transplant. Such mode of transplants have several advantages like, tumour formation at the site of interest can be studied (breast cancer in breast tissue), metastatic and tumour-initiating properties of a given number of cancer cells can be studied for a particular type of cancer and clinically relevant studies can be performed.

In vitro Clonogenicity Assays

Though *in vivo* transplantation experiments are the best way to assay stem cells, it takes long periods for such assays to be conclusive, they are often very time consuming, cumbersome and not feasible. Researchers have thus also devised a number of *in vitro* functional assays by which the attributes of a stem cell can be tested. *In vitro* colony forming assays are routinely used and widely accepted as an authentic test for self-renewal of stem cells. An additional proof to the stemness of these cells is their ability to be serially passaged in *in vitro* cultures demonstrating their self-renewal potential. Many reports have shown that when cells are cultured in three-dimensional (3D) clonogenic cultures, not all of these cells are able to give rise to colonies. This suggests that the cell populations are heterogeneous and most likely the colonies that arise are due to division and self-renewal of stem cells.

These 3D assays can be performed by various methods as described below:

3D culture system using matrix. Single cells isolated from the tissues are embedded in a 3D matrix scaffold, either as single cells or as cellular aggregates. 3D scaffolds can be generated from purified biomolecules, like Collagen I; native extra-cellular matrices or synthetic polymers, like agar and methyl cellulose. The use of semisolid synthetic scaffolds, like agar and methylcellulose has proven to be extremely useful since single cells embedded in such scaffolds are unable to aggregate to form clumps [75]. Thus, any colony that arises is due to authentic division and self-renewal of stem cells and not due to clumping.

A cell which possesses tumour-initiating property, when embedded in such a scaffold proliferates and gives rise to a colony. This typically is referred to as the "colony forming assay", which was initially exploited to discover and characterize leukemic stem cells. Here single cells derived from patients suffering from Acute Myeloid Leukemia (AML) were seeded in methylcellulose and scored for colony formation. The cells which gave rise to colonies in methylcellulose were termed as "Colony Forming Units" (CFUs). The number of CFUs represented the number of primitive leukemic stem cells present in the AML sample, which were the only cells capable of giving rise to the colonies in methylcellulose [76].

Suspension culture or sphere culture system. Another type of 3-dimensional model, which came up a decade back in the field of stem cells, is the suspension culture or sphere culture system. This system has proven to be extremely useful to study stem cells of the epithelial tissues and tumour-initiating cells in solid tumours. In this system, single cell suspensions are seeded in a "defined" serum free media, containing a cocktail of growth factors and mitogens, like EGF, bFGF, insulin etc. The cells are seeded in low attachment plates, where most of the cells undergo anoikis, since a "substratum" is an absolute necessity for epithelial cells. Only the stem cells grow as spherical colonies of cells in suspension. The suspension culture system was initially used to enrich and expand normal neural stem cells *in vitro*, where a small population of neural stem cells were seen to survive and expand as spherical colonies termed "neurospheres" [77]. It was confirmed that neurospheres were enriched in neural stem cells. Similar conditions were used to culture and enrich brain tumour stem cells from Glioblastoma Multiforme (GBM) patient samples [31]. The neurospheres formed in this case had self-renewal potential and could differentiate into all the cell types found in GBM. Using the same approach, Dontu et al developed the mammosphere system to enrich for mammary stem cells in an undifferentiated state by culturing cells under anchorage independent conditions [37]. These mammosphere derived cells were then showed to be enriched for stem and progenitor cells. The suspension culture of mammospheres was then later also used to culture and propagate breast CSCs from primary breast tumour samples [78] (Fig. 24.19). The "breast tumour spheres" so formed were shown to have the potential to self-renew, undergo extensive proliferation and gave rise to all the cell types found in breast tumours. This concept of culturing stem cells, both normal and cancer, as "spheres" in suspension has been shown to be successful in multiple other systems. For example prostate stem cells, when cultured as "spheroids" show long term self-renewal ability [79]. Stem cells from thyroid and cells from the corresponding cancer tissue were shown to grow as spheres in suspension [80]. It is therefore

Plate 65

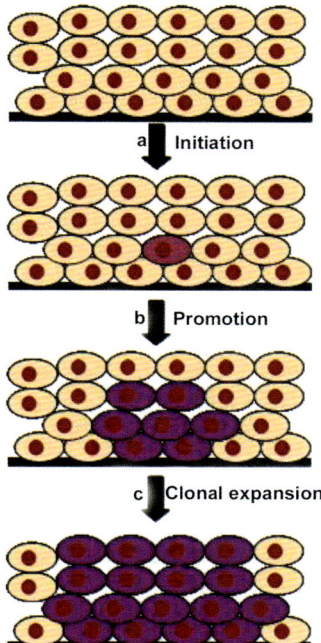

➤ *Fig. 24.2 Clonal theory. Cancer arises from the clonal expansion of a single cell, in which a genetic change (mutation) had "initiated" tumorigenesis. Over a period, accumulation of multiple genetic and epigenetic changes results in "promotion" of tumorigenesis.*

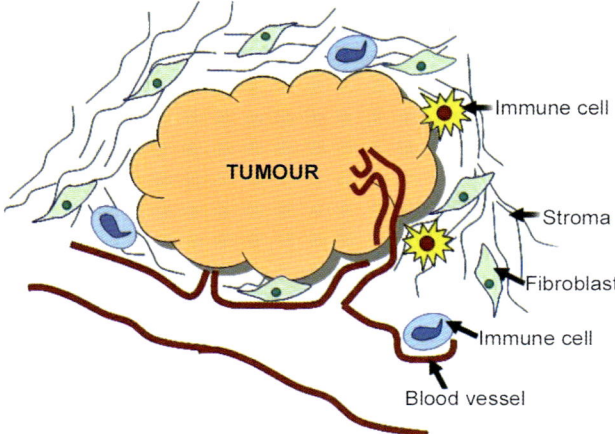

➤ *Fig. 24.4 A cartoon to represent a tumour in vivo. Within the body, a tumour doesn't exist in isolation. It is present in association with other components, like blood vessels, which nourish the tumour by supplying nutrients and oxygen. Stroma, which forms the extracellular matrix, supports the tumour and stromal cells such as fibroblasts are closely associated with the tumour. Cells of the immune system such as neutrophils, dendritic cells and WBCs are also known to be present in the vicinity of a tumour.*

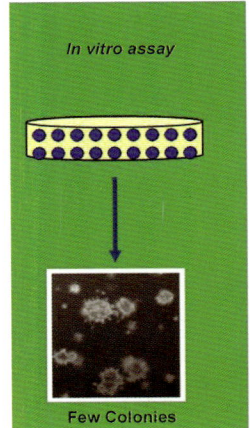
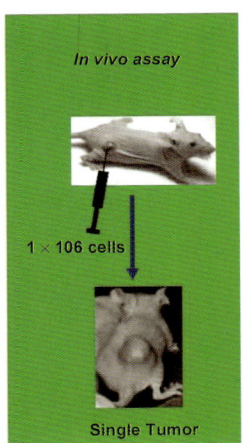

➤ *Fig. 24.5 Functional heterogeneity within a tumour. The 'function' of a tumour cell is assessed by its ability to give rise to a tumour. Two techniques are routinely used to assess the tumourigenic potential of cells. First, an in vitro method, the 'soft agar' assay, which scores for formation of colonies by single cells seeded in a 3D matrix like agar. The second is an in vivo assay which scores for tumour formation when cells are injected in nude mice. It was observed that not every cell which was seeded in an in vitro matrix could give rise to a colony and cell numbers as high as $1 \infty 10^6$ tumour cells had to be injected for the formation of a tumour, indicating a functional heterogeneity within tumours.*

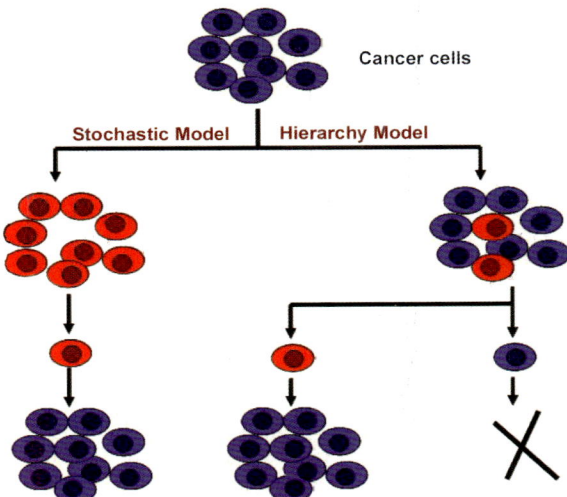

➤ *Fig. 24.6 Stochastic and Hierarchy Models. Two hypotheses had been suggested to explain the functional heterogeneity as seen in tumours. First, that every cell within the tumour has the ability to give rise to a tumour (all red cells) and it is chance that determines which cell becomes tumourigenic (Stochastic Model). In the second hypothesis only few cells within the tumour have an inherent potential to give rise to the tumour (the red cells) while most of the cells cannot give rise to a tumour (Hierarchy Model).*

Plate 67

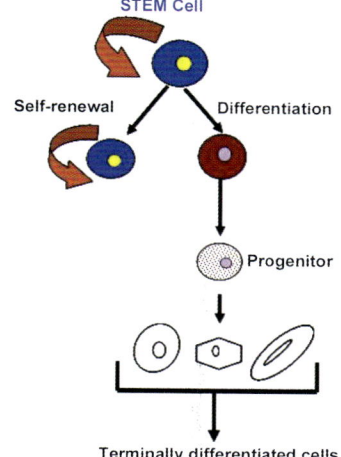

▶ *Fig. 24.7 Stem cells and hierarchy. In normal body, there exists a functional hierarchy within every tissue/organ and the cells present at the top of the hierarchy comprises of a small population of cells, known as the stem cells. These cells have two characteristics, which set them aside from all other somatic cells. First, their potential for "self-renewal" by which they are capable of giving rise to a cell of their own kind and second, multilineage differentiation potential, whereby a stem cell can give rise to all the other cell types of the particular tissue/organ via a pool of its progeny, the "progenitors."*

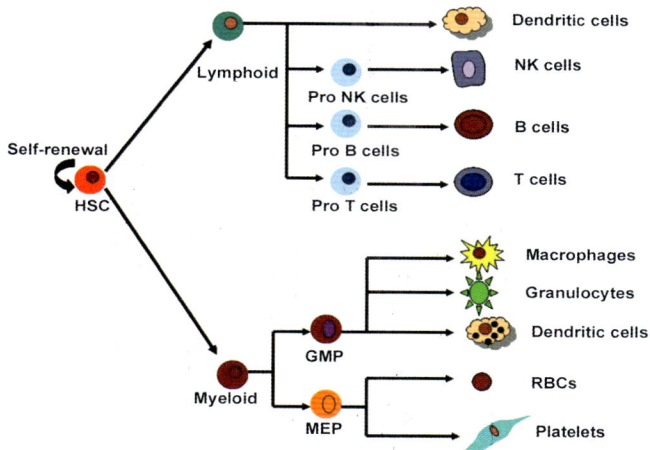

Courtesy: Adapted from Reya et al. Nature 2001, Vol. 414

▶ *Fig. 24.8 Hierarchy in the haematopoeitic system. One of the best examples of stem cell hierarchy can be explained through the haematopoeitic system, where cells at every stage of the hierarchy have been well characterised. As the Haematopoetic Stem Cell (HSC) renew itself, also gives rise to all the cell types of the blood, via two early progenitors, the lymphoid progenitor, which gives rise to the B and T lymphcytes, dendritic and NK cells, and the myeloid progenitor, which gives rise to the macrophages and granulocytes via the Granulocyte Macrophage Precursor (GMP) and the RBCs and platelets via the Megakaryocte Erythroid Precursor (MEP). A single HSC can, thus, give rise to all the cell types of the blood.*

Plate 68

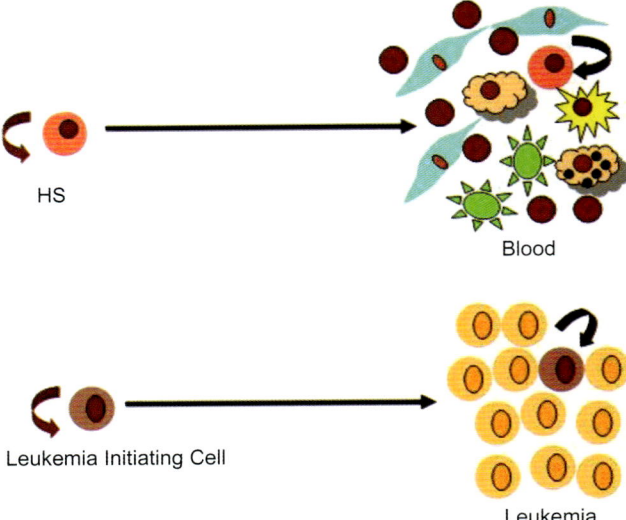

▶ *Fig. 24.9 Schematic representation of the similarity in hierarchy between normal HSC and leukemia-initiating cell. A normal HSC sustains the haematopoeitic system, by renewing itself on the one hand and giving rise to all the cell types of the blood on the other hand, via progressively restricted progenitors. On similar lines, the Leukemia-Initiating Cell (LIC) sustains leukemia, by giving rise to all the cell types comprising leukemia and renewing itself at the same time to maintain the pool of LICs.*

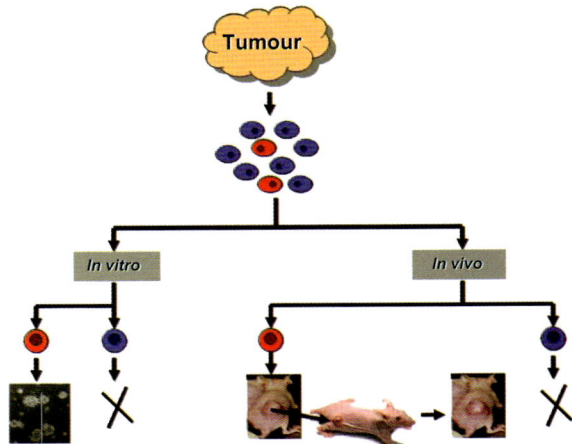

▶ *Fig. 24.10 Proof of the principle of Hierarchy Model: the cancer stem cells (CSCs). Small populations of tumour-initiating cells have been discovered in multiple cancers by now so that as few as 100 of these cells (red cells) can give rise to colonies in a 3D matrix, like soft agar in vitro and a tumour in vivo. Furthermore, cells isolated from the tumour formed by these 100 cells could further seed a fresh tumour when transplanted in another animal, demonstrating the ability of self-renewal of this small population. Given the distinct similarities between the tumour-initiating population and the normal stem cells, this small population of cells in multiple cancers was termed as "Cancer Stem Cells".*

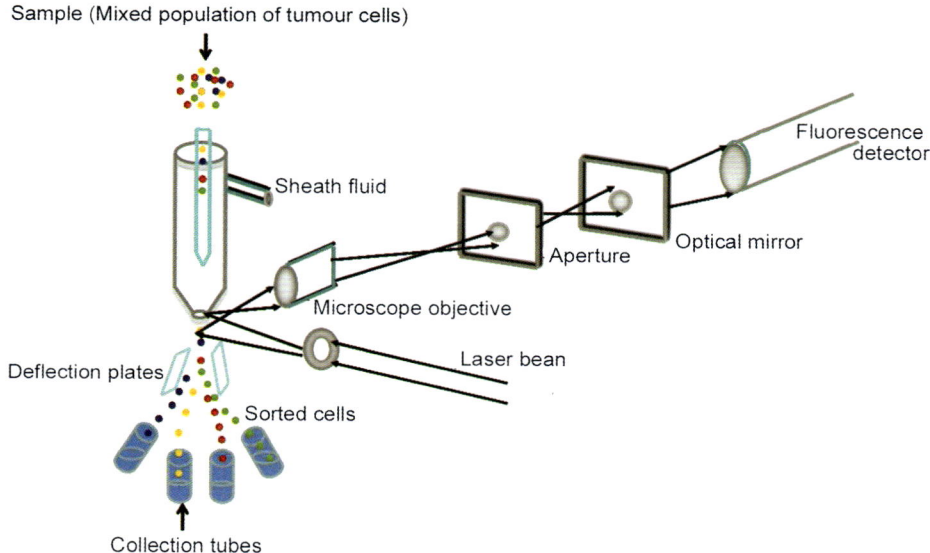

▶ *Fig. 24.11 Principle of flow cytometry to analyse and sort cells. The suspension of heterogenous cell populations is introduced in the centre of a narrow, rapidly flowing stream of sheath fluid, such that these cells move in a single file. A vibrating mechanism causes the stream of cells to break into individual droplets at a nozzle, each droplet carrying with it a single cell. Laser of a particular wavelength is focused onto each droplet. When a cell labelled with a fluorescent antibody is probed with the laser, it gets excited and the fluorescent moiety emits light at a longer wavelength. The "unlabelled" cells are not detected. They pass down the stream into a waste collector. The light emitted by the "fluorescent" cells is detected by a number of detectors. Based on the fluorescence signal, a positive or a negative electric charge is assigned to each drop, which then gets deflected into respective tubes. The desired cells are thus sorted into different collection tubes by an electrostatic deflection system.*

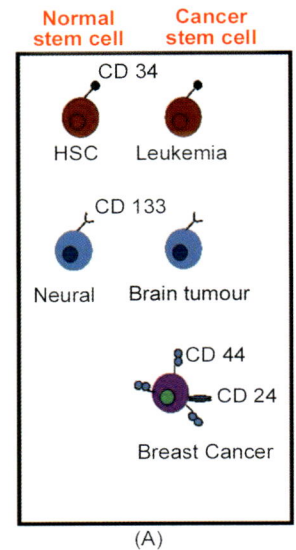
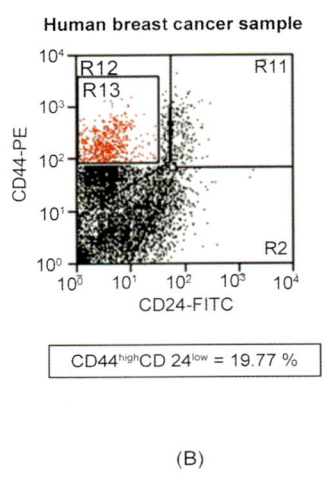

(A)

(B)

Courtesy: Nidhi Lal, IISc, Bangalore (unpublished data)

▶ *Fig. 24.12 Cell surface marker based isolation of cancer stem cells.*
(A) Normal and cancer stem cells are isolated on the basis of cell surface markers. In Leukemia, leukemic stem cells were isolated on the basis of CD34+CD38− expression, which is also the marker of the HSC. Similarly, CD133, expression of which marks stem cells of the brain is also used for isolation of brain cancer stem cells. In breast cancer CD44^{high} CD24^{low} population has been found to be a marker of breast cancer stem cells. (B) Identification of breast cancer stem cells from human breast cancer samples. Primary human breast tumour cells were stained with CD44-PE and CD24-FITC and analysed by flow cytometry. Along the X axis of the FACS plot is represented the CD24 profile, while the Y axis represents CD44 profile. R2, R11, R12 represent various quadrants of the two marker combinations. In this case, CD44^{high} CD24^{low} population (Quadrant R13) was found to comprise 19.77% of the total population of cells.

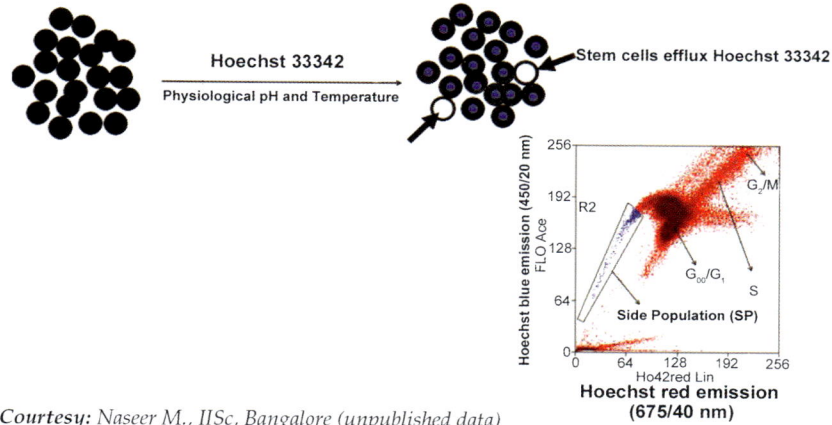

Courtesy: Naseer M., IISc, Bangalore (unpublished data)

▶ *Fig. 24.13 Side population based isolation of cancer stem cells. Cancer stem cells have the ability to efflux, the lipohilic DNA binding dye, Hoechst 33342. Typically, Hoechst 33342 is excited in a flow cytometer equipped with a UV laser, when excited at a wavelength of 352nm. This dye emits in two wavelengths, Hoechst blue (450/20 nm) and Hoechst red (670/40 nm). SP stands out as a distinct and small population of cells as compared to the rest of the cells and has low Hoechst emission characteristics, which indicates low levels of the dye within these cells. It has been identified that ATP transporters like P-glycoprotein or ABCG2 which are preferentially expressed in cancer stem cells causes the Hoechst to efflux mediating the phenotype of Side population.*

From a SP profile, one can also get an idea of the cell cycle profile of the cells being analyzed. The distinct regions of the SP profile mark different phases of the cell cycle, as shown in the profile (G_0/G_1, S, G_2/M).

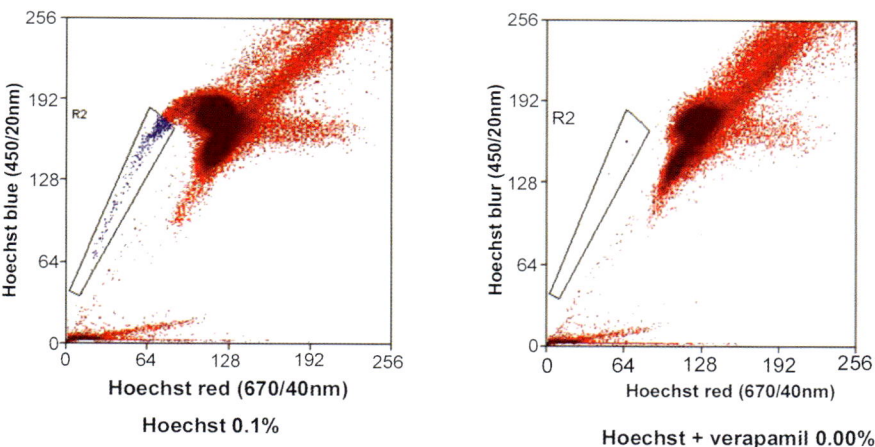

Courtesy: Naseer M., IISc, Bangalore (unpublished data)

▶ *Fig. 24.14 Identification of side population in mouse bone marrow. Mouse bone marrow cells were isolated from C57/BL6 stain of mouse and stained with Hoechst 33342 for identification of SP cells by flow cytometry. 0.1% of the total population of bone marrow cells was found to be the "SP" (the boxed blue population). When in a control set, the cells were incubated with verapamil, a p-glycoprotein inhibitor, the efflux was completely blocked. As a consequence, the SP vanished completely (0.00%).*

Plate 72

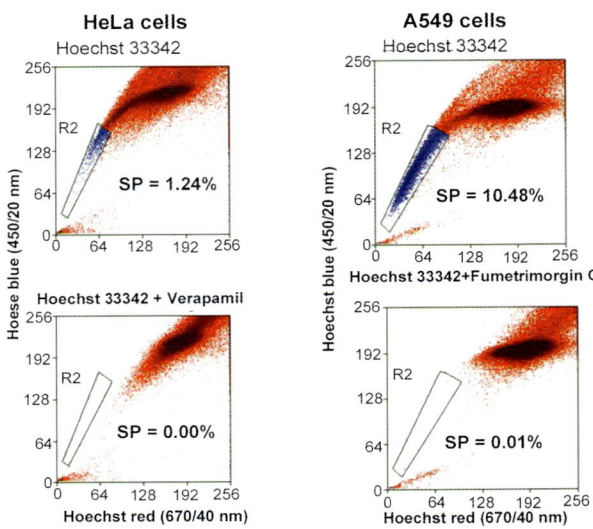

Courtesy: Naseer M., IISc, Bangalore (unpublished data)

▶ *Fig. 24.15 Identification of side population in cancer cell lines (HeLa and A549 cells).*

Single cells of HeLa, a cervical cancer cell line and A549, a lung cancer cell line were stained with Hoechst 33342 for identification of SP cells. In HeLa and A549 cells, SP was found to be 1.24% and 10.48%, respectively. When cells were incubated with verapamil, SP vanished completely (0.00%). By a similar approach, when A549 cells were incubated with Fumetrimorgin C which is a ABCG2 specific inhibitor, SP was seen to vanish completely (0.01%).

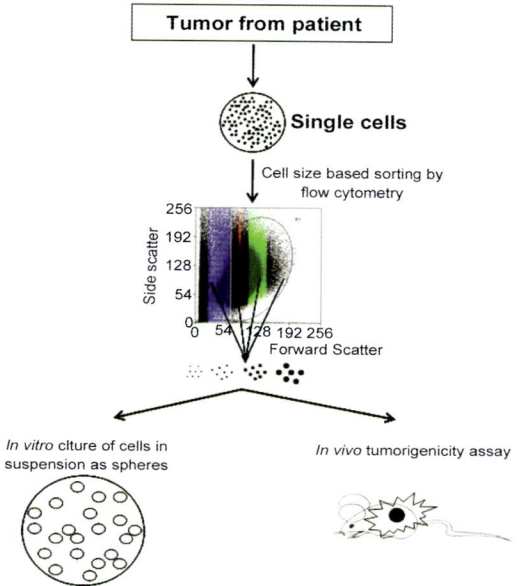

Courtesy: Nidhi Lal, IISc Bangalore (unpublished data)

➤ *Fig. 24.16 Cell size based isolation of cancer stem cells.*

Cancer tissue is mechanically and enzymatically digested to yield single cells. Single cells are subsequently analysed by flow cytometry to compare the relative cell sizes based on forward scatter of light. The X axis represents the forward scatter and the Y axis represents the "side scatter", which is a measure of the granularity of cells. In this particular experiment, low cell size has been indicated by the blue population; medium cells are represented as the red and green populations while the grey population marks the large cells. These populations are sorted out on a flow cytometer, and subsequently, their clonogenic ability is studied in vitro *and* in vivo.

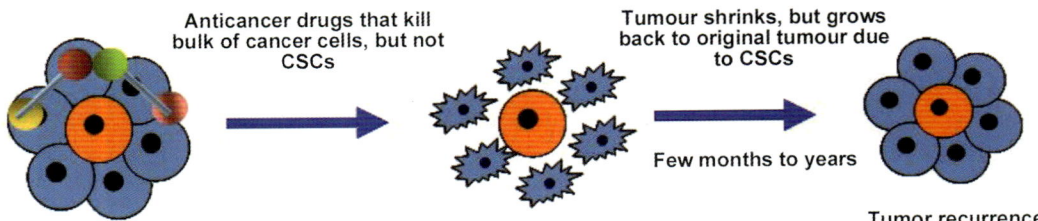
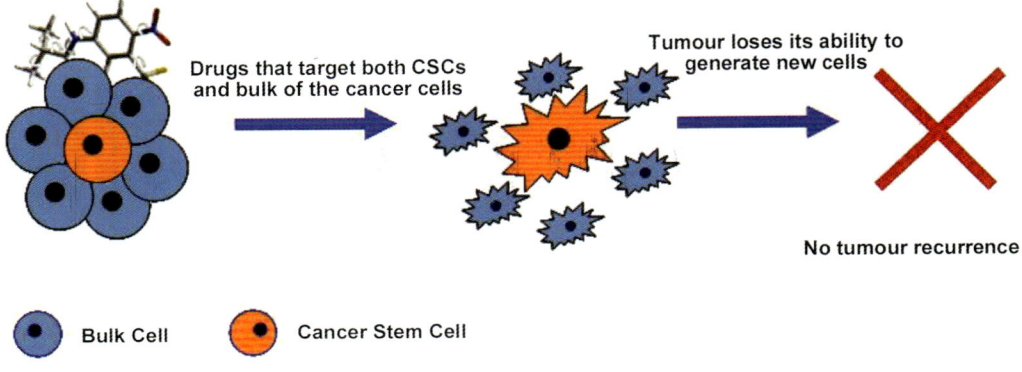

➤ *Fig. 24.20 Therapeutic implications of cancer stem cells.*
(A) Conventional chemotherapy is effective in killing rapidly dividing cells. On account of their low numbers and their drug efflux property, cancer stem cells are less sensitive to these therapies. Hence, while the bulk of cancer cells is killed, leading to an "apparent" cure of the cancer, the CSCs remain viable and active even after the therapy and contribute to re-growth of the entire tumour leading to recurrence of the cancer after a certain period. (B) Therapies which can target both cancer stem cells and bulk would be more effective in eliminating cancer at the root. CSC specific drugs, which might target the efflux pumps and some of the CSC specific survival pathways, in combination with the conventional drugs that kill the bulk would be the ideal treatment regime for cancers. Thus, targeted cancer stem cell therapy leads to remission of the disease at the root.

Plate 75

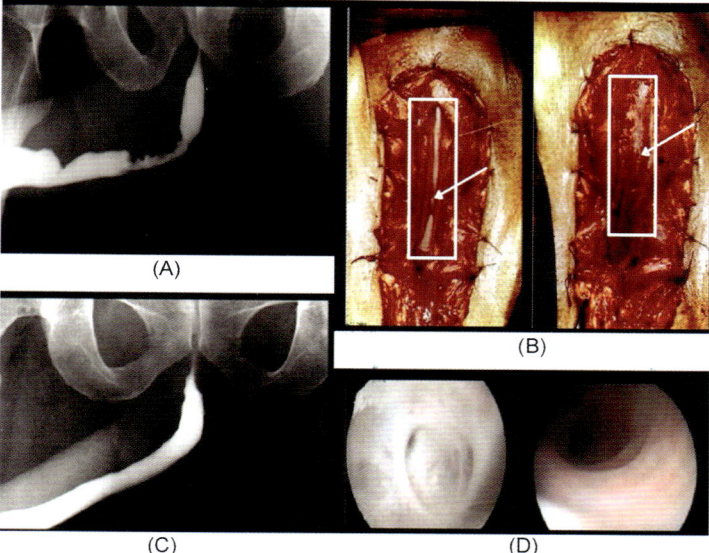

► Fig. 26.2 Urethral repair using a collagen matrix. (A) Representative case of a patient with a bulbar stricture. (B) During surgery, strictured tissue is excised, preserving the urethral plate on the left hand side, and the matrix is anastamosed to the urethral plate in an onlay fashion on the right. The boxes in both photos indicate the area of interest, including the urethra, which appears white in the left hand side photograph. In the left hand photograph, the arrow indicates the area of stricture in the urethra. On the right hand, the arrow indicates the repaired stricture. Note that the engineered tissue now obscures the native white urethral tissue in an onlay fashion in the right photograph. (C) Urethrogram 6 months after repair. (D) Cystoscopic view of urethra before surgery on the left hand side, and 4 months after repair on the right hand side.

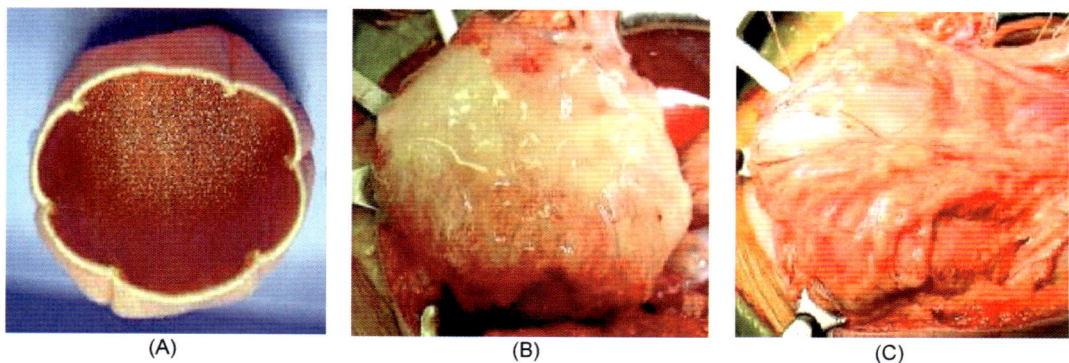

► Fig. 26.3 Construction of engineered bladder. (A) Scaffold material seeded with cells for use in bladder repair. (B) The seeded scaffold is anastamosed to native bladder with running 4-0 polyglycolic sutures. (C) Implant covered with fibrin glue and omentum.

Plate 76

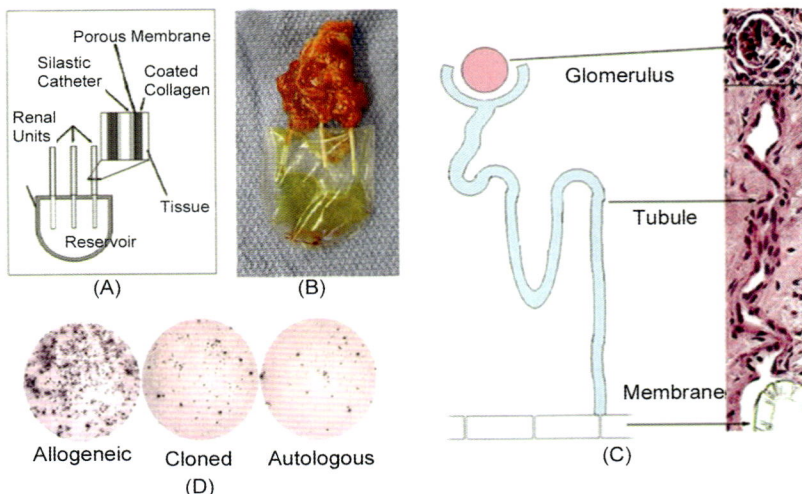

▶ *Fig. 26.5 Combining therapeutic cloning and tissue engineering to produce kidney tissue. (A) Illustration of the tissue-engineered renal unit. (B) Renal unit seeded with cloned cells, three months after implantation, showing the accumulation of urine-like fluid. (C) Clear unidirectional continuity between the mature glomeruli, their tubules and Silastic catheter. (D) ELISPOT analyses of the frequencies of T cells that secrete IFN-γ after stimulation with allogeneic renal cells, cloned renal cells or nuclear donor fibroblasts. Cloned renal cells produce fewer IFN-γ spots than the allogeneic cells, indicating that the rejection to cloned cells is diminished. The presented wells are single representatives of duplicate wells.*

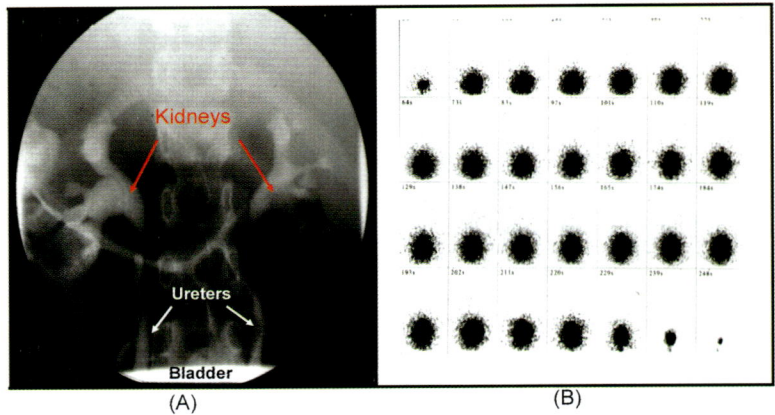

▶ *Fig. 26.6 Autologous chondrocytes for the treatment of vesicoureteral reflux. (A) Pre-operative voiding cystourethrogram of a patient with bilateral reflux. A catheter was inserted into the bladder via the urethra and contrast material was instilled intravesically. Here, contrast material can be seen within both ureters and within kidneys, indicating that reflux is present. (B) Post-operative radionuclide cystogram of the same patient 6 months after injection of autologous chondrocytes. A catheter was inserted into the bladder via the urethra and a radioactive solution was inserted into the bladder. The bladder was scanned during filling and emptying phases. This panel includes sequential images of the bladder as it was filled and emptied. This shows a normal, round bladder that fills and empties properly. If reflux had been present, ureters would have been visible in the scan above the round bladder.*

Plate 77

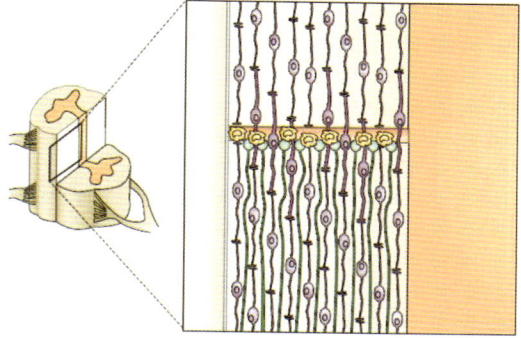
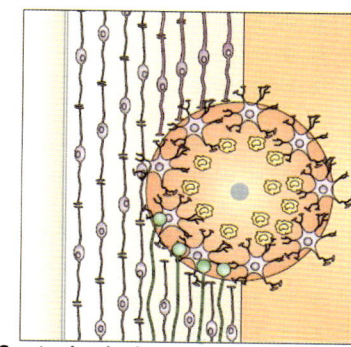

(A) Microlesion
- Blood-brain barrier is minimally disrupted.
- Astrocytes maintain normal alignment but produce CSPGs and KSPGs along the injury tract.
- Axons cannot regenerate beyond the lesion.
- Macrophages invade the lesion site.

(B) Contusive lesion
- Blood-brain barrier is disrupted, but meninges are intact.
- Cavitation occurs at the lesion epicentre.
- Astrocyte alignment is altered at the lesion site.
- Astrocytes produce CSPGs and KSPGs in a gradient, increasing from the penumbra towards the centre of the lesion.
- No fibroblast invasion or lesion core and therefore no fibroblast-expressed inhibitors are present (see below).
- Macrophages invade the lesion and its core.
- Dytrophic axons approach the lesion before growth ceases.

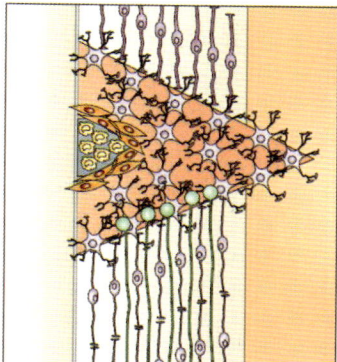
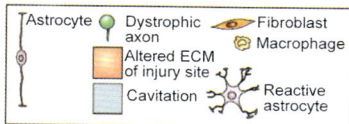

(C) Large stab lesion
- Blood-brain barrier is disrupted.
- Cavitation occurs at the lesion centre.
- Astrocyte alignement is altered at the lesion site.
- Astrocytes produce SCPG and KSPG in a gradient increasing towards the lesion.
- TGF, ephrin-B2 and Slit protein expression increases in reactive astrocytes adjacent to fibroblasts.
- Fibroblasts invade the lesion and express SEMA3 and the EPHB2 receptor.
- Macrophages invade the lesion and release inflammatory cytokines.
- Dystrophic neurons are highly repeled by the lesion core and express neuropilin 1.

➤ *Fig. 27.1 Schematic representation of three stereotypical CNS lesions. In all examples, macrophages invade the lesion, and both chondroitin sulphate proteoglycans (CSPGs) and keratan sulphate proteoglycans (KSPGs) are upregulated. (A) Microlesion in which astrocyte alignment is not altered by the injury process, but axons are unable to regenerate past the lesion site. (B) Contusive injury that does not disrupt the meninges, but produces cavitation and proteoglycan deposition. Again, axons are unable to regenerate beyond the lesion, but spared axons can be found distal to the injury site. (C) Stab lesion that penetrates the meninges and allows fibroblast invasion in addition to macrophages. Axons are highly repulsed by the increasing gradient of CSPGs and KSPGs. Several other inhibitory molecules are also made in this type of injury and are especially prevalent in the core of the lesion. ECM, extracellular matrix; TGF, transforming growth factor.*

Plate 78

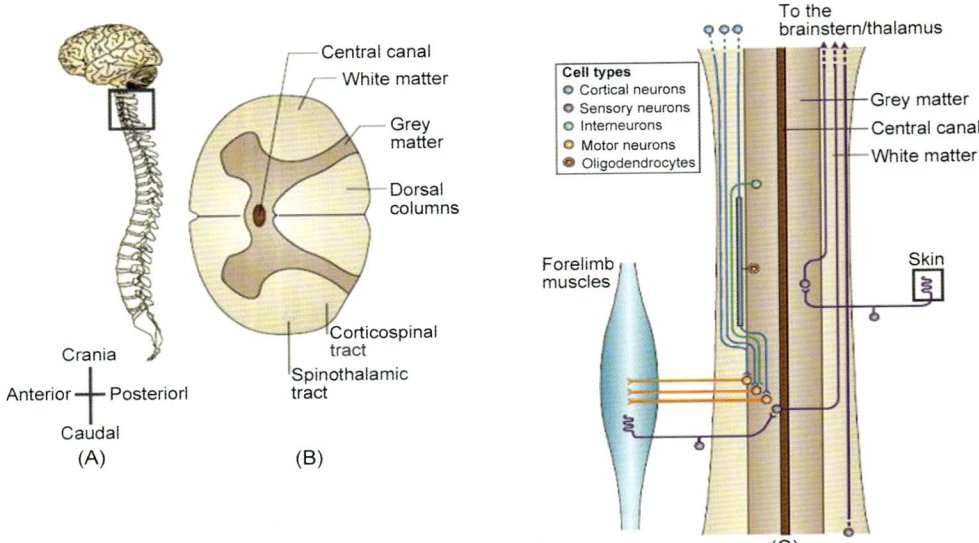

▶ *Fig. 27.2 Intact spinal cord. (A) Schematic diagram showing a sagittal view through the human CNS. (B) Transverse section through human spinal cord showing the relationship between axonal tracts and grey matter. (C) Cortical, brainstem and spinal axons project to motor neurons in spinal cord grey matter, which in turn send axons through the PNS to target organs, including muscle. Primary sensory axons send axons through the PNS to second order sensory neurons in the CNS grey matter, which, in turn, send axons through white matter in the dorsal columns to supraspinal regions. Oligodendrocytes myelinate ascending and descending axons.*

Plate 79

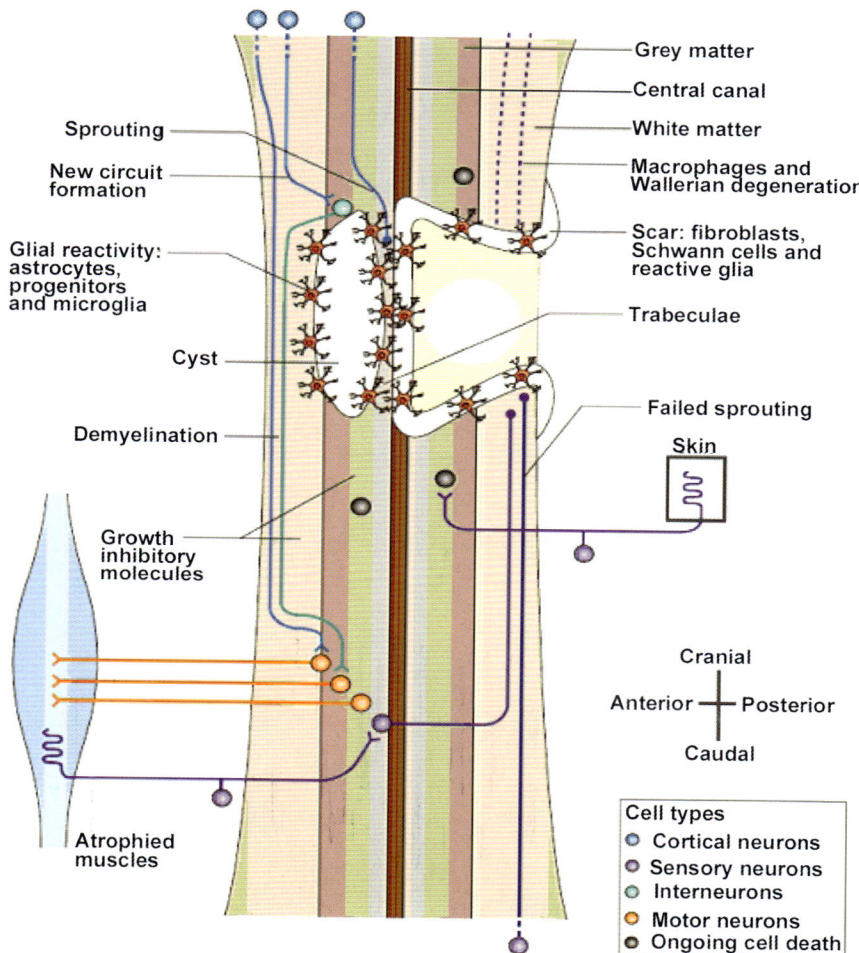

▶ *Fig. 27.3 Spinal cord after injury. Schematic diagram showing a sagittal view through a region of cervical SCI, depicting a combination of features from different types of injury. Many cells die immediately as well as progressively after SCI. Cysts usually form after contusion injury. After penetrating the injury site, cells from the PNS often invade the injury site to form a connective tissue scar that incorporates astrocytes, progenitor cells and microglia. Many ascending and descending axons are interrupted and fail to regenerate over long distances. Some axons form new circuits with motor neurons via interneurons. At the site of cyst formation, axons can sprout into trabeculae that are formed from ependymal cells. Disconnected myelinated axon segments are phagocytosed by macrophages. Some spontaneous remyelination occurs, largely by PNS Schwann cells, whereas denervated (non-spastic) muscles atrophy.*

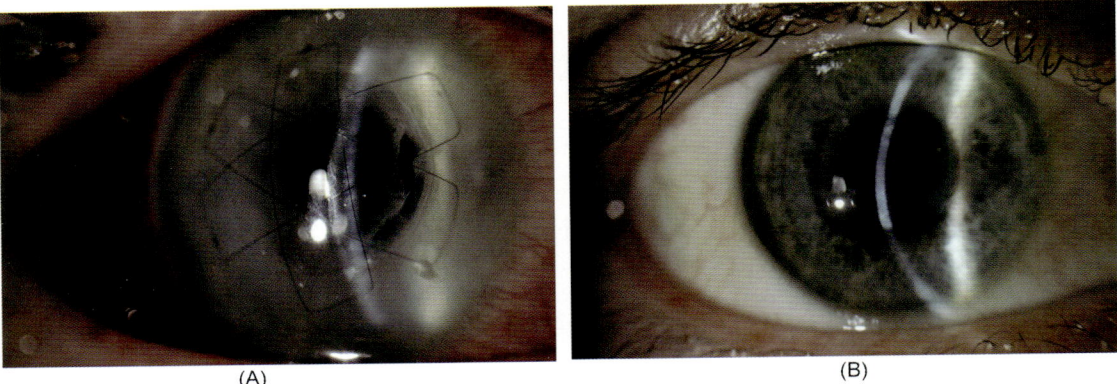

➤ *Fig. 27.4 Corneal transplantation with RHC corneas, as deep lamellar grafts, are currently in Phase I human clinical trials. (A) A recently implanted artificial cornea (sutures visible). (B) The success of the implant, 9 months after the operation.*

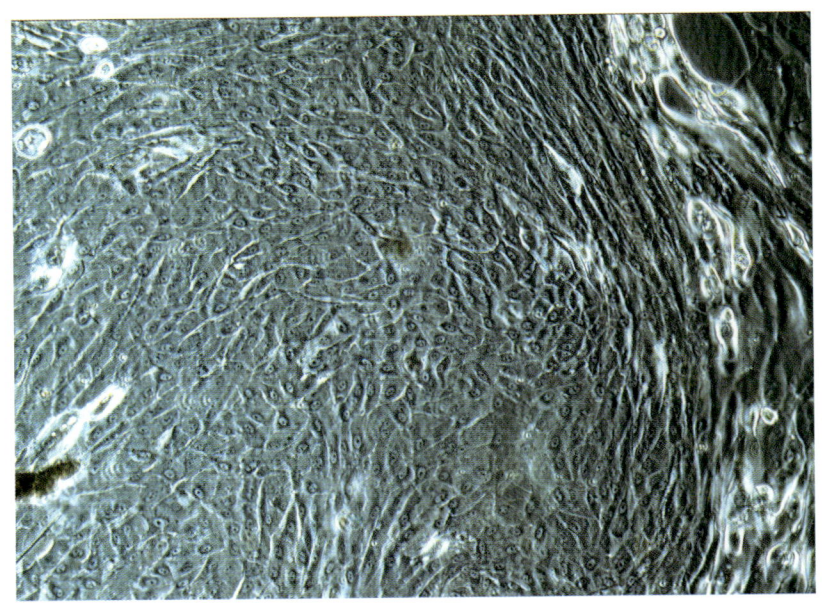

➤ *Fig. 27.5 Human limbal stem cells explanted onto a type III recombinant human collagen (RHC-III) scaffold, after 3 days of growth. Visualised using phase contrast microscopy.*

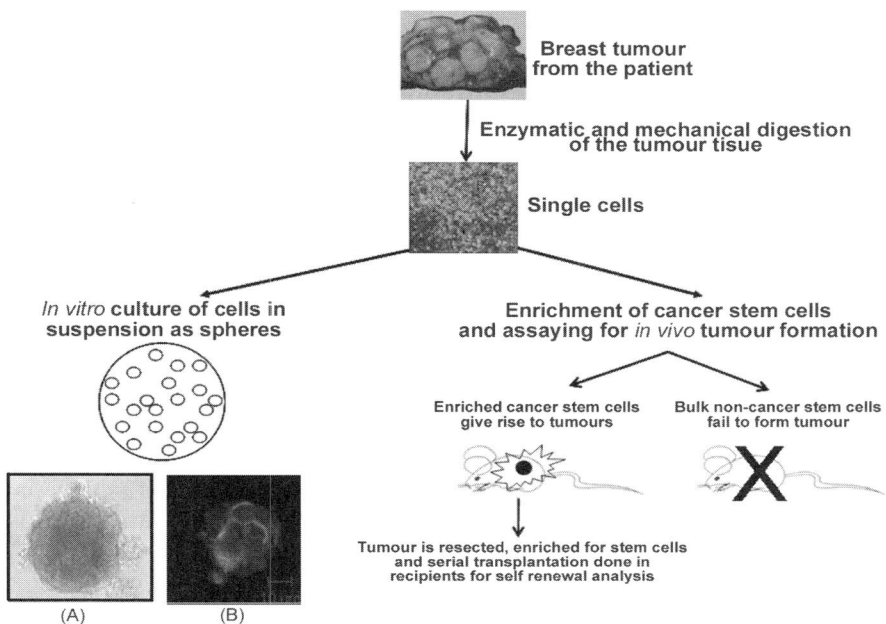

Courtesy: Devaveena Dey., IISc Bangalore (Unpublished data)

▶ **Fig. 24.19** *Prospective Identification and isolation of breast cancer stem cells.*

Breast cancer tissue is mechanically and enzymatically digested to yield single cells. These cells are subsequently seeded in non-adherent conditions in serum free medium with defined growth factors, like EGF, bFGF, insulin etc. In such suspension cultures, spherical colonies are formed, termed as breast cancer "mammospheres" (A), which are enriched in breast cancer stem cells. Detailed phenotypic analysis of these spheres indicate they are cellular structures (B), as seen by staining with a nuclear stain, propidium iodide (red). They also stain positive for progenitor specific markers, like cytokeratin 19 in the case of breast tissue (green; FITC), which is a cytoplasmic structural protein. Subsequent passaging of these mammospheres is done to assess the self-renewal property of the breast CSCs in vitro. Simultaneously, single cells isolated from breast cancer tissue are enriched for cancer stem cells by sorting out the $CD44^{high}$ $CD24^{low}$ population by flow cytometry, following which these cells are injected subcutaneously into NOD/SCID mice to assess tumour-initiating property. Enriched breast cancer stem cells fraction forms palpable tumours in mice while the bulk cells do not. Subcutaneous serial transplantation of the tumours from one recipient to another is carried out to study the self-renewal property of the breast cancer stem cells.

evident that there has been rapid development of multiple *in vitro* systems for culturing, maintaining and enriching both normal and CSCs, both in case of leukemias and solid tumours.

However, the "sphere" system has also been associated with some controversy since it does not strictly adhere to the concept of "clonality". Even though some studies confirmed clonality by retroviral tagging, cell–cell aggregation and clumping cannot be completely over ruled.

IDENTIFICATION AND ISOLATION OF CSCs FROM SOLID TUMOURS

As mentioned earlier, the discovery of leukemic stem cells was soon followed by discovery of a small population of stem-like cancer cells within solid tumours as well. Some of the earlier discoveries were made in case of breast and brain cancers. Again in these cases, the identity of this small population of cells was definitively established as CSCs due to the ability of as few as 100 s of these tumour cells to give rise to a tumour when injected *in vivo*, while most of the other tumour cells could not [30, 31]. These tumours comprised of the same heterogeneity which is very typical of solid tumours. Like in case of LSCs, the use of cell surface markers has enabled identification of this small population of cells. In many cases, this was based on the knowledge of marker profile of normal stem cells. Let us now discuss in brief how CSCs were identified in different cancers.

Breast Cancer

Nine breast cancer samples obtained from primary or metastatic sites were engrafted in the NOD/SCID mice subcutaneously and scored for tumour formation. Some of the tumours were passaged in mice once or twice [30]. The tumours formed in all cases were then analyzed for a set of cell surface markers. Three markers were selected for flow cytometry based analysis, which are known to be heterogeneously expressed on breast cancer cells. These were two cell adhesion molecules, CD24 and CD44 and an ovarian/breast cancer specific marker, B38.1. Cells were probed for these markers using antibodies specific for these surface proteins; populations positive or negative for each marker were sorted out and injected into mice to score for their tumour forming potential. It was seen that only the CD24–/low, CD44+ and B38.1+ cells could give rise to detectable tumours. They next went ahead do a detailed flow cytometric analysis of these sub populations by staining the cells for the other two markers. This was followed by sorting the dual stained subpopulations and injecting each subpopulation into mice. It was seen that only the $CD24^{-/low}CD44^{high}$ subpopulation of tumour cells could give rise to tumours in mice. This population comprised 11–35% in various tumours analyzed. Use of additional markers demonstrated that the isolation of ESA+ subset of the $CD24^{-/low}CD44^{high}$ population enhanced the tumourigenic potential of this population. When these tumours were analyzed for the expression of these markers, it was seen that they had recapitulated the same heterogeneity of expression as seen in the original tumour. Besides, these tumours could be further passaged in mice, reflecting the self-renewal potential of these cells.

As described under "Techniques of CSC isolation", 3D suspension cultures have recently been developed to enrich normal human mammary stem cells as cellular spheres, termed "mammospheres". The same approach was successfully used to enrich breast CSCs *in vitro*, where cells isolated from breast cancer lesions and some breast cancer cell lines could form spheres in culture [78]. It was also demonstrated that these spheres were enriched in CD24$^{-/low}$CD44^{high} population and the sphere derived cells could self renew and proliferate extensively. As few as 1000 cells derived from these spheres could give rise to a tumour in SCID mice. This study is another demonstration of the distinct similarities of normal and CSCs, in this case the breast.

Brain Tumour

In study of brain cancer, clonogenicity of cells is often tested *in vitro* as the ability to generate spherical colonies in suspension cultures, termed "neurospheres" [77] (For technical details, refer to Section VIIIB). It was observed that in brain tumours, neurosphere formation and self-renewal potential was exclusive to a minor subpopulation of tumour cells expressing the human neural stem cell markers CD133 and nestin [31]. These cells not only had the capacity to differentiate *in vitro* into cells with neural and glial phenotypes in proportions resembling the original tumour, but their proliferative capacity was also proportional to the aggressiveness of the original tumour. Injection of as few as 100 CD133+ cells produced tumours in NOD/SCID mice; these brain tumours could be serially transplanted and were identical to the patient's original tumour. In contrast, injections of up to $1 \infty 10^5$ CD133-negative cells failed to produce tumours. Even in glioblastoma multiforme, a particularly aggressive malignant adult human CNS tumour, only a small fraction of CD133+ cells was capable of neurosphere generation and multipotential differentiation, with as few as 5,000 of these cells being able to form tumours in nude mice [81]. In this tumour type, CD133+ cells are remarkably resistant to irradiation, probably because of an enhanced DNA damage response [82].

Pancreatic, Prostate and Ovarian Cancers

In human pancreatic cancer, a cell population with a surface marker phenotype closely related to breast CSCs, i.e. CD44+ CD24+ ESA+ was shown to be greatly enriched for tumour-initiating ability [33] making up only 0.2–0.8% of all cells. As few as 100 of these cells could produce tumours in 50% of mice. Very recently, in ovarian cancers, CD44, together with the expression of a key stem cell associated molecule, c-kit, has been shown to be present on a subset of ovarian tumour cells, which could initiate the tumour, and could be enriched by culturing as *in vitro* spheroids. These cells were identified as CD44+c kit−, such that as few as 100 cells of this population could serially propagate the original ovarian tumours [83]. In this population, the stemness markers were upregulated and the cells showed higher chemosensitivity. The molecule, CD44 is also expressed by a subpopulation of cells in prostate cancer. This subpopulation with the signature CD44$^+\alpha 2\beta 1^{high}$CD133$^+$ has no correlation with tumour grade but is substantially enriched for *in vitro* colony-forming ability [32]. CD44 has been confirmed as a useful prostate CSC marker, since compared with CD44- cells, CD44+ cells are more clonogenic and tumourigenic and express high levels of several "stemness genes" like Oct3/4, Bmi1, â-catenin and Smo [84].

> **Box 24.1. Genes associated with cancer**
>
> Two important classes of genes which are universally found to have acquired mutations in almost all cancers are **"oncogenes"** and **"tumour suppressor genes."**
>
> **Oncogenes:** The genes for which a gain-of-function mutation drives a cell towards cancer are called proto-oncogenes; their mutant, hyperactive forms are called oncogenes. Most of the oncogene encoded proteins, function as elements of the signaling pathways that regulate cell proliferation and survival in response to growth factor stimulation. A mutant form of this protein usually functions in a growth factor independent manner. These proteins include polypeptide growth factors, growth factor receptors, elements of intracellular signaling pathways, and transcription factors. **Examples** of some proto-oncogenes frequently implicated in cancer are growth factors like EGF, growth factor receptors like ErbB, intracellular signaling molecules like Ras and Raf and transcription factors like myc.
>
> **Tumour suppressor genes:** These are genes for which a loss-of-function mutation drives a cell towards cancer. The proteins encoded by most **tumour suppressor genes** inhibit cell proliferation or survival and also activate DNA repair pathways. Inactivation of tumour suppressor genes therefore leads to tumour development by eliminating the regulatory proteins. **Examples** of some tumour suppressor genes commonly invoked in cancer include transcriptional regulatory protein pRb, cycle progression and apoptosis regulator p53 and PTEN (A lipid phosphatase that dephosphorylates PIP_3).
>
> Recently the definition of oncogenes and tumour suppressor genes has also been extended to micro RNA (miRNA) class of regulatory molecules.

Colon and Colorectal Cancers

In 2006 [85] it was shown that tumourigenic cells in colon cancer were included in the high density CD133+ population, which comprised of about 2.5% of the tumour cells. Subcutaneous injection of colon cancer CD133+ cells was shown to readily reproduce the original tumour in immunodeficient mice, whereas CD133– cells did not. Such tumours were serially transplanted for several generations, in each of which progressively faster tumour growth was observed without significant phenotypic alterations. Unlike CD133– cells, CD133+ colon cancer cells were shown to grow exponentially for more than one year *in vitro* as undifferentiated tumour spheres in serum-free medium, maintaining the ability to engraft and reproduce the same morphological and antigenic pattern of the original tumour. Very recently, it was shown [86] that the ability of cells from colorectal cancers to engraft *in vivo* in immunodeficient mice was restricted to a minority subpopulation of Epithelial cell adhesion molecule ($EpCAM^{high}$)/$CD44^+$ epithelial cells. Tumours which originated from $EpCAM^{high}$/$CD44^+$ cells maintained a differentiated phenotype and reproduced the full morphologic and phenotypic heterogeneity of their parental lesions. Analysis of the surface molecule repertoire of $EpCAM^{high}$/$CD44^+$ cells led to the identification of CD166 as an additional differentially expressed marker, useful for CSC isolation in colorectal cancers. Another class of carcinomas where a similar small population of CSCs was discovered, marked by CD44+

cells is the head and neck squamous cell carcinoma. Less than 10% of these cells could give rise to tumours in mice when serially propagated while the CD44– cells could not [87].

CSCs in Cell Lines

In addition to identification and characterisation of CSCs from primary tumours obtained from patients, presence of CSCs has also been investigated in the well-established cell lines which are present in culture for several years now, have indefinite proliferation potential; form tumours like the original tumour from which these were derived, when transplanted *in vivo* [88]. Since many of these cancer cell lines are clones of a single cancer cell, and their properties are very similar to CSCs (indefinite division potential and formation of tumours), they serve as attractive models for study of CSCs. In fact, recently in 2004, it was shown that four cancer cell lines contain a distinct "Side Population" (SP), which has been shown to be enriched in stem cells, as explained earlier [89]. These cell lines were derived from different cancers; namely breast cancer, glioma, neuroblastoma and cervical cancer. It was seen in the glioma cell lines that only the SP could generate both SP and non-SP in culture and it was this fraction which was responsible for malignancy of this tumour *in vivo*. Since these cell lines are in culture for decades and are well adapted to *in vitro* growth conditions, this can serve as a simple system to carry out preliminary studies of CSCs, before moving on to the more complex primary tumours.

CSCs AND METASTASIS

Metastasis is a multistep process which involves specific cellular and molecular mechanisms. For a cell to be capable of migrating to a distant site and establishing a tumour there, it needs to acquire specific mutations and molecular changes which will equip it to successfully survive and thrive in a "foreign" microenvironment [6]. According to the clonal selection model, mutations acquired in the late stages of cancer progression underlies metastasis. However, recent studies indicates that metastatic capacity is predetermined by genetic changes acquired in the initial stages of cancer, when a tumour-initiating cell arises from one of the normal cells by acquiring genetic and epigenetic changes. Today the identity of the cells possessing metastatic potential and capable of surviving in a foreign "niche" is still unknown. CSCs have been proposed to be strong candidates possessing these abilities for multiple reasons [90]. First, CSCs are the exclusive "tumour-initiating" population, which would also enable them to initiate a tumour in a foreign site. The plasticity of stem cells and their increased responsiveness to signals from the microenvironment helps them adapt to a distant new site, away from the original tumour. Besides, the genetic instability of CSCs might also help them adapt to a new environment by acquiring additional mutations. Since tumour-initiating capacity is required for metastasis at any site, with the creation of the tumour niche, CSCs might be the metastasis-initiating cells.

Box 24.2. Normal adult stem cells: The "friend" within

- A small population of cells (<1%), residing in each tissue/ organ, which possess the property of "self-renewal," whereby it divides to give rise to a cell identical to it.
- These cells possess "multilineage differentiation" potential, whereby they can give rise to all the cell types of that particular tissue/organ.
- These cells repopulate the tissue/organ in case of an injury or disease.
- They divide infrequently and have tightly regulated "cell cycle checkpoints." Due to their "slow cycling" nature, a DNA binding label, like BrdU, when taken up by these cells, can be detected within them for a prolonged period, and hence these cells are often termed as "label retaining cells" (LRCs).
- The division of a normal stem cell is known to be "asymmetric," under most conditions, such that one daughter cell is a "progenitor", while another is an identical stem cell, retaining one of the parental strands. This is to ensure maintenance of the normal stem cell pool. When the stem cell "need" increases, like in injury and disease, they may undergo "symetric" division to increase stem cell numbers.
- Some critical pathways underlie their self-renewal potential. These are the Notch, Wnt and Hedgehog pathways.

CSCs and Invasive Gene Signature (IGS)

In one of the gene expression profiling studies done on samples of CD44+CD24–/low tumourigenic breast CSCs, in comparison with that of normal breast epithelium, a 186-"invasive" gene signature was arrived at [91]. This gene signature when evaluated for its correlation with metastasis and overall survival of patients, showed a strong association between the IGS and overall and metastasis free survival of patients. The IGS was successfully used for prognosis of high risk breast cancer patients and strong correlation was found with clinical outcomes, such that among patients predicted with a good prognosis using IGS, 10-year rate of metastasis free survival was seen to be 81% clinically. Though the IGS was originally developed for breast cancer, a strong correlation with prognosis was also seen for other cancers, like medulloblastoma, lung and prostate cancer. This strong correlation observed between a CSC marker phenotype and clinical outcome is an encouraging step in taking the CSC concept and findings from the "bench" to "bedside".

Migratory CSCs

A model system which has been extensively utilized to understand the concept of metastasis at the cellular and molecular level is colorectal cancer where all the steps of tumourigenesis, right from benign adenomas to metastatic carcinomas have been clearly defined. The conversion from adenomas to metastatic carcinomas is marked by upregulation of the "Epithelial Mesenchymal Transition" (EMT) related genes, by acquisition of additional mutations by a few adenoma cells. EMT endows migratory potential to epithelial cells, by downregulaing the cell–cell and cell-matrix adhesion molecules such that these cells can now move out of the primary tumour to seed tumours in a distant location.

When sections of colorectal cancer samples taken from different stages of progression were stained with EMT markers, like SLUG, L1CAM etc [92], it was found that only the cells located at the tumour-host interface in metastatic specimens stained positive for these markers. No positive cells were seen in the normal colonic crypts or benign adenomas. Interestingly, the EMT positive cells also stained positively for stemness markers, like survivin [93]. Cells positive for stemness markers could be detected in the normal colonic crypts as well as in all stages of progression of colorectal cancer. In all stages of cancer progression, these CSCs were scattered throughout the tumour. In the malignant form, a subset of these CSCs, which were also positive for the EMT markers were found only along the tumour-host interface [94].

This expression pattern of EMT and stemness markers led to the hypothesis that CSCs exist as two subsets during tumour progression: "stationary" CSCs (SCS) and "mobile" CSCs (MCS) [94]. The stationary CSCs are active in the benign lesions and are found throughout the tumour, in all stages of progression. However, these cells cannot migrate out of the tumour. Mobile CSCs, on the other hand are derived from the SCS but have acquired additional mutations that confer motility to these cells and they are found only at the tumour-host interface. Hence, the understanding the biology of MSCs, which possesses the combination of two traits: stemness and motility will help one understand the phenomenon of metastasis at its root.

CELL OF ORIGIN OF CSC

Even though the existence of small, subpopulation of CSCs within several cancers has been demonstrated beyond doubt, the cell of origin of cancer and CSCs remains a debatable issue [95]. Before the knowledge of the existence of stem cells within each tissue, it was predicted that mutations that occur in differentiated cells enabled them to regain their proliferative potential. However, differentiated cells seldom divide. Then how do they acquire mutations in the first place? Also, the life span of differentiated cells within several tissues is limited. Then how do they gain several mutations that are required to render them cancerous?

A more recent hypothesis takes into account both these issues, and predicts that normal stem cells (NSCs) endowed with an extensive life span and proliferative potential accumulates multiple mutations and gives rise to CSCs [17]. Given the fact that there already exist similarities between normal and CSCs of a given tissue, this is quite possible. Also, as mentioned earlier there are common signaling pathways, critical for both normal stem cell self-renewal and in carcinogenesis [17]. In fact, in leukemia and brain tumour, it has already been shown that the cancer arises in normal stem cells [25]. Yet another possibility is the progenitor origin of CSCs [96]. According to this, the rapidly proliferating progenitors or transit amplifying cells might accumulate mutations which confer on them an aberrant self-renewal property giving rise to CSCs. This could be a possibility given the fact that there is a high chance of mutations occurring during rapid proliferation of the progenitors. In fact, cancer arising in progenitors has been demonstrated recently in the haematological cancer AML [97]. The cell of origin of solid tumours, however, still remains an enigma.

THERAPEUTIC IMPLICATIONS OF CSCs
Conventional Chemotherapy and its Limitations

Chemotherapy, in context of cancer refers to the use of cytotoxic drugs to attack the cancer cells, most often exploiting the property of rapid proliferation of these cells. One of the inherent limitations of this approach of treatment is that these drugs also kill those normal cells that have a rapid proliferation rate. A second limitation of long term chemotherapeutic treatment is the development of drug resistance; a common universal phenomenon observed right from the antibiotic resistance seen in prokaryotes to resistance of insects and pests to commonly used cytotoxic chemicals. In the context of cancer cells, while some of the cancer cells are intrinsically resistant to various anticancer drugs, the others "develop" resistance to multiple chemotherapeutic drug upon treatment, a phenomenon known as "Multi Drug Resistance" (MDR). In those cancer cells which are inherently resistant to multiple anticancer agents, on prolonged administration of the drugs, there is an increase in expression of different classes of proteins involved in drug metabolism, detoxification and active drug efflux in addition to alteration of intracellular drug targets, like certain proteins to which the drug binds. All these mechanisms alter the effective drug concentration within a cell. These mechanisms are also known to be activated in the cancer cells which "acquire" drug resistance.

One of the immediate measures the cell utilizes towards drug resistance is active efflux of the drug so as to minimize the number of molecules of the drug entering the cell. This is usually a cell membrane based mechanism, where a family of plasma membrane efflux transporters are overexpressed. These include the most common ATP-binding cassette (ABC) transporters, which have received extensive attention [98]. These transmembrane protein transporters actively extrude the cytotoxic drugs by utilizing ATP. There are several sub classes of these transporters, the most well characterized in cancer being ABCB1 or MDR1 (P-glycoprotein), ABCC1 (MRP1), and ABCG2 (also known as BCRP as it was discovered in a breast cancer cell line) [99]. These transporters are not specific to a particular class of molecules and show a wide spectrum of substrate recognition and binding. The substrates these bind to vary from ions, bile salts, anticancer drugs, microbial toxins and hydrophobic compounds, like those derived from plants.

Box 24.3. Cancer stem cells: A closer look at the "enemy"

- CSCs comprises of a small population (<10%) within a tumour, which are the only cells capable of initiating a tumour.
- They have been found in all types of tumours, starting from leukemias to most of the solid tumours.

- Like normal stem cells, they can self-renew, thereby maintaining their pool so as to sustain the tumour.

- They have a functional **telomerase**, an enzyme which maintains the length of telomeres, the ends of the chromosomes, through subsequent rounds of cell division. It is the catalytic component of telomerase, "Tert" which makes the enzyme functional. Most of the somatic cells do not encode Tert. Only two types of cells have been shown to be positive for the tert component. The first includes those cells which have a prolonged lifespan (Stem cells) and the second type of cells are those which have a high division rate (Cancer cells). Given the "End Replication Problem", whereby a few base pairs at the ends of chromosomes are lost after each round of cell division, functional telomerase is critical for these cells.

- They have aberrant (usually upregulated) self-renewal pathways, like Notch, Wnt, and Sonic Hedgehog, which endows them with the property of indefinite self-renewal.

- They express some of the markers of metastatic cells (like Snail, Slug, Vimentin), making them potential "metastasis-initiating cells."

- They show an upregulation of membrane transporters, which bind to and efflux chemotherapeutic drugs out of the cell. This poses a major challenge in targeting this enemy within.

- How a CSC arises within a normal tissue/organ is still an enigma. A normal stem cell is the most potential candidate as the cell of origin of CSC, given the large number of similarities between the two. Along with the identification of the cell of origin of a CSC, the trigger and possible mechanism of conversion of a normal cell to a CSC are also critical questions waiting to be addressed.

ABC Transporters

P-glycoprotein or ABCB1 (MDR1)

Major mechanism of multidrug resistance in cultured cancer cells was shown to be due to the overexpression of an energy-dependent drug efflux pump, known as P-glycoprotein (P-gp) [100]. This efflux pump, the product of the *MDR*1 gene in the human [101] and the product of two different related genes, *mdr*1a and *mdr*1b in the mouse [102, 103] was one of the first members to be described of a large family of ATP-dependent transporters known as the ATP-binding cassette (ABC) family [104]. P-glycoprotein (P-gp) is highly expressed in bone marrow, on the cells lining the intestine (where the function of absorption that these cells have to carry out makes them highly susceptible to toxins), at the blood brain barrier (where they play a crucial role in protecting the brain) [105], in placental trophoblasts, and several other organs which are either directly exposed to environmental toxins, like the mucosal lining along the oral cavity or whose functions involve absorption and excretion of cellular wastes, like kidney [106].

P-gp can detect and bind a large variety of hydrophobic natural-product drugs as they enter the plasma membrane. These drugs include many of the commonly used natural product anticancer drugs such as doxorubicin and daunorubicin, vinblastine and vincristine, and taxol, as well as many commonly used pharmaceuticals ranging from antiarrhythmics and antihistamines to cholesterol-lowering statins [107] and HIV protease inhibitors [108].

Levels of expression of P-gp in many different tumours are high enough to confer significant drug resistance and the presence of P-gp correlates with drug resistance in several different cancers [109]. It was observed that aacquisition of drug resistance after chemotherapy is associated with increased P-gp levels [109]. One of the direct evidences of the link between this transporter and efflux of a chemotherapeutic agent is the acute induction of P-gp in human tumours following *in vivo* exposure to doxorubicin, a commonly used chemotherapeutic drug [110]. Clinically, it has been seen that expression of P-gp in some tumours predicts poor response to chemotherapy [111].

MRP1 (ABCC1)

Over a period of time, researchers realized that not all multidrug-resistant cells overexpress P-glycoprotein. This led to the discovery of other ABC transporters causing multidrug resistance in cancer. One of the ABC transporters that came up in this direction was the multidrug-resistance-associated protein 1 (MRP1, or ABCC1) [112]. MRP1 physiologically serves as a cellular defense mechanism because it is located in basolateral side of epithelial membrane [113]. It has been shown that MRP1 may protect the testicular tubules [114], cerebrospinal fluid [115] and bone marrow precursor cells [116].

MRP1 recognizes neutral and anionic hydrophobic natural products, and transports glutathione and other conjugates of these drugs, or, in some cases such as for vincristine co-transports unconjugated glutathione [117, 118]. MRP1 and MRP3 have been recently shown to transport methotrexate, a well-known chemotherapeutic drug extending the range of compounds potentially involved in the multidrug resistance phenotype [119, 120]. The discovery of MRP1 led to a search for other members of this family, resulting in the discovery of a total of 9 or 10 MRP genes, at least 6 of which have been characterized enough to indicate that they transport anticancer and antiviral compounds (MRP 1-6) [121].

ABCG2 (BCRP)

Selection for mitoxantrone resistance results in multidrug-resistant cells that produce a more distant member of the ABC transporter family, ABCG2 (mitoxantrone-resistance gene) also known as BCRP (breast cancer resistance protein) or ABC-P (ABC transporter in placenta) [122–124]. ABCG2 is a half transporter and hence unlike MDR1 and the MRP family members, ABCG2 protein is presumed to function as a dimer [99].

It is mainly expressed at the apical membranes of placental cells, in the mammary gland, intestine, colon, enterocytes, hepatocytes, erythrocytes and the human brain microvessel endothelium [125,126]. It was reported that ABCG2 is involved in transport of folate [127] and endogenous porphyrins, thereby protecting the cells against hypoxia induced by toxic porphyrins [128]. It has a renal and hepatic secretory function as it is involved in transport of organic sulfates and bile acids respectively [129,130]. ABCG2 can actively efflux a substantial variety of compounds ranging from fluorescent dyes to both anionic and cationic drugs. Reported drug substrates for this transporter mainly include mitoxantone, doxorubicin, topotecan, etoposide, prazosin, flavopiridol, Hoechst 33342, and anthracyclines [131]. The active efflux of these different classes of chemotherapeutic drugs by ABCG2 ultimately results in development of MDR.

Other Drug Transporters

Other ABC family members that have been associated with drug resistance include MDR2 gene product [132] encoding for the bile salt export protein (BSEP, ABCB11), first reported as the "sister of P-gp". This protein is expressed at high levels in liver cells, and it confers low level resistance to paclitaxel [133]. Another ATP transporter reported in cancer is MDR3, a phosphatidylcholine flippase that is closely related to P-gp, normally transports phospholipids into bile, but can transport paclitaxel and vinblastine out of the cell, albeit inefficiently, unless it is mutated [134]. It has also been shown that Lung resistance protein (LRP) is expressed at high levels in drug-resistant cell lines and some tumours [135]. Many of these transporters are expressed in the normal liver and are likely to be involved in drug disposition [136].

It is therefore evident that drug resistance is a serious barrier in the conventional chemotherapy which is used today as the primary mode of cancer treatment. The discovery of MDR as a widespread phenomenon occurring in cancer cells, followed by understanding its mechanism led to some important questions about the identity of the cells within a tumour which are multidrug resistant. Questions that had been asked decades back about the identity of tumour- "initiating" cells were now being asked for the "multidrug resistant" cells.

Clinical Implications of Cancer Stem Cells

It has been evident in haematopoietic malignancy and solid tumour that only a rare population of cells with unique self-renewal and survival mechanisms called CSCs drive and sustain the tumour. CSCs are biologically distinct from other cells in the tumour and are the only cells capable of initiating and sustaining tumour growth *in vivo*, whereas the bulk cells are not. With the discovery of CSCs, the ability to separate out the CSCs from the bulk and with the findings of distinct differences between CSCs and the bulk in multiple cancers, the property of drug efflux was also looked into for these distinct cell populations.

Current cancer treatments treat cancer as a whole, without taking into account the phenotypic and functional heterogeneity that exists within a tumour. With the discovery of distinct subpopulations like CSCs and bulk within multiple cancers, this conventional treatment regime needs to be reanalysed. Today it is known that the cell underlying at the root of cancer is the CSC, the only cell within a tumour which can initiate and sustain a tumour. Hence to cure cancer at the root, one needs to target the CSCs.

When the property of drug resistance was studied in dissociated tumours, it was found that a small population of tumour cells (1% or lesser) were resistant to commonly used chemotherapeutic drugs. These resistant cells could survive and proliferate in the presence of the drug, while most of the other cells died. This was observed in many cancers starting from blood cancer to solid tumours. When the drug resistant population was analysed with respect to their phenotype and function, they were found to be the CSCs. Further studies indicated that the CSCs overexpressed many of the transporters described above, unlike the bulk cells.

Targeted Cancer Stem Cell Therapy

It has been demonstrated that CD34+CD38− leukemic cells are less sensitive to daunorubicin than the more committed CD34−CD38− cells [137]. The ability of leukemic progenitors to efflux mitoxantrone and daunorubicin, two agents commonly used in treatment of AML, has also been shown [138]. The ability of leukemic progenitors to efflux mitoxantrone and daunorubicin, two agents commonly used in treatment of AML, has also been shown [139]. In CML patients, a commonly used drug is Gleevec, which targets the ATP-binding domain of the Bcr/Abl-kinase, the fusion protein which is a mark of CML and is present in all the CML cells. Most patients treated with Gleevec experience a clinically complete response. However on follow up of these patients, majority of the patients were seen to be positive for the presence of the fusion transcript [140]. This clearly indicated the presence of dormant CML cells in the body even after attaining clinical normalcy. This is another example of drug resistance of leukemia-initiating cells which enables these cells to survive even after the bulk of the tumour cells are killed. This small number of cells cannot be detected clinically. Besides, there are no clinical features of cancer at this stage. This stresses the need to preferentially target the tumour-initiating population.

Conventional cancer treatment regimes, including chemotherapy and radiation therapy focus on rapidly proliferating cells, which forms the bulk of a tumour. This often results in an initial "cure" of the cancer but in most cases, the tumour relapses, which more often than not is aggressive than the primary tumour. One of the major reasons for this could be the escape of the CSCs from the initial chemotherapeutic treatment (Fig. 24.20).

Challenges to eliminate CSCs

There are two primary challenges in the development of CSC-specific drugs; one is the small number of this population, which makes it difficult to target them amidst the bulk and second is the problem of drug efflux. Several agents effective in inhibiting the ATP-binding cassette transporters have been studied and found to have limited clinical efficacy. The reason for this is the similar level of expression of these transporters in normal stem cells, making them equally susceptible to the inhibitors [141]. As such, strategies directed at pathways that specifically regulate CSCs survival would probably be more fruitful.

It has been demonstrated that the transcription factor nuclear factor ?B (NF-?B) is constitutively activated in LSCs but not in normal HSCs [142]. This was followed by addition of NF-KB inhibitors to antileukemic agents such as idarubicin in experimental models [138, 142]. It has been observed that parthenolide, a potent inhibitor of NF-kB, could induce apoptosis in AML and CML LSCs and progenitors while sparing normal HSCs. Notably, parthenolide was much more selective in eliminating LSCs and sparing normal haematopoietic cells than the standard chemotherapy agent cytarabine [143]. The proteasome-inhibitor MG-132, which inhibits NF-kB activation through stabilisation of its cellular inhibitor IkB, induced apoptosis in CD34+CD38− AML cells while sparing normal primitive progenitors [138]. Similarly, constitutive activation of the phosphatidylinositide-3 kinase is also necessary for the survival of LSCs via the mTOR pathway. Its pharmacological inhibition leads to a dose-dependent decrease in survival [144]. The kinases Akt and mTOR are activated in human AML blasts and treatment with inhibitors of PI3K or mTOR, in combination with cytarabine or etoposide, induced apoptosis in AML cells and decreased the

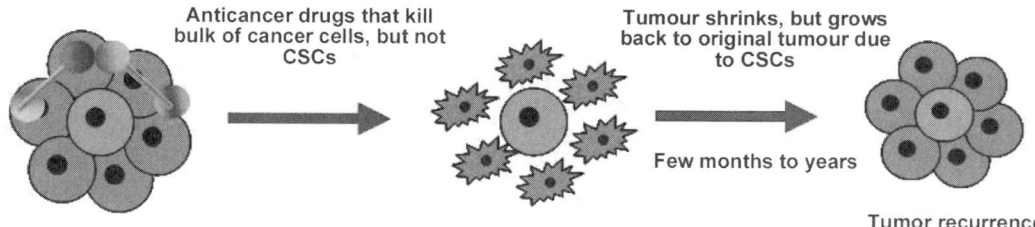

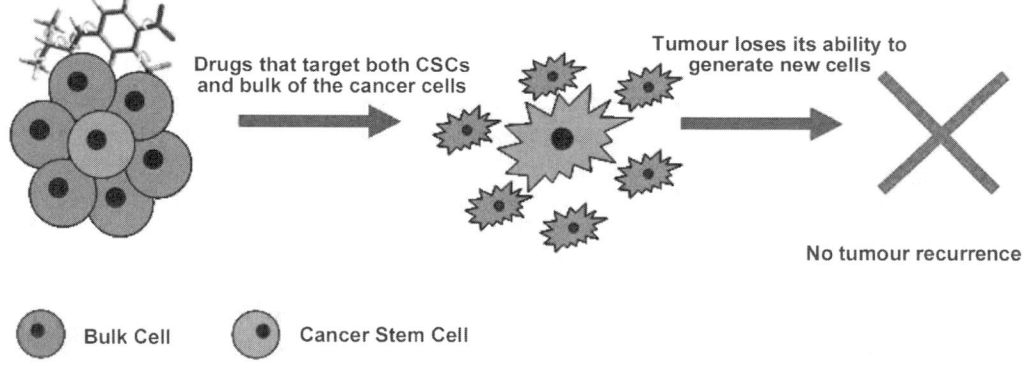

▶ Fig. 24.20 *Therapeutic implications of cancer stem cells.*
(A) Conventional chemotherapy is effective in killing rapidly dividing cells. On account of their low numbers and their drug efflux property, cancer stem cells are less sensitive to these therapies. Hence, while the bulk of cancer cells is killed, leading to an "apparent" cure of the cancer, the CSCs remain viable and active even after the therapy and contribute to re-growth of the entire tumour leading to recurrence of the cancer after a certain period. (B) Therapies which can target both cancer stem cells and bulk would be more effective in eliminating cancer at the root. CSC specific drugs, which might target the efflux pumps and some of the CSC specific survival pathways, in combination with the conventional drugs that kill the bulk would be the ideal treatment regime for cancers. Thus, targeted cancer stem cell therapy leads to remission of the disease at the root. (for colour figure see Plate 74).

abundance of LSCs [144,145]. It has also been demonstrated that NF-kB inhibitors like parthenolide (PTL), pyrrolidinedithiocarbamate (PDTC) and its analog diethyldithiocarbamate (DETC) could preferentially inhibit proliferation of a commonly used breast cancer cell line, MCF7, when these cells were cultured as 3D spheres, in which the breast CSCs were enriched. Such inhibition was not seen when these cells were cultured as single cells in a 2D system. Such preferential inhibition effect of these compounds can be explained due to their specific effects on CSCs and not all cells. Studies *in vitro* have indicated that the breast CSCs, cultured as spheres, are specifically killed when such compounds are used in combination with chemotherapy [146].

Another molecule, a tumour suppressor, which also functions through mTOR, has been shown to be to be essential for normal HSC function, while it is lost from the leukemic stem cells. This is the Pten protein (The Phosphatase and Tensin homologue), which negatively regulates signaling through the PI3 (K)-Akt pathway. While Pten is commonly deleted or inactivated in multiple cancers, its deletion from normal HSCs causes depletion of these normal stem cells. Hence, while Pten is critical for the survival and maintenance of normal HSCs, its loss converts the HSCs into LSCs. In this case, a therapeutic strategy would be the exogenous addition of Pten to leukemic cells. It was seen that on activating the mTOR pathway in LSCs, which is downstream of Pten, not only were the leukemia-initiating cells depleted, normal HSC function was also restored in them. [147].

The observations above have brought the PI3(K)-Akt pathway, mediated by Pten and mTOR at the forefront of CSC biology. It might be worthwhile to dissect out the role of this pathway in multiple cancers and compare it with the scenario in the normal stem cell counterparts, to devise drugs which specifically target the "evil" CSCs, sparing the "good" normal stem cells.

FUTURE DIRECTIONS

Based on the recent studies cancer can be viewed as a disease of unregulated self-renewal whose growth is sustained by a rare sub population of cells called the CSCs. These CSCs are biologically distinct from the other cells in the tumour and are able to initiate and sustain tumour growth *in vivo*, where the bulk cells cannot. Thus the biological function of cancer pathways might be fundamentally different in the rare CSC compartment compared with bulk of the cancer cells. Therefore, isolation and characterisation of these CSCs, followed by the ability to get rid of them could lead to improved therapeutics.

Pitfalls in the Field

The various strategies which have been discussed for the isolation and long term culture of CSCs currently allow mere "enrichment" of the CSC population and not the propagation of a pure CSC population. Thus further refinements in methods of isolation and long term propagation of pure CSC population are needed. Furthermore, in several instances, two techniques do not identify the same population. For example, even though SP has been found to house the stem cells in some tissues, in several other tissues SP and marker based analysis do not always narrow down to the same population. Moreover, not all cancers contain SP, but they may harbor real CSCs. The latest technique of ALDH assay also reveals that ALDH positive cells do not always show up as the SP.

More recently it has been argued that stem cells are not rigid "entities", and that stemness is a "state", as defined by the stem cell microenvironment, or the "niche" [148]. Disrupting this environment, which is done by isolating the cells from the body and culturing them under *in vitro* conditions may render a stem cell into a non-stem cell, and vice versa. This would lead to changes in both phenotypic and functional properties of these cells, making it impossible to use marker or SP based techniques. This issue is being dealt with by modifying culture conditions, so as to mimic the *in vivo* microenvironment as closely as possible. This concept has led to development of 3D matrices and specialized scaffolds, utilizing purified ECM molecules. The mechanical and biochemical properties of such scaffolds are often close mimics of the *in vivo* stem cell niche, thereby enabling isolation and maintenance of pure stem cell populations. The development of such scaffolds has been enabled by technical advances in the field of bioengineering.

The other critical issue which is still a grey area in this field is the identity of the cell of origin of cancer. It is only in leukemia that it has been definitively proven that LSC arises from normal HSC. In the same system it was also shown that deletion of three critical genes from progenitors, which serve as tumour suppressors, can convert progenitors into LSCs. In most of the other systems, the origin of CSC is still a black box. Finding the cell of origin of a CSC in multiple cancers would help one understand the molecular basis of conversion of a normal cell to a CSC, which in turn would help design target specific anti cancer drugs.

From the point of view of cancer therapy, as discussed earlier, the challenge today is to specifically target CSCs within a tumour to handle the disease at the root. A holistic mode of anti cancer therapy would be a combinatorial approach, involving use of the conventional chemotherapeutic drugs in combination with CSC specific drugs. Such mode of treatment would on one hand target the highly proliferating mutated progeny of the CSCs and on the other hand, target the root of cancer, the CSCs themselves. For patients with high risk of cancer development, stem cell specific therapies can be used, like antibodies directed against CSCs or agents that inhibit stem cell growth, like some of the inhibitors of the PI3 (K)-Akt pathway, as described earlier.

In conclusion, the identification of CSCs within cancers has changed the way we look at cancer from a disease of unlimited proliferation, to a disease of unlimited self-renewal. Even though similarities exist between cancer and normal stem cells, there is enough evidence to suggest that several pathways might be differentially regulated between these cell types. Thus, the ability to specifically target CSCs is likely to revolutionize the way cancer is treated.

DEFINITIONS OF IMPORTANT TERMS

Tumour An abnormal growth of tissue resulting from uncontrolled, progressive proliferation of a group of cells.

Metastasis The phenomenon by which cancer cells migrate to new sites in the body and initiate secondary tumours.

Carcinoma Cancers of the epithelial tissue

Chemotherapy Use of cytotoxic drugs to kill cancer cells.

Hayflick Limit The finite number of mitotic divisions that a somatic cell can undergo before attaining senescence. This limit appears to be a "programmed", fixed number for each normal cell type.

Epigenetics Epigenetics refers to heritable modifications to DNA that alter gene expression but do not affect coding sequences.

CFU Colony Forming Units; a term used to assess the number of single cells which could give rise to visible "colonies" when seeded in soft agar, or highly viscous matrix, like methylcellulose. An assay commonly used to test "clonogencity" of tumour cells.

Clonogenicity The property of a single cell to initiate tumour formation *in vivo* and colony formation *in vitro*.

Tumourigenicity The ability of a cell to give rise to a tumour, when injected into mice. This is a hallmark property of a CSC.

Soft Agar A viscous gel-like matrix which prevents cell–cell clumping. Single cells are embedded in this to check for their "clonogenic" potential, by scoring for formation of 3D colonies.

Stem Cells A small group of cells, which have the properties of: (1) Self-renewal and (2) differentiation. There are two types of stem cells. (a) embryonic stem cells, which are pluripotent; (b) adult stem cells (Tissue specific stem cells), which are multipotent.

Self Renewal The property of a cell to give rise to another genetically identical cell.

Haematopoeitic Of/ Dealing with blood.

Cancer Stem Cells (CSCs) The small number of cells within a tumour, which are the only cells capable of initiating a new tumour.

Niche The microenvironment around a stem cell, which comprises of other cells, ECM, blood cells, immune cells.

FACS (Fluorescence Activated Cell Sorting) A technique employed for identification, separation and sorting of a heterogeneous population of cells, marked with an antibody tagged with a small fluorescent molecule, such that the antibody binds to the antigen of interest. On probing with a single wavelength laser, the emission of the fluorescent molecule is detected by the electronic detection system, based on which "decisions" for analysis and sorting are made.

Cluster of Differentiation (CD) Markers These include many cell surface markers having huge practical applications for the identification of different cells types in multiple tissues, since each cell type has a specific "signature" of these markers. These proteins usually play an important role in cell–cell and cell-matrix adhesion and in cell signaling.

Side Population (SP) A small population of cells having high drug efflux, as measured by efflux of the DNA binding fluorescent dye, Hoechst 33342. These cells appear as a distinct low fluorescing population on a FACS plot. "SP" have been shown to mark stem cells.

Nude Mice A lab mouse called so because of their hairless phenotype. These mice lack a functional thymus and are thus immunodeficient. Used extensively for transplant and xenograft experiments.

SCID Mice Severely immunodeficient lab mice due to non-functional B and T lymphocytes. Being severely immunodeficient these mice do not reject tumours or transplants and thus serve as useful models in biology.

NOD/SCID Mice A variant of the SCID mice wherein the scid mutation has been transferred onto a diabetes-susceptible NOD (Non-Obese Diabetic) background.

Multidrug Resistance (MDR) Intrinsic or acquired resistance of cells to multiple classes of drugs.

ABC Transporters ATP Binding Cassette group of membrane proteins, which are involved in effluxing drugs out of a cell. Highly expressed in stem cells and CSCs.

Subcutaneous Injection Injection given in the fat layer between the skin and the muscle (just below the epidermis and dermis). The cells injected can enter the blood vessels and into circulation, for a systemic effect.

Orthotopic Transplant Transplantation of donor cells or graft into the host at the same site as that occupied by the original organ.

Suspension Culture An *in vitro* method of culture system for normal and CSCs. Cells are seeded in a non-adherent dish in liquid media, containing defined growth factors, following which only the stem cells form cellular "spheres" in suspension.

ALDH (Aldehyde Dehydrogenase) This is a cytosolic enzyme that is able to oxidize intracellular aldehydes.

REFERENCES

1. Weinberg R.A. (1996). 'How cancer arises'. *Sci. Am.* **275**(3): 62–70.
2. Widschwendter M., et. al. (2007). 'Epigenetic stem cell signature in cancer'. *Nat. Genet.* **39**(2): 157–158.
3. Weisenberger D.J., et. al. (2006). 'CpG island methylator phenotype underlies sporadic microsatellite instability and is tightly associated with BRAF mutation in colorectal cancer'. *Nat Genet.* **38**(7): 787–793.
4. Finkel, T., M. Serrano and M.A. Blasco (2007). 'The common biology of cancer and ageing'. *Nature.* **448**(7155): 767–74.
5. Hanahan D. and R.A. Weinberg (2000). 'The hallmarks of cancer'. *Cell* **100**(1): 57–70.
6. Fidler I.J. (2003). 'The pathogenesis of cancer metastasis: the 'seed and soil' hypothesis revisited'. *Nat. Rev. Cancer.* **3**(6): 453–458.
7. Kornblau S.M., et. al. (2006). 'Studying the right cell in acute myelogenous leukemia: dynamic changes of apoptosis and signal transduction pathway protein expression in chemotherapy resistant ex-vivo selected "survivor cells"'. *Cell Cycle.* **5**(23): 2769–2777.

8. Schätzlein A.G. (2006). 'Delivering cancer stem cell therapies—a role for nanomedicines'? *Euro. J. Cancer.* **42**: 1309–1315.
9. Von D.D. Hoff, et. al. (1981). 'Combination chemotherapy with cisplatin, bleomycin and methotrexate in patients with advanced head and neck cancer'. *Cancer. Clin. Trials.* **4**(2): 215–218.
10. Nowell P.C. (1976). 'The clonal evolution of tumour cell populations'. *Science.* **194**(4260): 23–28.
11. Fialkow, P.J., S.M. Gartler and A. Yoshida (1967). 'Clonal origin of chronic myelocytic leukemia in man'. *Proc. Natl. Acad. Sci. USA.* **58**(4): 1468–1471.
12. Nowell, P., et. al. (1976). 'T cell variant of chronic lymphocytic leukaemia with chromosome abnormality and defective response to mitogens'. *Br. J. Haematol.* **33**(4): 459–468.
13. Heppner, G.H. (1984). 'Tumour heterogeneity'. *Cancer. Res.* **44**(6): 2259–2265.
14. Shapiro, J.R., W.K. Yung and W.R. Shapiro (1981). 'Isolation, karyotype and clonal growth of heterogeneous subpopulations of human malignant gliomas'. *Cancer. Res.* **41**(6): 2349–2359.
15. Gatenby, R.A. and R.J. Gillies (2008). 'A microenvironmental model of carcinogenesis'. *Nat. Rev. Cancer.* **8**(1): 56–61.
16. Bergsagel, D.E. and F.A. Valeriote (1968). 'Growth characteristics of a mouse plasma cell tumour'. *Cancer. Res.* **28**(11): 2187–2196.
17. Reya, T., et. al. (2001). 'Stem cells, cancer, and cancer stem cells'. *Nature.* **414**: 105–111.
18. Huntly, B.J.P. and D.G. Gilliland (2005). 'Leukaemia stem cells and the evolution of cancer-stem-cell research'. *Nat. Rev.* **5**: 311–321.
19. Smalley, M. and A. Ashworth (2003). 'Stem cells and breast cancer: A field in transit'. *Nature.* **3**: 832–844.
20. Dontu, G., et. al. (2004). 'Role of Notch signaling in cell-fate determination of human mammary stem/progenitor cells'. *Br. Cancer. Res.* **6**(6): R605–R615.
21. Reya, T. and H. Clevers (2005). 'Wnt signalling in stem cells and cancer'. *Nature.* **434**: 843–850.
22. Lapidot, T., et. al. (1994). 'A cell initiating human acute myeloid leukaemia after transplantation into SCID mice'. *Nature.* **367**(6464): 645–648.
23. Kondo, M., et. al. (2003). 'Biology of haematopoietic stem cells and progenitors: implications for clinical application'. *Ann. Rev. Immunol.* **21**: 759–806.
24. Ibrahim, S.F. and G. van den Engh (2003). 'High-speed cell sorting: fundamentals and recent advances'. *Curr. Opin. Biotechnol.* **14**(1): 5–12.
25. Bonnet, D. and J.E. Dick (1997). 'Human acute myeloid leukemia is organized as a hierarchy that originates from a primitive haematopoietic cell'. *Nat. Med.* **3**(7): 730–737.
26. Behbod, F. and J.M. Rosen (2004). 'Will cancer stem cells provide new therapeutic targets'? *Carcinogenesis.* **26**(4): 703–711.
27. Jordan, C.T., M.L. Guzman and M. Noble (2006). 'Cancer Stem Cells'. *N. Engl. J. Med.* **355**: 1253–1261.
28. Wu, X.Z. (2008). 'Origin of cancer stem cells: the role of self-renewal and differentiation'. *Ann. Surg. Oncol.* **15**(2): 407–414.
29. Stingl, J., et. al. (2006), 'Purification and unique properties of mammary epithelial stem cells'. *Nature.* **439**(7079): 993–997.
30. Al-Hajj, M., et. al. (2003). 'Prospective identification of tumourigenic breast cancer cells'. *PNAS.* **100**(7): 3983–3988.

31. Singh, S.K., et. al. (2003). 'Identification of a Cancer Stem Cell in Human Brain Tumours'. *Cancer Res*. **63**: 5821–5828.
32. Collins, A.T., et. al. (2005). 'Prospective identification of tumourigenic prostate cancer stem cells'. *Cancer Res*. **65**(23): 10946–51.
33. Li, C., et. al. (2007). 'Identification of pancreatic cancer stem cells'. *Cancer. Res*. **67**(3): 1030–1037.
34. Goodell, M.A., et. al. (1996). 'Isolation and functional properties of murine haematopoietic stem cells that are replicating *in vivo*'. *J. Exp. Med*. **183**(4): 1797–1806.
35. Chaudhary, P.M. and I.B. Roninson (1991). 'Expression and activity of P-glycoprotein, a multidrug efflux pump, in human haematopoietic stem cells'. *Cell*. **66**(1): 85–94.
36. Zhou, S., et. al. (2001). 'The ABC transporter Bcrp1/ABCG2 is expressed in a wide variety of stem cells and is a molecular determinant of the side-population phenotype'. *Nat. Med*. 7(9): 1028–1034.
37. Dontu, G., et. al. (2003). '*In vitro* propagation and transcriptional profiling of human mammary stem/progenitor cells'. *Genes. Dev*. **17**(10): 1253–1270.
38. Summer, R., et. al. (2003). 'Side population cells and Bcrp1 expression in lung'. *Am. J. Physiol. Lung Cell Mol. Physiol*. **285**(1): L97–104.
39. Meeson, A.P., et. al. (2004). 'Cellular and molecular regulation of skeletal muscle side population cells'. *Stem. Cells*. **22**(7): 1305–1320.
40. Hussain, S.Z., et. al. (2005). 'Side population cells derived from adult human liver generate hepatocyte-like cells *in vitro*'. *Dig. Dis. Sci*. **50**(10): 1755–1763.
41. Yano, S., et. al. (2005). 'Characterisation and localisation of side population cells in mouse skin'. *Stem. Cells*. **23**(6): 834–841.
42. Kondo, T., T. Setoguchi and T. Taga (2004). 'Persistence of a small subpopulation of cancer stem-like cells in the C6 glioma cell line'. *Proc. Natl. Acad. Sci. USA*. **101**(3): 781–786.
43. Hirschmann-Jax, C., et. al. (2004). 'A distinct "side population" of cells with high drug efflux capacity in human tumour cells'. *Proc. Natl. Acad. Sci. USA*. **101**(39): 14228–14233.
44. Wu, C., et. al. (2007). 'Side population cells isolated from mesenchymal neoplasms have tumour initiating potential'. *Cancer. Res*. **67**(17): 8216–8222.
45. Chiba, T., et. al. (2006). 'Side population purified from hepatocellular carcinoma cells harbors cancer stem cell-like properties'. *Hepatology*. **44**(1): 240–251.
46. Ho, M.M., et. al. (2007). 'Side population in human lung cancer cell lines and tumours is enriched with stem-like cancer cells'. *Cancer. Res*. **67**(10): 4827–4833.
47. Haraguchi, N., et. al. (2006). 'Characterisation of a side population of cancer cells from human gastrointestinal system'. *Stem. Cells*. **24**(3): 506–513.
48. Mitsutake, N., et. al. (2007). 'Characterisation of side population in thyroid cancer cell lines: cancer stem-like cells are enriched partly but not exclusively'. *Endocrinology*. **148**(4): 1797–1803.
49. Wang, J., et. al. (2007). 'Identification of cancer stem cell-like side population cells in human nasopharyngeal carcinoma cell line'. *Cancer Res*. **67**(8): 3716–3724.
50. Zhou, S., et. al. (2002). 'Bcrp1 gene expression is required for normal numbers of side population stem cells in mice, and confers relative protection to mitoxantrone in haematopoietic cells in vivo'. *Proc. Natl. Acad. Sci. USA*. **99**(19): 12339–12344.
51. Zhou, J., et. al. (2007). 'Activation of the PTEN/mTOR/STAT3 pathway in breast cancer stem-like cells is required for viability and maintenance'. *Proc. Natl. Acad. Sci. USA*. **104**(41): 16158–16163.

52. Morita, Y., et. al. (2006). 'Non-side-population haematopoietic stem cells in mouse bone marrow'. *Blood.* **108**(8): 2850–2856.
53. Shackleton, M., et. al. (2006). 'Generation of a functional mammary gland from a single stem cell'. *Nature.* **439**(7072): 84–88.
54. Barrandon, Y. and H. Green (1985). 'Cell size as a determinant of the clone-forming ability of human keratinocytes'. *Proc. Natl. Acad. Sci. USA.* **82**(16): 5390–5394.
55. Kim, D.S., et. al. (2004). 'Isolation of human epidermal stem cells by adherence and the reconstruction of skin equivalents'. *Cell Mol. Life Sci.* **61**(21): 2774–2781.
56. Li, A., et. al. (2004). 'Extensive tissue-regenerative capacity of neonatal human keratinocyte stem cells and their progeny'. *J. Clin. Invest.* **113**(3): 390–400.
57. Webb, A., A. Li and P. Kaur (2004). 'Location and phenotype of human adult keratinocyte stem cells of the skin'. *Differentiation.* **72**(8): 387–395.
58. Youn, S.W., et. al. (2004). 'Cellular senescence induced loss of stem cell proportion in the skin *in vitro*'. *J. Dermatol. Sci.* **35**(2): 113–123.
59. Kang, M.K., et. al. (2000). '*In vitro* replication and differentiation of normal human oral keratinocytes'. *Exp. Cell Res.* **258**(2): 288–297.
60. Gao, F.B. and M. Raff (1997). 'Cell size control and a cell-intrinsic maturation program in proliferating oligodendrocyte precursor cells'. *J. Cell Biol.* **138**(6): 1367–1377.
61. Dazard, J.E., et. al. (2000). 'Switch from p53 to MDM2 as differentiating human keratinocytes lose their proliferative potential and increase in cellular size'. *Oncogene.* **19**(33): 3693–3705.
62. De Paiva, C.S., S.C. Pflugfelder and D.Q. Li (2006). 'Cell size correlates with phenotype and proliferative capacity in human corneal epithelial cells'. *Stem Cells.* **24**(2): 368–375.
63. Grichnik, J.M., et. al. (2006). 'Melanoma, a tumour based on a mutant stem cell'? *J. Invest. Dermatol.* **126**(1): 142–153.
64. Russo, J.E., D. Hauguitz and J. Hilton (1988). 'Inhibition of mouse cytosolic aldehyde dehydrogenase by 4-(diethylamino)benzaldehyde'. *Biochem. Pharmacol.* **37**(8): 1639–1642.
65. Kastan, M.B., et. al. (1990). 'Direct demonstration of elevated aldehyde dehydrogenase in human haematopoietic progenitor cells'. *Blood.* **75**(10): 1947–1950.
66. Russo, J.E. and J. Hilton (1988). 'Characterisation of cytosolic aldehyde dehydrogenase from cyclophosphamide resistant L1210 cells'. *Cancer Res.* **48**(11): 2963–2968.
67. Fallon, P., et. al. (2003). 'Mobilized peripheral blood SSCloALDHbr cells have the phenotypic and functional properties of primitive haematopoietic cells and their number correlates with engraftment following autologous transplantation'. *Br. J. Haematol.* **122**(1): 99–108.
68. Russo, J.E., J. Hilton and O.M. Colvin (1989). 'The role of aldehyde dehydrogenase isozymes in cellular resistance to the alkylating agent cyclophosphamide'. *Prog. Clin. Biol. Res.* **290**: 65–79.
69. Croker, A.K., et. al. (2008). 'High aldehyde dehydrogenase and expression of cancer stem cell markers selects for breast cancer cells with enhanced malignant and metastatic ability'. *J. Cell. Mol. Med.*
70. Ma, S., et. al. (2008). 'Aldehyde dehydrogenase discriminates the CD133 liver cancer stem cell populations'. *Mol. Cancer Res.* **6**(7): 1146–1153.
71. Pearce, D.J., et. al. (2005). 'Characterisation of cells with a high aldehyde dehydrogenase activity from cord blood and acute myeloid leukemia samples'. *Stem Cells.* **23**(6): 752–760.

72. Till, J.E. and C.E. Mc (1961). 'A direct measurement of the radiation sensitivity of normal mouse bone marrow cells'. *Radiat. Res.* **14**: 213–222.
73. Harrison, D.E. (1980). 'Competitive repopulation: a new assay for long-term stem cell functional capacity'. *Blood.* **55**(1): 77–81.
74. Bock, T.A., et. al. (1995). 'Improved engraftment of human haematopoietic cells in severe combined immunodeficient (SCID) mice carrying human cytokine transgenes'. *J. Exp. Med.* **182**(6): 2037–2043.
75. Buick, R.N., et. al. (1979). 'Development of an agar-methyl cellulose clonogenic assay for cells in transitional cell carcinoma of the human bladder'. *Cancer Res.* **39**(12): 5051–5056.
76. Guan, Y. and D.E. Hogge (2000). 'Proliferative status of primitive haematopoietic progenitors from patients with acute myelogenous leukemia (AML)'. *Leukemia.* **14**(12): 2135–2141.
77. Reynolds, B.A. and S. Weiss (1992). 'Generation of neurons and astrocytes from isolated cells of the adult mammalian central nervous system'. *Science.* **255**(5052): 1707–1710.
78. Ponti, D., et. al. (2005). 'Isolation and *In vitro* Propagation of Tumourigenic Breast Cancer Cells with Stem/Progenitor Cell Properties'. *Cancer Res.* **65**(13): 5506–5511.
79. Xin, L., et. al. (2007). 'Self-renewal and multilineage differentiation *in vitro* from murine prostate stem cells'. *Stem Cells.* **25**(11): 2760–2769.
80. Fierabracci, A., et. al. (2008). 'Identification of an adult stem/progenitor cell-like population in the human thyroid'. *J. Endocrinol.* **198**(3): 471–487.
81. Galli, R., et. al. (2004). 'Isolation and characterisation of tumourigenic, stem-like neural precursors from human glioblastoma'. *Cancer Res.* **64**(19): 7011–7021.
82. Bao, S., et. al. (2006). 'Glioma stem cells promote radioresistance by preferential activation of the DNA damage response'. *Nature.* **444**(7120): 756–760.
83. Zhang, S., et. al. (2008). 'Identification and characterisation of ovarian cancer-initiating cells from primary human tumours'. *Cancer Res.* **68**(11): 4311–4320.
84. Patrawala, L., et. al. (2007). 'Hierarchical organisation of prostate cancer cells in xenograft tumours: the CD44+alpha2beta1+ cell population is enriched in tumour-initiating cells'. *Cancer Res.* **67**(14): 6796–6805.
85. Ricci-Vitiani, L., et. al. (2007). 'Identification and expansion of human colon-cancer-initiating cells'. *Nature.* **445**: 111–115.
86. Dalerba, P., et. al. (2007). 'Phenotypic characterisation of human colorectal cancer stem cells'. *PNAS.* **104**(24): 10158–10163.
87. Prince, M.E., et. al. (2007). 'Identification of a subpopulation of cells with cancer stem cell properties in head and neck squamous cell carcinoma. *PNAS.* **104**(3): 973–978.
88. Locke, M., et. al. (2005). 'Retention of intrinsic stem cell hierarchies in carcinoma-derived cell lines'. *Cancer Res.* **65**(19): 8944–8950.
89. Setoguchi, T., T. Taga and T. Kondo (2004). 'Cancer stem cells persist in many cancer cell lines'. *Cell Cycle.* **3**(4): 414–415.
90. Li, F., et. al. (2007). 'Beyond tumourigenesis: cancer stem cells in metastasis'. *Cell Res* **17**(1): 3–14.
91. Liu, R., et. al. (2007). 'The Prognostic Role of a Gene Signature from Tumourigenic Breast-Cancer Cells'. *N. Engl. J. Med.* **356**(3): 217–226.
92. Huber, M.A., N. Kraut and H. Beug (2005). 'Molecular requirements for epithelial-mesenchymal transition during tumour progression'. *Curr. Opin. Cell Biol.* **17**(5): 548–558.

93. Tetsu, O. and F. McCormick (1999). 'Beta-catenin regulates expression of cyclin D1 in colon carcinoma cells'. *Nature.* **398**(6726): 422–426.
94. Brabletz, T., et. al. (2005). 'Opinion: migrating cancer stem cells—an integrated concept of malignant tumour progression'. *Nat. Rev. Cancer.* **5**(9): 744–749.
95. Bjerkvig, R., et. al. (2005). 'The origin of the cancer stem cell: current controversies and new insights'. *Nat. Rev.* **5**: 899–904.
96. Wang, J.C.Y. and J.E. Dick. 'Cancer stem cells: lessons from leukemia'. *Trends Cell Biol.* **15**: 494–501.
97. Akala, O.O., et. al. (2008). 'Long-term haematopoietic reconstitution by Trp53-/-p16Ink4a-/-p19Arf-/- multipotent progenitors'. *Nature.* **453**(7192): 228–232.
98. Litman, T., et. al. (2001). 'From MDR to MXR: new understanding of multidrug resistance systems, their properties and clinical significance'. *Cell Mol. Life Sci.* **58**(7): 931–959.
99. Gottesman, M.M. (2002). 'Mechanisms of cancer drug resistance'. *Ann. Rev. Med.* **53**: 615–627.
100. Juliano, R.L. and V. Ling (1976). 'A surface glycoprotein modulating drug permeability in Chinese hamster ovary cell mutants'. *Biochim. Biophys. Acta.* **455**(1): 152–162.
101. Chen, C.J., et. al. (1986). 'Internal duplication and homology with bacterial transport proteins in the mdr1 (P-glycoprotein) gene from multidrug-resistant human cells'. *Cell.* **47**(3): 381–389.
102. Croop, J.M., et. al. (1989). 'The three mouse multidrug resistance (mdr) genes are expressed in a tissue-specific manner in normal mouse tissues'. *Mol. Cell Biol.* **9**(3): 1346–1350.
103. Lothstein, L., et. al. (1989). 'Alternate overexpression of two P-glycoprotein [corrected] genes is associated with changes in multidrug resistance in a J774.2 cell line'. *J. Biol. Chem.* **264**(27): 16054–16058.
104. Higgins, C.F. (1992). 'ABC transporters: from microorganisms to man'. *Ann. Rev. Cell Biol.* **8**: 67–113.
105. Schinkel, A.H. (1999). 'P-Glycoprotein, a gatekeeper in the blood-brain barrier'. *Adv. Drug Deliv. Rev.* **36**(2-3): 179–194.
106. Sparreboom, A., et. al. (1997). 'Limited oral bioavailability and active epithelial excretion of paclitaxel (Taxol) caused by P-glycoprotein in the intestine'. *Proc. Natl. Acad. Sci. USA.* **94**(5): 2031–2035.
107. Bogman, K., et. al. (2001). 'HMG-CoA reductase inhibitors and P-glycoprotein modulation'. *Br J Pharmacol.* **132**(6): 1183–1192.
108. Lee, C.G. and M.M. Gottesman (1998). 'HIV-1 protease inhibitors and the MDR1 multidrug transporter'. *J Clin Invest.* **101**(2): 287–288.
109. Goldstein, L.J., et. al. (1989). 'Expression of a multidrug resistance gene in human cancers'. *J Natl Cancer Inst.* 81(2): 116–124.
110. Abolhoda, A., et. al. (1999). 'Rapid activation of MDR1 gene expression in human metastatic sarcoma after in vivo exposure to doxorubicin'. *Clin Cancer Res.* **5**(11): 3352–3356.
111. Chan, H.S., et. al. (1991). 'P-glycoprotein expression as a predictor of the outcome of therapy for neuroblastoma'. *N Engl J. Med.* **325**(23): 1608–1614.
112. Cole, S.P., et. al. (1992). 'Overexpression of a transporter gene in a multidrug-resistant human lung cancer cell line'. *Science.* **258**(5088): 1650–1654.

113. Wijnholds, J., et. al. (1997). 'Increased sensitivity to anticancer drugs and decreased inflammatory response in mice lacking the multidrug resistance-associated protein'. *Nat. Med.* **3**(11): 1275–1279.
114. Evers, R., et. al. (1996). 'Basolateral localisation and export activity of the human multidrug resistance-associated protein in polarized pig kidney cells'. *J. Clin. Invest.* **97**(5): 1211–1218.
115. Wijnholds, J., et. al. (2000). 'Multidrug resistance protein 1 protects the choroid plexus epithelium and contributes to the blood-cerebrospinal fluid barrier'. *J. Clin. Invest.* **105**(3): 279–285.
116. Wijnholds, J., et. al. (1998). 'Multidrug resistance protein 1 protects the oropharyngeal mucosal layer and the testicular tubules against drug-induced damage'. *J. Exp. Med.* **188**(5): 797–808.
117. Jedlitschky, G., et. al. (1996). 'Transport of glutathione, glucuronate, and sulfate conjugates by the MRP gene-encoded conjugate export pump'. *Cancer. Res.* **56**(5): 988–994.
118. Muller, M., et. al. (1994). 'Overexpression of the gene encoding the multidrug resistance-associated protein results in increased ATP-dependent glutathione S-conjugate transport'. *Proc Natl Acad Sci. USA.* **91**(26): 13033–13037.
119. Kool, M., et. al. (1999). 'MRP3, an organic anion transporter able to transport anti-cancer drugs'. *Proc. Natl. Acad. Sci. USA.* **96**(12): 6914–6919.
120. Zeng, H., et. al. (1999). 'Expression of multidrug resistance protein-3 (multispecific organic anion transporter-D) in human embryonic kidney 293 cells confers resistance to anticancer agents'. *Cancer. Res.* **59**(23): 5964–5967.
121. Borst, P., et. al. (1999). 'The multidrug resistance protein family'. *Biochim Biophys Acta.* **1461**(2): 347–357.
122. Miyake, K., et. al. (1999). 'Molecular cloning of cDNAs which are highly overexpressed in mitoxantrone-resistant cells: demonstration of homology to ABC transport genes'. *Cancer Res.* **59**(1): 8–13.
123. Doyle, L.A., et. al. (1998). 'A multidrug resistance transporter from human MCF-7 breast cancer cells'. *Proc. Natl. Acad. Sci. USA.* **95**(26): 15665–15670.
124. Allikmets, R., et. al. (1998). 'A human placenta-specific ATP-binding cassette gene (ABCP) on chromosome 4q22 that is involved in multidrug resistance'. *Cancer Res.* **58**(23): 5337–5339.
125. Maliepaard, M., et. al. (2001). 'Circumvention of breast cancer resistance protein (BCRP)-mediated resistance to camptothecins *in vitro* using non-substrate drugs or the BCRP inhibitor GF120918'. *Clin Cancer Res.* **7**(4): 935–941.
126. Cooray, H.C., et. al. (2002). 'Localisation of breast cancer resistance protein in microvessel endothelium of human brain'. *Neuroreport.* **13**(16): 2059–2063.
127. Ifergan, I., et. al. (2004). 'Folate deprivation results in the loss of breast cancer resistance protein (BCRP/ABCG2) expression'. *A role for BCRP in cellular folate homeostasis. J. Biol. Chem.* **279**(24): 25527–25534.
128. Krishnamurthy, P., et. al. (2004). 'The stem cell marker Bcrp/ABCG2 enhances hypoxic cell survival through interactions with heme'. *J. Biol. Chem.* **279**(23): 24218–24225.
129. Mizuno, N., et. al. (2004). 'Impaired renal excretion of 6-hydroxy-5,7-dimethyl-2-methylamino-4-(3-pyridylmethyl) benzothiazole (E3040) sulfate in breast cancer resistance protein (BCRP1/ABCG2) knockout mice'. *Drug. Metab. Dispos.* **32**(9): 898–901.
130. van Herwaarden, A.E., et. al. (2003). 'The breast cancer resistance protein (Bcrp1/Abcg2) restricts exposure to the dietary carcinogen 2-amino-1-methyl-6-phenylimidazo[4,5-b]pyridine'. *Cancer Res.* **63**(19): 6447–6452.

131. Sarkadi, B., et. al. (2006). 'Human multidrug resistance ABCB and ABCG transporters: participation in a chemoimmunity defense system'. *Physiol Rev.* **86**(4): 1179–1236.
132. Borst, P., N. Zelcer and A. van Helvoort (2000). 'ABC transporters in lipid transport'. *Biochim Biophys Acta.* **1486**(1): 128–144.
133. Childs, S., et. al. (1998). 'Taxol resistance mediated by transfection of the liver-specific sister gene of P-glycoprotein'. *Cancer Res.* **58**(18): 4160–417.
134. Smit, J.J., et. al. (1993). 'Homozygous disruption of the murine mdr2 P-glycoprotein gene leads to a complete absence of phospholipid from bile and to liver disease'. *Cell.* **75**(3): 451–462.
135. Scheffer, G.L., et. al. (2000), 'Lung resistance-related protein/major vault protein and vaults in multidrug-resistant cancer'. *Curr Opin Oncol.* **12**(6): 550–556.
136. Bates, S.E., et. al. (2001). 'The role of half-transporters in multidrug resistance'. *J Bioenerg Biomembr.* **33**(6): 503–511.
137. Costello, R.T., et. al. (2000). 'Human acute myeloid leukemia CD34+/CD38- progenitor cells have decreased sensitivity to chemotherapy and Fas-induced apoptosis, reduced immunogenicity, and impaired dendritic cell transformation capacities'. *Cancer Res.* **60**(16): 4403–4411.
138. Guzman, M.L., et. al. (2002). 'Preferential induction of apoptosis for primary human leukemic stem cells'. *Proc. Natl. Acad. Sci. USA.* **99**(25): 16220–16225.
139. Wulf, G.G., et. al. (2001). 'A leukemic stem cell with intrinsic drug efflux capacity in acute myeloid leukemia'. *Blood.* **98**(4): 1166–1173.
140. Hughes, T. and S. Branford (2003). 'Molecular monitoring of chronic myeloid leukemia'. *Semin Haematol.* **40**(2 Suppl 2): 62–68.
141. Dean, M., T. Fojo and S. Bates (2005). 'Tumour stem cells and drug resistance'. *Nat Rev Cancer.* **5**(4): 275–284.
142. Guzman, M.L., et. al. (2001). 'Nuclear factor-kappaB is constitutively activated in primitive human acute myelogenous leukemia cells'. *Blood.* **98**(8): 2301–2307.
143. Guzman, M.L., et. al. (2005). 'The sesquiterpene lactone parthenolide induces apoptosis of human acute myelogenous leukemia stem and progenitor cells'. *Blood.* **105**(11): 4163–4169.
144. Xu, Q., et. al. (2003). 'Survival of acute myeloid leukemia cells requires PI3 kinase activation'. *Blood* **102**(3): 972–980.
145. Xu, Q., J.E. Thompson and M. Carroll (2005). 'mTOR regulates cell survival after etoposide treatment in primary AML cells'. *Blood.* **106**(13): 4261–4268.
146. Zhou, J., et. al. (2008). 'Cancer stem/progenitor cell active compound 8-quinolinol in combination with paclitaxel achieves an improved cure of breast cancer in the mouse model'. *Br. Cancer Res. Treat.*
147. Yilmaz, O.H., et. al. (2006). 'Pten dependence distinguishes haematopoietic stem cells from leukaemia-initiating cells'. *Nature.* **441**(7092): 475–482.
148. Zipori, D. (2004). 'The nature of stem cells: state rather than entity'. *Nat. Rev* **5**: 873–878.

25

EXPLORING THE POTENTIAL OF ADULT STEM CELLS: WHY UMBILICAL CORD BLOOD?

Viviane Abreu Nunes*
Stem Cells Laboratory, Human Genome Research Centre, University of São Paulo, São Paulo, Brazil.

INTRODUCTION

The study and knowledge of the generation of blood cell lines throughout life has provided conceptual, experimental, and therapeutic possibilities to researchers and stem cell biologists. From a clinical perspective, no other area of stem cell biology has been applied as successfully as transplantation of bone marrow (BM) and cord blood for the treatment of haematological diseases. In the last few years, research in stem cell biology has expanded rapidly, engendering new perspectives concerning the identity, origin, and full therapeutic potential of tissue-specific stem cells. This chapter will focus on the use of cord blood stem cells for reconstitution of BM, looking at the biology of these stem cells in the context of cell plasticity and their therapeutic potential for repair of different tissues and organs.

UMBILICAL CORD BLOOD IS AN IMPORTANT SOURCE OF STEM CELLS

Umbilical cord blood (UCB) is the blood collected from the umbilical cord of a newborn child. Cord blood contains all the regular elements found in blood such as red and white blood cells, platelets, and plasma, but is a special source of primitive haematopoietic stem cells (HSC) and progenitor cells that have been used to reconstitute the haematopoietic system in patients with malignant and

---

* *vanunes@ib.usp.br*

non-malignant haematological disorders [1]. More recently, these adult stem cells are shown to differentiate into neural, cardiac, epithelial, hepatocytic, and dermal tissue.

HSC can be defined by two essential properties: self-renewal and multilineage haematopoietic differentiation. These combined HSC properties allow them to differentiate into all blood cell types (multilineage) in a sustained manner for the lifetime of the animal, which requires their ability to make cellular copies of themselves (self-renewal). These features can be tested by transplantation from donor to recipient and provide a functional basis to define and identify HSC. Currently, human BM, mobilized peripheral blood, and UCB represent the major sources of transplantable HSC, but their availability for use is limited by both quantity and compatibility.

Until recently, blood that remained in the umbilical cord and placenta after delivery was routinely discarded. However, at present this blood is known to contain both HSCs and pluripotent mesenchymal cells, which have been explored in clinical and research investigation and regenerative medicine [2].

UMBILICAL CORD BLOOD STEM CELLS AND TRANSPLANTS

An estimated 45,000–50,000 haematopoietic cell transplants (BM, peripheral blood stem cells, or cord blood) are performed annually worldwide to treat patients with life-threatening malignant and non-malignant diseases [3]. Due to advances in transplantation, long-term experience and current clinical trials, patient eligibility for transplant continues to expand and the list of indications for which transplant may be a standard treatment option continues to develop.

A list of diseases for which autologous or allogeneic haematopoietic cell transplant may be used includes leukemias and lymphomas, multiple myeloma, severe aplastic anemia, SCID and other inherited immune system disorders, beta thalassemia major, sickle cell disease, inherited metabolic disorders, refractory anemia (all types), and other malignancies.

After more than one decade of clinical experience, it is currently accepted that hUCB transplants, related and unrelated, are equivalent to or might be comparable with BM transplants, especially in children. Studies of long-term survival in children with both malignant and non-malignant haematological disorders, who were transplanted with hUCB from a sibling donor, demonstrated comparable or superior survival to children who received BM transplantation. In addition, using hUBC presents several advantages, including prompt availability, decreased risk of transmissible viral infections, and graft-versus-host disease (GVHD) in both human leukocyte antigen (HLA)-matched and HLA-mismatched stem cell transplants [4] and ease of collection with little risk to the mother or newborn. On the other hand, potential limitations of UCB transplantation include insufficient stem cell dose that may be harvested from umbilical cord to reliably treat larger children and adult recipients, which may result in a slower engraftment and higher transplant-related mortality, mainly due to the long aplasia period after transplantation.

Despite prolonged periods of aplasia, the apparent reduction in the incidence and severity of GVHD can in turn underline comparable rates of survival in comparison to other adult-donor sources. The immature nature of hUCB lymphocytes may explain the lower incidence and severity of GVHD compared to the allogeneic BM transplant. The susceptibility to viral and fungal infections during the aplasia period is also an inconvenience of transplantation using hUCB.

hUCB transplantation does not seem to be associated with increased rates of disease relapse. The available data suggest that nucleated cell dose should be the primary criterion for donor selection and *in vitro* studies have showed that single collections of cord blood probably do contain sufficient cells to engraft the haematopoietic system of an adult [5]. Engraftment will depend not only on the number of stem progenitors, but also on the quality of these cells for proliferation and self-renewal.

To accelerate the engraftment rate, several strategies such as multiple hUCB transplants and *ex vivo* expansion of HSC have been assayed. The current strategies are focused on the development of much more efficient technologies for *ex vivo* production of progenitor cells, but whether expansion will speed engraftment and improve outcome after hUCB remains to be determined. However, long-term cultures have suggested that UCB contains more early cells than BM with high proliferative potential and results have shown that cord blood cells can be extensively expanded in suspension culture with different combinations of cytokines [6]. Xiao et al. [7] have demonstrated that *ex vivo* expansion can also be established from single primitive cells. Thus, immature HSC and progenitor cells appear to be of good quality in cord blood.

GENERAL AND ETHICAL TOPICS ABOUT UCB BANKING

Using hUCB as a source of HSC and the fact that the cord blood exists in almost limitless supply have led to the establishment of cord blood banks worldwide. The standards for collecting and processing of hUCB to be used as a source of stem cells for transplantation were still not well-established; however, the American Association of Blood Banks (AABB) and the Foundation for Accreditation of Cellular Therapy (FACT-NETCORD) have created guidelines pertaining to collection, testing, processing, and banking of UCB for transplantation and provide the accreditation of banks.

In general, cord blood units are collected for allogeneic unrelated and related HSC transplantation. In unrelated cord blood banks, donated hUCB units are collected and stored for allogeneic use in patients who do not have an identified HLA-matched relative. For this purpose, hUCB banks report available units to national and international donor registries. The second model of hUCB banking is referred as family banking, where hUCB is stored for the benefit of the donor or their family members.

Cord blood is collected at the time of delivery by two methods: *intra* or *extra uterus* collection. *Intra uterus* collection is done while the placenta still remains in the uterus. Larger unit volumes and higher total nucleated cell counts may result by this method. Conversely, bacterial contamination may occur due to the proximity to the perineum. Additionally, in the case of cesarean section, *intra uterus* collection may result in increased operative time and make removal of the uterus more difficult once collection has occurred. *Extra uterus* collection is most common and usually performed by trained personnel outside the delivery room. Macroscopic clot formation may occur with the prolonged handling times necessary for this type of collection.

The main collection procedure involves a relatively simple venipuncture, followed by gravity drainage into a standard sterile anti-coagulant-filled blood bag, using a closed system, similar to the one utilized on whole blood collection. After aliquots have been removed for routine testing, the collected UCB is then sent to a specified blood bank for processing and storage. The units are, then, cryopreserved and stored in liquid nitrogen.

Nevertheless, the standards for collection and processing of hUCB to be used as a source of stem cells for transplantation were still not well-established; the discussion continues regarding the private banking of autologous blood for biologic insurance versus public banking for access by the general population.

In fact, to implement the widespread clinical use of cord blood, it is necessary to establish both autologous and allogeneic cord blood banks from which stem cells can be obtained. These cord blood banks must be large in size, reflect the ethnic diversity of the patient populations, and be financial and logistically feasible. Also, there are many unresolved ethical issues related to UCB banking, particularly related to the rapid growth of private cord blood banks offering long-term storage for potential future autologous or related allogeneic transplantation. However, directed donation of cord blood should be considered when there is a specific diagnosis of a disease within a family known to be amenable to stem cell transplantation. An examination of these issues is needed to inform public policy and to raise the awareness of prospective parents, clinicians, and researchers.

To study some aspects of UCB banking, persons with expertise in anthropology, blood banking, BM transplantation, ethics, law, obstetrics, pediatrics, and the social sciences were invited to join the Working Group on Ethical Issues in Umbilical Cord Blood Banking. The conclusions of this work were that UCB technology is promising but requires a very accurate investigational phase to assure secure linkage of stored hUCB to the identity of the donor. Also, hUCB banking for autologous use is controversial than banking for allogeneic use, and consequently hUCB banking in the private segment needs a lot more attention. At last, they pointed out to the process of obtaining informed consent for collection of hUCB, which should begin before labor and delivery as described by Sugarman et al. [8].

Focusing on UCB cells, Madlambayan and Rogers [9] discussed some of the challenges that stem cell therapy faces, including obtaining clinically relevant numbers of stem cells and the ability of stem cells to provide permanent engraftment of multiple tissue types. They suggest *in vitro* stem cell expansion as a possible solution to these problems, for example. In addition, these pluripotent or, perhaps, multipotent stem cells have been proposed as suitable elements for cellular therapy and regenerative medicine. till date, there are no conclusive data regarding these possibilities, but preliminary *in vitro* and animal studies in the field of tissue regeneration suggest some degree of plasticity and/or transdifferentiation. hUCB cells are showing their unique qualities and potential, and consequently hUCB banks might dramatically increase the scope of their clinical application.

UMBILICAL CORD BLOOD AS A SOURCE OF MESENCHYMAL STEM CELLS

Mesenchymal stem cells (MSCs) comprise a well-characterized population of adult stem cells. These cells, also found in the BM, can form a variety of cells, including adipocytes, cartilage, bone, tendon and ligaments, muscles and skin cells, and even neurons. MSCs have been studied in great detail and researchers have advanced knowledge about how to grow these cells in culture.

Unlike most other human adult stem cells, MSCs can be obtained in appropriate quantities for clinical applications, making them good candidates for use in cellular therapy and tissue repair. Techniques for isolation and amplification of MSCs in culture have been established and the cells can be maintained and propagated in culture for long periods, without compromising their capacity to form all the above cell types.

Although most therapeutic applications have focused on adult BM-derived MSCs, increasing evidence suggests that MSCs are present within a wide range of tissues. hUCB has been proven to be a valuable source of HSCs, but its therapeutic potential is now shown to extend beyond the haematopoietic component, suggesting regenerative potential in solid organs as well. There is evidence that other stem or progenitor populations, such as MSCs, exist in hUCB, which might be responsible for these effects [10]. Nevertheless, the isolation of HSCs from hUCB has been well-established; attempts to isolate MSC from UCB of full-term deliveries have previously either failed or been characterized by a low yield. Yet the isolation and characterisation of MSCs from hUCB is still being constantly established and evaluated.

Bieback et al. [11], using optimized isolation and culture conditions, were able to isolate, in low-volume hUCB units, cells showing a characteristic mesenchymal morphology and immune phenotype (MSC-like cells), which were similar to control MSCs from adult BM. In addition, osteogenic and chondrogenic differentiation demonstrated a multilineage capacity comparable with BM MSCs. However, in contrast to MSCs, MSC-like cells showed a reduced sensitivity to undergo adipogenic differentiation. Crucial points to isolate MSC-like cells from hUCB were a time from collection to isolation of less than 15 h, a net volume of more than 33 mL, and an MNC count of more than 1×10^8 MNCs. Because MSC-like cells could be isolated at high efficacy from full-term hUCB donations, hUCB plays the role of an additional stem cell source for experimental and potentially clinical purposes.

Several reports have showed that MSC-like cells could be consistently derived from hUCB and have the capabilities for self-renewal and differentiation into various cells types, including bone, neurons, cartilage [12], and skeletal muscle cells [13].

In the study performed by Gang et al., they isolated MSC from hUCB and induced them to differentiate into skeletal muscle cells. During cell expansion, UCB-derived mononuclear cells produced adherent layers of fibroblast-like cells expressing MSC-related antigens such as SH2, SH3, alpha-smooth muscle actin, CD13, CD29, and CD49. Also, when these hUCB-derived MSCs were incubated in myogenic conditions for up to 6 weeks, they expressed myogenic markers in accordance with myogenic differentiation pattern. Similarly, Nunes et al. [14] have found a myogenic potential in the mononuclear fraction of hUCB, which includes mesenchymal subpopulation. These stem cells were capable of acquiring a muscle phenotype when in contact with the *in vivo* environment leading to the expression of muscle proteins. On the other hand, this study raised the possibility that hUCB might ultimately represent a tool in future cell therapy of degenerative disorders of muscle such as the muscular dystrophies.

Zwart et al. [15] investigated the neurogenic potential of full-term hUCB-derived multipotent MSsC in response to neural induction media or coculture with rat neural cells. They used phenotypic and functional changes by immunocytochemistry, RT-PCR, and whole-cell patch-clamp recordings. Results showed that immature MSC expressed both mesodermal and ectodermal markers prior to neural induction. Exposure to retinoic acid, basic fibroblast growth factor, or cyclic adenosine monophosphate (cAMP) did not stimulate neural morphology, whereas exposure to dibutyryl cAMP and 3-isobutyl-1-methylxanthine stimulated a neuron-like morphology; however, it appeared to be cytotoxic. All protocols stimulated the expression of the neural precursor marker nestin, but expression of mature neuronal or glial markers MAP2 and GFAP was not observed. Nevertheless, MSC possessed no ionic currents typical of neurons before or after neural induction protocols. In addition, coculture of hUCB-derived MSCs and rat neural cells induced some MSCs to assume an astrocyte-like morphology and express GFAP protein and mRNA. These data suggested that UCB-derived MSCs were not able to transdifferentiate into mature functioning neurons in response to the above neurogenic protocols; however, coculture with rat neural cells led to a minority of cells to adopt an astrocyte-like phenotype.

UMBILICAL CORD DERIVED STEM CELLS FOR TISSUE THERAPY: CURRENT STATUS AND PERSPECTIVES

Since transplantation of hUCB cells was successfully performed for the first time in 1988 to treat Fanconi's anemia and other blood-related diseases [16–18], there has been extensive experience in the use of hUCB cells to reconstitute the haematopoietic system in humans [19].

UCB-derived stem cells offer multiple advantages over adult stem cells, including their immaturity, which may play a significant role in reduced rejection after transplantation. While research with embryonic stem cells continues to generate considerable controversy, human umbilical stem cells provide an alternative cell source that has been more ethically acceptable and appears to have widespread public support.

Recently it was observed that cells derived from this source have the ability to express markers and morphologies of other cell types within the same mesodermal germ layer such as bone [20], fat [21], smooth muscle [22], and skeletal muscle [14, 23]. Recent studies also documented that subpopulations of hUCB cells can transdifferentiate into hepatic [24] and endothelial cells [25]. More surprisingly, exposing hUCB cells to various experimental conditions showed that their progeny could also reveal properties typical of neuroectoderm-derived cells [26–29], suggesting that they may, therefore, be useful for numerous cell-based therapies requiring either the replacement of individual cell types or substitution/replenishment of missing substances [30].

It has been reported that UCB cells can be used as a source of therapeutically effective substances with the ability to improve functional outcomes after stroke [31–33] and delay both onset of symptoms and death of animals with amyotrophic lateral sclerosis [34, 35]. Furthermore, these cells were shown to be useful in models of traumatic brain injury [36, 37] and spinal cord injury [38, 39]. In the Sanfilippo animal model of mucopolysacharidosis IIIB, histopathological improvements were also found after hUCB administration [40]. We have to consider that all these studies rely more on the trophic effect of the mononuclear hUCB cells than on the actual cellular replacement. Only a few transplantation studies have addressed the possibility that the mononuclear fraction may contain a small number of non-differentiated cells that may, under specific circumstances, such as stroke or placement into a favorable neurogenic environment, give rise to neural-like cells [15].

Similarly, knowing that overlapping genetic programs for haematopoiesis and neuropoiesis exist, Chen et al. [41] explored the possibility that stem/progenitor cells in the mononuclear hUCB fraction can produce progeny that expresses neural antigens when grown as adherent or nonadherent cultures. They used culture conditions commonly used for neural tissue and examined through immunocytochemical stainings and Western blot analysis the typical antigens for neural lineages. They found that there were a significant number of stem cell antigens expressed on cells in the floating population as well as neural antigens. Fully 90% of the adherent hUCB cells expressed the common leukocyte antigen CD45. In addition, there were few cells in this fraction that transiently expressed stem/precursor antigens (CD133 and CD117). More cells that expressed early stem cell antigens were found in the floating population of hUCB cells and not the adherent cell subpopulation. This was consistent with the hypothesis that adult stem cells exist in a quiescent state in the blood until an injury or disease state triggers the stem cells to proliferate or differentiate through the production of specific factors [42].

Another stem cell antigen, CD15+ or SSEA-1, was found in the hUCB cell cultures. This antigen is expressed on embryonic stem cells, primitive HSCs, and also within the developing human brain [43, 44]. The expression of CD15 antigen in both haematopoietic and neural lineages suggests that there may be a set of overlapping genes controlling genesis in both lineages. They also examined the expression of neural proteins in hUCB cultures. All of the neural antigens chosen such as nestin, vimentin, and neurotrophin receptors were observed within the hUCB subpopulations. The expression of neurotrophin receptors in hUCB cells certainly suggests that the hUCB cells could be driven to a neural fate. By binding to their appropriate receptors, neurotrophins play important roles in differentiation and survival of neural cells during CNS development.

The coexpression of two different lineage markers could be indicative of cell fusion with the host cells *in vivo*, but alternative explanations are being elaborated to support these interesting findings. Lu et al. [45], for example, suggested that transdifferentiation of MSC from BM into neural cells is more likely a function of cellular toxicity resulting in cytoskeleton changes and not a change in the cellular differentiation pathway. Clearly the best demonstration that the cells have attained a neural phenotype will be through functional assays showing not only a neural morphology and protein expression but also synapse formation, neural electrophysiological activity, and neurotransmitter release. Further studies are necessary to better understand how to best use these cells in the development of cell therapies to restore brain function.

Recently, UCB has also been shown as a rich source of endothelial colony forming cells (ECFC), which are endothelial progenitor cells (EPCs) used for angiogenic therapies and as biomarkers of cardiovascular disease [46]. The authors first compared the amount of haematopoietic progenitor cells in premature UCB and term controls, showing that these cells are increased in premature UCB.

They found that 33–36 week gestational age UCB yielded predominantly ECFC colonies at equivalent numbers to term infants. UCB from 24–28 weeks gestational age infants had significantly fewer ECFC, and surprisingly, they yielded predominantly MSC colonies, capable of differentiating into adipocytes, chondrocytes, and osteocytes. Moreover, MSC were rarely identified in 37–40 weeks UCB. These studies demonstrate that circulating MSC and ECFC appear at different gestational age in the hUCB, and that premature UCB may be a novel source of MSC for therapeutic use in human diseases.

Recent advances in stem cell research have provided an exciting and potentially new approach to treat type 1 diabetes and many other clinical disorders. As an evidence of the potential of stem cells to treat diabetes, BM transplantation has been shown to prevent autoimmune insulitis and diabetes in NOD mice [47]. In addition, recent research in immunodeficient mice with chemically induced pancreatic damage has shown that BM-derived stem cells may have the capacity to initiate ?cell regeneration [48]. However, the mechanism involved in pancreatic regeneration may be somewhat contrary to the classic concept of direct stem cell differentiation into cells of the desired target tissue. Based on this line of evidence, Hess et al. determined that transplanted BM-derived stem cells travel preferentially to damaged organs and initiate tissue regeneration via the organ's hold stem cell population. As such, it may be that the role of stem cells in ameliorating diabetes is to protect remaining ? cells from further destruction or stimulate remaining tissue to regenerate rather than participating directly in production of *de novo* stem cell derived islets.

For human application, autologous stem cell transplantation is the most common and potentially safest form of stem cell transplantation. Autologous BM transplants have been used successfully for patients undergoing high-dose chaemotherapy, and for the treatment of many forms of cancer [49].

Voltarelli et al. [50] recently published the first efforts to determine the safety and efficacy of autologous transplantation in new onset type 1 diabetes patients. In this study, 14 of 15 subjects were able to discontinue insulin injections for at least 1 month post-therapy with the majority being able to remain off insulin for more than 6 months. Fortunately, no mortality was observed in that study. Nevertheless, morbidity was still an issue.

As such, researchers remain cautiously optimistic that stem cell therapies can be safely modified and applied to the treatment of diabetes.

More recently, a group has focused on the potential use of autologous UCB transfusion in children with type 1 diabetes. The first possible explanation for use of cord blood relies on the fact that stem cell therapies stirs controversy, while use of UCB is attractive because it avoids much of the debate surrounding this delicate issue. In addition, UCB offers additional major advantages over other ethically acceptable stem cell sources. As mentioned earlier, when compared to BM and peripherally mobilized stem cells, UCB is preferable because of its immediate availability, absence of risk to the donor (and, if autologous, to the recipient as well), low risk of GVHD, and increased capacity for expansion.

As further evidence of the potential use of UCB in type 1 diabetes therapies, UCB stem cells have successfully been directed to *in vitro* differentiate into insulin and c-peptide–producing cells [51].

Because cord blood contains a large population of immature unprimed functional regulatory T-lymphocytes, this may be the most important reason for exploring therapeutic applications of UCB in type 1 diabetes. This population of regulatory T-cells may function to decrease the inflammatory cytokine response of effector T-cells, which play a key role in the cellular-mediated autoimmune process [52].

Practically, the lack of disease-reversal trials for children with type 1 diabetes younger than the age of 8 years (due to safety concerns with the immunosuppressive regimens being tested) also makes use of UCB particularly appealing. As rates of UCB storage continue to increase exponentially, the number of potential subjects for autologous UCB-based clinical trials will continue to grow.

Based on available data and the agreement that infusion of minimally manipulated autologous cord blood cells was likely to be safe, Haller et al. [53] began to design and implement an unblinded observational pilot study to determine if autologous UCB infusion could ameliorate the type 1 diabetes autoimmune process and offer to the patients preservation of remaining endogenous insulin production. They hypothesized that autologous UCB transfusion may help mitigate the autoimmune process by different mechanisms. First, UCB stem cells may migrate to the damaged pancreas, where they will differentiate into insulin-producing β cells. Alternatively, UCB stem cells may promote proliferation of new islets from remaining viable tissue and/or UCB regulatory T-cells may facilitate suppression of effector T-cells or allow restoration of tolerance by their inhibitory effects on multiple cell types [54].

In terms of providing evidence of a mechanism for the action of UCB cells, researchers have focused on the frequency and function of regulatory T-cell in the peripheral blood of UCB-infused subjects, since while poor access to the pancreas limits the ability to dissect and study the potential mechanisms. Perhaps the most important finding at this time is that they had no significant adverse events associated to the UCB infusion.

Although the potential of UCB to participate in the future of therapies for type 1 diabetes, it is consensus that multiple therapeutic tentative will need to be explored and combined in order to achieve a safety protocol and permanently reversing type 1 diabetes.

Regarding stem cell therapy to muscular disorders, results are not so very optimistic, sometimes controversial, and it remains unclear if these cells are able to differentiate into muscle cells. Koponen et al. [55] have showed that rather than directly participating in angiogenesis or skeletal myogenesis, UCB-derived progenitor cells indirectly enhance the regenerative capacity of skeletal muscle after acute ischemic injury. Also, VEGF-D-transfected UCB cells did not improve the therapeutic effect in ischemic muscles.

Nishiyama et al. [56] tested the cardiomyogenic potential of the UCB-derived MSCs. For this purpose, the green fluorescent protein (GFP)-labelled UCB were cocultured with foetal murine cardiomyocytes. On day 5 of cocultivation, approximately half of the UCB mesenchymal cells contracted rhythmically and synchronously, suggesting the presence of electrical communication between them. Their results showed that approximately half of the UCB cells were successfully transdifferentiated into cardiomyocytes *in vitro*; however, the authors disclose the potential conflicts of interest about the reported data.

Nunes et al. [14] have showed that stem cells from hUCB did not differentiate into myotubes or express dystrophin when cultured in muscle-conditioned medium or with human muscle cells. However, delivery of GFP-transduced mononucleated cells from hUCB, which comprises both haematopoietic and mesenchymal populations, into quadriceps muscle of *mdx* (mouse dystrophy X-chromosome linked) mice resulted in the expression of human myogenic markers. After recovery of these cells from *mdx* muscle and *in vitro* cultivation, they were able to fuse and form GFP-positive myotubes that expressed dystrophin. These results indicate that chemical factors and cell-to-cell contact provided by *in vitro* conditions were not enough to trigger the differentiation of stem cells into muscle cells. Nevertheless, it was showed that the hUCB-derived stem cells were capable of acquiring a muscle phenotype after exposure to an *in vivo* muscle environment, which was required to activate the differentiation program.

In order to assess the therapeutic effect of UCB stem cells transplantation in muscular regeneration and dystrophin expression in Duchenne muscular dystrophy (DMD) patients, Zhang et al. [57] performed UCB transplantation by intravenous injection in a 12-year-old DMD boy, confirmed by gene analysis. The biochemistry profile including serum creatine kinase (CK), reconstruction of blood making, deletion exon of *DMD* gene, regenerating muscles and dystrophin protein expression, and the locomotive function of the DMD boy were regularly evaluated. Results showed that at different times after transplantation, peripheral blood DNA showed that his genotype was completely the same as the donor's. At 60 days, *DMD* gene analysis by PCR showed that the defected *DMD* gene (containing a deletion in the exon 19) had been corrected by the umbilical cord stem cells transplantation. At 75 days, the biopsy of calf muscle showed there were myoblast cells and muscular tubes growing. The dystrophin expression was weak, but a few of them were strong. DNA analysis showed that the donor's gene DNA accounted for 1–13%. The serum CK of the boy declined from 5735 U/L to 274 U/L. Finally, at 100 days, physical examination revealed improvement in his arms and legs. In summary, they conclude that therapy of DMD with allogeneic UCB HSC transplantation may reset up the blood-making function, decrease the serum CK level, restore the dystrophin in muscles, and improve the locomotive function of the DMD boy. These data suggest that the allogeneic UCB HSC transplantation may benefit the DMD boys.

Similarly, UCB have been tested in limb girdle muscular dystrophy type 2B form (LGMD-2B) and Miyoshi myopathy (MM). These muscle disorders are both caused by mutations in the dysferlin (*dysf*) gene. In one study conducted by Kong et al. [58], they used dysferlin-deficient *sjl* mice as a mouse model. A single-blind study evaluated the therapeutic potential of hUCB as a source of myogenic progenitor stem cells. Cells were administrated intravenously to each immunosuppressed animal, which were euthanized at different time points (1–12 weeks) after transplantation. Immunohistochemical analysis revealed that a small number of human cells from the whole hUCB and LIN(-)CD34(+/-)-enriched hUCB subgroups engraft in the recipient muscle to express both dysferlin and human-specific dystrophin at 12 weeks after transplantation. So, they conclude that myogenic progenitor cells are present in the hUCB that they can disseminate into muscle after intravenous administration and they are capable of myogenic differentiation in host muscle.

All these presented data demonstrate that UCB stem cells possess a potential of myogenic differentiation and also imply that these cells could be a suitable source for skeletal muscle repair and a useful tool of muscle-related tissue engineering.

In summary, this chapter showed that stem cells from UCB have been used world-wide to treat malignant or nonmalignant blood diseases. More recently, they have been shown to differentiate into many cell types in response to distinct stimuli. Clinical applications of UCB stem cells have potential in the regeneration of blood cells, skin, bone, cartilage, and muscle, and may have potential in degenerative diseases. However, the potential of these stem cells for generating human tissues is the subject of ongoing public debate, and, of course, stem cells must be used in standardised and controlled conditions in order to guarantee the best safety conditions for the patients.

REFERENCES

1. Bojaniæ I., B.G. Cepuliæ (2006). 'Umbilical cord blood as a source of stem cells'. *Acta. Med. Croatica.* **60**(3): 215–225.
2. Moise K. (2005). 'Umbilical cord stem cells'. *Obstet. Gynecol.* **106**(6): 1393–1407.
3. Laughlin M.J., M. Eapen, P. Rubinstein, J.E. Wagner, M.J. Zhang, R.E. Champlin, C. Stevens, J.N. Barker, R.P. Gale, H.M. Lazarus, D.I. Marks, J.J. van Rood, A. Scaradavou, M.M. Horowitz (2004). 'Outcomes after transplantation of cord blood or bone marrow from unrelated donors in adults with leukemia'. *N. Engl. J. Med.* **351**(22): 2265–2275.
4. Jaing T.H. (2007). 'Umbilical cord blood transplantation: Application in pediatric patients'. *Acta. Paediatr. Taiwan.* **48**(3): 107–111.
5. Broxmeyer H.E., G. Hangoc, S. Cooper et al. (1992). 'Growth characteristics and expansion of human umbilical cord blood and estimation of its potential for transplantation in adults'. *Proc. Natl. Acad. Sci. USA.* **89**: 4109–4113.
6. Ruggieri L., S. Heimfeld, H.E. Broxmeyer (1994). 'Cytokine-dependent *ex vivo* expansion of early subsets of CD34+ cord blood myeloid progenitors is enhanced by cord blood plasma, but expansion of the more mature subsets of progenitors is favored'. *Blood. Cells.* **20**(2–3):436–454.
7. Xiao M., H.E. Broxmeyer, M. Horie, S. Grigsby, L. Lu (1994). 'Extensive proliferative capacity of single isolated CD34+ human cord blood cells in suspension culture'. *Blood. Cells.* **20**: 455–467.
8. Sugarman J., V. Kaalund, E. Kodish, M.F. Marshall, E.G. Reisner, B.S. Wilfond, P.R. Wolpe (1997). 'Ethical issues in umbilical cord blood banking'. *Working Group on Ethical Issues in Umbilical Cord Blood Banking. JAMA.* **278**(11): 938–943.
9. Madlambayan G., I. Rogers (2006). 'Umbilical cord-derived stem cells for tissue therapy: Current and future uses'. *Regen. Med.* **1**(6): 777–787.
10. Bieback K., H. Klüter (2007). 'Mesenchymal stromal cells from umbilical cord blood'. *Curr. Stem. Cell. Res. Ther.* **2**(4): 310–323.
11. Bieback K., S. Kern, H. Klüter, H. Eichler (2004). 'Critical parameters for the isolation of mesenchymal stem cells from umbilical cord blood'. *Stem Cells.* **22**: 625–634.
12. Park K.S., Y.S. Lee, K.S. Kang (2006). '*In vitro* neuronal and osteogenic differentiation of mesenchymal stem cells from human umbilical cord blood'. *J. Vet. Sci.* **7**(4): 343–348.
13. Gang E.J., J.A. Jeong, S.H. Hong, S.H. Hwang, S.W. Kim, I.H. Yang, C. Ahn, H. Han, H. Kim (2004). 'Skeletal myogenic differentiation of mesenchymal stem cells isolated from human umbilical cord blood'. *Stem. Cells.* **22**(4): 617–624.

14. Nunes V.A., N. Cavaçana, M. Canovas, B.E. Strauss, M. Zatz. (2007). 'Stem cells from umbilical cord blood differentiate into myotubes and express dystrophin *in vitro* only after exposure to *in vivo* muscle environment'. *Biol. Cell.* **99**(4): 185–196.
15. Zwart I., A.J. Hill, J. Girdlestone, M.F. Manca, R. Navarrete, C. Navarrete, L.S. Jen (2008). 'Analysis of neural potential of human umbilical cord blood-derived multipotent mesenchymal stem cells in response to a range of neurogenic stimuli'. *J. Neurosci. Res.* **86**(9): 1902–1915.
16. de Medeiros C.R., L.M. Silva, R. Pasquini (2001). 'Unrelated cord blood transplantation in a Fanconi anemia patient using fludarabine-based conditioning'. *Bone Marrow Transplant.* **28**: 110–112.
17. Ooi J., T. Iseki, H. Nagayama et al. (2001). 'Unrelated cord blood transplantation for adult patients with myelodysplastic syndrome-related secondary acute myeloid leukaemia'. *Br. J. Haematol.* **114**: 834–836.
18. Tezuka K., H. Nakayama, K. Honda et al. (2002). 'Treatment of a child with myeloid/ NK cell precursor acute leukemia with L-asparaginase and unrelated cord blood transplantation'. *Int. J. Haematol.* **75**: 201–206.
19. Tse W., M.J. Laughlin (2005). 'Umbilical cord blood transplantation: A new alternative option'. *Haematology Am. Soc. Haematol. Educ. Program.* 377–383.
20. Rosada C., J. Justesen, D. Melsvik et al. (2003). 'The human umbilical cord blood: A potential source for osteoblast progenitor cells'. *Calcif. Tissue Int.* **72**: 135–142.
21. Goodwin H.S., A.R. Bicknese, S.N. Chien et al. (2001). 'Multilineage differentiation activity by cells isolated from umbilical cord blood: Expression of bone, fat, and neural markers'. *Biol. Blood Marrow. Transplant.* **7**: 581–588.
22. Simper D., P.G. Stalboerger, C.J. Panetta et al. (2002). 'Smooth muscle progenitor cells in human blood'. *Circulation.* **106**: 1199–1204.
23. Pesce M., A. Orlandi, M.G. Iachininoto et al. (2003). 'Myoendothelial differentiation of human umbilical cord blood–derived stem cells in ischemic limb tissues'. *Circ. Res.* **93**: 51–62.
24. Newsome P.N., I. Johannessen, S. Boyle, E. Dalakas, K.A. McAulay, K. Samuel, F. Rae, L. Forrester, M.L. Turner, P.C. Hayes, D.J. Harrison, W.A. Bickmore, J.N. Plevris (2003). 'Human cord blood-derived cells can differentiate into hepatocytes in the mouse liver with no evidence of cellular fusion'. *Gastroenterology.* **124**(7): 1891–900.
25. Igreja C., R. Fragoso, F. Caiado, N. Clode, A. Henriques, L. Camargo, E.M. Reis, S. Dias (2008). 'Detailed molecular characterisation of cord blood-derived endothelial progenitors'. *Exp. Haematol.* **36**(2): 193–203.
26. Sanchez-Ramos J.R. (2002). 'Neural cells derived from adult bone marrow and umbilical cord blood'. *J. Neurosci. Res.* **69**: 880–893.
27. Bicknese A.R., H.S. Goodwin, C.O. Quinn, et al. (2002). 'Human umbilical cord blood cells can be induced to express markers for neurons and glia'. *Cell Transplant.* **11**: 261–264.
28. Zigova T., S. Song, A.E. Willing, et al. (2002). 'Human umbilical cord blood cells express neural antigens after transplantation into the developing rat brain'. *Cell Transplant.* **11**: 265–274.
29. Walczak P., N. Chen, J.E. Hudson, et al. (2004). 'Do haematopoietic cells exposed to a neurogenic environment mimic properties of endogenous neural precursors'? *J. Neurosci. Res.* **76**: 244–254.
30. van de Ven C., D. Collins, M.B. Bradley, E. Morris, M.S. Cairo (2007). 'The potential of umbilical cord blood multipotent stem cells for nonhaematopoietic tissue and cell regeneration'. *Exp. Haematol.* **35**(12): 1753–1765.

31. Chen J., P.R. Sanberg, Y. Li, L. Wang, M. Lu, A.E. Willing, J. Sanchez-Ramos, M. Chopp (2001). 'Intravenous administration of human umbilical cord blood reduces behavioural deficits after stroke in rats'. *Stroke.* **32**(11): 2682–2688.

32. Willing A.E., J. Lixian, M. Milliken, et al. (2003). 'Intravenous versus intrastriatal cord blood administration in a rodent model of stroke'. *J. Neurosci. Res.* **73**: 296–307.

33. Willing A.E., M. Vendrame, J. Mallery, et al. (2003). 'Mobilized peripheral blood cells administered intravenously produce functional recovery in stroke'. *Cell Transplant.* **12**: 449–454.

34. Chen R., N. Ende. (2000). 'The potential for the use of mononuclear cells from human umbilical cord blood in the treatment of amyotrophic lateral sclerosis in SOD1 mice'. *J. Med.* **31**: 21–30.

35. Garbuzova-Davis S., A.E. Willing, T. Zigova, et al. (2003). 'Intravenous administration of human umbilical cord blood cells in a mouse model of amyotrophic lateral sclerosis: distribution, migration, and differentiation'. *J. Haematother. Stem Cell Res.* **12**: 255–270.

36. D. Lu, P.R. Sanberg, A. Mahmood, et al. (2002), 'Intravenous administration of human umbilical cord blood reduces neurological deficit in the rat after traumatic brain injury'. *Cell Transplant.* **11**: 275–281.

37. Cairns K., S.P. Finklestein. (2003). 'Growth factors and stem cells as treatments for stroke recovery. Phys. Med. Rehabil'. *Clin. N. Am.* **14**: S135–S142.

38. Saporta S., J.J. Kim, A.E. Willing, et al. (2003). 'Human umbilical cord blood stem cells infusion in spinal cord injury: engraftment and beneficial influence on behaviour'. *J. Haematother. Stem Cell Res.* **12**: 271–278.

39. Zhao Z.M., H.J. Li, H.Y. Liu (2004). 'Intraspinal transplantation of CD34+ human umbilical cord blood cells after spinal cord hemisection injury improves functional recovery in adult rats'. *Cell Transplant.* **13**: 113–122.

40. Sanberg R.R., A.E. Willing, L.A. Austin et al. (2003). 'Cerebral intraventricular transplantation of human umbilical cord blood cells as a potential treatment of Sanfilippo syndrome type B'. *Exp. Neurol.* **181**:104.

41. Chen J., P.R. Sanberg, Y. Li, et al. (2001). 'Intravenous administration of human umbilical cord blood reduces behavioural deficits after stroke in rats'. *Stroke.* **32**: 2682–2688.

42. Fortunel N., P. Batard, A. Hatzfeld, et al. (1998). 'High proliferative potential-quiescent cells: a working model to study primitive quiescent haematopoietic cells'. *J. Cell Sci.* **111**: 1867–1875.

43. Mai J.K., S. Krajewski, G. Reifenberger, et al. (1999). 'Spatiotemporal expression gradients of the carbohydrate antigen (CD15) (Lewis X) during development of the human basal ganglia'. *Neuroscience.* **88**: 847–858.

44. Mai J.K., R. Winking, K.W. Ashwell (1999). 'Transient CD15 expression reflects stages of differentiation and maturation in the human subcortical central auditory pathway'. *J. Comp. Neurol.* **404**: 197–211.

45. Lu P., A. Blesch, M.H. Tuszynski (2004). 'Induction of bone marrow stromal cells to neurons: differentiation, transdifferentiation, or artefact'? *J. Neurosci. Res.* **77**: 174–191.

46. Javed M.J., L.E. Mead, D. Prater, W.K. Bessler, D. Foster, J. Case, W.S. Goebel, M.C. Yoder, L.S. Haneline, D.A. Ingram (2008). 'Endothelial colony forming cells and mesenchymal stem cells are enriched at different gestational ages in human umbilical cord blood'. *Pediatr. Res.* **64**(1): 68–73.

47. Beilhack G.F., Y.C. Scheffold, I.L. Weissman, et al. (2003). 'Purified allogeneic haematopoietic stem cell transplantation blocks diabetes pathogenesis in NOD mice'. *Diabetes.* **52**: 59–68.

48. Hess D., L. Li, M. Martin, S. Sakano, D. Hill, B. Strutt, S. Thyssen, D.A. Grey, M. Bhatia (2003). 'Bone marrow-derived stem cells initiate pancreatic regeneration'. *Nat. Biotechnol.* 21(7): 763–770.
49. Lennard A.L., G.H. Jackson (2000). 'Stem cell transplantation'. *BMJ.* **321**: 433–437.
50. Voltarelli J.C., C.E.B. Couri, A.B.P.L. Stracieri, et al. (2007). 'Autologous nonmyeloablative haematopoietic stem cell transplantation in newly diagnosed type 1 diabetes mellitus'. *JAMA.* **297**(14): 1568–1576.
51. Denner L., Y. Bodenburg, J.G. Zhao, et al. (2007). 'Directed engineering of umbilical cord blood stem cells to produce C-peptide and insulin'. *Cell Prolif.* **40**: 367–380.
52. Han P., G. Hodge, C. Story, X. Xu (1995). 'Phenotypic analysis of functional T-lymphocyte subtypes and natural killer cells in human cord blood: relevance to umbilical cord blood transplantation'. *Br. J. Haematol.* **89**: 733–740.
53. Haller M.J., H.L. Viener, C. Wasserfall, T. Brusko, M.A. Atkinson, D.A. Schatz (2008). 'Autologous umbilical cord blood infusion for type 1 diabetes'. *Exp. Haematol.* **36**(6): 710–715.
54. Brusko T., M. Atkinson (2007). 'Treg in type 1 diabetes'. *Cell Biochem. Biophys.* **48**: 165–175.
55. Koponen J.K., T. Kekarainen, E. Heinonen, A. Laitinen, J. Nystedt, J. Laine, S. Ylä-Herttuala (2007). 'Umbilical cord blood–derived progenitor cells enhance muscle regeneration in mouse hindlimb ischemia model'. *Mol. Ther.* **15**(12): 2172–2177.
56. Nishiyama N., S. Miyoshi, N. Hida, T. Uyama, K. Okamoto, Y. Ikegami, K. Miyado, K. Segawa, M. Terai, M. Sakamoto, S. Ogawa, A. Umezawa (2007). 'The significant cardiomyogenic potential of human umbilical cord blood-derived mesenchymal stem cells *in vitro*'. *Stem Cells.* **25**(8): 2017–2024.
57. Zhang C., H.Y. Feng, S.L. Huang, J.P. Fang, L.L. Xiao, X.L. Yao, C. Chen, X. Ye, Y. Zeng, X.L. Lu, J.M. Wen, W.X. Zhang, Z. Li, S.W. Feng, H.G. Xu, K. Huang, D.H. Zhou, W. Chen, Y.M. Xie, J. Xi, M. Zhang, Y. Li, Y. Liu (2005). 'Therapy of Duchenne muscular dystrophy with umbilical cord blood stem cell transplantation'. *Zhonghua Yi Xue Yi Chuan Xue Za Zhi.* **22**(4): 399–405.
58. Kong K.Y., J. Ren, M. Kraus, S.P. Finklestein, R.H. Brown, Jr. (2004). 'Human umbilical cord blood cells differentiate into muscle in sjl muscular dystrophy mice'. *Stem Cells.* **22**(6): 981–993.

26

CELL THERAPY AND TISSUE ENGINEERING STRATEGIES IN REGENERATIVE MEDICINE

Anthony Atala[*]

The W. Boyce Professor and Chair, Department of Urology, Director, Wake Forest Institute for Regenerative Medicine, Wake Forest University School of Medicine, Winston-Salem, NC 27157, USA

INTRODUCTION

Patients suffering from diseased and injured organs are often treated with transplants; however, there is a severe shortage of donor organs that is worsening yearly. As modern medicine increases the human lifespan, the aging population grows, and the need for organs grows with it. Scientists in the field of regenerative medicine and tissue engineering apply principles of cell transplantation, material science and bioengineering to construct biological substitutes that can prolong life by eliminating the need for transplant surgery.

Tissue engineering, one of the major components of regenerative medicine, follows principles of cell transplantation, materials science and engineering toward the development of biological substitutes that can restore and maintain normal organ function. Tissue engineering strategies generally fall into two categories: the use of acellular matrices, which depend on the body's natural ability to use its own cells to regenerate, for proper orientation and direction of new tissue growth, and the use of matrices with cells. Acellular tissue matrices are usually prepared by manufacturing artificial scaffolds, or by removing cellular components from tissues via mechanical and chemical manipulation to produce collagen-rich matrices [1–4]. These matrices tend to slowly degrade on implantation and are generally replaced by the extracellular matrix (ECM) proteins that are secreted by the ingrowing cells. Cells can also be used for therapy via injection, either with carriers such as hydrogels, or alone.

---

[*] *aatala@wfubmc.edu*

When cells are used for tissue engineering, a small piece of donor tissue is dissociated into individual cells. These cells are either implanted directly into the host, or are expanded in culture, attached to a support matrix, and then reimplanted into the host after expansion. The source of donor tissue can be heterologous (such as bovine), allogeneic (same species, different individual) or autologous. Ideally, both structural and functional tissue replacement occur with minimal complications. The preferred cells to use are autologous cells, where a biopsy of tissue is obtained from the host, cells are dissociated and expanded in culture and the expanded cells are implanted into the same host [2, 5–12]. The use of autologous cells, although it may cause an inflammatory response, avoids rejection and thus the deleterious side effects of immunosuppressive medications can be avoided.

Most current strategies for tissue engineering depend upon a sample of autologous cells from the diseased organ of the host. However, for many patients with extensive end-stage organ failure, a tissue biopsy may not yield enough normal cells for expansion and transplantation. In other instances, primary autologous human cells cannot be expanded from a particular organ, such as the pancreas. In these situations, stem cells are envisioned as being an alternative source of cells from which the desired tissue can be derived. Stem cells can be derived from discarded human embryos (human embryonic stem cells (hESCs)), from foetal tissue or from adult sources (bone marrow, fat, skin). However, there are ethical issues involved in the use of ESCs, and many applications are currently banned in the United States. Despite this, the field of stem cell biology is advancing rapidly and cutting-edge techniques such as therapeutic cloning and somatic cell reprogramming circumvent some of the ethical questions and offer a potentially limitless source of these cells for tissue engineering applications.

This chapter reviews recent advances that have occurred in regenerative medicine and describes applications of these new technologies that offer the promise of a longer life.

THE BASIC COMPONENTS OF TISSUE ENGINEERING

Cells

Native cells

One of the limitations of applying cell-based regenerative medicine techniques to organ replacement has been the inherent difficulty of growing specific cell types in large quantities. Even when some organs, such as the liver, have a high regenerative capacity *in vivo*, cell growth and expansion *in vitro* can be difficult. By studying the privileged sites for committed precursor cells in specific organs, as well as exploring conditions that promote differentiation and/or self-renewal of these cells, it may be possible to overcome obstacles that limit cell expansion *in vitro*. One example is the urothelial cell. Urothelial cells could be grown in the laboratory setting in the past, but only with limited success. Several protocols were developed over the past two decades that identify the undifferentiated cells in a mixed culture, and keep them undifferentiated during their growth phase [11, 13–16]. Using these methods of cell culture, it is now possible to expand a urothelial culture that initially covered a surface area of 1 cm^2 to one covering a surface area of 4202 m^2 (the equivalent of one football field) within 8 weeks [11]. These studies indicated that it should be possible to collect autologous bladder cells from human patients, expand them in culture and return them to the donor in sufficient quantities for reconstructive purposes [11, 14–19]. Major advances have been achieved within the past decade regarding the expansion of a variety of primary human cells, with specific techniques that make the use of autologous cells possible for clinical application.

Embryonic stem cells

According to data from the Centre for Disease Control, as many as 1 million Americans die every year from diseases that, in the future, may be treatable with tissues derived from stem cells [20]. Diseases that might benefit from ESC-based therapies included diabetes, heart disease, cerebrovascular disease, liver and renal failure, spinal cord injuries and Parkinson's disease.

In 1981, pluripotent cells were found in the inner cell mass of the human embryo, and the term 'human embryonic stem cell' was coined [21]. These cells are able to differentiate into all cells of the human body, excluding placental cells (only cells from the morula are totipotent; that is, able to develop into all cells of the human body). These cells have great therapeutic potential; however, their use is limited by several factors, both biological and ethical.

The political controversy surrounding stem cells began in 1998 with the creation of hESCs derived from discarded, non-transferred human embryos. hESCs were isolated from the inner cell mass of a blastocyst (an embryo 5 days post-fertilisation) using an immunosurgical technique. Using this technique, the blastocyst was incubated with antibodies specific to the trophectoderm. Complement proteins then resulted in lysis of the trophectoderm so that the only surviving cells were the inner cell mass [22, 23]. Given that some cells cannot be expanded *ex vivo*, ESCs could be the ideal resource for tissue engineering because of their fundamental properties: the ability to self-renew indefinitely and the ability to differentiate into cells from all three embryonic germ layers. These cells have demonstrated longevity in culture and can maintain their undifferentiated state for at least 80 passages when grown using current published protocols [24, 25].

hESCs have been shown to differentiate into cells from all three embryonic germ layers *in vitro*. Skin and neurons have been formed, indicating ectodermal differentiation [26–29]. Blood, cardiac cells, cartilage, endothelial cells and muscle have been formed, indicating mesodermal differentiation [30–32]. Pancreatic cells have been formed, indicating endodermal differentiation [33]. In addition, as further evidence of their pluripotency, ESCs can form embryoid bodies, which are cell aggregations that contain all three embryonic germ layers while in culture, and can form teratomas *in vivo* [34].

However, in addition to the ethical dilemma surrounding the use of ESCs, their clinical application is limited because they represent an allogenic resource and thus have the potential to evoke an immune response. New stem cell technologies (such as somatic cell nuclear transfer and reprogramming) promise to overcome this limitation.

Therapeutic cloning (somatic cell nuclear transfer)

Somatic cell nuclear transfer (SCNT) entails the removal of an oocyte nucleus in culture, followed by its replacement with a nucleus derived from a somatic cell obtained from a patient. Activation with chemicals or electricity stimulates cell division up to the blastocyst stage. At this time the inner cell mass is isolated and cultured, resulting in ESCs that are genetically identical to the patient. It has been shown that nuclear transferred ESCs derived from fibroblasts, lymphocytes and olfactory neurons are pluripotent and generate live pups after tetraploid blastocyst complementation, showing the same developmental potential as fertilized blastocysts [35–38]. The resulting ESCs are perfectly matched to the patient's immune system and no immunosuppressants would be required to prevent rejection.

Although ESCs derived from SCNT contain the nuclear genome of the donor cells, mitochondrial DNA (mtDNA) contained in the oocyte could lead to immunogenicity after transplantation. To assess the histocompatibility of nuclear transfer-generated tissue, Lanza et al. microinjected the nucleus of a bovine skin fibroblast into an enucleated oocyte [39]. Although the blastocyst was implanted (reproductive cloning), the purpose was to generate renal, cardiac and skeletal muscle cells, which were then harvested, expanded *in vitro* and seeded onto biodegradable scaffolds. These scaffolds were then implanted into the donor from whom cells were cloned to determine if cells were histocompatible. Analysis revealed that cloned renal cells showed no evidence of T-cell response, suggesting that rejection will not necessarily occur in the presence of oocyte-derived mtDNA. This finding represents a step forward in overcoming the histocompatibility problem of stem cell therapy.

Although promising, SCNT has certain limitations that require further improvement before its clinical application, in addition to ethical considerations regarding the potential of the resulting embryos to develop into cloned embryos if implanted into a uterus. Many animal studies have shown that blastocysts generated from SCNT can give rise to a liveborn infant that is a clone of the donor when implanted into a uterus. In 1997, for example, a sheep named Dolly was derived from an adult somatic cell using nuclear transfer [40]. This is known as reproductive cloning, which is banned in most countries for human applications. In contrast, therapeutic cloning is used to generate only ESC lines whose genetic material is identical to that of their source, and the generated blastocysts are never implanted into a uterus. In this case blastocysts are only allowed to grow in culture until they reach a 100 cell-stage from which ESCs can be obtained.

In addition, this technique has not been shown to work in humans. The initial failures and fraudulent reports of nuclear transfer in humans have reduced enthusiasm for human applications [41–43]. However, it was recently reported that non-human primate ESC lines were generated by SCNT of nuclei from adult skin fibroblasts [44, 45]. A total of 304 oocytes yielded 35 blastocysts, from which two ESC lines were derived. Both lines demonstrated typical ESC morphology. They also demonstrated self-renewal and expressed the stem cell markers *OCT4, SSEA4, LEFTYA, TDGF, TRA1-60* and *TRA1-80*. To test their differentiation potential, cells were exposed to cardiac and neural differentiation conditions and these experiments resulted in cells that expressed markers of the specified lineages. When injected into SCID mice, the SCNT-derived non-human primate ESCs induced teratomas that contained differentiated cell types from all three embryonic germ layers.

Before SCNT-derived ESCs can be used as clinical therapy, careful assessment of quality of the lines must be determined. For example, some cell lines generated by SCNT have contained chromosomal translocations and it is not known whether these abnormalities originated from aneuploid embryos or if they occurred during ESC isolation and culture. In addition, the low efficiency of SNCT (0.7%) and the inadequate supply of human oocytes further hinder the therapeutic potential of this technique. Although promising, SCNT has limitations that require further improvement before clinical application, including ethical considerations regarding the potential of the generated blastocysts to develop into cloned embryos if implanted into a uterus. The destruction of embryos is not an ethically acceptable means to generate pluripotent stem cells. On the other hand, these studies renew the hope that ESC lines could one day be generated from human cells to produce patient-specific stem cells with the potential to cure many human diseases that are currently untreatable (see Fig. 26.1).

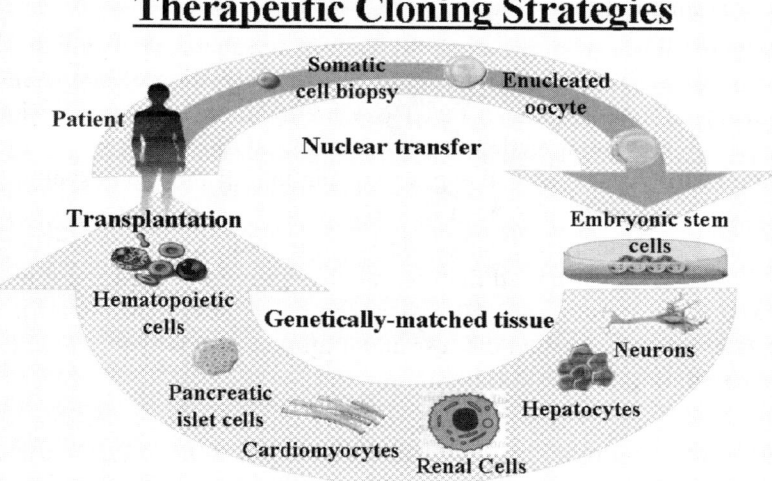

▶ **Fig. 26.1** *Strategies for therapeutic cloning in regenerative medicine.*

Reprogrammed somatic cells

Recently, exciting reports of the successful transformation of adult cells into pluripotent stem cells through a type of genetic 'reprogramming' have been published. Reprogramming is a technique that involves de-differentiation of adult somatic cells to produce patient-specific pluripotent stem cells, eliminating the need to create embryos. Cells generated by reprogramming would be genetically identical to the somatic cells (and thus, the patient who donated these cells) and would not be rejected. Yamanaka was the first to discover that mouse embryonic fibroblasts (MEFs) and adult mouse fibroblasts could be reprogrammed into an 'induced pluripotent state' (iPS) [46]. This group used MEFs engineered to express a neomycin resistance gene from the *Fbx15* locus, a gene expressed only in ESCs. They examined 24 genes that were thought to be important for ESCs and identified four key genes that, when introduced into the reporter fibroblasts, resulted in drug-resistant cells. These were *Oct3/4*, *Sox2*, *c-Myc* and *Klf4*. This experiment indicated that expression of the four genes in these transgenic MEFs led to expression of a gene specific for ESCs. Mouse embryonic fibroblasts and adult fibroblasts were co-transduced with retroviral vectors, each carrying one of the four genes, and transduced cells were selected via drug resistance. The resultant iPS cells possessed the immortal growth characteristics of self-renewing ESCs, expressed genes specific for ESCs and generated embryoid bodies *in vitro* and teratomas *in vivo*. When iPS cells were injected into mouse blastocysts, they contributed to a variety of cell types. However, although iPS cells selected in this way were pluripotent, they were not identical to ESCs. Unlike ESCs, chimeras made from iPS cells did not result in full-term pregnancies. Gene expression profiles of the iPS cells showed that they possessed a distinct gene expression signature that was different from that of ESCs. In addition, the epigenetic state of the iPS cells was somewhere between that found in somatic cells and that found in ESCs, suggesting that the reprogramming was incomplete.

These results were improved significantly by Wernig and Jaenisch in July 2007 [47]. Fibroblasts were infected with retroviral vectors and selected for the activation of endogenous *Oct4* or *Nanog* genes. Results from this study showed that DNA methylation, gene expression profiles and the chromatin state of the reprogrammed cells were similar to those of ESCs. Teratomas induced by these cells contained differentiated cell types representing all three embryonic germ layers. Most importantly, the reprogrammed cells from this experiment were able to form viable chimeras and contribute to the germline like ESCs, suggesting that these iPS cells were completely reprogrammed. Wernig et al. observed that the number of reprogrammed colonies increased when drug selection was initiated later (day 20 rather than day 3 post-transduction). This suggests that reprogramming is a slow and gradual process and might explain why previous attempts resulted in incomplete reprogramming.

It has recently been shown that reprogramming of human cells is possible [48, 49]. Yamanaka showed that retrovirus-mediated transfection of *OCT3/4*, *SOX2*, *KLF4* and *c-MYC* generates human iPS cells that are similar to hESCs in terms of morphology, proliferation, gene expression, surface markers and teratoma formation. Thompson's group showed that retroviral transduction of *OCT4*, *SOX2*, *NANOG* and *LIN28* could generate pluripotent stem cells without introducing any oncogenes (*c-MYC*). Both studies showed that human iPS were similar but not identical to hESCs.

Another concern is that these iPS cells contain three to six retroviral integrations (one for each factor) which might increase the risk of tumourigenesis. Yamanaka et al. studied the tumour formation in chimeric mice generated from Nanog-iPS cells and found 20% of the offspring developed tumours due to the retroviral expression of c-myc [50]. An alternative approach would be to use a transient expression method, such as adenovirus-mediated system, since both Jaenisch and Yamanaka showed strong silencing of the viral-controlled transcripts in iPS cells [50, 51]. This indicates that these viral genes are only required for the induction, not the maintenance, of pluripotency.

Another concern is the use of transgenic donor cells for reprogrammed cells in the mouse studies. Both studies used donor cells from transgenic mice harboring a drug resistance gene driven by *Fbx15*, *Oct3/4* or *Nanog* promoters so that, if these ESC-specific genes were activated, the resulting cells could be easily selected using neomycin. However, the use of genetically modified donors hinders clinical applicability for humans. To assess whether iPS cells can be derived from non-transgenic donor cells, wild type MEF and adult skin cells were retrovirally transduced with Oct3/4, Sox2, c-Myc, and Klf4 and ES-like colonies were isolated by morphology alone, without the use of drug selection for *Oct4* or *Nanog* [51]. iPS cells from wild type donor cells formed teratomas and generated live chimeras. This study suggests that transgenic donor cells are not necessary to generate iPS cells.

Although this is an exciting phenomenon, it is unclear why reprogramming adult fibroblasts and mesenchymal stromal cells have similar efficiencies [46]. It would seem that cells that are already multipotent could be reprogrammed with greater efficiency, since the more undifferentiated the donor nucleus the better SCNT performs [52]. This further emphasizes our limited understanding of the mechanism of reprogramming, yet the potential for this area of study is exciting.

Placental and amniotic fluid stem cells

Alternate sources of stem cells can be the amniotic fluid and placenta. Both amniotic fluid and placenta are known to contain multiple partially differentiated cell types derived from the developing foetus. We isolated stem cell populations from these sources, called amniotic fluid and placental stem cells (AFPSC) that express embryonic and adult stem cell markers [53]. The undifferentiated stem cells expand extensively without a feeder cell layer and double every 36 h. Unlike hESCs, the AFPSC do not form tumours *in vivo*. Lines maintained for over 250 population doublings retained long telomeres and a normal karyotype. Amniotic fluid stem (AFS) cells are broadly multipotent. Clonal human lines verified by retroviral marking that can be induced to differentiate into cell types representing each embryonic germ layer, including cells of adipogenic, osteogenic, myogenic, endothelial, neuronal and hepatic lineages. In this respect, they meet a commonly accepted criterion for pluripotent stem cells, without implying that they can generate every adult tissue. Examples of differentiated cells derived from AFS cells and displaying specialized functions include neuronal lineage secreting the neurotransmitter L-glutamate or expressing G-protein-gated inwardly rectifying potassium (GIRK) channels, hepatic lineage cells producing urea and osteogenic lineage cells forming tissue-engineered bone. The cells could be obtained either from amniocentesis or chorionic villous sampling in the developing foetus, or from the placenta at the time of birth. The cells could be preserved for self use, and used without rejection, or they could be banked. A bank of 100 000 specimens could potentially supply 99 percent of the US population with a perfect genetic match for transplantation. Such a bank may be easier to create than with other cell sources, since there are approximately 4.5 million births per year in the USA [53].

Biomaterials

In the most common tissue engineering procedures, isolated cells are seeded onto a scaffold composed of an appropriate biomaterial. These biomaterials replicate the biological and mechanical function of the native ECM found in tissues in the body by serving as an artificial ECM. Biomaterials provide a three-dimensional (3D) space for cells to develop into new tissues with appropriate structure and function. They can also allow delivery of appropriate bioactive factors (e.g., cell adhesion peptides, growth factors) to the developing tissue [54] to help regulate cellular function. As the majority of mammalian cell types are anchorage-dependent and die if no cell adhesion substrate is available, biomaterials provide this substrate that can deliver cells to specific sites in the body with high loading efficiency. Biomaterials can also provide mechanical support against *in vivo* forces so that the predefined 3D structure of the engineered implant is maintained during tissue development.

The ideal biomaterial should be biodegradable, bioresorbable and support the replacement of normal tissue without inducing inflammation. Incompatible materials are destined for an inflammatory or foreign-body response that eventually leads to rejection and/or necrosis. Degradation products, if produced, should be removed from the body via metabolic pathways at an adequate rate so that the concentration of these degradation products in tissues remains at a tolerable level [55]. The biomaterial should also provide an environment in which appropriate regulation of cell behaviour (adhesion, proliferation, migration and differentiation) can occur. Cell behaviour in the newly formed tissue has been shown to be regulated by multiple interactions of cells with their microenvironment, including interactions with cell adhesion ligands [56] and with soluble growth factors. Since biomaterials provide temporary mechanical support while the cells undergo spatial reorganisation into tissue, the properly chosen biomaterial should allow the engineered tissue to maintain sufficient mechanical integrity to support itself in early development, while in late development, it should begin to degrade so that it does not hinder further tissue growth [54].

Generally, three classes of biomaterials have been utilized for engineering tissues: naturally derived materials (e.g., collagen and alginate), acellular tissue matrices (e.g., bladder submucosa and small intestinal submucosa) and synthetic polymers such as polyglycolic acid (PGA), polylactic acid (PLA) and poly(lactic-co-glycolic acid) (PLGA). These classes of biomaterials have been tested in respect to their biocompatibility [57, 58]. Naturally derived materials and acellular tissue matrices have the potential advantage of biological recognition. However, synthetic polymers can be produced reproducibly on a large scale with controlled properties such as strength, degradation rate and microstructure, which would aid in the preparation of easily used "off-the-shelf" scaffold material.

Naturally derived materials

Collagen is the most abundant and ubiquitous structural protein in the body, and may be readily purified from both animal and human tissues with an enzyme treatment and salt/acid extraction [59]. Collagen implants, under normal conditions, degrade through a process involving phagocytosis of collagen fibrils by fibroblasts [60]. This is followed by sequential attack by lysosomal enzymes including cathepsins B1 and D. Under inflammatory conditions, implants can be rapidly degraded largely by matrix metalloproteins (MMPs) and collagenases [60]. However, the *in vivo* resorption rate of a collagen implant can be regulated by controlling the density of the implant and the extent of intermolecular crosslinking—the lower the density, the greater the space between collagen fibers and the larger the pores for cell infiltration, leading to a higher rate of implant degradation. Collagen contains cell adhesion domain sequences (e.g., RGD) that might assist to retain the phenotype and activity of many types of cells, including fibroblasts [61] and chondrocytes [62].

Alginate, a polysaccharide isolated from seaweed, has been used as an injectable cell delivery vehicle [63] and a cell immobilisation matrix [64] owing to its gentle gelling properties in the presence of divalent ions such as calcium. Alginate is relatively biocompatible and is approved by the Food and Drug Administration (FDA) for human use as wound dressing material. Alginate is a family of copolymers of D-mannuronate and L-glucuronate. The physical and mechanical properties of alginate gel are strongly correlated with the proportion and length of polygluronic block in the alginate chains [63].

Acellular tissue matrices

Acellular tissue matrices are collagen-rich matrices prepared by removing cellular components from tissues. The matrices are often prepared by mechanical and chemical manipulation of a segment of tissue [1–4]. These matrices slowly degrade upon implantation and are replaced and remodeled by ECM proteins synthesized and secreted by transplanted or in growing cells.

Synthetic polymers

Polyesters of naturally occurring a-hydroxy acids, including PGA, PLA and PLGA, are widely used in tissue engineering. These polymers are FDA-approved for a variety of applications, including sutures [65]. The ester bonds in these polymers are hydrolytically labile, and they degrade by non-enzymatic hydrolysis. The degradation products of PGA, PLA and PLGA are non-toxic natural metabolites and are eventually eliminated from the body in the form of carbon dioxide and water [65]. The degradation rate of these polymers can be tailored to the application by altering crystallinity, initial molecular weight and the copolymer ratio of lactic to glycolic acid. Generally, the optimal degradation time ranges from several weeks to several years. Since these polymers are thermoplastics, they can be easily formed into a 3D scaffold with a desired microstructure, gross shape and dimension by various techniques, including molding, extrusion, solvent casting [66], phase separation techniques and gas foaming techniques [67]. Many applications in tissue engineering often require a scaffold with high porosity and ratio of surface area to volume. Other biodegradable synthetic polymers, including poly(anhydrides) and poly(ortho-esters), can also be used to fabricate scaffolds for tissue engineering with controlled properties [68].

TISSUE ENGINEERING OF SPECIFIC TISSUES AND ORGANS

Investigators around the world, including our laboratory, have been working towards the development of several cell types and tissues and organs for clinical application. The following sections will describe this research in detail.

Liver

The liver can sustain a variety of insults, including viral infection, alcohol abuse, surgical resection of tumours and acute drug-induced hepatic failure. The current therapy for liver failure is liver transplantation. However, this therapy is limited by the shortage of donors and the need for lifelong immunosuppressive therapy. Cell transplantation has been proposed as a potential solution for liver failure. The liver has enormous regenerative potential *in vivo,* and this suggests that in the right environment, it may be possible to expand liver cells *in vitro* in sufficient quantities for tissue engineering [69]. While expansion of these cells has not been as straightforward as originally thought, many approaches have been used for this purpose, including development of specialised media, co-culture with other cell types, identification and use of growth factors that have proliferative effects on these cells and culture on 3D scaffolds within bioreactors that simulate *in vivo* environments [69].

Extracorporeal bioartificial liver devices that use porcine hepatocytes have been designed and applied. These devices are designed to filter and purify the patient's blood in place of the patient's own liver, and then blood is returned to the patient in a manner similar to kidney dialysis. Another cell-based approach is the injection of liver cell suspensions. This has been performed in animal models. Intraportal hepatocyte injection has also been used in patients with Crigler–Najjar syndrome Type 1 [70]; however, complications such as portal vein thrombosis and pulmonary embolism are major concerns, especially when large cell numbers are used [71]. Finally, cells including stem cells, oval progenitor cells and mature hepatocytes have been seeded onto liver-shaped biocompatible matrices to engineer artificial and implantable livers. These have been tested in various animal models [72, 73]; however, the transplantation efficiency as well as the functionality of these constructs must be improved substantially before the technology can be moved into the clinic.

Heart

In the United States, over 5 million people currently live with some form of heart disease, and this number increases yearly. While many medications have been developed to assist the ailing heart, transplantation still remains the only treatment for end-stage heart failure. Unfortunately, as with other organs, donor hearts are in short supply, and even when a transplant can be performed, the patient must endure effects of lifelong immunosuppressive therapy. Thus, alternatives are desperately needed, and the development of novel methods to regenerate or replace damaged heart muscle using tissue engineering and regenerative medicine techniques is an attractive option.

Cell therapy for infracted areas of the heart is attractive, as this sort of treatment requires a rather simple injection into the damaged area of a patient's heart rather than a rigorous surgical procedure. To create such a therapy, various types of stem cells have been investigated for their potential to regenerate damaged or dead heart tissue. Skeletal muscle cells, bone marrow stem cells (both mesenchymal and haematopoietic), AFS cells and ESCs have all been used for this purpose. In this technique, cells are suspended in a biocompatible matrix that can range from simple normal saline to complex yet biocompatible hydrogels, depending on the part of the heart involved, the type of damage it sustained and the type of injection to be performed. The cells are either injected into the damaged area of the heart itself, or they are injected into the coronary circulation with the hope that they will home to the damaged area, take up residence there and begin to repair the tissue. Unfortunately, injectable therapies for cardiac damage have been shown to be relatively inefficient, and cell loss is quite substantial. Newer methods of tissue engineering include the development of engineered 'patches,' which are comprised of cells adhered to a biomaterial. These patches could theoretically be used to replace the damaged area of the heart. These techniques have promise, but require further research to determine the optimal cell types and biomaterials for this purpose before they can be used extensively in the clinic (see [74] for an excellent review of these methods).

However, methods described above could only be used in cases where a relatively small section of heart muscle was damaged. In cases where a large area or even the whole heart has become non-functional, a more radical approach would be required. Here, the use of a bioartificial heart would be ideal, as rejection would be avoided and problems associated with a mechanical heart (such as thromboembolus formation) would be abolished. To this end, Ott et al. recently developed a novel heart construct *in vitro* using decellularized cadaveric hearts. A specialized decellularisation process was used to remove cellular material from these cadaveric hearts, leaving behind an acellular tissue matrix. These matrices were reseeded with various types of healthy cells that make up a heart (cardiomyocytes, smooth muscle cells, endothelial cells and fibrocytes) and the resulting construct was cultured in a bioreactor system designed to mimic physiologic conditions. This group produced a heart construct that could generate pump function on its own [75]. This study suggests that production of bioartificial hearts may one day be possible.

Blood Vessels

Xenogenic or synthetic materials have been used as replacement blood vessels for complex cardiovascular lesions. However, these materials typically lack growth potential, and might place the recipient at risk for complications such as stenosis, thromboembolisation or infection [76]. Tissue-engineered vascular grafts have been constructed using autologous cells and biodegradable scaffolds and have been applied in dog and lamb models [77–80]. The key advantage of using these autografts is that they degrade *in vivo*, and thus allow the new tissue to form without the long-term presence of foreign material [76]. Application of these techniques from the laboratory to the clinical setting has begun, with autologous vascular cells harvested, expanded and seeded onto a biodegradable scaffold [81]. The resultant autologous construct was used to replace a stenosed pulmonary artery that had been previously repaired. Seven months after implantation, no evidence of graft occlusion or aneurysmal changes was noted in the recipient.

Urethra

Various biomaterials without cells, such as PGA and acellular collagen-based matrices from small intestine and bladder, have been used experimentally (in animal models) for the regeneration of urethral tissue [4, 82–86]. Some of these biomaterials, like acellular collagen matrices derived from bladder submucosa, have also been seeded with autologous cells for urethral reconstruction. Our laboratory has been able to replace tubularized urethral segments with cell-seeded collagen matrices [87, 88].

Acellular collagen matrices derived from bladder submucosa by our laboratory have been used experimentally and clinically. In animal studies, segments of the urethra were resected and replaced with acellular matrix grafts in an onlay fashion. Histological examination showed complete epithelialisation and progressive vessel and muscle infiltration, and the animals were able to void through the neo-urethras [4]. These results were confirmed in a clinical study of patients with hypospadias and urethral stricture disease [89] (Fig. 26.2). Decellularized cadaveric bladder submucosa was used as an onlay matrix for urethral repair in patients with stricture disease and hypospadias (Fig. 26.2). Patent, functional neo-urethras were noted in these patients with up to a

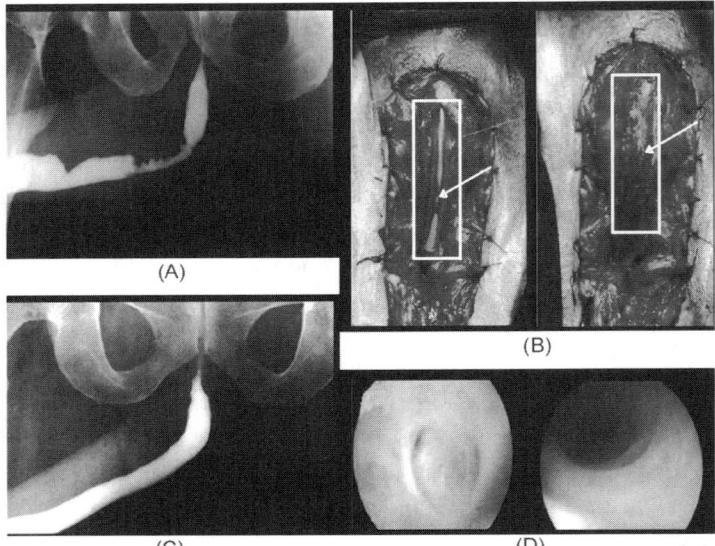

▶ **Fig. 26.2** *Urethral repair using a collagen matrix. (A) Representative case of a patient with a bulbar stricture. (B) During surgery, strictured tissue is excised, preserving the urethral plate on the left hand side, and the matrix is anastamosed to the urethral plate in an onlay fashion on the right. The boxes in both photos indicate the area of interest, including the urethra, which appears white in the left hand side photograph. In the left hand photograph, the arrow indicates the area of stricture in the urethra. On the right hand, the arrow indicates the repaired stricture. Note that the engineered tissue now obscures the native white urethral tissue in an onlay fashion in the right photograph. (C) Urethrogram 6 months after repair. (D) Cystoscopic view of urethra before surgery on the left hand side, and 4 months after repair on the right hand side. (for colour figure see Plate 75).*

7-year follow-up. The use of an off-the-shelf matrix appears to be beneficial for patients with abnormal urethral conditions and obviates the need for obtaining autologous grafts, thus decreasing operative time and eliminating donor site morbidity. Unfortunately, the above techniques are not applicable for tubularized urethral repairs. The collagen matrices are able to replace urethral segments only when used in an onlay fashion.

However, if a tubularized repair is needed, the collagen matrices should be seeded with autologous cells to avoid the risk of stricture formation and poor tissue development [87]. Therefore, tubularized collagen matrices seeded with autologous cells can be used successfully for total penile urethra replacement.

Bladder

Currently, gastrointestinal segments are commonly used as tissues for bladder replacement or repair. However, gastrointestinal tissues are designed to absorb specific solutes, whereas bladder tissue is designed for the excretion of solutes. Due to problems encountered with the use of gastrointestinal segments, numerous investigators have attempted alternative materials and tissues for bladder replacement or repair.

The success of cell transplantation strategies for bladder reconstruction depends on the ability to use donor tissue efficiently and to provide the right conditions for long-term survival, differentiation and growth. Urothelial and muscle cells can be expanded *in vitro*, seeded onto polymer scaffolds and allowed to attach and form sheets of cells [90]. These principles were applied in the creation of tissue engineered bladders in an animal model that required a subtotal cystectomy with subsequent replacement with a tissue-engineered organ in beagle dogs [12]. Urothelial and muscle cells were separately expanded from an autologous bladder biopsy, and seeded onto a bladder-shaped biodegradable polymer scaffold. The results from this study showed that it is possible to tissue engineer bladders that are anatomically and functionally normal. Clinical trials for the application of this technology are currently being conducted.

A clinical experience involving engineered bladder tissue for cystoplasty reconstruction was conducted starting in 1999. A small pilot study of seven patients was reported, using a collagen scaffold seeded with cells either with or without omentum coverage, or a combined PGA-collagen scaffold seeded with cells and omental coverage (Fig. 26.3). The patients reconstructed with the engineered bladder tissue created with the PGA-collagen cell-seeded scaffolds showed increased compliance, decreased end-filling pressures, increased capacities and longer dry periods [91] (Fig. 26.4). Although the experience is promising in terms of showing that engineered tissues can be implanted safely, it is just a start in terms of accomplishing the goal of engineering fully functional bladders. Further experimental and clinical work is being conducted.

Kidney

We applied the principles of both tissue engineering and therapeutic cloning in an effort to produce genetically identical renal tissue in a large animal model, the cow (*Bos taurus*) [92]. Bovine skin

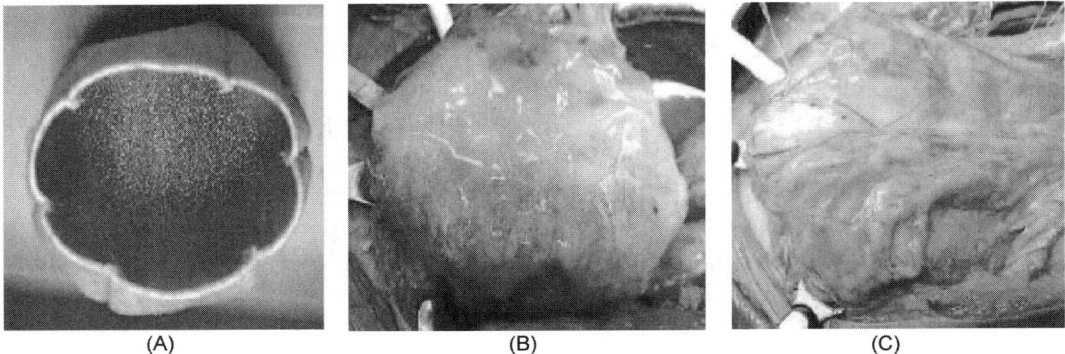

▶ **Fig. 26.3** *Construction of engineered bladder. (A) Scaffold material seeded with cells for use in bladder repair. (B) The seeded scaffold is anastamosed to native bladder with running 4-0 polyglycolic sutures. (C) Implant covered with fibrin glue and omentum. (for colour figure see Plate 75).*

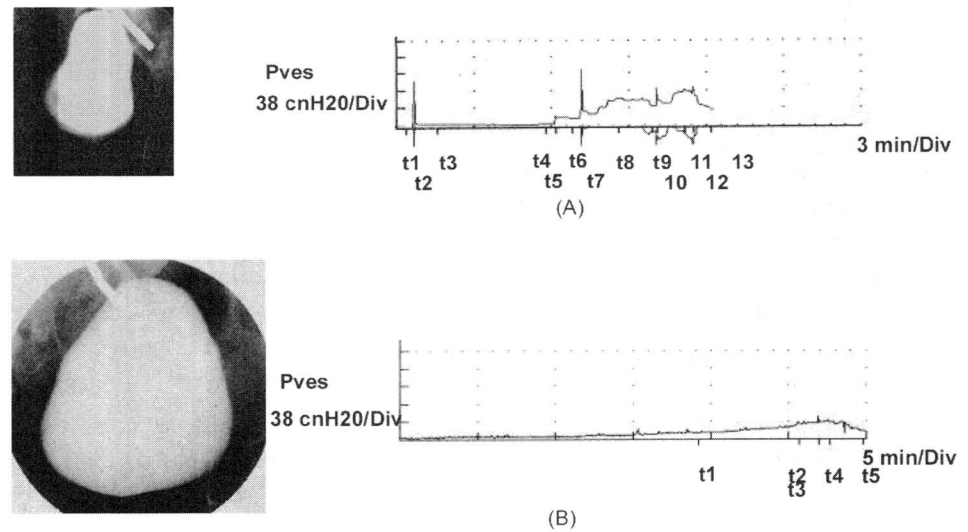

▶ **Fig. 26.4** *Cystograms and urodynamic studies of a patient before and after implantation of the tissue-engineered bladder. (A) Preoperative results indicate an irregular bladder in the cystogram and abnormal bladder pressures as the bladder is filled via urodynamic study. (B) Postoperatively, findings are significantly improved.*

fibroblasts from adult Holstein steers were obtained by ear notch, and single donor cells were isolated and microinjected into the perivitelline space of donor enucleated oocytes (nuclear transfer). The resulting blastocysts were implanted into progestin-synchronized recipients to allow for further *in vivo* growth. After 12 weeks, cloned renal cells were harvested, expanded *in vitro*, then seeded onto biodegradable scaffolds. The constructs, which consisted of the cells and scaffolds, were then implanted into the subcutaneous space of the same steer from which cells were cloned to allow for tissue growth.

The kidney is a complex organ with multiple cell types and a complex functional anatomy that renders it one of the most difficult to reconstruct [5, 93]. Previous efforts in tissue engineering of the kidney have been directed toward the development of extracorporeal renal support systems made of biological and synthetic components [94–102] and *ex vivo* renal replacement devices are known to be life-sustaining. However, there would be obvious benefits for patients with end-stage kidney disease if these devices could be implanted long term without the need for an extracorporeal perfusion circuit or immunosuppressive drugs.

Cloned renal cells were seeded on scaffolds consisting of three collagen-coated cylindrical Silastic catheters (Fig. 26.5A). The ends of the three membranes of each scaffold were connected to catheters that terminated into a collecting reservoir. This created a renal neo-organ with a mechanism for collecting the excreted urinary fluid (Fig. 26.5B). These scaffolds with the collecting devices were transplanted subcutaneously into the same steer from which the genetic material originated, and then retrieved 12 weeks after implantation.

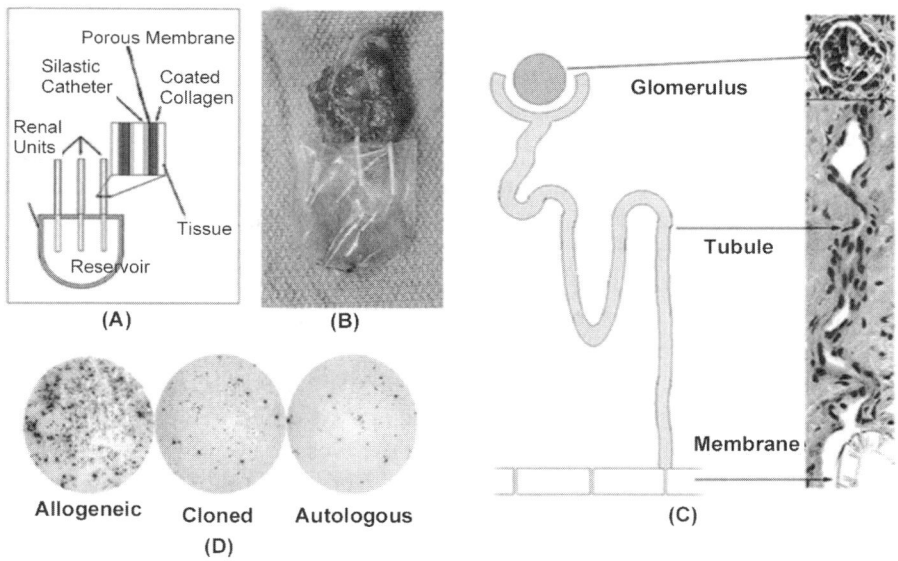

▶ **Fig. 26.5** *Combining therapeutic cloning and tissue engineering to produce kidney tissue. (A) Illustration of the tissue-engineered renal unit. (B) Renal unit seeded with cloned cells, three months after implantation, showing the accumulation of urine-like fluid. (C) Clear unidirectional continuity between the mature glomeruli, their tubules and Silastic catheter. (D) ELISPOT analyses of the frequencies of T cells that secrete IFN-γ after stimulation with allogeneic renal cells, cloned renal cells or nuclear donor fibroblasts. Cloned renal cells produce fewer IFN-γ spots than the allogeneic cells, indicating that the rejection to cloned cells is diminished. The presented wells are single representatives of duplicate wells. (for colour figure see Plate 76).*

Chemical analysis of the collected urine-like fluid, including urea nitrogen and creatinine levels, electrolyte levels, specific gravity and glucose concentration revealed that the implanted renal cells possessed filtration, reabsorption and secretory capabilities.

Histological examination of the retrieved implants revealed extensive vascularisation and self-organisation of the cells into glomeruli and tubule-like structures. A clear continuity between the glomeruli, tubules and the Silastic catheter was noted that allowed the passage of urine into the collecting reservoir (Fig. 26.5C). Immunohistochemical analysis with renal-specific antibodies revealed the presence of renal proteins, RT-PCR analysis confirmed the transcription of renal-specific RNA in the cloned specimens and Western blot analysis confirmed the presence of elevated renal-specific protein levels.

Since previous studies have shown that bovine clones harbor the oocyte mitochondrial DNA [103–105], the donor egg's mitochondrial DNA (mtDNA) was thought to be a potential source of immunologic incompatibility. Differences in mtDNA-encoded proteins expressed by cloned cells could stimulate a T-cell response specific for mtDNA-encoded minor histocompatibility antigens when the cloned cells are implanted back into the original nuclear donor [106]. We used nucleotide sequencing of the mtDNA genomes of the clone and fibroblast nuclear donor to identify potential antigens in the muscle constructs. Only two amino acid substitutions were noted to distinguish the clone and the nuclear donor and, as a result, a maximum of two minor histocompatibility antigens could be defined. Given the lack of knowledge regarding peptide-binding motifs for bovine MHC class I molecules, there is no reliable method to predict the impact of these amino acid substitutions on bovine histocompatibility.

Oocyte-derived mtDNA was also thought to be a potential source of immunologic incompatibility in the cloned renal cells. Maternally transmitted minor histocompatibility antigens in mice have been shown to stimulate both skin allograft rejection *in vivo* and cytotoxic T lymphocytes expansion *in vitro* [106] that could prevent the use of these cloned constructs in patients with chronic rejection of major histocompatibility matched human renal transplants [107, 108]. We tested for a possible T-cell response to the cloned renal devices using delayed-type hypersensitivity testing *in vivo* and ELISPOT analysis of interferon-gamma secreting T-cells *in vitro*. Both analyses revealed that the cloned renal cells showed no evidence of a T-cell response, suggesting that rejection will not necessarily occur in the presence of oocyte-derived mtDNA (Fig. 26.5D). This finding might represent a step forward in overcoming the histocompatibility problem of stem cell therapy [108].

These studies demonstrated that cells derived from nuclear transfer can be successfully harvested, expanded in culture and transplanted *in vivo* with the use of biodegradable scaffolds on which the single suspended cells can organize into tissue structures that are genetically identical to that of the host. These studies were the first demonstration of the use of therapeutic cloning for regeneration of tissues *in vivo*.

Genital Tissues

Reconstructive surgery is required for a wide variety of pathologic penile conditions, such as penile carcinoma, trauma, severe erectile dysfunction and congenital conditions such as ambiguous genitalia, hypospadias and epispadias. One of the major limitations of phallic reconstructive surgery is the scarcity of sufficient autologous tissue.

The major components of the phallus are corporal smooth muscle and endothelial cells. The creation of autologous functional and structural corporal tissue *de novo* would be beneficial. Autologous cavernosal smooth muscle and endothelial cells were harvested, expanded and seeded on acellular collagen matrices and implanted in a rabbit model [109, 110]. Histologic examination confirmed the appropriate organisation of penile tissue phenotypes, and structural and functional studies, including cavernosography, cavernosometry and mating studies, demonstrated that it is possible to engineer autologous functional penile tissue. Our laboratory is currently working on increasing the size of the engineered constructs.

Congenital malformations of the uterus may have profound implications clinically. Patients with cloacal exstrophy and intersex disorders may not have sufficient uterine tissue present for future reproduction. We investigated the possibility of engineering functional uterine tissue using autologous cells [111]. Autologous rabbit uterine smooth muscle and epithelial cells were harvested, then grown and expanded in culture. These cells were seeded onto preconfigured uterine-shaped biodegradable polymer scaffolds, which were then used for subtotal uterine tissue replacement in the corresponding autologous animals. Upon retrieval 6 months after implantation, histological, immunocytochemical and Western blot analyses confirmed the presence of normal uterine tissue components. Biomechanical analyses and organ bath studies showed that the functional characteristics of these tissues were similar to those of normal uterine tissue. Breeding studies using these engineered uteri are currently being performed.

Similarly, several pathologic conditions, including congenital malformations and malignancy, can adversely affect normal vaginal development or anatomy. Vaginal reconstruction has traditionally been challenging due to the paucity of available native tissue. The feasibility of engineering vaginal tissue *in vivo* was investigated [112]. Vaginal epithelial and smooth muscle cells of female rabbits were harvested, grown and expanded in culture. These cells were seeded onto biodegradable polymer scaffolds and the cell-seeded constructs were then implanted into nude mice for up to 6 weeks. Immunocytochemical, histological and Western blot analyses confirmed the presence of vaginal tissue phenotypes. Electrical field stimulation studies in the tissue-engineered constructs showed similar functional properties to those of normal vaginal tissue. When these constructs were used for autologous total vaginal replacement, patent vaginal structures were noted in the tissue-engineered specimens, while the non-cell-seeded structures were noted to be stenotic [112].

OTHER EMERGING TECHNOLOGIES
Injectable Therapies

Both urinary incontinence and vesicoureteral reflux (a condition in which urine flows backwards out of the bladder towards kidneys) are common conditions affecting the genitourinary system. Both conditions are usually the result of damage to or malformation of a specific sphincter muscle. Currently, injection of bulking agents around the defective sphincter can be used clinically for these conditions; however, biocompatibility of current synthetic bulking agents is a concern. The ideal substance for endoscopic treatment of reflux and incontinence should be injectable, non-antigenic, non-migratory, volume stable and safe for human use. Animal studies have shown that chondrocytes (cartilage cells) can be easily harvested and combined with alginate *in vitro* and the resulting suspension can be easily injected cystoscopically. A similar technique using muscle and muscle precursor cells with the hope of repairing the defective sphincter muscle has been studied. These technologies have been applied in humans for the correction of vesicoureteral reflux in children and for urinary incontinence in adults [113, 114].

Autologous chondrocytes as bulking agents

Injectable bulking agents can be endoscopically used in the treatment of both urinary incontinence and vesicoureteral reflux. The advantages in treating urinary incontinence and vesicoureteral reflux with this minimally invasive approach include the simplicity of this quick outpatient procedure and the low morbidity associated with it. Several investigators are seeking alternative implant materials that would be safe for human use [115].

The ideal substance for the endoscopic treatment of reflux and incontinence should be injectable, non-antigenic, non-migratory, volume stable and safe for human use. Toward this goal long-term studies were conducted to determine the effect of injectable chondrocytes *in vivo* [116]. It was initially determined that alginate, a liquid solution of gluronic and mannuronic acid, embedded with chondrocytes, could serve as a synthetic substrate for the injectable delivery and maintenance of cartilage architecture *in vivo*. Alginate undergoes hydrolytic biodegradation and its degradation time can be varied depending on the concentration of each of the polysaccharides. The use of autologous cartilage for the treatment of vesicoureteral reflux in humans would satisfy all requirements for an ideal injectable substance.

Chondrocytes derived from an ear biopsy can be readily grown and expanded in culture. Neocartilage formation can be achieved *in vitro* and *in vivo* using chondrocytes cultured on synthetic biodegradable polymers. In these experiments, the cartilage matrix replaced the alginate as the polysaccharide polymer underwent biodegradation. This system was adapted for the treatment of vesicoureteral reflux in a porcine model [117]. These studies showed that chondrocytes can be easily harvested and combined with alginate *in vitro*, the suspension can be easily injected cystoscopically and the elastic cartilage tissue formed is able to correct vesicoureteral reflux without any evidence of obstruction.

Two multicentre clinical trials were conducted using this engineered chondrocyte technology. Patients with vesicoureteral reflux were treated at 10 centres throughout the US. The patients had a similar success rate as with other injectable substances in terms of cure (Fig. 26.6). Chondrocyte formation was not noted in patients who had treatment failure. It is supposed that patients who were cured have a biocompatible region of engineered autologous tissue present, rather than a foreign material [113]. Patients with urinary incontinence were also treated endoscopically with injected chondrocytes at three different medical centres. Phase 1 trials showed an approximate success rate of 80% at follow-up, 3 and 12 months postoperatively [114]. Several of the clinical trials involving bioengineered products have been placed on hold because of the costs involved with the specific technology. With a bioengineered product, costs are usually high because of the biological nature of the therapies involved. As with any therapy, the cost that the medical health care system can allow for a specific technology is limited. Therefore, costs of bioengineered products have to be lowered for them to have an impact clinically. This is currently being addressed for multiple tissue-engineered technologies. As technologies advance over time, and the volume of the application is considered, costs naturally decrease.

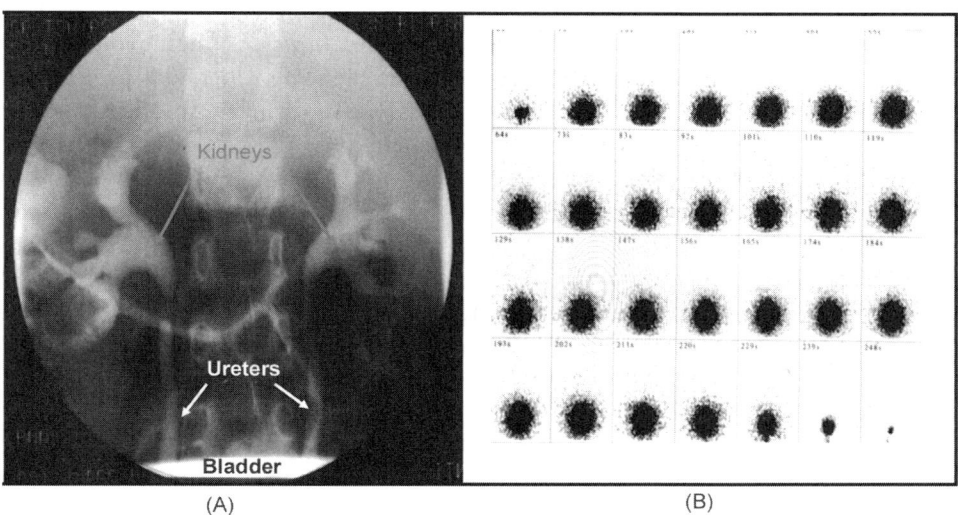

▶ **Fig. 26.6** *Autologous chondrocytes for the treatment of vesicoureteral reflux. (A) Pre-operative voiding cystourethrogram of a patient with bilateral reflux. A catheter was inserted into the bladder via the urethra and contrast material was instilled intravesically. Here, contrast material can be seen within both ureters and within kidneys, indicating that reflux is present. (B) Post-operative radionuclide cystogram of the same patient 6 months after injection of autologous chondrocytes. A catheter was inserted into the bladder via the urethra and a radioactive solution was inserted into the bladder. The bladder was scanned during filling and emptying phases. This panel includes sequential images of the bladder as it was filled and emptied. This shows a normal, round bladder that fills and empties properly. If reflux had been present, ureters would have been visible in the scan above the round bladder. (for colour figure see Plate 76).*

Injectable muscle cells

The potential use of injectable cultured myoblasts for the treatment of stress urinary incontinence has been investigated [118, 119]. Myoblasts were labelled with fluorescent latex microspheres (FLM) in order to track them after injection. Labelled myoblasts were directly injected into the proximal urethra and lateral bladder walls of nude mice with a microsyringe in an open surgical procedure. Tissue harvested up to 35 days post-injection contained the labelled myoblasts, as well as evidence of differentiation of the labelled myoblasts into regenerative myofibers. The authors reported that a significant portion of the injected myoblast population persisted *in vivo*. Similar techniques of sphincteric-derived muscle cells have been used for the treatment of urinary incontinence in a pig model [120]. The fact that myoblasts can be labelled and can survive after injection and begin the process of myogenic differentiation further supports the feasibility of using cultured cells of muscular origin as an injectable bioimplant.

The use of injectable muscle precursor cells has also been investigated for use in the treatment of urinary incontinence due to irreversible urethral sphincter injury or maldevelopment. Muscle precursor cells are the quiescent satellite cells found in each myofiber that proliferate to form myoblasts and eventually myotubes and new muscle tissue. Intrinsic muscle precursor cells have previously been shown to play an active role in the regeneration of injured striated urethral sphincter [121]. In a subsequent study, autologous muscle precursor cells were injected into a rat model of urethral sphincter injury, and both replacement of mature myotubes as well as restoration of functional motor units was noted in the regenerating sphincteric muscle tissue [122]. This is the first demonstration of the replacement of both sphincter muscle tissue and its innervation by the injection of muscle precursor cells. As a result, muscle precursor cells may be a minimally invasive solution for urinary incontinence in patients with irreversible urinary sphincter muscle insufficiency.

Foetal Tissue Engineering

The prenatal diagnosis of foetal abnormalities is now more common and more accurate. Improvements in prenatal diagnosis have led to demand for novel interventions designed to reverse potentially life-threatening processes before birth. Having a ready supply of urologic-associated tissue for immediate surgical reconstruction of congenital malformations at birth may be advantageous. Theoretically, once the diagnosis of the pathologic condition is confirmed prenatally, a small tissue biopsy could then be obtained under ultrasound guidance. These biopsy materials could then be processed expanded *in vitro*. Using tissue engineering techniques, *in vitro*-reconstituted structures could then be readily available at the time of birth for reconstruction.

CONCLUSION

Regenerative medicine efforts are currently underway experimentally for virtually every type of tissue and organ within the human body. As regenerative medicine incorporates fields of tissue engineering, cell biology, nuclear transfer and materials science, personnel who have mastered techniques of cell harvest, culture, expansion, transplantation, as well as polymer design are essential for the successful application of these technologies to extend human life. Various tissues are at different stages of development, with some already being used clinically, a few in preclinical trials and some in the discovery stage. Recent progress suggests that engineered tissues may have an expanded clinical applicability in the future and might represent a viable therapeutic option for those who would benefit from the life-extending benefits of tissue replacement or repair.

ACKNOWLEDGEMENT

The author wishes to thank Jennifer L. Olson, Ph.D., for editorial assistance with this manuscript.

REFERENCES

1. Dahms S.E., H.J. Piechota, R. Dahiya, T.F. Lue, E.A. Tanagho (1998). 'Composition and biomechanical properties of the bladder acellular matrix graft: Comparative analysis in rat, pig and human'. *Br. J. Urol.* **82**(3): 411–419.
2. Yoo J.J., J. Meng, F. Oberpenning, A. Atala (1998). 'Bladder augmentation using allogenic bladder submucosa seeded with cells'. *Urology.* **51**(2): 221–225.
3. Piechota H.J., S.E. Dahms, L.S. Nunes, R. Dahiya, T.F. Lue, E.A. Tanagho (1998). '*In vitro* functional properties of the rat bladder regenerated by the bladder acellular matrix graft'. *J. Urol.* **159**(5): 1717–1724.
4. Chen F., J.J. Yoo, A. Atala (1999). 'Acellular collagen matrix as a possible "off the shelf" biomaterial for urethral repair'. *Urology.* **54**(3): 407–410.
5. Amiel G.E., A. Atala (1999). 'Current and future modalities for functional renal replacement'. *Urol. Clin. North Am.* **26**(1): 235–246.
6. Amiel G.E., M. Komura, O. Shapira, J.J. Yoo, S. Yazdani, J. Berry, S. Kaushal, J. Bischoff, A. Atala, S. Soker (2006). 'Engineering of blood vessels from acellular collagen matrices coated with human endothelial cells'. *Tissue. Eng.* **12**(8): 2355–2365.
7. Yoo J.J., H.J. Park, I. Lee, A. Atala (1999). 'Autologous engineered cartilage rods for penile reconstruction'. *J. Urol.* **162**(3 Pt 2): 1119–1121.
8. Atala A. (1998). 'Autologous cell transplantation for urologic reconstruction'. *J. Urol.* **159**(1): 2–3.
9. Atala A. (2001). 'Bladder regeneration by tissue engineering'.[see comment]. *BJU Int.* **88**(7): 765–770.
10. Atala A. (1999). 'Creation of bladder tissue *in vitro* and *in vivo*. A system for organ replacement'. *Adv. Exp. Med. Biol.* 46231–46242.
11. Cilento B.G., M.R. Freeman, F.X. Schneck, A.B. Retik, A. Atala (1994)' 'Phenotypic and cytogenetic characterisation of human bladder urothelia expanded *in vitro*'. *J Urol.* **152**(2 Pt 2): 665–670.
12. Oberpenning F., J. Meng, J.J. Yoo, A. Atala (1999). 'De novo reconstitution of a functional mammalian urinary bladder by tissue engineering'.[see comment]. *Nat Biotechnol.* **17**(2): 149–155.
13. Scriven S.D., C. Booth, D.F. Thomas, L.K. Trejdosiewicz, J. Southgate (1997). 'Reconstitution of human urothelium from monolayer cultures'. *J Urol.* **158**(3 Pt 2): 1147–1152.
14. Liebert M., A. Hubbel, M. Chung, G. Wedemeyer, M.I. Lomax, A. Hegeman, T.Y. Yuan, M. Brozovich, M.J. Wheelock, H.B. Grossman (1997). 'Expression of mal is associated with urothelial differentiation *in vitro*: Identification by differential display reverse-transcriptase polymerase chain reaction'. *Differentiation.* **61**(3): 177–185.
15. Liebert M., G. Wedemeyer, L.V. Abruzzo, S.L. Kunkel, C. Hammerberg, K.D. Cooper, H.B. Grossman (1991). 'Stimulated urothelial cells produce cytokines and express an activated cell surface antigenic phenotype'. *Semin. Urol.* **9**(2): 124–130.
16. Puthenveettil J.A., M.S. Burger, C.A. Reznikoff (1999). 'Replicative senescence in human uroepithelial cells'. *Adv. Exp. Med. Biol.* 46283–46291.

17. Freeman M.R., J.J. Yoo, G. Raab, S. Soker, R.M. Adam, F.X. Schneck, A.A. Renshaw, M. Klagsbrun, A. Atala (1997). 'Heparin-binding EGF-like growth factor is an autocrine growth factor for human urothelial cells and is synthesized by epithelial and smooth muscle cells in the human bladder'. *J. Clin. Invest.* **99**(5): 1028–1036.
18. Nguyen H.T., J.M. Park, C.A. Peters, R.M. Adam, A. Orsola, A. Atala, M.R. Freeman (1999). 'Cell-specific activation of the HB-EGF and ErbB1 genes by stretch in primary human bladder cells'. *In Vitro Cell Dev. Biol. Anim.* **35**(7): 371–375.
19. Harriss D.R. (1995) Smooth muscle cell culture:, 'A new approach to the study of human detrusor physiology and pathophysiology'. *Br. J. Urol. 75 Suppl.* 118–126.
20. Minino A. (2004). 'In National Vital Statistics Reports'.
21. Martin G.R. (1981). 'Isolation of a pluripotent cell line from early mouse embryos cultured in medium conditioned by teratocarcinoma stem cells'. *Proc. Natl. Acad. Sci. USA***78**(12): 7634–7638.
22. Brivanlou A.H., F.H. Gage, R. Jaenisch, T. Jessell, D. Melton, J. Rossant (2003). 'Stem cells. Setting standards for human embryonic stem cells'. [see comment]. *Science.* **300**(5621): 913–916.
23. Solter D., B.B. Knowles (1975). 'Immunosurgery of mouse blastocyst'. *Proc. Natl. Acad. Sci.USA.* **72**(12): 5099–5102.
24. Reubinoff B.E., M.F. Pera, C.Y. Fong, A. Trounson, A. Bongso (2000). 'Embryonic stem cell lines from human blastocysts: Somatic differentiation *in vitro.*'[see comment][erratum appears in Nat Biotechnol 2000 May;18(5):559]. *Nat. Biotechnol.* **18**(4): 399–404.
25. Thomson J.A., J. Itskovitz-Eldor, S.S. Shapiro, M.A. Waknitz, J.J. Swiergiel, V.S. Marshall, J.M. Jones (1998). 'Embryonic stem cell lines derived from human blastocysts'.[see comment][erratum appears in Science 1998 Dec 4;282(5395):1827]. *Science.* **282**(5391): 1145–1147.
26. Reubinoff B.E., P. Itsykson, T. Turetsky, M.F. Pera, E. Reinhartz, A. Itzik, T. Ben-Hur (2001). 'Neural progenitors from human embryonic stem cells'.[see comment]. *Nat. Biotechnol.* **19**(12): 1134–1140.
27. Schuldiner M., R. Eiges, A. Eden, O. Yanuka, J. Itskovitz-Eldor, R.S. Goldstein, N. Benvenisty (2001). 'Induced neuronal differentiation of human embryonic stem cells'. *Brain. Res.* **913**(2): 201–205.
28. Schuldiner M., O. Yanuka, J. Itskovitz-Eldor, D.A. Melton, N. Benvenisty (2000). 'Effects of eight growth factors on the differentiation of cells derived from human embryonic stem cells'. *Proc. Natl. Acad. Sci. USA.* **97**(21): 11307–11312.
29. Zhang S.C., M. Wernig, I.D. Duncan, O. Brustle, J.A. Thomson (2001). '*In vitro* differentiation of transplantable neural precursors from human embryonic stem cells'.[see comment]. *Nat. Biotechnol.* **19**(12): 1129–1133.
30. Kaufman D.S., E.T. Hanson, R.L. Lewis, R. Auerbach, J.A. Thomson (2001). 'Haematopoietic colony-forming cells derived from human embryonic stem cells'. *Proc. Natl. Acad. Sci. USA.* **98**(19): 10716–10721.
31. Kehat I., D. Kenyagin-Karsenti, M. Snir, H. Segev, M. Amit, A. Gepstein, E. Livne, O. Binah, J. Itskovitz-Eldor, L. Gepstein (2001). 'Human embryonic stem cells can differentiate into myocytes with structural and functional properties of cardiomyocytes'.[see comment]. *J. Clin Invest.* **108**(3): 407–414.
32. Levenberg S., J.S. Golub, M. Amit, J. Itskovitz-Eldor, R. Langer (2002). 'Endothelial cells derived from human embryonic stem cells'. *Proc Natl. Acad. Sci. USA.* **99**(7): 4391–4396.

33. Assady S., G. Maor, M. Amit, J. Itskovitz-Eldor, K.L. Skorecki, M. Tzukerman (2001). 'Insulin production by human embryonic stem cells'. *Diabetes*. **50**(8): 1691–1697.
34. Itskovitz-Eldor J., M. Schuldiner, D. Karsenti, A. Eden, O. Yanuka, M. Amit, H. Soreq, N. Benvenisty (2000)' 'Differentiation of human embryonic stem cells into embryoid bodies compromising the three embryonic germ layers'. *Mol Med*. **6**(2): 88–95.
35. Brambrink T., K. Hochedlinger, G. Bell, R. Jaenisch (2006). 'ES cells derived from cloned and fertilized blastocysts are transcriptionally and functionally indistinguishable 1,. *Proc. Natl. Acad. Sci. USA*. **103**(4): 933–938.
36. Rideout W.M., III, K. Hochedlinger, M. Kyba, G.Q. Daley, R. Jaenisch (2002). 'Correction of a genetic defect by nuclear transplantation and combined cell and gene therapy'. *Cell*. **109**(1): 17–27.
37. Hochedlinger K., R. Jaenisch (2002). 'Monoclonal mice generated by nuclear transfer from mature B and T donor cells'. *Nature*. **415**(6875): 1035–1038.
38. Eggan K., K. Baldwin, M. Tackett, J. Osborne, J. Gogos, A. Chess, R. Axel, R. Jaenisch (2004). 'Mice cloned from olfactory sensory neurons 1'. *Nature*. **428**(6978): 44–49.
39. Lanza R.P., H.Y. Chung, J.J. Yoo, P.J. Wettstein, C. Blackwell, N. Borson, E. Hofmeister, G. Schuch, S. Soker, C.T. Moraes et al. (2002)' 'Generation of histocompatible tissues using nuclear transplantation 1'. *Nat.Biotechnol*. **20**(7): 689–696.
40. I. Wilmut A.E. Schnieke, J. McWhir, A.J. Kind, K.H. Campbell (1997). 'Viable offspring derived from foetal and adult mammalian cells'.[see comment][erratum appears in Nature 1997 Mar 13;386(6621):200]. *Nature*. **385**(6619): 810–813.
41. Hwang W.S., S.I. Roh, B.C. Lee, S.K. Kang, D.K. Kwon, S. Kim, S.J. Kim, S.W. Park, H.S. Kwon, C.K. Lee et al. (2005). 'Patient-specific embryonic stem cells derived from human SCNT blastocysts 1'. *Science*. **308**(5729): 1777–1783.
42. Simerly C., T. Dominko, C. Navara, C. Payne, S. Capuano, G. Gosman, K.Y. Chong, D. Takahashi, C. Chace, D. Compton et al. (2003). 'Molecular correlates of primate nuclear transfer failures 2'. *Science*. **300**(5617): 297.
43. Hwang W.S., Y.J. Ryu, J.H. Park, E.S. Park, E.G. Lee, J.M. Koo, H.Y. Jeon, B.C. Lee, S.K. Kang, S.J. Kim et al. (2004). 'Evidence of a pluripotent human embryonic stem cell line derived from a cloned blastocyst 1'. *Science*. **303**(5664): 1669–1674.
44. Byrne J., D. Pedersen, L. Clepper, M. Nelson, W. Sanger, S. Gokhale, D. Wolf, S. Mitalipov (2007). 'Producing primate embryonic stem cells by somatic cell nuclear transfer'. *Nature*. **450**(7169): 497–502.
45. Mitalipov S. (2007). 'Reprogramming following somatic cell nuclear transfer in primates is dependent upon nuclear remodeling'. *Hum. Reprod*. 222232–222242.
46. Takahashi K., S. Yamanaka (2006). 'Induction of pluripotent stem cells from mouse embryonic and adult fibroblast cultures by defined factors 2'. *Cell*. **126**(4): 663–676.
47. Wernig M., A. Meissner, R. Foreman, T. Brambrink, M. Ku, K. Hochedlinger, B.E. Bernstein, R. Jaenisch (2007). '*In vitro* reprogramming of fibroblasts into a pluripotent ES-cell-like state 1'. *Nature*. **448**(7151): 318–324.
48. Takahashi K., K. Tanabe, M. Ohnuki, M. Narita, T. Ichisaka, K. Tomoda, S. Yamanaka (2007). 'Induction of Pluripotent Stem Cells from Adult Human Fibroblasts by Defined Factors'. *Cell*. **131**(5): 861–872.

49. Yu J., M.A. Vodyanik, K. Smuga-Otto, J. Antosiewicz-Bourget, J.L. Frane, S. Tian, J. Nie, G.A. Jonsdottir, V. Ruotti, R. Stewart et al. (2007). 'Induced Pluripotent Stem Cell Lines Derived from Human Somatic Cells'. *Science.* **318**(5858): 1917–1920.
50. Okita K., T. Ichisaka, S. Yamanaka (2007). 'Generation of germline-competent induced pluripotent stem cells 1'. *Nature.* **448**(7151): 313–317.
51. Meissner A., M. Wernig, R. Jaenisch (2007). 'Direct reprogramming of genetically unmodified fibroblasts into pluripotent stem cells 1'. *Nat.Biotechnol.* **25**(10): 1177–1181.
52. Blelloch R., Z. Wang, A. Meissner, S. Pollard, A. Smith, R. Jaenisch (2006). 'Reprogramming efficiency following somatic cell nuclear transfer is influenced by the differentiation and methylation state of the donor nucleus 1'. *Stem Cells.* **24**(9): 2007–2013.
53. De Coppi P., G. Bartsch, Jr., M.M. Siddiqui, T. Xu, C.C. Santos, L. Perin, G. Mostoslavsky, A.C. Serre, E.Y. Snyder, J.J. Yoo et al. (2007). 'Isolation of amniotic stem cell lines with potential for therapy'.[see comment]. *Nat Biotechnol.* **25**(1): 100–106.
54. Kim B.S., D.J. Mooney (1998). 'Development of biocompatible synthetic extracellular matrices for tissue engineering'. *Trends Biotechnol.* **16**(5): 224–230.
55. Bergsma J.E., F.R. Rozema, R.R. Bos, G. Boering, W.C. de Bruijn, A.J. Pennings (1995). '*In vivo* degradation and biocompatibility study of *in vitro* pre-degraded as-polymerized polyactide particles.[see comment]'. *Biomaterials.* **16**(4): 267–274.
56. Hynes R.O. (1992) Integrins:, 'versatility, modulation, and signaling in cell adhesion'. *Cell.* **69**(1): 11–25.
57. Pariente J.L., B.S. Kim, A. Atala (2001). '*In vitro* biocompatibility assessment of naturally derived and synthetic biomaterials using normal human urothelial cells'. *J Biomed Mater Res.* **55**(1): 33–39.
58. Pariente J.L., B.S. Kim, A. Atala (2002). '*In vitro* biocompatibility evaluation of naturally derived and synthetic biomaterials using normal human bladder smooth muscle cells'. *J. Urol.* **167**(4): 1867–1871.
59. Li S.T. (1995) Biologic biomaterials:, 'Tissue derived biomaterials (collagen)'. *In: The Biomedical Engineering Handbook* (J.D. Bronzino., ed.). 627–647.CRS Press, Boca Raton, FL.
60. Arora P.D., M.F. Manolson, G.P. Downey, J. Sodek, C.A.G. McCulloch (2000). 'A Novel Model System for Characterisation of Phagosomal Maturation, Acidification, and Intracellular Collagen Degradation in Fibroblasts'. *J. Biol. Chem.* **275**(45): 35432–35441.
61. Silver F.H., G. Pins (1992). 'Cell growth on collagen: A review of tissue engineering using scaffolds containing extracellular matrix'. *J. Long. Term. Eff. Med. Implants.* **2**(1): 67–80.
62. Sams A.E., A.J. Nixon (1995). 'Chondrocyte-laden collagen scaffolds for resurfacing extensive articular cartilage defects'. *Osteoarthritis Cartilage.* **3**(1): 47–59.
63. Smidsrod O., G. Skjak-Braek (1990). 'Alginate as immobilisation matrix for cells'. *Trends Biotechnol.* **8**(3): 71–78.
64. Lim F., A.M. Sun (1980). 'Microencapsulated islets as bioartificial endocrine pancreas'. *Science.* **210**(4472): 908–910.
65. Gilding D. (1981). 'Biodegradable Polymers. In:, Biocompatibility of Clinical Implant Materials (D. Williams, ed.)', 209–232. *CRC Press, Boca Raton, FL.*
66. Mikos A.G., M.D. Lyman, L.E. Freed, R. Langer (1994). 'Wetting of poly(L-lactic acid) and poly(DL-lactic-*co*-glycolic acid) foams for tissue culture'. *Biomaterials.* **15**(1): 55–58.

67. Harris L.D., B.S. Kim, D.J. Mooney (1998). 'Open pore biodegradable matrices formed with gas foaming'. *J Biomed Mater Res*. **42**(3): 396–402.
68. Peppas N.A., R. Langer (1994). 'New challenges in biomaterials'.[see comment]. *Science*. **263**(5154): 1715–1720.
69. Bhandari R.N., L.A. Riccalton, A.L. Lewis, J.R. Fry, A.H. Hammond, S.J. Tendler, K.M. Shakesheff (2001). 'Liver tissue engineering: A role for co-culture systems in modifying hepatocyte function and viability'. *Tissue Eng*. **7**(3): 345–357.
70. Fox I.J., J.R. Chowdhury, S.S. Kaufman, T.C. Goertzen, N.R. Chowdhury, P.I. Warkentin, K. Dorko, B.V. Sauter, S.C. Strom (1998). 'Treatment of the Crigler-Najjar syndrome type I with hepatocyte transplantation'.[see comment]. *N. Engl. J. Med*. **338**(20): 1422–1426.
71. Nieto J.A., J. Escandon, C. Betancor, J. Ramos, T. Canton, V. Cuervas-Mons (1989). 'Evidence that temporary complete occlusion of splenic vessels prevents massive embolisation and sudden death associated with intrasplenic hepatocellular transplantation'. *Transplantation*. **47**(3): 449–450.
72. Kaufmann P.M., U. Kneser, H.C. Fiegel, D. Kluth, H. Herbst, X. Rogiers (1999). 'Long-term hepatocyte transplantation using three-dimensional matrices'. *Transplant Proc*. **31**(4): 1928–1929.
73. Gilbert J.C., T. Takada, J.E. Stein, R. Langer, J.P. Vacanti (1993). 'Cell transplantation of genetically altered cells on biodegradable polymer scaffolds in syngeneic rats'. *Transplantation*. **56**(2): 423–427.
74. Jawad H., N.N. Ali, A.R. Lyon, Q.Z. Chen, S.E. Harding, A.R. Boccaccini (2007). 'Myocardial tissue engineering: A review'. *J Tissue Eng Regenerat Med*. **1**(5): 332–342.
75. Ott H.C., T.S. Matthiesen, S.K. Goh, L.D. Black, S.M. Kren, T.I. Netoff, D.A. Taylor (2008). 'Perfusion-decellularized matrix: Using nature's platform to engineer a bioartificial heart'. *Nat Med*. **14**: 213–221.
76. Matsumura G., S. Miyagawa-Tomita, T. Shin'oka, Y. Ikada, H. Kurosawa (2003). 'First evidence that bone marrow cells contribute to the construction of tissue-engineered vascular autografts *in vivo*'. *Circulation*. **108**(14): 1729–1734.
77. Watanabe M., T. Shin'oka, S. Tohyama, N. Hibino, T. Konuma, G. Matsumura, Y. Kosaka, T. Ishida, Y. Imai, M. Yamakawa et al. (2001). 'Tissue-engineered vascular autograft: Inferior vena cava replacement in a dog model'. *Tissue Eng*. **7**(4): 429–439.
78. Shinoka T., C.K. Breuer, R.E. Tanel, G. Zund, T. Miura, P.X. Ma, R. Langer, J.P. Vacanti, J.E. Mayer, Jr. (1995). 'Tissue engineering heart valves: Valve leaflet replacement study in a lamb model'. *Ann Thorac Surg*. **60**(6 Suppl): S513–516.
79. Shinoka T., D. Shum-Tim, P.X. Ma, R.E. Tanel, N. Isogai, R. Langer, J.P. Vacanti, J.E. Mayer, Jr. (1998). 'Creation of viable pulmonary artery autografts through tissue engineering'. *J Thorac Cardiovasc Surg* **115**(3): 536–545; *discussion*. 545–536.
80. Shinoka T., D. Shum-Tim, P.X. Ma, R.E. Tanel, R. Langer, J.P. Vacanti, J.E. Mayer, Jr. (1997). 'Tissue-engineered heart valve leaflets: Does cell origin affect outcome'? *Circulation*. 96(9 Suppl): **II**: 102–107.
81. Shin'oka T., Y. Imai, Y. Ikada (2001). 'Transplantation of a tissue-engineered pulmonary artery'. *N Engl. J. Med*. **344**(7): 532–533.
82. Chen F., J.J. Yoo, A. Atala (2000). 'Experimental and clinical experience using tissue regeneration for urethral reconstruction'. *World. J. Urol*. **18**(1): 67–70.

83. Atala A., J.P. Vacanti, C.A. Peters, J. Mandell, A.B. Retik, M.R. Freeman (1992). 'Formation of urothelial structures *in vivo* from dissociated cells attached to biodegradable polymer scaffolds *in vitro*'. *J. Urol.* **148**(2 Pt 2): 658–662.
84. Olsen L., S. Bowald, C. Busch, J. Carlsten, I. Eriksson (1992). 'Urethral reconstruction with a new synthetic absorbable device'. *An experimental study. Scand J. Urol. Nephrol.* **26**(4): 323–326.
85. Kropp B.P., J.K. Ludlow, D. Spicer, M.K. Rippy, S.F. Badylak, M.C. Adams, M.A. Keating, R.C. Rink, R. Birhle, K.B. Thor (1998). 'Rabbit urethral regeneration using small intestinal submucosa onlay grafts'. *Urology.* **52**(1): 138–142.
86. K.D. Sievert, M.E. Bakircioglu, L. Nunes, R. Tu, R. Dahiya, E.A. Tanagho (2000). 'Homologous acellular matrix graft for urethral reconstruction in the rabbit: Histological and functional evaluation'. *J. Urol.* **163**(6): 1958–1965.
87. Filippo R.E. De, J.J. Yoo, A. Atala (2002). 'Urethral replacement using cell seeded tubularized collagen matrices'. *J. Urol.* **168**(4 Pt 2): 1789–1792; *discussion.* 1792–1783.
88. Atala A. (2002). 'Experimental and clinical experience with tissue engineering techniques for urethral reconstruction'. *Urol. Clin. North. Am.* **29**(2): 485–492.
89. El-Kassaby A.W., A.B. Retik, J.J. Yoo, A. Atala (2003). 'Urethral stricture repair with an off-the-shelf collagen matrix'. *J. Urol.* **169**(1): 170–173; discussion 173.
90. Atala A., M.R. Freeman, J.P. Vacanti, J. Shepard, A.B. Retik (1993). 'Implantation *in vivo* and retrieval of artificial structures consisting of rabbit and human urothelium and human bladder muscle'. *J. Urol.* **150**(2 Pt 2): 608–612.
91. Atala A., S.B. Bauer, S. Soker, J.J. Yoo, A.B. Retik (2006). 'Tissue-engineered autologous bladders for patients needing cystoplasty'. *Lancet.* **367**(9518): 1241–1246.
92. Lanza R.P., H.Y. Chung, J.J. Yoo, P.J. Wettstein, C. Blackwell, N. Borson, E. Hofmeister, G. Schuch, S. Soker, C.T. Moraes et al. (2002). 'Generation of histocompatible tissues using nuclear transplantation'.[see comment]. *Nat. Biotechnol.* **20**(7): 689–696.
93. Auchincloss H., J.V. Bonventre (2002). 'Transplanting cloned cells into therapeutic promise'.[comment]. *Nat. Biotechnol.* **20**(7): 665–666.
94. Aebischer P., T.K. Ip, L. Miracoli, P.M. Galletti (1987). 'Renal epithelial cells grown on semipermeable hollow fibers as a potential ultrafiltrate processor'. *ASAIO. Trans.* **33**(3): 96–102.
95. Aebischer P., T.K. Ip, G. Panol, P.M. Galletti (1987). 'The bioartificial kidney: Progress towards an ultrafiltration device with renal epithelial cells processing'. *Life Support. Syst.* **5**(2): 159–168.
96. Amiel G.E., J.J. Yoo, A. Atala (2000). 'Renal therapy using tissue-engineered constructs and gene delivery'. *World. J. Urol.* **18**(1): 71–79.
97. Humes H.D., D.A. Buffington, S.M. MacKay, A.J. Funke, W.F. Weitzel (1999). 'Replacement of renal function in uremic animals with a tissue-engineered kidney'.[see comment]. *Nat. Biotechnol.* **17**(5): 451–455.
98. Humes H.D., S.M. MacKay, A.J. Funke, D.A. Buffington (1999). 'Tissue engineering of a bioartificial renal tubule assist device: *In vitro* transport and metabolic characteristics'. *Kidney Int.* **55**(6): 2502–2514.
99. Ip T.K., P. Aebischer, P.M. Galletti (1988). 'Cellular control of membrane permeability. Implications for a bioartificial renal tubule'. *ASAIO Trans.* **34**(3): 351–355.

100. Joki T., M. Machluf, A. Atala, J. Zhu, N.T. Seyfried, I.F. Dunn, T. Abe, R.S. Carroll, P.M. Black (2001). 'Continuous release of endostatin from microencapsulated engineered cells for tumour therapy'.[see comment]. *Nat Biotechnol.* **19**(1): 35–39.

101. Lanza R.P., J.L. Hayes, W.L. Chick (1996). 'Encapsulated cell technology'. *Nat Biotechnol.* **14**(9): 11071111.

102. MacKay S.M., A.J. Funke, D.A. Buffington, H.D. Humes (1998). 'Tissue engineering of a bioartificial renal tubule'. *ASAIO. J.* **44**(3): 179–183.

103. Evans M.J., C. Gurer, J.D. Loike, I. Wilmut, A.E. Schnieke, E.A. Schon (1999). 'Mitochondrial DNA genotypes in nuclear transfer-derived cloned sheep'. *Nat. Genet.* **23**(1): 90–93.

104. Hiendleder S., S.M. Schmutz, G. Erhardt, R.D. Green, Y. Plante (1999). 'Transmitochondrial differences and varying levels of heteroplasmy in nuclear transfer cloned cattle'. *Mol Reprod Dev.* **54**(1): 24–31.

105. R., P. Schinogl, V. Zakhartchenko, R. Achmann, W. Schernthaner, M. Stojkovic, E. Wolf, M. Muller, G. Brem (2000). 'Mitochondrial DNA heteroplasmy in cloned cattle produced by foetal and adult cell cloning'. *Nat. Genet.* **25**(3): 255–257.

106. Fischer Lindahl K., E. Hermel, B.E. Loveland, C.R. Wang (1991). 'Maternally transmitted antigen of mice: A model transplantation antigen'. *Annu. Rev. Immunol.* 9351–9372.

107. Hadley G.A., B. Linders, T. Mohanakumar (1992). 'Immunogenicity of MHC class I alloantigens expressed on parenchymal cells in the human kidney'. *Transplantation.* **54**(3): 537–542.

108. Yard B.A., M. Kooymans-Couthino, T. Reterink, P. van den Elsen, M.E. Paape, J.A. Bruyn, L.A. van Es, M.R. Daha, F.J. van der Woude (1993). 'Analysis of T cell lines from rejecting renal allografts'. *Kidney International—Supplement.* **39**: S133–138.

109. Kershen R.T., J.J. Yoo, R.B. Moreland, R.J. Krane, A. Atala (2002). 'Reconstitution of human corpus cavernosum smooth muscle *in vitro* and *in vivo*'. *Tissue. Eng.* **8**(3): 515–524.

110. Kwon T.G., J.J. Yoo, A. Atala (2002). 'Autologous penile corpora cavernosa replacement using tissue engineering techniques'. *J. Urol.* **168**(4 Pt 2): 1754–1758.

111. Wang T., C. Koh, J.J. Yoo (2003). 'American Academy of Pediatrics Section on Urology'. *New Orleans, LA*'.

112. De Filippo R.E., J.J. Yoo, A. Atala (2003). 'Engineering of vaginal tissue *in vivo*'. *Tissue Eng.* **9**(2): 301–306.

113. Diamond D.A., A.A. Caldamone (1999). 'Endoscopic correction of vesicoureteral reflux in children using autologous chondrocytes: Preliminary results'. *J. Urol.* **162**(3 Pt 2): 1185–1188.

114. Bent A.E., R.T. Tutrone, M.T. McLennan, L.K. Lloyd, M.J. Kennelly, G. Badlani (2001). 'Treatment of intrinsic sphincter deficiency using autologous ear chondrocytes as a bulking agent'. *Neurourol Urodyn.* **20**(2): 157–165.

115. Kershen R.T., A. Atala (1999). 'New advances in injectable therapies for the treatment of incontinence and vesicoureteral reflux'. *Urol Clin North. Am.* **26**(1): 81–94.

116. Atala A., L.G. Cima, W. Kim, K.T. Paige, J.P. Vacanti, A.B. Retik, C.A. Vacanti (1993). 'Injectable alginate seeded with chondrocytes as a potential treatment for vesicoureteral reflux'. *J. Urol.* **150**(2 Pt 2): 745–747.

117. Atala A., W. Kim, K.T. Paige, C.A. Vacanti, A.B. Retik (1994). 'Endoscopic treatment of vesicoureteral reflux with a chondrocyte-alginate suspension'. *J. Urol.* **152**(2 Pt 2): 641–643; *discussion* 644.

118. Yokoyama T., J. Huard, M.B. Chancellor (2000). 'Myoblast therapy for stress urinary incontinence and bladder dysfunction'. *World J. Urol.* **18**(1): 56–61.
119. Chancellor M.B., T. Yokoyama, S. Tirney, C.E. Mattes, H. Ozawa, N. Yoshimura, W.C. de Groat, J. Huard (2000). 'Preliminary results of myoblast injection into the urethra and bladder wall: A possible method for the treatment of stress urinary incontinence and impaired detrusor contractility'. *Neurourol Urodyn.* **19**(3): 279–287.
120. Strasser H., S. Berjukow, R. Marksteiner, E. Margreiter, S. Hering, G. Bartsch, S. Hering (2004). 'Stem cell therapy for urinary stress incontinence'. *Exp Gerontol.* **39**(9): 1259–1265.
121. Yiou R., J.P. Lefaucheur, A. Atala (2003). 'The regeneration process of the striated urethral sphincter involves activation of intrinsic satellite cells'. *Anat Embryol* (Berl). **206**(6): 429–435.
122. Yiou R., J.J. Yoo, A. Atala (2003). 'Restoration of functional motor units in a rat model of sphincter injury by muscle precursor cell autografts'. *Transplantation.* **76**(7): 1053–1060.

27

BIOMATERIALS FOR ENHANCING CORNEAS AND SPINAL CORD REGENERATION

Joanne M. Hackett[1,2], *Benu Sethi*[3], *Xudong Cao*[3], *Mehrdad A. Rafat*[1], *May Griffith*[1,2]*

[1]*University of Ottawa Eye Inst, 501 Smyth Road, Ottawa, ON K1H 8L6, Canada*
[2]*Department of Cellular and Molecular Medicine, University of Ottawa, 451 Smyth Road, Ottawa, ON K1H 8M5, Canada*
[3]*Department of Chemical and Biological Engineering, University of Ottawa, 161 Louis-Pasteur, Ottawa, ON K1N 6N5, Canada*

INTRODUCTION

Repairing damaged or diseased tissues or organs to prevent failure using natural or bioengineered materials, is the essence of regenerative medicine. Technologies can involve the use of stem cells, gene therapy, tissue engineering, and the use of artificial organs. To date, researchers have been recreating a range of tissues and organs in vitro, with varying degrees of success. The tools to create scaffolds and structures are limited, and fine control over mechanical and environmental variables is complex. Also, there is a limited means of diagnosing at the molecular, cellular, and tissue levels what is actually happening within the constructs. In this review, we focus on efforts employed to build *in vitro* models of the nervous system, in particular, the peripheral nervous system and the related visual system. We examine the development of novel biomaterials that serve as the building blocks for the fabrication of scaffolds in engineered tissues.

Scaffolds must be fabricated from biologically compatible materials to be used as cellular supports for an engineered tissue or organ. Cells must be able to proliferate and differentiate into the appropriate target tissue or organ, when a chosen 3D scaffold is used. Engineered tissues need to mimic morphological, physiological and biochemical properties of the natural tissue as closely as possible. Thus, construction requirements are rigorous and demanding. The cornea and peripheral

---

* *mgriffith@ohri.ca*

nervous system (PNS) are currently experiencing advances in tissue engineering efforts, for both transplantation and *in vitro* testing.

Biomaterials Used in Regenerative Medicine
Synthetic materials

Natural or synthetic materials can be used in the development and fabrication of scaffolds. Synthetic materials, in particular polymers, have superior strength when compared to natural ones. Due to this, prostheses and implants are often composed of synthetic materials, although a material lacking biological function can not be an appropriate replacement for human tissues and organs. Using synthetic materials does have advantages, such as the consistency in reproducing the product, and the ease of manufacturing. Poly-methyl-methacrylate (PMMA) was used in the development of the first artificial corneas, known as keratoprostheses (KPro) [1–3]. KPros utilize materials that allow oxygen permeation and nutrient permeability, and are non-immunogenic and non-mutagenic [4]. Great advances have been made in the composition of KPros, although polymer technology is prominently utilized. Recombinant human collagen (RHC) is a synthetic version of human collagen and is used to support limbal epithelial proliferation and differentiation [5]. Some of the synthetic non-biodegradable polymers used in spinal cord injury (SCI) research for fabricating tubular conduits are acrylic polymers such as poly [*N*-(2-hydroxypropyl) methacrylamide] (PHPMA), poly (2-hydroxyethyl methacrylate-*co*-methyl methacrylate) (PHEMA), and poly (2-hydroxyethyl methacrylate-*co*-methyl methacrylate) (P (HEMA-*co*-MMA). Poly (lactic-*co*-glycolic acid) (PLGA) is an aliphatic polyester that has received extensive attention due to its availability, ease of processing, approval by FDA, low inflammatory response and modifiable degradation rates. PLGA has a long and successful history for use as a suture material, and recently it has also been studied as localized and sustained molecule delivery vehicle in the PNS and central nervous system (CNS). In PNS, PLGA and its degradation products do not appear to result in an untoward inflammatory response. In CNS, aliphatic polyesters have been used mainly for drug delivery applications in a microsphere form and also as guidance scaffolds for neuronal regeneration with reports of biocompatibility. Thus, PLGA devices used for axonal regeneration after SCI should have no adverse effects [6–8].

Natural and biomimetic materials

Modifying existing synthetic molecules by hybridizing them with naturally occurring bioactive peptides or growth factors, is one way to address the need for new materials with biological function. By using appropriate biomaterials as scaffolds or tissue templates, the native environment is being replicated. The environment that directed tissue or organ development is being synthetically manufactured in a way that encourages regeneration of tissues or organs. This allows the properties of synthetic materials to be utilized, while the biological compatibility of the material is no longer a concern. Optimal cellular function requires a complement of extracellular matrix (ECM) macromolecules. Structural proteins from the mammalian ECM, such as collagen, fibrin and proteoglycans, are often used in tissue engineered scaffolds. With continued efforts to improve KPros, scientists have grafted collagen, fibronectin, laminin, and other cell adhesive peptides such as Ile-Lys-Val-Ala-Val (IKVAV), Tyr-Ile-Gly-Ser-Arg (YIGSR) and Arg-Gly-Asp (RGD) onto the synthetic medium. By modifying the surface with multiple peptides, there was a synergistic effect of corneal epithelial cell attachment compared to single peptides on their own [9]. Corneal epithelial cell growth rate and adhesion were tested to see what would occur when bioactive factors (including fibronectin, laminin, substance P (SP), insulin-like growth factor 1 (IGF-1) and RGD) were directly coated on the surface of polymethacrylic acid-co-2-hydroxyethyl methacrylate (PHEMA/MAA) or tethered through poly(ethylene glycol) (PEG) spacers. The spacer molecules were found to provide the correct microenvironment for the epithelial cells to reach confluence, compared with little to no epithelial growth on the surfaces that were only coated with the bioactive factors [10]. These studies show that biologically derived glycoproteins, or peptide sequences, can be used effectively to promote biointeraction with cells. The Stanford keratoprosthesis consists of a double network of PEG and poly(acrylic acid) (PAA) as its central optic, as this allows for overgrowth of corneal epithelial cells. A double network of polymers increases the mechanical strength, optical transparency and biocompatibility of the polymer. Microperforations on the outer layer allow further integration with the host tissue. Also, a photochemical surface modification strategy has been utilized to tether bioactive factors to specific sites of the polymer to promote cell adhesion [3, 11, 12]. In SCI research, use of semipermeable polyacrylonitrile: polyvinylchloride (PAN/PVC) channels filled with suspension of Schwann cells facilitated the regeneration of propriospinal, and some long-tract axons, when implanted into a transected rat spinal cord [13]. PAN/PVC by itself has not been shown to have any beneficial properties toward the regeneration as compared to biodegradable materials and hence can be easily substituted [7]. Even though collagen as a material is a naturally available biodegradable polymer, it has been shown by experimental studies that it may be a component of the inhibiting glial scar after SCI. This casts a doubt on its effectiveness and usage [14].

Chitosan

Chitosan is a natural polymer that is derived from chitin, which the major component of crustacean shells. Chitosan scaffolds have been used to fabricate tissues, vascular grafts, cartilage, skin and to provide substrates for neural tissues. Chitosan is biocompatible, promotes wound healing, and has antibacterial properties [15–17]. Tubular SCI implants made from chitosan were more permeable than PLGA fibers, but it was found difficult to control pore size of chitosan tubular scaffolds [18]. Chitosan fibers have been shown to swell, which could be related to the difficulty in controlling pore size [19]. A drug delivery system (DDS) can be designed to provide growth factors at the lesion site after SCI. Goraltchouk et al. [20] in an effort to enhance the nerve regeneration after SCI suggested localized delivery of therapeutic molecules, by incorporating protein encapsulated microspheres into the nerve guidance channel wall. PLGA 50/50 microspheres were physically entrapped into the annulus of two concentric tubes; the inner one fabricated from chitosan and the outer one from chitin. The inner layer provided a cell adhesive surface, advantageous for cell interaction and guided regeneration, and the outer tube provided a non-adhesive cell surface to reduce cell and tissue infiltration at the outer surface.

Alginate

Alginate is a linear polysaccharide that is extracted from seaweed [21]. Alginate hydrogels are used in tissue engineering applications, such as drug delivery and cell encapsulation. To be useful in scaffold fabrication, the hydrogels require covalent modification or coupling with bioactive peptides or other ECM molecules. For ocular applications, a collagen hydrogel containing protein encapsulated alginate microspheres was developed. The composite hydrogel was able to support human corneal epithelial cell growth, had adequate mechanical strength, and excellent optical clarity. Thus, the composite hydrogel has the possibility of being used as a therapeutic lens for drug delivery, and/or used as a corneal substitute for transplantation into patients who have corneal diseases [22].

Collagen

Collagen is a fibrous protein used to provide tensile strength and elasticity to matrices supporting body cells and tissues. Collagen is found in multicellular organisms and accounts for approximately 30% of all body proteins [23]. Natural crosslinks ensure that native collagen is robust *in vivo*; however, recombinant and animal-derived collagens rapidly degrade and lack the mechanical toughness and elasticity required for implantation. Collagen molecules can be reinforced using synthetic crosslinkers and polymeric agents. Once reinforced, collagen can be fashioned into various shapes and sizes. *In vitro* corneal equivalents reconstructed from immortalized human cell lines have been developed from collagen-based materials [60]. Also, corneal scaffolds based on the copolymer poly(*N*-isopropylacrylamide-*co*-acrylic acid-*co*-acryloxylsuccinimide), and simple cross-linkers such as 1-ethyl-3-(3-dimethylaminopropyl) carbodiimide (EDC), and *N*-hydroxysuccinimide (NHS), have also been designed [62, 64]. These scaffolds allowed regeneration of corneal cells and nerves, when implanted as lamellar grafts. Although these materials lacked the optimal toughness and elasticity required, research is currently aimed at improving these collagen lamellar grafts [24].

MODELS OF THE NERVOUS SYSTEM

Cornea

The human cornea forms a transparent window through which light is transmitted to the retina, enabling one to see. A unique property of the cornea is its optical clarity, which accounts for over 70% of the light that is transmitted to the retina for vision. Corneal clarity is now believed to result from a combination of refractive index matching, and the presence of structural components that are well below the wavelength of visible light [25–27]. The cornea is avascular, as well as immune privileged; however it contains a dense network of nerves. Central thickness is about 500 ìm, with a total periphery of about 750 ìm [28]. It can be viewed as a largely collagenous "stromal" matrix containing cells sandwiched between an outer epithelium and an inner endothelium. The corneal epithelium consists of stratified, non-keratinizing epithelial cells, which is approximately 50 ìm thick. Biochemically, the stroma consists of 13.6% type I collagen, 0.9% glycosaminoglycans, and

80% water, making the stroma resemble a hydrogel. Stromal cells, or keratocytes, are localized in lamellar networks within this hydrogel. The basal layer of the epithelium proliferates to replace the superficial cells lost at the anterior surface [29], which are subsequently replenished by a population of corneal stem cells that reside within the corneal/scleral limbus [30]. Several antiinflammatory and antimicrobial factors are secreted by the epithelium, as an insoluble mucous layer, which aids in stabilizing the tear film [31]. Corneal nerves are especially important for the health of the epithelial cells they innervate [32]. Loss of corneal innervation decreases epithelial cell proliferation and migration, which compromises the epithelial barrier function and slows corneal wound healing [33, 34]. The 10 ìm single cell inner endothelium functions to regulate the hydration level of the corneal stroma, which helps to maintain central corneal thickness. Sodium/potassium ATPase pumps, within the endothelium, circulate aqueous humor between the anterior chamber and stroma [35].

The function of the cornea is compromised when it loses its transparency. Loss of clarity, due to disease or damage, results in vision loss and where irreversible, results in corneal blindness. According to the World Health Organization (WHO), there are approximately 50 million people worldwide who possess bilateral blindness, as well as at least 150 million people who have impaired vision in both eyes [36, 37]. Furthermore, it is estimated that corneal ulceration and ocular trauma results in 1.5–2 million new cases of corneal blindness annually [37]. At present, transplantation of matched human donor tissue is the only widely acceptable treatment.

Depending on the nature of the damage, replacement of one or more corneal layers is possible, using corneal allografts or progenitor cell grafts (e.g. limbal stem cell grafts). When the stromal matrix is damaged or weakened (e.g. keratoconus), attempts to strengthen the matrix using *in situ* crosslinking have been attempted [38]. Corneal transplants are the most successful organ transplants, but graft survival rates are only moderate: 86% at one year postoperative, 73% at five years, 62% at ten years, and 55% at fifteen years. Also, no improvements in graft survival rates have been seen throughout the fifteen years of grafting [39]. The demand for good quality donor corneas is expected to rise beyond the available supply, with all countries except for the USA already experiencing a shortage. Life expectancy is longer now than before, there is a higher incidence of infectious diseases (HIV, hepatitis, Creutzfeldt-Jakob disease), and the growing popularity of refractive surgery renders the surgically treated corneas too thin to be acceptable as donor tissues [40–42].

As alternatives to donor tissues, bioengineered corneas are designed to replace some or all of a damaged or diseased cornea. They range from prosthetic devices that solely address replacement of the cornea's function, to tissue engineered hydrogels that permit regeneration of host tissues. If corneal stem cells have been depleted by injury or disease, tissue engineered lamellar implants reconstructed with stem cells have been transplanted. *In situ* methods using ultraviolet A (UVA) crosslinking have also been developed to strengthen weakened corneas. In addition to the clinical need, bioengineered corneas are also rapidly gaining importance in the area of *in vitro* toxicology. In Europe, there is currently a ban on consumer product testing using animals (European Union Directive 76/768/EEC) that is expected to expand worldwide. Complex, fully innervated, physiologically active, 3D organotypic corneal models are currently being developed and tested.

Spinal Cord

Normal architecture of the spinal cord can be radically disrupted by injury [43]. Worldwide, an estimated 2.5 million people live with a SCI, with more than 130,000 new injuries reported each year [44]. SCI can result from contusion, compression, penetration or maceration and transection of spinal cord (Fig. 27.1). It leads to death of cells, including neurons, oligodendrocytes, astrocytes and precursor cells that exist within the spinal cord. Even though circumferential white matter is spared during injury, cavities and cysts resulting from SCI interrupt the descending and ascending axonal tracts. After initial injury, additional structure and functional loss occurs through active secondary processes. This tissue damage in the CNS leads to glial scarring that involves reactive astrocytes, glial progenitors, microglia, macrophages, and secreted and transmembrane molecular inhibitors of axon growth that creates a barrier for regeneration (Figs. 27.2 and 27.3).

The glial scar stabilizes the fragile CNS tissue after injury, prevents an overwhelming inflammatory response and limits cellular degeneration. Unfortunately, this interferes with the long distance regenerative capacity of axons, and hence leads to functional degeneration. Astrocytes and the glial scar inhibit remyelination of axons within lesions. Proteoglycans are additional inhibitory molecules produced by astrocytes. There are four of these ECM molecules: heparin sulphate proteoglycan (HSPG), dermatan sulphate proteoglycan (DSPG), keratin sulphate proteoglycan (KSPG) and chondroitin sulphate proteoglycan (CSPG). They are implicated as barriers in the axonal regeneration, and repel embryonic as well as adult axons. In mature mammals, CSPG are secreted after injury and can persist for many months. Over time the glial scar develops to become a rubbery, tenacious, growth that blocks the membrane. Progressive expansion of injury across more than one segment, in the form of cysts, known as syringomyelia, can occur over months or years and can be fatal. The physical structure of the glial scar is comprised of tightly interlaced processes of astrocytes connected by gap junctions and interwoven with an extensive extracellular matrix [45].

A tubular structure is needed to bridge spinal cord gaps so that axons can extend from the inhibitory boundary of the glial scar to a permissive environment of uninjured CNS [46]. Both the PNS and CNS require chaemotactic cues, and appropriate guidance for axonal regeneration. The PNS is often considered to be a model for regeneration, since it has a much more permissive environment that encourages growth after injury.

Studies suggest that the initial attempt, after SCI, should be to reduce secondary tissue loss through neuroprotective, antiinflammatory or immunomodulatory interventions [47]. Then an attempt should be made to promote regrowth of axons and restore function. Studies using cellular therapies, scaffolds and tubular structured implants have been carried out with the goal to restore functional recovery after SCI. None of these therapeutic strategies, by themselves, seem to be sufficient to achieve the final goal. However, combinatorial approaches may aid in functional restoration after SCI.

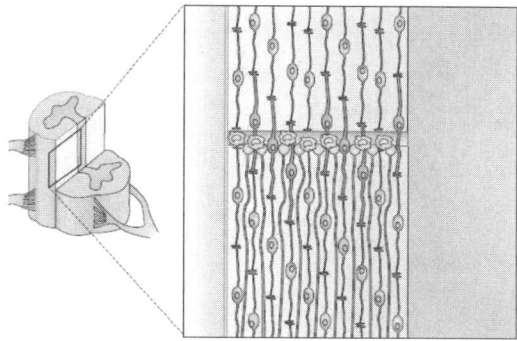
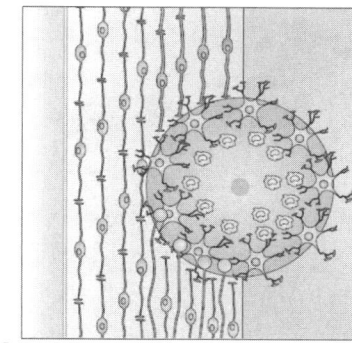

(A) **Microlesion**
• Blood-brain barrier is minimally disrupted.
• Astrocytes maintain normal alignment but produce CSPGs and KSPGs along the injury tract.
• Axons cannot regenerate beyond the lesion.
• Macrophages invade the lesion site.

(B) **Contusive lesion**
• Blood-brain barrier is disrupted, but meninges are intact.
• Cavitation occurs at the lesion epicentre.
• Astrocyte alignment is altered at the lesion site.
• Astrocytes produce CSPGs and KSPGs in a gradient, increasing from the penumbra towards the centre of the lesion.
• No fibroblast invasion or lesion core and therefore no fibroblast-expressed inhibitors are present (see below).
• Macrophages invade the lesion and its core.
• Dytrophic axons approach the lesion before growth ceases.

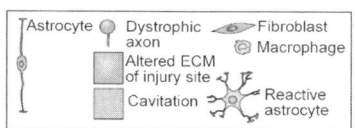

(C) **Large stab lesion**
• Blood-brain barrier is disrupted.
• Cavitation occurs at the lesion centre.
• Astrocyte alignement is altered at the lesion site.
• Astrocytes produce SCPG and KSPG in a gradient increasing towards the lesion.
• TGF, ephrin-B2 and Slit protein expression increases in reactive astrocytes adjacent to fibroblasts.
• Fibroblasts invade the lesion and express SEMA3 and the EPHB2 receptor.
• Macrophages invade the lesion and release inflammatory cytokines.
• Dystrophic neurons are highly repeled by the lesion core and express neuropilin 1.

▶ **Fig. 27.1** *Schematic representation of three stereotypical CNS lesions. In all examples, macrophages invade the lesion, and both chondroitin sulphate proteoglycans (CSPGs) and keratan sulphate proteoglycans (KSPGs) are upregulated. (A) Microlesion in which astrocyte alignment is not altered by the injury process, but axons are unable to regenerate past the lesion site. (B) Contusive injury that does not disrupt the meninges, but produces cavitation and proteoglycan deposition. Again, axons are unable to regenerate beyond the lesion, but spared axons can be found distal to the injury site. (C) Stab lesion that penetrates the meninges and allows fibroblast invasion in addition to macrophages. Axons are highly repulsed by the increasing gradient of CSPGs and KSPGs. Several other inhibitory molecules are also made in this type of injury and are especially prevalent in the core of the lesion. ECM, extracellular matrix; TGF, transforming growth factor. (for colour figure see Plate 77).*

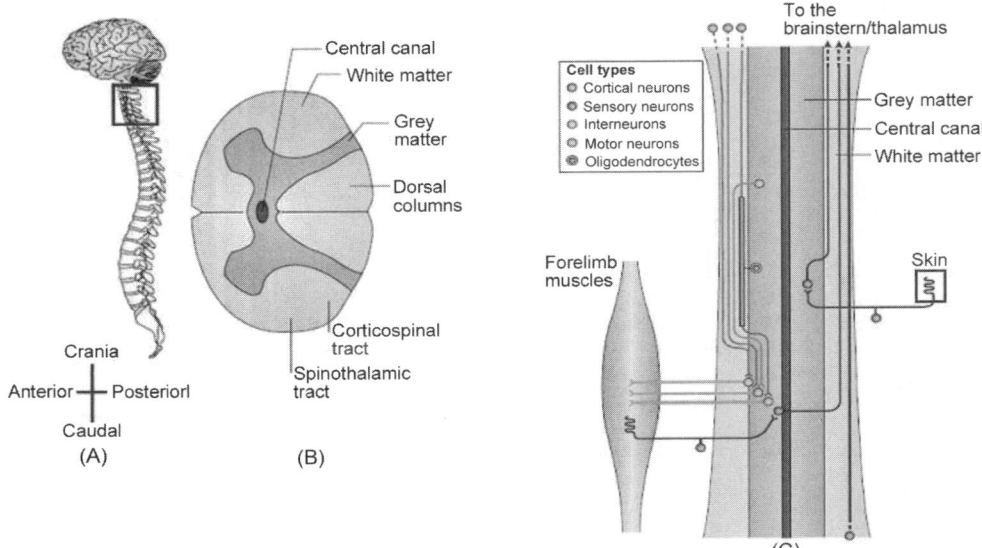

▶ **Fig. 27.2** *Intact spinal cord. (A) Schematic diagram showing a sagittal view through the human CNS. (B) Transverse section through human spinal cord showing the relationship between axonal tracts and grey matter. (C) Cortical, brainstem and spinal axons project to motor neurons in spinal cord grey matter, which in turn send axons through the PNS to target organs, including muscle. Primary sensory axons send axons through the PNS to second order sensory neurons in the CNS grey matter, which, in turn, send axons through white matter in the dorsal columns to supraspinal regions. Oligodendrocytes myelinate ascending and descending axons. (for colour figure see Plate 78).*

CORNEAL SUBSTITUTES
Corneal Substitutes for Transplantation
Keratoprostheses

KPros are synthetic implants designed to replace the central portion of an opaque cornea. Most designs are based upon a transparent central optic surrounded by a skirt which provides anchorage through integration into the host tissue. The microporous skirt is designed to allow cellular integration of the host tissue which promotes ECM (mainly collagen) deposition to provide firm anchorage. The material comprising the skirt must maintaining proper strength to allow for suture placement. The anterior surface of the device should promote epithelialization, as well as inhibit downgrowth of epithelial cells into the stroma/implant interface. The posterior surface can be modified to inhibit cellular attachment and proliferation, to prevent retroprosthetic membrane formation which leads to corneal opacification. The materials used in KPros should also promote oxygen and nutrient permeation to maintain the cellular components [4]. Finally, the synthetic

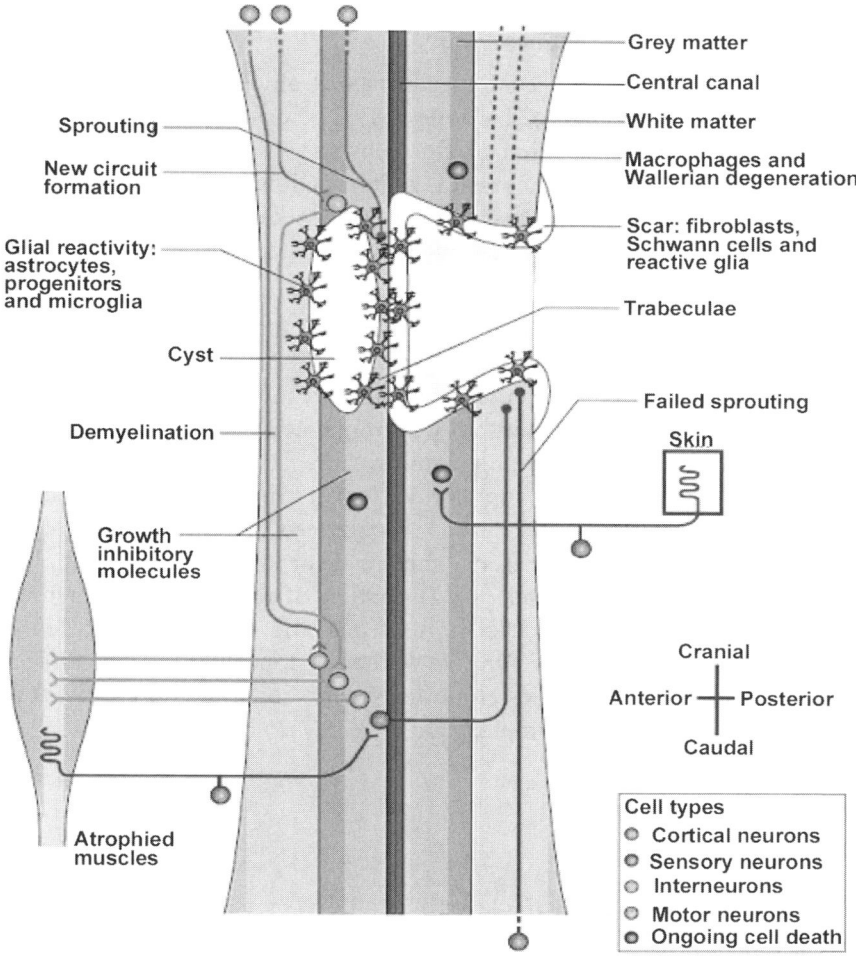

▶ **Fig. 27.3** *Spinal cord after injury. Schematic diagram showing a sagittal view through a region of cervical SCI, depicting a combination of features from different types of injury. Many cells die immediately as well as progressively after SCI. Cysts usually form after contusion injury. After penetrating the injury site, cells from the PNS often invade the injury site to form a connective tissue scar that incorporates astrocytes, progenitor cells and microglia. Many ascending and descending axons are interrupted and fail to regenerate over long distances. Some axons form new circuits with motor neurons via interneurons. At the site of cyst formation, axons can sprout into trabeculae that are formed from ependymal cells. Disconnected myelinated axon segments are phagocytosed by macrophages. Some spontaneous remyelination occurs, largely by PNS Schwann cells, whereas denervated (non-spastic) muscles atrophy. (for colour figure see Plate 79).*

materials used in the device should have non-immunogenic nor mutagenic properties, allowing for graft acceptance. A few specific examples follow.

The osteo-odonto keratoprosthesis (OOKP) consists of autologous tissue derived from tooth and bone, which surrounds a central PMMA optic [48]. The osteodental skirt is preimplanted into the buccal mucosa, allowing colonization of fibroblasts to support integration when implanted ocularly. This KPro is one of the most successful, as it has a low extrusion rate due to the excellent integration of the skirt material with the host tissue [49]. Associated complications with the OOKP include retroprosthetic membrane formation, glaucoma and decentration of the central optic, due to absorption of the osteodental skirt [50].

The AlphaCor™ KPro is fabricated as a one-piece device that comprises a transparent core and an opaque porous skirt, both made of PHEMA [51]. This KPro is unified by an interpenetrating network of polymers, which differ only in water content, resulting in a device that lacks glued or mechanical junctions [52]. The porous skirt allows stromal cells to grow into it, thereby anchoring the device stably within the host eye. Implantation of the device involves a two-stage surgical procedure. This device has regulatory approval in North America, Australia and Europe. The targeted clinical use is for patients with scarred, vascularized or diseased corneal tissues who are either not eligible for conventional donor tissue transplants or who have had multiple previous graft failures. Associated complications include the formation of retroprosthetic membranes, corneal melt, retained lenticular material, optic depositions and rare cases of device extrusion. Contraindications include abnormal tear film, as well as uncontrollable high intraocular pressure. Topical administration of medroxyprogesterone has been shown to limit corneal melting of the device. It should be noted that herpes simplex virus-1 (HSV-1) infection is no longer considered a contraindication for implantation [51].

Keratoprostheses with cell coverage

The regrowth of an intact corneal epithelial layer over a KPro may help to stabilize the tear film, and prevent extrusion and infection [53], which are complications observed with more traditional KPros. Several KPros have naturally occurring ECM proteins grafted onto them, and some of these are discussed in Section "Hybrid materials."

Self-assembled corneal substitutes

Scientists at the Laboratoire d'Organogenèse Expérimentale (LOEX) have developed a construct that uses a self-assembly approach [54]. Ascorbic acid is used to stimulate the production of collagen and other ECM molecules by dermal fibroblast cells. These matrix glycoproteins are elaborated into sheets which get stacked together, allowed further integration in culture. An epithelium is then seeded on top of the stack. These constructs show excellent corneal morphology, and cells express appropriate tissue-specific markers. The primary drawback is the time needed to produce the amount of material required for transplantation [55]. An improved self-assembled model has been developed that is comprised of a stroma consisting of human corneal and dermal fibroblasts [56]. The combination of corneal and dermal fibroblasts is more conducive to the formation of a well-differentiated epithelium, which shows higher re-epithelialization rates than just corneal fibroblasts alone. This model reproduces the microanatomy of the native human cornea. In addition, this model was able to reproduce a mechanistically accurate wound healing process, rendering it useful for studying wound healing, screening bioactive factors that could modulate wound healing, or as a prescreen prior to animal testing [56]. The ECM macromolecules deposited by primary human corneal fibroblasts have been characterized in this model [57]. These constructs were highly cellular and were morphologically similar to the stroma of mammalian corneas, with multiple, parallel layers of cells and small fibrillar ECM arrays.

Biomimetic corneal implants promoting regeneration

Type I collagen is the dominant biopolymer found in the human cornea, comprising 70% of its dry weight. It is amenable to modification and has been investigated for use as a scaffold for artificial cornea construction [58, 59]. Numerous collagen-based corneal substitutes have been implanted successfully into mice, rabbits, guinea pigs, dogs and pigs as either deep lamellar grafts, or full thickness implants [60–64]. These gels can be fabricated to the appropriate dimensions and curvatures, which allow for transmission of 90% or higher of white light. Corneas harvested from implanted eyes mimic normal pig corneas in key mechanical properties. Generally, mechanical properties of implants were found to be either similar or superior to the ones for control pig corneas. Implants fabricated with RHC showed very similar results [65]. Using RHC is advantageous, as the risk of disease transmission and the possibility of an immune response is avoided.

Lenticules have been implanted in RHC artificial corneas as inlays under the Bowman's membrane of a patient. Corneal transplantation with the RHC corneas, as deep lamellar grafts, is currently in Phase I human clinical trials (Fig. 27.4). Early postoperative results show regeneration of cornea epithelium and stroma in the patients with vision loss due only to keratoconus or central scarring [66]. Six month postoperative results show the presence of regenerating nerves [67]. Such implants have not been shown to cause adverse reactions, and are therefore suitable as temporary grafts or patches.

Synthetic materials have also been hybridized with collagen based corneal alternatives, as a way of enhancing interaction with the host cornea [62]. Finally, it has been shown that corneal substitutes can be fabricated to incorporate micro- or nanoparticles that release a drug [22].

Scaffold Substitutes for Cell-based Therapies

Often, only one corneal layer is damaged. Usually it is the outermost epithelial layer, as it is exposed to the environment and is a target for accidental exposure to irritating and toxic substances. The stem and progenitor cells that normally respond to the injury may also be destroyed. As such, there have been various attempts to repopulate the cornea.

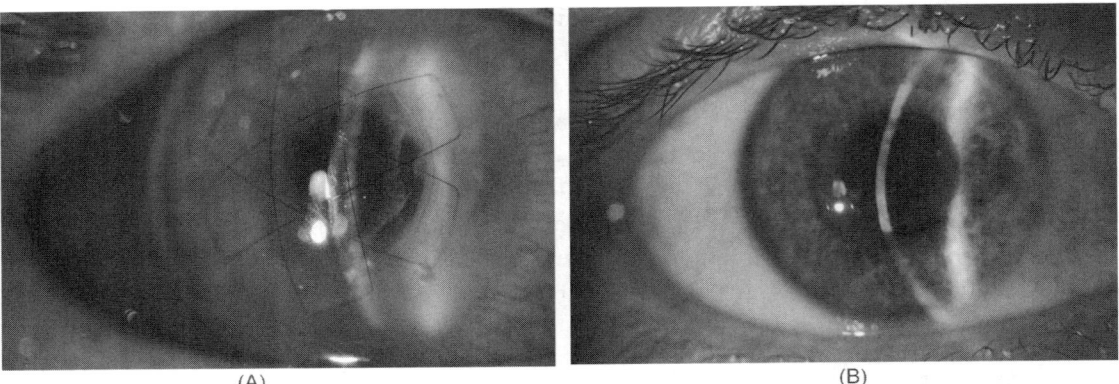

▶ **Fig. 27.4** *Corneal transplantation with RHC corneas, as deep lamellar grafts, are currently in Phase I human clinical trials. (A) A recently implanted artificial cornea (sutures visible). (B) The success of the implant, 9 months after the operation. (for colour figure see Plate 80).*

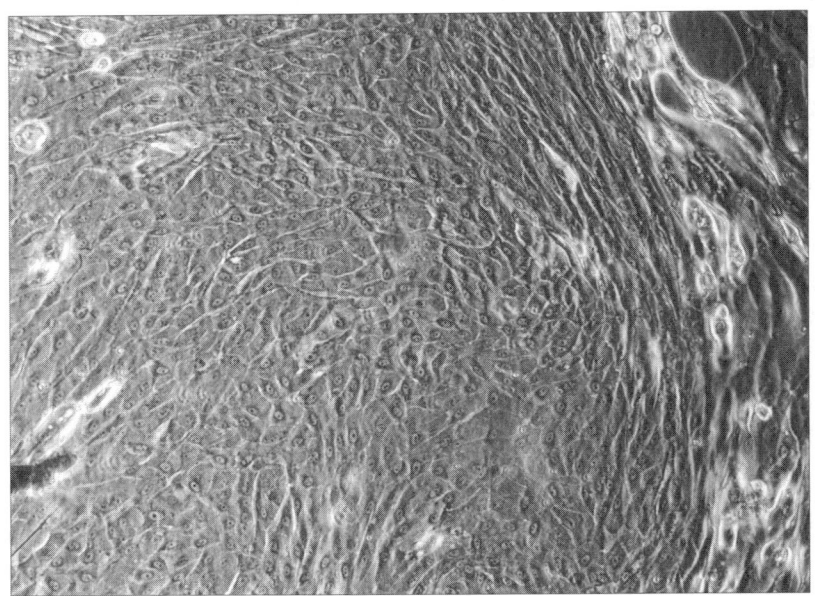

▶ **Fig. 27.5** *Human limbal stem cells explanted onto a type III recombinant human collagen (RHC-III) scaffold, after 3 days of growth. Visualised using phase contrast microscopy. (for colour figure see Plate 80).*

Tissue engineered scaffolds

Repopulating the epithelium involves using corneal stem cells from the surrounding limbus, either from the undamaged contralateral eye (autograft) or from allogeneic sources. Explants are often seeded on human amniotic membranes [67] or fibrin substrates [68], and outgrowing cells will form sheets that are transplanted onto the damaged eye. Corneal limbal cells can also be seeded onto fully synthetic, crosslinked human recombinant collagen substrates, as fully stratified corneal epithelia can be reconstituted on these (Fig. 27.5) [5].

When both corneal surfaces have been depleted of stem cells, autologous reconstruction of the corneal surface by transdifferentiation of oral mucosal epithelium has been performed [69]. The oral mucosal cells are cultured on human amniotic membranes before transplantation. It has been suggested that transdifferentiation by autologous epithelial precursor cells may be safer for ocular resurfacing than with allogeneic grafts [69].

A common reason for transplantation in patients with congenital or age-related conditions, is failure of the monolayered corneal endothelium. Transplants of tissue engineered endothelial sheets have been tested in animals, mainly the rabbit model, even though the endothelium is able to regenerate readily, unlike in humans [70]. Human corneal endothelial cells have also been grown successfully *in vitro* on a range of supports, including Descemet's membrane [70], amniotic membrane and crosslinked type I RHC.

Corneal Substitutes for Toxicology Testing

The Draize or Low Volume Eye Test (LVET) [71] has been used to observe changes in optical clarity, in rabbits, after the cornea has been exposed to test substances. The Bovine Cornea Opacity and Permeability (BCOP) test is an *in vitro* method used to assess ocular irritancy or toxicity [72]. Corneas are exposed to test substances and then measured for a loss in transparency and changes in permeability, indicating the potential toxicity of a substance. Researchers are investigating tissue engineered *in vitro* corneal alternatives, as a way to eliminate animal research in the field of ocular toxicity testing.

Three dimensional models

A human corneal epithelial cell line (HCE-T) has been developed for *in vitro* test protocols. The cells are grown on a simplified collagen membrane and induced to stratify at an air liquid interface. This allows for the formation of tight junctions and a true epithelial barrier [73]. A commercially available corneal epithelium model is the EpiOcular model, designed by MatTek Corporation [74]. Human foreskin keratinocytes are allowed to stratify in a specialized medium at an air liquid interface, and then can be purchased for cellular viability testing. Although these models are gaining acceptance, they are not able to predict responses to chemicals that may affect the other cell layers of the cornea, or that are dependent upon the interactions of the epithelial cells with other cell types within the cornea.

Full thickness models

A fabricated full-thickness cornea was developed by utilizing a stroma comprised of neutralized type I collagen blended with rabbit stromal fibroblasts, and seeded with rabbit corneal epithelial cells. Immortalized mouse endothelial cells were seeded on the reverse side, making a true three dimensional corneal construct [75]. An improved full-thickness cornea was later developed that used an immortalized human corneal epithelium, endothelium, and keratocyte cell lines [60]. The tissue engineered artificial corneas were fabricated around a type I collagen/ chondroitin sulphate stroma and crosslinked with glutaraldehyde to provide a more physiologically relevant and robust structure. The epithelium was allowed to stratify at an air liquid interface, before constructs were tested for opacity changes in response to irritating chemicals, as well as for changes in the expression of wound healing genes [60]. These models lack innervation, which shows the limitations, as no indication of pain can be extrapolated.

Innervated corneal model

A sensory nerve supply is crucial for optimal tissue function [76]. To develop an appropriately engineered organ or tissue substitute, the model should be innervated. Using a sensory nerve source, a functionally innervated *in vitro* corneal model has been described [77]. Using glutaraldehyde crosslinking, laminin and nerve growth factor (NGF) were crosslinked into type I rat-tail collagen and chondroitin sulphate hydrogel scaffolds, as laminin and NGF have been shown to promote differentiation and guidance of nerves [78, 79]. The sensory neuron source originated from isolated chicken dorsal root ganglia explants (DRG), and were implanted within the hydrogel scaffold prior to gelling. Retinoic acid, known to induce neurite outgrowth from embryonic mouse dorsal root ganglia explants, was added to media to promote neurite extension [80]. In this model, structural morphology, similar to what is found in the human cornea, was observed [81, 82]. Also using this model, a more physiologically relevant toxicological assay was developed [83]. This assay confirms the hypothesis that proper epithelium homeostasis can be maintained in the presence of functional corneal nerves, *in vivo* [84]. Based on this design, ocular irritancy can be tested using the innervated artificial cornea, *in vitro*. Upon further refinement of this model, a corneal–scleral model has been developed [85]. Innervating the sclera and cornea, and having nerves within the sclera form a ring around the cornea with branches penetrating into the cornea, gives a design that is of great likeness to that found in the normal cornea [86].

SPINAL CORD REPAIR AND REGENERATION
Schwann Cell Grafts

Schwann cell grafts, when positioned between two transected stumps of spinal cord, can serve as a bridge to promote axonal regeneration from both rostral and caudal stumps [13]. PAN/PVC copolymer channels were fabricated with a smaller Schwann cell cable suspended in DMEM:Matrigel (70:30). Schwann cells survive inside the channels causing propriospinal and dorsal root ganglion axons to grow. When implanted within rats, the Schwann cell cable leads to growth of myelinated and unmyelinated axons. Fibers regenerate from both caudal and rostral stump into channels. Although there was a Schwann cell cable within the entubulated PAN/PVC channels, no regrowth of brainstem or cortical neurons was observed. Also, regenerated axons did not exit the Schwann cell environment to enter host cord tissue. Therefore, regeneration of axons only occurred within the graft, meaning no synaptic connections will form and a functional recovery will not occur [13]. In a hemisected model of spinal cord, Schwann cell tissue cable led to regrowth of brainstem axons from many regions [87]. Even though this model produced positive results, using a hemisection model to obtain growth of different axons is not realistic. A hemisection model takes only a very specific SCI model into consideration, and such injuries may or may not exist in reality. A transection model is considered to be a more suitable model for SCI studies, as it considers the adverse scenarios and maximum possible damage that can occur as a result of SCI. Genetically modified Schwann cells secreting human brain-derived neurotrophic factor (BDNF), have been shown to regenerate brainstem neurons within Schwann cell grafts implanted after transection of spinal cords in rats [88]. As the rate of release of BDNF cannot be controlled and only a particular section of neurons get targeted, this technique by itself is not sufficient.

To see if adult human Schwann cells would aid in remyelinating injured CNS axons, poly (L-lactic acid) (PLLA) conduits were employed [89]. Results showed that seeding PLLA conduits with Schwann cells did not functionally enhance peripheral nerve regeneration. Schwann cells were only able to proliferate when in contact with neurons and axons [90].

Tubular Strategies

Combining a cellular therapy with a tubular scaffold is important for axonal regeneration [91–93]. During the early phase of growth response after SCI, the entubulation limits formation of scar tissue by allowing accumulation of neurite growth promoting factors. Combining a peripheral nerve grafts with a growth factor has also shown axonal regeneration after SCI. Fibroblast growth factor (FGF-1) helps reduce inhibitory proteoglycans, affects neuroprotection and encourages neurite growth. The presence of a peripheral graft improves generation, due to the presence of Schwann cells [94]. The combined effect of a Schwann cell bridge and olfactory ensheathing glia (OEG) was studied in adult rats with a spinal cord transection [95]. Hindlimb and forelimb coupling was improved, an increased number of myelinated axons in the Schwann cell bridge were observed, and serotonergic fibers grew through the bridge into caudal spinal cord. Unfortunately prominent descending tracts, such as the corticospinal tract, did not successfully regenerate.

Tubular implants

An ideal engineered artificial neural implant would enclose the rostral and caudal ends of the injured spinal cord, as well as guide and help axons grow and regenerate across the glial scar [96]. The presence of a tubular implant should prevent formation of scar tissue, allow for accumulation of axonal growth promoting molecules, and serve as a protective casing with appropriate mechanical strength and flexibility [44]. Tubular implants, also referred to as guidance conduits, nerve guidance channels or hollow fiber membranes, must be fabricated from biomaterials that are non-cytotoxic, non-carcinogenic, non-immunogenic, non-mutagenic, and cause no irritation or allergic response a local or systemic level *in vivo*. An ideal biomaterial should be able to be modified into a desired shape, be simple to handle during preparation and transplantation, have a controlled porous architecture to allow for cell growth, tissue regeneration, and vascularization, serve as a scaffold for growth factors, and degrade after serving its purpose *in vivo* [97, 98].

3-D printing fabrication technology has been used to design implants from PLGA [7], however only standard shapes can be generated at a given time. An implant modelled after an intact spinal cord, consisting of a multicomponent polymer scaffold seeded with neural stem cells, has been fabricated [99]. The inner surface of the implant was fabricated from a salt leaching process, while the outer surface was made from a solid–liquid phase separation technique using a blend of PLGA 50 : 50 and polylysine. The outer layer inhibited cellular ingrowth and impeded scarring and cyst formation. Implantation of the scaffold with neural stem cells (NSCs) improved functional recovery and reduced the scar tissue. In a hemisection model, it is not clear whether the regeneration is occurring because of the infiltration of cells from the adjacent section of the intact spinal cord.

PLGA 10 : 90 fibers have been used to fabricate tubular scaffolds using a microbraiding technique [100]. There was no inflammatory response, a thin tissue capsule was formed on the outer surface of the conduit, and the conduits were flexible, permeable and displayed no swelling. One month after implantation the majority of rats showed successful nerve regeneration. Channels fabricated by the same methodology were compared when composed of different materials, PLGA 90 : 10 and chitosan fibers. PLGA conduits showed negligible swelling whereas chitosan exhibited 60% swelling, however cell viability was shown in both the cases [19].

A linearly oriented nerve guidance scaffold was fabricated from agarose, by using a freeze drying process [101]. These scaffolds had uniaxial linear pores that extended the full length of the scaffold, which helped physically align and restrict the direction of growing axons. Pore size was large enough to allow vascularization and infiltration of cells that support regeneration. The scaffold was able to physically guide the linear growth of axons across a site of injury, as well as provide neurotrophic and cellular support. This aids in retaining the native organization of regenerating axons across the lesion site and into distal host tissue, and could increase the probability of achieving functional recovery.

The concept of hapto-axis was implemented into 3-Dl neural based scaffolds, creating PLLA conduits that possess multiple intralumenal walls [102]. These conduits have a surface area up to eight times larger than a comparably sized implant, providing an adsorption area for endogenous adhesion proteins. Physical parameters such as number of lumens, conduit length and diameter are able to be controlled. The choice of biomaterial used to fabricate the conduits is capable of being custom tailored, so that reabsorption can occur in parallel with the healing process. This study highly appreciates the structural attributes of the developed tubular scaffolds, however further studies *in vitro* and *in vivo* are required to establish their efficacy.

Collagen filaments have been implanted in the young rat model after SCI [103]. Complete recovery was elusive, and has not been reported so far in adult mammals. An agarose scaffold that gels *in situ* at 37 °C has been used to fill the irregular cavity that results after SCI in adult rats [104]. This scaffold also serves as a carrier of trophic factors, using BDNF to encourage neurite growth into scaffolds. A novel DDS was developed to provide localized release of growth factors from an injectable gel blend of hyaluronan and methylcellulose (HAMC) [105]. It did not bring about adverse reactions, and displayed a low inflammatory response. Other materials, like silk fibroin matrices, are also being considered for use as scaffolds for SCI [106]. Biocompatibility, slow biodegradability and good mechanical properties are all desired. Adequate research is needed to establish the utility of these materials to be used as a scaffold, or to be formulated into a tubular structure for use in regeneration after SCI.

In many recent studies, the concept of a multi-layered scaffold has been investigated and shown to be effective after SCI [107–110]. Layers of stiffer and harder materials have been successful in increasing mechanical strength of scaffolds. Various factors, including pore size, morphology of the layers, and their surface interaction, play a crucial role in determining the final mechanical strength of scaffold.

CONCLUSION

Significant advances have been made in the development of biomaterials as scaffolds and other support structures for enhancing stem cell based therapies in regenerative medicine. They can be used as substrates for cell delivery to a particular layer, or as acellular, biointeractive scaffolds that serve as templates for host regeneration.

In the cornea, regeneration of the host cornea could overcome the rejection problems and other postoperative complications from donor tissue transplantation and KPros. In addition, corneal implants that allow nerve regeneration could also circumvent problems resulting from the lack of nerve regeneration after surgery, found both in human donor tissue and in synthetic KPros. At the current pace of product development and testing, viable alternatives to donor corneas for transplantation are not far off.

There are substantial roadblocks and challenges to achieve functional recovery after SCI, as the adult human spinal cord boasts of three million axons that project rostrally and caudally. A combination of precisely timed events is required to develop axonal elongation and growth. Over the past twenty years, several repair strategies have been devised that allow regeneration of the central axons within the spinal cord. None of these strategies, on their own, have been able to result in a complete functional recovery after SCI. Thus, a combination of strategies may be required to improve the extent of functional recovery, in order to achieve a suitable application for humans. The most reliable approach, as of now, is to combine a bridge like tubular implant with neurotrophic factors. A systematic and combined approach which uses tubular implants and growth factors, along with the presence of extracellular molecules and myelin neutralization compounds, could be what allows for a functional recovery after SCI.

REFERENCES

1. Sheardown H., M. Griffith (2008). 'Regenerative medicine in the cornea. In: Principles of Regenerative Medicine' A. Atala, R. Lanza, J. Thompson, R. Nerem, eds. *Elseiver, Boston, USA*. 1060–1071.
2. Griffith M., P. Fagerholm, W. Liu, et. al. (2008). 'Corneal regenerative medicine: Corneal substitutes for transplantation'. 'In: Cornea and External Disease, Essentials in Ophthalmology' T. Reinhard, F. Larkin, eds. *Springer, Heidelberg, Germany*.
3. Myung D., N. Farooqui, L. Zheng, et. al. (2008). 'Bioactive interpenetrating polymer network hydrogels that support corneal epithelial wound healing'. *J. Biomed. Mater. ResA., E-Pub.* 15 May.
4. Sweeney D.F., R.Z. Xie, D.J. O'Leary, et. al. (1998). 'Nutritional requirements of the corneal epithelium and anterior stroma: Clinical findings'. *Invest. Ophthalmol. Vis. Sci.* **39**: 284–291.
5. Dravida S., S. Gaddipati, M. Griffith, et. al. (2008). 'A biomimetic scaffold for culturing limbal stem cells: Promising alternative for clinical transplantation'. *J. Tissue Eng. Regen.* **2**: 263–271.
6. Sundback C., T. Hadlock, M. Cheney, J. Vacanti (2003). 'Manufacture of porous polymer nerve conduits by a novel low-pressure injection molding process'. *Biomaterials.* **24**: 819–830.
7. Friedman J.A., A.J. Windebank, M.J. Moore, et. al. (2002). 'Biodegradable polymer grafts for surgical repair of the injured spinal cord'. *Neurosurgery.* **51**: 742–752.
8. Wen X., P.A. Tresco (2006). 'Fabrication and characterization of permeable degradable poly (DL-lactide-*co*-glycolide) (PLGA) hollow fiber phase inversion membranes for use as nerve tract guidance channels'. *Biomaterials.* **27**: 3800–3809.
9. Aucoin L., C.M. Griffith, G. Pleizier, et. al. (2002). 'Interactions of corneal epithelial cells and surfaces modified with cell adhesion peptide combinations'. *J. Biomater. Sci. Polym. Ed.* **13**: 447–462.
10. Jacob J.T., J.R. Rochefort, J. Bi, B.M. Gebhardt (2005). 'Corneal epithelial cell growth over tethered-protein/peptide surface-modified hydrogels'. *J. Biomed. Mater. ResB- Appl. Biomater.* 72: 198–205.
11. Myung D., W. Koh, A. Bakri, et. al. (2007). 'Design and fabrication of an artificial cornea based on a photolithographically patterned hydrogel construct'. *Biomed. Micro.* **9**: 911–922.
12. Myung D., D. Waters, M. Wiseman, et. al. (2008). 'Progress in the development of interpenetrating polymer network hydrogels'. *Poly. Adv. Tech.* **19**: 647–657
13. Xu X.M., A.Q. Chen, V. Guenard, et. al. (1997). 'Bridging Schwann cell transplants promote axonal regeneration from both the rostral and caudal stumps of transacted adult rat spinal cord'. *J. Neurocytol.* **26**: 1–16.
14. Spilker M.H., I.V. Yannas, S.K. Kostyk, et. al. (2001). 'The effects of tubulation on healing and scar formation after transection of the adult rat spinal cord'. *Restor. Neurol. Neurosci.* **18**: 23–38.
15. Mi F.L., S.S. Shyu, Y.B. Wu, et. al. (2001). 'Fabrication and characterization of a sponge-like asymmetric chitosan membrane as a wound dressing'. *Biomaterials.* **22**: 165–173.
16. Suzuki M., S. Itoh, I. Yamaguchi, et. al. (2003). 'Tendon chitosan tubes covalently coupled with synthesized laminin peptides facilitate nerve regeneration *in vivo*'. *J. Neurosci. Res.* **72**: 646-659.
17. Yamaguchi I., S. Itoh, M. Suzuki, et. al. (2003) 'The chitosan prepared from crab tendons:II'. 'The chitosan/apatite composites and their application to nerve regeneration'. *Biomaterials.* **24**: 3285–3292.

18. Huang Y., Y. Huang, C. Huang and H. Liu (2005). 'Manufacture of porous polymer nerve conduits through a lyophilizing and wire heating process'. *J. Biomed. Mat. ResB-Appl. Biomater.* **74**: 659–664.
19. Bini T.B., S. Gao, S. Wang and S. Ramakrishna (2005). 'Development of fibrous biodegradable polymer conduits for guided nerve regeneration'. *J. Mat. Sci. Mat. In Med.* **16**: 367–375.
20. Goraltchouk A., V. Scanga, C.M. Morshead and M.S. Shoichet. (2006). 'Incorporation of protein-eluting microspheres into biodegradable nerve guidance channels for controlled release'. *J. Cont. Release.* **110**: 400–407.
21. Bouhadir K.H., D.J. Mooney (2002). 'Synthesis of hydrogels: Alginate hydrogels'. *In: Methods of Tissue Engineering* (A. Atala, R.P. Lanza, eds.) *Academic Press, San Diego. USA.*
22. Liu W., M. Griffith and F. Li (2008). 'Alginate microsphere-collagen composite hydrogel for ocular drug delivery and implantation'. *J. Mat. Sci. Mat. In Med.* **19**: 3365–3371.
23. Nimni M.E., D. Cheung, B. Strates, et. al. (1987). 'Chemically modified collagen: a natural biomaterial for tissue replacement'. *J. Biomed. Mat. Res.* **21**: 741–771.
24. Rafat M., F. Li, P. Fagerholm, et. al. (2008). 'PEG-Stabilized Carbodiimide Crosslinked Collagen-Chitosan Hydrogels for Corneal Tissue Engineering'. *Biomaterials.* **29**: 3960–3972.
25. Meller D., K. Peters and K. Meller (1997). 'Human cornea and sclera studied by atomnic force microscopy'. *Cell Tissue Res.* **288**: 111–118.
26. Freegard T.J (1997). 'The physical basis of transparency of the normal cornea'. *Eye.* **11**: 465–471.
27. Niederkorn J.Y. 'Immune privilege and immune regulation in the eye'. *Adv Immunol.* **48**: 191–226.
28. Jonas J.B. and L. Holbach (2005). 'Central corneal thickness and thickness of the lamina cribrosa in human eyes'. *Invest. Ophthalmol. Vis. Sci.* **46**: 1275–1279.
29. Ren H. and G. Wilson (1996). 'Apoptosis in the corneal epithelium'. *Invest. Ophthalmol. Vis. Sci.* **37**: 1017–1025.
30. Dua H.S. and A. Azuara-Blanco (2000). 'Limbal stem cells of the corneal epithelium'. *Surv. Ophthalmol.* **44**: 415–425.
31. Sack R.A., I. Nunes, A. Beaton and C. Morris (2001). 'Host-defense mechanism of the ocular surfaces'. *Biosci. Rep.* **21**: 463–480.
32. Baker K.S., S.C. Anderson, E.G. Romanowski, et. al. (1993). 'Trigeminal ganglion neurons affect corneal epithelial phenotype'. *Influence on type VII collagen expression in vitro. Invest. Ophthalmol. Vis. Sci.* **34**: 137–144.
33. Beuerman R.W. and B. Schimmelpfennig (1980). 'Sensory denervation of the rabbit cornea affects epithelial properties'. *Exp. Neurol.* **69**: 196–201.
34. Garcia-Hirschfeld J., L.G. Lopez-Briones and C. Belmonte (1994). 'Neurotrophic influences on corneal epithelial cells'. *Exp. Eye Res.* **59**: 597–605.
35. Nishida T. (1997). Cornea, 'Fundamentals of cornea and external disease'. In: Cornea (J.J. Krachmer, M.J. Mannis, E.J. Holland (eds.). *Mosby-Year Book Inc., St. Louis. USA.*
36. Foster A. (2003). 'Vision 2020-the Right to Sight'. *Trop. Doct.* **33**: 193–194.
37. Whitcher J.P., M. Srinivasan and M.P. Upadhyay (2001). 'Corneal blindness: a global perspective'. *Bull World Health Organ.* **79**: 214–221.
38. Wollensak G., E. Spoerl and T. Seiler (2003). 'Riboflavin/ultraviolet-a-induced collagen crosslinking for the treatment of keratoconus'. *Am. J. Ophthalmol.* **135**: 620-627.

39. Williams K.A., A.J. Esterman, C. Bartlett, et. al. (2006). 'How effective is penetrating corneal transplantation? Factors influencing long-term outcome in multivariate analysis'. *Transplantation.* **81**: 896–901
40. Duffy P., J. Wolf, G. Collins, et. al. (1974). 'Letter: Possible person-to-person transmission of Creutzfeldt-Jakob disease'. *N. Engl. J. Med.* **290**: 692–693.
41. Srinivasan A., E.C. Burton, M.J. Kuehnert, et. al. (2005). 'Transmission of Rabies Virus from an Organ Donor to Four Transplant Recipients'. *N. Engl. J. Med.* **352**: 1103–1111.
42. Schwarz A., F. Hoffmann, J. L'Age-Stehr, et. al. (1987). 'Human immunodeficiency virus transmission by organ donation outcome in cornea and kidney recipients'. *Transplantation* **44**: 21–24.
43. Thuret. S., L.D.F. Moon and F.H. Gage, et. al. (2006). 'Therapeutic interventions after spinal cord injury'. *Nat. Rev. Neurosci.* **7**: 628–643.
44. Patist C.M., M.B. Mulder, S.E Gautier et. al. (2004). 'Freeze-dried poly (D, L-lactic acid) macroporous guidance scaffolds impregnated with brain-derived neurotrophic factor in the transected adult rat thoracic spinal cord'. *Biomaterials.* **25**: 1569–1582.
45. Carmen J., T. Magnus, R. Cassiani-Ingoni, et. al. (2007). 'Revisiting the astrocyte-oligodendrocyte relationship in the adult CNS'. *Prog. Neurobiol.* **82**: 151–162.
46. Novikova L.N., J. Pettersson, M. Brohlin, et. al. (2003). 'Biodegradable poly-β-hydroxybutyrate scaffold seeded with Schwann cells to promote spinal cord repair'. *Biomaterials.* **29**: 1198–1206.
47. Bunge M.B. and D.D. Pearse (2003). 'Transplantation strategies to promote repair of the injured spinal cord'. *J. Rehab. Res. Develop.* **40**: 55–62.
48. Strampelli B. (1963). 'Osteo-Odontokeratoprosthesis'. *Ann. Ophthamol. Clin. Ocul.* **89**: 1039–1044.
49. Mehta J.S., C.E. Futter, S.R. Sandeman, et. al. (2005). 'Hydroxyapatite promotes superior keratocyte adhesion and proliferation in comparison with current keratoprosthesis skirt materials'. *Br. J. Ophthalmol.* **89**: 1356-1362.
50. Falcinelli G., B. Falsini, M. Taloni, et. al. (2005). 'Modified osteo-odonto-keratoprosthesis for treatment of corneal blindness: long-term anatomical and functional outcomes in 181 cases'. *Arch. Ophthalmol.* **123**: 1319–1329.
51. Hicks C.R., G.J. Crawford, J.K. Dart, et. al. (2006). 'AlphaCor: Clinical outcomes'. *Cornea.* **25**: 1034–1042.
52. Chirila T.V., S. Vijayasekaran, R. Horne, et. al. (1994). 'Interpenetrating polymer network (IPN) as a permanent joint between the elements of a new type of artificial cornea'. *J. Biomed. Mater. Res.* **28**: 745–753.
53. Trinkaus-Randall V. (2000). 'Cornea. In: Principles of Tissue Engineering'. Eds: R. Lanza, R. Langer and J.P. Vacanti'. *Academic Press, New York, USA.*
54. Gaudreault M., P. Carrier, K. Larouche, et. al. (2003). 'Influence of sp1/sp3 expression on corneal epithelial cells proliferation and differentiation properties in reconstructed tissues'. *Invest. Ophthalmol. Vis. Sci.* **44**: 1447–1457.
55. Auger F.A., M. Remy-Zolghadri, G. Grenier, et. al. (2002). 'A truly new approach for tissue engineering: the LOEX self-assembly technique'. *Ernst. Schering. Res. Found. Workshop.* **35**: 73–88.
56. Carrier P., A. Deschambeault, M. Talbot, et. al. (2008). 'Characterization of wound reepithelialization using a new human tissue-engineered corneal wound healing model'. *Invest. Ophthalmol. Vis. Sci.* **49**: 13761385.

57. Guo X., A.E.K Hutcheon, S.A. Melotti, et. al. (2007). 'Morphologic characterization of organized extracellular matrix deposition by ascorbic acid-stimulated human corneal fibroblasts'. *Invest. Ophthalmol. Vis. Sci.* **48**: 4050–4060.
58. Borene M.L., V.H. Barocas and A. Hubel (2004). 'Mechanical and cellular changes during compaction of a collagen-sponge-based corneal stromal equivalent'. *Ann. Biomed. Eng.* **32**: 274–283.
59. Crabb R.A., E.P. Chau, M.C. Evans, et. al. (2006). 'Biomechanical and microstructural characteristics of a collagen film-based corneal stroma equivalent'. *Tissue Eng.* **12**: 1565–1575.
60. Griffith M., R. Osborne, R. Munger, et. al. (1999). 'Functional human corneal equivalents constructed from cell lines'. *Science*. **286**: 2169–2172.
61. Griffith M., M. Hakim, S. Shimmura, et. al. (2002). 'Artificial human corneas: Scaffolds for transplantation and host regeneration'. *Cornea*. **21**: S54–S61.
62. Li F., D. Carlsson, C. Lohmann, et. al. (2003). 'Cellular and nerve regeneration within a biosynthetic extracellular matrix for corneal transplantation'. *Proc. Natl. Acad. Sci. USA*. **100**: 15346–15351.
63. Li F., M. Griffith, S. Li, et. al. (2005). 'Recruitment of multiple cell lines by collagen-synthetic copolymer matrices in corneal regeneration'. *Biomaterials*. **16**: 3093–3104.
64. Liu Y., L. Gan, D.J. Carlsson, et. al. (2006). 'A simple, cross-linked collagen tissue substitute for corneal implantation'. *Invest. Ophthalmol. Vis. Sci.* **47**: 1869–1875.
65. Merrett K., P. Fagerholm, C.R. McLaughlin, et. al. (2008). 'Tissue engineered recombinant human collagen-based corneal substitutes for implantation: performance of type I versus type III collagen'. *Invest. Ophthalmol. Vis. Sci.* **49**: 3887–3894.
66. McLaughlin C.R., P. Fagerholm, L. Muzakare, et. al. (2008). 'Regeneration of corneal cells and nerves in an implanted collagen corneal substitute'. *Cornea*. **27**: 580–589.
67. Nakamura T., T. Inatomi, C. Sotozono, et. al. (2006). 'Transplantation of autologous serum-derived cultivated corneal epithelial equivalents for the treatment of severe ocular surface disease'. *Ophthalmology*. **113**: 1765–1772.
68. Rama P., S. Bonini, A. Lambiase, et. al. (2001). 'Autologous fibrin-cultured limbal stem cells permanently restore the corneal surface of patients with total limbal stem cell deficiency'. *Transplantation*. **72**: 1478–1485.
69. Inatomi T., T. Nakamura, M. Kojyo, et. al. (2006). 'Ocular surface reconstruction with combination of cultivated autologous oral mucosal epithelial transplantation and penetrating keratoplasty'. *Am. J. Ophthalmol.* **142**: 757–764.
70. Hsiue G.H., J.Y. Lai, K.H. Chen and W.M. Hsu (2006). 'A novel strategy for corneal endothelial reconstruction with a bioengineered cell sheet'. *Transplantation*. **81**: 473–476.
71. Daston G.P. and F.E. Freeberg (1991). 'Ocular irritation testing'. In: Dermal and Ocular Toxicology'. *Ed: D.W. Hobson. CRC Press, Florida, USA.*
72. Gautheron P., M. Dukic, D. Alix and J.F. Sina (1992). 'Bovine corneal opacity and permeability test: An *in vitro* assay of ocular irritancy'. *Fundam. Appl. Toxicol.* **18**: 442–449.
73. Wertz P.W. (2006). 'Biochemistry of human stratum corneum lipids'. 'In:, 'Skin Barrier' (P.M. Elias, K.R. Feingold)'. *Taylor & Francis, New York, USA.*
74. Klausner M., P.J. Hayden, B.A. Breyfogle, et. al. (2003). 'The EpiOcular prediction model: A reproducible *in vitro* means of assessing ocular irritancy'. *In: Alternative Toxicological Methods* (H. Salem, S.A. Katz, eds.). *CRC Press, FL, USA.*

75. Zieske J.D., V.S. Mason, M.E. Wasson, et. al. (1994). 'Basement membrane assembly and differentiation of cultured corneal cells: importance of culture environment and endothelial cell interaction'. *Exp. Cell. Res.* **214**: 621–633.
76. Holzer P. (1988). 'Local effector functions of capsaicin-sensitive sensory nerve endings: involvement of tachykinins, calcitonin gene-related peptide and other neuropeptides'. *Neuroscience.* **24**: 739–768.
77. Suuronen E.J., M. Nakamura, M.A. Watsky, et. al. (2004). 'Innervated human corneal equivalents as *in vitro* models for nerve-target. cell interactions'. *Faseb. J.* **18**: 170–172.
78. Chan K.Y. and R.H. Haschke (1982). 'Isolation and culture of corneal cells and their interactions with dissociated trigeminal neurons'. *Exp. Eye Res.* **35**: 137–156.
79. Riggott M.J. and S.A. Moody (1987). 'Distribution of laminin and fibronectin along peripheral trigeminal axon pathways in the developing chick'. *J. Comp. Neurol.* **258**: 580–596.
80. Corcoran J., B. Shroot, J. Pizzey & M. Maden (2000). 'The role of retinoic acid receptors in neurite outgrowth from different populations of embryonic mouse dorsal root ganglia'. *J. Cell. Sci.* **113**: 2567–2574.
81. Muller L.J., L. Pels and G.F. Vrensen (1996). 'Ultrastructural organization of human corneal nerves. Invest'. *Ophthalmol. Vis. Sci.* **37**: 476–488
82. Muller L.J., G.F. Vrensen, L.Pels, et. al. (1997). 'Architecture of human corneal nerves'. *Invest. Ophthalmol. Vis. Sci.* **38**: 985–994.
83. Suuronen E.J., C.R. McLaughlin, P.K. Stys, et. al. (2004). 'Functional innervation in tissue engineered models for *in vitro* study and testing purposes'. *Toxicol. Sci.* **82**: 525–533
84. Araki-Sasaki K., S. Aizawa, M. Hiramoto, et. al. (2000). Substance P-induced cadherin expression and its signal transduction in a cloned human corneal epithelial cell line'. *J. Cell. Physiol.* **182**: 189–195.
85. Suuronen E.J., L. Muzakare, C.J. Doillon, et. al. (2006). 'Promotion of angiogenesis in tissue engineering: developing multicellular matrices with multiple capacities'. *Int. J. Artif. Organs.* **29**: 1148–1157.
86. Bee J.A. (1982). 'The development and pattern of innervation of the avian cornea'. *Dev. Biol.* **92**: 5–5.
87. Xu X.M., S. Zhang, H. Li, et. al. (1999). 'growth of axons into the distal spinal cord through a Schwann-cell-seeded mini-channel implanted into hemisected adult rat spinal cord'.*Eur. J. Neurosci.* **11**: 1723–1740.
88. Menei P., C. Montero-Menei, S.R. Whittemore, et. al. (1998). 'Schwann cells genetically modified to secrete human BDNF promote enhanced axonal regrowth across transected adult rat spinal cord'. *Eur. J. Neurosci.* **10**: 607–21.
89. Evans G.R.D., K. Brandt, S. Katz, et. al. (2002). 'Bioactive poly (L-lactic acid) conduits seeded with Schwann cells for peripheral nerve regeneration'. *Biomaterials.* **23**: 841–848.
90. Qiu J., D. Cai, H. Dai, et. al. (2002). 'Spinal Axon regeneration induced by elevation of cyclic AMP'. *Neuron.* **34**: 895–903.
91. Oudega M., S.E. Gautier, P. Chapon, et. al. (2001). 'Axonal regeneration into Schwann cell grafts within resorbable poly (alpha-hydroxyacid) guidance channels in the adult rat spinal cord'. *Biomaterials.* **22**: 1125–1136.
92. Radtke C., M. Sasaki, K.L. Lankford, et. al. (2008). 'Potential of olfactory ensheathing cells for cell-based therapy in spinal cord injury'. *J. Rehab. Res. Dev.* **45**: 141–152.

93. Wen Z. and J.Q. Zheng (2006). 'Directional guidance of nerve growth cones'. *Curr. Opin. Neurobiol.* **16**: 52–58
94. Lee M., C.J. Chen, C. Cheng, et. al. (2008). 'Combined treatment using peripheral nerve graft and FGF-1: Changes to the glial environment and differential macrophage reaction in a complete transected spinal cord'. *Neurosci. Letters.* **433**: 163–169.
95. Fouad K., L. Schnell, M.B. Bunge, et. al. (2005). 'Combining Schwann cell bridges and olfactory-ensheathing glia grafts with chondroitinase promotes locomotor recovery after complete transection of the spinal cord'. *J. Neurosci.* **25**: 1169–1178.
96. Cai J, K.S. Ziemba, G.M. Smith and Y. Jin (2007). 'Evaluation of cellular organization and axonal regeneration through linear PLA foam implants in acute and chronic spinal cord injury'. *J. Biomed. Mat. Res A.* **83A**: 512–520.
97. Yang Y., L.D. Laporte, C.B. Rives, et. al. (2005). 'Neurotrophin releasing single and multiple lumen nerve conduits'. *J. Cont. Release.* **104**: 433–46.
98. Dunnen W.F.A., M.F. Meek, D.W. Grijpma, et. al. (2000). '*In vivo* and *in vitro* degradation of poly [$^{50/}_{50}$ ($^{85}/_{15}$ $^{L}/_{D}$) LA/å-CL], and the implications for the use in nerve reconstruction'. *J. Biomed. Mat. Res.* **51**: 575–585.
99. Teng Y.D., E.B. Lavik, X. Qu, et. al. (2002). 'Functional recovery following traumatic spinal cord injury mediated by a unique polymer scaffold seeded with neural stem cells'. *Proc. Natl. Acad. Sci. USA.* **99**: 3024–3029.
100. Bini T.B., S. Gao, X. Xu, et. al. (2004). 'Peripheral nerve regeneration by microbraided poly (L-lactide-*co*-glycolide) biodegradable polymer fibers'. *J. Biomed. Mat. ResA.* **68**: 286–295.
101. Stokols S. and M.H. Tuszynski (2004). 'The fabrication and characterization of linearly oriented nerve guidance scaffolds for spinal cord injury'. *Biomaterials.* **25**: 5839–5846.
102. Li J. and R. Shi (2007). 'Fabrication of patterned multi-walled poly-L-lactic acid conduits for nerve regeneration'. *J. Neurosci. Meth.* **165**: 257–264.
103. Yoshii S., M. Oka, M. Shima, et. al. (2003). 'A. Bridging a Spinal Cord Defect Using Collagen Filament'. *SPINE.* **28**: 2346–2351.
104. Jain A., Y. Kim, R.J. McKeon and R.V. Bellamkonda (2006). 'In situ gelling hydrogels for conformal repair of spinal cord defects, and local delivery of BDNF after spinal cord injury'. *Biomaterials.* **27**: 497–504.
105. Gupta D., C.H. Tator and M.S. Shoichet (2006). 'Fast gelling injectable blend of hyaluronan and methylcellulose for intrathecal, localized delivery to the injured spinal cord'. *Biomaterials.* **27**: 2370–2379.
106. Uebersax L., M. Attotti, M. Papaloizos, et. al. (2007). 'Silk fibroin matrices for controlled release of nerve growth factor (NGF)'. *Biomaterials.* **28**: 4449–4460.
107. Miao X., L-P. Ta, L-S. Tan and X. Huang (2007). 'Porous calcium phosphate ceramics modified with PLGA-bioactive glass'. *Mat. Sci. Eng.* **C27**: 74–79.
108. Wu L. and J. Ding (2004). '*In vitro* degradation of 3D porous poly (D, L-lactide-*co*-glycolide) scaffolds for tissue engineering'. *Biomaterials.* **25**: 5821–5830.
109. Rowlands A.S., S.A. Lim, D. Martin and J.J. Cooper-White (2007). 'Polyurethane/poly (lactic-*co*-glycolic) acid composite scaffolds fabricated by thermally induced phase separation'. *Biomaterials.* **28**: 2109–2121.
110. Ma P.X. and J. Choi (2001). 'Biodegradable polymer scaffolds with well defined interconnected spherical pore network'. *Tiss. Eng.* **7**: 23–33.

Index

Symbols

(HNF4a)   182, 184
2′,3′-cyclic nucleotide-3′-phosphodiesterase (CNPa)   434
293T fibroblasts   234
46, XX karyotype   227
5-VpreB1   157
5/6 preimplantation embryo   120
6-DMAP   227
67/548 EEC   294

A

α-globin   237
α-glycerophosphate   362
α-MEM   433
α-SM actin   430
α-smooth muscle actin   362
α1,3GalT   218
ABC transporters   327, 587
ABCG2   323, 571, 586, 588, 589
Aberrant spindles   230
Acinar cells   169
Activin   58, 140, 180, 185
Activin B   186
Acute myocardial infarction   403
AD-MSC   400

Adherence   360
Adhesion   201
Adipocyte   364
Adipogenesis   362
Adipogenic   281, 362
Adipogenic differentiation   413
Adipose   366
ADSCs adipose tissue-derived stem cells   338
Adult stem cells (ASCs)   426
Advanced coronary artery disease   408
AFP   180
Albumin, a-1-antitrypsin   304
ALCAM   431
Alkaline phosphatase activity   26
Alloantigens   218
Allogeneic   207, 216
    bone grafts   367
    embryonic stem cells   224
    MSC   223
ALS   417
ALU   435
Alzheimer   417
Amcyte   3
AMD 3100 (Plerixafor)   328
Ameloblasts   427
Amelogenin   428

AML3 428
Amnion
    epithelial stem cell 7
    stem cells 223
Amyotrophic lateral sclerosis 403
Aneuploidy 126, 230
Aneuploid cell 84
Ang-1 402
Angiogenesis 281, 332
Animal origin free (AOF) 144
ANP 302
Anterior–posterior 177
Anti
    a5b1 203
    CD25 219
    proliferative 263
Antioxidant 402
Antler 402
Antral follicle 269
Aphaeresis 328
Apligraf 3
Apomorphine 415
Apoptosis 323
Area
    opaca 59
    pellucida 275
Artificial Gametes 275
Ascorbic acid 2-phosphate 362
Aspirate 328
Assisted reproductive technology (ART) 125, 262,
Asthma 338
Astrocytes 369, 333
Astrocytic 376
Astrotactin 381
Asymmetry 455, 474
Ataxia telangiectasia (ATM) 253
Atherosclerotic lesions 330
ATP-Binding cassette (ABC) 567, 321
Atrial-like 302
Auto-cell transplantation therapy 417
Auto-reactive T-cells 104

Autocrine 480, 378
Autograft 480, 378
Autoimmune 169
Autologous 216, 421, 618
Autologous fat transplantation 366

B

β-cell 218, 328
β-Cell Replacement Therapies 172
β-cell Surrogates 191
β-galactosidase 235
β-unsaturated aldehydes 305
b3 tubulin 434
Bacterial artificial chromosomes (BAC) 89
Bacteriophage P1 95
BALBc 287
Banking 433
Banks 226
Basal
    cell carcinomas (BCCs) 336
    ganglia 368
    lamina 134
Basic fibroblast growth factor (bFGF) 41, 608
BCR-ABL⁺ leukemic stem cells (LSCs) 329
BCRP 323, 586
BDNF 370, 421, 659, 661
Betacellulin 187
BFGF 378
BG01v-Oct4-GFP 139
Bgeo cassette 235
Bi-potent 324
Bicyclam 328
Bilaminar disc 131
Bioceramics 198, 367
Biohybrid 3
Biotransformation 305
Bipolar islets 178
Bipronuclear (2PN) 125
Birefringence 230
Birth defects 299
Bivalent modification 156

Blastocysts 121, 230, 275, 630
Blastodermal cells (BCs) 59
Blastula-germ-ring stage 25
Blimp1 63, 263, 335
Blood glucose levels 170
Blood-compatible 209
Bloom syndrome (BLM) 253
Blue–purple colour 362
BM-MSC 430
Bone morphogenic protein (BMP) 80, 446
    antagonist 186
Bmp15 269
BMP4 63, 263, 272, 388
BMP8 263
BMP8b 63
Body mass index (BMI) 122
Bone marrow transplantation 219
Brachyury 60, 180
Brain-derived neurotrophic factor (BDNF) 378, 659
Breast carcinoma 327
Breast mastectomies 366
Bronchioalveolar stem cells 338
Btkxid mice 225

C

c-KIT (CD117) 47, 321, 431
    MYC 102, 252, 236
    peptide 175, 178
C3H 10T1/2 and BALB-3T3/A31 80
C57BL6 79, 287
Ca^{2+} channels, Na^+ and K^+ channels 302
Ca^{2++} 178
Californian Institute for Regenerative Medicine (h) 2
Cambial layer 402
Cancer stem 320, 322
Cardiac precursor 302
Cardiac stem/progenitor cells (CSCs) 331
Cardiogenic 281
Cardiogenol C 103
Cardiomyocytes 80, 295, 301, 331
Carotid body 333

Cbfa1 428
CD 105 361, 437
CD 106 361
CD 117 362
CD 11b 362, 399
CD 13 431, 607
CD 133 321, 322
CD 14 362, 399
CD 166 (activated leukocyte cell adhesion) 362, 431
CD 19 399
CD 29 361, 430, 431, 437, 607
CD 3/CD28 401
CD 31 353, 462, 431, 566
CD 33 362
CD 34 362, 399, 430, 431, 562
CD 3e 328
CD 4^+ 217
CD 4^+ helper T-cells 170
CD 44 321, 322, 361, 566
CD 45 362, 399, 430, 609
CD 54 361
CD 55 361
CD 73 222, 399, 437
CD 79a 399
CD 8 217
CD 8 cytotoxic cells 170
CD 80 207, 225
CD 86 207, 225
CD 90 222, 361
CD 95 224
CDx2 58
Cell cycle S phase 266
Cell fusion 377
Cell replacement 417
Cellastim™ 398
Cementoblasts 428
Cerebellum 414
CF (cystic fibrosis) 250
CF transmembrane conductance regulator (CTFR) 291

CGG repeat   250
Chaperon   402
Chemical
    mutagenesis   88
    mutagens   88
Chemotaxis   203
Chemotherapy   329
Chicken ESCs (cESC)   59
Chimera   44, 282
Chimerism   229
Chloramphenicol acetyltransferase (CAT)   50
CHO   398
Cholangiocytes   307
Cholesterol (Chol)   210
Chondorogenic   434
Chorions   24
Christiaan Barnard   2
Chromosomal
    anomaly   250
    fusions   327
Chromosome 17   216
Chromosome 6   216
Chromosome elimination cassette (CEC)   233
Chromosome 11   233
Chromosome 12   233
Chronic bronchitis   338
    heart failure   403
    Inflammatory Bowel Disorders (IBDs)   329
cis-acting   284
cis-regulatory elements   158
CK5/14   335
Cleavage   262
Clonogenic   431
CNS   367, 380, 417
CNTF   60, 205, 378
Co-culture   86
Co-culture of ESCs   189
Cohnheim   392
Collagen   143, 624, 646, 648, 661
    type I   430

Collagenase IV   296
Collagenase type 1   433
Colony-forming unit-fibroblasts (CFU-F)   359
Colony-forming-unit-fibroblast assay (CFU-F)   428
Colony-forming-units   411
Complement 5a receptor (C5aR)   97
Complete culture medium (CCM)   360
Conditional Knockouts   94
Conditioned media (CM)   381
Connexin 43/45   302
Conrad Waddington   152
Contralateral   420
Controlled ligand supplementation (CLS)   208
Corpus striatum   368
Corticosteroids   219
CpG "islands,"   154
CPouV   67
Cre   289
    recombinase   95, 289
Crest of the pelvis   360
Cristae   130
CSF   204
CXCR4   203, 321, 416, 446
CXXC finger protein 1   158
Cyclic AMP   178
Cyclosporine/FK-506   219
Cyp26b1   265
CYP3A4   306
CYPs   306
Cystic EB   86
Cystic fibrosis   291, 338
Cytochalasin B   227
Cytochrome P450   305
Cytokeratin 19   304, 335
Cytosol   217
Cytotoxic T-cells   217
Cytotoxicity   300

D

Danio rerio sequence project   22
DAPT   187

Dax1   58
DAZ   264
DAZL   47, 58, 265, 273
De-differentiated   6, 130, 221, 251
Defined culture medium   141
Definitive
    endoderm   176, 305
    pancreas   177
Delayed blastocysts   78
Deleted in-azoospermia (DAZ)   47
Dendritic   225
Denominated human immature dental pulp cells (hIDP   431
Dense thin solid films (TSF)   209
Dental
    crown   432
    follicle   428
    lamina   427
    papilla   427
    Pulp   427
    Pulp Extraction   432
    pulp stem cells (DPSCs)   428
Dentate gyrus   333, 380
Dentin   427, 428, 432
    Specific Protein (DSPP)   434
Dentinogenesis   437
Derivation of hESC   125
    medaka ES Cell   41
    medaka spermatogonial stem cells   47
    zebrafish ESC   25
    zebrafish PGC   31
Dermal papilla   336
Dermatopathies   336
Desmosomes   130
Desquamation   335
Developmental toxicity   299
Dexamethasone   362
Diabetes   610
Diabetes mellitus   169
Dicer   98

Dictyate   268
Differentiation   178
Digital differential display   67
Digyny   125
Diploid pESCs   227
Diplotene   268
Dispase   296, 433
Dispermy   125
Dl′2   428
Dmc1   271
DMEM-   433
    F12   397
    KO   397
DMSO (diméthylsulfoxide)   61
DNA fingerprinting   230
DNA methyl-transferases (DNMTs)   246
DNAM-1   207
DND   33
Dolly   226
Donors   217
Dopamine   333
Dopaminergic   334, 376, 419
Double-strand break   98
Down syndrome (DS).   250
Doxycycline (Dox)   100
Drug Screening   257, 294
Duct cells   169
Ductal   177
Dysmyelinated   334
Dysplasia   336

E

E-cadherin (E-cad)   81
E2   120
E4.5–E5   65
E7.5   65
Ecat (ESC-associated transcripts)   67
ECAT1   236
Ectoderm   58, 76, 120, 132, 252
Ectomesenchymal cells   427
ECVAM   299, 300

Edmonton Protocol   172
EED protein   158
EGFR   321, 327
Electroporation   28, 98
EMA-1   63
Embryo   262
    assessment   125
    transfer   121
Embryoid bodies   32, 81, 174, 232, 272, 294
Embryonal carcinoma (EC)   231, 535
    cells (ECC)   77, 269
Embryonic germ cells (EGC)   79, 263, 269, 487
Embryonic lethal   255
Embryonic stem cell test (EST)   299
Embryotoxic   299
EMEA medium   397
Emphysema   338
Enamel   427, 432
    knot   427
Endocrine pancreas   169
Endocytic vesicles   217
Endoderm   58, 60, 76, 120, 132, 252
Endodontic   433
Endoglin   361
Endosteum   328
Endothelial progenitor cells (EPC)   50, 206, 609
Engraftment   369
Eomes   58
EPCs   330
Epiblast   57, 132, 177, 263, 506
Epigenetic   152, 255, 326
Epigenome   153
EpiSC (post-implantation epiblast-derived stem cell)   58, 78
Epoxides   307
ES cells   224
ES-somatic-cell   234
ESC-Derived Gametes   276
ESM1   42
ESM2   42
ESM3   42
ESM4   42
Ets   291
Euploid   81
Exendin-4   186
Exocrine pancreas   169
Expression of Fas   224
Extra-cellular matrix (ECM)   295
Extra-embryonic ectoderm   57, 263
EZH2   159, 246, 532

F

FA (FANC genes)   253
Facial reconstructions   366
Family gene   284
FANCD2   249, 253
Fanconi anemia (FA)   249
FasL or CD95L   224
FBS   397
Fbx15   235, 236
Fbx15–bgeo   235
Feeder cells   21
Feeder-free   296
Female Germ Cells   268
Femur   360, 412
Fertilisation   123
FGF   205, 402, 427
FGF 10   182, 185
FGF 18   186
FGF 2   58, 185
Fibroblasts   230, 411, 428, 630, 659
    feeders   78
Fibronectin   143
Figla   269
Flavin monooxygenases   306
Flp recombinases   289
Fluorescent in situ hybridization (FISH)   250
FMR1   250, 255
Follicle-stimulating hormone (FSH)   122, 475
Follicles   122
Forebrain   333
Forskolin   187

FoxA2 179, 180, 183, 188
FoxA2, HNF-1/4/6 304
FoxD3 (forkhead-box D3) 85
Fractures 367
Fragile X 253
Fragilis 263, 273
Fragmentation 128
FRAX (Fragile X) 250
FRAXA 250
Free radical 402
Friedenstein 359
frt 289
Frt-Flp systems 95

G

G1 cell cycle 81
G418 50, 235
Gain-of-function 282
Galactocerebroside (GalC) 210, 382
Gametes 270
Ganciclovir 287
Gap junctions 130
GASSI strain 26
Gastrointestinal 324
GATA-1 289
GATA-4 302, 304
Gdf9 269, 273
GDNF 205, 482
Gelatin 86
Gelatin-coated 78
Geltrex™ 143
Gene
    therapy 256
    trapping 90, 284
Genetic modification 88
Genetically modified 300
Genomic instability 254
Genomics 294
Germ
    cells (GCs) 262
    line 282
    ring stage 23

GFAP 376, 434, 608
Ghrelin 188
Giemsa (G)-banding 84
Gli1 428
Glial cells 333
    line-derived neurotrophic factor (GDNF) 473, 481
Glucagon 188
Glucose transporter 2 (Glut2) 174
Glucose
    dependent 177
    responsive 188
Glucurunosyltransferases 306
Glutamax 397
Glycogen synthase 171
Glycolipid 210
GM-CSF 328, 446
Gonad 64, 263, 265
Gonadotrophins 266
Gonadotropin-releasing hormone agonist (GnRH agonist) 122
Gonocytes 265
Goosecoid 60
Gout-like 249
Gp130 57, 60
GPA 60
Gpbox 269
Graft rejection 297
Graft Versus Host Disease (GVHD) 3, 219, 220, 366, 604
Graft-versus-leukaemia 220
Granulose 268
GST 305
GVHD 7, 221

H

H19 232
H2A, H2B, H3, and H4 155
H3 Lys-27 155
H3 Lys-27 methylation 156
H3 Lys-4 155
H3 Lys-4 methylation 156

H3 Lys-9   155
H3 Lys-9 acetylation   156
H3-K27   63
H3-K9   63
H3-K9 methylation   255
H33342   221
H3K27me3   160
H4 acetylation   155
Haematopoietic   381, 319, 361, 392, 413
    chimeras   5
    stem cells (HSCs)   221, 328
Hair follicle   334, 335
Ham's F12   433
Hand1   58
Hanging drops   85
Haploid   126, 273, 275
Heat shock protein   402
Hedgehog   427, 455
    signalling   182
HEK293   374
HEK293T   234
Hematopoietic   442, 447, 451
Hematopoietic stem cells (HSCs)   532
HePAR1   304
Hepatocarcinoma   304
Hepatocytes   295, 304
    growth factor   187
    nuclear factor 1b (HNF1b)   182
Hepatotoxicity   304
HepG2   304, 306, 569
Hercules   2
Heregulin   140
hERG potassium channels   303
Herpes Simplex Virus thymidine kinase (HSV-tk)   92
hESCs   245
hes1   185
hEScGRO   142
HGF   378
hIDPSC   434
High-throughput screening (HTS)   293
Hippocampus   333, 380

Histone deacetylation   255
Histone
    H1   155
    methylation   236
    modification patterns   155, 161
HLA   5
    A   226
    B   226
    DR   226, 399
HLXB9   183
HNF1b   184
HNF3   60
HNF4a   182, 184
HNF6   183, 188
HNI   41
Hoescht   230
Homeobox genes   246
Homeostasis   401
Homing   203
Homologous recombination (HR)   58
Hormones-insulin   188
Host neural cells   370
Hox   157
HPRT   284
Human
    amniotic epithelial cell (hAEC)   223
    embryonic stem cells (hES cells)   4, 120, 215
    foreskin fibroblast feeder layers   138
    islet transplantation   172
    leukocyte antigen (HLA)   216
    Menopausal Gonadotrophin (HMG)   122
    neural stem cells (hNSCs)   203
    serum   143
    tubal fluid (HTF)   124
Humanised substrate CELLstart™   143
Huntigton's disease   252, 368
HY gene   219
Hybridoma   232
Hydroperoxides   307
hyg   50
Hygromycin   50

resistant   232
Hyperacute   219
Hyperuricemia   249
Hypoblast   132, 177
Hypoglycaemia   188
Hypoxanthine phosphoribosyl transferase (HPRT)   19, 232
Hypoxanthine phosphoribosyl transferase 1 (HPRT1)   248
Hypoxia   393

I

ICH M3(R1)   299, 303
ICH S5(R2)   299
ICH S7B   303
ICR19   88
Igf2R   64
Igf2r   232
IL-1, IFN-g, TNF-b, and TNF-a   206
IL-11   60, 378
IL-15   207, 378
IL-1a   420
IL-2-   207
IL-3   398
IL-4   207
IL-5   207
IL-6   68, 378, 398, 481
IL-6, -7, -8, -11, -12, -14 and 15   369
Immune privileged sites   224
Immunodeficient   296
Immunosuppressive   216
Imprinting   153
Imprints   231, 275
In vitro fertilization (IVF)   247
In vitro 'Trans'-Differentiation   372, 375
Indocyanine green   305
Indolamine dioxygenase (IDO)   401
Induced pluripotent stem (iPS) cells   6, 79, 235, 621
Inducible systems   100
Induction models   63
Infarcted   420
Infertile   489

Inner cell mass (ICM)   57, 263, 506
iNOS   222
Ins   174
Insulin growth factor 1   59
Insulin I   175
Insulin II   175
Insulin-
   dependent   169
   like growth factor   181
   like growth factor 2 receptor (Igf2r)   270
   like growth factor I   187
   like growth factor II   187
   producing   188
   producing β-cell   337
   like growth factor II (IGF-II)   283
   receptor   171
Integrin b1   431
Integrin
   -mediated   202
   a2   203
   a4   201
   a5   201
   a6   201
   b1   201
Intercalating agent   88
Interfollicular   334
Interleukin 11 (IL11)   59
Internal ribosome entry sites (IRES)   90
International
   Gene Trap Consortium   90
   Mouse Knockout Consortium   94
   Stem Cell Initiative   140
Interstitial lung diseases   334
Intracytoplasmic sperm injection   121
Ipsilateral   420
Irx3   156
Ischemic   332, 379
Ischemia   416
Islet-specific Glucokinase (GCK)   174
Islet1 transcription factor Sca-1   331
IVF laboratories   120

J

JAK/Stat3 80
Jurkat cell line 234

K

KAAD-cyclopamine 182
Karyoplast 233
Karyotype 81
Keratinocyte stem cells (KSCs) 335
KGF 185
KIT 322, 328
Kitlga 32
Klf2 99
Klf4 97, 104, 234, 237, 252, 464
Klf5 99
Knockins 95, 235, 289
Knockout 92, 283, 289
Knockout™ Serum Replacement (KSR) 140
Krebs citric acid cycle 178
Kuppffer cells, stellate cells 307

L

Label retaining cells (LRCs) 322
LacO/LacIR 101
Laminin 143, 647
Laminin A/C 236
LamininB1, Dab2 68
Lateral ventricle 333
LDF medium 24
Lesch-Nyhan 249
Leukaemia 220, 329, 463
Leukemia inhibitory factor (LIF) 57, 79, 60, 481
Leydig 267
Leydig
  cells 266
Limitations in Non-Ontogeny Based
  Differentiation 175
Lin-28 237, 252, 622
Lineage-control 161
  genes 156, 157
Lipid droplets 364
Lipophilic Nile red 264
Loss-of-function 282
LoxP 289
LoxP-Cre 95
Lrp6 286
Lung disorder 338
Luteinizing hormone (LH) 266
Lymphoid organs 225
Lymphomas 226
Lystbg mice 225

M

M. Capecchi 57
Machado-Joseph, Kennedy's disease (SBMA) 254
Major Histocompatibility Complex 216
Male chimera 87
Malformations 299
Malignant 327
MAPC 392
Mario Capecchi 77
Martin Evans 77
Mash1 157
Mater 269
Matrigel™ 138, 143, 296
Maturation promoting factor (MPF) 229
Mature islets 178
MCAo 379
Myotonic Dystrophy (MD) 250
MDR-1 323
Medaka 40
  mid-blastula embryo (MBE) 42
Medicult's SSR media 398
MEF-2C 302
Meiotic 266, 271
Membrane, phosphatidylcholine (PC) 210
Merkel cells 336
MES1 42
MES2 42
MES3 42
Mesenchymal stem cell 7, 222
Mesendoderm 60
Mesoderm 58, 76, 120, 132, 252

Metabolic competence 305
Metaphase II 123
Metaphases 83
Methanesulfonic acid ethyl ester (EMS) 88
Methylcellulose-containing media 85
Methyltransferases (DNMTs) 154
MHC 5
    Class I 366
    Class II 366
    Class III 217
MHC-I 217, 225
MHC-II 217, 225
MICA 207
Micro Electrode Arrays (MEA) 303
Microinjection 282
MicroRNA (miRNA) 99, 514
Microvascular pericytes 393
Microvasculature 328
Migration 369
Mixed lymphocyte reaction (MLR) 400
MIXL1 181
MMPs 428
Mobilisation 328, 445, 393
Modifier genes 94
Monkey oocytes 230
Morula 133
Mos 269
Mosaicism 126
Mouse embryonic fibroblasts (MEF) 57, 80, 235
Mouse embryonic stem cells (mESCs) 57, 76, 281
Mouse tail tip fibroblasts (TTFs) 236
MPF 230
MS5 86
MSC-CM 381
MSCs 412
Msx1 156, 246, 427, 532
Msx2 246
mTert 81
mTeSR1™ 141
Multiple sclerosis 368

Multipotent 324, 358
Multipotent epithelial stem cells (bESCs) 335
Muscular dystrophy 256
Musculoskeletal 339
Mutagenesis 283
Mutations 282, 457
Mvh (mouse vasa homologue) 264, 271
Myelin basic protein (MBP) 382
Myeloablation 7
Myf5 157
MyoD 157
Myogenic 281
Mz-Hep-1 304

N

N-ethyl-N-nitrosourea (ENU) 88
NADH 178
Nanog 57, 67, 85, 237, 252, 431
Nanos 33
Nanotechnology 3
NCCIT 234
NCX 302
Neo 50
Neomycin resistance genes 235
Nephropathy (renal failure) 170
Nerve growth factor (NGF) 333, 378
Nestin 175
    positive cells 173
Netrin-4 and reticulon-4 381
NeuN 434
Neural
    crest 431
    rosettes 133
Neural-like cells 365
Neurectoderm 60
Neuregulins 333
Neurite outgrowth 374
Neurotrophic factors 205, 378
Neurodegenerative 368, 414
Neurofilament 376
Neurogenic 281, 413
    effects 379

Neurogenin 3 (Ngn3)   174
Neuroimmunomodulation   418
Neuromuscular   281
Neuron-specific enolase (NSE)   372
Neuropathy (loss of feeling)   170
Neuroprotective   379
Neurotrophic   418
Neurotrophins   378
Newt   56
Nfic   428
NFM   434, 435
Ngn3   184, 186
Niche   320, 325, 368, 392, 394, 583
Niemann-Pick disease type C (NP-C)   414
NIH3T3 fibroblasts   234
Nijmegen breakage syndrome (Nbs1), Ataxia telangie   253
Ninjurin-1/2   381
Nitric oxide synthase   206
NK cells   225
NKG2D   207
NKX homeodomain   67
Nkx2.2   156, 184, 186, 532
Nkx2.5   302
Nkx6.1   186, 188
No tail gene (ntl)   27
Nobox   269
Nodal   180
Nodes of Ranvier   334
Noggin   140, 186
Non-hematopoietic   411
Non-human primates   230
Non-insulin dependent   169
Non-ischemic heart   332
Non-neurotrophic   381
  factors   378
Non-Ontogeny-Based Differentiation   173
Normoglycaemic   188, 337
Notch   182, 187, 335, 455, 559
Novocell   3
npMSC-CM   382

Nr6a1   269
NT-ESC   79, 105

O

O. Smithies   57
Obox   269
OCT-3/4-null embryos   246
Oct-4/GFP   301
Oct4   57, 84, 104, 233, 234, 236, 252, 271, 431, 532, 533, 620, 622
  promoter   138
Oct4-GFP hESC Reporter Line   145
Odontoblasts   427
OECD-program   294
Oil Red O   363
Oil red-0   413
OLES1   41
Oligodentrocytes   333
Oliver Smithies   77
Ontogeny-based differentiation   178
Oocytes   122
Oogonia   268
OP9   86
Oryzias latipes   41
Osgood   3
Osseous   366
  defects   367
Osteo-progenitor   393
Osteoblasts   428
Osteogenesis   437
  imperfecta   330
Osteogenic differentiation   413
Osteonectin   430
Osteoporosis   359
Outgrowth   78, 295
Ovulation   122
Oxytocin receptor (Oxtr)   95

P

P-zero   413
p16INK4a   326

P450   304
p53   50, 326
PA6   86
Pac   50
Paired box 4 (Pax4)   185
Pancreas   337
Pancreatic and duodenal homeobox 1 (Pdx-1)   174
Pancreatic and duodenal homeobox factor-1   337
Pancreatic development   176
Pancreatic stem/progenitor cells (PSCs)   337
Pancreatic transcription factor Pdx-1   175
Paracrine   378, 480
Parietal endoderm   68
Parkinson   368, 376
Parkinson's disease (PD)   414
Parthenogenesis   227, 226
Parthenogenetic ESC (pESC)   78, 271
Patched1   428
Patched2   428
Pax   157
Pax3   156
Pax4   189
Pax6   60
Pax9   427
PC-12   374
pCAGG   301
pCMV   301
PDGFs   333
Pdx-1   96, 181, 182, 186, 188, 189, 273
PDZ-domain   286
PEBP2A1   428
Pedicle periosteum   402
Peg1/Mest   231
Pericytes   394
Periodontal ligament   392
Periodontium   428
Periosteum   392, 402
Perivascular   360
Permeabilisation   234
pESC   227

PGC   270, 272
PHA (phytohaemagglutinin)   431
Phase 1   306
PhiC31   89, 147
Phosphatidylethanolamine (PE)   210
Phosphatidylinositol-3 Kinase   181
Phospholipid   210
PINs   327
Piwi   47
PiwiL2   58, 273
Plasma   398
Plastic is   360
Plastic-adherent precursor/stem cells (pMSCs)   417
PLGA-gelatin/chondroitin/hyaluronate   205
Pluripotency, Pluripotent   4, 40, 30, 153, 226, 232, 296, 531
Pluripotin   103
PNS   367, 646
Poland syndrome   366
Polarized   230
Polyadenylation signal (pA)   284
Polycomb group protein complexes (PcG)   157
Polycomb repressive complexes   158
Polyethylene glycol (PEG)   231
Polyglutamine   254
Polymorphisms   217
Polyploid   233
Polyploidy   126,
Porcine endogenous retrovirus (PERV)   218
Porcine islets   172
Post-meiotic   266, 485
Post-natal   391
PP   188
pR26   301
PRC1   158, 522
PRC2   158, 522
Premature chromosome condensation (PCC)   229
Primate   229
Primitive Gut Tube   182
Primordial follicles   268

Primordial germ cell (PGC)   21, 63, 263
Prion   397
Prm1   273
Progesterone   267
Prometheus   2
Promoter demethylation   237
Prophase   268
Prospermatogonia   265
Protein kinase C   178
pUbiC   301
Purine   249
Purkinje-like   302
Puromycin   50

Q
QT   303
Quiescent   426
Quinines   307

R
Rag–/–   256
Rag2–/–   229
Rainbow trout spleen cell line (RTS34st)   21
Raldh2   183
Random integration   88
RAR   183
Rat   412
RB   326
Receptor-a   328
Recombinase-mediated cassette exchange (RMCE)   89
Refractory cancers   329
Rejection   216
RepeatMasker program   97
Reproductive cloning   229
Reprogramming   226, 234, 251, 621
Reticulated   130
Retinaldehyde dehydrogenase (RALDH)   183
Retinoic acid (RA)   265, 608
Retinoid X receptors (RXR)   183
Retinopathy (blindness)   170
Retroviral   283

Retroviruses   237
Revascularisation   332
Rex1   68, 235, 431
Rfp14   269
Rhesus macaque   251
Richard Feynman   3
RISC   99
RNAi   98
RNAi libraries   103
RNase III   98
RNase L   99
ROCK-inhibitor   296
Romberg's disease   366
RTG   49
RTS34st   25
RtTA   100
Runx2   427, 428
RXR   183
RyR   302

S
Saccharomyces cerevisiae   95, 286
Scavengers   402
SCF   264, 322, 398, 482
Sciatic function index (SFI)   380
Sciatic nerve   379
SCID   225, 459
SDF-1   204, 445, 446
SDF-1/CXC   321
Second polar body   227
Secretome   378
Sedative   299
Self-renewal   40, 454, 559
Self-tolerance   104
Sendai viruses   231
SERCA   302
Sertoli   267
Serum-free   138
Set1 H3Lys-4 methyltransferase   158
Severe combined immunodeficient (SCID)   86
SG3   40
SH2   361, 431, 607

SH3   361, 431, 607
SH4   361, 431
SHED   435
Shh   185, 427, 428
Shroom   286
Sialophosphoprotein   434
Side population "SP"   221, 323, 567
Single X chromosome   256
Sinus nodal-like   302
Sir M. Evans   57
siRNAs   249
Skin cells   251
Smoothened   428
Sodium Butyrate and Bmp4   181
Soft tissue defects   366
Soma   374
Somatic cell nuclear transfer (SCNT)   79, 226, 251
Somatostatin   188
Sonic hedgehog (Shh)   185
Sox1   58, 156, 532
Sox17   67, 179, 180
Sox2 (sex determining region Y-box 2)   57, 60, 68, 85, 104, 236, 252, 464
Sox-2-GFP   299
Sox3   68
Sperm immobilization   123
Spermatogenesis   26, 473
Spermatogonia   266
Spermatogonial Stem Cells   46, 64, 266
Spermatozoa   266
Spg/Pou2   66
Sphere-forming   437
Spinal cord   417, 650
Spinocerebellar ataxias (SCAs)   254
Splice acceptor (SA)   284
Splice donor (SD)   285
Spontaneous differentiation   296
Squamous cell carcinomas   336
SRY   179
SRY box containing gene 7 (SOX7)   180
SRY-box 2   223

SRY-related HMG   68
SSEA-1   232, 273, 609
SSEA-3   431
SSEA-4   223, 431
Stage 13–15 (H&H)   64
Stage 17 (H&H)   64
Stage 28 (H&H)   64
Stage X (EG & K)   64
Stage X embryo   59
Stage-specific embryonic antigen-1 antigen (SSEA-1)   81
Stella   58, 263
Stem cell factor (SCF)   59, 264, 481
Stem cell maintenance   159
StemPro   398
StemPro® hESC-SFM   141
STO   235
Stored-umbilical cord (UC)   329
Stra8   58, 265, 275
Striatum   415
STRO-1   339
Stroke   420, 421
Stromal   326
Stromal cell derived factor 1 (Sdf1)   32
Stromal cells from human dental pulp (SBP-DPSC)   435
Subgerminal cavity   59
Subjcent ectomesenchymal   427
Subventricular   333
Super-numerary   120
Suppressor of zeste-12 (SUZ12)   157
SUZ12   246
SYCP3   273
Sycp3   271, 273
Symmetric   325
Synaptogenesis   281
Synaptogenic   379
Synaptotagmin-4   157
Syncitiotrophoblast   128
Syngeneic   214
Synth-a-freeze   139, 144

T

T-helper 1 207
T-helper 2 207
T-lymphocyte 401
T-cells 218, 431
　　receptor 217
T-regulatory cells 104
Taqman Low Density Arrays (TLDA) 300
Target-induced local lesions in genomes (TILLING) 22
Tbx5 302
TCR 227
Telomerase (TERT) 81, 237, 294, 323
Tendon 392
Teratocarcinomas 58, 77
Teratogenesis 205
Teratoma 4, 191, 223, 225, 296
Testis 265
Testosterone 267, 475, 479
Tet-on system 100
Tetraploid embryos 88, 229
Tetraploid ES cells 233
TGF-b 180, 186, 205, 224, 402
Th0 207
Thalidomide 299
THBD 180
Therapeutic cloning 229, 230
Thin film 200
Thin solid films (TSF) 200
Thoracic and lumbar spine 360
Thy-1 361
Thymic epithelial cells (TEC) 104
Thymic epithelial progenitor cells (TEPC) 104
Thymidine kinase (TK) 287
Thymocyte 232
Thymus 104
Tibia 360
Tissue non-specific alkaline phosphatase (TNAP) 263
TNF-a 203, 210
Tolerance 226
Totipotency 358
TRA-1-60, TRA-1-81(tumour recognition antigen) 431
TRA-1-61 232
TRA-2-3 233
Trans-differentiate 221, 252, 325, 326, 365, 412, 419, 606
Transcription factors 252
Transgene 282
Transgenesis 283
Transgenic mice 282
Transit-amplifying (TA) 325
Transmigration 203
Transplantation 216, 489, 574
Trapping vectors 284
Traumatic brain injury 421
Triple-helix forming oligonucleotide (TFO) 98
Triploid embryos 125
Tripronuclear 125
Trisomy 21 251
Trisomy 8 83
Trithorax group (trxG) proteins 158, 160
Trophectoderm 57, 619
Trophic factors 370, 402
　　delivery 418
Tropomyosin, MLC-2A/2V 302
Troponins 302
TrypleTM Select 296
Trypsin 296
tTA (for the tet-off system) 100
TTAGGG 323
TTFs 237
Tumour resection 366
TWS119 103
Type 1 diabetes 169, 610
Type 2 diabetes 169
Type IV collagen 86
Tyroxine hydroxylase 415

U

ULBPs 207
Ulcers 336
Umbilical cord blood (UCB) 221, 329
Undifferentiated 295
Unipotent 358
Up-scaling 395

V

Vasa 33, 46, 47
Vasa-GFP 33
Vascular 281
    endothelial growth factor (VEGF) 203
    smooth-muscle cells (SMCs) 331
Vasculogenesis 325
VCAM-1 203, 362, 430
VEGF 206, 333, 378, 402, 446
Ventricular repolarisation 303
Ventricular-like 302
Viral
    delivery 90
    transduction 235
Vitamin A 265, 487
VitroGro® 143
von Kossa 362

W

Werner syndrome (WRN) 253
Whole pancreas transplantation 171
William Heseltine 2
Wnt 189, 427, 446, 559
Wnt act 80
Wnt/b-catenin 569
Wnt3a 181

X

X chromosome 232
    inactivation 154, 282, 509
X-linked 289
X-linked FMR1 255
XEN 58
Xeno-free 138, 296
Xeno-transplantation 369
Xenoantigen 220
Xenobiotic 139, 307
Xenogenic 216
    islet cells 172
Xenografting 296
Xenopus 66
Xenopus oocyte 234
Xenorejection 225
Xenotransplantation 220
XO mice 256
Xoct25 66
Xoct60 66
Xoct91 66
XX 265
XY 265

Y

Y-chromosome 264

Z

Zar1 269
Zebra fish FGF 142
Zebrafish embryonic stem 21
Zebrafish kit ligand a (Kitlga) 21
Zebrafish, SDF1/Sdf1 serves 32
Zeste-2 (EZH2) 158
Zfx [51] 65
Zinc finger nucleases (ZFN) 22, 98
Zona pellucids (ZP) 269
Zoonoses 397
ZP1, Zp2, ZP3 269, 273